T0387129

SUSTAINABLE MARITIME TRANSPORTATION AND EXPLOITATION OF SEA RESOURCES

PROCEEDINGS OF THE 14TH INTERNATIONAL CONGRESS OF THE INTERNATIONAL MARTIME ASSOCIATION OF THE MEDITERRANEAN (IMAM), GENOVA, ITALY, 13–16 SEPTEMBER, 2011

Sustainable Maritime Transportation and Exploitation of Sea Resources

Editors

Enrico Rizzuto

University of Genoa – DICAT, Genoa, Italy

Carlos Guedes Soares

Centre for Marine Technology and Engineering (CENTEC)
Technical University of Lisbon, Lisboa, Portugal

VOLUME 1

CRC Press is an imprint of the
Taylor & Francis Group, an **informa** business

A BALKEMA BOOK

Cover photo credit (front and back cover): 'Views of the port of Genoa', courtesy of the Genoa Port Authority

CRC Press/Balkema is an imprint of the Taylor & Francis Group, an informa business

Typeset by Vikatan Publishing Solutions (P) Ltd., Chennai, India
Printed and bound by CPI Group (UK) Ltd, Croydon, CRO 4YY

Published by: CRC Press/Balkema
P.O. Box 447, 2300 AK Leiden, The Netherlands
e-mail: Pub.NL@taylorandfrancis.com
www.crcpress.com – www.taylorandfrancis.co.uk – www.balkema.nl

ISBN: 978-0-415-62081-9 (set of 2 volumes + CD-ROM)
ISBN: 978-0-415-62082-6 (Vol 1)
ISBN: 978-0-415-68393-7 (Vol 2)
ISBN: 978-0-203-13033-9 (eBook)

Sustainable Maritime Transportation and Exploitation of Sea Resources – Rizzuto & Guedes Soares (eds)
© 2012 Taylor & Francis Group, London, ISBN 978-0-415-62081-9

Table of contents

3 *Structures & materials*

3.1 *Analysis of local structures*

3.2 *Analysis of the global structures*

3.3 *Static and dynamic structural assessment*

3.4 *Materials*

3.5 *Degradation/Defects in structures*

VOLUME II

5 *Propulsion plants*

5.1 *Prime movers*

5.2 *Machinery*

6 *Safety & operation*

6.1 *Safety*

Sustainable Maritime Transportation and Exploitation of Sea Resources – Rizzuto & Guedes Soares (eds)
© 2012 Taylor & Francis Group, London, ISBN 978-0-415-62081-9

Preface

This book presents the proceedings of the XIV Conference of the International Maritime Association of the Mediterranean (IMAM), held in Genoa on September 13th to 16th 2011 under the theme 'Sustainable Maritime Transportation and Exploitation of Sea Resources'.

The book covers the most updated aspects of maritime transports and of coastal and sea resources exploitation, with focus on the Mediterranean area. Vessels for transportation are analysed from the viewpoint of ship design in terms of hydrodynamic, structural and plant optimisation, as well as from the perspective of construction, maintenance, operation and logistics. The exploitation of marine and coastal resources is covered in terms of fishing, aquaculture and renewable energy production as well as of subsea resources extraction.

The characterisation of the marine environment is seen under the twofold perspective of providing reference loads and conditions for the design of means for the resources exploitation, but also of setting limits to the design in order to preserve the natural ambient and minimise the impact of anthropogenic activities related to both transportation and exploitation.

Efficiency, reliability, safety and sustainability of sea- and Mediterranean-related human activities are the focus throughout the book. The text is mainly devoted to technical operators of the various field involved, coming from shipbuilding and ship-owner companies, research organisations, Universities, certifying bodies, but it represents an updated reference text of interest also for government Agencies and other institutional and educational bodies.

ABOUT THE INTERNATIONAL MARITIME ASSOCIATION OF THE MEDITERRANEAN (IMAM)

The Association was established in 1974 with the acronym of IMAEM (International Maritime Association of East Mediterranean), initially including institutions from six countries (Bulgaria, Egypt, Greece, Italy, Turkey and Yugoslavia). The membership was later progressively enlarged to most of the Mediterranean countries and neighbouring areas and, since the 1990 conference, the acronym was changed to IMAM, dropping the reference to the eastern part of the basin.

The International Maritime Association of the Mediterranean (IMAM) is proud of its more than thirty-year-long history committed to the enhancement and dissemination of technical knowledge related to study, design, construction and lifetime operation of ships and other marine structures. The focus of the Association is on the development of marine transports and on the exploitation of sea resources in the Mediterranean area, in line with the principles of a sustainable growth.

The IMAM Congresses are privileged forums for the maritime technical community of the Mediterranean. They have been hosted in the following key locations:

Istanbul (1978)	Trieste (1981)	Athens (1984)	Varna (1987)
Athens (1990)	Varna (1993)	Dubrovnik (1995)	Istanbul (1997)
Ischia (2000)	Crete (2002)	Lisbon (2005)	Varna (2007)
Istanbul (2009)	Genoa (2011)		

The IMAM conferences are traditionally attended by qualified representatives from the Academia and from Professional and Technical Associations of the various fields involved. The conferences represent therefore a window for areas of high potential growth such as the Balkans, Eastern Europe, the Middle and the Near East as well as North Africa.

 ISBN 978-0-415-62081-9

IMAM Organisation

IMAM EXECUTIVE COMMITTEE

Prof. C. Guedes Soares, *Instituto Superior Técnico, Portugal (President)*
Eng. K. Lapa, *Vlora University, Albania*
Prof. P. Kolev, *Technical University of Varna, Bulgaria*
Prof. V. Zanic, *University of Zagreb, Croatia*
Prof. M. Gaafary, *Suez Canal University, Egypt*
Prof. B. Molin, *École Superieure d'Ingenieurs de Marseille, France*
Prof. C. Spyrou, *Hellenic Institute of Marine Techn, Greece*
Prof. E. Rizzuto, *University of Genova, Italy*
Prof. I. Chirica, *"Dunarea de Jos" University, Romania*
Prof. O. Goren, *Istanbul Technical University, Turkey*
Prof. Y. Vorobyov, *Odessa State Maritime University, Ukraine*

IMAM TECHNICAL COMMITTEES

Hydrodynamics

Stefano Brizzolara	University of Genoa	IT
Dario Bruzzone	University of Genoa	IT
Pierre Ferrant	Ecole Centrale – Nantes	FR
Nuno Fonseca	Instituto Superior Técnico-Lisbon	PT
Omer Goren	Istanbul Technical Univ.	TR
Gregory Grigoropoulos	National Techn. Univ. Athens	GR
Atilla Incecik	University of Strathclyde	UK
Fernando López Pena	University of La Coruna	ES
Adolfo Maron	CEHIPAR-Madrid	ES
Touvia Miloh	Tel Aviv University	IL
Bernard Molin	Ecole Centrale – Marseille	FR
Marcelo Neves	Federal University of Rio de Janeiro	BR
Jasna Prpic-Orsic	University of Rijeka	HR
Constantinos Spyrou	National Technical Univ. Athens	GR
Penny Temarel	University of Southampton	UK
Leszek Wilczynski	CTO – Gdańsk	PL
Kostadin Yossifov	Bulgarian Ship Hydrodynamics Centre	BG

Marine structures

Dino Cervetto	RINA	IT
Matteo Codda	CETENA-Genoa	IT
Ionel Chirica	University of Galati	RO
Leonard Domnisoru	University of Galati	RO
Yordan Garbatov	Instituto Superior Técnico-Lisbon	PT
Reza M. Khedmati	Amirkabir University of Technology	IR
Mario Maestro	University of Trieste	IT
Josko Parunov	University of Zagreb	HR
Cesare Rizzo	University of Genoa	IT

Ports & transports systems

Makoto Arai	Yokhohama University	JP
Carlos Botter Rui	University of S. Paulo	BR
Dimitrios Lyridis	Nat. Technical University of Athens	GR
Ovidius Mamut Eden	University of Constantza	RO
Nikitas Nikitakos	Aegean University	GR
Harilaos Psaraftis	Nat. Technical University of Athens	GR
Giovanni Solari	University of Genoa	IT

Off-shore & coastal development

Francisco Taveira Pinto	University of Porto	PT
Mohamed Chagdali	University Ben M'Sik Casablanca	MR
Inigo Losada	University of Cantabria	ES
Spyros Mavrakos	Nat. Technical University of Athens	GR
Vicent Rey	University of Toulon	FR
Leonardo Brunori	RINA	IT

Aquaculture & fishing

Aida Campos	IPIMAR	PT
Teresa Dinis	Algarve University	PT
Rajko Grubisic	University of Zagreb	HR
Barry O'neil	MARLAB	UK
Daniel Priour	IFREMER	FR
Antonello Sala	ISMAR	IT
Emma Tomaselli	RINA	IT

Small and pleasure crafts

Carlo Bertorello	University of Naples	IT
Dario Boote	University of Genoa	IT
Izvor Grubisic	University of Zagreb	HR
Massimo Musio-Sale	University of Genoa	IT
Lorenzo Pollicardo	Federagenti Yacht	IT
Antonio Scamardella	Parthenope University (Naples)	IT

Sustainable Maritime Transportation and Exploitation of Sea Resources – Rizzuto & Guedes Soares (eds)
© 2012 Taylor & Francis Group, London, ISBN 978-0-415-62081-9

XIV IMAM conference organisation

The conference was organised by the University of Genoa in the seat of the Faculty of Engineering, Villa Cambiaso Pallavicini in Genoa.

CONFERENCE COMMITTEE

Enrico Rizzuto (Chairman) – *University of Genoa*
Carlo Podenzana Bonvino – *University of Genoa*
Bruno Della Loggia – *Atena*

CONFERENCE SECRETARIAT

Isa Traverso – *University of Genoa*
Michela Tizzani – *Promoest*
Elisabetta Gembillo – *University of Genoa*

1 *Propulsion & resistance*

1.1 *Resistance*

Sustainable Maritime Transportation and Exploitation of Sea Resources – Rizzuto & Guedes Soares (eds)
© 2012 Taylor & Francis Group, London, ISBN 978-0-415-62081-9

Theoretical and experimental investigation on the total resistance of an underwater ROV remotely operating vehicle

D. Obreja & L. Domnisoru
"Dunarea de Jos" University of Galati, Galati, Romania

ABSTRACT: The present study focuses on the theoretical and experimental hull resistance analysis of an underwater ROV remotely operating vehicle, small in mass and dimensions, used for maritime and offshore operations survey at 30 m design depth. In order to evaluate the mini-ROV resistance, with and without free surface influence, both experimental and statistical methods were applied. An ellipsoidal body shape was selected, whose main dimensions were 500 × 350 × 250 mm, having horizontal and vertical propeller tubes in the external shell structure, placed at both sides and ends. In the experimental tests four different immersion cases are analysed, corresponding to the speed domain 1 to 2 m/s. The 1:1 scale experimental ROV model tests were performed at the Towing Tank of the Naval Architecture Faculty, from the "Dunarea de Jos" University of Galati. The comparison between the experimental results and statistical predictions evinced the overestimation of the mini-ROV resistance based on statistical relations.

1 INTRODUCTION

The present day construction and exploitation of the maritime and offshore devices require periodic underwater survey. One of the practical solutions is the survey based on underwater ROV remotely operating vehicles (Christ & Wernli, 2007; Griffiths, 2003).

As a rule, a modern ROV has to carry out different tasks. A wide diversity of technical solutions, in terms of shape, general arrangement, propulsion system and onboard devices can be obtained (Valencia et al., 2008; Ross, 2006; Domnisoru et al., 2010).

The experimental resistance tests of the mini-ROV is carried out on a 1:1 scale built ROV model, attached to the towing tank carriage by means of a hydrodynamically-shaped support. The ROV model resistance is obtained from the difference between the total resistance of the ROV-profile support system and the resistance of the hydrodynamic profile support only. In order to point out the free surface influence on the resistance force, the experimental tests include four different immersion cases, the test speed ranging from 1 to 2 m/s. Section 3 illustrates the experimental tests performed at the Naval Architecture Faculty Towing Tank, at the "Dunarea de Jos" University of Galati, using a new Cussons Technology Ltd. measuring system (Cussons, 2009).

The study includes also under Section 4 an evaluation of the mini-ROV model resistance, based on fluid mechanics statistical relations (Blevins, 2003; Munson et al., 2004; Batchelor, 2000), applied to an ellipsoidal body shape, without the free surface influence.

The statistical values represent the comparing reference data base for the experimental test results.

2 THE MINI ROV HULL CHARACTERISTICS

In the present study the experimental resistance tests are developed for a concept of our own design of a mini-ROV model (Domnisoru et al., 2010), used for general purpose applications, whose main dimensions are shown in Table 1, and hull structural layout in Figure 1a.

The mini-ROV design has only one external watertight shell, ellipsoidal in shape. In order to obtain a compact mini-ROV design, the propulsion systems for horizontal and vertical motions are mounted symmetrically, at the sides as well as the ends, in the external shell propulsion tubes (Fig. 1a).

Table 1. The main dimensions of the mini-ROV ellipsoidal hull model.

$2a$ [mm]	500	e_1 [mm]	215
$2b$ [mm]	350	f_1 [mm]	35
$2c$ [mm]	250	e_2 [mm]	140
d [mm]	50	f_2 [mm]	35
Δ [kg]	21.3	M_{hull}[kg]	5.2
$M_{devices}$[kg]	16.2	v [Knots]	3

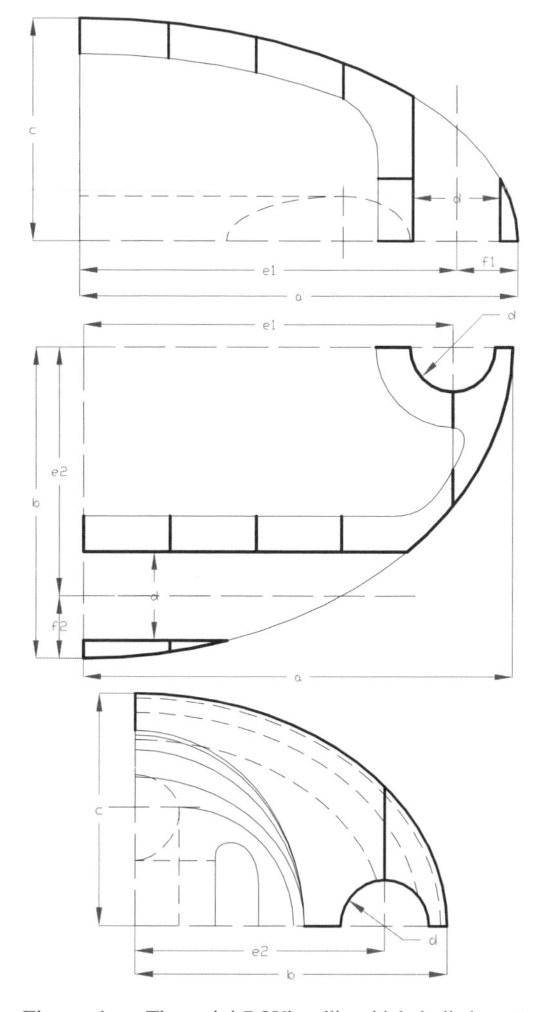

Figure 1a. The mini-ROV's ellipsoidal hull layout design model.

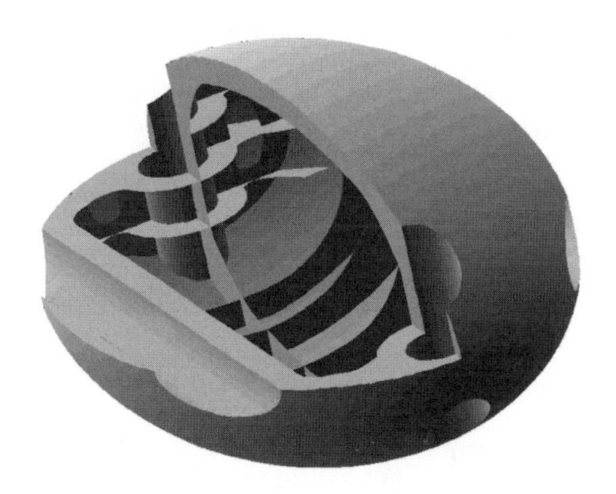

Figure 1b. The mini-ROV's ellipsoidal hull 3D CAD design model.

Figure 1b shows our own 3D CAD design concept for the experimental model of the mini-ROV.

The ROV hull model is made of composite fibreglass material (GL, 2011), with structural strengths corresponding to the design operation depth of 30 m.

The mass of the hull structure is 5.2 kg, resulting in a 16.2 kg onboard devices carrying capacity.

The design speed is 3 Knots (1.54 m/s), so that the experimental testing speeds range is set to 1÷2 m/s.

3 EXPERIMENTAL INVESTIGATION ON THE TOTAL RESISTANCE OF THE MINI-ROV

3.1 *The experimental mini-ROV model and the set-up of the towing tank test facilities*

The experimental tests used to measure the ROV model resistance in a forward motion were carried out in the Towing Tank of the Faculty of Naval Architecture of the "Dunarea de Jos" University of Galati (Fig. 2).

The Towing Tank, 45 × 4 × 3 meters in size, is fitted with an automatic carriage able to tow experimental models at a maximum speed of 4 m/s, built by the British Company Cussons Technology, operating since 2009 (Cussons, 2009).

In order to measure the resistance in forward motion, the forward motion resistance dynamometer R35 was used, the measuring range being up to 200 N and the measuring error 0.2%. By applying the standard procedure, the calibration constant of the dynamometer was determined (1.05268 V/N).

The ROV experimental model was built at a 1:1 scale on the basis of the shape plan generated by the "Dunarea de Jos" University of Galati. In order to determine the resistance in forward motion, the submerged ROV model was coupled to the carriage by means of a support with a symmetrical hydrodynamic profile, of segment type

Figure 2. Towing Tank of the "Dunarea de Jos" University of Galati, Naval Architecture Faculty.

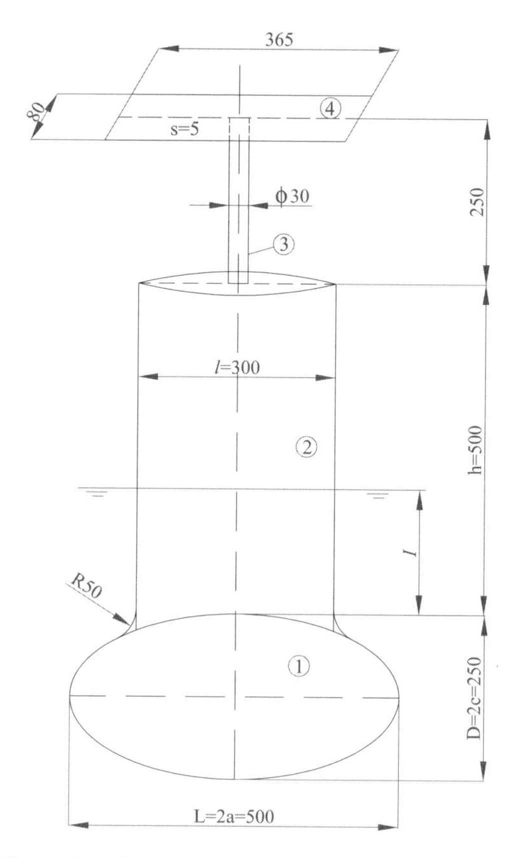

Figure 3. The mini-ROV-hydrodynamic profile support system components: 1- the ROV hull model; 2- the hydrodynamic profile support; 3-beam support; 4-plate support mounted on the resistance dynamometer.

(Fig. 3), generating its own waves of medium amplitudes. The geometric characteristics of the hydrodynamic profile are shown in Table 2.

Measurements were performed both for the forward motion resistance of the system made up of the ROV and the hydrodynamic support, and of the hydrodynamic support by itself. In the hypothesis of overlapping effects, the forward motion resistance of the ROV model is the difference between the resistance of the ROV-hydrodynamic system and the hydrodynamic support resistance.

The automatic system of experimental data acquisition and analysis used for processing the results of experimental tests was developed by the Cussons company. The electric signals transmitted by the forward motion resistance translator were automatically acquired and transformed into physical dimensions by applying the calibration constant. The sampling time step was 0.1 s, and the sample number depended on the carriage speed (390 samples at the minimum speed 1 m/s and 220 samples at the maximum speed of 2 m/s).

Table 2. Characteristics of the hydrodynamic support.

Profile thickness, $t = 0.036$ m	Aspect ratio, $h/l = 1.667$
Profile length, $l = 0.300$ m	Relative thickness, $t/l = 0.12$

The average values of the experimental results were calculated for the stationary flow range. The measurement error of the forward motion resistance tests was about 2%.

The serial of experimental tests included sets of trials for 4 different immersions of the hydrodynamic support (0.05 m, 0.35 m, 0.45 m, 0.55 m) in order to determine the influence of the free surface upon the forward motion resistance. Also, keeping into account the designed value of the speed, the evolution of the forward motion resistance was analysed within a range between 1 m/s and 2 m/s, at a step of 0.25 m/s.

3.2 *Experimental resistance tests on the hydrodynamic profile support*

The experimental results of forward motion resistance tests of the hydrodynamic profile support are shown in Table 3. Figure 4 illustrates the diagram of the forward motion resistance R_{Ts} function of the speed v, for the four immersion cases analysed.

Generally speaking, the forward motion resistance of the hydrodynamic support increases with speed. Only in the case of minimal immersion, $I = 0.05$ m, the forward motion resistance does not significantly depend on speed within the analysed range.

Figure 5 shows the diagram of the forward motion resistance of the hydrodynamic support, R_{tS} in relation to immersion I, for constant speeds. Naturally, there is an increase of the forward motion resistance of the hydrodynamic support with the immersion depth.

Figures 6a, b, c show the flow around the hydro-dynamic support at maximum immersion $I = 0.55$ m, at speeds of 1 m/s, 1.5 m/s and 2 m/s respectively. The Figures 6 analysis leads to the following observations:

- the segment profile attack board generates its own wave of moderate height;
- at the bow of the flight board there is increased wake at the speed of 1 m/s, with vortices and broken waves;
- due to the segment profile flight board, the wake width significantly decreases with the increase of the speed.

Figures 7a, b, c show the flow around the hydrodynamic support at the speed of 1.5 m/s, at

Table 3. Resistance of the hydrodynamic support, R_{tS} [N].

Immersion I [m]	Speed, v [m/s]				
	1	1.25	1.5	1.75	2
0.05	0.57	0.60	0.53	0.56	0.68
0.35	1.54	1.91	1.97	2.18	2.44
0.45	1.99	2.24	2.43	2.64	2.90
0.55	2.07	2.54	2.79	3.01	3.55

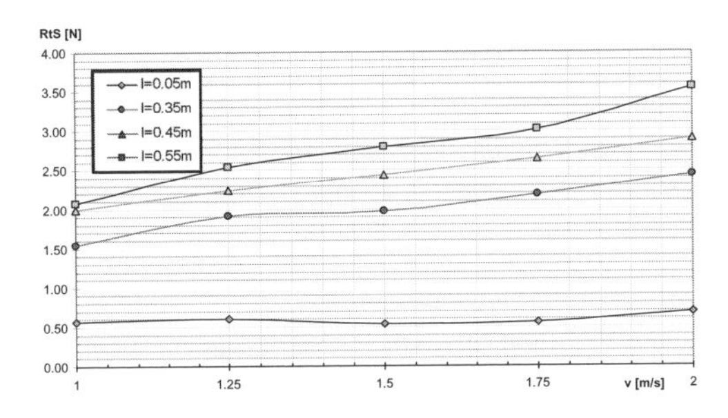

Figure 4. Resistance of the hydrodynamic support function of speed, at constant immersion (0.05, 0.35, 0.45, 0.55 m).

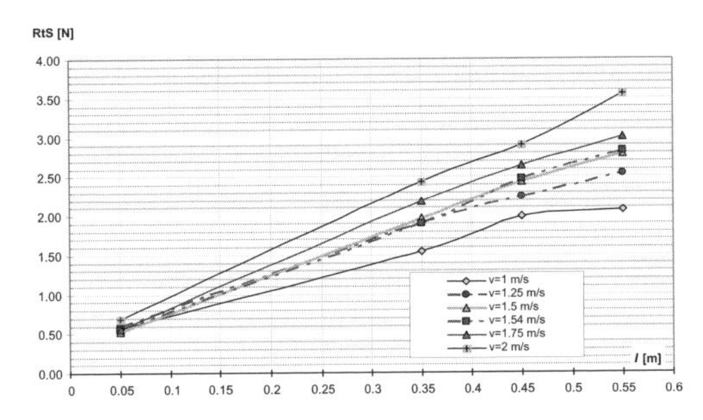

Figure 5. Resistance of the hydrodynamic support function of immersion, at constant speed (1, 1.25, 1.5, 1.75, 2 m/s).

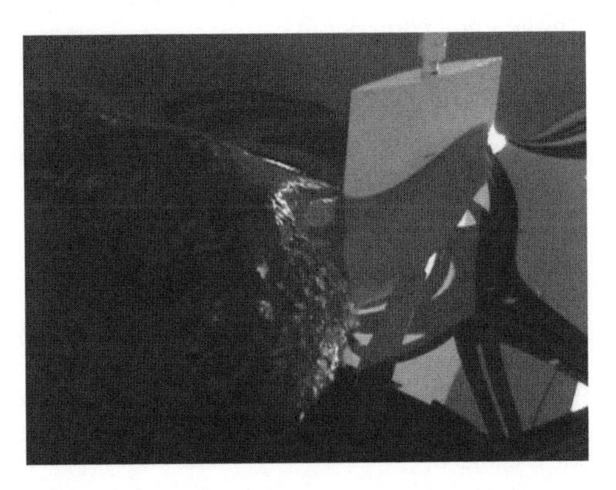

Figure 6a. Hydrodynamic support, immersion $I = 0.55$ m and speed $v = 1$ m/s.

Figure 6b. Hydrodynamic support, immersion $I = 0.55$ m and speed $v = 1.5$ m/s.

Figure 6c. Hydrodynamic support, immersion $I = 0.55$ m and speed $v = 2$ m/s.

Figure 7a. Hydrodynamic support, immersion $I = 0.05$ m and speed $v = 1.5$ m/s.

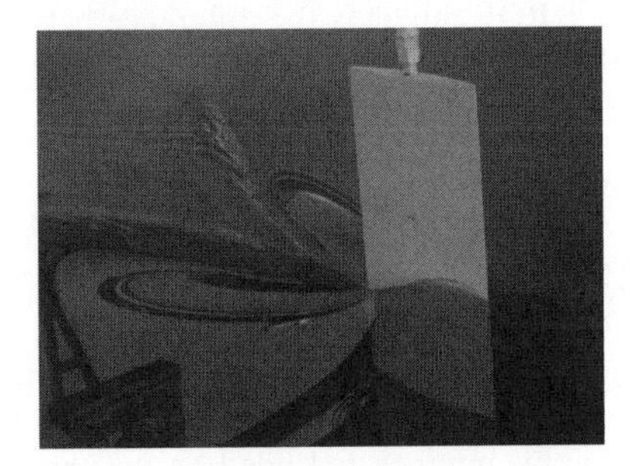

Figure 7b. Hydrodynamic support, immersion $I = 0.35$ m and speed $v = 1.5$ m/s.

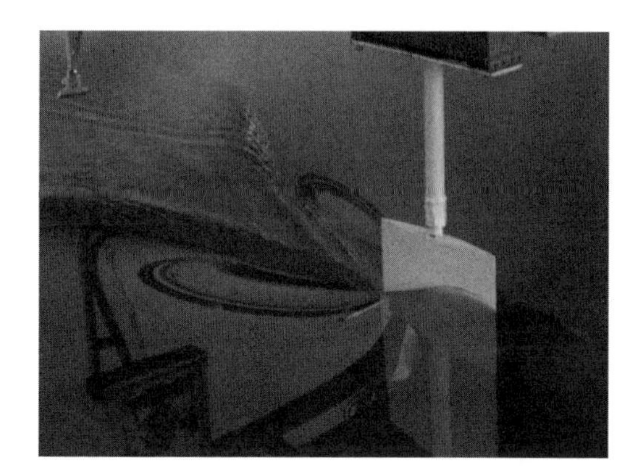

Figure 7c. Hydrodynamic support, immersion $I = 0.55$ m and speed $v = 1.5$ m/s.

immersions of 0.05 m, 0.35 m and 0.55 m. Except for the minimum immersion case, no noticeable differences occur in the hydrodynamic spectre of the flow around the hydrodynamic profile. The wake field contains vortices and broken waves.

3.3 *Experimental resistance tests on the mini-ROV hull coupled to the hydrodynamic profile support*

The second stage of the experimental tests contained the measurements of the forward motion resistance for the system consisting of the ROV and the hydrodynamic profile support, for the same range of speeds and immersions.

The experimental results are shown in Table 4. Figure 8 shows the diagram of the forward motion resistance R_{tA} function of the speed v, for the 4 immersions I tested, and Figure 9 shows the diagram of the forward motion resistance of the ROV- hydrodynamic support, R_{tA} in relation to the immersion I, at the 5 speeds under consideration.

The following observations were derived from the experimental data:

- the forward motion resistance of the ROV-hydrodynamic support system increases with speed, except for the minimum immersion $I = 0.05$ m, where there is a significant decrease of the forward motion resistance starting from speeds higher than 1.5 m/s (Fig. 8), which was confirmed by repeated tests;
- the forward motion resistance of the ROV-hydrodynamic support system decreases at minimum immersion (I = 0.05 m) for speeds under 1.5 m/s, which may be explained by the considerable calming of the flow's hydrodynamic spectre within this speed range, as seen in Figures 10a, b, c;

Table 4. Resistance of the ROV-hydrodynamic support, R_{tA}[N].

Immersion I[m]	Speed, v [m/s]				
	1	1.25	1.5	1.75	2
0.05	21.19	34.23	43.87	36.39	29.65
0.35	8.07	12.73	16.44	20.47	25.44
0.45	7.78	11.73	16.36	20.28	25.34
0.55	7.77	11.37	16.21	20.19	24.44

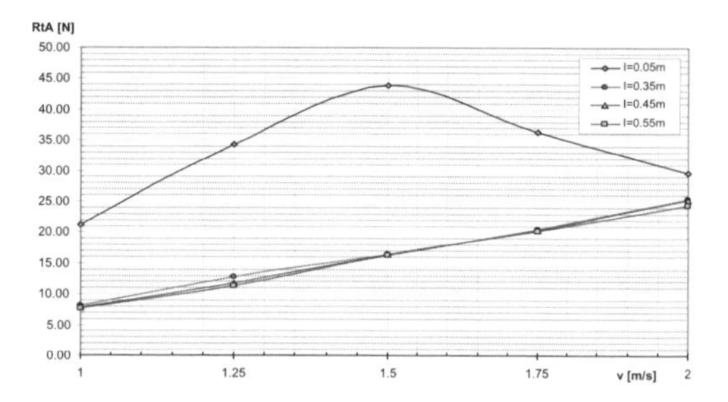

Figure 8. Resistance of the ROV-hydrodynamic support system function of speed, at constant immersion (0.05, 0.35, 0.45, 0.55 m).

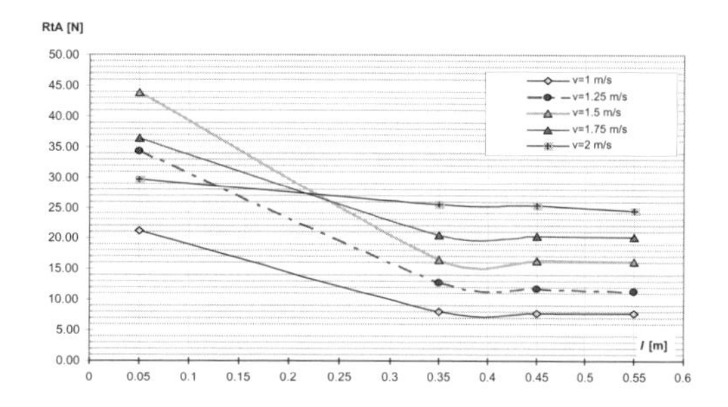

Figure 9. Resistance of the ROV-hydrodynamic support system function of immersion, at constant speed (1, 1.25, 1.5, 1.75, 2 m/s).

Figure 10a. ROV-hydrodynamic support system, immersion $I = 0.05$ m and speed $v = 1.5$ m/s.

Figure 10b. ROV-hydrodynamic support system, immersion $I = 0.05$ m and speed $v = 1.75$ m/s.

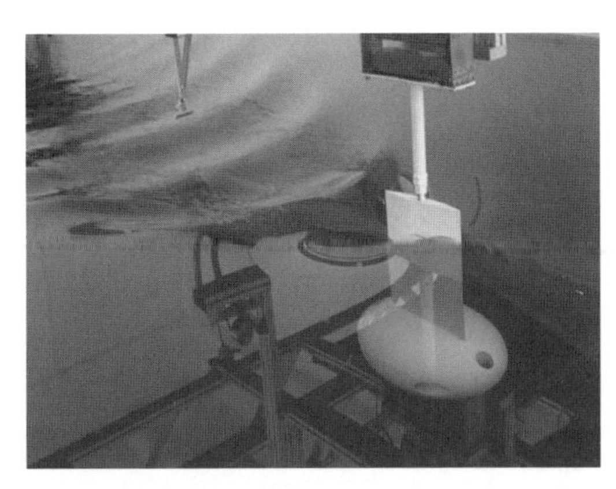

Figure 11a. ROV-hydrodynamic support system, immersion $I = 0.55$ m and speed $v = 1$ m/s.

Figure 10c. ROV-hydrodynamic support system, immersion $I = 0.05$ m and speed $v = 2$ m/s.

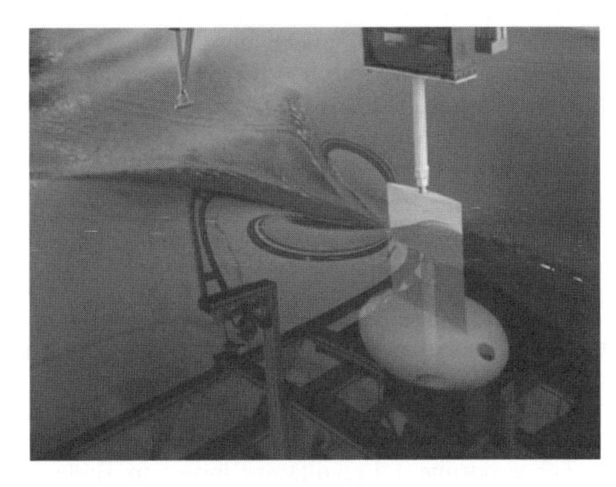

Figure 11b. ROV-hydrodynamic support system, immersion $I = 0.55$ m and speed $v = 1.5$ m/s.

- the values of the forward motion resistance at the minimum immersion I = 0.05 m are much higher than for the other immersions tested, due to the negative effect of the free surface;
- in the case of the immersions I = 0.45 m and I = 0.55 m there are very small differences between the values of the forward motion resistance of the ROV- hydrodynamic support system; the increase of the forward motion resistance due to the increased immersion of the ROV- hydrodynamic support system is practically compensated for by the decrease in the forward motion resistance resulting from the decreased effect of the free surface.

Similarly, Figures 11a, b, c illustrate the flow around the ROV- hydrodynamic support system,

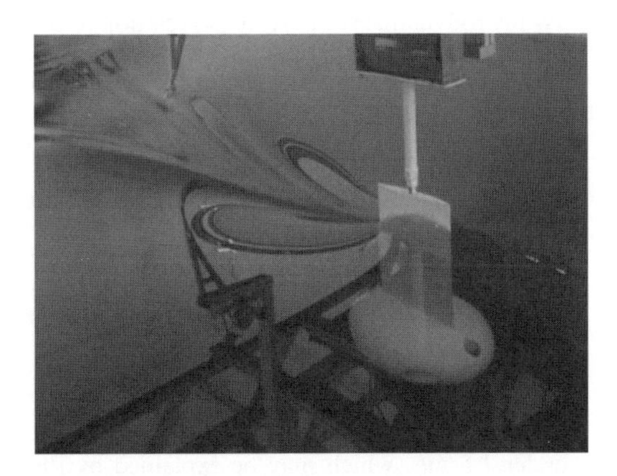

Figure 11c. ROV-hydrodynamic support system, immersion $I = 0.55$ m and speed $v = 2$ m/s.

at the maximum immersion $I = 0.55$ m, for speeds equal to 1 m/s, 1.5 m/s and 2 m/s, respectively.

Figures 12a, b, c show the flow around the ROV- hydrodynamic support system, at the speed of 1.5 m/s, for the immersions 0.05 m, 0.35 m and 0.55 m.

The analysis of the hydrodynamic spectres of the ROV- hydrodynamic support system allows the following observations:

- the height of the own wave is above the one in the hydrodynamic support tests, and the length and height of the stern waves increases with the current speed;
- the wake width significantly decreases when the current speed increases;
- at the minimum immersion $I = 0.05$ m, the wake of the ROV- hydrodynamic support system is more noticeable than in the case of the hydrodynamic support alone, generating broken waves and strong vortices; the influence of the ROV hull, located close to the free surface, on the aspect of hydrodynamic flow is very important;
- for the other cases of immersion, there are no noticeable differences in the hydrodynamic spectre of the flow at the same speed; a strong uneven wake is generated at bow, with vortices and broken waves at the ends of the wake field, similar to the wake created by the hydrodynamic support.

Figure 12a. ROV-hydrodynamic support system, immersion $I = 0.05$ m and speed $v = 1.5$ m/s.

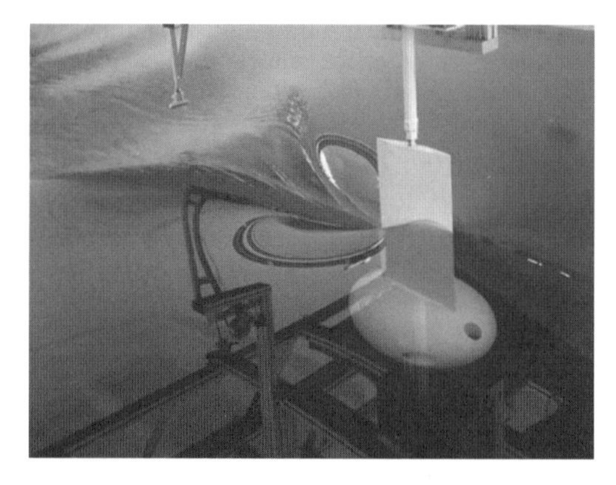

Figure 12b. ROV-hydrodynamic support system, immersion $I = 0.35$ m and speed $v = 1.5$ m/s.

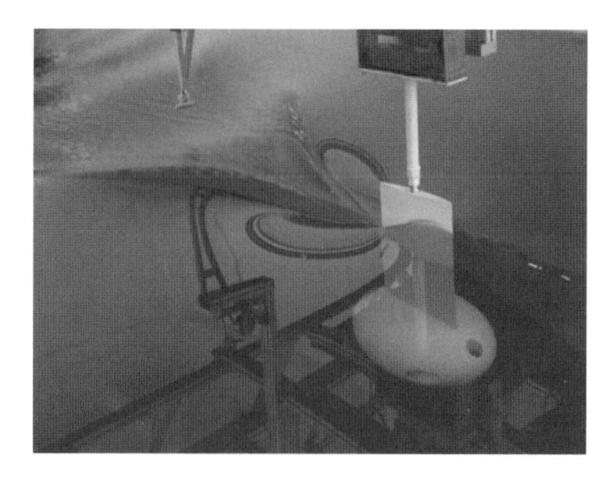

Figure 12c. ROV-hydrodynamic support system, immersion $I = 0.55$ m and speed $v = 1.5$ m/s.

3.4 *The mini-ROV resistance analysis*

Based on the comparative analysis between the eigen wave patterns generated at the maximum immersion by the hydrodynamic support (Figs. 6a, b, c) and by the mini-ROV hull and hydrodynamic support system (Figs. 11a, b, c), it results that for the tested speed domain the influence of the immerged hull is very reduced in compare to the hydrodynamic support. These justifies the use of the linear hypothesis, so that the forward motion resistance of the ROV model is determined as the difference between the resistance of the ROV—hydrodynamic support system and the hydrodynamic support resistance:

$$R_{trOV} = R_{tA} - R_{tS} \qquad (1)$$

The results of the calculations are provided in Table 5. Figure 13 shows the diagram of the forward motion resistance R_{tROV} versus speed v, at the 4 immersions I tested, while Figure 14 shows the diagram of the ROV forward motion resistance, R_{tROV} in relation to the immersion I, for the 5 speed cases.

Result analysis leads to the following observations:

- the ROV forward motion resistance increases with speed, except for the minimum immersion

$I = 0.05$ m case, where a significant decrease of the forward motion resistance is seen at speeds over 1.5 m/s (Fig. 13);
- the values of the forward motion resistance at the minimum immersion $I = 0.05$ m are much higher than at the other immersions tested, which proves the negative effect of the free surface;
- in the cases of immersions ranging from 0.45 m to 0.55 m, the influence of the free surface decreases, yet maximum reductions of about 7% in the forward motion resistance are noticed with the increase of immersion, within the speed range analysed.

Considering by approximation that the variation diagram of the ROV forward motion resistance is the diagram corresponding to the maximum immersion, $I = 0.55$ m, the actual towing power P_{EROV} at v speed was calculated by means of the relation (see Table 6):

$$R_{EROV} = R_{tROV} \cdot v \tag{2}$$

Table 5. ROV resistance, R_{tROV}[N].

Immersion I[m]	Speed, v [m/s]				
	1	1.25	1.5	1.75	2
0.05	20.62	33.63	43.34	35.83	28.97
0.35	6.53	10.82	14.47	18.29	23.0
0.45	5.79	9.49	13.93	17.64	22.44
0.55	5.70	8.83	13.42	17.18	20.89

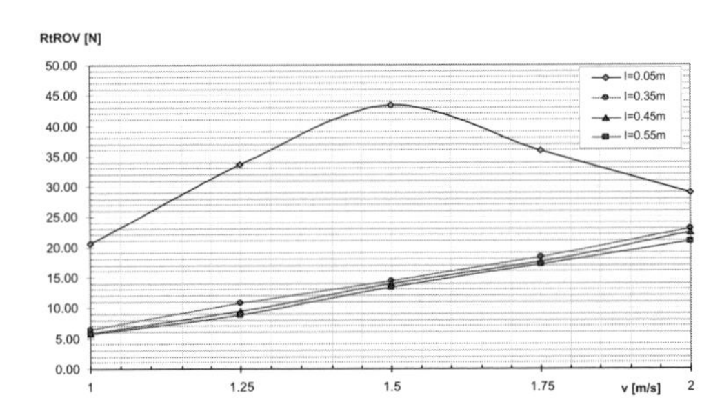

Figure 13. ROV resistance function of speed, at constant immersion (0.05, 0.35, 0.45, 0.55 m).

Table 6. ROV resistance and effective power versus speed.

v [m/s]	1	1.25	1.5	1.75	2
R_{tROV} [N]	5.70	8.83	13.42	17.18	20.89
P_{EROV} [W]	5.70	11.04	20.13	30.07	41.78

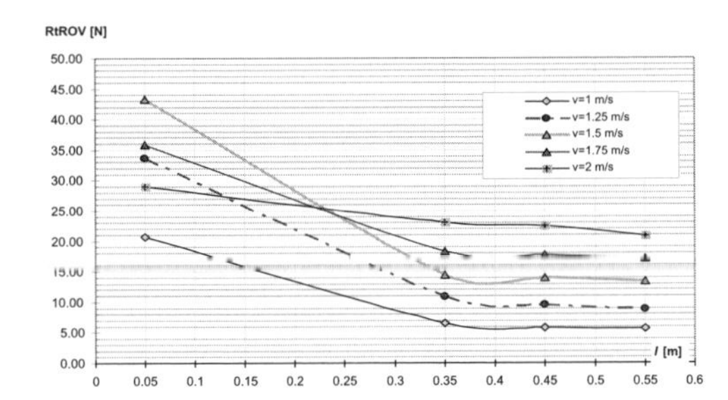

Figure 14. ROV resistance depending of the immersion, at constant speed (1, 1.25, 1.5, 1.75, 2 m/s).

4 THE MINI-ROV DRAG FORCE PREDICTION BASED ON STATISTICAL RELATIONS

In order to obtain preliminary reference values for the towing tank experimental data, this section focuses on the mini-ROV hull model drag force and power evaluation based on fluid mechanics statistical relations (Blevins, 2003; Munson et al., 2004).

The statistical approach for the min-ROV drag force and power evaluation is applied according to the following hypothesis:

- the immerse ellipsoidal body has the same dimensions and speed as the ROV vehicle (Table 1);
- the ellipsoidal shape is smooth and complete, without taking into account the ROV horizontal and vertical propulsion tubes ($d = 0$);
- the free surface influence is neglected, considering the immersion at full operation depth 30 m.

The statistical drag force and power are obtained with Equation 3, with C_x statistical coefficient from Table 7 (Blevins, 2003; Munson et al., 2004).

Table 8 shows the resulting statistical drag force and power values for the min-ROV ellipsoidal hull model, without propulsion tubes and free surface influence.

$$R_{tROV\,st} = C_x \rho_a \frac{v^2}{2} A_E [\text{N}];\; P_{EROV\,st} = R_{tROV\,st}\, v[\text{W}] \tag{3}$$

$$C_x = f\left(R_{eE}\right); R_{eE} = \frac{vL}{\upsilon}; \upsilon = \frac{\mu}{\rho_a}; L = 2a; A_E = \pi bc$$

where: $R_{tROV\,st}$ [N], $P_{EROV\,st}$ [W] are the full immerse ellipsoidal hull statistical drag force and power; C_x is the non-dimensional drag force coefficient; R_{eE} is the Reynolds number; ρ_a = 998.2 kg/m^3, $\mu = 1.002\ 10^{-3}$ Pa · s, $\upsilon = 1.004\ 10^{-6}$ m^2/s are the water density, dynamic and cinematic viscosity, for

Table 7. Statistic drag force non-dimensional coefficient for full immerse ellipsoidal body (Blevins, 2003; Munson et al., 2004).

R_{eE}	$\log_{10}(R_{eE})$	C_x	R_{eE}	$\log_{10}(R_{eE})$	C_x
$1.00\ 10^1$	1.00	8.00	$1.00\ 10^5$	5.00	0.47
$1.00\ 10^2$	2.00	1.00	$5.00\ 10^5$	5.70	0.29
$5.00\ 10^2$	2.70	0.50	$1.00\ 10^6$	6.00	0.15
$1.00\ 10^3$	3.00	0.47	$1.00\ 10^7$	7.00	0.10

Table 8. ROV resistance and effective power of the immerse ellipsoidal ROV hull, based on statistic values for the C_x drag force coefficient.

v[m/s]	R_{eE}	$\log_{10}$ (R_{eE})	C_x	R_{tROVst} [N]	P_{EROVst} [W]
1.00	4.98E + 05	5.70	0.29	9.79	9.79
1.25	6.23E + 05	5.79	0.24	12.97	16.22
1.50	7.47E + 05	5.87	0.21	15.95	23.92
1.75	8.72E + 05	5.94	0.18	18.56	32.48
2.00	9.96E + 05	6.00	0.15	20.68	41.36

t = 20°C reference tests temperature; v = 1÷2 m/s is the model speed testing range; L = 0.500 m is the model reference length; A_E = 0.06872234 [m^2] is the frontal reference area of the immerse ellipsoidal body; a, b, c [m] are the hull model main dimensions (Table 1).

5 CONCLUSIONS

Based on the experimental tests data in Section 3 and the statistic values in Section 4, the following conclusions were drawn for the mini-ROV resistance analysis:

1. The mini-ROV resistance, based on experimental tests, increases with speed for all immersions, except for the minimum immersion case when a significant decrease of the resistance is recorded for speeds higher than 1.5 m/s (Table 5, Fig. 13).
2. The decrease of the resistance recorded for the minimum immersion may be accounted for on the basis of a flow around the hull without vortices or its own braking waves, for speeds higher than 1.5 m/s (Figs. 10a, b, c).
3. The negative effect of the free surface on the mini-ROV hull resistance, determined by experimental tests, decreases when the immersion of the ROV-hydrodynamic profile system increases. The free surface effect explains the highest ROV resistance for the minimum immersion case (I = 0.05 m).

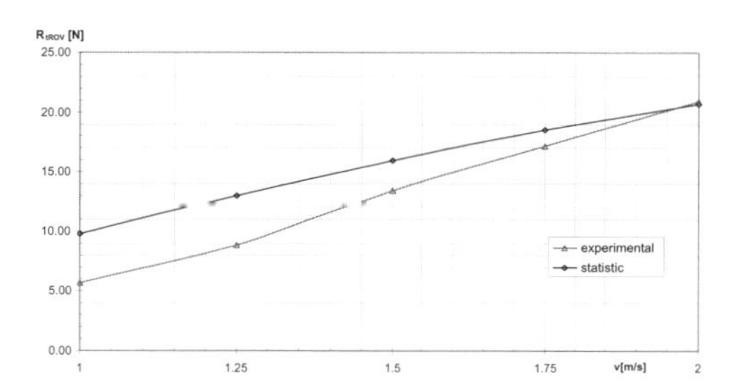

Figure 15. Statistical and experimental diagrams of the mini-ROV resistance, at maximum immersion.

4. The mini-ROV resistance based on experimental tests (Table 6) is 14.13 N for the design speed of 1.54 m/s (3 Knots). The effective power calculated at design speed is 21.76 W.
5. The mini-ROV resistance estimations, based on fluid mechanical statistical relations of an ellipsoidal immerse body (Blevins, 2003; Munson et al., 2004), without propeller tubes and free surface influence (Table 8) is 16.43 N for the design speed of 1.54 m/s (3 Knots). The statistically effective power at design speed is 25.35 W.
6. Comparing the experimental results and the statistical predictions (Fig. 15) results in the fact that the mini-ROV resistance based on statistical relations is overestimated by 16.28% for the reference design speed of 1.54 m/s (3 Knots).
7. For 2 m/s speed the ROV experimental and statistical values are almost de same, but the resistance versus speed curves slope are different (Fig. 15).
8. The extreme eigen wave pattern generated at the hydrodynamic support stern in low speed cases induces a significant increase of the support initial resistance (Figs. 6 & 11). Based on the superposition hypothesis, it results that the ROV hull resistance has a significant decrease for lower speed cases. These results are justifying the high differences between experimental and statistical resistance values for the lower speed domain (Fig. 15).

ACKNOWLEDGEMENTS

The work has been performed within the scope of the projects for mini-ROV submerged vehicles design and fluid flow analysis on hydrodynamic profiles financed by the Romanian Education and Research Ministry, under contracts CNMP-PNII-P4/3401/12-116/2008-2011 and CNCSIS PNII-ID-790/2008-2011.

REFERENCES

Batchelor, G. 2000. *An introduction to fluid dynamics*, Cambridge: Cambridge University Press.

Blevins, Robert, D. 2003. *Applied fluid dynamics handbook*. Malabar, Florida: Krieger Publishing Company.

Christ Robert, D., Wernli, Sr. & Robert, L. 2007. *The ROV manual. A user guide to observation-class remotely operated vehicles.* Oxford: Butterworth Heinemann.

Cussons. 2009. Marine hydrodynamic research. Cussons Technology Ltd. Manchester.

Domnisoru, L., Dumitru, D. & Mocanu, C. 2010. *Initial structural design of an submerged vehicle, made of composite materials*. Galati: Galati University Press.

GL. 2011. Germanischer Lloyd's Rules. Non-metallic materials. Remotely operated underwater vehicles. Hamburg.

Griffiths, G. 2003. *Technology and applications of autonomous underwater vehicles*. London: Taylor & Francis.

Munson, Bruce, R., Young, Donald, F. Young & Okishi, Theodore, H. 2004. *Fundamentals of fluid mechanics*. New York: John Wiley & Sons Publications.

Ross, C. 2006. *A conceptual design of an underwater vehicle*. Ocean Engineering, Vol. 33, pp. 2087–2104, 2006.

Valencia, R.A., Ramirez, J.A., Gutierrez, L.B. & Garcia, M.J. 2008. Modelling and simulation of an underwater remotely operated vehicle for surveillance and inspection of port facilities using CFD tools. Proceedings of the ASME 27th International Conference OMAE 2008, Estoril.

Sustainable Maritime Transportation and Exploitation of Sea Resources – Rizzuto & Guedes Soares (eds)
© 2012 Taylor & Francis Group, London, ISBN 978-0-415-62081-9

Numerical optimization of a hull form with bulbous bow

D. Matulja & R. Dejhalla
Department of Naval Architecture and Ocean Engineering, University of Rijeka, Rijeka, Croatia

ABSTRACT: A genetic algorithm technique has been applied to deal with the hydrodynamic optimization of a hull form with bulbous bow. A non interactive procedure has been developed to minimize the objective function, related to the flow past a ship moving with steady forward speed in calm water. The paper discusses the application of the genetic algorithm coupled with a potential flow solver to the optimization of a hull form with bulbous bow from the hydrodynamic point of view. A Bézier surface, defined by several bulb geometrical parameters, has been used to allow the forebody modifications. According to the geometrical constraints, a new optimized hull form has been automatically obtained. The obtained ship hull form modification has been presented and analyzed in order to demonstrate the effectiveness and validity of the developed procedure.

1 INTRODUCTION

The prediction of ship resistance is normally based on results from model tests at towing tanks. Even with accurate CFD simulations which are increasingly performed in ship design, these model tests are still a very important method in determining and verifying the ship resistance and power requirements. However, different powerful computational tools can be used for preliminary selection of variants before testing, as well as to study flow details to gain insight into how a ship hull form can be improved.

The ship hull form design is a complex process where compromises must be made among various and often conflicting requirements. The problem can be formulated as determination of a set of design variables subjected to certain relations between variables and restrictions of these variables. In general, many factors must be considered and not all of them are hydrodynamic in nature. But, from the hydrodynamic point of view, the most interesting optimization contribution is to minimize the ship resistance, or the ship resistance components. Many interesting works on hull form optimization have been presented through the years (Hsiung 1981, Janson & Larsson 1996, Peri et al., 2000, Percival et al., 2001, Campana et al., 2006, Wilson et al., 2011). In addition to that, different approaches to hull form optimization using genetic algorithm have been considered (Day & Doctors 1997, Dejhalla et al., 2001, Dejhalla et al., 2002).

Numerical optimization is a well established mathematical field and there are numerous references on the theory and application of numerical optimization tools. Such optimization is usually much faster and cheaper than experiments and it offers more insight into flow details.

2 THE OPTIMIZATION PROBLEM FORMULATION

Optimization is a procedure which allows finding the best solution within a limited or unlimited number of choices. Many optimization problems may be generally formulated as the problems of minimizing an objective function $f(\varphi)$, of a number of variables ξ_1, ξ_2, ξ_3, ..., ξ_n subject to a group of constraints that can be formulated as equalities or inequalities.

The solution of the optimization problem calls for the formulation of a suitable optimization procedure. Therefore, the three-dimensional potential flow solver (Dejhalla 1999) and the genetic algorithm (Goodman 1996) have been coupled to build a procedure for the bulbous bow optimization. The linear potential flow solver is an in-house made computer code that has been comprehensively validated. The FORTRAN program is based upon the boundary element method, i.e., a well known Rankine source method and basically it is the Dawson's method (Dawson 1977). The forces are computed by integration of pressure forces over the hull. Sinkage and trim effects have not been included in the procedure. Since a potential flow solver has been used as computation tool, the wave resistance has been selected as the objective function for the presented study. The wave resistance is one of the most important resistance components, generally contributing from 20 to 80% of the total resistance of a ship sailing in still water. Experience has shown that this resistance component is sensitive to modifications in the hull form design, and reductions of the wave resistance can often be obtained without any important sacrifice in displacement volume.

The optimization procedure starts from a specified basis hull form, by calculating the flow past the hull form and evaluating the wave resistance. The genetic algorithm creates an initial population of specified number of individuals (hull shapes), randomly generated within upper and lower bounds for the design variables. The bow geometry modification algorithm is an integral component of the optimization procedure. It allows to obtain the changes of the bulb shape and to automatically remesh the fore part of the hull form for each design case.

The optimization of a ship hull form, from a hydrodynamic point of view, forms a non-linear optimization problem. The objective function and the constraints are non-linear functions of the design variables. The optimization problem may include multiple minima and the general optimization method cannot be expected to find a global minimum for this case. The genetic algorithm is very effective at finding optimal or near optimal solutions to a wide variety of problems because it does not impose many of the limitations required by traditional methods (Goldberg 1989). Genetic algorithms process populations rather than individual solutions and as a consequence, genetic algorithms are extremely desirable if the search space is too large to be exhaustively searched, or has multiple local optima with the topology unfamiliar to the user. Since genetic algorithms do not rely on gradient information, they are unaffected by discontinuities and are not as likely to become trapped at local optima.

Therefore, the ship hull form optimization problem has been identified as one that can be successfully solved using the genetic algorithm. In every generation, each individual is evaluated using a fitness function and assigned a fitness value. The fitness of an individual is determined calling the potential flow solver. Based on their relative fitness values, individuals in the current population are selected for reproduction. Based upon genetic and evolutionary principles, the genetic algorithm repeatedly modifies the population of artificial individuals. Generating a new generation, individuals in its current population are improved by performing genetic algorithm operators. The process continues until the specified number of generations is attained and acceptable or the best possible solution evolves. The developed procedure is fully automatic and no user interference is needed during the optimization.

The three most important aspects of using genetic algorithm are:

- definition of the objective function,
- definition and implementation of the genetic representation,
- definition and implementation of the genetic operators.

For the presented optimization problem the real coding has been chosen.

To transform a minimization problem to a maximization problem needed for the genetic algorithm procedure, it is necessary to map the objective function to a fitness function form (so called “raw fitness”) through one or more mappings. The following transformation has been used:

$$\text{Fitness} = \begin{cases} C_{\max} - f(\xi) & \text{when } f(\xi) < C_{\max} \\ 0 & \text{otherwise} \end{cases}. \quad (1)$$

The value of parameter $C_{\max}$ is taken as input coefficient to avoid negative fitness values and its value should be greater than the expected largest value of objective function in the simulation. Often, the raw fitness must be scaled in order to help the genetic algorithm maintain diversity between very similar individuals.

The results presented in the study have been carried out by means of genetic algorithm employing:

- the linear scaling of the raw fitness
- the stochastic uniform sampling as selection operator
- the two-point crossover as crossover operator
- the multi-bit mutation as mutation operator.

In addition, the following genetic algorithm parameters have been adopted:

- String length = 7
- Crossover probability $p_c = 0.5$
- Mutation probability $p_m = 0.3$
- Population size = 40
- Number of generations = 50

Through experimentation with the optimization procedure it has been decided to keep this very complex problem as simple as possible. As geometrical constraints, the design waterline and the stem profile were kept the same as in the basis form, while the shape of the fore part was allowed to change only by altering the y—coordinates of the chosen points.

In this manner, $y_1, y_2, y_3, \ldots, y_n$ form the set of design variables, taken as variables $\xi_1, \xi_2, \xi_3, \ldots, \xi_n$ introduced previously. The number of design variables n treated in the optimization procedure must remain within some reasonable range. On the other hand, the grid used in computation must capture the ship geometry appropriately in order to resolve changes in the flow with sufficient resolution.

3 BÉZIER SURFACE

In order to allow the modifications of the bulbous bow defined by several parameters (Kracht 1978), the fore part has been modeled with a Bézier surface because of its possibility to create a grid of desired density from a relatively low number of points. The fore part has been defined as the part between the foremost part of the bow and section at 10% L_{PP} aft of the fore perpendicular. The Bézier surface is a species of mathematical spline used in computer graphicsand computer-aided design (Salomon 2006). It consists of a patch defined by a set of control points, and it enables to generate a mesh of quadrilaterals of any required density. The Bézier surface passes through the four points which define the vertices of the patch, while it does not generally pass through the other control points. It is rather attracted by those points, creating a smooth surface which can adequatly match the hull form surface. The obtained surface is mathematically convinient since it provides any required number of quadrilaterals and keeps good continuity properties.

A Bézier surface of order (n,m) needs a grid of $(n+1)(m+1)$ control points $P(i,j)$. If $p(u,v)$ is a function of the parametric coordinates (u,v), then the Bézier surface can be defined as a parametric surface given by:

$$p(u,v) = \sum_{i=0}^{n} \sum_{j=0}^{m} B_i^n(u) B_j^m(v) \cdot p_{i,j} \tag{2}$$

In the above expression the parts $B_i^n(u)$ and $B_j^m(v)$ are Bernstein polynomials, expressed by:

$$B_i^n(u) = \binom{n}{i} u^i (1-u)^{n-1} \tag{3}$$

and

$$B_j^m(v) = \binom{m}{j} v^j (1-v)^{m-1} \tag{4}$$

where:

$$\binom{n}{i} = \frac{n!}{i!(n-i)!}, \tag{5}$$

and

$$\binom{m}{j} = \frac{m!}{j!(m-j)!} \tag{6}$$

are the binomial coefficients.

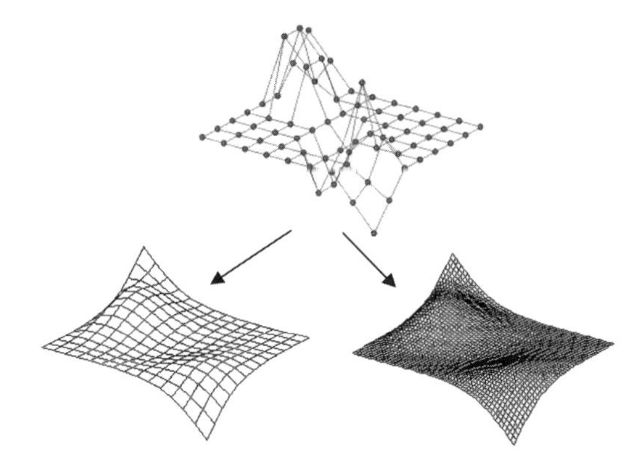

Figure 1. The Bézier surface grid.

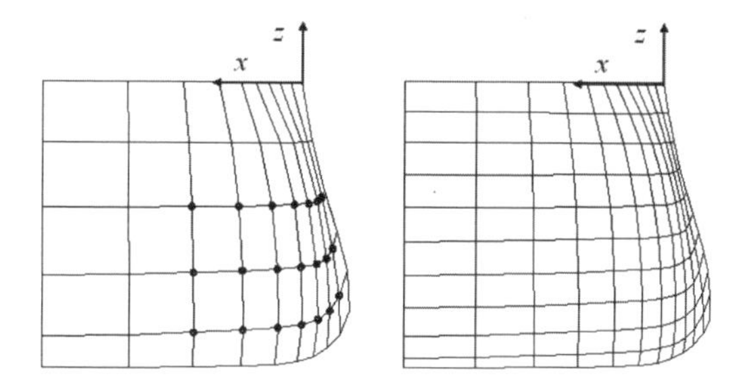

Figure 2. The control points and the fore part meshed by the Bézier surface.

The density of the mesh is defined by changing the parameters n and m.

In Figure 1 (top) the initial set of control points ($n = 9$, $m = 8$) is given.

On the left, the Bézier surface is defined by $n = 15$ and $m = 14$, while on the right the parameters are $n = 55$ and $m = 54$. In the given case, the ship fore part has been defined by 21 control points taken as optimization parameters, Figure 2.

To keep the stem profile and waterline unchanged, the points on the keel line and the points on the design waterline were kept fixed, as well as the points on the stem.

4 RESULTS

The optimization procedure has been applied to the S175 containership hull form taken as the basis hull form. The principal ship particulars are given in Table 1. The fore part of the hull form has been optimized for a single speed corresponding to the Froude number 0.22. The Froude number is based on L_{PP}. The wave resistance R_W is calculated as:

$$R_W = 0.5\rho C_W U^2 S, \tag{7}$$

using the wave resistance coefficient C_W obtained by integrating the x-components of the pressure forces acting on the submerged portion of the hull form. In (7) ρ represents the water density in kg/m^3, U *is* the ship speed in m/s and S is the wetted surface in m^2.

The computational domain has been represented by a regular grid with 469 panels on the hull surface, and 1980 panels on the free surface, Figure 3.

The extension of the free surface has included the area from $0{,}3 \cdot L_{WL}$ upstream, $0{,}8 \cdot L_{WL}$ downstream and $0{,}80 \cdot L_{WL}$ in the transverse direction.

The evolution history of wave resistance for the evaluated hull forms is presented in Figure 4.

Table 1. Principal particulars of the basis hull form (S175 containership).

Length between perpendiculars	175.0 m
Breadth	25.4 m
Draught	9.5 m
Displacement	24779.7 t
Block coefficient	0.572
Midship coefficient	0.97

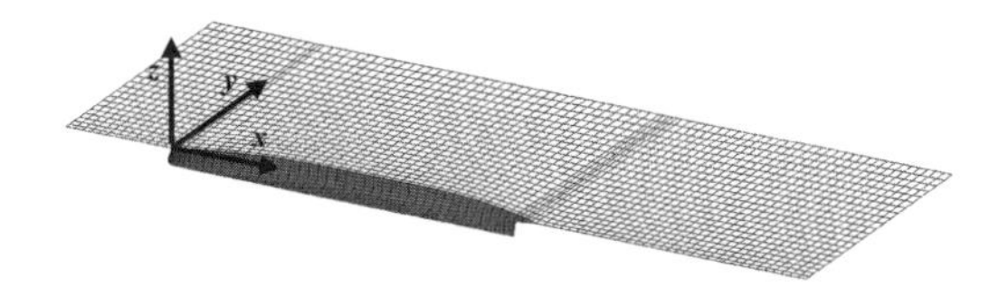

Figure 3. Computational domain.

In the figure each dot represents the absolute value of the wave resistance for each individual (hull shape) evolved during the optimization. Having 40 individuals through 50 generations, a total of 2000 evaluation cycles have been executed before the optimal hull form emerged. The solid line represents the optimal solutions front over the generations. It is formed by minimal values of the wave resistance obtained in the simulations. As demonstrated, generation by generation a set of solutions has converged as the given number of generations is gained.

The optimal solution i.e. the hull form with the lowest value of the wave resistance has been identified as the 37th individual in the 47th generation.

To demonstrate the effectiveness of developed procedure, the maximum, average and minimum fitness values over generations from the genetic algorithm statistical report are presented in Figure 5.

The maximum fitness represents the raw fitness of the best evolved individual (hull shape), while the minimum fitness is the raw fitness of the worst evolved individual in the generation. The average fitness is obtained as an arithmetic mean of the raw fitness for individuals in generation. Shown fitness function values are dimensionless, normalized by the fitness of the best evolved individual, giving the extreme value of 1.0 for the hull form with the lowest wave resistance. The fitness function values show a standard genetic algorithm behavior—faster improvement of the best individual and steady increase of the average fitness.

For the minimum fitness values, certain oscillations due to mutations and crossovers are

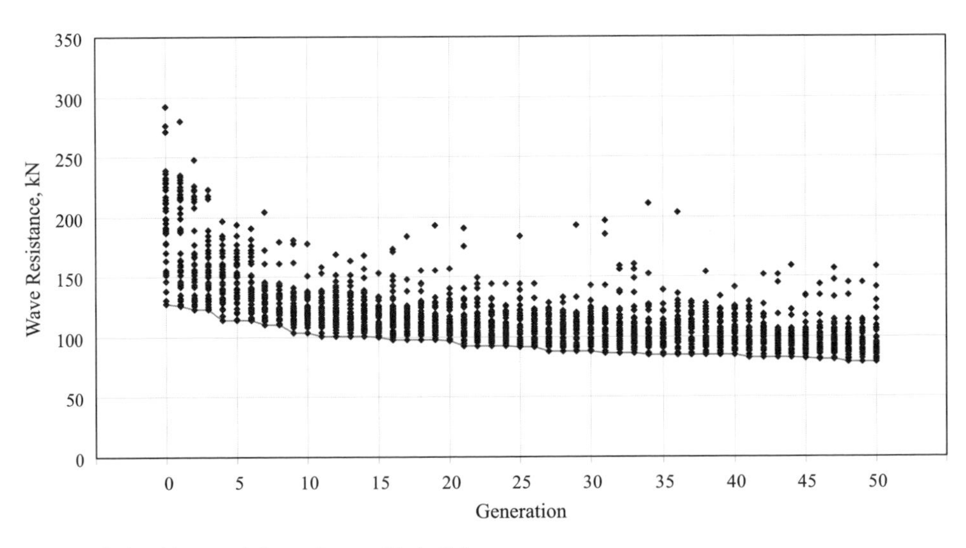

Figure 4. Evolution history of the optimum ship hull form.

pronounced. Despite the random choice of an initial population, once the genetic algorithm starts, it finds relatively quickly a good performance.

As the run continues, a further improvement is found in both maximum and average fitness. Toward the end of the run, a form of convergence is observed.

The optimized bulb shape is compared to the initial bow shape in Figures 6 and 7. The resulting bow shape is entirely dictated by the hydrodynamic behavior associated with the changes in the bow sections shape. The results of the optimization procedure are evident.

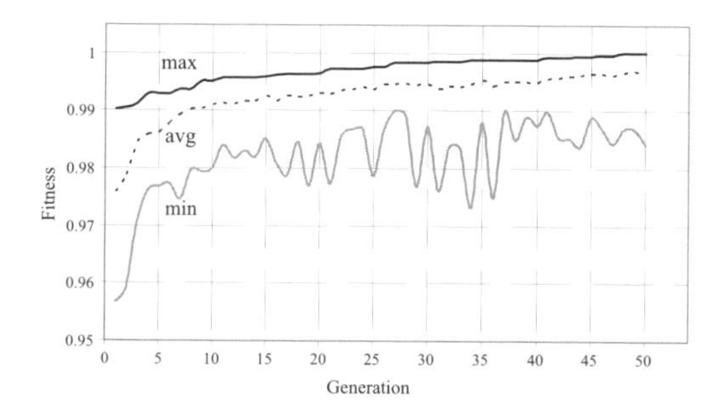

Figure 5. The fitness value over the generations.

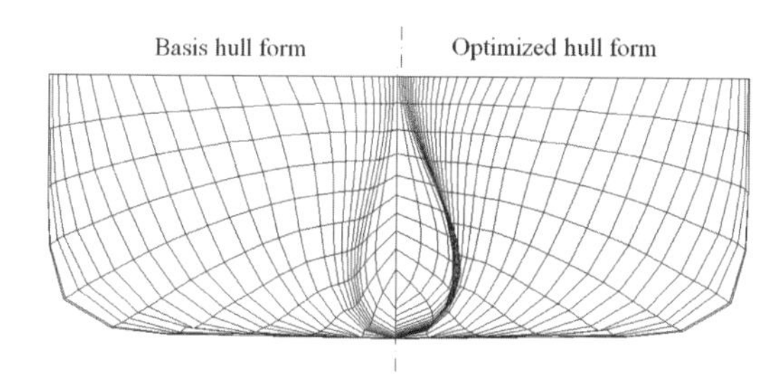

Figure 6. Panelized basis and optimized hull grid.

Figure 8 presents the comparison for the wave profile for the basis hull form and optimized hull form advancing at $Fr = 0.220$. Both the hull length and the wave elevations are normalized by $L_{PP}/2$.

Wave elevations η are plotted for the collocation points of the panels next to the centerline ($y/L_{PP} = 0.0$), when $2x/L_{PP} < 0.0$ or $2x/L_{PP} > 2.0$, and for the panels next to the hull when $0 < 2x/L_{PP} < 2.0$. A certain wave elevation reduction can be noticed near the fore perpendicular. Since only the fore part has been changed, the wave profile does not show any changes aft of the midship.

Finally, although the problem is treated as a single point design problem, the additional numerical tests have been performed with obtained hull form for Froude numbers ranging from 0.175 to 0.28.

The predicted wave resistance coefficients are summarized in Table 2 and Figure 9.

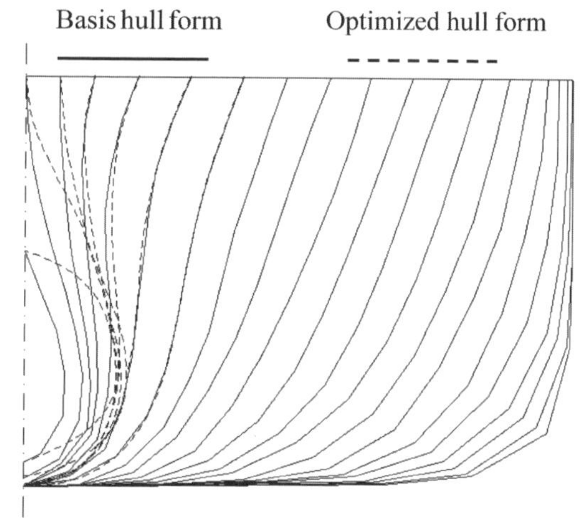

Figure 7. Comparison of the original and optimized sections.

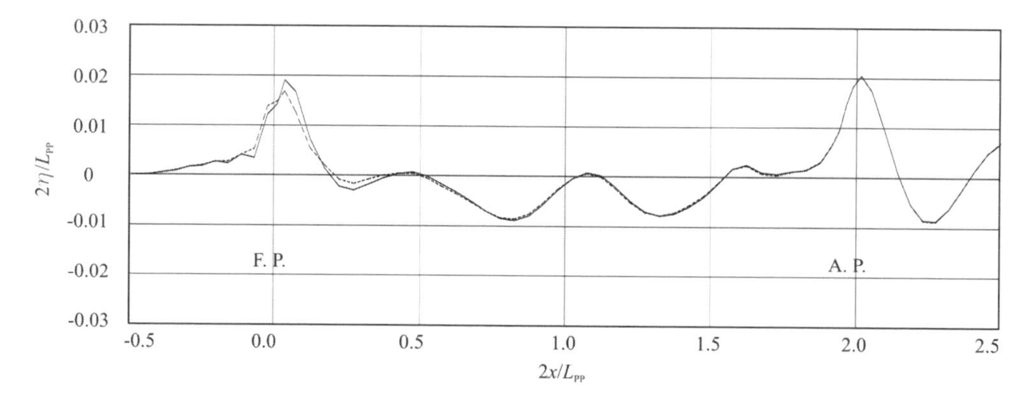

Figure 8. The wave profile at $Fr = 0.22$. Solid line—basis hull form. Dashed line—optimized hull form.

Table 2. Comparison of the results.

Fr	Basis hull form		Optimized hull form	
	C_W	R_W, kN	C_W	R_W, kN
0.175	$0.062 \cdot 10^{-3}$	9.300	$0.106 \cdot 10^{-3}$	15.920
0.200	$0.112 \cdot 10^{-3}$	21.810	$0.082 \cdot 10^{-3}$	16.118
0.220	$0.405 \cdot 10^{-3}$	95.739	$0.329 \cdot 10^{-3}$	77.798
0.240	$0.461 \cdot 10^{-3}$	129.739	$0.379 \cdot 10^{-3}$	106.670
0.260	$0.623 \cdot 10^{-3}$	205.882	$0.538 \cdot 10^{-3}$	177.982
0.280	$0.921 \cdot 10^{-3}$	353.063	$0.829 \cdot 10^{-3}$	317.829

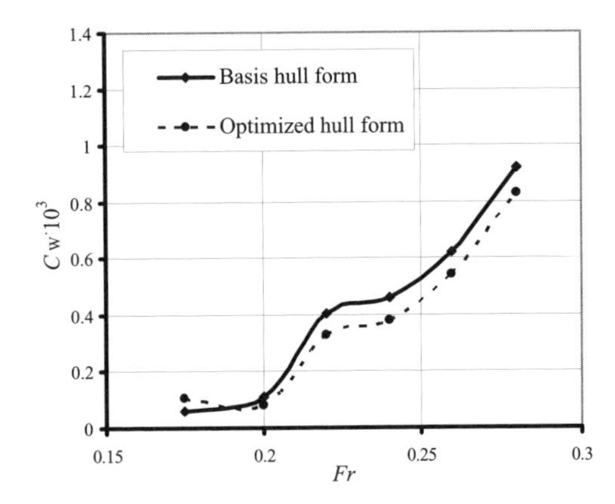

Figure 9. Comparison of wave resistance coefficients versus Froude number.

The wave resistance coefficients are calculated for the panelized version of the hull forms. It is evident that the reduction of the wave resistance coefficient has been achieved over a wide range of Froude numbers.

5 CONCLUSION

At present time, the most common application of potential flow solvers is the analysis of fore part of the ship hull form. The potential flow solvers are additionally attractive due to their low computational costs.

A numerical procedure for the optimization of a ship hull form with the bulbous bow from a hydrodynamic point of view is established. The procedure is based on the genetic algorithm and a linearized potential flow solver. The method uses wave resistance as a single objective function to find the optimal hull.

The study has shown that the developed optimization procedure can be successfully applied to the optimization of the fore part of the ship hull and used as a valuable method to favorably modify a ship hull.

An additional work is intended to be done for the application of the developed optimization procedure. In such further work, instead of wave resistance, other objective functions, as well as design variables and geometrical constraints, which inevitably arise from resistance or other requirements, will be included.

REFERENCES

Campana, E.F., Peri, D., Tahara, Y. & Stern, F. 2006. Shape optimization in ship hydrodynamics using computational fluid dynamics. *Computer methods in applied mechanics and engineering* 196: 634–651. Elsevier.

Dawson, C.W. 1977. A Practical Computer Method for Solving Ship-Wave Problems. *Proceedings of the 2nd International Conference on Numerical Ship Hydrodynamics*: 30–38, Berkeley.

Day, A.H. & Doctors, L.J. 1997. Resistance optimization of displacement vessels on the basis of principal parameters, *Journal of Ship Research* 41: 249–259, Jersey City.

Dejhalla, R. 1999. *Numerical modeling of flow around the ship hull*, Ph.D. Thesis. Rijeka: Faculty of Engineering, University of Rijeka.

Dejhalla, R., Mrša Z. & Vuković, S. 2001. Application of Genetic Algorithm for Ship Hull Form Optimization, *International Shipbuilding Progress* 48: 117–133, Delft.

Dejhalla, R., Mrša, Z. & Vuković, S. 2002. A Genetic Algorithm Approach to the Problem of Minimum Ship Wave Resistance, *Marine Technology* 39: 187–195, New Jersey.

Goldberg, E.D. 1989. *Genetic Algorithms in Search, Optimization, and Machine Learning*. New York,: Addison -Wesley Publishing Co.

Goodman, E.D. 1996. *An Introduction to GALLOPS - The "Genetic Algorithm Optimized for Portability and Parallelism" System, Release 3.2*, Technical Report #96-07-01, East Lansing: Michigan State University.

Hsiung, C.C. 1981. Optimal Ship Forms for Minimum Wave Resistance, *Journal of Ship Research* 25: 95–116, Jersey City.

Janson, C.E. & Larsson, L. 1996. A Method for the Optimization of Ship Hulls from a Resistance Point of View. *21st Symposium on Naval Hydrodynamic.* Trondheim.

Kracht, A.M. 1978. Design of bulbous bows. *SNAME Transactions* 86: 197–271, Jersey City.

Percival, S., Hendrix, D. & Noblesse, F. 2001. Hydrodynamic Optimization of Ship Hull Forms, *Applied Ocean Research* 23: 337–355. Elsevier.

Peri, D., Rossetti, M. & Campana, F. 2000. Improving the hydrodynamic characteristics of ship hulls via numerical optimization techniques. *Proceedings of IX Congress IMAM 2000*: 16–24. Ischia.

Salomon, D. 2006. *Curves and surfaces for computer graphics*. Northridge: Springer Science+Business Media, Inc.

Wilson, W., Hendrix, D. & Gorski, J. 2011. Hull Form Optimization for Early Stage Ship Design, *Naval Engineers Journal* 122: 53–65. Alexandria.

Sustainable Maritime Transportation and Exploitation of Sea Resources – Rizzuto & Guedes Soares (eds)
 ISBN 978-0-415-62081-9

Numerical prediction of the resistance characteristics of catamaran cargo ships

G.D. Tzabiras, G.N. Zaraphonitis & S.P. Polyzos
Laboratory for Ship and Marine Hydrodynamics, National Technical University of Athens, Greece

ABSTRACT: The present work is concerned with the calculation of the resistance of catamaran cargo ships by applying CFD tools that have been developed at LSMH of NTUA. A non-linear potential flow code is applied to calculate the free-surface geometry, while a RANS solver calculates the viscous flow underneath this surface. Both methods are extended to calculate the flow past catamaran vessels. The non-linear potential solver adopts an iterative method for the calculation of the free-surface. The viscous flow method solves the RANS equations by applying the finite volume method in a computational domain where the boundary conditions are calculated from the potential flow solution. The computational domain comprises three blocks while a successive grid refinement technique is applied. Turbulence is simulated be means of the k-ε model. In order to validate the method, numerical resistance results at model scale, are compared to experiments conducted in the towing tank of LSMH.

1 INTRODUCTION

The development of Computational Fluid Dynamics (CFD) codes provides nowadays a powerful tool for the design of effective hull forms. In complex flow situations they may offer valuable information which cannot be attained by experiments. A typical example is the design of appendages, where the flow is mostly affected by viscous effects and solutions at full scale are needed in order to optimize the local geometry. However, in such difficult cases, the common problem of existing methods is the high computing time required to achieve accurate solutions, e.g., (Carrica et al., 2010). A significant part of the effort is devoted to the construction of the grids while the solver of Reynolds Averaged Navier-Stokes (RANS) equations has also to calculate the unknown free-surface boundary.

On the other hand, during the preliminary design stage, a naval architect has usually to examine a variety of hull forms in order to decide about the optimum. Therefore, in this case, a fast tool is needed in order to perform resistance calculations with as far as possible accurate results. In this respect, the development of the so-called hybrid methods, (Huan & Huang 2007), (Raven & Starke 2002), (Tzabiras & Kontogiannis 2010), may be considered as a strong alternative with regard to full viscous solutions. These methods use a potential flow solver to compute the free-surface geometry and then, they solve the RANS equations underneath a specified boundary. Consequently, the computing time is drastically reduced, provided also that a restricted computational domain is examined with boundary conditions taken from the potential solution. Besides, the relevant procedures may apply very fine grid resolutions which allow for obtaining accurate results.

Naturally, the efficiency of a hybrid method depends mainly on the accurate calculation of the free-surface. The results are influenced when strong viscous phenomena affect the wave formation about the hull. Characteristic cases are the wave breaking and the flow recirculation around a partially submerged transom stern. While the former seems not to cause significant problems in many cases, the submergence of a transom stern may affect seriously the resistance calculations because the potential flow cannot simulate separation. The purpose of the present work is to study two alternative ways in order to model a related situation: the treatment of the stern as 'dry' or its numerical substitution with a cruiser-type stern by changing slightly the local geometry. The case of a catamaran is examined and numerical results for the total resistance at model scales are compared to experimental data near the design speed of the vessel.

2 METHOD

The numerical method that is employed to calculate the flow about catamarans is based on two codes. Both of them have been developed in the

Laboratory for Ship and marine Hydrodynamics (LSMH) of the National Technical University of Athens (NTUA). The first one solves the non-linear potential flow about the body and calculates the free-surface iteratively on which both the dynamic and kinematic conditions are satisfied, Tzabiras (2008). This surface is introduced as a fixed boundary to calculate the viscous flow underneath by the second code which solves the RANS equations.

In the potential code, the free-surface and the solid boundary are covered by quadrilateral elements to solve the Laplace equation according to the classical Hess & Smith (1968) method. Since the non-linear problem is faced, an iterative Lagrangian procedure is adopted in conjunction with an Eulerian solution of the vertical momentum equation. Originally it was developed to handle cruiser-type and dry transom sterns as well as to take into account sinkage and trim corrections, (Tzabiras et al., 2009). In the present investigation, it has been extended to compute the flow about a catamaran at steady forward speed, (Fig. 1). Since the flow has one symmetry plane only one hull is taken into account and the computational domain is extended up to the ship's center plane. In addition, the case of a partially submerged transom stern is also considered by introducing panels on the transom up to the free-surface line and following the same fundamental procedures.

The numerical solution of the RANS equations is performed according to the finite volume approach. An orthogonal curvilinear co-ordinate system is employed for all transport equations which are solved in a partially orthogonal grid, Tzabiras (2004). The grid generation is based on the conformal mapping of the ship transverse sections onto the unit circle, e.g., Tzabiras & Kontogiannis (2010). Turbulence is simulated by two-equation

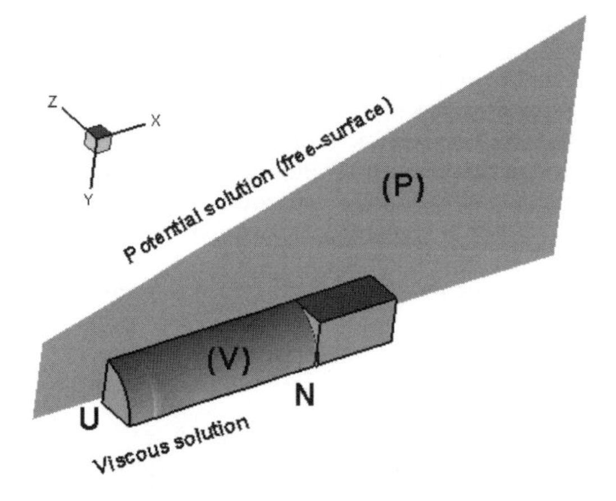

Figure 1. Computational domain for the potential (P) and RANS (V) calculations.

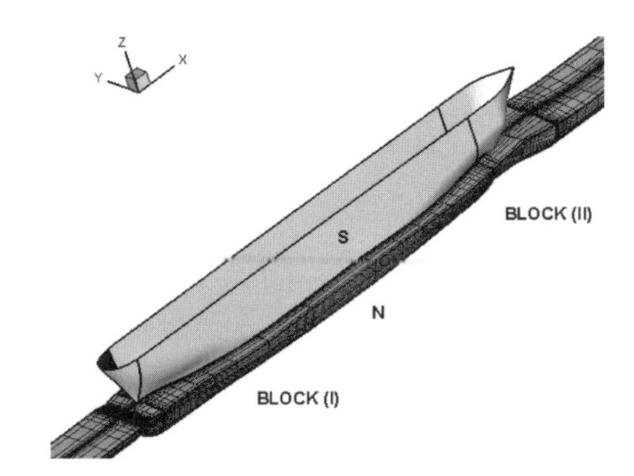

Figure 2. Two block grid arrangement for viscous calculations about a hull with cruiser-type stern.

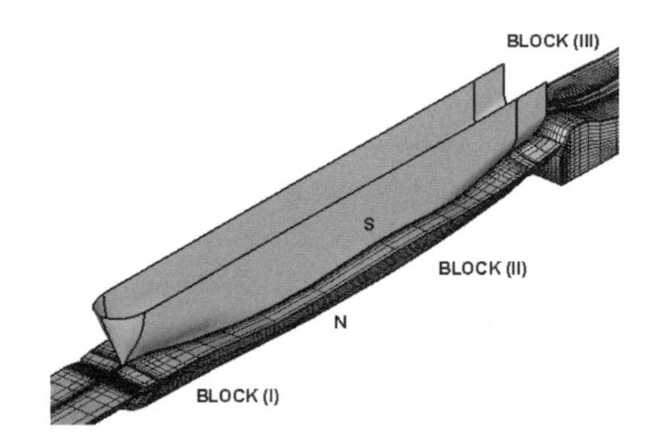

Figure 3. Three block grid arrangement for viscous calculations about a hull with transom stern.

models, i.e., the classical k-ε model of Launder & Spalding (1974) or the SST model of Menter (1993). The convective terms are approximated by first or second order TVD schemes. In order to apply fine grid resolutions the computational domain is separated in two or three blocks, (Figs. 2, 3), corresponding to cruiser or dry transom sterns. In each block a marching solution of the momentum and turbulence model equations is followed, while the pressure field is calculated by solving a fully 3D pressure correction equation according to a SIMPLE variant, Tzabiras (2004). Since the domain is restricted about the hull, the velocity components and the pressure values on the inlet and the external boundary N are evaluated through the potential flow solution. On the solid boundary S the method of wall functions or a direct solution can be applied, e.g., Tzabiras & Prifti (2001).

3 TEST CASES

3.1 *Experiments*

In order to validate the numerical method, a series of experiments were conducted and the results were compared to those of the corresponding numerical tests. All experiments were performed in the towing tank of the LMSH. The dimensions of the towing tank are 91 m (effective length), 4.56 m (width), and 3.00 m (depth). The towing tank is equipped with a running carriage that can achieve a maximum speed of 5.2 m/s.

The catamaran hull under consideration features a bulbous bow and a transom stern with a stern bulb and a flat seating for a podded propeller. Its main particulars are given in Table 1. A body plan of one demi-hull is presented in Figure 4 while a perspective view of the stern is given in Figure 5.

For the experiments scaled models of both demi-hulls were constructed at a scale ratio of 1/24. The tests were conducted in calm, fresh water, while its temperature was measured to be 15.4°C. Following Froude's assumption, experiments were conducted keeping the Froude number (Fn) equal between model and full scale. This resulted in a speed range of 1.05 to 2.10 m/s at model scale, corresponding to speeds of 10 to 20 kn at full scale. During the experiments the model was free to heave and pitch and the corresponding values were measured along with the total resistance. The total resistance coefficient C_T was then calculated:

$$C_T = \frac{R_T}{\frac{1}{2}\cdot \rho \cdot WS \cdot V^2} \tag{1}$$

where R_T = total resistance; ρ = water density; WS = wetted surface; V = model's speed.

Table 1. Main particulars.

			Ship	Model
Length between perpendiculars	L_{BP}	m	75.200	3.133
Waterline length	L_{WL}	m	78.155	3.257
Demi-hull beam	B_{Hull}	m	7.150	0.298
Overall beam	B_{OA}	m	21.018	0.876
Distance between demi-hulls	S	m	13.868	0.578
Design draught	T	m	4.500	0.188
Overall wetted surface	WS	m^2	1839.32	3.193
Overall volume of displacement	∇	m^3	2908.03	0.210
Overall displacement	Δ	tn	2981	0.210

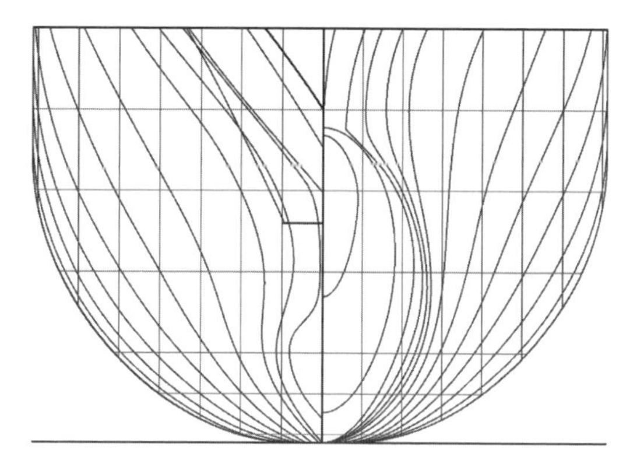

Figure 4. Body plan of the hull tested.

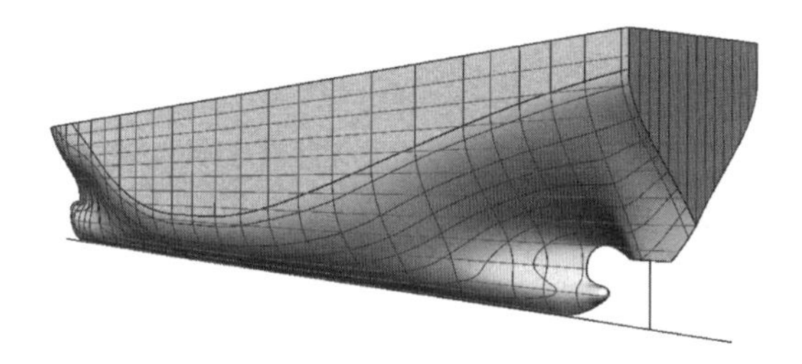

Figure 5. 3D rendering of the stern of the tested hull.

The total resistance is then decomposed into the skin friction resistance C_F and the residual resistance C_R:

$$C_T = C_F + C_R \tag{2}$$

For the calculation of C_F, the ITTC-57 formula is employed:

$$C_{F,ITTC} = \frac{0.075}{\left(\log_{10} Rn - 2\right)^2} \tag{3}$$

where Rn = Reynold's number. In equation (1) the wetted surface is considered equal to the wetted surface at rest, for the specific loading condition. The resulting values for C_T and C_R are presented in Figure 6. During the experiments the transom was observed to be partially submerged, even at the highest speed.

3.2 *Potential calculations*

A number of numerical tests were conducted for the above mentioned hull, using the potential flow solver described in paragraph 2. The draught was set equal to the one used in the experiments and the dynamic heave and trim were calculated by the method. The numerical experiments were

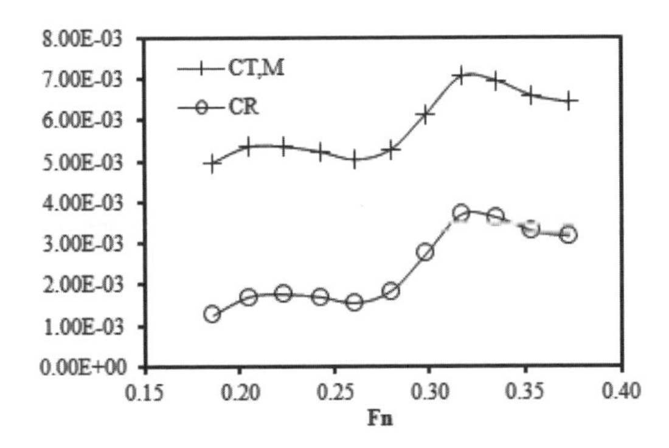

Figure 6. Experimental measurements for the total (CT, M) and residual (CR) resistance coefficients at model scale.

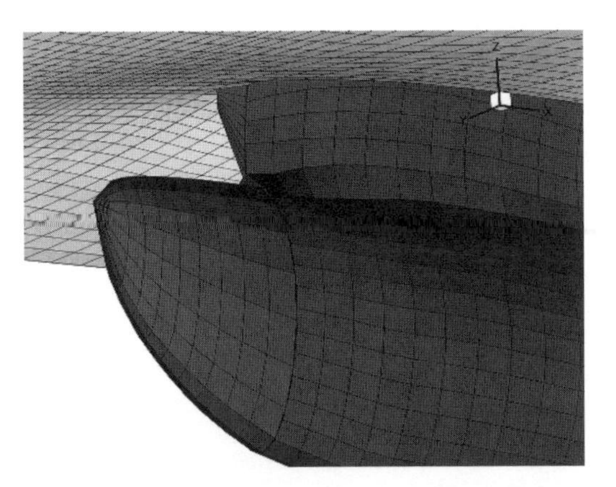

Figure 7. Panel arrangement at the bow of the hull.

conducted at two values of Fn corresponding to ship's speeds of 14 and 15 kn.

Since the transom during the experiments was always partially submerged it was decided to use three different configurations for the stern. First the stern was extended abaft the transom in order to form a cruiser-type stern (Figs. 8, 11). Then a dry transom configuration was tested, were the free surface panels are forced to follow the shape of the transom (Figs. 9, 12) and finally a partially submerged transom configuration were panels on the transom are added and the free-surface is left free to rise until it satisfies the kinematic and dynamic conditions (Figs. 10, 13). The number of panels used for each of the above cases is presented in Table 2.

In all cases, the computational domain extended $1.6 \times L_{BP}$ upsrteam the bow and $2.5 \times L_{BP}$ downstream the stern. In the transverse direction it extended from the ship's centerline up to $1.6 \times L_{BP}$ at the upsrteam end and at the $2 \times L_{BP}$ downstream end.

In Figure 7, the panels at the bulbous bow of a demi-hull are presented, while the panels at the stern for each case are presented in Figures 8–10. In Figures 11–13, the panels on the free surface, abaft the stern are presented for $Fn = 0.280$ corresponding to a ship's speed of 15 kn, for the transom stern case. It should be pointed out that the Froude number for the cruiser-type is slightly less since the waterline length is longer.

Table 2. Number of panels for the potential calculations.

	Cruiser-type	Dry transom	Wet transom
Hull	7346	9322	9341
Free surface	14621	16597	16616
Total	21967	25919	25957

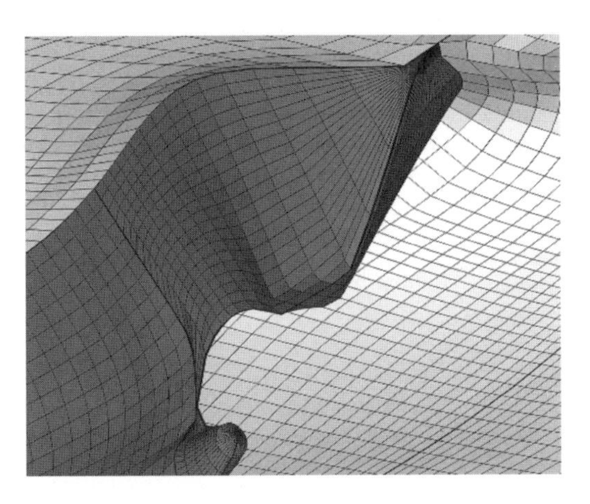

Figure 8. Panels at the stern of the hull for the case of cruiser-type stern.

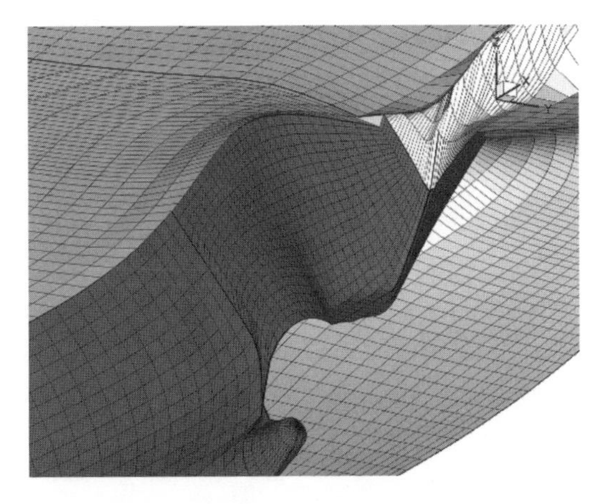

Figure 9. Panels at the stern of the hull for the case of the dry transom. The first row of free surface panels is adjacent to the edge of the transom.

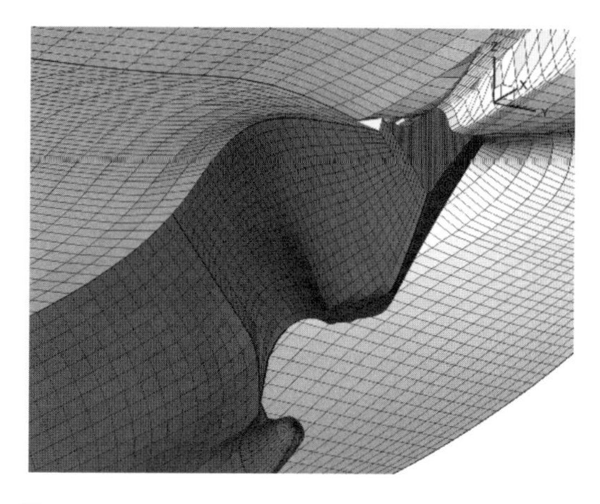

Figure 10. Panels at the stern of the hull for the case of the partially submerged transom. Panels are introduced on the transom up to the free surface line, the height of which is now calculated.

Figure 11. Panels at the free-surface about the cruiser-type stern, $V_S = 15$ kn, $Fn = 0.280$.

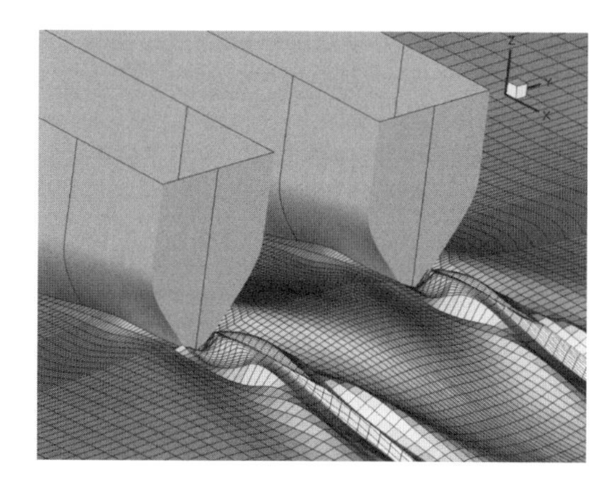

Figure 12. Panels at the free-surface about the dry transom stern, $V_S = 15$ kn, $Fn = 0.280$.

Figure 13. Panels at the free-surface about the partially submerged transom stern, $V_S = 15$ kn, $Fn = 0.280$.

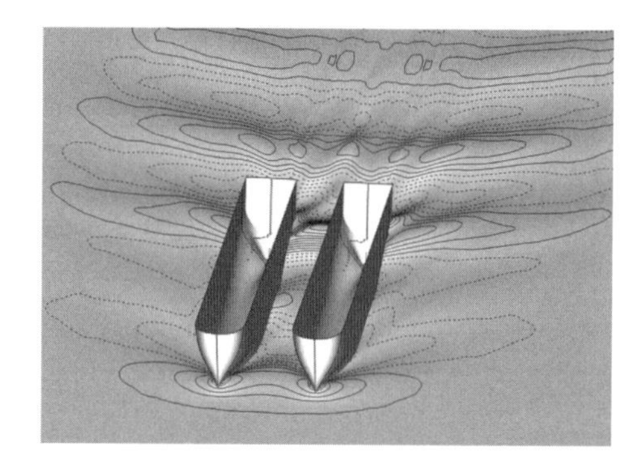

Figure 14. Perspective view of the wave elevation contours for the case of the partially submerged transom stern, $V_S = 15$ kn, $Fn = 0.280$.

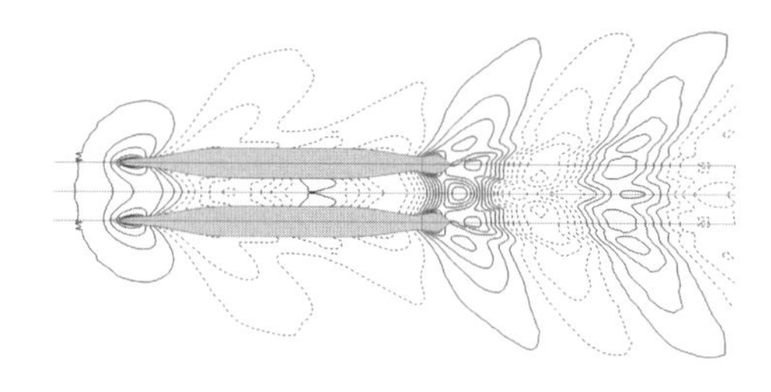

Figure 15. Wave elevation contours for the case of the cruiser-type stern, $V_S = 15$ kn, $Fn = 0.280$.

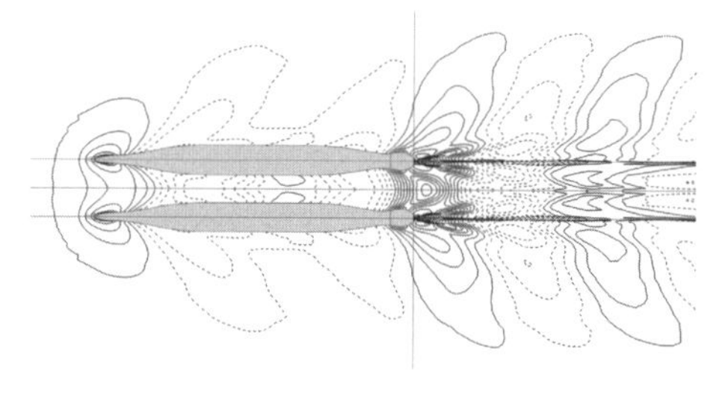

Figure 16. Wave elevation contours for the case of the dry transom stern, $V_S = 15$ kn, $Fn = 0.280$.

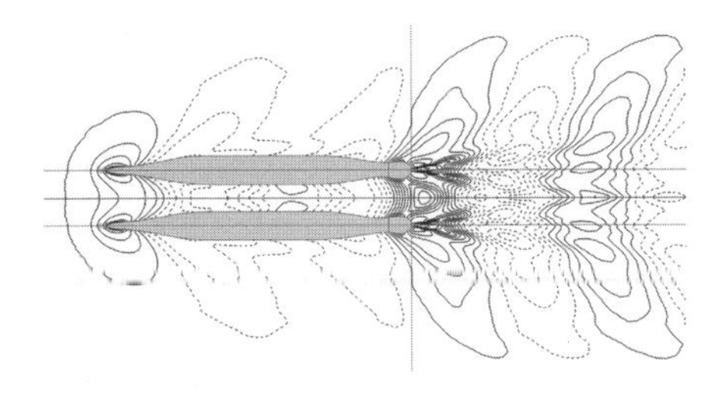

Figure 17. Wave elevation contours for the case of the partially submerged transom stern, V_S = 15 kn, $Fn = 0.280$.

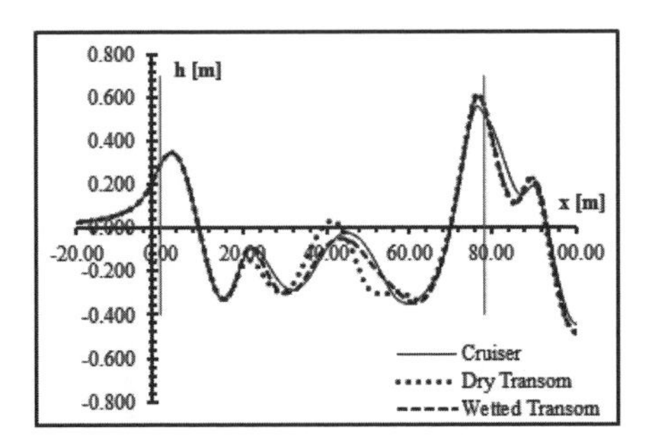

Figure 18. Longitudinal wave-cuts, at a distance of 0.5 cm (at model scale) from the hull beam, V_S = 14 kn, $Fn = 0.261$.

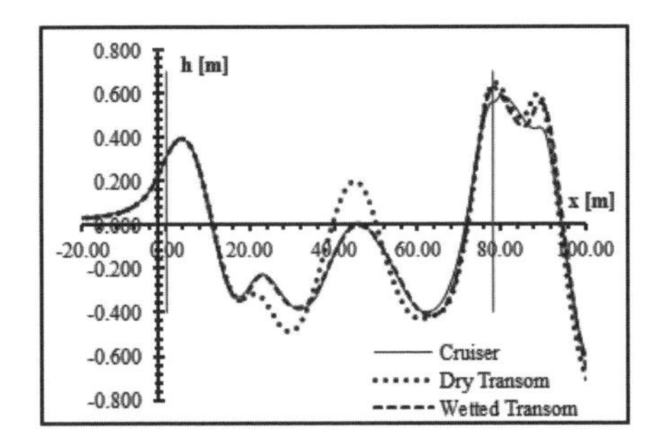

Figure 19. Longitudinal wave-cuts, at a distance of 0.5 cm (at model scale) from the hull beam, V_S = 15 kn, $Fn = 0.280$.

Figure 14 depicts a perspective view of the wave elevation and the corresponding contours for the partially submerged transom case, V_S = 15 kn, $Fn = 0.280$. Figures 15–17, depict the wave elevation contour for the three cases, for V_S = 15 kn and $Fn = 0.280$. In Figures 18 and 19, longitudinal wave cuts at a distance from the demi-hull beam of 0.5 cm at model scale are presented for each case, for V_S = 14 kn, $Fn = 0.261$ and V_S = 15 kn, $Fn = 0.280$ respectively.

From the above figures it is evident that the way the stern is modeled affects the formation of the waves along the length of the hull and abaft. The use of a cruiser-type stern and partially submerged transom result in similar waves along the length of the hull while both types of transom generate similar waves abaft the hull. This is an important observation since in combined methods such as the one presented here, the surface generated by the potential solver is used as a fixed boundary for RANS calculations.

Furthermore the wave making component of the resistance is also affected. The calculated values for the wave making resistance coefficient, C_W, are presented in Table 3. It is evident that the dry transom case results in far higher values for C_W, while the results of the other two methods are relatively close. This is the result of the loss of the positive effect of pressure at the aft section of the hull in the case of the dry transom. On the other hand the effect of viscosity is considerable at the stern of a ship, hence none of the above tests can provide with an accurate prediction for C_W.

3.3 *Viscous calculations*

The free-surface calculated with the potential solver for the cruiser-type and dry transom cases, was used as a fixed boundary for the RANS calculations. In all tests, the computational domain extended $1 \times L_{BP}$ upsrteam the bow and $0.5 \times L_{BP}$ downstream the stern. In the transverse direction it extended ether side a length equal to half the distance of the demi-hulls hence the inner-most points lay on the ship's center plane. In the cruiser-type stern case, the numerical grid comprised two blocks while in the dry transom case, the numerical grid comprised three blocks. Both stern configurations were tested at model scale and at two values of Fn, corresponding to ship speeds of 14 and 15 kn respectively.

In order to study the sensitivity of the calculated total resistance with respect to the grid resolution, grid dependence tests were conducted for the cruiser-type stern at both speeds, by applying the mesh sequencing technique of Tzabiras & Prifti (2001). The grid sizes for grid blocks (I) and (II), as well as the resulting value for the resistance coefficient C_T, are presented in Tables 4 and 5, for V_S = 14 and 15 kn respectively. The grid size is denoted by the numbers NI, NJ and NK, where NI is the number of grid points in the normal direction, NJ circumferentially

Table 3. Potential calculations for the wave making resistance coefficient C_W.

Ship speed kn	Froude number	Cruiser-type	Dry transom	Wetted transom
14.00	0.279	0.377×10^{-3}	1.022×10^{-3}	0.403×10^{-3}
15.00	0.260	0.793×10^{-3}	1.866×10^{-3}	0.843×10^{-3}

Table 4. Grid dependence tests for the cruiser-type stern at model scale, $V_S = 14$ kn, $Fn = 0.261$, $Rn = 4.24 \times 10^6$.

	NI × NJ × NK	Resistance coefficient C_T
Block (I)	80 × 54 × 194	3.921×10^{-2}
	120 × 80 × 286	3.914×10^{-2}
Block (II)	82 × 54 × 230	5.557×10^{-3}
	120 × 80 × 316	5.318×10^{-3}

Table 5. Grid dependence tests for the cruiser-type stern at model scale, $V_S = 15$ kn, $Fn = 0.280$, $Rn = 4.55 \times 10^6$.

	NI × NJ × NK	Resistance coefficient C_T
Block (I)	80 × 54 × 230	3.438×10^{-2}
	120 × 80 × 315	3.427×10^{-2}
Block (II)	80 × 54 × 230	5.894×10^{-3}
	120 × 80 × 280	5.676×10^{-3}

Table 6. Number of grid points for the transom stern.

Block	NI × NJ × NK
(I)	120 × 80 × 286
(II)	120 × 80 × 196
(III)	61 × 120 × 135

and NK in the longitudinal direction. The resistance coefficient in the second and third lines refers to block (I) while the coefficient in the fourth and fifth lines is the total coefficient for both blocks. Since the values of the resistance coefficient differ little (less than 4.5%) between the two grid sizes, it was concluded that the finer grid was fine enough to generate accurate solutions.

Based on the results of the grid independence tests for the cruiser-type stern, a grid size of similar fineness was adopted for the three blocks for the transom stern. The grid size for each block is given in Table 6.

By solving the viscous resistance problem using the grid configuration presented in Tables, 4–6, the resistance is calculated for each case and both speeds. The total resistance R_T, is calculated as:

$$R_T = R_P + R_F \tag{4}$$

were R_P is the pressure resistance and R_F is the friction resistance. R_P and R_F are calculated by integrating on the wetted portion of the hull, the dynamic pressure and the viscous shear stresses respectively:

$$R_P = \iint_{WS} (p^* - \rho \cdot g \cdot h)(\mathbf{n} \cdot \mathbf{i}) ds \tag{5}$$

$$R_F = \iint_{WS} \tau_w (\mathbf{n} \cdot \mathbf{i}) ds \tag{6}$$

where g = gravity; h = vertical distance from a reference level; $\mathbf{n}$ = unit vector normal to the hull surface in the outward direction; $\mathbf{i}$ = unit vector parallel to the longitudinal axis; and τ_w = shear stress on the hull surface.

The corresponding coefficients C_T, C_P and C_F are then calculated:

$$C_P = \frac{R_P}{1/2 \cdot \rho \cdot WS \cdot V^2}, C_P = \frac{R_P}{1/2 \cdot \rho \cdot WS \cdot V^2} \tag{7a}$$

$$C_T = \frac{R_T}{1/2 \cdot \rho \cdot WS \cdot V^2} = C_P + C_F \tag{7b}$$

were WS = calculated wetted surface at speed, as opposed to the experimentally calculated value of C_T were the wetted surface at rest is used.

The resulting values for C_P, C_F, C_T and R_T are presented in Tables 7 and 8 for $V_S = 14$ kn and $V_S = 15$ kn respectively. The ITTC-57 (Equation 3) value for C_F is also included for comparative reasons. In the last column of the above tables the corresponding experimental results are presented for C_R, $C_{F,ITTC}$, C_T and R_T. In the last row the difference between the numerically calculated and the experimentally measured value of R_T is presented.

From the results of Tables 7 and 8, it is evident that the most accurate results for R_T are obtained when the free-surface calculated for the cruiser-type stern is employed. This is due to the fact that during the experiments the transom was partially submerged, hence forcing the flow to detach from the transom results in less accurate calculations. On the other hand as speed increases, the tendency of the flow to detach also increases. As a consequence, the cruiser-type stern configuration becomes less accurate while the transom stern one appears to become more accurate. Consequently the decision to use a dry transom configuration is not a straight-forward one and should be backed-up be experimental evidences, while the cruiser-type stern is a valid alternative when the transom is partially submerged.

In any case the effect of the viscosity on the free-surface elevation is neglected in the present method, compromising somewhat the overall accuracy. This otherwise small effect is magnified in the present case where the transom is partially submerged as a result of the flow re-attaching to the transom.

Regarding the friction resistance, the calculated values for C_F are higher than those provided by the ITTC-57 formula. This is an important observation for resistance calculations on catamaran vessels, since this commonly used formula result in underestimation of the resistance. The observed

Table 7. Resistance calculations at model scale, $V_S = 14$ kn, $Fn = 0.261$, $Rn = 4.24 \times 10^6$.

		Cruiser-type stern	Dry transom stern	Experimental
C_P, C_R		1.324×10^{-3}	1.895×10^{-3}	1.554×10^{-3}
C_F		3.990×10^{-3}	3.993×10^{-3}	–
$C_{F, ITTC}$		3.502×10^{-3}	3.510×10^{-3}	3.486×10^{-3}
C_T		5.314×10^{-3}	5.888×10^{-3}	5.039×10^{-3}
R	kp	1.840	2.046	1.771
δR	%	3.90	15.53	

Table 8. Resistance calculations at model scale, $V_S = 15$ kn, $Fn = 0.280$, $Rn = 4.55 \times 10^6$.

		Cruiser-type stern	Dry transom stern	Experimental
C_P, C_R		1.735×10^{-3}	1.904×10^{-3}	1.833×10^{-3}
C_F		3.940×10^{-3}	3.953×10^{-3}	–
$C_{F, ITTC}$		3.457×10^{-3}	3.465×10^{-3}	3.411×10^{-3}
C_T		5.675×10^{-3}	5.857×10^{-3}	5.274×10^{-3}
R	kp	2.268	2.324	2.128
δR	%	6.58	9.21	

increase of C_F is the consequence of increased velocities between the two demi-hulls due to them interacting, resulting a channeling of the flow.

4 CONCLUSIONS

The performed numerical experiments have shown that the employed hybrid method predicts with satisfactory accuracy the total resistance of catamaran cargo ships at model scale and can be used in the design process for optimizing similar hull forms. The ITTC-57 formula for the friction resistance in the case of catamaran vessels should be used with precaution since the actual resistance component may be significantly higher due to demi-hull interactions.

Regarding the modeling of a partially submerged transom stern, the use of a cruiser-type extension of the stern may be preferable, instead of forcing the detachment of the flow, since the viscous calculations using the cruiser-type configuration are in better agreement with the experiments.

ACKNOWLEGEMENTS

The authors are grateful to the European Community, since part of the present work has been supported under the Cargo-XPRESS project: Developing a high-tech vessel for EU-cargo-transport.

They also acknowledge the most helpful assistance of Mr. D. Sinetos, Mr. J. Trachanas, Mr. D. Triperinas and Mr. G. Milonas in conducting the experiments in the towing tank of NTUA.

The authors' would like to thank Lloyd's Register Educational Trust (LRET), since Mr. Polyzos' Phd studies are supported by LRET.

The Lloyd's Register Educational Trust (LRET) is an independent charity working to achieve advances in transportation, science, engineering and technology education, training and research worldwide for the benefit of all.

REFERENCES

Carrica, P.M., Castro, A.M. & Stern, F. 2010. Self-propulsion computations using a speed controller and a discretized propeller with dynamic overset grids. *J. Marine science and technology*.15: 4: 316–330.

Hess, J.L. & Smith, A.M.O. 1996. Calculation of potential flow about arbitrary bodies. *Prog. In Aeronaut. Sci.* 8: 1–138.

Huan, J.C. & Huang, T.T. 2007. Surface ship total resistance prediction based on a nonlinear free surface potential flow solver and a Reynolds-averaged Navier-Stokes viscous correction. *Journal of Ship Research*, March: 47–64.

Launder, B.E. & Spalding, D.B. 1974. The numerical computation of turbulent flows. *Computation Method in Applied Mechanics and Enginnering.* 3: 269–289.

Menter, F.R. 1993. Zonal two equation k-ω turbulence models for aerodynamic flows. *Proc. 24th Fluid dynamics conference.*

Raven, H.C. & Starke, B. 2002. Efficient Methods to Compute Steady Ship Viscous Flows with Free Surface. *24th Symposium on Naval Hydrodynamics, Fukuoka, Japan.*

Tzabiras, G. 2008. A method for predicting the influence of an additive bulb on ship resistance. *Proc. 8th International Conference on Hydrodynamics, Nantes 2008*, 53–60.

Tzabiras, G.D. 2004. Resistance and Self-propulsion simulations for a Series-60, CB = 0.6 hull at model and full scale. *Ship Technology Research* 51: 21–34.

Tzabiras, G.D. & Kontogiannis, K. 2010. An integrated method for predicting the hydrodynamic resistance of low-Cb ships. *JCAD* 1568: 1–16.

Tzabiras, G.D., Kontogiannis, K.D. & Papakonstantinoy, V.K. 2009. Numerical prediction of the resistance and self-propulsion characteristics of Passenger-Ferry Ships. *Proc.* 13th Congress of Int. Maritime Assoc. *of Mediterranean, Istanbul 2009*. 1: 291–298.

Tzabiras, G.D. (ed.) & Prifti, A. 2001. Numerical simulation of the separated, turbulent flow past the stern of traditional fishing vessels. *Advances in Fluid Mech. series, Special volume: Calculation of Complex Turbulent Flows.* Southampton: CMEM: 131–166.

Sustainable Maritime Transportation and Exploitation of Sea Resources – Rizzuto & Guedes Soares (eds)
© 2012 Taylor & Francis Group, London, ISBN 978-0-415-62081-9

A preliminary study for the numerical prediction of the behavior of air bubbles in the design of ACS

A. Cristelli, F. Cucinotta, E. Guglielmino, V. Ruggiero & V. Russo
D.C.I.I.M., Naval Engineering University of Messina, Contrada di Dio (S. Agata) Messina

ABSTRACT: Air-Cavity Ships (ACS) are advanced marine vehicles that use air injection under hull to improve the vessel's hydrodynamic characteristics. Although the concept of drag reduction by supplying gas under the ship's bottom was proposed in the 19th century by Froude and Laval, at this time there are not many systematic studies on this subject. This paper is a preliminary work with the purpose of being a basic tool for the design of the ACS with computational fluid dynamic methods. The study aims to conduct a series of computational tests to compare the numerical models of bubble with experimental data. The first step of this study was to investigate the behavior of free bubble in water, considering as parameters the critical mass of air, the rising speed and aspect ratio of the bubble. Then it is evaluated the interaction bubble-flat plate in order to obtain a reliable prediction of the behavior of air bubbles under the hull.

1 HULL'S VENTILATION IN EXPERIMENTAL TESTS

In planing hulls, a phenomenon of great importance is the ventilation. Thanks to it, it's possible to reduce the friction component of resistance.

Ventilation, in planing hull, is an extremely complex phenomenon of great importance. It's derived by the establishment of air channels under the hull that serve the dual purpose of reducing frictional resistance, due to lower viscosity of the air than water, and reduce the low precession region aft, and generally downstream of each discontinuity in the hull, thereby decreasing the component of form drag.

Ventilation is both exploited in a conventional manner, i.e., through geometries that facilitate the generation of these air channels (examples of this are the hulls with spray rails and steps) and, ultimately, by forced insufflations of air compressed by special nozzles (e.g., ACS).

Moreover, the study of other biphasic phenomena, in which is present an air-water mixing, such as the spray, which often implies, in planing hulls, a 20% share of resistance, follows the same laws.

Studies in the past were based on experimental experience, which trying to reproduce the phenomenon in scale (on a model in tank) and then estimate the results in one to one scale. This procedure, widely used in the naval field, has proved unsuitable to investigate a phenomenon whose implications are manifold. In fact, the parameters governing the biphasic phenomenon, under

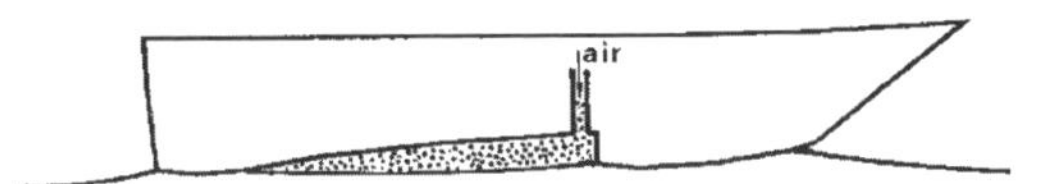

Figure 1. ACS scheme.

the hypothesis of negligible thermal phenomena, are free flow speed V, linear dimension of body d, acceleration of gravity g, density ρ, pressure p, surface tension σ, viscosity μ.

$$V = f(d^{\alpha}, g^{\beta}, \rho^{\gamma}, p^{\delta}, \sigma^{\varepsilon}, \mu^{\varsigma})$$

Therefore, from dimensional analysis, we obtain:

$$\left(\frac{L}{T}\right) = (L)^{\alpha}\left(\frac{L}{T^2}\right)^{\beta}\left(\frac{FT^2}{L^4}\right)^{\gamma}\left(\frac{F}{L^2}\right)^{\delta}\left(\frac{F}{L}\right)^{\varepsilon}\left(\frac{FT}{L^2}\right)^{\varsigma}$$

Which leads:

$$\frac{V}{\sqrt{gd}} \propto \left(\frac{p_a}{\rho V^2}; \frac{V\mathrm{d}}{\nu}; \frac{\rho d V^2}{\sigma}\right)$$

Then, with an adimensionalisation of the equations of motion, we obtain the following independent parameters:

$$Fr \equiv \left[\frac{Inertial\ forces}{Gravitational\ forces}\right]^{\frac{1}{2}} \propto \left[\frac{\rho V^2/_d}{\rho g}\right]^{\frac{1}{2}} = \frac{V}{\sqrt{gd}}$$

$$Eu \equiv \frac{Pressure\ forces}{Inertial\ forces} = \frac{p_a}{\rho V^2}$$

$$\mathrm{Re} \equiv \frac{Inertial\ forces}{Viscous\ forces} \propto \frac{\rho^u \partial u / \partial \chi}{\mu \partial^2 u / \partial \chi^2} \propto \frac{\rho V^2 / d}{\mu V / d^2} = \frac{Vd}{v}$$

$$We \equiv \frac{Inertial\ forces}{Surface\ tension\ forces} = \frac{\rho d V^2}{\sigma}$$

As is evident, in a model test in Froude similarity, it is impossible to take into account the similarity of the other parameters unless using surfactants to modify the properties of surface tension of water. Generally, the similarity of Euler and Weber are neglected in model tank tests because ventilation phenomena are not important. In this case the geometry of the cavity, the air flow and pressure of the bubble are the parameters that affect the generation and development of a stable cavity. The cushion of air under the hull also interacts with the wave generated from the boat and its wake.

So, for ventilated hulls, it's extremely important to use very high scales factor or alternative methods. Computational Fluid Dynamic methods are a good alternative in order to test virtual models on a one to one scale.

In this model are neglected the elastic effects due to fluid compressibility (Mach number).

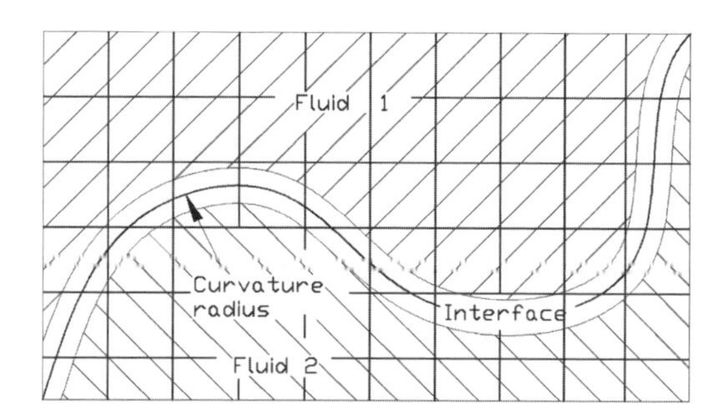

Figure 2. Mesh near the interface.

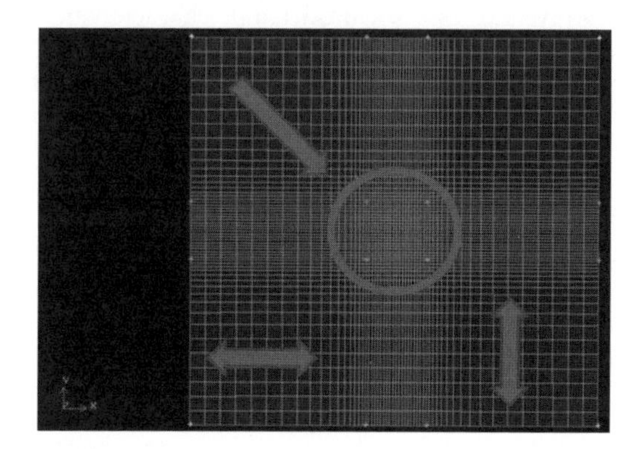
Figure 3. Mesh size function.

2 CFD'S MODEL OF VENTILATION

To take into account the viscous phenomena, which are necessary to evaluate the frictional resistance, RANSe (Reynolds Averaged Navier Stokes equations) methods have been used with a k-epsilon model for the study of turbulence.

The surface tension phenomena, those in which the We <<1, like bubbles and sprays, need a mesh for calculating well enough to describe the geometry of the interface. The surface is obtained with the model proposed by Brackbill et al. (1992) that add a source term in the momentum equation considering surface tension constant along the interface. This term involves a pressure drop across the surface depends upon the surface tension coefficient, and the surface curvature as measured by two radii in orthogonal directions:

$$\Delta p = \sigma \left(\frac{1}{R_1} + \frac{1}{R_2} \right)$$

Since the surface tension forces are proportional to the radius of curvature of interface, it's important to create a mesh that has a size of between 5% and 10% of the local radius of curvature of interface, otherwise it's difficult to achieve convergence of the result and good results (see Figs. 2–3). It is therefore essential to have a mesh small enough in all areas of the hull involved in spray and ventilation.

When an air bubble or a drop of water flows on the surface of the hull, it tends to adhere to the wall because the balance of forces that is established at the meeting of the three different phases (air, water, solid wall). This means that the hull tends to drag the spray (with resulting increase in resistance). Similarly, the air channels under the hull tends to cover a surface of the hull below that which would be expected if there were no surface tension (air does not tend to wet the solid wall). Rather than impose this boundary condition at the wall itself, the contact angle that the fluid is assumed to make with the wall is used to adjust the surface normal in cells near the wall. This so-called dynamic boundary condition results in the adjustment of the curvature of the surface near the wall.

3 BUBBLE'S MODEL VALIDATION

Bubbles dynamics is a rather complex phenomenon. When the bubble is at rest, it tends to assume a spherical shape, due to the forces of surface tension. But when the bubble is moving, the shape is altered due to the flow field. In particular, we can detect the simultaneous presence of:

1. Force due to hydrostatic pressure.
2. Inertial force due to buoyancy.
3. Force of surface tension.
4. Viscous drag force.
5. Centrifugal force due to internal circulation of gas.

These forces alter significantly the geometry of bubbles as a function of Weber and Reynolds. Hill's model describes the internal circulation of gas (Figs. 4–5). It generates a centrifugal force acting in opposition to the force of surface tension. These two forces, alternately prevail over one another, generating most of the dynamic phenomena seen during the ascent of the bubbles (distortion, asymmetry, pulse). Because of these forces the bubbles below 6 mm in diameter tend to maintain a spherical shape, while those above 14 mm in diameter take the so-called cap shape (Fig. 6). The bubbles above 200 mm instead break due to the centrifugal forces.

The resultant of these forces causes distortion in bubble's shape that assume, during motion,

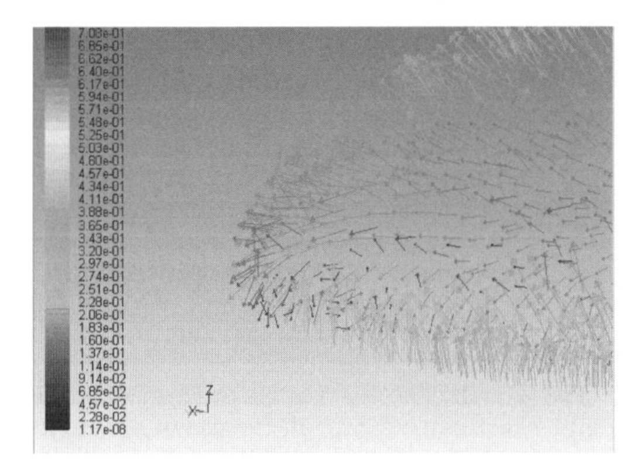

Figure 4. Gas velocity vectors inside bubble.

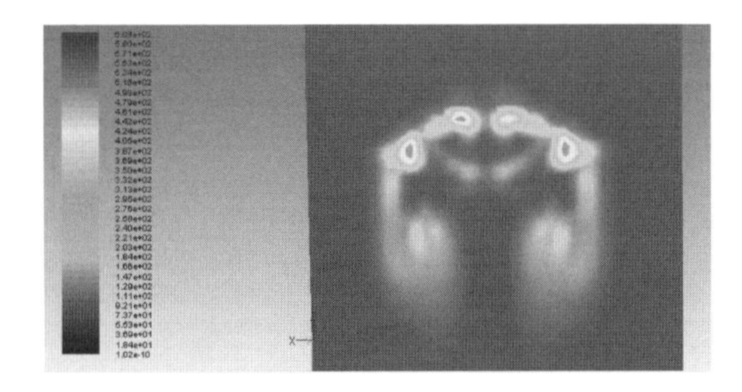

Figure 5. Vorticity inside and below the bubble.

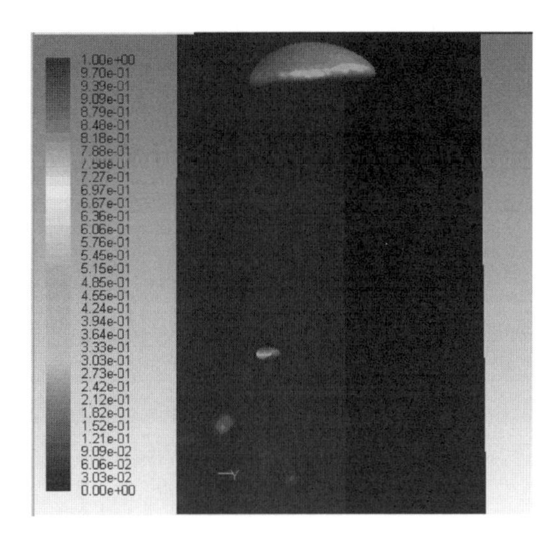

Figure 6. Cap-shape bubble.

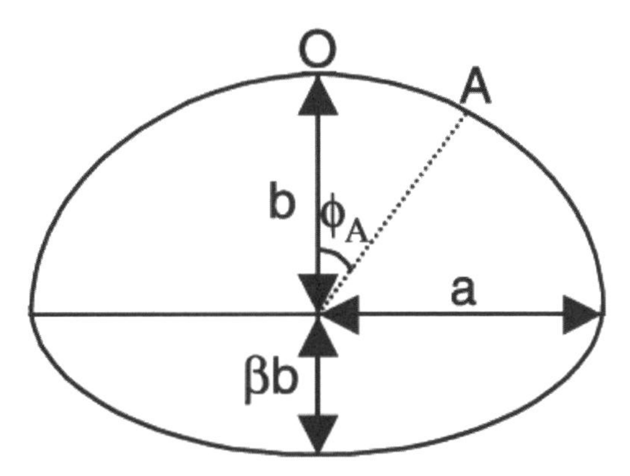

Figure 7. Ellipsoidal shape of bubble.

a pulsating ellipsoidal shape. This shape can be described by the aspect ratio E and the distortion factor γ, defined by Tomayama (2001) as:

$$E = \frac{b + \beta b}{2a}$$

$$\gamma = \frac{2}{1+\beta}$$

In which a is the major axis and the minor axes are b and βb like in Fig. 7.

Results were compared with experimental results obtained by Davies and Taylor (1950).

The graph shows the evolution of bubble velocity along z axis (vertical) and y (horizontal). The evolution of Z velocity tends, after a period of transition, to stabilize around the average experimental value. Pulsations are due to the alternation of the

Table 1. Shape parameters and rise velocity (R.v.) of bubbles. R.v. is compared with Davies & Taylor formulation.

∅	E	γ	R. v. CFD	R.v. D & T	Velocity difference
mm			m/s	m/s	%
6	0.49	1.02	0.117	0.114	2.5%
10	0.44	1.26	0.140	0.148	5.2%
14	0.38	1.42	0.160	0.170	5.8%

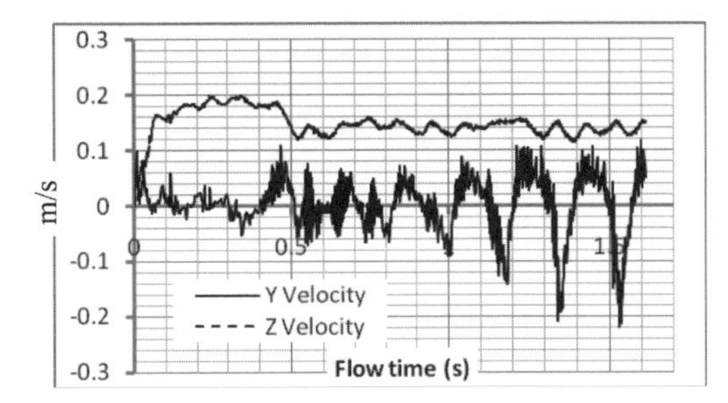

Figure 8. Velocity components of the bubb.

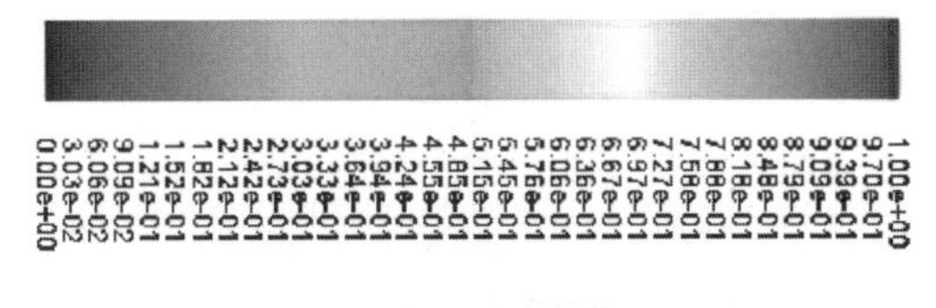

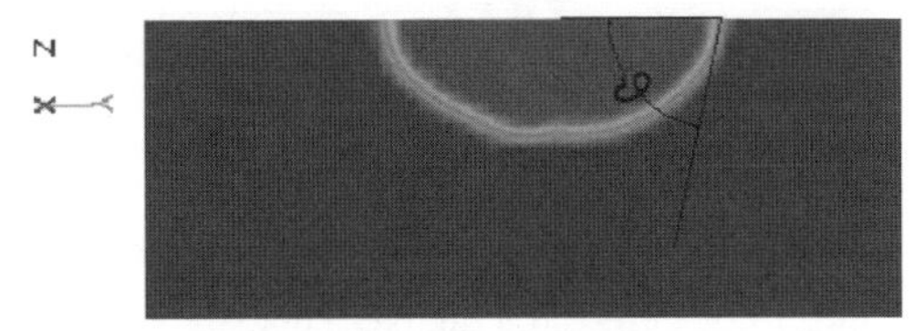

Figure 9. Static contact angle on solid wall.

dominance of the hydrodynamic forces (which speed up the bubble and flattening it) and viscous forces (which slows it down and create a sharp front). The Y velocity, instead, oscillates around zero because of the helical motions that are established. It's interesting to note that the transverse velocity has the same order of magnitude as the vertical (see Tomiyama et al., Talaia). Accordance between experimental and CFD results is very good (under 6% in velocity magnitude).

The contact angle is obtained by properly setting the wall adhesion model. It's really important to create a structured mesh with a max size of 5% of bubble's radius, for free bubbles, and 5% of curvature radius of contact angle for bubble on wall.

4 FLAT PLATE'S MODEL

The fundamental requirements for meshing are mainly two:

1. It should be structured hexahedral.
2. It should be easily editable.

The first one is to the needs of the Brackbill's model simulation of surface tension. This model also imposes a maximum size of the cell, in order to obtain convergence of the solution in the simulations with boundary. The second one is to the need to carry out a very extensive campaign of simulations, varying some control parameters. The system used is based on a calculation journal, obtained with pages of text containing command lines in sequential order for the processing software. Journal is written in function of the parameters of interest, using the language of its own software.

From a practical point of view, the most critical parameters for the structured mesh are: Skew EquiSize and Aspect Ratio. Both were established, through the experiences of the simulations, as:

- EquiSize Skew <0.2
- Aspect Ratio <18

Beyond these limits, in two-phase turbulent flow, it is difficult to achieve convergence of the calculation.

Flatures plates tested had a square base of side 1 m × 1 m and a thickness of 0.1 m. The volume control is also a square base of side 10 m and height of 3 m. The size of the mesh is designed so as to achieve higher definition in the areas of interest. In our case the most important area is the occurrence of forced ventilation, i.e., the bottom of the plate. In correspondence to that area it was possible to obtain cells with maximum size of 2.5 mm. In order to avoid having an excessive number of cells, were used hanging nodes.

The growth of cells along the three directions x, y and z is obtained through the use of size function, created to allow the adjustment of the size

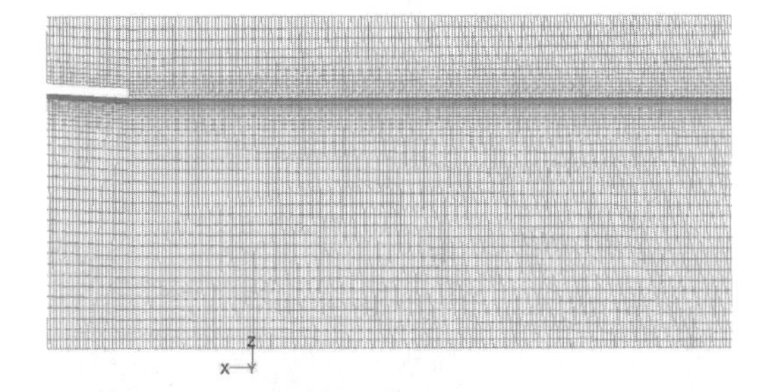

Figure 10. Mesh on symmetry plane. It's possible to see the thickening of the mesh on the free surface and under the plate.

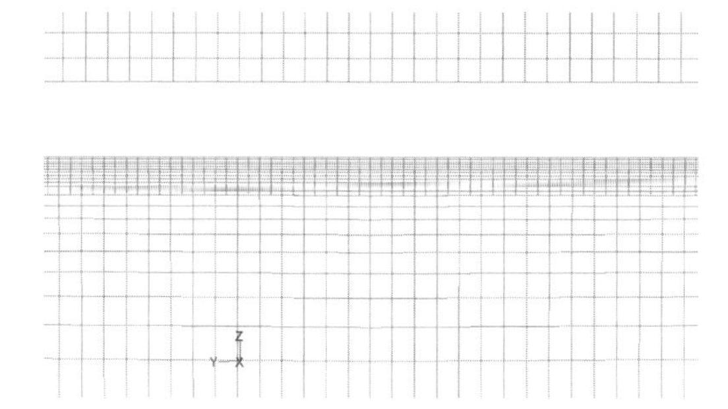

Figure 11. Particular of the mesh under the plate.

of the single cell. This growth was designed in order to not lead to the excessive increase of the parameter of Aspect Ratio on the periphery of the volume control. For this reason, the growth factor has been set to a maximum value of 1.15, i.e., an increase of up to 15% from cell to the next.

The computational model used is the transient VOF model with k-epsilon turbulence. The free surface was treated with a function of geometric reconstruction. In the geometric reconstruction approach, the standard interpolation schemes are used to obtain the face fluxes whenever a cell is completely filled with one phase or another. The geometric reconstruction scheme is used only for the cells near the interface between two phases. This scheme represents the interface between fluids using a piecewise-linear approach and, in this way, allows to obtain a free surface very accurate, with a very little diffusion of water into the air only limited to the thickness of a cell. The scheme used is that of Youngs (1982). It assumes that the interface between two fluids has a linear slope within each cell, and uses this linear shape for calculation of the advection of fluid through the cell faces. The first step in this reconstruction scheme is calculating the position of the linear interface relative to the center of each partially-filled cell, based on information about the volume fraction and its derivatives in the cell. The second step is calculating the advecting amount of fluid through each face using the computed linear interface representation and information about the normal and tangential velocity distribution on the face. The third step is calculating the volume fraction in each cell using the balance of fluxes calculated during the previous step.

The nozzle is positioned in the middle of the plate, along its entire width. The boundary condition imposed is a velocity inlet. The air flow is the same in all simulations and is 1.2 m/s. Many testes are effectuated varying vector orientation. Best results are obtained with a vector orientated at 33°. For larger angles, the flow tends to not adhere

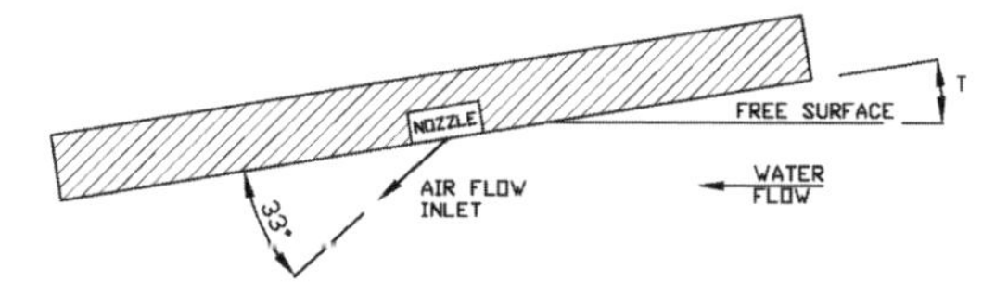

Figure 12. Flat plate schem.

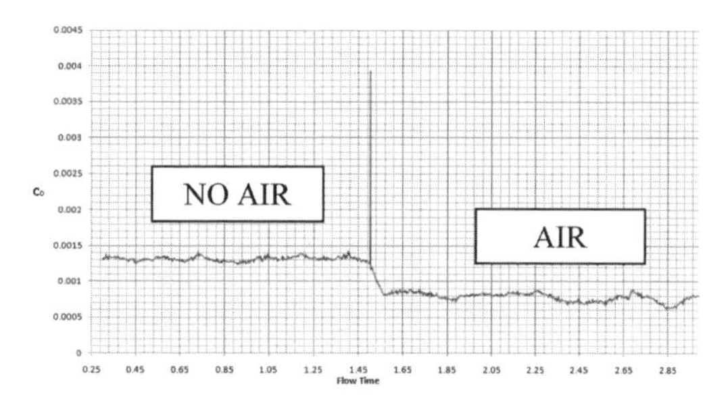

Figure 13. C_D vs. flow time. It's possible to note the two phases with and without ventilation.

to the solid wall creating turbulence and without covering all the surface correctly.

A too little angle is not a good condition, as it is essential to create a cavity to accommodate nozzles. But in this model, the nozzles are flush, so that the surface is fully lubricated by an air layer without altering the flow with geometrical discontinuity.

The simulations are carried out by setting an initial flow, with speed depending on the Froude of interest, in absence of ventilation. Once the flow field becomes stable, with resistance values that oscillate around a mean value with a small deviation, journal activates the ventilation on the bottom of the plate. In this step transients arise, mainly due to the sudden change in the pressure range and speed. At the end of the transitory, the phenomenon tends to evolve towards a new situation of regime. As can be seen in Fig., the trend of the C_D has two flat areas corresponding to the period without ventilation and the period with ventilation.

The system is based on the following cycle of operations:

1. Choice of simulation parameters (τ, C_V)
2. Structuring the mesh by journal
3. Setting solver by journal
4. Numerical analysis
5. Extrapolation of results
6. PostProcessing

5 RESULTS

Model validation has allowed a series of tests on a ventilated flat plate. This is because the basic

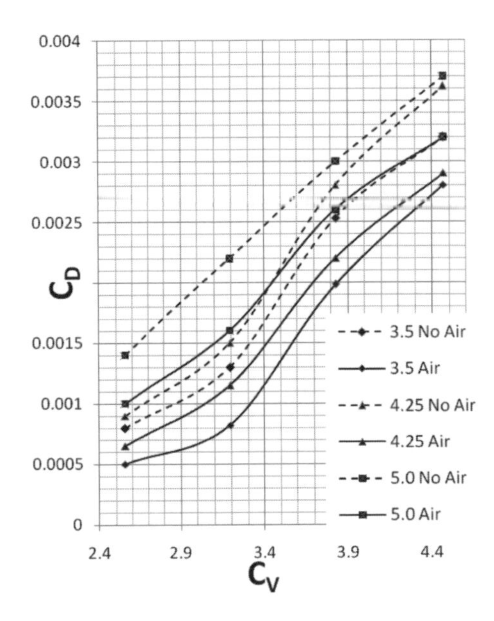

Figure 14. C_D vs. C_V varying trim angle. Continuous and dashed lines are, respectively, with and without ventilation.

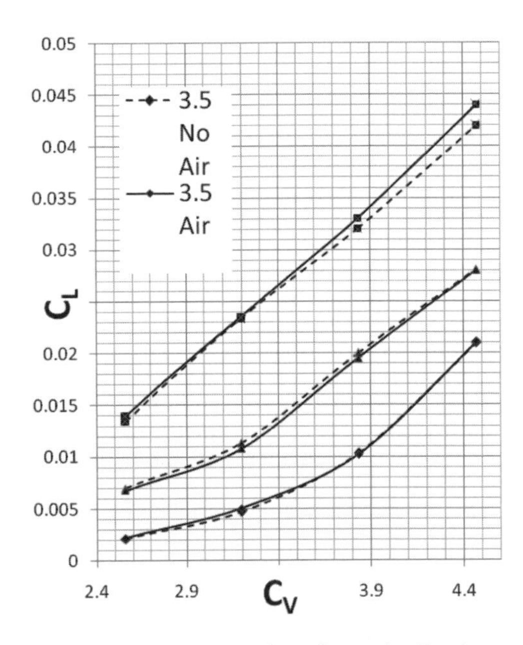

Figure 15. C_L vs. C_V varying trim angle. Continuous and dashed lines are, respectively, with and without ventilation.

knowledge of the phenomena is essential for the proper design of ACS. The campaign was conducted by varying the flow velocity and angle of attack, the flat plates were not free to trim, heel and heave.

Table 2. Drag coefficient difference between ventilated and unventilated plates.

C_D		τ	
Cv	3.5	4.25	5
2.55	37.5%	–27.8%	–28.6%
3.19	–36.9%	–23.3%	–27.3%
3.83	–21.7%	–21.4%	–13.3%
4.47	–12.5%	–19.9%	–13.5%

Table 3. Lift coefficient difference between ventilated and unventilated plates.

C_L		τ	
Cv	3.5	4.25	5
2.55	4,8%	–2,9%	3,7%
3.19	6,4%	–4,6%	0,7%
3.83	–1,0%	–2,5%	3,1%
4.47	0,5%	–0,4%	4,8%

The results were included in graphs according to dimensionless parameters that are a speed, a drag and a lift coefficient:.

$$Cv = \frac{v}{\sqrt{gb}}$$

$$C_D = \frac{D}{\frac{1}{2}\rho b^2 v^2}$$

$$C_L = \frac{L}{\frac{1}{2}\rho b^2 v^2}$$

Curves were plotted as a function of trim angle of the plate, the continuous curves are those obtained with ventilation, while the dashed ones are obtained without ventilation.

The following tables summarize the percentage gain obtained on the total drag and lift, in the case of ventilated plate compared to the unventilated one.

6 CONCLUSIONS

The series of tests carried out led to the following conclusions:

- CFD allows to overcome the problems of scale, becoming an indispensable tool in all applications where natural or forced ventilation is important. In fact the CDF simulations are made in one to one scale.

- The error returned, in free rising bubbles, has never exceeded the threshold of 6% compared to Davies & Taylor theoretical approach.
- CFD is an instrument suitable to the study of highly complex flows, provided they have the necessary theoretical knowledge for the optimal use of available empirical models for the description of phenomena and in order to construct a suitable mesh to obtain the convergence of differential equations.
- Automating the process, obtained by the exploitation of journals, has proven to be essential for the competitiveness of the method for the drastic computation time reduction.
- Curves obtained may be a good tool for preliminary design for the study of ACS.
- Air lubrication contributes to reduced resistance flat plate up to 40%. The advantage is reduced significantly for high C_V for the increase of the wave phenomena.
- Air lubrication don't change significantly lift that does not vary more than 7%.

REFERENCES

Batchelor G.K., (1967). *An Introduction to Fluid Dynamics.* Cambridge University Press, Cambridge, pp. 235–238.

Brackbill J.U., Kothe D.B. Zemach c. (1992). *A continuum method for modeling surface tension.* Journal of computational physics 100, 335–354.

Brennen C.E., (1995). *Cavitation and Bubble Dynamics.* Oxford University Press, Inc.

Chen C.-J., Jaw S.-Y. (1998). *Fondamentals of Turbulence Modeling.* Taylor and Francis.

Clift R., Grace J.R., Weber M.E., (1978). *Bubbles, Drops, and Particles.* Academic Press, New York.

Davies R.M., Taylor G.I. (1950). *The mechanics of large bubbles rising through extended liquid in tubes.* Proc. R. Soc. Ser. A 200, 375–390.

Dijkhuizen W., Van Den Hengel E.I.V., Deen N.G., Van Sint Annaland M., Kuipers J.A.M. (2005). *Numerical investigation of closures for interface forces acting on single air-bubbles inwater using Volume of Fluid and Front Tracking models.* Chemical Engineering Science 60 (2005) 6169–6175.

Hoffman A.H. (2000). *Computational Fluid Dynamics Vol. III,* 4th ed., Engineering Education System.

Hoffmann A.C., Van Den Bogaard H.A. (1995). *A Numerical Investigation of Bubbles Rising at Intermediate Reynolds and Large Weber Numbers.* World Academy of Science, Engineering and Technology 28 2007.

Lamb H. (1932). *Hydrodynamics.* sixth ed. Cambridge University Press, Cambridge.

Landau L.D., Lifshitz (1959) E.M. *Fluid Mechanics,* Pergamon Press, London.

Landahl M.T. Mollo-Christensen E. (1992) *Turbulence and Random Processes in Fluid Mechanics,* Second Edition, Cambridge University Press.

Larsson L. Raven H.C. (2010). *The principles of naval architecture series: Ship resistance and flow.* J. Randolph Pauling, Editor.

Lewis, E.V. (Editor), (1989). *Principles of Naval Architecture.* Volume II and III. The Society of Naval Architects and Marine Engineers.

Martìnez-Bazan C., Libby P.A. (1996) *Introduction to Turbulence,* Taylor and Francis.

Montanes J.L., Lasheras J.C. (2002). *Statistical description of the bubble cloud resulting from the injection of air into a turbulent water jet.* International Journal of Multiphase Flow 28 (2002) 597–615.

Patro R., Leifer I., Bowyer P. (1995). *Better Bubble Process Modeling: Improved Bubble Hydrodynamics Parameterization.*

Polonsky S., Shemer L, Barnea D. (1999). *The relation between the Taylor bubble motion and the velocity field ahead of it.* International Journal of Multiphase Flow 25 (1999) 957–975.

Prandtl L., Tietjens O.G. (1934). *Fundamentals of hydro and aeromechanics.* Dover Publications Inc. New York.

Prosperetti A., Tryggvason G. (2007). *Computational methods for multiphase flow.* Cambridge University Press.

Pushkarova R.A., Horn R.G. (2005). *Surface forces measured between an air bubble and a solid surface in water.* Colloids and Surfaces A: Physicochem. Eng. Aspects 261 (2005) 147–152.

Razzaque M.M., Afacan A., Liu S., Nandakumar K., Masliyah J., Sanders S.R. (2003). *Bubble size in coalescence dominant regime of turbulent air–water flow through horizontal pipes.*

Savitsky D. (1964). *Hydrodinamic design of planing hulls.*

Schlichting, H. (1979). *Boundary Layer Theory,* 7th ed., New York, McGraw-Hill.

Spalart P.R., Allmaras S.R. (1992). *A One-Equation Turbulence Model for Aerodynamics Flows,* AIAA-92–0439.

Talaia M.A.R. (2007). *Terminal Velocity of a Bubble Rise in a Liquid Column.* World Academy of Science, Engineering and Technology 28 2007.

Tomiyama A., Celata G.P., Hosokawa S., Yoshida S., (2002). *Terminal velocity of single bubbles in surface tension force dominant regime.* International Journal of Multiphase Flow 28 (2002) 1497–1519.

Tomiyama A., Kataoka, I., Zun, I., Sakaguchi, T. (1998). *Drag coefficients of single bubbles under normal and micro gravity conditions.* JSME Int. J. Ser. B. 41 (2), 472–479.

Wen-Ching Y., (2003). *Handbook of fluidization and fluid-particle systems.* Marcel Dekker Inc. 2003.

White F.M. (1974) *Viscous Fluid Flow.* McGraw-Hill Book Company.

Wilcox, D.C. (1993). *Turbulence modelling for CFD.* DCW Industries, Inc.

Wu M., Gharib M. (2002). *Experimental studies on the shape and path of small air bubbles rising in clean water.* Physics of fluids volume 14, number 7 july 2002.

Youngs D.L. (1982). *Time-Dependent Multi-Material Flow with Large Fluid Distortion.* In K.W. Morton and M.J. Baines, editors, Numerical Methods for Fluid Dynamics. Academic Press, 1982.

1.2 *Propulsion*

Sustainable Maritime Transportation and Exploitation of Sea Resources – Rizzuto & Guedes Soares (eds)
© 2012 Taylor & Francis Group, London, ISBN 978-0-415-62081-9

Numerical and experimental optimization of a CP propeller at different pitch settings

D. Bertetta, S. Brizzolara, S. Gaggero & M. Viviani
Department of Naval Architecture, Marine Engineering and Electrical Engineering (DINAEL), Genoa University, Genoa, Italy

L. Savio
Rolls-Royce University Technology Center 'Performance in a Seaway' Department of Marine Technology, Norwegian University of Science and Technology, (formerly DINAEL) Norway

ABSTRACT: Propeller design is an activity which nowadays presents ever increasing challenges to the designer, involving not only the usual mechanical characteristics and cavitation erosion avoidance, but also other side effects, such as radiated noise and/or pressure pulses. Moreover, in some cases propeller characteristics have to be optimized in correspondence to very different functioning points, including considerably off-design conditions, which are hardly captured by conventional design methods. In present paper, a recently presented method, based on the coupling between a multiobjective optimization algorithm and a panel code, is applied to the design of a CP propeller at different pitch settings, with the aim of reducing cavitating phenomena and, consequently, resultant radiated noise. Numerical results are validated by means of an experimental campaign at cavitation tunnel, showing the capability of the method to assess propeller functioning characteristics, thus representing a very useful tool for the designer in correspondence of challenging problems.

1 INTRODUCTION

Propeller design has evolved significantly in last years, with the introduction of numerical methods which can provide an ever improving assessment of propeller characteristics, considering propeller non stationary functioning and cavitating behavior, not only in correspondence to the usual design conditions, but also to off-design conditions. This assessment has become rather usual, with numerical methods being able to predict propeller characteristic curves (and cavitating behavior) in correspondence to a wide range of advance coefficients. Modern propeller requirements involve many different characteristics, not limiting only to maximum efficiency, but considering also propeller cavitating behavior and, more and more, its side effects, in terms of radiated noise and pressure pulses. This is evident with the ever increasing demand for improvement of comfort onboard and discussions about radiated noise problems, especially in proximity of protected areas.

Traditional design methods, using lifting line and lifting surface codes, are very well known, and their utilization has been established for a long time. Nevertheless, these methods cannot be used directly for a propeller which needs to be designed for very different working conditions (e.g., a CPP propeller operating at constant RPM in a wide range of ship speeds at different blade pitch angles). From this point of view, coupling of panel codes (usually adopted for propeller analysis) with optimization algorithm can represent an efficient alternative to the classical approach for the designer, as presented in (Gaggero et al., 2009). A theoretical background of the methods adopted is reported in section 2.

In present paper, a multiobjective optimization of a CP propeller at two very different working conditions is presented, in order to test the capabilities of the design procedure mentioned above. In particular, the usual design condition at maximum speed is considered, together with a very reduced speed, obtained at constant RPM and reduced pitch. This results in very different cavitating behavior, with presence of back sheet cavitation and tip vortex at maximum speed condition, while in correspondence to reduced pitch cavitating behavior is dominated by face related phenomena (mainly vortex from sheet face and sheet face cavitation). This, in turn, results in noise and vibration problems in correspondence to this operating condition, as remarked by the shipowner. As a consequence, scope of the optimisation activity was a new design with main attention given to the reduced pitch, with the aim of reducing propeller radiated noise, trying contemporarily to keep cavitation extent as low as possible at maximum speed and maintaining propeller mechanical characteristics. The optimization activity is summarized in section 3.

The two propellers (original and newly designed) have been tested at towing tank and cavitation tunnel, in order to validate the design code, thus considering not only mechanical characteristics but also cavitation extent. Moreover, in order to have an insight into the above mentioned side effects, a series of measurements of radiated noise have been carried out in order to verify their interrelation with cavitation and confirm the design assumptions. Measurements are reported in section 4 and compared with numerical results.

2 THEORETICAL BACKGROUND

2.1 *Propeller analysis/design codes*

For the analysis of marine propellers performances, several theoretical and numerical approaches are, nowadays, available. From the 1960's lifting line method, suitable even for the design than for the analysis, to the most recent RANS solvers, different codes have been developed on the basis of different theoretical assumption for the flow field around the propeller. Each of these methods has its own advantages and its own drawback. Lifting line and lifting surface codes, based on the hypothesis of inviscid, irrotational and incompressible fluid, are still the better established methods for the design of a propeller. Potential panel methods, with their capabilities to capture thickness effects and to take into account nonlinear effects like cavitation (at least sheet cavitation) could be employed for a first tentative analysis in steady and unsteady condition, being essentially an analysis approach for an already established geometry. In the framework of potential theory the flowfield around a solid body is modeled by means of a scalar function, the perturbation potential $\phi(x, t)$, whose spatial derivatives represent the component of the perturbation velocity vector. Irrotationality, incompressibility and absence of viscosity are the hypothesis needed in order to write the more general continuity and momentum equations as a Laplace equation for the perturbation potential itself:

$$\nabla^2\phi(x,t)=0 \tag{1}$$

Green's third identity allows to solve the three dimensional differential problem as a simpler integral problem written for the surfaces that bound the domain. The solution is found as the intensity of a series of mathematical singularities (sources and dipoles) whose superposition models the inviscid cavitating flow on and around the body. Assuming that the cavity bubble thickness is small with respect to the profile chord, singularities that model cavity bubble can be placed on the blade surface instead than on the real cavity surface, leading to an integral equation in which the subscript q corresponds to the variable point in the integration, $\boldsymbol{n}$ is the unit normal to the boundary surfaces and $\boldsymbol{r}_{pq}$ is the distance between points p and q, S_B is the fully wetted surface, S_W is the wake surface and S_{CB} is the projected cavitating surface on the solid boundaries:

$$\begin{aligned}2\pi\phi(x_p,t)= &\int_{S_B+S_{CB}} \phi(x_q,t)\frac{\partial}{\partial \boldsymbol{n}_q}\frac{1}{\boldsymbol{r}_{pq}}dS \\ &-\int_{S_B+S_{CB}} \frac{\partial\phi(x_q,t)}{\partial \boldsymbol{n}_q}\frac{1}{\boldsymbol{r}_{pq}}dS \\ &+\int_{S_W} \Delta\phi(x_q,t)\frac{\partial}{\partial \boldsymbol{n}_q}\frac{1}{\boldsymbol{r}_{pq}}dS\end{aligned} \tag{2}$$

The set of required boundary condition is:

- Kinematic boundary condition on the wetted solid boundaries:

$$\frac{\partial\phi(x_p,t)}{\partial \boldsymbol{n}_p}=-Vinflow(xp,t)\cdot\boldsymbol{n}_p \tag{3}$$

- Kutta condition at blade trailing edge:

$$\Delta\phi_{TE}(x_p,t)=\phi_{TE}^U(x_p,t)-\phi_{TE}^L(x_p,t) \tag{4}$$

- Kelvin's theorem to drive the unsteadiness of the problem:

$$\frac{D}{Dt}(\Delta\phi^W(x,t))=0 \tag{5}$$

- Dynamic boundary condition on the cavitating surfaces:

$$p=p_{vap}\ on\ S_{CB} \tag{6}$$

- Kinematic boundary condition on the cavitating surfaces (where $\boldsymbol{n}$ is the normal vector and th is the local cavity bubble thickness):

$$n(x)-th(x,t)=0 \tag{7}$$

- Cavity closure condition at cavity bubble trailing edge.

The numerical solution consists in an inner iterative scheme, delegated to solve the nonlinearities

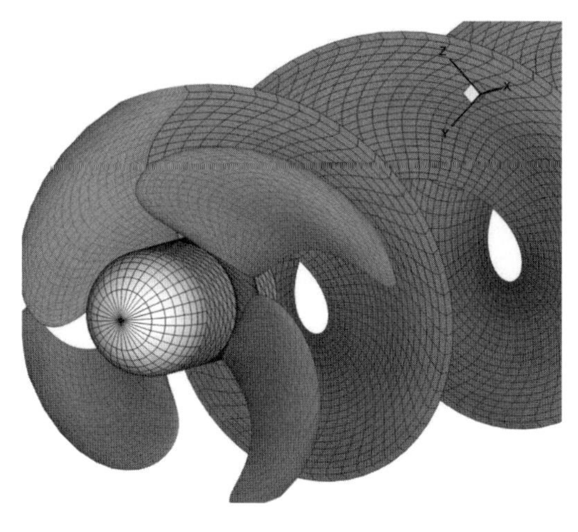

Figure 1. Panel representation of the propeller and of its trailing wake.

connected with the Kutta, the dynamic and the kinematic boundary conditions on the unknown cavity surfaces and an outer iterative cycle stepping the time. The discretized surface mesh consists of 1500 hyperboloidal panels for each blade. The trailing vortical wake extends for six complete revolutions (needed, for the unsteady case, to reach a periodic solution) and it is discretized with a time equivalent angular step of 6°, as in Figure 1.

2.2 *Multiobjective optimization*

The first step in order to apply an analysis code (like a panel method) into a design procedure through optimization is to obtain a robust parametric representation of the propeller geometry (Gaggero et al., 2009, Brizzolara et al., 2009).

The classical design table, obtained by traditional design codes, from which the original propeller has been built, is, inherently, a parametric description of the geometry itself. All the main dimensions that defines propeller geometry, like pitch, camber and chord distribution along the radius, represent main parameters that can easily be fitted with B-Spline parametric curves, whose control points turn into the free variables of the optimization procedure, as in Figure 2.

For what regard the profile shape, instead of adopting standard NACA or Eppler types, with the same parametric approach it is possible to describe only with few control points thickness and camber distribution along the chord for a certain number of radial sections (or, more consistently, to adopt a B-Surface representation of the mean nondimensional propeller surface) and include also profiles in the optimization routine (Figure 3).

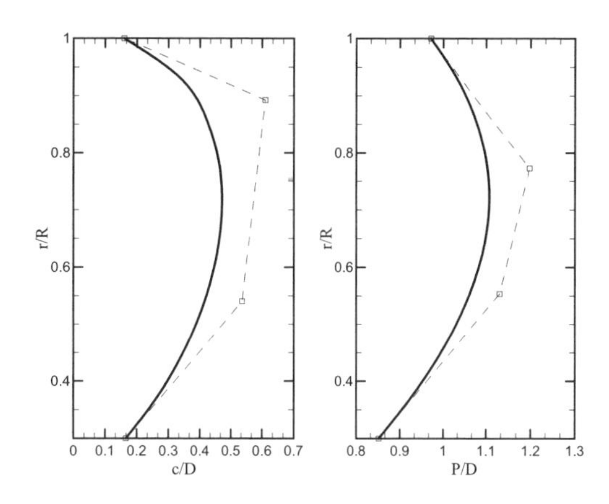

Figure 2. B-Spline representation of radial distributions of chord and pitch.

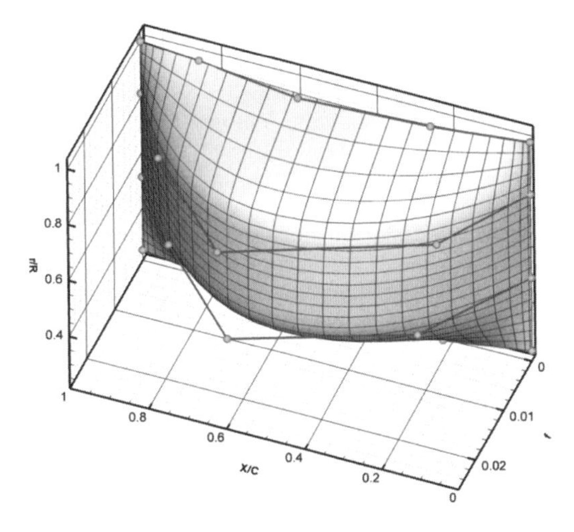

Figure 3. Parametric representation of propeller nondimensional mean surface.

The adopted optimization algorithm, built into the Mode FRONTIER environment is of genetic type: from an initial population (whose each member is randomly created from the original geometry, altering the parameter values within a prescribed range), successive generations are created via crossover and mutation. The members of the new generations arise from the best members of the previous one that satisfy all the imposed constraints (thrust identity, for instance) and grant better values for the objectives. To improve the convergence of the algorithm and speed up the entire procedure, a certain tolerance (within few percent points) has been allowed for the constraints, letting the inclusion of some more different geometries in the optimization loop. Each member of the initial population is analyzed via the potential code. Results, in terms of thrust, torque, efficiency and cavity area/volume (on the back and on the face) are collected together with the values of the parameters that describe that given geometry. The optimization algorithm, through these data, identifies the "direction" to be followed in order to satisfy the constraints and improve the objectives until convergence (or pareto convergence in the case that more than one objectives are addressed).

3 OPTIMIZATION ACTIVITY

3.1 *Original propeller characteristics*

The propeller considered for present study is a conventional 4-bladed CPP for a twin screw ship, whose main characteristics are reported in following Table 1, where D is propeller diameter, $P_{0.7}$ is pitch at 70% radial position, d_{hub} is hub diameter, A_E and A_O are propeller expanded area and disc area, Z is the number of blades.

As anticipated, the propeller is operating at constant revolution rate (about 180 RPM) in correspondence to very different ship speeds by means of blade pitch angle variation. In particular, in present work two considerably different speeds have been considered during optimisation activities, i.e., about 24 and 11 kn (referred to as "reference pitch" and "reduced pitch" respectively in the following). Pitch setting reported in table 1 is referred to the higher speed, while a reduced pitch ($P_{0.7}/D = 0.47$) corresponds to the lower speed.

Design activities have been carried out considering the nominal hull wake, which is reproduced in Figure 4, together with a shaft inclination of about 9°.

3.2 *Optimization*

Propeller optimization has been carried out in order to obtain a new geometry able to reduce back cavitation (at the design pitch) and face cavitation (at the reduced pitch) with the same numerical delivered thrust (within a range of ±2.5% to speed up the convergence) of the original propeller. At the reduced pitch, in fact, numerical predictions of thrust and torque for the original propeller

Table 1. Propeller characteristics.

Propeller characteristics	
D [m]	4.60
$P_{0.7}/D$	1.08
d_{hub}/D	0.30
A_E/A_O	0.72
Z	4

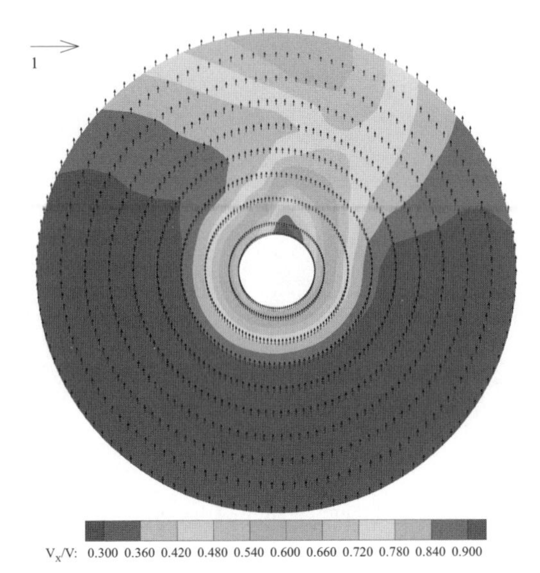

Figure 4. Nominal inflow hull wake seen from aft.

show some differences with respect to the available experimental measures. For the optimization it has been assumed that these differences, ascribed to the numerical approach, remain the same also for the newly designed propellers, let the numerical predictions for the original propeller be reference point of the optimization procedure. Also efficiency has been monitored and constrained not to be lower than that of the original geometry.

A fully unsteady optimization, although carried out with a panel method, is excessively time expensive. As presented in Gaggero (2009), unsteady performances are approximated, with a quasi-steady approach, as the mean (or the sum) of the steady performances evaluated in "N" angular wake sectors, whose mean flow characteristics (axial, radial and tangential velocity distributions along the radius) are taken as the mean radial inflow for a steady computation:

$$\overline{K_T} = \frac{1}{N}\sum_{i=1}^{N} K_{Ti}$$
$$\overline{K_Q} = \frac{1}{N}\sum_{i=1}^{N} K_{Qi} \qquad (8)$$
$$A_{cav} = \sum_{i=1}^{N} A_{cavi}$$

For the design of the new propeller three different optimization approaches have been applied. In the first case only global parameters have been investigated. Chord, maximum thickness, maximum camber and pitch distribution along the radius have been taken as free variables, maintaining the original propeller profile shapes. Chord and maximum thickness have been further constrained in order to achieve the same blade robustness of the original design. In the second case camber distribution along the chord has been added to the set of free parameters, while in the third case thickness distribution along the chord has been considered as a free parameter, adopting the better chord and maximum thickness distribution along the radius from the previous cases, with the original profile camber line. The number of free parameters included into the optimization process varies between 21 to 30. In all the cases an initial population of 300 members has been considered and let evolved for 100 generations, for a total number of 35 thousands different geometries for each optimization strategy.

Figure 5, for instance, shows a typical plot of the characteristics, in terms of cavity extension, of all the designed propellers obtained during the optimization having, as free variables, the global parameters plus the profile mean line. The main objectives of the optimization, back cavity area at the design pitch and face cavity area at the reduced pitch respectively, are reported on x and y axis, while bubble radius and colour monitor back cavity area at the reduced pitch (not evident for the original propeller and thus to be avoided) and face cavity area at the design pitch (not evident for the original propeller and thus to be avoided).

The analysis of the pareto designs shows some common trends for all the optimization strategies. Chord seems not to play an important role in the optimization process: its shape changes from the original one, but remains almost constant along the pareto frontier for all the three different approaches selected. Pitch and camber,

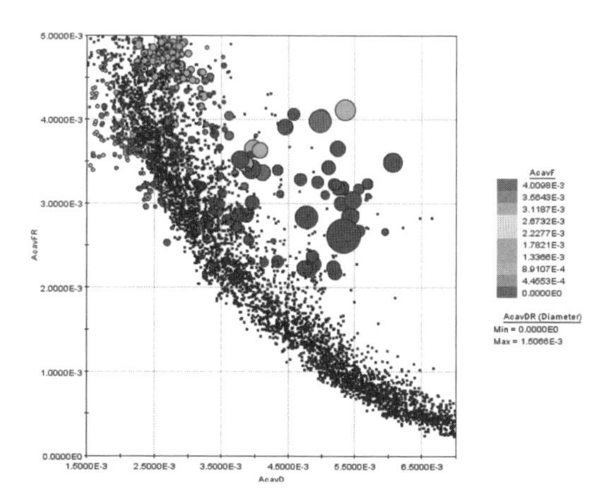

Figure 5. Pareto designs for the global parameters plus mean line optimization.

as obvious, are the main parameters from which cavity extension depends and they change mutually to maintain the required thrust. As expected, moving from geometries with reduced back cavitation (with respect to the original design) to geometries with reduced face cavitation, pitch increases at tip (to avoid opposite angles of attack) and it is balanced by a reduction of maximum camber. At the same time mean camber line is flattened at the leading edge, in order to reduce once more the risk of face cavitation.

In the light of the results of the three different optimization approaches, the optimal geometry has been selected among the pareto designs that satisfy the thrust constraints and grant zero face cavitation at the design condition and zero back cavitation at the reduced pitch condition.

In Figure 6 the pareto designs obtained with the three optimization approaches are compared. With respect to the original geometry all the pareto designs allow to sensibly reduce face cavitation, only with a minor reduction of back cavitation, showing the fact that the original propeller design was centered on maximum speed condition. With respect to the global parameters optimization, the inclusion also of the mean camber line distribution along the chord into the design procedure produces lower values of face cavity area. On the contrary, the modification of the thickness distribution is more important for the reduction of the back cavitation. The new propeller has been selected, among the pareto solutions, as a compromise between back and face cavitation, having in mind also the side effects of cavitation (in terms of radiated noise).

With respect to the original propeller, face cavitation of the selected optimum propeller is numerically reduced of about 50% while back cavitation is the 35% greater than the original one. Computed quasi steady delivered thrust is 1.67% lower at the reference pitch and 2.5% higher at the reduced pitch. This choice that increases back cavitation is justified by the fact that main noise and vibration problems were experienced by the original propeller

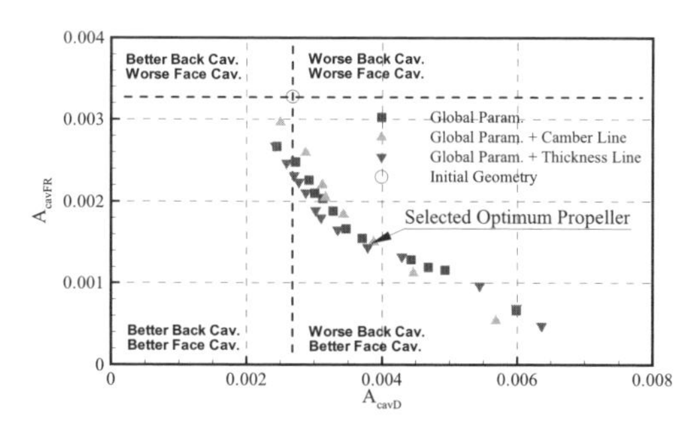

Figure 6. Comparison between the pareto designs.

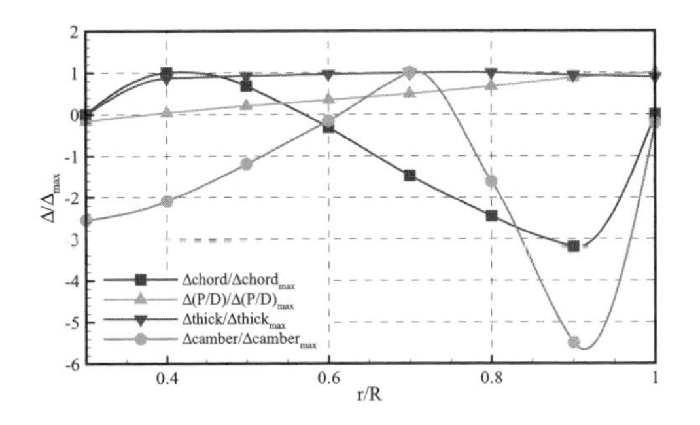

Figure 7. Nondimensionalized variations of chord, pitch, camber and thickness of the optimum propeller with respect to the original geometry.

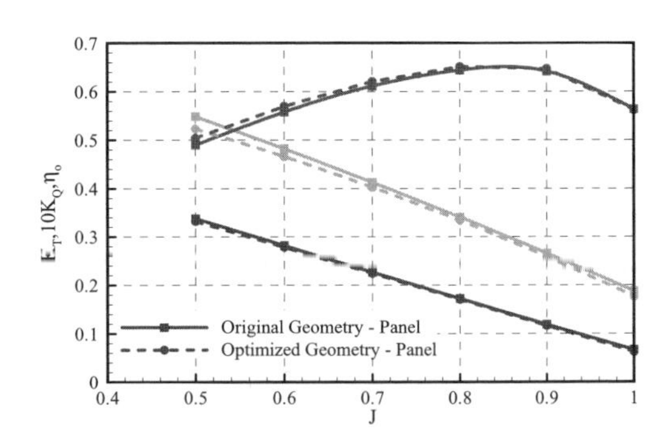

Figure 8. Comparison between original and optimized propeller—numerical open water tests—reference pitch.

at the reduced pitch condition. The choice of an optimum propeller strongly favoring performances at reduced pitch condition is also related to a better possibility to test it at the cavitation tunnel in the framework of this activity, as reported later. Comparison of main propeller characteristics in nondimensional form is reported in Figure 7. The nondimensional variations (with respect to the maximum variation) of the optimized propeller geometry clearly show how the optimization modified the original design. The lowering in chord at midspan is balanced with an increase of blade thickness necessary to maintain the same blade robustness. Pitch is increased, almost linearly, towards the tip with a stronger reduction of profile maximum camber that favors the reduction of face cavitation and slightly worsens back cavity extension.

3.3 *Analysis of original and optimized propeller*

The quasi steady optimization requires to be validated, first numerically, comparing the original and the optimized propeller in fully cavitating unsteady conditions for both the reference and the reduced pitches and, finally, as reported in the following sections, experimentally through an extensive campaign at the towing tank and in the cavitation tunnel.

In Figure 8, for instance, the numerical open water characteristics of the two designs, at the reference pitch, are compared.

Propellers are numerically equivalent in terms of working points. Thrust curves, from which ship speed, at a fixed propeller rate of revolution, depends, are overlapped within the complete range of advance coefficient. Moreover, as constrained in the optimization loop, efficiency has not been worsened: on the contrary a slightly increase (also evident in the experimental measures of Figures 13 and 14) has been achieved. At the reduced pitch conditions similar open water results have been obtained, with some major differences if compared with the experimental measures.

From the unsteady point of view, Figures 9 and 10 compare, at three different angular position inside the nominal hull wake, the back and face cavity extension at the reference and at the nominal pitches. As expected from the quasi steady optimization, back cavitation of the optimized propeller at the reference pitch is stronger. The new geometry, in fact, has been chosen to privilege a strong reduction of face cavitation at reduced pitch (as it is clear from Figure 10) instead of being a compromise that lightly reduces both back and face cavitation (for which a pareto design on the left lower side of the pareto plot of Figure 6 should have been chosen). In terms of performances (mean unsteady thrust) the optimized propeller delivers, at the reference pitch, a K_T 3% lower than that of the original design. At the reduced pitch the optimized propeller delivered thrust is 1.8% lower than the original one.

As wished fully unsteady characteristics are very close to quasi unsteady computations (at the reference pitch, for the optimized propeller, thrust coefficient varies from quasi steady conditions to fully unsteady of about −2%), giving a posteriori validation of the proposed optimization approach.

An equivalent unsteady analysis has been carried out also for the propellers operating inside the tunnel wake of Figure 16. The most significant results are reported in the following section, together with the comparison with the observed unsteady cavity extension. From a qualitative point of view it is worth to note that the higher pitch of the optimized propeller determines back cavitation at the reference pitch also in a more uniform and less decelerated wake as the tunnel one, while the original geometry is not subjected to this phenomena if working in the same inflow conditions. At the reduced pitch, instead, differences in face cavity extensions are less evident.

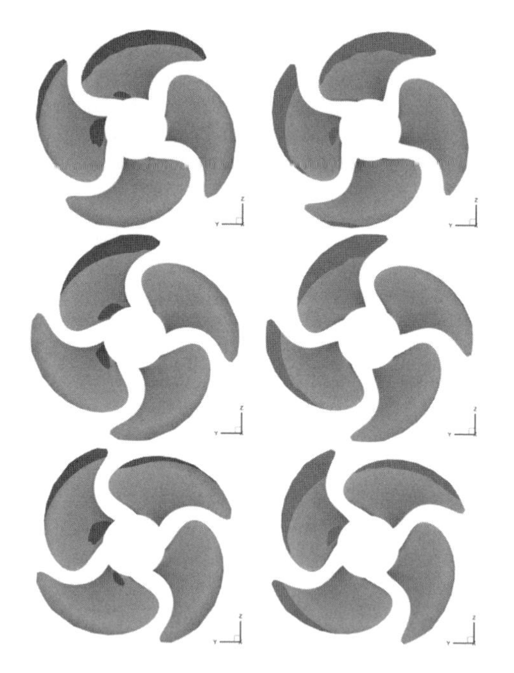

Figure 9. Comparison of predicted unsteady back cavity extension (0°–30°–60°) between original (left, blue) and optimized (right, red) propeller at reference pitch. Nominal inflow.

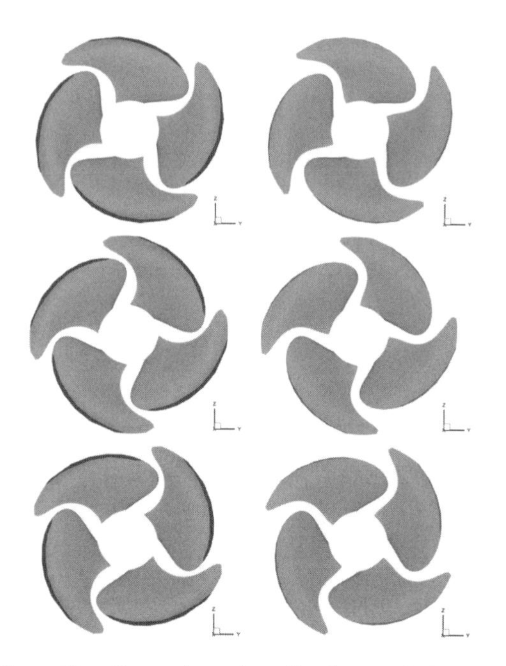

Figure 10. Comparison of predicted unsteady face cavity extension (0°–30°–60°) between original (left, blue) and optimized (right, red) propeller at reduced pitch. Nominal inflow.

For face cavitation the wake high speed regions, that are similar for the nominal hull and the tunnel wake, dominates the phenomena. Consequently numerical and experimental results in hull and in tunnel wake are comparable: face cavitation developed by the original propeller inside the nominal wake is almost the same of the face cavitation developed in tunnel wake conditions. The improvements obtained by the new propeller optimized for the nominal wake, as a consequence, can be appreciated, numerically and, more significantly, experimentally, also for the inflow conditions reproduced at the cavitation tunnel.

4 EXPERIMENTAL CAMPAIGN

4.1 *Open water tests*

As a first step in the experimental campaign, propeller open water tests have been carried out at CEHIPAR towing tank for both propellers and pitches.

In following Figures 11 and 12, comparison of numerical results and measurements at both pitches for the optimised propeller are reported.

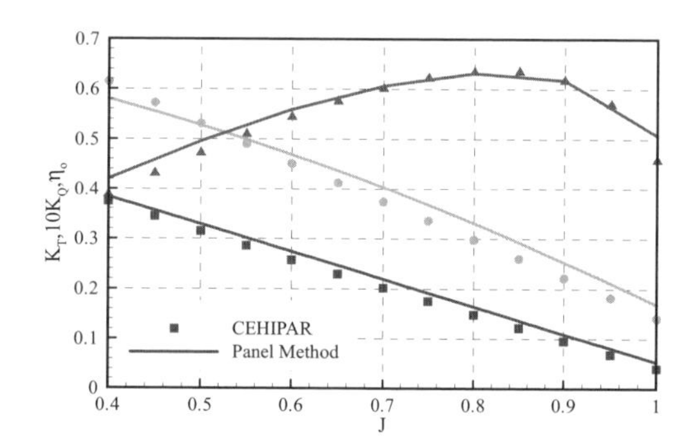

Figure 11. Comparison between experiments and numerical results, reference pitch, optimized propeller.

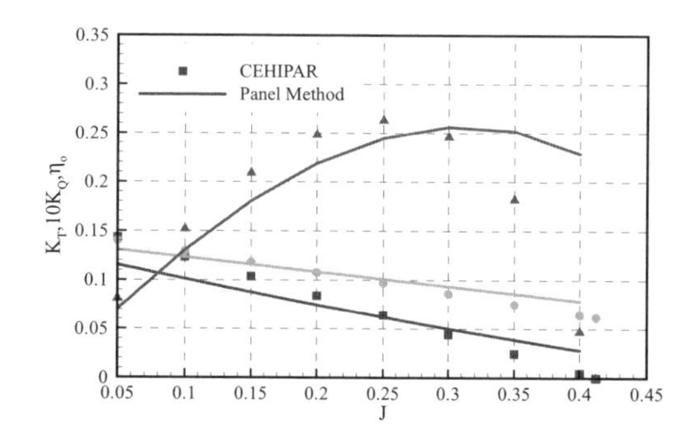

Figure 12. Comparison between experiments and numerical results, reduced pitch, optimized propeller.

In correspondence to reference pitch the overall agreement between measures and the numerical code is satisfactory, with a slight overestimation of both thrust and torque in a large range of advance coefficient values. Only for lower values numerical torque curve presents a change in slope, related to leading edge separation.

In correspondence to reduced pitch differences are higher, computed thrust and torque curves present a different slope with respect to the measured ones, which results in rather marked errors at higher and lower values of advance coefficient.

Results for original propeller (not reported for the sake of brevity) present a similar trend. Notwithstanding the problem related to low pitch, the method was successful in predicting propeller cavitating behaviour and, more in general, to provide a design with desired characteristics (constant thrust and reduced face cavitation).

In following Figures 13 and 14, open water test results of the two propellers are compared.

It is clear from previous figures that propellers are equivalent in terms of functioning point at reference pitch (same thrust at same advance coefficient, thus leading to same speed at same RPM), as requested and numerically verified in the design loop; moreover, the optimized propeller presents a higher efficiency. Similar trend (even if with slightly different thrust curves) was obtained also at reduced pitch.

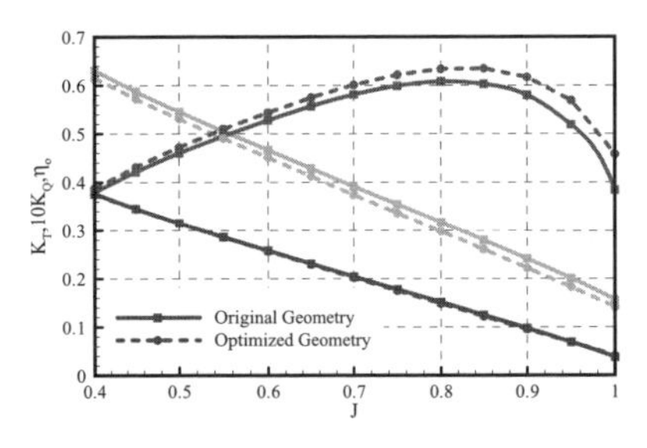

Figure 13. Comparison between original and optimized propeller—open water tests—reference pitch.

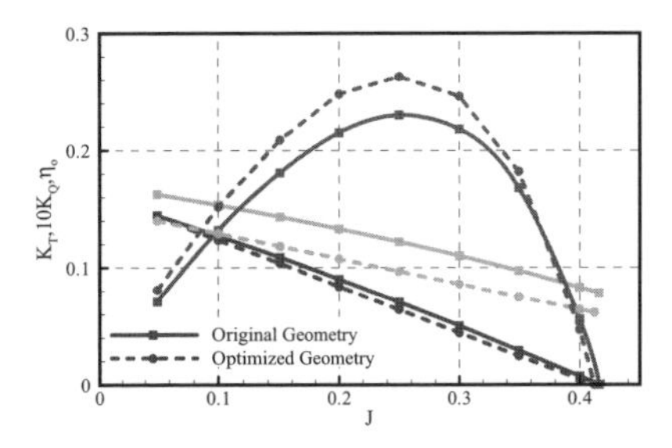

Figure 14. Comparison between original and optimized propeller—open water tests—reduced pitch.

4.2 *Cavitation tests*

Cavitation tests were performed at DINAEL Cavitation Tunnel, where inception points of various cavitation phenomena have been measured, and cavitation extent observations have been carried out in correspondence to different functioning points. Thrust equivalence (respectively 0.18 for the reference pitch, 0.032 for the reduced pitch at a cavitation index of 1.35) has been forced, changing the undisturbed inflow speed in the numerical computations to match the experimental measured values. In present paragraph, cavitation observation results for the design conditions of both the propellers are reported.

As anticipated, tests were not carried out in correspondence to the nominal wake considered in the design activity. Reason for this is to avoid adoption of a wake screen, which could provide a disturbance for the successive noise measurements. As a consequence, in testing condition inclined shaft was adopted together with two shaft brackets, as in the original ship. Resulting wake is reported in Figure 15, showing a clear discrepancy with original one, and in particular the absence of a rather marked low speed region in correspondence to 0° position.

Nevertheless, it is believed that tests are already indicative of the capability of codes adopted to capture cavitation extent in correspondence to different geometries and pitches, since calculations were carried out also considering "tunnel wake". Moreover, it has to be remarked that worst condition for original propeller was pointed out to be the reduced pitch. Numerical simulation showed that the wake has limited influence on the cavitating behaviour at this pitch setting, which is dominated by off-design functioning, thus the difference was accepted.

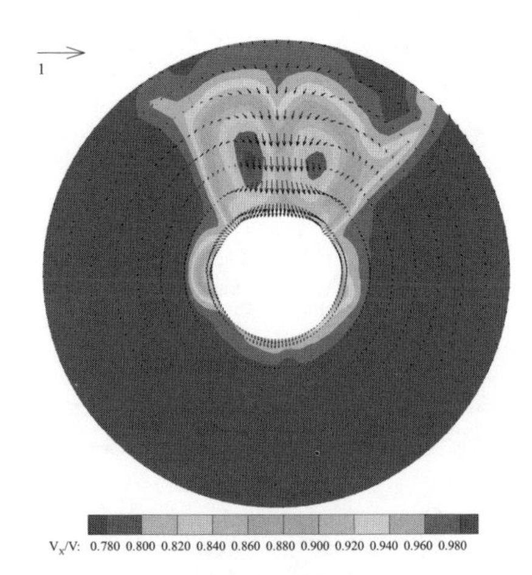

Figure 15. Cavitation tunnel wake.

In following Figures 16–19, numerical and experimental results for both propellers at the two pitches are reported; as it can be seen, a very good agreement is obtained in correspondence to all conditions, confirming code capability to capture with good accuracy cavitation extension. Moreover, results from optimization activity are confirmed, with worse performances in correspondence to reference pitch and better performances in correspondence to reduced pitch. It has to be noted that differences in correspondence to reference pitch are certainly amplified with adopted wake with respect to design condition. As it can be seen, original propeller presents almost no cavitation in this condition (only at 90° angular position), thus showing a rather high difference with respect to optimized one. In the effective wake, the effect of the low speed region tends to increase cavitation extent for both propellers, with the result of diminishing differences between them.

4.3 *Radiated noise measurements*

In order to carry out radiated noise measurements, one hydrophone was mounted on a small cylindrical support protruding from a fin; the hydrophone is located downstream the propeller, outside direct propeller slipstream, and its position relative to the propeller is reported in following Figure 20.

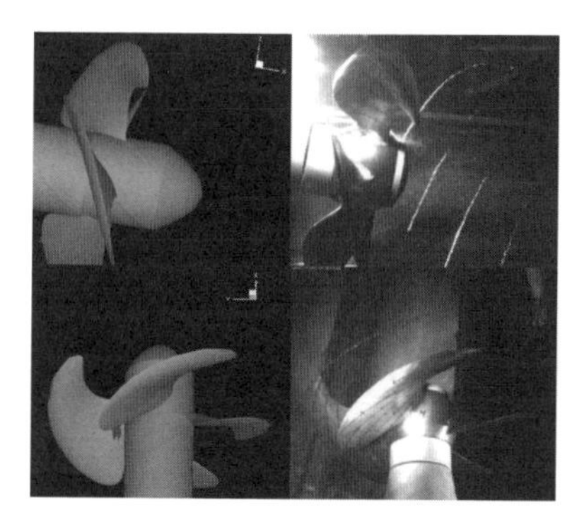

Figure 16. Observed and predicted cavitation extent—original propeller at reference pitch.

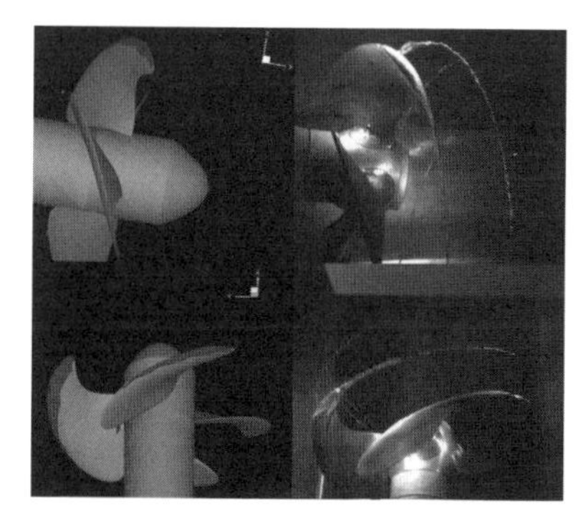

Figure 17. Observed and predicted cavitation extent—optimized propeller at reference pitch.

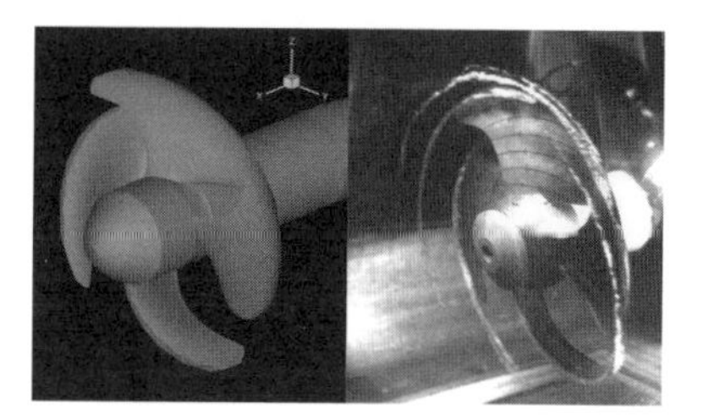

Figure 18. Observed and predicted cavitation extent—original propeller at reduced pitch.

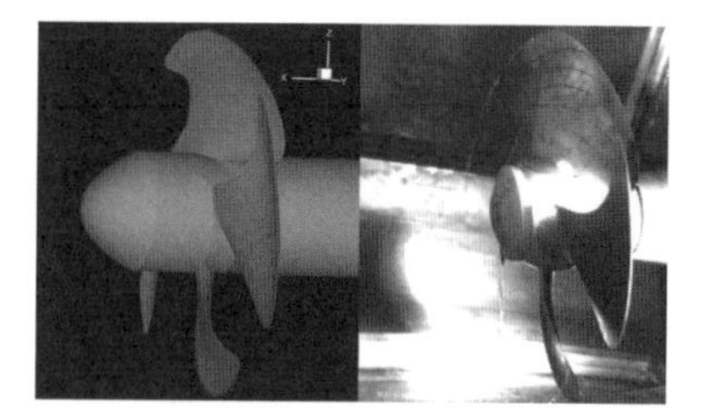

Figure 19. Observed and predicted cavitation extent—optimized propeller reduced pitch.

All tests were carried out with propeller working at 25 Hz. In correspondence to each functioning point, two successive measurements were carried out, i.e., with propeller and without propeller (substituting it with a dummy hub), in order to measure background noise. From the comparison between background and propeller noise, it was possible to verify that acquired signal with propeller functioning is considerably higher than background in the complete frequency range considered (up to 30 kHz).

For the two propellers, a considerable number of different functioning points have been tested, in terms of both K_T and σ_N values (Bertetta et al., 2011). In present work, only measurements in correspondence to the two design conditions at reference and reduced pitch are reported in following Figures 21 and 22. Numerical values have been omitted from graphs for industrial reasons.

Results are presented in terms of 1/3 octave spectrum scaled at 1m; in particular, for each band, nondimensional value K_p is evaluated as follows, together with the corresponding level, following Bark (1986):

$$K_p = \frac{prms}{\rho n^2 D^2}$$
$$L_p(K_p) = 20\log_{10}\left(\frac{K_p}{10^{-6}}\right) \tag{9}$$

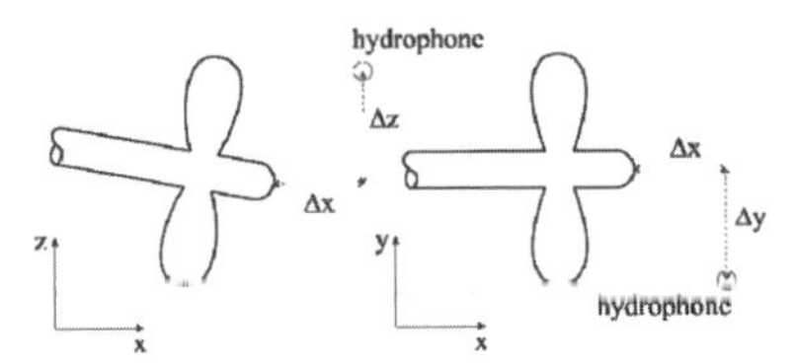

Figure 20. Hydrophone/propeller relative positions.

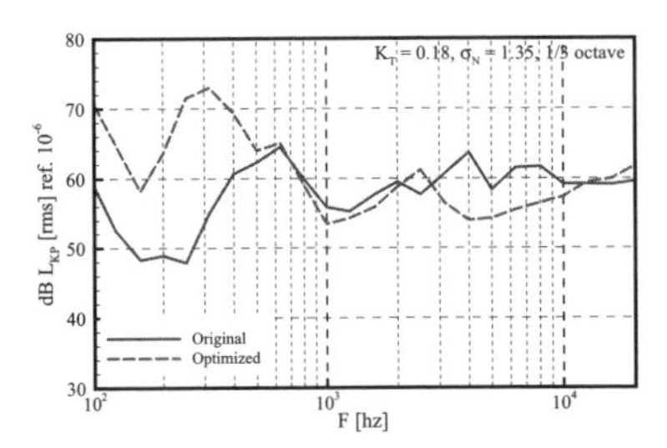

Figure 21. Radiated noise measurements at reference pitch.

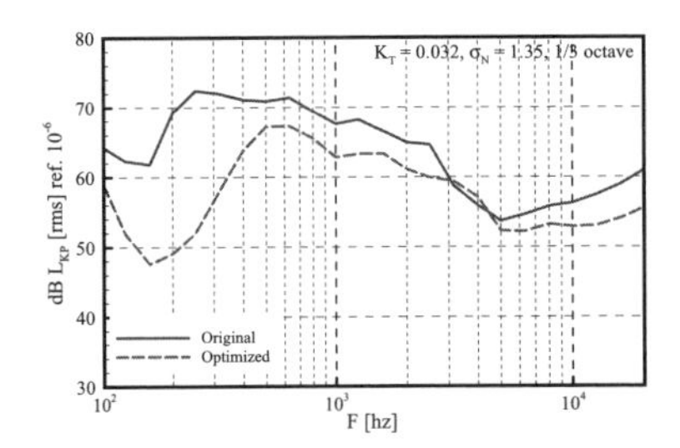

Figure 22. Radiated noise measurements at reduced pitch.

where p_{rms} is the root mean square value of each spectrum component.

Main difference between the two propellers at reference pitch setting is due to the tip vortex related phenomena, which are amplified in case of the optimized propeller, consistently with the design assumptions. In Figure 21 the higher peak of the tip vortex is clearly visible. Also in correspondence to reduced pitch condition main difference is due to the vortex related phenomena (in this case face vortex and vortex from sheet face), with opposite effect with respect to design pitch condition. In particular, it is clear that the delay of the vortex from sheet face phenomenon results in a considerable reduction of the noise spectrum of optimized propeller with respect to the original propeller.

5 CONCLUSIONS

In present work, design of a CPP propeller has been carried out, utilising a cavitating panel code for propeller analysis coupled to an optimisation tool. A particularly difficult task has been considered, i.e., optimisation at two very different design points with different pitch settings at constant RPM. In particular, attention has been given to reduced pitch condition, since worse problems were experienced in correspondence to this condition, with strong propeller induced vibrations on the hull. As a consequence, it was voluntarily decided to accept slightly worse behaviour of propeller at design condition in order to improve behaviour at reduced pitch.

Results from the experimental campaign carried out at towing tank and cavitation tunnel showed a very good agreement with numerical calculations; propeller noise and pressure pulses were considerably reduced in correspondence to reduced pitch condition (eliminating almost completely the original problem), with some worsening in correspondence to design pitch.

The optimisation tool proved to be an efficient mean to improve propeller characteristics, even if in a such problematic case it is difficult to obtain very large improvement at both operating conditions with conventional pitch propellers if operated at constant RPM. In case the same tool is applied to a single operating point, it is expected that much larger improvements may be obtained.

ACKNOWLEDGEMENTS

Part of the activities presented in this work have been carried and funded by SILENV—European Collaborative Project n° 234182.

REFERENCES

Bark, G. 1986. Prediction of Propeller Cavitation Noise from Model Tests and Its Comparison with Full Scale Data, *SSPA Report No. 103.*

Bertetta, D., Savio, L. & Viviani, M. 2011. Experimental characterization of two CP propellers at different pitch settings, considering cavitating behaviour and related noise phenomena, *Proc. Second International Symposium on Marine Propulsors, Hamburg, Germany, June 2011.*

Brizzolara, S., Gaggero, S. & Grasso, A. 2009. Parametric Optimization of Open and Ducted Propellers, *Proc. Propeller/Shafting Symposium, Williamsburg, Virginia, USA, September 2009.*

Gaggero, S. & Brizzolara, S. 2009. Parametric CFD Optimization of Fast Marine Propellers, *Proc. 10th International Conference on Fast Sea Transportation, Athens, Greece, October 2009.*

SILENV—Ship Oriented Innovative Solutions to reduce Noise and Vibrations. FP7 Collaborative Project 234182. Work Package 3, Subtask 3.1.1: Propellers.

Sustainable Maritime Transportation and Exploitation of Sea Resources – Rizzuto & Guedes Soares (eds)
 ISBN 978-0-415-62081-9

Latest experiences with Contracted and Loaded Tip (CLT) propellers

Juan Gonzalez Adalid
SISTEMAR s.a, Madrid, Spain

Giulio Gennaro
SINM s.r.l., Genoa, Italy

ABSTRACT: Since 1976 tip loaded concept has been evolved continuously. Presently SISTEMAR's CLT propellers are installed on about 280 vessels world wide. Despite this fact the use of CLT propellers has been curtailed due to the lack of specific knowledge of this type of propeller and due to the fact that CLT propellers need a different model-to-full-scale extrapolation methodology in respect with conventional propellers.

The conclusions that can be drawn from more than 30 years of model and full scale experiences are the following:

- CLT propellers are a fully developed technology;
- CLT propellers grant several significant advantages over conventional propellers, the most important being:
  - 5 to 8% higher efficiency over the entire operational range (i.e., 5 to 8% fuel saving and 5 to 8% reduced emissions);
  - lower induced noise and vibrations;
  - Improved ship maneuverability characteristics;
- CLT propellers are extremely indicated for new buildings and very attractive for ships in service.

1 INTRODUCTION

The first claims about the potential advantages of tip loaded propellers (TVF propellers, Tip Vortex Free propellers) were published in October 1976 in "Ingeniería Naval" [Ref. 1].

Since 1976 tip loaded concept has been evolved continuously. After the first few years the design of tip loaded propellers was improved by accounting for the contraction of the fluid vein crossing the propeller disk. In this manner the new generation of SISTEMAR's tip loaded propellers was devised and, consequentially, it was named CLT propeller: Contracted and Loaded Tip Propeller.

Presently CLT propellers are the most efficient and widespread type of unconventional propeller, being currently installed on about 280 vessels worldwide.

In 1980 the "Lifting Line Theory" was generalized and the first designs were produced. From 1983 onward the use of folded tip propeller became systematic. Later on Prof. Gonzalo Perez Gomez and Juan González-Adalid completely revisited the screw propulsion theory; in 1993 they published the "New Momentum Theory", which corrects the conventional "Momentum Theory"; in 1995 they published the "New Cascade Theory", which takes into account the three dimensional cascade effects differentiating a propeller blade from an isolated wing profile.

From the beginning, the full scale experiences with CLT propellers showed remarkable agreement with the design calculation, while it became apparent the ITTC extrapolation procedures were not adequate for CLT propellers. This was, and it still is, a major obstacle for the installation of CLT propellers on new-buildings.

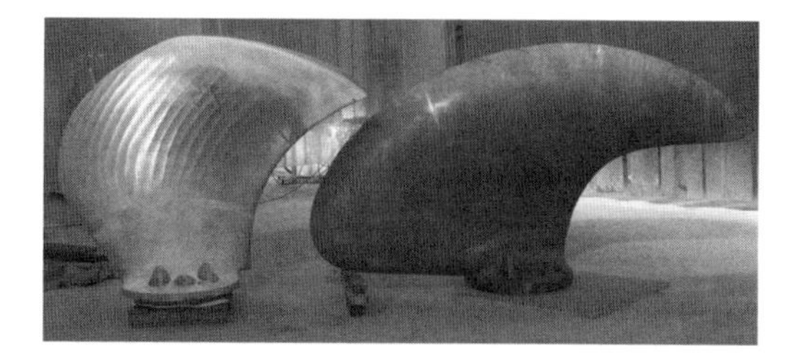

Figure 1. The striking difference between a CLT and a state-of-the-art high skew conventional propeller blade designed for the same Ro-Pax.

For this reason a series of R&D activities were carried out by SISTEMAR, in cooperation with CEHIPAR and NAVANTIA, between 1996 and 2006, in order to devise suitable model test procedures and extrapolations capable of predicting the full scale performance of CLT propellers with a similar level of confidence as in the case of conventional propellers.

Since 2007 SISTEMAR has been actively participating in different R&D projects of EU convocatories and has performed a very extensive R&D project with the A.P. Moeller Maersk Group.

2 GENERAL DESCRIPTION OF CLT PROPELLERS

CLT propellers are characterized by the following:

- The blade tip generates a substantial thrust.
- The pitch increases from the root to the tip of the blades.
- The chord at the tip is finite.
- End plates are fitted at the blade tips, toward pressure side; they are adapted to the fluid vein contraction to reduce as much as possible their viscous resistance.

The end plates operate as a barriers, avoiding the communication of water between the pressure and the suction side of the blades, allowing to establish a finite load at the tip of the blades. At the same time the presence of longer chords at the tip determines the decrease of the local loading per square meter, thereby helping in controlling the cavitation.

The fundamental goal of the CLT propeller is to improve the propeller open water efficiency by reducing the hydrodynamic pitch angle through the reduction of the magnitudes of induced velocities at the propeller disk.

In the new momentum theory the parameter ε is defined; ε is the ratio between the suction in front of the propeller disk and the pressure jump across the propeller disk. In other words ε defines how the propeller thrust is obtained by combining the under-pressure existing at the suction side of the propeller blades:

$$(p_o - \varepsilon \Delta p)$$

with the over-pressure existing at the pressure side of the propeller blades:

$$(p_o + (1 - \varepsilon)\Delta p).$$

In accordance with the new momentum theory, to reduce the magnitude of the induced velocities at the propeller disk it is necessary to reduce the value of ε, which means to reduce the suction for the same pressure jump across the propeller disk.

The non dimensional propeller specific load coefficient is defined as follows:

$$C_{TH} = T / (0.5\ \rho\ A\ V^2)$$

The ideal propeller efficiency can be expressed as a function of the non dimensional propeller specific load; according to the classic momentum theory the formulation is as follows:

$$\eta_0 = 2 / (1 + (1 + C_{TH})^{0.5})$$

The expression of the ideal propeller efficiency according to the New Momentum Theory is rather different:

$$\eta_0 = 1 / (1 + \varepsilon\ C_{TH})^{0.5}$$

In Figure 3 both formulations are plotted; the following comments can be made:

- According to the new momentum theory η_0 increases when ε decreases;
- The new momentum theory allows for greater ideal efficiency than classic momentum theory in case of ε parameter having a low value.

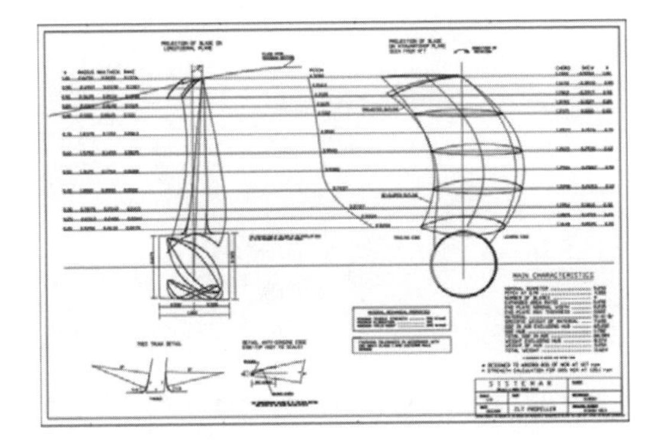

Figure 2. Typical drawing of a FP CLT propeller.

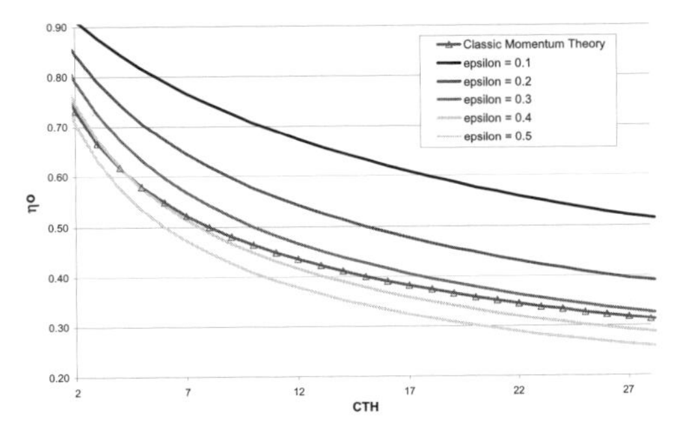

Figure 3. Ideal efficiency according to either the classic and the new momentum theory.

Table 1. Open water efficiency vs thrust loading coefficient for some recent CLT propeller designs.

Case	C_{Th}	η_o
1	3.077	0.596
2	2.973	0.576
3	1.249	0.726
4	1.090	0.709
5	0.712	0.729

The ε coefficient depends on the type of propeller, its main characteristics (diameter, number of blades, blade area ratio, etc.) and on the radial load distribution.

For conventional propeller ε is in the range of 0.4, for a CLT propeller ε is in the range of 0.1.

In the following table the full scale open water efficiency of recent CLT propellers is given in relation to their thrust loading coefficient.

3 ADVANTAGES OF CLT PROPELLERS

Up to date CLT propellers, both of fix and controllable pitch type, have been successfully installed on more than 280 vessels, of very different types:

Tankers, Product carriers, Chemical carriers, Bulk carriers, Cement carriers, General cargoes, Container ships, Reefers, Ro-Ro, Ro-Pax, Fishing vessels, Trawlers, Catamarans, Hydrofoils, Patrol boat, Landing crafts, Oceanographic ships, Yachts.

The application range has been extremely wide:

- Up to 300,000 DWT
- Up to 22 MW per propeller
- Up to 36 knots.

The advantages of CLT propellers over conventional propellers resulting from full scale installations and from several comparative full scale trials and long term observation are the following:

- Higher efficiency (between 5 and 8%)
  - Fuel saving
  - Reduced emissions
  - Saving on MM/EE maintenance
  - Higher top speed
  - Greater range
- Inhibition of cavitation and of the tip vortex
  - Less noise
  - Less vibrations
  - Lower pressure pulses
  - Lower area ratio
- Greater thrust
  - Smaller propeller optimum diameter
  - Better maneuverability.

It should be remarked that the advantages offered by CLT propellers in terms of reduced emissions and fuel consumption add to what achieved by other means (e.g., hull form optimization, hull maintenance, slow steaming, exhaust gas treatment …).

The percentage of efficiency improvement over an alternative conventional propeller and hence the fuel saving achieved depends on the type of vessel, being higher for slow vessels with high block coefficient as tankers, bulk carriers, etc.

CLT propellers can be applied both for newbuildings and ships in service, either in FP or CP type. The boss for FP applications and the blade flange for CP are interchangeable with the ones of the alternative conventional propeller/blades and the inertia is almost the same, therefore the installation of CLT propeller/blades does not introduce any modification in the shaft line neither for newbuildings nor retrofittings.

In case of CP applications there are additional advantages for CLT blades operating in off-design conditions at constant rpm derived from its special radial pitch distribution. For conventional blades the radial pitch distribution at design pitch setting is unloaded at the blade tip with the aim to reduce the risk of high pressure pulses; such unloading becomes excessive in off-design low-pitch conditions so that the outer sections of the blades provide a negative thrust while the inner sections provide a positive thrust, as a consequence the propeller efficiency decreases while the level of pressure pulses increases because of the existence of a broad band spectra.

This is not the case for CLT blades because in off-design conditions the blade tip is a little bit unloaded but it still produces a positive thrust for a wide range of pitch settings and therefore the propeller efficiency is high also in off-design conditions and the broad band spectra of pressure pulses is not generated.

Figure 4. FP CLT propellers installed on an hydrofoil.

The effective reduction of pressure pulses, for CLT propellers, is due to both the higher clearances between propeller and stern post contour (thanks to the lower optimum diameter) and to the lower amplitude of higher orders harmonics (thanks to the more stable sheet cavitation developed on the suction side of the propeller). Model and full scale measurements demonstrated that the amplitude of the first harmonic may be similar to than the one of the alternative conventional propeller, or even higher in some case, but the higher order harmonics are much lower and therefore the total excitation for the CLT propeller is lower than for an alternative conventional propeller.

The amelioration in the maneuverability is due to the higher overpressure downstream of the CLT propeller and to the flow concentration produced by the end plates which can be roughly compared with the effect of a nozzle. Due to the higher pressure and the flow concentration acting on the rudder the action of the rudder is more effective in combination with a CLT propeller than with an equivalent conventional propeller. Significant reductions in tactical diameter and in the crash stop distance have been measured at full scale with different vessels alternatively fitted with conventional and CLT propellers.

4 PAST R&D PROJECTS ON CLT PROPELLERS

In the past the following R&D activities were carried out on CLT propellers.

1997–2000 "Optimization of ship propulsion by means of innovative solutions including tip plate propellers." CEHIPAR, NAVANTIA, SISTEMAR. This R&D project resulted in the development of an ad hoc extrapolation procedure for open water tests of CLT propeller. The extrapolation is based on the ITTC-78 method adapted for CLT propellers by considering the presence of the end plates and the scale effects on lift forces.

During 1999 a new type of mean lines has been developed by SISTEMAR with the aim to improve further the efficiency of CLT propellers by reducing the under-pressure on the suction side and increasing the overpressure on the pressure side. These mean lines are characterized by a higher slope at the trailing edge compared to standard NACA mean lines.

2001–2003 "Research on the cavitation performance of CLT propellers, on the influence of new types of propeller blades annular sections and the potential application to POD's" CEHIPAR, NAVANTIA, SISTEMAR. This R&D project resulted in the development of a new procedure for cavitation tests and pressure

Figure 5. A CP CLT propeller installed in a modern Ro-Pax.

fluctuation measurements with CLT propeller at model scale.

2003–2005 "Research on the performance of high loaded propellers for high speed conventional ferries" CEHIPAR, NAVANTIA, SISTEMAR, TRASMEDITERRANEA, TSI. The aim of this research was the full scale application of CLT propeller blades to a large and modern conventional Ro-Pax and a complete full scale measurement campaign, aimed at comparing the CLT propeller blades with state-of-the-art high skew conventional propeller blades.

2005–2008 "SUPERPROP: Superior Life Time Operation of Ship Propeller" an EU sponsored R&D project aimed at studying the influence of different maintenance policies on the hydrodynamic performance of tugs and trawlers.

Within this project a CLT propeller was successfully retrofitted on a trawler.

5 ONGOING R&D PROJECTS ON CLT PROPELLERS

Several R&D projects on CLT propellers are currently being conducted.

In 2009 SISTEMAR has been invited by CEHIPAR and VTT to participate as subcontractor to the SILENV project implemented under the Seventh Framework Program of the EC, with the main objective of establishing a "green label" for vessels achieving low levels of noise and vibration on board as well as defining design guidelines to achieve said levels. CLT propellers will be analyzed by means of CFD calculations and model tests as one of the potential resources to decrease the noise and vibration levels on board.

The project "Triple Energy Saving by use of CRP, CLT and PODed propulsion" (TRIPOD) has been approved within the FP7 of the EU. The project has started on 1st November 2010 with the participation of A.P. Moeller Maersk, ABB, VTT, CEHIPAR, CINTRANAVAL DEFCAR and SISTEMAR. The main goal of this project is the development and validation of a new propulsion concept for improved energy efficiency of ships through the advance combination of three existing propulsion technologies: podded propulsion, CLT propellers and counter-rotating propeller (CRP) principle.

The ship selected for this R&D project is the 8.500 TEU's container vessel "Gudrun Maersk."

6 THE ROY MAERSK R&D PROJECT

In 2006 SISTEMAR entered into talks with A.P. Moeller Maersk who, at that time, was conducting an internal evaluation of energy saving devices. The CLT propellers were selected as the single most promising device and a joint R&D campaign was launched with the aim of conducting comparative model and full scale tests with CLT propellers.

CLT propellers were designed for a 2,500 TEU container vessel, a 35,000 DWT product tanker and a VLCC. Subsequently all three CLT propellers were tested at model scale at HSVA, Hamburg. The CLT propeller for the 35,000 DWT product tanker was also tested at CEHIPAR.

It was finally decided to proceed to a full scale on the 35,000 DWT product tanker Roy Maersk. At the end of October 2009 she was retrofitted with a CLT propeller.

Figure 6. 35,000 DWT product tanker "Roy Maersk", new CLT propeller and pre-existing WED.

Table 2. Main characteristics of M/V Roy Maersk.

L_{PP}	162.0	m
B	27.40	m
T	9.75	m
Δ	35,300	t

Table 3. Main characteristics of the propellers.

	Conventional	CLT	
D	5.65	5.25	m
Z	4	4	–
a_E	0.563	0.490	–
P @ 0.7 r	3.685	4.050	m

A long measurement campaign has been conducted at full scale on the Roy Maersk, with two main goals:

- to compare the model scale extrapolations with the full scale measurements;
- to ascertain the advantages of the CLT propeller over the conventional propeller.

At the time of writing the present paper we were not yet authorized to disclose fully the results of the Roy Maersk R&D project, which will be the subject of a paper to be published in the near future.

In any case we can confirm that both goals were satisfactorily achieved:

- the full scale pressure pulses showed a good agreement with the ones measured at model scale;
- the cavitation patterns observed at full scale showed very good agreement with the ones observed in HSVA's HYCAT cavitation tunnel;
- the CLT propeller open water test extrapolation was checked by means of a KQ analysis which gave satisfactory results, uncertainties taken in due care;
- the gain in efficiency was comparable to the expected one, uncertainties taken in due care.

7 CLT PROPELLERS AND CFD

CFD technology has being applied to CLT propellers by several reputable institutions, such as CEHIPAR and VTT, in order to try to gain insight to the scale effects peculiar of CLT propellers.

The results of these calculations have been contradictory, in general the CFD have confirmed the larger scale effects on KT, in comparison with conventional propellers, but they have been inconclusive in respect with the scale effects on KQ.

In particular very large deviations have been observed between measurements and calculations performed at model scale, with errors up to about 6.5% on KT and 16.5% on KQ. This has so far prevented the use of CFD other than for qualitative purposes.

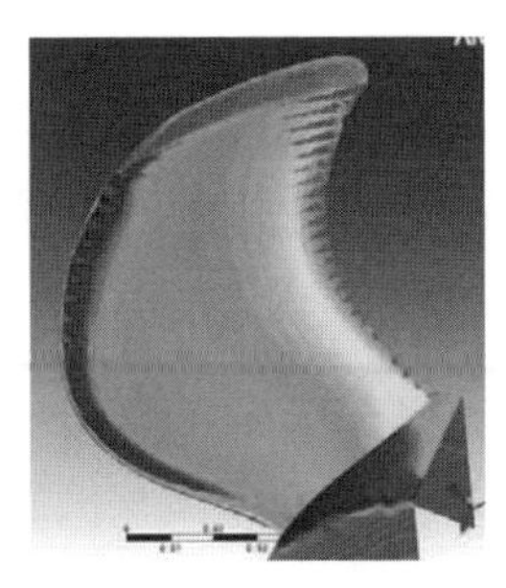

Figure 7. Pressure distribution on the pressures side (left) and on the suction side (right) of a CLT propeller blade computed by means of CFD.

Lately CEHIPAR and the University of Genoa have reported very good agreement between measurements and CFD calculations of CLT propellers. In addition Lloyd Register has reported good agreement between observed and calculated cavitation patterns.

In conclusion it appears that today CFD is not yet sufficiently reliable for calculating CLT propellers but that it will be in a matter of years.

8 LATEST CLT PROPELLER INSTALLATIONS

In the last few years several noticeable CLT propeller installations have been completed.

In October 2009 the sea trials of the m/v Cantabria (Buque de Aprovisionamiento en Combate, BAC Class), the new replenishment vessel of the Spanish Navy, were carried out.

The m/v Cantabria is equipped with the largest and most powerful CP CLT propeller manufactured to date (single screw, 5 blades, diameter 5.7 meter, MCR 21.8 MW).

The full scale performance of the CLT propeller was in line with both design calculation and model test predictions.

The Spanish Navy has programmed a series of 14 corvettes (Buques de Acción Marítima, BAM class) to be built by NAVANTIA. All the units of the BAM class will be equipped with CP CLT propellers (twin screw, 4 blades, diameter 3.45 meter, MCR 2 × 4.5 MW).

All the 4 units of the first batch have been already launched and the sea trials of the lead unit were performed in March 2011 with satisfactory results.

The next batch of 5 units has been already announced by the Spanish Ministry of Defense and it is expected to be signed by summer 2011.

The Brazilian company EMPRESA DE NAVEGAÇAO ELCANO has signed a new building contract with ESTALEIROS ITAJAI shipyard for three 7,500 m³ LPGs to be chartered to PETROBRAS. The propulsion system will consist of a two stroke Diesel engine, MCR 4,400 kW, driving a 3.9 m CP CLT propeller. Model tests have been satisfactorily carried out at CEHIPAR, Madrid, in early 2011.

Figure 8. The replenishment vessel A15 Cantabria.

Figure 9. The launching of P41 Meteor, the lead unit of the BAM class.

9 CONCLUSIONS

The technology for the design of CLT propellers is fully developed as well as the technology related with their model tests. This allows to estimate the performance of a CLT propeller at full scale with the same degree of accuracy of a conventional propeller.

The merits claimed for the CLT propeller (higher efficiency, lower noise an vibration levels and better maneuverability characteristics) have been demonstrated in more than 280 full scale applications for very different ship types.

The percentage of efficiency increase is in the range of 5–8%, in general being higher for slow vessels with high block coefficient as tankers, bulk carriers, etc ... Such increase in efficiency translates directly into comparable fuel saving and emission reduction.

SISTEMAR is currently involved in the application of CLT propellers to podded propulsors as well as in the application CFD codes to CLT propellers.

REFERENCES

A. Sanchez Caja; T.P. Sipila & J.V. Pylkkanen. "Simulation of the Incompressible Viscous Flow around an Endplate Propeller Using a RANSE Solver" 26th Symposium on Naval Hydrodynamics Rome, Italy, 17–22 September 2006.

B. Cerup.Simonsen; J.O. De Kat; O.G. Jakoben; L.R. Pedesen; J.B. Petersen & T. Posborg. "An integrated approach towards cost-effective operation of ships with reduced GHG emissions". SNAME, 2010.

"CLT: A Proven Propeller for Efficient Ships". Supplement of the Naval Architect, July/August 2005 issue.

G. Perez Gomez. "Una innovación en el proyecto de hélices". Ingeniería Naval, October 1976.

G. Gennaro; J.G. Aadalid & R. Folso. "Contracted and loaded tip (CLT) propellers. Latest installations and experiences". 16th International Conference of Ship and Shipping Research, NAV 2009. Messina, Italy. November 2009.

G. Perez Gomez & J. Gonzalez-Adalid. "Detailed Design of Ship Propellers". Book edited by FEIN (Fondo Editorial de Ingeniería Naval), Madrid 1998.

G. Perez Gomez & J. Gonzalez-Adalid. "Nuevo procedimiento para definir la geometría de las líneas medias de las secciones anulares de las palas de una hélice". Ingeniería Naval, September and December 1999.

G. Perez Gomez & J. Gonzalez-Adalid. "Optimisation of the propulsión system of a ship using the Generalised New Momentum Theory". WEMT Golden Metal Awards, 1997.

G. Perez Gomez & J. Gonzalez-Adalid. "Scale effects in the performance of a CLT propeller". The Naval Architect, July/August 2000.

G. Perez Gomez & J. Gonzalez-Adalid. "Tip Loaded Propellers (CLT). Justification of their advantages over high skewed propellers using the New Momentum Theory". SNAME—New York Metropolitan Section, Fiftieth Anniversary 1942–1992, 11th February 1993 and also at International Shipbuilding Progress nº 429, 1995.

G. Perez Gomez; M. Perez Sobrino; J. Gonzalez-Adalid; A. Garcia Gomez; J. Masip Hidalgo; R. Quereda Lavina; E. Minguito Caderna & P. Beltran Palomo. "Un hito español en la propulsión naval: rentabilidad de un amplio programa de I+D+i". Ingeniería Naval, June 2006.

H. Haimow; J. Vicario & J. Del Corral. "Ranse Code Application for ducted and endplate propellers in open water". 2nd International Symposium on Marine Propulsors SMP'11, Hamburg, June 2011.

M. Perez Sobrino; J.A. Alaez Zazurca, A. GarciaGomez; G. Perez Gomez & J. Gonzalez-Adalid "Optimización de la propulsión de buques. Un proyecto español de I+D", XXXVI Technical Sessions of Ingeniería Naval. Cartagena, 25th and 26th of November 1999.

M. Perez Sobrino; E. Minguito Cardena; A. GarciaGomez; J. Masip Hidalgo; R. Quereda Lavina; L. Pangusion Cidales; G. Perez Gomez & J. Gonzalez-Adalid. "Scale Effects in Model Tests with CLT Propellers". 27th Motor Ship Marine Propulsión Conference. Bilbao, Spain, 27th–28th January 2005.

M. Perez Sobrino; E. Minguito Cardena; A. GarciaGomez; J. Masip Hidalgo; R. Quereda Lavina; L. Pangusion: G. Perez Gomez & J. Gonzalez-Adalid. "Scale Effects in Model Tests with CLT Propellers"., C. Galindo, Full scale comparison of a superferry performance fitted with both High Skew and CLT blades". WMTC Conference. London, U.K., 2006.

Sustainable Maritime Transportation and Exploitation of Sea Resources – Rizzuto & Guedes Soares (eds)
© 2012 Taylor & Francis Group, London, ISBN 978-0-415-62081-9

Endplate effect propellers: A numerical overview

Stefano Gaggero & Stefano Brizzolara
Department of Naval Architecture, Marine Engineering and Electrical Engineering (DINAEL), Genoa University, Genoa, Italy

ABSTRACT: Energy saving is a primary objective, historically the first and, probably still now, the most important one, in the design of marine propellers. Modern design approaches, like fully numerical lifting line/lifting surface codes and optimization applied to potential panel methods satisfy this objective and allow to design conventional propellers with maximum efficiency for a given operating point. On the other hand nonconventional propellers, like CLT and Kappel like geometries, represent a further opportunity to increase efficiency and reduce the risk of cavitation. In the present work a numerical analysis of unconventional propellers will be carried out. Two different numerical approaches, a potential panel method and a RANS solver will be employed. The analysis will highlights the peculiarities of these kind of propellers, the possibility to increase efficiency and reduce cavitation risk, in order to exploit the design approaches already well proven for conventional propellers also in the case of these unconventional geometries.

1 INTRODUCTION

Energy saving is a primary objective in the design of marine propellers. The constant increase of oil price, the more strict regulations in terms of air pollution and the limits for NO_X and SO_X emissions require more and more efficient designs. At the same time, also requirements in term of radiated noise and vibration emissions became more strict: avoid negative effects on marine life and reduce the risk of hydro-acoustic signature are the primary aims of new commercial and navy constructions. Unconventional propellers, like Contracted and Loaded Tip (CLT) propellers, represent a valid answer to all these demands that can be fulfilled without the employment of completely different propulsive solutions, like contra- and co-rotating propellers or by the adoption of ducts, stators or wake regularizers.

The first concept of tip loaded propellers goes back to late seventies: Tip Vortex Free propellers were the first application of the Loaded Tip concept, that quickly evolved toward the CLT solution when also contraction of the fluid vein has been taken into account for the definition of the optimal geometry (www.sistemar.com).

CLT propellers are characterized by a monotonic increase of pitch from blade root to tip, a finite chord at tip, moderate values of skew and an endplate at the outermost radial edge of the blade towards the pressure side. Full scale installations and observations, together with model scale measures and theoretical studies, as reported by SISTEMAR (wws.sistemar.com) identified, as the main advantages of CLT propellers, higher values of efficiency (thus lower fuel consumption, air polluting lower emissions), higher value of thrust per unit area (thus higher ship speed and lower optimum diameter), lower noise and vibration levels with better margin for face cavitation and cavitation inception speed.

The gain in efficiency is obtained by the displacement of the maximum load towards the tip, that is made possible, without high noise and energy losses (typical of tip loaded conventional propellers) by the presence of the endplate. In fact the outer radial sections of the blade contribute more efficiently to the generation of thrust: velocities are higher and, geometrically, local pitch angle is lower, i.e., the local lift is more "aligned" with the axial propeller direction. A way to achieve high efficiency is, thus, to produce the most part of the required propeller thrust in this region of the blade. The additional span (but not in the radial direction!) provided by the endplate lets to locate the maximum load near the blade tip with a gradual and smooth reduction of the loading curve. In this way it is possible to avoid the presence of a strong tip vortex and higher values of induced velocities on the propeller plane, whose effect on the hydrodynamic pitch is fundamental in achieving high values of efficiency.

The presence of the endplate itself, that increases the pressure difference on the tip region, and lets to adopt a finite chord at tip, produces higher value of thrust per unit area and a local unload of the sections. The resulting smaller optimum diameter lets the propeller to operate in a more uniform hull wake, with wider propeller sections passing through areas where the change in local wake can be strong, achieving more stable cavity bubbles and less vibrations.

However several problems, connected with the peculiarities of this kind of propellers, still need to be investigated. The shape and position of the endplate poses some issues regarding local strength, the overall influence on propeller mechanical characteristics and the higher risk of "double" tip vortex cavitation at endplate root and tip. Scale effects, in addition, represent another challenging task for this kind of unconventional propellers, for which a deeper investigation is needed (Dyne, 2005). From a numerical point of view a lot of interest has been dedicated to the analysis and development of new propulsion concepts (Sanchez-Caja, 2006) involving CLT propellers. The TRIPOD European Project, for instance, is an example of the application of numerical tools, like potential flow based methods and RANS codes, in order to reduce the number of propeller design iterations for new propulsive solutions, in which unconventional Contracted and Loaded Tip propeller are adopted in contra-rotating and POD configurations.

In present work both the numerical approaches, a potential panel method and a RANS code, will be applied for the analysis of open water CLT performances, including prediction of steady cavitation extent and scale effects. While the former approach can be considered the best compromise between accuracy and computational time in the initial design stage to have an initial estimation of forces and cavitation extent, the latter represents a reliable tool to analyze the effects of viscosity that, especially in off design conditions, where potential approaches generally fails, are the leading aspects. The reliability of both the approaches will be investigated and their application limits for unconventional geometries highlighted.

2 NUMERICAL METHODS

The numerical modeling of Contracted and Tip Loaded Propellers, from and hydrodynamic point of view is equivalent to the conventional propellers case. Once the geometry has been defined, paying special attention to the modeling of the endplate, the hydrodynamic characteristics of CLT propellers can be computed straightforward applying both potential and RANS solvers.

2.1 *Panel method*

Panel/boundary elements methods model the flowfield around a solid body by means of a scalar function, the perturbation potential $\phi\ (x)$, whose spatial derivatives represent the component of the perturbation velocity vector. Irrotationality, incompressibility an absence of viscosity are the hypothesis needed in order to write the more general continuity and momentum equations as a Laplace equation for the perturbation potential itself:

$$\nabla^2\phi(x) = 0 \tag{1}$$

For the more general problem of cavitating flow, Green's third identity allows to solve the three dimensional differential problem as a simpler integral problem written for the surfaces that bound the domain. The solution is found as the intensity of a series of mathematical singularities (sources and dipoles) whose superposition models the inviscid cavitating flow on and around the body.

$$\begin{aligned} 2\pi\phi(x_p) = & \int_{S_B+S_{CB}} \phi(x_q)\frac{\partial}{\partial \boldsymbol{n}_q}\frac{1}{\boldsymbol{r}_{pq}}dS \\ & - \int_{S_B+S_{CB}} \frac{\partial\phi(x_q)}{\partial \boldsymbol{n}_q}\frac{1}{\boldsymbol{r}_{pq}}dS \\ & + \int_{S_W} \Delta\phi(x_q)\frac{\partial}{\partial \boldsymbol{n}_q}\frac{1}{\boldsymbol{r}_{pq}}dS \end{aligned} \tag{2}$$

Neglecting the supercavitating case (computation is stopped when the cavity bubble reaches the blade trailing edge) and assuming that the cavity bubble thickness is small with respect to the profile chord, (Gaggero, 2009) singularities that model cavity bubble can be placed on the blade surface instead than on the real cavity surface, leading to an integral equation in which the subscript q corresponds to the variable point in the integration, $\boldsymbol{n}$ is the unit normal to the boundary surfaces and $\boldsymbol{r}_{pq}$ is the distance between points p and q, S_B is the fully wetted surface, S_W is the wake surface and S_{CB} is the projected cavitating surface on the solid boundaries. This approach can be addressed as a partial nonlinear approach that takes into account the weakly nonlinearity of the boundary conditions (the dynamic boundary condition on the cavitating part of the blade and the closure condition at its trailing edge) without the need to collocate the singularities on the effective cavity surface. The set of required boundary conditions for the steady problem is:

- Kinematic boundary condition on the wetted solid boundaries:

$$\frac{\partial\phi(x_p)}{\partial \boldsymbol{n}_p} = -\boldsymbol{V}_{inflow}(x_p)\cdot\boldsymbol{n}_p \tag{3}$$

- Kutta condition at blade trailing edge:

$$\Delta\phi_{TE}(x_p) = \phi_{TE}^U(\boldsymbol{x_p}) - \phi_{TE}^L(\boldsymbol{x_p}) \tag{4}$$

- Dynamic boundary condition on the cavitating surfaces:

$$p = p_{vap} \, on \, S_{CB} \tag{5}$$

- Kinematic boundary condition on the cavitating surfaces (where $\boldsymbol{n}$ is the normal vector and th is the local cavity bubble thickness):

$$\boldsymbol{n}(\boldsymbol{x}) - th(x,t) = 0 \tag{6}$$

- Cavity closure condition at cavity bubble trailing edge.

Arbitrary detachment line, on the back and/or on the face sides of the blade can be found, iteratively, applying a criteria equivalent, in two dimensions, to the Villat-Brillouin cavity detachment condition, as in Mueller (1999). Starting from a detachment line obtained from the initial wetted solution (and identified as the line that separates zones with pressures higher than the vapor tension from zones subjected to pressure equal or lower pressures) or an imposed one (typically the leading edge), the detachment line is iteratively moved according to:

- If the cavity at that position has negative thickness, the detachment location is moved toward the trailing edge of the blade.
- If the pressure at a position upstream the actual detachment line is below vapor pressure, then the detachment location is moved toward the leading edge of the blade.

The numerical solution consists in an iterative scheme delegated to solve the nonlinearities connected with the Kutta, the dynamic and the kinematic boundary conditions on the unknown cavity surfaces until the cavity closure condition has been satisfied.

Viscous forces, neglected by the potential approach, can be computed following two different approaches. In the first case, as proposed by Hufford (1992) and Gaggero (2010), a thin boundary layer solver can be coupled, through transpiration velocities, to the inviscid solution, in order to obtain a local estimation of the frictional coefficient computed in accordance to the integral approach of Curle (1967) (for the laminar boundary layer) and Nash (1969) (for the turbulent boundary layer). This approach, though being applied successfully for the analysis of conventional propellers, poses some problems of convergence in very off design conditions and suffers from the tip influence on streamlines on which the boundary layer calculation is performed. As a consequence in the present work a local estimation of frictional coefficient has been carried out applying a standard frictional line. In particular, the VanOossanen formulation, based on local chord and thickness/chord ratio, has been employed:

$$C_n(r) = \frac{1}{2} a \left(1 + b \frac{t(r)}{c(r)}\right) Re_N^K(r) \tag{7}$$

The discretized surface mesh consists of 1800 hyperboloidal panels for each blade. The trailing vortical wake extends for five complete revolutions and it is discretized with a time equivalent angular step of 6°, as in Figure 1.

2.2 *RANS solver*

Viscous analysis of open water propeller characteristics has been carried out through StarCCM+, a commercial finite volume RANS solver (CD-Adapco, 2010). Continuity and momentum equation, for an incompressible flow, are expressed by:

$$\begin{cases} \nabla \cdot \boldsymbol{U} = 0 \\ \rho \dot{\boldsymbol{U}} = -\nabla p + \mu \nabla^2 \boldsymbol{U} + \nabla \cdot \boldsymbol{T}_{Re} + \boldsymbol{S}_M \end{cases} \tag{8}$$

in which $\boldsymbol{U}$ is the averaged velocity vector, p is the averaged pressure field, μ is the dynamic viscosity, $\boldsymbol{S}_M$ is the momentum sources vector and $\boldsymbol{T}_{Re}$ is the tensor of Reynolds stresses computed in according to the *Realizable* $k - \varepsilon$ turbulence closure equations.

The StarCCM+ computational domain, by means of the symmetries, is represented by an angular sector of amplitude $2\pi/Z$ around a single blade, discretized with an unstructured mesh of polyhedral cells of about 800 k elements (as in Figure 2). This discretization level has been assumed valid for all the computations after numerical convergence was checked. All the simulations have been carried out as steady, using the Moving Reference Frame

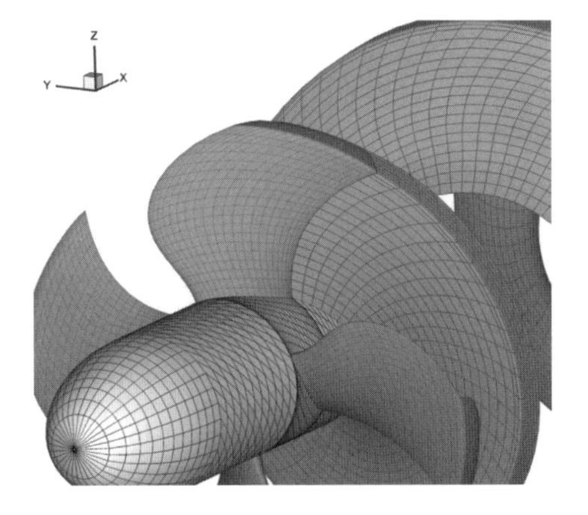

Figure 1. CLT propeller structured mesh for potential solver.

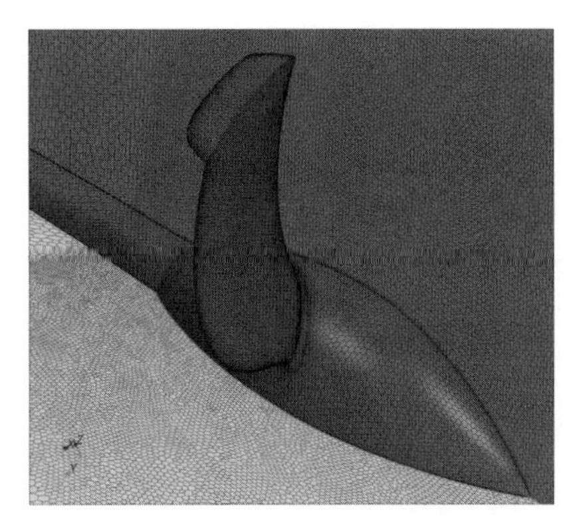

Figure 2. CLT propeller unstructured mesh based on polyhedral cells.

approach, whit SIMPLE algorithm to link pressure with velocity fields.

3 RESULTS

CLT propellers are claimed to be unconventional propellers able to grant some advantages with respect to conventional solutions. In particular, especially in full scale, these advantages results in higher efficiency, reduced cavitation and tip vortex, lower noise and vibrations and, in general, smaller propeller optimum diameter due to the high delivered thrust. The numerical analyses, consequently, have been carried out having in mind these primary aspects: prediction of propeller performances in open water, prediction of the steady cavity extension (only through the boundary element method) and prediction of full scale propeller characteristics. Objectives of this study are the validation of the numerical codes both in non cavitating and in cavitating conditions, the analysis of the capabilities of the panel method in predicting main features of the cavity bubble also for unconventional propellers and, finally, the possibility in predicting scale effects through the proposed numerical methods.

3.1 *Open water—model scale*

Model scale open water computations, compared with measures carried out at CEHIPAR towing tank (SILENV Subtask 3.1.1, 2011) are reported in figure 3. Both the numerical predictions (Panel Method and RANS solver) are very close to the experimental values for a wide range of advance coefficient. In mean the difference between computed and measured values is less than 3%: Panel Method tends to overestimate both thrust and torque, while RANS results are slightly under predicted.

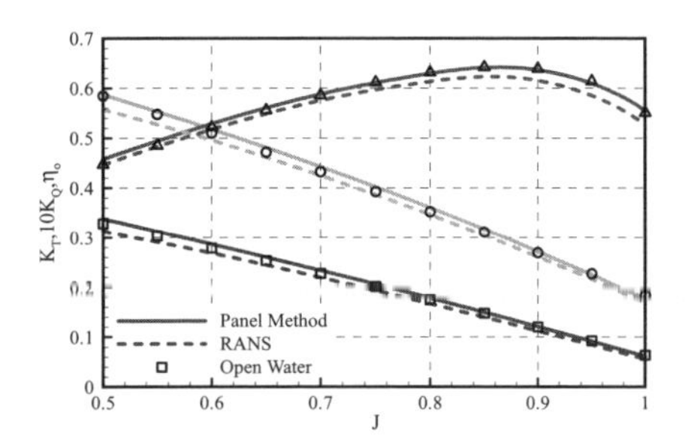

Figure 3. Open water propeller characteristics. RANS and panel method in model scale.

A more detailed analysis of the flow characteristics is presented in Figures 4 and 5 that show predicted pressure coefficient (in non-dimensional form with respect to the rate of rotation $C_p = (p - p_0)/0.5\rho N^2D^2$) on the CLT propeller blade in correspondence of the higher (J = 0.5) and lower (J = 1.0) load conditions of diagram 3. The main features of the flow are similarly computed by the two methodologies. When the propeller is highly loaded suction side is clearly distinguishable from the pressure side: they almost correspond to back and face of the blade. Instead, at lower load, the leading edge pressure peak moves towards the back side due to the local negative flow incidence. Both the numerical approaches, although with some differences, captures the influence of the endplate on the local pressure field. The increase of pressure on the face side is clear for the two loading conditions. However, while for the panel method this influence is more spread around the endplate leading edge, RANS computations show a very concentrated (more evident at higher advance values) pressure peak. These differences, obviously, can be attributed to the very different discretization level adopted for the two cases: the size of the cells around the endplate for the viscous computations are, at least, an order of magnitude smaller than the panels adopted for the inviscid computation. Anyhow this is not the only, and probably not the most important, reason to explain the different pressure distribution. The analysis of flow streamlines on the propeller blade (Figure 6), shows, in fact, a very different behavior (due to the viscous terms) of the velocity field in proximity of the blade tip. Especially on the pressure side, while in the RANS solution the streamline "hit" the endplate (producing the pressure peak) and it is deviated downward, in the panel method the streamline has a smooth trajectory (that is responsible of a smooth pressure distribution) and almost follow the endplate shape.

Finally, for the open water case, blade bounded circulations, computed by the panel method, of the CLT propeller are compared (at identical

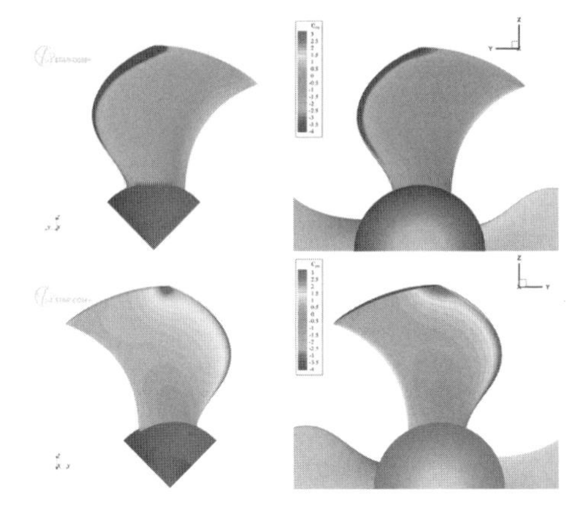

Figure 4. Propeller back (up) and face (bottom) pressure distribution. RANS (left) versus Panel (right). J = 0.5.

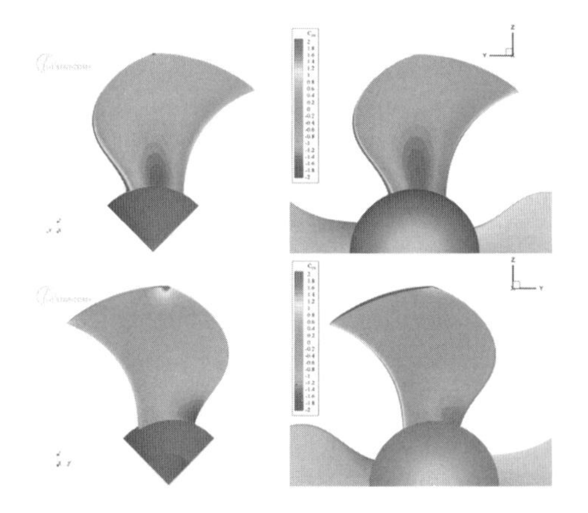

Figure 5. Propeller back (up) and face (bottom) pressure distribution. RANS (left) versus Panel (right). J = 1.

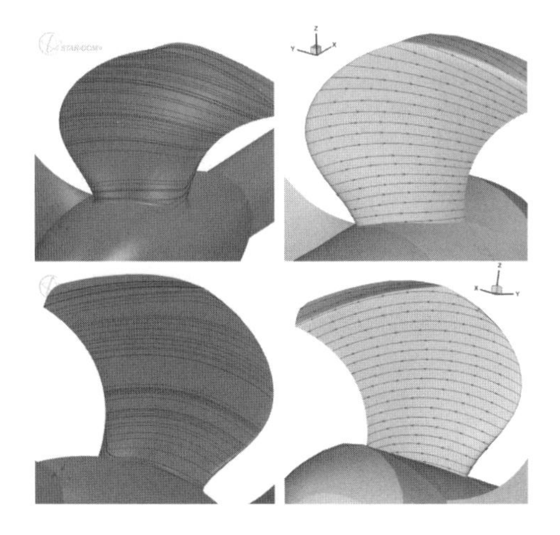

Figure 6. Streamlines at J = 0.8. RANS (left) vs. Panel Method (right).

thrust coefficients) to those of an high skewed conventional propeller designed, in the framework of the SILENV Project, by the University of Genova (Bertetta et al., 2011 and Bertetta et al., 2011) for the same working conditions. The conventional propeller, as presented in Bertetta et al. (2011) has been designed via the same developed potential code (adopted in this work only for analysis purposes) coupled with an optimization algorithm through a parametric geometry representation. Reduction of cavity extension (at both the design and the reduced pitch) and increase of efficiency with constrained values of delivered thrust were the objectives of the design.

Circulation, within the potential approaches, can be considered, in fact, as an integral parameter that measures load distribution of the blade along the radius and whose derivative in radial direction is proportional to the strength of the tip vortex. The comparison of Figure 7 demonstrates how endplated propellers work and justifies the advantages of these kind of propellers.

While for conventional propellers the maximum load is close to 0.7 r/R and this optimum location is a compromise between efficiency (that could be increased moving the load towards the outer sections) and vortical losses (that could be kept low if circulation decreases smoothly from its maxim value to the tip), the presence of the endplate (and the related increase in blade span) on CLT propellers lets to locate the maximum load closer to the tip, maintaining a smooth and gradually decreasing shape of the circulation curve. If for the conventional propeller the location of the maximum load varies from 0.65 r/R to 0.7, as a function of the load itself, the CLT solution has its maximum between 0.75 and 0.85 r/R, with a considerably lower slope of the circulation curve at the propeller tip. The viscous analysis of Figures 8 and 9, in which an isosurface of the Q-factor (defined as the sum of the symmetric and anti-symmetric part of the

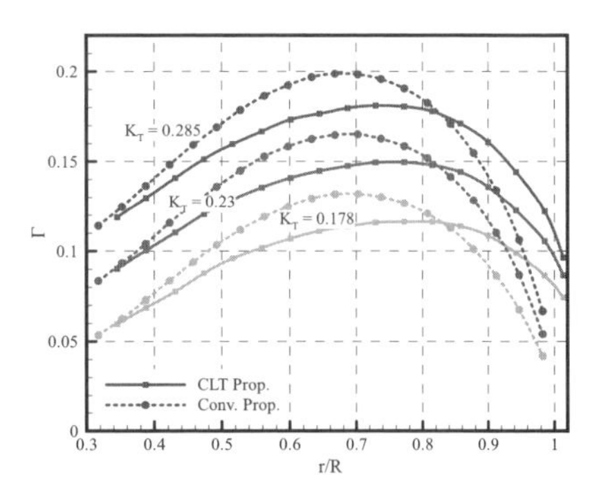

Figure 7. Comparison of circulation distribution for three different thrust coefficient between CLT and conventional propeller.

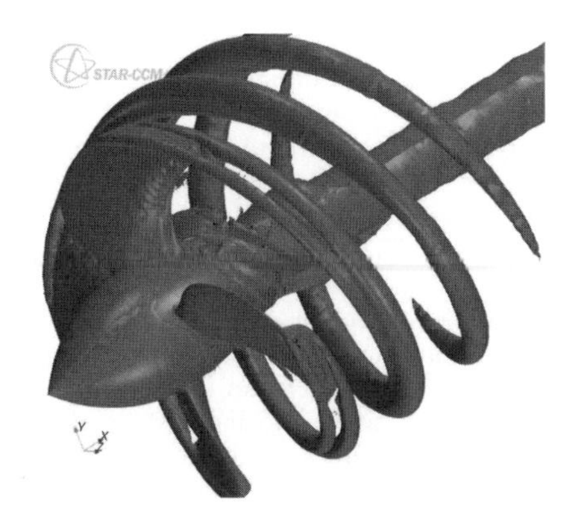

Figure 8. Q-factor isosurface for the CLT propeller. $K_T = 0.178$.

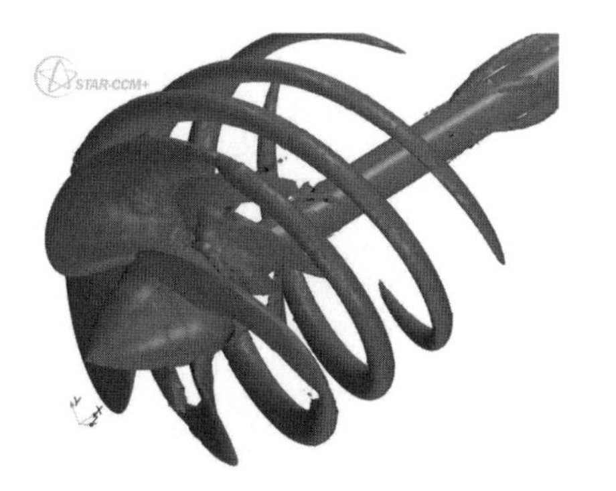

Figure 9. Q-factor isosurface for the conventional propeller. $K_T = 0.178$.

velocity gradient tensor) is presented, highlights once more how the tip vortexes from the endplate (one from the root and one from the tip with different pitches) develop and merge themselves while the conventional propeller (comparison at same thrust coefficient) of Figure 9 is characterized as usual by a single and stronger tip vortex.

3.2 *Cavitating condition—model scale*

The prediction of cavity extension is another key aspect for CLT unconventional propellers. The higher load near the tip, if cavitation cannot be avoided, could stabilize cavitation itself, especially when the propeller operates in a spatial non uniform wake (behind the hull, for instance). Avoiding cavity growing and collapse by a larger, more stable, cavity bubble results in reduction of vibrations and propagated noise. On the contrary, the longer chord at the propeller tip could locally reduce the load per unit area, reducing the risk of cavitation and the bubble extension. Moreover, the influence on the pressure field of the endplate, that, with respect to a conventional geometry, increase the overpressure on the pressure side (as in Figure 4) and reduce the pressure drop on the suction side, helps to reduce the back cavity extension. Anyhow, the presence of the endplate increases the risk of cavity inception (at its root and tip) that should be taken into account for a better propeller design.

Prediction of cavity extension is, therefore, a necessary step in understanding the flow dynamics around CLT propellers. A steady analysis performed with the panel method, that has already proven to be a reliable and sufficiently accurate approach to compute steady and unsteady cavitating flow on conventional geometries (Gaggero, 2009 and Gaggero, 2010), could provide a preliminary insight into these phenomena.

Two different advance coefficients, respectively 0.8 and 0.9, at a cavitation index $\sigma_N = 1.65$ (that corresponds to the tested conditions of the propeller), have been analyzed. Experimental measures, as reported in SILENV Subtask 3.1.1 (2011), have been carried out in unsteady flow. Nevertheless the unsteady mean advance coefficient for thrust identity is something between 0.8 and 0.9 and a preliminary steady analysis at these conditions could, in mean, describe the cavitating characteristics of the propeller.

Figures 10 and 11 show the steady cavity extension on the back side of the propeller.

The main features of the cavity bubble, identified at the cavitation tunnel for the SILENV Subtask 3.1.1 (2011), are captured quite well. In addition to the sheet cavitation on the suction side of the blade, the numerical computations predict the development of a smaller sheet cavity bubble (from the leading edge) on the outer side of the endplate side that is qualitatively in agreement with the experimental observations. Prediction of tip cavitating vortexes is beyond the capabilities of the developed panel method, that is limited to the analysis of supercavitating sections. However the thicker sheet cavity bubble at the blade trailing edge can be considered as a symptom of probable tip vortex cavitation. Especially at the lower advance coefficient the prediction of sheet cavity bubbles up to the trailing edge agrees well with the two observed cavitating vortexes from the endplate tip and root.

3.3 *Open water—ship scale*

Another important issue that it is worth to be investigated regards the scale effects that, as evidenced in literature (Perez Gomez, Sanchez-Caja, 2006), influence unconventional CLT propellers strongly than conventional designs. Also in this case a preliminary analysis has been carried out, considering only one propeller geometry and one grid configuration for the RANS computations,

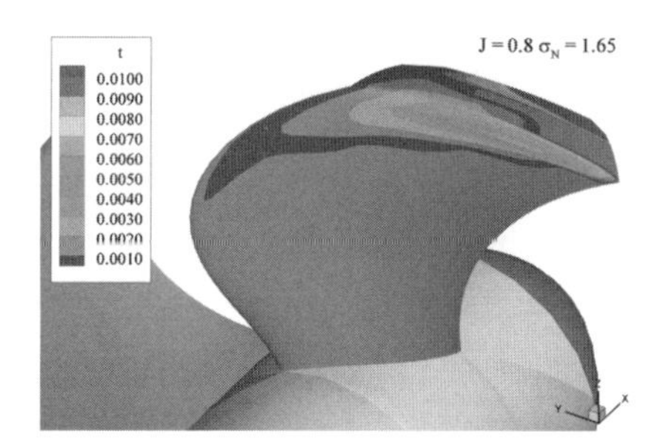

Figure 10. Cavity extension prediction. J = 0.8 at $\sigma_N = 1.65$.

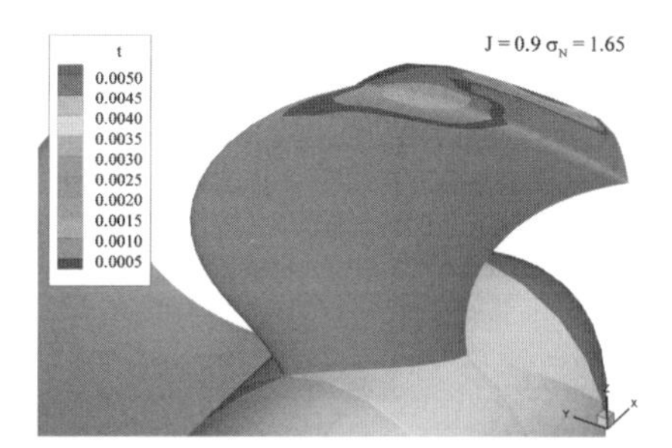

Figure 11. Cavity extension prediction. J = 0.9 at $\sigma_N = 1.65$.

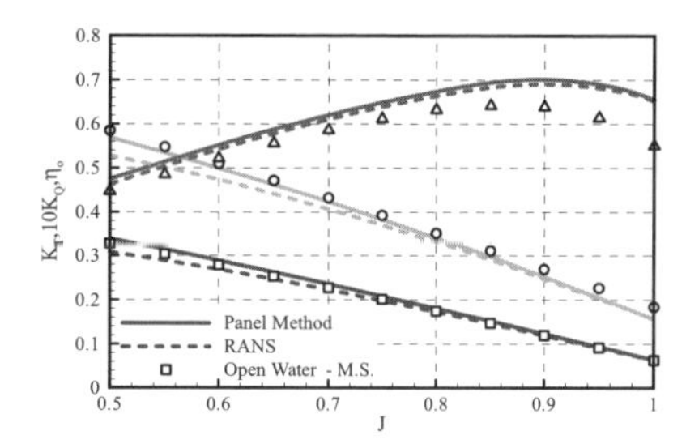

Figure 12. Open water propeller characteristics. RANS and panel method in ship scale.

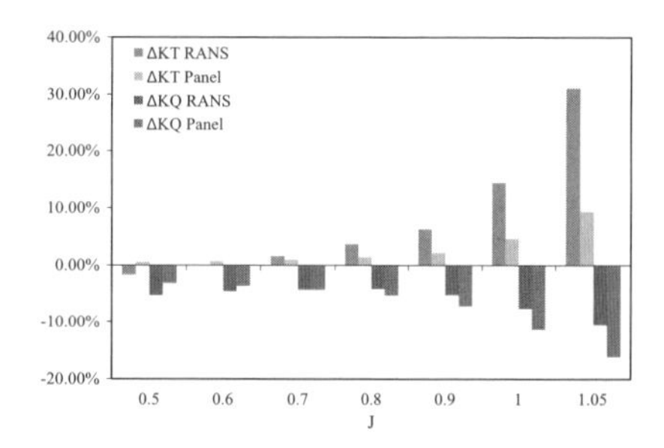

Figure 13. Variations of thrust and torque from model to ship scale.

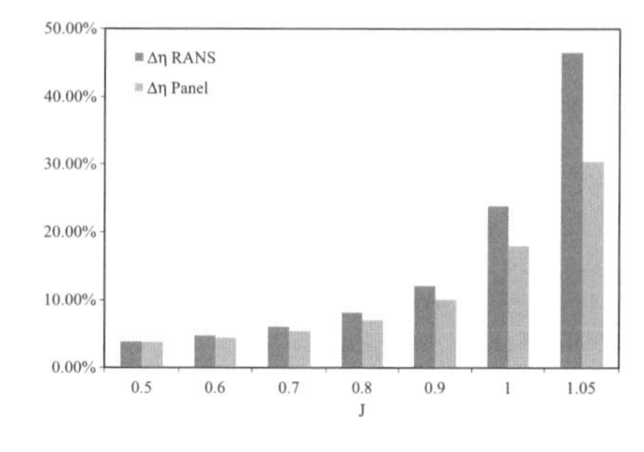

Figure 14. Variations of open water efficiency from model to ship scale.

simply obtained by scaling to ship dimensions the model scale mesh and rearranging the prims layer according to the new Reynolds number. A real deep knowledge of the phenomena will require a systematic analysis of all the aspects that could affect scale effects. The differences in the scaling factors between CLT and conventional propellers are mainly due to the interactions between viscous effects (different from model to full scale) and the overpressure, produced by the endplate, on the pressure side, than, in model scale could increase the region subjected to laminar flow and thus, to laminar separation (in full scale, with a turbulent inflow this risk is highly reduced). A first step to obtain reliable and accurate results, consequently, should require the investigation on the influence of turbulence models and mesh density around the region subjected to separation and overpressure.

As in the previous computations, also for the analysis of the scale effects the aim of the work is only to identify some major peculiarities of the adopted numerical methods and their applications limits on the light of their adoption in large design/optimization procedures. A simple analysis, consequently, could be considered adequate.

The comparison between model scale (previous figure 3) and full scale open water computations of figure 12 is further analyzed in Figures 13 and 14, in which the relative deviations ((*ship scale – model scale*)/*ship scale*) for different advance coefficients is computed both with the viscous RANS solver and the potential Panel Method. Unexpectedly the scale effect, strongly connected with the viscous influence, is qualitatively predicted also by the Panel Method. The application of the viscous correction of eq. 7, that, even if in an approximate way, depends on the Reynolds number and on the local shape of the profile, lets to partially compute model/ship scale differences in a better way than the application of a raw, constant, frictional coefficient (as in Lee, 1987).

The gain in efficiency is obtained, as in conventional propellers, by a simultaneous increase of thrust and decrease of torque. For the viscous computations, up to J = 0.8 the decrease of torque is the leading factor in increasing the efficiency.

For higher values of advance, instead, the increase of thrust in ship scale with respect to model scale is greater than the reduction of torque. Panel method computations, instead, are characterized by a reduction of torque always greater than the increase of thrust, with an overall underestimation of the performance variations computed with the RANS, that reflects in lower values of gained "panel method" efficiency in full scale (Figure 14).

With respect to conventional propellers, however, the gain in efficiency achieved, numerically, (with the panel method and with the RANS) in ship scale for the CLT propeller is generally higher than that achieved, numerically again, for conventional geometries. Also the agreement with traditional scaling procedures is qualitatively acceptable.

4 CONCLUSIONS

Two numerical approaches, a potential panel method and a RANS solver, have been successfully applied for the analysis of CLT propellers. Model scale open water characteristics have been numerically computed by the two methodologies with a discrepancy lower than 3% for the entire advance coefficient range. Full scale computations give an insight into scale effects and, without pretending to be a state of the art application of viscous solver for the prediction of scale effects (as mentioned above not all the aspects, like turbulence models and grid dependency, have been taken into account), suggest that the application of RANS solvers can contribute to a better knowledge of these phenomena. Finally a preliminary, steady and inviscid computations of the cavitating flow demonstrates that panel method could be employed, at least in a preliminary phase, and with sufficient accuracy, also for the analysis of the cavitating behavior of unconventional Tip Loaded Propellers. The key feature of the cavity bubble (of sheet type at the endplate leading edge, cavitating tip vortexes at its edges) are predicted qualitatively and quantitatively quite well. This suggests the application of the potential panel method (which computational time is a fraction of that required by a viscous RANS solver for the same flow conditions), in conjunction with a parametric geometry description and optimization algorithms, in an inverse design approach, in order to prescind from the simpler, well established, but strongly approximated (especially for this kind of propellers) design codes based on lifting line/surface.

ACKNOWLEDGEMENTS

This work was developed in the frame of the collaborative project SILENV—Ships oriented Innovative soLutions to rEduce Noise & Vibrations, funded by the E.U. within the Call FP7-SST-2008-RTD-1 Grant Agreement SCP8-GA-2009–234182. The authors would also like to thank SISTEMAR and Dr. Enrique Haimov from CEHIPAR for their support and suggestions in carrying out these analysis.

REFERENCES

Bertetta, D., Brizzolara, S., Gaggero, S., Savio, L. and Viviani, M. 2011. Numerical and Experimental characterization of a CP propeller unsteady cavitation at different pitch settings, *Proc. Second International Symposium on Marine Propulsors, Hamburg, Germany, June 2011.*

Bertetta, D., Brizzolara, S., Gaggero, S., Savio, L. and Viviani, M. 2011. Numerical and Experimental optimization of a CP propeller at different pitch settings, *Proc. International Maritime Association of the Mediterranean, Genoa, Italy, September 2011.*

CD-Adapco 2010. StarCCM+ v5 User's Manual, *2010.*

Curle, H. 1967. A two Parameter Method for Calculating the two dimensional incompressible Laminar Boundary Layer, *J.R. Aero Soc., 1967.*

Dyne, G. 2005. On the principles of Propellers with endplates, *International Journal of Maritime Engineering, RINA, 2005.*

Gaggero, S. 2010. Development of a Potential Panel Method for the Analysis of Marine Cavitating and Supercavitating Propellers. *Ph.D. Thesis, University of Genova, Italy, 2010.*

Gaggero, S. and Brizzolara, S. 2009. A Panel Method for Trans-Cavitating Marine Propellers, *7th Int. Symposium on Cavitation, Ann Arbor, Michigan, USA, 2009.*

Hufford, G.S. 1992. Viscous Flow Around Marine Propellers using Boundary Layer Strip Theory, *Ph.D. Thesis, Massachusetts Institute of Technology, USA, 1992.*

Lee, J.T. 1987. A Potential based Panel Method for the Analysis of Marine Propellers in steady flow, *Ph.D. Thesis, Massachussetts Institute of Technology, USA, 1987.*

Mueller, A.C. and Kinnas, S.A. 1999. Propeller sheet cavitation prediction using a panel method, *Journal of Fluid Engineering, 1999.*

Nash, J.F. and Hicks, J.G. 1969. An Integral Method including the effect of Upstream History on the turbulent shear stress for the computation of turbulent boundary layer, *Proc. A FOSR-IFT Stanford Conference, Stanfor University Press, 1969.*

Perez Gomez, G. and Gonzalez-Adalid, J. Scale Effects in the Performance of a CLT Propeller, *www.sistemar.com.*

Sanchez-Caja, A., Sipila, T. and Pylkkanen, J. 2006, Simulation of the Incompressible Viscous Flow Around an Endplate Propeller using a RANSE *solver. Proc. 26th Symposium on Naval Hydrodynamics, Rome Italy, 2006.*

SILENV—Ship Oriented Innovative Solutions to reduce Noise and Vibrations. FP7 Collaborative Project 234182. *Work Package 3, Subtask 3.1.1: Propellers, 2011.*

Sustainable Maritime Transportation and Exploitation of Sea Resources – Rizzuto & Guedes Soares (eds)

3D measurements of the effective propeller jet flow behind the rudder

T. Abramowicz-Gerigk
Gdynia Maritime University, Poland

ABSTRACT: The paper presents 3D measurements of the effective velocity field generated by a controllable pitch propeller of a twin propeller twin rudder ferry model in 1:16 scale. The bollard pull tests were performed in the experimental setup situated on the lake. The distribution of the propeller jet velocity behind the rudder, in the zone of flow establishment was investigated. The use of the large physical model allowed studying the effective propeller jet characteristics. The influence of the propeller pitch, rotational speed, aft body form of the ship hull and shallow water effect on the velocity field are discussed.

1 INTRODUCTION

The formation and diffusion characteristics of the propeller jet are important to predict the scouring of seabed and banks in the navigational areas, mainly at near quay wall and in channels. The results of ship's propeller wash investigations are used for design purposes in civil engineering and for determination of allowable operational conditions for the vessels.

The major concern is the downstream propeller jet flow which lasts for the distance of several propeller diameters (Lam 2006) and has the axial velocity components whose magnitudes can exceed 10 m/s (Wilson et al., 2006).

The maximum permitted current speed due to the seabed scouring is about 0.5 m/s for sandy bottom and 1.2 m/s for the firm clay therefore the protection systems with different types of current-resistant top layers are installed to counteract this effects.

Propeller jets can be the reason of the stone riprap and concrete plates instability and displacement, damage of geotextile mattresses, drift of sediments and their deposition into the protected areas.

The formation of a propeller jet is a complex phenomenon. In practical applications the local jet velocities are calculated on the basis of the axial propeller efflux velocity (Chin & Li 2002, Lee et al., 2004) and the jet diameter in the efflux plane, assumed as 0.7 D (propeller diameter). The efflux velocity is usually approximated by the function of the applied engine power.

Due to the obstruction effects of the propeller hub the two zones of flow pattern, depending on flow character, can be distinguished: zone of flow establishment and zone of established flow. At the distance of $3D$ the propeller outflow stream is fully established.

The axial velocities have the highest magnitudes (Lam et al., 2006), however for the big self-manoeuvring vessels, operating in shallow water conditions the radial and tangential components at the exit of propeller jet are not neglectful and can induce considerable scouring effects in shallow water (Chin & Li 2002).

2 3D VELOCITY FIELD OF PROPELLER JET

The 3D velocity field comprises of axial, tangential and radial components. The axial component is parallel to the axis of propeller rotation, the tangential and radial components are perpendicular and coincident with the radius.

In the zone of flow establishment the axial velocities at 0.5 D are approximately symmetrical about the propeller axis, the maximum velocities occur at the radial distance of 0.238 D from the axis of the propeller (Chin & Li 2002). The maximum velocities at the efflux plane occur at the radial distance from the axis of the propeller equal to 0.27 – 0.38 D (Lam et al., 2006), 0.263 D (Lam et al., 2010).

The propeller hub reduces the axial velocities in the center of the propeller jet, the rudder situated behind the propeller splits the jet into to parts directed upwards and downwards (Yuksel et al., 2005).

The distribution of jet velocities is influenced by the shape of ship stern. The slipstream deflects to the nearest boundary surface. The flow in the stern region of a fully appended hull was analyzed by Muscari et al. (2011) by both computational and experimental fluid dynamics. The study was focused on the velocity field induced by the rotating propellers. Measurements have been performed by laser Doppler velocimetry (LDV) on the vertical midplane of the rudder and in two transversal planes behind the propeller and behind the rudder.

The time-averaged velocity components at the propeller jet were presented by Kee et al. (2006) and Lam et al. (2010). The propellers used in those investigations were 4-bladed 0.092 m in diameter, operated over a range of rotational speeds from 8 rps to 21 rps and 3-bladed 0.076 m in diameter, operated over a range of rotational speed 16.67 rps, respectively. Three-dimensional velocity measurements of the jets produced by the propeller attached onto a test rig via a rotating shaft at about mid-depth of a free-surface tank, large enough to allow the unhindered expansion of the rotating propeller jets were conducted in bollard pull conditions, using a Laser Doppler Anemometer.

The maximum velocities in tangential direction can be about 40% (Chin & Li 2002), 62% (Kee et al., 2006) or 82% (Lam et al., 2010) of axial flow. It can induce considerable effects under the propeller in shallow water.

The tangential components in deep water conditions observed for the ferry model showed values of about 10% of the axial velocities magnitude. The horizontal and vertical component magnitudes are 10% and 22% of the axial velocity, both the tangential and radial are 10%. (Abramowicz-Gerigk 2010). Shallow water effect results in the similar shape of the area of higher velocities. However the area is widen in both horizontal and vertical directions and slightly shifted towards the ship centre line and base plane accordingly.

3 MODEL TESTS OF JET FLOW GENERATED BY PROPELLER OF TWIN PROPELLER TWIN RUDDER FERRY

3.1 *Experimental setup*

Model tests were carried out in bollard-pull conditions in the open water experimental setup (Fig. 1) constructed in the lake Silm at Ship Research and Training Centre of the Foundation for Safety of Navigation and Environment Protection in Ilawa-Kamionka, Poland.

The main particulars of the man manned model of car passenger ferry used in the tests are presented in Table 1. The model was equipped with two four blade, controllable pitch propellers of inward direction of revolution and two rudder.

The measurements were conducted using the multi-hole pressure probe. The three-dimensional velocity field was measured behind the rudder in the distance of 0.13 m (0.41 D) from the rudder trailing edge equal to 0.81 D form the propeller disk. The three dimensional velocities were measured at discrete points, The square measurement area 0.304 × 0.304 m was scanned with 0.001 m steps to obtain the array of values.

The limits of the measurements of the turbulent flow were in the range of velocity vectors slant +/–60° with uncertainty 3% FS.

The coordination system used for the measurements and the measurement area are presented in Figure 2.

The main parameters of the propeller are presented in Table 2. The view of the aft body of the car—passenger ferry model are presented in Figure 3.

Figure 1. Experimental setup.

Table 1. Main particulars of the model.

Displacement [m³]	V	4.89
Length over all [m]	LOA	10.98
Length between perpendiculars [m]	L_{PP}	9.64
Breadth [m]	B_{WL}	1.78
Draft [m]	T	0.42
Block coefficient [–]	C_B	0.687
Model scale	λ	1:16

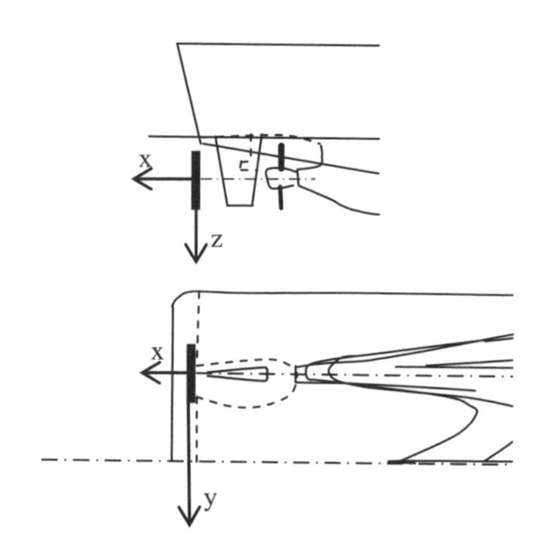

Figure 2. Coordination system.

Table 2. Main parameters of the propeller model.

Propeller diameter [m]	D	0.319
Number of blades	z	4
Hub diameter [m]	d_h	0.107
Max pitch ratio	p	1.081
Blade Area ratio	A_D/A_0	0.67
Rake [°]	Θ	0
Max rps	n	17.083
Direction of revolution		inward

Figure 3. Aft body of the model.

3.2 *Program of model tests*

The program of model test was developed to obtain the data for further CFD computation of propeller jet in deep and shallow water conditions. The velocity field was measured behind the port rudder, for the propeller settings presented in Table 3, for rudder angles 0° and 35° to port and two depth to draft ratios, assumed as $h/T = 1.2$ for shallow water and $h/T = 3$ for deep water.

The propeller settings and the corresponding propeller pitch angle θ, rate of turn n and thrust coefficient K_T (Equation 1) are presented in Table 3.

$$K_T = \frac{T}{\rho \cdot n^2 \cdot D^4} \tag{1}$$

where: T = propeller thrust [N]; ρ = water density [kg/m³]; n = propeller revolutions [rps]; D = propeller diameter [m].

3.3 *Model test results*

Visualization of 3D measurements of the velocity field behind the rudder in deep water is presented in Figures 4 and 5.

Visualization of measurements in shallow water conditions is presented in Figures 6 and 7.

The influence of water depth to draft ratio on the velocity field measured behind the rudder can

Table 3. Propeller settings.

Propeller setting	θ [°]	n [rps]	K_T [–]
Dead slow ahead	5	13.63	0.012
Slow ahead	10	15.88	0.032
Half ahead	15	17.08	0.079
Full ahead	19	17.08	0.114

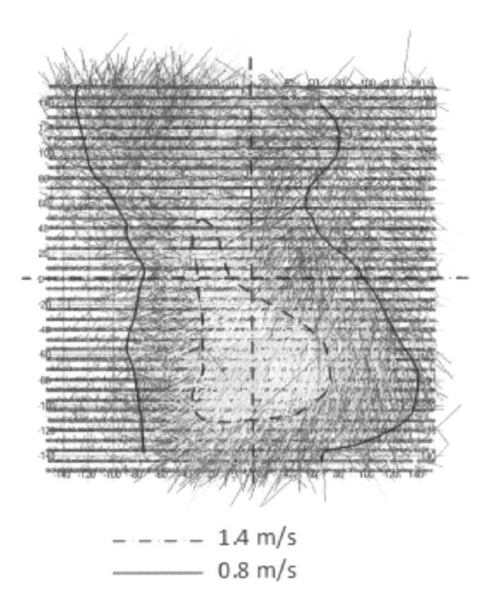

Figure 4. Propeller's jet velocity field behind the rudder in deep water, visualisation of 3D measurements: Full Ahead $H = 19°$, $n = 17.08$ rps, axial velocity component is marked in grey scale.

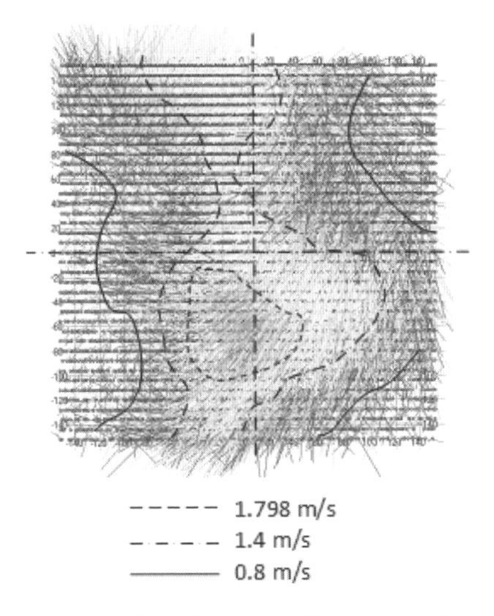

Figure 5. Propeller's jet velocity field behind the rudder in shallow water, visualisation of 3D measurements: Full Ahead $H = 19°$, $n = 17.08$ rps, axial velocity component is marked in grey scale.

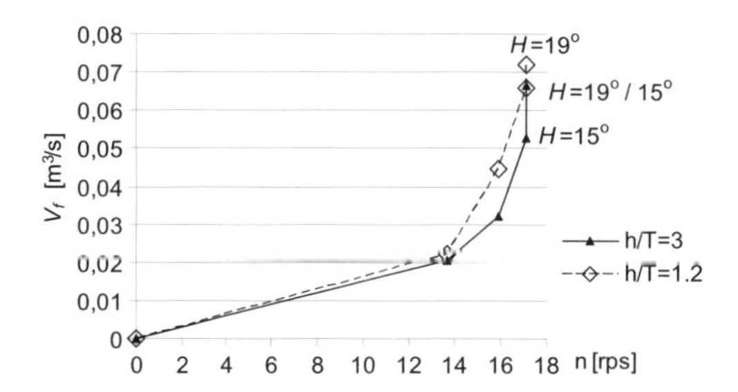

Figures 6. Volume of flow rate of propeller jet behind the rudder, in dependence of water depth to drft ratio, propeller rate of turn and corresponding pitch (Table 3).

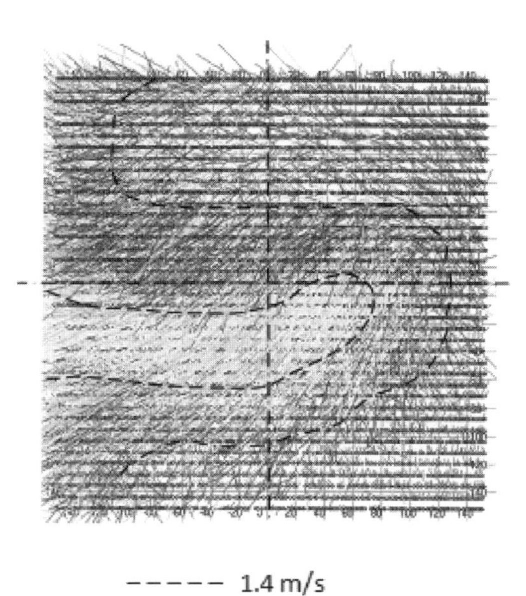

Figure 7. Propeller jet velocity field behind the rudder, visualisation of 3D measurements, rudder 35° to port side, Half Ahead p = 15°, n = 17.08 rps in deep water, axial velocity component is marked in grey scale.

be observed for higher propeller settings. It is confirmed that the rotation speed of the propeller is the most important factor in formation of the propeller jet flow (Chin & Li 2002).

The distribution of the axial velocity is not axisymmetrical. It is influenced by the rudder, aft body shape, free surface and seabed modelled by a wooden flat plate.

Volume of flow rate of propeller jet behind the rudder, in dependence of water depth to draft ratio, propeller rate of turn and corresponding pitch is presented in Figure 6.

The flow rate V_f results from the integration of flow speed over the circle area 0.304 m in diameter, equal to the maximum range of the pressure probe used for the measurements.

There is the 18% percent increase of the volume of flow rate V_f in shallow water which corresponds to 40% increase of the forces and moments generated on the rudder.

The mean axial velocity v calculated for deep and shallow water, compared with efflux velocity values calculated using the semi-empirical equation based on the actuator disc theory, proposed by Hamil (Kee et al., 2006) are presented in Table 4.

$$v_0 = 1.33 \cdot n \cdot D \cdot \sqrt{K_T} \quad (2)$$

where: n = propeller revolution [rps]; D = propeller diameter [m], K_T = thrust coefficient;

The axial component of velocity show acceleration up to 0.5 D from the propeller disk and then

Table 4. Mean axial velocities.

Propeller settings		v [m/s]		v_o [m/s]
θ	n	h/T = 3	h/T = 1.2	open water propeller
5	13.63	0,284	0,307	0.627
10	15.88	0,444	0,616	1.197
15	17.08	0,725	0,907	2.036
19	17.08	0,917	0,992	2.450

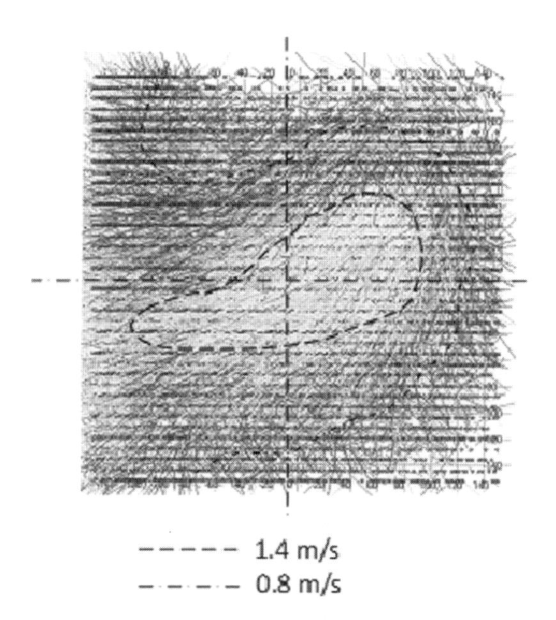

Figure 8. Propeller jet velocity field behind the rudder, visualisation of 3D measurements, rudder 35° to port side, Half Ahead p = 15°, n = 17.08 rps in shallow water, axial velocity component is marked in grey scale.

Table 5. Mean axial velocities of propeller jet behind the rudder for Half Ahead for rudder angle 0° and 35°.

	rudder angle	v
h/t	[°]	[m/s]
3	0	0,725
	35	0,700
1.2	0	0,907
	35	0,811

the velocity gradually decreases (Kee et al., 2006), therefore it can be assumed that for the open water propeller at the distance of 0.81 D they are equal to the efflux velocity.

The measured mean values are about 50% less than the maximum values. They are about 50% less than the mean values calculated using Formula 2. The influence of rudder helm hard to port on the velocity field is presented in Figures 7 and 8.

Mean axial velocities of propeller jet behind the rudder in deep and shallow water conditions are presented in Table 5.

The effect of 35° helm to port is propeller jet deflection to the port side and smaller area of higher velocities observed within the axisymmetrical measuring plane.

4 CONCLUSIONS

The presented results of 3D measurements of the effective velocity field generated by propeller of twin propeller twin rudder ferry show the under estimation of the values of mean axial velocity calculated on the basis of measurements in comparison with the maximum measured and calculated by a formula often used in practical applications.

The axial velocity was confirmed to be the dominating component, however the other components especially tangential and vertical should be considered in shallow water conditions.

In the calculations of the propeller wash only 20–25% of the maximum installed engine power used per propeller is assumed with respect to operational restrictions in ports. In practice the amount of power in different weather or ice conditions, used during manoeuvres, can be much greater. In shallow water conditions the tangential and vertical velocity components can be the reason of propeller scouring under the vessel in the propeller plane.

The results of model tests presented in the paper can be used to model the velocity field generated by propellers around the vessel using CFD methods.

REFERENCES

Abramowicz-Gerigk, T. 2010. Distribution of flow velocity generated by propellers of twin propeller vessel. *Scientific Journals Maritime University of Szczecin, No. 20/92*, 5–13.

Chin, C.O. & Li, W. 2002. Propeller jet flow. Index of /e-library/beijing_proceedings/Theme_E.http://www.iahr.org/ e-library/beijing_proceedings/Theme_E.

Kee, C., Hamill, G, Lam, W. & Wilson, P. 2006. Investigation of the Velocity Distributions within a Ship's Propeller Wash. *Proc. of the Sixteenth Int. Offshore and Polar Engineering Conference, San Francisco, California, USA, 2006*, 451–456.

Lam, W., Robinson, D.J. Hamill, G.A., Raghunathan, S. & Kee, C. 2006. Simulations of a Ship's Propeller Wash. *Proc. of the Sixteenth (2006) Int. Offshore and Polar Engineering Conference, San Francisco, California, USA.* Copyright © 2006 by The International Society of Offshore and Polar Engineers, 1380–1388.

Lam, W., Hamill, G., Robinson, D. & Raghunathan, K. 2010. Observations of the initial 3D flow from a ship's propeller. Ocean Engineering vol. 37, 2010. 1380–1388.

Lee, S.J., Paik, B.G. Yoon, J.H. & Lee, Ch. M. 2004. Three-component velocity field measurements of propeller wake using a stereoscopic PIV technique. Experiments in Fluids 36 (2004) 575–585.

Muscari, R., Felli, M. & Mascio, A. 2011. Analysis of the flow past a fully appended hull with propellers by computational and xperimental fluid dynamics. *Journal of Fluids Engineering,* vol. 133, Issue 6.

Wilson, P.R., Hamill, G.A. Johnston, H.T. & Kee, C. 2006. Influence of a Horizontal Boundary on a Marine Propeller Wash. *Proc. of the Sixteenth (2006) International Offshore and Polar Engineering Conference, San Francisco, California, USA, 2006* Copyright © 2006 by The International Society of Offshore and Polar Engineers.

Yuksel, A., Celikoglu, Y., Cevik, E. & Yuksel, Y. 2005. Jet scour around vertical piles and pile groups. *Ocean Engineering* No. 32.

2 *Marine vehicles dynamics*

2.1 *Water & air flows*

Sustainable Maritime Transportation and Exploitation of Sea Resources – Rizzuto & Guedes Soares (eds)

Experimental investigation of the wave pattern around a sailing yacht model

K. Sfakianaki, D. Liarokapis & G.D. Tzabiras
National Technical University of Athens (NTUA), Athens, Greece

ABSTRACT: The significance of towing tank testing in the evaluation of the performance of ships both in calm and rough waters has been recognised by many authors. For instance, on a competitive sailing yacht design the study of the free surface is of great importance. In towing tank measurements on sailing yachts, the keel is acting as lifting surface at high yaw angles which affect considerable all resistance parameters as well as the free surface.

The flow field around a sailing yacht is investigated in this paper. A ¼ scaled model of a 50-ft modern sailing yacht has been tested. The experimental results referring to the drag, the side force, the dynamic C.G. rise and the dynamic trim, as well as the flow pattern around the model are presented. Furthermore, the performance of the model in calm water was evaluated, both with and without the keel for a grid of heeling and leeway angles. The procedure used for the alignment, the calibration procedure as well as issues of great importance will be discussed in detail. Useful conclusions are drawn following the discussion of the experimental results.

Furthermore, the experimental results were compared with numerical calculations, in order to evaluate the efficiency of the numerical results. The codes were generated by Prof. G. Tzabiras and K. Sfakianaki. Finally, both the experimental results and the evaluation of the numerical analysis are presented in this paper.

1 INTRODUCTION

The investigation of the flow around racing yachts is an important issue, especially in the final design stage, where optimization of the hull form results in a competitive design. A cerium section of this investigation is the free surface study, which is particularly useful for wave resistance calculations [Dumez & Cordier, 1997] and for validation of numerical prediction codes [Brizzolara, Bruzzone, Cassela, Scamardella and Zotti, 1999].

Kirkman [Kirkman, 1979] discussed the evolving role of the towing tank in providing assistance to the designers and the appropriate means of using model tests in light of the contemporary understanding of scale effects. Nowadays, although the role of the numerical methods in the design of sailing yachts has significantly increased, the experimental methods have also been considerably refined since Davidson's memorable towing tank investigation [Davinson, 1936].

Especially for sailing yachts, balancing under the combined effect of aerodynamic and hydrodynamic forces and sailing in most of the cases in an inclined and yawed condition, the contribution of the experimental evidence to the prediction of their behavior is invaluable.

Furthermore, racing yachts compete in long races where the winner is only a few seconds faster than the other participants. In such cases the incorporation of high technology and the adoption of innovative solutions can make the difference.

2 TURBULENCE STIMULATORS

Ship model tests use turbulence stimulators to compensate for the violation of Reynolds similarity and enforce laminar-turbulent transition in models roughly at the same location as in full scale. Joubert and Matheson [Joubert & Matheson, 1970] studied the effect of stimulators on the boundary layer characteristics past a model in a wind tunnel. It is common to use empirical rules to correlate the size of the stimulator with its position from the forward end of the body as well as with the model speed.

In conventional resistance towing tank tests, the results from this procedure can be assumed reliable, as the size of the stimulator affects only a small area near the bow. The forced transition leads to the recovery of the desired turbulent flow and influences mainly the total skin friction component.

There are cases, however, where this standard procedure is not applicable because the turbulence stimulators disturb drastically the flow-field. For instance, in towing tank measurements on sailing yachts, where the keel, acting as lifting surface at high yaw angles, affects considerable all resistance parameters.

2.1 *Turbulence stimulator selection*

There are two conditions that must be satisfied by the turbulence stimulation devices in order to have the desired effects. Firstly, their geometrical properties should be selected in such a way that steady turbulence is stimulated. For the case of trip wires, the flow may be considered equivalent to the one around an infinite cylinder. This criterion leads to a lower limit in the Reynolds number, where the flow is laminar. Secontly, their height should not be greater than the thickness of the boundary layer at the location they are positioned. If this is the case, they present high parasitic drag, leading to overestimation of the resistance force. Combination of these two conditions results in a range

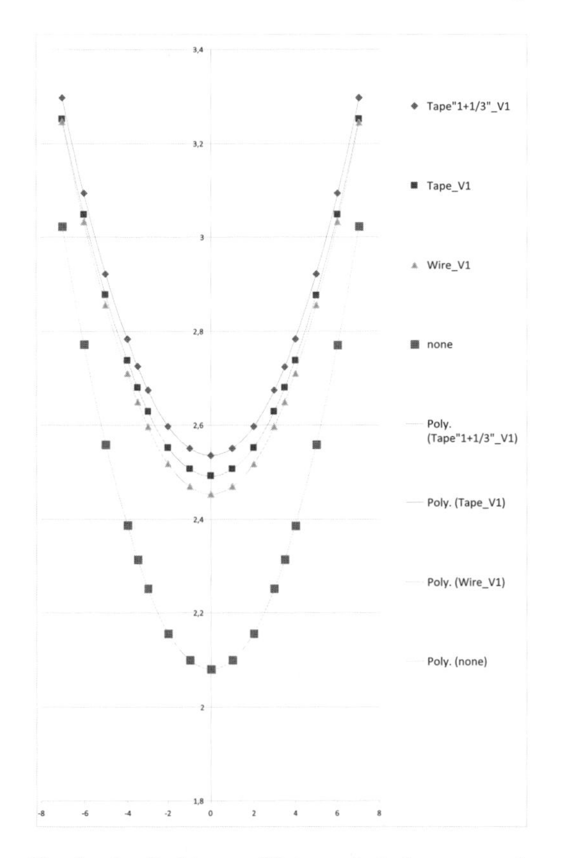

Graph 1. Resistance (Kp) against leeway angle Vm = 2 m/s.

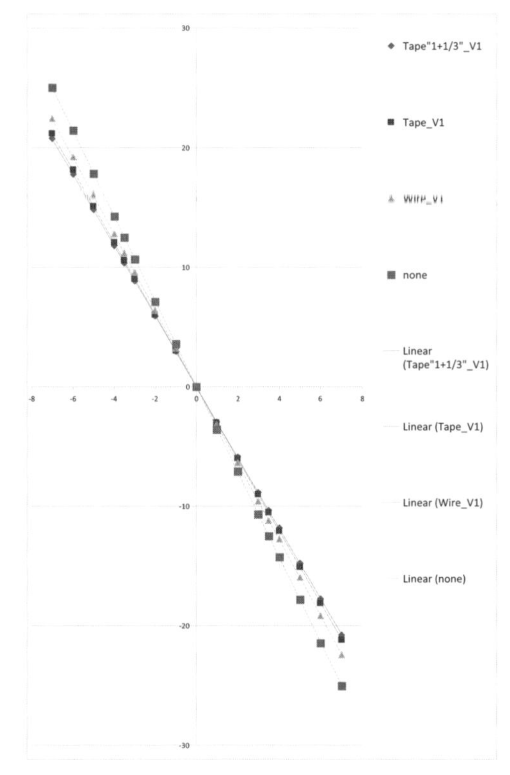

Graph 2. Side force (Kp) against leeway angle Vm = 2 m/s.

of geometrical values that can be selected for the stimulation, when a speed range is considered for testing.

Trip wires, which are commonly used in Laboratory for Ship and Marine Hydrodynamic (LSMH) of NTUA, have low parasitic drag and work sufficiently for a wide range of ship model types. However, they are unreliable in cases where lifting surfaces exist, such as the keel of a sailing yacht. In fact, the rapid change of the pressure in front of the trip wire affects the local pressure and changes the flow deviating from the full scale phenomenon.

Sand strips on the other hand, although they induce more parasitic drag, they are expected to form the turbulence with lower local picks of the pressure.

Because, it is generally difficult to estimate the parasitic drag, it is desirable for their drag to be kept to a minimum value [Mishkevich, 1995].

From previous experimental investigation on the effect of the turbulence stimulators on sailing yacht model we concluded to use sand strips on the keel and trip wire on the canoe hull of the model (Graph 1, 2). [Liarokapis, Sfakianaki, Perissakis and Tzabiras, 2010]

3 EXPERIMENTAL APPARATUS

The LSMH of NTUA posseses a four-component balance yacht dynamometer specially designed for its towing tank (Fig. 1) by Wolfson Unit. The dynamometer is capable of measuring drag, side-force, yaw moment, roll moment, roll, pitch and heave.

The model was attached to the dynamometer at the LCG, via a pivot, which allowed the vertical motion (heaving) and the rotation around the lateral axis through the attaching point (pitching). The model was restrained in surge, sway, yaw and heel. The drag along the yacht's track down the tank, the side force, vertical to the drag, the yaw moment and the heel moment were recorded. At the same time the vertical motion of the centre of gravity and the trimming angle of the model were measured from the attitude of the model relative to the dynamometer. Although these data have no direct use for any performance estimate, as explained by Campbell and Claughton [3], they can provide some qualitative insight into the hydrodynamic behaviour of the yacht. Furthermore, the model was restrained at preset angles of heel and leeway, selected for performance prediction.

Sailing yacht measurements resolve two critical issues on the experimental procedure. The first issue is the accuracy of the dynamometer positioning relatively to the water surface.

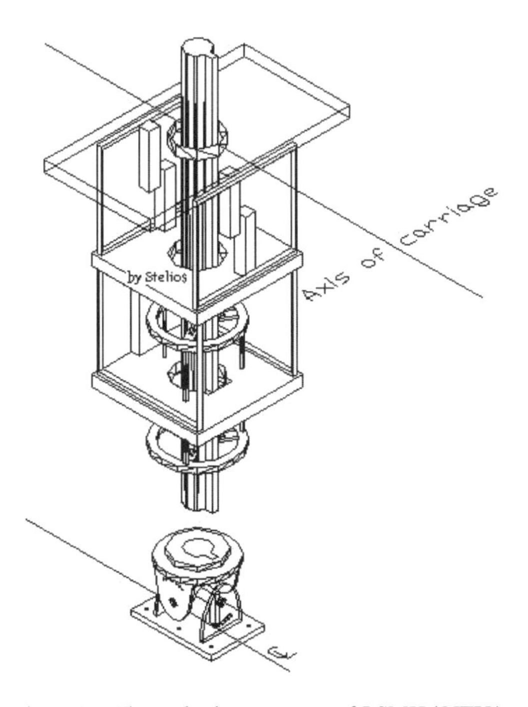

Figure 1. The yacht dynamometer of LSMH / NTUA.

The yacht dynamometer is attached to the tank carriage through two parallel rails. A fully adjustable rig connecting the dynamometer with the towing tank carriage was devised. The constructed rig allows for 6-degrees of freedom adjustments and by using modern measuring techniques e.g., (laser, etc) the experimentalist can accurately align the dynamometer parallel to the water surface.

The second issue associated with the position of yacht model relatively to the longitudinal axis of the towing tank. This misalignment affects the measurement axes of resistance and side force. The alignment procedure proposed by the manufacturer suggested rotating the model till both the side force and drag is minimized (resistance versus side force squared diagram). From the results it can be noticed the fairly small misalignments (less than 1 degree) from the carriages track direction results to substantial high side forces, which leads to a considerable misrepresentation of the lifting phenomenon. Another source of misalignment is inherent to the model as a result of the construction asymmetries mainly of the keel and its installation to the hull but also of the hull itself. Thus, contemporary analytical method to calculate the exact upright position was developed.

The basic principle of the software is as follows: Referring to model asymmetries, and provided that these asymmetries are small, the physical zero leeway angle could be arbitrary defined at zero Side Force. However a second correction should take place, considering the angle between the main axis of the dynamometer and the axis of the motion of the carriage. Both problems are encountered by assuming an overall correction angle of the position of the dynamometer and recalculating the Resistance and Side Force according to the new corrected axis. The assumed correction angle should satisfy the following rule of thumb: the recalculated resistance must be minimum and at the same time the recalculated side force must be zero.

The wave pattern was measured by the well established wave probes. Wave probe is the most common instrument for this kind of measurements. It records the time history of the free surface elevation at a specific location, which represent a wave cut of the wave pattern. For our measurements we used two kinds of wave probes, the resistance and the acoustic type.

- The resistance wave probe, which although is a low cost, reliable and relatively accurate instrument, has considerable limitations mainly due to the fact that it is an intrusive method. Thus, it is not feasible to position these wave probes on the path of the model.
- The acoustic wave probe, which measures the distance to the wave surface by sound propagation.

Steep and very fast moving waves can be measured with a relative velocity of 15 m/s, and frequency response of 100 Hz. The acoustic probes, due to their non intrusive nature, where used for measuring wave elevation near the hull.

4 TEST PROCEDURE

A 1:4 scaled wooden model of a 50-ft modern sailing yacht, designed by Mortain and Mavrikios according to the British Oxygen Corporation (BOC) regulations has been extensively tested in the towing tank of the Laboratory for Ship and Marine Hydrodynamics (LSMH) of the National Technical University of Athens (NTUA).

Both, the canoe body and the keel are made of wood to ensure precise representation of the hull form (Fig. 2). In order to achieve a light model construction to enable testing at light displacements, water-resistant plywood was used to shape the transverse frames and wooden strip planks which formed the shell. The main particulars of the yacht are given in Table 1.

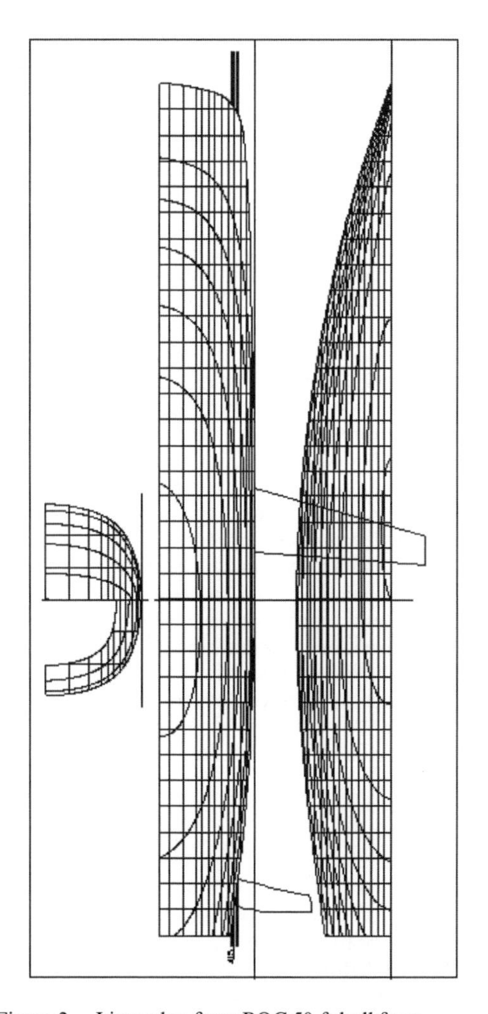

Figure 2. Lines plan from BOC 50-ft hull form.

Table 1. Main particulars of the racing yacht tested.

Main particulars	BOC 50-ft yacht
Length at waterline	14.87 m
Breadth at waterline	2.66 m
Design draft (canoe body)	0.415 m
Design draft (maximum)	1.065 m
Design displacement	7.175 mt
Appendage displacement	0.569 mt
Trim	Even keel
Sectional foils of Keel	NACA 64 A015
Rigging	Ketch
Sailing area	130 m^2

The performance of the model in calm water was evaluated, both with and without the keel for upright position and in variety of leeway angles. More specific, the model was tested at three speeds (0.5, 1 and 2 m/s) and at three leeway angles (0, 3.5, 7 degrees) on both tacks. Two turbulent stimulators are used, one for the keel and one for the canoe body.

Follow up, a trip wire (with a mean diameter of 1.8 mm) was fitted on the canoe body at a distance of 23 cm from the bow. In sequel, a sand strip was fitted on the keel. The stimulator was fitted at a distance of 2 cm from the leading edge of the keel. Trip wire diameter fitted on the hull was chosen according to the Fn, and the position of the installation.

5 ANALYSIS AND PRESENTATION OF THE RESULTS

The experimental results referring to the model of BOC 50-ft sailing yacht, are presented in the following graphs (Graphs 3, 4):

In the above two graphs, the resistance and the side force is plotted against the leeway angle for model speed of 1 and 2 m/s, with and without the keel. The results regarding the resistance and the side force are presented in the tables below:

It can be easily noticed that all the curves are symmetrical to leeway angle axis. This indicates that the dynamometer operates satisfactory on both tacks and the software developed by Prof. Tzabiras calculates the upstream position with less than 0.1 degrees accuracy. In both cases, the drag increases with the keel fitted on. In the upright position, at the speed of 1 m/s, we can observe that the resistance increased by 41.7% when the keel was

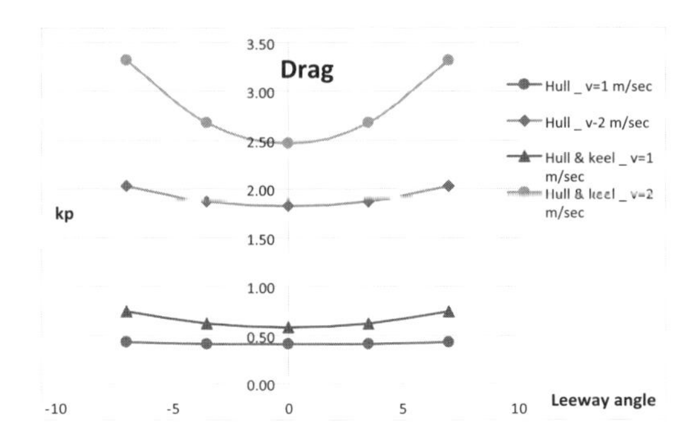

Graph 3. Resistance against leeway angle.

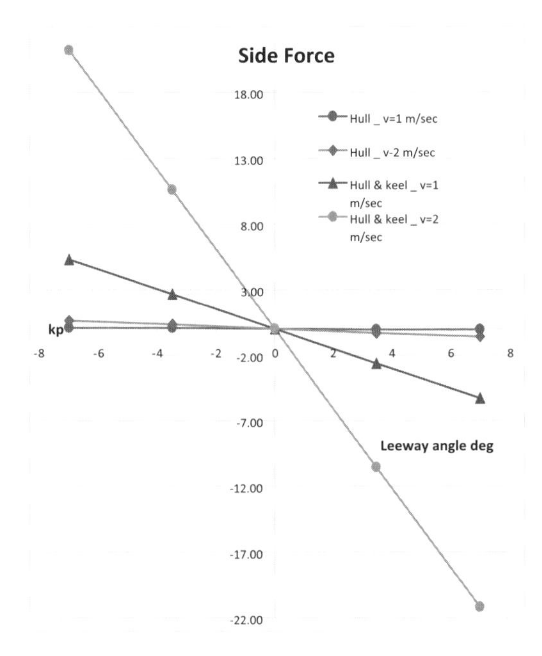

Graph 4. Side force against leeway angle.

fitted on the body, while this percentage increased further to 49.7% and 72.4% for the leeway angle of 3.5 and 7 degrees respectively. Moreover, at the speed of 2 m/s, the resistance increased by 35.5% when the keel was fitted on, while this percentage increased further to 43% and 63.7% for the leeway angle of 3.5 and 7 degrees respectively. Furthermore, we observe that the side force reduces by 98% at both leeway angles and at both speeds when we removed the keel.

In the next graph (Graph 5) the CG rise is plotted against the velocity both with and without the keel. It can be assumed that the difference between the two curves is negligible.

In addition, the effect of the trim on the resistance was investigated. From the following graphs

Table 2. Side force against leeway angle.

	Side force				
Hull	Leeway angle				
Speed	−7	−3.5	0	3.5	7
0.5	0.026	0.013	0	0.013	0.026
1	0.076	0.038	0	0.038	0.076
2	0.626	0.312	0	0.312	0.626
Hull & keel	Leeway angle				
Speed	−7	−3.5	0	3.5	7
0.5	1.575	0.785	0	−0.785	−1.575
1	5.311	2.645	0	−2.647	−5.313
2	21.198	10.558	0	−10.563	−21.202
	% Side force reduction				
	Leeway angle				
Speed	−7	−3.5	0	3.5	7
0.5	98.32%	98.32%	0	98.32%	98.32%
1	98.57%	98.57%	0	98.57%	98.57%
2	97.05%	97.05%	0	97.05%	97.05%

Table 3. Resistance against leeway angle.

	Resistance				
Hull	Leeway angle				
Speed	−7	−3.5	0	3.5	7
0.5	0.109	0.110	0.112	0.111	0.109
1	0.430	0.413	0.408	0.413	0.076
2	2.027	1.873	0	1.873	2.027
Hull & keel	Leeway angle				
Speed	−7	−3.5	0	3.5	7
0.5	0.222	0.172	0.156	0.172	0.222
1	0.741	0.619	0.578	0.619	0.741
2	3.318	2.679	2.468	2.679	3.319
	% Side force difference				
	Leeway angle				
Speed	−7	−3.5	0	3.5	7
0.5	103%	56.6%	39.3%	56.6%	103%
1	72.4%	49.7%	41.7%	49.7%	72.4%
2	63.7%	43%	35.5%	43%	63.7%

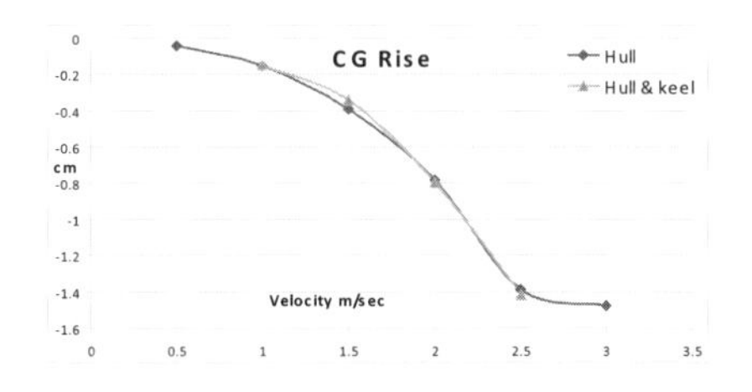

Graph 5. Side heave against velocity.

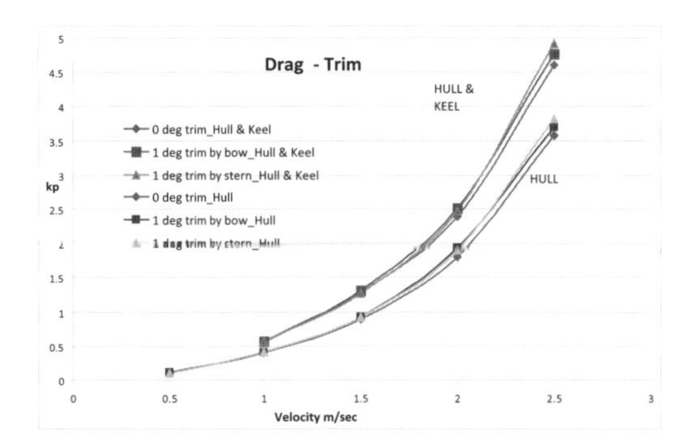

Graph 6. Drag against trim.

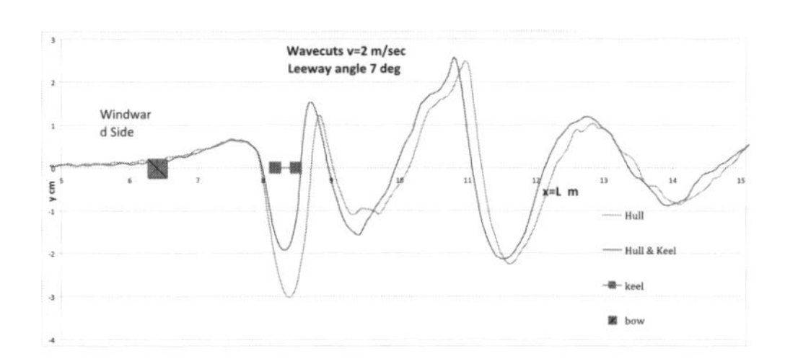

Graph 7. Wavecut at windward side Vm = 2 m/s.

Graph 8. Wavecut at leeward side Vm = 2 m/s.

we can observed that a small increase at the resistance is noticeable at the speed of 2 m/s. The minimum values of Drag both with and without the keel were obtained at even kill.

5.1 *Wave pattern*

The results regarding the wave pattern around the model derived using both resistance wave probes at the side of the towing tank and acoustic wave probes measuring close to the hull. Again, the model was tested at three speeds (0.5, 1 and 2 m/s) and at three leeway angles (0, 3.5, 7 degrees) on both tacks.

In the graphs 7, 8, and 9, the wave elevation is plotted against the distance for leeway angle 0 and 7 degrees and model speed of 2 m/s, both with and without the keel. We can observe that at the upward condition, the keel reduces the pressure

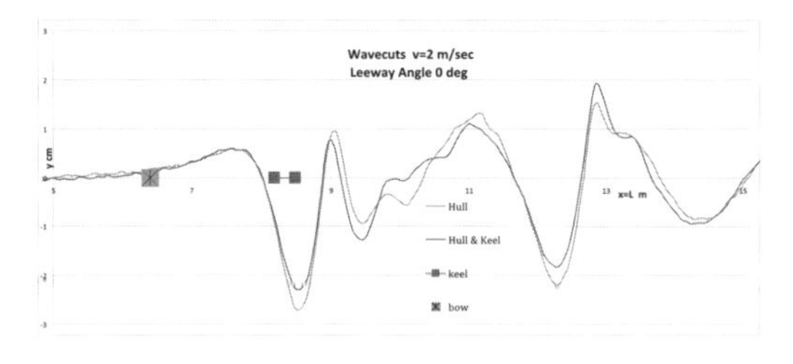

Graph 9. Wavecut at upstream position Vm = 2 m/s.

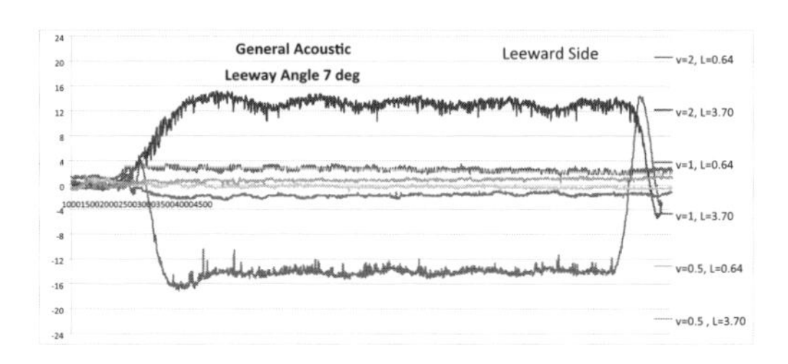

Graph 10. Wave elevation against velocity.

locally, thus we notice a reduction on the howe and the crest. At a leeway angle of 7 deg. on the windward side we notice a big reduction of the howe at the area of the keel, while the crest seems to increase at the end of the keel. At the same angle on the leeward side, a big howe is created after the keel and then a big crest immediately afterwards.

Next, we measured the wave elevation at a leeway angle of 7 degrees with the keel fitted on. From the acoustic sensor near the hull we notice a significant wave elevation only at the speed of 2 m/s. More analytical, the elevation at two m/s was more than 14 mm at various lengths while at 0.5 the wave elevation is less about 1 mm.

6 DATA EVALUATION

To evaluate the data from the sailing yacht dynamometer we carried out repetitive tests with the static dynamometer. Ineluctably, all the tests carried out at the upstream position, both with and without the keel. The results confirmed that both dynamometers produce the same output. In the following graph the drag is plotted against the velocity both with and without the keel for the two dynamometers. It can be noticed that the data from the static dynamometer are lying at the curves obtained from the sailing yacht dynamometer.

Moreover, for evaluating the results from the wave probes we carried five repetitive tests and plotted the results on the same graph. Phase shift was synchronized by using time cross-correlation

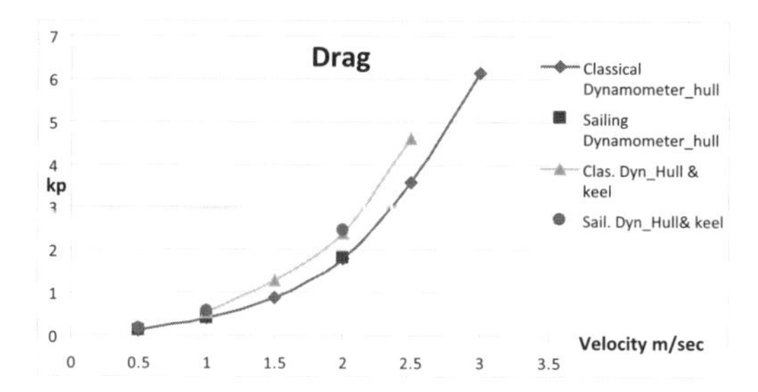

Graph 11. Drag against velocity.

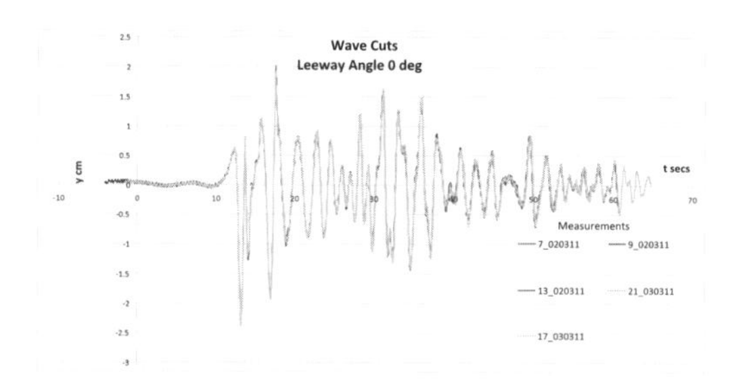

Graph 12. Wave cuts leeway angle 0 Deg.

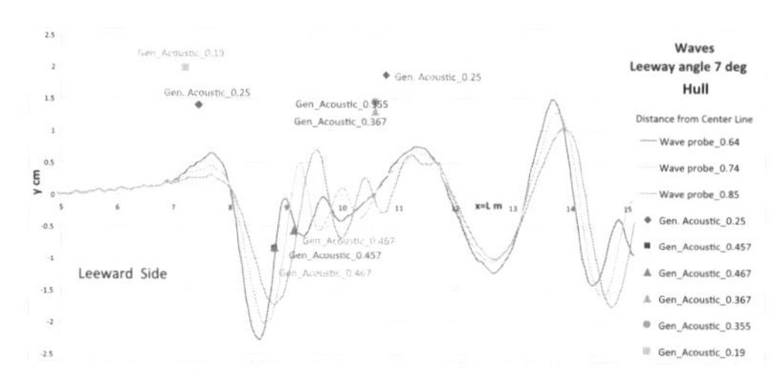

Graph 13. Wave elevations and probes position.

function. The five different curves are in good agreement with variation of less than 3 mm.

Finally, at the above graph the wave elevations as well as the positions of the wave probes are plotted relative to the distance from the model. To synchronize the model passage with the actual measurements we used an opto-coupled devise that sends a pulse to the data acquisition when the tip of the model reaches the wave probes rake.

7 DISCUSSION

The Laboratory for Ship and Marine Hydrodynamic of National Technical University of Athens (NTUA), has recently built a rig to accurate align the sailboat dynamometer. From the results it can easily be noticed that all the curves are symmetrical to leeway angle axis. This indicates that the dynamometer operates satisfactory on both tacks and the software developed for the alignment calculates the upstream position with less than 0.1 degrees accuracy.

During the installation of the dynamometer and the model in the towing carriage, there are some unavoidable sources of misalignment. It was observed that even a small fraction of a degree in the alignment of the dynamometer, may induce reasonable errors.

The resistance, the side force, CG rise were plotted against the leeway angle for model speed of 1 and 2 m/s, with and without the keel. In both cases, the drag increases with the keel fitted on. Furthermore, we observe that the side force is reduced by 98% at both leeway angles and at both speeds when we removed the keel.

The increase of the resistance relative to the leeway angle was expected due to the shape of the keel. On the other hand, we can observe that the percentage of increment encouraged between the two speeds is slightly lower at the speed of 2 m/s mainly due to the fact that the wave pattern, influence the wetted surface, thus reducing the drag.

This is in general agreement with the results obtained by the wave probes. As it was mentioned earlier the howe reduces both at the leeward and windward angle at the area of the keel. At upright position there is a noticeable decrease of the wave pattern at the same area. Results obtained by analytical calculations indicating this behavior. At speed of 1 m/s, the wave elevation was about 3 mm, and for 0.5 m/s the wave elevation was less than 1 mm. In sequel, the wave pattern does not affect the resistance friction at low speeds.

8 CONCLUSION

The scope of the work was to investigate the wave pattern around a sailing yacht model. A trip wire and a sand strip were fitted at the hull and the keel of the model. The experimental results referring to the drag, the side force, the GS rise, the dynamic trim and the wave pattern, for three model speeds were presented. Furthermore, the performance of the model in calm water was evaluated for a grid of leeway angles.

In general, the experimental results derived in the towing tank of the Laboratory for Ship and Marine Hydrodynamics (LSMH) of the National Technical University of Athens (NTUA) are in satisfactory agreement with published experimental results. In any case, the provision of the measured free surface significantly increases the efficiency of the analytical codes, at least for simple hull geometries, keeping at the same time the required computer power to a minimum.

REFERENCES

Brizzolara, S., Bruzzone, D., Cassela, P, Scamardella, A. & Zotti, I., 1999. 'Wave Resistance and Wave Patterns for High-Speed Crafts; Validation of Numerical Results by Model Test', 22nd Symposium on Naval Hydrodynamics.

Brodeur, R. & Dam, C.P. 2001 "Transition prediction for a two-dimensional Reynolds-averaged Navier-Stokes method applied to wind turbine airfoils, Wind Engineering 4, pp. 61–75.

Campbell, I. & Claughton, A. 1987. "The interpretation of results from tank tests on 12-m yachts", Proc. 8th Chesapeake Sailing Yacht Symposium, SNAME, pp. 91–107.

Davinson, K.S.M. 1936 "Some experimental studies of the sailing yatch", Trans. SNAME, Vol. 44, pp. 288–334.

Djeridi, H., Sarraf, F. & Billard, J.Y. 2007. "Thickness effect of NACA symmetric hydrofoils on hydrodynamic behaviour and boundary layer states, IUTAM Symp. Unsteady Separated Flows and their Control, Corfu.

Dumez, X.F. & Cordier, S. 1997. 'Accuracy of Wave Pattern Analysis Methods in Towing Tanks', 21st Symposium on Naval Hydrodynamics.

Garofallidis, D.A. June 1996. "Experimental and numerical investigation of the flow around a ship model at various Froude Numbers" Phd Thesis, NTUA.

Hughes, J.F. Allan, 1951. "Turbulence stimulation on ship models", Trans. SNAME, Vol. 59.

Joubert, P.N. & Matheson, N. 1970. "Wind tunnel tests of two Lucy Ashton reflex geosims", Journal of Ship Research, Part 4.

Kirkman, K.L. Jan 1979. "The evolving role of the towing tank", Proc. 4th Chesapeake Sailing Yacht Symposium, SNAME, pp. 129–155.

Liarokapis, D., Sfakianaki, D., Perissakis, & Tzabiras, G. 2010. "Experimental Investigation of the Turbulence Stimulator on a Sailing Yacht Model".

Mishkevich, G. 1995. "Scale and roughness effects in ship performance from the designer's point of view", Marine Technology, pp. 126–131, vol. 32, No. 2.

Sustainable Maritime Transportation and Exploitation of Sea Resources – Rizzuto & Guedes Soares (eds)
© 2012 Taylor & Francis Group, London, ISBN 978-0-415-62081-9

A panel method based on vorticity distribution for the calculation of free surface flows around ship hull configurations with lifting bodies

K.A. Belibassakis
School of Naval Architecture & Marine Engineering, NTU of Athens, Greece

ABSTRACT: In the present paper a panel method is developed for the calculation of free surface flows around ship hull configurations including lifting bodies. The method is based on source distribution on the wetted surface of the hull and the free-surface, in conjunction with vorticity distribution of the lifting bodies and trailing vortex sheets. A free wake analysis has been applied in order to locate the position of the vortex sheets. The non-linear boundary conditions on the free surface are satisfied by iterations, starting from a linearization. Detailed numerical results are presented, including initial comparisons against other methods and measured data, illustrating the efficiency and usefulness of the present approach.

1 INTRODUCTION

Performance of many hull types is dependent on hydrodynamic properties of their keels, rudders, hydrofoils and similar appendages operating as lifting bodies. Interaction of free-surface flows with lifting surfaces constitutes therefore an interesting problem, finding applications in the design of yachts and sailing boats and the performance of stabilizers, hydrofoils and similar devices. Moreover, due to its specific importance in ship powering prediction and optimisation of ship hulls, the investigation of steady ship motion in calm water is a significant problem.

Many mathematical formulations and corresponding numerical techniques are available today for the wave resistance problem of a steadily translated ship including lifting bodies in calm water, ranging from solvers based on RANSE (see, e.g., Carrica et al., 2007, Tzabiras 2004, Larsson et al., 2003), to fully nonlinear or partially linearized potential flow solvers; see, e.g., Janson (1997), Bal (2008); recent advances are also presented in the reports by ITTC (see ITTC 2005, 2008 and the references cited there). In addition, methods based on potential flow provide useful information for the intialization of CFD methods and geometry optimization at the initial stages. The most popular way to reduce problems involving the Laplace equation to equivalent boundary integral equations are the method of potentials and the direct method, which is based on Green's representation formula. This reduction is by no means unique, therefore, a constantly growing number of equivalent formulations is available today for the approximate solution of the problem under consideration. The application of three dimensional potential flow theory to the steady ship motion problem results in an essentially non-linear boundary value problem (due to the non-linear character of the boundary condition at the free surface), from which the unknown velocity potential and the free-surface disturbance are calculated. Boundary Element Method (BEM, see, e.g., Paris & Canas 1997) using the simple Rankine source (the fundamental solution of Laplace equation) as the elementary singularity is widely used to solve potential flow problems in marine hydrodynamics and especially the wave resistance and the ideal wave pattern of ships advancing with steady forward speed. The method was first presented by Dawson (1977) and since then it has been widely applied as a practical method to predict wave resistance. Many improvements have also been made to account for non-linear effects; see, e.g., Nakos & Sclavounos (1990), Raven (1996), Bertram (2000), Bal (2008). In a similar way, it is possible to reduce the present boundary value problem, formulated with respect to the unknown disturbace velocity field, to boundary integral equations. In the present work low-order a panel method is developed for the calculation of free surface flows around ship hull configurations including lifting bodies. The method is based on source distribution on the wetted surface of the hull and the free-surface, in conjunction with vorticity distribution on the lifting bodies and trailing vortex sheets. A free wake analysis has been applied in order to locate the position of the vortex sheets. The non-linear boundary conditions on the free surface are satisfied by iterations, starting from a linearization (see, e.g., Raven 1988, Janson 1997). Numerical results are presented, including first comparisons against other methods and measured data, illustrating the efficiency and usefulness of the present approach.

2 FORMULATION

We consider inviscid, incompressible flow around a geometrical configuration (Fig. 1) moving at constant forward speed $\mathbf{U} = -\mathbf{U}_\infty$, where $\mathbf{U}_\infty$ denotes the parallel incoming flow. A Cartesian coordinate system is introduced in the body-fixed frame of reference, with its origin in the center of flotation of the body. The longitidinal and transverse axes are x_1 and x_2, respectively, and the x_3-axis is pointing upwards. The total flow velocity is decomposed as $\mathbf{w} = \mathbf{U}_\infty + \mathbf{u}$, where $\mathbf{u} = \nabla\varphi(x_1, x_2, x_3)$ denotes a potential disturbance-flow component.

We denote the flow domain by $D = D^+$ and its boundary by $\partial D = S_B \cup S_F \cup S_L \cup S_W$, where S_B is the non-lifting part of the wetted surface of the body, S_F is the free surface, $S_L = \cup S_{LK}$ denotes the various lifting parts (as, e.g., wings and keels) and $S_W = \cup S_{Wk}$ the associated trailing vortex sheets. Let also $D^- = \mathbb{R}^3 \backslash D^+$ be the complementary domain, and $\mathbf{n}$ the unit normal vector on ∂D directing inwards of D^+; see Fig. 1. It is assumed that the boundary ∂D can be locally represented through a regular parametric representation of the form $\mathbf{x} = \mathbf{x}(\xi^\alpha)$, mapping an open parametric domain $\{\xi^1, \xi^2\} \in E \subseteq \mathbb{R}^2$ bijectively onto a surface patch of ∂D. The vectors

$$\mathbf{e}_\alpha = \partial\mathbf{x}\left(\xi^1,\xi^2\right)/\partial\xi^\alpha, \qquad \alpha = 1, 2, \tag{1}$$

forming the covariant local base on the surface are linearly independent at each point of each surface patch. Direction of the parametric curves is appropriately selected, in order to obtain the unit normal vector on the boundary ∂D by the following relation,

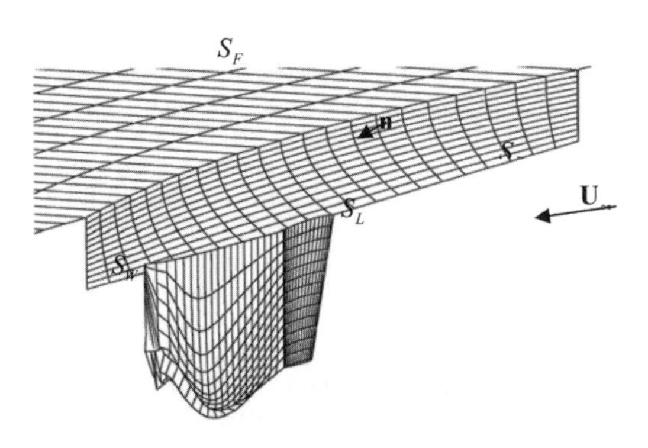

Figure 1. Geometrical configuration and notation.

$$\mathbf{n} = \left(\mathbf{e}_1 \times \mathbf{e}_2\right)/\sqrt{a} \tag{2}$$

where $a = |\mathbf{e}_1 \times \mathbf{e}_2|^2$ the metric tensor determinant. The no-entrance boundary condition at the solid boundaries in the moving frame of reference requires

$$\mathbf{u}\cdot\mathbf{n} = -\mathbf{U}_\infty\cdot\mathbf{n} = f\left(\mathbf{x}\right), \qquad \mathbf{x} \in S_B \cup S_L, \tag{3}$$

where $f(\mathbf{x})$ denotes the boundary data.

Following Hess (1972) and many authors since then (see, e.g., Katz & Potkin 1991), our method is based on using simple source distributions (σ) with support on the non-lifting parts of the wetted surface of the body (S_B), as well as on the free surface (S_F), and surface vorticity distributions ($\boldsymbol{\gamma}$) with support on the lifting parts (S_L) and the corresponding trailing vortex sheets (S_W). Thus, the disturbance velocity is

$$\mathbf{u}(P) = \frac{1}{4\pi}\left\{\int_{S_B \cup S_F} \sigma\frac{\mathbf{r}}{r^3}dS + \int_{S_L \cup S_W} \gamma\times\frac{\mathbf{r}}{r^3}dS\right\}, \tag{4}$$
$$P \in D,$$

where $\mathbf{r} = \mathbf{x}(P)-\mathbf{x}(Q)$ denotes the distance between the field point (P) and integration point (Q) and $r = |\mathbf{r}|$. For the disturbance velocity field, as represented by the above equation, to be irrotational the surface vorticity must fulfill a continuity condition on its support

$$\nabla_\Sigma\gamma = 0, \qquad \text{on} \quad S_L \cup S_W, \tag{5}$$

where ∇_Σ denotes divergence of a surface vector field. The above equation ensures the existence of an equivalent normal dipole distribution (μ) on the surface such that

$$\gamma = -\mathbf{n}\times\nabla_\Sigma\mu. \tag{6}$$

For points on the boundary, $P \in \partial D$, the same expression (Eq. 2, with the integrals considered as their principal Cauchy values) provides the mean value of the velocity field,

$$\left[\mathbf{u}(P)\right] = \frac{1}{2}\left(\mathbf{u}^+(P) - \mathbf{u}^-(P)\right), \quad P \in \partial D, \tag{7}$$

while its discontinuity across the boundary is given by

$$<\mathbf{u}> = \mathbf{u}^+ - \mathbf{u}^- = \sigma\,\mathbf{n} + \gamma\times\mathbf{n}, \quad P \in \partial D. \tag{8}$$

Use of Eqs. (2), (3) and (4) in Eq. (1) results to the following (singular) boundary integral equations, for points on the body

$$\frac{\sigma}{2}+\frac{\mathbf{n}(P)}{4\pi}\left\{\int_{S_B \cup S_F}\sigma\frac{\mathbf{r}}{r^3}dS+\int_{S_L \cup S_W}\gamma\times\frac{\mathbf{r}}{r^3}dS\right\}=g,$$

$$P\in S_B, \tag{9a}$$

and for points on the lifting pars and vortex wakes

$$\frac{\mathbf{n}(P)}{4\pi}\left\{\int_{S_B \cup S_F}\sigma\frac{\mathbf{r}}{r^3}dS+\int_{S_L \cup S_W}\gamma\times\frac{\mathbf{r}}{r^3}dS\right\}=g,$$

$$P\in S_L \cup S_W. \tag{9b}$$

Pressure is calculated from velocity through Bernoulli's theorem, which in the present case reads

$$p=p_\infty+\frac{\rho}{2}\left(U_\infty^2-w^2\right)-\rho g x_3, \tag{10}$$

where ρ stands for the fluid density and g is the gravitational acceleration, p_∞ denotes the ambient (atmospheric) pressure and $U_\infty = |\mathbf{U}_\infty| = U$. On the free-surface $x_3 = \eta(x_1, x_2)$ the kinematic condition requires

$$\mathbf{n}\cdot\mathbf{w}=0,\ P\in S_F,\text{or}\ w_1\frac{\partial\eta}{\partial x_1}+w_2\frac{\partial\eta}{\partial x_1}-w_3=0, \tag{11}$$

and the corresponding dynamic condition is

$$p=p_\infty \text{ on } S_F,\text{ or } \eta=-\frac{1}{2g}\left(w^2-U_\infty^2\right)=0. \tag{12}$$

Assuming that the free-surface elevation and the disturbance velocity are small quantities, the free-surface boundary conditions (11), (12), can be linearised by separating the total flow field and the free surface elevation into two parts $\mathbf{w}_b + \mathbf{w}$ and $\eta = \eta_b + \eta$, respectively. One option leading to enhanced linear formulations is based on using as background field ($\mathbf{w}_b$, η_b) the one associated with the double-body flow, obtained by imposing a homogeneous Neumann condition on the undisturbed water surface ($x_3 = 0$); see, e.g., Janson (1997, Sec. 2.5). A more simple formulation, appropriate for thin hull forms, is obtained by selecting $\eta_b = 0$, $\mathbf{w}_b = \mathbf{U}_\infty$ (the parallel incoming flow), leading to Neumann-Kelvin boundary conditions (see, e.g., Baar & Price 1988)

$$\frac{\partial u_1}{\partial x_1}+ku_3=0 \text{ and } \eta=U_\infty u_1/g,\text{ on } x_3=0, \tag{13}$$

where $k=g/U_\infty^2$ is the characteristic wavenumber, controlling the wavelength of the transverse ship wave system (and is directly connected with the square inverse of the corresponding Froude number $F=U/\sqrt{gL}$, with L denoting the length of the body). The linearised version of boundary condition, Eq. (13), is used in the present work to obtain a starting solution. Subsequently, the non-linear problem is tracked iteratively, by using Eq. (11b) to update the free-surface boundary from the previous iteration, remeshing the free surface and satisfying the kinematic free surface boundary condition Eq. (11a) at the new position (see, e.g., Raven 1988).

To discretize the present system of boundary integral equations (9) and (11) or (13a), we use 4-node quadrilateral (hyperboloid) elements (see, e.g., Katz & Plotkin 1991) to approximate the boundary surfaces, ensuring C^0-continuity. The source distribution σ has been approximated by a piecewise constant scheme on each element, and the corresponding induced velocities and potential from each source element are calculated by semi-analytical integration, consistently treating the singular part. The vorticity distribution is approximated by means of a piecewise constant scheme of the associated dipole distribution, which is equivalent to a collection of vortex ring elements, with vorticity concentrated along the element edges (see, e.g., Hess 1972, Politis 2005). The corresponding induced velocities and potential are analytically calculated by repetitive application of the Biot-Savart rule for each of the 4 straight vortex filaments surrounding the quadrilateral element (Katz & Plotkin 1991). In this sense, the vorticity surfaces ($S_L \cup S_W$) are approximated by means of a vortex-lattice, consisted of an arrangement of non co-planar, closed vortex loops (vortex rings) of constant intensity. The integral equations are satisfied at a set of collocation points $\{P_j\}$, defined as the centroids of each element of the mesh on $S_B \cup S_F$ and $S_L \cup S_W$.

For obtaining the starting solution, Eq. (13a) has been also discretized on the undisturbed free-surface ($x_3 = 0$), by using the four-point, upwind finite difference scheme by Dawson (1977) for approximating the horizontal derivative. A minimum number of 15 elements per wavelength is used in discretizing S_F, in order to eliminate errors due to damping and dispersion characteristics of the above discrete scheme (see also, Sclavounos & Nakos 1988 and Janson 1997).

3 MODELLING OF TRAILING VORTEX SHEETS

The present formulation requires the knowledge of the position and geometry of the trailing vortex sheets, extended far downstream the lifting parts of the geometrical configuration. In the case when $S_L = \cup S_{LK}$ are located relatively far from the free-surface, we may assume weak interaction between free-surface and trailing vorticity dynamics. This permits us to approximately determine the position of the trailing vortex sheets by solving an auxiliary problem concerning the flow evolution around the present steadily translating geometrical configuration at constant forward speed ($\mathbf{U} = -\mathbf{U}_\infty$), using a homogeneous Neumann condition on undisturbed free surface $x_3 = 0$, and starting from rest. For the description of the kinematical characteristics of the system and the induced flow dynamics two reference systems are used (see Katz & Plotkin 1991, Sec. 13): (i) the stationary, inertial coordinate system (X_1, X_2, X_3) and (ii) the body-fixed, moving coordinate system (x_1, x_2, x_3) with constant velocity $\mathbf{U} = (U, 0, 0)$ with respect to the former,

$$X_1 = Ut + x_1,\ \ X_2 = x_2,\ \ X_3 = x_3. \tag{14}$$

As each lifting component is steadily translated, the vorticity created in the boundary layers of its upper and lower surfaces is continuously shed into its wake. Furthermore, this vorticity, which is subsequently convected away from the lifting parts with the local velocity, is assumed to be concentrated into vortex sheets of infinitesimal thickness, constituting the trailing vortex wakes, and flow outside these vortical regions is assumed irrotational. The disturbance velocity is represented by the corresponding time-dependent potential $\psi(X_i;t)$ as $\nabla_X\psi$, and satisfies the no-entrance boundary condition $\mathbf{n} \cdot \nabla\psi = \mathbf{U} \cdot \mathbf{n}$ on all solid boundaries and at the Neumann plane ($X_3 = 0$). Exploiting the fact that, for incompressible flows, the influence of the momentary boundary condition is immediately radiated across the whole fluid domain, steady-state techniques can be implemented to solve the boundary value problem on ψ, satisfying the boundary condition expressed in the body-fixed frame of reference, using a similar boundary integral formulation method as described in the previous section. However, the wake shape of the lifting component depends on the time history of the motion (which is assumed starting from rest) and is calculated by applying the laws governing the evolution of trailing vortex sheets in the stationary frame of reference. The latter force-free material surfaces and the vorticity field (concentrated on these surfaces) must be divergence-free, see Eq. (5). The numerical solution is obtained by modeling the trailing vortex sheets by a collection of joint-together vortex rings of constant intensity, which is equaivalent to piesewise constant dipole distribution on $S_W(t)$; see, e.g., Katz & Plotkin (1991, Sec. 13.12).

Shed vorticity is convected with the local velocities, as calculated at the nodal points of each free vortex ring. By this technique, Kelvin's theorem,

$$\frac{D\Gamma}{Dt} = \left[\frac{\partial}{\partial t} + \nabla\psi \cdot \nabla\right] \oint_{c(t)} (\nabla\psi) \cdot d\ell = 0, \tag{15}$$

Concerning the conservation of circulation Γ around any closed material circuit $c(t)$, is automatically fulfilled. By applying Bernoulli's theorem in unsteady flow,

$$\frac{p - p_\infty}{\rho} = -\frac{\partial\psi}{\partial t} + \frac{1}{2}(\nabla\psi)^2, \tag{16}$$

where p_∞ stands for for the ambient pressure, it can be seen that Eq. (15) is, practically, equivalent to the satisfaction of the dynamical condition on the free-vortex sheet, that requires equal limit pressures at points on these surfaces as approached from either side.

Furthermore, the present inviscid formulation is supplemented by the Kutta condition applied along the trailing edge of the lifting sections (here the wing-keel below the hull). Among the alternative approximate forms of the Kutta condition in unsteady flow, in the present work this condition is implemented by enforcing the instantaneous vorticity generated at the trailing edge (and the tip) of the wing to shed with the local velocity. By applying Kelvin's theorem, Eq. (15), in conjunction with Bernoulli's theorem, Eq. (16), one can derive that this postulate is asymptotically (for very low time increments, $\delta t \to 0$) equivalent to the requirement of equal limit pressures at points on the upper and lower surfaces of the wing (or the vortex wake), as these flow separation lines are approached.The flow evolution problem is treated by means of a time marching algorithm, as e.g., described in Katz & Plotkin (1991, Sec. 13). This scheme permits us to calculate the space-time evolution of the trailing vortex surfaces $S_W(t)$, in the stationary coordinate system (X_1, X_2, X_3), by using the fact that these material surfaces should evolve while moving with the local velocities

$$d\mathbf{X}/dt = [\nabla\psi], \quad \text{for } \mathbf{X} \in S_W(t), \tag{17}$$

where $[\nabla\psi]$ denotes the mean perturbation velocity on the trailing vortex sheets $S_W(t)$. The above formulation suggests that dipole strengths associated with each element, which is equivalent to the intensity of the vortex rings in the wake, should remain invariant during the motion with the mean velocity. This property, permits us to obtain successive instances of the mesh on $S_W(t)$ by free-wake analysis. We note here that the same practically approach has been used in Belibassakis et al. (1997) for calculating the propulsive performance of oscillating wing tails, and in Politis (2005, 2011) for unsteady marine propeller analysis.

As an example, we present in Figs. 2, 3 and 4, the evolution of the wake downstream a trapezoidal wing, used as keel below the parabolic Wigley hull, as calculated by the present method. In this case, the ratios of the main dimensions of the hull are: $L/B = 10$, $L/T = 16$, $B/T = 1.6$. The above hull is joined together with a wing of trapezoidal planform and symmetric NACA0012 sections extended below its keel. In this case, the leading edge of the wing's root-chord is located at the midship section of the hull, on the base line. Furthermore, the span of the wing-keel is taken $s/L = 0.12$ the root chord $c_1/L = 0.1$, the tip chord $c_2/L = 0.086$, and the leading and trailing edge swept angles are $\Lambda_1 = 70$ deg and $\Lambda_2 = 90$ deg, respectively. Thus, the wing aspect ratio is relatively small, $AR = 1.53$, and its wetted surface is $S_W/L^2 = 0.0191$, while the wetted surface of the bare hull is $S_B/L^2 = 0.1484$.

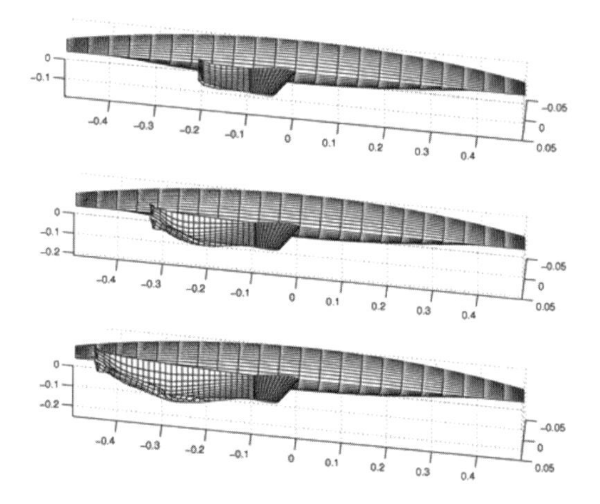

Figure 2. Spatial evolution of the vortex wake of the wing-keel below the parabolic Wigley hull, at a = 5 deg incidence angle.

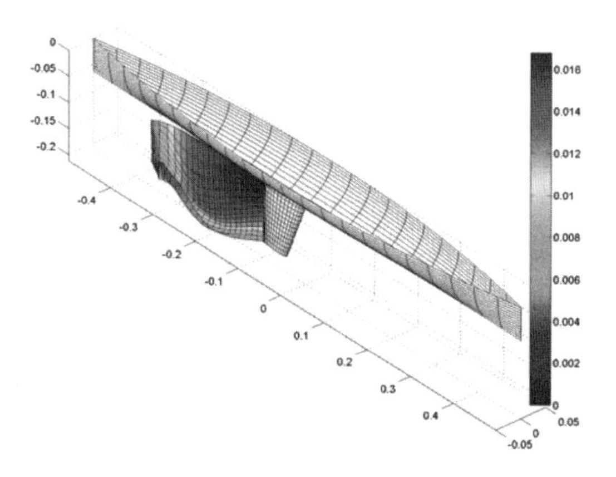

Figure 3. Development of trailing vortex sheet associated with trapezoidal wing below Wigley hull.

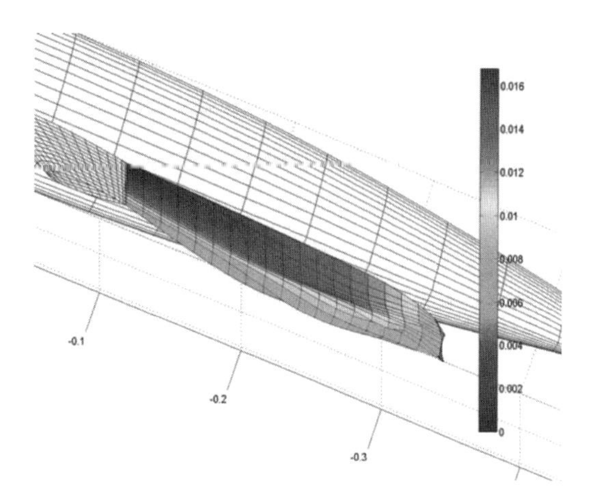

Figure 4. Same as in Fig. 3, zooming on the region of the tip of the wing-keel tip and illustrating the rollup of the vortex wake. Color is used to indicate the calculated vortex-ring intensity which equals to the local dipole strength distribution, and trailing vorticity is proportional to the surface gradient of the latter field (see Eq. 6).

In the results shown in the above figures, the geometrical configuration is steadily translated at a = 5 deg angle of attack (flow is incident from the left side of the hull), and speed corresponding to Froude number $F = U/\sqrt{gL} = 0.316$. We observe in Fig. 2 the development of the wing's wake, including the roll-up of tip vortices, as calculated by the present method, at three instances $\tau = tU/L = 0.09$, 0.18, 0.36, respectively. Moreover, in Figs. 3 and 4 the intensity of vortex rings (equivalent to the doublet strength) in the wake of the wing is plotted using color at the instance $\tau = tU/L = 0.18$. We recall here that the vorticity distribution is proportional to the surface gradient of the dipole strength and is maximized at the tip of the wing-keel, as expected. The slow variation of color in the middle part and near the root of the wing's wake is directly connected with the low intensity of vorticity in these regions.

4 NUMERICAL RESULTS AND DISCUSSION

At the next stage, the developed wake surface S_W of the wing-keel is used to fix the position of the trailing vortex sheet and numerically treat the flow interaction problem, including the free-surface efects, by means of the present steady-state technique, as described in the section 2. To illustrate the present approach, in the following we consider the previous Wigley parabolic hull, with and without wing keel. Results for the geometrical configuration including the keel and advancing at a = 5 deg angle of attack are presented in Figs. 5–10. Numerical results shown in these figures have been obtained by using the following discretization of the various boundary surfaces: (i) hull surface: 22 panels (in the longitudinal direction) by 24 (sectionwise), (ii) wing-keel: 13 panels (spanwise) by 32 (chordwise), (iii) trailing vortex surface: 13 panels spanwise by 30 (in the downstream direction) and (iv) the left and right parts of the free surface, extended in $-5 < x_1/L < 0.8$ and $-1.2 < x_2/L < 1.2$, using 56 panels (in the long direction) by 14 (in the tranverse direction) for each part. Thus, the total number of boundary elements are 2759. Ship velocities have been considered corresponding to Froude numbers $Fn = U/\sqrt{gL}$ in the interval $0.25 < Fn < 0.4$. Consequently, the wavelength ranges in $0.45 < \lambda/L < 1$ and the number of panels on S_F per wavelength is in the range 14–20, ensuring small errors due to dispersion and dissipation of the present discrete scheme (see, e.g., Sclavounos & Nakos 1988, Janson 1997). Further improvement can be achieved by horizontally shifting the collocation points on the free-surface and raising the corresponsing panels (see, e.g., Janson 1997).

The calculated surface velocities on the hull and keel and the distribution of the pressure coefficient on the wing-keel are presented in Figs. 5, 6 and 7, in the case of and $Fn = 0.316$ a = 5 deg angle of attack. We observe in these figures that the predicted velocity field is reasonable, and does not exhibit irregularities especially in the junction area between the hull and wing-keel. The pressure coefficient is defined as follows

$$C_p = \frac{p - p_\infty + \rho g x_3}{0.5\rho U_\infty^2} = 1 - \left(w/U_\infty\right)^2, \tag{18}$$

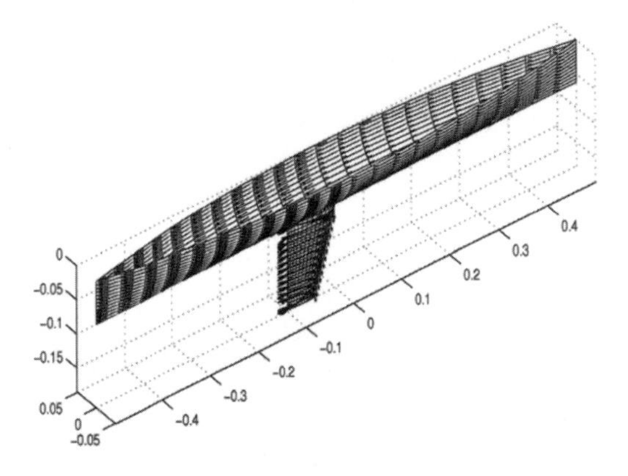

Figure 5. Calculated velocity field on the surface of Wigley hull with keel at $Fn = 0.316$ and a = 5 deg incidence angle.

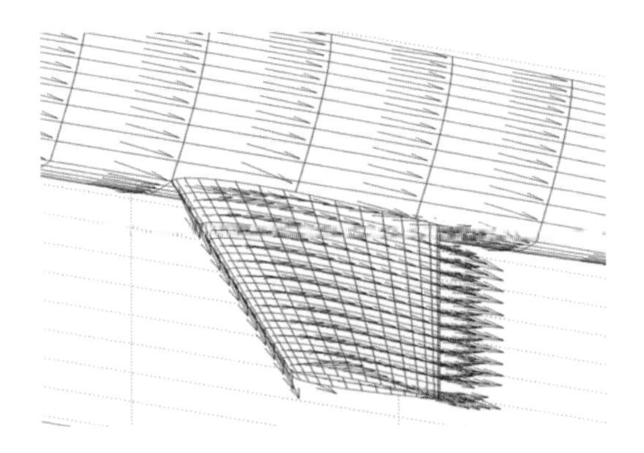

Figure 6. Same as in Fig. 5, viewed from the pressure side of the wing-keel. Froude no $Fn = 0.316$.

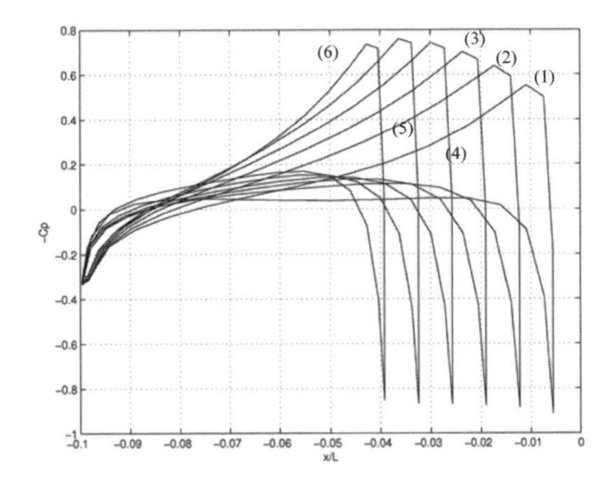

Figure 7. Chordwise pressure distributions at various sections of the wing-keel at Froude number $Fn = 0.316$, as calculated by the present method. (1) wing section at $x_3/L = -0.076$, (2) $x_3/L = -0.094$, (3) $x_3/L = -0.113$, (4) $x_3/L = -0.131$, (5) $x_3/L = -0.150$ (6) $x_3/L = -0.168$.

and chordwise plots are presented in Fig. 7, at various sections ranging from the root to the tip of the wing-keel. In the examined case, the lift and drag coefficients of the wing at a = 5 deg are obtained by pressure integration on its surface and the calculated values are: $C_L = L/\left(0.5\rho U_\infty^2 S_L\right) = 0.102$ and $C_D = D/\left(0.5\rho U_\infty^2 S_L\right) = 0.0015$, and are found to be practically constant for velocities in the interval $0.25 < Fn < 0.4$, being essentially dependent on the angle of attack. Moreover, for $Fn = 0.316$, the position of the pressure center is calculated at $x_3/L = -0.118$ (a distance 5.6% L below base line) and 2.84% aft the hull midsection. Colorplots of the pressure distributions on the hull surface for Froude number $Fn = 0.316$ and a = 5 deg are presented in Figs. 8 and 9, respectively, as obtained by the present method. We see in these figures that side forces are practically generated by the wing lift, while the pressure forces on the hull mainly

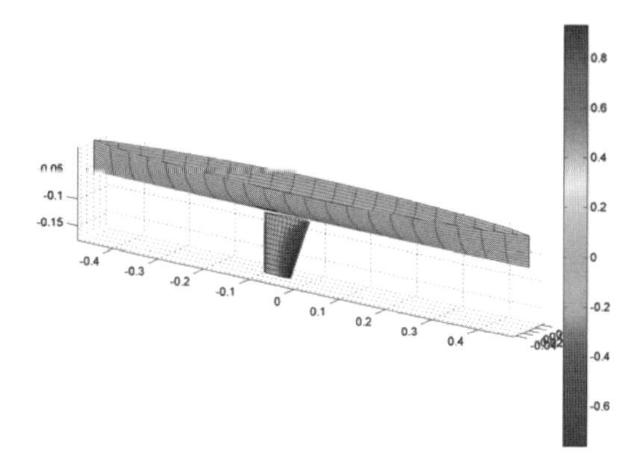

Figure 8. Calculated Cp distribution on the geo-metric configuration, for Froude number $Fn = 0.316$, advancing at a = 5 deg incidence angle (suction side).

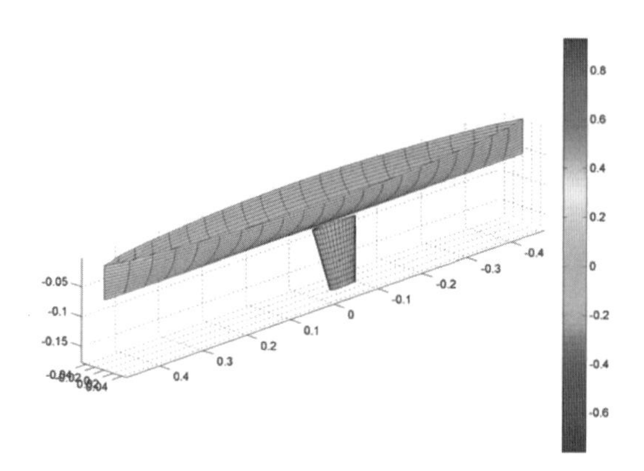

Figure 9. Calculated Cp distribution on the geometric configuration, for $Fn = 0.316$, advancing at a = 5 deg incidence angle (view from pressure side).

produce yaw moment. Finally, in the same case and conditions, the calculated wave pattern is shown in Fig. 10 and Fig. 11, as viewed form the right (suction) and left (pressure) sides, respectively. In these plots the free-surace elevation is two times exaggerated for clarity.

Finally, the wave resistance (R_W) at various speeds is obtained by pressure integration on the surface of the submerged body,

$$C_W = \frac{R_W}{0.5\rho U^2 S} = S^{-1} \int_{S_B \cup S_W} C_P n_x dS, \tag{19}$$

where $S = S_L + S_W$ is the total wetted surface (hull plus wing-keel) of the configuration. Results obtained by our method are plotted in Fig. 12, where, the blue line corresponds to the wave resistance of the bare hull, advancing at a = 0 deg, and compares well with the range of experimental data

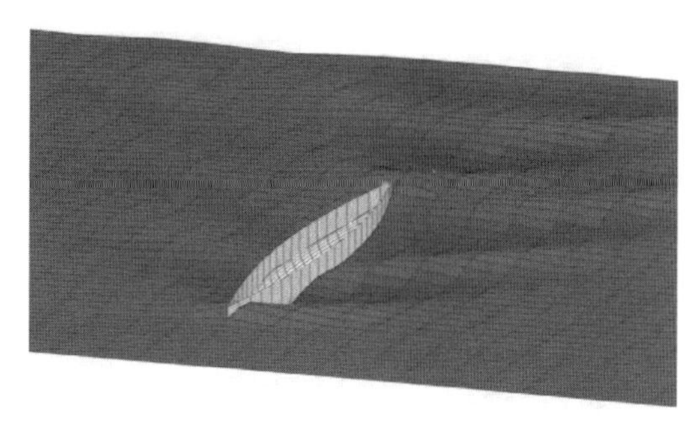

Figure 10. Calculated free-surface elevation for the hull & keel in the upward position steadily advancing at a = 5 deg incidence angle.

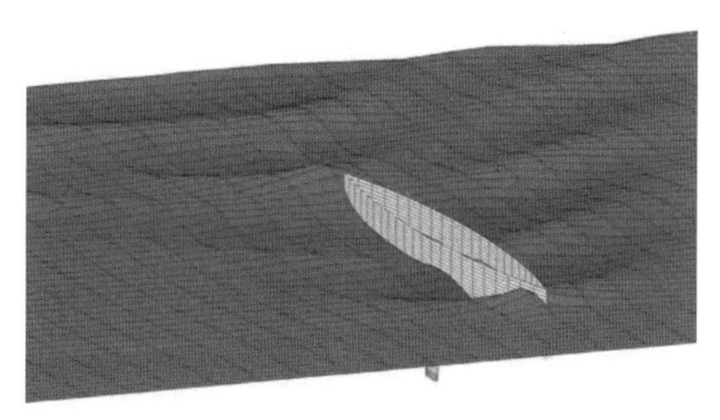

Figure 11. Same as in Fig. 10 viewed from the suction (right) side. The free surface elevation has been exaggerated for clarity.

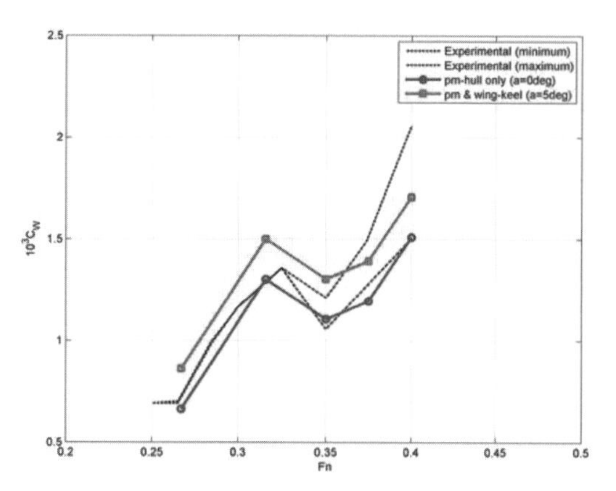

Figure 12. Wave resistance coefficient (10^3Cw) of Wigley parabolic hull vs. the Froude number.

for this hull and predictions by other boundary element methods shown by using dashed lines (see e.g., Bal 2008). The resistance curve shown by using red lines in Fig. 12 is for the same hull together with wing keel, advancing at a = 5 deg, and the notable increase is mainly due to the induced drag of the lifting body.

5 CONCLUSIONS

A low-order panel method based on surface singularity distributions is applied to the solution of the boundary integral equation associated with steadily translating surface piercing bodies or geometrical configurations, including both lifting and non-lifting parts. The applicability of the present method is illustrated in the case of Wigley hull equipped with a trapezoidal keel characterised by symmetrical sections. Future work is planned towards the detailed investigation of accuracy of the present method predictions in the case of realistic hulls, including flow asymmetry due to inclination, and its applications to hull-form optimization with respect to performance.

REFERENCES

Bal, S. 2008. Prediction of wave pattern and wave resistance of surface piercing bodies by a boundary element method. *International Journal for Numerical Methods in Fluids* 56(3), pp. 305–329.

Belibassakis, K.A., Politis, G.K. & Triantafyllou, M.S. 1997. Application of the Vortex Lattice Method to the propulsive performance of a pair of oscillating wing-tails. Proc. 8th Inter. *Conf. on Comput. Methods and Experim.l Measurements*, CMEM'97, Rhodes, Greece.

Bertram, V. 2000. *Practical Ship Hydrodynamics*. Butterworth-Heinemann, Oxford.

Carrica, P.M., Wilson, R.V. & Stern, F. 2007. An Unsteady Single-Phase Level Set Method for Viscous Free Surface Flows. *International Journal Numerical Methods Fluids*, 53, No. 2, 229–256.

Dawson, C.W. 1977. A practical computer method for solving ship-wave problems, Proc. 2nd Int. *Conf. on Numerical Ship Hydrodynamics*, Berkeley, USA.

Hess, J.L., 1972. Calculation of potential flow around arbitrary three dimensional lifting bodies. Report no. MDC J5679–01, McDonnell Douglas Corporation.

ITTC 2005. Report of the Resistance Committee, 24th Inter. Towing Tank Conf., Edinburgh, U.K.

ITTC 2008. Report of the Resistance Committee, 24th Inter. Towing Tank Conf. Fukuoka, Japan.

Janson, C.E. 1977. Potential flow panel methods for the calculation of free-surface flows with lift, Ph.D thesis, Chalmers Univ of Technology, Goeteborg 1997.

Katz, J. & Plotkin, 1991. *Low Speed Aerodynamics*. McGraw Hill, New York.

Larsson, L., Stern, F. & Bertram, V. 2003. Benchmarking of Computational Fluid Dynamics for Ship Flows: The Gothenburg 2000 Workshop, *Journal of Ship Research*, 47, No. 1, 63–81.

Nakos, D. & Sclavounos, P. 1990. On steady and unsteady ship wave patterns, *Journ. Fluid Mech*, 215, 263–288.

Paris, F. & Canas, J. 1997. *Boundary Element Methods*. Oxford University Press.

Politis, G. 2005. Simulation of unsteady motion of a propeller in a fluid including free-wake modelling. *Engineering Analysis Boundary Elements* 28, 633–653.

Politis, G. 2011. Application of a BEM time stepping algorithm in understanding complex unsteady propulsion hydrodynamic phenomena. *Ocean Engineering* 38, 699–711.

Raven, H.C. 1988. Variations on a theme by Dawson, Proc. 17th Symp Naval Hydrodynamics, Seoul, Korea.

Raven, H.C. 1996. A solution method for the nonlinear ship wave resistance problem. PhD-thesis, Delft University of Technology.

Sclavounos, P. & Nakos, D. 1988. Stability analysis of panel methods for free surface flows with forward speed. Proc. 17th *Symp. Naval Hydrodynamics*, The Hague, Netherlands.

Tzabiras, G. 2004. Resistance and self-propulsion calculations for a series 60, Cb060 hull at model and full scale. Ship Technology Research, 51, 21–34.

Sustainable Maritime Transportation and Exploitation of Sea Resources – Rizzuto & Guedes Soares (eds)
© 2012 Taylor & Francis Group, London, ISBN 978-0-415-62081-9

Numerical and experimental analysis of the flow field around a surface combatant ship

S. Zaghi & R. Broglia
CNR-INSEAN, The Italian Ship Model Basin, Rome, Italy

D. Guadalupi
ITALIAN NAVY, Ship Design Office, Rome, Italy

ABSTRACT: In this paper an analysis of the flow field around a surface combatant ship is carried out. The test case considered is the flow around a model of the US naval combatant DDG51; three different speeds of advance are analyzed, corresponding to $Fr = 0.28$, 0.35 and 0.41, with Reynolds numbers equal to $1.200\ 10^7$, $1.499\ 10^7$ and $1.756\ 10^7$ respectively, the scale of the model being $\Lambda = 28.83$. Numerical simulations have been carried out by means of the unsteady Reynolds-Averaged Navier-Stokes equations solver *χnavis*, developed at CNR-INSEAN. The computations are conducted on a sequence of grids in order to perform an accurate analysis of the grid convergence property of the solver. Validation of the results are provided by comparisons with experimental data.

1 INTRODUCTION

This paper describes the work that has been done in the framework of the WP5 of the Research Project EUROPA ERG1 RTP N° 110.067 "Development and Validation of Tools for the Prediction of Hydrodynamics Signatures", DALIDA, financially supported by the Italian and the French Navies, throughout the European Defence Agency. The main goal of this work package is to develop and evaluate numerical solutions in order to simulate wave field, wave breaking, risk of air capturing and convection, and wake around a ship. Several numerical tools among those at DGA Hydrodynamic (already, Bassin d'Essais des Carènes) and INSEAN-CNR disposal will be evaluated. In particular in the present paper the simulations that have been performed at CNR-INSEAN with the in-house unsteady Reynolds-Averaged Navier-Stokes Equations (RANSE) solver *χnavis* will be presented. The test case considered is the steady advancement of the DDG51 ship model; three different speeds will be investigated. The choice of this test is ascribed to two main reasons: first of all, the availability of a large experimental data base; second, the different regimes at which those tests were realized, ranging from low speed ($Fr = 0.28$) to high speed ($Fr = 0.41$) at which strong breaking wave phenomena have been observed.

The paper is organized as follows: in the next section, the mathematical and he numerical models with the proper boundary and initial conditions will be briefly recalled. Numerical results will follow; first a description of the numerical grid employed will be given, and then an analysis of the computed flow fields will be reported. Wave and velocity fields will be investigated, comparison with experimental data will be provided. The analysis of the resistance, with verification and validation of the results, will be presented in the last section.

2 MATHEMATICAL AND NUMERICAL MODELS

The governing equations for the unsteady motion of an incompressible viscous fluid can be written in integral form as:

$$\oint_{S(V)} \mathbf{U}\cdot\mathbf{n}\mathrm{d}S = 0$$
$$\frac{\partial}{\partial t}\int_V \mathbf{U}\mathrm{d}V + \oint_{S(V)} (F_C - F_D)\cdot\mathbf{n}\mathrm{d}S = 0 \qquad (1)$$

where V is a control volume, $S(V)$ its boundary, and $\mathbf{n}$ the outward unit normal. In the general formulation, the equations are written in an inertial frame of reference, in order to take into account the possibility of block motion to follow possible moving boundaries. The equations are made non-dimensional with a reference velocity U_∞ and a reference length L and the water density ρ.

In equation (1), F_c and F_d represent inviscid (advection and pressure) and diffusive fluxes, respectively:

$$F_C = p\mathbf{I} + (\mathbf{U} - \mathbf{V})\mathbf{U}$$
$$F_D = \left(\frac{1}{Re} + \nu_t\right)\left(\nabla\mathbf{U} + \nabla\mathbf{U}^{\mathrm{T}}\right) \tag{2}$$

In the previous equation, $p = P + z/Fr^2$ is the hydrodynamic pressure (i.e., the difference between the total P and the hydrostatic pressure $-z/Fr^2$, $Fr = U_\infty/(gL)^{1/2}$ being the Froude number and g the acceleration of gravity parallel to the vertical axis z, positive upward). $\mathbf{V}$ is the local velocity of the control volume boundary, $Re = U_\infty L/\nu$ the Reynolds number, ν the kinematic viscosity, and ν_t the non-dimensional turbulent viscosity; in the present work, the turbulent viscosity has been calculated by means of the Spalart-Allmaras (1994) one-equation model. In what follows, u_i is the i-th Cartesian component of the velocity vector (the Cartesian components of the velocity will be also denoted with u, v, and w).

The problem is closed by enforcing appropriate conditions at physical and computational boundaries. On solid walls, the relative velocity is set to zero (whereas no condition on the pressure is required); at the (fictitious) inflow boundary, velocity is set to the undisturbed flow value, and pressure is extrapolated from inside; on the contrary, pressure is set to zero at the outflow, whereas velocity is extrapolated from inner points.

At the free surface, whose location is one of the unknowns of the problem, the dynamic boundary condition requires continuity of stresses across the surface; if the presence of the air is neglected, the dynamic boundary condition reads:

$$p = \tau_{ij}n_i n_j + \frac{z}{Fn^2} + \frac{\kappa}{We^2}$$
$$\tau_{ij}n_i t_j^1 = 0$$
$$\tau_{ij}n_i t_j^2 = 0 \tag{3}$$

where τ_{ij} is the stress tensor, κ is the average curvature, $We = (\rho U_\infty^2 L/\sigma)^{1/2}$ is the Weber number (σ being the surface tension coefficient), whereas $\boldsymbol{n}$, $\boldsymbol{t}^1$ and $\boldsymbol{t}^2$ are the surface normal and two tangential unit vectors, respectively.

The actual position of the free surface $F(x, y, z, t) = 0$ is computed from the kinematic condition:

$$\frac{DF(x, y, z, t)}{Dt} = 0 \tag{4}$$

Initial conditions have to be specified for the velocity field and for the free surface configuration:

$$u_i(x, y, z, 0) = \overline{u}_i(x, y, z) \qquad i = 1, 2, 3$$
$$F(x, y, z, 0) = \overline{F}(x, y, z) \tag{5}$$

The numerical solution of the governing equations (1) is computed by means of a simulation code developed at CNR-INSEAN; the code yields the numerical solution of the Unsteady Reynolds averaged Navier Stokes equations with proper boundary and initial conditions. The algorithm is formulated as a finite volume scheme, with variable co-located at cell centres. Turbulent stresses are taken into account by the Boussinesq hypothesis, with several turbulence models (both algebraic and differential) implemented. Free surface effects are taken into account by a single phase level-set algorithm. Complex geometries and multiple bodies in relative motion are handled by a suitable dynamical overlapping grid approach. High performance computing is achieved by an efficient shared and distributed memory parallelization. For more details, the interested reader is referred to Di Mascio *et al.* (2001), Di Mascio *et al.* (2006), Di Mascio *et al.* (2007), Di Mascio *et al.* (2009) and Broglia *et al.* (2007) for details.

3 GEOMETRY AND COMPUTATIONAL PARAMETERS

The calculations were performed around the model (model scale $\Lambda = 24.824$) of the US naval combatant DDG51 (Figure 1) whose main dimensions are given in Table 1. This ship has been taken as benchmark for an international collaboration between three institutes (namely the David Taylor Model Basin, the Iowa Institute of Hydraulic Research and the Italian Ship Model Basin, CNR-INSEAN), and therefore, extensive experimental results are available (see Olivieri *et al.* (2001), Olivieri *et al.* (2004) and Stern *et al.* (2000)); moreover, in the last two workshops on CFD for ship hydrodynamics

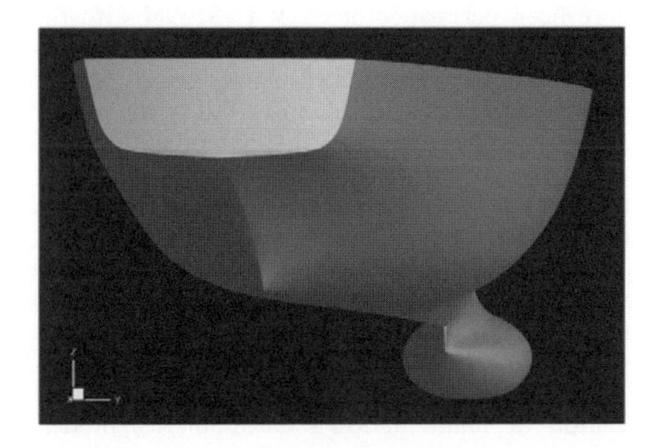

Figure 1. View of the DDG51 (bare hull).

Table 1. DDG51 main dimensions.

Dimension	Full scale	Model
$L_{pp}(m)$	142.000	5.719
λ	1.000	28.830
$T_m(m)$	6.150	0.248
$B(m)$	19.000	0.768
$\Delta(m^3)$	8424.400	0.554
C_B	0.507	0.507

Table 2. Computational parameters.

Speed (m/s)	Froude	Reynolds	Weber	Trim (deg)	Sinkage
2.097	0.28	1.200×10^7	586.56	0.108	1.18×10^{-3} L_{pp}
2.621	0.35	1.499×10^7	733.20	0.069	3.18×10^{-3} L_{pp}
3.071	0.41	1.756×10^7	858.89	−0.421	4.70×10^{-3} L_{pp}

held in Gothenburg in 2000 (Larsson *et al.* (2000)) and in Tokyo in 2005 (Hino (2005)), it has been used as representative of modern ship hull forms for a surface naval combatant. In the computations which follow, the model is considered in bare hull configuration.

The frame of reference which has been considered has the longitudinal axis aligned with the free stream velocity, positive backward; the z-axis is vertical, positive upward; y-axis completes a right hand system of reference. The origin is placed on the undisturbed free surface, amidships.

Simulations were carried out at three different speeds; the attitude of the ship is fixed at the dynamical position, with values of trim and sinkage taken from the experiments carried out at CNR-INSEAN (see Olivieri *et al.* (2001) and Olivieri *et al.* (2004)). Computational parameters ($Fr = U_\infty/(gL_{pp})^{1/2}$, $Re = U_\infty L_{pp}/\upsilon$, $We = \rho U_\infty^2 L_{pp}/\sigma$) are summarized in Table 2.

The trim is positive when the ship rotates the bow downward, the sinkage is positive if the LCG moves downward.

4 COMPUTATIONAL GRID

The volume grid around the hull has been built by means of a standard grid generator, namely ANSYS ICEM CFD. The computational grid exploits the Chimera-type topology capability (i.e., domains can overlap) available in the CNR-INSEAN RANS code. The grid topology is chosen in order to match the requirements of the numerical algorithms, i.e. grid clustering toward solid walls is determined by the estimated boundary layer thickness and refinements are considered

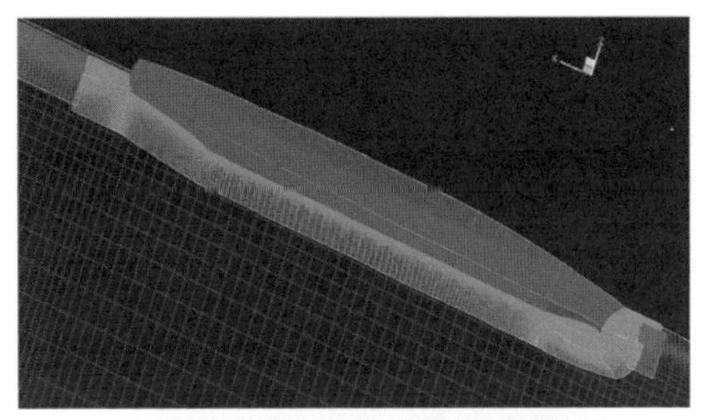

Figure 2. Overview of the overlapping grid.

around the free surface as well as in regions were strong gradient are expected.

The design of the grid focused on geometrical details of the hull, trying to keep as orthogonal as possible the cell faces in order to avoid degradation in the accuracy of the solution. The use of an overlapping grid approach easily allows the treatment of complex geometries, keeping the quality of the mesh, in terms of orthogonality, expansion ratio and grid refinement properties, satisfactory high.

A view of the computational mesh is given in Figure 2, where an overview of the surface mesh on the hull and on the plane of symmetry and a particular of the grid around the bulbous bow are shown. By the overlapping grid approach the time required for the generation of the mesh is highly reduced, even in the case of bodies with complex shapes. Moreover, since each region of the domain is discretized with a group of blocks, each independent of the other, any modification of the grid can be easily handled.

The entire computational mesh is made by a total of 26 blocks, the total number of volumes is around 6,000,000. In the following table the number of volumes used for the discretization of different regions are summarized; it has to be note that only one of the two sides has been discretized, the geometry being symmetrical about the vertical plane. Medium and coarse grids are obtained by removing every other point from the finer mesh.

5 RESULTS

In this section the results obtained for the three cases are presented; the wave field first and then the velocity field will be analyzed. Results will be compared with available experimental data. In the final section an analysis of the computed resistance coefficients will be shown; the verification of the results, as well as the assessment of the numerical uncertainty and the comparison with experimental values will be carried out.

5.1 *Wave fields*

An overview of the computed free surface at Froude number equal to 0.28 and 0.41 are presented in Figure 3; from these figures the capability of the single phase level set approach developed at CNR-INSEAN in dealing with free surface waves in non breaking and breaking conditions can be clearly inferred. This test case is very severe to check the capabilities of free surface simulation algorithms in dealing with complex three dimensional wave breaking phenomena, because of the formation of a water sheet on the hull at the bow, the formation of multiple jets, impingements and splash-ups at least at medium and high Froude numbers.

The wave patterns obtained for the three different Froude numbers are presented in Figure 4; results obtained on finest grid are compared with the measurements taken at INSEAN (Olivieri *et al.* (2001), Olivieri *et al.* (2004)). As expected, at $Fr = 0.41$ larger wave elevations can be observed, minimum and maximum wave heights being $h/Lpp = -0.015,\ 0.026$; for the lower Froude number, extreme values are almost halved: $h/Lpp = -0.007,\ 0.015$. At the medium Froude number the wave elevation range is $h/Lpp = -0.012,\ 0.021$. In these figures, the prediction of the shoulder wave, and the wave system at the dry transom, composed by both the transverse and the divergent front waves can be clearly observed. For the medium and the higher Froude number,

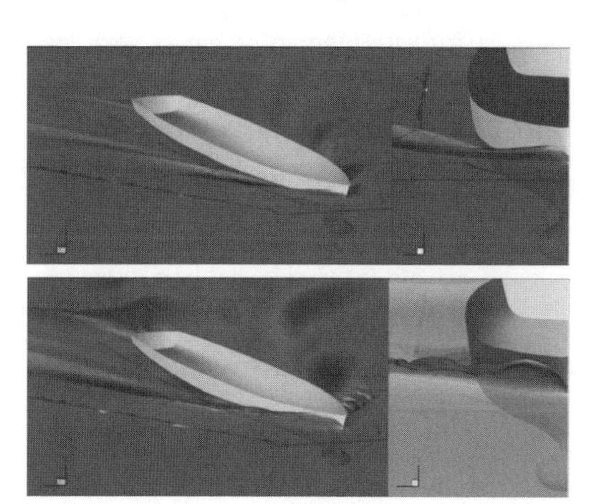

Figure 3. Overview of the wave field: $Fr = 0{,}28$ and $0{,}41$.

it is interesting to note the divergent wave system downstream the bow breaking wave, whose crests are almost aligned with the free stream, and the so-called rooster tail system at the transom stern. In those figures, it can be noticed that some interesting features of the breaking waves are well resolved by the numerical simulation (see the works by Dong *et al.* (1997) and Waniewski *et al.* (2002)): as the flow impinges on the bow, a strong vertical velocity component induces a motion which creates a liquid sheet along the wall; this rise up increases with the Froude number. At the medium and the higher speed the formation of a jet that impinges on the free surface follows. A second and even a third jet due to the splash-up and its plunging can be clearly observed in Figure 4, followed by the reconnection phase.

The qualitative agreement between numerical results and measurements are well satisfactory. In the experiments surface tension effects seem more pronounced; moreover, differently from

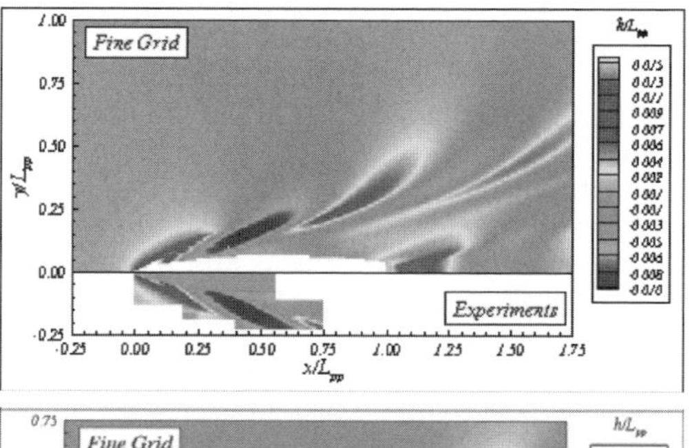

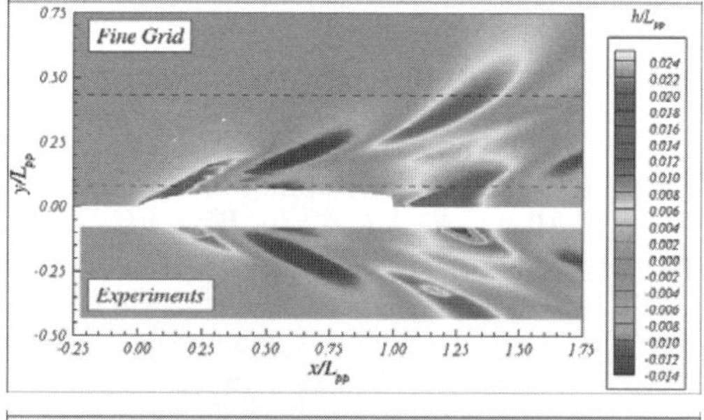

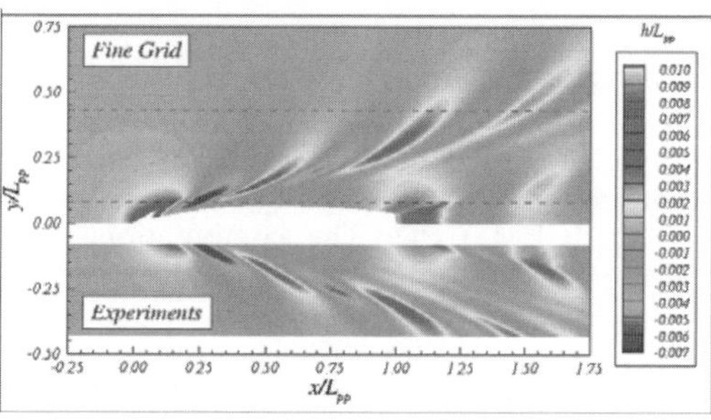

Figure 4. Numerical and experimental wave patterns: from top to bottom $Fr = 0.28$, 0.35 and $0{,}41$.

experimental observations, breaking waves at the transom stern did not appear in the simulation, because of lack in the grid resolution.

In Figure 5, the computed wave profiles are compared (only for the lower and the higher Froude numbers) with the experimental data. At the lower Froude number the agreement is satisfactory, in spite of the large deviation at the first crest; such disagreement has been already observed in all the calculations in Larsson *et al.* (2000), and it should be due to the way the experimental wave profile is taken. In fact, a photographic and digitizing system was used; therefore, if the maximum wave height is not attained on the hull surface, the result in the photo does not show the profile on the hull but the envelope of the maximum wave heights. This conjecture is supported by the DTMB measurements of the wave profile on a similar model, where a waterproof pencil is used and the first crest of the wave profile is somewhat lower (see Fig. 5 in Stern *et al.* (2000)). However, the agreement with experimental data for the crest at the bow is well within the experimental uncertainties (Stern *et al.* (2005)); moreover, for the shoulder and stern waves, the agreement is very satisfactory. For the higher speed, the agreement with the experimental data is very good along the whole hull surface, whereas for the medium Froude number experimental data are not available. For all the speeds a monotonic convergence is observable.

5.2 *Velocity fields*

An overview of the computed axial velocity fields is given in Figure 6 for the $Fr = 0.28$ and 0.41 test cases respectively; this figure gives a global view of the flow around the model and highlight interesting flow features. The axial velocity component for the speed corresponding to $Fr = 0.28$ is shown for cross sections around the fore perpendicular ($x/Lpp = 0.00$), close to the trailing edge of the bulbous bow ($x/Lpp = 0.10$) and in its wake ($x/Lpp = 0.20$ and $x/Lpp = 0.40$), around midships ($x/Lpp = 0.60$), in the region where the hull

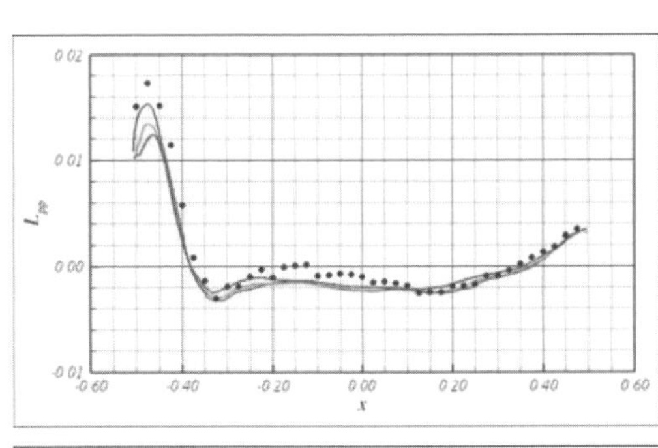

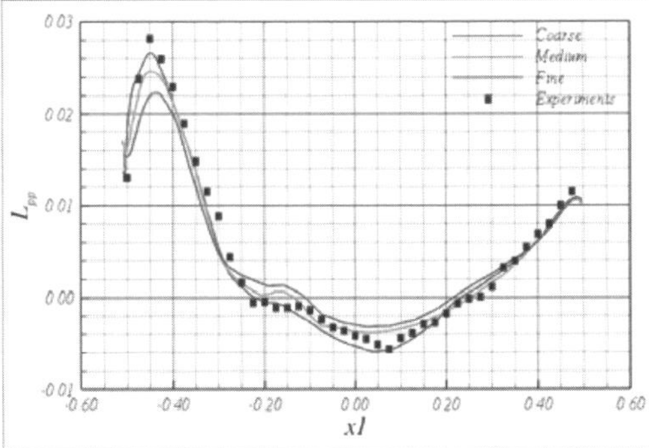

Figure 5. Numerical and experimental wave profile: top Fr = 0.28, bottom 0.41.

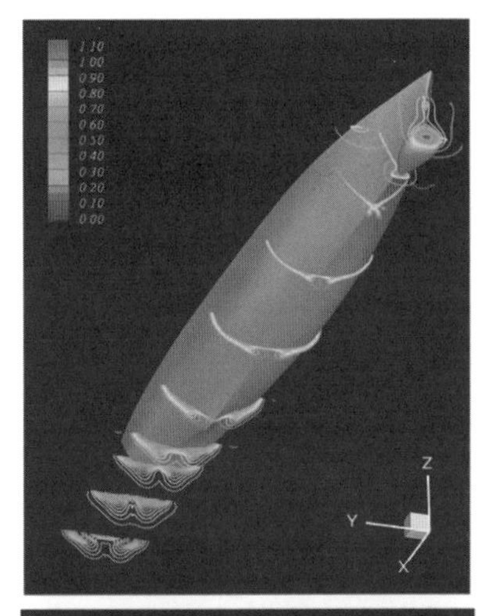

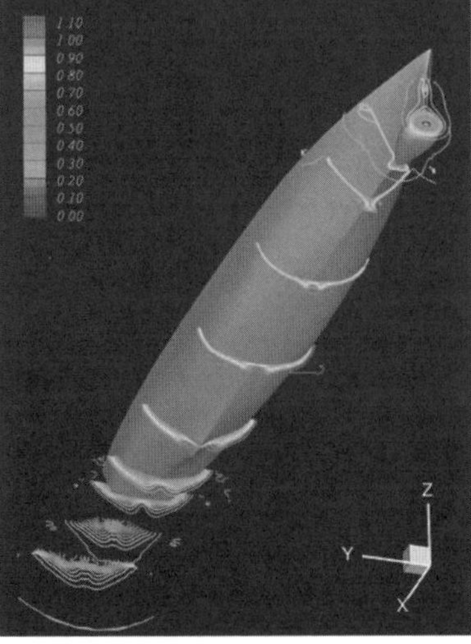

Figure 6. Overview of the axial velocity contour: top $Fr = 0.28$, bottom 0.41.

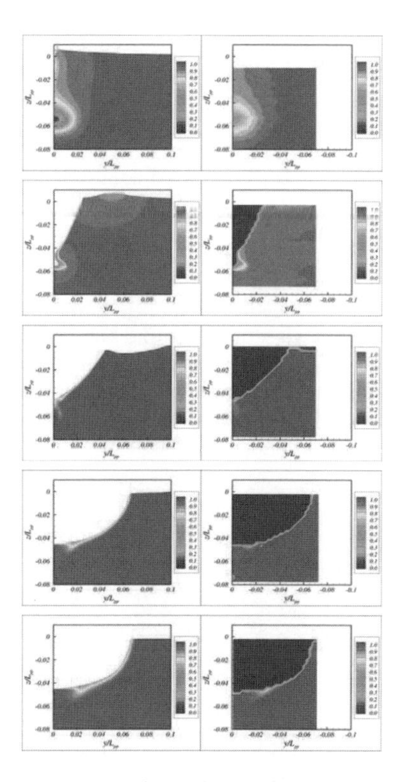

Figure 7. Cross sections of the axial velocity contour: $Fr = 0.28$. From top to bottom $x/L_{pp} = 0.00$, 0.10, 0.20, 0.40, 0.60.

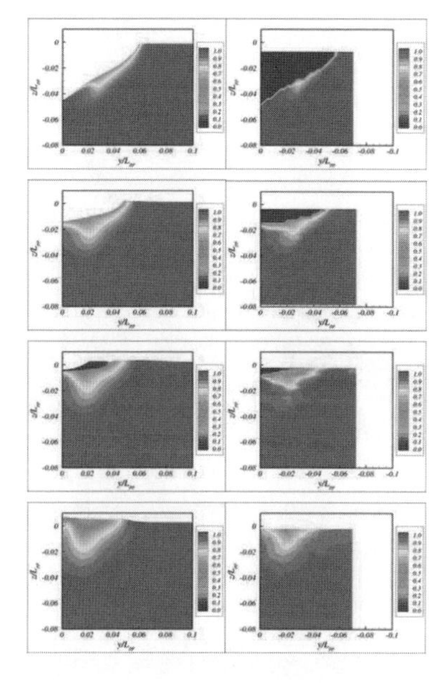

Figure 8. Cross sections of the axial velocity contour: $Fr = 0.28$. From top to bottom $x/L_{pp} = 0.80$, 0.9346, 1.00, 1,10.

surface narrows ($x/Lpp = 0.80$), at the propeller plane ($x/Lpp = 0.9643$) and in the wake of the hull ($x/Lpp = 1.00$, $x/Lpp = 1.10$) in Figure 7 and Figure 8. In these figures the comparison with the measurements taken at CNR-INSEAN (Olivieri *et al.* (2001)) is also presented.

In the cross section at the fore perpendicular, a stagnation point in correspondence of the leading edge of the bulb is evident; the peculiar shape of the bulb, due to the sonar dome, determines the generation of a pair of streamwise vortices, which are shed one from the upper region of the bulb and one from the edge of the bulb itself. These two vortices merge at the trailing edge of the bulb, forming a pair (port and a starboard side) of contra rotating bow bilge vortices.

The presence of the bow bilge vortex at port side is clearly revealed by the iso-lines of the axial velocity in Figure 7, related to the section located $x = 0.20Lpp$ downstream the fore perpendicular. The position and the strength of the bow bilge vortex clearly depend on the Froude number. At sections $x/Lpp = 0.40$, $x/Lpp = 0.60$ and $x/Lpp = 0.80$ (around amidships, and in the stern region), the bow bilge vortex is still present. By the action of the bow bilge vortex, high momentum fluid is convected toward the hull at the center plane, whereas low moment fluids is convected from the boundary layer region toward the far field on the side of the surface hull; consequentially a significant decrease of the boundary layer along the keel and a considerable growth of the boundary layer along the side wall are well highlighted in the figures. At section $x/Lpp = 0.6$, a second vortex is observed, having the same rotation of the first one. This second vortex, known as the stern bilge vortex, is attributable to the convergence of the limiting streamlines in the after body flow, due to the adverse pressure gradient (Tanaka (1988)).

The growth of the boundary layer thickness along the ship hull is well captured by the numerical simulations. The wake ($x/Lpp = 0.9346$) is characterized by the presence of a very thick boundary layer and it is dominated (as the whole flow in the stern region) by the vertical motion due to the hull shape. The main characteristics of the wake are well represented by the numerical results up to the last cross section presented ($x/Lpp = 1.10$); the dimension and the shape of the wake clearly depend on the speed of advance, with a strong interaction with the wave system at the transom of the ship.

The comparison between the computed and the measured velocity at different cross planes for the lowest Froude number case, clearly highlights the reliability of the numerical simulations, the agreement being well satisfactory.

Table 3. Verification analysis: Total resistance coefficient.

Fr	$10^3(C_T)_3$	$10^3(C_T)_2$	$10^3(C_T)_1$	$10^3(C_T)_{RE}$	σ	U_{SN}
0.28	5,056	4,602	4,525	4,499	2,56	0,57%
0.35	5,729	5,077	4,976	4,942	2.69	0,68%
0.41	7,147	6,882	6,816	6,794	2.00	0,32%

Table 4. Comparison with experimental data.

Fr	$10^3(C_T)_{RE}$	D	E%
0.28	$4{,}499 \times 10^{-3}$	$4{,}291 \times 10^{-3}$	4,84%
0.35	$4{,}942 \times 10^{-3}$	$4{,}906 \times 10^{-3}$	0,74%
0.41	$6{,}794 \times 10^{-3}$	$6{,}765 \times 10^{-3}$	0,44%

5.3 *Resistance*

In this section an analysis of the total resistance coefficient computed at the three speeds is reported; both verification (the evaluation of the order of convergence and the assessment of numerical uncertainty) and validation (the comparison with the experimental data) analysis of the results are carried out following the classical approach by Roache (1997). The analysis of the grid convergence for the total resistance coefficient (defined as $C_T = X/(0.5\rho\, U^2 S_0)$, where X is the longitudinal force, S_0 the wetted surface in ballast condition, ρ the density of the water and U the speed of advance) is reported in Table 3.

In the previous table values computed on the coarse, medium and fine grid, denoted with the subscript 3, 2 and 1 respectively, are reported; the extrapolated value is computed following the classical Richardson's extrapolation procedure, i.e., considering that the medium and the fine computation are in the asymptotic range and therefore the theoretical order of convergence (2 in this case) has been reached. The extrapolated value is defined as:

$$(C_T)_{RE} = (C_T)_1 + \frac{1}{3}\left((C_T)_1 - (C_T)_2\right) \tag{6}$$

The actual order of convergence is computed as:

$$\sigma = \frac{\ln\left(\dfrac{(C_T)_2 - (C_T)_3}{(C_T)_1 - (C_T)_2}\right)}{\ln(2)} \tag{7}$$

The grid uncertainty is computed as $U_G = ((C_T)_1 - (C_T)_2)/3$ and is expressed as percentage of the extrapolated value; the iterative convergence uncertainty being negligible, grid uncertainty is the only contribution to the numerical uncertainty.

At all the Froude number, the estimated order of accuracy is close to the theoretical value of two; moreover, for the numerical uncertainty is very small. This indicates that the grid converged results are provided. The comparison with available experimental data (Oliveiri *et al.* (2001)) is reported in Table 4.

The agreement well satisfactory, the error (in percentage of the experimental data) being less than *1%* at medium and higher Froude number, and around *5%* at the lower speed.

The higher value of the error at the lower Froude number seems indicate a lack in the resolution of the free surface in the longitudinal direction.

6 CONCLUSIONS

The simulations of the flow around the US naval combatant DDG51 model have been carried out by means of the numerical solutions of the RANS equations. Three different speed of advance have been analyzed; plunging wave breaking phenomena of the bow wave have been observed at the medium ($Fr = 0.35$) and the highest ($Fr = 0.41$) Froude numbers. Spilling breaker develops at the lower Froude ($Fr = 0.28$). The Navier Sotkes solver *χnavis* developed at INSEAN has proved to provide accurate results in both breaking and non breaking conditions; by comparison with experimental data, a well satisfactory agreement for both the wave and the velocity fields has been observed. The wave profiles on the surface hull is well reproduced at both low and high Froude number, i.e., regardless the presence of a breaking wave.

The numerical uncertainty for the total resistance coefficient is rather small (less than 1%); the measured order of convergence is close to the theoretical value of two for the whole Froude number range investigated. Comparison with available experimental data has been also performed, the agreement is well satisfactory; the error, with respect to the measured values, is less than 5% at the lower Froude number, and less than 1% for the medium and the higher speed.

ACKNOWLEDGEMENTS

This work has been done in the framework of Research Project EUROPA ERG1 RTP

N°110.067 *"Development and Validation of Tools for the Prediction of Hydrodynamics Signatures"*, DALIDA, financially supported by the Italian and the French Navies, throughout the European Defence Agency, for which activities include experimental tests in both calm water and in waves, as well as complementary numerical simulations.

Numerical computations presented here have been performed on the parallel machines of CASPUR Supercomputing Center (Rome); their support is gratefully acknowledged.

REFERENCES

Broglia, R., Di Mascio, A. & Amati, G. (2007). "A Parallel Unsteady RANS Code for the Numerical Simulations of Free Surface Flows", 2nd International Conference on Marine Research and Transportation, Ischia (NA), Italy, June.

Di Mascio, A., Broglia, R. & Favini, B. (2001). A Second Order Godunov-Type Scheme for Naval Hydrodynamics, Kluwer Academic/Plenum Publishers, pp. 253–261.

Di Mascio, A., Broglia, R. & Muscari, R. (2007). "On the application of the one-phase level set method to naval hydrodynamics flows", Computers and Fluids 36, 868–886.

Di Mascio, A., Broglia, R. & Muscari, R. (2009). "Prediction of Hydrodynamic Coefficients of Ship Hulls by High-Order Godunov-Type Methods", J. Marine Sci. Tech., Vol. 14, N. 1, pp. 19–29.

Di Mascio, A., Muscari, R. & Broglia, R. (2006). "An Overlapping Grids Approach for Moving Bodies Problems", In Proc. of 16th Int. Offshore and Polar Engineering Conference, San Francisco, California (USA).

Dong, R.R., Katz, J. & Huang, T.Y.T. (1997). On the Structure of Bow Waves on a Ship Model, J. Fluid Mech. 347, 77–115.

Hino, T. (2005). Proc. of TOKYO 2005 Workshop, in: A Workshop on CFD in Ship Hydrodynamics, Tokyo, Japan.

Larsson, L., Bertram, V. & Stern, F. (2000). Proc. of GOTHENBURG 2000, in: A Workshop on CFD in Ship Hydrodynamics, Gothenburg, Sweden.

Olivieri, A., Pistani, F., Avanzini, G., Stern, F. & Penna, R. (2001). Tank Experiments of Resistance, Sinkage and Trim, Boundary Layer, Wake and Free Surface Flow Around a Naval Combatant INSEAN 2340 Model, IIHR Report, No. 421.

Olivieri, A., Pistani, F., Wilson, R., Benedetti, L., La Gala, F., Campana, E.F. & Stern, F. (2004). Froude Number and Scale Effects and Froude Number 0.35 Wave Elevations and Mean Velocity Measurements for Bow and Shoulder Wave Breaking of Surface Combatant DTMB5415, IIHR Report, No. 441.

Roache, J. (1997). Quantification of uncertainty in computational fluid dynamics, Ann. Rev. Fluid Mech., vol. 60, pp. 29–123.

Spalart, P.R. & Allmaras, S.R. (1994). A One-Equation Turbulence Model for Aerodynamic Flows, La Recherche Aerospatiale 1, 5–21.

Stern, F., Longo, J., Penna, R., Olivieri, A., Ratcliffe, T. & Coleman, H. (2000). International Collaboration on Benchmark CFD Validation Data for Surface Combatant DTMB Model 5415, in: Proc. of 23th Symposium on Naval Hydrodynamics, Val de Reuil, France.

Stern, F., Olivieri, A., Shao, J., Longo, J. & Ratcliffe, T. (2005). Statistical approach for estimating intervals of certification or biases of facilities or measurement systems including uncertainties, J Fluid Eng., vol. 127(3), pp. 604–610.

Tanaka, I. (1988). "Three-dimensional ship boundary layer and wake", Advances in Applied Mechanics, vol. 26, 1988.

Waniewski, T.A., Brennen, C.E. & Raichlen, F. (2002). Bow Wave Dynamics, J. Ship Research 46 (1), 1–15.

Sustainable Maritime Transportation and Exploitation of Sea Resources – Rizzuto & Guedes Soares (eds)
© 2012 Taylor & Francis Group, London, ISBN 978-0-415-62081-9

Investigation of 2D nonlinear free surface flows by SPH method

M. Ozbulut & O. Goren
Istanbul Technical University, Maslak Istanbul, Turkey

ABSTRACT: Smoothed Particle Hydrodynamics (SPH) arises as a useful tool in analyzing violent fluid flow problems with a free surface. The present study treats 2D nonlinear fluid motions with a free surface by means of SPH method. As a Lagrangian based method, SPH is employed first for the solution of the 2-D dam break problem, which can be used as a benchmark study, using Euler's equation and continuity equation as the governing differential equations. While the usual mathematical treatments of SPH are adopted, the assumptions or empirical schemes and parameters used by previous researchers are revisited to see the effect of these parameters—such as ε in "XSPH" variant velocity—on the resultant fluid flow. It is observed that such parameters, which find widespread acceptance, have crucial effects on the evolution of the flow. 2D sloshing problem of a swaying partially filled rectangular tank is subsequently investigated. It is understood from the numerical investigations that it is possible to obtain acceptable, valid results by tuning the basic parameters (such as ε in "XSPH") without further developing new techniques such as density re-initialization and particle refinement to improve flow characteristics.

1 INTRODUCTION

The complex nature of highly nonlinear free surface flows is a challenging problem for people working on hydrodynamics. Eulerian methods, especially in violent flows like dam break, sloshing or overturning and breaking ship bow waves have limited capability. The present work studies a Lagrangian based, meshless, particle method called "Smoothed Particle Hydrodynamics" (SPH) to handle 2D violent fluid flows with a free surface. SPH is still a developing numerical method that represents the continuum domain with a set of particles which bears all the fluid properties like density, pressure and velocity. It was first developed simultaneously by Lucy (1977) and Monaghan & Gingold (1977) for solving three dimensional astrophysics problems.

As the method is based on defining the continuum of the fluid domain by particles, it has a great compatibility with the nonlinear fluid flow problems. After the pioneering works of Lucy (1977) and Monaghan & Gingold (1977), the method has been applied to various type of problems like shock (Monaghan, 1983), under-water explosion (Swegle & Attaway, 1995) and impact simulation (Liu, 2003). The method started to be employed in hydrodynamic problems by the beginning of 1990's. The first free surface simulation was made by Monaghan (1994) and then it has been applied to several fluid flow problems like waves on beaches, sloshing tank and bow waves produced by certain ship hulls (Monaghan, 2005).

The present study focuses on the usual numerical treatments of SPH method which receive a general acceptance among SPH researchers. For example, parameter ε in the correction term for the rate of change of particle position is suggested as 0.5 in most cases (Monaghan, 1994 and Pakozdi, 2008). The present study systematically investigates the effect of such parameters. To show the effect of such parameters in the correction algorithms, two particular nonlinear free surface flow cases are considered, namely dam break (Liu & Liu, 2003 and Marrone et al., 2011) problem which is one of the popular benchmark studies in SPH and sloshing problem (Pakozdi, 2008 and Iglesias, 2004) of a rectangular tank partially filled with water. Pakozdi's (2008) results are used for comparison. Although the present work does not propose any new correction algorithms, like density re-initialization and/or modification of kernel functions, but discloses the sensitivity of the results to some parameters and points out that it is possible to obtain acceptable results without the use of novel numerical refinements such as—locally implemented—particle refinement procedure.

2 GOVERNING EQUATIONS AND SPH APPROXIMATION

2.1 *Governing equations*

During the solution process, continuity equation and Euler's equation are used, respectively, as in the following:

$$\frac{d\rho}{dt} = -\rho \nabla \cdot \boldsymbol{u} \tag{1}$$

$$\frac{d\boldsymbol{u}}{dt} = -\frac{1}{\rho}\nabla p + \boldsymbol{g} \tag{2}$$

where, ρ is the density. In 2D case, $\mathbf{u} = (u_x, u_y)$ is the velocity, p is the pressure of the fluid and $\mathbf{g}$ is the gravitational acceleration.

2.2 *SPH approximation*

SPH is basically an interpolation method which can let any function to be expressed by the value of a group of disordered particles (or points) and weighted by a kernel function at these particles (Haiuha & Ha, 2010). According to the SPH approximation, the kernel function $W(x-x',h)$ is an approximate substitute for the Dirac Delta function where *"h"* is the smoothing length and so we get the general representation of the function $A(x)$ as:

$$\langle A(x) \rangle = \int_{\Omega} A(x')W(x-x',h)dx' \tag{3}$$

In the SPH terminology the bracket symbols "<>" shows the SPH approximation. As a kernel function, the 5th order *"Quintic"* kernel is used in this study:

$$W(R,h) = \alpha_d \begin{cases} (3-R)^5 - 6(2-R)^5 + 15(1-R)^5, & 0 \le R < 1 \\ (3-R)^5 - 6(2-R)^5, & 1 \le R < 2 \\ (3-R)^5, & 2 \le R < 3 \\ 0, & R \ge 3 \end{cases} \tag{4}$$

where, $R = |x - x'|/h$ and α_d depends on the dimension of the problem, such that 120/h, $7/(478\Pi h^2)$ and $3/(359\Pi h^3)$ for one, two and three dimensions, respectively, (Liu, 2010). Because of the relatively higher numerical stability as mentioned in the literature (Yildiz et al., 2008 and Bian et al., 2010), the quintic kernel function is chosen in the present study.

Determining the smoothing length parameter "*h*" in the kernel function is a critical issue because it effects the influence area of the smoothing function and decides the number of particles which can interact with each other. It can be said that it is the equivalent of the width of a grid-cell in finite difference method (Schlatter, 1997).

There are two general approaches to determine the pressure gradient which appears in the RHS of Equation (2). The first one uses an equation of state expression including the speed of sound. It can be said that this approach is a result of the first application of SPH method to the compressible fluid flow problems. In order to solve incompressible fluid flows, this approach is used by the inspiration of the idea that all fluids can weakly be compressed theoretically. This approach is generally called *"Weakly Compressible Smoothed Particle Method"* in the SPH literature (Cummins & Rudman, 1999). The second approach for determining the pressure gradient is the *"Fully Incompressible Method"*. The main reason for searching a fully incompressible solution is that working with sonic time scales in weakly compressible approach requires an extremely small time step in order to satisfy the Courant condition (Ellero et al., 2007, Yildiz et al., 2008). Additionally, weakly compressible methods have problems in high Reynolds number flows and sound wave reflections near boundaries (Cummins & Rudman, 1999).

Weakly compressible smoothed particle method approach is employed in this study. In this approach (Monaghan, 1994), the governing equations are approximated as in the following:

$$\frac{Du_i}{Dt} = -\sum_{j=1}^{N} m_j \left(\frac{p_i}{\rho_i^2} + \frac{p_j}{\rho_j^2} + \Pi_{ij} \right) \nabla_i W_{ij} \tag{5}$$

here, m_j is the jth particle mass, $W_{ij} = W(\boldsymbol{x}_{ij}, h)$ is the kernel where $\boldsymbol{x}_{ij} = \boldsymbol{x}_i - \boldsymbol{x}_j$. Π_{ij} is the artificial viscosity term which is explained below.

In Equation (5), the pressure and density are defined as follows:

$$\langle \frac{D\rho}{Dt} \rangle_i = \rho_i \sum_{j=1}^{N} \frac{m_j}{\rho_j} (u_i - u_j) \nabla_i \cdot W_{ij} \tag{6}$$

$$p = \frac{\rho_0 c_0^2}{\gamma} \left[\left(\frac{\rho}{\rho_0} \right)^{\gamma} - 1 \right] + p_0 \tag{7}$$

where, c_0 is the reference sound speed, p_0 is the atmospheric pressure and ρ_0 is the reference density which is equal to 1000 [kg/m^3] for fresh water. The value of reference sound speed is obtained by the help of Mach number (M). It is required that $M \sim 0.1$ to keep density fluctuations around %1 (Monaghan, 1999).

Monaghan (1999) defined the artificial viscosity term in the following form:

$$\Pi_{ij} = \begin{cases} -\alpha \mu_{ij} \dfrac{c_i + c_j}{\rho_i + \rho_j}, & u_{ij} \cdot x_{ij} < 0 \\ 0, & u_{ij} \cdot x_{ij} \ge 0 \end{cases} \tag{8}$$

$$\mu_{ij} = h \frac{(\boldsymbol{u}_i - \boldsymbol{u}_j)(x_i - x_j)}{\| x_i - x_j \|^2 + \beta h^2} \tag{9}$$

where, β is taken as 0.01. As is known, the purpose of adding artificial viscosity into the pressure addition term is to increase the stability of the numerical code. The artificial viscosity parameter, α, may be obtained by:

$$\alpha = \frac{8v}{hc_0} \tag{10}$$

where ν is the artificial kinematic viscosity (Monaghan & Kos, 1999). The calculation of local speed of sound is done according to:

$$c(\boldsymbol{x}) = c_0 \left(\frac{\rho(\boldsymbol{x})}{\rho_0} \right)^{\frac{\gamma-1}{2}} \tag{11}$$

where, γ is the specific heat-ratio of water and is equal to 7.

The rate of change of particle position is calculated by:

$$\frac{d\boldsymbol{x}_i}{dt} = \boldsymbol{u}_i \tag{12}$$

Monaghan (1994) adds a correction term $\Delta \mathbf{u}_i$ to the right hand side of Equation (12):

$$\Delta u_i = \epsilon \sum_{j=1}^{N} \frac{m_j (\boldsymbol{u_i} - \boldsymbol{u_i})}{\frac{\rho_i + \rho_j}{2}} W_{ij} \tag{13}$$

The reason for adding this correction term is to keep the particles more ordered and to prevent penetration of one particle by another in high speed flows. This corrected velocity is called *"XSPH"* variant velocity. The typical value for ε in the literature is 0.5 (Monaghan, 1994 and Monaghan & Kos, 1999). However, the present work systematically tried smaller values which will be expressed in the numerical investigation section.

In order to help the reader, the artificial viscosity parameters for both cases considered in the present study are given in Table 1.

2.3 *Boundary conditions*

For the interaction force between fluid particles and solid boundary particles, the most common expression which is given by Monaghan (1994) is used:

$$f(r) = \begin{cases} D \frac{r}{r^2} \left[\left(\frac{r_0}{r} \right)^{p1} - \left(\frac{r_0}{r} \right)^{p2} \right], & r > r_0 \\ 0, & r \leq r_0 \end{cases} \tag{14}$$

The boundary force (14) exerted by solid boundary particles to the fluid particles near them, pushes the fluid particles and not allowed them to penetrate into the boundaries of the tank. In the above equation, *"r"* represents the distance between boundary particle and the fluid particle, and *"r_0"* represents the cut-off distance. Cut-off distance is important for deciding the influence area of the boundary particles and generally taken as the initial particle spacing (Liu, 2003). Parameter *"D"* is in the same order with the square of maximum velocity and it depends on the physical arrangement of the system (Monaghan, 1994). This parameter is taken as 2.25 in the solution process of this study. Finally, the parameters *"p_1"* and *"p_2"* are generally taken as 12 and 6, respectively, (Schlatter, 1997).

The solid boundary particles are settled as three parallel rows instead of single line in the present work. In addition, particles in the middle row of these three rows are shifted by a half particle spacing as shown in Figure 1. The aim of this configuration is to get a balanced and more uniform force which is exerted from boundary particles to fluid particles. Our experience with channel flow shows that the use of multi-line particle collocations for modelling solid boundaries, gives more reasonable results as compared to that of a single line modelling. Repulsive force approach is adopted here in modelling solid boundary following the recommendations in Monaghan (1994).

For the free surface boundary condition; the pressure of the particles close to free surface is

Table 1. Artificial viscosity parameters.

Test case	c_0	α	β
Dam-break	15	0.0075	0.01
Sway sloshing	15	0.0075	0.01

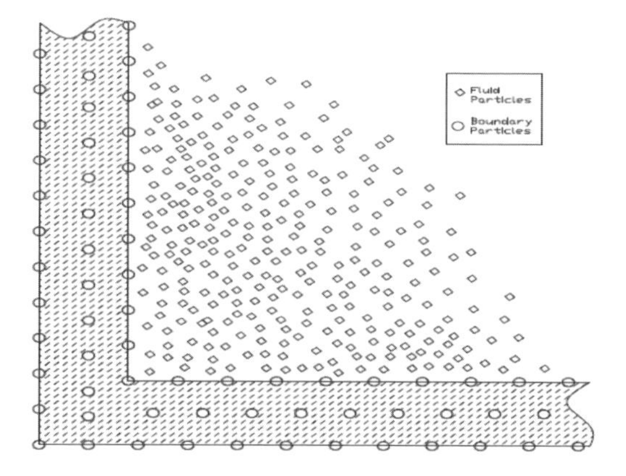

Figure 1. Distribution of boundary particles.

set to the atmospheric pressure, p_0, by taking the densities of these particles as reference density. The decision whether a particle could be regarded as a free surface particle depends on the number of neighbouring particles of that particle. During the simulations of the present work, the average amount of neighbouring particles is around 40–45. We considered a particle as a free surface particle if the amount of neighbouring particles of that particle is less than 35.

2.4 *Time stepping*

Time-stepping is carried out by leapfrog time integration scheme where the particle velocities and positions offset by a half time step while integrating the equation of motion. For simplicity, time step is taken as constant during the integration. The value of time step is determined by the Courant-Friedrichs-Lewy (CFL) condition where the recommended time step is; $\Delta t \leq C_{CFL} h_{ij,\min} / (c_i + v_{\max})$ (Rodriguez, 2005). Here; $h_{ij} = 0.5(h_i + h_j)$ and $h_{ij,\min}$ is the minimum smoothing length among all i–j pairs which is a constant in this work. C_{CFL} number should be; $0 \leq C_{CFL} \leq 1$. The present work used 0.4 and 0.5 for dam-break and sway sloshing problems, respectively. Finally, $v_{\max}$ is the expected maximum velocity in the problem. It is obvious that as the value of speed of sound decreases, time step value decreases, too. The speed of sound is taken as 15 for both dam-break and sway sloshing problems in the numerical investigation part.

3 NUMERICAL INVESTIGATION

3.1 *Dam-break problem*

Dam-break problem is a very famous benchmark case for SPH researchers and a good rehearsal for sway sloshing motion of a partially filled rectangular tank. In order to check the code, we compare our free surface profiles at the same instant with those of Pakozdi (2008). The geometry of the problem is given in Figure 2.

The initial particle spacing is taken as 0.01[m], the total number of particles in the problem domain is N = 8968. The smoothing length parameter is 0.0133 and artificial viscosity parameter, α, is equal to 0.0075. The particles are in rest and have hydrostatic pressure initially. The time step value for dam break problem is 0.00032 [s]. The comparison of free surface profiles predicted by Pakozdi and our simulation is given in Figure 3.

The result of the present study plotted in Figure 3 is obtained by a systematic search on the parameter ε in equation (13) and on the kinematic viscosity. Typical value of 0.5 proposed for **ε** does

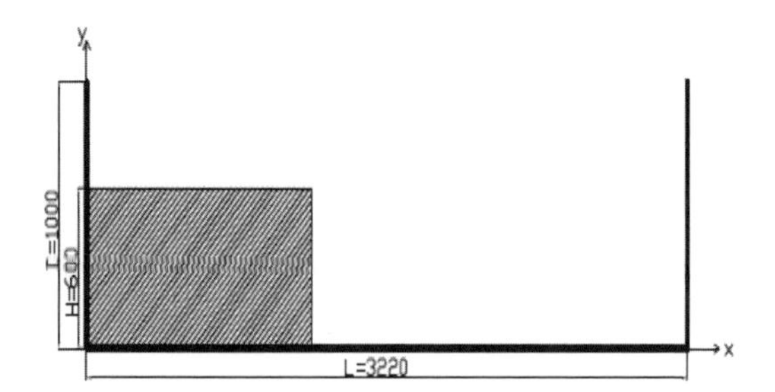

Figure 2. Geometry of the dam-break problem.

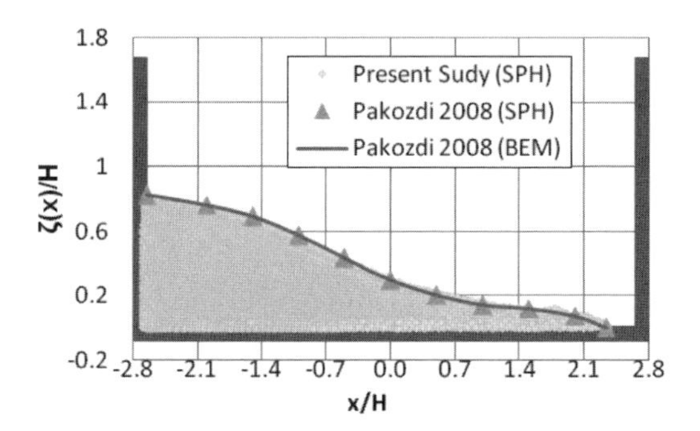

Figure 3. Free surface profile at $t = 2.2\,(H/g)^{0.5}$.

not work for dam-break problem within the frame of the present numerical model, and according to the present investigation; ε = 0.0125 gives the most compatible result whereas higher values cause the flow damped and smaller values cause the flow blow out. Even radical changes in kinematic viscosity do not have a significant effect on the results.

It should be noted, as well, that SPH simulation is slightly slower in the evolution of the flow; a fact which is also observed by Pakozdi (2008) and Colagrossi (2004). Pakozdi (2008) gave a formula of $0.08–0.11(H/g)^{0.5}$ for the delay mentioned, and the present simulation has a similar latency around 0.075 [s] when the fluid flow hits the wall and makes the maximum run-up as depicted in Figure 4. The comparison of the free surface profiles presents a clear indication of the relative compatibility of the three studies.

3.2 *Sway sloshing problem*

As a second implementation, sway sloshing motion of a partially filled rectangular tank is investigated. The geometry of the problem is given in Figure 5.

The initial particle spacing is taken as 0.01 [m] and accordingly the total number of particles in the problem domain is N = 6564. The smoothing length parameter, h, is 0.0133 [m] and artificial

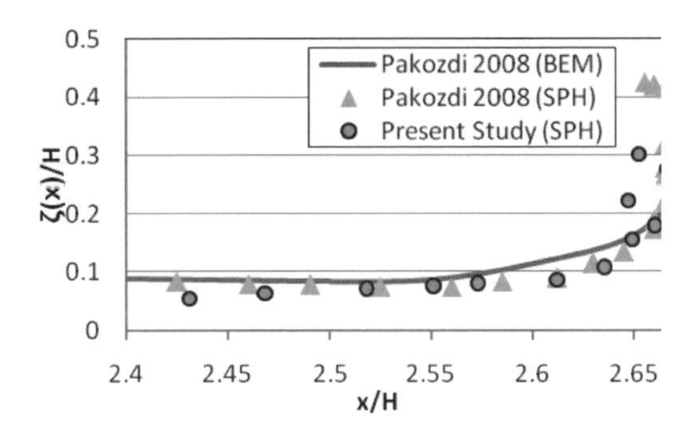

Figure 4. Free surface boundary particles at the right hand side of the tank at $t = 2.61(H/g)^{0.5}$.

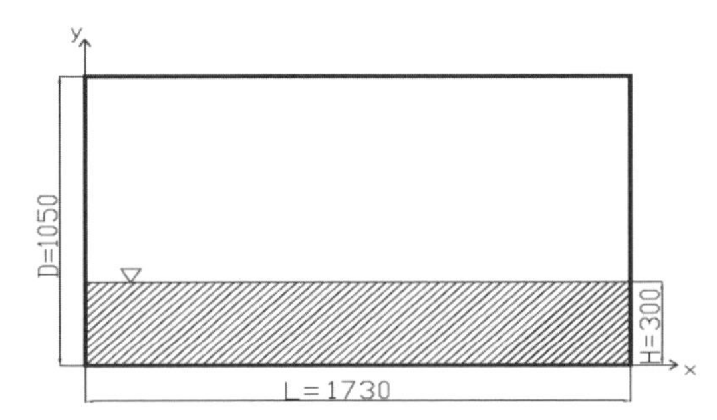

Figure 5. The geometry of the sway sloshing problem.

viscosity parameter, α, is taken as 0.0075. The same initial conditions of the dam-break problem are also valid for this problem and the time step for the time integration procedure is 0.0004 [s]. It takes approximately 16 hours for 20 second simulation on a computer with a single processor.

The sway motion of the tank is prescribed as:

$$x(t) = A\sin(\omega t) \tag{15}$$

where A is the amplitude and ω is the circular frequency of the motion. In the present numerical investigations amplitude of the sway motion and the circular frequency is taken as 0.08 [m] and 3.488 [rad/s], respectively. Sway motion directly starts with the equation (15) at t = 0 and no other treatment is made to decrease the acceleration at the beginning of the numerical procedure.

Our dam-break study shows that the *"XSPH"* correction on velocity has a significant effect on the motion of fluid particles. It is observed that the fluid particles move more freely as parameter ε in the correction expression (13) decreases and they move in a more restricted manner as parameter ε increases. This effect of ε on the evolution of the fluid flow is explained in Figure 6 where

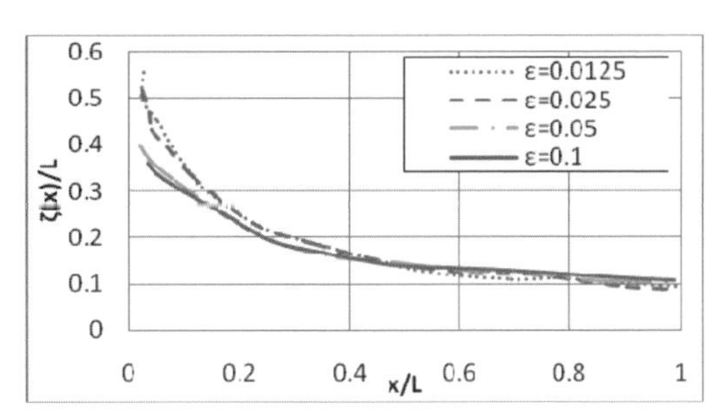

Figure 6. Comparison of free surface deformations with different ε parameters at $t = 1.34/(gL)^{0.5}$.

Table 2. Consecutive extremums (ζ_m) of the free surface elevations at x = 0 (at the left side wall).

	Experiment pakozdi (2008) (ζ_m/L)	Pakozdi (2008) (SPH) (ζ_m/L)	Present study (SPH) (ζ_m/L)
3rd	0.272 −0.086	0.266 −0.081	0.221 −0.080
4th	0.270 −0.085	0.246 −0.071	0.271 −0.078
5th	0.270 −0.085	0.255 −0.101	0.266 −0.077
6th	0.230 −0.081	0.219 −0.076	0.269 −0.751
7th	0.196 −0.081	0.219 −0.076	0.245 −0.067
8th	0.176 −0.076	0.224 −0.071	0.226 −0.073

instantaneous free surface deformations for four different ε values are compared.

As understood from Figure 6, smaller ε values in the sloshing problem do not adequately damp the fluid motion whereas ε values greater than 0.05 over-dampen the fluid motion. Thus ε = 0.05 is found to be appropriate for realistic results. Since digital time series results of the other researchers are not available at present, only extremums of the free surface elevations with respect to still water level are compared at x = 0 (at the left side wall of the tank) after the fluid motion gains momentum (see Table1). Although slight discrepancies are observed as the time proceeds, an overall general agreement is seen from the comparison.

Since digital time series results of the other researchers are not available at present, only extremums of the free surface elevations with respect to still water level are compared at x = 0 (at the left side wall of the tank) after the fluid motion gains momentum (see Table 2).

4 FUTURE WORK AND CONCLUDING REMARKS

2D nonlinear fluid flows with a free surface is treated by means of SPH method. It is known that SPH is very powerful method in modelling violent flows on the one hand but still requires adjustments to the problems tackled on the other. Two fluid flow cases are taken into consideration to reveal the sensitivity of the solutions to some of the parameters used in the numerical model employed in the present work. Weak compressibility is assumed in the flow governed by Euler's and continuity equations. Accordingly, artificial viscosity is introduced to preserve the stability of the numerical solution. Among other parameters, the present study particularly focuses on *XSPH* variant velocity and its parameter, ε. Interestingly, artificial viscosity parameter, α, is found not to be effective on the results but parameter ε in *XSPH* variant velocity, that is; numerical algorithm is stable and is not affected by changes to the artificial viscosity but sensitive to the parameter, ε. A search is made to disclose the most appropriate values of ε and proposed as 0.0125 for dam-break problem and 0.05 for sway sloshing problem. The differences of the proposed numerical parameters for different problems as well as differences among the findings of the researchers pinpoint the importance of the selected SPH algorithm best fit to the nature of the problem tackled.

As a future work, the dynamics of the fluid motion which requires the analysis of time history data of the particles will be analysed in order to determine pressures/forces exerted on the walls of the tank.

ACKNOWLEDGEMENTS

The authors are thankful to the referee for his/her valuable comments and suggestions.

REFERENCES

Bian, X., Ellero, M. & Adams, N.A. 2010. Resolution Study on Smoothed Particle Hydrodynamics with Mesoscopic Thermal Fluctuations, *Proceedings of the 5th International SPHERIC Workshop, Manchester, U.K.*

Colagrossi, A. 2004. A Meshless Lagrangian Method for Free-Surface and Interface Flows with Fregmentation. PhD Thesis, Universita di Roma La Sapienza.

Cummins, S.J. & Rudman, M. 1999. An SPH Projection Method, *Journal of Computational Physics,* 152, 584–607.

Ellero, M., Serrano, M. & Espanol, P. 2007. Incompressible smoothed particle hydrodynamics. *Journal of Computational Physics*, 226, 1731–1752.

Haihua, X. & Ha, D. 2010. A SPH Model with C1 Particle Consistency, *Proceedings of the 5th International SPHERIC Workshop*, 194–200.

Iglesias, S.A., Rojas, L.P. & Rodriguez R.Z. 2004. Simulation of Anti-roll Tanks and Sloshing Type Problems with Smoothed Particle Hydrodynamics, *Ocean Engineering* 31, 1169–1192.

Liu, G.R. & Liu, M.B. 2003. *Mesh-free Methods: Moving Beyond The Finite Element Method*, CRC Press, New York.

Liu, M.B. & Liu, G.R. 2010. Smoothed Particle Hydrodynamics (SPH): An Overview and Recent Developments, *Arch. Comput. Methods Eng.* 17, 25–76.

Lucy, L.B. 1977. Numerical Approach to Testing The Fission Hypothesis, *Astronomical Journal*, 82, 1013–1024.

Marrone, S., Antuono, M., Colagrossi, A., Colicchio, G., Le Touze, D. & Graziani, G. 2011. δ-SPH Model for Simulating Violant Impact Flows, *Computer Methods in Applied Mechanics and Engineering.*

Monaghan, J.J. 1994. Simulating Free Surface Flow with SPH, *Journal of Computational Physics,* 110, 399–406.

Monaghan, J.J. 2005. Smoothed Particle Hydrodynamics, *Reports on Progress in Physics*, 68, 1703–1759.

Monaghan, J.J. & Gingold, R.A. 1977. Smoothed Particle Hydrodynamics: Theory and Application to Non-Spherical Stars, *Monthly Notices of The Royal Astronomical Society*, 181, 375–389.

Monaghan, J.J. & Gingold, R.A. 1983, Shock Simulation by the Particle Method SPH, *Journal of Computational Physics*, 52, 374–389.

Monaghan, J.J. & Kos, A. 1999. Solitary Waves on a Cretan Beach, *Journal of Waterway Port Coastal and Ocean Engineering-Asce*, 125(3), 145–154.

Pakozdi, C. 2008. A Smoothed Particle Hydrodynamics Study of Two-dimensional Nonlinear Sloshing in Rectangular Tanks, PhD Thesis, *Norwegian University of Science and Technology*.

Rodriquez, P.M. & Bonet, J.A. 2005. A Corrected Smoothed Particle Hydrodynamics Formulation of the Shallow Water Equations, *Computers and Structures*, 83, 1396–1410.

Schlatter, B. 1997. A Pedagogical Tool Using Smoothed Particle Hydrodynamics To Model Fluid Flow Past A System Of Cylinders, *Dual MS Project*, Oregon State University.

Swegle, J.W. & Attaway, S.W. 1995. On the Feasibility of Using SPH, for underwater explosion calculations, *Computational Mechanics*, 17, 151–168.

Yildiz, M., R.A. Rook, & A. Suleman, 2008. SPH with the multiple boundary tangent method, *International Journal for Numerical Methods in Engineering*, 77, 1416–1438.

Sustainable Maritime Transportation and Exploitation of Sea Resources – Rizzuto & Guedes Soares (eds)
 ISBN 978-0-415-62081-9

Parametric nonlinear sloshing in a 2D rectangular tank with finite liquid depth

Christos C. Spandonidis & Kostas J. Spyrou
School of Naval Architecture and Marine Engineering, National Technical University of Athens, Iroon Polytechneiou, Zographos, Athens, Greece

ABSTRACT: An investigation of parametrically-excited sloshing in a two-dimensional (2D) rectangular tank with finite liquid depth is described. The analysis is based on the adaptive multimodal approach of nonlinear sloshing in a rectangular tank, described in Faltinsen and Timokha (2001). Following the standard adaptive mode ordering, a finite-dimensional system of ordinary differential equations is obtained. Third-order polynomial nonlinearities are retained. The external vertical forcing of the tank is assumed to be of "sufficiently" small amplitude. The novel part of the work lies in the advanced investigation of the nonlinear free surface oscillations that brought to light a new (in parameters space) region of liquid surface instability that was up to now considered as quiescent.

1 INTRODUCTION

Parametrically excited engineering systems can often exhibit dangerous behaviour (Ibrahim 1985). Parametric sloshing is the motion of a liquid's free surface due to an excitation perpendicular to the plane of the undisturbed free surface. Such vertical excitation of ship tanks could be produced by the heaving motion in a seaway. In our case the heave is considered as a harmonic function. This arises physically in combination with excitations in other modes of ship motion; for example in pitch and in roll. In the current work we have considered however the heave excitation alone in order to clarify the free surface dynamics under pure parametric excitation. The issue is not new, yet it is less often considered as important compared to cases of directly excited sloshing (Dodge 1966). In an investigation of parametric sloshing the nonlinear treatment of the free surface is very essential; otherwise unrealistic infinite free surface displacements will be uniformly predicted inside the instability region.

A concise overview of research concerning the nonlinear behaviour of liquids contained in tanks of various shapes and subjected to parametric excitation can be found for example, in the book of Ibrahim (2005). Standing waves generated in vertically oscillating tanks were firstly studied experimentally by Faraday (1831), Mathiessen (1868, 1870) and Lord Rayleigh (1883a & b, 1887). The same problem was investigated theoretically by Lewis (1950), Taylor (1950), Benjamin & Ursell (1954), Konstantinov et al. (1978), Nevolin (1985), Feng & Sethna (1989), Simoneli & Gollub (1989), Henderson & Miles (1990), Nagata (1991), Miles (1994), Perlin & Schultz (2000). Different numerical approaches to sloshing prediction have been reported by Telste (1985), Chen et al. (1996), Takizawa & Kondo (1995), Chern et al. (1999), Pawell (1997), Turnbull et al. (2003), Wu et al. (1998, 2001 and 2007), Fradnsen (2003), Y. Kim et al. (2001, 2007) treating the moving free surface either by using Lagrangian tracking of free surface nodes with regrinding; or by mapping. Both have advantages and disadvantages; however a common drawback is that they are not the most appropriate for long time simulations.

In the current work we have adopted a well-known semi-analytical method developed by Faltinsen and Timokha (2001) that is based on modal analysis. However, as our objective here is not to advance the modelling of sloshing but to deepen into the character of the nonlinear oscillations exhibited by the free surface of a liquid in a two-dimensional (2D) rectangular tank with finite liquid depth, a direct method for capturing nonlinear steady dynamics is coupled to the hydrodynamic model. More specifically, we couple the model with a "continuation analysis" algorithm of nonlinear dynamics in order to predict (in a single run) the amplitudes of steady liquid surface oscillations as the frequency and/or the amplitude of excitation are varied, without performing multiple simulations (which would not capture the unstable oscillations anyway). The focus is on determining the region of parameters' values where free surface activity takes place (termed here as "region of instability"). For the model that we use, this region of instability is predicted to be more extended than the one associated with linear parametrically excited systems (as described for example in Benjamin & Ursell 1954). The unique feature of the newly discovered region is that it is initial-conditions dependent i.e., one may obtain a stable surface wave or a flat surface depending on the exact state of the free surface when the harmonic excitation was firstly applied.

2 FORMULATION OF THE PROBLEM

We consider a mobile, rectangular, smooth and rigid tank, filled partly by an inviscid, incompressible fluid. Liquid depth is finite; but the tank top is high enough so that it is never reached by the moving liquid. The flow is two-dimensional and irrotational. The origin of the coordinate system is placed at the middle of the mean free surface (Figure 1).

The problem of sloshing of an incompressible fluid with irrotational flow when part of the boundary (the free surface) is free to move is formulated in the standard manner in terms of the Laplace equation in the fluid, with suitable boundary conditions:

$$\Delta\Phi = 0 \text{ in } Q(t),$$
$$\frac{\partial\Phi}{\partial n} = \vec{u}_0 \cdot \vec{n} \text{ on } S(t),$$
$$\frac{\partial\Phi}{\partial n} = \vec{u}_0 \cdot \vec{n} - \frac{\partial Z/\partial t}{|\nabla Z|} \text{ and}$$
$$\frac{\partial\Phi}{\partial t} - \nabla\Phi\cdot\vec{u}_0 + \frac{1}{2}(\nabla\Phi)^2 + U_g = 0 \text{ on } \Sigma(t) \quad (1)$$

$Q(t)$ is the fluid volume, $\Sigma(t)$ is the free surface which is associated with the equation $Z(y, z, t) = 0$, $S(t)$ is the tank surface below $\Sigma(t)$, t is time, $\Phi(y, z, t)$ is the velocity potential in the reference frame, and $\vec{n}$ is the unit vector that is normal to $S(t)$. Tank's velocity is $\vec{u}_0(t) = \dot{n}_3 \cdot \vec{e}_3$, where n_3 is the magnitude of external excitation and $\vec{e}_3$ is the unit vector in the z- axis.

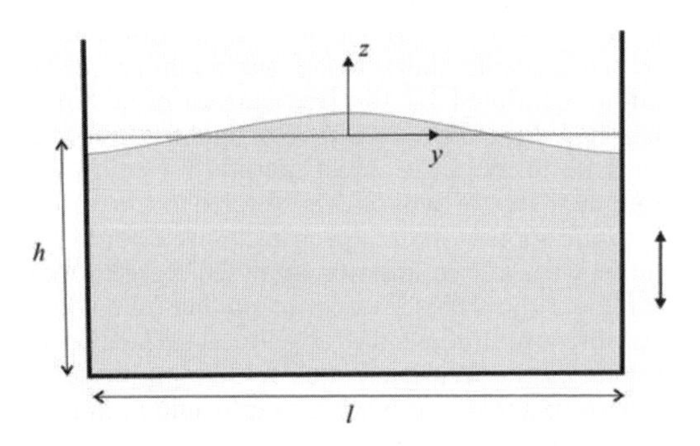

Figure 1. The origin of the coordinate system is placed at the middle of the mean free surface. Tank length and height are denoted by l, h respectively. Tank is free to move in the z-axis.

3 NONLINEAR ASYMPTOTIC ADAPTIVE MODAL ANALYSIS

The multimodal method uses a Fourier series representation of the solution with time-dependent unknown coefficients. The sloshing problem is expressed by means of two functions, describing the free-surface elevation and the velocity potential. Faltinsen et al. (2000) postulated their Fourier series representations as follows:

$$\zeta(y,t) = \sum_{i=1}^{\infty} \beta_i(t) f_i(y),$$
$$\Phi(y, z, t) = \sum_{i=1}^{\infty} R_i(t)\varphi_i(y, z) \quad (2)$$

The modal representation (2) is based upon the functions $\{f_i\}$ and $\{\varphi_i\}$ which must provide "complete" sets on the mean free surface and the whole tank domain, respectively. The most common choice for the basis $\{\varphi_i\}$ is the set of linear natural modes. However, the natural modes are theoretically defined only in the unperturbed hydrostatic domain. In view of this problem, we interpret the natural modes as an *asymptotic* basis, assuming that the free surface is, to some extent, asymptotically close to its unperturbed state (Faltinsen & Timokha, 2009). In our case the modal basis $f_i(y)$ and the set of functions $\varphi_i(y, z)$ coincide with the linear natural sloshing modes (3) as derived by Faltinsen & Timokha (2002).

$$\varphi_i(y, z) = \cos\left(\frac{\pi i}{l}(y + \frac{1}{2}l)\right) \times \frac{\cosh(\frac{\pi i(z)}{l})}{\cosh(\frac{\pi i h}{l})}$$
$$f_i(y) = \cos\frac{\pi i(y + \frac{1}{2}l)}{l}, \quad i \geq 1 \quad (3)$$

By using the asymptotic non-linear modal theory we obtain an infinite-dimensional system of nonlinear differential equations (modal system). Following the adaptive approach proposed by Faltinsen & Timokha (2001), this modal system could be asymptotically reduced to an infinite-dimensional system of ODEs. An important fact is that asymptotically truncated systems may use natural modes because the procedure needs only the completeness of $\varphi_n(y, z)$ in the unperturbed liquid domain (Faltinsen & Timokha, 2002). If, in the first instance, nonlinear terms are kept only up to third-order, the considered system comes to the following form in the case of vertical excitation.

$$\sum_{i=1}^{p}\ddot{\beta}_i(\delta_{p\mu}+d_{p,q}^{1,\mu}\sum_{i=1}^{q}\beta_i+d_{p,q,r}^{2,\mu}\sum_{i=1}^{q}\beta_i\sum_{i=1}^{r}\beta_i)+$$

$$\sum_{i=1}^{p}\dot{\beta}_i\sum_{i=1}^{q}\beta_i(t_{p,q}^{0,\mu})+\sum_{i=1}^{p}\dot{\beta}_i\sum_{i=1}^{q}\dot{\beta}_i\sum_{i=1}^{r}\beta_i(t_{p,q,r}^{1,\mu})+$$

$$[\sigma_\mu^2+\frac{\ddot{n}_3}{l}\pi\mu\tanh(\frac{\pi\mu h}{l})]\beta_\mu=0,\ \mu=1,2\ldots \quad (4)$$

where δ stands for Kronecker's delta and p, q, r are upper summation limits. The d and t coefficients can be expressed as functions of the ratio of liquid depth to tank breadth (such analytical expressions can be found in Faltinsen and Timokha 2001). σ_μ represents the μth natural frequency and is given (Faltinsen & Timokha, 2009) by the equation:

$$\sigma_\mu=\sqrt{g\left(\frac{\pi\mu}{l}\right)\tanh\left(\frac{\pi\mu}{lh}\right)},\ \mu=1,2\ldots \quad (5)$$

Furthermore, by using the condition $\beta_n=O(\varepsilon^{1/3})$, $\beta_i=O(\varepsilon)$, $i\geq n+1$ we could introduce more than one dominant mode (in contrast with a Moiseev-like ordering). This leads to a finite-dimensional nonlinear modal system that will be called from here on "Model-k", where the integer k denotes the number of dominant modes. The coefficient $\varepsilon=n_{3\alpha}/l$ is an indicator of the smallness of excitation. Here it is assumed that $\varepsilon<<1$.

4 LINEAR MODEL

Linearization of the modal system (Equation 4) leads to the following set of uncoupled modal equations:

$$\ddot{\beta}_\mu+\left[\sigma_\mu^2+\ddot{n}_3 l^{-1}\pi\mu\tanh\left(\frac{\pi\mu h}{l}\right)\right]\beta_\mu=0,\mu=1,\ldots \quad (6)$$

By assuming harmonic excitation $n_3=n_{3\alpha}\cos(\sigma t)$, Equation 6 comes to the form of a set of Mathieu-type equations. Let us now restrict our investigation to the case of vertical excitation with relatively small amplitude, with excitation frequency in the vicinity of the principal parametric resonance of the first mode; i.e. $(\sigma_1/\sigma)^2\approx 1/4$.

We also choose the tank's height-to-breadth-ratio to be larger than the critical depth ($h/l=0.03368$) restricting our investigation to *finite liquid depth*. We do this because, according to Fulzt (1962), in the vicinity of critical depth strong changes and amplifications in the liquid behaviour occur.

It is remarked that Faltinsen and Timokha (2009) have introduced to Equation 6 an empirical linear damping term, represented by the damping ratio ζ_1. From a physical perspective, this damping term could empirically account for boundary–layer damping. Incorporating such damping, the linear modal equation for the dominant mode β_1 obtains the following form:

$$\ddot{\beta}_1+2\sigma_1\zeta_1\dot{\beta}_1+\sigma_1^2\left[1-\frac{n_{3a}\sigma^2}{g}\cdot\cos(\sigma\cdot t)\right]\beta_1=0 \quad (7)$$

Benjamin and Ursell (1954) had investigated the free surface elevation under similar forcing, using Mathieu functions for expressing analytically the solution. As well known, depending on parameters' values, a system described by Equation 7 could exhibit stable as well as unstable behaviour.

Frandsen (2004) considered the case of a rectangular tank under vertical forcing and checked the stability of the free surface by using two-dimensional CFD simulations. She has shown that the prediction of the stable regions is in good agreement with Benjamin and Ursell's (1954) predictions when the forcing parameter is small. If the excitation amplitude is raised, nonlinearities due to intermodal interaction have to be considered.

In Figure 2 is shown the well-known stability chart that corresponds to Equation (7), for damping ratio $\zeta_1=0.02$. The linear model entails that, to the interior of the curve a solution is unstable and diverges to infinity; whereas every point outside the curve corresponds to a stable steady solution, in the sense of having the liquid surface maintained flat and horizontal.

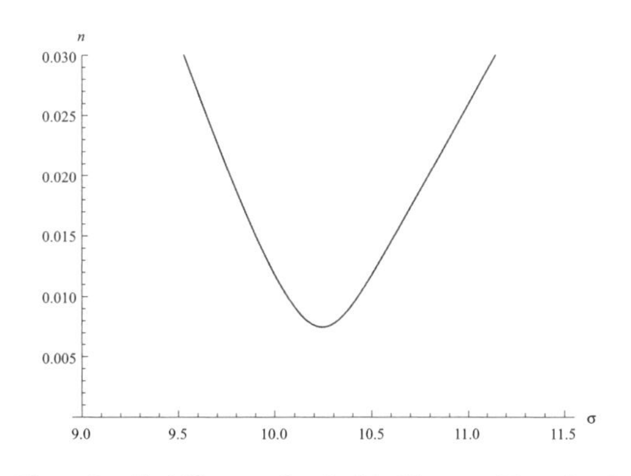

Figure 2. Stability map for the Mathieu type Equation 7, for damping ratio $\zeta_l=0.02$.

As realised, β_1 stands for the time-dependant response of the free surface. In what follows we track the elevation of free surface at $y = -l/2$; i.e., at its interface with the left tank wall. To obtain the elevation for all other points $(y, 0)$ of the liquid surface β_1 should be multiplied by $\cos(\pi(y + l/2)/l)$.

5 NON-LINEAR MODEL (MODEL-1)

For mode ordering $\beta_1 = O(\varepsilon^{1/3})$, $\beta_\mu = O(\varepsilon)$, $\mu > 1$, and by retaining only β_1 following the same thinking like before, the system of Equations 4 generates the following form ("Model-1"):

$$\ddot{\beta}_1 + 2\zeta_1\sigma_1\dot{\beta}_1 + (\sigma_1^2\beta_1 + Q_1\ddot{n}_3\beta_1) + d_2(\ddot{\beta}_1\beta_1^2 + \dot{\beta}_1^2\beta_1) = 0 \quad (8)$$

where: $d_2 = \frac{\pi^2}{4}\left[1 - 2\cdot\tanh(\pi\frac{h}{l})\cdot\tanh(2\pi\frac{h}{l})\right]$ and

$$Q_1 = \pi\tanh(\frac{\pi h}{l}).$$

As in the linear model case, a damping term has been introduced. Assuming harmonic external excitation $n_3 = n_{3\alpha}\cos(\sigma t)$ with small amplitude, Equation (8) becomes:

$$\ddot{\beta}_1 + 2\zeta_1\sigma_1\dot{\beta}_1 + \sigma_1^2(1 - \frac{n_{3a}\sigma^2}{g}\cdot\cos(\sigma\cdot t))\beta_1$$
$$+ d_2(\ddot{\beta}_1\beta_1^2 + \dot{\beta}_1^2\beta_1) = 0 \quad (9)$$

As observed, the above non-linear system is the same as the linear plus two non-linear terms that feature products of the surface elevation with the corresponding acceleration and with the corresponding velocity. This makes this problem somehow different from several others that are also modeled through a Mathieu-type equation [e.g., for ships' parametric rolling investigations we usually include nonlinearity only n the stiffness term, Spyrou et al. (2008). It is remarked that all other nonlinear terms of Equation (4) are not present because their coefficients, according to the current approximation, are equal to zero. The only non-zero nonlinear term parameter is d_2.

The nonlinear model of Equation (9) was derived by following the adaptive approach. With the restrictions described above, the same equation could have been produced from the single dominant mode approach, also introduced by Faltinsen et al. (2000). If large-amplitude response occurs; or if the tank-height-to depth-ratio is equal or smaller than the critical depth; or lastly if secondary resonance occurs, the assumption of lowest order dominant mode collapses. In such cases, the adaptive approach is the way to follow for model derivation (Faltinsen and Timokha 2002). We have thus selected to follow the adaptive method, in order to maintain the prospect of comparison of our results against those produced from higher order models that we are concurrently investigating (not included in this paper).

6 CONTINUATION ANALYSIS FOR THE NON-LINEAR MODEL

"Continuation" algorithms usually accept the mathematical model in the so called "autonomous canonical form":

$$\frac{dx}{dt} = f(x;b) \quad (10)$$

Here x and b are, respectively, the state and control parameters' vectors of our problem. Variation of one or more components of the control vector b creates, through solution of the above vector differential equation, branches of steady-state (in our case periodic) solutions. These branches constitute the "spine" of the dynamical response of the system. It is thus imperative to be able to trace such branches of steady-states efficiently, even when nonlinearities are strong and multiplicity of steady solutions arises. For such types of investigation "continuation" is a truly indispensable tool. The specific algorithm implemented here is "MATCONT". For mathematical details Dhooge et al. (2003) can be consulted.

The basic mathematical model expressed through Equation (8) is characterised by explicit time-dependence in the restoring term. Thus, it is not in the autonomous form of Equation (10) entailed by the continuation algorithm. To overcome this, a suitable additional pair of differential equations is introduced with respect to the dummy variables $x = \sin(\sigma t)$, $y = \cos(\sigma t)$. Thereafter, Equation (9) can be converted into the following system of four 1st-order ordinary differential equations that, whilst being equivalent to Equation (10), is not characterised by explicit time-dependence (Spyrou & Tigkas 2011):

$$\dot{\beta}_1 = \phi_1$$
$$\dot{\phi}_1 = (-2\zeta_1\sigma_1\phi - \sigma_1^2(1 - \frac{n_{3a}\sigma^2}{g}\cdot y)\beta_1 - \phi_1{}^2\beta_1)/(1 + d_2\beta_1{}^2)$$
$$\dot{x} = x + \sigma\cdot y - x\cdot(x^2 + y^2)$$
$$\dot{y} = y - \sigma\cdot x - y\cdot(x^2 + y^2) \quad (11)$$

Due to the cubic nonlinearity of Model-1, inside the unstable region one should expect a stable limit cycle. Selecting such limit cycle as initial state (this is easily captured through simulation) we can use the continuation method to produce the dependence of that "cycle" from the excitation amplitude and/or frequency, while restricting ourselves to the region of small amplitudes so that we don't violate the assumption of relatively small surface elevation that is intrinscic to the model. Such a continuation analysis result is presented in Figure 3. It is observed that there is an excitation region where two different limit cycles coexist. The limits of this region is determined by two bifurcation points. When the excitation amplitude increases from zero and until the first bifurcation is met ("limit point of cycles"), there isn't any limit cycle. This is the area where a stable flat condition is possible for the water surface.

As the amplitude continues to increase, two coexisting limit cycles are suddenly met, one stable and one unstable. The stable one gets gradually larger as the excitation is increased. The other limit cycle shrinks till the second bifurcation point (that is of "subcritical" type) where it disappears. On the basis of Figure 3 we have extracted Figure 4 that

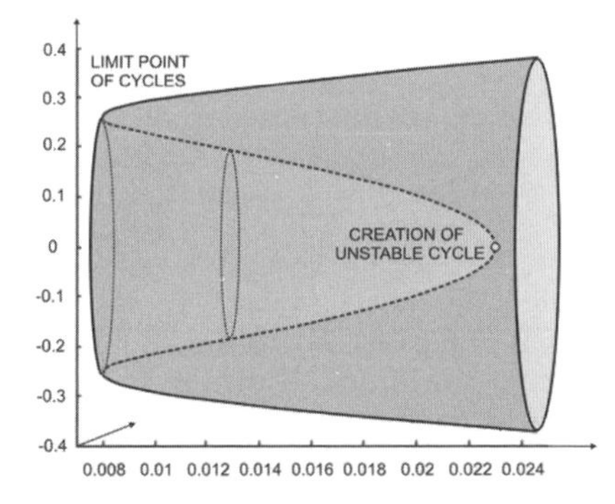

Figure 3. Limit-cycle dependence on excitation amplitude ($n_{3\alpha}$, β_1), for $\sigma = 9.7$ rad/s. There is an excitation region determined by the point of creation of limit cycles and the fold (limit point) of cycles (the two Branch points of Cycles) where two different limit-cycles coexist (one unstable).

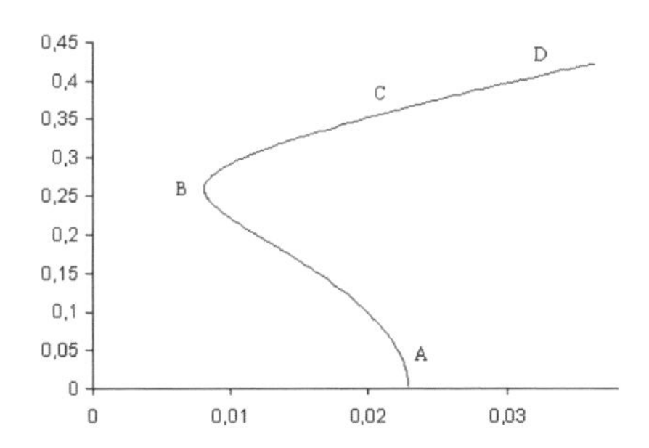

Figure 4. Dependence of the response amplitude (y-axis) on the excitation amplitude $n_{3\alpha}$ (x-axis). Dependence has a qualitatively similar form with that extracted by Ibrahim (2005).

summarizes the dependence of surface elevation upon the excitation amplitude $n_{3\alpha}$. It is remarked that subcritical bifurcations are associated with hysteretic behaviour as explained next:

As the excitation amplitude increases from the zero value to the bifurcation point A, the liquid-free-surface amplitude will jump in a fast dynamic transition to point C right after A is reached. As the excitation level is further increased, the amplitude also increases monotonically along the solid curve CD. If, on the other hand, the excitation begins to decrease while the liquid free surface is in a state in the neighbourhood of point C, the surface amplitude will decrease along the curve DCB until the point B is reached. Figure 4 has a qualitatively similar form with one by Ibrahim (2005) who had targeted an approximation of the solution according to a perturbation technique.

To investigate the stability of these periodic solutions the formal choice is to perform calculation of Floquet multipliers. But we can also get some strong indication about the stability of periodic solutions by simply selecting some forcing level that lies in between the two Branch points of Cycles (BPC); and then integrating the nonlinear system under different initial conditions. As expected from nonlinear dynamics, one of the limit cycles (the outer) is stable and the other one (the inner) is unstable. In Figure 5 is presented the time history of β_1 for two slightly different initial conditions. Initial elevation $\beta_1 = 0.091$ m (with zero initial vertical velocity of the surface) leads to a stable (zero) point; i.e., there is attraction towards the stationary state.

On the other hand, an initial elevation $\beta_1 = 0.095$ m leads to a stable periodic pattern.

Working in a similar manner for more than one frequency we are able to demonstrate the

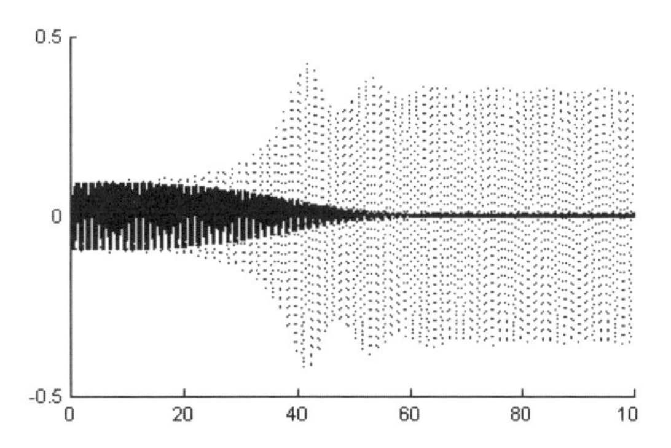

Figure 5. Time history of β_1 for $\sigma = 9.7$ rad/s and $n_{3\alpha} = 0.02$. Initial elevation $\beta_1 = 0.091$ m (solid line) leads to a stable (zero) point and $\beta_1 = 0.095$ m (dot line) leads to a stable periodic pattern.

steady dynamic behavior of the autonomous nonlinear system represented by Equation (11), in the frequency region that surrounds $\sigma = 2\sigma_1$. In Figure 6 is summarised this behaviour. As one expects, for small frequency every excitation ends up to a stable point. Increase of the excitation frequency leads, through a fold of limit-cycles bifurcation, to an initial-conditions-dependent area (Figure 6-Left) where a stable and an unstable limit-cycle coexist. Further increase of frequency leads, through the subcritical bifurcation mentioned earlier, to the "classical" area of instability associated with principal parametric resonance (Figure 6-Right), where the unstable limit-cycle disappears and the stable trivial solution becomes unstable. Lastly, a supercritical ("smooth") bifurcation locus represents in fact the boundary with the stability area to the right of the "classical" linear instability area.

Incorporating the above observations into the stability chart of the linear Mathieu-type system we obtain the diagram of Figure 7. If we compare Figure 7 against the similar result of Benjamin and Ursell we see that a new initial-condition-dependent area (area C) has been added. As a result, the forcing-versus-frequency parameters' plane is divided into three areas that are identified in Figure 7. Area A is the stable one where every external excitation (with the fitting specification) leads invariably to a flat liquid surface. Area B is the classical area of instability where external excitation generates periodic oscillations of the free surface. Area C is the initial-condition-dependent area, (in terms of β_l and φ_1) where the same external

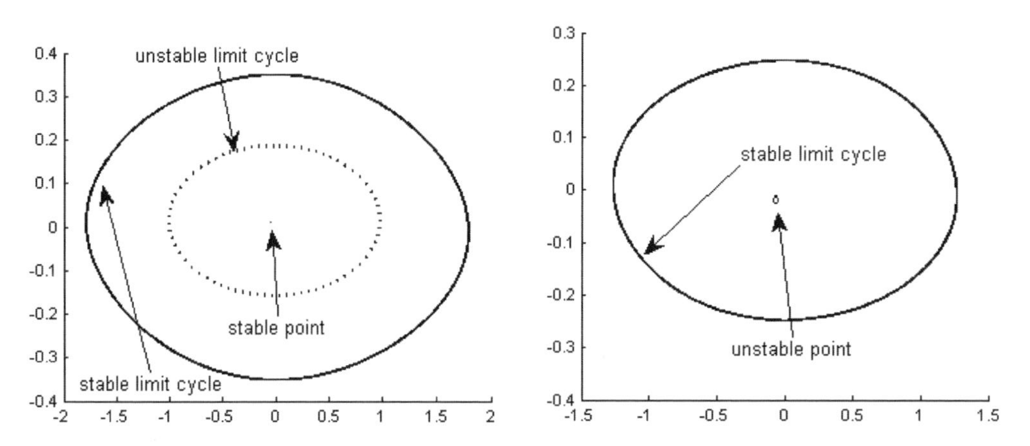

Figure 6. Phase diagram (φ_1, β_1) for different excitation frequencies, $\zeta_1 = 0.02$ m and $n_{3a} = 0.02$ m. Left: initial-condition-dependent area for excitation frequency σ = 9.7 rad/sec. The unstable limit cycle drives the behaviour of the system either to a stable fixed point or to a stable limit cycle, Right: typical behavior in the classical unstable area for excitation frequency σ = 10.2 rad/sec. Every excitation ends to a stable limit-cycle.

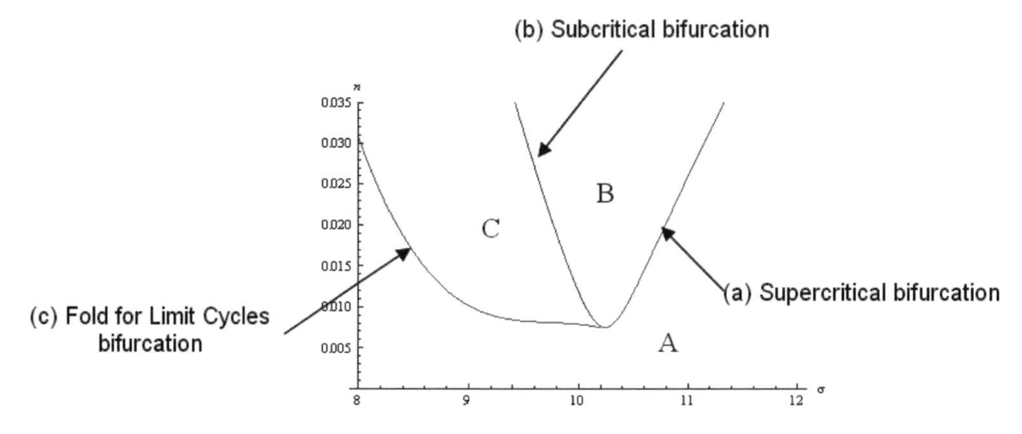

Figure 7. Stability chart for the nonlinear Model-1 system ($\zeta_1 = 0.02$). A new initial-condition-depended area (indicated by C) is added to the linear stability chart. Inside C, one may obtain a stable surface wave or a flat surface depending on how the free surface looked like when the harmonic excitation was firstly applied.

excitation leads either to a quiescent surface or to a wavy one. The phase space structure of the considered dynamical system is quite a common one. The domains of attraction of the two competing stable patterns occupy certain complementary regions of phase space. They are separated by a surface defined by the incoming (stable) manifold of the unstable periodic solution.

7 CONCLUSIONS

2-D liquid sloshing in a rectangular and vertically excited tank has been investigated. A non-linear model has been used, based on modal modelling and according to the adaptive analysis of Faltinsen and Timokha (2001). The work was limited to finite liquid depth, corresponding to a tank-height-to-depth-ratio of 0.4. A global picture of liquid surface dynamics was obtained, befitting to model nonlinearities retained up to third-order. A new area of bi-stability should be added to the stability chart of free surface oscillations. Investigation of the influence of the damping term on the size of that area is currently in progress.

A first step towards confirming the validity of results will be the investigation of the dynamic behaviour associated with the immediately higher order non-linear model. That should allow elicitation of the liquid surface dynamics and the associated stability properties without severe limitations on the excitation amplitude or frequency. Areas of secondary resonance (higher excitation frequency) and also the chaotic areas (higher excitation amplitude) discussed by Ibrahim (2005) are also very interesting topics of further research. These could not be considered here, partly due to time limitations and partly due to limitations of the model.

Of course, experimental reproduction of the identified types of numerical solutions will be required before these are considered as established patterns of the behaviour of the physical system under consideration.

REFERENCES

Abramson, H.N. 1966. *The Dynamics of Liquids in Moving Containers*. NASA Report SP 106.

Benjamin, T.B. & Ursell, F. 1954. The stability of the plane free surface of a liquid in a vertical periodic motion. *Proceedings of the Royal Society* A225: 505–15.

Bredmose, H., Brocchini, M., Peregrine, D.H. & Thais, L. 2003. Experimental investigation and numerical modelling of steep forced water waves. *Journal of Fluid Mechanics*, 490: 217–49.

Chen, W., Haroun, M.A. & Liu, F. 1996. Large amplitude liquid sloshing in seismically excited tanks. Earthquake Engineering and Structural Dynamics 25: 653–669.

Chern, M.J., Borthwick, A.G.L. & Taylor, R. 1999. A pseudospectral s-transformation model of 2-D nonlinear waves. Journal of Fluids & Structures 13: 607–630.

Dhooge, A., Govaerts, W., Kuznetsov, Y.A., Mestrom, W., Riet, A.M. & Sautois, B. 2003. MATCONT and CL_MATCONT: Continuation Toolboxes for MATLAB. Report of Gent (Belgium) and Utrecht (Netherlands) Universities.

Dodge, F.T. 1966. Verticall Excitation of Propellant Tanks. In the book: *The Dynamics of Liquids in Moving Containers*. NASA Report SP 106.

Faltinsen, O.M. & Rognebakke, O.F. 2000. Sloshing. Keynote lecture. *Proceedings* International Conference on Ship and Shipping Research, NAV2000, Venice, 19–22 September, Italy.

Faltinsen, O.M. & Timokha, A.N. 2001. Adaptive multimodal approach to nonlinear sloshing in a rectangular rank. *Journal of Fluid Mechanics*, 432: 167–200.

Faltinsen, O.M. & Timokha, A.N. 2002. Asymptotic modal approximation of nonlinear resonant sloshing in a rectangular tank with small fluid depth, *Journal of Fluid Mechanics* 470 (2002): 319–357.

Faltinsen, O.M. & Timokha, A.N. 2009. *Sloshing*. New York: Cambridge University Press. ISBN: 978-0-521-88111-1.

Faltinsen, O.M., Rognebakke, O.F., Lukovsky, I.A. & Timokha, A.N. 2000. Multidimensional modal analysis of nonlinear sloshing in a rectangular tank with finite water depth. *Journal of Fluid Mechanics*, 407: 201–234.

Faraday, M. 1831. On a peculiar class of acoustical figures, and on certain forms assumed by groups of particles upon vibrating elastic surfaces. Phil Trans R Soc Lond, 121:299–340.

Feng, Z.C. & Sethna, P.R. 1989. Symmetry-breaking bifurcations in resonance surface waves. *Journal of Fluid Mechanics* 199: 495–518.

Frandsen, J.B. Sloshing motions in the excited tanks (2004) *Journal of Computational Physics*, 196, pp. 53–87.

Frandsen, J.B. & Borthwick, A.G.L. 2003. Simulation of sloshing motions in fixed and vertically excited containers using a 2-D inviscid rtransformed finite difference solver. J. Fluids Struct. 18 (2): 197–214.

Fultz, D. 1962. An experimental note on finite-amplitude standing gravity waves. *Journal of Fluid Mechanics* 13: 193–212.

Henderson, D.M. & Miles, J.W. 1990. Single mode Faraday waves in small cylinders. *Journal of Fluid Mechanics* 213: 95–109.

Ibrahim, R.A. 1985. *Parametric Random Vibration*. New Jersey: Wiley-Interscience. ISBN: 086380-032-7.

Ibrahim, R.A. 2005. *Lliquid Sloshing Dynamics*. New York: Cambridge University Press. ISBN: 978-0-521-83885-6.

Ibrahim, R.A., Pilipchuk, V.N. & Ikeda, T. 2001. Recent advances in liquid sloshing dynamics. *Applied Mechanics Research*, 54(2): 133–199.

Kim, Y. 2001. Numerical simulation of sloshing flows with impact load. *Applied Ocean Research* 23: 53–62.

Kim, Y., Nam, B.W., Kim, D.W. & Kim, Y.S. 2007. Study oncoupling effects of ship motion and sloshing. *Ocean Engineering* 34:2176–2187.

Konstantinov Akh., Mikityuk YuI & Pal'ko, L.S. 1978. Study of the dynamic stability of a liquid in a cylindrical cavity, in *Dynamics of Elastic Systems with Continuous-Discrete Parameters*. Naukova Dumka 69–72.

Lewis, D.J. 1950. The instability of liquid surfaces when accelerated in a direction perpendicular to their planes. *Proceedings of the Royal Society* A 202: 81–96.

Mathiessen, L. 1868. Akustische versuche, die klieinsten transversalivellen der flussigkeiten betreffend. *Annalen der Physik* 134: 107–117.

Mathiessen, L. 1870. Uber die transversal-schwingungen tonender tropharer und elastisher flussigkeiten. *Annalen der Physik* 141: 375–393.

Miles, J.W. 1994. Faraday waves: rolls versus squares. *Journal of Fluid Mechanics* 269: 353–371.

Nagata, M. 1991. Behavior of parametrically excited surface waves in square *geometry. European Journal of Mechanics B/Fluids* 10(2): 61–66.

Nevolin, V.G. 1985. Parametric excitation of surface waves. *Journal of Engineering Physics* 49: 1482–1494.

Pawell, A. 1997. Free surface waves in a wave tank. *International Series Numerical Mathematics* 124: 311–320.

Perlin, M. & Schultz, W.W. 1996. On the boundary conditions at an oscillating contact-line: a physical/numerical experimental program, *Proceedings* NASA 3rd Microgravity Fluid Physics Conference, Cleveland, OH, 615–620.

Rayleigh, J.W.S. 1883a. On the crispations of fluid resting upon a vibrating support. *Philosophical Magazine* 15: 229–235.

Rayleigh, J.W.S. 1883b. On maintained vibrations. *Philosophical Magazine* 15: 229–235.

Rayleigh, J.W.S. 1887. On the maintenance of vibrations by forces of double frequency and on the propagation of waves through a medium endowed with a periodic structure. *Philosophical Magazine* 24: 145–159.

Simonelli, F. & Gollub, J.P. 1989. Surface wave mode interactions: effects of symmetry and degeneracy. *Journal of Fluid Mechanics* 199: 471–494.

Spandonidis, C. 2010. *Parametric Sloshing in A 2D Rectangular Tank*. Postgraduate Thesis, National Technical University of Athens.

Spyrou, K.J. & Tigkas, I. 2011. Nonlinear surge dynamics of a ship in astern seas: "Continuation analysis" of periodic states with hydrodynamic memory. *Journal of Ship Research* 55: 19–28.

Spyrou, K.J., Tigkas, I., Scanferla, G., Pallikaropoulos, N. & Themelis, N. 2008. Prediction potential of the parametric rolling behaviour of a post-panamax containership, *Ocean Engineering* 35: 1235–1244.

Takizawa, A. & Kondo, S. (1995), Computer discovery of the mechanism of flow-induced sloshing. Fluid-Sloshing and Fluid–Structure Interaction, *Proceedings*, ASME Pressure Vessel Piping Conference, PVP-314, 153–158.

Taylor, G.I. 1950. The instability of liquid surfaces when accelerated in a direction perpendicular to their planes. *Proceedings of the Royal Society* A 201: 192–196.

Telste, J.G. 1985. Calculation of fluid motion resulting from large amplitude forced heave motion of a two-dimensional cylinder in a free surface. *Proceedings*, The Fourth International Conference on Numerical Ship Hydrodynamics, Washington, USA, pp. 81–93.

Turnbull, M.S., Borthwick, A.G.L. & Eatock Taylor, R. 2003. Numerical wave tank based on a s-transformed finite element inviscid flow solver. *International Journal for Numerical Methods in Fluids* 42: 641–663.

Wu, G.X. 2007. Second—order resonance of sloshing in a tank. *Ocean Engineering* 34: 2345–2349.

Wu, G.X., Eatock Taylor, R. & Greaves, D.M. 2001. The effect of viscosity on the transient free-surface waves in a two-dimensional tank. *Journal of Engineering Mathematics* 40: 77–90.

Wu, G.X., Ma, Q.A. & Eatock Taylor, R. 1998. Numerical simulation of sloshing waves in a 3D tank based on a finite element method. *Applied Ocean Research* 20: 337–355.

Sustainable Maritime Transportation and Exploitation of Sea Resources – Rizzuto & Guedes Soares (eds)
© 2012 Taylor & Francis Group, London, ISBN 978-0-415-62081-9

A numerical investigation of exhaust smoke-superstructure interaction on a naval ship

S. Ergin, Y. Paralı & E. Dobrucalı
Istanbul Technical University, Department of Naval Architecture and Marine Engineering, Maslak, Istanbul, Turkey

ABSTRACT: The exhaust smoke-superstructure interaction for a generic frigate is investigated numerically. The frigate was driven by a CODOG system. The k-ε model is adopted for turbulent closure, and the governing equations in three dimensions are solved using a finite volume technique. The computations were performed for different yaw angles, efflux velocities and temperatures of the exhaust smoke. The cases with diesel engines and gas turbines are considered. The calculated streamlines, temperature contours and smoke concentrations are presented and discussed. Furthermore, the detailed predictions are compared with the available experimental measurements. A good agreement between the predictions and experiments is obtained. The study has demonstrated that computational fluid dynamics is a powerful tool to study the problem of exhaust smoke–superstructure interaction on ships.

1 INTRODUCTION

Nowadays, the understanding of the exhaust behavior of ship stacks has become quite important in the naval ship design due to advances in superstructure electronics (radar and antenna) and heat seeking missiles. The predictions of velocity and temperature fields of the ship exhaust plume from the stack are of vital importance for positioning and arranging of various superstructure electronics, weapons, gas turbine intake and ventilation intakes in the superstructure with the minimum interference with the hot exhaust smoke from the stack. However, all modern naval ships tend to favor short stacks and tall mast to house electronics. This makes them prone to the problem of smoke nuisance where hot exhaust gas gets entrapped into turbulent wake of the superstructure and deteriorates the performance of electronics, weapons, sensors, increase the ships infrared signature and hampers the internal ventilation and gas turbine intakes. This is also problem for the operation of helicopters from ships (see, for example, Park et al., 2011, Mishra et al., 2010, Ergin et al., 2010, Syms 2008, Kulkarni et al., 2005, 2007, Fitzgerald, 1986).

The dispersion of exhaust smoke is affected by a large number of parameters such as efflux velocity and temperature of smoke, level of turbulence, wind velocity and direction, geometry of the structures on ship's deck etc. (see for example, Baham et al., 1977; Fitzgerald, 1986, Jin et al., 2001, Park et al., 2011, Heywood, 1988). Therefore, the prediction of exhaust smoke dispersion from the ship stacks is extremely complicated task. Recently, a comprehensive review on the numerical modeling of exhaust smoke dispersion from ships is given by Kulkarni et al. (2011) and Mishra et al. (2010). Traditionally, the exhaust smoke dispersion has been investigated using scale models in wind tunnel (see, for example, Nolan 1946, Acker 1952, Isyunov et al., 1979 & Kulkarni et al., 2005). However, these experimental studies are expensive, lengthy and time consuming.

In this paper, the exhaust smoke dispersion from a generic frigate and its interaction with rest of the superstructure are investigated numerically. The perspective view of the frigate model is shown in Figure 1. The frigate is driven by a combined diesel or gas turbine, CODOG system. The k-ε model is adopted for turbulent closure, and the

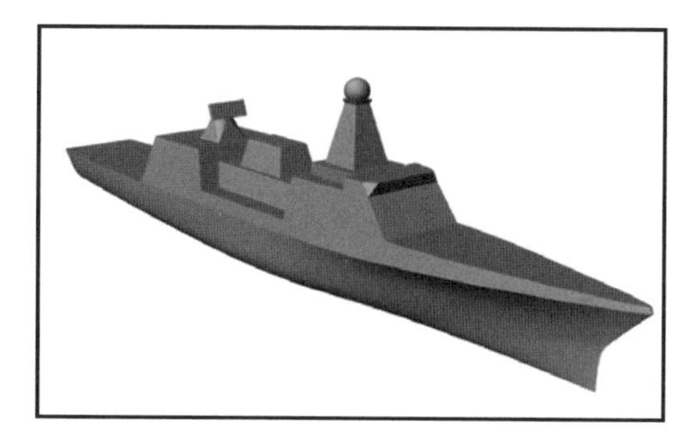

Figure 1. Perspective view of the frigate model.

governing equations in three dimensions are solved using a finite volume technique. The computations were performed for different yaw angles, efflux velocities and temperatures of the exhaust smoke. However, the location, height and shape of the stack are kept the same. The calculated streamlines, temperature contours and smoke concentrations are presented and discussed. Furthermore, the predictions are compared with the available experimental measurements. A good agreement between the predictions and experiments is obtained. The study has demonstrated that computational fluid dynamics is a powerful tool to study the problem of exhaust smoke–superstructure interaction on ships and is capable of providing a means of visualising the dispersion of the exhaust smoke under different operating conditions very early in the design spiral of a ship.

2 PHYSICAL MODEL

The exhaust smoke-superstructure interaction for a generic frigate is investigated numerically. Figure 1 shows the perspective view of the frigate model. The frigate is driven by a combined diesel or gas turbine, CODOG system. The over-all length, breadth and height of the frigate are 138 m, 18.5 m and 10 m, respectively. As seen from Figure 1, there is an APAR (Active Phased Array Radar) tower on the fore and 3-D radar on the aft part of the superstructure. A helipad is located at the back end of the superstructure. The exhaust stack is placed at about the centre of the superstructure.

3 MATHEMATICAL MODEL

The numerical analysis is based on the time-averaged equations describing the conservation of mass, momentum and energy, and because of the turbulence model, the equations governing the transport of turbulence kinetic energy k and its dissipation rate ε. It is assumed that the fluid is incompressible with constant thermophysical properties. The governing equations for three-dimensional and turbulent flow, in a Cartesian coordinate system and using tensor notation, can be written as

Continuity equation:

$$\frac{\partial}{\partial x_j}(\rho u_j) = 0 \tag{1}$$

Momentum equation:

$$\frac{\partial}{\partial x_j}(\rho u_j u_i) = \frac{\partial}{\partial x_j}\left(\mu_{eff}\frac{\partial u_i}{\partial x_j}\right) - \frac{\partial p}{\partial x_i} \tag{2}$$

Energy equation:

$$\frac{\partial}{\partial x_j}(\rho u_j T) = \frac{\partial}{\partial x_j}\left(\Gamma_{T,eff}\frac{\partial T}{\partial x_j}\right) + \rho g_i \tag{3}$$

Turbulent kinetic energy:

$$\frac{\partial}{\partial x_j}(\rho u_j k) = \frac{\partial}{\partial x_j}\left(\Gamma_{k,eff}\frac{\partial k}{\partial x_j}\right) + P - \rho\varepsilon \tag{4}$$

Dissipation rate:

$$\frac{\partial}{\partial x_j}(\rho u_j \varepsilon) = \frac{\partial}{\partial x_j}\left(\Gamma_{\varepsilon,eff}\frac{\partial \varepsilon}{\partial x_j}\right) + C_1 P\frac{\varepsilon}{k} - C_2\rho\frac{\varepsilon^2}{k} \tag{5}$$

The effective viscosity in the above equations are defined as

$$\mu_{eff} = \mu + \mu_t \tag{6}$$

Figure 2. Structure of the computational grid.

a)

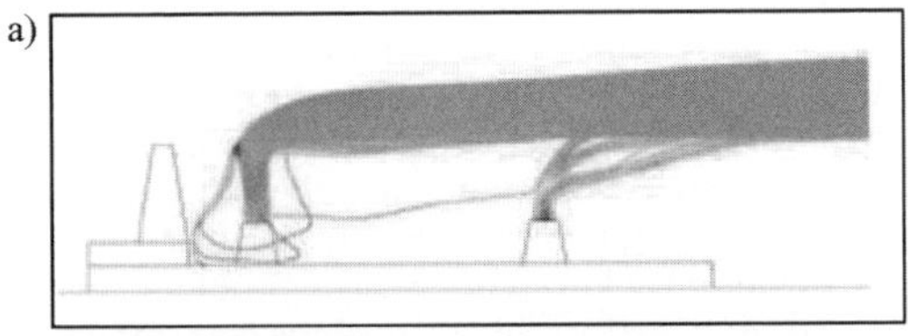

b)

Figure 3. Comparison of the calculations with experiments for K = 2. a) Calculated streamlines, b) Experimental results (Kulkarni et al., 2005).

The turbulent viscosity is obtained from

$$\mu_t = C_\mu \rho \frac{k^2}{\varepsilon} \quad (7)$$

where C_μ is a constant. In equations (3), (4) and (5), Γ is the diffusion coefficient given by μ_{eff}/σ_T, μ_{eff}/σ_k and $\mu_{eff}/\sigma_\varepsilon$, respectively. The values of the constants appearing in the above equations are $C_1 = 1.44$, $C_2 = 1.92$, $C_\mu = 0.09$, $\sigma_T = 1.0$, $\sigma_k = 1.0$, and $\sigma_\varepsilon = 1.3$. These values have been used for many forced convection studies. The term P in equations (4) and (5) represents the generation of turbulence energy given by

$$P = \mu_t \frac{\partial u_i}{\partial x_j}\left(\frac{\partial u_i}{\partial x_j} + \frac{\partial u_j}{\partial x_i}\right) \quad (8)$$

4 NUMERICAL MODEL AND BOUNDARY CONDITIONS

The finite volume method is employed to obtain the numerical solution of the governing equations with using unstructured grid. The computational domain is a box shape volume that includes ships body above waterline. The computational domain is divided into a set of tetrahedral cells. Figure 2 shows the typical structure of the grid used in the computations. The Drichlet boundary condition is applied on the inlet of the box. While on the exit boundary, the Neumann boundary condition, i.e., the derivative of the solution was applied. The computational domain was large enough to minimize the inlet boundary influence on the mixing where the model ship placed. The calculations were carried out with 714700 cells. The grid points were distributed non-uniformly with higher concentration of grids closer to the walls, see Figure 2. The effect of grid size on the results was investigated for each case.

Convective terms of momentum equations are discretized using the hybrid differencing scheme. Diffusive terms are discretized by the central differencing scheme. The derivation of pressure is based on the SIMPLE algorithm. The details of the discretisation and solution procedures are given in Malalasekara (1995), Patankar (1980), Ergin et al. (2005, 2010) and Ref. (ANYSYS 12.1, CFX),

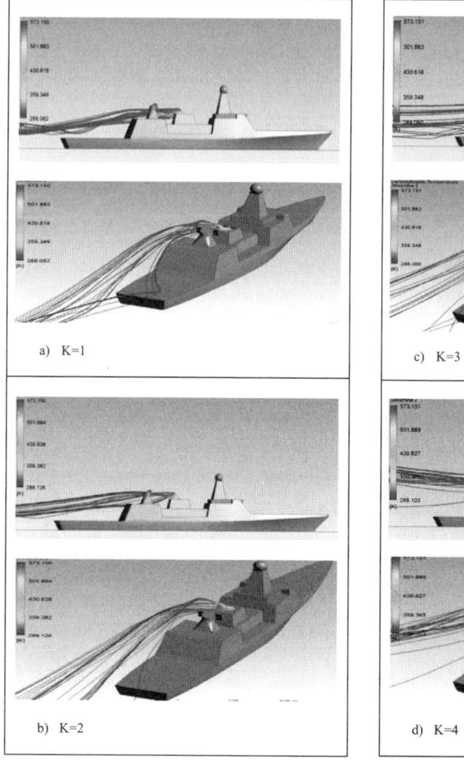

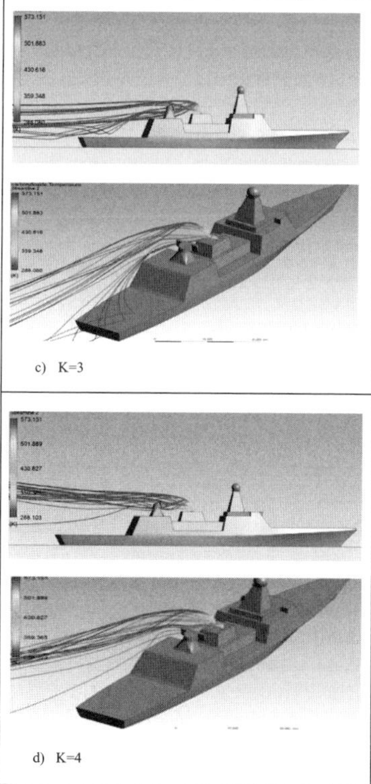

Figure 4. Streamlines and temperature distribution for the case with $T_{exh} = 300$ K, $\Psi = 10°$.

which also describes the computer code used in the present work.

In the computations, four different velocity ratios (ratio of exhaust velocity to relative wind velocity), K = 1, 2, 3 and 4 are considered. At the exit of the stack, the constant exhaust gas temperatures of T_{exh} = 15°C, 200°C, 300°C and 400°C are used as the boundary conditions. Six different yaw angles (direction of upcoming air flow), Ψ = 0°, 5°, 10°, 15°, 20° and 30° are considered. Furthermore, the cases of diesel engines on and gas turbines on are considered.

The wind tunnel experiments of Kulkarni et al. (2005) were used to validate the numerical model used in the study. As can be seen from Figure 3, the agreement between the simulations and experiments are good.

5 RESULTS AND DISCUSSIONS

The results of the numerical study are presented in Figures 4–8 for different velocity ratios (K), yaw angles (Ψ) and exhaust gas temperatures at the exit of the stack (T_{exh}).

Figure 4a-d show the effects of velocity ratio, K on the dispersion of exhaust gases for T_{exh} = 300°C and Ψ = 10°, when the main diesel engines are on. As can be seen from the streamlines in Figure 4a-c, the plume is directed towards the helicopter deck at the back end of the superstructure. Since the exhaust gas cannot rise with its momentum and buoyancy forces, and it cannot escape from the turbulent zone around superstructure. Therefore, the downwash phenomena occur. The results in Figure 4a-d show that the exhaust plume rises more as the velocity ratio increases.

Figure 5 shows the effect of yaw angle, Ψ on the dispersion of the exhaust gases. The results show that the yaw angle up to Ψ = 10°, the exhaust gases can rise (see, Figure 4) and no downwash phenomena occurs. However, when the yaw angle Ψ = 10°, the exhaust gases directed to the back part of the superstructure, and, consequently, the downwash phenomena occurs. As can be seen from Figure 4 and 5, the exhaust plume rises to higher levels while the yaw angle increases from Ψ = 10° to 30°.

Figures 6–8 show the isosurfaces of the exhaust gas concentrations for different velocity ratios (K), yaw angles (Ψ) and exhaust gas temperatures at the exit of the stack, T_{exh}. The isosurface presents% 1 concentration value of the exhaust gas given at the stack exit. The concentration values are presented in Figure 6 for the velocity ratio values, K = 1–4 with T_{exh} = 300 and Ψ = 10°. As the velocity ratio, K increases, the isosurface area increases. It means that the higher concentration values are on the

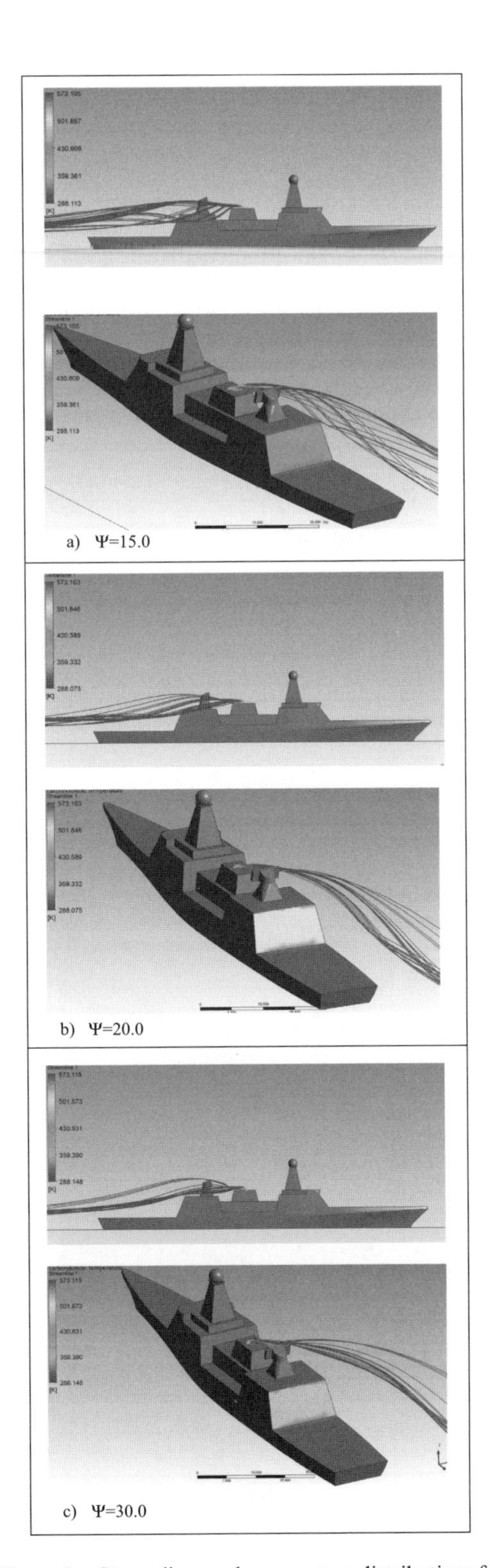

Figure 5. Streamlines and temperature distributions for the case with K = 1 and T_{exh} = 300.

back end of the superstructure including helicopter deck.

When the yaw angle increases, the area occupied by the high concentration values first increase, and then decrease (see, Figures 6 and 7).

The effect of exhaust gas temperatures at the exit of the stack on the plume rise and concentration distribution can be seen in Figure 8. The results in the Figure are presented for the yaw angle, $\Psi = 0°$ and velocity ratio K = 1. It can be seen that the plume and high concentration region rise more with increasing temperature. On the other hand, the size of the high concentration region decreases with increasing temperature.

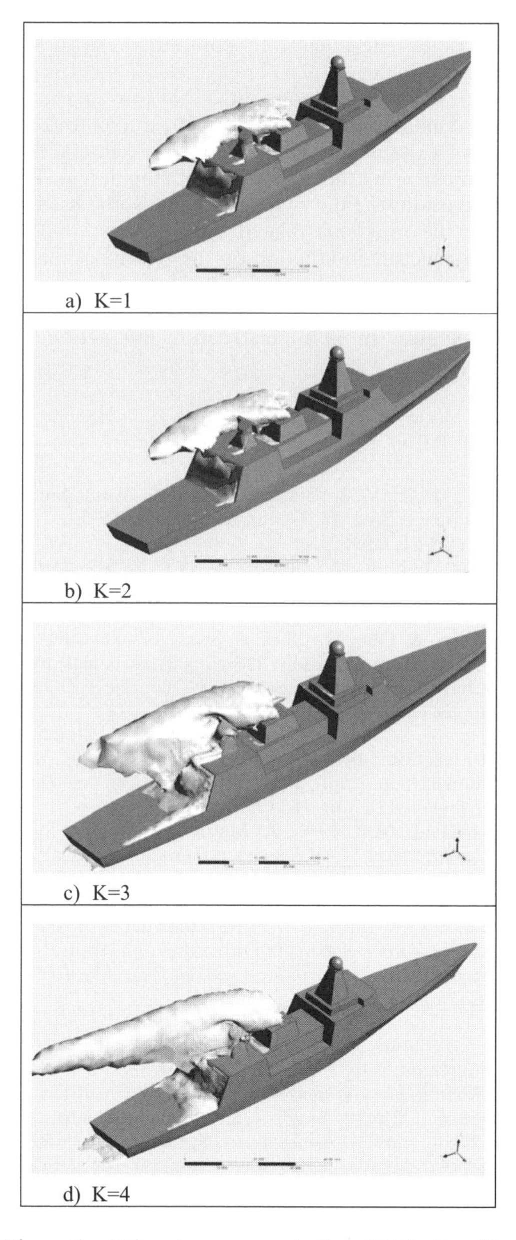

Figure 6. Exhaust gas concentration distribution (the isosurface presents % 1 concentration of the plume at the stack outlet) with $T_{exh} = 300$ and $\Psi = 10°$.

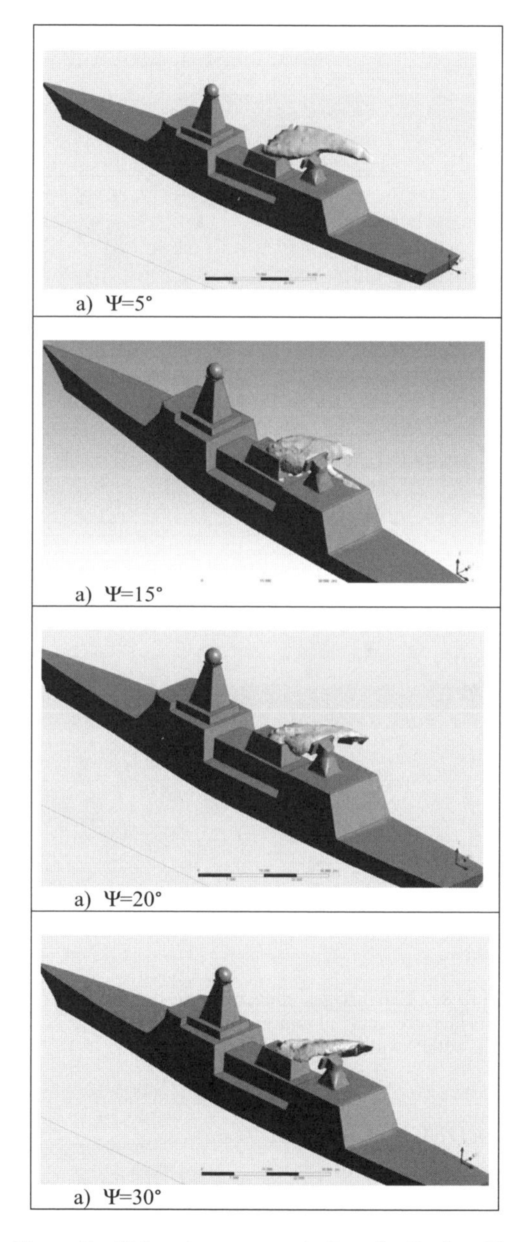

Figure 7. Exhaust gas concentration distribution (the isosurface presents % 1 concentration of the plume at the stack outlet) with $T_{exh} = 300$ and K = 1.

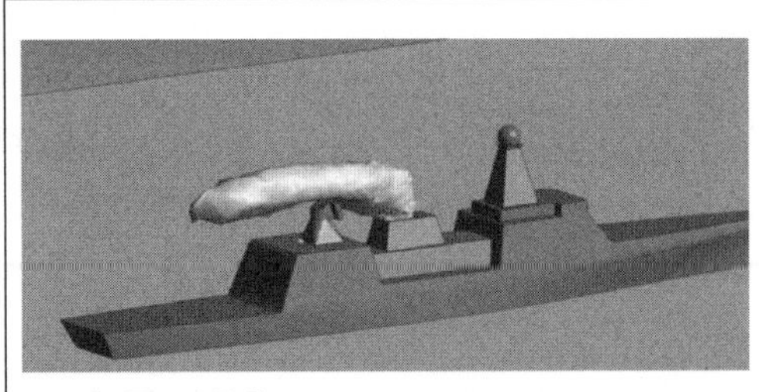

a) T = 15°C

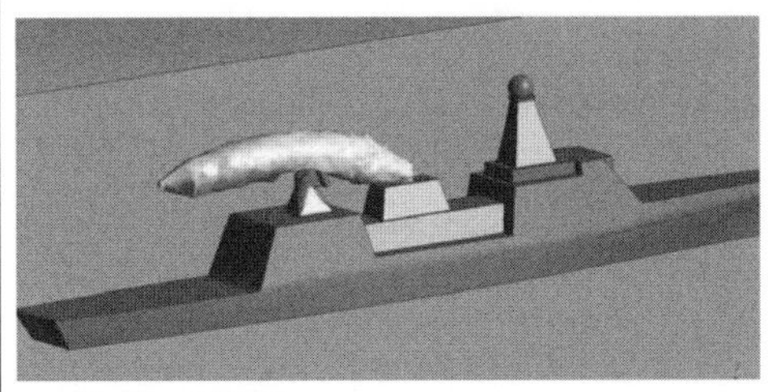

b) T = 200°C

c) T = 300°C

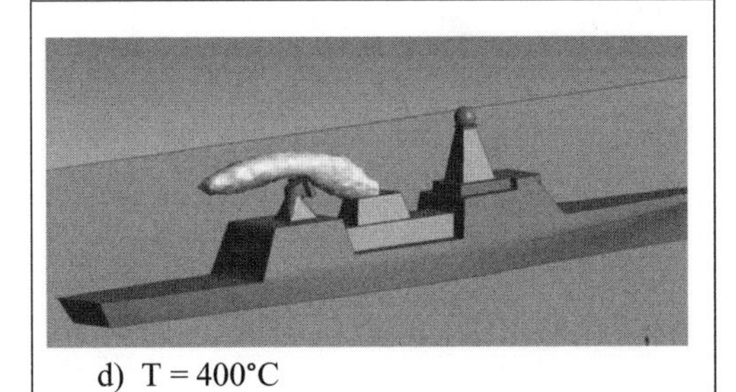

d) T = 400°C

Figure 8. Exhaust gas concentration distribution (the isosurface presents % 1 concentration of the plume at the stack outlet) with $\Psi = 0$ and $K = 1$.

6 CONCLUSIONS

The exhaust smoke-superstructure interaction for a generic frigate is investigated through several calculations. The conservation of mass and momentum equations have been solved numerically with k-ε turbulence model using finite volume method. It is showed that the dispersion of smoke is affected by efflux velocity, temperature, and turbulence, wind velocity and direction and geometry of the superstructure. The results show that the yaw angle larger than $\Psi = 10°$ and the velocity ratio $K = 2$ should be maintained to avoid downwash phenomena. It is also found that the effect of the buoyancy forces on the plume rise is less significant in comparison with the momentum.

The study has demonstrated that computational fluid dynamics is a powerful tool to study the problem of exhaust smoke–superstructure interaction on ships during the early stages of ship design. The computational fluid dynamics can also be used to solve the problems related with exhaust smoke-superstructure interaction on the already available ships.

The study provides a further understanding of the exhaust smoke-superstructure interaction for naval ships.

REFERENCES

Acker, H.C. 1952. Stack Design to Avoid Smoke Nuisance, *SNAME Transactions*, 60, 566–593.

ANSYS 12.1, CFX.

Baham, G.J. & McCallum, D. 1977. Stack Design Technology For Naval and Merchant Ships, *SNAME Transactions,* Vol. 85, 324–349.

Ergin, S. & Ota, M. 2005. A Study of the Effect of Duct Width on Fully Developed Turbulent Flow Characteristics in a Corrugated Duct, *Heat Transfer Engineering*, Vol. 26, No.2, s. 54–62.

Ergin, S. & Paralı, Y. 2010. Numerical Study of Interaction between the Exhaust Gases and Superstructure of a Naval Ship, *Gemi ve Deniz Teknolojisi Dergisi*, 185 (in Turkish).

Fitzegerald, M.P. 1986. A Method to Predict Stack Performance, *Naval Engineers Journal*, 5–41.

Heywood, J.B. 1988. *Internal Combustion Engine Fundamentals*, McGraw-Hill.

Isyumov, N. & Tanaka, H. 1979. Wind-tunnel Modelling of Stack Gas Dispersion Difficulties and Approximations, *Proc. of the fifth Int. Conf. on Wind Engineering*, Colorado, USA.

Jin, E., Yoon, J. & Kim, Y. 2001. A CFD Based Parametric Study on the Smoke Behavior of a Typical Merchant Ship, *Practical Design of Ships and Other Floating Structures*, 459–465.

Kulkarni, P.R., Singh, S.N. & Seshadri, V. 2005. Flow Visualisation Studies of Exhaust Smoke–Superstructure Interaction on Naval Ships, *Naval Eng J*, ASNE, 117, 41–56.

Kulkarni, P.R., Singh, S.N. & Seshadri, V. 2007. Parametric Studies of Exhaust Smoke-Superstructure

Interaction on a Naval Ship Using CFD, *Computers & Fluids* 36, 794–816.

Malalasekera, W. & Versteeg, H.K. 1995. *An Introduction to Computational Fluid Dynamics the Finite Volume Method*, Longman Scientific and Technical.

Mishra, D.P. & Dash, S.K. 2010. Numerical Investigation of Air Suction Through the Louvers of a Funnel due to High Velocity Air Jets, *Computers & Fluids*, 39, 1597–1608.

Nolan, R.W. 1946. Design of Stacks to Minimize Smoke Nuisance. *Trans SNAME*, 54, 42–84.

Park, S., Heo, J., Yu, B.S., Rhee & S.H. 2011. Computational analysis of ship's exhaust-gas flow and its application for antenna location, *Applied Thermal Engineering*, doi:10.1016/j.applthermaleng. 2011.02.011 (in press).

Patankar, S.V. *Numerical Heat Transfer and Fluid Flow*, McGraw-Hill, New York, 1980.

Syms, G.F. 2008. Simulation of simplified-frigate airwakes using lattice-Boltzman Method, Journal of Wind Engineering and Industrial Aerodynamics, 96, 1197–1206.

Sustainable Maritime Transportation and Exploitation of Sea Resources – Rizzuto & Guedes Soares (eds)
© 2012 Taylor & Francis Group, London, ISBN 978-0-415-62081-9

Numerical analysis of the shadow effect of an LNG floating platform on an LNG carrier under wind conditions

A.D. Wnęk & C. Guedes Soares
Centre for Marine Technology and Engineering (CENTEC), Instituto Superior Técnico, Technical University of Lisbon, Lisbon, Portugal

ABSTRACT: The analysis of aerodynamic forces acting on a floating LNG platform and an LNG carrier is the main concern of this paper. The investigation focuses on an important issue in the approaching maneuver of the ship to the platform, which is the effect of platform shadow on the vessel during operation under wind conditions. Two particular cases of models position have been taken into consideration, LNG arrival and departure from the floating platform, in which the wind attacks prior the platform. Results have been obtained using commercial CFD code and compared with experimental measurements performed in a wind tunnel. Numerical and experimental results, presented in a form of coefficients of the drag, lift components and yaw moment, reached approximate agreement.

1 INTRODUCTION

The effect of wind acting on ships can become critical in several situations of close proximity maneuvers of the ships. As one of the environmental loads, wind can vary randomly in direction and velocity, which can decrease the maneuvering capability of the ship and effectively affect her planned maneuvers. One of the examples, in which the wind becomes an important factor, is the maneuvering situation at ship arrival and departure from another vessel because in those situations the speed is very small and thus the effect of rudder is limited and wind forces can become relatively large. The example is the carrier with liquefied natural gas approaching floating platform which is presented in this work. The effect of platform's shadow on an LNG under wind conditions can become crucial and this is the main interest of this work.

The situation is similar to the case of shuttle tankers approaching an FPSO, which was considered recently by Tannuri et al. (2010).

Wind forces acting on an LNG carrier have already been analyzed by Wnęk et al. (2010) using the commercial RANS code Ansys CFX. The ship model was rotated at full incidence of 360° with step of 10 degrees. Results in the form of coefficients of the drag, lift force components and yaw moment were compared with experimental measurements giving approximate agreement.

Another model, which is a floating platform, has already been studied too (Wnęk et al., 2009). Using the same commercial code, the model was analyzed in 19 different angles of wind attack. According to the symmetry of the hull in relation to its Y axis, the incidence was decreased to 180°. Results were compared with experimental tests and with two simplified methods of Isherwood (1972), who presented coefficients expressions that are coming from multiple regression analysis of the previous published experimental results and Blendermann (1996), who used the collection of wind load data for coefficients expressions.

These simplified methods were also used by Haddara & Guedes Soares (1999) together with other two methods (OCIMF, Gould) in an investigation of wind loads as an important factor of ships maneuverability.

The paper presents the set of two models (a floating platform and an LNG carrier) and the effect of platform shadow on an LNG under wind conditions in particular relative positions using the CFD tool and wind tunnel. Two main cases are investigated: wind loads on the LNG at arrival and departure from the floating platform. Each case is considered for two different angles of LNG. Results in the form of dimensionless force components of the drag, lift and yaw moment are compared with measurements from the wind tunnel.

2 EXPERIMENTAL MEASUREMENTS

Experimental tests have been performed in open jet wind tunnel with rectangular cross section of measurement zone, which was 3 meters in length, 2 m in width and 1.3 m in height. Two ship models (Fig. 1) made of wood in a scale of 1:400 and with characteristics given in Table 1 were used for measurements. A strain gauge load cell was used to

Figure 1. Physical models of floating platform and LNG carrier.

Table 1. Models' characteristics.

Model →		Platform	LNG carrier
Characteristics	↓	m, m²	m, m²
Length overall	L_{OA}	0.916	0.725
Breadth	B	0.15	0.115
Frontal projected wind area	A_F	0.00683	0.01051
Lateral projected wind area	A_L	0.05275	0.05011

measure the forces and moment of the LNG carrier and was set right under the model. The maximum wind speed used in the tests was 15 m/s. In order to decrease uncertainty of results, experimental tests have been performed three times for each case of models position.

Several particular cases of models position took place in the experimental measurements, in which the LNG carrier was fully or at least partly located in the platform shadow. The LNG model was mounted to the load cell and was rotating on its axis Z so as to simulate an arrival or departure from the floating platform. Wind was acting on the side of LNG, which was always shielded by the floating platform. The platform was moving forward and backward in order to expose the part of LNG carrier, the detailed description of the models distance is given in the next section of this paper.

The measurements of the wind loads on LNG carrier are presented in the form of coefficients of the drag, lift force components and yaw moment:

$$c_x = \frac{F_x}{\frac{1}{2}\rho A_F U^2} \tag{1}$$

$$c_y = \frac{F_y}{\frac{1}{2}\rho A_L U^2} \tag{2}$$

$$c_N = \frac{N}{\frac{1}{2}\rho A_L L_{BP} U^2} \tag{3}$$

where F_x, F_y (parallel and perpendicular to LNG's lateral area) are the force components of the drag, lift measured by the strain gauge load cell; and N is the yaw moment of LNG model with positive down Z axis. Air density ρ is equal to 1.2 kg/m³ and A_F, A_L, L_{BP} are the models' characteristics of the LNG (Table 1).

3 NUMERICAL ANALYSIS

3.1 *Methodology*

Virtual models of the floating platform and the LNG carrier (Fig. 2) have been created in Rhinoceros 4.0 software and they are an identical to the physical models used in the wind tunnel (Table 1).

Computational domain in the form of rectangle with dimensions: $l = 7\ L_{Plat}$, $b = 7\ L_{Plat}$, $h = 1.5\ L_{Plat}$ have been generated in Ansys ICEM CFD software and discretised by a mesh of 6,200,000 tetrahedral elements (Fig. 3). Ten prism layers have been created around the LNG carrier where the dimensionless wall distance y+ was close to 1.

Computations have been performed in RANS code Ansys CFX-13 using 8 CPUs, Intel® Xeon® CPU E5420 @2.50 GHz, 16GB of RAM. Shear stress transport SST viscous model was used to predict the flow around models. At the inlet wall, the uniform wind velocity of U = 10 m/s, V = 0 m/s, W = 0 m/s was distributed, as taken from the experimental tests, zero pressure at the outlet, free

Figure 2. Virtual models of floating platform and LNG carrier.

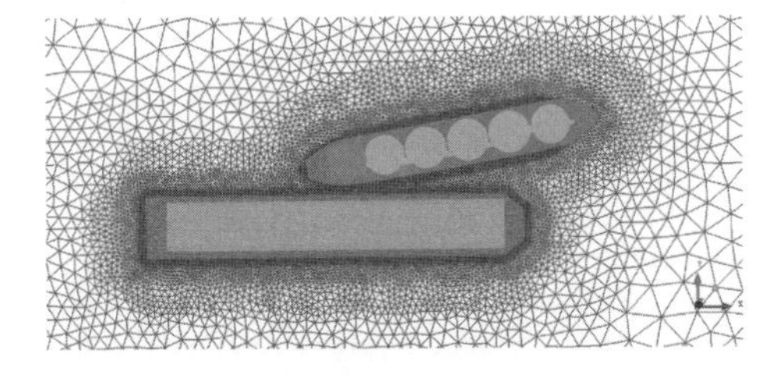

Figure 3. Tetrahedral mesh.

slip condition on the bottom and the side walls, no slip condition on model surfaces has been implemented and top was treated as a symmetry plane.

3.2 *Measurements*

Four cases of models position were taken into consideration:

- LNG approaching floating platform at the angle of 10°—ARRIVAL 10°
- LNG departing from the floating platform at the angle of 10°—DEPARTURE 10°
- LNG approaching floating platform at the angle of 20°—ARRIVAL 20°
- LNG departing from the floating platform at the angle of 20°—DEPARTURE 20°

In each case, the floating platform was exposed directly to the wind forces. The LNG carrier was located behind the platform at varying angles of its location. The platform was also moving forward or backward making the same two particular distances between models amidships x_{mid} = 0.2 m, 0.28 m (Fig. 4). The minimum distance between sides of the models was 0.025 m.

In order to obtain the most precise results, one particular case of models position which is arrival at 10°, x_{mid} = 0 m has been studied several times. The sensitivity of results to the code settings is presented in Table 2.

Simulation has been performed in two states: steady and transient. Two most universal turbulence models (SST, k-e) have been implemented. Turbulence has been considered at two intensities: low (1%) and high (10%). Prediction of the wind flow has been made using mesh without prism layers around LNG model and with 6 and 10 prism layers. Results closest to the experimental measurements were obtained using steady state simulation, SST model, low turbulence intensity and 10 prism layers around LNG model.

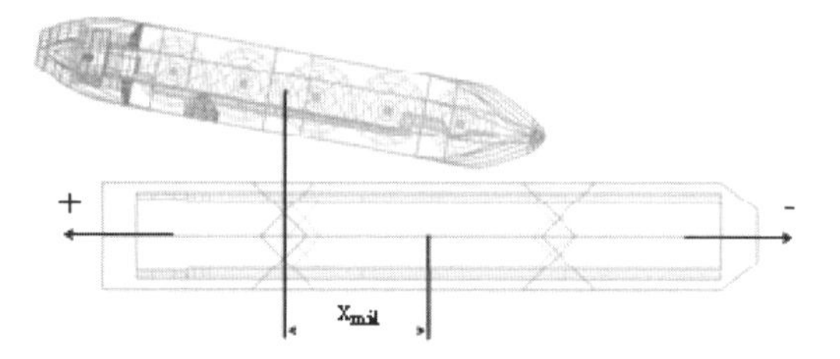

Figure 4. LNG approaching platform at the angle 10°, x_{mid} = 0.2 m.

Table 2. Results sensitivity to the code settings.

State	Model	Intensity	Prism	c_x	c_y	c_N
Steady	SST	1%	6 lay	−0.01	0.09	0.02
Steady	SST	10%	6 lay	0.025	0.1	0.009
Transient	SST	1%	6 lay	0.03	0.1	0.01
Steady	SST	1%	–	0.03	0.1	−0.005
Steady	k-ε	1%	6 lay	−0.1	0.08	0.01
Steady	k-ε	1%	10 lay	−0.03	−0.16	0.005
Steady	SST	1%	10 lay	−0.06	−0.2	0.003
EXP				−0.1	−0.32	0.007

4 RESULTS

Results are presented in the form of coefficients of the drag, lift force components and yaw moment: longitudinal force coefficient c_x, lateral force coefficient c_y and yaw moment coefficient c_N. Coefficients are related to the wind loads on the LNG carrier only. The floating platform was directly exposed to the wind forces, what made the shadow on the LNG and caused smaller wind loads (drag and lift) on the model (Figs. 5, 6).

Figure 7 presents the longitudinal force coefficient c_x of the LNG carrier in two cases: arrival and departure at the angle of 10°. For each case three positions of LNG (x_{mid} = 0, 0.2, 0.28 m) have been considered. The biggest force occurred in departure case, when the fore part of the ship was

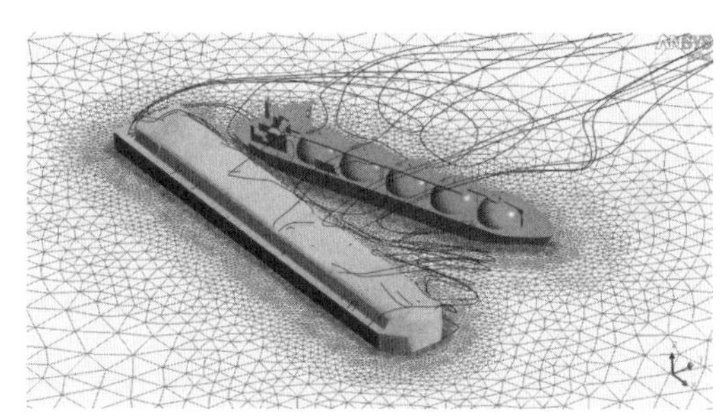

Figure 5. Streamlines and pressure on LNG and floating platform/LNG arrival at the angle 20°, x_{mid} = 0 m.

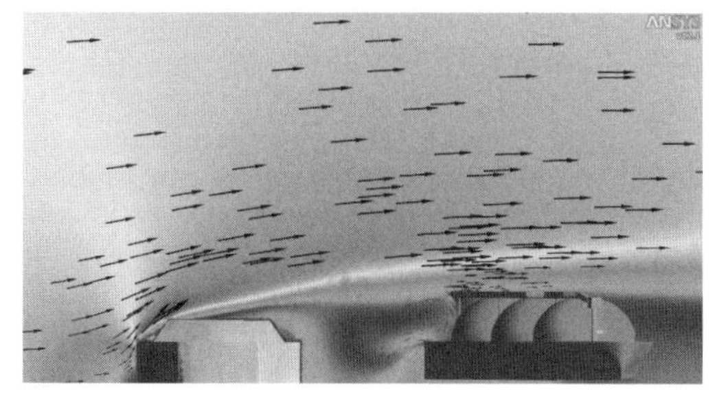

Figure 6. Wind forces acting on LNG and floating platform/LNG arrival at the angle 20°, x_{mid} = 0 m.

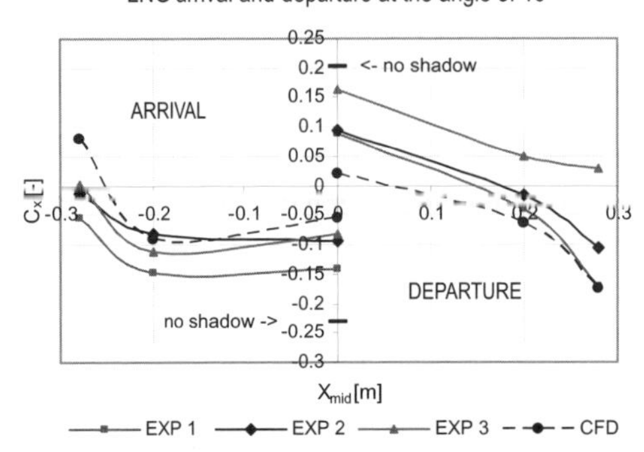

Figure 7. Longitudinal force coefficient for LNG carrier / arrival and departure at 10°, $x_{mid} = 0, 0.2, 0.28$ m.

partly exposed to the wind forces ($x_{mid} = 0.28$ m). However one can observe significant discrepancy between all results. Reliability of experimental measurements (EXP 1–3) is not constant, what makes CFD and EXP results difficult to compare. The points marked by "no shadow" indicate the force coefficients for LNG carrier without the platform shadow.

Figure 8 presents the lateral force coefficient c_y at the same angle of wind attack. In this case, the forces increase almost linearly with the growth of the distance between amidships x_{mid}.

One can observe the similarity in magnitude of the forces between arrival and departure. Significant discrepancy between CFD and experimental results could be caused by numerical errors. Although, considering small forces on the LNG model, instrumental error of the load cell could also be the reason of results variance.

Figure 9 shows yaw moment coefficient c_N (LNG arrival, departure at 10°).

Very good agreement between numerical and experimental measurements has been observed. Small discrepancy occurred only in the departure case at zero distance of the midships. Moment grows similarly in magnitude in both cases: arrival and departure. One can also observe significant growth of the moment on the LNG when it is only partly exposed to the wind.

Figures 10–12 present force coefficients for the LNG at the angle of 20°. Longitudinal forces are almost two times bigger than in the case of LNG at the angle of 10°. Numerical and experimental results agree well except one significant discrepancy at LNG arrival, where $x_{mid} = 0$ m. Similar to previous case (angle of 10°) three experimental measurements are different from each other. In this case the difference is smaller, due to bigger distance between ships (angle of 20°) and more exposed LNG model to the wind forces.

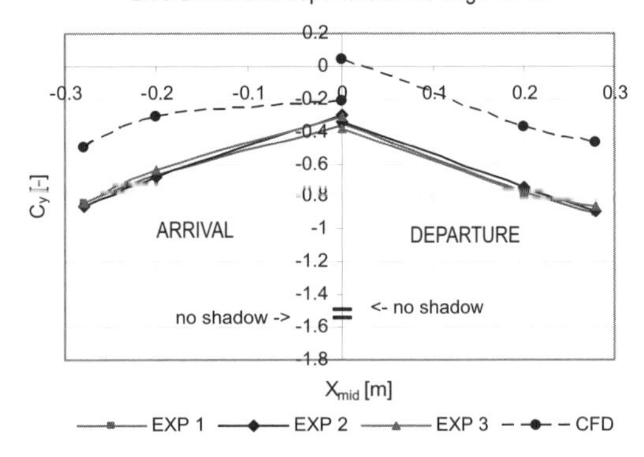

Figure 8. Lateral force coefficient for LNG carrier/ arrival and departure at 10°, $x_{mid} = 0, 0.2, 0.28$ m.

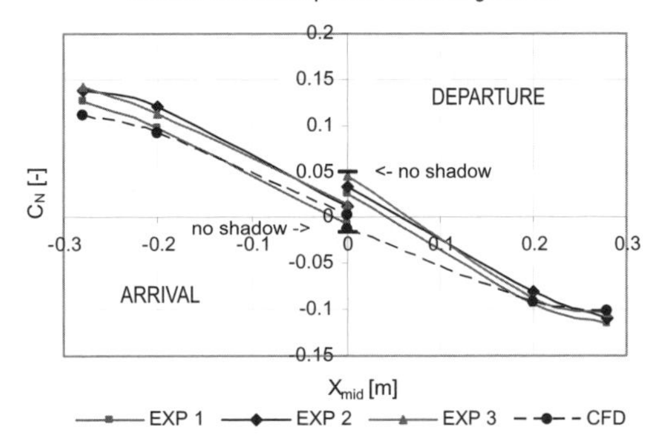

Figure 9. Yaw moment coefficient for LNG carrier/ arrival and departure at 10°, $x_{mid} = 0, 0.2, 0.28$ m.

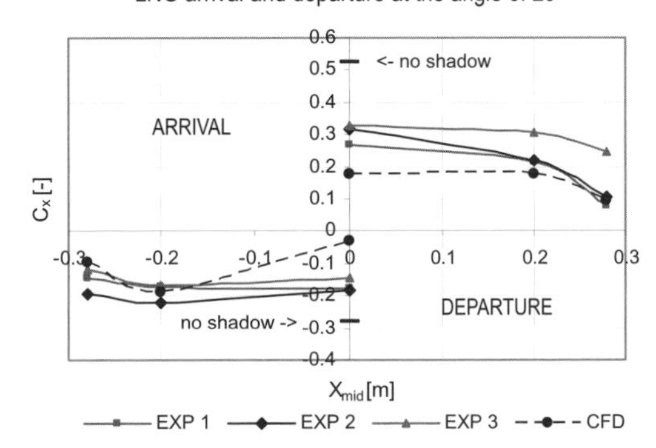

Figure 10. Longitudinal force coefficient for LNG carrier/arrival and departure at 20°, $x_{mid} = 0, 0.2, 0.28$ m.

Significant divergence is observed for the lateral force coefficients as was observed in previous case for angle of 10°. CFD calculations underpredict experimental measurements by about 50%. Both

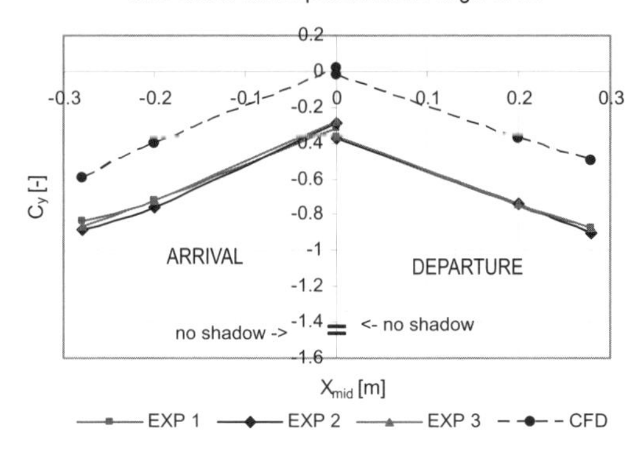

Figure 11. Lateral force coefficient for LNG carrier/ arrival and departure at 20°, x_{mid} = 0, 0.2, 0.28 m.

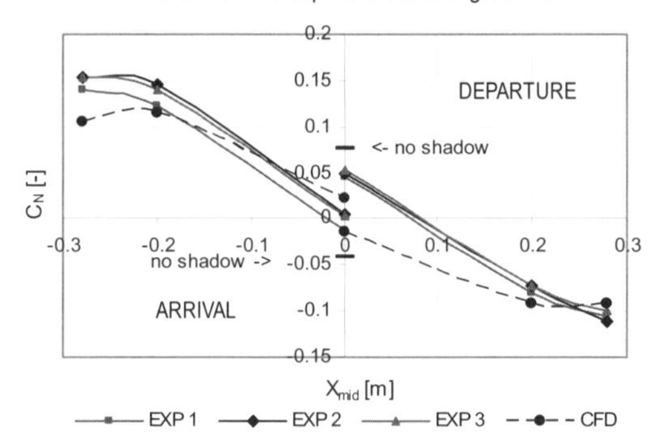

Figure 12. Yaw moment coefficient for LNG carrier/ arrival and departure at 20°, x_{mid} = 0, 0.2, 0.28 m.

results (CFD, EXP) increase linearly with the growth of x_{mid}.

Another interesting observation is the ratio E/C of experimental measurements to the numerical results in both cases (angle of 10° and 20°). Table 3 presents the ratio between lateral force coefficients for the LNG obtained experimentally and numerically for two positions (x_{mid} = 0.2 m, 0.28 m) of the models. The negative distance between midships determines the LNG arrival and the positive is the LNG departure.

The average ratio of EXP to CFD lateral force coefficients is 2, which means that the experimental measurements are about two times bigger than the forces obtained by CFD.

In the case of yaw moment coefficients, numerical results and measurements from the wind tunnel are in very good agreement. The biggest discrepancy occurred at the LNG departure for x_{min} = 0 m, as it was already observed in the previous case (angle of 10°). The significant growth of the moment occurred also in the case of wind loads on the LNG with platform shadow and x_{mid} = 0.2, 0.28 m at arrival and departure.

Table 3. The ratio E/C between lateral force coefficients obtained in the wind tunnel and CFD code.

x_{mid}	deg	EXP*	CFD*	E/C	deg	EXP*	CFD*	E/C
0.28	10	−0.84	−0.5	1.7	20	−0.86	−0.59	1.5
−0.2	10	−0.66	−0.31	2.1	20	−0.73	−0.4	1.8
0.2	10	−0.76	−0.38	2.0	20	−0.74	−0.38	2.0
0.28	10	−0.88	−0.47	1.9	20	−0.88	−0.5	1.8

* Lateral force coefficient c_y.

5 CONCLUSIONS

This work has presented an analysis of wind loads on an LNG carrier located in a shadow of floating platform under wind conditions. Using a commercial CFD code and the wind tunnel, two main cases of relative positions of the LNG have been analyzed: arrival and departure of the LNG carrier from the floating platform. Wind force coefficients of the LNG carrier at angles of 10° and 20° have been presented.

In general a good agreement between numerical and experimental measurements has been observed. In some cases results almost perfectly match each other as it is mainly observed for yaw moment coefficients. The biggest discrepancy occurred for lateral forces, where CFD underpredicts by about 50% the experimental results. In this investigation, the LNG carrier is not directly exposed to the wind forces and the wind loads on the model are generally small, as compared to the LNG fully exposed to the wind without the platform shadow. The discrepancies between the three repeated experimental measurements of longitudinal forces can suggest some instrumental errors of the load cell, justifying some differences with respect to the numerical results.

The biggest difference between the results for the angles of 10° and 20° has been observed in the magnitude of the longitudinal forces, in the latter case, forces are almost two times bigger.

ACKNOWLEDGEMENTS

This paper has been prepared within the project "SAFEOFFLOAD—Safe Offloading from Floating LNG Platforms", which has been funded by the European Union through the GROWTH program under contract TST4-CT-2005-012560.

The first author has been financed by The Portuguese Foundation for Science and Technology (*Fundação para a Ciência e a Tecnologia*) under contract number SFRH/BD/67070/2009.

REFERENCES

Blendermann, W. 1996. *Wind loading of ships—Collected data from wind tunnel tests in uniform flow.* Institut fur Schiffbau der Universitat Hamburg, Bericht No. 574, p. 62.

Haddara, M.R. & Guedes Soares, C. 1999. Wind loads on marine structures. *Marine Structures 12*, pp. 199–209.

Isherwood, R.M. 1972. Wind resistance of merchant ships. *Trans. of RINA*, 114:327–38.

Tannuri, E.A., Fucatu, C.H., Rossin, B.D., Montagnini, R.C.B. & Ferreira, M.D. 2010. Wind shielding effects on DP system of a shuttle tanker, *Proceedings of the 29th International Conference on Ocean, Offshore and Arctic Engineering*, OMAE2010, pp. 143–149.

Wnęk, A.D., Paço, A., Zhou, X.-Q. & Guedes Soares, C. 2009. Numerical and experimental analysis of the wind forces acting on a floating LNG platform, *Proceedings of the 13th International Congress*, IMAM2009, pp. 697–702.

Wnęk, A.D., Paço, A., Zhou, X.-Q. & Guedes Soares, C. 2010. Numerical and experimental analysis of the wind forces acting on LNG carrier, *Proceedings of the 5th European Conference on Computational Fluid Dynamics*, ECCOMAS CFD2010.

2.2 *Dynamic stability*

Sustainable Maritime Transportation and Exploitation of Sea Resources – Rizzuto & Guedes Soares (eds)
© 2012 Taylor & Francis Group, London, ISBN 978-0-415-62081-9

Dynamic instabilities in following seas caused by parametric rolling of C11 class containership

A. Turk & J. Prpić-Oršić
Department of Naval Architecture and Ocean Engineering, Faculty of Engineering, Rijeka, Croatia

S. Ribeiro e Silva & C. Guedes Soares
Centre for Marine Technology and Engineering (CENTEC), Technical University of Lisbon, Instituto Superior Técnico, Lisboa, Portugal

ABSTRACT: Parametric roll resonance is known as one of dangerous modes of ship motions in waves. This resonance occurs as a result of roll restoring energy variations in astern or head seas. Post-Panamax container ships seem to be particularly susceptible to parametric rolling because the vessels tend to feature wide beam and large bow flares in order to carry more containers on deck while at the same time minimizing the resistance and improve propeller performance with the streamlined underwater hull. This paper focuses on the parametric resonance observed in following seas. A 6 degree-of-freedom model has been used to explain the parametric roll resonance of the ship taking into account coupling between roll and vertical motions. Parametric rolling resonance in following waves was also investigated experimentally. Several experiments using a scale model of a post-Panamax containership were carried out in regular following waves at the CEHIPAR basin in Spain. In these experiments the wavelength, the wave height, the model speed and the encounter angle were widely varied to clarify overall property of parametric rolling resonance in following waves. As a result, conditions in which the parametric rolling resonance is likely to occur were determined. Also outlined in the paper are recommendations for additional research needed to better understand the influence of vessel design and operational considerations on the propensity of post-Panamax containerships towards parametric rolling.

1 INTRODUCTION

Following the disaster of APL China casualty in October 1998, attention on parametric rolling in head seas came to the forefront of research in academic (France et al., 2003, Shin et al., 2004), insurer (Roenbeck, 2003), and regulatory settings (ABS, 2004; Sweden, 2004; ITTC, 2006). Although naval architects have been aware of the problem mathematically for decades, the development of the so-called parametric resonance became an area that was starting to receive wider attention. As early as 2000, Ribeiro e Silva and Guedes Soares demonstrated that both linearised and nonlinear theories could be used to predict parametric rolling in regular head waves. The investigation of this casualty included theoretical computations of *GZ* variations, model experiments and numerical simulations as well as meteorological studies of the wind and sea conditions prevailing at the time.

In regular waves the authors confirmed a set of conditions for the occurrence of parametric rolling, namely: wave encounter frequency nearly twice the roll natural frequency, ship length of the same order as wave length, roll damping (which is speed dependent) below a certain threshold and wave height above a certain threshold.

Levadou and Gaillarde (2003) followed with numerical simulations for the C11 containership, using FREDYN, by investigating a range of speeds, headings (head and bow quartering seas), significant wave heights and load conditions. Their investigations showed that high sustained speeds reduce the risk of parametric rolling. Following seas were not investigated, but the authors noted the possibility of parametric rolling occurrence for such cases at low or zero speed.

Even though majority of the studies are presently concentrated on the head wave parametric rolling the phenomena was initially thought to be a phenomenon mainly limited to the following seas condition (Spyrou, 2000) and of significance for smaller, high-speed displacement vessels such as some fishing boats (Perez and Sanguinetti, 1995), and seagoing tugs.

The 6 DOF model has been utilized to check the inception of the parametric rolling for C11 class containership. By applying this "partly-nonlinear" time domain model with great computational efficiency, an analysis of parametric rolling in following seas was preformed paying attention to the validity of the numerical results and the consequences of the validity with respect to experiments in regular and polychromatic waves on a containership, sought as the initial step towards derivation of optimal experimental/numerical procedures for safety assessment in a realistic sea. Important parameters that are effective in roll resonance are pointed out. For this purpose, a study is conducted to determine susceptibility and severity criteria for unfavorable sailing conditions such as heading and speed, which directly depend on the environmental conditions. Numerical details of the procedure have been worked out and provided as well.

2 ASSESSMENT MODEL FOR PARAMETRIC ROLLING ON SHIPS

2.1 *Susceptibility criteria*

Following the structure proposed by ABS (2004) on longitudinal (following) waves, the analysis criteria is applied to determine if this particular vessel is vulnerable to parametric rolling (susceptibility criteria) for the attainable ship speed and heading. The most comprehensive mathematical method to assess susceptibility to parametric rolling is the solution of the Mathieu equation.

It enables us to discuss parametric roll resonance within the Mathieu equation, which then characterizes the rolling motion of a ship in longitudinal waves. Depending on the values of the coefficients, this equation yields a stable or unstable solution. It is well-known that the directly excited pitch can create growth of roll, which will influence back pitch and so forth.

To check if parametric resonance is possible, standard roll equation, must be transformed to the Mathieu type varying restoring coefficient of the form,

$$\frac{d^2\eta_4}{d\tau^2} + (p + q\cos\tau)\eta_4 = 0 \qquad (1)$$

where η_4 denotes roll amplitude, p is a function of the ratio of forcing and natural frequency, q the parameter that dictates the amplitude of parametric excitation (reflects the level of GM change in waves), and τ represents nondimensional time. The solution of the Mathieu equation can establish certain boundaries and if for instance (p, q) lie in an unstable region, an arbitrarily small initial disturbance will trigger an oscillatory motion that tends to increase indefinitely with time.

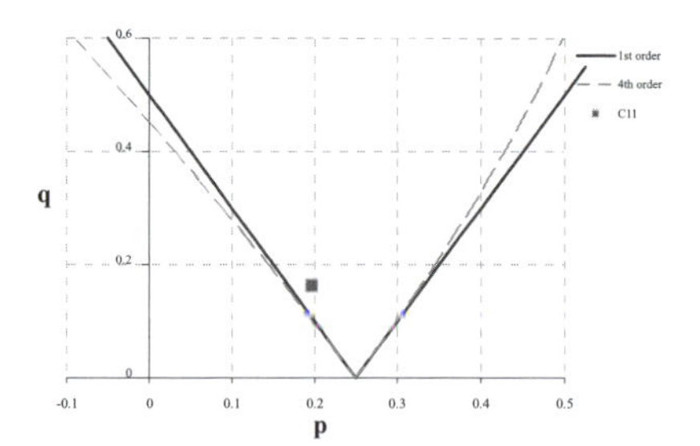

Figure 1. Basic verification of Susceptibility Criteria (wave amplitude 4 m, frequency 0.542 s^{-1}).

The loading condition adopted corresponds to a displacement of 76,056 t and a position of the centre of gravity of the vessel and its cargo which leads to a transverse metacentric height (GM_T) of 1.973 m (in still water). As shown in Figure 1, Ince-Strutt diagram for Mathieu's equation falls into zone of instability for the operational conditions in the principal parametric resonance which starts precisely at p = 0.25 and corresponds to a natural period exactly twice the period of parametric excitation.

2.2 *Prediction of amplitude of parametric rolling in regular waves (severity criteria)*

If the susceptibility to parametric rolling has been determined, the application of an arbitrary numerical procedure with different possible analytical methods has to be applied; see (Bulian et al., 2003, Umeda et al., 2004, Neves et al., 2003, Ribeiro e Silva et al., 2005, Ribeiro e Silva and Guedes Soares, 2009).

A numerical procedure (Ribeiro e Silva and Guedes Soares, 2009) is applied for the assessment of parametric rolling resonance for the following wave's conditions while evaluating the proneness of a C11 post-Panamax container carrier.

In terms of numerical simulations one begins by obtaining the solution in the frequency domain. The numerical code is based on the transfer of frequency domain data to time domain and the inclusion of so called Froude-Krylov nonlinear part of the loading in time domain. Considering the overall set of computations, this is a hybrid method, where some of these calculations are performed in the frequency domain and partly in the time domain. This procedure transfers diffraction and radiation forces, supposed linear, from frequency to time domain. The nonlinear Froude-Krylov and hydrostatic forces are computed considering the actual hull wetted surface at each time step.

With the model adopted for this case study, radiation and wave excitation forces are calculated at the equilibrium waterline using a standard strip theory (Salvesen et al., 1970), as the Frank's close fit method is used for computing the two-dimensional frequency-dependent coefficients of added mass and damping (Frank et al., 1975), while the Haskind-Newman relations are utilised to evaluate sectional diffraction forces.

The governing equations can be written in the abbreviated form, such as,

$$\sum_{J=1}^{6}\left\{\left(M_{kj}+A_{kj}\right)\ddot{\eta}_j+B_{kj}\dot{\eta}_j+C_{kj}(t)\eta_j\right\}=F_k e^{i\omega_e t} \quad (2)$$

combining all hydrodynamic forces with the mass forces, using subscript notations, making it is possible to obtain six linear coupled differential equations of motion where the subscripts k, j are associated with forces in the k -direction due to motions in the j-mode (k = 1, 2, 3 represent the surge, sway and heave directions, and 4, 5, 6 represent roll, pitch and yaw directions). M_{kj} are the components of the mass matrix for the ship, A_{kj} and B_{kj} are the added mass and damping coefficients, $C_{kj}(t)$ are the hydrostatic (time dependent) restoring coefficients, and F_k are the amplitudes of the exciting forces, where the forces are given by the real part of $F_k e^{i\omega_e t}$. Encounter frequency is denoted by ω_e.

After each instantaneous wetted cross section is set, the underwater part of the hull with its geometric, hydrostatic and hydrodynamic properties represented as follows,

$$\Phi = \Phi_U + \Phi_{I\,+}\, \Phi_D + \Phi_R \quad (3)$$

enables obtaining the coefficients of restoring hydrostatic forces, hydrodynamic radiation forces and excitation forces denoting Φ_U as the potential of the steady motion in still water, Φ_I as the incident wave potential, Φ_D as the diffraction potential while Φ_R represents the forced motion potential.

Owing to the complex integro-differential equations (2), interactions between the hull and ship generated waves can be referred to as a cyclic process leading to the solution of equations of motion in time. In each simulation cycle, the displacements and velocities of the ship are extrapolated for the next time instant. Next step is to calculate the distribution of nonlinear sectional forces along the length of the ship. These include the hydrostatic, Froude-Krylov and viscous contribution.

Subsequently, sectional forces are integrated over the length of the ship and used to derive global forces on the vessel to be inputted into the equations of motion. The forces and moments, calculated using these instantaneous properties, are used to derive the resulting translational and rotational motions for the next time instant governing the relative position of ship with respect to waves.

The results as stated depend on the geometry of the vessel, the position of the waves relative to the hull but also on the ship speed. Finally, non-linear restoring coefficients in heave, roll, and pitch motions in waves within the quasi-static approach are calculated over the instantaneous waterline using the pressure integration technique (Schalck and Baatrup (1990)) along each segment of each transverse section of the ship hull under either a regular or irregular wave profile.

Special attention was given to damping evaluation. The modified Ikeda's method was introduced. Based on this very comprehensive approach an upgrade of the existing commonly used platform for the damping assessment was designed especially for large vessels like C11 while the reliability of the implemented model in simulating damping behaviour was validated against experimental data. The proposed methodology very much enables scrutinizing the damping components that contribute to it. However, the problem of viscous roll damping prediction probably will not be fully solved in the near future. The details of this procedure are omitted from this paper.

3 IMPLEMENTATION OF THE NUMERICAL MODEL FOR THE ASSESSMENT OF PARAMETRIC ROLLING

3.1 *Simulated instability in regular following seas*

As a part of a HydraLab III project there is an ongoing experimental investigation that took place in CEHIPAR (Canal de Experiencias Hidrodinámicas del Pardo), Spain, where the numerical results can be evaluated and verified. In order to validate such capability, the criteria should be applied for the case when existence of parametric roll is certain, such as conditions from France, et al., (2003). It is well documented that a post-Panamax, C11 class containership, (see Table 1), encountered extreme

Table 1. Main particulars of container vessel.

Length between perpendiculars (L_{PP})	262.00 m
Depth at main deck (D)	24.40 m
Breadth, design waterline (B_{DWL})	40.00 m
Displacement, design waterline (Δ_{DWL})	76056 t
Block coefficient (C_B)	0.66
Draught at amidships (T_{DWL})	12.34 m
Transverse metacentric height, still water ($GM1_T$)	1.973 m
Natural roll period, linearised in waves (T_{44})	22.78 s
Maximum speed (U)	20 kn

weather and sustained extensive loss and damage to deck stowed containers.

It is imperative that before the experiment itself, such conditions are predicted in order to calibrate test instruments and devices, wave generation etc. for all the measurements that will be taking place. It coincides with the effort to get the exact speed of the ship which will most likely cause parametric rolling for a given condition. That speed is in a way a "design speed" for parametric rolling and thus can be taken as a starting point.

With the above given particulars the numerical model briefly explained under section 2 is applied. To emphasise, this procedure implies the re-meshing of the hydrodynamic model up to the intersection of the ship with the incident wave elevation at each time instant. Since the parametric roll phenomenon is caused by time variation of transverse stability, the numerical simulation method must be able to adequately model the changes of geometry of the immersed part of the hull due to large waves and ship motions.

For this specific case the chosen wave length to ship ratio varies from 0.8 to 1.4. The investigation was carried out for wave heights ranging from 4 m to 10 m. Finally, several speed surveys have been conducted for each meteocean scenario referred above, starting from zero forward speed up to the maximum achievable speed. More specifically, forward speeds were considered up to 20 knots with successive increments of 1 knot. By varying each of these parameters, maximum steady roll amplitude is analyzed and recorded.

As it can be seen from Figure 2, starting from the case of zero knots the resonant effect only occurs for the wave length to ship's length ratio of 0.8, corresponding to the wave period of $T_W = 11.59$ s.

The resonant roll conditions can start to occur at the wave amplitude of 3.5 m, where the maximum steady roll amplitude is $\eta_4 = 12°$, but then increases up to $\eta_4 = 38°$ at the wave amplitude of 5 m, where it reaches the saturated value. Similarly to zero speed scenario, Figure 3 shows that at the start of the simulation the ship was pitching to angles of about 2°, with negligible roll response. A small initial roll angle, which may be caused by, for instance, rudder movement, is introduced to disturb the initial roll equilibrium and acts as trigger that results in parametric rolling.

The simulation records are systemized and presented in Table 2. As it can be seen, there is a clear trend on parametric rolling responses in following waves which can be traced to the speed of 0 and 2 knots, where it starts at the wave amplitude of 3.5 m and dies out at 5 m.

However, the ship speed of 1 knot reveals that the highest amplification of roll motion is shifted towards higher wave heights. This actually supports the theory stating that such an effect cannot be sustained for a prolonged range of ahead speeds and it usually vanishes at certain threshold wave heights due to the energy balance between damping and change of stability.

The energy put into the pitch and heave motion by the wave excitations may be partially transferred into the roll motions by means of nonlinear coupling among these modes; consequently roll motion can be indirectly excited which leads to excessive resonant rolling and stability problems. Actually, as compared to heave and pitch motions, which are

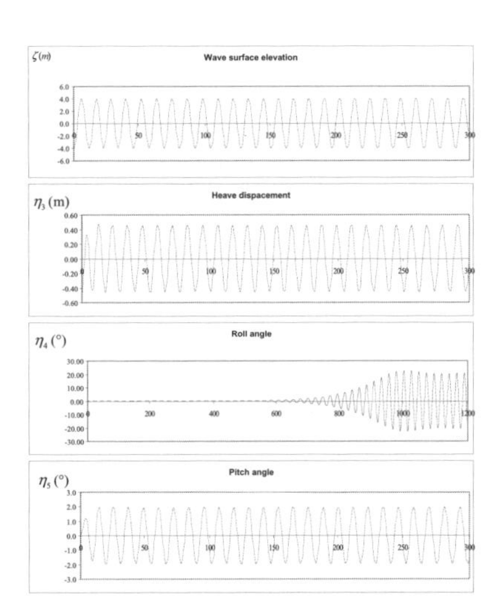

Figure 2. Development of parametric rolling in following regular waves ($H_W = 8$ m, $T_W = 11.59$ s, speed 0 kn).

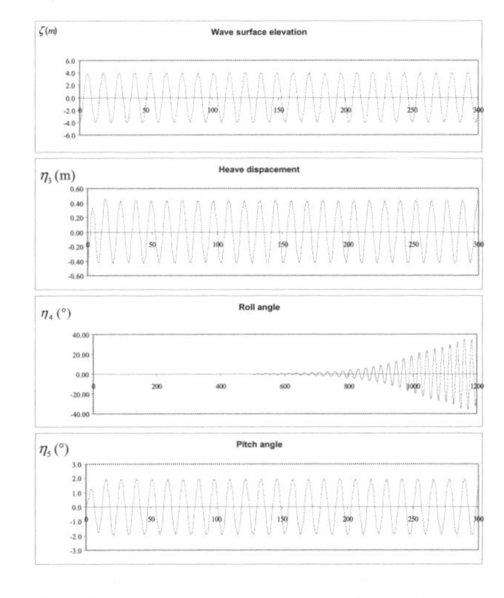

Figure 3. Development of parametric rolling in following regular waves ($H_W = 8$ m, $T_W = 11.59$ s, speed 1 kn).

Table 2. Numerical simulations in following seas for 0, 1 and 2 knots.

Speed (kts)	Heading (°)	λ/Lpp	Height (m)	Roll amplitude (°)
0.00	0.00	0.80	4.00	0.00
0.00	0.00	0.80	5.00	0.00
0.00	0.00	0.80	6.00	0.00
0.00	0.00	0.80	7.00	13.50
0.00	0.00	0.80	8.00	21.81
0.00	0.00	0.80	9.00	39.30
0.00	0.00	0.80	10.00	38.42
1.00	0.00	0.80	4.00	0.00
1.00	0.00	0.80	5.00	0.14
1.00	0.00	0.80	6.00	0.06
1.00	0.00	0.80	7.00	2.19
1.00	0.00	0.80	8.00	37.18
1.00	0.00	0.80	9.00	36.42
1.00	0.00	0.80	10.00	27.96
2.00	0.00	0.80	4.00	0.02
2.00	0.00	0.80	5.00	0.44
2.00	0.00	0.80	6.00	0.00
2.00	0.00	0.80	7.00	14.55
2.00	0.00	0.80	8.00	32.91
2.00	0.00	0.80	9.00	26.44
2.00	0.00	0.80	10.00	0.44

reaching stabilised solutions for a small number of oscillations, the induced roll motion has firstly a transitory phase and then the rolling amplitude slowly increases until the steady state is reached.

Figure 4 shows an expanded view of the roll and pitch response in regular waves.

Positive pitch values mean the vessel is pitched down by the bow. Also, it can be seen from Figure 5 that there are two pitch cycles for each roll cycle, and that the model is always pitched down by the bow at maximum roll. Throughout the test program, this relationship between pitch and roll motions existed whenever parametric rolling was induced.

In the Figure 5 below time traces are given for the heave, roll and pitch motion and wave height during a run in regular following waves at a different wave period $T_W = 12.95$ s.

When the vessel encounters a sequence of wave components of a certain period and height, parametric rolling is initiated if the resonant conditions are met, which is not the case in here. When the wave period changes or the wave height diminishes, the parametric rolling response quickly dissipates. The instability domains are a result of multiple interdependence of physical parameters such as: the energy partially transferred to the roll mode of motion, the excitation coefficient depending on the *GM* variation on following waves, the natural roll period, the ship speed, the wave length, the heading angle, the encountering period, etc.

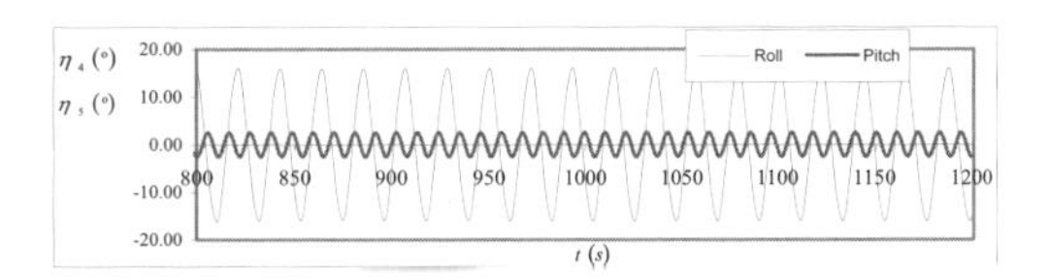

Figure 4. Expanded view of roll and pitch motions at largest roll ($H_W = 7$ m, $T_W = 11.59$ s, speed 2 kn).

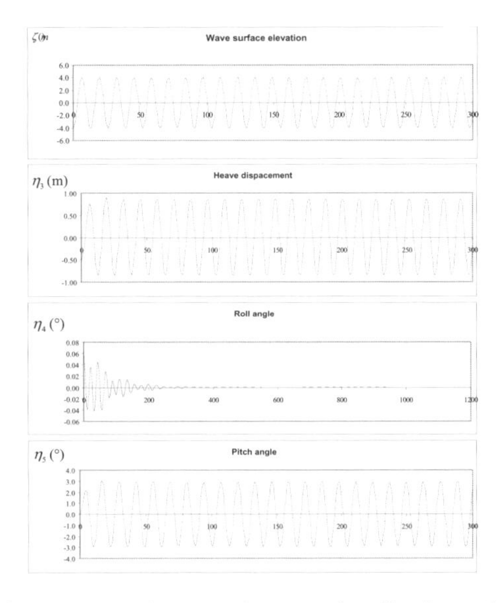

Figure 5. Development of parametric rolling in regular waves ($H_W = 8$ m, $T_W = 12.95$ s, speed 0 kn).

The same outcome has been kept for the regular following waves simulations at higher frequencies. Mention should be made that the time traces given in Figures 2 to 5 suggest the correlation between parametric resonance condition and the relation between the encountering period and the incident wave period.

3.2 *Experiments in regular following sea*

The search of the "critical speed" for parametric rolling in following waves was taken as the starting point for the model tests, which was initially set as zero knots. The parametric rolling was observed at the wave frequency of $\omega_0 = 0.542$ rad/s. The results obtained from the regular following waves are presented in scope of this section.

Although the design speed for following seas parametric rolling experimental assessment was correctly set, these groups of tests were operated at a distinct metacentric height (see Table 2, note subsequent change of *GM*1 to *GM*2) for the purpose of evaluating the contribution of loading condition on the resonance effect. Although the effects of range of roughly possible loading conditions that are available to the designer are not a primary task related to this work, it could be useful to have at hand a methodology for checking the relative level of vulnerability of a ship given those conditions.

Therefore, these simulations were recalculated using the appropriate geometrical set up (consequent change of displacement, *VCG* and *GM*). A total of 8 tests for $GM2 = 0.990$ m were carried out for the above mentioned loading condition and the complete list of results is summarized in Table 3.

During the execution of the designed experiments it was found that parametric rolling is sustained for all the wave frequencies tested. This is hardly surprising, since as explained during the previous section, the theoretical background which sets a sequence of components for parametric rolling condition to initiate is very complex. The results for the $H_W = 10$ m were unable to be interpreted since the model suffered water ingress due to the green water effect.

The shaded selection corresponding to test No. 314 is also presented on Figure 6 (obtained from experimental data) and on Figure 7 (obtained from numerical simulations). Figure 8 depicts the roll response from both.

Table 3. Experiments in following seas for heading 0°, different frequencies.

Test N°	λ/Lpp	Height (m)	T_W (m)	Speed (kts)	Roll amplitude (°)
309	0.8	6	11.59	0.00	35.79
310	1	6	12.95	0.00	35.79
311	1.2	6	14.19	0.00	31.53
312	1.4	6	15.33	0.00	25.96
313	0.8	8	11.59	0.00	35.74
314	1	8	12.95	0.00	38.20
315	1.2	8	14.19	0.00	34.07
316	1.4	8	15.33	0.00	26.35

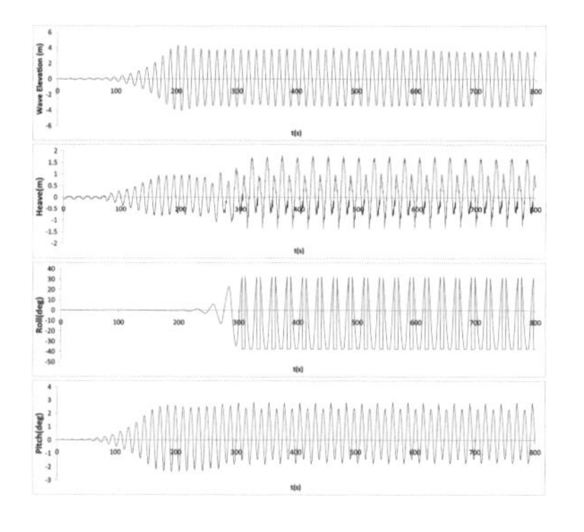

Figure 6. Experimental development of parametric rolling in regular following waves ($H_W = 8$ m, $T_W = 12.95$ s, speed 0 knots).

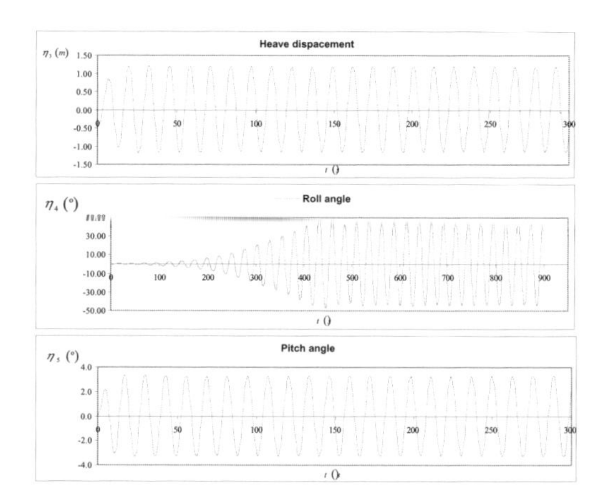

Figure 7. Numerical development (*GM2*) of parametric rolling with regular following waves ($H_W = 8$ m, $T_W = 12.95$ s, speed 0 knots).

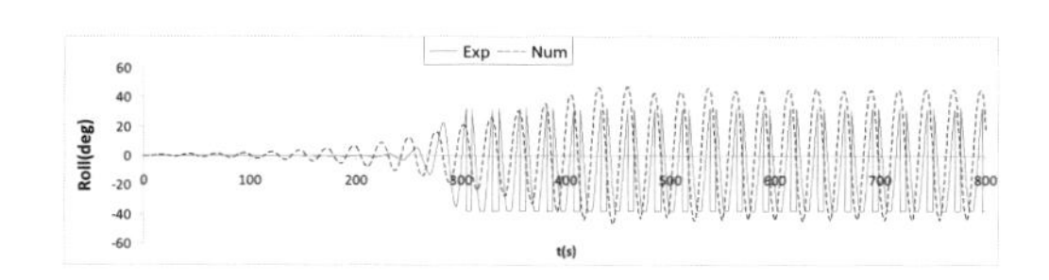

Figure 8. Roll history (*GM2*) of parametric rolling with regular following waves ($H_W = 8$ m, $T_W = 12.95$ s, speed 0 knots).

There is a difference in the maximum steady roll amplitude of about $\eta_4 \approx 7°$, which is in general a discrepancy range, and which represents the case where the numerical prediction overestimates the experimental result. A change in the *GM* value affects the natural roll period, which is one of the main parameters that will influence the risk of parametric roll (together with the encounter wave period, wave length, speed and roll damping). More specifically, the increase of *GM* value reduces the natural roll period, and as the significant wave height is also one of the governing parameters that will unleash parametric rolling. Therefore, adopting a *GM* value that will yield lower natural roll period can be a rather efficient way to reduce the risk of encountering unfavourable wave conditions.

3.3 *Stern quartering simulations*

Systematic model tests have clearly demonstrated that the instability problems also occurs in stern quartering seas, and low values of initial stability, especially when the vessel has sufficient time to stay on the crest (Krüger et al., 2004).

After extensive simulations have been conducted, the method developed by Ribeiro e Silva and Guedes Soares, 2009, seems very capable of predicting the events of parametric rolling. Even though there is only a very small scope of work presented in here, the validation attempt has shown very good agreement with the experimental results for roll both in the experiments where parametric roll resonance occurred and in the experiments where it did not occur.

Therefore, the concept has been applied for the stern waves (330–30°) set up. The propensity and effects of parametric rolling have been considered from all the various aspects that it contribute both on the qualitative and quantitative properties of solutions. The simulations were preformed ranging from:

- Headings 330°–30°, with increments of 10°,
- Wave heights 4–10 m, with increments of 1 m,
- Wave to ship length ratio λ/L_{pp}, of 0.8–1.4, with increments of 0.2,
- Ship speeds 0–20 knots, with increments of 1 knot,

which makes roughly 2400 simulations in total. The operated range of headings is taken as such where parametric rolling has a dominant presence.

The following list of Figures (9 to 12) presents a set of polar plots of the results utilized with

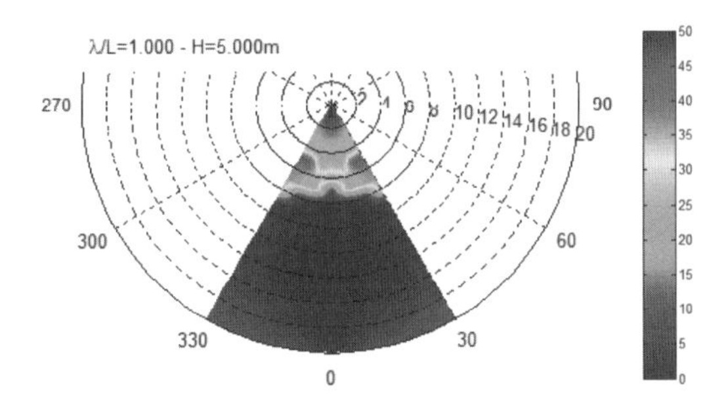

Figure 9. Numerical evidence with polar plots ($GM2$) of parametric rolling with regular following waves ($H_W = 5$ m, $T_W = 12.95$ s).

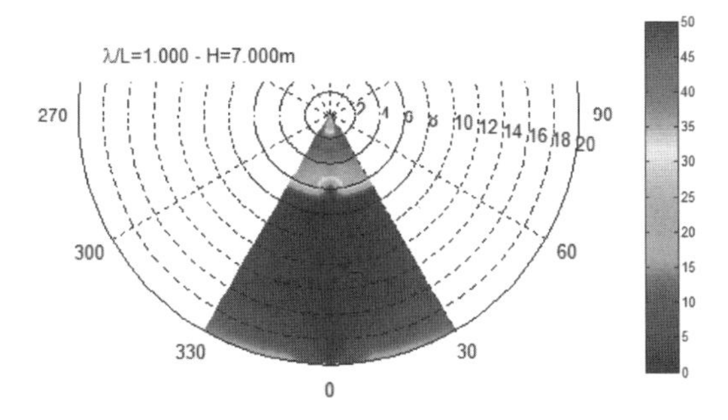

Figure 10. Numerical evidence with polar plots ($GM2$) of parametric rolling with regular following waves ($H_W = 7$ m, $T_W = 12.95$ s).

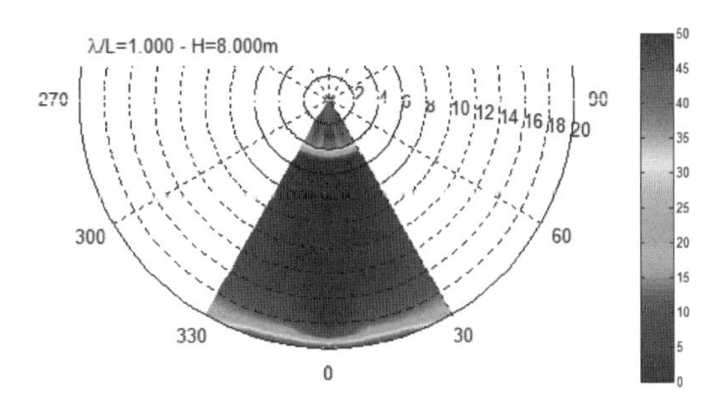

Figure 11. Numerical evidence with polar plots ($GM2$) of parametric rolling with regular following waves ($H_W = 8$ m, $T_W = 12.95$ s).

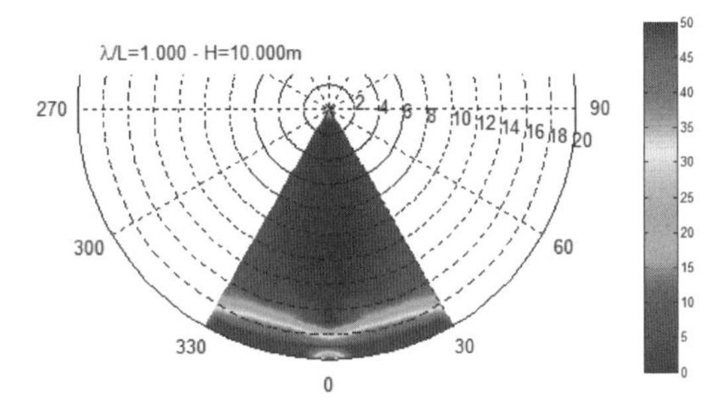

Figure 12. Numerical evidence with polar plots ($GM2$) of parametric rolling with regular following waves ($H_W = 10$ m, $T_W = 12.95$ s).

the above proposed methodology which is deliberately associated with the $\lambda/L_{pp} = 1$ typically pointed out as the most critical condition where severe parametric rolling condition is most likely to occur.

Parametric rolling usually occurs within the enclosed range of the encounter periods, wave heights and encounter angles, and because of it, this specific case shown in Figures 9–12 is referred as embarkation of "regional" occurrence of parametric rolling (in terms of shaded parts of polar plots).

Because a (container) vessel will navigate in different wave conditions at different speeds with different loading conditions the amount of information which is needed for a reliable operational guidance is rather large and possibly even infeasible. However, by employing the results from regular wave's excitation and being aware of the different physical phenomena that contribute to large rolling motions, there is a possibility to use reliable numerical model to identify the risk zones.

Following Figures (13 to 16) summarize all the above presented as 3D polar plots for all the range of frequencies and wave heights.

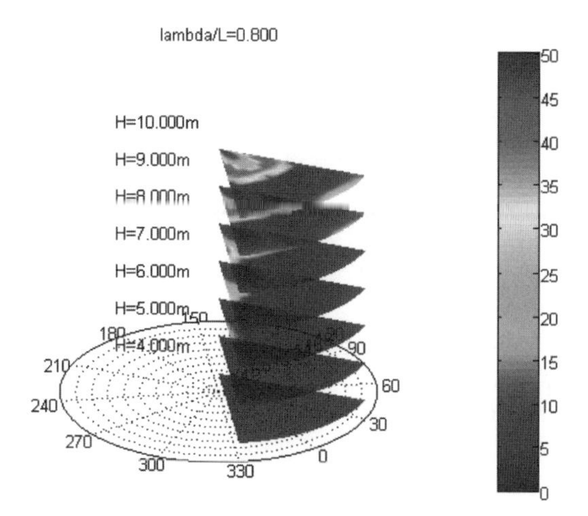

Figure 13. Numerical evidence with 3D polar plots ($GM2$) of parametric rolling with regular following waves ($T_W = 11.59$ s).

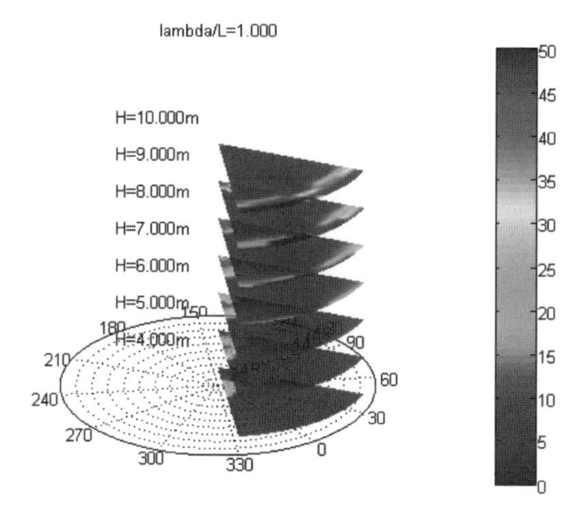

Figure 14. Numerical evidence with 3D polar plots ($GM2$) of parametric rolling with regular following waves ($T_W = 12.95$ s).

The physical explanations that govern these results will be explained in the aftermath. Even so, with the preliminary visual observation the results are in perfect compliance with the reported from Levadou and Gaillarde (2003) for the stern quartering seas which in fact was not the part of the main investigation rather a surplus.

Within this work the procedures were optimized in a manner that a presentation as such, being demonstrated above, is feasible within couple of hours. This itself strengthens the applicability of method given its desirable CPU effort especially while investigating more complex irregular wave excitations.

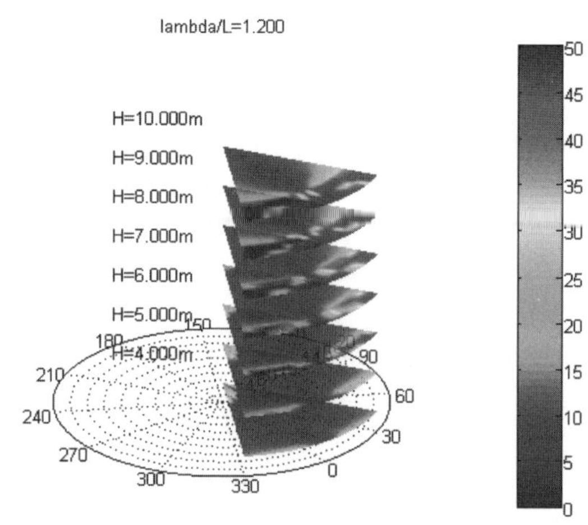

Figure 15. Numerical evidence with 3D polar plots ($GM2$) of parametric rolling with regular following waves ($T_W = 14.19$ s).

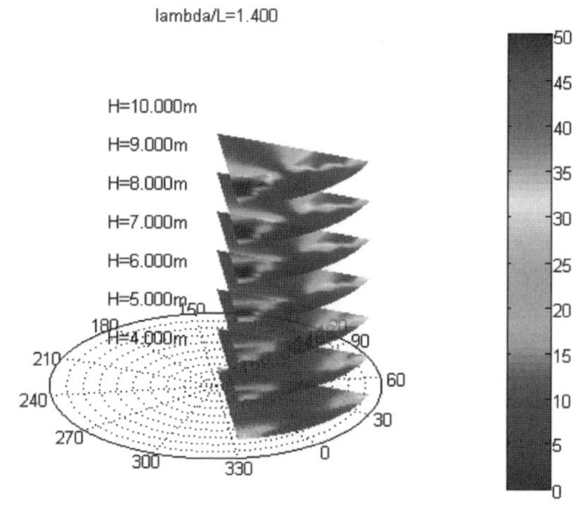

Figure 16. Numerical evidence with 3D polar plots ($GM2$) of parametric rolling with regular following waves ($T_W = 15.33$ s).

4 CONCLUSION

A numerical procedure for parametric rolling inception is presented and applied to a post-Panamax, C11 class containership, for a regular following wave's excitation as well as some results from an experimental program conducted on the same vessel.

Comparisons between numerical results and model tests showed a very good agreement and as such emphasized the usefulness of the numerical model as a tool to study the influence of vessel operational restrictions on the tendency of containerships towards parametric rolling contained within the intuitive graphical presentation.

ACKNOWLEDGEMENTS

The experimental results reported here were obtained at CEHIPAR, the Canal de Experiências Hidrodinâmicas del Pardo, Spain, in the scope of the Hydralab III Project, which has been partially financed by the European Commission.

REFERENCES

ABS, 2004, *Guide for the Assessment of Parametric Roll Resonance in the Design of Container Carriers*, American Bureau of Shipping, Houston, USA.

Belenky, V.L., Weems, K.M., Lin, W.M. and Paulling, J.R. 2003. Probabilistic Analysis of Roll Parametric Resonance in Head Seas. *Proc. 8th Int. Conf. on Stability of Ships and Ocean Vehicles STAB*, 325–340.

Bulian, G. and Francescutto, A., 2003, On the nonlinear modeling of parametric rolling in regular and irregular waves. *Proc. 8th International Conference on the Stability of Ship and Ocean Vehicles, Madrid, pp. 305–323.*

France, W.N., Levadou, M., Treakle, T.W., Paulling, J.R., Michel, R.K. and Moore, C. 2003. An investigation of headsea parametric rolling and its influence on container lashing systems. *Marine Technology*, 40:1 1–19.

Frank, W. and Salvesen, N. 1975. *The Frank close-fit ship-motion computer program*. NSRDC Report.

Ikeda, Y., Himeno, Y. and Tanaka, N. 1978. *A prediction method for ship roll damping.* Report No. 405, of Department of Naval Architecture, University of Osaka.

ITTC, 2006. *Recommended Procedures and Guidelines; Predicting the Occurrence and Magnitude of Parametric Rolling*.

Krüger, S., Hinrichs, R. and Cramer, H.; 2004. Performance Based Approaches for the Evaluation of Intact Stability Problem, susceptibility *Roll Damping*, Technical Report 6136-74-80, NAVSPEC.

Neves, M., Pérez, N., Lorca, O. and Rodriguez, C. 2003. Hull design considerations for improved stability of fishing vessels in waves. *Proc. of STAB'03 8th International Conference on Stability of Ships and Ocean Vehicles.*

Perez, N. & Sanguinetti, C. 1995. Experimental results of parametric resonance phenomenon of roll motion in longitudinal waves for small fishing vessels. *International Shipbuilding Progress 42 (431)*, 221–234.

Ribeiro e Silva, S. and Guedes Soares, C. 2000. Time Domain Simulation of Parametrically Excited Roll in Head Seas. *Proceedings of the 7th International Conference on Stability of Ships and Ocean Vehicles (STAB'2000); Renilson, M. (ed) Launceston, Tasmania, Australia. Pages* 652–664.

Ribeiro e Silva, S., Santos, T.A. and Guedes Soares, C. 2005. Parametrically excited roll in regular and irregular head seas. *Int. Shipbuild. Progr.*, 52: 29–56.

Ribeiro e Silva, S. and Guedes Soares, C. 2008, Non-Linear Time Domain Simulation of Dynamic Instabilities in Longitudinal Waves, *Proceedings of the 27th International Conference on Offshore Mechanics and Arctic Engineering;(OMAE)*, Estoril, Portugal, Paper OMAE2008-57973.

Ribeiro S. e Silva and Guedes Soares, C. (2009): "Parametric Rolling of a Container Vessel in Longitudinal Waves", Stability of Marine Structures and Ocean Vehicles—STAB 2009 Conference, St. Petersburg (Russia).

Salvesen, N., Tuck, E.O. and Faltinsen, O. 1970. Ship Motions and Sea Loads, *Transactions of SNAME*, Vol. 8: 250–287.

Schalck, S. and Baatrup, J. 1990, "Hydrostatic Stability Calculations by Pressure Integration", *Ocean Engineering*, Vol. 17, 155–169.

Shin, Y., Belenky, V.L., Paulling, J.R., Weems, K.M. and Lin, W.M. 2004. Criteria for parametric roll of large containerships in head seas. *Transactions of SNAME*, Vol. 42.

Spyrou, K.J. 2000. Designing against parametric instability in following seas. *Ocean Engineering 27 625–653*.

Turk, A., Ribeiro e Silva, S. Guedes Soares, C. Prpić-Oršić, J. 2009. An investigation of dynamic instabilities caused by parametric rolling of C11 class containership // *The 13th Congress IMAM 2009 / Istanbul* 167–175.

Umeda, N., Hashimoto, H., Vassalos, D., Urano, S. and Okou, K. 2004, Nonlinear Dynamics on Parametric Roll Resonance with Realistic Numerical Modelling, *International Shipbuilding Progress*, 51:2/3, 205–220.

Sustainable Maritime Transportation and Exploitation of Sea Resources – Rizzuto & Guedes Soares (eds)
© 2012 Taylor & Francis Group, London, ISBN 978-0-415-62081-9

Influence of righting moment curve on parametric roll motion in regular longitudinal waves

E. Pesman & M. Taylan
Istanbul Technical University, Istanbul, Turkey

ABSTRACT: Parametric roll is a phenomenon which typically occurs in longitudinal waves. Although its theoretical existence has been known for decades, parametric roll attracted a great deal of interest in recent years. Stability accidents result from parametric roll make the topic more important because of the potential risks to money and human lives. In the present study, parametrically excited roll motion is modeled as a single degree of freedom system incorporating heave and pitch effects by means of restoring moment variations. The nonlinear roll model is solved by the averaging technique to avoid viable correlation between the responses and area under restoring moment curves. Nonlinear effects like discontinuous bifurcations were also introduced into the results. As a result, risk free regions, risky regions, risky regions due to discontinue bifurcations and extremely risky regions were determined.

1 INTRODUCTION

The phenomenon of parametrically excited roll motion has been known for a long time that the restoring moment of a ship depends on position relatively to the waves (Kempf, 1938). The stability variations of a ship moving in longitudinal waves have been studied by a number of researchers including Graff, Heckscher (1941), Kerwin (1955), Paulling and Rosenberg (1959). The first experimental observation of parametric roll was done by Paulling et al. (1972) in San-Francisco bay. Although its theoretical existence has been known for a long time, parametric roll attracted a great deal of interest in recent years because of the incidents that resulted in loss of life and money. In October, 1998, a post-Panamax, C11 class containership encountered extreme weather and sustained extensive loss and damage to deck stowed containers. This incident encouraged researchers to make investigations through a series of model tests and numerical analyses. Results of these studies showed that post-panamax containerships tend to experience parametric roll motion in head sea (France et al., 2003). These casualties urged designers, researchers and regulatory authorities to initiate further researches and investigations. In further studies, researchers such as Spyrou (2000), Neves, Rodrigues (2006) and Bulian et al. (2004) focused on nonlinear aspects and effect of changing tuning factors on parametric roll motion, such as frequency range. In another aspect, some researchers focused on probabilistic properties of parametric roll (Shin et al., 2004; Belenky, 2004; Hashimoto et al., 2006; Bulian et al., 2006).

Studies related with probabilistic properties indicate that non-ergodicity is stronger if nonlinearity is stronger and it is especially noticeable for parametric roll motion. Beside researchers, regulatory and classification societies have also been studying on parametric roll. Current IMO intact stability code has been mainly formed with the static and quasi-static stability rules. The need to develop stability criteria with the latest achievements in the research of dynamic intact stability physical mechanisms has been indicated. The IMO working group proposed the framework for the development of the new criteria on the meetings in 2007 (SLF 50/WP.2, 2007). A revision of the intact stability criteria has been discussed towards including roll restoring arm variation problems in e.g., the document submitted to SLF 51 (IMO, SLF/4/3, 2008). A very detailed explanation of the physics behind parametric roll has been given in the ABS assessment of parametric roll in the container ship design (Shin et al., 2004). The state of the art in methodology development and regulations in assessment of ship intact stability can be found in Francescutto (2007).

This paper examines the influence of righting moment curve between the extreme values; wave trough and wave crest conditions on parametric roll motion in regular longitudinal waves. For this purpose, various hull forms having different geometries were investigated for numerous wave and damping conditions. In the present study, parametrically excited roll motion is modeled as a single degree of freedom system incorporating heave and pitch effects by means of restoring moment variations. Bulian (2006) approximated GZ surface with polynomial coefficients and Fourier series. The model presented in this study may be considered as simplified version of the earlier work. Restoring moment variations in waves with respect to time and instantaneous roll angle was modeled analytically between the righting moment curves for the wave crest and wave trough conditions. Simply, the polynomial coefficients of wave crest and wave trough GZ curves and sinusoidal function were utilized unlike the previous approach. The nonlinear roll model is solved by the averaging technique in frequency domain to avoid viable correlation between the responses and area under restoring moment curves. Nonlinear effects like discontinuous bifurcations were also introduced into the evaluation. Roll responses were non-dimensionalized by dividing them with the angle of vanishing stability to obtain more common conclusions.

2 MATHEMATICAL MODEL

In general, the equation of roll motion in regular longitudinal waves can be written as follows:

$$(I_{xx} + \delta I_{\mathbf{xx}})\ddot{\phi} + B(\dot{\phi},\phi) + \Delta GZ(\phi,t) = 0 \tag{1}$$

where:

$(I_{xx} + \delta I_{xx})$: Moment of inertia,
ϕ : Roll angle,
$B(\dot{\phi},\phi)$: Damping function,
$\Delta GZ(\phi, t)$: Restoring function.

Eq. (1) may be re-written as;

$$\ddot{\phi} + b(\dot{\phi}, \phi) + \frac{\omega_0^{\ 2}}{GM_0} GZ(\phi, t) = 0 \tag{2}$$

In the above equation, $GZ(\phi,t)$ may be approximated by the following expression neglecting surge and Froude-Krylov forces. In this study, the restoring moment variation is modeled by using only wave crest and wave trough restoring moment curves.

$$GZ(\phi, t) = \sum_{n=1}^{N} \left(m_{2n-1} + k_{2n-1}\cos(\omega_e t)\right)\phi^{2n-1} \tag{3}$$

The coefficients "m" and "k" in Eq. (3) are obtained from righting lever curves in wave crest and wave trough conditions.

$$m_{2n-1} = \frac{c_{2n-1,\, trough} + c_{2n-1,\, crest}}{2} \tag{4}$$

$$k_{2n-1} = \frac{c_{2n-1,\, trough} - c_{2n-1,\, crest}}{2} \tag{5}$$

$$b(\phi,\dot{\phi}) = 2\mu\dot{\phi} \tag{6}$$

In Eqs. (4) and (5), "$c_{2n-1,\, crest}$ and $c_{2n-1,\, trough}$" show the coefficients of polynomials fitted to restoring lever curves in wave trough and wave crest conditions. When the ship is in upright position, she is usually symmetrically loaded, so the even terms of the polynomial approximation of the righting lever curve disappear. In this work, seventh degree polynomials are utilized for developing the restoring lever surfaces. The linear damping term is used and is given in Eq. (6). Coefficients of the damping function were addressed by Ikeda, Himeno and Tanaka (1978).

Substitution of Eq. (6) and Eq. (3) in Eq. (2) leads to the following differential equation:

$$\ddot{\phi} + 2\mu\dot{\phi} + \omega_0^{\ 2}\frac{\left(\sum_{n=1}^{N}\left(m_{2n-1} + k_{2n-1}\cos(\omega_e t)\right)\phi^{2n-1}\right)}{GM_0} = 0 \tag{7}$$

Eq. (7) can be expressed in a non-dimensional form by the following transformations.

$$\tau = \omega_0 t, \qquad \gamma_{2n-1} = \frac{m_{2n-1}}{GM_0}, \qquad \zeta_{2n-1} = \frac{k_{2n-1}}{GM_0},$$
$$\Lambda = \frac{\omega_e}{\omega_0}, \qquad \nu = \frac{\mu}{\omega_0} \tag{8}$$

After the necessary manipulations, the nonlinear equation of roll motion in longitudinal waves may be represented by Eq. (9). The prime denotes derivation with respect to the non-dimensional time variable τ.

$$\phi'' + 2\nu\phi' + \left(\sum_{n=1}^{4}\left(\gamma_{2n-1} + \zeta_{2n-1}\cos(\Lambda\tau)\right)\phi^{2n-1}\right) = 0 \tag{9}$$

In this study, Eq. (9) was solved in frequency domain by using averaging method.

The following transformations are introduced in order to convert the system in the standard Lagrange form for averaging (Bogoliubov & Mitropolsky, 1961).

$$\left\{\begin{aligned} &\phi(\tau) = A\cos(Q) && \{A = A(\tau)\} \\ &\phi'(\tau) = -A\frac{\Lambda}{2}\sin(Q) && \left\{Q = \frac{\Lambda}{2}\tau + \frac{\psi(\tau)}{2}\right\} \\ &\phi''(\tau) = -A'\frac{\Lambda}{2}\sin(Q) - A\frac{\Lambda^2}{4}\cos(Q) \\ &\qquad - A\frac{\Lambda}{2}\frac{\psi'}{2}\cos(Q) \end{aligned}\right\} \tag{10}$$

In Eq. (10), A is amplitude and ψ is phase. After substituting Eq. (10) into Eq. (9) the following system is obtained:

$$A'\frac{\Lambda}{2}=\begin{bmatrix}-A\frac{\Lambda^2}{4}\cos(Q)+b(\phi')\\+r(\phi)+\delta r(\phi,\tau)\end{bmatrix}\sin(Q) \tag{11}$$

$$A\frac{\Lambda}{2}\frac{\psi'}{2}=\begin{bmatrix}-A\frac{\Lambda^2}{4}\cos(Q)+b(\phi')\\+r(\phi)+\delta r(\phi,\tau)\end{bmatrix}\cos(Q) \tag{12}$$

$$b(\phi')=2\nu\phi' \tag{13}$$

$$r(\phi)=\sum_{n=1}^{N}\gamma_{2n-1}\phi^{2n-1} \tag{14}$$

$$\delta r(\phi,\tau)=\sum_{n=1}^{N}\xi_{2n-1}\cos(\Lambda\tau)\phi^{2n-1} \tag{15}$$

As it can be seen from Eq. (11) and Eq. (12), "$-A(1/4)\Lambda^2\cos(Q)+b(\phi')+r(\phi)+\delta r(\phi,\tau)$" is a periodic function of Q. By using the properties of slowly varying A and ψ, periodic averages can be used in the right hand side of Eq. (11) and Eq. (12) as follows (Hayashi, 1964; Nayfeh & Mook, 1979; Nayfeh, 1985; Bullian, 2006);

$$A'\frac{\Lambda}{2}=\frac{1}{\pi}\int_0^{2\pi}\begin{bmatrix}-A\frac{\Lambda^2}{4}\cos(Q)+b(\phi')\\+r(\phi)+\delta r(\phi,\tau)\end{bmatrix}\sin(Q)dQ \tag{16}$$

$$A\frac{\Lambda}{2}\frac{\psi'}{2}=\frac{1}{\pi}\int_0^{2\pi}\begin{bmatrix}-A\frac{\Lambda^2}{4}\cos(Q)+b(\phi')\\+r(\phi)+\delta r(\phi,\tau)\end{bmatrix}\cos(Q)dQ \tag{17}$$

After averaging over "$Q-\pi$, $Q+\pi$", and letting $\tau^*\approx\tau/\Lambda$, the following approximate autonomous system is being found:

$$\frac{d\psi}{d\tau^*}=2\left[M_{11}\cos(\psi)-B_{11}\right]=F_1(\psi,A,\Lambda) \tag{18}$$

$$\frac{dA}{d\tau^*}=A\left[M_{22}\sin(\psi)-B_{21}\right]=F_2(\psi,A,\Lambda) \tag{19}$$

Coefficients M_{ik} and B_{ik} are determined as follows:

$$B_{21}=\nu\Lambda+\frac{2}{3\pi}\beta\Lambda^2A+\frac{3}{32}d\Lambda^3A^2 \tag{20}$$

$$B_{11}=-\left(-\frac{\Lambda^2}{4}+\gamma_1+\frac{3}{4}\gamma_3A^2+\frac{5}{8}\gamma_5A^4+\frac{35}{64}\gamma_7A^6\right) \tag{21}$$

$$M_{11}=\frac{1}{2}\zeta_1+\frac{1}{2}\zeta_3A^2+\frac{15}{32}\zeta_5A^4+\frac{7}{16}\zeta_7A^6 \tag{22}$$

$$M_{22}=\frac{1}{2}\zeta_1+\frac{1}{4}\zeta_3A^2+\frac{5}{32}\zeta_5A^4+\frac{7}{64}\zeta_7A^6 \tag{23}$$

Steady state solution of Eq. (18) and Eq. (19) can be easily obtained by setting time derivative to zero with the assumption, $A>0$.

$$\begin{pmatrix}\left[M_{11}\cos(\psi)-B_{11}\right]=0\\\left[M_{22}\sin(\psi)-B_{21}\right]=0\end{pmatrix} \tag{24}$$

Solution of Eq. (24) yields the Eq. (25). Expanding Eq. (25) and collecting coefficients of the same powers of A, a 24th degree polynomial, whose roots are the steady state amplitudes, can be obtained. Only positive real roots of the polynomial are considered meaningful due to the assumption $A>0$. The corresponding phases can be obtained for all amplitudes by using Eq. (24).

$$B_{11}{}^2M_{22}{}^2+B_{21}{}^2M_{11}{}^2-M_{11}{}^2M_{22}{}^2=0 \tag{25}$$

Stability of trivial and nontrivial solutions determined by utilizing known characteristics of Mathieu equation and Routh-Hurtwitz criterion respectively (Nayfeh & Mook, 1979).

3 AREAS UNDER GZ CURVES IN WAVE TROUGH AND WAVE CREST

In this study, restoring moment variations in waves with respect to time and instantaneous roll angle was modeled analytically between the righting moment curves for the wave crest and wave trough conditions. Variation of restoring moment has a significant effect on parametric roll motion. It may be further concluded that the primary cause of parametric roll motion is the difference between wave trough and wave crest GZ curves.

In this paper, area under GZ curves in wave trough and wave crest are named as "A_{trough}" and "A_{crest}" respectively and calculated between zero to angle of maximum GZ as follows (Figure 1). "A_{crest}" is divided by "A_{trough}" to obtain non-dimensional "area ratio".

$$A_{crest}=\left(\int_0^{\phi_{GZmaks.}}GZ(\phi)d\phi\right)_{crest} \tag{26}$$

$$A_{trough}=\left(\int_0^{\phi_{GZmaks.}}GZ(\phi)d\phi\right)_{trough} \tag{27}$$

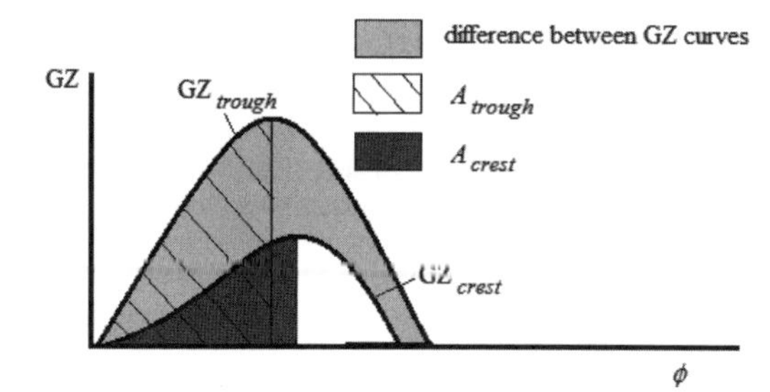

Figure 1. Areas under GZ curves in wave trough and wave crest.

4 SAMPLE SHIPS

In this paper, 8 different ship forms were used. Ships were named as C1, RR1, RR2, F1, D2, D1F, D1 W and D1T. C1 is a Post-Panamax C11-class containership (France et al., 2001). RR1 is a Ro-Ro ship whose experimental tests were carried out at the towing tank of DINMA (Bullian, 2006) and INSEAN Laboratory (Bullian, 2006). RR2 is also a Ro-Ro ship smaller than RR1. Experimental tests of RR2 were carried out at the towing tank of DINMA (Bullian, 2006). F1 is a frigate and her experimental tests were carried out at the towing tank of DINMA (Bullian, 2006). D2 is a destroyer and her experimental tests were carried out at the towing tank of DINMA (Bullian, 2006). D1F, D1 W and D1T are destroyers whose underbodies are similar in size and characteristics as DDG51 but above-water shapes are different. D1F has 10° flare, D1 W has a parallel body and D1T has 10° tumblehome above-water forms (McCue et al., 2007). Main characteristics and forms of sample ships are given in Table 1 and Figure 2, respectively.

5 ROLL RESPONSE ANALYSIS BASED ON THE AREA UNDER GZ CURVES

The analyses were carried out for 9 different wave slopes (1/20, 1/30 … 1/100) and 6 different damping ($\mu = 0.01, 0.02 \ldots 0.06$) conditions for 8 sample ship forms. Roll response curves were calculated with respect to ship speed by using averaging method and were divided into two parts as given in Figure 3. The lower part is called the "main resonance region" and the upper part is called the "bifurcated resonance region".

Maximum roll amplitudes in the main resonance and bifurcated regions were non-dimensionalized by dividing with the angle of vanishing stability. The non-dimensional amplitudes were plotted for various damping values in Figure 4 and Figure 5.

The results given with Figure 4 and Figure 5 show that the roll amplitudes linearly increase with difference between areas under wave trough and wave crest restoring moment curves. Furthermore, it is observed that roll motion is somewhat damped with increasing damping coefficient "μ" and area ratio "A_{crest}/A_{trough}". Therefore, first degree polynomials were fitted to the results and polynomials are plotted in Figure 6 and 7 with respect to damping coefficient and area ratio for the main resonance and bifurcated resonance regions.

It is observed that roll amplitudes increase with decreasing area ratio and damping coefficient. The results also showed that roll amplitudes rapidly increase due to discontinuous bifurcations. In the light of these results, it may be concluded that the risk of parametric roll motion (risk free zone, risky zone, very risky zone and extremely risky zone) can be determined based on the damping coefficient and area ratio as shown in Figure 8.

Linear polynomials were fitted to the boundaries of risk zones to determine the correlation between damping and area ratio. Equations of damping coefficients related to area ratio were defined as follows for risk zones.

Risk free zone ($\phi < 0.1\ \phi_v$):

$$\mu > -0{,}0925\frac{A_{crest}}{A_{trough}} + 0{,}1025 \tag{28}$$

Risky zone (large roll amplitudes $\phi < \phi_v$):

$$\begin{Bmatrix} \mu < -0{,}0925\dfrac{A_{crest}}{A_{trough}} + 0{,}1025 \\ \mu > -0{,}357\dfrac{A_{crest}}{A_{trough}} + 0{,}06714 \end{Bmatrix} \tag{29}$$

Very risky zone (large roll amplitudes and possibility of capsizing due to the discontinuous bifurcations $\phi \le \phi_v$):

$$\begin{Bmatrix} \mu < -0{,}357\dfrac{A_{crest}}{A_{trough}} + 0{,}06714 \\ \mu > -0{,}0980\dfrac{A_{crest}}{A_{trough}} + 0{,}0786 \end{Bmatrix} \tag{30}$$

Extremely risky zone (capsizing zone $\varphi > \varphi_v$):

$$\mu < -0{,}0980\frac{A_{crest}}{A_{trough}} + 0{,}0786 \tag{31}$$

Table 1. Main characteristic of sample ships.

	Type	LBP	B	T	KG
C1	Container	262.00 m	40.00 m	12.360 m	17.550 m
D1F	Destroyer flared	154.00 m	18.80 m	5.500 m	8.200 m
D1 W	Destroyer wall-sided	154.00 m	18.80 m	5.500 m	8.200 m
D1T	Destroyer tumblehome	154.00 m	18.80 m	5.500 m	8.200 m
D2	Destroyer	126.60 m	13.65 m	3.985 m	5.800 m
F1	Frigate	120.00 m	14.25 m	4.060 m	6.557 m
RR1	Ro-Ro	132.22 m	19.00 m	5.875 m	8.660 m
RR2	Ro-Ro	52.55 m	10.00 m	2.100 m	4.558 m

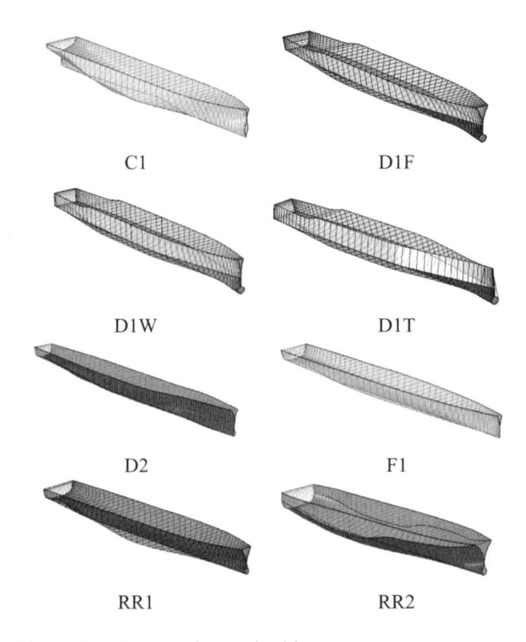

Figure 2. Forms of sample ships.

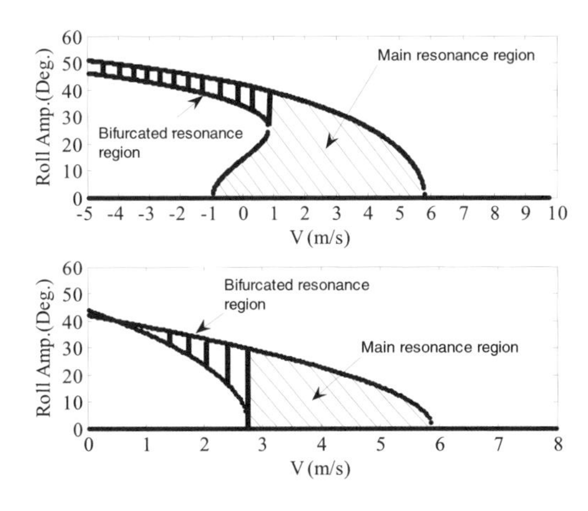

Figure 3. Roll response curves of F1 and D1T ship forms.

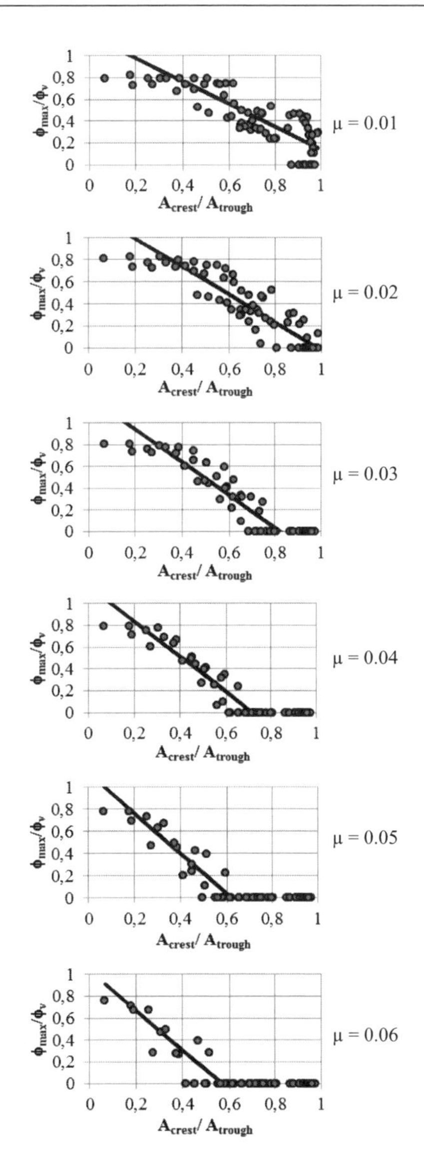

Figure 4. Maximum roll responses in main resonance region.

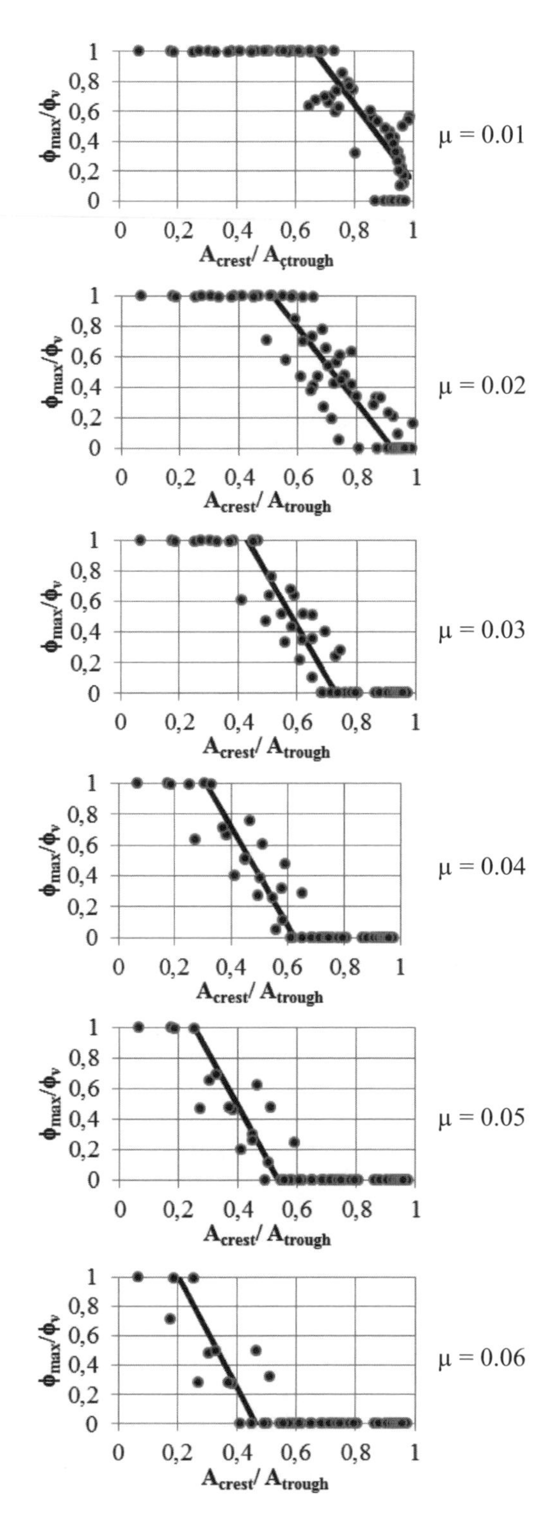

Figure 5. Maximum roll responses in resonance region where discontinues bifurcations were observed.

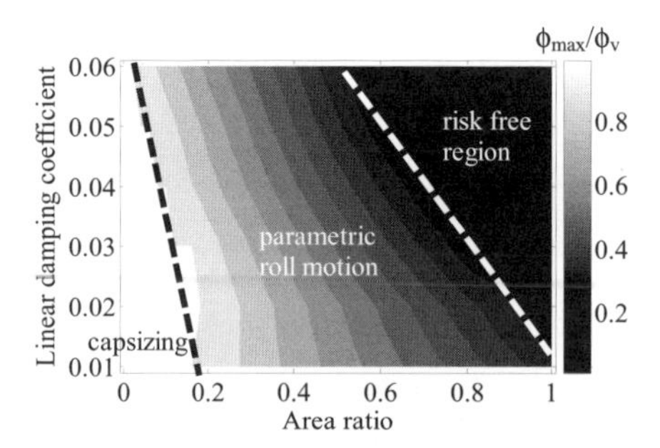

Figure 6. Maximum roll responses with respect to damping coefficient and area ratio in main resonance region.

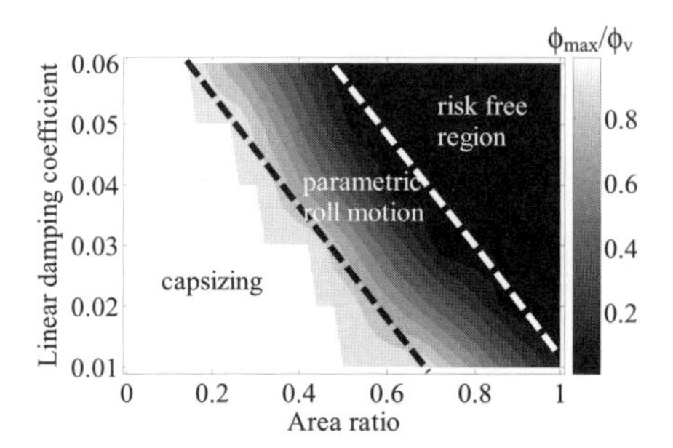

Figure 7. Maximum roll responses with respect to damping coefficient and area ratio in bifurcated resonance region.

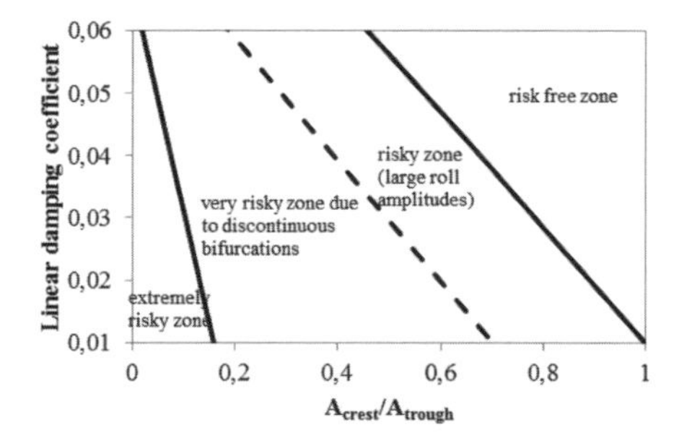

Figure 8. Risk zones related to area ration and damping coefficient.

6 CONCLUSIONS

Parametric roll motion is a phenomenon that has to be considered in the preliminary design stage. In the preliminary design stage, following precautions can be taken to avoid large roll amplitudes caused by parametric excitation. These precautions are form optimization, increasing damping capability (bilge keels, etc.) and limiting sea state that ship serviced in. These parameters govern the area ratio and damping coefficients.

In the present study, potential risk regions of parametric roll motion were determined based on the relationship between damping coefficient and area ratio. As a conclusion, this tool may be utilized as a measure for parametric roll determination in the preliminary design stage.

Finally, using the procedure outlined in this paper may help designers to optimize hull geometry and damping devices against extreme roll amplitudes due to the parametric roll in the preliminary design stage.

REFERENCES

Belenky, V.L. 2004. On Risk Evaluation at Extreme Seas, *Proc. of the 7th Int. Stability Workshop*, 188–202, Shanghai, China.

Bogoliubov, N.N. & Mitropolsky, Y.A. 1961. Asymptotic Methods in the Theory of Non-Linear Oscillations, *Hindustan Publishing Corp*, Delhi.

Bullian, G. 2006. Development of Analytical Nonlinear Models for Parametric Roll and Hydrostatic Restoring Variations in Regular and Irregular Waves. *University of Trieste Ph.D. Thesis*, Trieste.

Bullian, G., Francescutto, A. & Lugni, C. 2004. On the nonlinear modeling of parametric rolling in regular and irregular waves, *International Shipbuilding Progress*, 51: 205–220.

Bullian, G., Francescutto, A. & Lugni, C. 2006. Theoretical, Numerical and Experimental Study on the Problem of Ergodicity and 'Practical Ergodicity' with an Application to Parametric Roll in Longitudianl Long Crested Irregular Sea, *Ocean Engineering*, 33: 1007–1043.

France, W.N., Levaduo, M., Treakle, T.W., Paulling, J.R., Michel, R.K. & Moore, C. 2003. An Investigation of Head-Sea Parametric Rolling and its influence on Container Lashing Systems. *Marine Techn,*. 40(1): 1–19.

Francescutto, A. 2007. Intact Stability of Ships Recent Developments and Trends, *Proc. of 10th International Symposium on Practical Design of Ships and Other Floating Structures PRADS'07*, 1: 487–496, Houston.

Graff, W. & Heckscher, E. 1941. Widerstand und Stabilität Versuche mit Drei Fischdampfer Modellen, *Werft Reederei Hafen*. 22: 115–120.

Hashimoto, H., Umeda, N. & Matsuda, A. 2006. Experimental and Numerical Study on Parametric Roll of a Post-Panamax Container Ship in Irregual Wave, *Proc. of STAB'06 9th Int. Conf. on Stability of Ships and Ocean Vechicles*, 181–190, Rio de Janeiro, Brazil.

Hayashi, C. 1964. Nonlinear Oscillations in Physical Systems, *McGraw Hill*, New York.

Ikeda, Y., Himeno, Y. & Tanaka, N. 1978. A Prediction Method for Ship Roll Damping, *Report No. 00405 of Department of Naval Architecture, University of Osaka Prefecture*.

IMO SLF 50/4/12, 2007. Review of the IS Code.

IMO SLF 51, 2008. Revision of the Intact Stability Code. Report of the Working Group. (Part 1), London 2008.

Kempf, G. 1938. Die Stabilität Beanspruchung der Schiffe Durch Wellen und Schwingungen, *Werft Reederei Hafen*, 19: 200–202.

Kerwin, J.E. 1955. Not on Rolling in Longitudinal Waves. *Int. Shipbuilding Progress*, 2(16): 597–614.

McCue, L.S., Campbell, B.L. & Belknap, W.F., 2007. On the Parametric Resonance of Tumblehome Hullforms in a Longitudinal Seaway, *American Society of Naval Engineers Journal*, 3: 35–44.

Nayfeh, A.H. 1985. Problems in Perturbation, *John Wiley & Sons, Inc.*

Nayfeh, A.H. & Mook, D.T. 1979. Nonlinear Oscillations, *John Wiley & Sons, Inc.*

Neves, M.A.S. & Rodriguez, C.A., 2006. Influence of non-linearities on the limits of stability of ships rolling in head seas, *Ocean Engineering*, 34: 1618–1630.

Paulling, J.R., Kastner, S. & Schaffran, S. 1972. Experimental Studies of Capsizing of Intact Ships in Heavy Seas. *U.S. Coast Guard Technical Report* (also IMO Doc. STAB/7, 1973).

Paulling, J.R. & Rosenberg, R.M. 1959. On unstable ship motions resulting from nonlinear coupling, *Journal of Ship Research*, 3: 36–46.

Shin, Y.S., Belenky, V.L., Paulling, J.R., Weems, K.M. & Lin, W.M., 2004. Assesment of Parametric Roll Resonance in the Design of Container Carriers. *ABS*, 41–62.

Spyrou, K.J. 2000. Designing against parametric instability in following seas, *Ocean Engineering*, 27: 625–653.

Sustainable Maritime Transportation and Exploitation of Sea Resources – Rizzuto & Guedes Soares (eds)
© 2012 Taylor & Francis Group, London, ISBN 978-0-415-62081-9

Investigation about wave profile effects on ship stability

A. Coraddu, P. Gualeni & D. Villa
University of Genoa, Italy

ABSTRACT: The influence of wave profile on ship righting arm is among the possible stability failure issues addressed by the recent activity at IMO. The present paper is mainly focused on the so called "pure loss of stability", a dangerous phenomenon that can occur for specific conditions in following seas.

In the paper, the development of a computational tool is described, able to evaluate the influence of wave profile on ship metacentric height and righting arms, giving the possibility to investigate different ship geometries and loading conditions in relation with wave profiles in terms of different lengths and steepness. The tool is applied to several ship typologies and a special attention in the investigations is given to transversal metacentric height GM and to the maximum righting arm GZmax since, in some proposals, these are the parameters assumed as indicative while performing first and second level vulnerability analysis for pure loss of stability.

1 INTRODUCTION

1.1 *The rulemaking activity at IMO*

In the recent IMO renewal process about Intact Stability Code (IMO, 2008), an accent is posed on awareness that some ships are more at risk of encountering critical situations in waves. Righting lever variation, resonant roll in dead ship condition, broaching and other maneuvering related phenomena are identified as possible source of such dangerous conditions, involving large roll angles and/or accelerations.

At present an item named "Second generation intact stability criteria" is constantly put on agenda of SLF sub-committee and an inter-sessional correspondence group is regularly established to carry on the activity.

One of the most important technical implications is the development of the so called vulnerability criteria i.e., procedures able to assess the susceptibility to a particular stability failure mode. The modality of interaction among IS Code, vulnerability levels, direct stability assessment and operational guidance are available in SLF 52/WP.1, SLF 53/3/5 and SLF 53/WP.4, where in particular a possible scheme summarizing the whole approach is presented.

In this paper a special attention is paid to ship righting lever variation due to wave profile that, in principle, could induce parametric roll or pure loss of stability or their possible combinations.

A computational tool for analyzing the influence of wave profile is developed and applied. Investigations on several ship typologies are presented, at upright conditions as well as at large heel angles: a modern destroyer-type hull form with tumblehome topside, a Ro-Ro passenger ferry and a container ship are analysed. For comparison purposes also in principle not vulnerable ships like a bulk carrier and a tanker are assessed. Focus of investigation are the transversal metaentric height GM and the maximum righting arm GZmax. In fact, for the loss of stability vulnerability assessment, these are the significant elements of the proposed criteria, at first and second level (SLF 53/WP.4, SLF 53/INF.10).

1.2 *New exigencies for the ship design process*

The ship susceptibility to stability failures in waves is of course dependant on the sea environmental context but a great importance is to be recognized to geometrical properties of the hull. Necessary precautionary provisions may need to be taken during design to address the severity of such phenomena. The new generation intact stability criteria is based on a multi-tiered assessment approach: for a given ship design, each stability failure mode will be evaluated using two level vulnerability criteria. A ship which fails to comply with two levels of vulnerability criteria is requested to be examined with a direct assessment procedure—the third level criteria. The third level criteria should be as close to physics as practically possible, taking into account limitations (validation, verification, organizational and economical burdens) of accessible tools (SLF 53/3/5). The outcome of the direct assessment procedure could be changes in the ship design or development of an operational guidance supplemented with crew training, if necessary.

For several practical reasons it is foreseeable that the ship design team is willing to avoid the burden of a direct assessment. Therefore, in case of ships sensitive to the problem, during the design phase, an iterative process to improve ship performance under the second level of vulnerability is likely to happen.

Whichever will be the final procedure for vulnerability assessment, in terms of "loss of stability", an agile and versatile tool able to support design team during the design process is of outstanding importance, bearing in mind that among the different possible modes of stability failures, some can be avoided by accurate design (this is the case of loss of stability) and some others have to strongly rely on an appropriate operational profile in charge of the training and the experience of the ship crew.

2 THE WAVE PROFILE EFFECT

During the last decades some new ship geometries have developed, as the result of a compromise between the necessity of speed performance and a large amount of volume, devoted mainly to modular cargo like containers or vehicles. This has created a sort of strong differentiation of geometry between the dead-work and the quick-work that influences a lot the actual water plan area and the underwater volume distribution of the ship when a crest of wave of length comparable to the ship length is located at amidships section.

The wave effect should be analyzed as the results of surface integration of hydrodynamic and hydrostatic pressures instantaneously acting on the submerged hull. The methodology applied for the computational tool described in this paper is based on hydrostatic pressures integration only.

As a last comment, it is mentioned that timing may be critical factor for pure loss of stability (Belenky & Sevastianov, 2007). It means that the most critical situation can be envisaged when the wave and ship length are comparable and at the same time, in case of following seas, the ship speed and wave celerity are comparable as well.

3 THE COMPUTATIONAL TOOL

An in-house software has been developed in order to evaluate the stability at upright condition and at large angles for a ship in still water and in longitudinal waves.

The geometry of the ship is defined by transversal sections in principles without any limitations in terms of points per sections and number of sections.

The free surface is defined by a cylindrical sinusoidal wave. The wave profile is defined in terms of wave length and wave height. The ship waterline is the result of geometrical intersection between the hull and undisturbed wave free surface.

The computational tool is able to evaluate any geometrical property of the hull at any longitudinal or transversal angle of inclination, with linear or wave waterline profile: by a numerical integration over the sections all the geometrical characteristics are evaluated, for example the sectional submerged area and the breadth of the waterline. Then with a longitudinal integration all global characteristics are computed, e.g., the center of buoyancy and the hull volume.

The second feature of the code calculates the righting arms GZ, given a fixed volume and a transversal angle. Calculations are carried out with the ship left free to trim.

Therefore this routine is able to find the righting arm (GZ) for any load condition, any imposed roll angle and any sinusoidal wave profile. In Figure 1 an example of ship underwater volume computation with a heel angle of 30 degrees is presented. The model is the tumblehome vessel (it is possible to see the sonar dome at the bow at the right) and it is heeled on portside. The sections number is kept very high in order to better appreciate the surface of the waved waterline.

4 APPLICATIONS

4.1 *General*

The computational tool has been applied to several ship typologies, some deemed to be rather sensitive to the righting arm variation due to the wave profile and some others that in principle should be immune from the problem: a tumblehome vessel, a 9000 TEU container ship, a 50000 GT Ro-Ro passenger vessel operating in the Mediterranean sea, a 41000 DWT tanker and a 54000 DWT bulk carrier are going to be examined. Beside the tumblehome vessel, all the other ships are completely independent from the usual database used by the IMO community representing therefore a possible source of increasing the investigation cases.

In the following, main particulars and geometry of the investigated ships are given. Volume and Vertical Center of Gravity (VCG) given in tables can be considered representatives of possible full load departure condition.

As far as the wave profile is concerned, in each application case the wave length is assumed equal to ship length, and the wave steepness ranges from

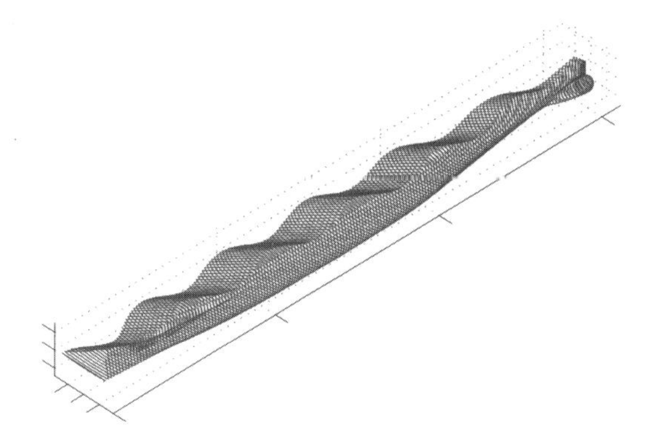

Figure 1. Tumblehome vessel with wave surface at 30 degrees of ship heel angle.

Table 1. Containership main data.

Container ship		
V	[m³]	163530
Lpp	[m]	317.21
B	[m]	45.32
D	[m]	24.60
T	[m]	14.50
VCG	[m]	19.50

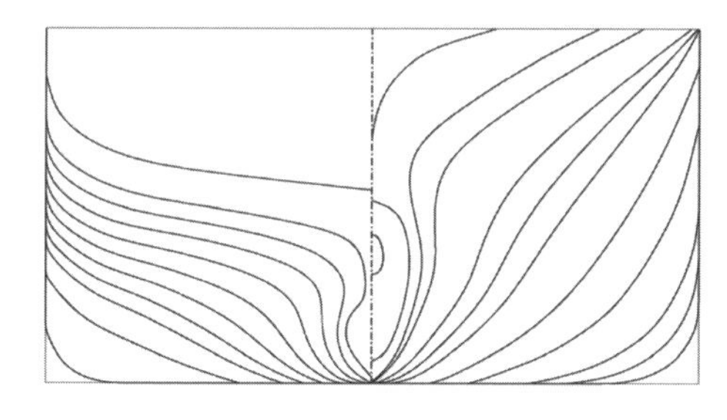

Figure 2. Body plan container ship.

0.01 to 0.1. Calculations are carried out for each ship at the design waterline with relevant vertical center of gravity. Then an inferior draft is defined corresponding to 50% of displacement derived from main data tables and the relevant even keel draft is assumed as initial condition; the VCG values are not changed in order to appreciate solely the stability variation due to the geometry effect. For the tanker, bulk carrier and container ship this can be believed as a standard (rather light) ballast condition. For other ship typologies this draft is to be interpreted just as significantly lower draft, in comparison with the full load departure but still with a reasonable possibility to be deemed a service condition.

Table 2. Ro-Ro ship main data.

RO-RO ship		
V	[m³]	25226
Lpp	[m]	186.21
B	[m]	30.40
D	[m]	15.50
T	[m]	7.45
VCG	[m]	13.38

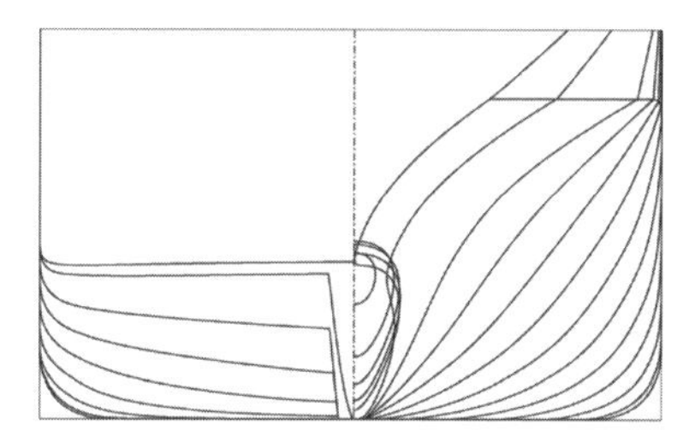

Figure 3. Body plan Ro-Ro ship.

Table 3. Tumblehome main data.

Tumblehome vessel		
V	[m³]	8552
Lpp	[m]	154.00
B	[m]	18.80
D	[m]	16.61
T	[m]	5.50
VCG	[m]	5.50

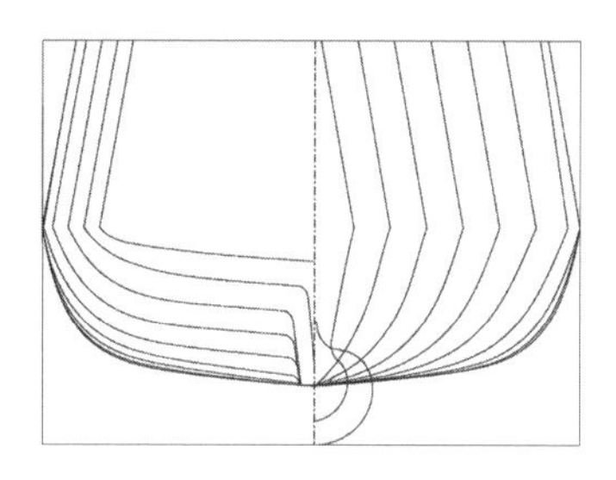

Figure 4. Body plan tumblehome vessel.

Table 4. Bulk carrier main data.

Bulk Carrier		
V	[m³]	62926
Lpp	[m]	194.50
B	[m]	32.25
D	[m]	17.00
T	[m]	12.40
VCG	[m]	9.67

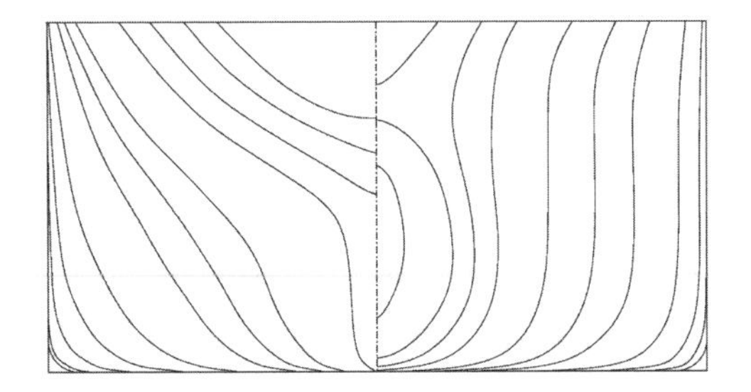

Figure 5. Body plan bulk carrier.

Table 5. Tanker main data.

Tanker		
V	[m³]	77346
Lpp	[m]	204.74
B	[m]	34.00
D	[m]	23.21
T	[m]	13.71
VCG	[m]	14.20

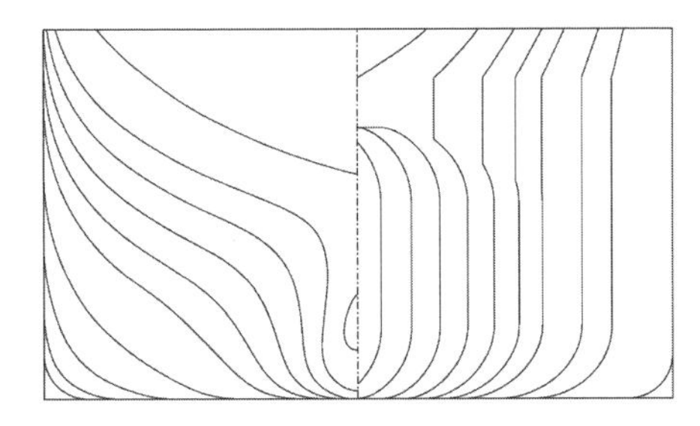

Figure 6. Body plan tanker.

5 COMMENTS TO RESULTS

All the investigated hulls are sensitive to the effect of a longitudinal sinusoidal wave profile, but with different emphasis.

GZ diagrams show differences in terms of initial.

Beside the full load departure condition, a lower draft has been considered with in mind the possible more pronounced geometrical sensitivity of the lower part of the hull to the wave profile effect. From results it is evident that the full load departure conditions is always the most important to be assessed: the increment of stability at lower draft is so important that the possible negative effect of wave profile is not significant.

The wave steepness ranges from 0.01 to 0.1 as suggested in the Japan proposal (SLF53/INF.10) related to the second level of vulnerability for pure loss of stability assessment.

In the analysis of the following results a special attention is going to be given to GM and to GZmax position and value.

Going into details for each specific hull typology, the tumblehome vessel (Figs. 7a, b) seems to be the less sensitive to the wave profile at both drafts.

The tanker (Figs. 8a, b), as expected, is rather imperturbable as well, with a light influence only on the maximum GZ value at the deepest draft. The situation nevertheless is very far to be critical.

The bulk carrier GZ diagrams (Figs. 9a, b), at the lower draft, present curves that are nearly not distinguishable up to 15 degrees of hell angle, i.e., the metacentric heights GM is nearly imperturbable, therefore the KM values. The position of the maximum GZ is nearly fixed at 40 degrees, for any wave steepness.

Ro-Ro (Figs. 10a, b) and container ships (Figs. 11a, b), as expected, are the most sensitive (vulnerable) to the wave profile effect.

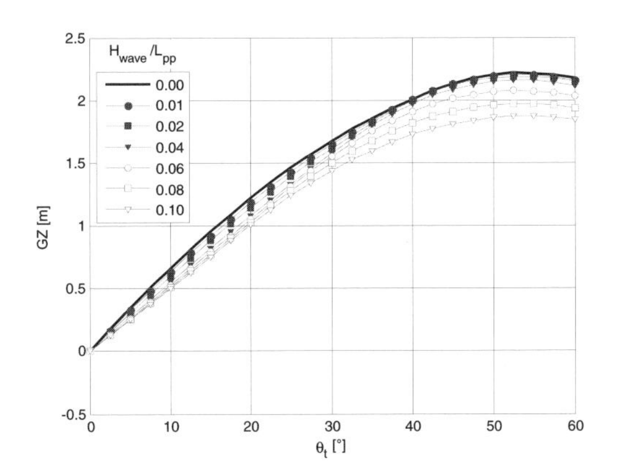

Figure 7a. Tumblehome vessel GZ curves at deeper draft.

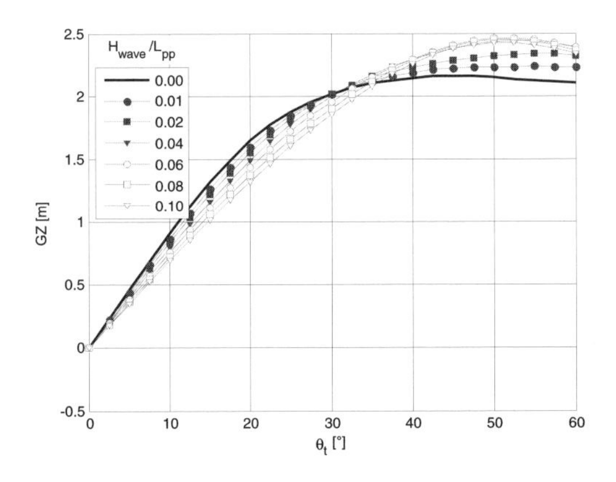

Figure 7b. Tumblehome vessel GZ curves at lighter draft.

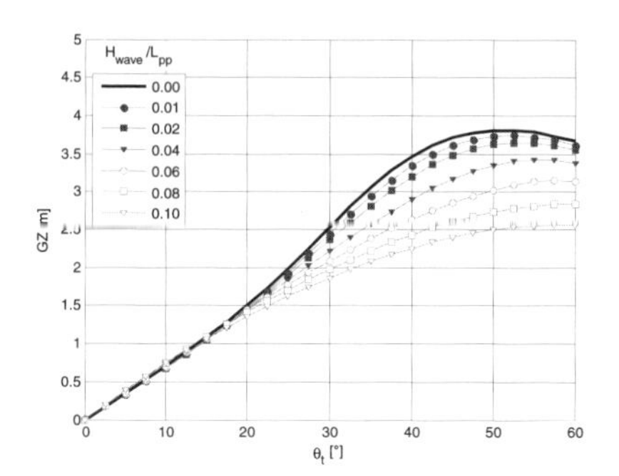

Figure 8a. Tanker ship GZ curves at deeper draft.

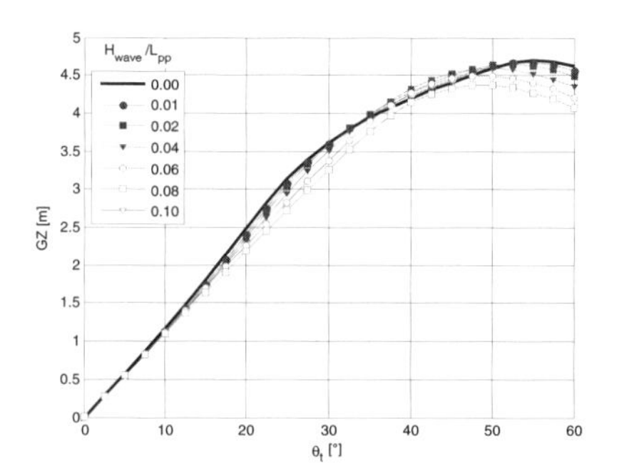

Figure 8b. Tanker ship GZ curves at lighter draft.

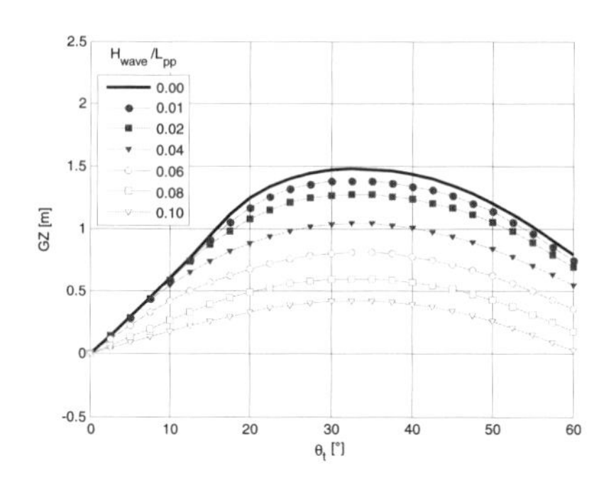

Figure 9a. Bulk carrier GZ curves at deepest draft.

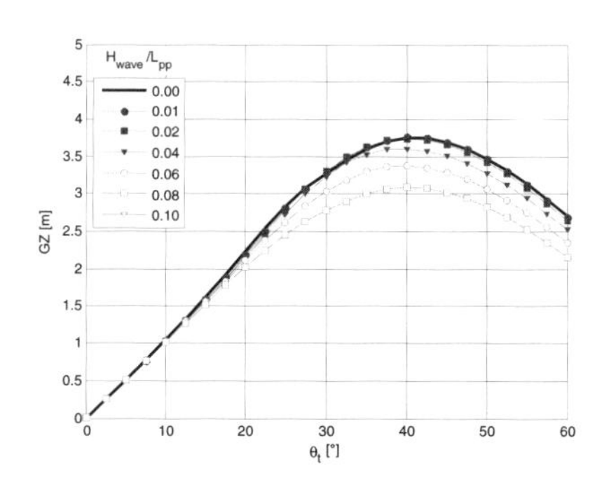

Figure 9b. Bulk carrier GZ curves at lighter draft.

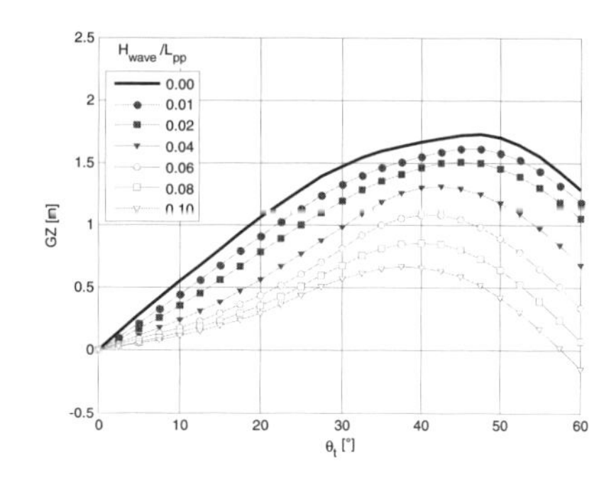

Figure 10a. Ro-Ro ship GZ curves at deeper draft.

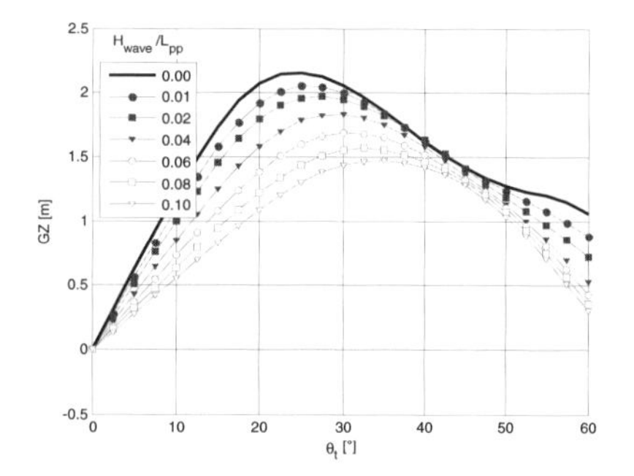

Figure 10b. Ro-Ro ship GZ curves at lighter draft.

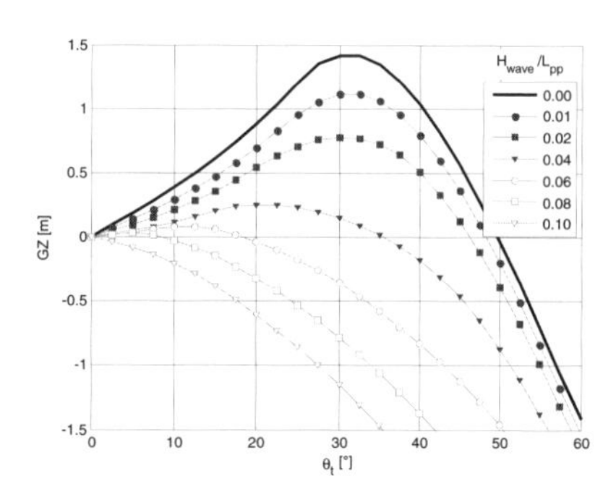

Figure 11a. Container ship GZ curves at deeper draft.

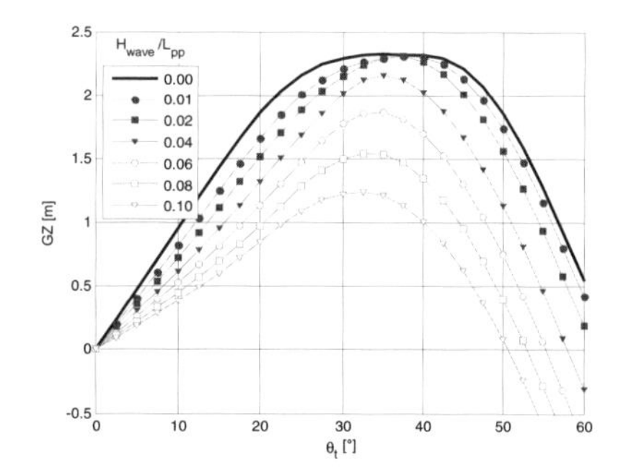

Figure 11b. Container ship GZ curves at lighter draft.

At the deepest draft the GZ curves of Ro-Ro show a significant diminution in terms of GM values. Also the maximum GZ arm is much affected and it reduces nearly to one third with the highest wave steepness. The position of the maximum GZ shifts of nearly ten degrees toward the origin. The same parameter is very much influenced in the lightest draft condition but with less influence on the arm values. The GM is very variable as well.

The investigated container ship, for the lower draft shows a variation in terms of GM ad of maximum GZ that become nearly one half, without any significant change in its position. At the deepest draft she presents the most critical results among the investigated cases. The GM reduction is so remarkable that the ship becomes unstable at the upright condition; the whole GZ diagrams are negatively influenced and from steepness 0.04 the situation appears to be unacceptable, from the safety point of view.

From the above mentioned results, it is confirmed that GM and GZmax values are the most important parameters and only the Ro-Ro ship has evidenced an important variation also of the GZmax position due to wave steepness increment.

For a further insight into the problem, an investigation about the maximum VCG, for each ship typology, has been carried out at the deepest draft, with and without wave profile. For this evaluation the wave profile has been selected in relation with draft i.e., wave height is equal to 0.25T.

Maximum VCG are evaluated in relation with two separate criteria: the GM not less than 0.15 m, the maximum GZ not less than 0.20 m.

In Figure 12 results are reported in terms of maximum VCG for each ship for the two above mentioned criteria. The influence of the wave profile has been already evidenced with previous figures and it is confirmed also in this one.

The aim this time is to investigate whether the GM criterion is more or less severe that the GZmax

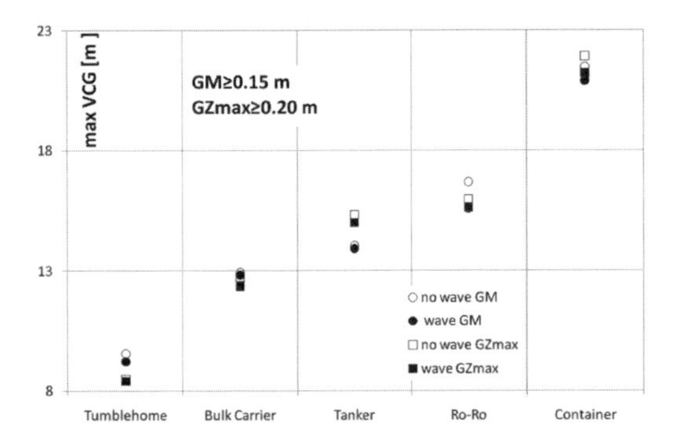

Figure 12. Maximum VCG from GM and GZmax criteria for all ships.

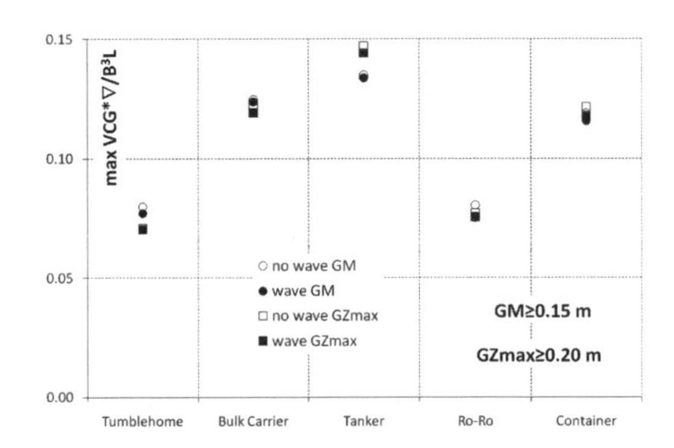

Figure 13. Maximum VCG from GM and GZmax criteria for all ships in non dimensional form.

one. Focusing comments only on sensitive ships, it seems that the GZmax criterion is more sever for the tumblehome, while for the container vessel the most severe constraint derives from the GM criterion. For the Ro-Ro ship the two criteria are nearly equivalent.

In Figure 13 the same results are reported in a non-dimensional form.

6 CONCLUSIONS

A computational tool has been developed inhouse to investigate the ship stability, at upright conditions and at large angles, influenced by wave profile. The necessity to perform this kind of studies and to develop the proper tools is raised by the recent IMO activity about the second generation intact stability criteria. In particular this paper has focused on the problem of pure loss of stability due to the longitudinal wave with crest at amidships.

Five ship typologies have been assessed at different drafts with sinusoidal wave profile at different steepness (but always wave length equal to ship length). Ships have shown different vulnerability to the phenomenon. As it was expected tanker and bulk carrier ships are not very sensitive; the tumblehome vessel is less perceptive to the problem than expected. Ro-Ro and container ship geometries are indeed the most susceptible to the wave profile effect with a really important effect on the container ship.

The investigation have been performed at two different drafts, the full load condition and a lower draft; the deepest draft has shown to be always the most significant one.

Some considerations in comment to results are given in relation to maximum VCG, evaluated separately with two criteria, i.e., GM and GZmax, in order to confirm their selection as significant parameter for respectively the first and the second level of vulnerability for pure loss of stability problem.

ACKNOWLEDGMENTS

Authors would like to express their huge gratitude to Mr Luca Bonfiglio for the friendly concession of the technical documentation relevant to the design of the 9000 TEU container ship he developed for his final graduation thesis.

We thank very much also Ms Alida Abbo and Mr Filippo Gozzi, for their initial work, during their final graduation thesis activity, on the computing tool we further developed and applied for this paper.

REFERENCES

Belenky, V.L. & Sevastianov, N.B., 2007. Stability and Safety of Ships. Bhattacharyya and Mc Cormic Editors—*The Society of Naval Architects and Marine Engineers.*

IMO, 2008. Adoption of the International Code On Intact Stability (2008 IS CODE), Resolution MSC.267(85) MSC 85/26/Add.1. London.

SLF 52/WP.1, 2010. Development Of New Generation Intact Stability Criteria—Report of the working group (part 1). *International Maritime Organization*, London.

SLF 53/3/5, 2010. Development Of New Generation Intact Stability Criteria—Comments on the structure of new generation intact stability criteria—(report of the correspondence group)—Submitted by Poland. *International Maritime Organization*, London.

SLF53/INF.10, 2010. Development Of New Generation Intact Stability Criteria—Information collected by the Correspondence Group on Intact Stability—Submitted by Japan. *International Maritime Organization*, London.

SLF 53/WP.4, 2011. Development Of Second Generation Intact Stability Criteria—Report of the working group (part 1). *International Maritime Organization*, London.

Sustainable Maritime Transportation and Exploitation of Sea Resources – Rizzuto & Guedes Soares (eds)
© 2012 Taylor & Francis Group, London, ISBN 978-0-415-62081-9

Examination of the transverse stability of fishing vessels by residual energy

E. Ücer
Istanbul Technical University, Istanbul, Turkey

ABSTRACT: In this study, the transverse stability of eight fishing vessels is analyzed by determining their residual energy for different sea and weather conditions. For this aim, the single degree of freedom rolling motion is modeled by a nonlinear differential equation including steady wind and gust moments. In the method of residual energy, firstly the nonlinear equation is integrated for a bounded area of initial conditions. Second, capsizing initial conditions and their energies are determined. Finally, the smallest energy of the capsizing initial conditions is chosen as residual energy which shows the width of safe domain of fishing vessel. If it is equal to zero, there is no safe domain of fishing vessel. From the results of this study, it can be concluded that the capsizing probability of the fishing vessels increases dramatically by increment of wind forces and decrement of beam.

1 INTRODUCTION

Oceans and seas have random, unpredictable and sometimes chaotic character in rough weather conditions. A sudden strong gust can cause capsizing of statically stable small fishing vessels instantly. Therefore, capsizing of small ships such as fishing vessels in rough weather conditions is a nonlinear phenomenon. Different analytical and numerical methods and also experimental studies to analyze nonlinear capsizing phenomenon of small fishing vessels and to improve the intact stability rules were developed or used by scientists and researchers.

The essential examples of experimental studies for analyzing capsize of the ships are the studies of Wright & Marshfield (1980), Grochowalski (1989) and Cotton & Spyrou (2001). In the study of Wright and Marshfield (1980), roll response and capsize behavior in beam seas is presented. In the experimental study of Grochowalski (1989), the mechanism of ship capsizing in quartering and beam waves for light and full load condition, danger created by bulwark submergence and accumulation of water on deck and also stability reduction on a wave crest are presented. In the experimental study of Cotton and Spyrou (2001), the results of a series of capsize tests based on a prismatic model ship are presented and a capsize boundary is derived.

Melnikov method, Lyapunov exponents and Lyapunov's Direct Method are the essential analytical methods used for analyzing the stability of ships in beam seas. Melnikov method is used to derive the capsizing criterion for ships in regular and irregular seas by Falzarano (1990), Hsieh et al. Troesch & Shaw (1994) and Jiang (1995). The method of Lyapunov exponents is used to analyze and predict capsize and other chaotic behavior of ships by McCue (2004). Lyapunov Direct method is used to analyze the stability of ships by Odabaşı (1978), Ozkan (1981) and Caldeira-Sariava (1986).

Numerical investigation is the other way to analyze the nonlinear capsizing of ships in beam seas. In rough seas, the capsizing of ships is a nonlinear phenomenon and it is highly dependent on the variation of initial conditions (Thompson, 1989). For that reason, the method of numeric safe basin which enables to show the effects of thousands of initial conditions on the stability of ships by using only one graphic is commonly used to analyze the stability of ships. This method is firstly used by Thompson (1989), Soliman & Thompson (1991) and Rainey & Thompson (1991) to analyze the stability of ships. In those studies, transient capsizing diagrams showing critical wave height against capsizing are also presented.

In the case of large amplitude rolling, rolling motion of a ship is stochastic. In 2004, large amplitude rolling in a realistic sea is examined by Francescutto & Naito. Calculation of capsizing rate of a ship in beam sea is presented by Hoon and Pyo (2006). An analytical method of capsizing probability in the time domain for ships in the random beam seas is presented by Liu, Tang & Li (2007). The survival probability is calculated by estimating erosion of "safe basin" during ship rolling motion by Monte Carlo simulations (Long, Lee & Kim, 2010).

In this study, the transverse stability of eight fishing vessel which have the same displacement, beam-draught ratio (B/T) and block coefficient (C_B) is analyzed considering regular sea by the method of residual energy. However, the method of residual energy is also applicable to examine stochastic rolling motion. Method of residual energy is the expansion of numeric safe basin concept developed by Thompson (1989). Residual energy method is based on the investigation of the existence of safe basin around origin. By using this method, the width of the domain of safe initial conditions is determined. If there is no safe basin, the ship definitely capsizes. The rapid reduction of residual energy indicates the beginning of claw type erosion (Thompson, 1989). The maximum allowable wave height (critical wave height) is assumed to be the wave height causing initiation of claw type erosion.

2 MATHEMATIC MODEL

In beam seas, rolling motion of the ship has a greater influence on ship stability rather than the other modes of ship motion. Due to the difficulty of accurately determining the complete hydrodynamic forces, a rolling model which decouples the six degrees of freedom is generally assumed. Generally, only roll and sway motions are considered for the purpose of ship stability analysis (Jiang, 1995). The roll and sway coupled model can be reduced to a 1-DOF rolling model if a virtual roll centre is introduced (Jiang 1995 and Balcer 2004). In this paper, while the rolling motion of the ship is being modeled, interactions between rolling and other modes of motion are ignored by defining virtual roll centre and the ship is considered to have a rigid body. Under these assumptions, rolling motion of a ship is written as in Eq. (1).

$$I\ddot{\phi} + D(\phi,\dot{\phi}) + M_R(\phi) = E(t) + M_{wind} \quad (1)$$

where ϕ is rolling angle with respect to calm sea surface (rad), $\dot{\phi}$ is roll angular velocity (rad/s), I is virtual moment of inertia corresponds to a virtual (physical) axis of rotation, located at the virtual ship mass centre (the mass centre of the ship along with the added mass in sway), as discussed by Balcer (2004).

$D(\phi)$ is Non-linear damping moment. It is amplitude dependant, normally obtained from free roll tests. In this study, it is determined by the semi empirical formulas of Himeno (1981) and modeled as shown below:

$$D(\dot{\theta}) = B_e\dot{\phi}$$

where B_e is the equivalent linear damping coefficient

$M_R(\phi)$ is righting moment in calm water and can be represented as follows:

$$M_R(\phi) = \Delta GZ(\phi)$$

where Δ is the buoyancy force and GZ is the righting arm as a function of the roll angle. GZ can be approximated by odd polynomials.

E(t), roll exciting moment is the hydrodynamic moment due to a regular wave. This moment is calculated according to the linear theory, in which the ship is in an upright position and can be approximated as follows (Senjanovic et al., 2000).

$$E(t) = \kappa K \zeta_0 \left[\left(\Delta GM - \omega^2 m_{44}\right)^2 + \left(\omega N_{44}\right)^2 \right]^{1/2} \times \cos\omega t$$

where κ is the reduction coefficient for the effective wave slope, GM is the initial metacentric height, $K = 2\pi/\lambda$ is the wave number, $N_1 \equiv N_{44}$ is linear damping moment, ($\zeta_0 = 1/2 \times h$) is the wave amplitude, and ω is the wave circular frequency. Assuming that ($\omega \times N_{44}$) is negligible in relation to ($\Delta k_1 - \omega^2 m_{44}$) and $\Delta GM = \omega_0^2 I$, the above equation takes the form

$$E(t) = E_0 \cos\omega t$$

where $E_0 = \kappa(\pi h/\lambda)\,|\omega_0^2 I - \omega^2 m_{44}|$ is the amplitude of wave excitation. The wave slope ($\pi h/\lambda$) is taken smaller than 12° in numerical computation.

M_{wind}, wind moment consists of two parts steady wind and gust moments. Steady wind moment (M_{sw}) is assumed to be dependent on roll angle (Wendel, 1967 and Odabasi, 1976) and defined as follows:

$$M_{sw} = M_0 \cos\phi$$

where M_0 represents the side wind moment on a ship in upright position and determined as follows:

$$M_0 = (1/2)\,\rho_A V_w^2\, A\, Z\, \varsigma_w$$

where ρ_a is the air density, V_w is the velocity of the wind, A is the lateral area including erections and rigging exposed to the wind (m^2), Z is the vertical distance between the centre of wind pressure and the centre of water pressure: in practice it is taken as the vertical distance between the centre of exposed area and a point at half draught (m) and ς_w is the wind pressure coefficient (Wendel, 1967).

Gust moment is also roll angle dependant and can be defined as follows (Ucer & Odabasi, 2008):

$$M_G = E_G\left[H(t-t_0)-H(t-t_0-\tau)\right]\cos\phi$$

where E_G is the magnitude of gust moment, t_0 is the initiation of gust, τ is duration of gust and H(t) is Heaviside function defined as follows:

$u > 0 \rightarrow H(u) = 1$ and $u < 0 \rightarrow H(u) = 0$

In the lights of these assumptions, Eq. (1) can be written as follows:

$$\begin{aligned} & I\ddot{\phi} + B_e\dot{\phi} + \Delta GM\left(\phi + c_3\phi^3 + c_5\phi^5 + c_7\phi^7\right) \\ & = E_0\cos\omega t \\ & \quad +(M_0 + E_G\left(H(t-t_0)\times H(t-t_0-\tau)\right)\cos\phi \end{aligned} \tag{2}$$

Dividing both sides of the equation by the virtual moment of inertia and substituting for the wave length $\lambda = 2\pi g/\omega^2$, the expression given in Eq. (3) is obtained.

$$\begin{aligned} & \ddot{\phi} + b_e\dot{\phi} + \omega_0^2\left(\phi + c_3\phi^3 + c_5\phi^5 + c_7\phi^7\right) \\ & = e_0\cos\omega t \\ & \quad +(m_0 + e_G\left(H(t-t_0)\times H(t-t_0-\tau)\right)\cos\phi \end{aligned} \tag{3}$$

where $e_0 = (h/2\,g)\,\kappa\,\omega^2\,|\omega_0^2 - \omega^2 m_{44}/I|$, $\omega_0^2 = \Delta GM/I$, $m_0 = 0.5\,I\,\rho_A V_w^2\,A\,Z\,\varsigma_w$, $b_e = B_e/I$ and $e_G = E_G/I$

3 FISHING VESSELS AND SEA STATES

Eight fishing vessels which have the same displacement (Δ), beam-draught ratio (B/T) and block coefficient are taken as sample ships. The main dimensions and stability characteristics of those fishing vessels are presented in Table 1 and Table 2 respectively.

In Table 2, GM is the initial metacentric height, GZ_{max} is the maximum righting arm, ϕ_m is the angle where GZ_{max} occurs, A_{30} and A_{40} are the area under $GZ - \phi$ curve till to 30° and 40° respectively.

Table 1. Main dimensions of fishing vessels (Yılmaz, 1998).

Ship No	L (m)	B (m)	T (m)	D (m)	Δ (ton)
1	21.4	6.75	2.70	3.780	200
2	22.6	6.57	2.63	3.679	200
3	23.3	6.47	2.59	3.624	200
4	24.3	6.34	2.53	3.548	200
5	25.2	6.22	2.49	3.484	200
6	26.1	6.11	2.44	3.423	200
7	28.4	5.86	2.34	3.283	200
8	29.0	5.80	2.32	3.248	200

Table 2. Static stability characteristics of fishing vessels (Yılmaz, 1998).

Ship No	GM (m)	GZ_{max} (m)	ϕ_m(rad)	A_{30} (m × rad)	A_{40} (m × rad)
1	0.334	0.229	0.660	0.053	0.092
2	0.336	0.230	0.666	0.053	0.092
3	0.339	0.231	0.679	0.053	0.091
4	0.337	0.230	0.672	0.053	0.091
5	0.344	0.233	0.677	0.053	0.092
6	0.349	0.237	0.681	0.054	0.094
7	0.351	0.237	0.688	0.054	0.093
8	0.350	0.238	0.691	0.053	0.093

Table 3. Wind and sea state (Sabuncu, 1993).

Beaufort wind force	Wind velocity (knot)	Wave height $H_{1/3}$ (m)	$H_{1/10}$ (m)	Characteristic wave period
4	12	0.67	0.85	1.0–7.0
	13.5	0.88	1.12	1.4–7.6
	14	1.01	1.28	1.5–7.8
	16	1.40	1.77	2.0–8.8
5	18	1.85	2.40	2.5–10.0
	19	2.10	2.65	2.8–10.6
	20	2.45	3.05	3.0–11.1
6	22	3.05	3.95	3.4–12.2
	24	3.65	4.90	3.7–13.5
	24.5	3.95	5.20	3.8–13.6
	26	4.60	6.10	4.0–14.5
7	28	5.50	7.00	4.5–15.5
	30	6.70	8.50	4.7–16.7
	32	7.95	10.0	5.0–17.5

In Table 3, wind and sea states are shown.

4 EXAMINATIONS OF STABILITY BY RESIDUAL ENERGY METHOD

Before beginning the stability analysis of fishing vessels by residual energy method, the bounded area of initial conditions should be defined firstly. The bounded area of initial conditions is expressed as follows:

$$A_B : \left\{ (\phi,\dot{\phi}) : -\phi_m \le \phi \le \phi_m, -\dot{\phi}_m \le \dot{\phi} \le \dot{\phi}_m \right\}$$

where A_B is divided into 201 × 201 points, and the lattice points are taken as the initial values for the solutions of Eq. (3) (Long, Lee & Kim 2010). θ_m (the maximum initial roll angle) is equal to the angle of deck immersion or the angle at which the bilge comes out from water or 0.6981 radians whichever is smaller. $\dot{\phi}_m$ is the maximum initial roll angular velocity and assumed to be equal to 0.65 rad/s.

Secondly, single degree of freedom nonlinear rolling motion equation, Eq. (3) is integrated by using DLSOAD sub program (Hindmarsch, 1982) for each initial condition. The integrations are continued until either roll angle exceeds a capsizing criterion (vanishing stability angle of ship in calm water), at which point the ship is assumed to be capsized or the maximum allowable excitation cycles (twenty excitation cycle) is reached, in which case it is assumed that the ship will remain upright under these condition (Soliman & Thompson, 1991). If the ship is capsized (roll angle exceeds capsize criterion), the initial condition is defined as unsafe initial condition whereas if the ship is not capsized, the initial condition is called as a safe initial condition.

Finally, residual energies of the fishing vessels for capsizing initial conditions are determined as follows:

$$\frac{\phi_i^2}{2}+\frac{\Delta}{I}\int_0^{\phi_i} GZ(\phi)\,d\phi = V_i \quad i=1,\dots n \tag{4}$$

where V_i is the energy and n is the number of unsafe initial conditions

$$V_{residual} = \mathrm{Min}[V] \tag{5}$$

The magnitude of the residual energy shows the width of the safe domain around the origin. By the increment of residual energy, the size of the safe basin increases or in other words the survivability of the ship increases. When the residual energy is equal to zero, there is no safe basin of the roll motion.

Normalized residual energy is obtained by dividing residual energy with the area under the GZ – ϕ curve.

When the wind force is Beaufort 5 and 6, the variation of normalized residual energies due to wave height and frequency are shown in Figure 1a-b for Ship 1, in Figure 2a-b for Ship 4 and in Figure 3a-b for Ship 8.

Ship 1 has a larger beam than the other ships with same block coefficient and beam-draught ratio so it has a higher damping moment and can stand more steep waves. For example, Ship 1 can operate safely when the wind force is Beaufort 6 and wave height is smaller than five meters whereas ship 8 can operate safely when the wave height is smaller than three meters.

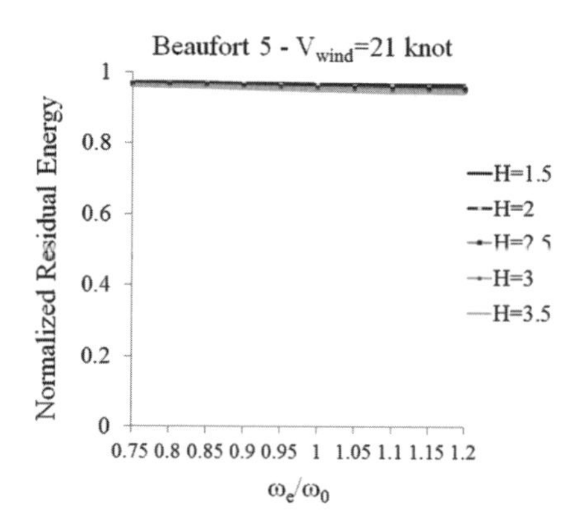

Figure 1a. Normalized residual energy of ship 1 (B_n5, $V_w = 21$).

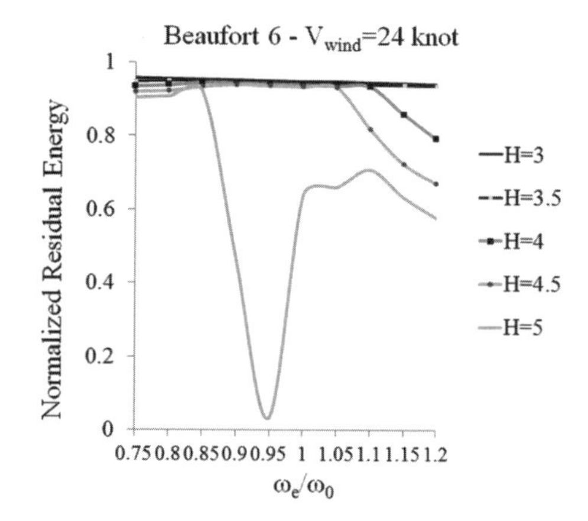

Figure 1b. Normalized residual energy of ship 1 (B_n6, $V_w = 24$).

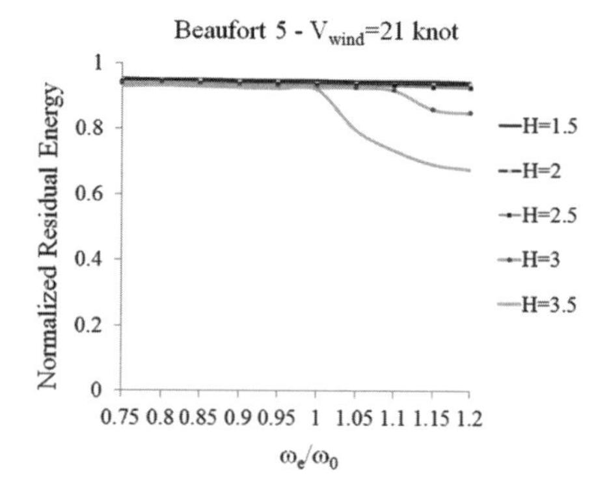

Figure 2a. Normalized residual energy of ship 4 (B_n5, $V_w = 21$).

As can be seen from Figure 1b, when the wave heights are equal to 4.5 m and 5 m, the corresponding normalized residual energy values are 0.936 and 0.034 respectively. That sudden

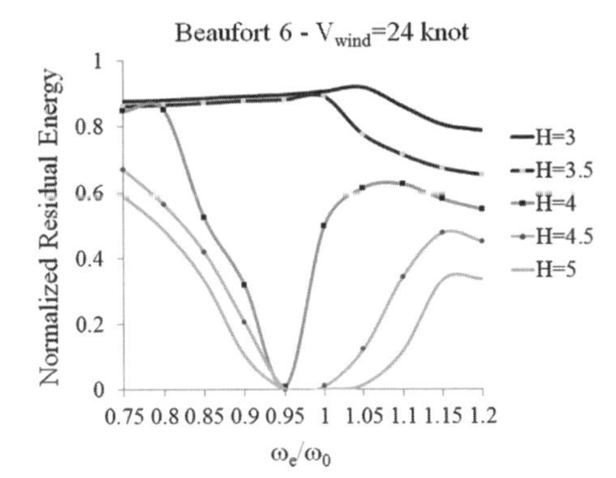

Figure 2b. Normalized residual energy of ship 4 (B_n6, V_w = 24).

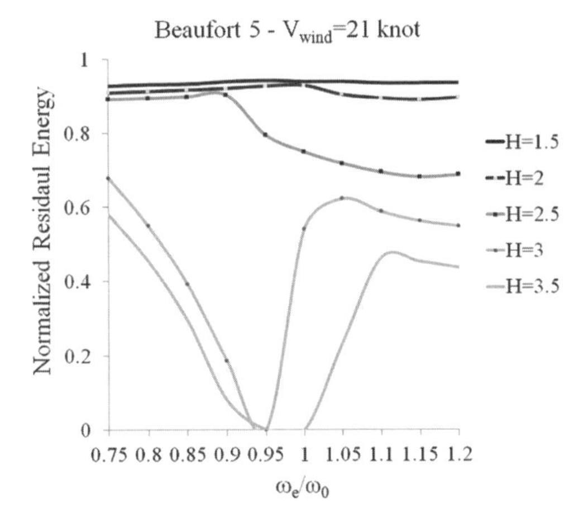

Figure 3a. Normalized residual energy of ship 8 (B_n5, V_w = 21).

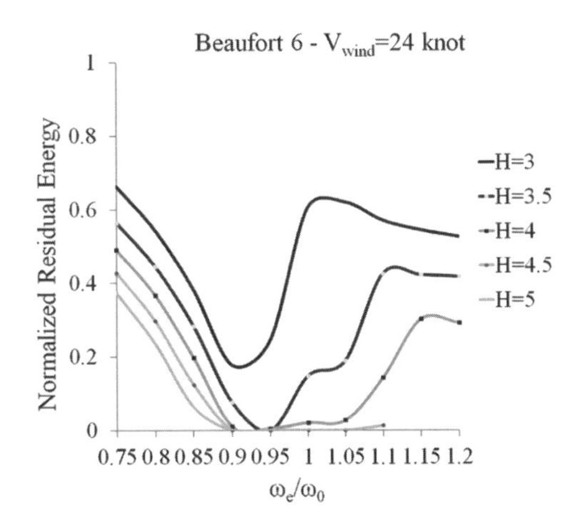

Figure 3b. Normalized residual energy of ship 8 (B_n6, V_w = 24).

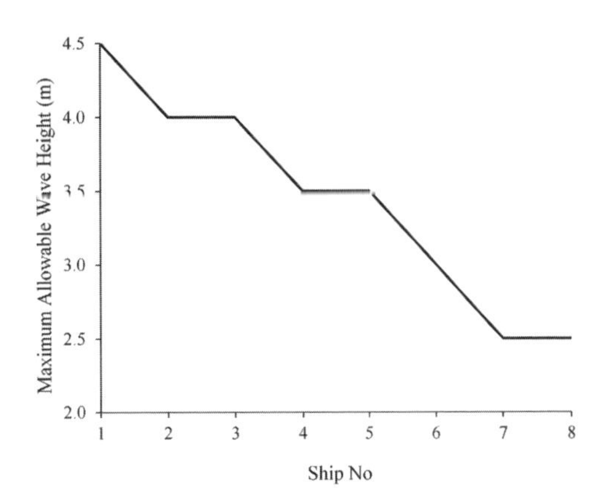

Figure 4. Maximum allowable wave heights for ships when the wind force is Beaufort 6.

reduction in residual energy indicates the claw type erosion. The wave height causing initiation of claw type erosion is chosen as maximum allowable wave height (critical wave height).

In Figure 4, the maximum allowable wave heights for each ship are presented when the wind force is Beaufort 6.

The safe operating conditions of other fishing vessel are determined by finding critical wave heights for each sea condition and summarized in Table 3. The classification of being safe, dangerous or etc. is as follows:

Safe: The maximum allowable wave height is greater than the observable wave heights and characteristic wave height of the wind force.

Dangerous: The maximum allowable wave height is greater than characteristic wave heights of the wind force but the observable wave heights can be greater than maximum allowable wave height.

Very dangerous: The maximum allowable wave height is equal to characteristic wave height of the wind force but the observable wave heights can be greater than maximum allowable wave height.

Table 3. Operation status of fishing vessels for different wind forces.

Ship No	Wind force beaufort 5	Wind force beaufort 6
1	Safe	Dangerous
2	Safe	Very dangerous
3	Safe	Very dangerous
4	Safe	Incapable of operate
5	Safe	Incapable of operate
6	Dangerous	Incapable of operate
7	Dangerous	Incapable of operate
8	Dangerous	Incapable of operate

Incapable of operate: The maximum allowable wave height is smaller than characteristic wave height of the wind force.

5 CONCLUDING REMARKS

The aim of this study is to determine the safe operating conditions of fishing vessels by using residual method. Outcomes of this work can be presented as follows:

- Residual energy method enables to find the width of safe basin around the origin in roll angle and roll angular phase space for each sea condition.
- The rapid reduction in residual energy indicates the occurrence of claw type erosion. The wave height causing initiation of claw type erosion is critical wave height.
- The decrement of beam-wave length ratio (B/λ) and wind velocity (V_w) enable the fishing vessels operate at more steep waves.
- The fishing vessels which have higher damping moments due to their larger beams can stand worse weather and sea conditions.
- Whole of the fishing vessels operate safely at Beaufort 4.
- Whole of the fishing vessels capsize at Beaufort 7.
- In beam seas, the most dangerous scenario for a fishing vessel occurs when the ratio of wave frequency to natural frequency is around one.

In rough weather and sea conditions, large amplitude rolling occurs. Therefore, for further studies, stochastic rolling motion should be considered and method of residual energy can be used in order to obtain better results.

REFERENCES

Balcer, L. 2004. Location of ship rolling axis. *Polish Maritime Research* 11(1): 3–7.

Caldeira-Sariava, F. 1986. The boundedness of solutions of a Leinard Equation arising in the theory of ship rolling. *IMA Journal of Applied Mathematics* 36: 126–139.

Cotton, B. & Spyrou, K.J. 2001. An experimental study of nonlinear behavior in roll and capsize. *International Shipbuilding Progress* 48(1): 5–18.

Falzarano, J.M. 1999. Predicting complicated dynamics leading to vessel capsize. *PhD. Thesis*, University of Michigan, Ann Arbor.

Francescutto, A. & Naito, S. 2004. Large amplitude rolling in realistic sea. *International Ship Building Progress* 51(2–3): 221–235.

Grochowalski, S. 1989. Investigation into the Physics of Ship Capsizing by Combined Captive and Free running Model Tests. *Trans. SNAME* 97: 169–212.

Himeno, Y. 1981. Prediction of Ship Roll Damping-State of the Art. Report 239, University of Michigan, Ann Arbor.

Hindmarsch, A.C. 1982. ODEPACK, A Systemized Collection of ODE Solvers. *Tech. Rept. UCRL–88007*. Lawrance Livermore National Lab.

Hoon, S.K. & Pyo, R.K. 2006. Calculation of a capsizing rate of a ship in stochastic beam seas. *Ocean Eng.* 33(3–4): 425–438.

Hsieh, S.R. et al. 1994. A nonlinear probabilistic method for predicting vessel capsizing in random beam seas. *Proceedings of Royal Society London* A446: 1–17.

Jiang, C. 1995. Highly Nonlinear Rolling Motion Leading to Capsize. *PhD. Thesis*, University of Michigan, Ann Arbor.

Liu, L., Tang, Y. & Li, H. 2007. Analytical method of capsizing probability in the time domain for ships in the random beam seas. *Frontiers of Architecture and Civil Engineering in China* 1(3): 361–366.

Long, Z.Z., Lee, S.K. & Kim, J.Y. 2010. Estimation of survival probability for a ship in beam seas using the safe basin. *Ocean Eng.* 37: 418–424.

McCue, L. 1995. Chaotic Vessel Motions and Capsize in Beam Seas. *PhD. Thesis*, University of Michigan, Ann Arbor.

Odabasi, A.Y. 1976. Ultimate stability of ships. *Trans. RINA* 118: 237–262.

Odabasi, A.Y. 1978. Conceptual understanding of the stability theory of ships. *Schiffstechnik* 25: 1–18.

Ozkan, I.R. 1981. Total (practical) Stability of ships. *Ocean Eng.* 8: 551–598.

Rainey, R.C.T. & Thompson, J.M.T. 1991. The transient capsize diagram-a new method of quantifying stability analysis. *Journal of Ship Research* 35(1): 58–92.

Sabuncu, T. 1993. *Ship Motions*. Istanbul Technical University Press, Istanbul (In Turkish).

Senjanovic, I., Cipric, G. & Parunov, J. 2000. Survival analysis of fishing vessels rolling in rough seas. *Phil. Trans.: Mathematical, Physical and Engineering Sciences* 358(1771): 1943–1965.

Soliman, M.S. & Thompson, J.M.T. 1991. Transient and steady state analysis of capsize phenomena. *Applied Ocean Research* 13(2): 82–92.

Thompson, J.M.T. 1989. Loss of engineering integrity due to the erosion of absolute and transient basin boundaries. *Proceedings of IUTAM Symposium on the Dynamics of Marine Vehicles and Structures in Waves*, 313–320.

Ucer, E. & Odabasi, A.Y. 2008. Significance of roll damping on weather criteria. *Trans. RINA Int. Journal of Maritime Eng.* 150(A1): 1–8.

Wendel, K. 1967. Safety from capsizing. In Traung, J.O. (ed.), *Fishing Boats of the World*: 496–504. New Jersey.

Wright, J.H.G. & Marshfield, W.B. 1980. Ship roll response and capsize behavior in beam seas. *Trans. RINA*. 122: 129–149.

Yılmaz, H. 1998. The examination of form parameters of fishing vessels in preliminary design stage for the practical ship stability criteria. *PhD. Thesis*, Yıldız Technical University, Istanbul, (In Turkish).

Sustainable Maritime Transportation and Exploitation of Sea Resources – Rizzuto & Guedes Soares (eds)
 ISBN 978-0-415-62081-9

An investigation on parametric rolling prediction using neural networks

J.-R. Bellec
ENSTA-Bretagne, Brest, France

C.A. Rodríguez & M.A.S. Neves
UFRJ, Rio de Janeiro, Brazil

ABSTRACT: Parametric rolling is a recently identified phenomenon that affects vessels and is characterized by large roll angles that lead to severe rolling of the vessel and even capsizing. Parametric rolling is caused by a periodic variation of the self-righting characteristics of the ship. Due to its short inception time it is necessary to have an on-board warning system that monitors the current state and provides an advance warning of possible onset of parametric rolling. Vessels mainly encounter parametric rolling in monochromatic head seas.

In this paper an application of artificial neural network technology to head seas parametric rolling prediction will be discussed. Neural network is an algorithm that imitates the mechanism of neurons in the brain. It can learn a function given by input-output pairs and return approximate outputs for inputs that were not given. Such algorithms are already used in naval architecture for approximation, control and classification.

Here neural networks will be used in a recursive manner with discrete time-series to predict three to five future natural rolling periods to allow time to react and to counteract the phenomenon. The model is then improved in a remarkable way to include pitch and frequency data. The authors developed a systematic methodology and validation method which include the use of multiple initial conditions to avoid biased data.

Experimental data were obtained in monochromatic head seas with a hull of a modern container vessel and a nonlinear numerical model using six degrees of freedom with terms defined up to third order derivatives Rodríguez (2010). This numerical model was shown to provide a good prediction of parametric rolling.

1 INTRODUCTION

Parametric Rolling is a rare phenomenon involving sudden large amplitude rolling oscillation of the vessel. It can cause container loss, crew injuries, or loss of the vessel in the most severe cases. How and when parametric rolling occurs in real sea is nowadays still being studied as warning systems are sometimes inefficient Palmquist & Nygren (2004).

Neural network is a widely spread tool that has the ability to simulate almost any function. Its main advantages is to offer real-time results and does not need to know the wave excitation, when one would run a mostly complicated inverse model to get the excitation.

Neural networks have been recently used with success in maritime engineering: controlling roll motion using rudder Alarçin & Gulez (2007), stabilizer fin Guo et al. (2003), and Li et al. (2005) or identifying rolling parameters in transverse seas Xing & McCue (2009).

Míguez et al. (2010) had some success in predicting the parametric rolling time-series of a fishing vessel.

Objective of the study was to predict parametric rolling of a container ship hull using neural networks. The study focuses on regular head seas for rolling angles smaller than 10° (in order to prevent larger movement) on a container ship, but also on adaptability to irregular seas and to other hulls.

The results are discussed in the context of different architectures of network. Direct neural implementation is compared to an algorithm which uses frequency information from the pitch mode. It is shown that this second models is improved in a remarkable way.

2 MATHEMATICAL MODEL AND DATA

2.1 *Mathematics model*

The general equation of motion of the vessel can be written as

$$I \cdot \ddot{X} + B(X,\dot{X}) \cdot \dot{X} + C(X) \cdot X = f(t) \quad (1)$$

For roll motion:

$$I_{\phi} \cdot \ddot{\phi} + B_{\phi}(X,\dot{X}) \cdot \dot{\phi} + C_{\phi}(X) \cdot \phi = f_{\phi}(t) \quad (2)$$

Travelling along the hull, the wave modifies the metacentric height (GM). In regular waves at first order and without coupling, one understands parametric rolling as a resonance in the damped Mathieu equation:

$$\ddot{\phi} + B \cdot \dot{\phi} + \omega^2 \cdot \phi \cdot (1 + \varepsilon \cdot \cos(\omega_e \cdot t + \varphi)) = 0 \quad (3)$$

This equation is a particular case of differential equations studied by Hsu & Bhatt (1966). Its areas of stability are well-known. However, this model is too simple and many more terms are to be taken into account such as terms of coupling and terms of higher order.

2.2 *Numerical model and vessel used*

The numerical model used here is based on a 3D model described in Rodríguez et al. (2007). The 3D model was later extended to six degrees of freedom Rodríguez (2010). It is a model with a strong coupling of all freedom degrees with all third-order derivatives. Its results are very close to experimental results especially for parametric rolling cases.

The vessel used is a container ship of which a brief description can be found in Figure 1.

The aim of the present study is to be able to predict is parametric rolling might possibly happen or not using this numerical model.

2.3 *Data generation*

The times series used in the study were generated using the numerical model. Different frequencies and wave heights were used as described in Figures 8 and 9. As the behavior is very different depending on the conditions (roll amplitude decrease, steady rolling response or capsizing of the boat), different initial conditions were used. Series which present a parametric rolling response start with a very smaller roll angle (0.2°). Series which does not present parametric rolling response start with a bigger roll angle (2°).

As plotted on Figures 2–3, to obtain more realistic time series, beginning were systematically deleted. Endings of the time series were generally deleted, especially when reaching steady state, as redundant data are not relevant for a neural network. This simplifies the problem and allows the neural network to converge faster.

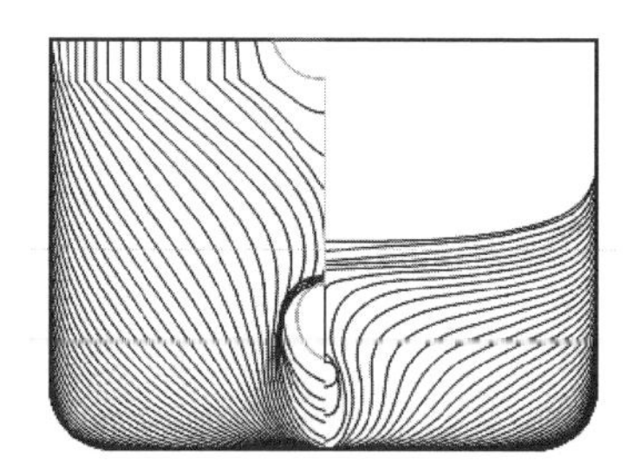

Length between perpendiculars	293.5 m
Breadth	32.3 m
Depth	11.7 m
Metacentricheight: GM	1.83 m
Displacement	74,600 t
Natural roll period	21.3 s
Natural pitch period	9.0 s
Froude number (for the study)	0.1022

Figure 1. The hull form and some vessel characteristics.

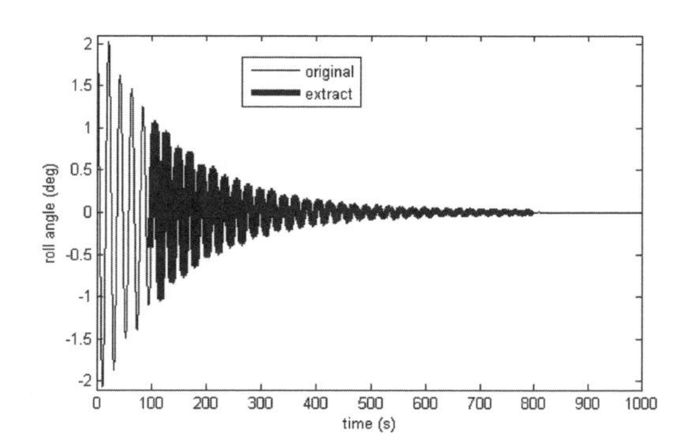

Figure 2. Extraction of data for a typical decay case: the bold line shows the data that will be used to train the neural network. The thin line represents the output of the numerical model, while the bold line is the extract chosen to train the network. $A_w = 0.5\,m, \omega_w = 0.530\,rad \cdot s^{-1}$.

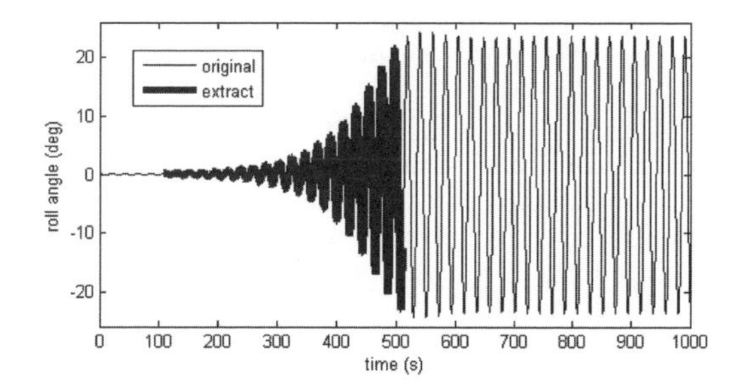

Figure 3. Extraction of data for a typical parametric rolling case. The thin line represents the output of the numerical model, while the bold line is the extract chosen to train the network. $A_w = 1.5\,m, \omega_w = 0.464\,rad \cdot s^{-1}$.

3 NEURAL NETWORKS

Neural Networks are more and more used as a tool in naval architecture for approximation, control, classification .... They are well known for their ability to reproduce real-time data. Here, feed-forward neural network are used as function approximation.

Cybenko (1989) proved that it was possible to simulate any continuous function with a feed-forward network as precisely as required if a sufficient number of neurons is used.

3.1 *A neural network layer*

One layer is described by the input $A=(a_i)_{1<i\leq N}$, the weights $W=(w_{ji})_{1<j\leq M,1<i\leq N+1}$, the activation function h, and the output $B=(b_j)_{1<j\leq M}$. The output of the layer is given by

$$b_j = h\left(\sum_{1<i\leq N} w_{ji}a_i + w_{jN+1}\right) \tag{4}$$

with $a_{N+1}=1$ which is the bias and $h=id$ (linear activation function) or $h(x)=tanh(x)$ (hyperbolic tangent function activation). Figure 4 summarizes in a diagram the notations used.

3.2 *Feed-forward neural networks*

A feed-forward neural network consists of connected layers. Each output of a layer is the input for the next layer.

In this study, we used feed-forward neural network with hyperbolic tangent function activation for hidden layers and linear function activation in the output layer as it is common in neural network literature. A usual way to describe it is sketched in Figure 5.

The output of the network is a non-linear function of the inputs: for a neural network with N inputs and one hidden layer of M neurons, we have:

$$B=\sum_j w_{kj}^{(2)}\cdot tanh\left(\sum_i w_{ji}^{(1)}A_i + w_{jN+1}^{(1)}\right)+w_{kM+1}^{(2)} \tag{5}$$

In this study, a simplified notation is used to describe the feed-forward neural networks used: $[i\, h_1\, h_2 \ldots h_k\, o]$ describes a network with i inputs, o outputs and k hidden layers with respectively $h_1, h_2, \ldots, h_k$ neurons. For instance, [25 40 30 1] is a network with 25 inputs, 1 output, a first hidden layer of 40 neurons and a second hidden layer of 30 neurons.

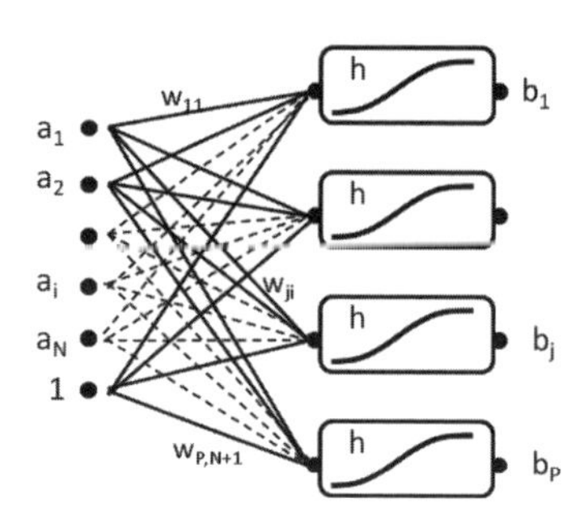

Figure 4. Layer diagram.

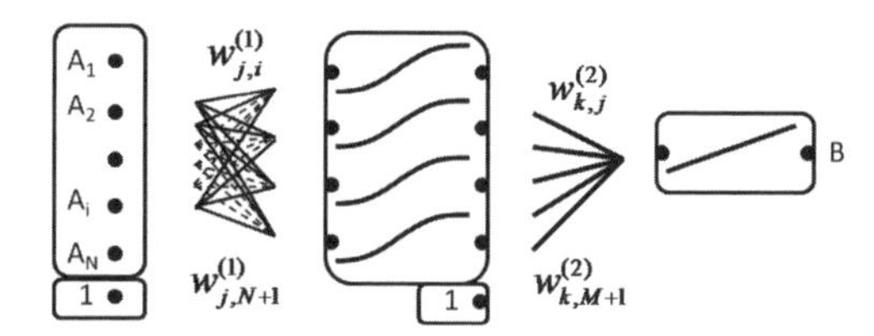

Figure 5. Neural network diagram.

3.3 *Learning algorithms*

The parameters of a neural network are the weights $W^{(i)}$. To be able to approximate a function, a network has to learn it. Learning is a crucial and complicated phase. There are a lot of parameters including the quality of input/output set.

Learning algorithms generally perform a gradient descent, and propose at each step new weights calculated in order to reduce the error between output of the training vectors and desired vectors. These algorithms can be separated in two classes: first-order algorithms and second-order algorithms. In this work, Levenberg-Macquardt algorithm was used, a second-order algorithm, known as very fast and accurate with networks that are not too big.

To avoid the problem of *over-fitting* (over-training), if enough data are available, one can reserve data (15–20%) to the validation. These data are not used to train the network but to check if the network has good generalization ability.

4 METHODS

In this work, two methods are compared: a direct neural implementation, and a second method that only differs in adding one input containing frequency information about pitch time-series.

4.1 *Direct neural implementation*

Figure 6 shows how roll time-series are predicted. The inputs are a discretization of the signal to be predicted and the output is the prediction one time step after. The network is then used recursively to predict as many time steps as needed.

A *s-seconds prediction*, means that the network was once ran recursively until reaching s seconds of prediction. Then, iteratively, the time-series are shifted and a new prediction of s seconds is done from which only the last point is used and added to the precedent curve. It allows to understand how accurately the prediction of s seconds is achieved.

4.2 *Using frequency information*

This time we add an input to the neural network as shown in Figure 7. This input is the mean value of the pitch windowed Fourier coefficient calculated over the entire signal as described in Equation (6). This input is kept constant while the network is ran recursively.

$$< \mathcal{F}_{\omega,S}(x) >_{\xi} = \frac{1}{n}\sum_{i=1}^{n}\left(\int_{\mathbb{R}}^{\cdot} e^{-i\omega t} w(t-\xi_i)x(t)dt\right) \quad (6)$$

4.3 *Times-series used*

Experiments were carried out using only time-series of roll and pitch. Three benches of data were generated: one at constant wave frequency, one at constant wave amplitude and the third one varying both frequency and amplitude. Data set 1 and 3 are represented in Figures 8 and 9. Data set 2 is done varying frequency only at a wave height of $A_w = 2.5m$. 9 training cases and 2 testing cases were generated for this data set.

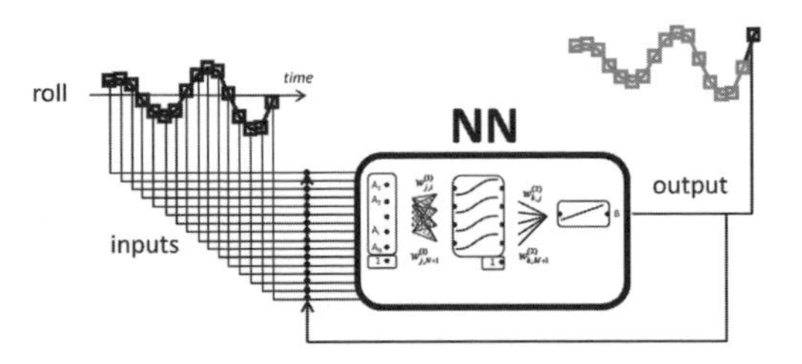

Figure 6. Neural network used to predict recursively time-series.

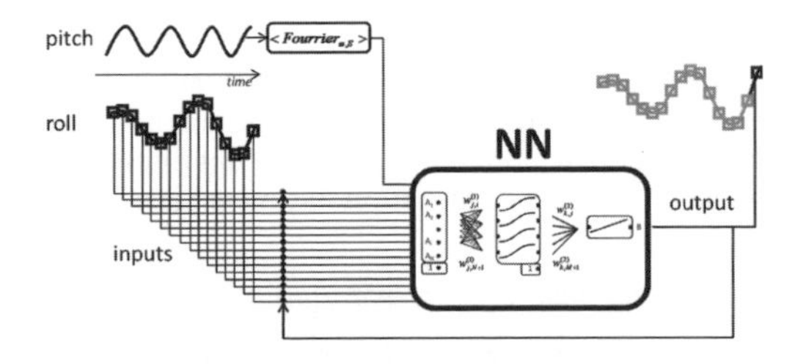

Figure 7. Neural network used to predict recursively time-series with one frequency information.

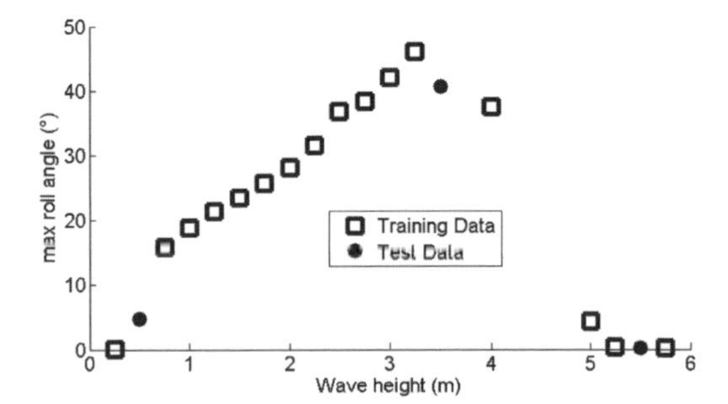

Firuer 8. Data set 1 (16 training +3 testing cases): $\omega_w = 0.46\ rad \cdot s^{-1}$ and $A_w = 0.25 \ldots 5.75m$.

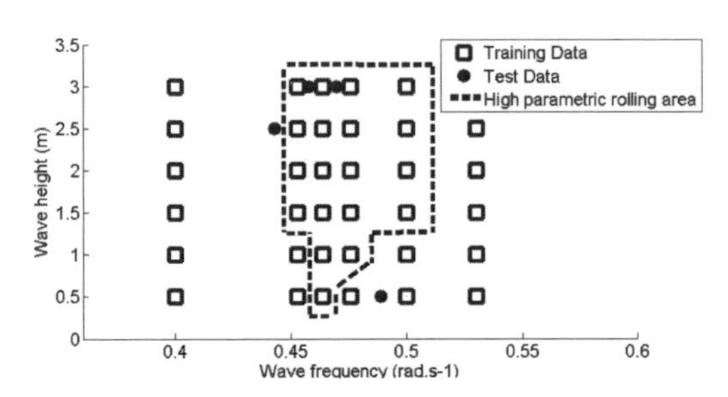

Figure 9. Data set 3 (36 training + 4 test cases): $\omega_w = 0.40 \ldots 0.53\ rad \cdot s^{-1}$ and $A_w = 0.5 \ldots 3.0\ m$.

Generated data were discarded to keep only the most relevant part as shown in Figures 2 and 3.

5 RESULTS

Results are shown for time prediction of 50 and 100 seconds and the different data sets. The neural network is trained and then tested using the testing cases which are different from the training cases. Figure 10 compares different time of prediction of a result for a typical parametric rolling test case of the data set 3.

Numerical results are shown in Table 1 for the *direct neural implementation* and in Table 2 when using the Fourier coefficient. The numbered data given are mean sum square of difference between original and prediction. Points that predict less than s seconds are not taken into account. There is also a comparison between neural networks of one hidden layer and neural networks of two hidden layers.

Results from data set 3 which compare the accuracy of both methods are displayed in Figures 11 and 12. Interesting in Figure 12 is the error of the prediction. The error is diminished when using the Fourier coefficient.

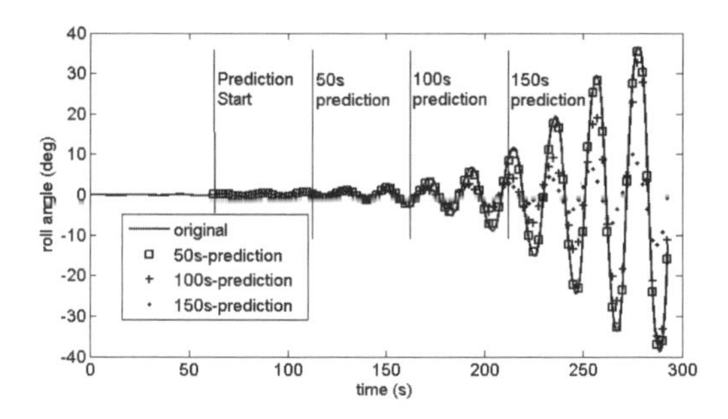

Figure 10. Example of results of the neural network prediction in a simple implementation (without Fourier coefficient). Prediction is done 50 s, 100 s, and 150 s in advance. 50 s prediction is very accurate while 150 s shows the prediction limit. $A_w = 3.0\,m$, $\omega_w = 0.47\,rad \cdot s^{-1}$.

Table 1. Direct neural implementation, mean prediction error for test cases of data set 1, 2, 3 with different architectures of network and prediction advances.

		50 s-prediction	100 s-prediction
data set 1	[15 10 10 1]	0,051	1,5
	[15 20 1]	0,045	1,4
data set 2	[25 40 30 1]	0,26	3,40
	[25 70 1]	0,40	5,7
data set 3	[25 40 30 1]	10	21
	[25 70 1]	0,61	17

Table 2. Mean prediction error for test cases of data set 2 and 3 with different network architecture using frequency information.

		50 s-prediction	100 s-prediction
data set 2	[26 40 30 1]	0,43	2,6
	[26 70 1]	0,32	0,51
data set 3	[26 40 30 1]	1,7	27
	[26 70 1]	0,45	4,8

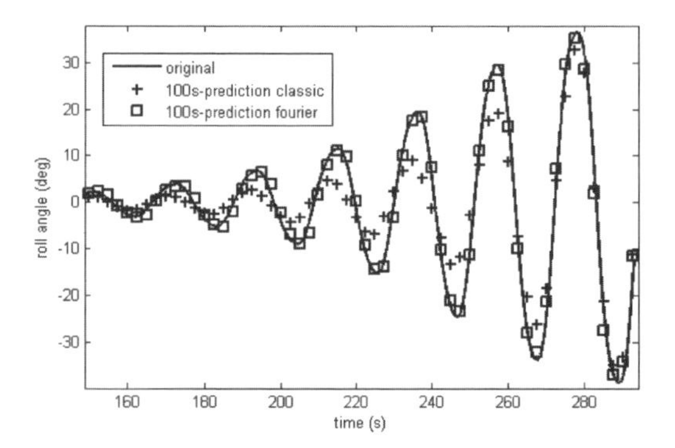

Figure 11. Comparing using or not the fourier coefficient for a test case with a prediction advance of 100 s, $A_w = 3.0\,m$, $\omega_w = 0.47\,rad \cdot s^{-1}$. Prediction is improved when using the fourier coefficient.

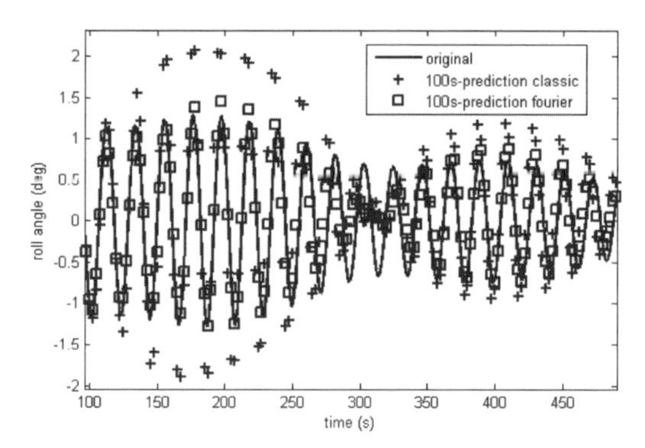

Figure 12. Comparing using or not the fourier coefficient for a test case with a prediction advance of 100 s, $A_w = 0.5\,m$, $\omega_w = 0.49\,rad \cdot s^{-1}$. Prediction is improved with the second method including the fourier coefficient.

6 DISCUSSION

6.1 *Why using a neural network?*

It is interesting here to recall some facts: parametric rolling develops within six to ten regular waves to critical angles. On the other side, the sea spectrum is diversified, irregular and depends on the time of the sample. This imposes the use of a real-time system to predict parametric rolling.

The goal of this study is to be able to provide relevant information to a captain or to an auto pilot of a ship to undertake quickly action if the ship is undergoing such waves. This is why the following times were chosen:

50 s of last movement "memory" (more or less 5 waves periods).

50 s-100 s prediction time for the future state of the ship (time to develop a hardly full parametric rolling case).

In case the prediction predicts a critical roll situation 50 s in advance, it would then let 5 to 10 seconds to undertake an action (modifying speed or heading) to avoid a dramatic situation.

The neural network permits to do an instantaneous analysis of the situation and this is its principal strength.

6.2 *About neural network*

A difficult problem with neural networks is to find the best architecture. Some authors believe that the number of hidden layers gives an indication of the degree of non-linearity of the function or an indication of the local/global character of the property researched. Finding the best neural network is a difficult and non-pertinent problem. However, one wants to have a neural network of sufficient accuracy, this should be kept in mind. Here are some clues to find a good neural network:

a. Trying to use networks of equivalent sizes but varying the number of hidden layers has to be done. Here, the networks [25 40 30 1] and [25 70 1] are compared. Although, the first network has 2301 parameters and the second only 1891, the second network was more successful in every data set. This difference principally comes from the learning algorithm. Learning algorithms are quite crucial for neural network abilities. Research in this area should improve the results.
b. It is important to identify the complete wave domain and to train the network with the whole data set. Using only parametric rolling cases to train the network will induce a network that predicts parametric rolling in any case and that will be inefficient for navigation.
c. Parametric rolling is sometimes delayed in the simulations until the phases of wave and roll match. It is better not to take into account this critical state of the boat waiting for the roll phase to match to start rolling dangerously. This is why beginning of such time series should be deleted in order to train the network. The neural network will predict a parametric rolling earlier than it would happen in fact, therefore it allows counteracting the phenomena earlier. The neural network will also gain in simplicity and robustness.
d. As mentioned before, neural networks are quite sensitive to many parameters. Some of them are the initial weights that are given before running the learning phase. To avoid this, it is necessary to compare networks with different initial weights. When approaching good abilities, the best networks reach approximately the same performances. In this study, for each case, ten different initial conditions were enough to get this state, but more initial conditions should be use for others functions.

Eventually, this study only considered one hull. Application on a different hull may need improvement as neural network tools are very dependent on the specific function it approximates.

6.3 *Future researching lines*

As the neural technique starts to be mature enough, networks should now be confronted with irregular seas. Additionally, future research should focus on towing tank data. Data of recorded parametric rolling at sea are available and should be considered Palmquist & Nygren (2004).

Tests with other hulls should be done in order to validate or invalidate the universality of this approach.

7 CONCLUSION

The use of neural networks to predict roll time series in regular heading seas in order to develop an alert system for parametric rolling has been studied and shows ability to prevent catastrophic situation. It is then shown that adding other degrees of freedom to the roll time-series and frequency information can improve the prediction performance.

The strong point of this method is that as many training data as wanted can be generated, basin results are only used to validate the numerical model. However, this process should then be validated with experimental results.

When working with neural networks, some considerations are to be taken into account, as sensitivity to many parameters: architecture of the network, learning functions, initial weights. In order to improve the robustness of the system, many different neural networks should be considered and may be used simultaneously.

Future work should concentrate on applying the neural networks to irregular seas. Another important line is improving the differential equation model testing other hulls.

ACKNOWLEDGEMENTS

The present investigation is supported in Brazil by CNPq within the STAB Projet (Nonlinear Stability of Ships) and in France by ENSTA-Bretagne. The UFRJ authors also acknowledge financial support from LabOceano, CAPES and FAPERJ.

NOTATIONS

A_W	wave height (m)
B	damping coefficient (6×6 matrix)
b, B	layer output, neural network output
C	restoring moment (6×6 matrix)
F_n	Froude number
f	excitation (wind, waves, ...)
ϕ	roll angle (*degree*)
GM	metacentric height (m)
g	gravity (9.81 $m\cdot s^{-2}$)
I	Inertia (kg)
L	vessel length (m)
$X(t)$	current state of the vessel ($6\times$ vector)
ω	natural roll frequency ($rad\cdot s^{-1}$)
ω_w	wave frequency ($rad\cdot s^{-1}$)

REFERENCES

Alarçin, F. & Gulez, K. (2007). Rudder roll stabilization for fishing vessel using neural network approach. *Ocean Engineering 34*(13), 1811–1817.

Cybenko, G. (1989). Approximation by superpositions of a sigmoidal function. *Mathematics of Control, Signals, and Systems (MCSS) 2*(4), 303–314.

Guo, C., Simaan, M. & Sun, Z. (2003). Neuro-fuzzy intelligent controller for ship roll motion stabilization. In *2003 IEEE International Symposium on Intelligent Control*, pp. 182–187.

Haddara, M. & Xu, J. (1998). On the identification of ship coupled heave-pitch motions using neural networks. *Ocean Engineering 26*(5), 381–400.

Hsu, C. & Bhatt, S. (1966). Stability charts for second-order dynamical systems with time lag. *Journal of applied mechanics 33*(1), 119–124.

Insperger, T. & Stépán, G. (2003). Stability of the damped Mathieu equation with time delay. *Journal of dynamic systems, measurement, and control 125*(2), 166–171.

Li, H., Guo, C. & Jin, H. (2005). Design of Adaptive Inverse Mode Wavelet Neural Network Controller of Fin Stabilizer. In *International Conference on Neural Networks and Brain*, pp. 1745–1748.

Lloyd, A. (1989). *Seakeeping: ship behaviour in rough weather*. Halsted Press.

Mahfouz, A. (2004). Identification of the nonlinear ship rolling motion equation using the measured response at sea. *Ocean Engineering 31*(17–18), 2139–2156.

Míguez, M., Peña, F., Casás, V. & Neves, M. (2010, September). An Artificial Neural Network Approach for Parametric Rolling Prediction. In *PRADS, Rio de Janeiro, Brazil*.

Neves, M.A.S., Rodríguez, C. & Vivanco, J. (2009). On the limits of stability of ships rolling in head seas. *Institution of Mechanical Engineers, Part M: Journal of Engineering for the Maritime Environment 223*(4), 517–528.

Palmquist, M. & Nygren, C. (2004). Recordings of head-sea parametric rolling on a PCTC. *submitted by Sweden to the Sub-Committee on Stability and Load Lines and on Fishing Vessels Safety, IMO*.

Roberts, J. & Vasta, M. (2000). Markov modelling and stochastic identification for nonlinear ship rolling in random waves. *Philosophical Transactions: Mathematical, Physical and Engineering Sciences 358*(1771), 1917–1941.

Rodríguez, C. (2010). *On the Nonlinear Dynamics of Parametric Rolling (in Portuguese)*. Ph.D. thesis, COPPE/UFRJ, Brazil.

Rodríguez, C.A., Holden, C. Perez, T., Drummen, I., Neves, M.A.S. & Fossen, T. (2007). Validation of a container ship model for parametric rolling. *9th International Ship Stability Workshop, Hamburg, Germany*.

Xing, Z. & McCue, L. (2009). Parameter identification for two nonlinear models of ship rolling using neural networks. In *10th International Conference on Stability of Ships and Ocean Vehicles*, pp. 421–428.

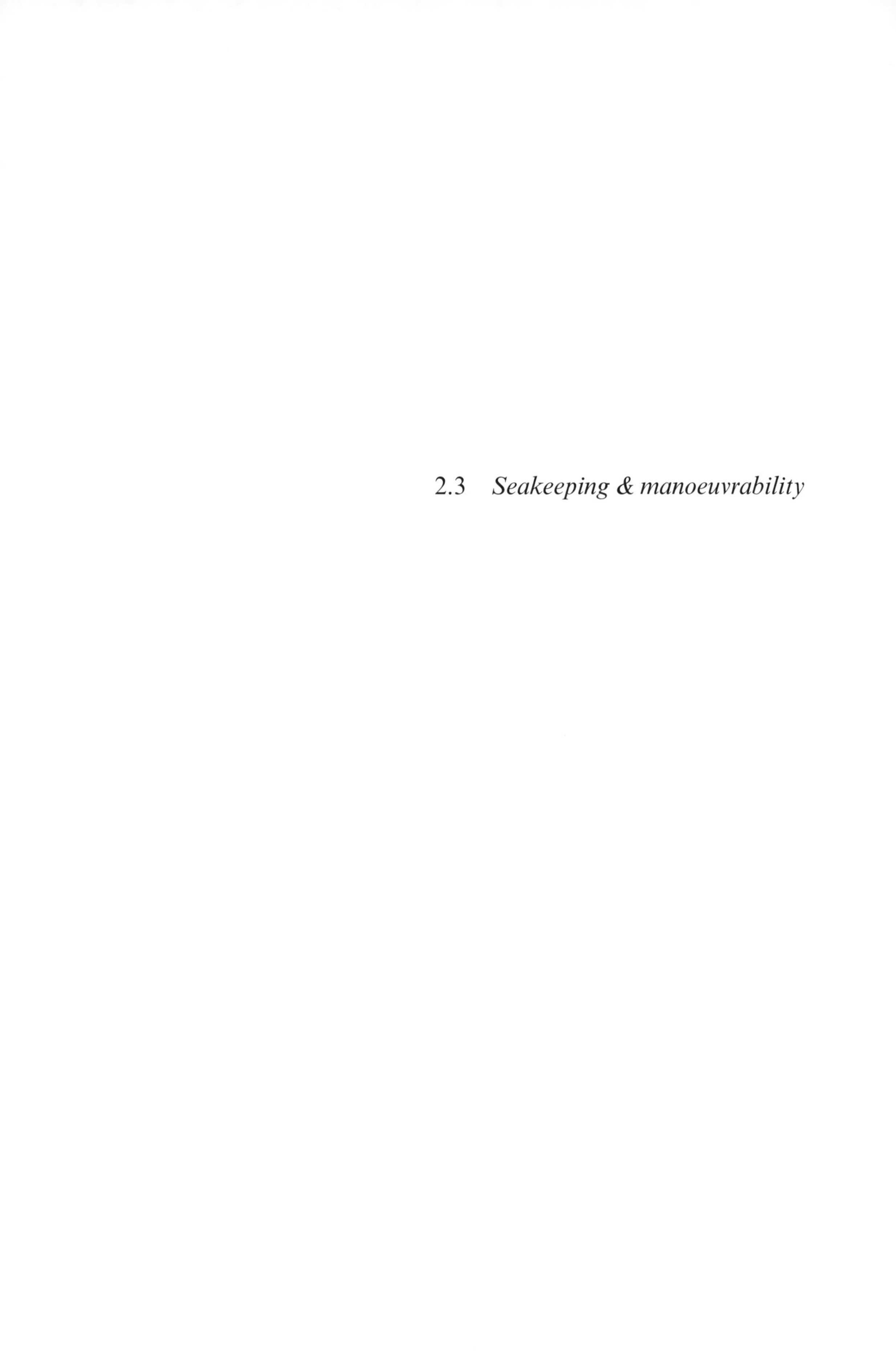

2.3 *Seakeeping & manoeuvrability*

Sustainable Maritime Transportation and Exploitation of Sea Resources – Rizzuto & Guedes Soares (eds)

Automatic trajectory planning and collision avoidance of ships in confined waterways

Y.Z. Xue & D.F. Han
College of Shipbuilding Engineering, Harbin Engineering University, Harbin, Hei Longjiang, China

S.Y. Tong
College of Power and Energy Engineering, Harbin Engineering University, Hei Longjiang, China

ABSTRACT: An automatic manoeuvring simulation is a cost-effective tool for minimising accidents and assisting design and operational planning which does not involve a human steersman but realistically emulates his/her performance. Automatic simulation of ship manoeuvring has to be able to con the ship automatically to reach a destination point whilst avoiding other ships and navigational hazards, keeping well clear of non-navigable areas, such as shallow water and shore line. One of the key problems in this task is automatic route planning and collision avoidance. This paper concentrates on developing a simple and practical method of automatic trajectory planning and collision avoidance based on an artificial potential field. The potential field method first applied in robotic research is used for ship trajectory planning and collision avoidance. The method has been adopted successfully for automatic navigation in busy, dynamic and confined seaways. In this work, 3 DOF ship manoeuvring model is used. The PID controller is designed to control the ship. An "extreme" case—the Strait of Istanbul—was chosen to investigate the performance of the system. The developed algorithm is fairly straightforward and simple to implement, and has been shown to be effective in decision support and automatic ship handling in the test case.

1 GENERAL INSTRUCTIONS

Due to increases in traffic, speeds and sizes of modern vessels, today's waterways and harbours are becoming busier and the navigation environment is becoming more complex. Increasing congestion in navigation channels has led to continued unacceptably high levels of occurrences of marine accidents despite considerable advances in navigational aids and equipment. It is without question that collision is one of the most severe marine accidents with potentially catastrophic consequences. Especially in confined waterways, collision avoidance takes on increased significance, and the increased threat of possible collisions gives navigators significantly more pressure and work load.

An automatic ship navigation system, probably used as an advisory tool to start with, would be an effective assistant to the crew contributing considerably to safe and efficient navigation. In such system, one of key elements is an intelligent decision-making capability. The problem of intelligent decision-making is connected with collision avoidance manoeuvres and route planning of vessels. In recent years, intensive research on automatic ship navigation has been carried out along with developments in computer hardware and software algorithms. Smierzchalski [Smierzchalski, 1996] adopted Evolutionary Algorithms to develop a ship guidance system for possible collision scenarios. Ito [Ito et al., 1999] employed Genetic Algorithms to compute the collision avoidance navigation path. Hwang [Hwang et al., 2001] developed a system for collision avoidance and track-keeping employing Fuzzy Logic. Liu and Shi [Liu & Shi, 2005] developed a Fuzzy-Neural Inference Network model for ship collision avoidance.

In general, numerous attempts have been made in the past using different approaches to develop an automatic navigation system. Most of the reported studies can generate collision-free paths for the vessel, but the work is by no means complete. For example, some methods are relatively complex and time-consuming and some systems either ignore collision prevention regulations (COLREGS) or are incapable of describing the complex encounter situations in detail.

This paper presents work in progress to solve the problems mentioned above and simulate realistic situations, such as when the ship navigates in the confined waterway which has narrow sections with many sharp turns, a new approach is required. Ideally, this method should be simple and effective whilst taking advantage of the state-of-the-art technology in this area of research.

2 MANOEUVRING MODEL USED

Accurate and safe navigation requires precise knowledge of the manoeuvring behaviour of the ship. It is well known that to represent a manoeuvring ship fully in space requires a mathematical model with six degrees of freedom. To simplify the problem, it is assumed that the steering of a ship can be regarded as a rigid-body motion on the horizontal plane, as is customary. Thus, the mathematical model is simplified to three degrees of freedom. With the global and ship coordinate systems shown in Figure 1, the equation of the motion can be written as [Fossen, 1994]

$$\text{Surge } m\left(\dot{u} - vr - x_G r^2\right) = X \tag{1}$$

$$\text{Sway } m\left(\dot{v} + ur + x_G r^2\right) = Y \tag{2}$$

$$\text{Yaw } I_Z \dot{r} + m x_G\left(\dot{v} + ur\right) = N \tag{3}$$

where m is the mass of the ship; u, v represent surge speed and sway speed respectively; $\dot{u}, \dot{v}$ represent surge and sway acceleration respectively; $r, \dot{r}$ are yaw rate and yaw acceleration; X is the force applied on the ship in x-direction; Y is the force applied on the ship in y-direction; I_Z is yaw moment of inertia of the ship; and N is yaw moment.

The forces X, Y and moment N can be expressed as functions of the state variables u, v, r, their time derivatives $\dot{u}, \dot{v}, \dot{r}$ and the rudder angle δ:

$$X = X\left(u, v, r, \dot{u}, \dot{v}, \dot{r}, \delta\right) \tag{4}$$

$$Y = Y\left(u, v, r, \dot{u}, \dot{v}, \dot{r}, \delta\right) \tag{5}$$

$$N = N\left(u, v, r, \dot{u}, \dot{v}, \dot{r}, \delta\right) \tag{6}$$

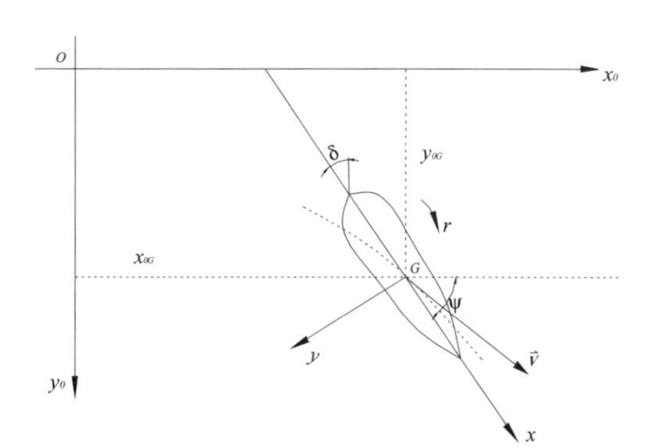

Figure 1. Global and ship coordinate systems.

This model of the ship that includes surge, sway and yaw will yield sufficient information to show the manoeuvring behaviour of the ship.

3 POTENTIAL FIELD METHOD APPLIED TO TRAJECTORY PLANNING AND COLLISION AVOIDANCE

The potential field method was first used by Khatib [Khatib, 1986] for robot path planning in the 1980s. In this method, a potential field is defined in the configuration space such that it has a minimum potential at the goal configuration. While the target is ideally at the minimum, all obstacles, or walls, are treated as high potential hills. In such a potential field, the robot is attracted to its goal position and repulsed from any obstacles. This method is particularly attractive because of its mathematical elegance and simplicity. It allows real-time robot operations in a complex environment and is currently widely used for path planning and collision avoidance of mobile robots.

The ship's trajectory planning and collision avoidance are, in a sense, similar to the mobile robots'. A ship sails from its starting position to its destination point. There is an obstacle in the way of a direct route between the two points. The shortest route for the ship to follow is shown in blue line ('Desired Track') in Figure 2. However, the actual safe route will be something like that shown as the 'Actual Track'. This actual track can be determined by applying the potential field method.

Since the ship is pulled towards the destination point, the potential energy responsible for it acts in a way reminiscent of gravitational potential energy. The existence of the obstacle with imaginary potential field energy can be denoted as $\vec{U}_{rep}$. Thus,

$$\vec{U}(\vec{p}) = \vec{U}_{att}(\vec{p}) + \vec{U}_{rep}(\vec{p}) \tag{7}$$

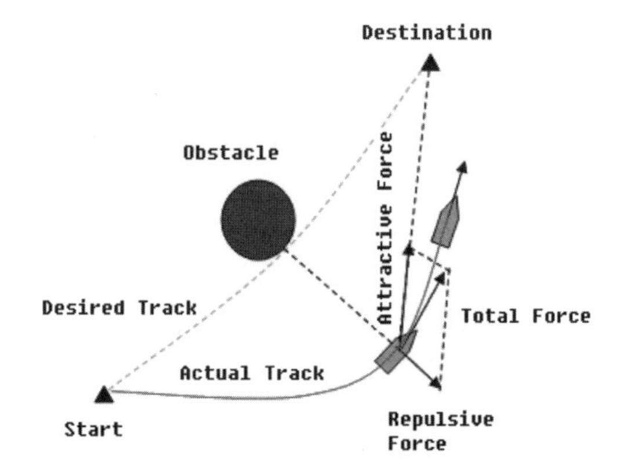

Figure 2. The potential field in ship's route finding.

where $\vec{U}(\vec{p})$ is the total potential energy; $\vec{U}_{att}(\vec{p})$ is the potential energy due to attraction towards destination point; $\vec{U}_{rep}(\vec{p})$ is the potential energy due to repulsion of the obstacle; $\vec{p}$ denotes a point on the water surface.

The ship then is subjected to a force which is derived from this total potential force as follows:

$$\vec{F} = \vec{F}_{att} + \vec{F}_{rep} \tag{8}$$

where

$$\vec{F}_{att} = -grad\left(\vec{U}_{att}(\vec{p})\right), \vec{F}_{rep} = -grad\left(\vec{U}_{rep}(\vec{p})\right)$$

$\vec{F}_{att}$ may be called the attractive force as it pulls the ship towards the destination; $\vec{F}_{rep}$ is repulsive force as it pushes the ship away from the obstacle thus avoiding collision. The feasible path now can be found by following the direction of the total force at any given position. More than one obstacle can be accounted for by summing all the repulsive forces due to the obstacles.

3.1 *Attractive potential function*

The attractive potential is defined as a function of the relative distance between the ship and the destination point. In this paper, the attractive potential function is presented as follows:

$$\vec{U}_{att}(\vec{p}) = \alpha\left\|\vec{p}_d - \vec{p}(t)\right\|^m \tag{9}$$

where $\vec{p}_d$ and $\vec{p}(t)$ denote the destination position and the position of ship at time t, respectively; $\left\|\vec{p}_d - \vec{p}(t)\right\|$ is the Euclidean distance between the ship at time t and the destination position; α is a scalar positive parameter; and m is a positive constant.

The corresponding virtual attractive force is defined as the negative gradient of the attractive potential

$$\vec{F}_{att}(\vec{p}) = -\nabla\vec{U}_{att}(\vec{p}) = -\frac{\partial\vec{U}_{att}(\vec{p})}{\partial\vec{p}} \tag{10}$$

Substituting (9) into (10), the equation is

$$\vec{F}_{att}(\vec{p}) = m\alpha\left\|\vec{p}_d - \vec{p}(t)\right\|^{m-1} \tag{11}$$

From Equation (9) and (11), we can modify the shape of attractive potential function by changing the value of m, and the effect of the attractive potential field by changing the value of α.

3.2 *Repulsive potential function*

To avoid the obstacle, the relative position between the ship and the obstacle is taken into account when constructing the repulsive potential function. If the ship is within the circle of a certain radius measured from the obstacle in question, the repulsive force exists. Otherwise, the repulsive force is zero. In this paper, the following form of repulsive potential function was used and was found to be satisfactory in this regard:

$$\vec{U}_{rep}(\vec{p}) = \begin{cases} \frac{1}{2}\eta\left(\frac{1}{p_s} - \frac{1}{p_o}\right)^2 \left\|\vec{p}(t) - \vec{p}_d\right\|^n & if \quad p_S \le p_o \\ 0 & if \quad p_S > p_o \end{cases} \tag{12}$$

where, $\vec{U}_{rep}(\vec{p})$ denotes the repulsive potential generated by the obstacle; η and n are constants; $\vec{p}_d$ and $\vec{p}(t)$ denote the destination and the ship position of ship at time t respectively; p_s is the shortest Euclidean distance between the ship and the obstacle surface; p_o is a positive constant describing the influence range of the obstacle.

Similar to the definition of the attractive force, the corresponding repulsive force is defined as the negative gradient of the repulsive potential in terms of position.

$$\vec{F}_{rep}(\vec{p}) = -\nabla\vec{U}_{rep}(\vec{p}) = -\frac{\partial\vec{U}_{rep}(\vec{p})}{\partial\vec{p}} \tag{13}$$

Substituting (12) into (13), the equation is

$$\vec{F}_{rep}(\vec{p}) = \begin{cases} \vec{F}_{rep1} + \vec{F}_{rep2} & if \quad p_s \le p_o \\ 0 & if \quad p_s > p_o \end{cases} \tag{14}$$

where

$$\begin{aligned} \vec{F}_{rep1} &= \eta\left(\frac{1}{p_s} - \frac{1}{p_o}\right)\frac{1}{p_s^2}\left\|\vec{p}(t) - \vec{p}_d\right\|^n \\ \vec{F}_{rep2} &= \frac{n}{2}\eta\left(\frac{1}{p_s} - \frac{1}{p_o}\right)^2\left\|\vec{p}(t) - \vec{p}_d\right\|^{n-1} \end{aligned} \tag{15}$$

3.3 *Total potential function*

After the calculation of the attractive and repulsive potential functions, the total potential can be obtained by

$$\vec{U}(\vec{p}) = \vec{U}_{att}(\vec{p}) + \vec{U}_{rep}(\vec{p}) \tag{16}$$

The total virtual force can be obtained by

$$\vec{F}_{total}(\vec{p}) = \vec{F}_{att}(\vec{p}) + \vec{F}_{rep}(\vec{p}) \tag{17}$$

where $\vec{F}_{att}(\vec{p})$ and $\vec{F}_{rep}(\vec{p})$ can be calculated through Equation (11) and (15).

For the case where there are multiple obstacles, the repulsive force is given by

$$\vec{F}_{rep}(\vec{p}) = \sum_{i=1}^{n_{obs}} (\vec{F}_{rep})_i \tag{18}$$

where n_{obs} is the number of obstacles and $(\vec{F}_{rep})_i$ is the repulsive force generated by the ith obstacle. The total virtual force $\vec{F}_{total}$ will be used for route finding.

4 PID HEADING CONTROLLER

In this paper, the PID heading controller is designed to control the ship. It can be designed as follows [Fossen, 2002]:

$$\delta_{PID}(t) = -K_p\tilde{\psi} - K_d\tilde{r} - K_i\int_0^t \tilde{\psi}(\tau)d\tau \tag{19}$$

where $\delta_{PID}(t)$ is the rudder angle required at time t; $\tilde{\psi} = \psi - \psi_i$ is the heading error; K_p is the proportional gain constant; K_d is the derivative gain constant; $\tilde{r} = r - r_i$ is the yaw rate error; K_i is the integral gain constant.

The controller gains can be found in terms of the design parameters ω_n and ξ, through:

$$\begin{aligned} K_P &= \frac{\omega_n^2 T}{K} \\ K_d &= \frac{2\xi\omega_n T - 1}{K} \\ K_i &= \frac{\omega_n^3 T}{10K} \end{aligned} \tag{20}$$

where ω_n is the natural frequency of the system and ξ is the relative damping ratio, T and K are time constant and gain constant respectively. The K_p term allows the controller to simulate the restoring action of a manual helmsman to a heading error. The K_d term simulates the anticipatory action of a well trained helmsman and finally the K_i term simulates the helmsman compensating for low frequency drift of the vessel heading.

5 THE CASE STUDY

In order to investigate the performance of the system, an "extreme" case—the Strait of Istanbul—was chosen for the next case study of navigation in a confined waterway. This straight presents one of the greatest challenges for ship navigation as it snakes through the heart of Asia Minor [Kose etc., 2003]. The part of this strait between two red lines as shown in Figure 3 (a) is used for this study. The model of the coastline in the simulation program is shown in Figure 3 (b).

In this case, the Mariner class ship was chosen and the speed was set at 7.5 kn, the starting point was at $(x_s, y_s) = (0, 0)$ (km) and the destination point was at $(x_d, y_d) = (8,10)$ (km). The coastline of the strait was represented by discrete obstacle points as shown in Figure 4 and the influence range of the obstacle p_o was 0.3 km.

Figure 5 shows the contour plot of the potential field. We can see that high potential is generated at the surface of the obstacle, and the coastline is covered by the potential, effectively building 'dykes' so

(a)

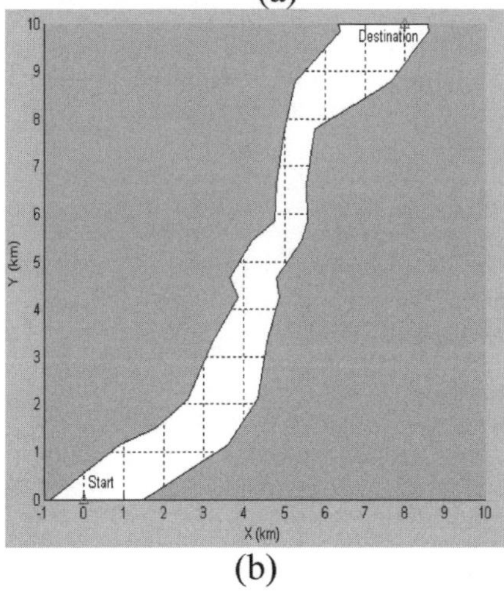

(b)

Figure 3. (a) The selected part of the strait (b) Simulation model.

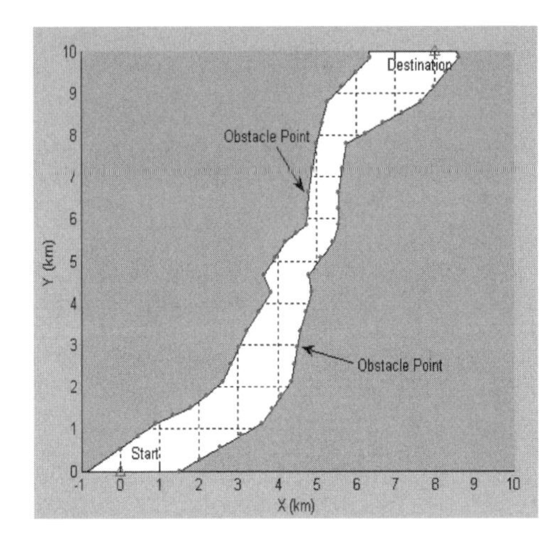

Figure 4. The representation of coastline.

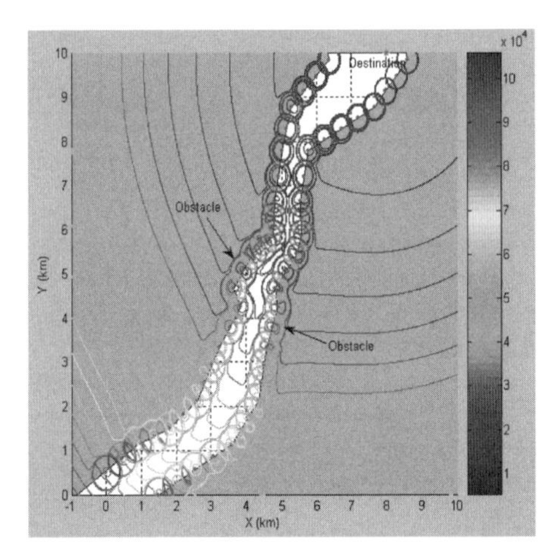

Figure 5. The contour plot of the potential field.

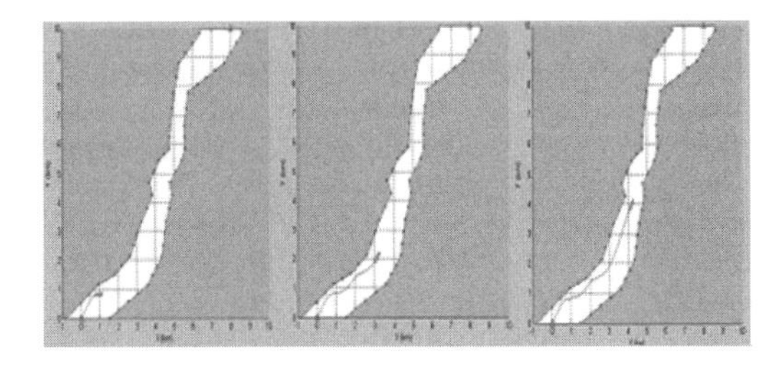

Figure 6. The simulation progress (1).

that the ship is prevented from 'leaking' out of the shipping lane.

Figure 6 and 7 show the progress of the simulation and the key manoeuvring parameters are shown against time in Figure 8.

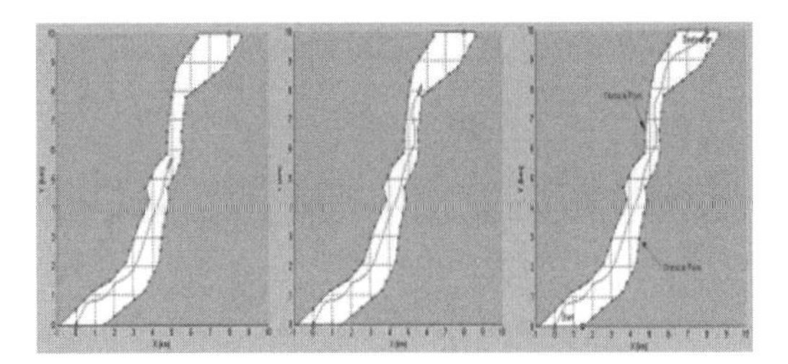

Figure 7. The simulation progress (2).

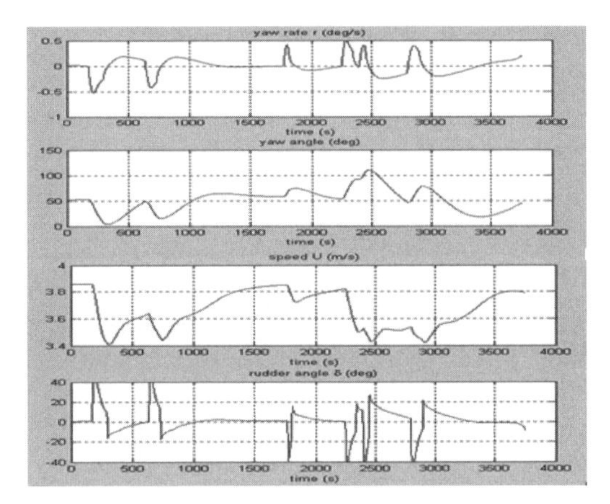

Figure 8. The ship's yaw rate, yaw angle, speed and rudder angle.

From the simulation results, we can see the ship can find its way and follow the gradient of the potential field to reach the desired location. This case demonstrates that the developed algorithm is capable of automatically navigating a ship though the confined waterway which has narrow sections with many sharp turns.

6 CONCLUDING REMARKS

This paper has presented a method for automatic route finding and collision avoidance for the ship in confined waterways. It has been shown that the method can also be used for a decision-aid tool for navigators in control of a ship. So far as can be ascertained through the case study, the method works well.

The method in its current form does have some deficiencies. For example, the weather conditions are not taken into consideration; only the Mariner class ships were considered in the simulation study. Nevertheless, the developed algorithm has been shown to be effective in decision support and automatic ship handling for ships involved in confined waterways.

REFERENCES

Fossen, T.I. (1994). *Guidance and control of ocean vehicles*. Wiley, Chichester.

Fossen, T.I. (2002). *Marine control systems: guidance, navigation and control of ships, rigs and underwater vehicles*. First edition, Marine Cybernetics, Trondheim, Norway.

Hwang, C.N., Yang, J.M. & Chiang, C.Y. (2001). The design of fuzzy collision avoidance expert system implemented by H-infinity autopilot. *Journal of Marine Science and Technology*, 9, 25–37.

Ito, M., Zhang, F. & Yoshida, N. (1999). Collision avoidance control of ship with genetic algorithms. *Proceedings of the 1999 IEEE International Conference on Control Applications*, Kohala Coast, Hawaii, USA, August, Vol. 2, pp. 1791–1796.

Khatib, O. (1986). Real-time obstacle avoidance for manipulators and mobile robots. *J. Robotics Res.*, 5(1), 90–98.

Kose, E., Basar, E., Demirci, E., Guneroglu, A. & Erkebay, S. (2003). Simulation of marine traffic in Istanbul strait. *Simulation Modelling Practice and Theory*, 11 (2003) 597–608.

Liu, Y. & Shi, C. (2005). A fuzzy-neural inference network for ship collision avoidance. *Proceedings of IEEE the Third International Conference on Machine Learning and Cybernetics ICMLC2005*, Guangzhou, China, pp. 4754–4759.

Smierzchalski, R. (1996). The decision support system to design the safe manoeuvre avoiding collision at sea. *14th International Conference Information Systems and Synthesis*, Orlando, USA.

Sustainable Maritime Transportation and Exploitation of Sea Resources – Rizzuto & Guedes Soares (eds)
 ISBN 978-0-415-62081-9

Improvement of steering efficiency for sailing boats

P. Vidmar & M. Perkovič
University of Ljubljana, Portorož, Slovenia

ABSTRACT: The research explained in this paper was carried out to investigate the efficiency of different steering systems on sailing yachts. The steering system of a sail yacht mostly includes a simple steering system and a hydrodynamic shaped single rudder or multiple rudders, depending on boat characteristics. One of the basic design guidelines for fast sailing yachts is to reduce wetted surface to minimum allowed by the dynamic stability and maintaining the sailing performances. Deficiencies of different steering systems are discussed and their influences on total drag and yacht manoeuvrability in different sailing directions is analysed. The discussion is focused on steering systems applicable in practice and accepted by the yacht-building industry, although several innovations could be found that remained on their development stage because of their complexity in construction, maintenance, use itself and reliability.

1 INTRODUCTION

Observing the evolution of racing sailing yachts in last three decades we can see a progress in using innovative construction materials and new design approach in hull design. Observing the latest racing sail yachts give a clear idea that those yachts have very few things common with yachts of the previous decade. Observing a VOR (Volvo Ocean Race) 60 feet class sail yachts that in 1995 reach the top sailing speed of 25 mph and a 70 feet class that now day exceed 40 mph, give the idea that an exponential step in technology and design have arisen.

The primary objective of the research is to provide design information as to the effect of sailing yacht hull and appendage characteristics and their interactions on the resistance and lift of the yacht. There is a wide field in yacht research and applications with and without influence of the free surface where the drag and lift on the appendages have since long been an area of extensive research.

In their earlier publications "Course keeping qualities and motions in waves of a sailing yacht" (Gerritsma, 1971) and later "Balnce of helm of sailing yachts, a ship hydromechanics approach on the problem" (Nomoto and Tatano, 1979) the authors have presented assessment methods for determining the force distribution and position of CLR (Center of Lateral Resistance) in yaw and sway over the hull and appendages in calm water and in waves. In this method the use was made of a so called: Extended Keel Method (EKM) as introduced by (Gerritsma, 1971) for calculating the side force on the keel and rudder (and hull) of a sailing yacht. EKM gave good results for the total side force of the hull, keel and rudder together in the upright condition, indicating that the mayor part of the side force is produced by the appendages, in particular for boats with average to high aspect ratio keels and rudders. The analyses of the yaw moment gave the results that the hull of the canoe body has a significant contribution that is not accounted with the EKM. A modified approach to the correction method as introduced by (Nomoto, 1979) yields good results for the yaw moment as well. With the development of Delft Systematic Yacht Hull Series (DSYHS) a large number of towing tank tests have been conducted on sailing yachts hulls and the results were used to develop improved methods for the calculation of lift and drag. Results were presented by Keuning in several publications for bare hull resistance (Keuning 1996 & Keuning 1997) and appendages (Keuning 1998) with particular attention on the interaction between hull and appendages and appendages themselves, keel-rudder interactions. Rodriguez (Rodriguez, 2008) has conducted his research on a series of experiments in the ETSIN towing tank focusing more on the influence of rudder, evaluating the distribution of forces in different conditions of navigation as well as for the interactions between hull and appendages. In recent Keuning publications he was focusing in determining the force distribution in yaw and sway over the hull, keel and rudder. In (Keuning, 1998) his method was used to deal with the yaw balance of a sailing yacht. In later publications the similar approach was used to determine forces and moments during manoeuvring of sailing yacht.

The interaction of keel and rudder has been studies by Keuning in (Keuning 2006) where the effect of downwash from keel to the rudder changes the effective angle of attack on rudder, changing the force distribution and increasing the resistance on rudder. Without leeway the resistance of the rudder is influenced by the wake of the keel and the wave forming around the stern of the ship.

1.1 *The step forward*

Changing the underwater geometry of a sailing yacht could change the stability and the hydrodynamic characteristics of the boat. A useful and competitive system is the canting keel that gave several advantages in sailing but also disadvantages. One of the main disadvantages is the need of additional foils or front rudder to reduce sideslip—dagger boards. If they are extractible has no influence on additional drag in downwind sailing, but if fixed it does. The second system is twin foil manoeuvring system that is usually combined with a canting keel. In this system the "working" rudder is preventing sideslip but the upwind rudder produces additional resistance. In any case the best manoeuvring efficiency of the rudder is in its upright position when the maximum lift is produced. The idea is the introduction of the bending rudder. This implies the system that keeps the rudder in upright position independently of the heel of the boat.

In the following the effect of the bending rudder on force distribution in upright and heeled condition is analysed. Further the downwash effects are observed where the hypothesis is that its effects are reduced.

2 SAILING BALANCE

During sailing the force of wind on sails is equalized by the forces acting on hull and forces that move the boat forward. If the keel is generating less lift than the sideways force from the rig the boat will slip across track more quickly. The increasing sideways speed increases the leeway, which increases the angle of attack and so increases the lift from the keel. Eventually the leeway builds up to the point where the lift from the keel balances the sideways force from the rig and the forces on the boat are back in balance. If the lift from the keel increases then the boat is accelerated decreasing the leeway and lift until everything is back in balance again. We can see that the leeway angle is determined by the sideways force from the rig balancing against the lift from the keel. For a given boat speed the boat will settle on a leeway angle that gives just the right amount of lift to balance the sideways force from the rig. The driving force from the rig accelerates the boat forwards and as the boat speed increases the drag caused by moving through the waters goes up. When the speed gets to the point where the drag balances the driving force the boat stops accelerating and settles at that speed. Of course a change in the driving force puts the forces out of balance and so changes the boat speed. In the same way a change in the drag changes the boat speed. (oceanis.co.uk)

2.1 *Effect of heel on sailing balance*

The lift that a foil generates is perpendicular to its surface, if our boat is upright any lift generated by the keel or rudder acts horizontally. When we're sailing it's unusual for the boat to be absolutely upright, as the boat heels the lift forces from the foils move away from the horizontal. We're interested in generating a horizontal force from the keel and rudder, to look at how these changes with heel angle we use the fact that a force at an angle can be represented as the combined effect of a horizontal force and a vertical force.

The lift from the rudder is used to turn the boat, and also to stop the boat from turning. This second point is important to remember when we're sailing to windward with some weather helm. The person on the helm will be steering to leeward to keep the boat running straight. The more weather helm the boat has the more force is required from the rudder to keep it on track, and the force from the rudder depends on the boat's speed and the angle of the tiller.

As the boat heels over the horizontal component of the rudder's lift is reduced. If the weather helm and boat speed are constant then we need to increase the rudder angle to generate more lift so that the horizontal component stays the same. At 25 degrees of heel the rudder has to generate about 10% more lift than it did when vertical to produce the same tuning force, if we push the boat to 40 degrees we're asking the rudder for 30% more lift.

Increasing the lift generated by the rudder also increases the rudder's drag. So as we heel the boat we get more drag from the rudder for a given tuning force. Drag slows the boat down, and slowing the boat down increases leeway. Slowing the boat down also means that we need a bigger rudder angle to generate the lift that we need. As the boat heals we've got three effects adding up to make life hard for the rudder; the increasing heal reduces the horizontal component of the rudder's lift, it increases weather helm which the rudder needs to produce more lift to counteract and the increased rudder angle increases drag, slowing the boat and requiring a bigger rudder angle to produce the same amount of lift. (oceansail.co.uk)

3 THE APPROACH

As stated by (Lin et al., 2010) the earlier used manoeuvring prediction methods are based almost entirely on empirical equations (Whicker & Fehlner 1958). Such methods give satisfying results for boats that are geometrically similar to models tested in towing tanks and which measurement results were used to derive coefficients of empirical equations. For new and unconventional ships and boats this empirical data are usually not available. The use of computer and advanced computational flow prediction numerical methods to predict the ship motion and steering capabilities allows the analysis of different hulls with different appendages configurations. (El Moctar, 1998) has used viscous flow methods to predict the rudder flow, and (Gaggero et al., 2008), like several other authors have used the panel method in a potential flow to compute forces in 2D and 3D profiles. Although several improvements have been introduced in a panel method the potential flow methods did not take into account the viscosity, turbulence, and flow separation. On other side viscous flow methods applied in time dependent calculations like ship movement and steering still remain technically difficult and computationally expensive. Therefore, considering limitations of potential flow methods, many practical flow problems are still solved by obtaining experimental data or computed by empirical methods, or by potential flow calculations (Lin et al., 2010).

The first presented analysis is based on the hull model which has been used for extensive measurements at Delft Ship hydromechanics Laboratory of the Delft University of Technology in late nineties. The DSYHS series model 3, named 366 is found to be well documented in several Keuning publications like (Keuning, 1998 & Keuning 2003) and (Verwerft, 2008) and is therefore used for the empirical model validation and the analysis of performance changes due to a bending rudder. The complete oversight of the hull shape parameters for model 366 is presented in Table 1.

Three different keel geometries have been used for this study, varying in aspect ratio and thickness/chord ratio, the forth is the keel of the tested model. The final test has been conducted for the measured keel of the tested model. The principal dimensions are presented in Table 2. Furthermore one rudder, and the rudder of the tested boat, of which the principal dimensions are also presented in Table 2, has been used in the calculation.

The line plane and the longitudinal profile of the model measured are presented on Figure 2. The model has a low beam/draft ratio and represents a typical racing sailboat from late nineties. The sailboat Moro di Venezia was an America's Cup class from 1992 to 1995.

3.1 *Side force computation methods for the hull and appendages*

The side force production of the hull and appendages is the key element in the sail yacht dynamic motion, because allow upwind sailing. Different models and empirical equations have been developed in years, mainly based on towing tank testing results. DSYHS hull series tests conducted by Keuning at Delft University have given new answers to open questions left by L.F. Whicker and late by J. Gerritsma and K. Nomoto on lift production in different conditions. Tests or in practice sailing conditions that have influence on hull lift are; yaw angle and heel angle. Depending on hull form and type of appendages the magnitude of lift and the resistance/lift ratio is varying from ship to ship often on sister ships too. The total side force of the hull and appendages and the separate contributions of hull, keel and rudder, are assessed differently in the upright and the heeled conditions. In the upright condition the so called Extended Keel Method, as derived by (Gerritsma, 1992), is used to calculate the side force on keel and rudder. The side force generated by the hull is accounted for by the virtually extended keel inside the canoe body to the waterline. The downwash angle on the rudder is approximated as 50% of the leeway angle and the water velocity over the rudder reduced by 10% to account for the wake of the keel.

The total side force is calculated as the sum of the force on extended keel and rudder.

$$Y_{tot} = Y_{ek} + Y_r \tag{1}$$

$$Y_{ek} = \frac{1}{2}\rho V_S^2 A_{ek}\left(\left(\frac{dC_L}{d\alpha}\right)_{ek}\beta\right) \tag{2}$$

$$Y_r = \frac{1}{2}\rho\left(0{,}9V_S^2\right)A_r\left(\left(\frac{dC_L}{d\alpha}\right)_r 0{,}4\beta\right) \tag{3}$$

Table 1. Hull form parameter of model 366.

Moro di venezia	Lwl/Bwl	Bwl/Tc	Lwl / VOLc$^{1/3}$	LCB %	LCF %	Cb	Cp	Cw	Cm	Aw/ VOLc$^{1/3}$
33	6.15	4.7	7.4338	−6.55	−8.73	0.413	0.549	0.659	0.6315	7.4178

Table 2. Geometry particulars for keels and rudder.

			Keel moro	Rudder moro
Lateral area	A_{lat}	[m²]	0.0651	0.0188
Wetted area	S	[m²]	0.1432	0.0413
Aspect ratio	AR	[–]	10.33	7.733
Span	b	[m]	0.620	0.290
Mean chord	c_{mean}	[m]	0.12	0.075
Sweepback angle	Λ	[°]	5	5
Volume	V_k	[m³]		
Thickness/chord ratio	t/c	[–]		

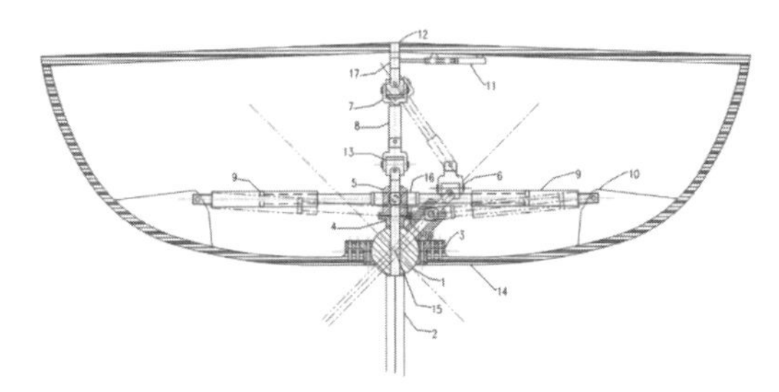

Figure 1. Device and system of bending rudder for sailing yachts.

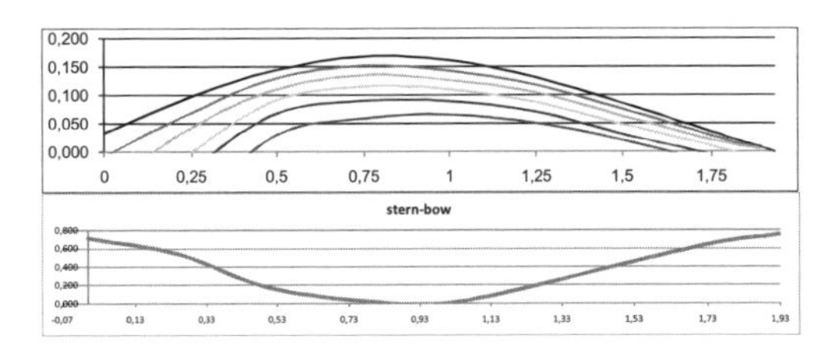

Figure 2. Line plan and longitudinal profile.

where

Y_{tot} = total side force in the horizontal plane [N]
Y_{ek} = side force generated by the extended keel [N]
Y_r = side force generated by the rudder [N]
A = lateral area of the foil [m²]
$(dC_L/d\alpha)$ = lift curve slope of the foil [deg⁻¹]

The extended keel method is often applied by yacht designers to make first approximations about forces and size of appendages. However under heel this procedure does not work. Therefore in these conditions the results of the side force polynomial as derived from the results of the DSYHS by (Keuning & Sonnenberg, 1998) are used. This polynomial accounts for effects of heel angle and forward speed on the total side force production.

$$Fh\cos(\varphi) = \left(b_1 \frac{T^2}{Sc} + b_2 \left(\frac{T^2}{Sc} \right)^2 + b_3 \frac{Tc}{T} + b_4 \frac{Tc}{T}\frac{T^2}{Sc} \right) \tfrac{1}{2}\rho V_S^2 Sc\left(\beta - \beta_{Fh:0}\right) \tag{4}$$

where

$$\beta_{Fh:0} = B_3 \varphi^2 Fn \text{ and}$$
$$B_3 = 0{,}0092 \cdot \frac{Bwl}{Tc} \cdot \frac{Tc}{T}$$

$Fh\cos(\phi)$ = side force in for horizontal plane [N]
T = total draft of hull with keel [m]
Tc = draft of the canoe body [m]
Sc = wetted surface of the canoe body [m²]
$\beta_{Fh:0}$ = zero lift drift angle [deg]
Fn = Froude Number

The coefficients b_1 to b_4 used in a presented function are obtained from DSYHS tests for the heeling angle between 0 and 30 degrees of heel and presented by (Keuning, 1998) in Table 3.

The use of this expression yields however no information on the contribution of the three different components, i.e., hull, keel and rudder and therefore no result for the yaw moment can be found. Verwert & Keuning, (2008) have developed a new formulation for keel and rudder lift calculation that takes in account the interaction effect of hull on the keel and the rudder. To overcome this problem the distribution over keel and rudder as found in the upright condition is used in the heeled condition also. The Munk moment on the hull is calculated taking the geometry of the heeled hull in account. This procedure is also described in (Keuning, 1997). Keuning, Katgert & Vermeulen, (2003) improved the prediction of the side force production for higher aspect ratio keels and the yaw moment under heel by taking the newly derived formulation for the influence of the downwash of the keel on the rudder into the calculations.

This situation of using two different approaches was considered undesirable and inconsistent. So in the framework of the present study a new method has been developed.

Table 3. Coefficients for the lift force polynom.

φ	0⁰	10⁰	20⁰	30⁰
b_1	2,025	1,989	1,980	1,762
b_2	9,551	6,729	0,633	–4,957
b_3	0,631	0,494	0,194	–0,087
b_4	–6,575	–4,745	–0,792	2,766

In this new method the side force generated by keel and rudder is calculated using the expression derived by Whicker and Fehlner (1958) for thin airfoils.

$$\frac{dC_L}{d\alpha} = \frac{a_0 AR_e}{\cos\Lambda\sqrt{\frac{AR_e^2}{\cos^4\Lambda} + 4 + \frac{57,3a_0}{\pi}}}, \tag{5}$$

where

AR_e: effective aspect ratio [m]
Λ: the sweepback of quarter-chord line [rad]
α: angle of attack [deg]
a_0: corrected section lift curve slope [–]

$$a_0 = 0,9\left(\frac{2\pi}{57,3}\right) \text{ per degree}$$

The aspect ratio is obtained from expression:

$$AR_e = 2\frac{b}{c_{mean}},$$

where b is the span of the foil and c_{mean} the mean geometric chord in meters.

In this calculation the keel is not extended to the free surface, but taken as it is. The effect of the hull is therefore calculated separately.

Another effect is also the lift carry over from keel to the hull that is expressed over the ratio between the entire lift of the appended hull and the lift generated by the keel and rudder computed from equation 5. This ratio is represented as hull influence coefficient c_{hull}. The formulation for the extended range of keels in upright conditions states:

$$c_{hull} = 1.8\frac{Tc}{bk} + 1, \tag{6}$$

where bk is the span of the keel.

The influence of the heel angle on the lift production is represented by the lift reduction of appendages and expressed by the hell influence coefficient c_{hell}. The second is the zero lift drift angle βo that originates from the asymmetry of the hull when heeled. This reduces the angle of attack on appendages and the effect increases with heel angle. As presented by Verwerft, (2008) a linear relation between lift reduction and heel angle is applied.

$$c_{hell} = 1 - b_0\varphi, \tag{7}$$

with $b_0 = 0.382$ and φ in radians.

The influence of hull asymmetry when heeled is represented by the zero lift drift angle obtained from measurements:

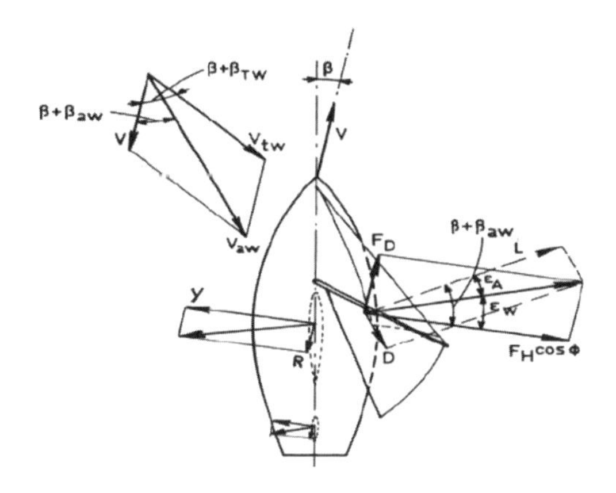

Figure 3. Forces acting on sailing yacht. [On the balance of large sailing yachts].

$$\beta_0 = \left(c_0\frac{B_{wl}}{Tc}\varphi\right)^2, \tag{8}$$

with $c_0 = 0.405$ and φ in radians.

The downwash angle of the keel on the rudder is calculated from the formulation of Keuning (2007):

$$\Phi = a_0\sqrt{\frac{C_{Lk}}{A\mathrm{Re}_k}}, \tag{9}$$

where

Φ : downwash angle at the rudder [rad]
$A\mathrm{Re}_k$: effective aspect ratio of the keel
C_{Lk} : lift coefficient of the keel
and $a_0 = 0,137$ for 15° heel angle.

The lift of the keel and the rudder is than calculated from:

$$Lc_{keel} = C_{L\,keel_{WF}} \cdot \alpha \cdot \frac{\rho \cdot v_{e\,keel}^2 \cdot A_{lat\,keel}}{2} \cdot c_{hull} \cdot c_{hell} \tag{10}$$

where $v_{e\,keel}$ is assumed to be equal to the velocity of the boat v_B.

$$Lc_r = C_{L\,r_{WF}} \cdot \alpha_r \cdot \frac{\rho \cdot v_{e\,r}^2 \cdot A_{lat\,r}}{2} \cdot c_{hull} \cdot c_{hell} \tag{11}$$

where v_{er} is assumed to be $0.9 \cdot v_B$.

The equilibrium obtained is in practice very thin and is controlled by helmsman acting on rudder.

The results obtained for the model of Moro di Venezia give a side forces produced by the keel and the rudder: Depending on the heel angle side forces

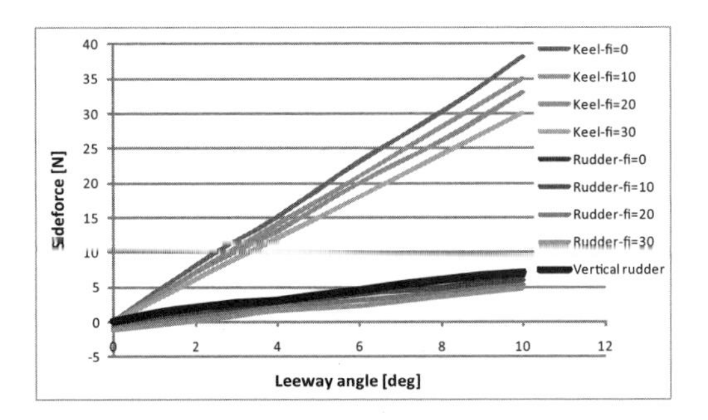

Figure 4. Side forces for keel and rudder depending on heel angle.

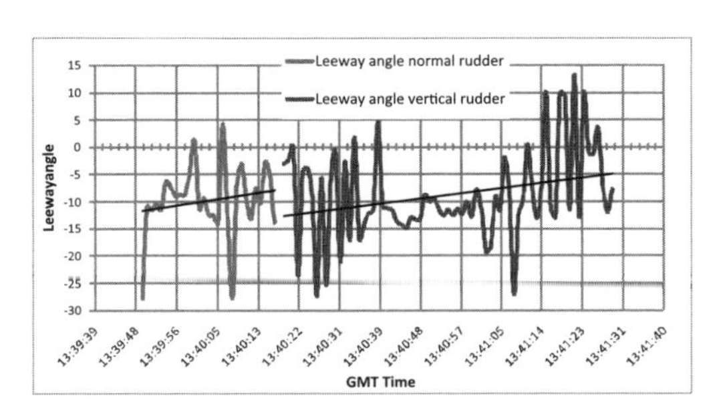

Figure 5. Leeway angle measured for booth rudder positions.

are presented on Figure 4. The velocity is taken as measured at sea, about 1 m/s.

The balance of underwater side forces controls the yaw and the drift of the boat. The reduction of the side forces on the appendages caused by heel angle is presented on Figure 4 as calculated by the above method. Applying the bending rudder that is kept vertical independently on the heel angle neglect some parameters in the calculation of rudder side force. This is the downwash angle Φ, that represent the influence of the keel leaving flow to the rudder and the zero drift lift angle β_0. The result is more side force on the rudder and the ability to reduce the leeway angle by pushing the boat upwind with a reducer drift.

3.2 *Way to reduce leeway angle*

Modern racing boats like open 60 s have wide, flat sterns. This style is beginning to appear in some cruiser-racer designs, particularly in smaller boats. With a single rudder this means that at large heeling angles some of the rudder is out of the water where it's not doing any good at all, so the force available from the rudder is reduced even more. Many boats of this design get around this problem by having twin rudders canted outwards a little. As the boat heels the windward rudder lifts out of the water but the leeward one is submerged perpendicular to the water. The disadvantage of two rudders is in more wetted surface in all other sailing conditions than windward sailing and in disturbances when the windward rudder is not completely out of water.

Instead of twin rudders the single bending rudder (Figure 7) does not influence the original underwater geometry of the boat and could be maintained perpendicular the water surface at any time.

Measurements conducted in water, as presented in the next paragraph, demonstrate the positive influence of the bended rudder in reducing the leeway angle, see Figure 5.

4 MEASUREMENT

The analysis of boat sailing and steering improvement was conducted on sailboat prototype of 2 m LOA. The boat is remote controlled applying a standard rig of mainsail and jib. Mail dimensions of the sailboat model are presented on Figure 7.

Measurements are conducted at the sea although of no controlled conditions. Non-controlled conditions up to a certain level, because boat movements in all tree directions is measured with accelerometer with 25 Hz and positions were measured with GPS at 4 Hz. At the same time the wind speed and direction is measured with 2D anemometer.

The main purpose of measurements is to find the difference of sail characteristics applying a classic rudder and a bending rudder.

The steering system mounted on a sailboat model allows rudder to bend 35 degree in each direction, without influence on underwater geometry of the hull or on rudder profile.

Figure 8 shows the path of the model test and the true wind directions and strengths. The point indicated by a single star is the point where the rudder position is changed from normal, perpendicular to the boat, to perpendicular to the water. At that point the wind has not changed its direction and strength.

When sailing the curse of the sailboat was keep as most as possible upwind regarding to the jib wool tickers mounted on the luff. The obvious conclusion is that the boat is going more upwind when the rudder is in vertical position. However this is still not an overall indication of a more efficient sailing.

Further review on wind conditions and boat speed in necessary. Figure 9 shows the true and apparent wind speed and the speed of the boat. The line at time 14:40:18 indicates when the rudder changes from normal to vertical position. At that time the true wind is slightly slowing down reducing the speed of the boat. There is a contradiction between Figure 8 and Figure 9 because at lower boat speed the drift increase, what has not happen this time contrariwise, the boat is sailing more upwind.

Just before the wind has changed direction, time around 14:41:10, the boat speed has touched the velocity of the true wind speed, what is quite good for a small sailboat model.

Next is the review of wind angle dynamics during measurements. Figure 10 shows the angles of the true and apparent wind that in average do not changes particularly up to the time 14:41:10. Up to that time the true wind angle is from about 250 degrees. The apparent wind angle is going to reduce at 14:40:18, because the wind is slowing down and consequently the speed of the boat.

The finding of the benefit of the vertically positioned rudder is found calculating the (V_{mg}) velocity made good for windward sailing. Applying the model described in [19] the V_{mg} is increasing after the change of rudder in vertical position. This is the result the authors were looking for to confirm the benefit of the bending rudder in windward sailing.

The second test regarded the resistance of the bending system holding the rudder, its stiffness and water sealing. The survey after several measurements and several hours in water in different weather and sea conditions shows that no one drop of water has entered through the mechanism.

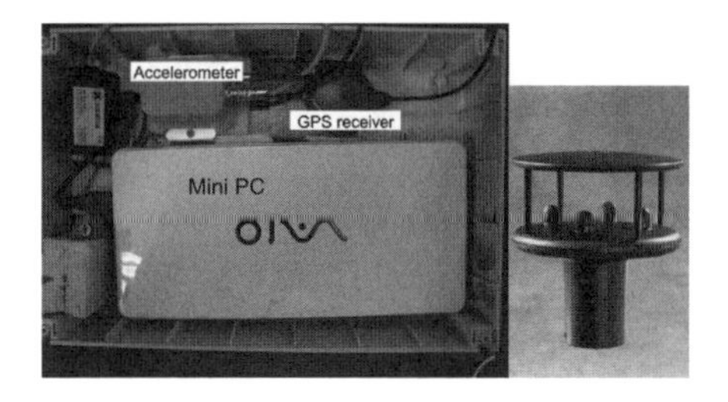

Figure 6. Measurement equipment and data collection.

Figure 7. Sailboat model applying bending rudder device.

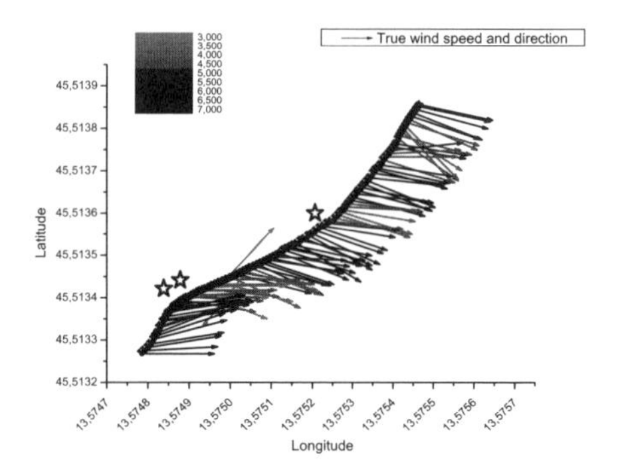

Figure 8. Sailing path and true wind characteristics.

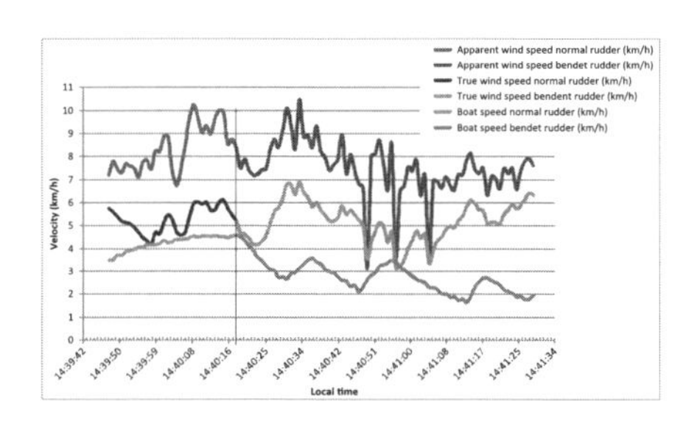

Figure 9. Apparent and true wind speed compared with sail boat speed.

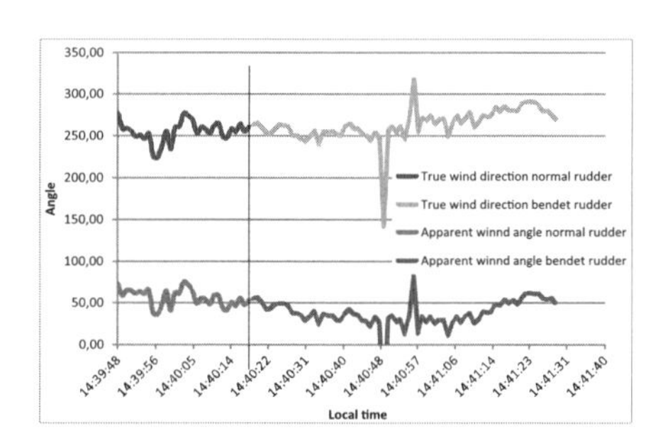

Figure 10. Wind direction and apparent wind angle for booth rudder positions.

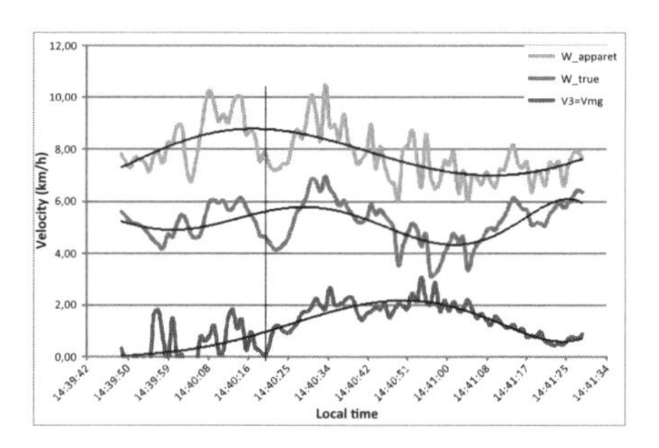

Figure 11. Apparent and true wind speed compared with velocity made good for windward sailing.

5 CONCLUSION

The bending rudder system and its holding mechanism has been installed in a 2 meter sailboat model and tested at the sea. The initial hypothesis was that the boat applying a bending rudder, could reach better sailing performances in windward sailing. During tests, the boat position, accelerations in three directions, wind speed and direction have been measured. The analysis of results demonstrates that the sailboat reduces the drift and is sailing at lower upwind angle. The result is a better V_{mg} for windward sailing. The bending rudder system was also tested for mechanical resistance and water sealing. Results were positive; the mechanism maintains his stiffness and water tightness for all the time of testing.

From the author point of view the analyses conducted and presented in this paper gives enough information and proves that the system could be applicable to larger racing and cruising sail yachts.

An additional performance that is quite funny for a banding rudder is the possibility of use as emergency propulsion. Bending the rudder continuously moves the boat forward and the rudder works like a long oar.

REFERENCES

B. Verwerft & J.A. Keuning. The Modification and Application of a Time Dependent Performance Prediction Model on the Dynamic Behaviour of a Sailing Yacht, 20th International HISWA Symposium on Yacht Design and Yacht Construction, Amsterdam, The Netherlands, 2008.

F. Fossati. Aero-hydrodinamics and the performance of sailing yachts, Politecnico di Milano, Milano, 2007.

J. Gerritsma. "Course keeping qualities and motions in waves of a sailing yacht" Technical Report, Delft University of Technology, May 1971.

J. Gerritsma & J.A. Keuning. "Sailing yacht performance in calm water and waves", 12th International HISWA Symposium on Yacht Design and Yacht Construction, The Netherlands,1992.

J. Meseguer Ruiz & A. Sanz Andrés. "Aerodinámica Básica". ETSI Aeronáuticos, Universidad Politécnica de Madrid 2005.

J.A. Keuning & B.J. Binkhorst. "Appendage Resistance of a Sailing Yacht Hull", the 13th Chesapeake Sailing Yacht Symposium, Annapolis, Maryland, USA, 1997.

J.A. Keuning & G.K. Kapsenberg. "Wing—Body Interaction on a Sailing Yacht", the 18th Chesapeake Sailing Yacht Symposium, Annapolis, Maryland, USA, 2005.

J.A. Keuning & K.J. Vermeulen. The yaw balance of sailing yachts upright and heeled, Chesapeake Sailing Yacht Symposium, 2003.

J.A. Keuning, M. Katgert & K.J. Vermeulen. "Keel-Rudder Interaction on a Sailing Yacht", 19th International HISWA Symposium on Yacht Design and Yacht Construction, Amsterdam, The Netherlands, November 2006.

J.A. Keuning, M. Katgert & K.J. Vermeulen. "Further Analysis of the Forces on Keel and Rudder of a Sailing Yacht", the 18th Chesapeake Sailing Yacht Symposium, Annapolis, Maryland, USA, March 2007.

J.A. Keuning, R. Onnink, A. Versluis & A.A.m. van Gulik. "The Bare Hull Resistance of the Delft Systematic Yacht Hull Series" 14th International HISWA Symposium on Yacht Design and Yacht Construction, Amsterdam, The Netherlands, 1996.

J.A. Keuning & U.B. Sonnenberg. "Approximation of the Calm Water Resistance on a Sailing Yacht Based on the Delft Systematic Yacht Hull Series", the 13th Chesapeake Sailing Yacht Symposium, Annapolis, Maryland, USA, 1997.

J.A. Keuning & U.B. Sonnenberg. Approximation of the Hydrodynamic forces on a sailing yacht based on the Delft Systematic Yacht Hull Series, HISWA Symposium on Yacht Design and Construction, 1998.

J.R. Binns, K. Klaka & A. Dovell. "Hull-Appendage Interaction of a Sailing Yacht, Investigated with Wave Cut Techniques", the 13th Chesapeake Sailing Yacht Symposium, Annapolis, Maryland, USA, 1997.

L.F. Whicker & L.F. Fehlner. Free-stream characteristics of a family of low-aspect-ratio, all-movable control surfaces for application to ship design, Technical report 933, David Taylor Model Basin, 1958.

K. Nomoto & Tatano, H. Balance of helm of sailing yachts, a shiphydromechanics approach on the problem, HISWA 1979.

O.M. El Moctar. (1998). Numerical determination of rudder forces, Euromech 374, Futurescope, Poitiers, 27–29 April 1998.

P. van Oossanen. "Predicting the Speed of Siling Yacht". SNAME transactions, Vol. 101, 1993, pp. 333–397.

R. Garret. The Symetry of sailing-the physics of sailing for yachtsman, Adlard Coles Ltd., London, 1987.

R.Q. Lin, M.l Hughes & T. Smith. Prediction of ship steering capabilities with a fully nonlinear ship motion model. Part 1: Maneuvering in calm water, Journal of Marine Science and Technology Volume 15, Number 2, 131–142, 2010.

R.Z. Rodríguez, J.I. Yerón & E.B. Vera. Sailing Yacht Rudder Behaviour, 47° Congreso de Ingeniería Naval e Industria Marítima, España 2008.

S. Gaggero & S. Brizzolara. A potential based panel method for the analysis of marine propellers in steady flow, 6th International conference on high-performance marine vehicles, Naples (Italy), 2008.

S.F. Hoerner. "Fluid Dinamic Drag", Hoerner Fluid Dynamics, Bricktown, N.J., 1965–1992.

http://www.oceansail.co.uk

Sustainable Maritime Transportation and Exploitation of Sea Resources – Rizzuto & Guedes Soares (eds)
 ISBN 978-0-415-62081-9

An overview of seakeeping tools for maritime applications

V. Bertram
FutureShip, Hamburg, Germany

P. Gualeni
University of Genoa, Italy

ABSTRACT: The paper discusses the background of various seakeeping tools, ranging from strip theory to sophisticated 3d Navier-Stokes solvers, in the perspective to sketch a sort of users guideline for efficient exploitation. In fact, there is no approach that is suited for all seakeeping applications, but their associated advantages and disadvantages should be understood and tools should be applied intelligently, often combining them to have appropriate response time and accuracy in predictions. After a description of approaches (and most prominent codes based on these approaches), recent applications are shown for unconventional craft and offshore/coastal structures involving strong nonlinearities, as demonstration of challenging cases at present.

1 INTRODUCTION

Since the fifties the evolution of computational power and applied hydrodynamics has permitted a continuous improvement in seakeeping performance evaluation, with increasing accuracy and reliability even for audacious applications in terms of geometry, speed, environmental conditions and implied non linearities of the problem. A significant amount of methodological approaches and tools are nowadays available and an interesting taxonomy of them is available in Beck & Reed (2001) and Bertram (2000). For a further comprehensive analysis of the matter a large amount of references can be found in ITTC (2005) and ITTC (2008), where, beside efficient state of the art review of methodologies and tools for seakeeping calculations, attention is also given to the very important aspects of validation and development of experimental techniques.

The present paper discusses the background of various seakeeping tools, in the perspective to create the awareness that a sort of road-map for ship design teams might be desirable. The consistency between the investigation target and the proper numerical tool to be applied is vital in order not to fail or misinterpret the objective: to obtain a satisfactory accuracy in predictions in a reasonable response time.

Moreover the education, the experience and the capabilities of the operator can make a significant difference, while using numerical tool for a seakeeping performance evaluation.

A brief description of approaches implemented in the most prominent codes is presented in the following paragraphs together with recent applications.

In relation with the peculiarities of the investigated cases, calculations are going to be performed with RANSE simulations or using the most advanced codes based on inviscid fluid flow, exploiting Green Function or based on time domain strip method. Some challenging examples will be presented. To this regard it is interesting to point out that any seakeeping investigation might be challenging not only for performances that have to be examined, but also for the marine vehicle/structure to be analyzed: this in fact can present unusual characteristics both for geometry and operational profile.

It is well-known that seakeeping performances might in principle cover a very wide range of issues; we can recall for example:

- Maximum speed in a seaway associated with "involuntary" speed reduction due to added resistance in waves and "voluntary" speed reduction to avoid excessive motions, loads, etc.
- Route optimization (routing), most frequently to minimize fuel consumption
- Structural design of the ship to withstand slamming loads in seaways
- Habitation comfort and safety of people on board
- Ship and cargo safety
- Operational limits for special purpose ships (e.g., offshore supply vessels)

What above mentioned implies evaluation of motions (absolute and/or relative) and accelerations at different locations on board. It might be necessary to estimate loads on structures, level of comfort onboard. Computations can regard a short term prediction or a long term prediction and some criteria can also be introduced to support a decision making process. Some performances can be satisfactorily captured with a linear approach some others need the implementation of non-linearities effects relevant for example to motions, geometry, environment and their simultaneous influence.

Another point worth mentioning is the recent evolution of seakeeping investigation, more and more in close relation with safety issues and safety rules based on performance assessment, for both the intact and the damage ship.

2 BASIC SIMULATION OPTIONS: LINEAR VS NON-LINEAR

For conventional ships, experience as for example documented in Classification Rules may suffice for practical engineering predictions. However, increasingly advanced engineering analyses are used for decisions involving seakeeping in design and operation. Beside predictions made by experimental model tests, several tools are in use to predict ship seakeeping performance:

- Computations in the frequency domain to determine ship reactions in harmonic waves of different wave lengths and wave directions
- Computations in the time domain (simulation in time) to yield forces on the ship for given point in time and, based on that information, motions at subsequent times
- Computations in the statistical domain to obtain statistically relevant seakeeping measures in natural (irregular) seaways, e.g., average frequencies and exceedence limits of events in a given seaway or ocean region.

Many seakeeping investigations comprise the following steps:

- Representation of the natural seaway as a superposition of a large number of regular (harmonic) waves,
- Computation of ship reactions of interest in these harmonic waves, and
- Addition of the reactions in these harmonic waves to obtain the total reaction.

This procedure requires that the reaction of one wave on the ship is not changed by the simultaneous occurrence of another wave. This assumption is valid for small wave heights and for almost all ship reactions with the exception of added resistance. The procedure is often applied also for severe seaways with large waves. However, in these seaways only approximate estimates requiring appropriate corrections can be obtained. One consequence of the assumed independence of individual wave reactions is that all reactions of the ship are proportional to wave height. This is called linearization with respect to wave height.

Wave loads and their effects cannot be predicted within standard linear seakeeping theory, especially not for ships with large bow flare and stern overhang. Vertical hull girder loads can be nonlinear although the motions are linear. For practical applications, codes based on linear theory may suffice to provide motion predictions for ships.

For weakly or moderately nonlinear cases, motions and loads of a ship in natural seaways can be approximated by superimposing responses in elementary waves of different frequency and direction. For better accuracy, semi-empirical corrections are introduced for viscous force contributions and other nonlinear effects. Repeating these computations for all wave amplitudes, frequencies and headings yields the desired nonlinearly corrected transfer functions. These pseudo transfer functions depend on wave amplitude. Strictly speaking, they are valid only for one wave amplitude; in practice, they suffice for a limited range of wave amplitudes.

For highly nonlinear motions and loads, only time-domain simulations (or model tests) are appropriate. Extreme motions appear for example in synchronous roll (ship natural period near encounter period in waves) and parametric roll (ship natural period near half the encounter period). Ships with large flare are susceptible to parametric roll in (oblique) following or head seas of considerable wave height. Occurrence of parametric roll can be adequately described in terms of a linear equation (so-called Mathieu equation), but prediction of actual roll angles is difficult because of the nonlinear stability curve and nonlinear roll damping. Nonlinear time-domain simulations seem to be the most promising and practical approach. Several simulation runs are needed to obtain useful statistical measures of the stochastic process for the nonlinear system.

3 METHODS

3.1 *Strip methods*

The essence of strip theory is to reduce the three-dimensional hydrodynamic problem to a series of two-dimensional problems that are easier to solve. The underwater part of the ship is divided into a number of strips (typically 20).

Thus, two-dimensional coefficients are calculated for each strip, implying that the fluid flow velocities in the cross-sectional plane are much greater than in the longitudinal direction. Although the underlying physical models are crude, strip methods are able to calculate most seakeeping properties of practical relevance accurately enough for displacement monohulls. Strip methods are generally applicable up to Froude numbers of 0.4. With some corrections, this range can be extended up to Froude numbers of 0.6. For displacement hulls at Froude numbers above 0.4, 2D + t methods (also called high-speed strip methods HSST) are fast and yield good results, Bertram & Iwashita (1996).

Special strip methods were developed for catamarans, taking into account the interaction of radiated and diffracted waves between the two hulls at forward speed, e.g., SEDOS by Söding (1999). These are relatively complex in coding and not widely available. Therefore, frequently 3d methods are applied to assess seakeeping of catamarans, e.g., Landrini & Bertram (2002).

For extreme motions (particularly extreme roll motion up to capsizing), a suitable tool can be in principle a nonlinear strip method like ROLLS, GLSIMBEL, Pereira (1988), or LAMP. Large amplitude rigid body motions of the ship in six degrees of freedom are computed by time integration of the motion equations, considering forces caused by gravity, incident wave, radiation and diffraction pressure, speed effects (resistance and maneuvering forces caused by the oblique forward motion), rudder and propeller actions, and wind. Forces caused by radiation and diffraction are deduced from the two-dimensional potential flow at each of the transverse ship sections (strips). For large amplitude motions, radiation and diffraction forces and moments are obtained by integrating the pressure over the instantaneously wetted surface. Typically, semi-empirical corrections account for viscous roll damping of hull and bilge keels. Today, these codes can also account for damaged hulls with water flooding into hull compartments. However, computational times can be considerable, ranging up to several hours, while linear strip methods give results within minutes.

Although strip methods are practical design tools to assess global wave-induced motions, they have important limitations:

- Strip theory is basically a high-frequency theory. Thus, it is more applicable in head and bow waves than in following and quartering waves for a ship with forward speed.
- Strip theory is a low Froude number theory. It does not properly account for the interaction between the steady wave system and the oscillatory effects of ship motions.
- Due to the assumed linearity between response and incident wave amplitude, it is questionable to apply the standard strip method to severe sea states and/or hull forms that have strong flare or stern overhang.

3.2 *Three-dimensional potential flow methods*

For offshore applications, global loads and motions in seakeeping can be computed quite well by three-dimensional Green function methods (GFM). These distribute panels on the ship's average wetted surface (usually the calm-water floating position). For zero speed, the steady wave system vanishes and various diffraction and radiation wave systems coincide. If the geometry of offshore structure and waves are of the same order of magnitude, GFMs can successfully capture three-dimensional effects and complex interactions. The employed codes determine forces and motions either in the time or the frequency domain. First-order forces and motions are calculated reliably and accurately. For practically required accuracy of first-order quantities, 1000–2000 elements are typically deemed sufficient. Commercial program packages (WAMIT, TIMIT, AQWA or DIODORE) are widely accepted and used for offshore applications.

For non-zero speed, GFM are less suited. They neglect generally some forward-speed effects, such as the dynamical trim and sinkage and the steady wave profile. The various methods differ primarily in the way the Green function is computed. This involves the numerical evaluation of complicated integrals from 0 to ∞ with highly oscillating integrands. Some GFM approaches formulate the boundary conditions on the ship under consideration of the forward speed, but evaluate the Green function only at zero speed. This saves a lot of computational effort, but cannot be justified physically. Nevertheless satisfactory results for are reported, at least for motions, for speeds up to Froude number 0.4. As an alternative to the standard frequency-domain approach, GFM may also be formulated in the time domain. This avoids the evaluation of highly oscillating integrands, but introduces other difficulties related to the proper treatment of time history of the flow in so-called convolution integrals. All GFMs are fundamentally restricted to simplifications in the treatment of the steady flow. Usually the disturbance of the steady flow is completely omitted which is questionable for usual ship hulls. It will introduce, especially in the bow region, larger errors in predicting local pressures.

The 1990s saw the advent of Rankine singularity methods (RSM) for seakeeping, e.g., the SWAN (Ship Wave Analysis) code. Bertram & Yasukawa (1996) give an extensive overview of these methods. The approaches are similar to those used widely in industry for the steady wave resistance problem, but failed to reach a similar level of acceptance. RSM, in principle, may capture the steady flow completely and allow also more complicated boundary conditions on the free surface and the hull. This comes at a price. Both ship hull and the free surface in the near field around the ship have to be discretized by panels. The steady flow problem has to be solved first, before solving the seakeeping problem, increasing computational time and complexity. Both RPM and GFM can guarantee satisfactory results but they have in common the inherent difficulties to represent the flow around immersed transom sterns.

3.3 *RANSE solver*

Computational fluid dynamics (CFD) methods capture all relevant nonlinearities (fluid dynamics equation, hull geometry, and water surface geometry). State-of-the-art codes solve the so-called Reynolds-averaged Navier-Stokes equations (RANSE), modeling turbulent fluctuations by semi-empirical turbulence models. As turbulence does not play a significant role in seakeeping, RANSE solvers capture all relevant physics directly in a simulation. Modern CFD codes usually allow representation of complex free-surface flows (spray, breaking waves, green water on deck, etc.), as that one in Figure 1, e.g., El Moctar et al., (2007). They can be useful tools for slamming and sloshing analyses.

High-speed craft (HSC) rules of classification societies include formulas for safe design loads, and these loads reflect service experience. However, for unconventional designs and for optimized lightweight designs, special simulations are appropriate to determine wave loads. To determine amplitudes of equivalent design waves for fast ships, special effects of slamming or green water on wave-induced global loads need to be considered, e.g., Schellin and Perez de Lucas (2004), Ziegler et al. (2006).

Seakeeping of planing hulls is another area where RANSE simulations are showing encouraging results and in principle represent the recommended choice. Rolla Research in Switzerland and MTG in Germany have presented convincing applications for real planing hull geometries, Caponnetto (2001), Caponnetto et al. (2003), Azcueta (2003).

CFD simulations require significant effort in grid generation and are computationally expensive, (Fig. 2). The proper grid generation is nevertheless a key point in order to obtain reliable results.

As already mentioned education and skills of the operator can make a significant difference when dealing with any computational tool, but especially with RANSE codes. This is true in general also when in principle the application is "simpler", dealing with only ship resistance for example.

Simulating several seconds in real time may take several hours on powerful computers (employing parallel computing). A simulation case is represented in Figure 3.

Figure 1. CFD simulation of platform in extreme waves.

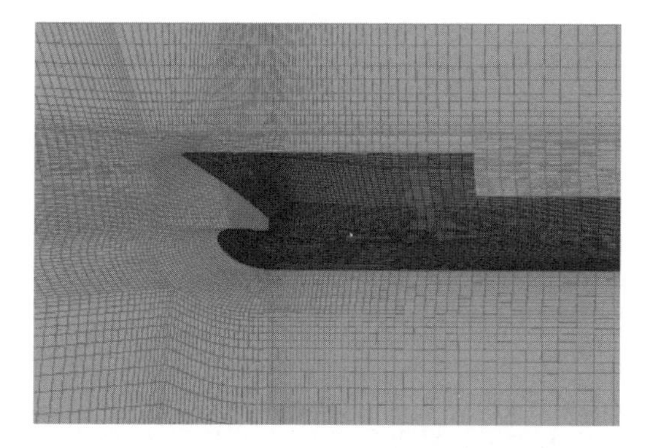

Figure 2. Sample of RANSE grid for slamming analysis.

Figure 3. Sample of a RANSE simulation with water on deck.

A final comment in this context is dedicated to a special problem: offshore structures. They typically feature a variety of relatively thin cylindrical structures, e.g., tubular elements to connect buoyancy components and platform, marine risers, or mooring cables. The hydrodynamic forces on these structures are dominated by viscous effects. Traditionally, the Morison formula has been applied to estimate the forces in seaways for these elements. Despite its undeniable usefulness, the Morison formula is not applicable to cases with background currents or interaction between several such elements. For such more complicated effects, experience based estimates are gradually replaced by CFD simulations.

It is worth mentioning that an ITTC (International Towing Tank Conference) Specialist Committee has been appointed on RANSE code matters named "CFD in Marine Hydrodynamics". Among the terms of references for 2008–2011 it can be read "It is inevitable that these methods will have an even larger role in the future as computer power increases and the application of such codes matures even further. However, it will still take considerable effort to have the confidence in these methods that currently exist with the same level as in model tests, since grid resolution, turbulence modeling and other sources of uncertainties are still major factors which affect the accuracy of solutions". As a specific task of the group it is indicated: "Identify CFD elements of importance to the ITTC from a user's point of view, including applicability, accuracy, reliability, time and cost".

4 SLAMMING: AN EXAMPLE FOR A VIRTUOUS TOOL COMBINATION

Wave-related impact (slamming) loads are larger than other wave loads. They may cause local damage or large-scale buckling of deck structures. Even if each impact load is relatively small, frequent impact loads accelerate fatigue failures of hulls, especially for high-speed ships. A fully satisfactory theoretical treatment of slamming has been prevented so far by the complexity of the problem:

- The process is random like other processes involving natural seaways.
- Slamming is strongly nonlinear and sensitive to relative motion and contact angle between body and surface.
- Since wave impact is short, hydro-elastic effects are important. Applying slamming peak pressures over ship structural plate fields is not appropriate for design purposes.
- Air trapping may lead to compressible flows with water-air interaction.
- Slamming is a three-dimensional process requiring three-dimensional analyses for most practical applications.

Slamming pressures depend on ship speed, wave height, and wave length. Some results showing the importance of these parameters are presented in Figures 4a, b, c, where pressures are normalized against maximum pressure of p0 = 270 kPa and times against wave encounter period of T0 = 6 s.

The highest pressures occur at large ship speeds, in high waves, and in waves of length approximately equal to ship length.

All classical slamming approaches are variations on the work of Von Karman and Wagner in the 1930s. With semi-empirical corrections, peak impact pressures are reasonably well predicted, albeit only for two-dimensional sections without flat bottom or large deadrise. Thus, these classical

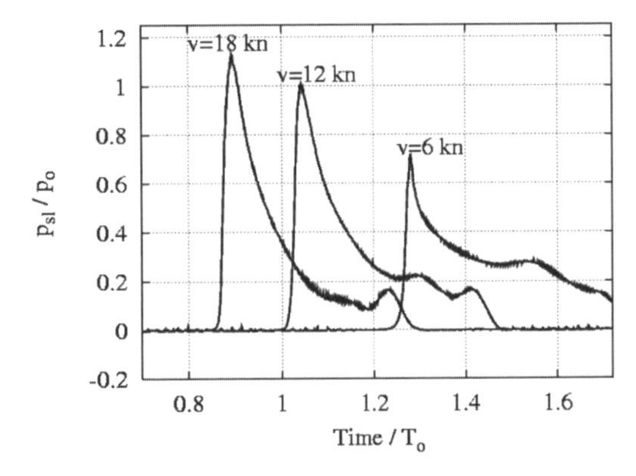

Figure 4a. Influence of ship speed on computed slamming pressure acting at a critical plate field under a flared bow.

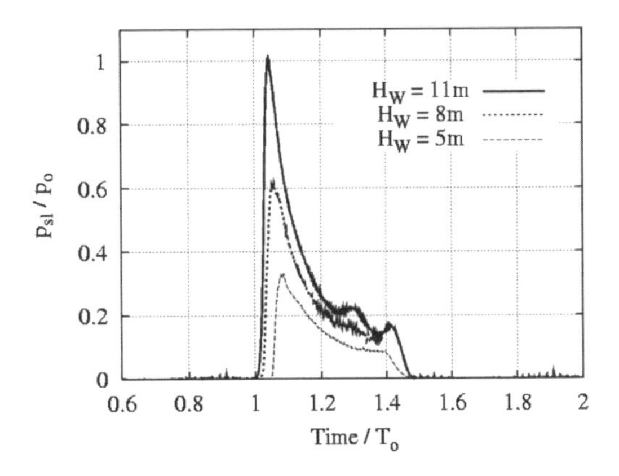

Figure 4b. Influence of wave height on computed slamming pressure acting at a critical plate field under a flared.

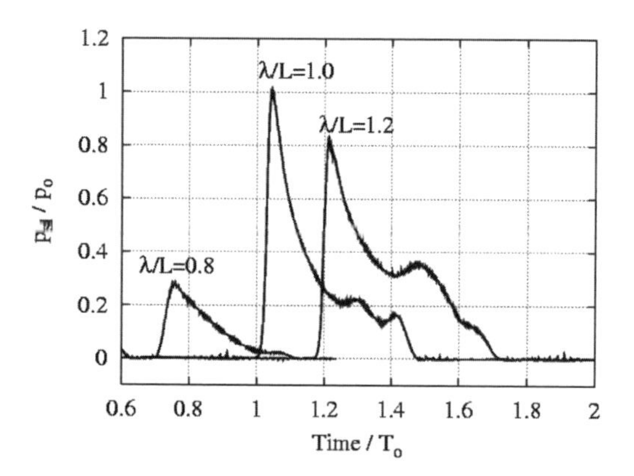

Figure 4c. Influence of wave length on computed slamming pressure acting at a critical plate field under a flared bow.

theories are hardly applicable for real ship hull geometries.

A suitable tool of choice today is a combination of fast potential flow methods with accurate RANSE methods, El Moctar et al. (2004). The approach can be summarize in to two phases. At first a linear, frequency-domain code computes ship responses in regular waves is applied. Wave frequency and wave heading are systematically varied to cover all possible combinations that are likely to cause slamming. Sometimes only head waves are investigated, because these conditions are rated most critical for slamming. Results are linearly extrapolated to obtain responses in wave heights that represent severe conditions, here characterized by steep waves close to breaking. Then free-surface RANSE computations enable to determine motions and loads for the identified critical parameter combinations.

The obtained average slamming pressures are applied as equivalent static design loads to determine scantlings of the ship structure. This multistage procedure allows predicting slamming loads suitable for design of monohulls and catamarans alike. It represents a good compromise between attainable accuracy and computational effort. Some comparison results between measured and computed pressures time histories, at a position shown in Figure 5a, are presented in Figures 5b. Figure 5c shows time histories of measured and computed vertical forces on bow section.

Although the employed design wave approach only roughly approximates real wave conditions when slamming, there is at present no alternative available for practical application.

Class Rules formulas for slamming pressures are suitable for local design of the fore and aft ship structure, but not for the global hull girder. For hydro-elastic analyses of the hull girder, RANSE

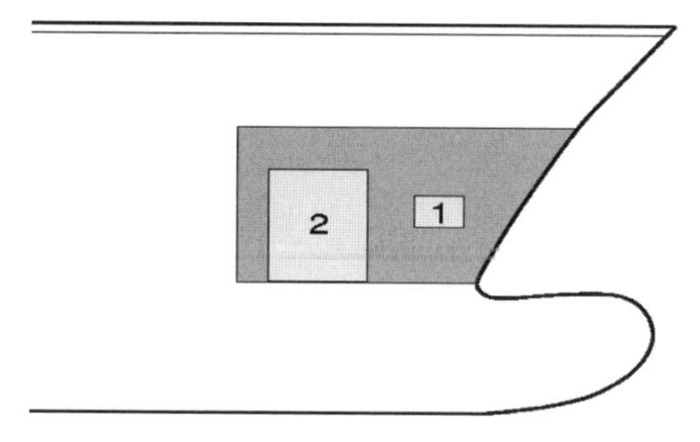

Figure 5a. Location of plate field 1 where time histories are measured (HSVA) and computed (GL).

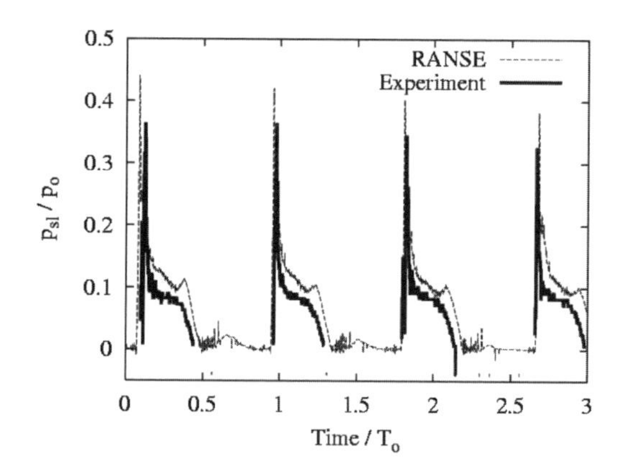

Figure 5b. Time histories of measured (HSVA) and computed (GL) pressures on plate field 1.

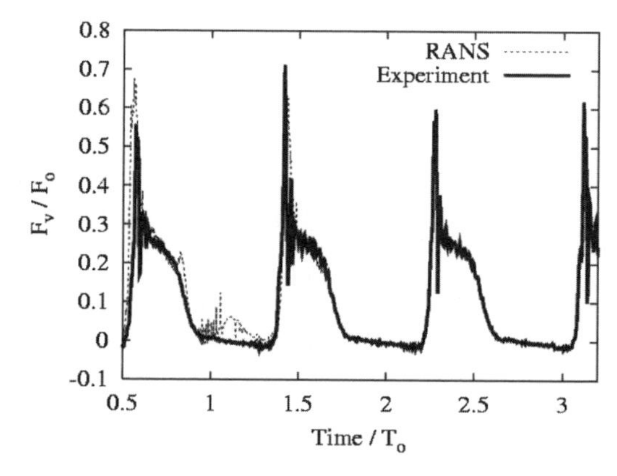

Figure 5c. Time histories of measured (HSVA) and computed (GL) vertical forces on bow section.

simulations of the freely moving ship coupled with finite-element analysis (FEA), Oberhagemann et al. (2008), Fig. 6, can reproduce whipping (high-frequency vertical vibrations of the hull girder) that may increase the wave-induced vertical bending

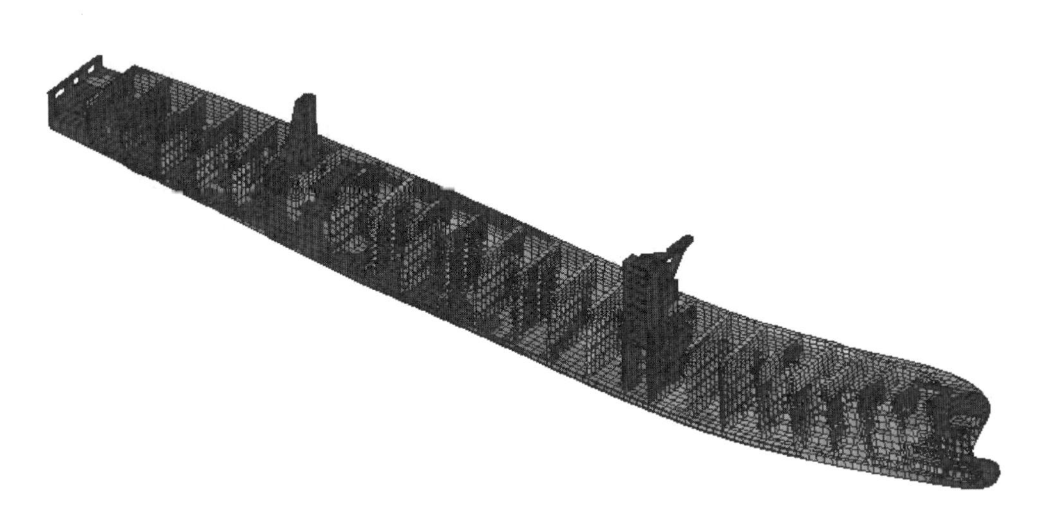

Figure 6. Time histories of measured (HSVA) and computed (GL) pressures on plate field 1.

moment (e.g., by 20% for a large containership). The FEA model comprises all major structural components (decks, bulkheads, walls, floors, web frames, etc.). The RANSE simulation determines slamming loads and the FE simulation the corresponding stresses and deformations. These, in turn, may be considered for the flow solution.

5 FINAL CONSIDERATIONS

It is possible to perform a ranking of different methodologies and tools for seakeeping performance evaluation with reference to different criteria. The following options can be pointed out: potential or viscous theory, 2D or 3D or something hybrid, linear or non linear, frequency domain or time domain, just to recall the most general issues among those mentioned in this paper.

It is more and more diffuse the awareness that unsteady RANSE codes with fully nonlinear free-surface boundary conditions are in principle the most theoretically accurate and elegant option. Nevertheless, results are very much dependant on the operator skills and a long activity of application and validation is still required to acquire the state of the art status. But the most important aspect is that their application still require an unacceptable computational effort.

On the complete opposite side there are potential strip methods, simple, fast and robust. On strip methods are based many valuable prediction tools which can provide (in some occasions inexplicably) satisfactory results, in very low time frame.

In relation with the kind of prediction, the proper tool can be selected in the conceptual space that has been identified and described in the paper. In some occasions, a combined approach can be advantageous and a successful example has been provided for the slamming problem.

Activity on seakeeping tools should be carried out on different levels, i.e. on the consolidation/combination of mature tools for new application cases and on the development and validation of innovative tools for very challenging problems, like prediction of ship behavior in extreme seas, with the proper treatment of the wind and wave scenario.

REFERENCES

Azcueta, R. 2003. Steady and unsteady RANSE simulations for planing crafts. *7th Conf. Fast Sea Transportation* (FAST), Ischia.

Beck, R.F. & Reed, A.M. 2001. Modern computational methods for ships in a seaway. *SNAME Transactions vol. 109.*

Bertram, V. 2000. Practical Ship Hydrodynamics. *Butterworth—Heinemann*, Oxford.

Bertram, V. & Iwashita, H. 1996. Comparative evaluation of various methods to predict seakeeping qualities of fast ships. *Schiff+Hafen 48/6.*

Caponnetto, M. 2001. Practical CFD simulations for planing hulls. *2nd Conf. High-Performance Marine Vehicles* (HIPER), Hamburg.

Caponnetto, M., Söding, H., Azcueta, R. 2003. Motion simulations for planing boats in waves. *Ship Technology Research vol. 50.*

El Moctar, O., Brehm, A. & Schellin, T.E. 2004. Prediction of slamming loads for ship structural design using potential flow and RANSE codes. *25th Symp. of Naval Hydrodynamics*. St. John's, Canada.

El Moctar, O.M., Schellin, T.E. & Peric, M. 2007. Wave load and structural analysis for a jack-up platform in freak waves, *26th Int. Conf. Offshore Mechanics and Arctic Engineering* (OMAE), San Diego, USA.

ITTC, 2005. "The Seakeeping Committee—Final Report and Recommendations to 24th ITTC." Proc. 24th ITTC, Edinburgh, UK.

ITTC, 2008. "The Seakeeping Committee—Final Report and Recommendations to 25th ITTC." Proc. 25th ITTC, Fukuoka, Japan.

Landrini, M. & Bertram, V. 2002. Three-dimensional simulation of ship seakeeping in time domain. *Jahrbuch der Schiffbautechnischen Gesellschaft*, Springer.

Oberhagemann, J., El Moctar, O., Holtmann, M., Schellin, T., Bertram, V. & Kim, D.W. 2008. Numerical simulation of stern slamming and whipping. *11th Numerical Towing Tank Symp.*, Brest.

Pereira, R. 1988. Simulation of nonlinear sea loads, *Ship Techology Research/Schiffstechnik. 3.*

Schellin, T.E. & Perez de Lucas, A. 2004. Longitudinal strength of the high-speed ferry. *Journal of Applied Ocean Research 26(6).*

Söding, H. 1999. Seakeeping of multihulls. *1st Int. Conf. High-Performance Marine Vehicles* (HIPER), Zevenwacht, South Africa.

Yasukawa, H. & Bertram, V. 1996. Rankine source methods for seakeeping problems *Jahrbuch der Schiffbautechnischen Gesellschaft, Springer.*

Ziegler, W., Fach, K., Hoffmeister, H. & El Moctar, O.M. 2006. Advanced analyses for the EARTHRACE project. *5th Conf. High-Performance Marine Vehicles* (HIPER), Launceston, Tasmania.

Sustainable Maritime Transportation and Exploitation of Sea Resources – Rizzuto & Guedes Soares (eds)
 ISBN 978-0-415-62081-9

On the interaction between random sea waves and a floating structure of rectangular cross section

G. Malara & F. Arena
"Mediterranea" University, Reggio Calabria, Italy

P.D. Spanos
Rice University, Houston, TX, US

ABSTRACT: This paper deals with the interaction of a floating structure of rectangular cross section with random sea waves. The structure can be regarded as a simple model of a barge or a ship. A matching technique involving an eigen-functions expansion is presented, in the context of a linear potential theory. It renders an estimate of the velocity potential and of the free surface displacement. It is shown that the wave field is the superposition of a scattered wave field and of a radiated one. The changes on the free surface are investigated. They are interpreted according to a variation parameter. It is shown that this parameter is approximately 2 at the head of the structure; therefore the mean wave height is 2 times the mean wave height of the incident wave field. Next, the interaction between a high wave and the structure is investigated by the Quasi-Determinism theory. Results are validated by a pertinent Monte Carlo simulation.

1 INTRODUCTION

The interaction between sea waves and floating bodies of rectangular cross section has been studied in the context of floating breakwaters, floating buoys, ice sheet modelling, and design of artificial islands and floating airports for several decades. Both numerical and analytical approaches have been used in these studies.

In numerical applications, the source function for fluids with free surface has been proposed, for example, by John (1950), Kim (1965) and Peter & Meylan (2004). Further, Yamamoto et al. (1980) applied the Green's identity formula to solve numerically the boundary value problem for a floating object in the context of two dimensional water waves. Also Mei & Black (1969) and Black et al. (1971) used a variational formulation to study the diffraction and the radiation of rectangular bodies.

Analytical solutions have been developed by matching techniques involving eigen-functions expansions. The technique was first proposed by Stoker (1957). Then, it was used by Drimer et al. (1992), for developing a simplified analytical model of a floating breakwater. Williams & McDougal (1996) investigated the behaviour of submerged and of floating breakwaters. They showed that solutions determined by matching techniques are in good agreement with data from small scale field experiments. Recently, the technique was also used by Abul-Azm & Gesraha (2000), Gesraha (2004) and Zheng et al. (2006). Further, the technique has been used in the context of random waves of a specified spectrum by Malara et al. (2010).

In this paper a matching technique involving an eigen-functions expansion is used for analysing the interaction between random sea waves and a structure of rectangular cross section. The structure can be regarded as a simple model of a barge or a ship. It is modelled a 3-D-O-F system in sway, heave and roll. The free surface modification is investigated. Then, the interaction of the wave field with the structure is analysed by the Quasi-Determinism theory. This theory allows the determination of the free surface when an exceptionally high wave occurs.

2 MATHEMATICAL BACKGROUND

2.1 *Governing equations*

Consider a Cartesian coordinate system $Oyxz$ (Figure 1), in which the origin is in the still water level, the z-axis is vertical and upward oriented, the y-axis and the x-axis are horizontal and orthogonal to each other. The fluid is assumed incompressible and inviscid. The flow is irrotational. In this context, the particle kinematics and the wave pressure

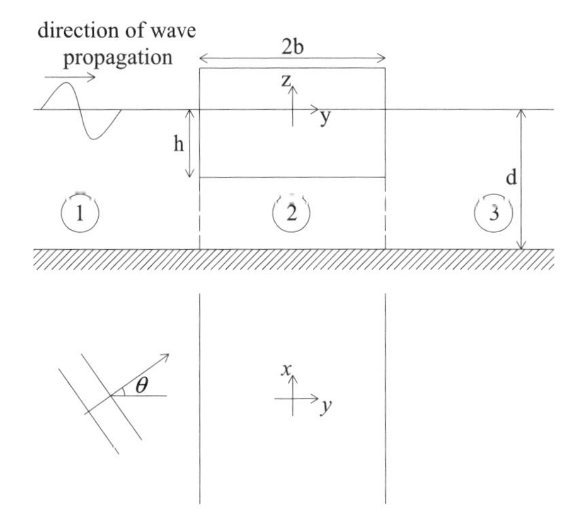

Figure 1. Reference sketch for the wave-structure interaction.

are estimated from the velocity potential. Next, consider a structure of rectangular cross section, rigid, infinitely long in the x-direction and with 3-D-O-Fs (sway, heave and roll). The characteristic dimensions of the structure in water of depth d are: the width $2b$, the draft h. The dimensions are such that the modification of the wave field due to the diffraction and radiation of waves is not negligible.

In this context, the velocity potential, to the first order in a Stokes' expansion, is the solution of the following boundary value problem

$$\frac{\partial^2 \Phi}{\partial y^2}+\frac{\partial^2 \Phi}{\partial x^2}+\frac{\partial^2 \Phi}{\partial z^2}=0, \quad (1)$$

$$\frac{\partial \Phi}{\partial z}-\frac{\omega^2}{g}\Phi=0, \ \text{at} \ z=\eta, \quad (2)$$

$$\left.\frac{\partial \Phi}{\partial z}\right|_{z=-d}=0, \quad (3)$$

$$\left.\frac{\partial \Phi}{\partial n}\right|_{\Omega}=V_n, \quad (4)$$

$$\lim_{|y|\to\infty}|y|^{1/2}\left[\frac{\partial}{\partial y}+ik\right]\left(\Phi-\Phi^{(I)}\right)=0, \quad (5)$$

where Φ is the velocity potential, g is the acceleration due to gravity and V_n is the component of the body velocity normal to the solid bound Ω. Eq. (1) expresses the mass conservation, eq. (2) is the free surface condition, eq. (3) is the condition on the seabed, eq. (4) is the condition on the solid bound and eq. (5) is the Sommerfield condition.

2.2 *Incident wave field*

Following the theory of wind-generated waves, formulated by Longuet-Higgins (1963) and Phillips (1967) a random sea state may be represented as a sum of a large number N of periodic components with infinitesimal amplitudes a_j, frequencies, ω_j, different from each other, and phase angles ζ_j uniformly distributed over the interval [0, 2π] and stochastically independent from each other.

Under these assumptions, the linear water surface and velocity potential processes are stationary Gaussian processes in the time domain. In this context, the velocity potential is given by the equation

$$\Phi^{(I)}=\text{Re}\left\{\sum_{j=1}^{N}\phi_j^{(I)}\exp\left[-k_{j0}x\sin\theta+i\left(\omega_j t+\zeta_j\right)\right]\right\}, \quad (6)$$

where

$$\phi_j^{(I)}(y,x,z)=iga_j\,\omega_j^{-1}\xi_{j0}^{(1)}\exp\left(-k_{j0}y\cos\theta_j\right), \quad (7)$$

and Re{·} denotes the real part of a complex quantity and k_{j0} is the wave number. That is, the positive imaginary solution of the dispersion relation

$$k_j\tan\left(k_j d\right)=-\omega_j^2/g. \quad (8)$$

$\xi_{j0}^{(1)}$ is a function of water depth and frequency

$$\xi_{j0}^{(1)}=\frac{\cos\left[k_{j0}(d+z)\right]}{\cos\left(k_{j0}d\right)}. \quad (9)$$

The free surface displacement is given by

$$\eta^{(I)}=\text{Re}\left\{\sum_{j=1}^{N}a_j\exp\left[-k_{j0}\left(y\cos\theta_j+x\sin\theta_j\right)+i\left(\omega_j t+\zeta_j\right)\right]\right\} \quad (10)$$

with variance

$$\sigma^2=\int_0^{\infty}\int_0^{2\pi}S(\omega,\theta)\,d\theta d\omega, \quad (11)$$

where $S(\omega,\theta)$ is the directional spectrum of the surface displacement in an undisturbed field. That is

$$S(\omega,\theta)\,\delta\theta\delta\omega=\frac{1}{2}\sum_i a_i^2 \quad \text{for } i \text{ such that } \begin{cases}\omega<\omega_i<\omega+\delta\omega\\ \theta<\theta_i<\theta+\delta\theta\end{cases} \quad (12)$$

In the following the Re symbol, the time dependence, and the x-dependence will be omitted for convenience.

2.3 *Diffracted and radiated potentials*

The boundary value problem (1)–(5) is solved by a matching technique involving an eigen-functions expansion. The fluid domain is divided in three regions (Figure 1). In each region the velocity potential is a solution of a scattered wave field and of a radiated wave field. The former is the superposition of the incident wave field (6) and the diffracted wave field. The latter is calculated by assuming forced oscillations of the structure. Wave fields are estimated independently from each other.

The diffracted potential is calculated by dividing the potential in a symmetrical part and an asymmetrical part. Thus, it is given by the equation

$$\phi^{(D_r)} = \sum_{j=1}^{N} \frac{1}{2}\phi_j^{(r,s)} + \frac{1}{2}\phi_j^{(r,a)}, \tag{13}$$

where r denotes the region, and the potentials are calculated by

$$\phi_j^{(1,s)} = \phi_j^{(1,a)} = \phi_j^{(I)} + iga_j\,\omega_j^{-1}\sum_{m=0}^{\infty} A_{jm}^{(s,a)}\,\xi_{jm}^{(1)}\,e^{\tilde{k}_{jm}(y+b)} \tag{14}$$

$$\phi_j^{(2,s)} = iga_j\,\omega_j^{-1}\sum_{m=0}^{\infty} B_{jm}^{(s)}\,\xi_m^{(2)}\frac{\cosh\left(\tilde{\alpha}_{jm}y\right)}{\cosh\left(\tilde{\alpha}_{jm}b\right)} \tag{15}$$

$$\phi_j^{(2,a)} = iga_j\,\omega_j^{-1}\sum_{m=0}^{\infty} B_{jm}^{(a)}\,\xi_m^{(2)}\frac{\sinh\left(\tilde{\alpha}_{jm}y\right)}{-\sinh\left(\tilde{\alpha}_{jm}b\right)} \tag{16}$$

$$\phi_j^{(3,s)} = \phi_j^{(3,a)} = iga_j\,\omega_j^{-1}\sum_{m=0}^{\infty} D_{jm}^{(s,a)}\,\xi_{jm}^{(1)}\,e^{\tilde{k}_{jm}(y+b)}. \tag{17}$$

In this equation

$$\xi_{jm}^{(1)} = \frac{\cos\left[k_{jm}(d+z)\right]}{\cos\left(k_{jm}d\right)}, \tag{18}$$

and

$$\xi_m^{(2)} = \cos\left[\alpha_m(d+z)\right], \tag{19}$$

in which k_{jm}, with $m > 0$, are positive real solutions of the dispersion relation (8), $\alpha_m = m\pi/(d\text{-}h)$, for $m = 0, 1, 2, \ldots, \infty$ and $\tilde{\alpha}_{jm}$, $\tilde{k}_{jm}$ are related to α_m and k_{jm} by

$$\tilde{\alpha}_{jm} = \sqrt{\alpha_m^2 - k_{j0}^2\sin^2\theta_j};$$
$$\tilde{k}_{jm} = \sqrt{k_{jm}^2 - k_{j0}^2\sin^2\theta_j}.$$

$A_{jm}^{(s,a)}$, $B_{jm}^{(s,a)}$, $D_{jm}^{(s,a)}$ are unknown complex constants to be determined. The use of even and odd functions in (14)-(17) allows reducing the number of unknowns to four sets, because $D_{jm}^{(s)} = A_{jm}^{(s)}$, $D_{jm}^{(a)} = -A_{jm}^{(a)}$, for $m = 0, 1, \ldots, \infty$ and $j = 1, 2, \ldots, N$.

For the estimation of the radiated field, it is assumed that the sway and the roll motions produce antisymmetrical waves, and that heave motion produces symmetrical waves (Gesraha 2004). The radiated potentials in region 1 are approximated as

$$\phi_j^{(X_l,1)} = \sum_{j=1}^{N} i\omega_j X_{lj} b_l \sum_{m=0}^{\infty} A_{jm}^{(X_l)}\,\xi_{jm}^{(1)}\,e^{\tilde{k}_{jm}(y+b)}. \tag{20}$$

In region 2, the radiated potential of the heave motion is approximated as

$$\phi_j^{(X_2,2)} = \sum_{j=1}^{N} i\omega_j X_{2j} b_l \sum_{m=0}^{\infty} B_{jm}^{(X_2)}\,\xi_m^{(2)}\frac{\cosh\left(\tilde{\alpha}_{jm}y\right)}{\cosh\left(\tilde{\alpha}_{jm}b\right)} + \Theta_j^{(X_2)}, \tag{21}$$

and the radiated potentials of sway and roll motions are determined by the equation

$$\phi_j^{(X_l,2)} = \sum_{j=1}^{N} i\omega_j X_{lj} b_l \sum_{m=0}^{\infty} B_{jm}^{(X_l)}\,\xi_m^{(2)}\frac{\sinh\left(\tilde{\alpha}_{jm}y\right)}{-\sinh\left(\tilde{\alpha}_{jm}b\right)} + \Theta_j^{(X_l)}. \tag{22}$$

In (20)–(22) b_l is b (for $l = 1, 2$) or b^2 (for $l = 3$), X_{lj} are the complex amplitudes of motion of the structure in sway ($l = 1$), heave ($l = 2$) and roll ($l = 3$) corresponding to a frequency ω_j. Functions $\Theta_j^{(X_l)}$ are given in the appendix.

In eqs. (20)–(22), $A_{jm}^{(X_l)}$ and $B_{jm}^{(X_l)}$ are complex constants to be determined by a matching technique (Gesraha 2004, 2006).

2.4 *Equation of motion*

The wave field is completely determined by connecting the scattered wave field to the radiated one. For this purpose, the equation of motion is used. It is estimated by assuming small oscillations of the structure around an equilibrium point. In matrix notation, it is given by (see e.g., Gesraha 2006)

$$\left[-\omega_j^2\left(\underline{\underline{M}} + \underline{\underline{A}}_j\right) - i\omega_j\underline{\underline{B}}_j + \underline{\underline{K}}_h\right]\underline{X}_j = \underline{F}_j, \tag{23}$$

for $j = 1, ..., N$. Eq. (23) is determined by considering forces per unit length. The symbols in eq. (23) denote: $\underline{\underline{M}}$ = mass matrix; $\underline{\underline{A}}_j$ = added mass matrix; $\underline{\underline{B}}_j$ = radiation damping matrix; $\underline{\underline{K}}_h$ = hydrostatic restoring force; $\underline{\underline{F}}_j$ = exciting force; $\underline{X}_j$ = complex amplitude of motion. The specific geometrical configuration allows the explicit determination of all terms appearing in Eq. (23). Indeed,

$$\underline{\underline{M}} = \begin{bmatrix} m & 0 & -mZ_G \\ 0 & m & 0 \\ -mZ_G & 0 & M_0 \end{bmatrix}, \tag{24}$$

$$\underline{\underline{K}}_h = \begin{bmatrix} 0 & 0 & 0 \\ 0 & 2\rho gbh & 0 \\ 0 & 0 & mg\overline{GM} \end{bmatrix}, \tag{25}$$

where m is the mass per unit length of the structure, Z_G is the z-coordinate of the centre of gravity, M_0 is the mass moment of inertia about the x-axis, GM is the arm of the hydrostatic restoring moment developed by small inclination.

The exciting force vector $\underline{F}$ is determined from the scattered wave field. Specifically, the pressure fluctuations are determined from the velocity potentials (14–17) by means of the linear Bernoulli equation ($\Delta p = -\rho\partial\Phi/\partial t$; being ρ water density). Then, the wave forces are calculated by integrating the wave pressure on the solid boundary. The wave force components are

$$F_y = \rho g \sum_{j=1}^{N} a_j \left[I_{j0}^{(1)} + \sum_{m=0}^{\infty} \left(A_{jm} - D_{jm} \right) I_{jm}^{(1)} \right], \tag{26}$$

$$F_z = 2\rho g \sum_{j=1}^{N} a_j \sum_{m=0}^{\infty} B_{jm}^{(s)} (-1)^m \frac{\tanh(\tilde{\alpha}_{jm} b)}{\tilde{\alpha}_{jm}}, \tag{27}$$

$$M = \rho g \sum_{j=1}^{N} a_j \left\{ I_{j0}^{(6)} + \sum_{m=0}^{\infty} \left(A_{jm} - D_{jm} \right) I_{jm}^{(6)} - 2 \sum_{m=0}^{\infty} B_{jm}^{(s)} (-1)^m \frac{\tanh(\tilde{\alpha}_{jm} b)}{\tilde{\alpha}_{jm}} \right\} \tag{28}$$

where $I_{jm}^{(1)}$ and $I_{jm}^{(6)}$ have been defined in the appendix.

The added mass matrix and the radiation damping matrix are determined from the radiated wave field. In this regard, the wave pressure is determined from the potentials (20–22) by means of the Bernoulli equation. Then, the reaction forces F_{ij} are calculated. Note that F_{ij} denotes the force in the i-th direction due to the j-th motion. The added mass and the radiation damping components are related to the reaction forces by the equation

$$F_{ij} = -\left(\omega^2 A_{ij} + i\omega B_{ij} \right). \tag{29}$$

Eq. (29) shows that the added mass is the component of the reaction force in phase with the acceleration, while the radiation damping is the component in phase with the velocity. The analytical expressions of the reaction forces are readily determined from the radiation potentials (20–22) (Gesraha 2004, 2006).

3 FREE SURFACE DISPLACEMENT PROCESS

3.1 *Variation parameter*

The free surface is derived from the velocity potential functions. It is related to the potential by the equation

$$\eta = -\frac{1}{g}\frac{\partial \Phi}{\partial t}, \tag{30}$$

The velocity potentials are the superposition of scattered and radiated waves. Therefore, the free surface is modified by these wave fields. In region 1 and in region 3 the analytical expression of the free surface displacements are given by

$$\eta_1 = \sum_{j=1}^{N} a_j e^{-\tilde{k}_{j0}(y+b)} + \sum_{j=1}^{N} a_j \sum_{m=0}^{\infty} A_{jm} e^{\tilde{k}_{jm}(y+b)} + \sum_{l=1}^{3} \sum_{j=1}^{N} X_{lj} \frac{\omega_j^2}{g} b_l \sum_{m=0}^{\infty} A_{jm}^{(X_l)} e^{\tilde{k}_{j0}(y+b)} \tag{31}$$

$$\eta_3 = \sum_{j=1}^{N} a_j \sum_{m=0}^{\infty} D_{jm} e^{-\tilde{k}_{jm}(y-b)} + \sum_{l=1}^{3} \sum_{j=1}^{N} X_{lj} \frac{\omega_j^2}{g} b_l \sum_{m=0}^{\infty} D_{jm}^{(X_l)} e^{-\tilde{k}_{jm}(y-b)} \tag{32}$$

The symbols D_{jm} and $D_{jm}^{(X_l)}$ denote the quantities

$$D_{jm} = \left(A_{jm}^{(s)} - A_{jm}^{(a)} \right) / 2, \tag{33}$$

and

$$D_{jm}^{(X_l)} = \begin{cases} A_{jm}^{(X_l)}; & for\ l = 2 \\ -A_{jm}^{(X_l)}; & for\ l = 1,3 \end{cases} \tag{34}$$

The free surface displacement is given by the superposition of the incident wave field, of the diffracted wave field and of the radiated wave field.

Its modification is investigated by means of a variation parameter. It is defined as the ratio of the standard deviation of the free surface displacement (31–32) over the standard deviation of the surface displacement (10). After some algebra, it can be proved that, in region 1 and region 3, it is given by

$$K_r = \frac{\sqrt{\int_0^\infty \int_0^{2\pi} S(\omega,\theta)\left|\Lambda_1(y;\omega,\theta)\right|^2 d\theta d\omega}}{\sqrt{\int_0^\infty \int_0^{2\pi} S(\omega,\theta) d\theta d\omega}}, \tag{35}$$

$$K_t = \frac{\sqrt{\int_0^\infty \int_0^{2\pi} S(\omega,\theta)\left|\Lambda_3(y;\omega,\theta)\right|^2 d\theta d\omega}}{\sqrt{\int_0^\infty \int_0^{2\pi} S(\omega,\theta) d\theta d\omega}}, \tag{36}$$

where the symbol $|\cdot|$ is the modulus of a complex quantity and $\Lambda_1(y;\omega,\theta)$ and $\Lambda_3(y;\omega,\theta)$ are given by the equations

$$\Lambda_1(y;\omega,\theta) = e^{\tilde{k}_{j0}(y+b)} + \sum_{m=0}^{\infty} A_{jm} e^{\tilde{k}_{jm}(y+b)} + \sum_{l=1}^{3}\sum_{j=1}^{N} X_{lj} \frac{\omega_j^2}{g} b_l \sum_{m=0}^{\infty} A_{jm}^{(X_l)} e^{\tilde{k}_{jm}(y+b)}, \tag{37}$$

and

$$\Lambda_3(y;\omega,\theta) = \sum_{m=0}^{\infty} D_{jm} e^{-\tilde{k}_{jm}(y-b)} + \sum_{l=1}^{3}\sum_{j=1}^{N} X_{lj} \frac{\omega_j^2 b_l}{g} \sum_{m=0}^{\infty} D_{jm}^{(X_l)} e^{-\tilde{k}_{jm}(y+b)}. \tag{38}$$

In Eq. (35) and Eq. (36), complex amplitudes X_{lj} are estimated from the equation of motion (23) up to the constant amplitudes a_j.

The parameters (35) and (36) render an estimate of the mean wave height modification in the space domain. If K_r (or K_t) is smaller than 1, there is a reduction of the mean wave height with respect to the mean wave height of the incident wave field. The vice versa occurs if K_r (or K_t) is larger than 1.

The variation parameters (35) and (36) are calculated for given spectral shape and geometrical configuration. The results of the calculations are shown in Figures 2–3–4. Specifically, figure 2 shows the influence of the spectral shape. In this regard, a JONSWAP (Hasselmann et al., 1973) and a Pierson-Moskowitz (Pierson et al., 1964) frequency spectrum have been assumed. In both cases a directional spreading function of Mitsuyasu et al. (1975) has been considered. It is shown that the coefficient is approximately 2 in front of the structure ($y = -b$) and is approximately 1.35 far from the structure ($y \to -\infty$). The spectrum modifies K_r behaviour in the space domain. Indeed, the narrower the spectrum is, the larger and the smaller the local maxima and local minima are, respectively.

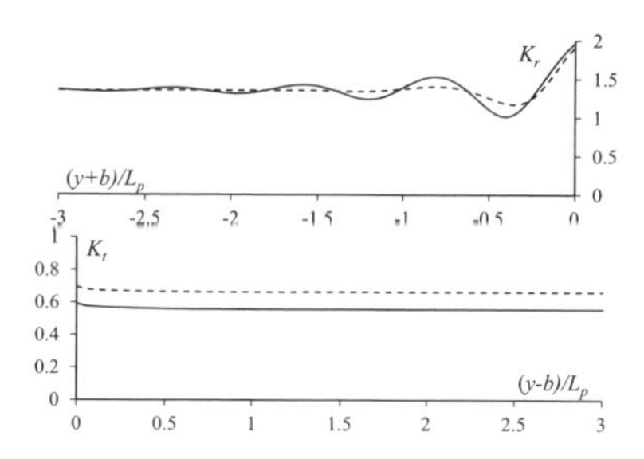

Figure 2. Variation parameter (35) (upper panel) and (36) (lower panel) estimated by assuming b/L_p = 0.1, h/L_p = 0.1, d/L_p = 0.5. A JONSWAP (continuous line) and a Pierson-Moskowitz (dotted line) frequency spectrum have been considered.

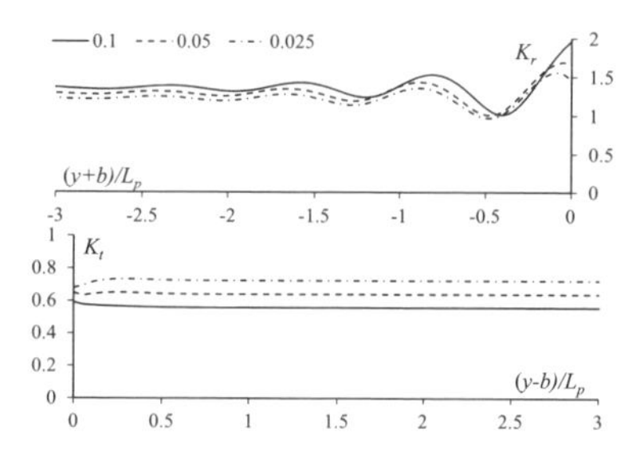

Figure 3. Variation parameter (35) (upper panel) and (36) (lower panel) calculated by assuming b/L_p = 0.1, d/L_p = 0.5, a JONSWAP frequency spectrum with directional spreading function of *Mitsuyasu* et al. Different values of h/L_p have been considered.

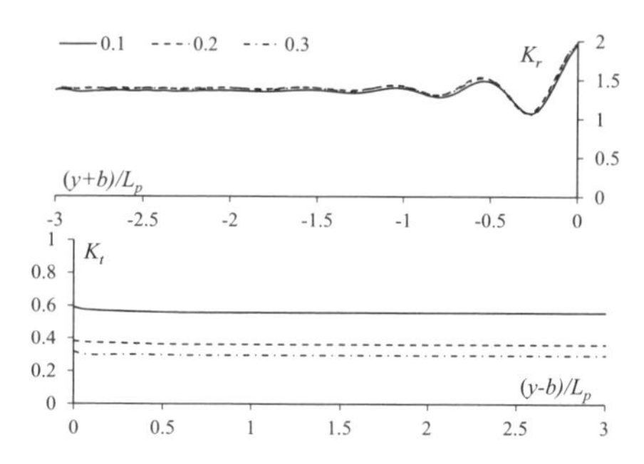

Figure 4. Variation parameter (35) (upper panel) and (36) (lower panel) calculated by assuming h/L_p = 0.1, d/L_p = 0.5, a JONSWAP frequency spectrum with directional spreading function of *Mitsuyasu* et al. Different values of b/L_p have been considered.

In region 3 the parameter is less than 1. It means that the mean wave height is reduced. It can be seen that, the wider the spectrum is, the larger the coefficient (36) is.

Figure 3 shows the influence of the thickness of the structure. In region 1, it is shown that the thicker the structure is, the smaller K_r is.

On the other hand, the thicker the structure is, the larger K_t is. This behaviour is well rendered at $y=-b$, where the coefficient decreases up to 1.5 for a quite small thickness.

Figure 4 shows the influence of the width b. This parameter does not modify the wave field in region 1 significantly. Indeed, for the configuration under study, there are only small variations of maxima and minima. This parameter influences the wave field in region 3. It is shown that the wider the structure is, the smaller K_t is.

4 RANDOM WAVE FIELD WHEN A HIGH WAVE CREST OCCURS

4.1 *Quasi-determinism theory*

The Quasi-Determinism theory has been developed in the 80's by Boccotti (1989, 1997) in the context of random waves. The theory allows the estimation of the free surface elevation and of the velocity potential given that a high wave crest occurs at a given time instant, and at a given point of the wave field. It shows that the wave crest is due to the passage of a wave group at the apex of its evolution. Its mechanics can be analytically described in the time-space domain. One of the most important results of the theory is that it holds irrespective the configuration of the solid boundary is. Indeed, the free surface and the velocity potential must be stationary processes; however they can be either homogeneous or inhomogeneous.

By focusing on the free surface, the analytical expression of the Quasi-Deterministic free surface is given by the equation

$$\bar{\eta}(y_0+Y,t_0+T)=H_c\frac{\langle\eta(y_0,t_0)\eta(y_0+Y,t_0+T)\rangle}{\langle\eta^2(y_0,t_0)\rangle},\tag{39}$$

The symbol <·> denotes the mathematical expectation operator. The right hand side of eq. (39) is explicitly determined from the eq. (31) and from the eq. (32). The free surface displacement in region 1 and in region 3 are given by

$$\begin{aligned}&\bar{\eta}_1(y_0+Y,t_0+T)\\&=\frac{H_c}{\int_0^\infty\int_0^{2\pi}S(\omega,\theta)\left|\Lambda_1(-b;\omega,\theta)\right|^2d\theta d\omega}\\&\times\left[\int_0^\infty\int_0^{2\pi}S(\omega,\theta)\left|\Lambda_1(-b;\omega,\theta)\right|\left|\Lambda_1(Y;\omega,\theta)\right|\right.\\&\left.\times\cos\left(\omega T+\arg\left(\Lambda_1(Y;\omega,\theta)\right)-\arg\left(\Lambda_1(-b;\omega,\theta)\right)\right)d\theta d\omega\right]\end{aligned}\tag{40}$$

and

$$\begin{aligned}&\bar{\eta}_3(y_0+Y,t_0+T)\\&=\frac{H_c}{\int_0^\infty\int_0^{2\pi}S(\omega,\theta)\left|\Lambda_1(-b;\omega,\theta)\right|^2d\theta d\omega}\\&\times\left[\int_0^\infty\int_0^{2\pi}S(\omega,\theta)\left|\Lambda_1(-b;\omega,\theta)\right|\left|\Lambda_3(Y;\omega,\theta)\right|\right.\\&\left.\times\cos\left(\omega T+\arg\left(\Lambda_3(Y;\omega,\theta)\right)-\arg\left(\Lambda_1(-b;\omega,\theta)\right)\right)d\theta d\omega\right],\end{aligned}\tag{41}$$

In eq. (40) and in eq. (41) arg (·) denotes the phase of a complex quantity.

The wave group evolution is shown in Figure 5. The maximum wave crest occurs at point y = $-b$, at the time instant $T/T_p=0$. At $T/T_p=-1$ and $T/T_p=-0.5$ there is a growing stage, in which the wave propagates along the positive y-axis. At $T/T_p=0.5$ and at $T/T_p=1$ there is a decaying stage. The wave is partially reflected and transmitted. The reflected wave propagates along the negative y-axis in region 1; the transmitted wave propagates along the positive y-axis in region 3.

The Quasi-Deterministic free-surface is shown in time domain in Figure 6. It is calculated at the head of the structure ($y=-b$).

It is a symmetric function in time domain. Further, minima and maxima, at $T/T_p\neq 0$ are smaller that H_c. The first minimum and the first maximum are the 23% and the 37% smaller than H_c, respectively. This is a bandwidth effect. Indeed, if a Pierson—Moskowitz spectrum is assumed, the reduction of the wave amplitudes are 29% and 52% respectively.

4.2 *Validation by monte carlo simulation*

The present stochastic model is validated by Monte Carlo simulations of a Gaussian wave field with a specified spectrum. The time history is synthesized by a moving average algorithm (Spanos & Zeldin, 1998). The algorithm estimates the current value of the time history as a combination of weighed white noise deviates. The algorithm is not "physically" consistent in terms of dynamic systems input-output consideration (it does not respect the principle of causality). It has been chosen for its efficiency in simulating a time history. Further, the design of the filter is straightforward even for "pathological" sea wave spectra (Spanos & Hansen, 1981).

For the present simulation, the incident wave field is defined by a mean JONSWAP spectrum with directional spreading function of Mitsuyasu et al. The sea state is characterized by a significant wave height $H_s=5$ m and a peak period $T_p=9.5$ s. The cut-off frequency is assumed $\omega_c=3\omega_p$. Then, the sampling period is $T_c=\pi/\omega_c$. The following structural characteristics have been assumed: $b=15$ m, $h=15$ m, $d=70$ m.

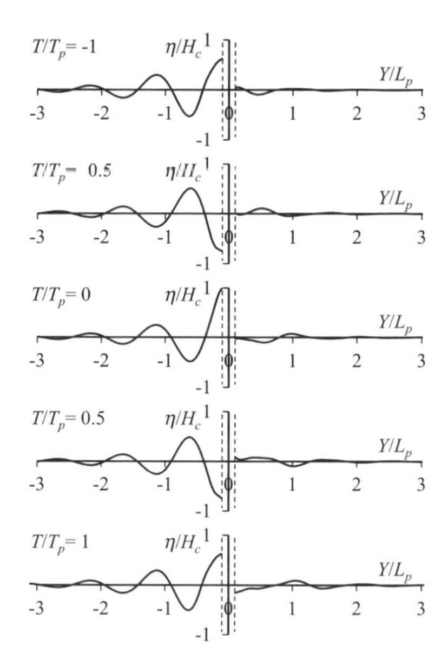

Figure 5. Free surface displacement in the space domain at given time instants. It has been assumed: $b/L_p = 0.1$, $h/L_p = 0.1$, $d/L_p = 0.5$, a JONSWAP frequency spectrum with directional spreading function of *Mitsuyasu* et al.

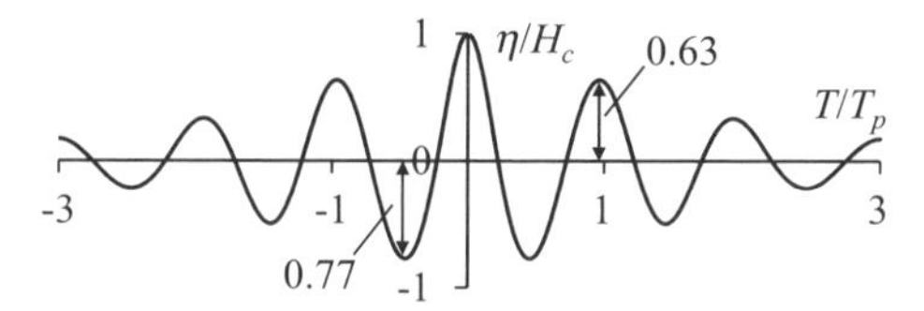

Figure 6. Free surface displacement in the time domain at the point (-*b*,0). It has been assumed: $b/L_p = 0.1$, $h/L_p = 0.1$, $d/L_p = 0.5$, a JONSWAP frequency spectrum with directional spreading function of *Mitsuyasu* et al.

The algorithm has been used to synthesize 200 sea states of 2000 waves. In each realization, the profile of the highest wave crest at the head of the structure has been calculated. Then, the mean profile of the free surface is estimated. The highest wave crest of the mean profile is $H_c = 9.30$ m. Note that, this wave crest amplitude is not unexpected, indeed the algorithm generates the time history of a scattered/radiated wave field at the head of the structure, where the standard deviation of the free surface is about 2 times the standard deviation of the incident free surface displacement.

The results obtained from the simulation have been compared with the Quasi-Deterministic free surface (40–41). In this case, the wave crest amplitude is an input datum, therefore it is assumed $H_c = 9.30$ m.

Figure 7 shows a comparison between the simulation and the Quasi-Determinism theory. It can be seen that the free-surface elevation is well estimated in the time domain. Indeed the structure of the wave group is well defined, and the local maxima and minima are well estimated.

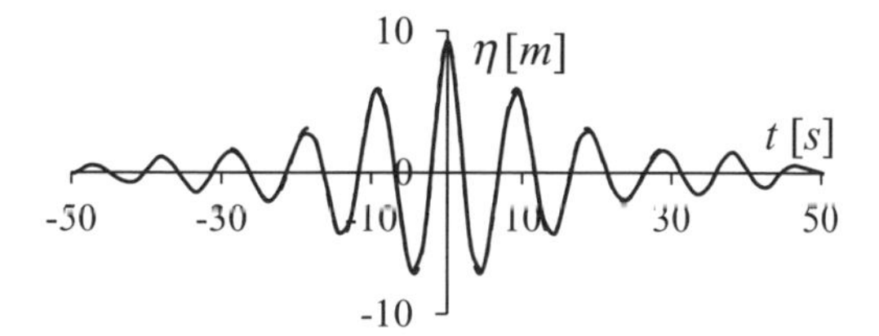

Figure 7. Comparison between the free surface given by the Quasi-Determinism theory (continuous line) and the free surface obtained by Monte Carlo simulations (dotted line).

5 CONCLUDING REMARKS

This paper has dealt with the interaction of a random incident wave field with a simple model of a ship. It has been modelled as a floating structure of rectangular cross section. An analytical approach for studying the wave field has been developed. It is based on a matching technique involving an eigen-functions expansion. By this technique, the free surface displacement process has been derived in the context of a first order potential theory. It has been shown that the free surface elevation is modified by the wave—structure interaction. The modifications are interpreted according to a variation parameter defined as the ratio of the standard deviation of the free surface elevation in the scattered-radiated wave field over the standard deviation of the free surface in the incident one. This parameter shows that there is a magnification of the mean wave height at the head of the structure and a reduction of the mean wave height behind it.

The interaction between a high wave and the structure is investigated by the Quasi-Determinism theory. The theory has allowed estimating the free surface displacement when an extreme wave crest occurs at the head of the structure. It has been used for investigating the wave field in the space domain and in the time domain. The results of the theory are in agreement with the results of Monte Carlo simulations.

ACKNOWLEDGEMENTS

The first author is grateful to the Rotary International District 2100 for the financial support in the context of "Progetto uniamo le 3T".

REFERENCES

Abul-Azm, A.G. & Gesraha, M.R. 2000. Approximation to the hydrodynamics of floating pontoons under oblique waves. *Ocean Engineering* 27: 365–384.

Black, J.L., Mei, C.C. & Bray, M.C.G. 1971. Radiation and scattering of water waves by rigid bodies. *J. Fluid Mech. 46*: 151–164.

Boccotti, P. 1989. On mechanics of irregular gravity waves. *Atti Acc. Naz. Lincei*, Memorie VIII: 111–170.

Boccotti, P. 1997. A general theory of three dimensional wave groups. *Ocean Engineering* 24:265–300.

Drimer, N., Agnon, Y. & Stiassnie, M. 1992. A simplified analytical model for a floating breakwater in water of finite depth. *Applied Ocean Research* 14: 33–41.

Gesraha, M.R. 2004. An eigenfunction expansion solution for extremely flexible floating pontoons in oblique waves. *Applied Ocean Research* 24: 171–182.

Gesraha, M.R. 2006. Analysis of Π shaped floating breakwater in oblique waves: impervious rigid wave boards. *Applied Ocean Research* 28: 327–338.

Hasselmann, K., Barnett, T.P., Bouws E., et al. 1973. Measurements of wind wave growth and swell decay during the Joint North Sea Wave Project (JONSWAP). *Deut. Hydrogr. Zeit* A8: 1–95.

John, F. 1950. On the motion of floating bodies. II. *Communications in Pure and Applied Mathematics* 3: 45–101.

Kim, W.D. 1965. On the harmonic oscillations of a rigid body on a free surface. *J. Fluid Mech. 21*: 427–451.

Longuet-Higgins, M.S. 1963. The effects of non-linearities on statistical distributions in the theory of sea waves. *J. Fluid Mech. 17*: 459–480.

Malara, G, Arena, F. & Spanos, P.D. 2010. Determination of random wave forces on a thick finite plate. *Proc. of the 6th Computational Stochastic Mechanics Conference*, Rodhos, Greece, 13–16 June.

Mei, C.C. & Black, J.L. 1969. Scattering of surface waves by rectangular obstacles in waters of finite depth. *J. Fluid Mech. 38*: 499–511.

Mitsuyasu, H., Tasai, F., Suhara, T. et al. 1975. Observation of directional spectrum of ocean waves using a clover-leaf buoy. *J. Phys. Oceanogr*. 5: 750–760.

Peter, M.A. & Meylan, M.H. 2004. The eigenfunction expansion of the infinite depth free surface Green function in three dimensions. *Wave Motion* 40: 1–11.

Phillips, O.M. 1967. The theory of wind-generated waves. *Advances in Hydroscience* 4: 119–149.

Pierson, O.M. & Moskowitz, L. 1964. A proposed spectral form for fully developed waves based on the similarity theory of S.A. Kitaigorodskii. *J. Geophys. Res.* 69, 5181–5190.

Spanos, P.D. & Hansen, J.E. 1981. Linear prediction theory for digital simulation of sea waves. *Journal of Energy Resources Technology* 103: 243–249.

Spanos, P.D. & Zeldin, B.A. 1998. Monte Carlo Treatment of Random Fields: A Broad Perspective. *Appl. Mech. Rev*., 51: 219–237.

Stoker, J.J. 1957. Water waves. New York: Interscience Publisher Inc.

Williams, A.N. & McDougal, W.G. 1996. A dynamic submerged breakwater. *J. Waterway, Port, Coastal, and Ocean Engineering* 122: 288–296.

Yamamoto, T., Yoshida, A. & Ijima, T., 1980. Dynamics of elastically moored floating objects. *Applied Ocean Research* 2: 85–92.

Zheng, Y.H., Shen, Y.M., You, Y.G., Wu, B.J. & Jie, D.S. 2006. Wave radiation by a floating rectangular structure in oblique seas. *Ocean Engineering* 33: 59–81.

APPENDIX

In this appendix the analytical expressions of the functions previously cited are given.

In Eqs. (20)–(22), $\Theta_j^{(Xj)}$ are functions necessary for satisfying the non-homogeneous boundary condition (4). They are given by the equations,

$$\Theta_j^{(X_1)} = 0, \tag{A1}$$

$$\Theta_j^{(X_2)} = i\omega_j X_{2j} b \sum_{m=0}^{\infty} F_{jn}^{(H)} \xi_m^{(2)} + i\omega_j X_{2j}(d-h)\frac{1}{2}\left(\frac{d+z}{d-h}\right)^2, \tag{A2}$$

and

$$\Theta_j^{(X_3)} = -i\omega_j X_{3j} b^2 \sum_{m=0}^{\infty} F_{jn}^{(R)}(y) \xi_m^{(2)} - i\omega_j X_{3j} b^2 \frac{1}{2}\frac{y}{b}\frac{d-h}{b}\left(\frac{d+z}{d-h}\right)^2, \tag{A3}$$

where

$$F_{jn}^{(H)} = \begin{cases} \dfrac{d-h}{b}\dfrac{1}{[\tilde{\alpha}_{jn}(d-h)]^2} + \dfrac{d-h}{b}\dfrac{k_{j0}^2 \sin^2\theta_j}{6\tilde{\alpha}_{jn}^2}; n=0 \\ \dfrac{d-h}{b}\dfrac{2k_{j0}^2(-1)^n \sin^2\theta_j}{n^2\pi^2\tilde{\alpha}_{jn}^2}; \quad n \neq 0 \end{cases} \tag{A4}$$

and

$$F_{jn}^{(R)}(y) = \frac{y}{b} F_{jn}^{(H)}. \tag{A5}$$

In eqs. (26)–(28) the following symbols have been used:

$$I_{jn}^{(1)} = \frac{\sin(k_{jn}d) - \sin[k_{jn}(d-h)]}{k_{jn}\cos(k_{jn}d)}, \tag{A6}$$

$$I_{jn}^{(6)} = \frac{1}{k_{jn}^2\cos(k_{jn}d)}\Big\{\cos(k_{jn}d) - \cos[k_{jn}(d-h)] + k_{jn}h\sin(k_{jn}d) - k_{jn}z_0\sin(k_{jn}d) + k_{jn}z_0\sin[k_{jn}(d-h)]\Big\}, \tag{A7}$$

with being z_0 the centre of rotation.

Sustainable Maritime Transportation and Exploitation of Sea Resources – Rizzuto & Guedes Soares (eds)
 ISBN 978-0-415-62081-9

Testing the impact of simplifications adopted in models describing wave loads on ship

P.A. Wroniszewski & J.A. Jankowski
Polish Register of Shipping, Gdańsk, Poland

ABSTRACT: The wave-ship system in naval architecture is generally assumed to be a linear one. This allows for the computation of a transfer function of ship response to waves including wave loads on ship. The transfer function and the wave spectral density function enable the development of the stochastic process of wave loads on ship, corresponding to the irregular wave generating that loads. If the mathematical models of wave load transfer functions were to be used in the designing process, they would require simple mathematical forms. The paper presents the attempt of developing a simplified models of wave loads on ships.

1 INTRODUCTION

In the presently binding rules of classification societies the requirements referring to the dynamic load components (wave bending moments and shear forces, external sea pressures, internal dynamic pressures, ship motion and accelerations) are given in the form of formulae. The formulae represent the amplitudes of the loads that can be exceeded with an assumed probability (normally 10^{-8}). The load combination factors, intended to be an approximation of the phase shifts between particular wave loads used to analyze the strength of ship structure, are given as tabulated values and are calculated by application of the equivalent design wave approach (IACS CSR, 2010).

Shigemi and Zhu 2003 and Zhu and Shigemi 2003 developed methods for the practical estimation of the design loads based on the following definitions:

- design sea state is the sea state that generates response value equivalent to the long-term prediction of the considered ship response to waves (e.g., stresses in structure member),
- design regular wave is the regular wave that generates response values equivalent to the response values generated in design irregular wave, and
- design load is the load generated by design regular wave used in designing ship hull structure.

The values of stresses estimated with the use of the proposed design loads are claimed to be equivalent to the long-term predictions of stresses for typical load cases.

In the design regular wave approach, the dominant load is determined for such wave-heading angle, wave period and height, at which the response is at its maximum. The dominant load is determined for each load case. Then the load combination factors, representing the relationship between the response to the dominant load and response to the secondary load, are determined (Shin et al., 2004). This regular wave approach has been widely used in local scantling and finite element analysis of ship structure.

However, it is questionable whether:

- the simple formula approximating the amplitudes of wave load components—with the assumed probability of exceeding thereof,
- the phase shifts between loads in form of factors, and
- the small number of load cases assumed in the simplified methods

can approximate the complex process of sea waving acting on the moving ship in waves.

The exact phase shifts between particular wave loads used to analyze the strength of the ship structure are naturally taken into account in the simulation (in time domain) of wave loads acting on the ship and the response (stresses) of the ship to the waves. However, due to their complexity the simulations of stresses are not used in design practice.

The wave-ship system is normally assumed to be a linear one in naval architecture analysis. This allows computing transfer functions of ship response to waves, including the transfer functions of wave loads on ship. The complex functions are normally used to determine the wave loads on ship as they include information on the amplitudes of: wave components, wave loads, and the responses of the ship to the wave loads; as well as information on phase shifts between them. Therefore, having the real and imaginary load components for different wave frequencies and having the wave spectral density function, the wave loads on ship in the time domain can be developed. Such an approach can be used in the strength analysis in designing process, provided the functions determining the wave loads have a simple mathematical form.

The paper presents a study on various simplifications in simulating wave loads on ships for design purposes. The method used is based on the linear approach in the determination of wave loads in the first step, and then on simulation of the wave loads in time domain, in the sense of discrete Fourier transform. The results of this method are compared with the simulation of vessel motions in waves based on numerical solutions of non-linear equations of motion (Jankowski, 2006).

2 THE LINEAR EQUATIONS OF SHIP MOTION IN REGULAR WAVES

The linear equations of ship motion in regular waves have the following form (e.g., Jankowski, 2006):

$$\left[-\left(\omega_E^2 M + C\right) + i\omega_E N\right]\xi_A = \left[Y_W + Y_D\right]_A, \tag{1}$$

where the matrices in the equation are equal to:

$$\begin{aligned} &M_{ii} = m + m_{ii}, \quad M_{i+3,\,i+3} = J_i + m_{i+3,\,i+3}, \quad i = 1,2,3 \\ &M_{ij} = m_{ij}, \quad \text{for remaining } i,j = 1,\dots,6, \\ &N_{ij} = n_{ij}, \quad i,j = 1,\dots,6; \end{aligned}$$

the restoring forces—elements of matrix C, are defined by the formulae:

$$\begin{aligned} &c_{33} = \rho g S_w, \quad c_{35} = c_{53} = -\rho g I_{S1}, \\ &c_{44} = \rho g\left(M_{V3} + J_{S2}\right), \quad c_{55} = \rho g\left(M_{V3} + J_{S1}\right), \end{aligned} \tag{2}$$

where ξ_A is the vector of ship motion amplitudes, m is the mass of the ship, J_j, $j = 1, \dots, 3$, are its moments of inertia, $|S_w|$ is the area of ship water plane, I_{S1} is the longitudinal static moment of water plane, J_{S1} and J_{S2} are the longitudinal and transverse moments of inertia of the water plane, respectively, and M_{V3} is the static moment of underwater ship volume in relation to the water plane, whereas, the added masses m_{ij}, $i, j = 1, \dots, 6$, and damping coefficients n_{ij}, $i, j = 1, \dots, 6$, are defined by the following integrals:

$$\begin{aligned} &m_{jk} = -\frac{\rho}{\omega_E}\int_{S_0}\left(\omega_E \varphi_{Rk}^R - u_0\frac{\partial \varphi_{Rk}^I}{\partial x_1}\right)n_j ds, \\ &n_{jk} = \rho\int_{S_0}\left(\omega_E \varphi_{Rk}^I + u_0\frac{\partial \varphi_{Rk}^R}{\partial x_1}\right)n_j ds, \; j,k = 1,\dots,6 \end{aligned} \tag{3}$$

where the encounter frequency is defined as follows:

$$\omega_E = \omega\left(1 - \frac{u_0\omega}{g}\cos\beta\right),$$

u_0 is the forward speed of the ship, ω is the wave frequency, S_0 is underwater part of a ship surface, β is the angle between the wave vector $\mathbf{k}$ and the x_1 axis (wave-heading angle), and $\mathbf{k} = (k \cos \beta, k \sin \beta, 0)$.

The exciting forces occurring in equations (1) are represented by the Froude-Krylov forces, determined by the following integrals:

$$\begin{aligned} &Y_{Wj}^R \approx y_{Wj}^R = -\rho g\zeta_A \int_{S_0} e^{kx_3}\cos(\mathbf{k}\cdot\overline{x})n_j ds, \\ &Y_{Wj}^I \approx y_{Wj}^I = \rho g\zeta_A \int_{S_0} e^{kx_3}\sin(\mathbf{k}\cdot\overline{x})n_j ds, \end{aligned} \tag{4}$$

where $j = 1, \dots, 6$, $k = \omega^2/g$, $\mathbf{k} = (k \cos \beta, k \sin \beta, 0)$, $\overline{x} = (x_1, x_2, 0)$; and diffraction forces:

$$\begin{aligned} &Y_{Dj}^R \approx y_{Dj}^R = -\rho\int_{S_0}\left[\omega_E\varphi_D^I + u_0\frac{\partial\varphi_D^R}{\partial x_1}\right]n_j ds, \\ &Y_{Dj}^I \approx y_{Dj}^I = \rho\int_{S_0}\left[\omega_E\varphi_D^R - u_0\frac{\partial\varphi_D^I}{\partial x_1}\right]n_j ds, \end{aligned} \tag{5}$$

where $j = 1, \dots, 6$

The radiation φ_R and diffraction φ_D potentials are determined by the following boundary-value problem (Haskind 1973):

- Laplace equation:

$$\Delta\varphi(x) = 0, \quad x \in R_3^- \setminus V, x = (x_1, x_2, x_3), \tag{6}$$

where V is a closed domain occupied by the ship, and by the following boundary conditions:

- on the free surface:

$$\begin{aligned} &-\nu\varphi(x) + 2i\tau\frac{\partial\varphi(x)}{\partial x_1} + \frac{1}{k_0}\frac{\partial^2\varphi(x)}{\partial x_1^2} + \frac{\partial\varphi(x)}{\partial x_3} = 0, \\ &x \in S_F = \{x : x_3 = 0\}\setminus V, \end{aligned}$$

where $\nu = \omega_E^2/g$, $\tau = \omega_E u_0/g$ and $k_0 = g/u_0^2$

- on the wetted surface of the ship:
  - for the diffraction potential:

$$\frac{\partial\varphi_D(x)}{\partial n} = -\frac{\partial\varphi_W(x)}{\partial n}, \quad x \in S_0,$$

where $\varphi_W(x)$ is the potential of the incident wave, and S_0 is the wetted surface of the ship;

– for the radiation potential:

$$\frac{\partial \varphi_{Ri}^{R}(x)}{\partial n} = n_i,$$
$$\frac{\partial \varphi_{Ri}^{I}(x)}{\partial n} = 0, \quad i = 1, \ldots, 6, \quad x \in S_0$$

where n_i are the components of the normal vectors to S_0: $n_4 = x_2 n_3 - x_3 n_2$, $n_5 = x_3 n_1 - x_1 n_3$, $n_6 = x_1 n_2 - x_2 n_1$;

- and appropriate conditions at infinity.

The formulae determining the internal forces in a given cross section $\bar{x}_1$ of the ship hull structure can be derived from equations (1), which yields (Jankowski, 2006):

$$\begin{aligned} &F_i(\bar{x}_1) = Y_i(\bar{x}_1) - \sum_{j=1}^{6}\left[\bar{m}_{ij}(\bar{x}_1) - i\bar{n}_{ij}(\bar{x}_1)\right]\xi_{Aj}, \\ &i = 2, \ldots, 6, \\ &M_s(\bar{x}_1) = F_4(\bar{x}_1) + dF_2(\bar{x}_1), \\ &M_V(\bar{x}_1) = F_5(\bar{x}_1) + \bar{x}_1 F_3(\bar{x}_1), \\ &M_H(\bar{x}_1) = F_6(\bar{x}_1) + \bar{x}_1 F_2(\bar{x}_1), \end{aligned} \tag{7}$$

where ξ_{Aj}, $j = 1, \ldots, 6$, are the components of ship motion computed from (1); $Y_i(\bar{x}_1)$ are the components of exciting forces vector $Y = Y_W + Y_D$; $F_k = F_k^R + iF_k^I$, $k = 2, \ldots, 6$, $F_{2,3}(\bar{x}_1)$ are respectively horizontal and vertical shear forces; $M_H(\bar{x}_1)$, $M_V(\bar{x}_1)$ and $M_s(\bar{x}_1)$ are respectively horizontal, vertical bending moments and torsional moment; d is the centre of torsion in the considered cross section $\bar{x}_1$.

Elements $\bar{m}_{ij}$ and $\bar{n}_{ij}$ of the matrices occurring in (7) are equal to:

$$\bar{m}_{ij} = -\left(\omega_E{}^2 m_{ij} + c_{ij}\right),$$
$$\bar{n}_{ij} = \omega_E n_{ij},$$

$Y_i(\bar{x}_1)$, $\bar{m}_{ij}(\bar{x}_1)$ and $\bar{n}_{ij}(\bar{x}_1)$ are computed for the part of the ship separated by the cross section $\bar{x}_1$, for example:

$$Y_{Dj}^{R}(\bar{x}_1) = -\rho \int_{S_0(\bar{x}_1)} \left[\omega_E \varphi_D^I + u_0 \frac{\partial \varphi_D^R}{\partial x_1}\right] n_j ds,$$

where $S_0(\bar{x}_1)$ denotes the wetted part of the hull up to the cross section $\bar{x}_1$.

3 FURTHER SIMPLIFICATIONS

The loads taken into account in the ship structure designing process occur in heavy seas in which the forward ship speed u_0 is normally reduced, therefore, this speed is usually neglected in the boundary-value problem and preserved only in the equations (1) and formulae (3), (5) and (10), in the encounter frequency ω_E. Such a simplified boundary-value problem can be solved numerically, enabling determination of diffraction forces, added masses and damping coefficients. However, this model is still too complex to be applied in ship design practice due to the prerequisite of solving boundary value problem (6) and further simplification need to be applied.

Some ideas have been given by (Haskind, 1973). He proposed, basing on the strict solution of the radiation problem for the ellipsoid by (Kotchin et al., 1963), the following form of the radiation potential:

$$\begin{aligned} &\varphi_{R1} = -C_1 x_1, \quad \varphi_{R2} = -C_2 x_2, \\ &\varphi_{R3} = -C_3 x_3, \quad \varphi_{R4} = -C_4 x_2 x_3, \\ &\varphi_{R5} = -C_5 x_1 x_3, \quad \varphi_{R6} = -C_6 x_1 x_2, \end{aligned} \tag{8}$$

where coefficients C_i, $i = 1, \ldots, 6$, can be determined for certain classes of ships. Functions (8) are real functions. Basing on the symmetry of added masses matrix, the coefficients satisfy the following relations (Jankowski, 2006):

$$\begin{aligned} &C_5 = -C_3, \quad C_6 = C_2, \\ &C_1 = C_5 \frac{1}{1 + \dfrac{I_{S1}}{(x_{V3} - x_{G3})|V|}} \\ &\quad = C_5 \frac{1}{1 + \dfrac{R_0}{(x_{V3} - x_{G3})}} \approx C_5 \frac{x_{V3} - x_{G3}}{R_0}, \\ &C_4 = -C_2\left(1 + \frac{I_{S2}}{(x_{V3} - x_{G3})|V|}\right) \\ &\quad = -C_2 \frac{MG}{x_{V3} - x_{G3}}, \end{aligned} \tag{9}$$

where R is large metacentric radius, MG is the metacentric height, x_{V3} is the third component of displacement centre, and x_{G3} is the third component of the mass centre.

Based on the approximations, the added masses can be computed according to (3). However, computations of damping coefficients according to (3) are impossible as φ_{Rj} in form of (8) are real. Again, (Haskind, 1973) developed the following formula determining these coefficients:

$$n_{ij} = \rho \frac{\omega_E \nu}{4\pi} \operatorname{Re} \int_{-\pi}^{\pi} H_i(\nu, \theta) \overline{H}_j(\nu, \theta) d\theta, \quad i, j = 1, \ldots, 6, \tag{10}$$

where

$$H_j(\nu,\theta)$$

$$= \int_{S_0} e^{\nu x_3 + i\nu\cdot\bar{x}} \left\{ \frac{\partial \varphi_{Rj}}{\partial n} - \varphi_{Rj}\left[\nu n_3 + i(n_1\nu_1 + n_2\nu_2)\right]\right\} ds,$$

$$\nu = (\nu_1, \nu_2, 0) = \nu(\cos\theta, \sin\theta, 0),\ \nu = \frac{\omega_E^{\ 2}}{g}.$$

Function $H_j(\nu, \theta)$, $j = 1, 2, \ldots, 6$, has the form of the so called Kotchin function.

To complete the simplified model, the diffraction force should be determined by a function and not by the numerical solution of the boundary-value problem (6). On the basis of the Green's identity and formula (5) the following form of the diffraction force function can be derived (Jankowski, 2006):

$$Y_{Dj} = \rho \int_{S_0} \varphi_{Rj}\varphi_w \left[kn_3 - i(k_1n_1 + k_2n_2)\right] ds, \tag{11}$$

By combining (4) and (11) the following function, determining the wave exciting function on ships, having the Kotchin form, is obtained:

$$Y = Y_W + Y_D = -\rho g \zeta_A \int_{S_0} e^{kx_3 + ik\cdot\bar{x}} \times \left\{ n_j - \varphi_{Rj}\left[kn_3 - i(k_1n_1 + k_2n_2)\right]\right\} ds \tag{12}$$

Radiation potential in formulae (10) and (12) is assumed to be a complex function.

4 COMPUTATIONAL MODELS AND RESULTS OF COMPUTATIONS

Methods determining the wave loads on ships for design purposes should be as simple as possible.

The computations of ship motions and loads were performed for four following ships:

- large container of block coefficient equal to 0.7, length 281 m:

- bulk carrier of panamax size of block coefficient equal to 0.8, length 225 m:

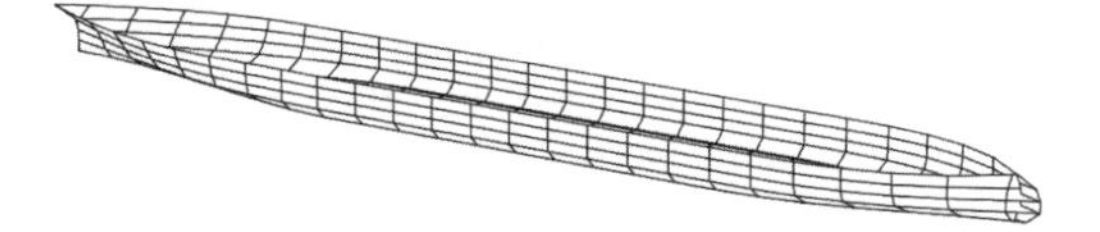

- ship of series 60, block coefficient equal to 0.7, length 123 m:

- cuboid, block coefficient equal to 1.0, length 123 m:

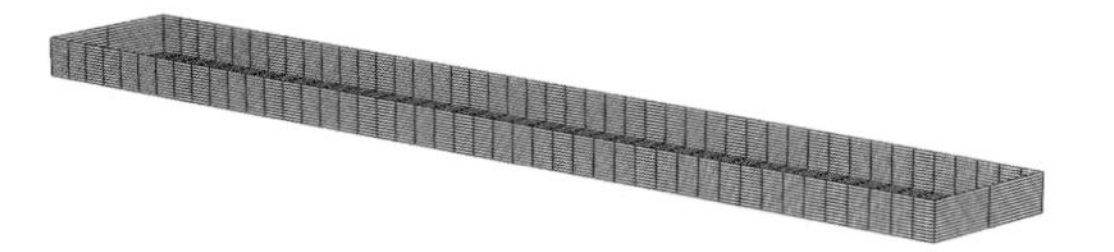

Therefore, an attempt was made to simplify the general linear method based on wave sources (boundary value model (6)) and in effect to simplify the computation of the loads.

The following models have been used to perform the computations:

Model 1. The model represents the direct approach—based on the ship motions and loads in waves—determined by motion equation (1), restoring forces (2), added masses and damping coefficients (3), Froude-Krylov forces (4), and diffraction forces (5). In this approach the radiation and diffraction potentials occurring in formulae (3) and (5) are obtained by solving the boundary-value problem (6), using 3D, zero speed wave source (Green's function).

Model 2. This model is based on equations as in *model* 1, with damping coefficients and the exciting forces computed on the basis of (10) and (12) and on radiation potential φ_{Rj}, $j = 1, 2, \ldots, 6$ obtained from the boundary-value problem (6). This approach is not a simplification of *model* 1 as solving the boundary value model (6) is required. *Model* 2 is only used to verify formulae (10) and (12).

Model 3. This model is based on equations as in *model 1* with damping coefficients and the exciting forces computed using (10) and (12) and radiation potential φ_{Rj} in simplified form (8).

Different ships were accounted for to investigate the affect of their shape and approximation of the radiation potential φ_{Rj} on the results of computations. The results of computations in the case of zero speed are presented in Figs. 1, 2, 3. Figure 4 presents the results of computation with forward speed of $u = 2.53\ m/s$ compared with experimental data.

The study shows that:

- The simplified *model* 3 can be used to compute the transfer function of ship motions in waves—Figs. 1 and 2
- *Model* 3 cannot be used to determine internal forces (e.g., bending moments) as formulae (10) and (12) are based on integral identities applied

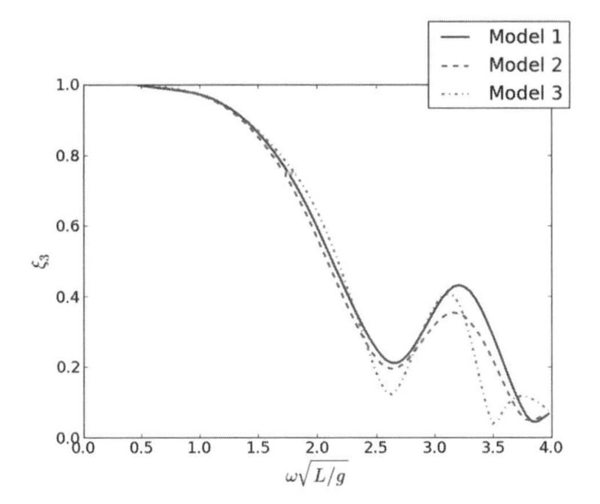

Figure 1. Response amplitude operator of bulk carrier heave for $\beta = 180°$, determined by different models.

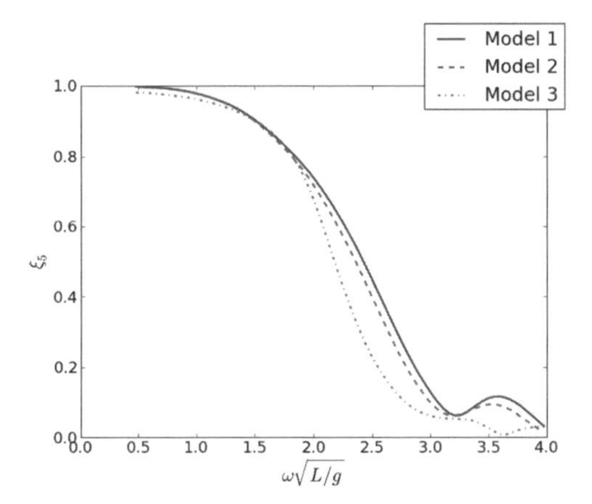

Figure 2. Response amplitude operator of bulk carrier pitch for $\beta = 180°$, determined by different models.

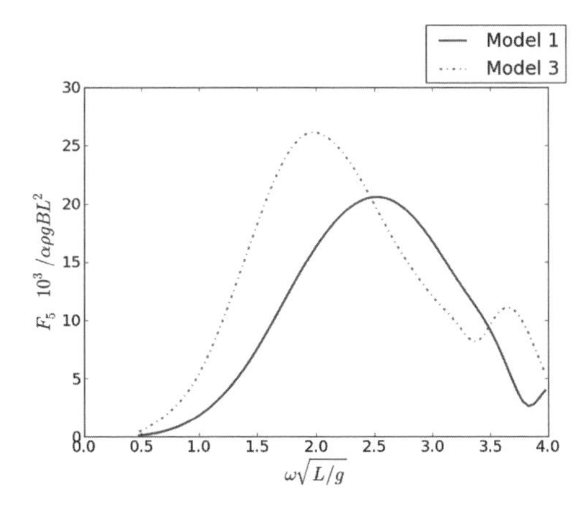

Figure 3. Response amplitude operator of bulk carrier vertical bending moment at midship cross section, for $\beta = 180°$, determined by different models.

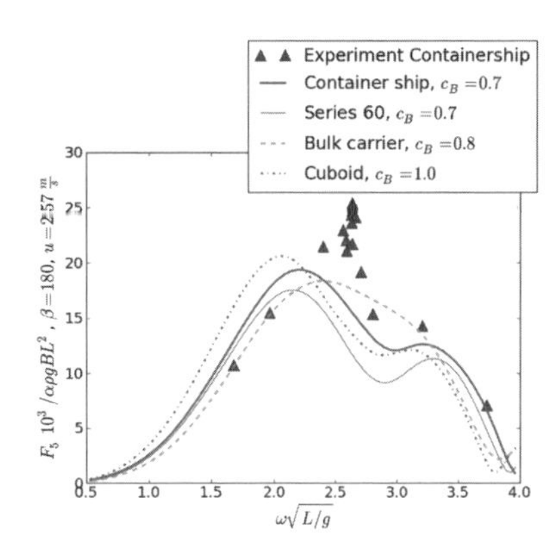

Figure 4. Response Amplitude Operator of vertical bending moments for $\beta = 180°$ and $u = 2.53\ m/s$ at midship cross section of different ships, obtained from the direct approach (*model* 1); experiment is presented in (Drummen, 2009).

to the whole ship and not to its part, as required in determination of internal forces in ship cross section $\bar{x}_1$ (Equation (7))—Fig. 3

- The internal forces (computed with the use of *model* 1) are very sensitive to the shape of the underwater part of the hull—Fig. 4.

5 SIMULATION OF WAVE LOADS ON SHIP

The simulation of vessel motions in waves is based on numerical solutions of non-linear equations of motion. The non-linear model used is presented in (Jankowski, 2006).

This method assumes that:

- Froude-Krylov forces are obtained by integrating the pressure caused by irregular waves, undisturbed by the presence of the ship, over the actual wetted ship surface;
- the diffraction forces are determined as a superposition of diffraction forces caused by the harmonic components of the irregular wave;
- the radiation forces are determined by added masses for infinite frequency and by the so-called memory functions given in the form of convolution.

The non-linear equations of motion are solved numerically.

The simulation based on the linear solution is carried out according to the formula:

$$F_j(\bar{x}_1, t) = \sum_{i=1}^{n} \left[F_{ji}^R \cos(\omega_i t + \varepsilon_i) - F_{ji}^I \sin(\omega_i t + \varepsilon_i) \right] a_i,$$
$$\xi_j(t) = \sum_{i=1}^{n} \left[\xi_{ji}^R \cos(\omega_i t + \varepsilon_i) - \xi_{ji}^I \sin(\omega_i t + \varepsilon_i) \right] a_i, \quad (13)$$

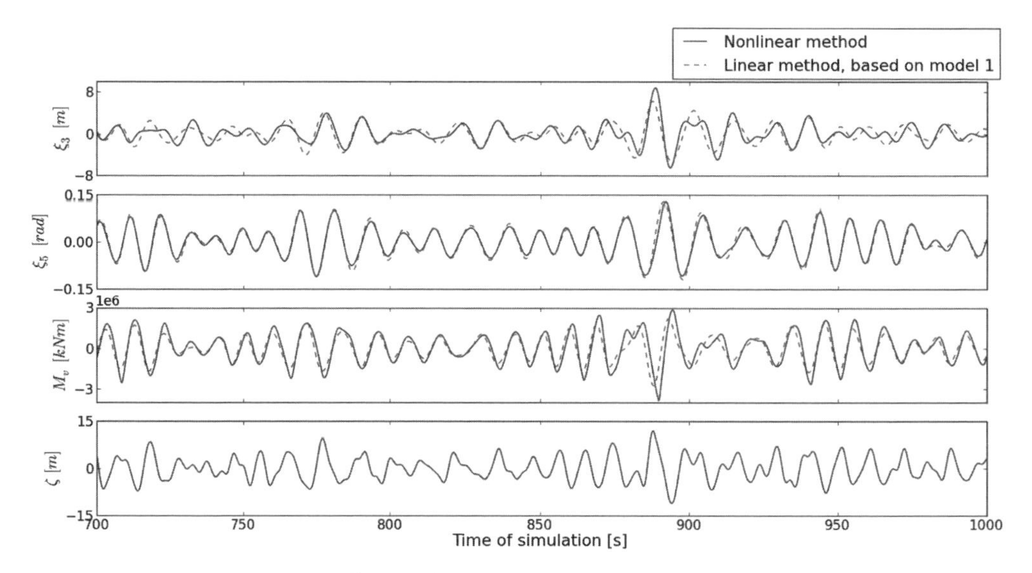

Figure 5. Panamax bulk carrier heave ξ_3, pitch ξ_5 and vertical bending moment M_V as a response to irregular wave, determined by: significant wave height H_s = 15 m, average zero-up-crossing period T_0 = 12 s, β = 180°; ζ is the wave elevation.

where $F_j(\bar{x}_1, t)$, j = 3, 5, are internal forces in the cross section $\bar{x}_1$ of the ship structure in the time instant t, F_{ji}^R and F_{ji}^I, j = 3, 5, are respectively real and imaginary part of the internal forces in the cross section $\bar{x}_1$ of the ship structure as a response to regular wave of frequency ω_i and unit amplitude, a_i is the regular wave amplitude obtained from the wave spectrum and ε_i is the random phase shift between the regular wave components i = 1, 2, …, n. The vertical bending moment in cross section $\bar{x}_1$ is computed according to formula (7).

The comparison of the computation results using non-linear and linear models is presented in Fig. 5.

6 CONCLUSIONS

The requirements in binding rules of classification societies referring to the dynamic loads are given in the form of:

- Formulae, determining amplitudes of the loads that can be exceeded with an assumed probability, and
- Tabulated values of load combination factors, intended to be an approximation of the phase shifts between particular wave loads used to analyze the strength of ship structure.

However, it is questionable whether such a simplified method can approximate the complex process of sea waving acting on the ship moving in waves.

The exact phase shifts between particular wave loads used to analyze the strength of the ship structure are naturally taken into account in the simulation (in time domain) of wave loads on ship and the response of the ship to the waves (ship motions in waves, wave loads on ships, stresses in the structure). However, due to complexity the simulation of loads and stresses in ship structure (ship response to irregular wave) is not used in design practice. Therefore, a simplified method should be developed.

The paper presents the attempt of developing such a simplified method, starting from the linear model based on the boundary-value problem (6) and then by adopting the following simplifications:

- in the first step, ship forward speed u_0 is neglected due to the fact that in high sea states, normally used as the design wave cases, ship speed u_0 is significantly reduced, simplifying the problem;
- in the second step, the attempt is made to approximate radiation potential φ_{Rj}, j = 1, …, 6, by formulae (8) and to compute the damping coefficients and the exciting forces using (10) and (12); this approach does not trigger the necessity to solve the boundary-value problem (6).

It turns out that:

1. Wave loads are very sensitive to the shape of the underwater part of the ship (see Fig. 5). Therefore, the models used to compute these loads (especially internal forces in the ship structure) should be based on precise description of the ship geometry.

2. Simplifications of the radiation potential φ_{Rj}, $j = 1, \ldots, 6$, can be applied only to the ship motions in waves (see Figs. 1 and 2). The proposed simplifications—*model* 3, can not be applied to determine internal forces in ship cross section $\bar{x}_1$ as formulae (10) and (12) are based on integral identities applied to whole ship and not to its part, as required in equation (7). Therefore, new ideas are necessary to develop the transfer functions of internal forces and then the simulation method for wave loads on ship.

The simulations with the use of formulae (13) and that based on the transfer function computed with the use of *model* 1, presented in Fig. 5, show that this approach is comparable to the non-linear method and can be applied in the designing processes, provided simpler than *model* 1 methods are developed for determining transfer function.

Therefore, necessary new ideas will be further developed to determine the transfer functions of internal forces of wave loads on ship with the use of *model* 3 in order to simulate them according to formulae (13).

REFERENCES

Drummen, I., Wu, M.K. & Moan, T. 2009. Marine Structures 22, *Experimental and numerical study of containership responses in severe head seas*, pp. 172–193.

IACS, 2010, *Common Structural Rules for Bulk Carriers*, International Association of Classification Societies.

Haskind, M.D. 1973. *Gosudarstvennoe Izdatelstvo fiziko—matematitcheskoi literatury*, Moscow, Hydrodynamic theory of vessel motion in waves (in Russian).

Jankowski, J.A. 2006. Technical Report No. 52, Polish Register of Shipping, *Ship facing waves*, (in Polish).

Kotchin, N.E. 1963. *Gosudarstvennoe Izdatelstvo fiziko—matematitcheskoi literatury*, Moscow, Theoretical hydrodynamics (in Russian).

Shigemi, T. & Zhu, T. 2003. Marine Structures 16, *Practical Methods of the design loads for primary structural members of tankers*, pp. 275–321.

Shin, Y.S., Kim, B. & Fyfe, A.J. 2004. OMAE, *Load Combination for Fatigue Analyses of Ship Structural.*

Zhu, T. & Shigemi, T. 2003. Marine Structures 16, *Practical Estimation Method of the Design Loads for Primary Structural Members of Bulk Carriers*, pp. 489–515.

Sustainable Maritime Transportation and Exploitation of Sea Resources – Rizzuto & Guedes Soares (eds)
 ISBN 978-0-415-62081-9

Motion calculations for fishing vessels with a time domain panel code

R. Datta, J.M. Rodrigues & C. Guedes Soares
Centre for Marine Technology and Engineering (CENTEC), Instituto Superior Técnico, Technical University of Lisbon, Portugal

ABSTRACT: In the present investigation, two different types of fishing vessels, which are larger than previously examined vessels, are considered and the motion response for various speed and heading angles are calculated with three different codes. For zero speed, the computed motion histories are computed with WAMIT, Strip Theory and a Linear Time Domain 3D Panel Method; while for the forward speed, results are compared between strip theory and time domain code. The objective of the present investigation is to analyse the co-relation of the motion results between three different theories applied to fishing vessels, which due to their size and shape are often in the borderline of applicability of important assumptions in the formulation of codes. It is observed that in spite of different level of approximations in the theories, the computed results show a consistent agreement among three different codes.

1 INTRODUCTION

The importance of developing reliable and practical tools for prediction of ship motions and sea loads requires no elaboration. It is therefore not surprising that such development efforts have been ongoing for over a century. Initially some analytical methods have been developed for simply shaped bodies such as sphere, spheroid etc, but due to the complexity of the theory, significant development in this direction only started since early 60's. A variety of two-dimensional strip theories were developed during the 1960's and 1970s, but the Salvesen et al. (1970) (STF) version of strip theory is one of the most popular and widely used by the industry for the practical ship motion and wave load calculations.

The main advantage of strip theory is that it is computationally not costly is also easy to implement. But at the same time it has limitations as it is a two dimensional theory. The assumption of slenderness of the hull, the low Froude number and the high frequency range are the basic limitations. Also the computation for following waves can be troublesome when the frequency of encounter approaches zero. Therefore the application of these theories to fishing vessels can be problematic.

Due to the advances in available computing power, the focus gradually shifted towards development of 3D solution methods since early 80's. Depending on the choice of the Green's function, one can classify the panel method in two types: frequency and time domain panel method. Frequency domain panel method is very popular in offshore industry such as WAMIT, but its application to forward speed ship motion problem is limited as computing frequency domain Green's function with forward speed is very difficult. Furthermore, frequency domain panel method doesn't account for any nonlinearity.

Therefore time domain methods are considered to be most suitable 3D methods for ship motion problem. There are many advantages of using time domain panel method in comparison to frequency domain methods such as introduction of forward speed which is less complicated. The time domain formulation allows some of the nonlinear effects to be investigated using the body-nonlinear approach such as non-linear hydrostatic, nonlinear Froude-Krylov force.

Some of the important developments in time domain panel methods are as follows: Liapis and Beck (1985) introduced the time-domain Green function based solution method for the 3D linear forward speed problem. Later Lin and Yue (1990), Korsemeyer and Bingham (1998) among others pursued variants of the same method for different class of 3D forward speed problems. Nakos and Sclavounos (1990) proposed Rankine panel method in frequency domain. Rankine source method was also applied in the time domain (Nakos et al., 1993).

The geometries of the fishing vessels are different from the classical ships. Normally the fishing vessels are smaller in length and having different aspect ratios between lengths to beam shows a significant effect on motion response. Also because of the large asymmetry between bow and stern part, the cross coupling coefficient become very important for fishing vessel. Hence it may be worthwhile to study and compare the motion of a fishing vessel with different theories. All the theories have different level of approximation and therefore it is interesting to see how the results co-relate to each other when the hull form is significantly different from the classical one.

The 3D time domain method proposed by Datta and Sen (2007) shows good agreement with other published results for large ships. But when it has been applied to fishing vessels, it has been found that there was a need to introduce some modifications in the scheme, to go beyond such restriction. Datta et al. (2011a) proposed some modifications with respect to the mesh generation, solving equation of motion and including cross coupling term in restoring force matrix. Such modifications resulted in more robust and efficient solution for the motion problem and led to more accurate results for fishing vessels.

Later Datta et al. (2011b) made an extensive study for three fishing vessels to check the robustness and the efficiency of the method. The results were calculated with three different speeds and with three different heading angles (following waves, quarterly head weaves and head waves) and compared with WAMIT (for zero speed case only) and strip theory code. The results show good correlation among all theories but at the service speed, agreement is not very good especially for quarterly head waves.

In the present study, two more fishing vessels are considered. The vessels taken here have different dimension (bigger in length, B/L ratio is different). Hence it is interesting to check the performance of the present time domain code for another kind of fishing vessel especially at the service speed.

Therefore in the present paper, results are computed for zero speed and service speed for different heading angles and compared with WAMIT for zero speed case and with strip theory for zero and forward speed case. The results show excellent agreement with other theories which confirm again the robustness and efficiency of the present linear time domain code and that difficulties with the theories appear only for very small vessels.

2 BRIEF DESCRIPTION OF THE TIME DOMAIN METHOD

In the present paper, three different theories are considered for calculation. Those are WAMIT, strip theory and time domain method proposed by Datta et al. (2011a). Since WAMIT and strip theory (STF code) are well known in the scientific community, a very brief discussion is given here for time domain panel method proposed in this paper. The detailed discussion on the same is available in many sources (Lin and Yue 1990, Datta and Sen 2007, Datta et al., 2011a).

In the present paper, a lower order panel method is selected as it is more adequate to represent the complicated ship hull and also the solution procedure is less complicated in comparison to higher order methods. The solution scheme is divided into the following parts:

a. Solving the hydrodynamic problem to get the hydrodynamic force
b. Solution for hydrostatic forces and Froude-Krylov forces
c. Coupling of equation of motion with hydrodynamic solution

2.1 *Brief discussion on hydrodynamic solution*

For the hydrodynamic solution, the time domain greens function can be divided into two parts:

$$G(p,t,q,\tau) = G^0 + G^f \quad \text{for} \quad p \neq q,\ t > \tau \tag{1}$$

in (1), G^0 is the Rankine part and G^f is the memory part of the Green's function. p and q is the source point at time step t and field point at time step τ. In the linear seakeeping problem, numerical instability occurs due to presence of G^f in the boundary integral equation. Since behaviour of G^f is very drastic near free surface, it leads to some kind of instability in the numerical solution when vessel's hull includes horizontal panels at the waterline.

A detailed presentation on the behaviour of the free surface Green's function and its derivative can be found in Sen (2002). Also from the work by Clement (1998), it may be noted that the behaviour of the free surface Green's function near the free surface is extremely diverging. It is observed that this nature of the free surface Green's function tends to some numerical instability for the seakeeping problem if the horizontal or nearly horizontal panel is presented in the free surface region.

Datta et al. (2011a) thoroughly studied this problem and suggested refinement of mesh if the ship contains horizontal or nearly horizontal panel near the waterline. To avoid such numerical instability, the body geometry of such vessels is modified near the waterline by introducing some small vertical panels there. It is assumed that, due to such small changes, the computed response of the ship is not affected significantly. Figure 1 illustrate this procedure, where the panel mesh has been modified accordingly. The detailed discussion regarding the refinement of mesh is given in Datta et al. (2011a) and hence not repeated here.

2.2 *Computation of hydrostatic co-efficient*

The hydrostatic restoring forces and moments $\left(\vec{F}_{Static}\left(\vec{F}_s, \vec{M}_s\right)\right)$ can be calculated as follows:

$$\vec{F}_{Static} = -c.\vec{x}(t) \tag{2}$$

where c is the restoring coefficient matrix, i.e.:

$$c = \begin{bmatrix} 0 & 0 & 0 & 0 & 0 & 0 \\ 0 & 0 & 0 & 0 & 0 & 0 \\ 0 & 0 & \rho g A_{wp} & 0 & -\rho g M_{wp} & 0 \\ 0 & 0 & 0 & \rho g \Delta \overline{GM}_T & 0 & 0 \\ 0 & 0 & -\rho g M_{wp} & 0 & \rho g \Delta \overline{GM}_L & 0 \\ 0 & 0 & 0 & 0 & 0 & 0 \end{bmatrix} \quad (3)$$

In the above expression, $g$is the acceleration of gravity, ρ is the density of water, A_{WP}, M_{wp} and Δ correspond to the waterplane area, the moment about the waterplane area and the displaced volume at the mean draft respectively. GM_T and GM_L are the transverse and longitudinal metacentric heights.

In the restoring coefficient matrix, the terms c_{35} and c_{53} are very important for fishing vessel since fishing vessels have large asymmetry between bow and stern. These two terms were absent in the work presented by Datta and Sen (2007), but for the calculation of the fishing vessel, one needs to include these two coefficient as contribution from this two terms are not negligible in case of fishing vessel.

2.3 *Solving the equation of motion*

Let F_T be the total force. Then F_T is consists of three following parts:

$$F_T = F_{Static} + F_K + F_{dynamic} \quad (4)$$

where F_K and $F_{dynamic}$ represent the Froude-Krylov and hydrodynamic force. The F_K force is obtained analytically from wave equation and $F_{dynamic}$ is calculated by integrating the hydrodynamic pressure. Once the total force is obtained in the present time step t, one needs to use some explicit/implicit scheme to predict the velocity and motion for the advanced time step $t + \Delta t$. In order to do that, an explicit rule can also be used for the integration which results in the value of the displacement. Detailed description of the scheme is presented in Datta et al. (2011), which is reformulation of the method proposed by Datta and Sen (2007). The reformulation has been done to avoid the numerical instability which occurs when fishing vessels are considered. The detailed discussion is given in Datta et al. (2011) and hence not repeated here.

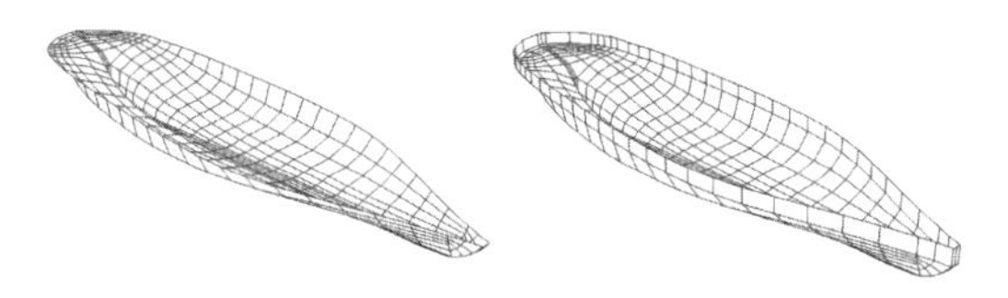

Figure 1. Illustration of modification of mesh. Left: Mesh of original Geometry, Right: mesh of modified geometry.

3 RESULTS AND DISCUSSIONS

In Datta et al. (2011b), an extensive study was made for three fishing vessel named FV1, FV2, FV3. In continuation to that, the vessels studied here are called FV4 and FV5. Table 1 presents the some basic characteristics of FV4 and FV5. In that table the basic characteristic of FV1, FV2 and FV3 vessels are also given to compare the difference of hull forms in terms of length, aspect ratios, block coefficient, B/L ratios with FV4 and FV5. The mesh of vessel FV5 needs to be modified because it has some nearly horizontal panel on the stern near free surface. Figure 2 represents the mesh for the FV4 vessel where as Figures 3a and 3b represent the original and modified mesh of vessel FV5.

Motion results for the unrestrained hull are presented. The results are presented in form of response amplitude operators (RAO). In the figure legends the acronyms W, ST and TD are defined as the results obtained using WAMIT (only for zero speed), strip theory and the present time domain scheme respectively.

The time domain transfer functions are obtained by averaging the peak to peak value of the steady part of the motion history. The frequency, heave and pitch motion are non-dimensionalized as follows:

$$\omega\sqrt{L_{bp}/g},\ x_3/a, \quad \text{and} \quad x_5 \quad (5)$$

For FV4 vessel, the motion calculation is carried out for zero speed, top speed of 16knot and for one intermediate speed (8 knot). In all speeds the computation is carried out for three different wave headings (head sea, i.e., for180 degree, following

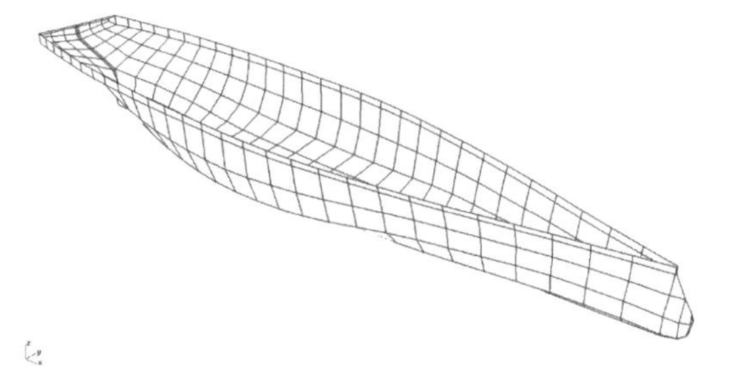

Figure 2. Mesh of FV4.

Table 1. Main characteristic of the set of fishing vessel.

	[m]	[m]	[m]	[mT]
Designation	LBP	B (moulded)	D (moulded)	Δ
FV1	15	5.4	2.4	104.9
FV2	22	5.8	1.6	94.7
FV3	20	7.4	3.4	302.9
FV4	63	13	6.17	3146.5
FV5	63	13.1	5.61	2923.4

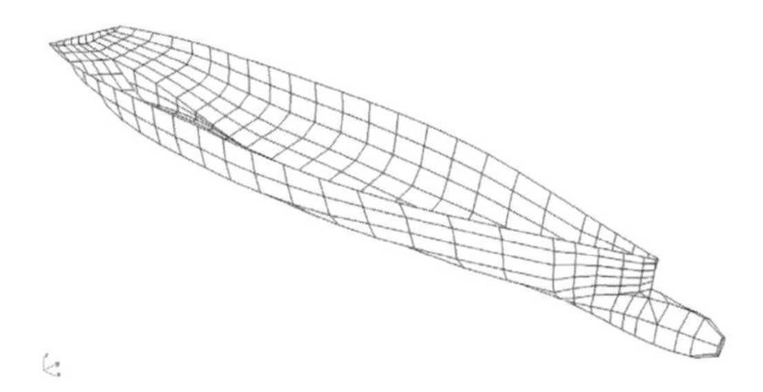

Figure 3a. Original mesh of FV5.

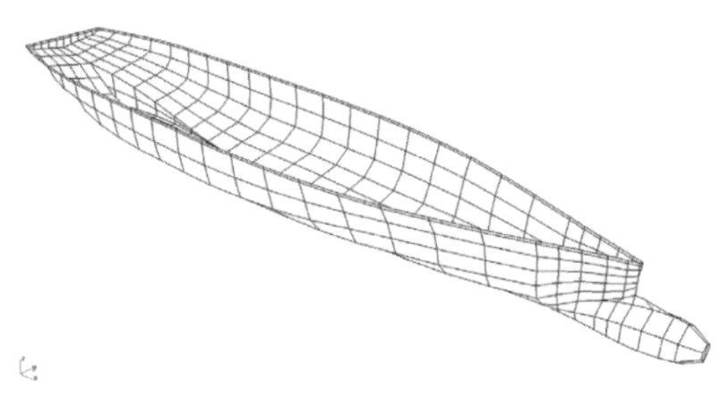

Figure 3b. Modified mesh for FV5.

sea, i.e. for 0 degree and quarterly head waves, i.e. for 135 degree).

Figure 4 represent the heave motion results for head waves. From the figure it may be noted that all the results agree excellently. In Datta et al. (2011b), the strip theory results showed some minor deviation with other two results for FV1 and FV3 vessel for zero speed case which decreased with speed. But in the present computation, zero speed result matches very well. From the Table 1, it may be noted that FV4 is much slender than FV1 and FV3 vessel which confirms again that slenderness is dominant factor when ship speed is low.

From Figure 4, it can be seen that the as speed increases, the correlation is good but some deviation can be found in the peak frequency region. Similar kind of nature in the motion result was observed for FV2 vessel in Datta et al. (2011b), where result are better for zero speed case with strip theory but had some disagreement for high speed case. This again confirms the argument that forward speed effect dominates the slenderness property for the strip theory. Also for heave, time domain results tends converges to a value that greater than1, this amplification may occur because of the diverging nature of the green's function, More accurate mesh can give better results. Since very low frequency range is out side of the domain of interest in present paper, therefore further modification of mesh is not incorporated here.

Figure 5 represents the results for pitch motion for head sea. The same phenomena can be seen for pitch results as well, results are excellent for zero speed case, but show some deviation at the peak frequency range.

Figures 6–9 are the FV4 results for different heading angle, Figures 6–7 are the results for quarterly head waves and 8–9 are the results for following waves.

These results are carried out to check the agreement between the results when different heading angles are considered. It is well know that time domain gives best results for head seas. Therefore it is interesting to check the robustness and efficiency of the proposed linear scheme when other heading angles are considered.

From the plots, it may be noted that, time domain results shows excellent agreement with WAMIT for zero speed which shows the efficiency of the time domain code. For the forward speed results, the co-relations between strip theory and time domain result is very good except for the higher frequency range, but overall behaviour is very consistent and represents the similar kind of phenomena.

The next set of results is shown for the vessel FV5. It may be noted that all the computation for the FV5 vessel is carried out using modified mesh, therefore from the figures, one can conclude that small geometric changes don't affect to the final response.

Figures 10–13 represents the heave and pitch RAO results for the FV5 vessel. The results are computed for 0 speed, and 16 knot speed which is service speed for the vessel. The computations are carried out for head waves quarterly head waves and following waves. Comparisons with WAMIT and strip theory results for different speed and for different angles are also presented.

From the plot, it can be observed that time domain results gives higher value in comparison to the strip theory results. Also like FV4; the agreement between these two theories is better for the zero speed case in comparison to forward speed case. It is obvious to say that the B/L ration of FV4

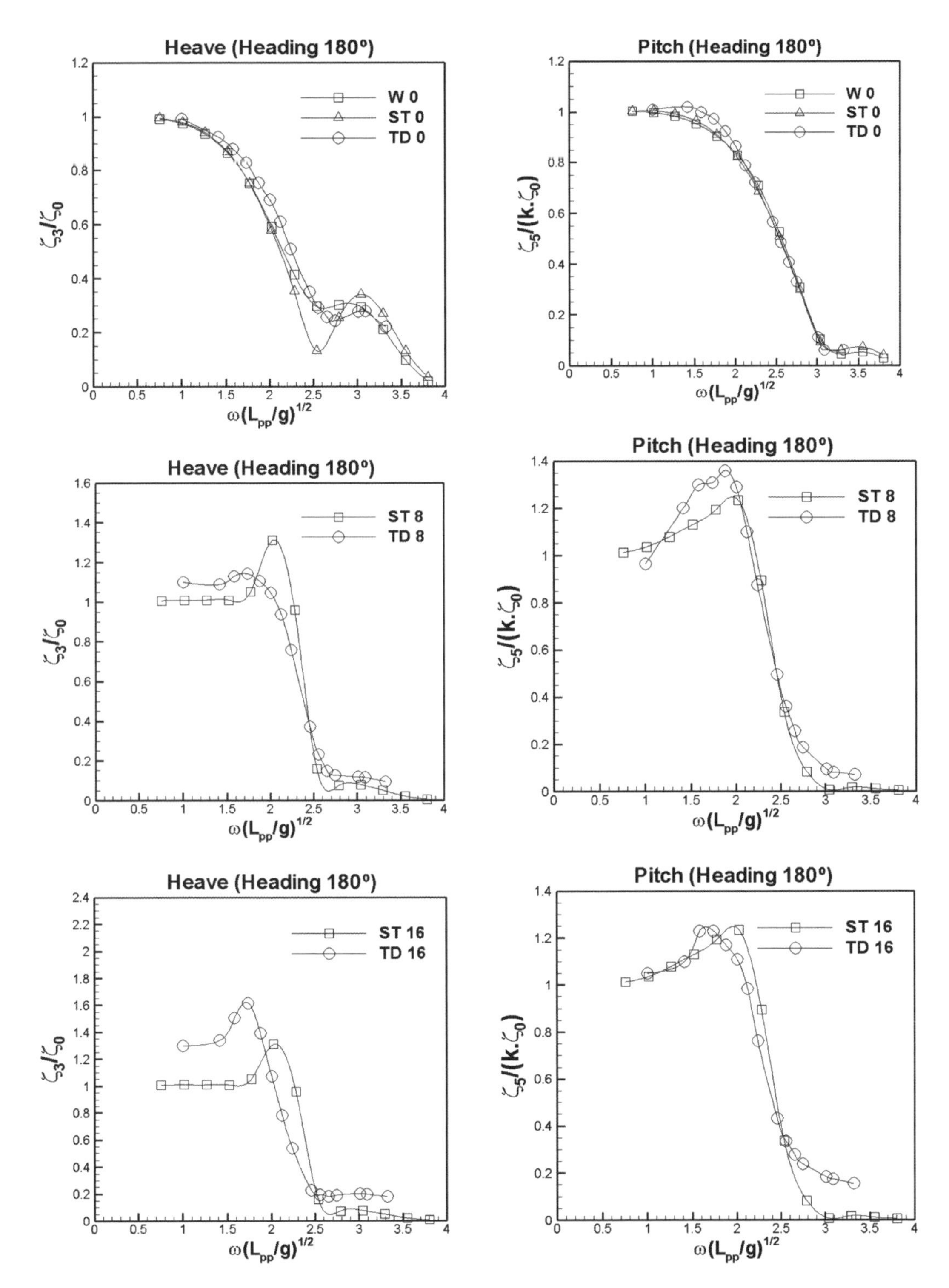

Figure 4. Heave response of FV4 vessel for different Speed. Wave heading: 180 degree. Top: 0 speed, middle: 8 knots, bottom: 16 knots.

Figure 5. Pitch response of FV4 vessel for different Speed. Wave heading: 180 degree. Top: 0 speed, middle: 8 knots, bottom: 16 knots.

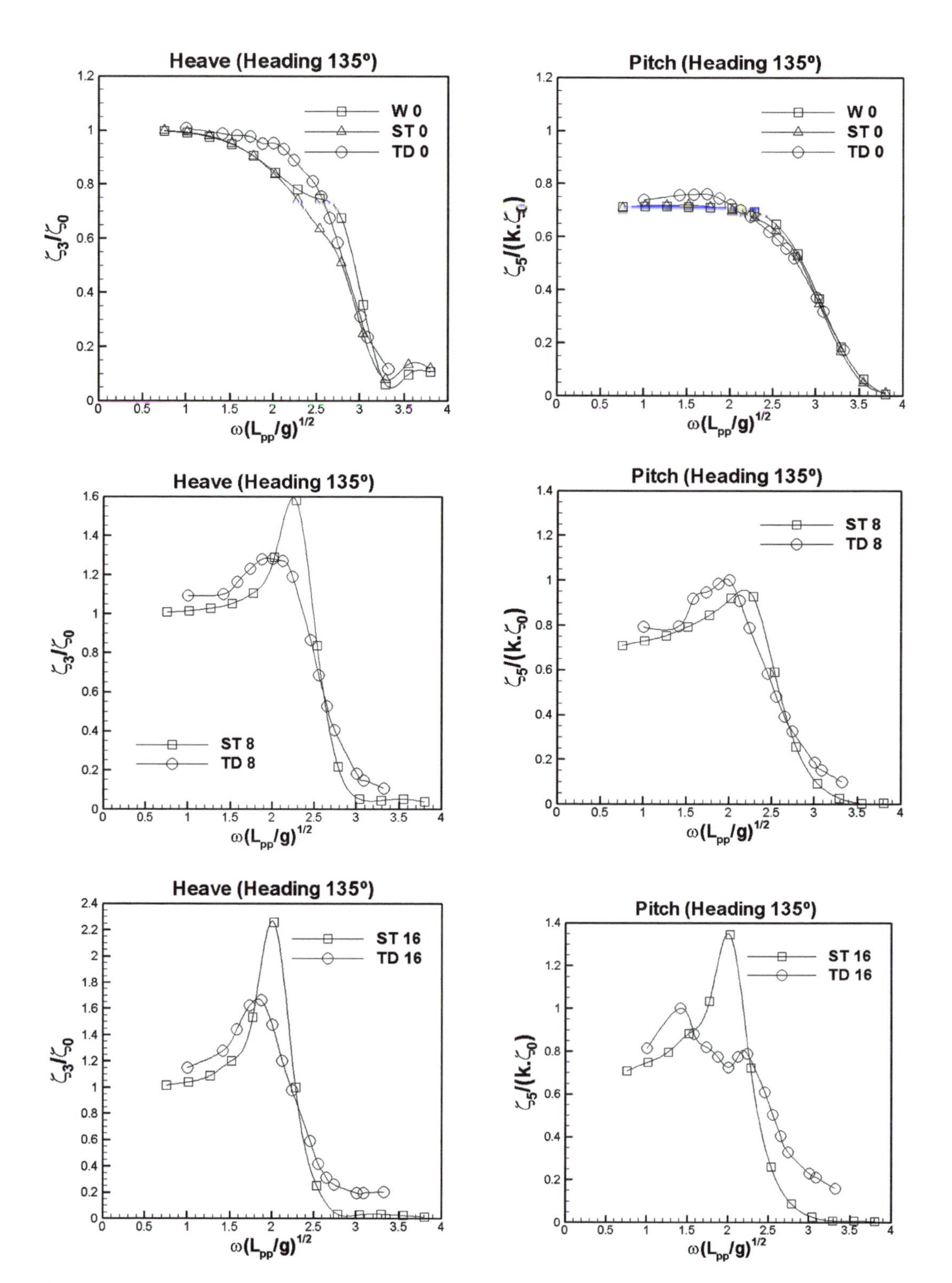

Figure 6. Heave response of FV4 vessel for different Speed. Wave heading: 135 degree. Top: 0 speed, middle: 8 knots, bottom: 16 knots.

Figure 7. Pitch response of FV4 vessel for different Speed. Wave heading: 135 degree. Top: 0 speed, middle: 8 knots, bottom: 16 knots.

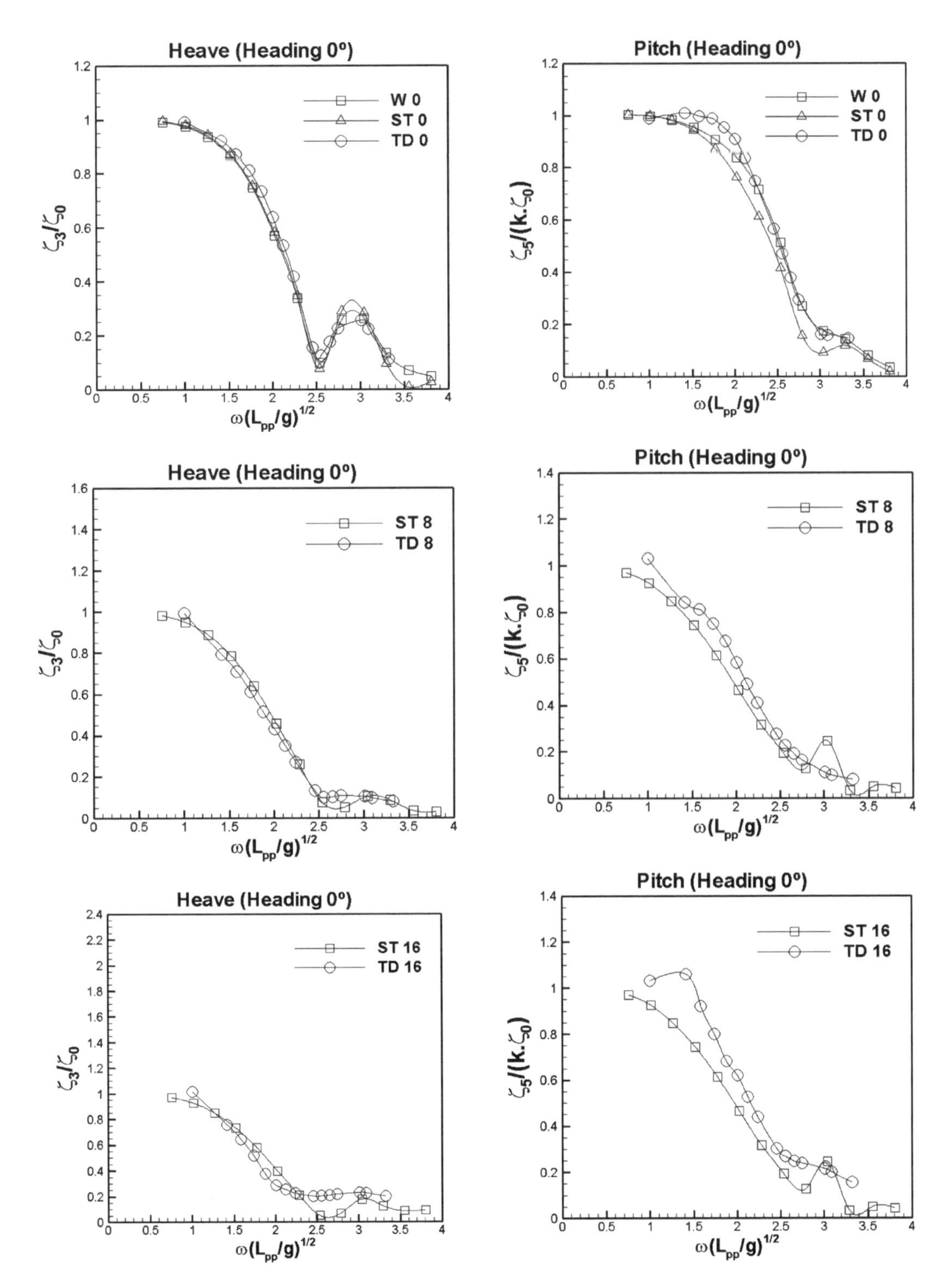

Figure 8. Heave response of FV4 vessel for different Speed. Wave heading: 0 degree. Top: 0 speed, middle: 8 knots, bottom: 16 knots.

Figure 9. Heave response of FV4 vessel for different Speed. Wave heading: 0 degree. Top: 0 speed, middle: 8 knots, bottom: 16 knots.

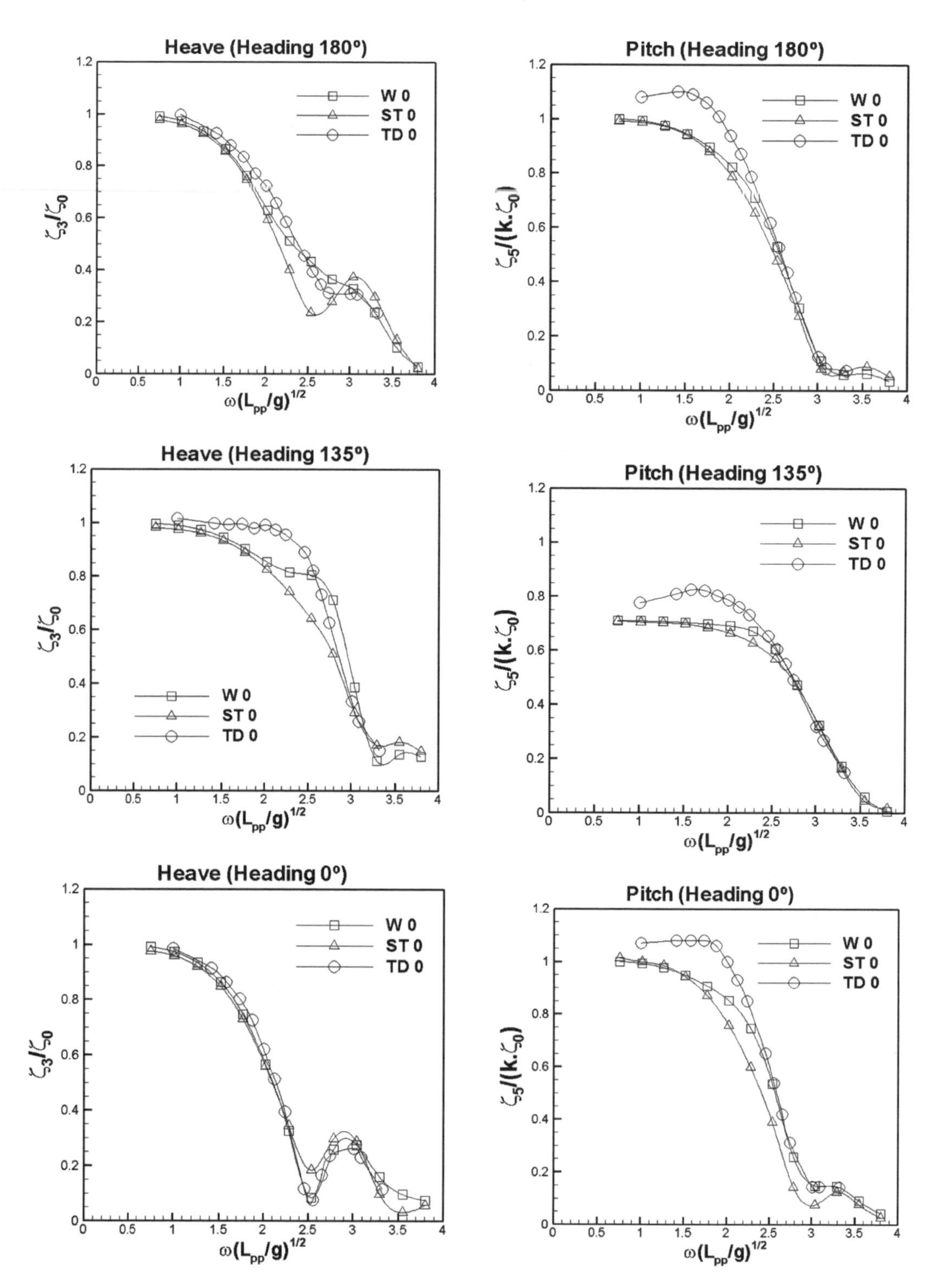

Figure 10. Heave response function of FV5 for different heading angles, Zero Speed case, Top: heading 180°, middle: heading 135°, bottom: heading 0°.

Figure 11. Pitch response function of FV5 for different heading angles, Zero Speed case, Top: heading 180°, middle: heading 135°, bottom: heading 0°.

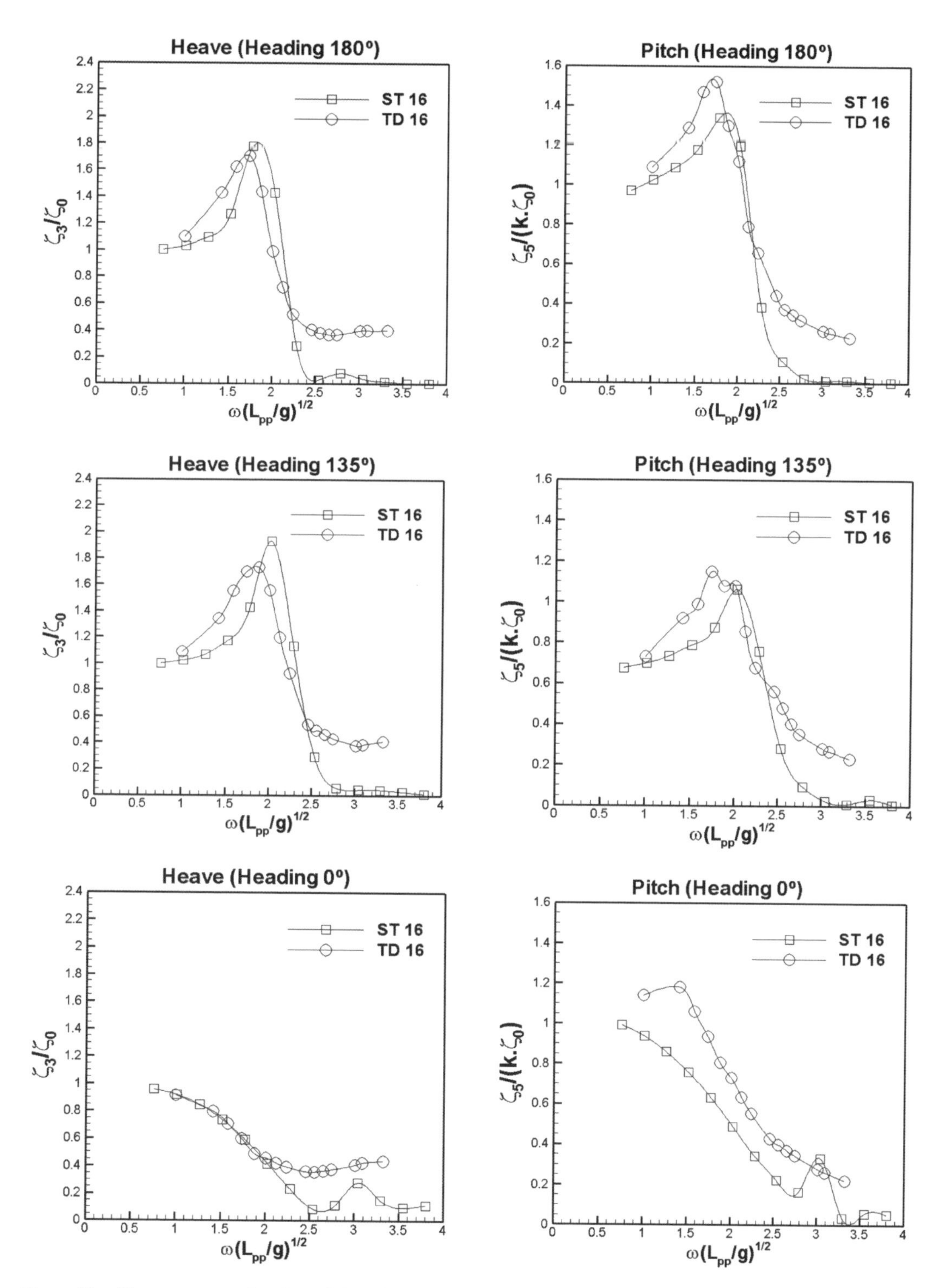

Figure 12. Heave response function of FV5 for different heading angles, 10 knot Speed, Top: heading 180°, middle: heading 135°, bottom: heading 0°.

Figure 13. Pitch response function of FV5 for different heading angles, 10 knot Speed, Top: heading 180°, middle: heading 135°, bottom: heading 0°.

and FV5 is very similar and therefore similar kind of behaviour in the motion result is expected

From the all results, it may be noted that the present time domain code is performing excellently for FV4 and FV5. Earlier the efficiency and robustness is verified for other kind of fishing vessel and found a satisfactory agreement with other commercially accepted theories/software. Therefore it can be concluded that the modification proposed by Datta et al. (2011a) is very significant and such changes increased the efficiency and robustness of present time domain code.

From the above results, it is also further verified that slenderness of the ship is important for the heave motions when zero speed is considered, but forward speed effect dominates the slenderness property.

4 CONCLUSION

In the present study, the time domain code is further verified with two more fishing vessels. The results compared very well with the industrially accepted and well validated WAMIT code for zero speed which again shows the efficiency and robustness of the proposed method. However for the forward speed, the present results deviate slightly from strip theory but that is due to the differences lies in the approximation of the theories. The differences are mainly observed in the peak frequency range. Otherwise overall behaviour is similar and shows the consistent pattern of the motion RAO. It is further verified that the slenderness property is important for the strip theory in the low speed situations, but gradually the forward speed effect becomes the dominated factor as speed increased.

ACKNOWLEDGEMENT

This work has been performed within the project "SADEP-Decision support system for the safety of fishing vessels subjected to waves", which has been financed by the Foundation for Science and Technology ("Fundação para a Ciência e a Tecnologia"), from the Portuguese Ministry of Science and Technology, under contract PTDC/EME-MFE/75233 /2006. The first and second authors have been funded by the Portuguese Foundation for Science and Technology under the contracts SFRH/BPD/ 66845/2009 and SFRH / BD / 64242 / 2009.

REFERENCES

Clement, A.H. 1998. An ordinary differential equation for green function of time-domain free-surface hydrodynamics, Journal of Engineering mathematics, vol. 33, pp. 201–217.

Datta, R. and Sen, D. 2007. The simulation of ship motion using a B-spline based panel method in time domain, Journal of Ship Research, vol. 51(3), pp. 267–284.

Datta, R, Rodrigues, J.M. and Guedes Soares, C. 2011a. Study of the Motions of Fishing Vessels by a Time Domain Panel Method, Ocean Engineering, vol. 38, pp. 782–792.

Datta, R, Rodrigues, J.M. and Guedes Soares, C. 2011b. Prediction of the motions of a fishing vessels using time domain 3D panel method, 1St MARTECH conference, IST, Lisbon.

Korsmeyer, F.T. and Bingham, H.B. 1998. The forward speed diffraction problem. Journal of Ship Research, vol. 42, no. 2, pp. 99–112.

Liapis, S.J. and Beck, R.F. 1985. Seakeeping computations using time domain analysis, Proceedings, 4th International Conference on Numerical Ship Hydrodynamics, National Academy of Sciences, Washington, D.C, pp. 34–54.

Lin, W.M. and Yue, D. 1990. Numerical solution for large amplitude ship motions in the time domain. Proceedings of 18th ONR Symposium on Naval Hydrodynamics, National Academy Press, Washington D.C, pp. 41–66.

Nakos, D.E., Kring, D.E. and Sclavounos, P.D. 1993. Rankine panel method for time-domain free surface flows., Proceedings, 6th International Conference on Numerical Ship Hydrodynamics., Iowa City, Iowa., University of Iowa.

Nakos, D.E. and Sclavounos, P.D. 1990. Ship motion by a three dimensional Rankine panel method, Proceedings of 8th ONR Symposium on Naval Hydrodynamics, National Academy Press, Washington D.C, pp. 21–40.

Salvesen, N., Tuck, E.O. and Faltinsen, O. 1970. Ship motions and sea loads. Trans. SNAME, vol. 78, 250–287.

Sen, D. 2002. Time domain computations of large amplitude 3D ship motions with forward speed, Ocean Engineering, vol. 29, pp. 973–1002.

Sustainable Maritime Transportation and Exploitation of Sea Resources – Rizzuto & Guedes Soares (eds)
© 2012 Taylor & Francis Group, London, ISBN 978-0-415-62081-9

Experimental evaluation of the dynamic responses of ship models moored in scaled ports

G.J. Grigoropoulos, D. Damala & D. Liarokapis
National Technical University of Athens (NTUA), Athens, Greece

ABSTRACT: A system to measure the linear and non-linear dynamic responses of ships and scaled models in waves using seven strap-down accelerometers mounted on the ship/model has been developed in the Laboratory for Ship and Marine Hydrodynamics (LSMH) of NTUA. In this paper it is used to evaluate experimentally the six degree of freedom dynamic responses of small models moored at piers of a 1:125-scaled model of the port of Piraeus. In addition, a modern optical system of four high-performance cameras tracking the motions of four positions on the model was used in the tests. Both of these user-friendly and reliable systems are presented. Their experimental results are compared providing useful guidelines to accomplish measurements on small-scaled ship models, including the modelling of the mooring system. The accelerometer-based system constitutes an inexpensive alternative to the optical tracking system.

1 INTRODUCTION

During the design of new ports or the extension of existing ones, extensive experimental investigations are carried out, in order to evaluate their capabilities. The latter are quantified via the measurement of the dynamic responses of vessels properly moored at the piers of ports by the bow, the stern or their sides. In all cases accounted for in this paper, the models are free to move in all six degrees of freedom (6 DOF), subject only to the constraints of the (flexible) mooring system.

The specific characteristic of this kind of measurements is that, since the harbors are quite extensive and the test facilities are relatively small. Thus, the facility at the Laboratory of Harbor Works (LHW) of the NTUA encompasses two rectangular tanks with dimensions 32 m × 26 m and 26 m × 24 m. The depth of the water in both of them varies from 1 to 100 cm. The first one was used for the tests described in this paper.

Based on the above characteristics the layouts of the ports are scaled down significantly in order to fit in the facility. This results inevitably to small-sized models, especially for ships of small to moderate size. Consequently, the models should be lightweight and stiff constructions, with limited margins to ballast them to the suitable draft and trim and to fit the associated measuring equipment onboard.

Thus, only limited methods are available for executing the recordings in the laboratory. In some of the major hydrodynamic facilities permanent optical systems have been installed, that are calibrated during their procurement and commissioning.

The Laboratory for Ship and Marine Hydrodynamics (LSMH) of NTUA was interested in mobile measuring systems to be used in its premises or in other facilities. Thus, a system of seven properly arranged strap-down accelerometers to estimate all six DOF motions of the model was developed (Grigoropoulos & Politis, 1999).

Furthermore, an optical system consisting of four high-precision cameras was procured and customized in the recordings of motions of freely moving models.

The aforementioned systems are described in the next section. Both of them were implemented in the measurements of three 1:125 scaled, moored models in the LHW of NTUA. The length of the models was 2.7, 1.7 and 0.4 m. Selected experimental results are presented, compared and discussed in the remaining sections of the paper.

2 DESCRIPTION OF THE SYSTEMS

2.1 *Accelerometer-based system*

In this system the seven strap-down accelerometers are mounted on the same horizontal plane of the model and their location and sensitivity axes are depicted in Figure 1. A pair of accelerometers is located at the stern, one for the vertical and one for the lateral acceleration. Another pair with the same directions is located at the bow region of the ship. Two vertically oriented accelerometers are fixed at both sides of the model (port and starboard) and a seventh accelerometer, in charge

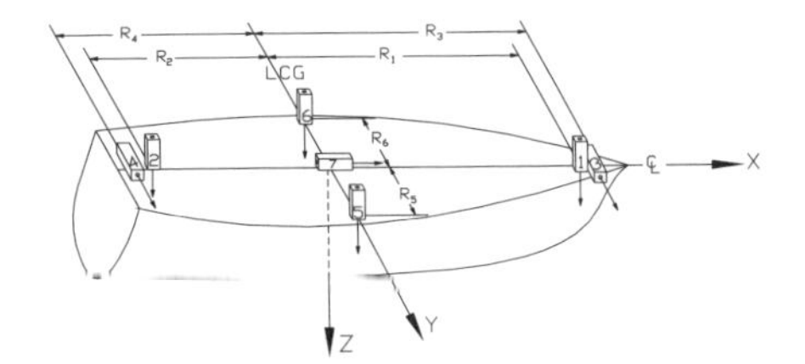

Figure 1. The arrangement of the strap-down accelerometers on the model and their direction.

of the longitudinal acceleration, is positioned at the centerline of the model. It should be noted that, although six accelerometers are sufficient to derive the equations of motions of the model, we use seven accelerometers for simplicity of the equations and for cross-checking of the measured accelerations. Although the calibration of the accelerometers is simple, the setup of the system on small models should be carried out carefully and it is time-consuming.

The theoretical background of the system is based on the derivation of the equations of motion of the model in terms of the measured accelerations. In this respect two coordinate systems are used:

- an inertial system moving with the constant forward speed of the body, and
- a body-fixed system, which coincides with the inertial one in the absence of the model oscillations.

Let $\vec{X}_0(P)$ be the position vector of an arbitrary point P with respect to the inertial system, and $\vec{X}(P)$ the corresponding position vector with respect to the body-fixed system. The following equation holds,

$$\vec{X}_0(P) = \vec{X}_0(C) + \mathbf{A}^{-1}\vec{X}(P), \tag{1}$$

where **A** is the transformation matrix transforming a vector from the inertial coordinate system to the body-fixed system given by Jeffers (1976):

$$\mathbf{A} = \begin{bmatrix} \cos\theta\cos\psi & \cos\theta\sin\psi & -\sin\theta \\ -\sin\psi\cos\varphi + \cos\psi\sin\varphi\sin\theta & \cos\varphi\cos\theta + \sin\theta\sin\varphi\sin\psi & \cos\theta\sin\varphi \\ \sin\psi\sin\varphi + \cos\psi\cos\varphi\sin\theta & -\cos\psi\sin\varphi + \sin\psi\sin\theta\cos\varphi & \cos\theta\cos\varphi \end{bmatrix} \tag{2}$$

In eq. (2) φ (angle of roll), θ (angle of pitch), ψ (angle of yaw) are the Euler angles, using the conventions provided in standard texts (e.g., SNAME 1952 and 15th I.T.T.C. 1978). Since **A** defines an orthogonal transformation, $\mathbf{A}^{-1}$, the inverse of **A**, defines an orthogonal transformation too and it is equal to the transpose of **A**.

Differentiating equation (1) twice with respect to the time and recalling that $\vec{X}(P)$ is constant, we readily find:

$$\ddot{\vec{X}}_0(P) = \ddot{\vec{X}}_0(C) + \ddot{\mathbf{A}}^{-1}\vec{X}(P) \tag{3}$$

Applying the above equation (3) for the positions of accelerometers 1,2,3,4 and 7 and taking the dot product with the unit vector associated to their sensitive axis, as well as appropriate corrections in order to take into account the gravitational effects of the inertial type accelerometers, we get the equation (Politis et al., 1995):

$$\ddot{\vec{X}}_0(P) = \mathbf{A}^{-1}\cdot\mathbf{P} + \ddot{\mathbf{A}}^{-1}\vec{X}(P) + \vec{g} \tag{4}$$

where

$$\mathbf{P} = \begin{bmatrix} a_7 \\ (a_3R_4 + a_4R_3)/(R_3 + R_4) \\ (a_1R_2 + a_2R_1)/(R_1 + R_2) \end{bmatrix} \tag{5}$$

a_i, i = 1, ...,7, is the output of the i-accelerometer and R_i, i = 1, ...,7, its radial distance with respect to the body-fixed system (Fig. 1).

Equation (4) can be considered as the equation of motion of the body in terms of the measured accelerations (the matrix **P**) and the transformation matrix **A**. Thus, if the matrix **A** can be calculated, equation (4) can be used to obtain the three translatory velocities and displacements of the arbitrary point P of the body.

Matrix **A** is calculated by applying equation (3) successively at accelerometer positions (i = 1, …, 6) and subtracting the outputs of each pair of accelerometers. Thus, we obtain the following non-linear system with respect to the angular velocities of the body resolved along the axes of the body-fixed system (Miles, 1986):

$$\begin{aligned} \dot{\omega}_x + \omega_y\omega_z &= F_1(t) \\ \dot{\omega}_y - \omega_x\omega_z &= F_2(t) \\ \dot{\omega}_z + \omega_x\omega_y &= F_3(t) \end{aligned} \tag{6}$$

where,

$$F_1(t) = \frac{a_5 - a_6}{R_5 + R_6}, \; F_2(t) = \frac{a_2 - a_1}{R_1 + R_2}, \; F_3(t) = \frac{a_3 - a_4}{R_3 + R_4}, \quad (7)$$

These are the fundamental equations of motions of the rigid body defining its angular velocity $\vec{\omega}$ in terms of the measured accelerations. It should be noted that the outputs of the six accelerometers (i = 1, …, 6) are non independent, e.g., the following relation holds:

$$a_6 = (a_1 R_2 + a_2 R_1) \cdot \frac{R_5 + R_6}{R_1 + R_2} - a_5 \frac{R_6}{R_5} \quad (8)$$

But in the present analysis we use the outputs of all accelerometers for the simplicity of the derived equations and for cross-checking of the measured accelerations.

Once the angular velocity $\vec{\omega}$ of the model has been calculated, the Euler angles, defining the transformation matrix **A** can be determined by solving the following non-linear system (SNAME, 1952):

$$\begin{aligned} \dot{\varphi} &= \omega_x + \tan\theta(\omega_y \sin\varphi + \omega_z \cos\varphi) \\ \dot{\theta} &= \omega_y \cos\varphi - \omega_z \sin\varphi \\ \dot{\psi} &= (\omega_y \sin\varphi + \omega_z \cos\varphi)/\cos\theta \end{aligned} \quad (9)$$

The numerical solution of the non-linear systems (6) and (9) is presented in detail by Grigoropoulos & Politis (1999).

The system can be used both for models and onboard physical ships.

2.2 *Optical system*

The optical system denoted as Optomatrix Motion Capture System (OMCS) consists of a set of four digital motion capture units (cameras) adapted to capture full-frame, full-speed, full-resolution high-speed video and a user-friendly Windows-based acquisition software (Qualisys Track Manager, QTM). In this configuration the cameras are equipped with a large buffer memory and a clear front glass to get the best possible performance out of the image capture. Retro-reflective markers, which reflect infrared light emitted by the cameras, are placed on the model.

The 2-D position of each marker is determined with high accuracy by the cameras internal signal processing algorithm. In a towing tank it could be enough with only two or four cameras depending on the configuration. The cameras can be put on the towing carriage or at fixed locations in wave flumes. It is essential that the model and the markers are visible during the whole recording time. For the calibration we used the wand-type technique. In more detail, a pre-fixed calibration table is placed at the water surface defining the local coordination system. All measurements taken refer to the above defined origin system. The calibration must be performed for every mooring arrangement and the time needed is less than one hour, depending on the physical limitation. The general layout of the optical system is depicted in Figure 2.

By combining 2-D data from the cameras, the QTM software calculates the 3-D position of the markers. Thus, one can get real-time 6-DOF data to analyze roll, pitch and yaw parameters of the model. The 6DOF tracking function uses the rigid body definition to compute P_{origin}, the positional vector of the origin of the local coordinate system in the global coordinate system (Fig. 3), and **R**, the rotation matrix which describes the rotation of the rigid body.

The rotation matrix **R** is similar to **A** with inversed signs where there are one or three signs in a product. It is used to transform a position P_{local} (e.g., x'_1, y'_1, z'_1) in the local coordinate system, which is translated and rotated, to a position P_{global}

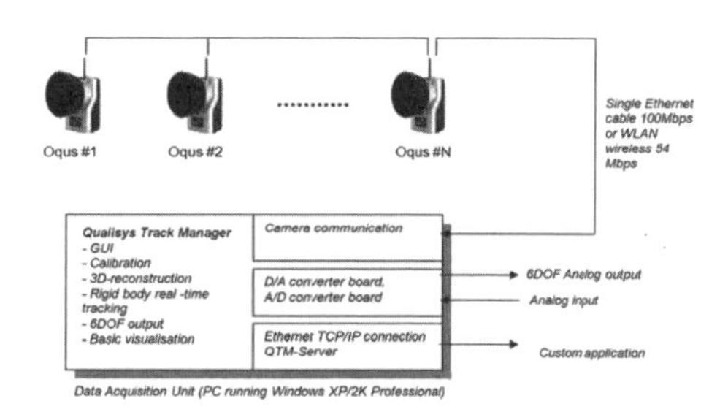

Figure 2. The layout of QMCS. In our case N = 4.

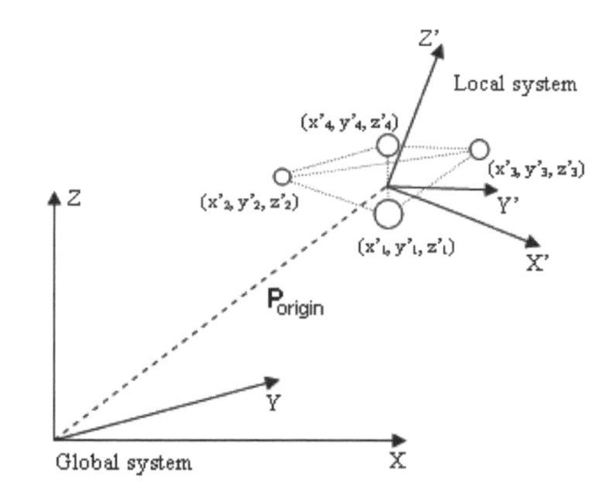

Figure 3. The global and the local coordinate systems. The former is used for the calibration.

(e.g., x'_1, y'_1, z'_1) in the global coordinate system. The following equation is used to transform a position:

$$P_{global} = \mathbf{R} \cdot P_{local} + P_{origin}$$

More detailed information on the system and its functionality can be found in the Marine Manual of Qualysis AB (2005).

2.3 *Additional setup arrangements*

In order to proceed to the measurements of the motions of moored ship models two further aspects should be taken into account, the effect of wind and the mooring systems.

Since there were no wind capabilities in the facility and models were constructed without their superstructures, the effect of the wind was numerically evaluated using the empirical method of Blendermann (1993) for each wind direction assuming that the directions of sea and wind coincide. The final arrangement was to impose an initial inclination on the model to represent the effect of the wind heeling angle on the transverse stability of the vessel.

The mooring systems were arranged using figures of the actual arrangements of the vessels, available in the internet (see Fig. 4 for FESTOS PALACE). Mooring cables with adequate elasticity and anchoring lines were properly chosen, to simulate the physical and the mechanical properties of the full-scale moored ships. The properties of the restrained mechanism were successfully assessed to simulate the restoring characteristics and the generalized spring constants of the mooring arrangement. To this end, when necessary, the elastic stiffness of the scaled down lines was calculated in accordance with the combined elasticity of the complex actual lines of the ship.

Figure 4. Mooring system of C/F FESTOS PALACE.

3 EXPERIMENTAL RESULTS

3.1 *Description of the tests*

During the test campaign, three vessels were experimentally evaluated. For two of them both methods were used. The smaller model, a 55-meter mega-yacht was just 0.40 m long and its displacement was less than 0.4 kgr. The dynamic performance of that model was evaluated by only using the accelerometer-based system. The optical system has a nominal displacement accuracy of 0.5 mm which is significant when displacements of a few mm are recorded. On the other hand, the sensitivity of the accelerometer-based system permits the capture of smaller displacements. Thus the optical system was not fitted on the smaller model. The results for the remaining two vessels only, where both systems were used, are presented in this paper.

The models correspond to the cruise liner QUEEN MARY II and the ferry FESTOS PALACE. The characteristics of the vessels are presented on Table 1.

3.2 *Regular waves*

Further to the standard tests, performed in specific random seaways, some additional runs have been carried out using regular waves in order to demonstrate more clearly the comparative capabilities of the two measuring systems under investigation.

In Figures 5 & 6 the roll and the absolute vertical acceleration at the bow of the model of QUEEN MARY II are plotted using both measuring systems for regular waves with frequencies 0.75 & 1.25 Hz. The results are in both Figures in excellent agreement in-between. The same holds true also for the absolute vertical acceleration at the stern for regular waves with frequency 1.50 Hz. The wave heading was 135° in all cases (180° correspond to head seas). The agreement is similar for the rest of the results recorded, although due to limited space a few are plotted in this paper.

Table 1. Main particulars of the tested models.

		QUEEN MARY II	FESTOS PALACE
Characteristic	Units	Model	Model
Length overall	m	2.76	1.71
Length at WL	m	2.53	1.63
Breadth	m	0.33	0.21
Draught	m	0.08	0.06
Trim by stern	m	0.00	0.004
Displacement	kgr	38.27	11.13
LCB fwd mid	m	–0.065	–0.063

It should be noted that we deliberately chose to compare the roll motion since it is derived directly on the basis of the markers movements in the Qualisys system, while it is calculated by solving a non-linear set of motion equations in the accelerometer-based system. On the other hand, the vertical acceleration is calculated by the second derivative of the motion at the same point in the former system while it is directly measured in the latter.

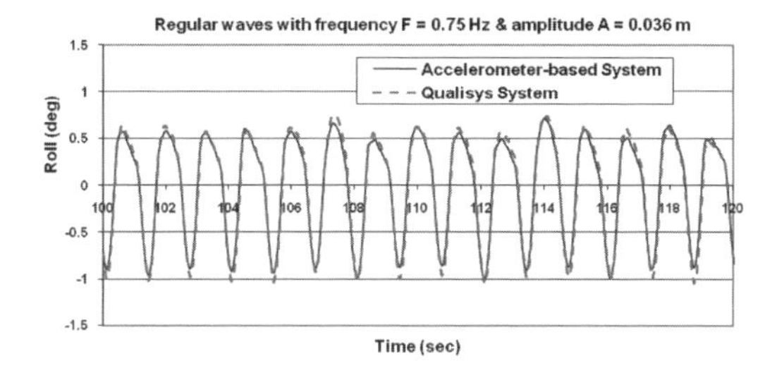

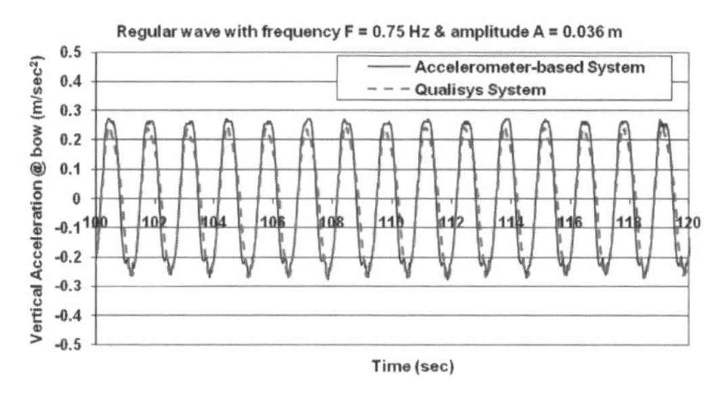

Figure 5a. Roll and vertical acceleration at the bow of the QUEEN MARY II model for wave frequency F = 0.75 Hz.

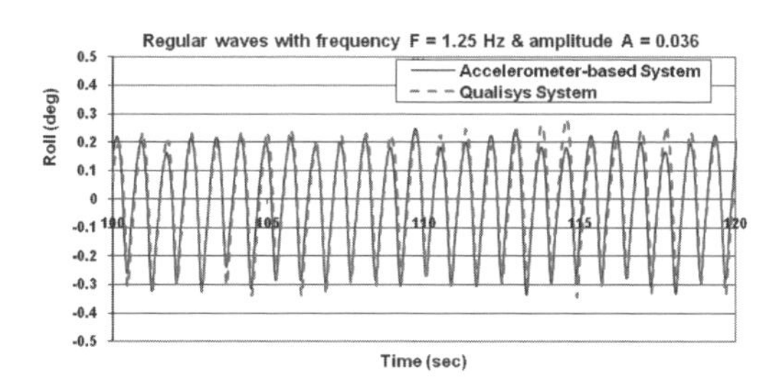

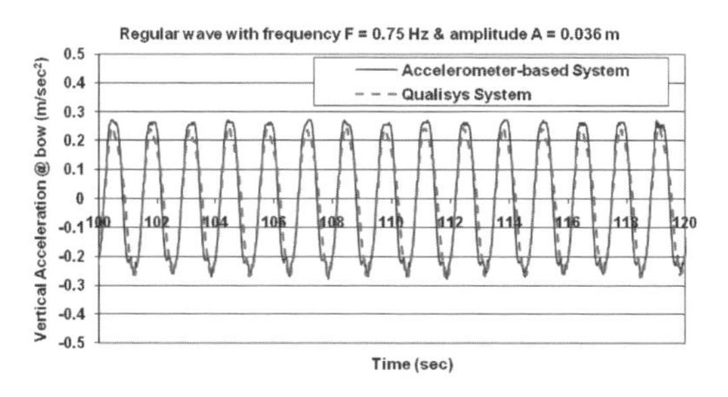

Figure 5b. Roll and Vertical acceleration at the bow of the QUEEN MARY II model for wave frequency F = 1.25 Hz.

3.3 *Random seaways*

In the case of random waves the agreement of the two measuring systems is very satisfactory. Again, we chose to test the worst case scenario. The wave direction was again 135°. The significant wave height and the respective modal period for each case is depicted on the graphs. In Figure 7, the time histories of roll and absolute vertical acceleration at the stern for FESTOS PALACE are plotted, as they derived by the two competitive systems under consideration.

In addition the RMS values are in excellent agreement, i.e., for the roll response of Figure 7 the

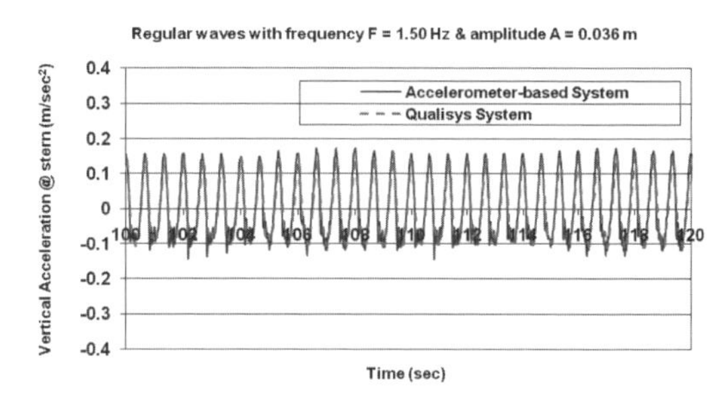

Figure 6. Vertical acceleration at the stern of the QUEEN MARY II model for wave frequency F = 1.50 Hz.

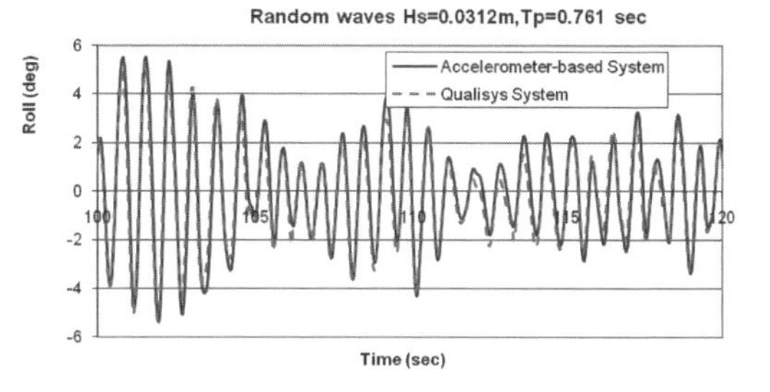

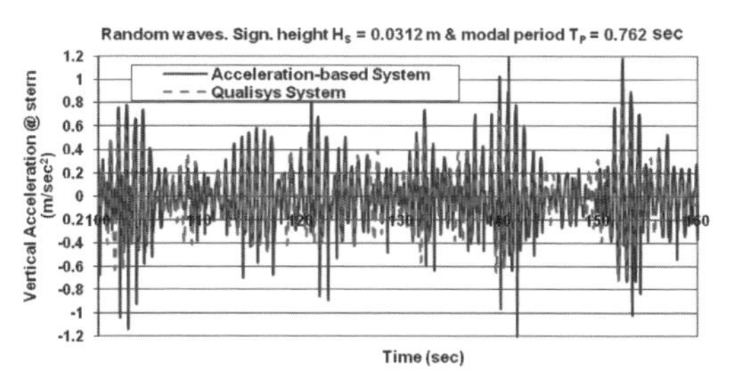

Figure 7. Roll and vertical acceleration at the stern of the FESTOS PALACE model for significant wave height $H_S = 0.031$ m and modal period $T_P = 0.762$ sec.

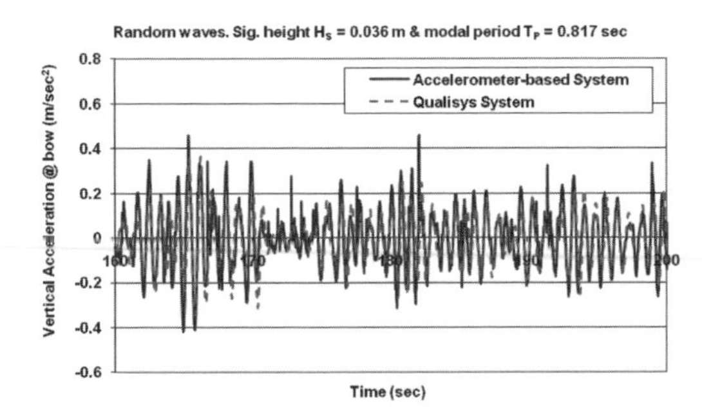

Figure 8. Vertical acceleration at the bow of the QUEEN MARY II model for significant wave height $H_S = 0.036$ m and modal period $T_P = 0.817$ sec.

Figure 9. Photo of the model and experimental set up during test measurements.

accelerometer-based system gives 1.810 deg, while the estimation of Qualisys system is 1.819 deg.

Similar results were derived for the QUEEN MARY II model, as it is shown in Figure 8 for the absolute vertical acceleration at the bow of the QUEEN MARY II model.

4 DISCUSSION & CONCLUSIONS

Two portable measuring systems were used for recording the dynamic responses of moored models in scaled ports. Both systems provide practically equivalent results.

The accelerometer-based system is relatively an inexpensive system which can be used both for models in the testing facility or at site, and onboard actual ships. Its setup is more tedious and time-consuming than the optical system. However, in cases where higher accuracy is required, this system is more competitive,

On the other hand, the Qualisys system is a fairly precise system to be used in the testing facility which readily derives the six DOF motions of the model and their derivatives.

REFERENCES

Blendermann, W. 1993. Parameter identification of wind loads on ships, *Rept. Universität Hamburg, Institut für Schiffbau*, Hamburg.

Grigoropoulos, G.J. & Politis, C.G. 1999. A measuring system of the six degrees of motions of a moving body, *ShipTechnology Research—Schiffstechnik Journal*, Vol. 45, No. 1, 1999 (with C. Politis).

Jeffers, M.F. Jr. 1976. Analytical methods for determining the motion of a rigid body equipped with internal motion-sensing transducers, *Rep. DTNSRDC, Ship Performance Dept., Bethesda*, Md. 20084, October.

Politis, C., Grigoropoulos, G.J. & Loukakis, T.A. 1995. A measuring system of the six D-O-F motions of a floating body, *Lab. for Ship & Marine Hydrodynamics of NTUA, Rep. No NAL-138-F-1995*, October.

Qualisys AB. (2005). Qualisys Track Manager, Marine Manual, *Qualisys Motion Campture Systems*, December 22, Sweden.

SNAME 1952. Nomenclature for treating the motion of a submerged body through a fluid, *SNAME T & R Bulletin 1–5*.

Van Berlekom, W.B. 1981. Wind forces on modern ship forms—Effects on Performance. *Rept. Swedish Maritime Research Centre-SSPA*, February.

Sustainable Maritime Transportation and Exploitation of Sea Resources – Rizzuto & Guedes Soares (eds)
© 2012 Taylor & Francis Group, London, ISBN 978-0-415-62081-9

Offshore wind generators dynamics

Matteo Masi
MSc Naval Architect and Marine Engineer, Italy

Stefano Brizzolara
Head of the Marine CFD group—University of Genoa, Italy

Stefano Vignolo
Assistant Professor in Mathematical Physics—University of Genoa, Italy

ABSTRACT: The paper presents the main features and application examples of a numerical method, created to simulate the 6DOF dynamics for an offshore floating wind turbine, subject to sea waves, wind and mooring lines forces, with the aim of evaluating many different design alternatives for the preliminary optimization of a concept design. The method presented is based on a parametric definition of the dimensions and characteristics of the main structural components. In fact, hull components can be represented by simple trunk cones, cylinders, prisms and concentrated masses. A set of arrays based on those parameters is used as input in the following numerical method where the dispositive is modeled as a rigid body, whose weight and matrix of inertia are calculated by a specific routine. This makes possible to study different kinds of offshore structures, with a rapid definition of different design solutions for a given design goal to be systematically analyzed.

The wind flowing over oceans can be considered a vector of energy, as it contains an amount of kinetic energy withdrawn from the sun radiation; such energy can be converted into electrical energy without emission of pollution using wind turbines. This vector has been used since epochs, e.g., for navigation or wind mills operation. The relatively recent use of wind turbines for electric energy conversion has received a partial success worldwide; however, the employment of wind turbines suffers from some unfavorable aspects such as the noise produced especially by the earlier turbines design and the visual impact of tall structures in open land. The present trend, aiming to overcome these difficulties and to increment the "green" fraction of energy flowing in the grid, is placing large turbines at sea, possibly far from human sight. Several wind farms have been installed in shallow waters but this can only partially solves the problem; in order to access sites in far and deeper waters, floating solutions are today proposed; these ones open a new range of problems that might be worth being considered. Floating wind turbines are quite unedited yet and their stability is simultaneously affected by hull hydrodynamics, mooring lines dynamics and rotor aerodynamics. The expected income for a single turbine is times lower than oil and gas dedicated structures, this reduces sensibly the cost margin of a machine and creates an interest in using simple and optimized structures, in which every element contributes to the functionality of the system. Probably only a specific engineering effort can cope with of this necessity.

The present study was born in March 2010, as an MSc thesis in Marine Engineering and Naval Architecture, at DINAEL—University of Genoa, with the aim to perform non linear seakeeping analysis for a specific concept design, based on both weight and shape stability. Since its beginning, a parametric and modular computer code has been developed to simulate the moored motion of a generic floating object composed by conical geometric figures. At present, the research is continuing at DIPTEM within a PhD program, with the aim to deepen specific features of the dynamic stability of wind turbines using adequate calculation techniques; to compose dedicated modular computer codes sustained by a theoretical background and to finally link homogeneously all the involved disciplines in a sufficiently complete model.

Offshore Wind Turbines create a link between offshore technologies (traditionally developed in the oil & gas industry) and wind technologies. Both rely on consolidated theoretical and empirical foundations, with a variety of dedicated design tools characterized by different levels of detail. Offshore platforms stability is usually studied by means of potential flow methods, giving the possibility to consider diffraction and refraction effects in wave motion. Some more simplified methods are available for slender vertical piles, where the effects mentioned above are not significant; on the contrary, more complicated methods, such as RANSE solvers, are employed in order to consider viscous and turbulent flows when needed. Mooring lines can be modeled using empirical or finite element methods. In the wind technologies field, horizontal axis wind turbines are usually designed using *aero-elastic* methods; they consist in applying aerodynamic loads, calculated by means of Blade Element Momentum theory, to elastic simplified structures. More detailed studies are carried out using RANSE methods for blade shape optimizing and FEM methods for detailed structures scantling, keeping as a reference aero-elastic global loads.

1 THE NUMERICAL METHOD

This paragraph briefly reports the key features of the numerical method developed in [Masi, 2011].

1.1 *Parametric approach*

The simplicity in the shapes of the parts composing the structure proposed by the designer, allows describing them by a set of solid trunk cones, prisms and concentrated masses, while the hull surface by conic and circular 3d surfaces. A set of parameters has been introduced to describe sizes, distances, orientations and density for each structural element. The value of each parameter is read as input by two scripts that provide the necessary input to the subsequent modules of the code. The scripts, together with the corresponding set of parameters, compose the structural input for the simulations. This feature allows the description of any design solution made of compatible elements and also a systematic optimization of a particular solution with respect to its proportions or weight. The use of a parametric approach was inspired by the necessity of modeling a floating turbine in the early stages of its design cycle with no reference to any existing exemplar, where the possibility to easily make large modifications in proportions might be very useful.

1.2 *Coordinate systems*

The turbine and the hull are represented in the numerical model as a set of *bodies*. Each body is described using its own coordinate system and all mechanical properties are expressed in that reference system. The complete system is thought as a *rigid body*. The motion of a *rigid body* can be studied using two different frames of reference: the *body fixed frame*, moving and rotating with the system; and the (pseudo) *inertial frame*, fixed with respect to the Earth's surface. A routine takes care of introducing, in the initialization phase of a simulation, the *body fixed frame* with its origin in the center of mass and to express the physical characteristics of the *rigid body*, composed by all its parts, with respect to this reference frame. Finally the *wave coordinate frame* has the same origin of the *inertial frame* but rotated of 180 deg around x axis.

1.3 *Euler's angles*

The orientation of the body fixed frame with respect to the inertial frame is described by 3 sequential axial rotations.

The entity of each rotation is defined by an Euler's angle.

Let us call $\{i\}$ the basis for the inertial frame and $\{b\}$ the basis for the body fixed frame; the sequence of rotations follows:

- Rotation around z-axis of basis {i} (pointing downwards) of an angle ψ, called yaw. Let us call the basis obtained by this rotation {i'}.
- Rotation around y-axis of basis {i'} of an angle θ, called pitch. Let us call the basis obtained by this rotation {i''}.
- Rotation around x-axis of basis {i''} of an angle Φ, called roll. The basis obtained by this rotation is finally {b}.

For further details about the mathematical aspects of these transformations see [Masi 2011, Fossen 1994].

1.4 *Rigid body 6DOF motion*

The whole structure is considered a *rigid body* under the hypothesis that the relative motions among its parts, due to the elasticity of the materials, are very small and do not significantly affects the motion of the whole structure. In this case, the structure can be physically described by means of a constant *mass matrix*, which is calculated by a dedicated structural module.

The motion is simulated solving the classical *rigid body dynamic (1) and kinematic (2) differential equations,* using a set of variables arrays, according to SNAME and [Fossen, 1994]:

$$[M]_{RB} \cdot \dot{v} + [C]_{RB} \cdot v = \tau_{\mathrm{RB}} \quad (1)$$

$$\dot{\eta} = [J] \cdot v \quad (2)$$

In these equations:

$\nu = [u, v, w, p, q, r]^T$ is the *velocity array*, $[u, v, w]^T$ are the components of the linear velocity of *body fixed frame* origin, evaluated with respect to the *inertial frame* and represented in the *body fixed basis* $\{b\}$; $[p, q, r]^T$ are the components of the angular velocity of *body fixed frame* with respect to *inertial frame*, represented in *body fixed basis* $\{b\}$.

$\tau = [X, Y, Z, K, M, N]^T$ is the *external forces array*, $[X, Y, Z]^T$ are the components of external forces and $[K, M, N]^T$ are the components of external moments both expressed in the $\{b\}$ basis.

$\eta = [x, y, z, \Phi, \theta, \psi]^T$ is the *position array*, $[x, y, z]^T$ are the coordinates of dispositive's center of gravity in the *inertial frame*, $[\Phi, \theta, \psi]^T$ are Euler's angles expressing orientation of the *body fixed frame* with respect to the *inertial frame*.

The "dot" over an array represents its time derivative.

$[M]_{RB}$ is the Matrix of Inertia of the rigid body; when the *body fixed frame* is chosen with its origin in the center of gravity of the rigid body, M_{RB} is composed by 2 sub-matrices.

$$M_{RB} = \begin{bmatrix} M_{RB1} & 0_{3x3} \\ 0_{3x3} & M_{RB2} \end{bmatrix} \quad (3)$$

$$M_{RB1} = [I]_{3x3} \cdot W_{RB} \quad (4)$$

is a diagonal matrix expressing the weight of the body.

$$M_{RB2} = \begin{bmatrix} I_x & -I_{xy} & -I_{xz} \\ -I_{yx} & I_y & -I_{yz} \\ -I_{zx} & -I_{zy} & I_z \end{bmatrix} \quad (5)$$

is the properly named *matrix of inertia*.

C_{RB} is the 6×6 Coriolis and Centripetal matrix, this has been compressed and *sym* indicates a symmetric term of the matrix with respect to its diagonal:

$$C_{RB} = \begin{bmatrix} 0_{3x3} & C_{RB1} \\ C_{RB1} & C_{RB2} \end{bmatrix} \quad (6)$$

$$C_{RB1} = \begin{bmatrix} 0 & m \cdot w & -m \cdot v \\ -m \cdot w & 0 & m \cdot u \\ m \cdot v & -m \cdot u & 0 \end{bmatrix} \quad (7)$$

$$C_{RB2} = \begin{bmatrix} 0 & -I_{yz}q - I_{xz}p + I_z r & I_{yz}r + I_{xy}p - I_y q \\ -sym & 0 & -I_{xz}r - I_{xy}q + I_x p \\ -sym & -sym & 0 \end{bmatrix} \quad (8)$$

$[J]$ represents the Kinematic transformation between the 2 frames of coordinate used.

$$J = \begin{bmatrix} J_1 & 0_{3x3} \\ 0_{3x3} & J_2 \end{bmatrix} \quad (9)$$

[J1] links the linear component of velocities expressed in $\{b\}$ to those expressed in $\{i\}$. Let us report the transformation column by column.

$$J_1 = [\mathrm{J}_{1:1}\ \mathrm{J}_{1:2}\ \mathrm{J}_{1:3}] \quad (10)$$

$$\mathrm{J}_{1:1} = \begin{bmatrix} \cos\psi\cos\theta \\ \sin\psi\cos\theta \\ -\sin\theta \end{bmatrix} \quad (11)$$

$$\mathrm{J}_{1:2} = \begin{bmatrix} -\sin\psi\cos\Phi + \cos\psi\sin\theta\sin\Phi \\ \cos\psi\cos\Phi + \sin\Phi\sin\theta\sin\psi \\ \cos\theta\sin\Phi \end{bmatrix} \quad (12)$$

$$\mathrm{J}_{1:3} = \begin{bmatrix} \sin\psi\sin\Phi + \cos\psi\cos\Phi\sin\theta \\ -\cos\psi\sin\Phi + \sin\theta\sin\psi\cos\Phi \\ \cos\theta\cos\Phi \end{bmatrix} \quad (13)$$

$$J_2 = \begin{bmatrix} 1 & \sin\Phi\tan\theta & \cos\Phi\tan\theta \\ 0 & \cos\Phi & -\sin\Phi \\ 0 & \dfrac{\sin\Phi}{\cos\theta} & \dfrac{\cos\Phi}{\cos\theta} \end{bmatrix} \quad (14)$$

The $[J_2]$ transformation (14) links the component of the angular of velocities expressed in $\{b\}$ to the time ratios of Euler's angles.

The system of equations is solved, at each time step, using the six components of $\dot{\nu}$ as unknowns. The instantaneous values are integrated by a second order scheme to solve velocities and displacements in the successive instant, using a constant time step.

1.5 *The inertia matrix*

This matrix contains information concerning the global inertia of the system. Equations (1) are expressed with respect to a coordinate system having its origin at the body's center of gravity and they are evaluated using the *structural arrays*. The inertia for each solid figure is analytically calculated with respect to its local coordinate system. The rigid body fixed frame is centered at the center of gravity of the system, calculated by the static momentum theorem and suitably orientated. The local inertia for each body are then rotated and transposed in order to be expressed with respect to the *body fixed frame*.

1.6 *External forces—Hydrodynamics*

Sea loads are generated by the waves striking the hull, the current, but also by the presence of the water itself; in this model they are calculated using Morison equation (15) together with Froude-Krilov forces on the hull (16).

Morison's equation expresses a force per unit of length acting on a transverse section of a cylindrical body immerged in a fluid in motion and it is solved in a finite number of hull sections; Morison's equation is supposed to be valid when the wave length is more than five times the diameter of the section. The terms composing the equation are functions of fluid acceleration (added mass) and the square of its velocity (drag). Both the kinematic quantities are the projection on the section plane of the un-disturbed water particles acceleration and velocity. The relative motion between the hull and the water is also considered.

$$\frac{f_M}{l} = \rho C_M \frac{\pi D^2}{4} \dot{u} + \frac{1}{2} \rho C_D \cdot D \cdot |u| u \tag{15}$$

These quantities are calculated at the center of the section:

- ρ is the density of the water.
- C_M is the added mass coefficient, its original expression according to (DNV) $C_M = 1 + C_A$ is modified into $C_M = C_A$ because the first term is calculated by the Froude-Krilov component of the dynamic forces.
- D is the diameter of the transverse section.
- $\dot{u}$ is the projection of the acceleration in the section plane.
- C_D is the drag coefficient.
- $|u|u$ is the square of the projection of the velocity on the plane preserving the sign.

Froude-Krilov forces are generated by the pressure of the water on the hull. Their contribution is calculated in a finite number of planar surfaces generated by a specific mesh of the hull surface.

$$f_{FK} = A \cdot p \cdot \boldsymbol{n} \tag{16}$$

where:

- A is the area of the planar surface.
- p is the pressure.
- n is the normal unit vector to the surface.

Water's kinematics is modeled using a linear Airy Potential function (expressed with respect to wave's frame of coordinates).

$$\Phi = \frac{ga}{\omega} e^{kz} \sin(kx - \omega t) \tag{17}$$

The pressure field, supposed to be undisturbed by the presence of the hull can be consequentially expressed as:

$$p = \rho \cdot g \cdot [a \cdot e^{k \cdot z} \cdot \cos(k \cdot x - \omega \cdot t) - z] \tag{18}$$

In case of still water (a = 0) the pressure field is hydrostatic.

As mentioned above, Equations (15) and (16) are solved in a finite number of rings and planar surfaces respectively. For this purpose two meshing models are used to decompose the hull surface. The contribution of each element is added to the others in order to generate the resultant external forces and moments array τ.

The method used for dynamic coupling between water and hull is based on the hypothesis that water particles' acceleration is prevailing with respect to structure's acceleration.

1.7 *RANSE comparison for hydrodynamic loads*

The hydrodynamic module has been preliminary validated using a commercial code (CD ADAPCO Star CCM+), based on Random Averaged Navier Stokes Equation Finite Volume Method. This tool represents a state of the art CFD solver for non linear turbulent flow of fluids with free surface.

In Figure (1) a comparison between global horizontal force and transversal moments reveals a tolerable margin between the two methods, which increases with time as the fluid in the RANSE simulation (green) evolves in the domain and loses part of its energy in *numerical viscosity*, while our method is characterized by an undisturbed water motion. Comparisons were performed on every volume composing the floating base, a larger error was found for those bodies partially submerged. This can be due both to the simplicity of our model and to the error margin in free surface modeling within the RANSE code. No comparison was possible in terms of motion yet.

1.8 *External forces—Mooring lines*

A floating body is hold inside its operational area by a group of *mooring lines*. These are made of

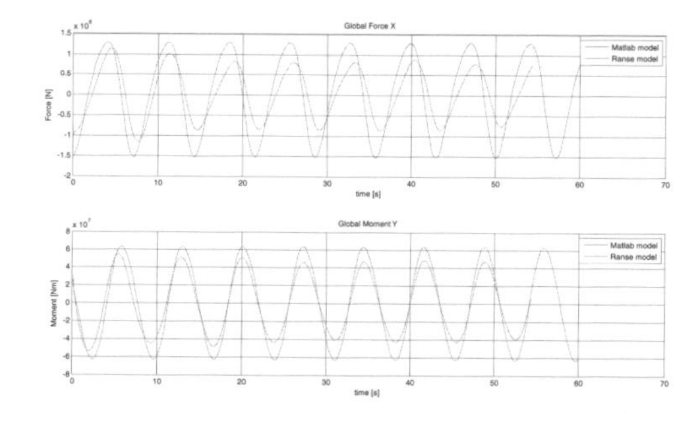

Figure 1. Global forces comparison.

chains or ropes (or both) and connect anchors to device's fairleads.

Several solutions for deep water offshore turbines are being studied and one of the main differences among them is actually in the mooring lines configuration.

In such structures, very tall and slender, forces due to mooring lines can significantly affect stability and platform motions. For this work, different models for mooring lines have been formulated: a quasi-static, finite element model with sea bottom and suction anchors interaction; a dynamic elastic one composed of two segments and finally, the model used in most of simulation is a rigid, fully coupled, segmented one. The main idea, of this mooring model, was to strongly link the dynamics of the rigid body to those of the mooring lines, in particular to the strain in the higher part of the line. In this case the motion is calculated by a greater system of differential equations, where a new variable and a new equation are introduced for each mooring line.

$$\begin{aligned} &\dot{v}\cdot\alpha_1+\dot{v}_2\cdot\alpha_2+\dot{v}_3\cdot\alpha_3+\dot{v}_4\alpha_4 \\ &+\dot{v}_5\alpha_5+\dot{v}_6\alpha_6+t_2\alpha_7+\alpha_0=0 \end{aligned} \quad (19)$$

Equation (19) is an example of the dynamic coupling between the floating body and one of the mooring lines; six body's kinematics unknowns, $(\dot{v}_i)$, are linked to mooring line's strain, t_2, considered as an unknown of the problem too. The coupled problem is characterized by a set of $3 + N_{lines}$ equations.

At present, a new Finite Element Elastic Method is under investigation.

1.9 *External forces—Wind loads*

Wind loads on the turbine could not be modeled in details having no competence about the subject at the time when the thesis was developed; extreme loads have been extracted from turbine manufacturer's data, and applied to tower top by a constant external force.

1.10 *Time integration scheme*

The time is discretized in a series of uniform time steps.

$$t_i = t_{i-1} + t_{step} \quad (20)$$

Velocity in a time instant is calculated trough a numerical integration, using velocity rate in the previous time instant and considering it constant in a time step.

$$v_i = v_{i-1} + \dot{v}_{i-1}\cdot t_{step} \quad (21)$$

The position and rotation of the rigid body in a time instant is calculated using previous time step transformation matrix J and actual velocity array in a numerical integration.

$$\eta_i = \eta_{i-1} + [J(\eta_{i-1})]\cdot v_{i-1}\cdot dt + \frac{d([J(\eta_{i-1})]\cdot v_{i-1})}{dt}\cdot\frac{dt}{2} \quad (22)$$

When velocity and displacement arrays are calculated for a time instant a linear system of equations is set up to solve "accelerations" and describe the time history of the motion.

$$M_{RB}\cdot\dot{v}_i + [C_{RBi}\cdot v_i - \tau_{RBi}] = 0 \quad (23)$$

1.11 *Initial conditions and hydrostatics*

The default starting condition for a simulation is characterized by still water equilibrium position. The space positioning of the floater and its mooring lines is performed by special routines. The hydrostatic equilibrium is calculated by an error cutting procedure which can be also used by another simple program that can perform stability calculations.

1.12 *Batch possibilities*

Most of computer codes give the opportunity to perform hundreds of different simulations in series or parallel. A parametric approach allows the computer to be programmed to act on parameters to obtain the satisfaction of given criteria.

A given model can be also tested in many load conditions for both wind and sea; an appropriate post processing analysis can give a wide spectrum of information about the model performances.

2 SOME RESULTS

In this paragraph some results of simulations performed during the study are presented. Please mind that those are not meant to be valid as "absolute results" because of the roughness of the model; our aim in this context is to show a possible use for a time domain simulation based on a parametrical approach, when the physical mathematic base is validated in its precision.

2.1 *Models used*

The former model was dimensioned on the basis of the initial idea of the designer, in the earlier stages of the work for hydrostatic stability purposes. The latter was produced in the last stage to verify the

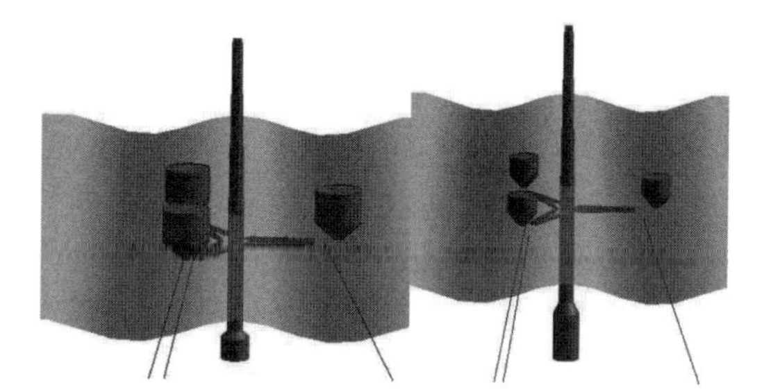

Figure 2. Models used.

Table 1. General data comparison.

	V5.1	V5	
Xg	0.00	0.00	[m]
Yg	0.00	0.00	[m]
Zg	30.95	27.04	[m]
T0	64.53	49.10	[m]
Weight	3904.58	3710.94	[Tons]
zAg	−33.58	−22.06	[m]

parametric feature of the code; its waterplane stability is reduced with respect to the former in order to reduce accelerations.

The parametric approach allows the immediate calculation of dynamic properties of the systems, which are reported here as an example.

Table (1) reports some features characterizing the two models used: the coordinates of the mass centre with respect to the *base point* coordinates frame, the hydrostatic draft, the weight and the initial depth of the centre of mass.

2.2 *V5 model results*

Each simulation produces a set of *time series* for the main variables involved in the problem and some derived from them, which cover a physical interest. The first set of results regards the V5 model in an extreme sea state, characterised by 7.5 m significant wave height with an *encountering wave angle* producing a non symmetrical motion.

Euler's angles express the rotations of the system.

An important physical parameter with respect to structural integrity is the acceleration of the nacelle which can be derived from the v and $\dot{v}$ arrays.

The parameter affecting dynamic forces is the acceleration amplitude (composition of the previous) rather than each single component.

Figure (6) shows clearly strong peaks of acceleration that will cause strong peaks in dynamic

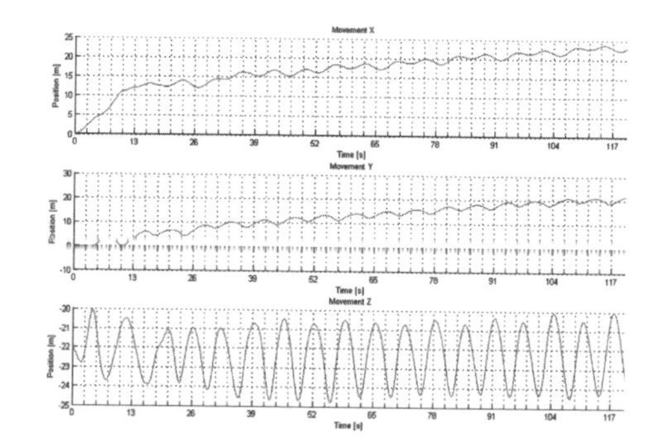

Figure 3. V5 model centre of mass path.

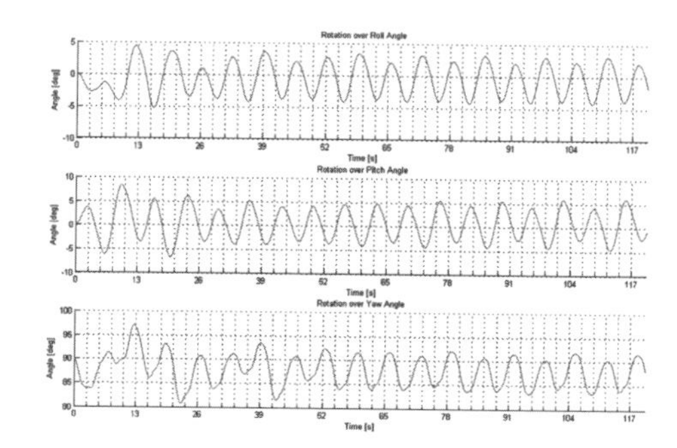

Figure 4. V5 euler's angles time series.

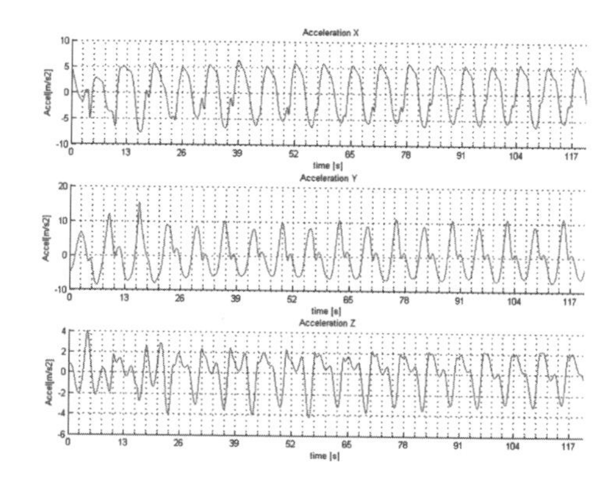

Figure 5. V5 nacelle accelerations (components).

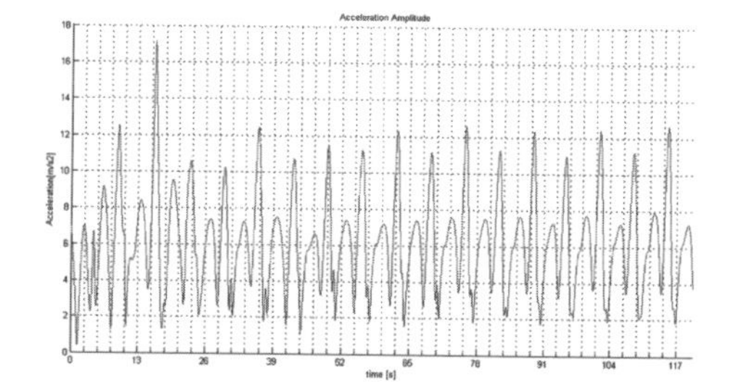

Figure 6. V5 nacelle accelerations (amplitude).

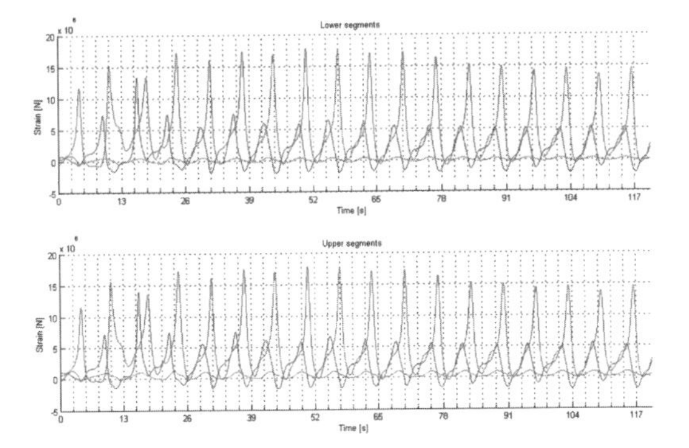

Figure 7. V5 mooring lines internal strain.

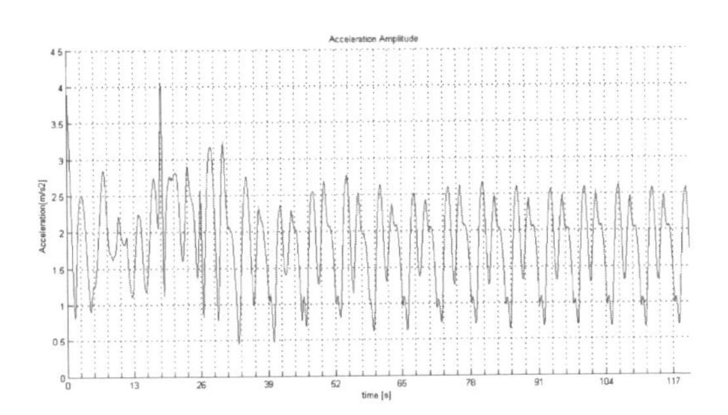

Figure 8. V5.1 nacelle accelerations (amplitude).

forces acting on nacelle components. These are only partially caused by the rigid mooring model, as the next case will demonstrate.

2.3 *V5.1 model results*

This version of the model was elaborated on the basis of the above results using the parametric approach to obtain a softer behaviour with respect to stability. To this aim, the water plane was reduced modifying some of the floaters' parameters. The keel was increased in its length and also the ballast area was modified. The advantage of the parametric approach is that those operations require a very small period and the effect of the modifications on the system can be analysed immediately.

The comparison between the nacelle accelerations for the 2 models shows the effectiveness of the parameters modification, having sensibly decreased their extreme values. Internal strains for mooring lines are also reduced.

The "softness" of the new configuration is transferred also to displacements and rotations.

The amplitude of motions both in translations than in rotations is also reduced with respect to previous configuration.

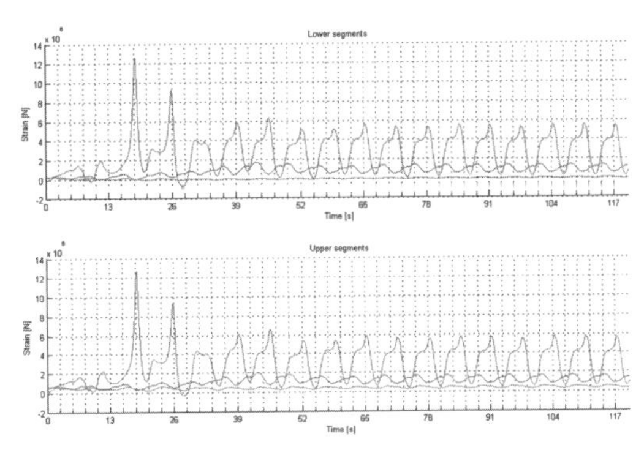

Figure 9. V 5.1 mooring lines internal strain.

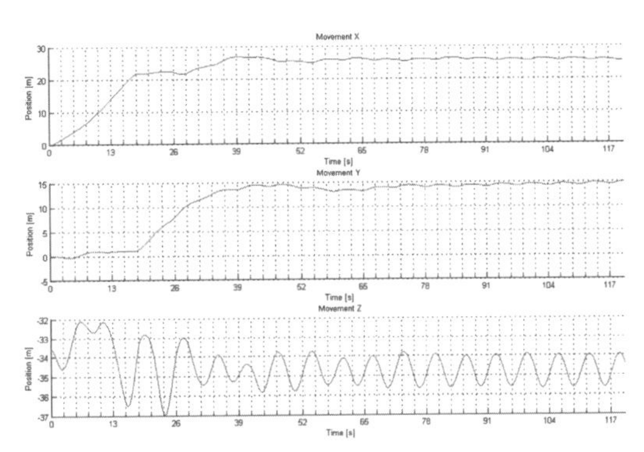

Figure 10. V5.1 centre of mass path.

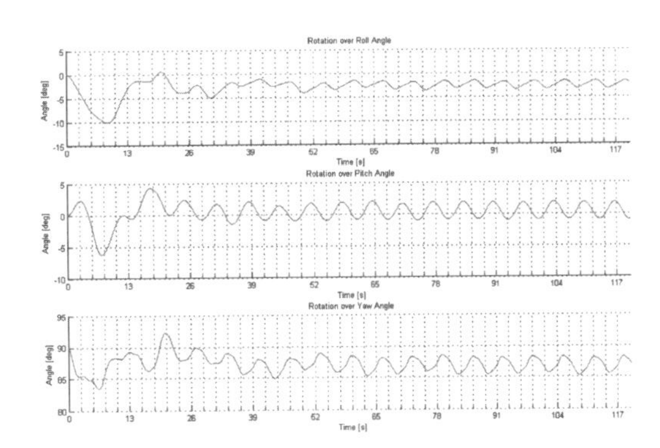

Figure 11. V5.1 euler's angles time series.

3 CONCLUSIONS & FUTURE DEVELOPMENTS

The numerical method presented in this paper, is an encouraging starting point for more sophisticate studies in the floating wind turbines field. The aim in the immediate future is to improve it within a PhD research program, as mentioned in the introduction.

For each issue will be developed a specific module of computer code, with the hope of building up a useful, scalable and flexible tool to face problems regarding floating wind turbines and not only.

The first improvement to be considered concerns mooring lines, with the introduction of a finite element dynamic method, coupled with the dynamics of the floating body. This effort aims giving, a more precise representation of forces that can considerably affect the stability of this family of marine structures. For this it is necessary modeling the highly non linear interaction of a floating body with its mooring lines, including seabed, wave and current effects for a general configuration (both chain than synthetic lines).

Also the hydrodynamic forces can be improved introducing hybrid panel methods which can take into account the diffraction and reflection effects on incoming waves, in order to decrease the error margin when calculating external hydrodynamic forces for all the bodies that do not respect Morison's hypothesis.

Aerodynamic interaction between the turbine and a given wind field can be modelled using the Blade Element Momentum theory [Manwell, McGowan, Rogers 2009]. This technique represents the common solution used for turbine design purposes. It is usually coupled with other empirical formulations that can take into account a series of features such as tip losses, misaligned flows and dynamic stall.

A complete modelling of a floating turbine will require also some considerations about the structural deformations.

The components of wind turbines are exposed to important dynamic loads in a wide range of frequencies. Both blades (in a greater measure) and the tower are flexible bodies and it is clear that their dynamic response can significantly affect the safety of the whole system.

Finally, in order to be efficient and safe, a turbine is equipped with a control system that manages some regulation of its geometry. The most evident is the orientation of the rotor with respect to wind direction, but also the pitch angle of blades is continuously regulated, adapting blades performances to the needs of electricity production system. The braking torque acted by the generator to extract energy is also controlled by the system. All these actions on the turbine modify a lot the forces acting on it and so are very important in a stability glance. The control system is then a fundamental part of the simulation process and its interaction with the turbines dynamics can't be neglected.

The coupling of the structural problems to control issues, as evaluation of hydrodynamic and aerodynamic forces seems worth being investigated.

Finally the design of Offshore Wind Turbines is characterized by very difficult issues.

The recent introduction of this technology causes a lack in proven design methods and experimental results analysis to validate them.

For this reason it is believed that a significant scientific effort is to be dedicated in this field, possibly melting different competences, from both the academic than industrial field.

ACKNOWLEDGMENTS

The work has been developed and based on the initial idea of the floating structure proposed by eng. Mongiardino. The authors wish to thank him.

REFERENCES

Det Norske Veritas, 2007. Recommended Practice DNV-RP-C205 Environmental Conditions and Environmental Loads.

Fossen, I. Thor. 1994. *Guidance and Control of Ocean Vehicles*. Chichester: John Wiley & sons Ltd.

Fossen, I. Thor. 2002. *Marine Control System*, Marine Cybernetics.

Manwell, J.F., McGowan, J.G. & Rogers, A.L. 2009. *Wind Energy Explained*. Chichester: John Wiley & sons Ltd.

Masi, M. 2011. Offshore Wind Generators Dynamics. MsC thesis in Naval Architecture, University of Genova.

Perez, T. & Fossen, I. Thor. 2007. *Kinematic Models for Manoeuvring and Seakeeping of Marine Vessels*, Modeling Identification and Control, Vol. 28, No. 1, pp. 19–30.

Sustainable Maritime Transportation and Exploitation of Sea Resources – Rizzuto & Guedes Soares (eds)

Hydrodynamics of a floating oscillating water column device

S.A. Mavrakos & D. Konispoliatis

Laboratory for Floating Bodies and Mooring Systems, School of Naval Architecture and Marine Engineering, National Technical University of Athens, Zografos, Greece

ABSTRACT: The geometric configuration of an Oscillating Water Column device (OWC) consists of a vertical cylinder party submerged as an open- bottom chamber in which air is trapped above the inner water surface. Forced by incident waves from any direction, the water surface inside the chamber pushes the dry air above through a Wells turbine system to generate power. In the present contribution, a formulation of the hydro-mechanic problem of a floating OWC structure is given, which is restrained in the presence of regular waves. Two types of first-order boundary value problems are investigated to represent the potential of the flow field inside the chamber. One diffraction problem with atmospheric pressure inside the OWC and one radiation problem, which is introduced to account for the oscillating pressure head above the interior water surface.

1 INTRODUCTION

Renewable energy technology is steadily gaining importance in the world energy market due to the limited nature of fossil fuel supplies, national requirements for security of supply, as well as political pressure toward the reduction of carbon emissions. In the last years considerable efforts and advances have been made world-wide for the wave energy exploitation and some technologies attaining a level of maturity with several commercial companies conducting sea trials with large- and full-scale prototypes. Among several classes of designs proposed for the wave energy conversion the Oscillating Water Column device (OWC) has received considerable theoretical attention (Evans, 1982; Sarmento et al., 1990; Evans and Porter, 1995, 1996 Falcao and Justino, 1999).

These kinds of wave power devices have been mostly installed near shore or onshore in order to maintain the energy storage and transmission cost low. Recently, some theoretical studies on OWC devices mounded at the tip of a breakwater have been made to examine the amount of absorbed wave energy (Martin-rivas and Mei, 2009)

In this article, a general formulation of the hydro-mechanical problem is presented for a three dimensions (3D) vertical OWC device restrained in regular waves. Numerical results will be given from the solution of two boundary–value problems, namely the diffraction problem (body fixed in wave, atmospheric pressure inside of the OWC) and the radiation problem resulting from an oscillating pressure head acting on the inner free- surface of the OWC. The developed method is based on matched axisymmetric eigenfunction expansions of the velocity potential in properly defined ring-shaped fluid regions around the device and could be considered as an extension of the methods employed by Miles and Gilbert (1968), Garrett (1971) and Yeung (1981) as far as a circular dock is concerned, by Mavrakos (1985, 1988) for bottomless cylinders, by Kokkinowrachos et al. (1986) for the hydrodynamic analysis (diffraction and radiation problem) of arbitrary shaped vertical bodies of revolution. In particular, the method extends the formulation given by Evans and Porter (1996) to account for finite wall thickness of the vertical bottomless cylindrical chamber and the related 2D formulation by Kokkinowrachos et al. (1987) to the case of vertical cylindrical ducts. Numerical results concerning the horizontal and vertical exciting wave forces along with the pressure inside the oscillating chamber and the absorbed power by the device are presented and compared with those of other investigators. Moreover, the volume flow through the turbine is being presented in order to calculate the pressure inside the chamber. These can be used for the evaluation and optimization of technical concepts for wave energy conversion.

2 FORMULATION OF THE PROBLEM

A vertical cylindrical OWC device is investigated considered restrained in the presence of a regular wave train. The sea depth, d, is assumed to be constant. A plane harmonic wave of amplitude $H/2$, frequency ω and wave number k, is incident on the device having internal and external radii equal to a and b, respectively. The device is immersed to a depth h (Fig. 1). We assume small amplitude, inviscid, incompressible and irrotational flow. Cylindrical coordinates (r,θ,z) are introduced with the vertical axis Oz directed upwards and the origin at the sea bed. For the OWC device, we expect the internal free surface of the device to be subjected to an oscillating pressure $P_{in}(t)$ having the same frequency, ω, as the incident wave.

The time harmonic complex velocity potential of the flow field around the structure can be expressed as:

$$\Phi(r,\theta,z;t) = \mathrm{Re}\left[\varphi(r,\theta,z)\cdot e^{-i\omega t}\right] \tag{1}$$

with

$$\varphi = \varphi_0 + \varphi_7 + p_{in0}\varphi_P = \varphi_D + p_{in0}\varphi_P \tag{2}$$

where, φ_0 is the potential of the undisturbed incident wave; φ_7 is the scattered potential for the body fixed in the wave with the duct open to the atmosphere, i.e. for a pressure in the chamber equal to the atmospheric one; φ_P is the radiation potential resulting from an oscillating pressure head P_{in} in the chamber for the body fixed in otherwise calm water, p_{in0} being the amplitude of the pressure head. Due to the time harmonic excitation, it holds $P_{in} = \mathrm{Re}\left(p_{in0}\cdot e^{-i\omega t}\right)$.

The undisturbed incident wave potential, ω_0, can be expressed in cylindrical co-ordinates as:

$$\varphi_0(r,\vartheta,z) = -i\omega\left(\frac{H}{2}\right)\sum_{m-\infty}^{\infty}\varepsilon_m i^m \Psi_{0,m}(r,z)\cos(m\vartheta) \tag{3}$$

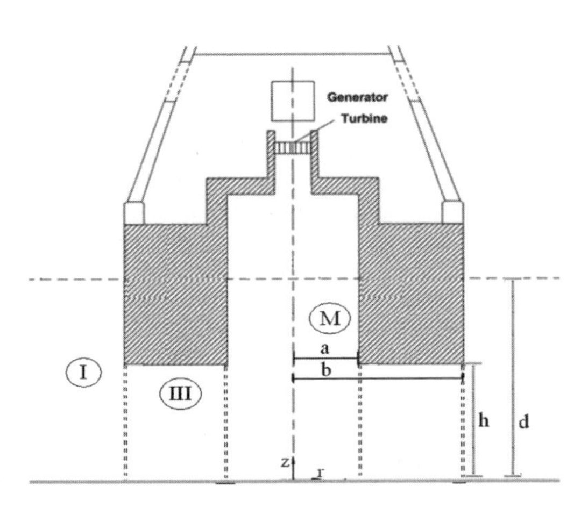

Figure 1. Typical floating oscillating water column device. Ring elements definition.

with:

$$\frac{1}{d}\Psi_{0,m}(r,z) = \frac{Z_0(z)}{dZ_0'(d)}J_m(kr) \tag{4}$$

Here, J_m is the m-th order Bessel function of the first kind, ε_m the Neumann's symbol defined as $\varepsilon_0 = 1$ and $\varepsilon_m = 2$ for $m \geq 1$ and $Z_0(z)$:

$$Z_0(z) = \left[\frac{1}{2}\left[1+\frac{\sinh(2kd)}{2kd}\right]\right]^{-1/2}\cosh(kz) \tag{5}$$

with $Z'_0(d)$ being its derivative at $z = d$. Frequency ω and wave number k are related by the dispersion equation:

$$\omega^2 = gk\tanh(kd) \tag{6}$$

In accordance to Equation 2, the total diffracted potential, φ_D, around the restrained structure can be written in the form:

$$\varphi_D(r,\theta,z) = -i\omega\frac{H}{2}\sum_{m=0}^{\infty}\varepsilon_m i^m \psi_{D,m}(r,z)\cos(m\theta) \tag{7}$$

As for the radiation potential φ_P, we note that due to the high sound speed in air and the low frequency of sea waves, the air pressure can be assumed to be spatially uniform throughout the axisymmetric chamber. Thus, since the forcing of the internal free surface is independent of θ, the radiation potential φ_P will be axially symmetric and so it will include only the $m=0$ angular mode, i.e., it holds:

$$\varphi_P(r,\vartheta,z) = \Psi_{P,0}(r,z) \tag{8}$$

The velocity potentials φ_j, $(j=0,7,p)$ have to satisfy the Laplace equation

$$\Delta\varphi = \frac{\partial^2\varphi}{\partial r^2}+\frac{1}{r}\frac{\partial\varphi}{\partial r}+\frac{1}{r}\frac{\partial^2\varphi}{\partial\theta^2}+\frac{\partial^2\varphi}{\partial z^2} = 0 \tag{9}$$

within the entire fluid domain and the linearized boundary condition at the outer and inner free sea surface (Evans and Porter, 1995, 1996), i.e.,

$$\omega^2\varphi - g\frac{\partial\varphi}{\partial z} = \begin{cases} 0 & for \quad r \geq b \\ i\omega/\rho & for \quad 0 \leq r \leq a \end{cases} \tag{10}$$

on the free surface $(z=d)$

Furthermore, the velocity potentials φ_j, $(j=0,7,p)$, have to satisfy following kinematic boundary conditions on the sea bed and on the wetted surface of the body, i.e.,

$$\frac{\partial\varphi}{\partial z} = 0 \text{ on the sea bed, } z = 0 \tag{11}$$

$$\frac{\partial\varphi}{\partial z} = 0 \text{ on the device's bottom, } z = h, a \le r \le b \quad (12)$$

$$\frac{\partial\varphi}{\partial r} = 0 \text{ on the walls, } r = a \text{ and } r = b, h \le z \le d \quad (13)$$

We also require the scattered potential to satisfy an appropriate radiation condition as $r \to \infty$, which has the form (Sommefeld, 1948):

$$\lim_{r\to\infty} \sqrt{r}\left(\frac{\partial}{\partial r} - ik\right)(\varphi - \varphi_0) = 0 \quad (14)$$

Moreover, both the velocity potential and its derivatives must be continuous at the boundaries $r = a$ and $r = b$ of the neighbouring ring elements shown in Fig. 1. This results in (for $j = D, P$):

$$\psi^{I}_{j,m}(b,z) = \psi^{III}_{j,m}(b,z) \text{ for } 0 \le z \le h \quad (15)$$

$$\left.\frac{\partial\psi^{I}_{j,m}}{\partial r}\right|_{r=b} = \left.\frac{\partial\psi^{III}_{j,m}}{\partial r}\right|_{r=b} \text{ for } 0 \le z \le h \quad (16)$$

$$\psi^{III}_{j,m}(a,z) = \psi^{M}_{j,m}(a,z) \text{ for } 0 \le z \le h \quad (17)$$

$$\left.\frac{\partial\psi^{III}_{j,m}}{\partial r}\right|_{r=a} = \left.\frac{\partial\psi^{M}_{j,m}}{\partial r}\right|_{r=a} \text{ for } 0 \le z \le h \quad (18)$$

The superscripts *I, III, M* imply quantities corresponding to respective types of ring-elements. Starting with the method of separation of variables for the Laplace differential equation appropriate expressions for the velocity potentials, $\Psi^{i}_{j,m}$, in each fluid domain (see Fig. 1; $i = I, III, M$) can be established (Garret, 1971; Mei, 1983; Mavrakos, 1985). These expressions satisfy the corresponding conditions at the horizontal boundaries of each fluid region and, in addition, the radiation condition at infinity in the outer fluid domain. As a result, the velocity potentials in each fluid domain fulfill a priori the kinematical boundary conditions at the horizontal walls of the bottomless cylindrical duct, the linearized condition on the free—surface, the kinematical one on the sea bed and the radiation condition at infinity.

The diffraction problem for the determination of φ_D in the case of an open vertical axisymmetric bottomless duct with finite wall thickness has been presented by Mavrakos (1985) and thus, it will be no further elaborated here.

Thus, in the following we will be dealing only with the radiation problem around the vertical cylindrical duct due to an imposed internal forcing pressure.

The series representations of the functions $\Psi^{i}_{P,0}$ (i = *I, III, M*) for the wave potentials in each fluid region around the vertical OWC device are given by:

For type element *I*:

$$\frac{1}{d}\psi^{I}_{P,0}(r,z) = F^{I}_{P,0}\frac{H_0(a_j r)}{H_0(a_j b)}Z_0(z) + \sum_{j=1}^{\infty} F^{I}_{P,j}\frac{K_0(a_j r)}{K_0(a_j b)}Z_j(z) \quad (19)$$

where H_0 and K_0 is the 0-th order Hankel function of the first kind and the modified Bessel function of the second type, respectively, and $F^{I}_{P,j}$ are unknown Fourier coefficients to be determined by the solution procedure. The first term in the series expansion 19 behaves as outgoing wave at large r satisfying the radiation condition, see Eq. 14. For $j \ge 1$, the remaining terms in Eq. 19 represent evanescent waves exponentially decaying at large r. Moreover, $Z_j(z)$ are orthonormal functions in [0, d] defined by Eq. 5 for $j = 0$ and:

$$Z_j(z) = \left\{\frac{1}{2}\left[1 + \frac{\sin(2\alpha_j d)}{2\alpha_j d}\right]\right\}^{-1/2} \cos(\alpha_j z), \quad j \ge 1 \quad (20)$$

The eigenvalues α_j are roots of the transcendental equation:

$$\omega^2 + g\alpha_j \tan(\alpha_j d) = 0 \quad (21)$$

which possesses one imaginary, $\alpha_0 = -ik$, $k > 0$, and infinite number of real roots. Substituting the value of α_0 in Equations 20 and 21, Equations 5 and 6 can directly be obtained.

For type element *III*:

$$\frac{1}{d}\psi^{III}_{P,0}(r,z) = \sum_{n=0}^{\infty} \varepsilon_n [R^{III}_{0n}(r)F^{III}_{P,n} + R^{*III}_{0n}(r)F^{*III}_{P,n}] \times \cos\left(\frac{n\pi z}{h}\right) \quad (22)$$

where $\varepsilon_0 = 1$, $\varepsilon_n = 2$, $n \ge 1$, in which

$$R^{III}_{0n}(r) = \frac{K_0(\frac{n\pi a}{h})I_0(\frac{n\pi r}{h}) - I_0(\frac{n\pi a}{h})K_0(\frac{n\pi r}{h})}{I_0(\frac{n\pi b}{h})K_0(\frac{n\pi a}{h}) - I_0(\frac{n\pi a}{h})K_0(\frac{n\pi b}{h})} \quad (23)$$

$$R^{III}_{00}(r) = \ln\left(\frac{r}{a}\right) \Big/ \ln\left(\frac{b}{a}\right), \quad n = 0 \quad (24)$$

$$R^{*III}_{0n}(r) = \frac{I_0(\frac{n\pi b}{h})K_0(\frac{n\pi r}{h}) - K_0(\frac{n\pi b}{h})I_0(\frac{n\pi r}{h})}{I_0(\frac{n\pi b}{h})K_0(\frac{n\pi a}{h}) - I_0(\frac{n\pi a}{h})K_0(\frac{n\pi b}{h})} \quad (25)$$

$$R_{00}^{*III}(r)=\ln\left(\frac{\mathrm{b}}{r}\right)\Big/\ln\left(\frac{\mathrm{b}}{\mathrm{a}}\right),\quad n=0 \tag{26}$$

For type element M:

$$\frac{1}{d}\psi_{P,0}^{M}(r,z)=\frac{p_{0m}}{\omega^2\rho}\left(1+\sum_{j=0}^{\infty}F_{P,j}^{M}\frac{I_0(a_j r)}{I_0(a_j \mathrm{a})}Z_j(z)\right) \tag{27}$$

Here, the orthonormal functions $Z_j(z)$, are defined by Eqs. 5 and 20, for $j=0$ and $j\geq 1$, respectively and α_j are given by Eq. 21. The potential functions $\Psi_{P,0}^{i}$, expressed through Equations 19, 22 and 27, have the advantage that on the boundaries $r=b$ and $r=\mathrm{a}$ can be described by simple Fourier series for all the types of ring elements. Moreover, the form of the selected solutions for the functions $\Psi_{P,0}^{i}$ is such that the boundary conditions at all the horizontal boundaries of the elements are satisfied.

The kinematic conditions at the pond's vertical walls, Equation 13, as well as the requirement for continuity of the potential and its first derivative, Equations 15–18, at the vertical boundaries of neighboring elements remain to be fulfilled. Expressing these conditions, the system of equations for the unknown Fourier coefficients is obtained in the next paragraph. Once these coefficients have been calculated, the functions $\Psi_{P,0}^{i}(r,z)$ $(i=I,III,M)$ and hence the velocity potential for all fluid regions can be obtained.

3 SOLUTION

The continuity requirements which must be fulfilled by the potential functions are described by Equations 17, 18, 13 and 15, 16, 13 for the boundaries $r=\mathrm{a}$ and $r=b$, respectively, and lead to the following relations:

Boundary $r=\mathrm{a}$

$$F_{P,n}^{*III}=\sum_{j=0}^{\infty}\left(R_n+L_{nj}F_{P,j}^{M}\right)\ \text{for}\ 0\leq z\leq h \tag{28}$$

$$\sum_{j=0}^{\infty}F_{P,j}^{M}\mathrm{A}_{0j}^{M}Z_j(z)=\sum_{n=0}^{\infty}\varepsilon_n\left[F_{P,n}^{III}D_{0n}^{III}+F_{P,n}^{*III}D_{0n}^{*III}\right]\times\cos\left(\frac{n\pi r}{h}\right)\ \text{for}\ 0\leq z\leq h \tag{29}$$

$$\sum_{j=0}^{\infty}F_{P,j}^{M}\mathrm{A}_{0j}^{M}Z_j(z)=0\ \text{for}\ h\leq z\leq d \tag{30}$$

Where

$$R_n=\frac{1}{h}\int_0^h\cos\left(\frac{n\pi z}{h}\right)dz=\begin{cases}0 & n=0\\ 1 & n\neq 0\end{cases} \tag{31}$$

$$L_{nj}=\frac{1}{h}\int_0^h Z_j(z)\cos\left(\frac{n\pi z}{h}\right)dz=(-1)^n N_j^{-1/2}\frac{a_j h}{a_j^2h^2-n^2\pi^2}\sin(a_j h) \tag{32}$$

when $j\geq 1$

$$L_{n0}=(-1)^n N_0^{-1/2}\frac{kh}{k^2h^2+n^2\pi^2}\sinh(kh) \tag{33}$$

when $j=0$

$$\mathrm{A}_{0j}^{M}=a_j\mathrm{a}\frac{I_0'(a_j\mathrm{a})}{I_0(a_j\mathrm{a})} \tag{34}$$

$$D_{0n}^{III}=\mathrm{a}\frac{\partial R_{0n}^{III}}{\partial r},\ \text{for}\ r=\mathrm{a} \tag{35}$$

$$D_{0n}^{*III}=\mathrm{a}\frac{\partial R_{0n}^{*III}}{\partial r},\ \text{for}\ r=\mathrm{a} \tag{36}$$

Boundary $r=\mathrm{b}$

$$F_{P,n}^{*III}=\sum_{i=0}^{\infty}L_{ni}F_{P,i}^{I}\ \text{for}\ 0\leq z\leq d \tag{37}$$

$$\sum_{n=0}^{\infty}\varepsilon_n\left[F_{P,n}^{III}A_{0n}^{III}+F_{P,n}^{*III}A_{0n}^{*III}\right]\cos\left(\frac{n\pi r}{h}\right)=B_0Z_0(z)+\sum_{i=0}^{\infty}F_{P,i}^{I}\mathrm{A}_{0i}^{I}Z_i(z)\ \text{for}\ 0\leq z\leq d \tag{38}$$

$$B_0Z_0(z)+\sum_{i=0}^{\infty}F_{P,i}^{I}\mathrm{A}_{0i}^{I}Z_i(z)=0\ \text{for}\ h\leq z\leq d \tag{39}$$

where

$$\mathrm{A}_{0i}^{I}=a_i\mathrm{b}\frac{K_0'(a_i\mathrm{b})}{K_0(a_i\mathrm{b})} \tag{40}$$

$$B_0=-\frac{2i}{\pi d H_0(kb)Z_0'(z)} \tag{41}$$

$$A_{0n}^{III}=\mathrm{b}\frac{\partial R_{0n}^{III}}{\partial r},\ \text{for}\ r=\mathrm{b} \tag{42}$$

$$A_{0n}^{*III}=\mathrm{b}\frac{\partial R_{0n}^{*III}}{\partial r},\ \text{for}\ r=\mathrm{b} \tag{43}$$

If we substitute Equation 28 in Equation 29, multiply both the resulting expression and Equation 30 by $1/d\,Z_p(z)$ integrate each over its region of validity and add, we obtain:

$$\sum_{j=0}^{\infty}F_{P,j}^{M}\left(\mathrm{A}_{0j}^{M}\delta_{jp}-\frac{h}{d}\sum_{n=0}^{\infty}\varepsilon_nL_{nj}D_{0n}^{*III}L_{np}\right)=\frac{h}{d}\sum_{n=0}^{\infty}\varepsilon_nF_{P,n}^{III}D_{0n}^{III}L_{np} \tag{44}$$

where

$$\delta_{jp} = \begin{cases} 1 & j = p \\ 0 & j \neq p \end{cases} \quad (45)$$

Following the same procedure with Equations 37–39 we can obtain:

$$B_0\delta_{i0} + \sum_{i=0}^{\infty} F_{P,i}^{I}\left(A_{0j}^{I}\delta_{ip} - \frac{h}{d}\sum_{n=0}^{\infty}\varepsilon_n L_{ni} A_{0n}^{III} L_{np} \right) = \frac{h}{d}\sum_{n=0}^{\infty}\varepsilon_n F_{P,n}^{*III} A_{0n}^{*III} L_{np} \quad (46)$$

The solution of the infinite system given by linear Equations 28, 37, 44 and 46 will provide the unknown Fourier coefficients for each type of ring element.

4 VOLUME FLOW

During the water oscillation inside the chamber the dry air above the free surface is being pushed through a Wells turbine. The volume flow produced by the oscillating internal water surface is given by:

$$Q(t) = \iint_S u_z dS = \iint_S u(r,\theta,z=h) r dr d\theta = \iint_S \frac{\partial \Phi}{\partial z} r dr d\theta \quad (47)$$

where $Q(t) = \mathrm{Re}\{q e^{-i\omega t}\}$ and S = the inner water surface.

It is convenient to decompose the total volume flow into three terms, as follows:

$$q = q_E + q_P - q_A \quad (48)$$

where q_E we have introduced the volume flow associated with the diffraction problem:

$$q_E = \iint_S \frac{\partial}{\partial z}(\varphi_0 + \varphi_7) dS \quad (49)$$

q_P is induced volume flow inside the chamber due to the difference between the atmospheric and the inner pressure:

$$q_P = p_{in_0}\iint_S \frac{\partial \varphi_P}{\partial z} dS \quad (50)$$

and q_A the volume flow through the turbine. It is assumed that the volume flow through the turbine is proportional to the inner pressure, i.e.,

$$q_A = g_T \cdot p_{in0} \quad (51)$$

where g_T is a turbine parameter. For a general definition of the turbine parameter g_T the reader is referred to the work by Sarmento and Falcao (1985). Here, it is assumed to be a real number.

The volume flow associated with the diffraction problem, from Equation 49, can be obtained:

$$q_E = (-i\omega)\frac{\omega^2}{g}\frac{H}{2} d 2\pi b\left(F_{D0,0}^{M}\frac{J_1(kb)}{kJ_0(kb)} N_0^{-1/2}\cosh(kh) + \sum_{j=1}^{\infty} F_{D0,j}^{M}\frac{I_1(a_j b)}{a_j I_0(a_j b)} N_j^{-1/2}\cos(a_j h) \right) \quad (52)$$

Where $F_{D0,j}^{M}$ are the Fourier coefficients for the M-th type of ring element, for the body fixed in the wave with the duct open to the atmosphere. These coefficients have been calculated by Mavrakos (1985).

The volume flow inside the chamber due to the difference between the atmospheric and the inner pressure, from Equation 50, is:

$$q_P = p_{in_0}(-i\omega)\frac{1}{g} 2\pi b\left(F_{P,0}^{M}\frac{J_1(kb)}{kJ_0(kb)} N_k^{-1/2}\cosh(kh) + \sum_{j=0}^{\infty} F_{P,j}^{M}\frac{I_1(a_j b)}{a_j I_0(a_j b)} N_j^{-1/2}\cos(a_j h) \right. \quad (53)$$

Where $F_{P,j}^{M}$ are the Fourier coefficients for the M-th type of ring element, for the body fixed in the wave with inter pressure different from the atmospheric one.

5 PRESSURE CALCULATION

The consideration of mass conservation in terms of volume flow can be written as

$$q_E + q_P - q_A + \frac{\partial V}{\partial t} = 0 \quad (54)$$

where V = the air volume in the chamber.

Here the assumption is made that the air inside the chamber behaves according to the adiabatic law, i.e.

$$PV^{\gamma} = P_0 V_0^{\gamma} = ct \quad (55)$$

where P_0 = the atmospheric pressure, V_0 = the initial air volume in the chamber and $\gamma = 1.4$ adiabatic constant.

The above Equation 55, after Taylor analysis, can be written as:

$$-\frac{p_{\text{in}_0}}{P_0} = -\gamma \frac{\dot{V}_{in}}{V_0} \tag{56}$$

If we substitute Equations 50, 51 and 54 in Equation 56, the complex amplitude of the internal pressure head p_{in_0} can be obtained as:

$$p_{in_0} = \frac{q_E}{-i\omega\left(\frac{V_0}{\gamma P_0}\right) - \iint_S \frac{\partial \varphi_P}{\partial z} dS + g_T} \tag{57}$$

Where the volume flow q_E is defined by Equation 52 and g_T is the turbine parameter, see Equation (51). The inner pressure p_{in_0} can be obtained as the solution of the Equation 57.

6 WAVE FORCES

The various forces on the oscillating water column device have been calculated from the pressure distribution given by the linearised Bernoulli's equation:

$$P(r,\theta,z;t) = -\rho\frac{\partial\Phi}{\partial t} = i\omega\rho\varphi\, e^{-i\omega t} \tag{58}$$

Where φ is the velocity potential in each fluid domain *I, III, M*.

The oscillating pressure inside the chamber does not affect the horizontal exciting force on the floating device since the radiation potential includes only the $m=0$ angular mode. So, the horizontal force on the OWC device is the same as on an open vertical axisymmetric bottomless duct with finite wall thickness and is given by (Mavrakos, 1985):

$$\begin{aligned}\frac{F_x}{B} = &-\frac{2ikd\tanh(kd)}{\text{b}}\left[\frac{g}{\omega^2}\frac{1}{\cosh(kd)}J_1(kb)\int_h^d \cosh(kz)dz\right.\\ &+N_k^{-1/2}\left(F_{D1,k}^I - \frac{J_1(k\text{b})}{dz'_{j=0}(d)}\right)\int_h^d \cosh(kz)dz\\ &\left.+\sum_{j=1}^{\infty} N_j^{-1/2}F_{D1,j}^I\int_h^d \cos(a_j z)dz\right]\\ &+\frac{kd\tanh(kd)\text{a}}{\text{b}^2}\left[2iF_{D1,k}^M N_k^{-1/2}\int_h^d \cosh(kz)\,dz\right.\\ &\left.+\sum_{j=0}^{\infty} 2iF_{D1j}^M N_j^{-1/2}\int_h^d \cos(a_j z)\,dz\right]\end{aligned} \tag{59}$$

Where $B = \pi\rho g\text{b}^2(H/2)$ and $F_{D1,j}^I, F_{D1j}^M$ are the Fourier coefficients for the *I*-th and *M*-th type of ring element, for the body fixed in the wave with the duct open to the atmosphere. These coefficients have been calculated by Mavrakos (1985). The oscillating pressure inside the chamber does affect the vertical force due to unit internal pressure head, on the floating device and is given by:

$$F_z = 2\pi\rho\omega^2 d\frac{H}{2}\left[\int_{\text{a}}^{b}\frac{1}{d}\psi_{P,0}^{III}(r,h)rdr\right] \tag{60}$$

Substituting Equation 22 in Equation 61 we have:

$$\begin{aligned}\frac{F_z}{B} = 2kd\tanh(kd)&\left\{\frac{1}{2}F_{P,0}^{III}\left(1-\frac{1-(\text{a/b})^2}{2\ln(\text{b/a})}\right)-\frac{1}{2}F_{P,0}^{*III}\right.\\ &\times\left(\frac{\text{a}^2}{\text{b}^2}-\frac{1-(\text{a/b})^2}{2\ln(\text{b/a})}\right)+2\sum_{n=1}^{\infty}(-1)^n\left(\frac{h}{n\pi\text{b}}\right)\\ &\left.\times\left[F_{P,n}^{III}(A_{0n}^{III}-D_{0n}^{III})+F_{P,n}^{*III}(A_{0n}^{*III}-D_{0n}^{*III})\right]\right\}\end{aligned} \tag{61}$$

where $B = \pi\rho g\text{b}^2(H/2)$ and D_{0n}^{III}, D_{0n}^{*III}, A_{0n}^{III}, A_{0n}^{*III} are defined in previous session. The $F_{P,n}^{III}$, $F_{P,n}^{*III}$ are the Fourier coefficients for the *III*-th due to different inter pressure, on the floating device.

The total vertical force on the OWC device is equal to the sum of the vertical force on the device when the duct is open to the atmosphere (diffraction load) and the vertical force due to unit internal pressure head, Equation 61, multiplied by the amplitude of the pressure head p_{in0}.

7 NUMERICAL RESULTS

The calculation of the Fourier coefficients $F_{P,mi}^I$, $F_{P,mn}^{III}, F_{P,mn}^{*III}, F_{P,mj}^M$ is the most significant part of the numerical procedure, because of their influence on the accuracy of solution. For the case of the first and the *M-th* ring element $i = j = 20$ terms were used, while for the third ring element $N = 50$, since it was found that the results obtained for those values, were correct to an accuracy of within 1%.

Since the pressure contribution inside the chamber affects only the vertical forces exerted on the device, the horizontal exciting force coincides with the one computed by Mavrakos (1985) for a device open to the atmosphere. So we examined a body with immersion $h/\text{a} = 3.46$ and water depth $d/\text{a} = 4$ for three cases of wall thickness $b/\text{a} = 4$, $b/\text{a} = 2$, $b/\text{a} = 1{,}5$. In Fig. 2 the horizontal exciting force on the OWC is being tested with the results of a cylindrical moon pool having the

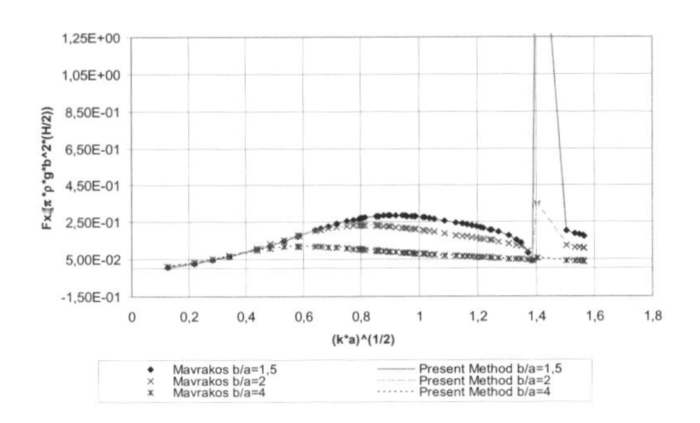

Figure 2. Horizontal exciting force on the OWC device for different cases of wall thickness.

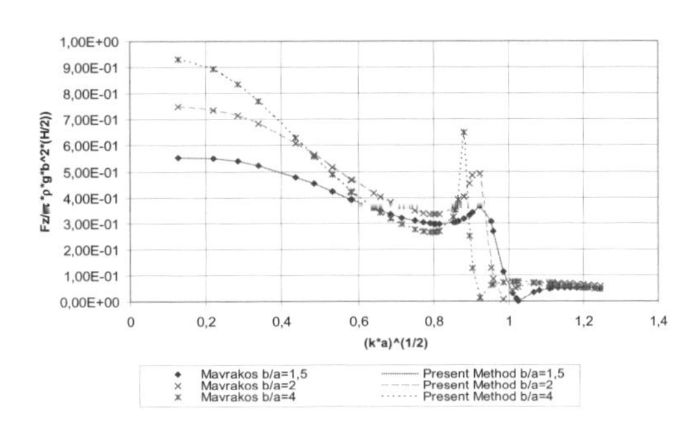

Figure 3. Vertical exciting force on the OWC device for different cases of wall thickness.

same dimensions as the OWC's ones. In Fig. 3 the vertical exciting force on the OWC device is compared with the results by Mavrakos (1985) for the moon pool body when the inner pressure is equal to the atmospheric.

Then, we investigated a three-dimensional oscillating water column device which is floating in a water depth $d = 40m$ with a chamber's high equal to $10m$ high. The OWC's internal and external radii are $15m$ and $20m$, respectively. The pressure head p_{in0} in the chamber differs from the atmospheric one and thus, a vertical force (modulus, phase) on the device due to the inner pressure is exerted. It is plotted in Figs.4 and 5 for a unit pressure head p_{in0}.

In Fig. 6, Fig. 7 and Fig. 8 results for the above OWC device are given for three values of turbine parameter $g_T = 1,\ 3, 6\ [m^4/(kN\,\mathrm{sec})]$. From Equation 58 we have assumed that $V_0/\gamma P_0 = 4.3m^4/kN$. In these figures the transfer functions of the pressure of the air trapped in the chamber (modulus and phase) and the time-averaged absorbed power output to the turbine, $\dot{\bar{E}} = 1/2\,\mathrm{Re}\{|p_{in0}^2| \cdot g_T\}$, are plotted.

In Figs. 9 and 10 the modulus and the phase of the total vertical force, consisting of the superposition of the diffraction and the radiation force on the device is given for a turbine parameter $g_T = 1$ $m^4/(kN\,\mathrm{sec})$.

Finally, we examined a three-dimensional oscillating water column device with very low wall thickness -5 cm- in order to compare the volume flow inside the chamber, with the results of Evans and Porter (1996). The device had a draught $(d-h)/d = 1/2$ and various duct radii $b/d = 1/8, 1/4, 1/2, 1$. In Figs. 11 and 12, the radiation conductance, ν, and susceptance μ, coefficients, defined by $\nu = -(1/\pi b^2)\mathrm{Im}\{q_P\}$, and $\mu = -(1/\pi b^2)\mathrm{Re}\{q_P\}$, respectively, are plotted in comparison with the results of Evans and Porter (1996) with a good coincidence.

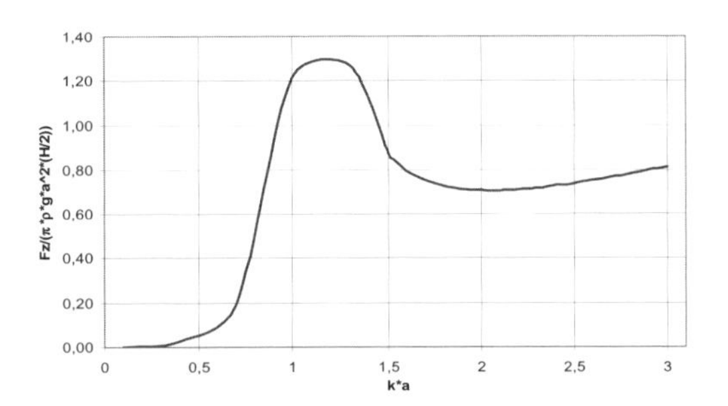

Figure 4. Modulus of the vertical exciting force on the OWC device due to internal pressure.

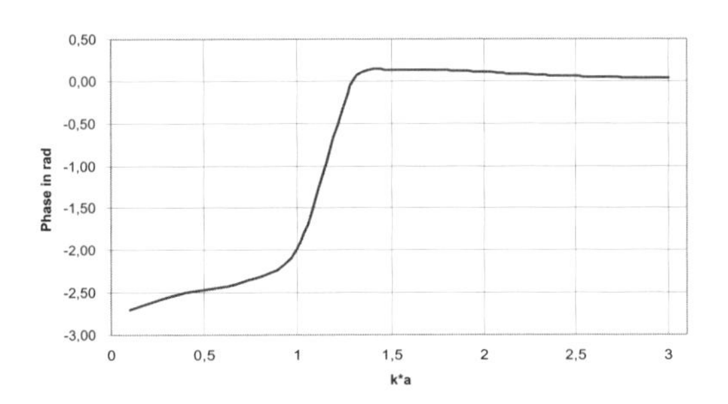

Figure 5. Phase of the vertical exciting force on the OWC device due to inter pressure.

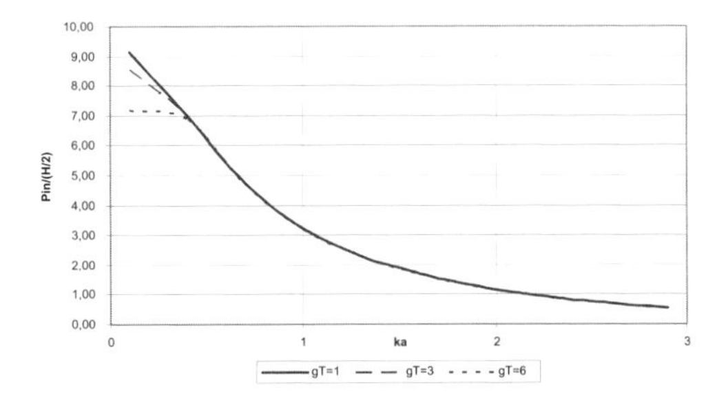

Figure 6. Inner pressure (modulus) for three values of the turbine parameter ($g_T = 1, 3, 6$).

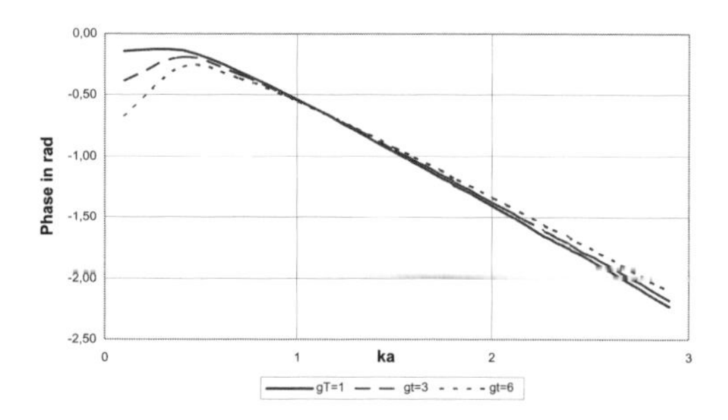

Figure 7. Inner pressure (phase) for three values of the turbine parameter (g_T = 1, 3, 6).

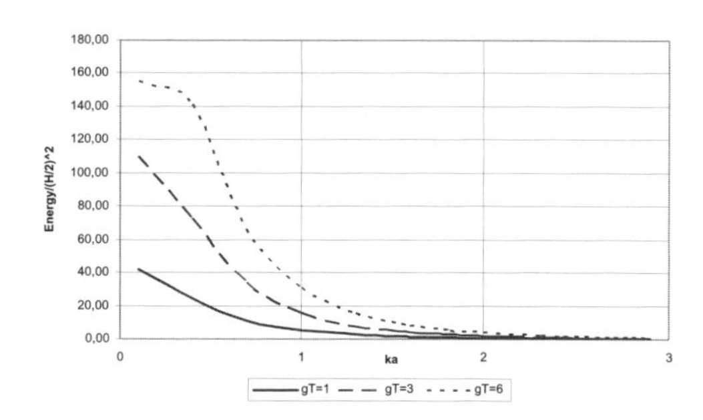

Figure 8. Absorbed power for three values of the turbine parameter (g_T = 1, 3, 6).

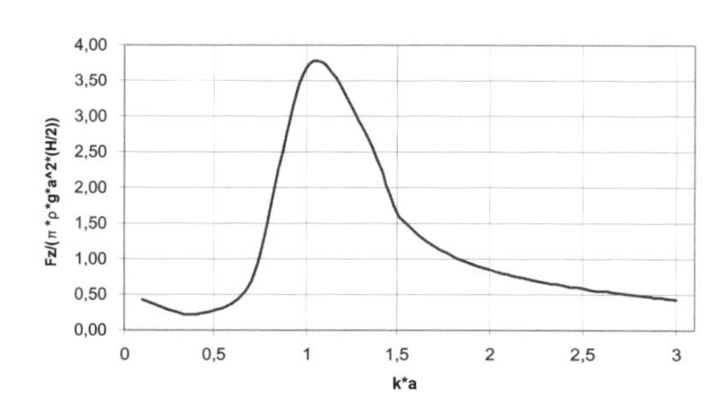

Figure 9. Modulus of the total vertical force on the OWC device for turbine parameter $g_T = 1$.

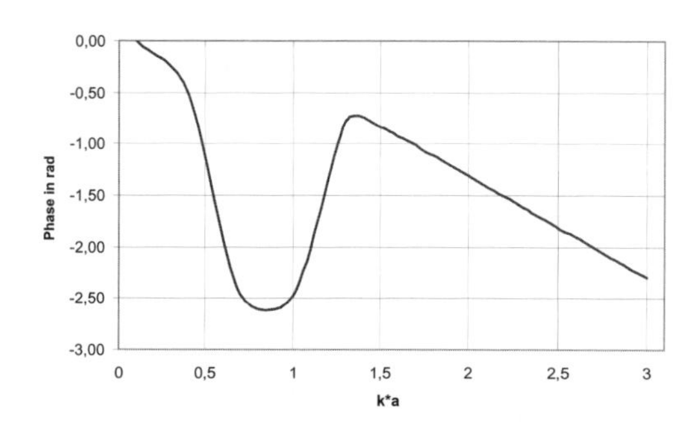

Figure 10. Phase of the total vertical excite force on the OWC device for turbine parameter $g_T = 1$.

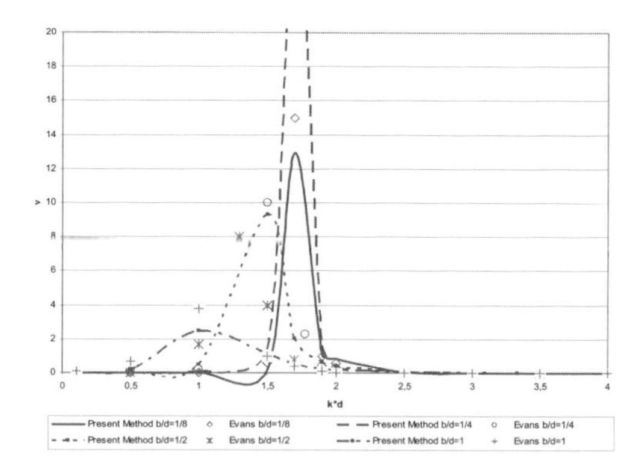

Figure 11. The radiation conductance (ν) against kd.

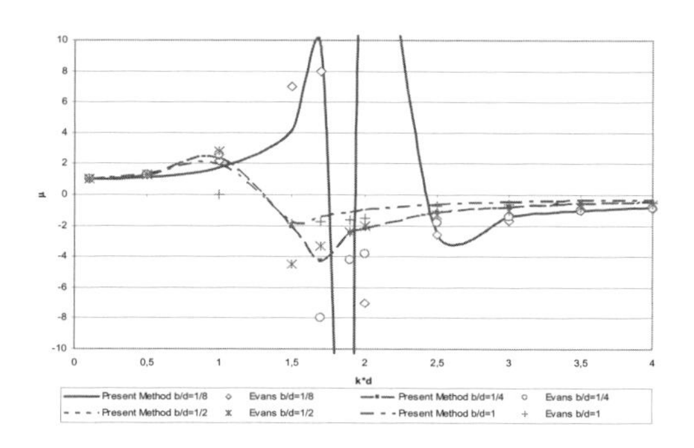

Figure 12. The radiation susceptance (μ) against kd.

8 CONCLUSIONS

The numerical method presented here provides an efficient tool for the complete hydromechanic analysis of an oscillating water column device. The results of this analysis are a fundamental importance for the design of these facilities. Moreover, the efficiency of the system depends on the selection of the turbine. The hydrodynamic parameters and characteristics of the turbine have to be combined in order to improve the wave energy conversion. Since the device is regarded as immovable the present theory can be straightforwardly modified for a nearshore oscillating water column device.

ACKNOWLEDGEMENT

This research has been co-financed by the European Union (European Social Fund—ESF) and Greek national funds through the Operational Program "Education and Life-long Learning" of the National Strategic Reference Framework (NSRF)—Research Funding

Program: Heracleitus II. Investing in knowledge society through the European Social Fund.

REFERENCES

Evans, D.V. 1982. Wave-power absorption by systems of oscillating surface pressure distributions, *J. Fluid. Mech.,* 114, 481–499.

Evans, D.V. & Porter, R. 1995. Hydrodynamic characteristics of an oscillating water column device, *Applied Ocean Research* 17, 155–164.

Evans, D.V. & Porter, R. 1996. Efficient calculation of hydrodynamic properties of O.W.C type devices, *OMAE*—Volume I—Part B.

Falcaõ, A.F. de O. & Justino, P.A.P. 1999. O.W.C wave energy devices with air flow control, *Ocean Engineering* 26, pp. 1275–1295.

Garrett, C.J.R. 1971. Wave forces on a circular dock, *J. Fluid Mech. 46* (I).

Kokkinowrachos, K., Mavrakos, S.A. & Asorakos, S. 1986. Behaviour of vertical bodies of revolution in waves, *Ocean Engineering*, 13(6), 505–538.

Kokkinowrachos, K., Thanos, I. & Zibell, H.G. 1987. Hydrodynamic analysis of some wave energy conversion systems, *Oceans' 87, Conf.*, Halifax, Canada.

Martins-rivas, H. & Mei, C.C. 2009. Wave power extraction from an oscillating water column at the tip of a breakwater, *Journal of Fluid Mechanics*, 626, 395–414.

Mavrakos, S.A. 1985. Wave loads on a stationary floating bottomless cylinder with finite wall thickness, *Applied Ocean Research*, 7(4), 213–224.

Mavrakos, S.A. 1988. Hydrodynamic Coefficients for a thick-walled bottomless cylindrical body floating in water of finite depth, *Ocean Engineering*, 15(3), 213–229.

Mei, C.C. 1983. The applied dynamics of ocean surface waves. John Wiley, New York.

Miles, J. & Gilbert, F. 1968. Scattering of gravity waves by a circular dock, *J. Fluid Mech. 34* (IV).

Miloh, T. 1983. Wave loads on a floating solar pond, Proc. *Int. Workshop on Ship and Platform Motions*, Dept of Nav. Arch. and Offshore Eng., University of California, Berkeley. California.

Sarmento, A.J.N.A., Falcaõ, A.F. & de O. 1985. Wave generation by an oscillating surface-pressure and its application in wave-energy extraction, *J. Fluid Mech. 1*50, 467–485.

Sarmento, A.J.N.A., Gato, L.M.C., Falcaõ, A.F. & de O. 1990. Turbine controlled wave energy absorption by oscillating water column devices, *Ocean Engineering* Vol. 17, No. 5, pp. 481–497.

Sommerfeld, A. 1948. *Vorlesugen uber theoretische Physik*, Bd. 6, Leipzig Akad, Verlagsgesellschaft.

Yeung, R.W. 1981. Added mass and damping of a vertical cylinder in finite depth waters, *Applied Ocean Reseach*, Vol. 3.

2.4 *Non-linear effects*

Sustainable Maritime Transportation and Exploitation of Sea Resources – Rizzuto & Guedes Soares (eds)

Evaluation of non linear vertical motions in head waves

D. Bruzzone, C. Gironi & A. Grasso
Department of Naval Architecture and Marine Engineering of University of Genoa, Italy

ABSTRACT: This paper deals with a 3D weakly non linear methodology for evaluating ship motions. It employs a Rankine- source boundary elements method for linear diffraction and radiation forces in the frequency domain and an iterative time-frequency domain solver to determine Froude-Krylov and hydrodynamic forces, in order to reduce the computational time respect to other methods based on the same assumptions. The procedure is divided into some consecutive steps. The first step consists in the evaluation of the steady flow and of the dynamic sinkage and trim in calm water. Then the linear sea keeping analysis is achieved to determine added mass and damping coefficients as well as diffraction forces. Finally, an iterative process is started in which Froude- Krylov and hydrostatic forces are evaluated in the time domain and transformed back in the frequency domain using Fourier transforms. The motions are solved in the frequency domain and transferred to the time domain for a new evaluation of the foregoing forces. This process proceeds until convergence. The applications presented here regard the DTMB 5415 hull. Numerical simulations have been performed in regular head waves at varying frequency and for different waves slopes in order to highlight nonlinear effects.

1 INTRODUCTION

The numerical description of the behavior of a ship hull in waves is a very important issue. To this aim, several methods have been developed in order to predict ship responses in a incident wave field. The linear strip theory, due to its simplicity and to the minimal computational effort required, is still the most employed model. Anyway, considering motions and loads in higher waves or sea states, non linear effects become not negligible and the linear approach does not give fairly accurate results. Different formulations have been proposed in literature and several methods have been developed in order to include non linear effects both in two and three dimensions; they are generally solved in time domain. Some of them combine linear with non linear terms, other apply fully non linear potential flow methods. Recently also applications of methods based on Reynolds equations have been reported. However, fully three dimensional potential flow methods still require considerable computational time and resources and may be affected by possible numerical instability and wave breaking problems. Even much more demanding are the methods based on RANSE equations. A comprehensive classification and review can be found in Beck and Reed (2000).

For these reasons and to have at disposal a tool to evaluate important non linear behaviours in reasonable computing times to be used systematically in ship design, "hybrid codes" have been developed.

These models, also called "blended methods", allow to partially consider non linearities; they generally evaluate in time domain, by pressure integration over the instantaneous wetted surface, hydrostatic and Froude-Krylov forces which are easy to compute in their intrinsic nonlinear form. Diffraction and radiation forces are instead obtained transforming in the time domain their frequency domain counterparts.

Into the classic hybrid approaches the computation of ship responses in the time domain is preceded by a frequency domain solution. This is done in order to evaluate the linear radiation and diffraction forces, from the solution of the unsteady hydrodynamic problem for a given number of meaningful arbitrary frequencies. The solution then proceeds solving the equations of motion in time domain, evaluating at each time step Froude-Krylov and hydrostatic forces by integration of the corresponding pressure over the actual hull wetted surface under the incident wave profile. Since added mass and damping are frequency dependent whereas ship motions are directly evaluated into the time domain and nonlinear effects are included, impulse theory is applied in order to transfer the radiation forces from frequency domain to time domain.

Methods based on this "blended" theory with the foregoing approach have been developed and tested also in our recent studies, with the aim to investigate about their capabilities and limits; an example can be found in Bruzzone and Grasso (2007).

Though based on the same hypothesis regarding the relevant forces, the methodology herein used exploits a different approach to determine the ship motions. It implies alternatively passing from the frequency to the time domain: for this reason it could be indicated as a "frequency-time-frequency" (FTF) method.

Therefore, in order to take into account nonlinearities, the blended method of the family in the foregoing description has been used in an alternative dual approach: Froude-Krylov and hydrostatic forces are evaluated in the time domain and the equation of motion are solved in the frequency domain (in their weakly nonlinear form) by an iterative procedure. The boundary value problems necessary to determine the hydrodynamic coefficients and the diffraction forces are solved by a Rankine panel method which allow to apply the complete methodology also to multi-hull marine vehicles (Bruzzone 2003). In this paper a short presentation of the basic theory and of the numerical method will be given at first, then the application to a cases study for which experimental data are available, including time histories of ship responses (Irvine, Longo, and Stern 2008), will be described and discussed.

2 THE MATHEMATICAL METHOD AND THE NUMERICAL SOLUTION

2.1 *Linear analysis in frequency domain*

Let define a right handed orthogonal coordinate system (x, y, z) advancing at the vessel speed U maintaining the xy plane coincident with the undisturbed free surface. x is the symmetry axis of the still waterplane and is assumed positive astern. The z-axis is positive upwards. Ship motions are defined by the instantaneous position of a body fixed reference system with respect to the previous system, that may be described by a vector $\eta_k(t)$, with $k = 1, \ldots, 6$. A regular incident wave $\eta(t) = \Re\{ae^{i\omega_e t}\}$ is assumed. Under the hypotesys of small amplitudes ship motions can be expressed as:

$$\eta(t) = \Re\{\zeta_k(\omega_e)e^{i\omega_e t}\} \tag{1}$$

where $\xi_k(\omega_e)$ is the complex amplitude of the kth motion component and ω_e the encounter frequency.

ζ_k can be determined solving the following complex system of linear equations:

$$\sum_{k=1}^{6}\left[-\omega_e^2(M_{jk} + A_{jk}(\omega_e)) + i\omega_e B_{jk}(\omega_e) + C_{jk}\right]\xi_k = F_j^D(\omega_e) + F_j^{FK}(\omega_e) \tag{2}$$

where $j = 1, 2, 3$ refer to the x, y, z force components respectively and $j = 4, 5, 6$ to the corresponding three moment components. M_{jk} and C_{jk} represent the mass and hydrostatic restoring matrix, A_{jk} and B_{jk} are the added-mass and the damping coefficients, F_j^D and F_j^{FK} are the the complex amplitudes of diffraction and Froude-Krylov forces.

As the problem is linear, superposition of the motions due to each frequency component of the incident wave pattern can be used for determining ship motions in irregular seas.

2.2 *The classical "blended" methodology*

Considering the ship as an unconstrained rigid body subjected to gravity, radiation, diffraction, Froude-Krylov and hydrostatic forces and applying the impulse theory (Cummins 1962) is possible to transfer the equation of motions from the frequency to the time domain, as:

$$\sum_{k=1}^{6}(M_{jk} + A_{jk}^{\infty})\ddot{\eta}_k(t) + \int_0^t h_{jk}(t-\tau)\dot{\eta}_k(\tau)d\tau = F_j^D(t) + F_j^{FK}(t) + F_j^H(t) \tag{3}$$

with $j = 1, \ldots, 6$ and $\dot{\eta}_k, \ddot{\eta}_k$ the first and the second time derivatives of η_k. A^{∞} and B^{∞} represent the mean infinite-frequency added mass and damping coefficients. $F_j^D(t)$, $F_j^{FK}(t)$ and $F_j^H(t)$ represent the diffraction, the Froude-Krylov and the hydrostatic forces (and moments) respectively. The terms h_{jk} are the impulse response functions and can be evaluated by the following releation:

$$h_{jk} = -\frac{2}{\pi}\int_0^{\infty}\omega_e(A_{jk} - A_{jk}^{\infty})\sin(\omega_e t)d\omega_e = \frac{2}{\pi}\int_0^{\infty}(B_{jk} - B_{jk}^{\infty})\cos(\omega_e t)d\omega_e \tag{4}$$

If the hydrostatic forces are considered linearly dependent on the motions and the Froude-Krylov and diffraction forces are supposed to be linear functions of the wave elevations only, system (3) represents a linear system of differential equations since both coefficients and exciting forces do not depend on motions and on theirs derivatives. So the impulse responses can be derived from the frequency dependent added-mass and damping coefficients and vice versa, since system (3) and system (2) are related by Fourier transforms. If, on the contrary, fully nonlinear hydrostatic and Froude-Krylov forces are introduced, they depend also on the instantaneous position of the hull, hence on the unknown motions and system (3) must be considered as nonlinear. Generally, in the so called blended methods, system (3) is solved in the time domain after the evaluation of the impulse response functions on the basis of the results of a previous linear calculation in the frequency domain. Fourier transforms allow to express the diffraction forces from the frequency to the time domain.

A time domain procedure can be found in Bruzzone and Grasso (2007), while in the following the methodology which exploits directly the frequency domain values of hydrodynamic coefficients and diffraction forces is briefly described.

2.3 *The FTF approach*

The process previously described is subjected to some inconveniences and uncertainties as regard the influence of the transient part and, overall, in the determination of the impulse response functions. In addition, there is some approximation in the fulfillment of the Kramers-Kronig relations from the numerical results. Even if the nonlinear forces must be evaluated in the time domain, the system of Equations (3) can be solved both in the time and in the frequency domain. The choice is related to the kind of analysis it is expected to be carried out. For this application the frequency domain has been preferred since it allows to avoid the initial transient phase and it is faster. The computational time is connected with the actual non-linearities and the time step is not constrained by time integration convergence requirements. A previous application is described in Bruzzone et al. (2011). Denoting with $\mathfrak{F}$ the Fourier transform and with $\mathfrak{F}^{-1}$ its inverse, $\eta_k(t)$ can be evaluated by:

$$\eta_k(t) = \mathfrak{F}^{-1}\{\xi_k(\omega_e)\} \tag{5}$$

The Fourier transform into the frequency domain of (3) can be written as:

$$\sum_{k=1}^{6}\{-\omega_e^2[M_{jk} + A_{jk}(\omega_e)] + i\omega_e B_{jk}(\omega_e)\}\xi_k(\omega_e) = \mathfrak{F}\{F_j^D(t)\}(\omega_e) + \mathfrak{F}\{F_j^H(t) + F_j^{FK}(t)\}(\omega_e) \tag{6}$$

The system in (6) cannot be solved in this form, because of the non-linear dependence of Froude-Krylov and hydrostatic forces on ship motions. An iterative procedure is hence adopted: at each iteration the time domain nonlinear forces are evaluated considering the motions obtained in the previous iteration. Froude-Krylov and hydrostatic forces are evaluated in the time domain, integrating hydrostatic and hydrodynamic pressure over the actual wetted surface under the incident wave profile. Then they are transformed from the time to the frequency domain. The linear solution is used as the first guess. The following formulation has been adopted:

$$\sum_{k=1}^{6}\left[-\omega_e^2\left(M_{jk} + A_{jk}(\omega_e)\right) + i\omega_e B_{jk}(\omega_e) + C_{jk}\right]\xi_k^{(p)}(\omega_e) = \mathfrak{F}\{F_j^D(t)\}(\omega_e) + \mathfrak{F}\{F_j^{H(p-1)}(t) + F_j^{FK(p-1)}(t)\}(\omega_e) + \sum_{k=1}^{6} Cjk\,\xi_k^{(p-1)}(\omega_e) \tag{7}$$

where p represents an iteration index and C_{jk} is the linear hydrostatic restoring matrix. It is considered in both sides of (7) in addition to the non linear hydrostatic forces to render the procedure more robust and as an aid to the convergence of the iterative procedure.

2.4 *Outline of the numerical method*

The method is based on the preliminary solution of a series of boundary value problems necessary to evaluate the dynamic attitude of the ship and the relevant terms in the previous equations, as added mass, damping and diffraction forces. Assuming non viscous irrotational flows, they can be stated in terms of a total velocity potential Φ which satisfies the Laplace equation in the fluid domain Ω and appropriate boundary conditions imposed over the boundaries $\partial\Omega$. The total potential Φ may be expressed as the sum of the potential of a steady base flow Φ_S and of a small unsteady perturbation potential Φ_{US}. In turn, the unsteady perturbation potential may be written as superposition of an incident wave potential Φ_I, of a diffraction potential Φ_D and of six radiation potentials.

$$\Phi = \Phi_S + \Phi_I + \Phi_D + \sum_{k=1}^{6}\Phi_k \tag{8}$$

Each of them should satisfy the relevant boundary conditions on the hull surface and on the free surface. The steady potential must be determined firstly since it will represent the "basis flow" for the subsequent determination of the radiation and diffraction potentials. In addition it allows to evaluate the still water sinkage and trim which represent the mean reference position for the seakeeping calculations. To solve all the involved boundary value problems, including the one relative to the steady flow, a three-dimensional Rankine panel method has been employed (more details can be found in Bruzzone 2003). Each of the unknown potentials is expressed in term of a distribution of Rankine sources on the hull and on the free surface. To this end the hull and a part of the free surface are approximated with quadrilateral panels, considering a uniform source strength on each. All the involved boundary value problems are hence solved in terms of these unknown source strengths. A suitable radiation condition is finally posed at the forward border of the computational domain. In the present method radiated and diffracted waves are considered not to propagate ahead the ship. Hence it can be applied only for $\omega_e U/g = 0.25$. Since the free surface computational domain is limited, its extension must be carefully considered in order to avoid wave reflections; moreover, the dimensions of the free surface panels should be chosen taking into account incident, radiated and diffracted wave lengths. After the solution of the foregoing boundary value problems, the relevant forces and coefficients can be determined for calculating the impulse response functions or for the iterative process frequency-time domain. To evaluate the nonlinear Froude-Krylov and hydrostatic forces the hull is described employing bi-cubic surfaces, depending on two normalized parameters u, v. At each time step the domain describing the wetted surface is evaluated as well as the pressure distribution on it. Forces and moments are then calculated by analytical integration of their distributions treated as bi-cubic function on the domain of the parameters u, v. As the methodology employed is based on the potential flow approach, viscous effects are neglected. In some cases, a viscous correction is required in order to avoid overestimation of the resonance peaks.

3 ANALYSIS OF COMPUTED RESULTS

The performance of these methods is to be validated by comparison with adequate and reliable experimental results, possibly for different ships and conditions. It is not so simple to find them in the open literature with adequate information on the geometry to which they are referred. In addition, information about their confidence interval and a description of their uncertainties are also important as discussed for instance by the seakeeping comitte of ITTC proceedings 2005.

The method here presented has been applied to different ships and marine vehicles, including catamarans and trimarans (Bruzzone, Grasso, and Zotti 2009) and on a container ship, the S175 hull, for which several experimental data was diffused in literature (see for inst., Fonseca 2004).

Generally, the data reported for regular waves are given in terms of response amplitude operators and higher harmonics. For a more complete validation it would be useful also to compare time histories that are not always available. Due to the possibility offered by cooperative programs and to the quick availability of data through the web, in this paper the DTMB5512 hull (a model of the DTMB5415 ship) has been considered, see Tab. 1. Many experimental data about pitch and heave motions in head waves are available from the towing tank of the University of Iowa which include all the foregoing quantities (Irvine, Longo, and Stern 2008).

First of all, for a method such the present one, the data coming from the frequency domain evaluation of the hydrodynamic coefficients and of the diffraction forces are an important issue because they are maintained unchanged for the whole process. The Rankine panel method used here has proven to give acceptable results and was tested in many cases for mono hull ships, catamarans and trimarans (Bruzzone 2003). Sometimes, from the comparison with experiments, overestimated values of heave and pitch resonance peaks are obtained, which was supposed to be due to an underestimation of damping. This is not the case in the present computation as it can be observed in Figs. 1–4. The heave and pitch response amplitude operators computed in the frequency domain compared with the experimental values at the lowest wave slope show a very similar trend thus indirectly indicating an adequate evaluation of the foregoing coefficients and forces. The results of the computations of the first harmonic for Froude numbers $F_N = 0.19$ (Figs. 5–8) and $F_N = 0.28$ (Figs. 9–12) are presented in this paper; the comparison with other experimental results is reported in Castagna (2011). For both Froude numbers significant differences in experiments due to the different wave slopes are not evidenced for pitch and heave. The same behavior is confirmed by numerical predictions. Time histories are represented in Fig. 13 to Fig. 16, the experimental data regarding heave have been here represented about they mean value. The the good

Table 1. Main dimensions of the selected ship.

DTMB 5512 main dimensions	
Length $L_{BP}[m]$	142.04
Beam $B[m]$	17.99
Draft $T[m]$	6.19
Wetted surface $S_W[m^2]$	2977
Block coefficient C_B	0.506
$LCG[m]$	71.61
$VCG[m]$	7.55
Radius of gyration $R_g[m]$	35.51

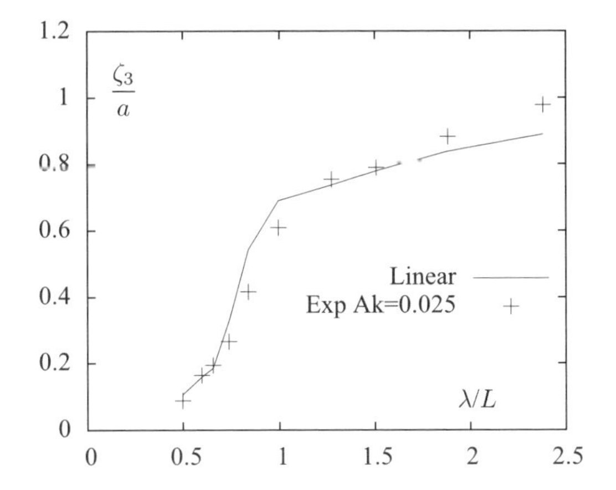

Figure 1. Linear (freq. domain) heave amplitude RAO, for Fn = 0.19.

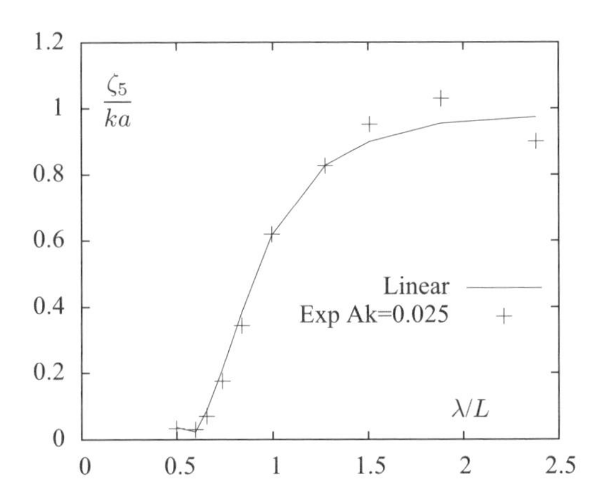

Figure 2. Linear (freq. domain) pitch amplitude RAO for Fn = 0.19.

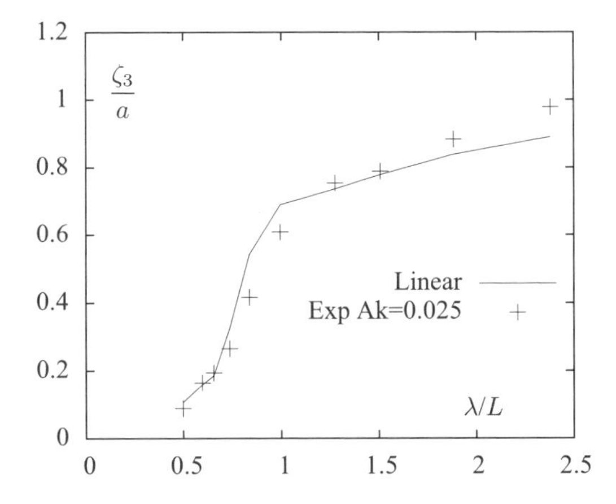

Figure 3. Linear (freq. domain) heave amplitude RAO for Fn = 0.28.

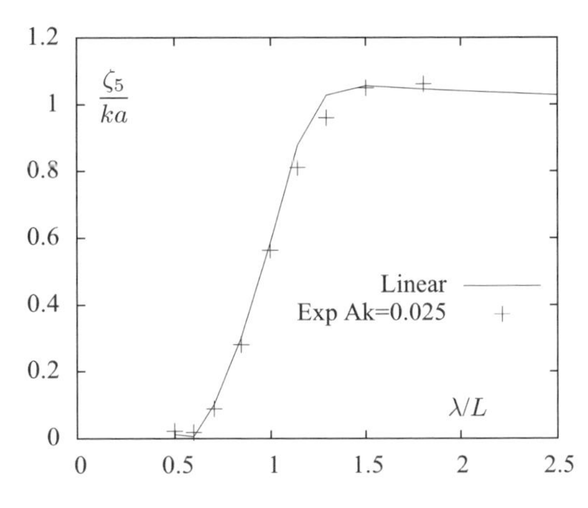

Figure 4. Linear (freq. domain) pitch amplitude RAO for Fn = 0.28.

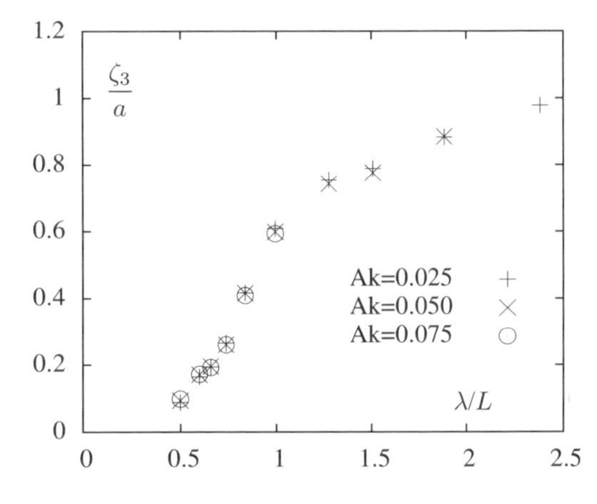

Figure 5. Experimental data of heave amplitude for Fn = 0.19.

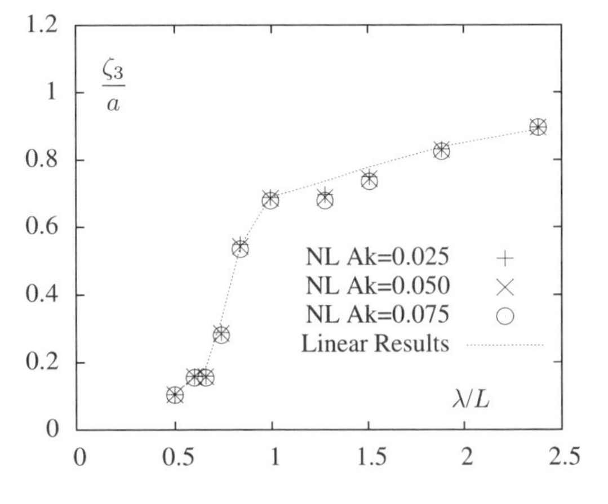

Figure 6. Numerical results of heave amplitude for Fn = 0.19.

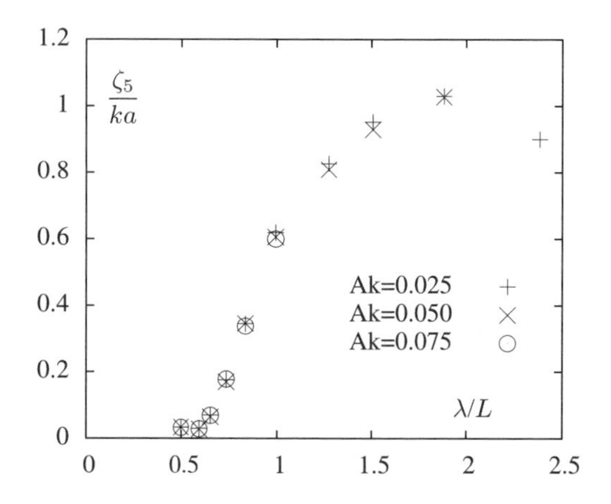

Figure 7. Experimental data of pitch amplitude for Fn = 0.19.

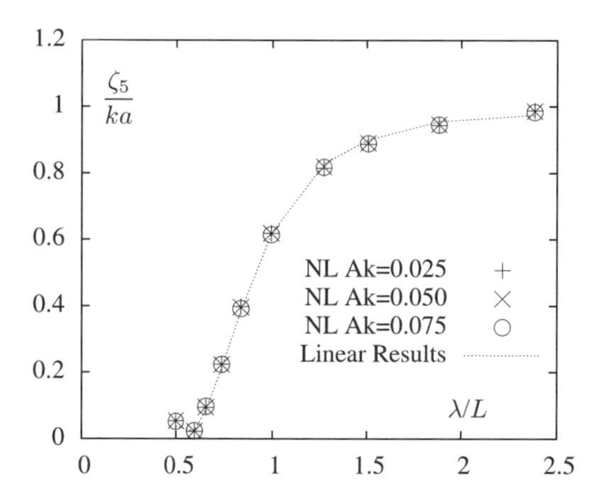

Figure 8. Numerical results of pitch amplitude for Fn = 0.19.

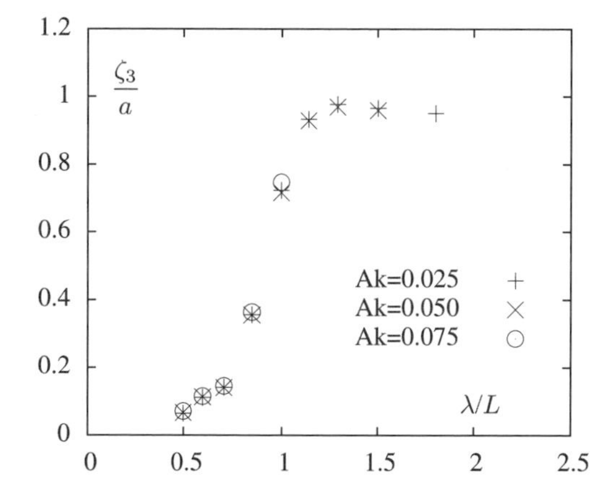

Figure 9. Experimental data of heave amplitude for Fn = 0.28.

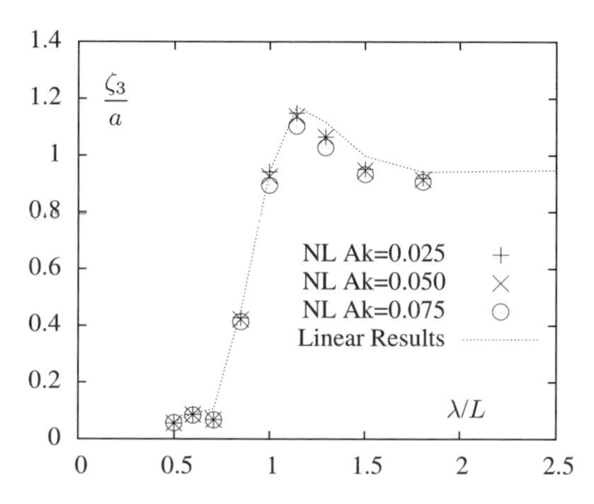

Figure 10. Numerical results of heave amplitude for Fn = 0.28.

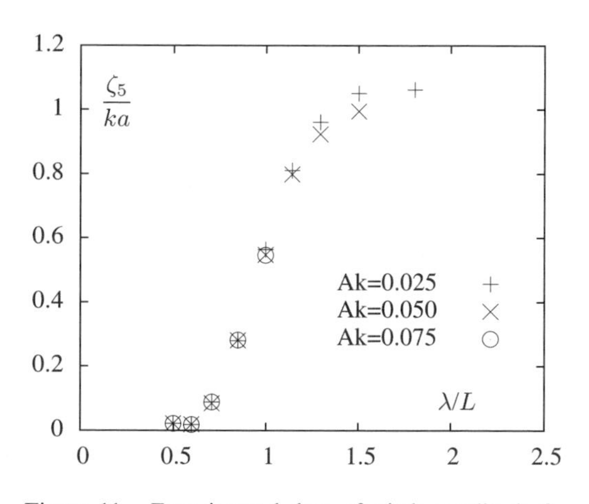

Figure 11. Experimental data of pitch amplitude for Fn = 0.28.

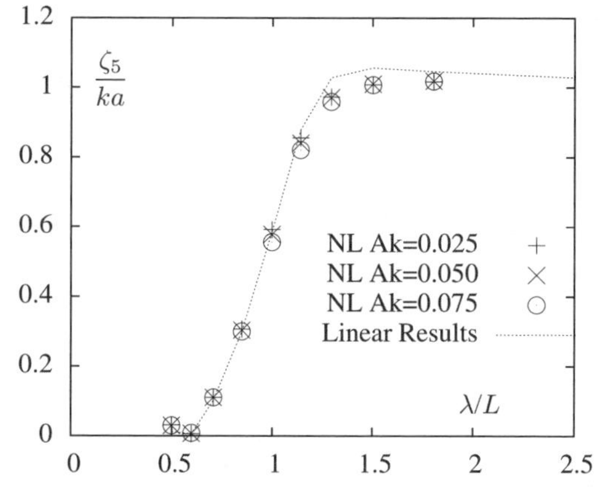

Figure 12. Numerical results of pitch amplitude for Fn = 0.28.

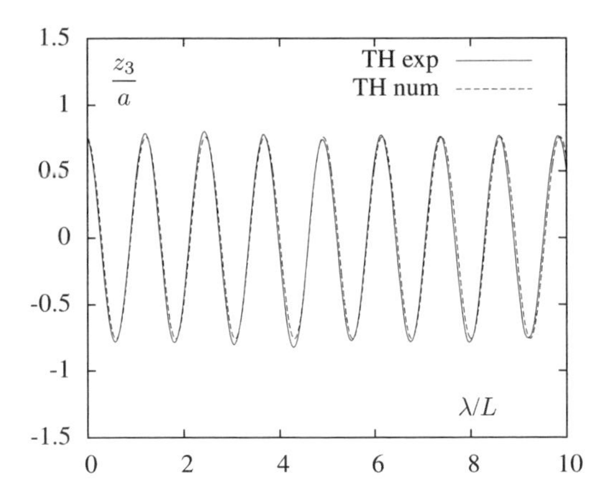

Figure 13. Heave time history for Fn = 0.19, Ak = 0.025 and $\lambda/L = 1$.

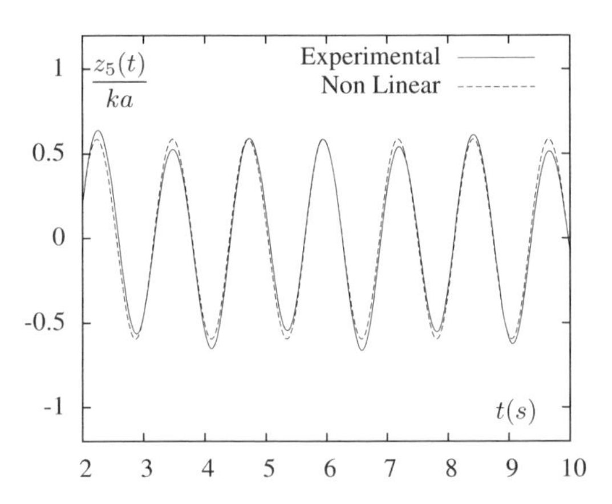

Figure 14. Pitch time history for Fn = 0.19, Ak = 0.025 and $\lambda/L = 1$.

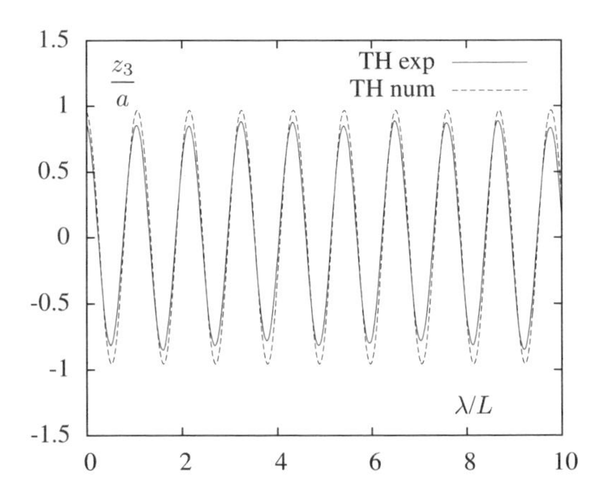

Figure 15. Heave time history for Fn = 0.28, Ak = 0.025 and $\lambda/L = 1$.

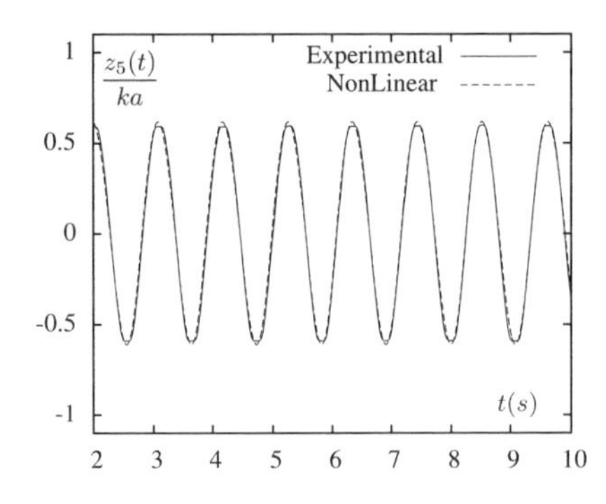

Figure 16. Pitch time history for Fn = 0.28, Ak = 0.025 and $\lambda/L = 1$.

correlation of computation with experiments is confirmed as already observed in the previous figures.

4 CONCLUSIONS

The paper presents a weakly nonlinear method for evaluating motions and loads in waves. Radiation and diffraction forces are assumed linear whereas Froude-Krylov and hydrostatic forces are evaluated in the time domain allowing for their non linearity. With respect to methodologies that adopt similar assumptions, the problem is approached in a different way. A procedure has been proposed that iteratively solves the equations of motions in the frequency domain, evaluating only the non linear forces in the time domain. This procedure allows to reduce computational time and does not require the evaluation of the initial transient phase.

The application presented in this study regards a surface combatant whose geometry is a test case for CFD computations. The experimental data concerning the behavior in regular head waves of the model DTMB5512, tested at the University of Iowa, have been available, including time histories of pitch and heave.

The obtained numerical results evidence a satisfactory correlation with the experiments. The trends of ship responses seem adequately predicted even if the tested cases and the example reported do not show significant non linear effects against wave slopes. The time histories of heave and pitch confirm the adequacy of computed results.

Further studies could be related to the improvement in the prediction of linear forces and to applications considering conditions where the effects of non linearity are more severe.

REFERENCES

Beck, R. and A. Reed, (2000). Modern seakeeping computations for ships. *Twenty-Third Symposium on Naval Hydrodinamics*. Val de Reuil, France.

Bruzzone, D. (2003). Application of a rankine source method to the evaluation of motions of high speed marine vehicles. *Proceedings of the 8th International Marine Design Conference, Athens 2*, 69–79.

Bruzzone, D., C. Gironi, and A. Grasso, (2011). Nonlinear effects on motions and loads using an iterative time-frequency solver. *International Journal of Naval Architecture and Ocean Engineering 3*(1), 20–26.

Bruzzone, D. and A. Grasso, (2007). Non linear time domain analysis of vertical ship motions. *Archives of Civil and Mechanical Engineering 7, N.4*, 27–37.

Bruzzone, D., A. Grasso, and I. Zotti, (2009). Experiments and computations of nonlinear effects on the motions of catamaran hulls. In *Proceedings of 10th International Conference on Fast Sea Transportation FAST 2009*, Volume I, pp. 349–360. Athens.

Castagna, A. (2011). Non linear analysis of vertical motions of a surface combatant. *Master Thesis*. Faculty of Engineering—University of Genova (in Italian).

Cummins, W. (1962). The impulse response function and ship motions. *Schiffstechnik 47*, 101–109.

Fonseca, N. and C. Guedes Soares, (2004). Experimental investigation of the nonlinear effects on the vertical motions and loads of a containership in regular waves. *Journal of Ship Research 48*(2), 118–147.

International Towing Tank Conference (2005). *Report of the Seakeeping Committee—Proceedings of the 24th Int. Towing Tank Conf.*, Volume I. International Towing Tank Conference. University of Newcastle.

Irvine, M.J., J. Longo, and F. Stern, (2008). Pitch and heave tests and uncertainty assessment for a surface combatant in regular head waves. *Journal of Ship Research 52*(2), 164–163.

Sustainable Maritime Transportation and Exploitation of Sea Resources – Rizzuto & Guedes Soares (eds)
 ISBN 978-0-415-62081-9

A nonlinear approach to the calculation of large amplitude ship motions and wave loads

G. Mortola, A. Incecik & O. Turan
Department of Naval Architecture and Marine Engineering, University of Strathclyde, Glasgow, UK

S.E. Hirdaris
Strategic Research Group, Lloyd's Register, London, UK

ABSTRACT: A two dimensional nonlinear time domain strip theory for the seakeeping analysis of ship behaviour under large amplitude waves is presented. Here the fluid actions are evaluated at each time step using the actual wetted hull portion and the exciting wave elevation as free surface. The radiation problem is partially solved in time domain with the employment of a convolution integral in the history of motion. The nonlinear system of equations of motion is numerically solved in time domain via a fourth order Runge-Kutta method.

The proposed nonlinear method is applied to the S-175 Container ship and the results are compared with the predictions for small amplitude waves obtained from the results published by other researchers. A study of the behaviour of the vessel under large amplitude waves is presented.

1 INTRODUCTION

The interest in non linear seakeeping methods has grown since the availability of fast computer allowed to introduce complex techniques in ship design. Usually the seakeeping analysis of ocean going vessels is carried out using linear methodologies; those procedures are well known and established in the naval architecture field. Linear methodologies are reliable to predict small amplitude motions, but the assumptions, on which they are based on, do not enable us to accurately model extreme ship motions and loads.

The response of a ship sailing in severe sea condition has a strong nonlinear behaviour. Motions and loads are not proportional to the exciting wave amplitude, they have a multi-harmonic spectrum and asymmetric responses appear. For these reasons the seakeeping analysis of vessel sailing under large amplitude waves should not be carried out using linear methods.

In the last twenty years several non linear methodologies have been presented. They differ from each other for their assumption and complexity (ISSC, 2009). Simple theories consider only the non linear effects generated by the real hull geometry (Fonseca N. and Guedes Soares C., 1998); other techniques introduce the nonlinearities associated with the radiation problem (Xia J. et al., 1998). The most complex methods are formulated without any linear simplification, modelling all the nonlinearities in the fluid (Beck R.F., 1994).

The simplest methodologies have a computational time that is comparable to the one required by linear theories; the most complex methods need longer time and for this reason their application during the first stages of the design is not possible nowadays.

The aim of this project is to develop a non linear seakeeping method that can be used in the preliminary design and optimisation. For these reasons the methodology used is a "quasi non linear" theory.

The "quasi non linear" or blended methods are a combination of linear and non linear assumptions. They permit to model the principal nonlinearities in the fluid using some linear hypothesis. Blended methodologies are not rigorous, but are an engineering compromise between accuracy of the model and the time of the calculation.

A comparison between the proposed method and the predictions of a linear strip theory and a three dimensional Green functions technique is presented for the S-175 container ship, in small and large amplitude regular waves.

2 THEORETICAL BACKGROUND

The proposed method is developed with the intention to model ship motions and loads under large amplitude waves. The hull is described as a rigid body with two degrees of freedom in heave and pitch motions respectively. In order to include all the main nonlinearities all the forces are calculated at each time step considering the actual wetted hull portion and the exciting wave elevation in a way of the free-surface as shown in Figure 1. The relative position and velocity between the vessel and the waves is considered in the computation of the radiation and hydrostatic forces.

The system of equations of motion is non linear. These equations are numerically solved in time domain using a fourth order Runge-Kutta method.

$$\begin{cases} M_{33}\ddot{\eta}_3(t) + M_{35}\ddot{\eta}_5(t) = \sum F_3^E \\ M_{55}\ddot{\eta}_5(t) + M_{53}\ddot{\eta}_3(t) = \sum F_5^E \end{cases} \tag{1}$$

where M_{ij} is the *i-th* and *j-th* element of the mass matrix of the vessel, η_j and $F^E{}_i$ are the displacement and the external force acting on the vessel respectively in the *j-th* and *i-th* mode of motion. The external forces are composed by the fluid forces generated by the motions and the exciting forces, produced by the incoming waves.

2.1 *Fluid forces*

The approach to the calculation of the fluid forces acting on the body is two dimensional. The hull is considered a sum of cylindrical segments with constant geometry. The hydrodynamic and hydrostatic forces are numerically evaluated at each time step along each section using the classical "strip theory" assumptions (Salvesen et al., 1970) and their global values are given by longitudinal integration.

The fluid is considered to be ideal, inviscid, irrotational, incompressible and uniform; thus it is possible to describe the velocity field inside the fluid using the potential flow analysis. Hydrodynamic actions on the body are evaluated via the formulation of the fluid momentum.

$$\frac{D}{Dt}\boldsymbol{M}(t) = -F_H(t) \tag{2}$$

$$\boldsymbol{M}(t) = \iint\limits_S \rho\varphi(y,z;t)\boldsymbol{n}ds \tag{3}$$

where $\boldsymbol{M}$ is the fluid momentum, $\boldsymbol{F}_H$ is the fluid force acting on the body and φ is the velocity potential. Equation 2 shows the definition of momentum inside the fluid and Equation 3 defines the fluid momentum considering the potential flow assumptions. S is the boundary of the two dimensional fluid domain with normal vector $\boldsymbol{n}$; it is composed, as described in Figure 2, by S_F, S_H, S_0 and S_∞. S_F is the portion of the boundary associated with the free-surface of the fluid; S_H is the hull section surface, S_0 are the two surfaces that limit the fluid volume along the longitudinal ship axis and S_∞ is the far away boundary that closes the domain at an infinite distance from the body.

Therefore the sectional fluid force is obtained.

$$\iint\limits_{S_H} p\boldsymbol{n}ds = -\rho\left(\frac{d}{dt} - U\frac{\partial}{\partial x}\right)\iint\limits_{S_H} \varphi\boldsymbol{n}ds - \iint\limits_{S_H} \rho gz\boldsymbol{n}ds \tag{4}$$

U is the vessel forward speed and S_H is the instantaneous wetted hull section portion. This surface changes in time domain as function of the ship motions and of the exciting wave pattern. The fluid action on a section is dived in two components, as shown in Equation 4.

The second term of the right hand side of Equation 4 describes the hydrostatic contribution to the fluid force. The first term is the radiation term related to the dynamics of the fluid. The radiation problem is partially solved in time domain and the definition of the fluid potential is formulated as proposed by Cummins (Cummins, 1962).

The velocity potential is divided in two components. One related to an instantaneous impulse of displacement and a second associated with the history of motions. Equation 5 describes the velocity potential.

$$\varphi(x;t) = \psi(x;t)V(t) + \int_0^t \chi(x;t-\tau)V(\tau)d\tau \tag{5}$$

The impulsive part is directly solved in time domain in analogy with the boundary values

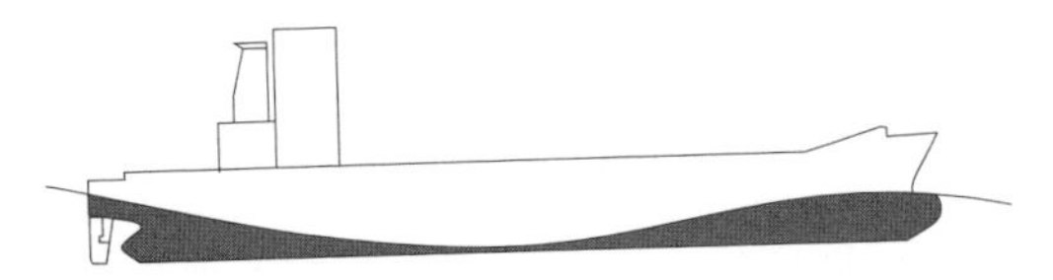

Figure 1. Example of the actual hull portion used in the simulation.

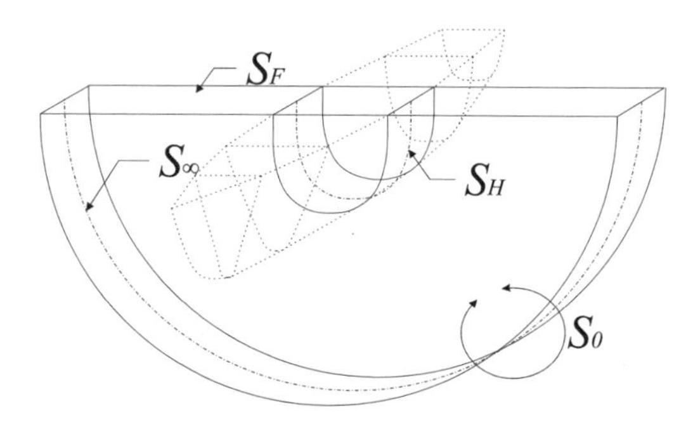

Figure 2. Components of the boundary surface S for the two dimensional radiation problem.

problem of a body oscillating at infinite frequency, as described in Equation 6. The second part of Equation 5 is solved using the Inverse Fourier Transform of the sectional damping coefficient in frequency domain. Equation 7.1 and Equation 7.2 are the formulations of the memory effect function and its kernel K_{ij}.

$$\rho \iint_{S_H} \psi_j n_i ds = a_{ij}^{\infty} \tag{6}$$

$$\rho \frac{D}{Dt} \iint_{S_H} \left[\int_{-\infty}^{t} \chi_i(x; t-\tau) V_j(x,\tau) d\tau \right] ds = \int_{-\infty}^{t} K_{ij}(x; t-\tau) V_j(x;\tau) d\tau \tag{7.1}$$

$$K_{ij} = \frac{2}{\pi} \int_0^{\infty} \left(b_{ij}(\omega) - b_{ij}(\infty) \right) \cos \omega t d\omega \tag{7.2}$$

Where a_{ij}^{∞} is the sectional added mass coefficient at infinite frequency and b_{ij} is the sectional damping coefficient in frequency domain.

The action of the fluid on the body is given by the integration of the fluid pressure, composed by a dynamic component and a hydrostatic one, along the hull surface. Equations (4), (5), (6) and (7) lead to the final formulation for the dynamic contribution of the fluid forces.

$$f_i^H = -\sum_j \{ a_{ij}^{\infty} \frac{D}{Dt} V_j - U \frac{\partial}{\partial x} a_{ij}^{\infty} V_j + \frac{\partial}{\partial z} a_{ij}^{\infty} V_j^2 + \int_{-\infty}^{t} K_{ij}(x; t-\tau) V_j(x,\tau)\, d\tau \} \tag{8}$$

The sectional hydrodynamic force is non linear and it is a function of the instantaneous wetted sectional surface and of the sectional motions. This nonlinear formulation is composed by four terms. The first three are related to the instantaneous impulse of displacement and are an inertial component a lift effect and an impulsive force; the last term is the memory effect function which is related to the history of motion.

2.2 *Exciting and restoring forces*

The exciting and the restoring forces are found using the actual wetted hull portion. At each time step the incoming wave surface and the related particle velocity are calculated via the linear waves theory. The ship sections are translated in accordance with the values of heave and pitch motions at the given time instant. The interception between the wave profile and the modified section is used to limit the wetted hull portion. The latter is divided into a predefined number of panels which are used to calculate exciting and restoring forces.

The exciting wave force is composed by the so called Froude-Krylov force and the diffraction forces. They are both calculated in time domain following the strip theory approach.

$$\begin{aligned} F_{30} &= \operatorname{Re} \int_L e^{-ikx\cos\mu} f_3 dx e^{i\omega_e t} \\ F_{50} &= -\operatorname{Re} \int_L x e^{-ikx\cos\mu} f_3 dx e^{i\omega_e t} \\ f_3 &= a_w \rho g \int_{S_H} e^{-iky\sin\mu} e^{kz} N_3 ds \end{aligned} \tag{9}$$

$$\begin{aligned} F_{37} &= \operatorname{Re} \int_L e^{-ikx\cos\mu} h_3 dx e^{i\omega_e t} \\ F_{57} &= -\operatorname{Re} \int_L \left(x + \frac{U}{i\omega_e} \right) e^{-ikx\cos\mu} h_3 dx e^{i\omega_e t} \\ h_3 &= a_w \rho \omega \int_{S_H} i N_3 e^{-iky\sin\mu} e^{kz} \psi_3 ds \end{aligned} \tag{10}$$

Equations 9 describes the exciting forces due to the incoming wave and Equations 10 shows the diffraction forces in heave and pitch respectively. The difference between the presented methodology and the classic methods is, in this case, the path for the numerical integration changes as a function of ship motions.

The restoring forces are a combination between the buoyancy and the weight of the vessel. They are the physical equivalent of the stiffness for a linear body. In the linear case the elastic component of the equation of motions is described as a linear increase or reduction of buoyancy, but in this nonlinear application the real wetted volume is calculated at each time step and its difference with the ship mass and its buoyancy gives the restoring forces and moment.

$$\begin{aligned} F_3^R &= gW - \int_{\nabla} \rho g z \boldsymbol{n} dv \\ F_5^R &= \int_{\nabla} x \rho g z \boldsymbol{n} dv - LCGgW \end{aligned} \tag{11}$$

The restoring vertical force and its moment are described in Equation 11.

3 RESULTS AND DISCUSSIONS

The proposed method is used to predict heave, pitch, vertical shear force and vertical bending moment for the S-175 container ship travelling in linear waves with and without forward speed. The body plan of the S175 container ship and the main particulars are shown in Table 1 and Figure 3.

Table 1. Main particulars of the S-175 container ship.

Length between perpendiculars—L_{BP} [m]	175.00
Beam—B [m]	25.40
Depth—D [m]	15.40
Draught—T [m]	9.50
Displacement—Δ [tonnes]	24792
LCG [m]	−2.57
Pitch radius of gyration—r_{yy} [m]	43.75

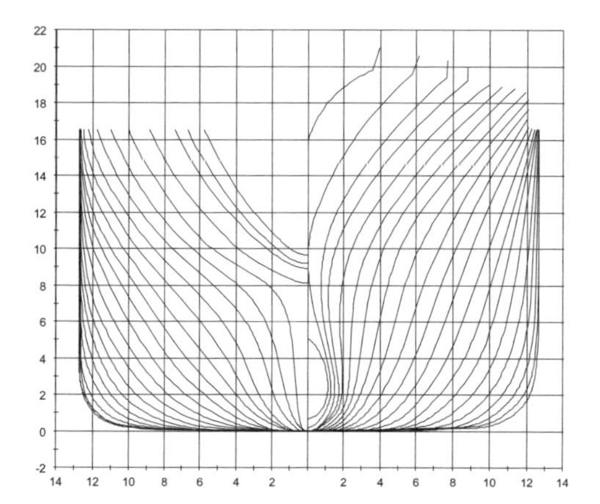

Figure 3. S-175 Container ship.

With the purpose of a first validation the results obtained by the proposed theory are compared with those obtained from commercial codes. The methodologies chosen for the comparison are a linear frequency domain strip theory and a three dimensional linear frequency domain Green's function method. In the latter method the hull is modelled using 789 pulsating source panels of maximum aspect ratio equal to 2:1. This idealisation is adequate to avoid any dependency of the hydrodynamic coefficients on the number of panels used.

The comparison between a linear and a non-linear methodology can be carried out only for exciting waves of small amplitude. Under the assumption of small amplitude waves the prediction of a nonlinear seakeeping method is not affected by any nonlinear behaviour. The motions and loads obtained for large amplitude waves are not compared with any method, because there is no availability of such a technique.

Several papers analyse the behaviour of the S-175 container ship under large amplitude waves.

Results for the S-175 container ship travelling in small and large amplitude waves are presented. Fonseca (Fonseca and Guedes Soares, 2004) published the result of experimental studies for the S-175. He presented the results for heave, pitch vertical shear force and vertical bending moment, at different combinations of forward speed, wave frequency and wave height.

Experimental trials show a reduction from the linear prediction for heave and pitch motions when wave amplitude rises. On the other hand, in the analysis of vertical loads it is visible an increase of shear force and bending moment from the values predicted using the assumption of linearity between response amplitude and exciting wave amplitude.

3.1 *Small amplitude results*

Simulation of the S-175 container ship travelling under small amplitude waves are carried out for different forward speed and wave frequencies in the head sea condition. The amplitude of the exciting wave in the nonlinear time domain model is equal to 0.05 m; the vessel immersion is 9.50 m. The simulations were conducted at zero forward speed and 10 knots ($F_N = 0.125$). The motions and loads obtained are compared with 2D and 3D linear frequency domain method, as previously described.

The non-dimensional heave and pitch responses at zero speed are presented in Figures 4 and 5. It is visible that in the zero speed case the solution of the non-linear method tends to converge to the response given by the 2D linear approach. This is due to the fact that for small amplitude waves the proposed method and the Strip Theory have a similar formulation.

Figure 6 shows the comparisons for the heave prediction for a forward speed of 10 knots ($F_N = 0.125$). The results for the pitch motion at the same velocity are displayed in Figure 7. The curve shows a good agreement for the heave motion also at forward speed condition. The pitch response given by the proposed formulation agrees with the

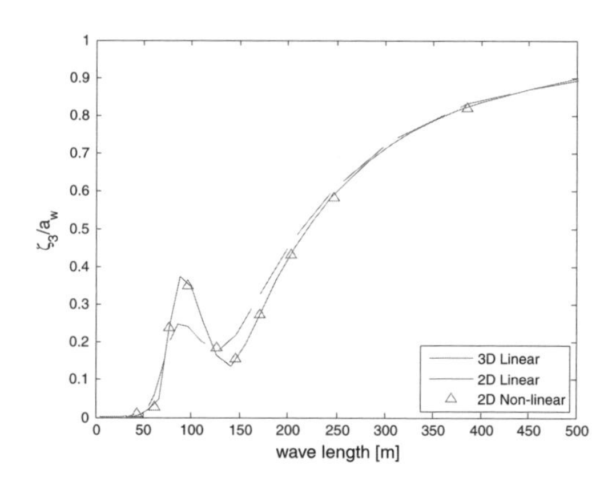

Figure 4. Non-dimensional heave response for small amplitude waves at zero speed.

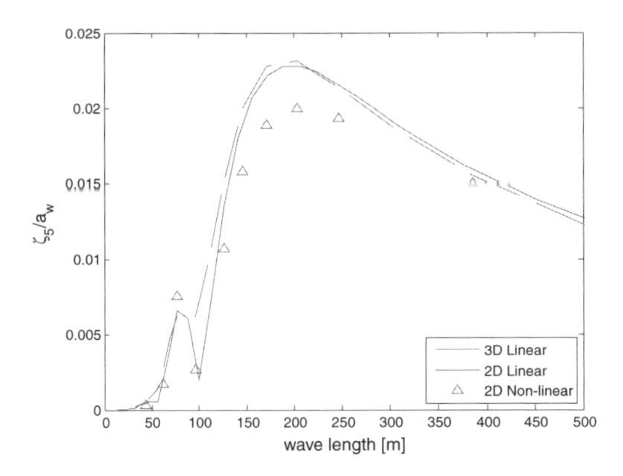

Figure 5. Non-dimensional pitch response for small amplitude waves at zero speed.

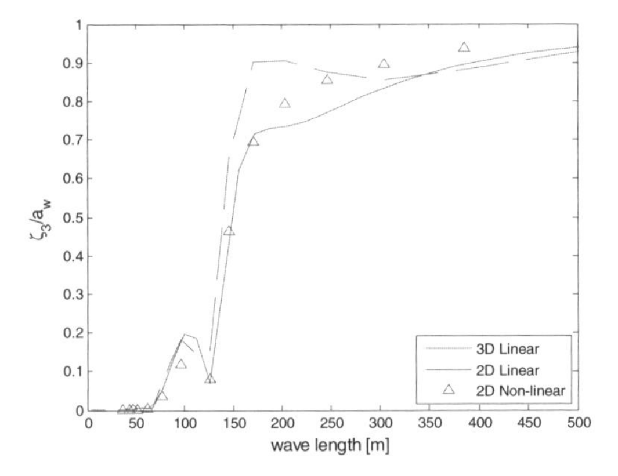

Figure 6. Non-dimensional heave response for small amplitude waves at $F_N = 0.125$.

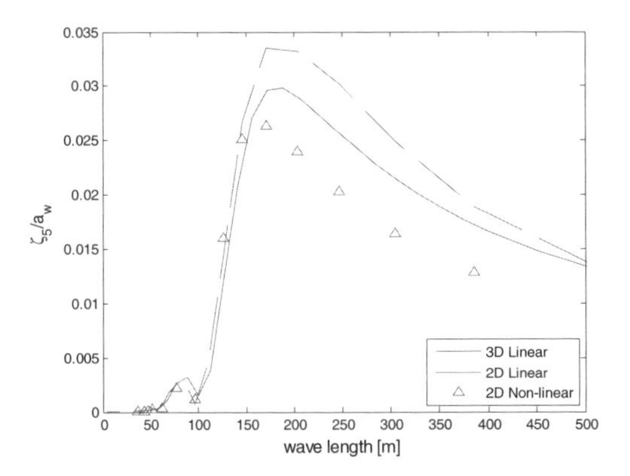

Figure 7. Non-dimensional pitch response for small amplitude waves at $F_N = 0.125$.

linear theories for the short waves; for longer waves the pitch is smaller than the one obtained from commercial software. This is due to the different formulation for the hydrodynamic coefficients and exciting forces. In the proposed method the forward speed correction for the sectional radiation and the diffraction problem is calculated without considering the contribution given by the aft-most section. The different formulation of the forward speed problem affects more the pitch motions because the forward speed terms have a larger magnitude in the equation of motion that models the pitch.

Figures 8 and 9 compare the results obtained for the vertical shear force at station 15 (131.25 m from the AP) and the vertical bending moment at station 10 (87.5 m from the AP), since at these sections the longitudinal distributions of both loads have their maximum values. The figures show a good agreement with the strip theory prediction. The results obtained by the three dimensional Green's function methods are different from the others. The pre-processing of the program modifies the mass distribution to obtain the balance between weight and buoyancy. The modified distribution is different from the ones used in the other calculations, and it reduces the load predictions.

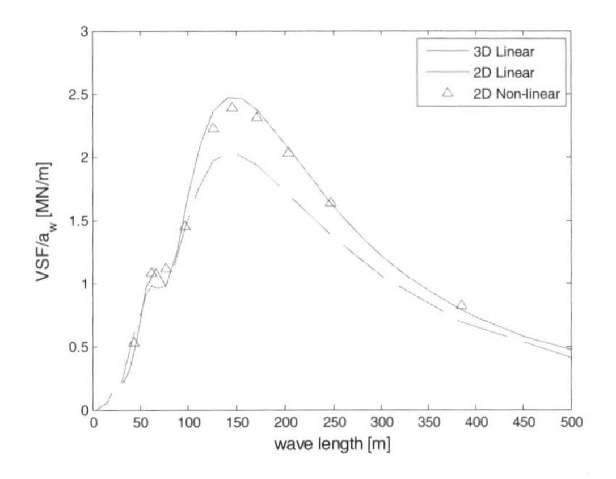

Figure 8. Non-dimensional vertical shear force at station 15 for small amplitude waves at zero speed.

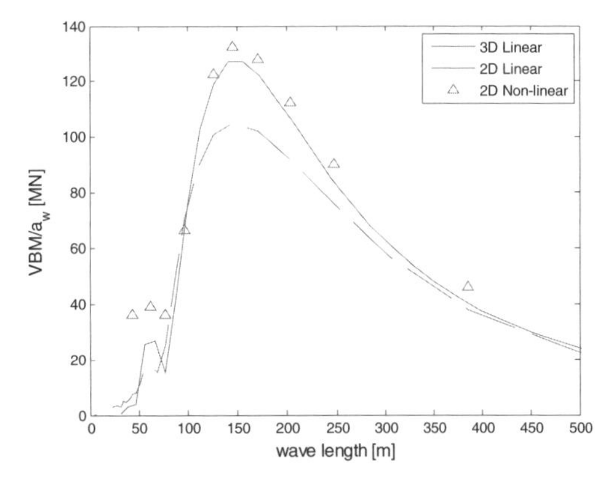

Figure 9. Non-dimensional vertical bending moment at station 10 for small amplitude waves at zero speed.

3.2 *Large amplitude results*

In order to study the behaviour of the vessel under large amplitude waves several simulations are carried out increasing the wave amplitude. The simulations are conducted in head sea condition at zero speed considering incoming waves with a length equal to 1.40 L_{BP}. The results are compared to the linear prediction. Linear methodologies are based on the assumption of linearity between the exciting force and the ship response. Following this hypothesis, the linear prediction at any wave amplitude is given by the product of the non-dimensional response obtained by the small amplitude analysis with the given exciting wave amplitude.

Figures 10 and 11 present the results for heave and pitch motions. The wave amplitude starts from 0.5 m and rises until 5 m. When the wave elevation raises the nonlinear predictions deviate from the linear response. Figure 12 shows the same analysis for the vertical bending moment at station 10.

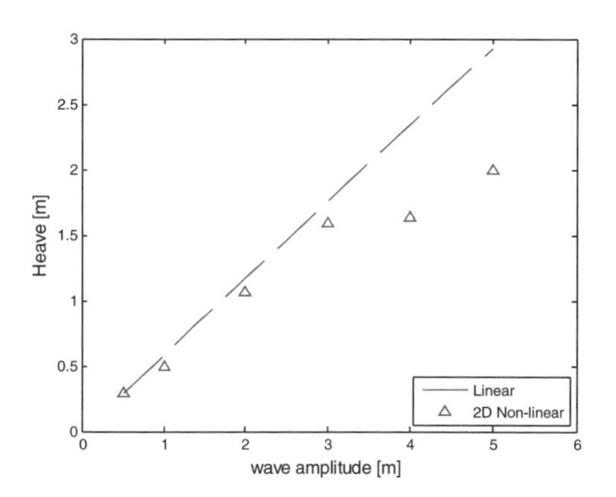

Figure 10. Heave response for different wave amplitudes at $\lambda = 1.40\ L_{BP}$.

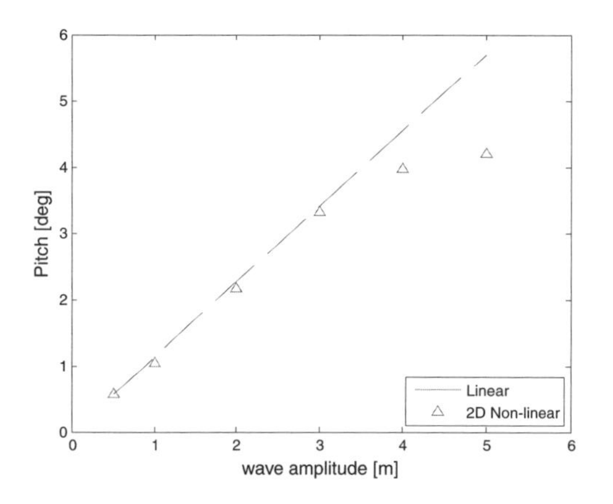

Figure 11. Pitch response for different wave amplitudes at $\lambda = 1.40\ L_{BP}$.

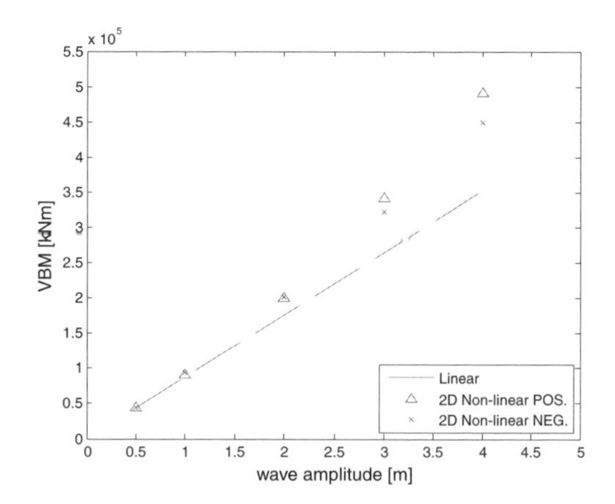

Figure 12. Vertical bending moment at Station 10 for different wave amplitudes at $\lambda = 1.40\ L_{BP}$.

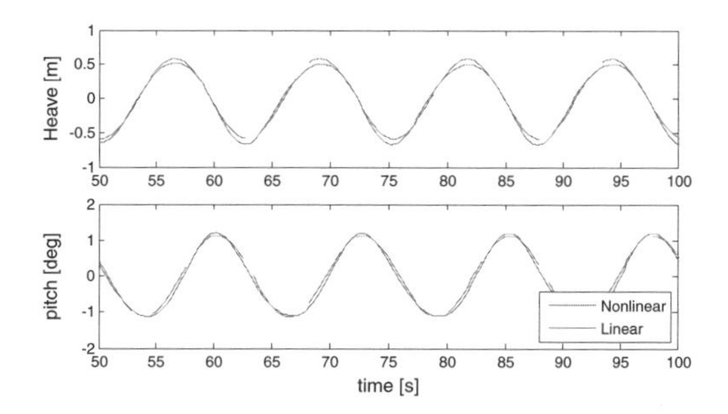

Figure 13. Heave and pitch for $a_W = 1.0\ m$ $\lambda = 1.40\ L_{BP}$.

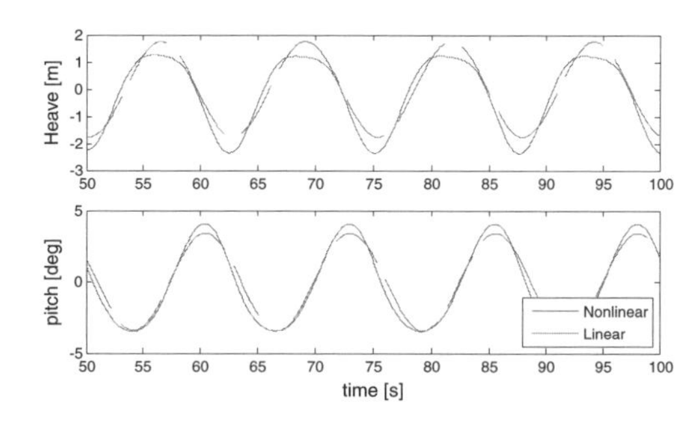

Figure 14. Heave and pitch for $a_W = 3.0\ m$ $\lambda = 1.40\ L_{BP}$.

In this case there is an increment of response at larger waves compared to the linear prediction. The non-linear vertical bending moment response is asymmetrical, i.e. the positive amplitude rises faster than the negative one. This analysis is conducted considering the maximum and the minimum value of the time histories. At higher amplitude waves the loads and motions are strongly nonlinear and a Fourier analysis should be conducted.

The loads and motions obtained by a linear calculation are sinusoidal. It can be accepted for small amplitude motions, but this is a strong simplification for larger motion. Figure 13 compares the heave and pitch time history obtained by linear and nonlinear calculations for a wave elevation of 1 m. The two responses are very similar and sinusoidal.

Figure 14 describes the same comparison for a wave with larger amplitude. In this case there is visible difference between the linear and the nonlinear prediction. The nonlinear response is non sinusoidal and asymmetrical especially for the heave motion.

4 CONCLUSIONS

A two dimensional time domain nonlinear "blended" method has been presented. The nonlinear radiation problem is solved partially in time domain through the convolution integral using the actual wetted hull portion and the exciting wave elevation as free surface at each time step. Restoring and exciting forces are nonlinear and numerically evaluated in time domain on the actual body. Results for the S-175 container ship have been presented.

Comparisons with linear techniques for small amplitude waves have been carried out. The results show a good agreement at zero speed. The forward speed prediction could be improved by introducing the contribution of the aft-most section for the radiation and diffraction problem. Loads analysis at zero speed has also been presented.

A first analysis of the ship response under large amplitude waves has been introduced. The results show the nonlinear behaviour of ship motions and loads. No comparison with other calculations or experiments has been presented.

To improve the prediction of motions and loads under large amplitude waves, other nonlinear phenomena, such as green water on the deck and water entry problem, should be modelled and considered in the formulation of equation of motions.

In order to validate the proposed methodology comparisons with experimental results will be conducted in the future, for the chosen vessel and for other hull forms.

ACKNOWLEDGEMENTS

The authors acknowledge the Lloyd's Register Marine Business and EPSRC that support this three year project on large amplitude motions analysis at the University of Strathclyde Naval Architecture and Marine Engineering department.

'Lloyd's Register, its affiliates and subsidiaries and their respective officers, employees or agents are, individually and collectively, referred to in this clause as the "Lloyd's Register Group". The Lloyd's Register Group assumes no responsibility and shall not be liable to any person for any loss, damage or expense caused by reliance on the information or advice in this document or howsoever provided, unless that person has signed a contract with the relevant Lloyd's Register Group entity for the provision of this information or advice and that in this case any responsibility or liability is exclusively on the terms and conditions set out in that contract'.

REFERENCES

Beck, R.F. 1994. Time-domain computations for floating bodies. *Applied Ocean Research* Vol. 16: 267–282.

Cummins, W.E. 1962. The impulse response function and ship motions. *DTMB Report 1661*.

Fonseca, N. & Guedes Soares, C. 1998. Time- domain analysis of large-amplitude vertical ship motions and wave loads. *Journal of Ship Research* Vol. 42, No. 2:139–153.

Fonseca, N. & Guedes Soares, C. 2004. Experimental investigation of the nonlinear effects on the vertical motions and loads of a containership in regular waves. *Journal of Ship Research* Vol. 48, No. 2: 118–147.

ISSC, 2009. Report of the ISSC Technical Committee I.2 on Loads. *17th International Ship and offshore Structures Congress (ISSC)*, ISBN: 9788995 473016, Seoul, Korea.

Salvesen, N. Tuck, E.O. & Faltinsen, O. 1970. Ship motions and sea loads. *Trans. SNAME 78*: 250–287.

Xia, J. Wang, Z. & Jensen, J.J. 1998. Non-linear wave loads and ship responses by a time-domain strip theory. *Marine structures 11*: 101–123.

Sustainable Maritime Transportation and Exploitation of Sea Resources – Rizzuto & Guedes Soares (eds)
© 2012 Taylor & Francis Group, London, ISBN 978-0-415-62081-9

A nonlinear time-domain hybrid method for simulating the motions of ships advancing in waves

S. Liu & A.D. Papanikolaou
Ship Design Laboratory, National Technical University of Athens, Athens, Greece

ABSTRACT: An efficient nonlinear time domain hybrid method for simulating motions of ships advancing in waves is presented. The method is a combination of a time-domain transient Green function method and a Rankine source method and enables the treatment of non-wall-sided bodies and shiplike forms. Radiation and diffraction forces (and moments) are calculated with respect to the mean wetted surface, whereas wave exciting and restoring forces (and moments) are taken over the actually wetted body surface. In the present paper, the application of the method to the simulation of the motions of the ITTC S175 container ship in a body-fixed coordinate system is presented, which greatly reduces computational time. Good agreement has been observed between the results of the present method, other numerical codes and experimental data.

1 INTRODUCTION

Accurate prediction of the seakeeping behavior of ships in heavy seas is of great importance for ship design and operation. Quasi 2D strip theory approaches were the first which delivered accurate enough results for practical applications to ship motions' prediction and they are widely applied even today. With the rapid advance of computer technology in the 70ties, various frequency domain 3D approaches were successfully developed. But due to some inherent limitations, its successful application is sometimes limited to a certain extent.

On the other hand, the time domain simulation methods become more and more popular as it enables the address of large amplitude ship motion problems which is very important for the design and the assessment of safe operation of modern ships operating in a variety of adverse environmental conditions. Following the pioneering work of Finkelstein (1957) and Cummins (1960), many researchers investigated seakeeping problems by different time domain approaches and showed promising results for both linear problem and nonlinear problems of different level. Lin & Yue (1990) showed the applicability of a time domain Green function method to large amplitude ship motions. Following this formulation, Singh et al. (2003) appeared to have obtained good results in some applications. However, this could not be confirmed in some other practical cases by other researchers, namely, when this method is applied to floating bodies with a flare at the waterline, which is common to modern ship designs, numerical problems may arise and computations fail. Duan & Dai (1999) found that the commonly used panel method employing the transient Green function for a non-wall-sided floating body does not satisfy the mean-value theorem of definite integrals for the near water surface panels and solved this problem by introducing an imaginary vertical surface, which encloses the hull surface in the fluid domain. This method works fine, unless the body has some bulb-like hull form, which exceeds the projection of the water plane. Zhang (1998) used a similar scheme, but the matching surface is placed some distance away from the body and moving at the same speed as the ship, so that there should be not any problem with body's shape. In this scheme, a part of the free-surface is included in the inner domain, which needs updating at each time step. Yasukawa (2003) and Kataoka & Iwashita (2004) also used similar schemes to solve the problem. The difference between these methods basically lies in the different ways of treatment of the boundary condition in the far field, in simulating the free surface and the numerical schemes. More recently, there is a trend of integrating CFD techniques into the hybrid method so as to study highly nonlinear phenomena. Iafrati & Campana (2003) presented a hybrid method combining a CFD scheme using conventional grids and the BEM for potential-flow free-surface problems. Sueyoshi et al. (2007) uses particle methods in the inner domain and a boundary element method in the outer domain to study various wave-free surface problems. Lin et al. (2009) presented recently a paper where they combined a viscous flow solver in the inner domain and potential flow solver in the outer domain. There is a slight overlap between the two introduced domains, which creates a matching domain.

In line with Lin & Yue's work (1990), the authors proceeded with the development of a Transient Green Function method (TDGF) and demonstrated its implementation by applications to some fundamental hydrodynamic problems, both of 1st order (Liu, et al. 2006) and of quasi 2nd order (Liu, et al. 2011). Following this development and in order to simulate the motions of practical shiplike forms, a time domain hybrid method was formulated and validated (Liu & Papanikolaou, 2009 & 2010). In this method, the fluid domain is decomposed into an inner and an outer part. The Rankine source method is applied in the inner domain to find the dominant equation for the velocity potential, while the transient Green function method is used in the outer domain to obtain a relationship for the velocity potential and its normal derivative, as necessary for the matching condition on the control surface between the inner and outer domain solution. This hybrid method works efficiently with a relatively small number of panels, compared to a pure Rankine source method, because the free surface panelization is restricted between the body boundary and the control surface. A double integration algorithm with respect to time, originally developed by Wang (2003), is adopted to simulate the free-surface condition. Especially, a *Chimera* grid method (Liu & Papanikolaou, 2011) has been adopted in the applications to forward speed problems, which is good for local geometry representation and allows small time step size. Figure 1 shows an example used in this scheme. Though good results have been obtained, as shown in Figure 2, the overall developed method demands too much CPU time for updating the influential matrices at every time step thus is not very satisfactory.

In order to reduce the simulation time, a new formulation, which is slightly different from the previous one, has been developed. In this new formulation, the hydrodynamic problem is solved in the body travelling coordinate system. The matching surface is set at certain distance away from the hull and travelling with the body, thus the free surface area (and corresponding paneling in the numerical solution) will remain unchanged, which is actually quite the same with the zero speed problem case; thus there is no update of the influence matrices necessary and the simulation time is greatly reduced.

In this paper, after a validation case study on the S175 ITTC container ship, the method will be applied to study the above water hull effect on ship motions, thus to show how it can be used in the above-water hull form optimization of ships.

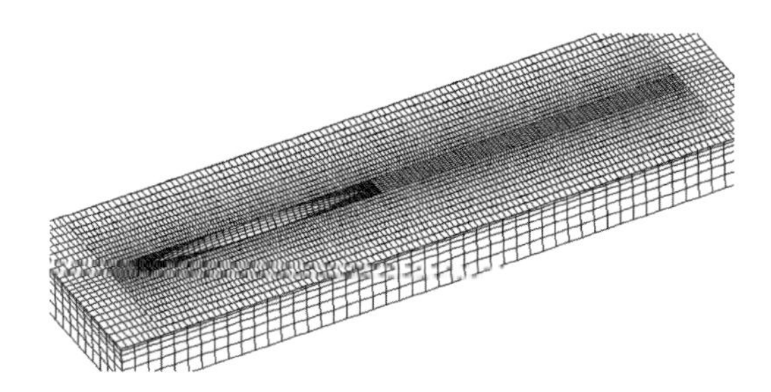

Figure 1. Example of runtime panelization of a Wigley hull.

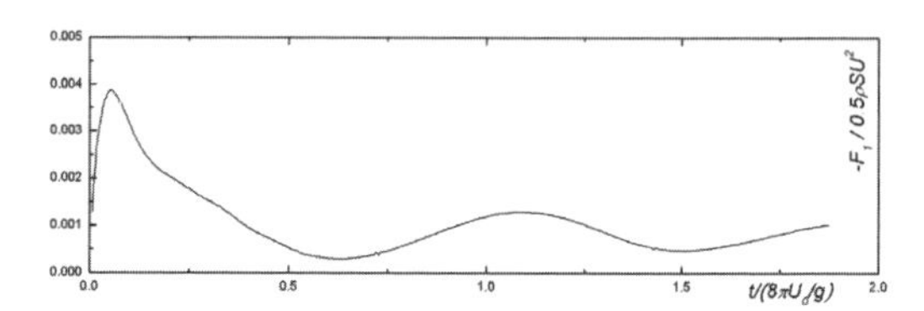

Figure 2. Wave making resistance of a Wigley hull, without sinkage/trim correction, Fn = 0.25.

2 FORMULATIONS

The ship motions are simulated by following Jan Otto de Kat's methodology (1990). The motions of the ship are determined by the orientation of body-fixed system $Gx'y'z'$ relative to $Oxyz$ system, which travels with the body. A total of six components are needed to define the motions, typically three translations, i.e. surge, sway and heave, and three rotations, i.e. roll, pitch and yaw. The six degrees of freedom motion of a rigid body in space is determined by the following two equations:

$$
\begin{aligned}
\frac{d\vec{P}}{dt} &= \Delta \cdot \frac{d\vec{v}}{dt} = \vec{F} = \vec{F}_{HS} + \vec{F}_I + \vec{F}_R + \vec{F}_D \\
\frac{d\vec{L}}{dt} &= I \cdot \frac{d\vec{\omega}}{dt} + \vec{\omega} \times (I \cdot \vec{\omega}) \\
&= \vec{M} = \vec{M}_{HS} + \vec{M}_I + \vec{M}_R + \vec{M}_D
\end{aligned}
\tag{1}
$$

where $\mathbf{P} = m\mathbf{v}$, $\mathbf{L} = \mathbf{I} \cdot \boldsymbol{\omega}$. $\mathbf{F}$ and $\mathbf{M}$ are the total force on the body and the total moment about mass center, Δ is the mass and $\mathbf{v}$ is the absolute velocity vector of the gravity center in the Oxyz system, and $\mathbf{I}$ and $\boldsymbol{\omega}$ are the inertia tensor and angular velocity about the rotating point, which is assumed to be ship's gravity center G. The moments and products of inertia in $\mathbf{I}$ are constants in the moving and rotating system Gx'y'z'. In this paper, the conservation of linear momentum equations will be solved in the earth-fixed reference system, while the conservation of angular momentum will be expressed in the local body-fixed system. The location of a rigid body in space is fully determined by the position of G in the fixed system O'xyz and the angular orientation of the Gx'y'z' system with respect to the earth fixed system.

On the right-hand side of the motion Equation (1), there are force components due to diffraction, radiation, incident wave, and restoring. Basically these forces/moments are calculated by integrating the pressure expressed by Bernoulli's equation on the body surface. Since an exact, fully nonlinear model is quite time-consuming and complicated for numerical computation, we restrict ourselves in this paper to the consideration of some of the more important and tractable nonlinear effects. In particular, for simulating large amplitude ship motions, the incident wave forces (Froude-Krylov) and restoring forces are taken over the actually wetted body surface and transferred into the motion equations.

For the hydrodynamic forces, both radiation and diffraction forces, are calculated up to mean wetted surface by using Bernoulli's equation. The corresponding velocity potentials are solved by applying the hybrid method as elaborated in following.

A matching surface is introduced to split the whole fluid domain into an inner domain and an outer domain. In the inner domain I, the Rankine source method is used to solve the flow field. The integral equation takes the form of:

$$\iint\limits_{S_c+S_b+S_f}\left[\Phi_I(q,t)\frac{\partial}{\partial n_q}\left(\frac{1}{r_{pq}}\right)-\frac{1}{r_{pq}}\frac{\partial}{\partial n_q}\Phi_I(q,t)\right]ds_q = -4\pi\,\Phi_I(p,t) \qquad (p\in I) \tag{2}$$

where **n** is the unit normal vector pointing outward of the inner domain. Let p approach the boundary then we will get:

$$\iint\limits_{S_c+S_b+S_f}\left[\Phi_I(q,t)\frac{\partial}{\partial n_q}\left(\frac{1}{r_{pq}}\right)-\frac{1}{r_{pq}}\frac{\partial}{\partial n_q}\Phi_I(q,t)\right]ds_q = -2\pi\Phi_I(p,t) \qquad (p\in S_I) \tag{3}$$

where $q(\xi,\eta,\zeta,\tau)$ is the source point, $p(x,y,z,t)$ is the field point; $\boldsymbol{r}_{pq} = (x-\xi)\boldsymbol{i} + (y-\eta)\boldsymbol{j} + (z-\zeta)\boldsymbol{k}$; the denotation S_b, S_c and S_f represent respectively the body surface, the control surface and the free surface.

In the outer domain II, the transient time-domain Green function is employed to solve the disturbed potential on S_c. The integral equation is expressed as:

$$\begin{aligned} &2\pi\Phi_{II}(p,t) \\ &+\iint\limits_{S_c}\left[\Phi_{II}(q,t)\frac{\partial}{\partial n_q}\left(\frac{1}{r_{pq}}-\frac{1}{r_{pq'}}\right)-\left(\frac{1}{r_{pq}}-\frac{1}{r_{pq'}}\right)\frac{\partial\Phi_{II}}{\partial n_q}\right]ds_q \\ &=\int_0^t d\tau\iint\limits_{S_c}\left(\tilde{G}\frac{\partial\Phi_{II}}{\partial n_q}-\Phi_{II}\frac{\partial\tilde{G}}{\partial n_q}\right)ds_q \\ &+\frac{1}{g}\int_0^t d\tau\int\limits_{wl(\tau)}V_N\left(\tilde{G}\frac{\partial\Phi_{II}}{\partial\tau}-\Phi_{II}\frac{\partial\tilde{G}}{\partial\tau}\right)dl_q \\ &\qquad (p\in S_c)\end{aligned} \tag{4}$$

2.1 *Free-surface condition*

The linearized free-surface condition can be expressed in the body-fixed coordinate system as:

$$\left(\frac{\partial}{\partial t}-U_0\frac{\partial}{\partial x}\right)^2\Phi_i+g\frac{\partial\Phi_i}{\partial z}=0 \qquad (i=1,2,\ldots,7) \tag{5}$$

This expression can be rewritten as:

$$\frac{\partial^2\Phi}{\partial t^2}-2U_0\frac{\partial^2\Phi}{\partial x\partial t}+U_0{}^2\frac{\partial^2\Phi}{\partial x^2}=-g\frac{\partial\Phi}{\partial z} \tag{6}$$

It has been shown by Wang (2003) that by using Laplace transform, Φ can be expressed as following:

$$\begin{aligned}\Phi(x,y,0,t)&=H\left(t-\frac{x_0-x}{u}\right)\\ &\times\left[\Phi\left(x_0,y,0,t-\frac{x_0-x}{u}\right)+\frac{x_0-x}{u}\frac{\partial}{\partial t}\Phi\left(x_0,y,0,t-\frac{x_0-x}{u}\right)\right.\\ &\left.-(x_0-x)\frac{\partial}{\partial x}\Phi\left(x_0,y,0,t-\frac{x_0-x}{u}\right)\right]+\frac{g}{u^2}\int_x^{x_0}(x-\xi)\frac{\partial}{\partial n}\\ &\Phi\left(\xi,y,0,t-\frac{\xi-x}{u}\right)H\left(t-\frac{\xi-x}{u}\right)d\xi\end{aligned} \tag{7}$$

where $H(t)$ is the *unit step function*. It should be noted that the initial conditions for panels located in the area that is free from disturbance/occupation of a sailing hull or not are different.

2.2 *Matching boundary condition*

On the control surface, the potential solutions from the inner domain model and the outer domain model should match each other. Considering the definitions of the normal vectors in different domains we have:

$$\Phi_I=\Phi_{II},\ \frac{\partial\Phi_I}{\partial n}=\frac{\partial\Phi_{II}}{\partial n}\ (on\ S_c) \tag{8}$$

2.3 *Body boundary condition*

The linearized body boundary condition is defined on the mean wetted surface:

$$\frac{\partial \Phi}{\partial n} = -\frac{\partial \Phi_0}{\partial n} + \sum_{j=1}^{6} \left[\dot{\eta}_j(t) n_j + U_0 \eta_j(t) m_j \right] \tag{9}$$

The m_j term is expressed as following:

$$\begin{bmatrix} m_1 \\ m_2 \\ m_3 \end{bmatrix} = -\frac{1}{U_0} \begin{bmatrix} n_1 \Phi_{sxx} + n_2 \Phi_{sxy} + n_3 \Phi_{sxz} \\ n_1 \Phi_{syx} + n_2 \Phi_{syy} + n_3 \Phi_{syz} \\ n_1 \Phi_{szx} + n_2 \Phi_{szy} + n_3 \Phi_{szz} \end{bmatrix}$$

$$\begin{bmatrix} m_4 \\ m_5 \\ m_6 \end{bmatrix} = \begin{bmatrix} y m_3 - z m_2 - \frac{n_2}{U_0} \Phi_{sz} + \frac{n_3}{U_0} \Phi_{sy} \\ z m_1 - x m_3 - \frac{n_3}{U_0} (\Phi_{sx} - U_0) + \frac{n_1}{U_0} \Phi_{sz} \\ x m_2 - y m_1 + \frac{n_2}{U_0} (\Phi_{sx} - U_0) - \frac{n_1}{U_0} \Phi_{sy} \end{bmatrix} \tag{10}$$

It is clear that during the procedure of linearizing the body boundary condition of radiation problems to the mean wetted surface, the contribution of the steady potential results in this m_j term which requires the computation of the second gradients of the basic steady potential. It has been shown by other researchers (Zhao & Faltinsen, 1989; Duan & Price, 2002) that it is difficult to obtain good results by the direct computations. In the present study, two different methods for m_j term calculation are implemented, as elaborated in the following.

2.3.1 *S.T.F. Simplification*

According to the Salvesen-Tuck-Faltinsen (S.T.F., 1970) method's simplification, the basic flow corresponds to an undisturbed stream $-U_0 x$, thus

$$(m_1, m_2, m_3) = -(\mathbf{n} \cdot \nabla)\nabla(-U_0 x)/U_0 = (0, 0, 0) \tag{11}$$

$$(m_4, m_5, m_6) = -(\mathbf{n} \cdot \nabla)[\mathbf{r} \times \nabla(-U_0 x)]/U_0 = (0, n_3, -n_2) \tag{12}$$

This is consistent with Neumann-Kelvin linearization and easy to implement.

2.3.2 *Direct calculation*

There are research groups who compute the m_j term on the basis of a double-body linearization (Chen et al., 2000). However, as extra integral equations will have to be solved, it is not adapted in the present study. Instead, the procedure which is proposed by Wu (1991) and followed by Chen & Malenica (1996) and Kim (2005) is implemented in the present hybrid method's framework.

The velocity potential in the inner domain can be expressed in the source distribution form as:

$$4\pi \Phi_s(p,t) = \iint_{S_c+S_b+S_f} \sigma(q,t) \frac{1}{r_{pq}} ds_q \tag{13}$$

Since we got Φ_s through the aforementioned hybrid solver, the source density σ can be evaluated by the above expression. Afterwards, the spatial derivatives can be evaluated as:

$$\frac{\partial \Phi_s(p,t)}{\partial x_k} = \frac{1}{4\pi} \iint_{S_c+S_b+S_f} \sigma(q,t) \frac{\partial}{\partial x_k} \frac{1}{r_{pq}} ds_q \tag{14}$$

The partial derivative of the velocity potential in the inner domain can also be expressed in the source distribution form as:

$$4\pi \frac{\partial \Phi_s(p,t)}{\partial x_k} = \iint_{S_c+S_b+S_f} \sigma_k(q,t) \frac{1}{r_{pq}} ds_q \tag{15}$$

Since $\partial \Phi_s(p,t)/\partial x_k$ has been obtained through the previous calculation, the source density σ_k can be evaluated by the above expression. Afterwards, the second order spatial derivatives can be evaluated as:

$$4\pi \frac{\partial^2 \Phi_s(p,t)}{\partial x_k \partial x_j} = \iint_{S_c+S_b+S_f} \sigma_k(q,t) \frac{\partial}{\partial x_j} \left(\frac{1}{r_{pq}} \right) ds_q \tag{16}$$

3 NUMERICAL SCHEMES

The constant panel method is used to numerically solve the defined BVP. Details have been given in papers published before (Liu & Papanikolaou, 2010). Here only the discretized form of the free surface condition is given:

$$\Phi(x,y,0,t) = \begin{cases} -g(\Delta t)^2 \sum_{j=1}^{n-1} j \frac{\partial}{\partial n} \Phi_{m-j}^{n-j} & n \le m \\ \Phi_0^{n-m} - m\Delta x \frac{\partial}{\partial x} \Phi_0^{n-m} + m\Delta t \frac{\partial}{\partial t} \Phi_0^{n-m} \\ -\frac{1}{2} g(\Delta t)^2 m \frac{\partial}{\partial n} \Phi_0^{n-m} \\ -g(\Delta t)^2 \sum_{j=1}^{m-1} j \frac{\partial}{\partial n} \Phi_{m-j}^{n-j} & n > m \end{cases} \tag{17}$$

where m is time-step index and n is the x-axis position index.

4 RESULTS AND DISCUSSION

In order to validate the above introduced time-domain hybrid method, some numerical tests have been carried out.

4.1 *Validation on S175 container ship*

Numerical experiments were carried out first to determine the optimal size and paneling of the free surface area. Based on our experience, we change the width, depth and length of the solution domain systematically and compare the resulting added masses and damping coefficients. Finally we set the width as max (3B, 0.5 λ), depth as max (5D, 0.4 λ), and length max (3L, 2λ+L). With this setting, the following results are obtained (denoted as HYBRID II), as shown in Figure 3.

Also plotted are the results from the 3D frequency domain code panel code NEWDRIFT (Papanikolaou, 1985; denoted by ND), a 3D body exact formulation (Zhang, et al., 2010; denoted by *3D-B.E.*) and the Hybrid method formulated in earth-fixed coordinate system (denoted as HYBRID I). Obvious deviations among different methods have been observed in this study.

The motions of the S175 container ship in head seas have also been studied by using the present method. For small amplitude motion case, we assume the gravity (mass) centre is on the calm waterplane at the mid-ship and all the forces are estimated about the assumed gravity center to obtain the motion prediction. For large amplitude motion modeling, we estimate the radiation and diffraction forces up to the mean wetted surface but the hydrostatic and Froude-Krylov part exactly about the actual gravity center to obtain the motion prediction. The m_j term is calculated either based on Neumann-Kelvin simplification or directly as elaborated in Section 2.3.

The numerical results on heave and pitch motions are shown in Figure 4 and Figure 5. During the simulation, the wave amplitude is set as constant A/L = 0.01. By using a small amplitude model, with either m_j term computation, the heave motion is overestimated in long wave range. By using the large amplitude model, which introduces the exact computation of Froude-Krylov force and restoring forces, the heave motion in long wave range is improved and gets closer to experimental data. For the m_j term, the effect is mainly around the peak range, which is also observed in others' computations (Bingham & Maniar, 1996). Computational results also suggest that the m_j term simplification according to the S.T.F. strip theory assumption leads to quite reasonable

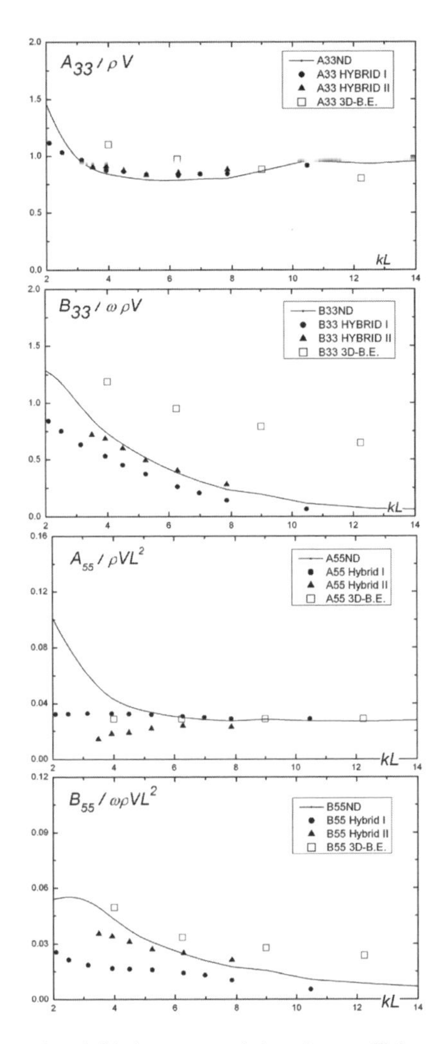

Figure 3. Added masses and damping coefficients due to forced heave and pitch motions, S175 ship, Fn = 0.275.

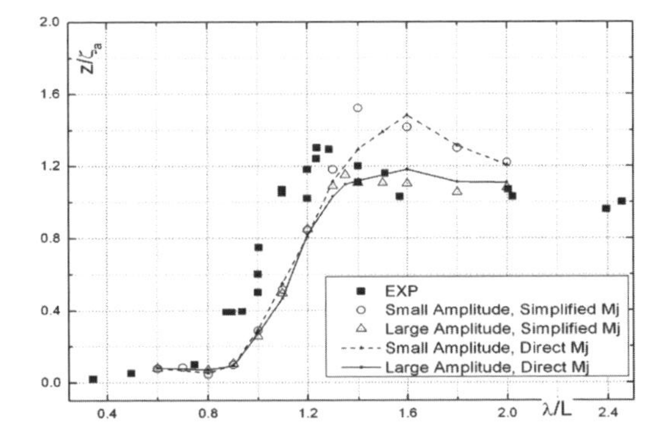

Figure 4. Heave motion amplitude calculation of S175 ship, Fn = 0.275.

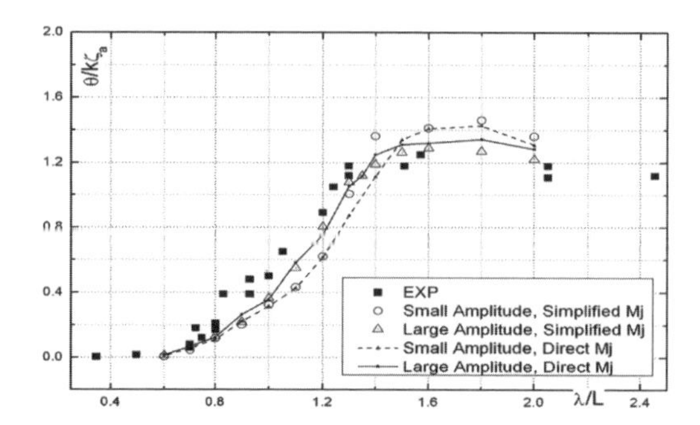

Figure 5. Pitch motion amplitude calculation of S175 ship, Fn = 0.275.

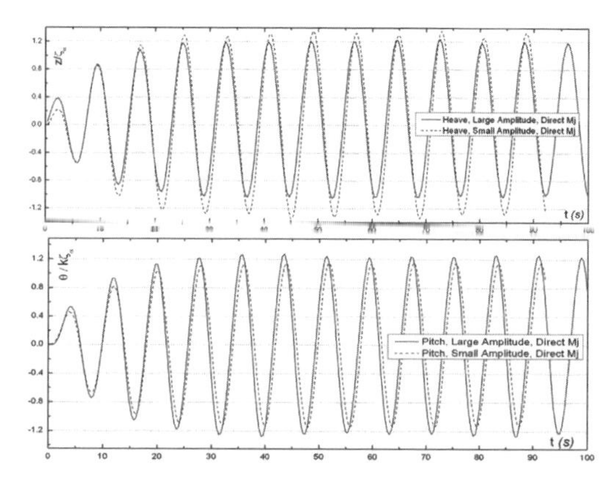

Figure 6. Heave and pitch motion histories calculation of S175 ship, λ/L = 1.4, Fn = 0.275.

results, except for some shift of the heave response curve for λ comparable to L (Figure 4).

For the pitch motion, when the small amplitude model is used, we observed some shift, when compared to the experimental data; it should be herein noted that the wave steepness (nonlinearity) in the physical model experiments is not available for more direct comparison. The different m_j term assumptions appear here to only affect the peak value. But by applying the large amplitude motion model, the pitch motion results were obviously improved, the shift becomes weak and the results in all the studied range closely match the experimental results. Furthermore, when the Froude-Krylov force and restoring forces are exactly calculated, it will introduce some serious differences, compared to the linear calculation in the small amplitude model, around bow and stern where the flare of sections is significant. This effect is more visible in the pitch response, thus it improved the prediction. Figure 6 shows the heave and pitch motion histories of the S175 ship with λ/L = 1.4, either with small amplitude simulation or with large amplitude simulation. Due to the fact that the wave amplitude is small, A/λ = 0.01, the nonlinearity is not strong. However, there are still obvious deviations between the two curves from different models. This indicates the importance of applying a more exact model even when the wave amplitude is small.

It is understood that around the resonance range, the hydrodynamic forces are comparable to the hydrostatic counterparts and the interaction phenomenon is quite complicated. When we drop the line-integral term in integral Equation (4), some error will obviously arise which cannot be estimated satisfactorily at the moment.

4.2 *The effect of the above waterplane hull form*

In the previous studies, it is shown that the developed hybrid method is capable of studying the effect of above waterplane hull shape changes; thus, it can be applied to the optimization of the above water hull shape of ships. In demonstrating this, we will apply the developed numerical method and computer code to the basic Wigley III hull and two modifications of the basic hull with respect to the above waterplane hull forms; it is defined by the following expression:

$$x = x_0(1+z) \quad (z \geq 0;\ \text{modified})$$

$$x = x_0(1+z) \quad (x \leq 0,\ z \geq 0;\ \text{modified V2})$$

$$y = \begin{cases} b\left(1-\left(\dfrac{2x_0}{L}\right)^2\right)\left(1-\left(\dfrac{z}{D}\right)^2\right) \\ \times\left(1+\left(\dfrac{2x_0}{L}\right)^2/5\right)\left(1+\left(\dfrac{2x_0 z}{L(D_0-D)}\right)^2\right) & z < 0 \\ b\left(1-\left(\dfrac{2x_0}{L}\right)^2\right)\left(1+\left(\dfrac{2x_0}{L}\right)^2/5\right) & z \geq 0 \end{cases}$$

Figure 7 shows the panelization of the modified Wigley hull V2, with a slight flare introduced in the bow/stern regions and an over-hang at the stern. The hybrid method is applied to predict the heave and pitch motion of this hull and results are shown in Figure 8 to Figure 11. Figures 8 and 9 show heave and pitch motions at small wave steepness, $A/L = 0.01$. At this condition, results of applying the large amplitude model are almost identical to the results of the small amplitude model. The results of both modified hulls slightly deviate from the results of the original hull, because the wave amplitude is small and the flare is also limited near the waterplane, so that the exact calculation of incident wave forces and restoring forces does not affect much the computation.

At a steeper wave condition, $A/L = 0.02$, more obvious deviations show up for both heave and

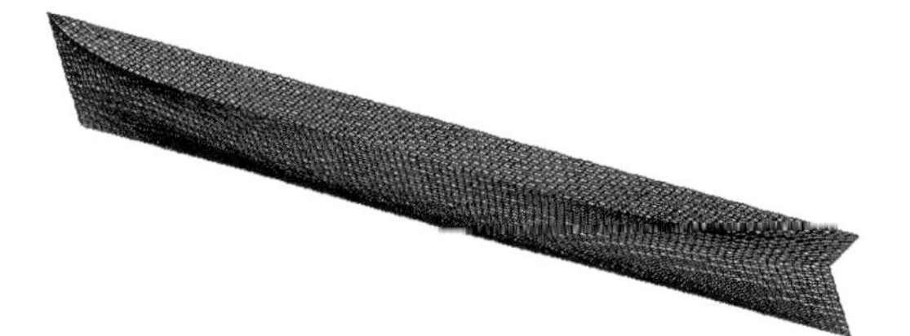

Figure 7. Panelization of the modified Wigley III hull V2.

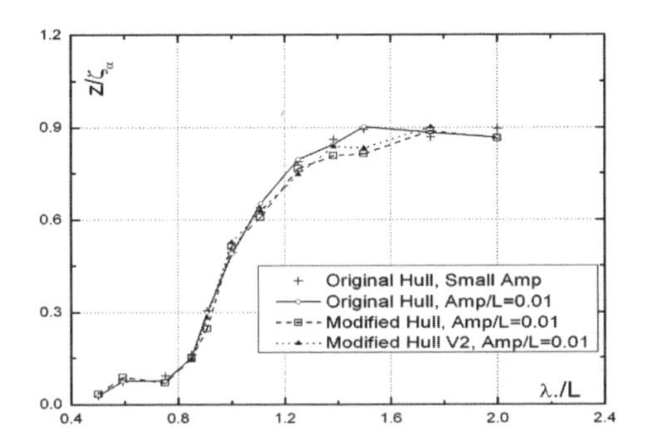

Figure 8. Heave amplitude of three different Wigley hulls, Fn = 0.2, Amp/L = 0.01.

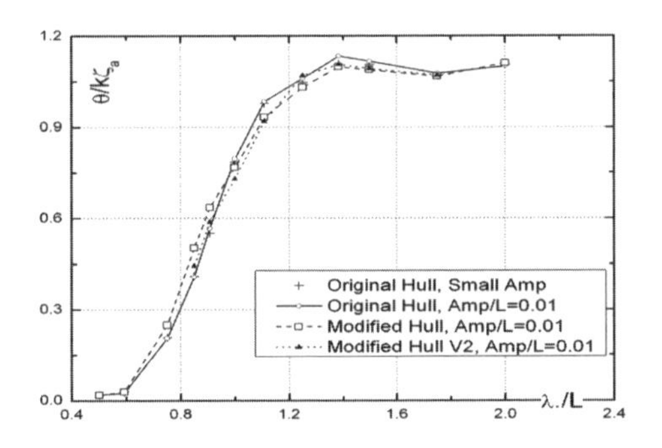

Figure 9. Pitch amplitude of three different Wigley hulls, Fn = 0.2, Amp/L = 0.01.

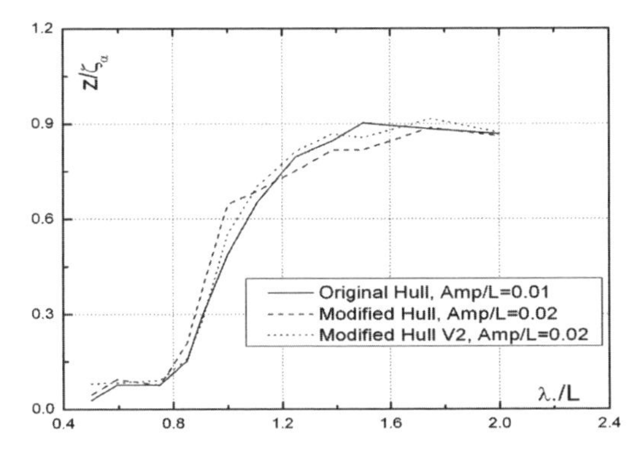

Figure 10. Heave amplitude of three different Wigley hulls, Fn = 0.2, Amp/L = 0.02.

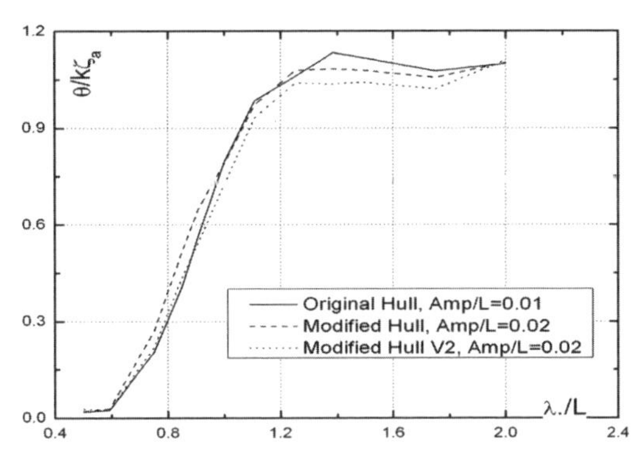

Figure 11. Pitch amplitude of three different Wigley hulls, Fn = 0.2, Amp/L = 0.02.

pitch motions, as shown in Figures 10 and 11. For heave motion, the amplitude of the modified hull has decreased compared to the other two hulls due to its large projected area on the vertical direction. On the other hand, the pitch motion of the modified hull V2 is obviously smaller than for the other two hulls in the long wave range. These calculations, offer some valuable information regarding the applicability of these methods in the preliminary design stage. When considering the actual sea conditions encountered in ship's service route, we may be able to determine the optimal route or even the optimal hull shape for specific routes. We should keep in mind that all the three Wigley hulls are actually very narrow, with L/B = 10, compared to actual shiplike forms and the introduced flares are also quite conservative compared to a real ship. Due to this reason, the mj term based on Neumann-Kelvin simplification is employed.

4.3 *Computational time*

An important aspect of the presented work refers to improving the computational efficiency of an early version of the developed hybrid method computer code, which is formulated in an earth-fixed system. For forward speed problem the hull will be moving, thus at every time step, the free surface near the ship will change, so that the influence matrices need to be updated at every time step. It takes about 30 seconds for preparing the matrices, depending on how complicated the problem is. After the *Chimera* grid concept is introduced, though the panels that are far away from the ship are fixed, thus do not need to be updated at every time step, there is still a considerable amount of data processing at each time step. For one simulation, it takes more than 5 hours for the forced motion problem or even more than 10 hours for more complicated large amplitude motion problems on a regular PC hardware with Intel Core 2 QUAD CPU (Q8200 2.33 GHz). As the method is essentially a potential flow solver, this is not satisfactory.

Solving the problem in the body-travelling coordinate system, which results in a panelization system that is quite similar to the zero speed prob lems, things greatly improve. For a typical motion simulation in head seas condition, it is possible to finish a run within 2 hours. But now the occurring problems are in two points:

1. the free surface condition update is complicate;
2. the matching surface condition update is not accurate due to the omission of the waterline integral term which appears in Equation (4).

The second point also affects the first point internally through the solved integral equations. This is reflected in the results, where some shift/bending shows up in the motions' RAOs. In this respect, recently other researchers argue on the effect of far field condition and propose some new type of far field condition (Lin & Kuang, 2009); also, methods for including the effect of the line integral terms have also been discussed, but not did reach yet maturity for practical applications (Delhommeau, et al., 2001).

5 CONCLUSIONS AND FUTURE WORK

An efficient time domain hybrid method is formulated and typical results of its numerical implementation are presented in this paper. As good agreement was obtained for a validation case study on S175 container ship, the method is applied to study the motion of a Wigley hull with different above water hull shape and show the effect on motion performance thus proves the method can be used as a fundamental tool for optimization purpose with decent computational efficiency.

The way ahead of this work is to explore alternative concepts for the matching surface conditions and further improve the efficiency of the numerical method by parallel programming in the new computer—cluster environment of NTUA-SDL.

ACKNOWLEDGEMENT

The presented work is part of the PhD work of the first author; the financial support of NTUA-SDL and continuous guidance by Professors A. Papanikolaou and W. Duan (Harbin Engineering University) is acknowledged.

REFERENCES

Bingham, H.B. & Maniar, H.D. 1996. Computing the double-body m-terms using a B-spline based panel method. *11th International Workshop on Water Waves and Floating Bodies*.

Chen, X.B. & Diebold, L. 2000. New Green-function method to predict wave induced ship motions and loads. *23rd Symposium on Naval Hydrodynamics*.

Chen, X.B. & Malenica, S. 1996. Nonlinear effects of the local steady flow on wave diffraction-radiation at low forward speed. *Int. Offshore and Polar Engineering Conference*.

Cummins, W.E. 1962. The impulsive response function and ship motions. *Schiffstechnik*, 9:124–135.

Delhommeau, G., Maury, C., Boin, J.-P., Guilbaud, M. & Ba, M. 2001. Influence of waterline integral and irregular frequencies in seakeeping computations. *4th Num. tank Symposium*, Hamburg.

Duan, W.Y. & Dai, Y.S. 1999. Time-domain calculation of hydrodynamic forces on ships with large flare. *International Shipbuilding Progress*, 46:223–232.

Duan, W.Y. & Price, W.G. 2002, A numerical method to solve the m-terms of a submerged body with forward speed. *Int. J. Num meth in Fluids*, Vol. 40.

Finkelstein, A. 1957. The initial value problem for transient water waves. *Pure App. Maths.*, 10: 511–522.

Iafrati, A. & Campana, E.F. 2003. A domain decomposition approach to compute wave breaking (wave breaking flows). *Int. J. Numer. Meth. Fluids*, Vol. 41, pp. 419–445.

Kat, J.O. 1990. The numerical modeling of ship motions and capsizing in severe seas. *Journal Ship Research*.

Kataoka, S. & Iwashita, H. 2004. Estimations of hydrodynamic forces acting on ships advancing in the calm water and waves by a time-domain hybrid method. *Journal of the Society of Naval Architects of Japan. 196:123–138.*

Kim, B. 2005. Some considerations on forward speed seakeeping calculations in frequency domain. *Int. J. Offshore and Polar Engineering*, Vol. 15.

Lin, R.Q. & Kuang, W.J. 2009. A nonlinear method for predicting motions of fast ships. *10th International Conference on Fast Sea Transportation*.

Lin, W.M. & Yue, D. 1990. Numerical Solutions for Large-Amplitude Ship Motions in the Time-Domain. *Proc. 18th Symposium on Naval Hydrodynamics*, pp. 41–66.

Lin, W.M., Chen, H. & Zhang, S. 2009. A hybrid numerical method for wet deck slamming on a high speed catamaran. *10th International Conference on Fast Sea Transportation*.

Liu, S.K. & Papanikolaou, A. 2009. A time-domain hybrid method for calculating hydrodynamic forces on ships in waves. *13th Inter. Congress of the Inter. Maritime Association of the Mediterranean*, Istanbul.

Liu, S.K. & Papanikolaou, A. 2010. Time-domain hybrid method for simulating large amplitude motions of ships advancing in waves. *Proceeding of ITTC benchmark study and workshop on seakeeping*, accepted for republication at the Journal of the Korean Society of Naval Architects and Ocean Engineers.

Liu, S.K. & Papanikolaou, A. 2011. Application of Chimera grid concept to simulation of the free-surface boundary condition. *26th International Workshop on Water Waves and Floating Bodies*.

Liu, S.K., Papanikolaou, A. & Duan W.Y. 2006. A time domain numerical simulation method for nonlinear ship motions. *Journal of Harbin Engineering University*, Vol. 27, pp. 177–185.

Liu, S.K., Papanikolaou, A. & Zaraphonitis, G. 2011. Prediction of added resistance of ships in waves. *Ocean Engineering*.

Papanikolaou, A. 1985. On Integral-Equation-Methods for the Evaluation of Motions and Loads of Arbitrary Bodies in Waves. *Journal Ingenieur—Archiv.*, 55:17–29.

Salvesen, N., Tuck, E.O. & Faltinsen, O. 1970. Ship motions and sea loads. *Trans. Soc. Nav. Arch. and Marine Eng.*, Vol. 78, pp. 250–287.

Singh, S.P., Sen, D. & Sarangdhar, D.G. 2003. 3D seakeeping computation for different hull forms for design evaluations. *15th International Conference on hydrodynamics in ship design, safety and operation*, Gdnask, 245–56.

Sueyoshi, M., Kihara, H. & Kashiwagi, M. 2007. A hybrid technique using particle and boundary-element methods for wave-body interaction problems. *9th International Conference on Numerical Ship Hydrodynamics*.

Wang, J.F. 2003. Numerical Simulation of the Linear Free-Surface Condition. Master Thesis, *Harbin Engineering University*.

Wu, G.X. 1991. A numerical scheme for calculating the mj-terms in wave-current-body interaction problem. *Applied Ocean Research*, Vol. 13.

Yasukawa, H. 2003. Application of a 3-D Time Domain Panel Method to Ship Seakeeping Problems. *24th Symposium on Naval Hydrodynamics*, 376–392.

Zhang, S. 1998. A hybrid boundary-element method for non-wall-sided bodies with or without forward speed. 13th International Workshop on Water Waves and Floating Bodies.

Zhang, X., Bandyk, P. & Beck, R.F. 2010. Time domain simulations of radiation and diffraction forces. *J. Ship Research*.

Zhao, R. & Faltinsen, O.M. 1989. A discussion on the mj-terms in the wave-current-body interaction problem. *Proc 4th International Workshop on Water Waves and Floating Bodies*.

Sustainable Maritime Transportation and Exploitation of Sea Resources – Rizzuto & Guedes Soares (eds)
 ISBN 978-0-415-62081-9

Loads and motions of offshore structures in extreme seas

G.F. Clauss & M. Dudek
Division of Naval Architecture and Ocean Engineering—Technical University Berlin, Berlin, Germany

ABSTRACT: During the design process of floating structures, different boundary conditions have to be taken into account. Besides the basic determination of the type of vessel, the range of application and the main dimensions at the initial stage, the reliability and the warranty of economical efficiency are an inevitable integral part of the design process. Model tests to evaluate the characteristics and the performance of the floating structure are an important milestone within this process. Therefore it is necessary to determine an adequate test procedure which covers all essential areas of interest. The focus lies on the limiting criteria of the design such as maximum global loads, maximum motions or relative motions between two or more vessels or maximum accelerations, at which the floating structure has to operate or to survive. These criteria are typically combined with a limiting characteristic sea state (H_s, T_p) or a rogue wave. However, the important question remains: What is the worst case scenario for each design parameter—the highest rogue wave or a wave group of certain frequency? And which sea states have to be taken into account for the experimental evaluation of the limiting criteria? As an approach to these questions, several critical wave sequences are characterized and analysed for different scenarios in order to identify the most important characteristics and to improve the efficiency of model tests.

1 INTRODUCTION

Productivity and survivability are critical parameters for offshore construction design. Modern computer-based analysis techniques allow calculation and simulation of these parameters in the design stage, but nevertheless model tests are indispensable for validating the respective approaches. The central problem that has to be solved in the design process of maritime structures is the choice of environmental design conditions to be considered:

- What are the maximum wave heights?
- What is the worst case scenario—the highest wave or a wave group with certain frequency characteristics?

Of course, this issue cannot be solved globally, i.e., different operating conditions (transit, operation, survival) and characteristics (body motions, local and global loads) will lead to individual results. In order to verify the calculated parameters sought, sophisticated model tests are required. Since complex sea states as wave groups have to be generated as exact as possible but with the least amount of time and money, only experienced test facilities are capable of conducting such experimental series.

Linear frequency-domain analysis, i.e., the calculation of RAOs (Response Amplitude Operators), is a basic but fast and elegant approach to investigate the motion characteristics of floating structures prior to model tests (Clauss et al., 1992), which is provided by various numerical tools available. By stochastic analysis procedures, the annual downtime of offshore production facilities can be calculated on basis of this method. Clauss and Birk (1996) proposed a procedure for optimizing buoyancy bodies of floating structures based on linear theory, where the geometry is generated automatically and subsequently analysed linearly and modified to find the minimum of a certain target function (annual downtime).

Extreme sea states and their consequences on body motions and loads can be investigated by transferring the linear RAOs into time-domain. Jacobsen and Clauss (2006) applied this method to analyse the seakeeping behaviour of multi-body systems. Motions and global loads of a FPSO in extreme sea states were numerically simulated and successfully validated by model tests (Clauss et al., 2004).

Impact loads and motions in severe seas and rogue waves are investigated with special emphasis on the influence of bow shape (Watanabe et al., 1989). Fonseca and Guedes Soares (2002) compared numerical and experimental results of wave-induced vertical ship motions and loads, revealing that the geometry of the bow flare in combination with wave steepness significantly influences the global loads and leads to higher sagging than hogging moments. This nonlinearity is mostly associated with the "nonlinear geometry" of the hull and increases with decreasing block coefficient. Furthermore, these nonlinear effects were approved and enhanced by Clauss et al. (2010, a) for different ship types in high, steep, regular waves as well as in irregular waves, confirming the influence of block coefficient, bow flare and freeboard height.

Besides the analysis of motion behaviour and global loads, the assessment of the maximum number of responses during a structure's lifetime also has to include local loads. Stansberg (2008) determined a parameter to assess the risks of slamming at a ship's bow due to waves and wave groups. As already mentioned above, the exact generation and reproduction, respectively, of sea states in a wave tank is an important component of the analysis of floating structures. An experimental optimization procedure for tailor-made wave sequence generation in a wave tank has been proposed by Clauss and Schmittner (2005), which enables the exact reproduction of wave or wave groups of desired characteristics.

This paper presents a new optimization approach for the determination of critical situations (i.e., wave sequences) combining the advantages of the previously mentioned methods:

- Step 1 discribes the process of developing tailored wave sequences—integrated into selected sea states,
- Step 2 presents the dedicated analysis of wave-structure interaction,
- Step 3 illustrates the response based design, i.e., the identification of critical waves or wave sequences leading to excessive response.

In conclusion the following scenarios are discussed:

- A single rogue wave hidden in a severe sea state is a highly dangerous event challenging the performance and safety of offshore structures.
- If a sea state contains wave groups which excite resonance motions, e.g., roll or parametric roll of cruising ships (depending on course and speed), the dynamic behaviour of the vessel (or structure) may exaggerate the wave height effects.
- The worst case scenario is occurring if wave sequences with relevant frequency characteristics are combining with extreme wave heights.

2 STEP 1: GENERATION OF TAILORED WAVE SEQUENCES

2.1 *Linear optimization of target wave trains*

The method for generating linear wave groups is based on the wave focussing technique of Davis and Zarnick (1964), and its significant development by Takezawa and Hirayama (1976).

Clauss and Bergmann (1986) recommended a special type of transient waves, i.e., Gaussian wave packets, which have the advantage that their propagation behaviour can be predicted analytically, later on completely performed numerically (Clauss and Kühnlein 1995). The generated wave trains are predictable at any instant and at any stationary or moving location. According to its high accuracy the technique is capable of generating special purpose transient waves.

In general, "rogue" waves or critical wave groups are rare events embedded in a random seaway. As long as linear wave theory is applied, the sea state can be regarded as superposition of independent harmonic "component" waves, each having a particular direction, amplitude, frequency and phase. For a given design variance spectrum of an unidirectional wave train, the phase spectrum is responsible for all local characteristics, e.g., the wave height and period distribution as well as the location of the highest wave crest in time and space.

Extreme wave conditions in a 100-year design storm arise from unfavourable superpositions of component waves which represent the severe sea spectrum. Freak waves have been registered in standard irregular seas when component waves accidentally superimpose in phase. Extensive random time-domain simulation of the ocean surface for obtaining statistics of the extremes, however, is very time consuming. Generally, when generating irregular seas in a wave tank the phase shift is supposed to be random, however, it can be fixed by the control program on the basis of a pseudo-random process: Consequently, the phase spectrum is also given as a deterministic quantity and can be optimized. Why should we wait for rare events if we can achieve freak wave conditions by intentionally selecting a suitable phase shift, and generate a deterministic sequence of waves converging at a preset concentration point? At this position all waves are superimposed without phase shift resulting in a single high wave peak. Assuming linear wave theory, the synthesis and up-stream transformation of appropriate wave packets is developed from this concentration point, and the Fourier transform of the wave train (phase spectrum) is transformed back to the upstream position at the wave board.

2.2 *Nonlinear transient wave description*

The generation of higher and steeper wave sequences requires a more sophisticated approach as propagation velocity increases with height. Kühnlein (1997) developed a semi-empirical procedure for the evolution of extremely high wave groups which is based on linear wave theory: The propagation of high and steep wave trains is calculated by iterative integration of coupled equations of particle positions. With this deterministic technique "freak" waves up to 3.2 m high have been generated in a wave tank (Clauss and Kühnlein 1997).

For the simulation of the nonlinear wave propagation a potential theory solver has been developed at Technical University Berlin (WAVETUB) (Clauss and Steinhagen 1999; Steinhagen 2001). Therefore, the two dimensional nonlinear free surface flow problem is solved in time domain—a complete description of this numerical wave tank is published by Steinhagen (2001). Tailored wave groups can be integrated into irregular seas using a Sequential Quadratic Programming (SQP) method and nonlinear wave theory (Hennig 2005).

Summarizing the achievements of the numerical wave tank we can generate deterministic high and steep wave sequences at a selected position, and integrate them into irregular seas with defined significant wave height and period.

2.3 *Self-validating procedure for optimizing tailored design wave sequences*

To further improve the process of the deterministic evolution of tailored wave sequences, a fully automized optimization technique with integrated validation has been developed. As shown in Figure 1 the wave tank is combined with a computer system representing a hybrid control loop for optimizing the selected wave train at a target position. Nonlinear free surface effects are automatically considered in the fitting process since the wave train propagates in the real wave tank, and thus the objective function is automatically validated (Clauss 2002). Based on deviations between the measured wave sequence and the design wave group at target location the control signal for generating the seaway is iteratively optimized in the fully automatic computer-controlled model test procedure (Fig. 1).

As an example, Figure 2(a) presents the evolution of the famous *Draupner* New Year Wave (Haver and Anderson 2000) (full scale data deduced from model tests at a scale of 1:70). The registrations show how the extremely high wave develops on its way to the target position: Two kilometers ahead of the concentration point we observe a wave group with three high waves between 700 s and 730 s. The slightly longer waves (right hand side waves of the registration) are travelling faster catching up with the leading wave of the group. Just 500 m ahead of the concentration point only two higher waves are visible, superimposing to the single New Year Wave at target. The registrations at 500 m and 1000 m behind the target document the dissipation of the wave sequence. Figure 2(b) presents the measured registration of the New Year Wave at target position. As compared to the registered New Year Wave the experimental simulation is quite satisfactory. Finally, Figure 2(c) shows a snapshot of the freak wave at target time.

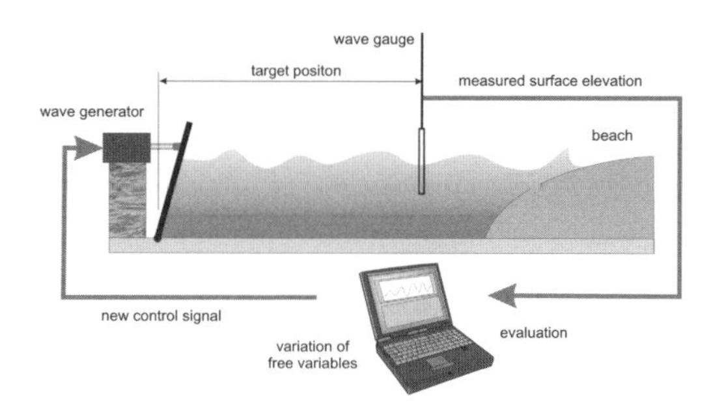

Figure 1. Computer controlled experimental optimization of tailored design wave sequences.

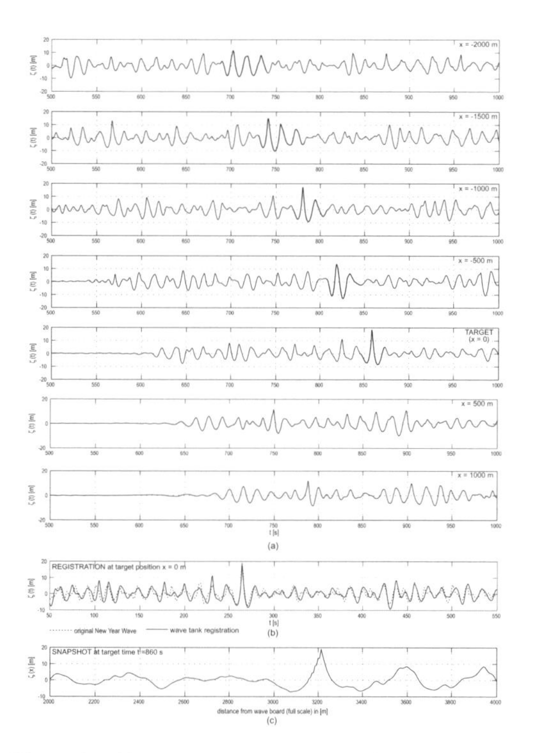

Figure 2. New Year Wave: (a) Evolution of the New Year Wave (scale 1:70), (b) Experimental simulation in comparison to the registered New Year Wave (full scale wave data) and (c) Snapshot of the experimental simulation at target time.

3 STEP 2: WAVE-STRUCTURE INTERACTION

3.1 *Modeling system behavior*

The performance assessment of offshore structures starts with computation of selected response amplitude operators of forces and motions. For hydrodynamically compact structures the well established 3D diffraction-radiation software package WAMIT (Wave Analysis Massachusetts Institute of Technology) is used (Department of Ocean Engineering, MIT 1994).

The Response Amplitude Operators (RAOs) of the body motions s_{la}/ζ_a follow from Newton's motion equation which constitute a system of linear equations if small harmonic motions $s_l = s_{la}e^{-i\omega t}$ are assumed:

$$[-\omega^2(M + A) + i\omega B + C]\frac{s}{\zeta_a} = \underline{f}. \tag{1}$$

Hydrodynamic analysis of the structure has to provide the system matrices for added masses A, potential damping B and hydrostatic restoring coefficients C as well as the exciting wave forces f.

Considering the well known boundary value problem of a rigid body in plane harmonic waves, the total velocity potential Φ has to satisfy the Laplace equation

$$\Delta\Phi = 0 \tag{2}$$

and can be separated into a superposition of

- the potential Φ_0 of the incident plane waves,
- the diffraction potential Φ_7, which expresses the interaction of the incident wave and the fixed body, and
- the six radiation potentials $\Phi_l = \dot{s}\phi_l$ which characterize the impact of body motions on the fluid. The local potential functions $\phi_l (l = 1, 2 ..., 6)$ are due to body motions.

Summation yields,

$$\Phi = \Phi_0 + \Phi_7 + \sum_{l=1}^{6} \dot{s}\phi_l. \tag{3}$$

In addition to the Laplace Equation (2), the solution must also satisfy the linearized boundary conditions on the ocean bottom (z = –d), on the free surface (z = 0), on the wetted body surface (mean position) and in the far field (radiation condition).

The diffraction-radiation program WAMIT solves the linear Fredholm integral equations for the six local potential functions ϕ_l and for the total diffraction potential $\phi_d = \phi_0 + \phi_7$. As a result, the integral equations are reduced to a set of seven systems of linear equations—the number of unknowns equals the number of panels N. The WAMIT algorithms for evaluating the coefficients of the linear equations with the Green function $G(X, \xi)$ are presented by Newman (1986). The accuracy of the solution of the linear systems is controlled by an iterative equation solver (Newman and Sclavounos 1988).

After calculating the velocity potentials we obtain the exciting forces and moments by integrating the dynamic pressure over the wetted surface of the body. Normalized by the wave amplitude ζ_a this gives the vector of transfer functions $f = (f_1, f_2, ..., f_6)^T$:

$$f_l = -\frac{i\omega\rho}{\zeta_a}\iint_{S_b} \phi_D n_l dS \tag{4}$$

The matrix of hydrodynamic masses $A = a_{lj}$ and the matrix of potential damping coefficients $B = b_{lj}$ are obtained from the well known relation

$$a_{lj} + \frac{i}{\omega}b_{lj} = \rho\iint_{S_b} n_l\phi_j dS. \tag{5}$$

The $[6 \times 6]$ matrix $C = c_{lj}$ contains the hydrostatic restoring coefficients (Newman 1977). Once the system matrices and the exciting force are known, the response amplitude operators can be calculated from Equation 1.

3.2 *Modeling environmental conditions*

In reality, the elevation of the ocean surface is irregular and of random nature. Hence, rational seakeeping criteria have to be based on a probabilistic description of random seas. Gaussian distribution of wave elevations and Rayleigh distribution of wave heights are assumed. The relevant parameters are significant wave height H_s and mean zero-upcrossing period T_0. The significant wave height is linked to the variance σ^2 of the random process by $H_s = 4\sqrt{\sigma^2}$ (Newland 1975). The probabilities of sea states are recorded in wave scatter diagrams (Hogben and Lumb 1967) which may be subdivided according to the region, season and direction of wave origin. For computing purposes the discrete data are approximated by an analytical joint probability density function $p(T_0, H_s)$ (Mathisen and Bitner-Gregerson 1990). Nonlinear regression yields the corresponding function parameters.

3.3 *Downtime prediction*

The application of spectral analysis in ship dynamics started with the fundamental publication of (Denis and Pierson 1953). Natural seaway is interpreted as a random superposition of a great number of harmonic long-crested waves of different amplitudes ζ_{an} and frequencies ω_n (Clauss et al., 1992).

Assuming linear behaviour of the system, the harmonic excitation of a structure due to a wave of frequency ω_n yields a phase shifted harmonic response of the same frequency. The seaway (input signal) as well as the response (output signal) are a superposition of an infinite number of harmonic elementary components. The complex ratio of output and input signals $s(\omega)$ and $\zeta(\omega)$ for each wave frequency ω constitutes the transfer function or response amplitude operator (RAO) which gives a complete description of the corresponding hydrodynamic characteristics.

$$H(\omega) = \frac{s(\omega)}{\zeta(\omega)} = \frac{s_a(\omega)}{\zeta_a(\omega)} e^{i\varepsilon}. \tag{6}$$

Each component wave contributes an amount of energy to the seaway proportional to its squared wave amplitude. The spectral density S represents the energy distribution as a function of wave frequency ω. Evidently two characteristic parameters define the shape of the energy distribution. The significant wave height

$$H_s = 4\sqrt{m_0} \tag{7}$$

and the zero-up-crossing period

$$T_0 = 2\pi\sqrt{m_0/m_2} \tag{8}$$

with $m_n = \int_0^\infty \omega^n S_{zz}(\omega) d\omega$. Corresponding to the wave spectrum $S_{zz}(\omega)$ of the seaway, the response spectrum $S_{ss}(\omega)$ represents the energy distribution of the output signal. Wave and response spectra are related by

$$S_{ss}(\omega) = |H_{s\zeta}(\omega)|^2 S_{zz}(\omega). \tag{9}$$

Similarly to the significant wave height H_s, a significant force or motion double amplitude $(2s)_s$ follows from the response spectrum:

$$(2s)_s = 4\sqrt{\int_0^\infty S_{ss}(\omega) d\omega}. \tag{10}$$

These significant double amplitudes characterize the behaviour of offshore structures in stationary sea states, and are appropriate hydrodynamic 'measure of merit' criteria for the evaluation procedure. Of course, any combination of significant double amplitudes of forces, motions, or other hydrodynamic parameters of interest may be used as objective functions. If limiting values of significant double amplitudes of forces and motions $(2s_a)_{s,limit}$ are known, the objective function (Eq. 10) defined above may be extended to include long term wave statistics.

4 STEP 3: RESPONSE BASED EVALUATION OF WAVE/STRUCTURE INTERACTIONS

In the previous sections the accurate generation of arbitrary extreme wave sequences embedded in irregular seas has been presented. Concerning wave/structure interactions, with respect to response based design the crucial question has to be answered: Is the highest wave with the steepest crest the most relevant design condition or should we identify wave sequences with critical frequency characteristics embedded in an irregular wave train? In addition to the global parameters H_S and T_P the wave effects on a structure depend on its dynamic behaviour as well as on superposition and interaction of wave components, i.e., on local wave characteristics.

If a cruising vessel is investigated the wave characteristics must be transformed to the moving reference frame, i.e., the characteristics of the encountering wave are relevant. Phase relations and nonlinear interactions are key parameters to specify the relevant surface profile at the (moving) structure. Only if wave kinematics and dynamics are known, cause-effect relationships can be detected.

The aiming target is the response of the structure which follows from the associated response amplitude operators (in frequency-domain) or impulse response functions (in time-domain).

4.1 *Offshore lift-off operation*

The first case study is a typical offshore problem, i.e., the lift-off operation of a heavy load from a barge by a semisubmersible crane vessel (Jacobsen and Clauss 2006). Figure 3 illustrates the proposed lift-off operation of such a multibody system, i.e., the crane vessel *Thialf* and a barge.

Three stationary states of the transfer of a 10000 t load are discussed.

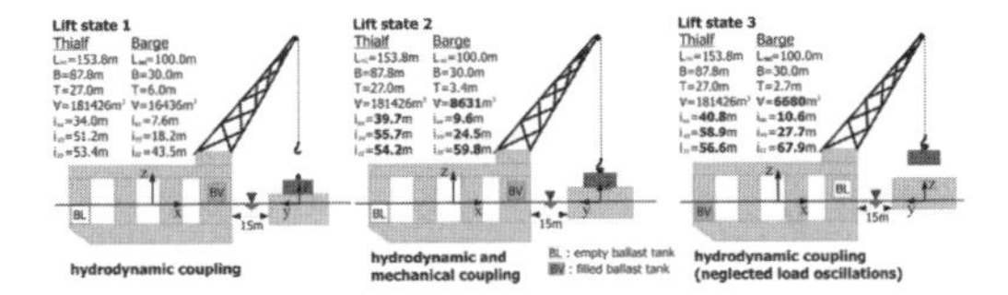

Figure 3. Sketch of the investigated multi-body systems—three states of the lift-off operation (load 10000 t).

- transport barge carrying the load, no mechanical connection, structures are coupled hydrodynamically only
- load is fastened, roap is pretensioned up to 80%, structures are coupled mechanically also
- load is hanging, fast lift up by usage of rapid ballast water system, hydrodynamically coupling again.

The most critical parameter of this process is the vertical relative motion between barge and load shortly after lift-off since this is the limiting criterion (Lift state 3). As the seaway excites both vessels the resulting motion depends on the incoming wave field as well as on the scattered and radiated wave fields of both structures, which are hydrodynamically coupled (Newman 2001). The hydrodynamic analysis in frequency-domain is performed by using the diffraction code WAMIT. The body surface is discretised to solve the system of integral equations numerically.

After evaluating the total potential the pressure is derived from the linearised, instationary Bernoulli equation. The forces and moments acting on the body are determined by integrating the pressure over the body surface, including hydrostatic as well as hydrodynamic forces and moments. Related to Newton's second law, the equation of motion for a stable, linear system is obtained:

$$(\underline{M}+\underline{A})\cdot\underline{\ddot{s}}+\underline{B}\cdot\underline{\dot{s}}+\underline{C}\cdot\underline{s}=\underline{F}_{err} \tag{11}$$

By assuming harmonic forces as well as motions due to harmonic waves the equation of motion becomes time-independent

$$\{-\omega^2(\underline{M}+\underline{A})-i\omega\underline{B}+\underline{C}\}\frac{s_a}{\zeta_a}e^{i\varepsilon}=\frac{F_{err,a}}{\zeta_a}e^{i\gamma} \tag{12}$$

and the associated response amplitude operator (Eq. 6) is obtained by relating motions and forces to the wave amplitude ζ_a. The decision, however, whether an operation is feasible needs information about the behaviour in natural seaways.

The motion behaviour in a natural seaway is obtained by spectral analysis. For engineering purposes standard spectra as the Pierson-Moskowitz spectrum are formulated. By multiplying the sea

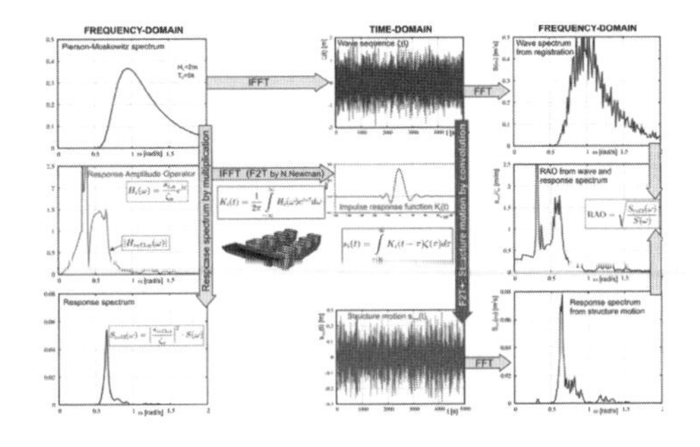

Figure 4. Response of structures in the sea—evaluation in frequency and in time-domain (relative motion between elevated load and barge during a lift-off operation with the crane semisub mersible *Thialf*).

spectrum with the squared RAO the response spectrum is obtained (Fig. 4). Relevant characteristics, as significant double amplitude $(2s_a)_s$ and zero-up crossing period $T_{o,s}$ of the response are derived from the associated response spectrum. The procedure is similar to the determination of significant wave height H_S and Zero-up crossing period T_0 from the sea spectrum. To derive assessment criteria in arbitrary seaways the procedure is repeated for several sea states with varied zero-upcrossing periods T_0. The resulting significant double amplitudes are normalized by the significant wave height. This leads to the significant response amplitude operator as a function of the zero-upcrossing period

$$\frac{(2s_a)_s}{H_S}=f(T_0), \tag{13}$$

which allows the comparison of different structures or configurations and serves as basis for operation decisions.

With frequency-domain results the motion behaviour of multi-body systems in waves are investigated very fast and efficiently, but the results can only be interpreted statistically. If cause-reaction effects are of interest and wave/structure interactions are evaluated in detail a time-domain analysis in deterministic wave trains is required. A sophisticated method of transforming frequency-domain results into time-domain enables the investigation of hydrodynamically coupled structures including memory effects. For this the response amplitude operators (calculated by WAMIT) are transformed into impulse response functions by Fourier transformation (Cummins 1962):

$$K_i(t)=\frac{1}{2\pi}\int_{-\infty}^{\infty}H_i(\omega)e^{i\omega t}d\omega \tag{14}$$

For this purpose a Fortran routine F2T by J.N. Newman has been provided to the authors and enhanced to an improved F2T+ procedure (Jacobsen 2005). With known impulse-response functions of the motions $K_i(t)$ the time dependent response in arbitrary wave trains $\zeta(t)$ is calculated by convolution (Fig. 4):

$$s_i(t) = \int_{\tau=-\infty}^{\infty} K_i(t-\tau)\zeta(\tau)d\tau \qquad (15)$$

The improved F2T+ procedure combines the

- Transformation of the complex frequency-domain RAOs into time-domain impulse-response functions.
- Convolution of the impulse response function with arbitrary wave sequences to determine the behaviour of a structure in time-domain.

From the rigid body motions, i.e., heave and pitch of the semisubmersible as well as heave of the barge, the response amplitude operator of the relative motion between the crane hook and the barge is determined by complex addition in order to include the phase angles correctly. For the barge positioned at the lee side of the floating crane the relative motion between the 10000 t-load and the barge is analysed for a sea state with a significant wave height of 2 m and a zero-upcrossing period of $T_0 = 5s$. Three wave sequences are chosen from a entire 1000 waves registration and compared in time—and frequency-domain. In the first sequence from 2650–2750 s the smallest relative motion occurs, while in the second the maximum relative motion is observed. The third includes the maximum wave height. The three ranges with the associated relative motions are shown in Figure 5. They illustrate the different motion behaviour:

- In the first sequence very small motions are observed. The zero-upcrossing period ($T_{0,rel} = 6.4s$) of the relative motion is very low. The basis of a safe lift is to start the operation in such a sequence with low motion.
- The second sequence is characterised by the maximum double amplitude of the relative motion. Though the exciting wave heights are quite similar to the waves of the first sequence (most of them are smaller than the significant wave height) the response is quite different. The associated zero-upcrossing period of the relative motions at 8.2 s is high. Such events should be avoided during lifting operations.
- During the third sequence, containing the highest wave, the response is something between the first and the second sequence. The zero-upcrossing period is ($T_{0,rel} = 7.13s$), located between the two others.

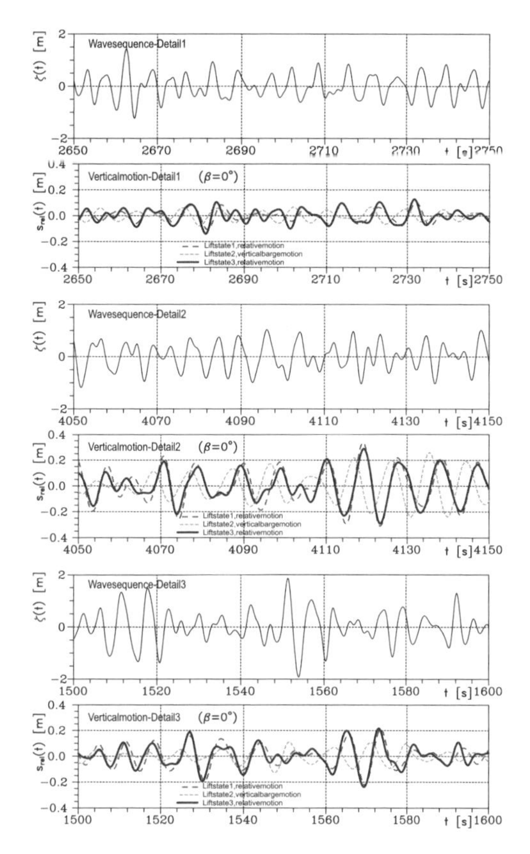

Figure 5. Zoomed registrations each of 100 s duration. Top: smallest relative motions. Middle: largest relative motions. Bottom: maximum wave height.

Apparently, the maximum wave exites a rather unspectacular reaction, while wave of medium height may lead to heavy responses.

If the analysis is focused on wave height only we cannot detect the effect of resonance-dominated wave sequences which excite temporally significant responses (Fig. 5-detail 2).

This observation is confirmed by spectral analysis of the respective wave sequences and motion response (cf. Jacobsen (2005)). The spectrum of wave sequence 1 (Fig. 5-top) contains relatively short waves, and hence the response remains low. In contrary, the wave sequence 2 (Fig. 5-middle) shows higher energies at lower frequencies. As the RAO is quite high in this region, we obtain a significant responses, and the associated relative motions are significant.

4.2 *Ships in encountering freak waves*

The second case study presents investigations of a RoRo-vessel (L_{pp} = 195 m, Δ = 44833 t, c_B = 0,71) encountering critical wave sequences with embedded freak waves such as there are:

- The "New Year Wave" (NYW), a giant single wave ($H_{max} = 25.63$ m) with a crest height of $H_c = 18.5$ m recorded during a storm on January 1, 1995 at the Draupner platform in the North Sea (Haver and Anderson 2000) in a surrounding sea state characterized by a significant wave height of $H_s = 11.92$ m ($H_{max}/H_s = 2.15$) at a water depth of $d = 70$ m.
- The "North Alwyn Wave" (NAW), recorded in a 5 day storm at the North Alwyn platform (Wolfram et al., 2000) on November 19th, 1997, i.e., a giant single wave ($H_{max} = 22.03$ m) in a surrounding sea state of $H_s = 8.64$ m ($H_{max}/H_s = 2.55$).

For the investigation of the influence of different wave characteristics on the vertical bending moment, the investigated vessel is chosen because it provides a pronounced horizontal asymmetry, due to its large flare at bow and stern. The wooden model is subdivided into three segments intersected at $1/2\,L_{pp}$ and $3/4\,L_{pp}$ (measured from the aft perpendicular) being connected with three force transducers at each cut. Based on the measured longitudinal forces and the given geometrical arrangement of the three force transducers, the resulting vertical wave bending moments and the longitudinal forces are obtained. On this basis, the superimposed vertical wave bending moment resulting from vertical and horizontal forces on the hull is determined.

For the investigations the model is fixed and towed with an elastic suspension system using a triangular towing arrangement pulling the model without inducing a moment. The longitudinal motions are restricted by a spring in front of and a counter weight behind the model. With this arrangement, heave and pitch motions as well as the measured forces and moments remain unrestrained.

The identification of the most critical wave sequence in an irregular sea state is not a single-edged problem. Loads and motions significantly depend on encountering speed and angle. To simplify this task the focus lays on long crested head seas as the worst case scenarios inducing the most critical vertical bending moments.

In the first part of the investigations, two target positions (at forward perpendicular and at midship) are analysed in the New Year Wave and compared to identify the more critical situation. The second part contains the investigations of two different time traces—a 20-minute time trace of the New Year Wave sequence and a 15-minute time trace of the North Alwyn Wave sequence, both at stationary condition.

5 TARGET POSITION

The first investigations are made to compare different encounter positions of extreme waves to identify the most critical target position. Therefore, the RoRo vessel is investigated in the New Year Wave with target position at forward perpendicular as well as at midship. Figure 6 illustrates the vessel position in the freak wave and the surrounding surface elevation for both target positions. The spatial development of the New Year Wave was measured in the seakeeping basin of the TU Berlin in a range from 2163 m (full scale) ahead of to 1470 m behind the target position by a total of 520 registrations (Clauss and Klein 2009). Figure 7 presents the vertical bending moments for both situations—the left column represents the target position at forward perpendicular, the right column shows the results of the target position midships. The first two rows compare the surface elevation at forward perpendicular and midships for both target positions. Due to minor local variations of the wave probes for the different test setups, the registrations at target location show some marginal differences, which can be neglected in the following. Row three and four provide the according vertical bending moments at $1/2\,L_{pp}$ and $3/4\,L_{pp}$. The last row shows the relative surface elevation, measured at the bow.

In contrast to expectations, global loads are higher in case of the midship target position (Fig. 7, right column). This becomes clearly apparent at the first sagging moment peak as well as at the maximum hogging moment. These higher loads at midship conditions are caused by the contour of the individual waves in the sequence, especially of the previous and following wave (cf. Fig. 7). Figure 6 illustrates that the wave trough in front of the rogue wave is much deeper than compared to the one in the forward perpendicular situation (Fig. 6, first row). Encountering the freak wave, the ship's bow is completely submerged in both cases and lifted up. The preceding wave trough at about

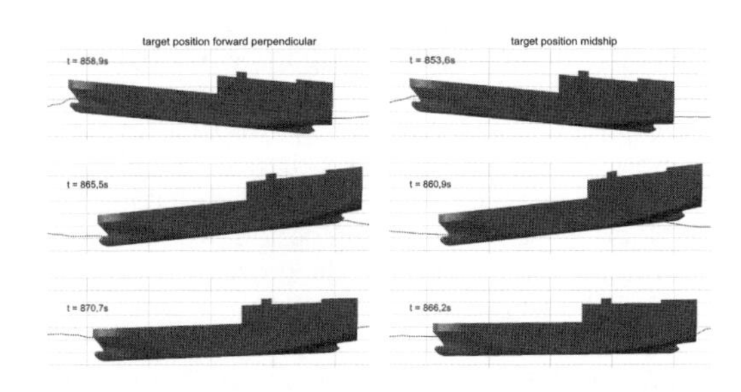

Figure 6. Visualisation of the ship motions (full scale) and surface elevation for target position forward perpendicular (left column) and midship (right column).

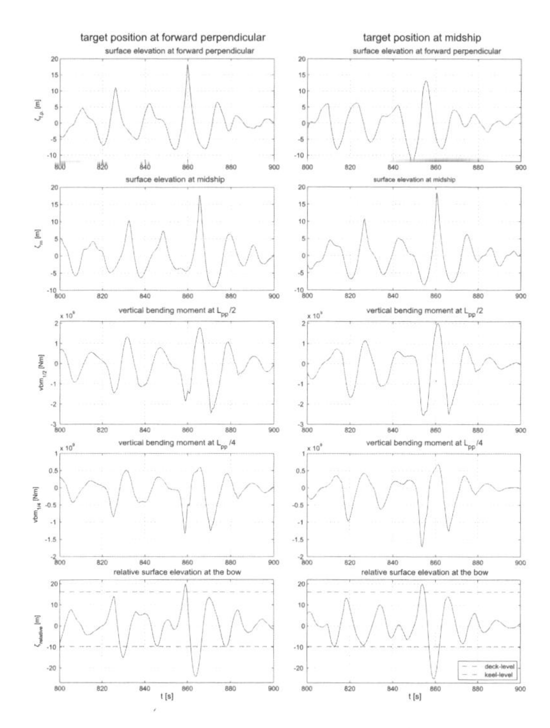

Figure 7. Comparison of model test results in the New Year Wave st different target positions—target position at forward perpendicular (left column) and target position at midship (right column).

midships is significantly deeper for the situation with midship target position than for the forward perpendicular one. This results in higher sagging loads. At the following hogging situation, the fully developed freak wave—in case of midship target position—induces higher hogging loads than the already broken and propagated freak wave (target position forward perpendicular) with a thereby decreased wave height. Submerging in the next wave crest, the second sagging moments are nearly the same in both conditions. With the restriction of investigating only global loads, target position midship presents a more critical situation for the RoRo vessel, and is chosen for the following investigations.

6 WAVE CHARACTERISTICS

For further investigations, two different time traces including freak waves are analysed, starting with the registration of the New Year Wave $(Fn=0)$, presented in Figure 9. Beside the surface elevation, three other wave characteristics are chosen to identify the critical potential of the different investigated waves—the crest front steepness, the horizontal asymmetry and the relative wave length. The first row presents the surface elevation ζ_a at midship (target position midship), the next two show the crest front steepness ε and the horizontal asymmetry μ,

$$\varepsilon = \frac{\eta'}{\frac{g}{2\pi} T \cdot T'} \text{ and } \mu = \frac{\eta'}{H}, \tag{16}$$

both defined by Kjeldsen (1983) and illustrated in Figure 8. The fourth row presents the relative wavelength, the ratio of wave length (L_ω) to ship length (L_{pp} = length between perpendiculars), which is an indicator for the critical wave length related to the bending moments. The last two rows show the vertical bending moments at $1/2\,L_{pp}$ and $3/4\,L_{pp}$.

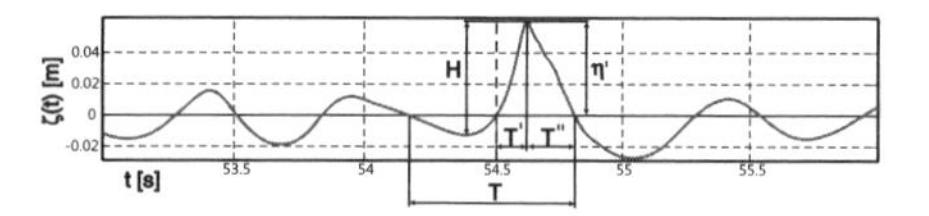

Figure 8. Wave parameters for the definition of the wave front steepness and the horizontal asymmetry as proposed by Kjeldsen.

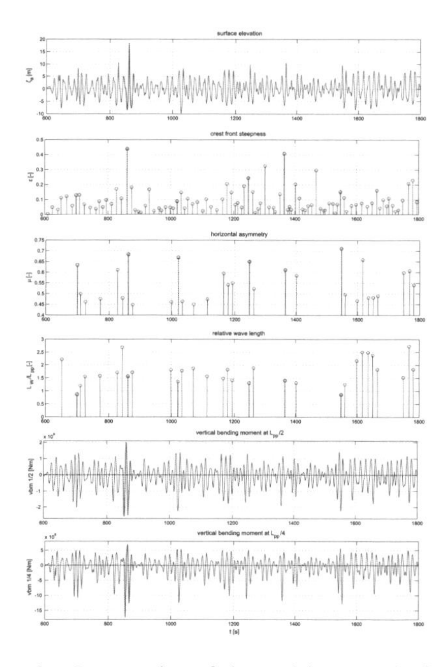

Figure 9. Presentation of the model test results for the New Year Wave: first row—surface elevation; second row—crest front steepness; third row—horizontal asymmetry; fourth row—relative wave length; fifth and sixth row—vertical bending moments at $1/2\,L_{pp}$ and $3/4\,L_{pp}$.

Previous investigations (Clauss et al., 2010) in regular waves revealed that the asymmetry of the vertical bending moment, induced by the horizontal asymmetry of the RoRo hull, increases for higher, steeper waves. This results in higher sagging and lowerhogging moments, which was also observed by Guedes Soares and Schellin (Soares and Schellin 1998). The global maximum of the Response Amplitude Operator (RAO) as well as the maximum asymmetry is about $L_{\omega}/L_{pp} \approx 1.1$.

7 NEW YEAR WAVE

Figure 9 presents a 20 minute time trace of the New Year Wave in stationary condition, target position midship. Regarding the vertical bending moments, extreme values are observed at $t \approx 860\,s$. Even though the wave length is not critical ($L_{\omega}/L_{pp} = 1.56$), the very deep preceding trough induces the first big sagging moment peak (cf. Fig. 6, right column). The extremely steep freak wave ($H_{max} = 25.63$ m, $H_c = 18.5$ m, $\varepsilon = 0.44$, $\mu = 0.68$, $H_{max}/H_s = 2.15$) induces a maximum hogging moment at target position midship with large emerged parts at bow and stern (cf. Fig. 6). Protruding from the New Year Waves crest and submerging in the following wave trough, a second critical sagging moment is induced. Beside this highest wave, some other critical wave sequences exist at $t \approx 700\,s$, $t \approx 1020\,s$, $t \approx 1250\,s$, $t \approx 1360\,s$ and $t \approx 1550\,s$. These smaller waves—or better to say the according wave groups—have a similar level of relative wave length $(0.87-1.4)$, which is about the maximum RAO region for the vertical bending moment ($L_{\omega}/L_{pp} \approx 1.1$). All wave groups show a distinct horizontal asymmetry $(0.61-0.71)$ but no extraordinary crest front steepness, with exception of the fifth wave ($t \approx 1360\,s$). This wave ($\varepsilon = 0.41$), as well as the New Year Wave itself ($\varepsilon = 0.44$), has a huge preceding wave trough. These troughs significantly increase the impact of the following giant wave and lead to a more critical vertical bending moment. The last critical wave sequence ($t \approx 1550\,s$) shows a following deep wave trough which results in a huge second sagging moment.

8 NORTH ALWYN WAVE

After investigating the time trace of the New Year Wave, another freak wave embedded in a 15 minute time trace, recorded in a five day storm at North Alwyn platform, is analysed and presented in Figure 10. Similarly to Figure 9, the first row shows the surface elevation, followed by diagrams of the crest front steepness, horizontal asymmetry, relative wave length and the vertical bending moments at $1/2\,L_{pp}$ and $3/4\,L_{pp}$.

First of all, the maximum moments are observed in the vicinity of the highest North Alwyn Wave ($t \approx 860\,s$). In relation to the New Year Wave and the surrounding sea state ($H_{s-NAW} = 8.64$ m *vs.* $H_{s-NYW} = 11.92$ m), this wave ($H_{max} = 22.03$ m, $H_c = 16.38$ m, $\varepsilon = 0.43$, $\mu = 0.72$, $H_{max}/H_s = 2.55$) is even more critical, in particular according to the relative wave length ($L_{\omega}/L_{pp} = 1.15$). This is obviously shown in similar bending moments—in comparison to the New Year Wave—which occur in this smaller freak wave. Regarding the whole time trace, there are a few other critical wave sequences, for example at $t \approx 550\,s$, $t \approx 620\,s$, $t \approx 1125\,s$ and $t \approx 1275\,s$. Again, all sequences have a similar ratio of relative wave length $(0.85-1.26)$, which is about the maximum RAO region for the vertical bending moment. However, neither a significant horizontal asymmetry nor an extraordinary crest front steepness are on hand, with an exception of the second wave group ($t \approx 620\,s$).

Comparing the individual situations in both investigations, the most critical parameter is the

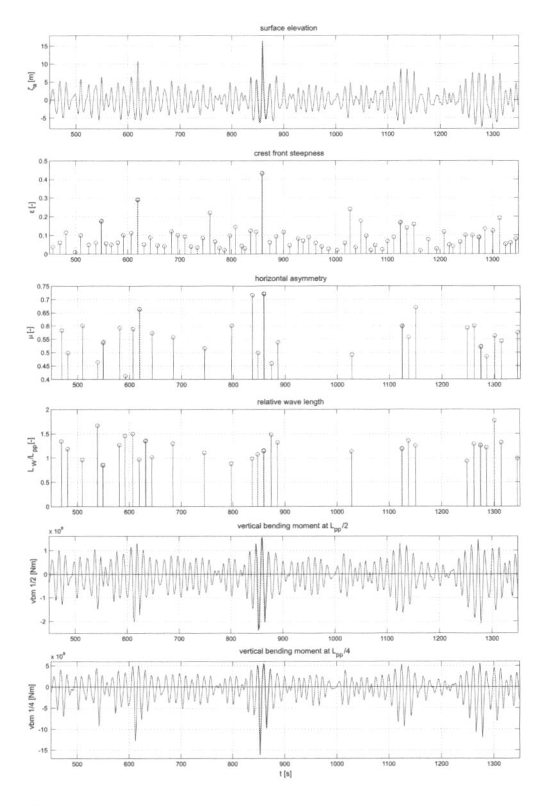

Figure 10. Presentation of the model test results for the North Alwyn Wave: first row—surface elevation; second row—crest front steepness; third row—horizontal asymmetry; fourth row—relative wave length; fifth and sixth row—vertical bending moments at $1/2\,L_{pp}$ and $3/4\,L_{pp}$.

relative wave length. Even with a smaller significant wave height (H_s =8.64 m), the time trace of the North Alwyn Wave provides comparable vertical bending moments in comparison to the New Year Wave (H_s =11.92 m). This fact is clearly caused by the smaller, more critical relative wave lengths. Another interesting result is the importance of the surrounding wave profile in terms of huge preceding or following wave troughs in combination with large crest front steepnesses. Situations like these can induce great sagging loads, which are comparatively big to those of the hogging condition. Nevertheless, freak waves like the New Year Wave can induce significant hogging moments and create very critical situations while lifting the vessel up at the crest and emerging large areas at bow and stern. This effect is amplified by high, asymmetrically and steep crests. Examining the characteristics of the three different time traces, all discussed criteria can be found at target position.

Regarding the procedure of a passing giant wave, the vessel is going down in the preceding wave trough. By entering the steep, giant wave front, the bow section submerges completely and hence induces maximum sagging loads. Floating through the New Year Waves crest with large emerged areas at bow and stern, maximum hogging loads are the consequence. Submerging heavily into the following wave front, a second critical sagging moment is induced.

9 CONCLUSIONS

This paper presents two different case studies for the identification of critical offshore situations. Therefore, a multibody system, e.g., an offshore lift operation, and ship in severe sea state, e.g., a RoRo-vessel encountering freak waves, are analysed. For the investigation of critical situations, the relative vertical motions is chosen for the lift of operation and the vertical bending moment for the RoRo-vessel.

First of all, the classical design evaluation, using random phase distributions, is compared to a phase optimization procedure, which is applied for a straightforward identification of the worst case scenario with regard to a selected sea state. It is shown that serious responses depend on critical wave length more than on the highest wave.

Furthermore, the influence of several wave parameters—such as target position, wave height, crest front steepness, horizontal asymmetry and relative wave length—on the occurring vertical bending moments are analysed for a RoRo-vessel. As mentioned in previous investigations (Clauss et al., 2010, a), one of the key parameter is the relative wave length. In combination with an extraordinary wave height and steep crest front, maximum critical loads can appear.

Investigations for both cases have shown that only a few parameters are responsible for the maximum responses. With the help of numerical optimization (Clauss et al., 2010, b), these parameters can be used to define tailored worst case sea state scenarios to determine limiting criteria for offshore operations or structural reliability. This procedure, prior to the investigations in the test facilities, can reduce the model test duration and make the investigations more efficiently.

ACKNOWLEDGEMENTS

The identification of critical wave scenarios and the attempt of a response based wave generation is investigated as a contribution to the project EXTREME SEAS, which is founded by the European Commission, under the Grant agreement no. 234175. The author highly acknowledges the support of this research project and wants to thank the project partners Det Norske Veritas AS, Instituto Superior Técnico, Germanischer Lloyd AG, Meteorological Institute, Universita di Torino, Institute of Applied Physics, Canal de Experiencias Hidrodinamicas de El Pardo, Meyer Werft, Estaleiros Navais Viana de Castelo and University Duisburg-Essen.

REFERENCES

Clauss, G. (2002). Dramas of the sea: episodic waves and their impact on offshore structures. In *Applied Ocean Research, Vol. 24(3), ISSN 0141-1187.*

Clauss, G. and Bergmann, J. (1986). Gaussian wave packets—a new approach to seakeeping tests of ocean structures. *Applied Ocean Research Vol. 8* (No. 4).

Clauss, G. and Birk, L. (1996, August). Hydrodynamic shape optimization of large offshore structures. *Applied Ocean Research, 18(4)*, 157–171.

Clauss, G. and Klein, M. (2009). The New Year Wave: Spatial Evolution of an Extreme Sea State. *JOMAE 2009—Journal of Offshore Mechanics and Arctic Engineering 131.*

Clauss, G., Klein, M. and Dudek, M. (2010). Influence of the Bow Shape on Loads in High and Steep Waves. In *OMAE 2010—29th International Conference on Offshore Mechanics and Arctic Engineeringn*, Shanghail, China.

Clauss, G., Klein, M., Sprenger, F. and Testa, D. (2010). Evaluation of Critical Conditions in Offshore Vessel Operation by Response Based Optimization Procedures. In *OMAE 2010—29th International Conference on Offshore Mechanics and Arctic Engineeringn*, Shanghail, China.

Clauss, G. and Kühnlein, W. (1995). Transient wave packets—an efficient technique for seakeeping tests of self-propelled models in oblique waves. In *Third International Conference on Fast Sea Transportation)*, Lübeck- Travemunde, Germany.

Clauss, G. and Kühnlein, W. (1997). Simulation of design storm wave conditions with tailored wave groups. In *7th International Offshore and Polar Engineering Conference (ISOPE)*, Honolulu,Hawaii, USA.

Clauss, G., Lehmann, E. and Östergaard, C. (1992). *Offshore Structures*, Volume 1: Conceptual Design and Hydrodynamics. Springer Verlag London.

Clauss, G.F. and Schmittner, C.E. (2005). Experimental Optimization of Extreme Wave Sequences for the Determinsitic Analysis of Wave/Structure Interaction. In *OMAE 2005—24th International Conference on Offshore Mechanics and Arctic Engineering*, Halkidiki, Greece. OMAE2005–67049.

Clauss, G.F., Schmittner, C.E., Hennig, J., Guedes Soares, C., Fonseca, N. and Pascoal, R. (2004). Bending Moments of an FPSO in Rogue Waves. In *OMAE 2004—23rd International Conference on Offshore Mechanics and Arctic Engineering*, Vancouver, Canada. OMAE2004–51504.

Clauss, G.F. and Steinhagen, U. (1999). Numerical simulation of nonlinear transient waves and its validation by laboratory data. In *Proceedings of 9th International Offshore and Polar Engineering Conference (ISOPE)*, Volume III, Brest, France, pp. 368–375.

Cummins, W. (1962, Juni). The impulse response function and ship motions. *Schiffstechnik Band 9* (Heft 47), 101–109.

Davis, M. and E. Zarnick (1964). Testing ship models in transient waves. In *5th Symposium on Naval Hydrodynamics*.

Denis, S. and Pierson, W. (1953). On the motions of ships in confused seas. Transactions, SNAME. 61.

Department of Ocean Engineering, MIT (1994). *WAMIT Version 5.1—A Radiation-Diffraction Panel Program For Wave-Body Interactions*. Department of Ocean Engineering, MIT. Userguide.

Fonseca, N. and Guedes Soares, C. (2002). Comparison of numerical and experimental results of nonlinear wave-induced vertical ship motions and loads. In *Journal of Marine Science and Technology*, pp. 193–204.

Haver, S. and Anderson, O.J. (2000). Freak Waves: Rare Realization of a Typical Population or Typical Realization of a Rare Population? In *Proceedings of the 10th International Offshore and Polar Engineering Conference (ISOPE)*, Seattle, USA, pp. 123–130.

Hennig, J. (2005). Generation and Analysis of Harsh Wave Environments. *Dissertation Technische Universität Berlin (D 83)*.

Hogben, N. and Lumb, F. (1967). *Ocean Wave Statistics*. London: Her Majesty's Stationery Office.

Jacobsen, K. (2005). Hydrodynamische Kopplung von Offshore-Strukturen im Seegang (Hydrodynamic Interaction of Multi-body Systems in Waves). *PhD Thesis Technische Universität Berlin (D 83)*.

Jacobsen, K. and Clauss, G.F. (2006). Time-Domain Simulations of Multi-Body Systems in Deterministic Wave Trains. In *OMAE 2006—25th International Conference on Offshore Mechanics and Arctic Engineering*, Hamburg, Germany. OMAE2006–92348.

Kjeldsen, S.P. (1983). Determination of severe wave conditions for ocean systems in a 3-dimensional irregular sea-way. In *Contribution to the VIII Congress of the Pan-American Institute of Naval Engineering*, pp. 1–35.

Kühnlein, W. (1997). Seegangsversuchstechnik mit transienter Systemanregung. *PhD Thesis Technische Universität Berlin (D 83)*.

Mathisen, J. and Bitner-Gregerson, E. (1990). Joint distributions for significant wave height and wave zero-upcrossing period. *Applied Ocean Research 12*(2), 93–103.

Newland, D. (1975). *Random vibration and spectral analysis*. New York: Longman Inc.

Newman, J. (1977). *Marine Hydrodynamics*. The MIT Press, Cambridge, Massachusetts.

Newman, J. (1986). Distributions of sources and dipoles over a quadrilateral panel. *Journal of Engineering Mathematics 20*, 113–126.

Newman, J. (2001). Wave effects on multi bodies in hydrodynamics in ship and ocean engineering. *RIAM, Kyushu University*, 3–26.

Newman, J. and Sclavounos, P. (1988). The computation of wave loads on large offshore structures. In *Proc. of Int. Conf. on Behaviour of Offshore Structures (BOSS '88)*, pp. 605–622. Trondheim, Norway.

Soares, C.G. and Schellin, T.E. (1998). Nonlinear effects on long-term distributions of wave-induced loads for tankers. *Journal of Offshore Mechanics and Arctic Engineering, 120(2)*, 65–70.

Stansberg, C.T. (2008). A Wave Impact Parameter. In *OMAE 2008—27th International Conference on Offshore Mechanics and Arctic Engineering*, Estoril, Portugal. OMAE2008–57801.

Steinhagen, U. (2001). Synthesizing Nonlinear Transient Gravity Waves in Random Seas. *Dissertation,* Technische Universität Berlin (D 83).

Takezawa, S. and Hirayama, T. (1976). Advanced experimental techniques for testing ship models in transient water waves. Part II: The controlled transient water waves for using in ship motion tests. In R. Bishop, A. Parkinson, and W. Price (Eds.), *Proceedings of the 11th Symposium on Naval Hydrodynamics: Unsteady Hydrodynamics of Marine Vehicles*, pp. 37–54.

Watanabe, I., Keno, M. and Sawada, H. (1989). Effect of bow flare on shape to wave loads of a container ship. *J Soc Nav Archit Jpn 166:259–266*.

Wolfram, J., Linfoot, B. and Stansell, P. (2000). Long- and short-term extreme Wave Statistics in the North Sea: 1994–1998. In *Rogue Waves 2000*, Brest, France, pp. 363–372.

3 *Structures & materials*

3.1 *Analysis of local structures*

Sustainable Maritime Transportation and Exploitation of Sea Resources – Rizzuto & Guedes Soares (eds)
© 2012 Taylor & Francis Group, London, ISBN 978-0-415-62081-9

On the *effective breadth* of plating

I.G. Tigkas & A. Theodoulides
Research & Rule Development Department, Hellenic Register of Shipping, Piraeus, Greece

ABSTRACT: Rules by Classification societies should continuously be evolved in order to address as accurately as possible the structural responses of ships. Along this way, a procedure for amending HRS Rules when calculating the effective breadth of the attached plating of primary and secondary stiffeners under normal bending loads, is clearly developed in this effort. This study shows how practical knowledge, first principles and FEA methods can be merged in an efficient manner in order to provide a reliable but thence easy-to-use Rule that can be widely used either by Class or more generally by structural engineers for calculating the *effective breadth* of a stiffened plate.

Various scenarios are investigated comprising stiffeners (including their attached plating) with different slenderness ratios and with boundary conditions of their ends either fixed or simply supported. More particularly, by employing FEA methods and least-square-fit techniques, the stress distributions at the transverse direction of the attached plating are modelled and calculated. Schade's (1951) well proven methodology is thenceforth used to compute the effective breadth of plating for each one of the investigated scenarios. From analysing the results, a practical formula can safely be derived for calculating the *effective breadth* of stiffened plating.

1 INTRODUCTION

The accurate assessment of the structural behaviour of ships from the Classification point of view, is crucial for ensuring ships with sound structural integrity throughout their life-span. To comply with this requirement, it is well known that Classification societies have been developing for many years Rules and Regulations. Nevertheless, due to their prescriptive nature, Rules may show deviations from the real structural responses of actual ship constructions. Such deviations may be proven detrimental when Rules under-predict stresses of structures that lie near the permissible limit range. Even if such may not be evident in relatively new constructions, the diminution of steel due to corrosion and other factors (although being within the acceptable limits required by Class) may yield the structures over their stress limits. In order therefore to close any inaccuracy gap, it is necessary to frequently provide interpretations or/and amendments that bring the prescriptive results of the Rules closer to the actual structural responses.

Accordingly, for the calculation of the section modulus of a primary or secondary member in bending, Classification Rules should contain a specific methodology for determining the effective breadth of the attached plate. A rule of thumb that may conveniently be used, is to equate the effective breadth of the attached plate with the frame spacing or more specific with the span of the plating between the midpoints of the two adjacent frames.

The validity of such a simplistic approach has not yet been assessed and hence a systematic and thorough methodology may be developed, that clearly outlines the development of Rules for calculating the effective breadth of the attached plating.

2 LITERATURE REVIEW

The contribution of plating attached to a stiffener that may also be part of a grillage arrangement, is mainly affected upon:

i. the direction and type of loading
ii. boundary conditions and end constraints
iii. geometrical properties of the section
iv. span between stiffeners and unsupported length of the stiffeners
v. connection arrangement (i.e., welding) of the stiffener with the attached plate
vi. joining or welding imperfections
vii. initial deformations and plate buckling
viii. residual stresses.

It should be noted from the above that the direction of loading is the only factor qualitatively altering the response behaviour of the structure and thus a different type of calculation analysis is required for instance when lateral or in-plane loads are applied. Accordingly, normal loads upon a plate will impose shear lag associated with stiffener/plate bending. On the contrary, axial compression will question the "stiffness" of the stiffener/plate arrangement, thus imposing local buckling. The effectiveness of plating in the former scenario is referred to as "effective breadth", where the effectiveness of the later as "effective width". Such nomenclature, still used today, was firstly introduced by Schade (1951), in order to properly distinguish the effectiveness of plating when lateral or axial loads are applied. It should not be omitted that real ship structures may in some cases experience a combination of loads that can simultaneously cause bending and buckling and thenceforth "effective breadth" is not totally remote from "effective width". Nevertheless, the effect of each one should be considered on separate when analytical calculation models are to be used. FEA tools and experiments on the other hand, may offer us the ability to depict structural responses in detail and draw conclusions even for more complicated loading combinations.

On the analytical and experimental forefronts, the effectiveness of plating has vastly been discussed by a plethora of research scientists. (*Effective breadth*: Papkovich 1923, Schade 1941, 1951, 1953, Vedeler 1950, Mansour 1970, 1971, Boote & Mascia 1991, Katsikadelis & Sapountzakis 2002, Belenkiy et al., 2007), (*Effective width*: von Karman 1924, Marguerre 1937, Murrey 1945, 1954, Cox 1946, Horne 1956, Bleich 1956). Moreover, a comprehensive and detail review by Faulkner (1975), discusses in addition further issues such as imperfections that are strongly related with plating effectiveness. From such complementary effort, it can be deduced that except the loading direction, all other precedently mentioned factors may deteriorate as appropriately the effectiveness but do not change qualitatively the structure's response behaviour.

3 THE USE OF FEA METHODS FOR THE CALCULATION OF THE EFFECTIVE BREADTH OF PLATING

Consider loading a reinforced by stiffeners plate in the lateral direction. It can thus be assumed that the load is entirely transmitted to the stiffeners. The web of the stiffeners on such a case, being normally aligned to the attached plate, exerts shear load on the plate at their connection. The stress developed is non-uniform, both along the longitudinal and the transverse direction of the attached plate. For example for a simply supported stiffener with free sides, the maximum stress of the attached plate is distributed longitudinally along the midspan of the plate and transversely at its intersection with the web. It can therefore be safely assumed that the boundary conditions of the edges and of the sides strongly affect the attached plate's stress distribution and thus such effects need to be investigated. The type of loading is also essential, but it can be assumed that it may be uniformly distributed, as this is a common type of loading on ship structures (e.g., from hydrostatic pressure).

When considering a uniformly loaded panel comprised of a plate with n supporting stiffeners spaced at a distance s, it can be assumed that this panel may behave as n individual stiffeners with attached plate having actual breadth equal to the distance s. The load acting on each stiffener is the original uniform distributed load divided by n.

A further simplification may be assumed when it is considered that that the load is fully resisted by the stiffeners. Thus the load should be uniformly distributed only upon the area of the plate located directly on top of the web. The sides of the attached plate of each stiffener can consequently be considered free without affecting the stress distribution along the transverse direction of the attached plate. This simplification evades irregularities imposed by the boundary conditions of the sides and additionally, it allows us to calculate without complexities the effective breadth of the attached plate of the stiffener.

The validity of most of the assumptions adopted will be tested later in this study. But firstly, it is of prime importance to define in-depth the meaning of the effective breadth of an assembly. That is the artificial breadth of the attached plate of a virtual stiffener where if the stress distribution is uniform and reaches a value up to the maximum stress, this virtual assembly behaves identical to the actual assembly having a non-uniform stress distribution.

Therefore according to Schade (1951), *effective breadth* can be expressed as following:

$$b_{eff} = \frac{\int_0^b \sigma_x dy}{\sigma_{\max}} \tag{1}$$

Along this way in order to calculate the effective breadth of plating for a stiffener at various slenderness ratios and at different end conditions, it is necessary to obtain stress distribution diagrams at the plate along its transverse direction. Such a procedure may be experimentally achieved by using strain gauges at different positions on the plating.

An alternative experimental mean to calculate indirectly the effective breadth is to measure deflections and from those measurements to deduce a value of breadth of plating that satisfies such deflections. More about the existing experimental methods for calculating the effective breadth can be found in Faulkner (1975).

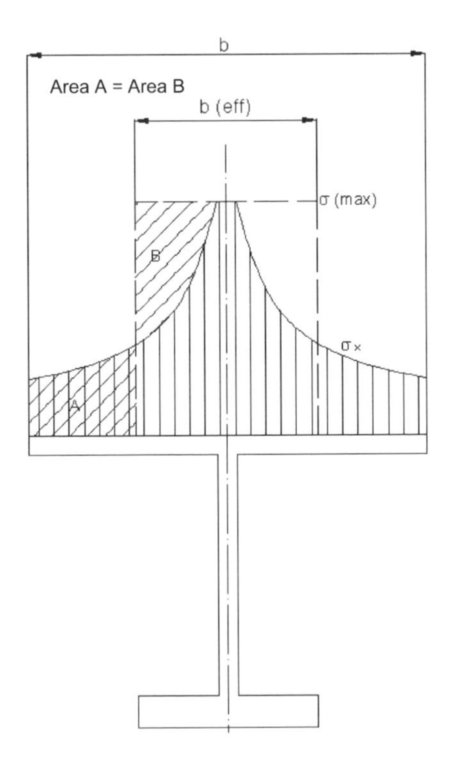

Figure 1. Definition of the effective breadth of plating.

Our procedure is not to conduct experiments but instead to use the numerical approach of FEA for calculating the stress distributions along the attached plate of a stiffener. Such indicative stress distributions are shown for example in Figure 2a-c.

Such reasoning is valid, since it may be safely assumed that FEA tools nowadays, can predict relatively accurately the structural behaviour at least of simple structures such as stiffened panels. We thus calculate the stress distributions of the attached plating at the midspan of stiffeners having increasing values of slenderness ratios, as seen in Figure 3. Calculating the stress distribution at the midspan is chosen since at this location:

i. The stress contour along the breadth of the associated plate does not vary abruptly in the longitudinal direction.
ii. The stress contour is not affected by local effects due to the boundary conditions.
iii. Stresses are maximised, at least for the simply supported case.

From the graphs (Fig. 3) shown below that are obtained with free sides, fixed supported edges and with uniform load distribution only on the part of the plating that is located above the web, it can be observed that as the slenderness ratio

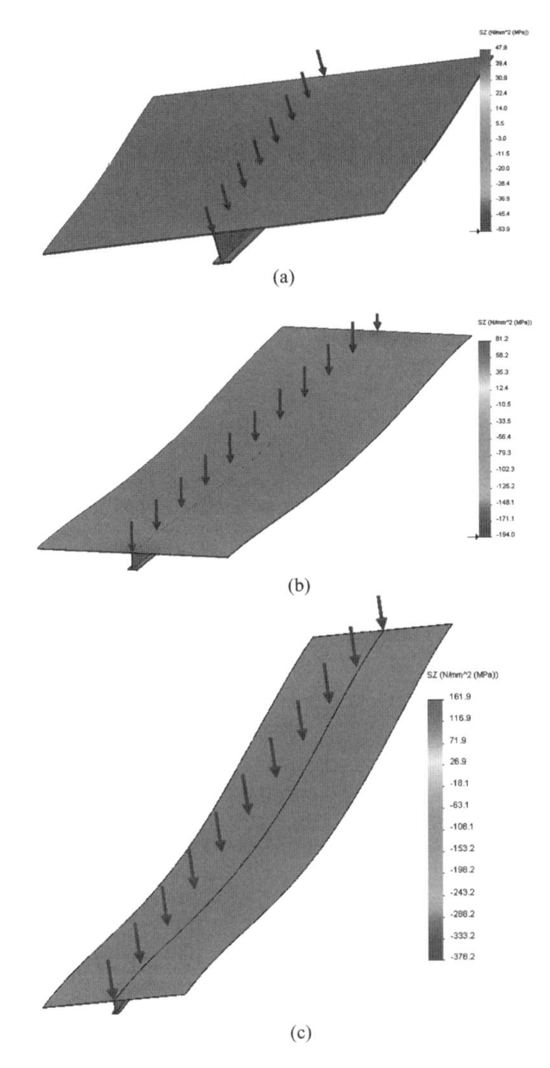

Figure 2. Indicative stress distributions for stiffened plates with fixed edges and slenderness ratios equal to 1 (L = 3000 mm), 5 (L = 15000 mm) and 10 (30000 mm) respectively. (Stiffener T 250 × 10 + 150 × 15 and plating: 3000 mm × 12 mm).

is increased, the slope of the stress distribution towards the centreline of the stiffener is being reduced. Furthermore, the difference between the maximum stress occurring at the centreline above the web with the minimum stress occurring at the sides deteriorates, as the slenderness ratio is once more being increased. It will be seen later that due to these facts, the effective breadth asymptotically converges to the actual breadth of the plate as the slenderness ratio reaches to infinity.

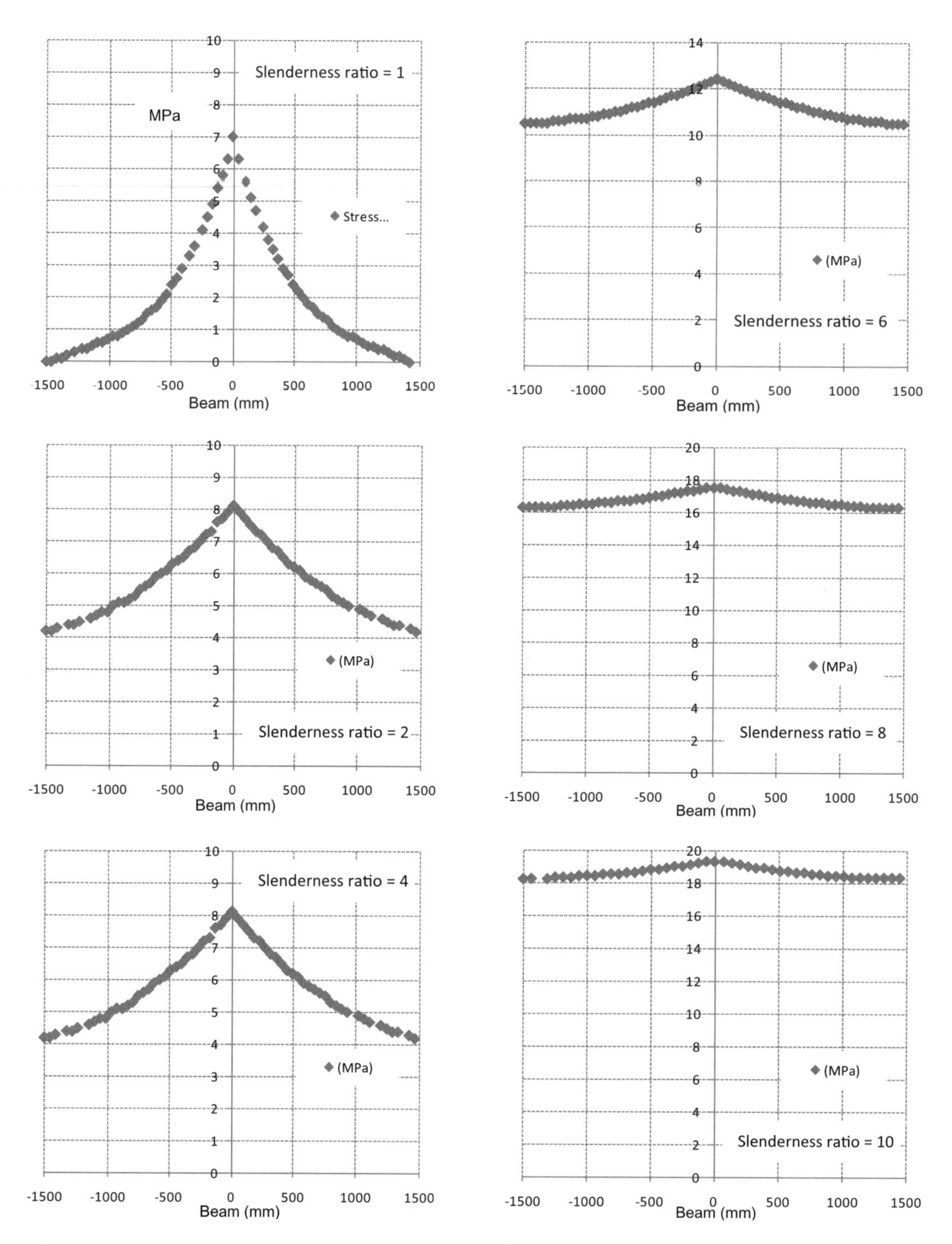

Figure 3. Stress distributions along the attached plating for slenderness ratios varying from 1 to 10. The distribution becomes more flat as the slenderness ratio is increased. (Stiffener T 250 × 10 + 150 × 15 and plating: 3000 mm × 12 mm).

Additional graphs, qualitative similar to Figure 3 are obtained when the boundary conditions at the edges are changed from fixed, to simply supported.

4 DEVELOPMENT OF NEW CRITERION

The procedure followed in each trial for calculating the effective breadth from the stress distributions, is to firstly discover a polynomial that best fits the stress distribution (Fig. 4), secondly to integrate

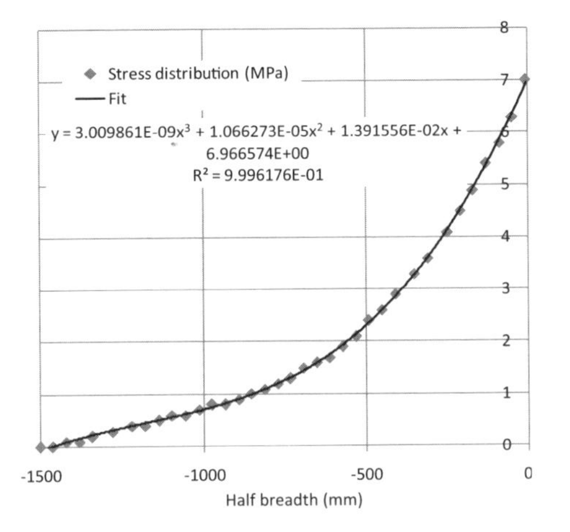

Figure 4. Indicative stress distribution at the attached plating that can accurately be fitted by a polynomial function. (e.g., Stiffener T 250 × 10 + 150 × 15 and plating: 3000 mm × 12 mm).

this polynomial along the transverse direction of the plate, then to evaluate the maximum stress incurred and finally to the use equation 1 to calculate the effective breadth.

In total 29 runs where conducted to comprise the different slenderness ratios, boundary conditions, sectional parameters and arrangements. Their results are consolidated into the graph shown in Figure 5.

From this figure it can be deduced that the minimum effective breadth is realised at low to medium slenderness ratios and accordingly, significant deviations from the actual breadth of the plating are evident at this region. Thus since the sum of the half breadths of the plating does not reflect the real case when calculating the section modulus of the stiffeners, it renders necessary the proposal of a new rule for the calculation of the effective breadth.

It is interesting to also note that in case simply supported stiffeners are being considered instead of fixed ones, their effective breadth follows a similar qualitative behaviour, but their differences from the actual breadth becomes less significant throughout almost all slenderness ratios. The results obtained in this case also are in an excellent convergence with the results obtained by the methodology of Shade (1951) when considering the same boundary conditions.

In order to expand the validity of our results we conduct a similar study, but we changed the geometrical properties and the shape of the investigated section (Fig. 6).

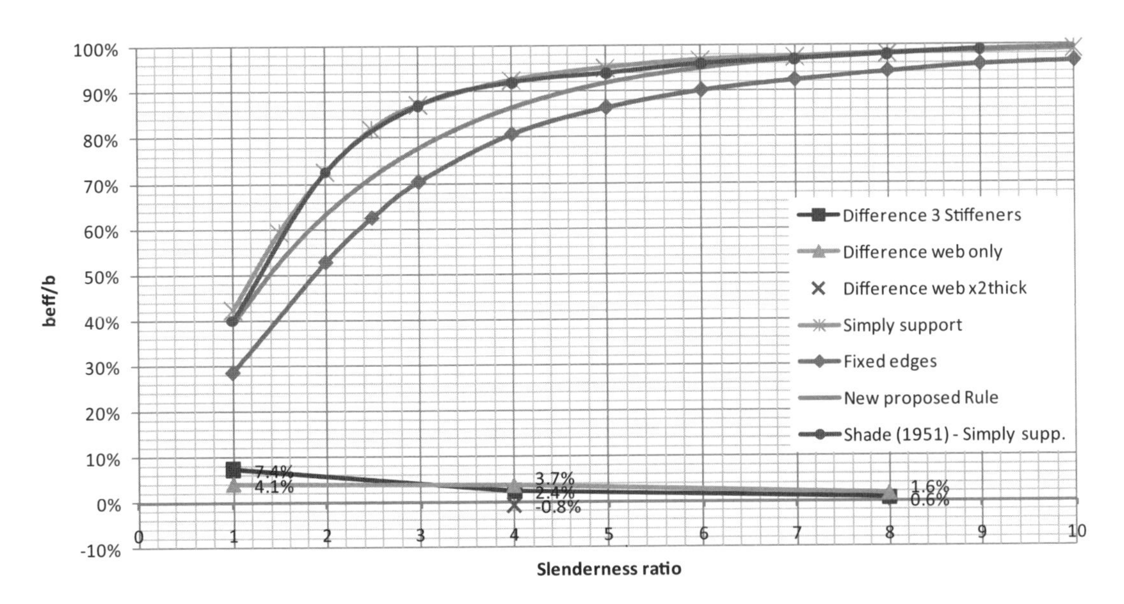

Figure 5. Effective breadth of attached plating as a function of slenderness ratio and new rule developed.

Accordingly, we firstly doubled the thickness of the web, keeping fixed ends, and we found that at slenderness ratio equal to 4, the difference in the value of the effective breadths is –0.8% (Shown in Figure 5 as the line named "Difference web × 2thick").

We secondly removed the lower flange of the stiffener (Fig. 7) and we observed that the changes of the effective breadths with and without the lower flange vary from +4.1% at unity slenderness ratio, to +1.6% at slenderness ratio equal to 8. (Shown in Figure 5 as the line named "Difference web only").

Lastly, since a loaded stiffener is rarely found on its own on ship structures but usually as a part of a structural assembly such as a grillage, it becomes necessary to investigate the behaviour of this stiffener when it is part of such an assembly. This investigation also simultaneously assesses the validity of the previous assumption where the sides of the stiffener are considered free. Therefore, we consider a grillage assembly comprised of 3 stiffeners that is loaded with a uniform load applied only on the areas of the plate that are located on top of the stiffeners (Fig. 8). When calculating the effective breadth at different slenderness ratios,

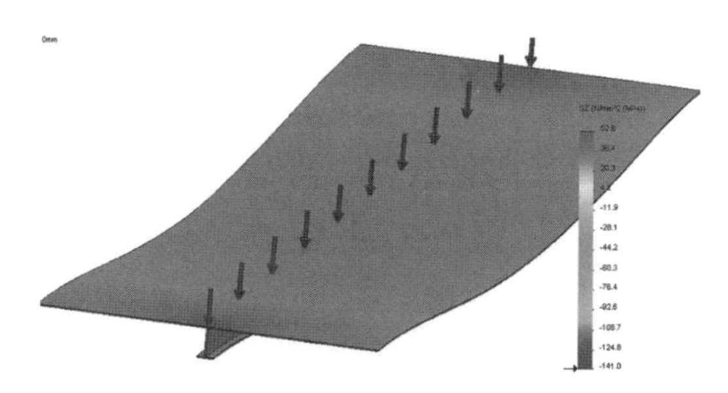

Figure 6. Plate and 'T' shape stiffener with slenderness ratio equal to 4 and fixed ends. The web thickness is doubled than previous cases. (Stiffener T 250 × 20 + 150 × 15 and plating: 3000 mm × 12 mm).

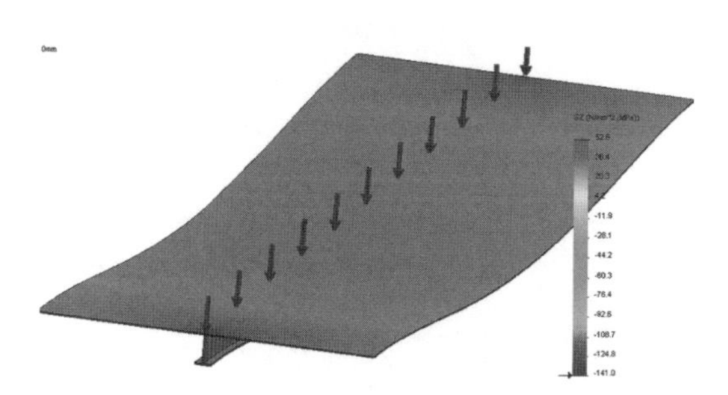

Figure 7. Plate and flat bar (web only) assembly with slenderness ratio equal to 4, experiencing bending. Fixed at the ends. (Stiffener F.B. 250 × 10 and plating: 3000 mm × 12 mm).

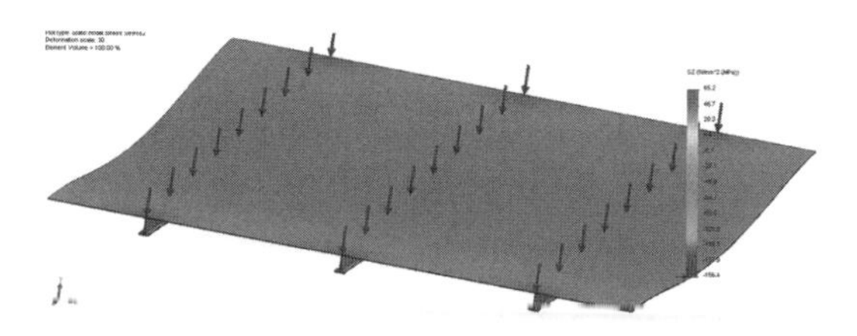

Figure 8. Grillage assembly comprised of plating supported by 3 stiffeners fixed at the edges.. (Stiffener 3pcs × T 250 × 10 + 150 × 15 and plating: 9000 mm × 12 mm).

it can be observed that the differences from the single stiffener case to the 3 stiffener case vary from +7.4% to +0.6% depending upon the slenderness ratio (Shown in Figure 5 as the line named "Difference 3 Stiffeners").

Such differences occurred between the different scenarios may be only partly attributed to numerical inaccuracies and to the methodology used. Nevertheless, since the qualitative behaviour is identical, they cannot really constitute a major hindrance for questioning the application of our results to a wide range of geometrical sections. Quantitative convergence is also feasible when an adequate factor of safety is adopted to tackle any possible deviation.

The new proposed Rule developed from the precedent graph and taking into account all such effects, is expressed in the following formula:

$$\frac{b_{eff}}{b} = \left(1 - e^{-0.4 \, l/b}\right) \tag{2}$$

where b = the actual width of the load-bearing plating, i.e., one-half of the sum of spacings between parallel adjacent members or equivalent supports; l = the overall length of the support member.

In order to observe the effectiveness of the formula, we calculate by using the numerical technique of FEA, the maximum tensile stress at the flange of a stiffener (T 250 × 10 + 150 × 12) with an attached plate of 12 mm thickness and an actual breadth of 3000 mm and we compare it with the stress derived from beam theory of a virtual stiffener that resembles the actual one, thus having identical properties except breadth of plating equal to the effective breadth (Fig. 9). By proceeding in this way for all range of slenderness ratios it may be noted that FEA method over-predict tensile stresses at the flange for more than 4%, when compared to Beam theory. By observing this difference at high slenderness ratios where $b_{eff} \approx b$, this difference should had been negligible. The fact that a difference of approximate 4% is evident in such slenderness ratio, can therefore be attributed to the

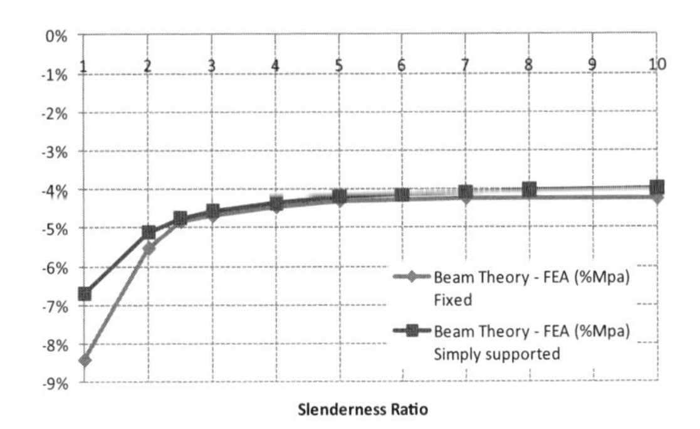

Figure 9. Difference in the tensile stresses developed at the lower flange of a beam with breadth b and a virtual beam with beam equal to the effective breadth b_{eff}. The former is calculated by FEA and the later by Beam theory.

slight over-prediction of stresses by the software used for FEA and not due to the method proposed in this effort. On the other hand, when slenderness ratio takes significant low values, as it was expected, the quantitative similarity deteriorates by the collapse of the beam theory's assumptions.

5 COMPARISON WITH RULES BY VARIOUS CLASSIFICATION SOCIETIES

In this case of study, it becomes also interesting to compare the results derived with the results obtained by different classification societies.

The following table (Table 1) summarises the applicable Rules of different classification societies for calculating the effective breadth of plating.

Accordingly, from Figure 10, it can be observed for the selected scenario that almost all Rules apart of CSR Rules, show similar qualitative behaviour. Despite that the convexity of the graphs is slightly different, all Rules apart of ABS reach $b_{eff} \approx b$ at a similar slenderness ratio (approx. 6). The only Rules that are qualitatively different are the CSR Rules for Bulk Carriers. These Rules at least for the primary stiffeners, do not distinguish the effective breadth of plating with the actual breadth of plating and it is assumed to be equal.

Furthermore, from Figure 11, it can be seen that the minimum section modulus is not significantly affected by changes in the effective breadth. It is evident that once the effective breadth is increased, the neutral axis moves towards the attached plate, consequently increasing the distance from the neutral axis and as a result, reducing the minimum section modulus. But concurrently, an increase in the effective breadth of the attached plating increases the moment of inertia and thus affects favourably the minimum section modulus. Therefore, it can be finally deduced that the change in the section modulus by an increase in the effective breadth, is a function of the changes of the moment of inertia and in the maximum distance from the neutral axis, both to the change of the effective breadth. Consequently, such effects in the breadth of plating especially for marginal cases should not be neglected.

Table 1. Rules by various Classification Societies for calculating the effective breadth of plating (b_{eff}).

Rules	
ABS	0.5 b or 0.33 l whichever is less
LR	$0.3(l/b)^{2/3}$ but is not to exceed 1.0
GL	Calculated according to a rigorous table, (see GL Rules Part 1, Sec. 3)
CSR Bulk (Primary)	b
HRS (new)	$\left(1-e^{-0.4 l/b}\right)b$

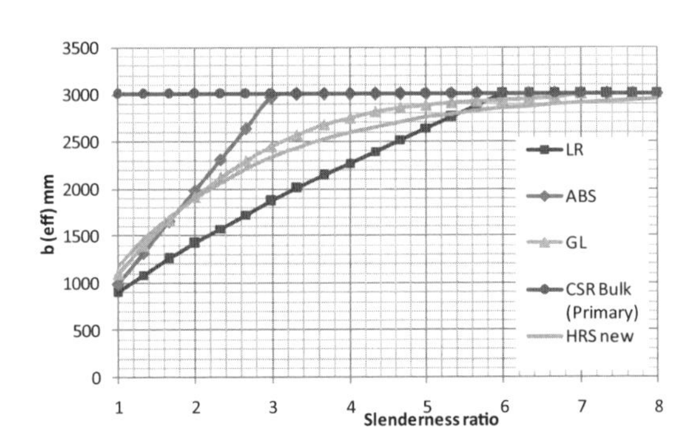

Figure 10. Breadth of attached plating for various Rules including new HRS Rule. (Primary stiffener T 250 × 10 + 150 × 15 and plating = 3000 × 12).

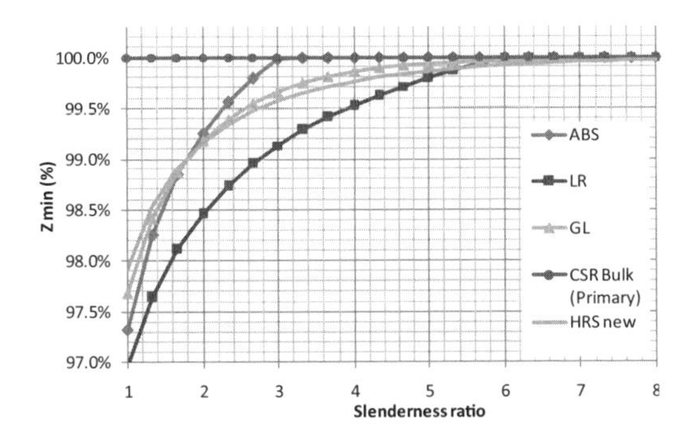

Figure 11. Section modulus of a stiffener for various Rules including new HRS Rule. (Primary stiffener T 250 × 10 + 150 × 15 and plating = 3000 × 12).

6 CONCLUSIONS

The current effort has practically attempted to investigate the effect of the breadth of the attached plating of a stiffener to the section modulus and hence the rigidity of an assembly. Hence the effect of different geometrical shapes and properties were investigated for a wide range of breadths. It may be deduced that despite that breadth of the plating is not a prime issue for the calculation of the section modulus, its importance especially at low slenderness ratios stiffened structures should not be omitted.

FEA analysis on 29 simple structures, yielded the non-uniform stress distributions at the plate imposed by shear from the web. By using Schade's formula (1951) the effective breadth at each case was extracted. Valuable diagrams for observing the change of the effective breadth of the plating by altering the slenderness ratio and the boundary conditions were developed.

Finally, a simple but useful Rule for the use of HRS was proposed that can be also widely used by structural engineers. Such Rule is able to calculate the effective breadth efficiently only as a function of slenderness ratio. This comes also in relative accordance with the existing Rules of other Classification societies for calculating the effective breadth of the attached plate of a stiffener (Fig. 10).

REFERENCES

ABS 2010. *Rules and Regulations*, Part 3, Chapter 1, Section 2–13.

Belenkiy L.M., Raskin Y.N. & Vuillemin J. 2007. Effective Plating in Elastic-Plastic Range of Primary Support Members in Double-Skin Ship Structures, *Marine Structures*, 20: 115–123.

Bleich H.H. 1956. *Notes on the Influence of Unfair Plating on Ship Failures by Brittle Fracture*, Ship Structure Committee Report SSC-96.

Boote D. & Mascia 1991. On the Effective Breadth of Stiffened Platings, *Ocean Engineering*, 16(6): 567–592.

Common Structural Rules 2010. *Bulk Carriers*, Chapter 3, Section 6, 5.4 & 4.3.

Cox H.L. 1946. *The Buckling of a Flat Rectangular Plate under Axial Compression and its Behaviour after Buckling*, ARC, R&M 2041.

Faulkner D. 1975. A Review of Effective Plating for Use in the Analysis of Stiffened Plating in Bending and Compression, *Journal of Ship Research*, 19(1): 1–17.

Germanischer Lloyd's 2010. *Rules and Regulations*, Part 1, Section 3, E & F Design Principles, Chapter 1.

Horne M.R. 1956. The Progressive Buckling of Plates Subjected to Cycles of Longitudinal Strain, *Transactions* of *RINA*, 98: 78.

Karman Th. V 1924. *Die Mittranende Breite*, Springer, Berlin.

Katsikadelis J.T. & Sapountzakis E.J. 2002. A Realistic Estimation of the Effective Breadth of Ribbed Structures, *International Journal of Solids and Structures*, 39: 897–910.

Lloyd's Register of Shipping 2010. *Rules and Regulations*, Ship Structures (General) – Structural Design – Structural Idealisation 3.2.

Mansour A.E. 1970. Effective Flange Breadth of Stiffened Plates under Axial Tensile Load or Uniform Bending Moment, *Journal of Ship Research*, 14(1).

Mansour A.E. 1971. On the Nonlinear Theory of Orthotropic Plates, *Journal of Ship Research*, (15)4.

Marguerre K. 1937. Die Mittrangende Breite der Gedruckten Platte, *Luftfahrt Forschung*, 14(3): 121.

Murrey J.M. 1945. Notes on Deflected Plating in Tension and Compression, *Transactions of RINA*, 87: 95.

Murrey J.M. 1954. Corrugation of Bottom Shell Plating, *Transactions of RINA*, 96: 229.

Papkovitch P.F. 1939. *Theory of Elasticity*, Leningrad-Moskva: Oborogiz.

Schade H.A. 1951. Design Curves for Cross-Stiffened Plating under Uniform Bending Load, *Transactions of SNAME*, 49: 403–430.

Schade H.A. 1951. The Effective Breadth of Stiffened Plating Under Bending Loads, *Transactions of SNAME*, 59: 154–182.

Schade H.A. 1953. The Effective Breadth Concept in Ship-Structure Design, *Transactions of SNAME*, 61: 410–430.

Vedeler G. 1950. Calculation of Beams, *Transaction of RINA*, 92: 30–58.

Sustainable Maritime Transportation and Exploitation of Sea Resources – Rizzuto & Guedes Soares (eds)
 ISBN 978-0-415-62081-9

Local and overall buckling of uniaxially compressed stiffened panels

V. Piscopo
Department of Applied Sciences, The University of Naples "Parthenope", Naples, Italy

ABSTRACT: This paper deals with the elastic buckling of longitudinally stiffened panels, for which overall and local plate buckling are two distinct failure modes. As for ship structures it is generally preferred that local buckling occurs firstly, it is necessary to define the minimum scantling of stiffeners that permits to prevent overall buckling from occurring before local plate buckling. In the following a systematic buckling analysis of one-bay stiffened panels under uniaxial compression is presented, taking into account both failure modes, and the convergence of solution, in terms of critical flexural rigidities of longitudinal stiffeners, is investigated. An improved expression is derived to identify the "crossover" panels, for which local and overall buckling stresses are almost the same, and subsenquently it is compared with those ones proposed by Klitchieff (1951), Hughes et al. (2004) and with the relevant results obtained by some FE eigenvalue buckling analyses.

1 INTRODUCTION

Ship structures can be regarded as constituted by multi-bay longitudinally stiffened panels, supported by transverse beams, having the following relevant buckling modes:

1. overall buckling: the multi-bay panel buckles overall;
2. partial buckling: the stiffened panel buckles between adjacent transverse beams;
3. local plate buckling: each single panel buckles locally;
4. longitudinal stiffeners' local buckling: each longitudinal stiffener buckles locally, due to web or tripping failure.

Even if these modes are not exclusive nor each other independent, the first one is generally preceeded by partial buckling of one-bay stiffened panels while the last one can be prevented if stiffeners have good proportions (see for example RINA Rules Rules Pt B, Ch. 7, Sec. 2, Par. 1.4). It follows that only overall and local plate buckling are two distinct failure modes, even if there are some specific dimensions of stiffeners' cross-section for which overall and local buckling stresses are almost the same.

Various authors presented approximate formulas to define the minimum scantling of stiffeners, necessary to prevent overall buckling from occurring before local plate buckling, as function of the stiffener flexural rigidity ratio γ, as defined by Bleich. Starting from the system of equations established by Timoshenko for panels reinforced by longitudinal ribs, Klitchieff (1951) derived a general solution $\gamma_{co,\,KL}$, valid for one-bay stiffened panels under uniaxial compression with any number of stiffeners, as function of the local panel aspect ratio α, the stiffeners' number n_s and area ratio δ:

$$\gamma_{co,\,KL} = 4\alpha^2\left[\delta + \frac{2}{\pi(C_1 - C_2)}\right] \tag{1}$$

with:

$$C_1 = \frac{1}{\sqrt{2\alpha+1}}\,\frac{sinh\left(\frac{\sqrt{2\alpha+1}}{\alpha}\pi\right)}{cosh\left(\frac{\sqrt{2\alpha+1}}{\alpha}\pi\right) - cos\left(\frac{\pi}{n_s+1}\right)} \tag{2}$$

and:

$$C_2 = \frac{1}{\sqrt{2\alpha-1}}\,\frac{sin\left(\frac{\sqrt{2\alpha-1}}{\alpha}\pi\right)}{cos\left(\frac{\sqrt{2\alpha-1}}{\alpha}\pi\right) - cos\left(\frac{\pi}{n_s+1}\right)} \tag{3}$$

Subsequently, Hughes et al. (2004) found an improved expression of the minimum flexural rigidity ratio $\gamma_{co,HU}$, taking into account the decrease in rotational restraint of the plating by the stiffeners, for local plate buckling, and considering, for overall buckling, a modified Euler expression that allows for the added deflection of an ideal column due to transverse shear. The local buckling stress is derived considering the plate, under uniaxial compression, as simply supported on the loaded edges and elastically restrained by the stiffeners on the unloaded edges. A formula by Paik and Thayamballi (2004), modified by means of a corrective factor C_r, that gave the best agreement with some FE eigenvalue buckling analyses of 55 crossover panels, was adopted. The overall buckling stress is evaluated applying the Euler formula, modified by Timoshenko to take into account the effect of transverse shear. The improved expression, proposed by Hughes et al. (2004), can be specialized for longitudinally stiffened panels as follows:

$$\gamma_{co,HU} = \alpha^2 \frac{A_t}{bt} \frac{k_{Cr}}{\left[\frac{A_w G}{A_w G + A_t \sigma_{E,c}}\right] \cdot \left[1 + 2\frac{H}{D_x} \frac{\alpha^2}{(n_s+1)^2}\right]} \tag{4}$$

having denoted by $\sigma_{E,c}$ the Euler buckling stress for an ideal column:

$$\sigma_{E,c} = \frac{\pi^2 E}{\left(\frac{a}{\rho}\right)^2} \tag{5}$$

and by k_{Cr} a function of ζ:

$$\begin{cases} k_{Cr} = 4 + 0.396\varsigma^3 - 1.974\varsigma^2 + 3.565\varsigma & ;0 \le \varsigma < 2 \\ k_{Cr} = 6.951 - \dfrac{0.881}{\varsigma - 0.4} & ;2 \le \varsigma < 20 \\ k_{Cr} = 7.025 & ;\varsigma \ge 20 \end{cases} \tag{6}$$

with:

$$\varsigma = \frac{\mu C_r}{2} \tag{7}$$

and:

$$C_r = \frac{1}{1 + 3.6\left(\frac{t}{t_w}\right)^3 \frac{h_w + 0.5(t + t_f)}{b}} \tag{8}$$

In the following, starting from the Timoshenko's system of equations for plates reinforced by longitudinal ribs, modified to take into account the stiffeners' torque and warping rigidities, a new expression $\gamma_{co,new}$ is derived. The new formula, that introduces in the Klitchieff one a corrective term, function of the stiffeners' torque and warping rigidities, has been obtained after an extensive eigenvalue buckling analysis of one-bay stiffened panels, carried out by a dedicated program developed in Mathworks Matlab. The convergence of solution, in terms of critical flexural rigidity ratio $\gamma_{co,new}$ is fully investigated, varying the number of harmonics of the vertical displacement field, developed into appropriate trigonometric series. Finally, to show the feasibility of the new expression, a numerical comparison with the formulas (1) and (4) and the relevant results obtained by some FE eigenvalue buckling analyses of different "crossover" stiffened panels, is presented.

2 THEORETICAL MODEL

Let us consider a one-bay stiffened panel (see Fig. 1) under uniaxial compression, comprised between adjacent primary supporting members. The stiffened panel, reinforced by n_s equally spaced longitudinal stiffeners, is simply supported on the four edges, so that the vertical displacement field $w(x, y)$ must verify the following boundary conditions:

$$\begin{cases} w(x=0,y) = w(x=a,y) = 0 \\ w(x,y=0) = w(x,y=B) = 0 \end{cases} \tag{9}$$

and:

$$\begin{cases} \left.\dfrac{\partial^2 w}{\partial x^2} + \nu \dfrac{\partial^2 w}{\partial y^2}\right|_{x=0,a} = 0 \\ \left.\dfrac{\partial^2 w}{\partial y^2} + \nu \dfrac{\partial^2 w}{\partial x^2}\right|_{y=0,B} = 0 \end{cases} \tag{10}$$

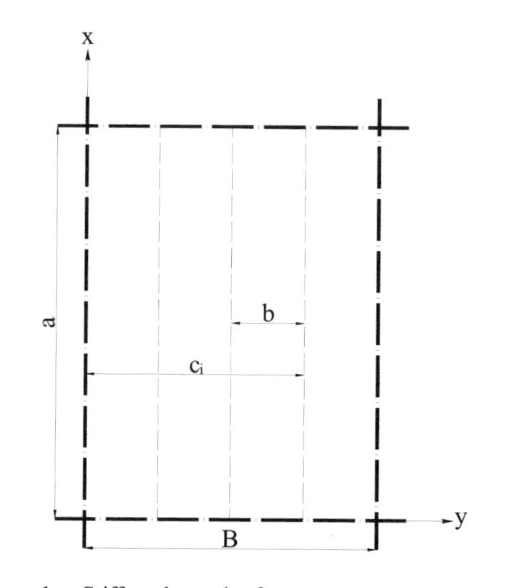

Figure 1. Stiffened panel reference system.

The displacement field may be developed intro appropriate double sine trigonometric series, satisfying the boundary conditions (9) and (10):

$$w(x,y)=\sum_{m=1}^{M}\sum_{n=1}^{N}w_{m,n}sin\left(m\pi\frac{x}{a}\right)sin\left(n\pi\frac{y}{B}\right) \quad (11)$$

To evaluate the Euler stress at which buckling occurs, the energy method can be adopted, assuming that the stiffened plate undergoes some small lateral bending, consistent with the given boundary conditions. Naturally, if the work done by the in-plane forces is smaller than the strain energy of plate and attached stiffeners, the equilibrium is stable, otherwise it is unstable and buckling occurs. So, the general equilibrium equation can be written as follows:

$$\Delta U_p+\sum_{i=1}^{n_s}\Delta U_{s,i}=\Delta T_p+\sum_{i=1}^{n_s}\Delta T_{s,i} \quad (12)$$

having denoted by ΔU_p ($\Delta U_{s,i}$) the strain energy due to plate bending (due to bending, torque and warping of the i-th stiffener) and by ΔT_p ($\Delta T_{s,i}$) the work done during buckling by the compressive forces acting in x-direction on the plate (on the i-th stiffener). The first term of Eq. (12) may be so developed:

$$\Delta U_p=\frac{D}{2}\int_0^a\int_0^B\left[\left(\frac{\partial^2 w}{\partial x^2}\right)^2+\left(\frac{\partial^2 w}{\partial y^2}\right)^2+2\frac{\partial^2 w}{\partial x^2}\frac{\partial^2 w}{\partial y^2}\right]dxdy \quad (13)$$

The i-th stiffener's strain energy due to bending, torque and warping is the sum of three terms:

$$\Delta U_{s,i}=\frac{1}{2}\int_0^a\left[\begin{array}{l}EI_i\left(\frac{\partial^2 w}{\partial x^2}\right)^2_{y=c_i}+GJ_i\left(\frac{\partial\phi_i}{\partial x}\right)^2_{y=c_i}\\ +EI_{w,i}\left(\frac{\partial^2\phi_i}{\partial x^2}\right)^2_{y=c_i}\end{array}\right]dx \quad (14)$$

having denoted, for the i-th stiffener, by c_i the relevant transverse distance from the edge $y=0$ and by $\phi_i(x, y)$ the rotation around the axis coinciding with the intersection of the stiffener's web and the inner plate surface. Imposing the continuity condition of deflection angle along the junction between the plate and the longitudinal stiffener, the following congruence condition must be verified:

$$\phi_i(x,c_i)=\left.\frac{\partial w}{\partial y}\right|_{y=c_i} \quad (15)$$

The work done by the compressive forces on the plate and the i-th stiffener can be so expressed:

$$\Delta T_p=\frac{N_x}{2}\int_0^a\int_0^B\left(\frac{\partial w}{\partial x}\right)^2 dxdy \quad (16)$$

$$\Delta T_{s,i}=-\frac{N_x A_i}{2t}\int_0^a\left(\frac{\partial w}{\partial x}\right)^2_{y=c_i}dxdy \quad (17)$$

Substituting Eqs. (13), (14), (16) and (17) into (12) the Euler buckling stress may be derived:

$$\sigma_E=\frac{\pi^2 E}{12(1-\nu^2)}\left(\frac{t}{B}\right)^2 k_{b,g}=\frac{\pi^2 E}{12(1-\nu^2)}\left(\frac{t}{b}\right)^2 k_{b,p} \quad (18)$$

having denoted by $k_{b,g}$ and $k_{b,p}$ the buckling coefficients of the one-bay stiffened panel and the local plate comprised between two adjacent longitudinal stiffeners, respectively. The two coefficients are related by the following expression:

$$k_{b,p}=\frac{k_{b,g}}{(n_s+1)^2} \quad (19)$$

As the coefficients of the series (11) must be chosen to make the equation (12) minimum, the following eigenvalue problem must be solved:

$$\frac{\partial}{\partial w_{j,k}}\left[\Delta U_p+\sum_{i=1}^{n_s}\Delta U_{s,i}-\Delta T_p-\sum_{i=1}^{n_s}\Delta T_{s,i}\right]=0 \quad (20)$$

finally becoming:

$$\frac{1}{k_{b,g}}\frac{\partial\rho_1}{\partial w_{j,k}}-\frac{\partial\rho_2}{\partial w_{j,k}}=0 \quad \forall\, j=1\ldots M; k=1\ldots N \quad (21)$$

with:

$$\frac{\partial\rho_1}{\partial w_{j,k}}=2w_{j,k}\left(\frac{j^2}{\alpha_g}+\alpha_g k^2\right)^2+2\sum_{q=1}^{N}w_{j,q}\sum_{i=1}^{n_s}[A_i+B_i+C_i] \quad (22)$$

$$\frac{\partial\rho_2}{\partial w_{j,k}}=2w_{j,k}j^2+\frac{4}{n_s+1}\sum_{q=1}^{N}w_{j,q}j^2\sum_{i=1}^{n_s}D_i \quad (23)$$

and:

$$\begin{cases} A_i = \dfrac{2}{\alpha_g^2(n_s+1)}\gamma_i j^4 \sin k\pi\chi_i \sin q\pi\chi_i \\ B_i = \dfrac{2\mu_i}{n_s+1} m^2 nq \cos k\pi\chi_i \cos q\pi\chi_i \\ C_i = \dfrac{2\pi^2}{\alpha_g^2(n_s+1)^3}\psi_i j^4 kq \cos k\pi\chi_i \cos q\pi\chi_i \\ D_i = \delta_i \sin k\pi\chi_i \sin q\pi\chi \end{cases} \tag{24}$$

having done the position: $\chi_i = c_i/B$. Assuming that all stiffeners have same scantlings, the system of equations (21) may be rewritten as follows:

$$\left[\frac{1}{k_{b,g}}\left(\mathrm{M}_1 + \gamma\mathrm{M}_2 + \mu\mathrm{M}_3 + \psi\mathrm{M}_4\right) - \mathrm{M}_5 + \delta\mathrm{M}_6\right]\mathrm{w} = 0 \tag{25}$$

so that, when γ, μ, ψ and δ are set, the buckling coefficient $k_{b,g}$ may be determined as the minimum eigenvalue of (25). Naturally, as for "crossover" stiffened panels the overall and local buckling coefficients are the same, it is possible to define the minimum stiffeners' flexural rigidity ratio $\gamma_{CO,new}$ that permits to prevent overall buckling from occurring before local plate buckling, proceeding as follows:

1. determine the minimum eigenvalue $k_{b,g}$ of (25) for fixed values of μ, ψ and δ, assuming $\gamma = \infty$;
2. determine the minimum eigenvalue $\gamma_{CO,new}$ of (25) for fixed values of μ, ψ and δ imposing for $k_{b,g}$ the value obtained at the first step.

To define a new expression of the minimum stiffener flexural rigidity ratio $\gamma_{CO,new}$, an extensive eigenvalue analysis has been carried out by a dedicated program developed in Mathworks Matlab, varying the involved non-dimensional parameters in the ranges: $n_s = 3, 4, 5, 6, 7$; $\alpha = 1 \ldots 6$; $\mu = 0 \ldots 4$; $\psi = 0 \ldots 4$; $\delta = 0 \ldots 1$. The large amount of data has been accurately fitted, so obtaining a corrective expression of the

Klitchieff formula:

$$\gamma_{co,new} = \gamma_{co,KL} f(\mu,\psi) \tag{26}$$

with:

$$f(\mu,\psi) = \left[k_1 e^{k_2\mu} + [1-k_1]e^{-k_3\mu}\right]\cdot\left[q_1 e^{q_2\psi} + [1-q_1]e^{-q_3\psi}\right] \tag{27}$$

having done the following positions:

$$\begin{cases} k_1 = 1.310 \\ k_2 = 0.00598 \\ k_3 = 1.320 \\ q_1 = 0.3227e^{-1.735\mu} + 1.070e^{-0.01089\mu} \\ q_2 = \dfrac{0.003989 + 0.00933\mu}{0.7604 + 0.4068\mu + \mu^2} \\ q_3 = \dfrac{15.83 + 7.889\mu + 2.075\mu^2}{1.5750 + 0.3628\mu + \mu^2} \end{cases} \tag{28}$$

In Figure 2 the corrective function $f(\mu, \psi)$ is shown: for $\mu = \psi = 0$, $f(0, 0) = 1$, while for positive

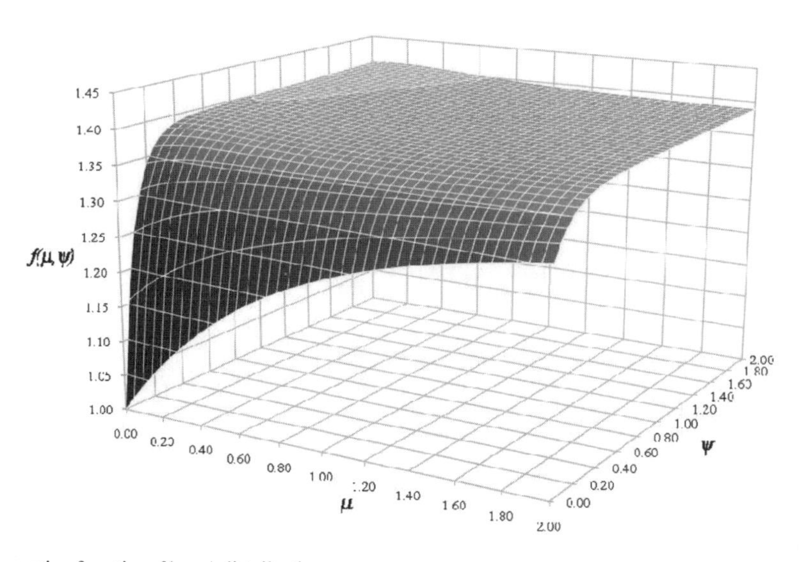

Figure 2. Corrective function $f(\mu, \psi)$ distribution.

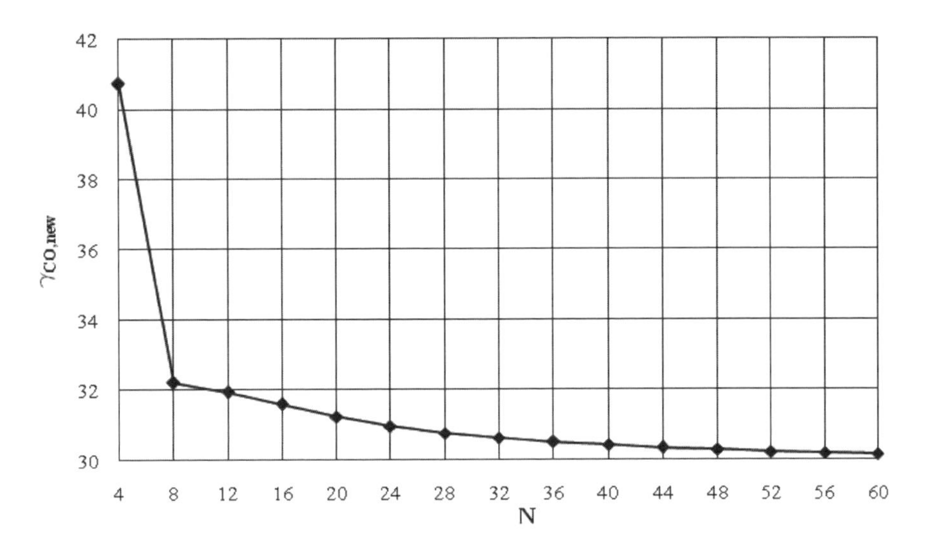

Figure 3. Panel 1—convergence of solution.

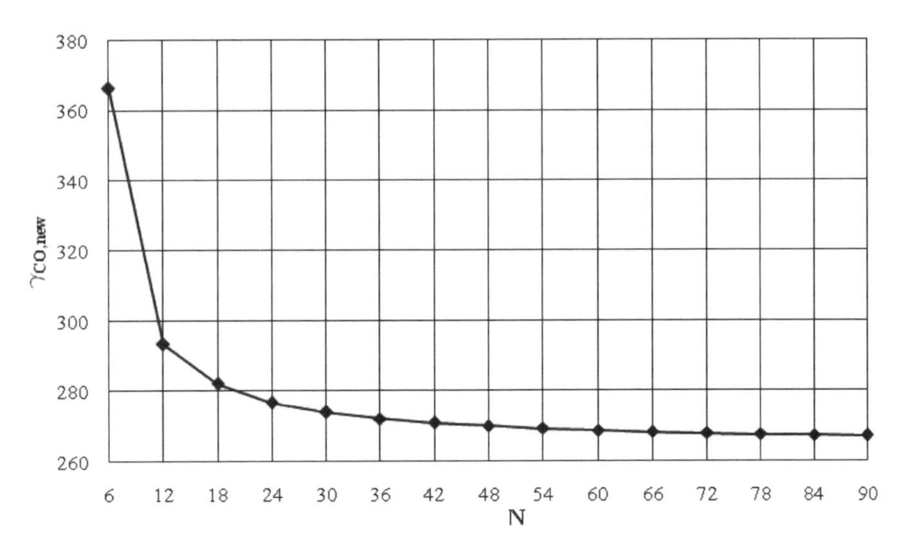

Figure 4. Panel 2—convergence of solution.

values of μ and ψ it is consistently greater than unit. As said, the convergence of data used to build the fitting surface, has been accurately investigated, varying the harmonics' number of the vertical displacement field in both x and y directions. It has been found that for one-bay stiffened panels $M = 7$ harmonics in x-direction and $N = 15(n_s+1)$ in y-direction are always sufficient to obtain consistent results. In Figures 3 and 4 the convergence of solution is shown for two different stiffened panels having the following scantlings:

1. panel 1: $n_s = 3$, $\alpha = 2$, $\mu = 0.50$, $\psi = 0.50$, $\delta = 0.20$;
2. panel 2: $n_s = 5$, $\alpha = 6$, $\mu = 0.50$, $\psi = 0.25$, $\delta = 0.40$.

3 NUMERICAL APPLICATIONS

To show the feasibility of the new improved expression, a numerical comparison with the relevant results obtained by Hughes et al. (2004) for some crossover panels with three and five stiffeners is presented. All panels are within practical proportions from a design point of view. Table 1 lists the scantlings of the crossover panels: all stiffeners have T-section in high strength steel with

Table 1. Geometric properties of crossover panels with three and five longitudinal stiffeners.

	a	B	n_s	t	h_w	t_w	b_f	t_f	I	A	J	I_w	δ	μ	ψ
N	mm	mm	–	mm	mm	mm	mm	mm	mm^4	mm^2	mm^4	mm^6	–	–	–
1	1800	3600	3	21	84	12	100	15	1.77E+07	2.51E+03	1.61E+05	8.82E+09	0.1327	0.0811	0.0143
2	1800	3600	3	16	56	12	100	15	8.01E+06	2.17E+03	1.45E+05	3.92E+09	0.1508	0.1649	0.0143
3	1800	3600	3	16	81	5	60	10	6.41E+06	1.01E+03	2.34E+04	1.18E+09	0.0698	0.0266	0.0043
4	1800	3600	3	10	41	5	60	10	1.67E+06	8.05E+02	2.17E+04	3.03E+08	0.0894	0.1013	0.0045
5	2640	3600	3	21	123	12	100	15	3.54E+07	2.98E+03	1.83E+05	1.89E+10	0.1575	0.0924	0.0306
6	2640	3600	3	16	84	12	100	15	1.59E+07	2.51E+03	1.61E+05	8.82E+09	0.1742	0.1833	0.0323
7	2640	3600	3	10	62	5	60	10	3.39E+06	9.10E+02	2.26E+04	6.92E+08	0.1011	0.1054	0.0104
8	3600	3600	3	21	112	20	200	30	9.19E+07	8.24E+03	2.10E+06	2.51E+11	0.4360	1.0575	0.4058
9	3600	3600	3	21	166	12	100	15	6.40E+07	3.49E+03	2.08E+05	3.45E+10	0.1848	0.1049	0.0557
10	3600	3600	3	16	120	12	100	15	3.10E+07	2.94E+03	1.82E+05	1.80E+10	0.2042	0.2069	0.0658
11	3600	3600	3	10	65	12	100	15	8.61E+06	2.28E+03	1.50E+05	5.28E+09	0.2533	0.6997	0.0791
12	3600	3600	3	10	86	5	60	10	6.26E+06	1.03E+03	2.36E+04	1.33E+09	0.1144	0.1101	0.0199
13	1800	3600	5	21	116	12	100	15	2.99E+07	2.89E+03	1.79E+05	1.68E+10	0.2295	0.1355	0.0918
14	1800	3600	5	21	93	10	160	20	3.51E+07	4.13E+03	4.58E+05	5.90E+10	0.3278	0.3459	0.3223
15	1800	3600	5	16	82	12	100	15	1.43E+07	2.48E+03	1.60E+05	8.41E+09	0.2588	0.2730	0.1037
16	1800	3600	5	10	56	5	60	10	2.70E+06	8.80E+02	2.23E+04	5.65E+08	0.1467	0.1563	0.0285
17	2640	3600	5	21	126	20	200	30	9.89E+07	8.52E+03	2.14E+06	3.18E+11	0.6762	1.6145	1.7333
18	2640	3600	5	21	168	12	100	15	6.15E+07	3.52E+03	2.09E+05	3.53E+10	0.2790	0.1582	0.1926
19	2640	3600	5	21	136	10	160	20	6.79E+07	4.56E+03	4.72E+05	1.26E+11	0.3619	0.3568	0.6893
20	2640	3600	5	16	120	12	100	15	2.89E+07	2.94E+03	1.82E+05	1.80E+10	0.3063	0.3104	0.2222
21	2640	3600	5	10	84	5	60	10	5.71E+06	1.02E+03	2.35E+04	1.27E+09	0.1700	0.1645	0.0642
22	3600	3600	5	21	174	20	200	30	1.75E+08	9.48E+03	2.26E+06	6.06E+11	0.7524	1.7113	3.3055
23	3600	3600	5	21	223	12	100	15	1.10E+08	4.18E+03	2.41E+05	6.22E+10	0.3314	0.1821	0.3393
24	3600	3600	5	21	185	10	160	20	1.20E+08	5.05E+03	4.88E+05	2.34E+11	0.4008	0.3691	1.2754
25	3600	3600	5	16	131	20	200	30	9.24E+07	8.62E+03	2.15E+06	3.43E+11	0.8979	3.6732	4.2362
26	3600	3600	5	16	164	12	100	15	5.37E+07	3.47E+03	2.07E+05	3.36E+10	0.3613	0.3537	0.4150
27	3600	3600	5	16	133	10	160	20	5.86E+07	4.53E+03	4.71E+05	1.21E+11	0.4719	0.8049	1.4905
28	3600	3600	5	10	95	12	100	15	1.58E+07	2.64E+03	1.67E+05	1.13E+10	0.4400	1.1705	0.5703

yield stress σ_y = 315 N/mm^2, E = 205.8 GPa and ν = 0.30.

In the analysis the Saint-Venant's moment of inertia of stiffeners J and the sectorial moment of inertia of stiffeners about the connection to the attached plating I_w have been evaluated applying the following formulas:

$$J = \frac{1}{3}\left(h_w t_w^3 + b_f t_f^3\right) \tag{29}$$

$$I_w = \frac{t_f b_f^3 h_w^2}{12} \tag{30}$$

In Table 2 the crossover values, obtained by Hughes et al. (2004) after some eigenvalue buckling analyses carried out by Abaqus, are compared with the relevant results obtained applying the formulas (1), (4) and the new expression (26).

The Klitchieff formula (1), normalized by $\gamma_{CO,FEA}$, has a mean of 0.812 with a COV of 8.19%, while the expression (4) a mean of 0.930 with a scatter of 5.18%. The new improved expression, normalized by $\gamma_{CO,FEA}$, has a mean of 1.018 with a COV of 7.69%.

4 CONCLUSIONS

In this paper the overall and local buckling problem of stiffened panels has been discussed with particular reference to the minimum stiffeners' flexural rigidity ratio that permits to prevent overall buckling from occurring before local plate buckling. Starting from the Timoshenko's system of equations for plates reinforced by ribs, a new formula, that takes into account the strain energy due to torque and warping of longitudinal stiffeners, has been obtained. To show the feasibility of the new improved formula, it has been compared with those ones proposed by Klitchieff (1951), Hughes et al. (2004) and with the relevant results obtained by some Abaqus FE eigenvalue buckling

Table 2. Comparison of crossover parameters.

N	$\gamma_{co,\,KL}$	$\gamma_{co,\,HU}$	$\gamma_{co,\,new}$	$\gamma_{co,\,FEA}$	$\gamma_{co,\,KL}/\gamma_{co,\,FEA}$	$\gamma_{co,\,HU}/\gamma_{co,\,FEA}$	$\gamma_{co,new}/\gamma_{co,\,FEA}$
	–	–	–	–	–	–	–
1	18.78	20.15	20.30	22.47	0.836	0.897	0.903
2	19.07	20.35	21.11	23.03	0.828	0.884	0.917
3	17.77	18.83	18.25	18.14	0.980	1.038	1.006
4	18.08	18.35	19.09	19.50	0.927	0.941	0.979
5	39.73	42.49	45.00	45.39	0.875	0.936	0.991
6	40.31	43.26	46.76	46.88	0.860	0.923	0.998
7	37.79	38.84	40.69	40.62	0.930	0.956	1.002
8	90.15	119.72	124.09	120.85	0.746	0.991	1.027
9	74.07	78.75	88.49	83.65	0.886	0.941	1.058
10	75.31	80.82	92.80	92.23	0.817	0.876	1.006
11	78.46	97.27	101.49	105.18	0.746	0.925	0.965
12	69.57	71.76	77.07	75.71	0.919	0.948	1.018
13	44.89	50.98	56.60	58.32	0.770	0.874	0.971
14	48.43	61.31	66.88	68.53	0.707	0.895	0.976
15	45.94	51.61	59.28	63.34	0.725	0.815	0.936
16	41.91	44.45	47.95	48.66	0.861	0.913	0.985
17	129.65	187.52	180.01	195.96	0.662	0.957	0.919
18	98.89	109.18	133.32	121.14	0.816	0.901	1.101
19	105.31	124.40	146.80	133.62	0.788	0.931	1.099
20	101.00	111.55	137.61	129.47	0.780	0.862	1.063
21	90.45	94.97	110.57	103.97	0.870	0.913	1.063
22	250.55	345.91	350.32	347.81	0.720	0.995	1.007
23	189.93	206.07	263.01	218.23	0.870	0.944	1.205
24	199.92	226.97	279.83	236.74	0.844	0.959	1.182
25	271.50	445.03	377.15	415.25	0.654	1.072	0.908
26	194.22	211.87	269.56	241.07	0.806	0.879	1.118
27	210.15	242.87	292.70	262.01	0.802	0.927	1.117
28	205.56	272.79	283.73	291.10	0.706	0.937	0.975
				Mean	**0.812**	**0.930**	**1.018**
				COV	**0.082**	**0.052**	**0.077**

analyses of different crossover stiffened panels. From the numerical comparison, it appears that the influence of stiffeners' warping and torque rigidities is noticeable and that the new improved formula furnishes results very close to those ones obtained by the relevant FE analyses.

5 LIST OF SYMBOLS

E	Young modulus
G	Coulomb modulus
v	Poisson modulus
n_s	Number of longitudinal stiffeners
a	Length of one-bay stiffened panel
B	Breadth of one-bay stiffened panel
b	Breadth of a local panel
α_g	Stiffened panel aspect ratio
α	Local panel aspect ratio
t	Plating thickness
t_f	Thickness of stiffener flange
h_w	Height of stiffener web
A	Sectional area of a single stiffener
A_t	Sectional area of a single stiffener with attached plating
A_w	Sectional area of stiffener web
I	Moment of inertia of a single stiffener with attached plating
J	Saint-Venant moment of inertia of a single stiffener
I_w	Sectorial moment of inertia of a single stiffener about its connection to the attached plating
$\rho = \sqrt{\frac{I}{A_t}}$	Radius of gyration of a single stiffener with attached plating
$D = \frac{Et^3}{12(1-v^2)}$	Flexural rigidity of isotropic plate

$D_x = \frac{EI}{b}$	Flexural rigidity of orthotropic plate
$H = \frac{G}{2}\left(\frac{J}{b} + \frac{t^3}{3}\right)$	Torque rigidity of orthotropic plate
$\delta = \frac{A}{bt}$	Stiffener area ratio
$\gamma = \frac{EI}{Db}$	Stiffener flexural rigidity ratio
$\mu = \frac{GJ}{Db}$	Stiffener torque rigidity ratio
$\psi = \frac{EI_w}{Db^3}$	Stiffener warping rigidity ratio

REFERENCES

Det Norske Veritas, 2002. Buckling strength of plated structures, Recommended Practice DNV-RP-C201.

G.M. Vörös. Buckling and vibration of stiffened plates, International Review of Mechanical Engineering (I.R.E.M.E.), Vol. 1, no. 1.

J.K. Paik & A.K. Thayamballi, 2000. Buckling strength of steel plating with elastically restrained edges, Thin-Walled Structures 37: 27–55.

J.M. Klitchieff, 1951. On the stability of plates reinforced by longitudinal ribs, Journal of Applied Mechanics 18(4): 364–366.

O.F. Hughes, 1988. Ship structural design: a Rationally-Based Computer Aided Optimization Approach, SNAME Edition.

O.F. Hughes, B. Gosh & Y. Chen, 2004. Improved prediction of simultaneous local and overall buckling of stiffened panels, Thin-Walled structures 42: 827–856.

RINA, 2011. Rules for the classification of ships.

S.P. Timoshenko & J.M. Gere, 1985. Theory of elastic stability, Mc-Graw-Hill International Book Company, 17th edition.

S.P. Timoshenko, J.N. Goodier, 1951. Theory of Elasticity, Mc-Graw-Hill International Book Company.

S.P. Timoshenko & S. Woinowsky-Krieger, 1959. Theory of Plates and Shells, Mc-Graw-Hill International Book Company.

Sustainable Maritime Transportation and Exploitation of Sea Resources – Rizzuto & Guedes Soares (eds)

Refined buckling analysis of plates under shear and uniaxial compression

V. Piscopo
The University of Naples "Parthenope", Department of Applied Sciences, Naples, Italy

ABSTRACT: In the classical buckling analysis of ship platings, Euler stresses are generally evaluated applying the classical thin plate theory and so neglecting the shear deformation, even if the plating thickness is not totally negligible as regards the panel's mean dimensions. In the paper the two variable refined theory proposed by Shimpi (2006) is applied to the buckling analysis of simply supported rectangular plates under combined shear and uniaxial stresses, in order to take into account the transverse shear effects. The coupled bending-shear governing equations, derived by the Hamilton's principle, are solved by the energy method, developing into appropriate double sine trigonometric series the displacement field. The convergence of solution, in terms of shear buckling coefficients, is fully investigated varying the number of harmonics. New relations, that permit to take into due consideration the plating thickness, are also obtained and compared with the classical design formulas and the relevant results, obtained by some FE eigenvalue buckling analyses of typical ship platings under the combined action of shear and uniaxial forces.

1 LIST OF SYMBOLS

E	Young modulus
G	Coulomb modulus
ν	Poisson modulus
a	Plating length
b	Plating breadth
α	Plating aspect ratio
t	Plating thickness
τ_E	Euler shear buckling stress
k_s	Shear buckling coefficient
$D=\dfrac{Et^3}{12(1-\nu^2)}$	Flexural rigidity of isotropic plate

2 INTRODUCTION

The buckling problem of simply supported rectangular plates, under the action of shearing forces uniformly distributed along the edges, was discussed by several authors, such as S. Bergmann and H. Reissner (1934), M. Stein and J. Neff (1947), Timoshenko (1963), always neglecting the transverse shear effects. Obviously, when the plating thickness is not totally negligible as regards the plate dimensions, the classical thin plate theory becomes not appropriate and a new formulation, that permits to take into account the shear effects, is required.

In the past different shear deformable theories were presented by several authors, such as Reissner (1945), Mindlin (1951), Levinson (1980), Reddy (1984), Shimpi (2006). In the following the Shimpi theory, based on two variable coupled governing equations for the bending and shear displacement fields, is applied to the buckling analysis of simply supported rectangular plates under the combined action of shear and uniaxial stresses. Its main advantage is that the governing equations are derived by the Hamilton's principle, so they are consistent with the assumed displacement field. To evaluate the Euler shear stresses at which buckling occurs, the energy method is adopted and the deflection surface of the buckled plate is expressed into appropriate double sine trigonometric series, whose terms satisfy the plate boundary conditions along all edges. The convergence of solution is fully investigated, varying the number of harmonics in both plate directions, too. A new relation, for the buckling coefficient of simply supported rectangular plates under pure shear, that permits to take into consideration the transverse shear effects, is presented and compared with the classical one:

$$\tau_E = \frac{\pi^2 E}{12(1-\nu^2)}\left(\frac{t}{b}\right)^2 k_s \tag{1}$$

in which the shear buckling coefficient my obtained by the following classical parabolic curves:

$$k_s = \begin{cases} 5.34 + \dfrac{4.00}{\alpha^2} & if \quad \alpha > 1 \\ 4.00 + \dfrac{5.34}{\alpha^2} & if \quad \alpha \le 1 \end{cases} \tag{2}$$

Furthermore, as in a lot of cases it is necessary to evaluate the buckling coefficients of plates under combined shear and uniaxial stresses, a new project formula that permits to take into account the presence of uniaxial compressive forces, is derived and discussed, too.

Finally, some numerical applications are presented for platings under shear and uniaxial compression, varying the thickness and the plating aspect ratio. The theoretical Euler shear forces per unit of length, obtained applying both the thin plate theory and the refined, are also compared with the relevant obtained by some FE eigenvalue analyses carried out by *ANSYS*.

3 THEORETICAL MODEL

Let us refer to the coordinate system of Fig. 1 with z axis having origin on the plate middle plane.

The basic assumptions of the two variable refined Shimpi theory are:

1. the displacements are small, if compared with the plating thickness;
2. the stress σ_z is negligible respect to the in-plane stresses σ_x and σ_y;
3. the vertical displacement $w(x, y)$, normal to the plate middle plane, is the sum of two components of bending and shear $w_b(x, y)$ and $w_s(x, y)$, respectively;
4. the in-plane displacements $u(x, y)$ and $v(x, y)$ along the x and y axes include two components of bending and shear and one component $u_0(x, y)$ and $v_0(x, y)$, due to the in-plane normal forces;
5. the bending components $u_b(x, y)$ and $v_b(x, y)$ are similar to those ones of the classical thin plate theory:

$$u_b(x,y) = -z\frac{\partial w_b}{\partial x};\ v_b(x,y) = -z\frac{\partial w_b}{\partial y} \backslash \quad (3)$$

6. the shear components $u_s(x, y)$ and $v_s(x, y)$ are related to the vertical shear displacement field w_s.

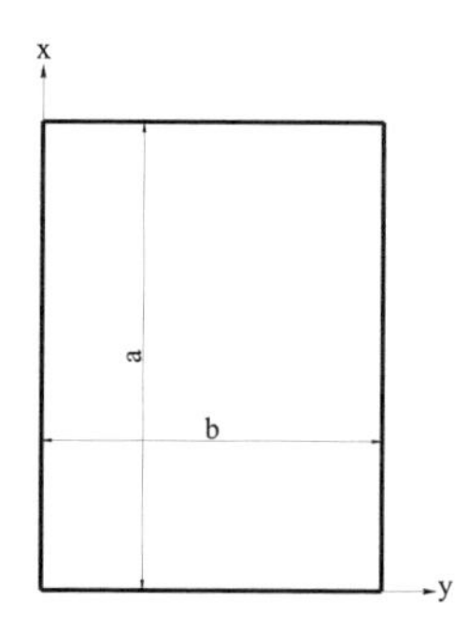

Figure 1. Plating reference system.

Starting from these basic assumptions, the displacement field may be so rewritten:

$$\begin{cases} u(x,y) = u_0(x,y) - z\dfrac{\partial w_b}{\partial x} + z\left[\dfrac{1}{4} - \dfrac{5}{3}\left(\dfrac{z}{t}\right)^2\right]\dfrac{\partial w_s}{\partial x} \\ v(x,y) = v_0(x,y) - z\dfrac{\partial w_b}{\partial y} + z\left[\dfrac{1}{4} - \dfrac{5}{3}\left(\dfrac{z}{t}\right)^2\right]\dfrac{\partial w_s}{\partial y} \\ w(x,y) = w_b(x,y) + w_s(x,y) \end{cases} \quad (4)$$

The Hamilton's principle is used to derive the equations of motion appropriate to the assumed displacement field (4), so imposing the following condition:

$$\int_0^T (\partial U + \partial V - \partial T)\,dt = 0 \quad (5)$$

having denoted by U the strain energy, V the work done by the applied forces and T the kinetic energy. Starting from (5), the governing equations are finally expressed as follows:

$$\begin{cases} D\nabla^4 w_b = p + N_x\dfrac{\partial^2 w}{\partial x^2} + N_y\dfrac{\partial^2 w}{\partial y^2} + 2N_{xy}\dfrac{\partial^2 w}{\partial x \partial y} \\ \dfrac{D}{84}\nabla^4 w_s = p + \dfrac{5}{6}Gt\nabla^2 w_s + N_x\dfrac{\partial^2 w}{\partial x^2} + N_y\dfrac{\partial^2 w}{\partial y^2} \\ \qquad +2N_{xy}\dfrac{\partial^2 w}{\partial x \partial y} \end{cases} \quad (6)$$

where N_x and N_y are the in-plane forces per unit of length, directed along the x and y axes, respectively, and N_{xy} is the shear in-plane force per unit of length. Assuming that these forces are constant throughout the plate, $N_y = 0$ and $N_x = \gamma N_{xy}$ with $-1 \leq \gamma \leq 0$, the Eq. (6) may be rewritten as follows, neglecting the bending due to the shear displacement field:

$$\begin{cases} D\nabla^4 w_b = N_{xy}\left(\gamma\dfrac{\partial^2 w}{\partial x^2} + 2\dfrac{\partial^2 w}{\partial x \partial y}\right) \\ \dfrac{5}{6}Gt\nabla^2 w_s = -N_{xy}\left(\gamma\dfrac{\partial^2 w}{\partial x^2} + 2\dfrac{\partial^2 w}{\partial x \partial y}\right) \end{cases} \quad (7)$$

As the plate is assumed simply supported along all edges, the relevant boundary conditions:

$$\begin{cases} w(x=0,y) = w(x=a,y) = 0 \\ w(x,y=0) = w(x,y=b) = 0 \end{cases} \quad (8)$$

and:

$$\begin{cases} \left.\dfrac{\partial^2 w}{\partial x^2}+\nu\dfrac{\partial^2 w}{\partial y^2}\right|_{x=0,a}=0 \\ \left.\dfrac{\partial^2 w}{\partial y^2}+\nu\dfrac{\partial^2 w}{\partial x^2}\right|_{y=0,b}=0 \end{cases} \tag{9}$$

are satisfied by taking for the deflection surface of the buckled plate the following double sine trigonometric series for the bending and shear components:

$$\begin{cases} w_b(x,y)=\displaystyle\sum_{m=1}^{m=\infty}\sum_{n=1}^{n=\infty} w_{m,n}^{(b)}\,sin\frac{m\pi x}{a}\,sin\frac{n\pi y}{b} \\ w_s(x,y)=\displaystyle\sum_{m=1}^{m=\infty}\sum_{n=1}^{n=\infty} w_{m,n}^{(s)}\,sin\frac{m\pi x}{a}\,sin\frac{n\pi y}{b} \end{cases} \tag{10}$$

Substituting the vertical displacement field (10) into the system of equation (7), it follows that for each harmonic the amplitude of the shear displacement field is related to the relevant bending one by the following relation:

$$w_{m,n}^{(s)}=k_{m,n}w_{m,n}^{(b)} \tag{11}$$

with:

$$k_{m,n}=\frac{\pi^2}{5(1-\nu)\alpha^2}\left(\frac{t}{b}\right)^2\left(m^2+\alpha^2 n^2\right) \tag{12}$$

so that the vertical displacement field may be finally rewritten as follows:

$$w(x,y)=\sum_{m=1}^{m=\infty}\sum_{n=1}^{n=\infty} w_{m,n}^{(b)}\left(1+k_{m,n}\right)sin\frac{m\pi x}{a}\,sin\frac{n\pi y}{b} \tag{13}$$

Obviously, for $k_{m,n}=0$ the well known results, obtained applying the classical thin plate theory, may be obtained again.

To evaluate the Euler shear stress τ_E at which buckling of plate occurs, the energy method is adopted, as there isn't a rigorous solution of Eq. (7). In applying this method, it is assumed the plate undergoes some small lateral bending, consistent with the given boundary conditions: obviously, if the work done by the in-plane forces is smaller than the strain energy of bending, for every possible shape of buckling, the equilibrium of plate is stable, otherwise it is unstable and buckling occurs.

The strain energy of the buckled plate is the sum of two terms due to bending and shear, respectively:

$$\Delta U_b=\frac{D}{2}\int_0^a\int_0^b\left[\left(\frac{\partial^2 w_b}{\partial x^2}\right)^2+\left(\frac{\partial^2 w_b}{\partial y^2}\right)^2+2\frac{\partial^2 w_b}{\partial x^2}\frac{\partial^2 w_b}{\partial y^2}\right]dxdy \tag{14}$$

$$\Delta U_s=\frac{5Et}{24(1+\nu)}\int_0^a\int_0^b\left[\left(\frac{\partial w_s}{\partial x}\right)^2+\left(\frac{\partial w_s}{\partial y}\right)^2\right]dxdy \tag{15}$$

finally becoming, considering the M and N partial sums of the series (10):

$$\Delta U=\frac{D\pi^4 b}{8a^3}\sum_{m=1}^{M}\sum_{n=1}^{N} w_{m,n}^{(b)2}\left(1+k_{m,n}\right)\left(m^2+\alpha^2 n^2\right)^2 \tag{16}$$

The work done by the external forces is:

$$\Delta T=-N_{xy}\int_0^a\int_0^b\left[\frac{\gamma}{2}\left(\frac{\partial w}{\partial x}\right)^2+\frac{\partial w}{\partial x}\frac{\partial w}{\partial y}\right]dxdy \tag{17}$$

finally becoming:

$$\Delta T=-4N_{xy}\left[\sum_{m=1}^{M}\sum_{n=1}^{N}\sum_{p=1}^{M}\sum_{q=1}^{N}\Delta T_{m,n}^{(1)}+\gamma\frac{\pi^2}{32\alpha}\sum_{m=1}^{M}\sum_{n=1}^{N}\Delta T_{m,n,p,q}^{(2)}\right] \tag{18}$$

with:

$$\gamma=\frac{N_x}{N_{xy}} \tag{19}$$

$$\Delta T_{m,n,p,q}^{(1)}=w_{m,n}^{(b)}w_{p,q}^{(b)}\left(1+k_{m,n}\right)\left(1+k_{p,q}\right)\times\frac{mnpq\chi_{mnpq}}{\left(m^2-p^2\right)\left(q^2-n^2\right)} \tag{20}$$

$$\Delta T_{m,n}^{(2)}=m^2 w_{m,n}^{(b)2}\left(1+k_{m,n}\right)^2 \tag{21}$$

and $\chi_{mnpq}=1$ if $m\pm p$ and $n\pm q$ are odd, $\chi_{mnpq}=0$ otherwise. Equating the work produced by the external forces to the strain energy, the coefficients $w_{m,n}^{(b)}$ of the series (10) may be chosen to make this expression a minimum, so obtaining the following system of equations:

$$\lambda\frac{\partial I_1}{\partial w_{m,n}^{(b)}}-\alpha^2\frac{\partial I_2}{\partial w_{m,n}^{(b)}}=0\ \forall m=1\ldots M;n=1\ldots N \tag{22}$$

with:

$$\lambda = -\frac{\pi^4 D}{32\alpha b^2 N_{xy}} \tag{23}$$

$$\frac{\partial I_1}{\partial w_{m,n}^{(b)}} = 2w_{m,n}^{(b)}\left(1+k_{m,n}\right)\left(m^2+\alpha^2 n^2\right)^2 \tag{24}$$

$$\frac{\partial I_2}{\partial w_{m,n}^{(b)}} = \gamma\frac{\pi^2}{16\alpha}\frac{\Delta T_{m,n}^{(2)}}{w_{m,n}^{(b)}} + 2\sum_{p=1}^{p=M}\sum_{q=1}^{n=N}\frac{\Delta T_{m,n,p,q}^{(1)}}{w_{m,n}^{(b)}} \tag{25}$$

Starting from eq. (23), the shear buckling coefficient k_s of eq. (1) may be immediately obtained:

$$k_s = -\frac{\pi^2}{32\alpha\lambda} \tag{26}$$

with λ defined as the minimum eigenvalue of (22).

The problem of finding the minimum value of λ that makes the determinant of (22) null, has been solved by a dedicated program developed in MATLAB. The solution is obtained varying the number of harmonics, to assure its convergence for very long narrow plates, too. The panel aspect ratio α has been varied, as well as the ratio γ between the shear and uniaxial stresses. For plates under pure shear the following expressions have been obtained for the buckling coefficient:

$$k_s = \begin{cases} 5.34 g_1\left(\frac{t}{b}\right) + \frac{4.00}{\alpha^2} f_1\left(\frac{t}{b}\right) & if \quad \alpha > 1 \\ 4.00 f_1\left(\frac{t}{b}\right) + \frac{5.34}{\alpha^2} g_1\left(\frac{t}{b}\right) & if \quad \alpha \le 1 \end{cases} \tag{27}$$

with:

$$f_1\left(\frac{t}{b}\right) = 1 - 26.870\left(\frac{t}{b}\right)^2; \; g_1\left(\frac{t}{b}\right) = 1 - 9.667\left(\frac{t}{b}\right)^2 \tag{28}$$

Obviously, for $t/b = 0$ the classical formulas may be obtained again. Similarly, for plates under shear and uniaxial compression the following relation has been derived for panels with aspect ratio $\alpha \geq 1$.

$$k_s = 5.34 g_1\left(\frac{t}{b}\right) g_2\left(\gamma\right) + \frac{4.00}{\alpha^2} f_1\left(\frac{t}{b}\right) f_2\left(\gamma\right) \tag{29}$$

with:

$$f_2(\gamma) = e^{1.6380\gamma}; \; g_2(\gamma) = e^{0.6317\gamma} \tag{30}$$

In Figures 2, 3 and 4 for plates under pure shear and shear plus uniaxial stresses with $\gamma = -0.50$ and $\gamma = -1.00$, the buckling coefficients k_s vs. α are shown for different values of the ratio $t/b = 0, 0.03, 0.04, 0.05$. The thick curves refer to the classical solution of the thin plate theory, while

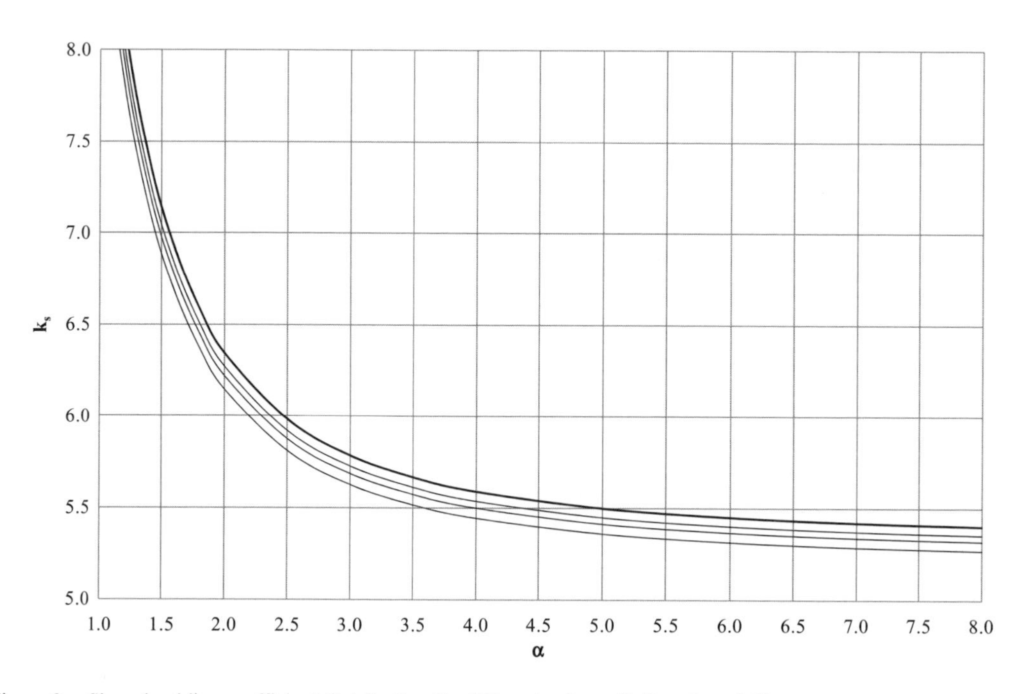

Figure 2. Shear buckling coefficient distribution for different values of t/b and $\gamma = 0.00$.

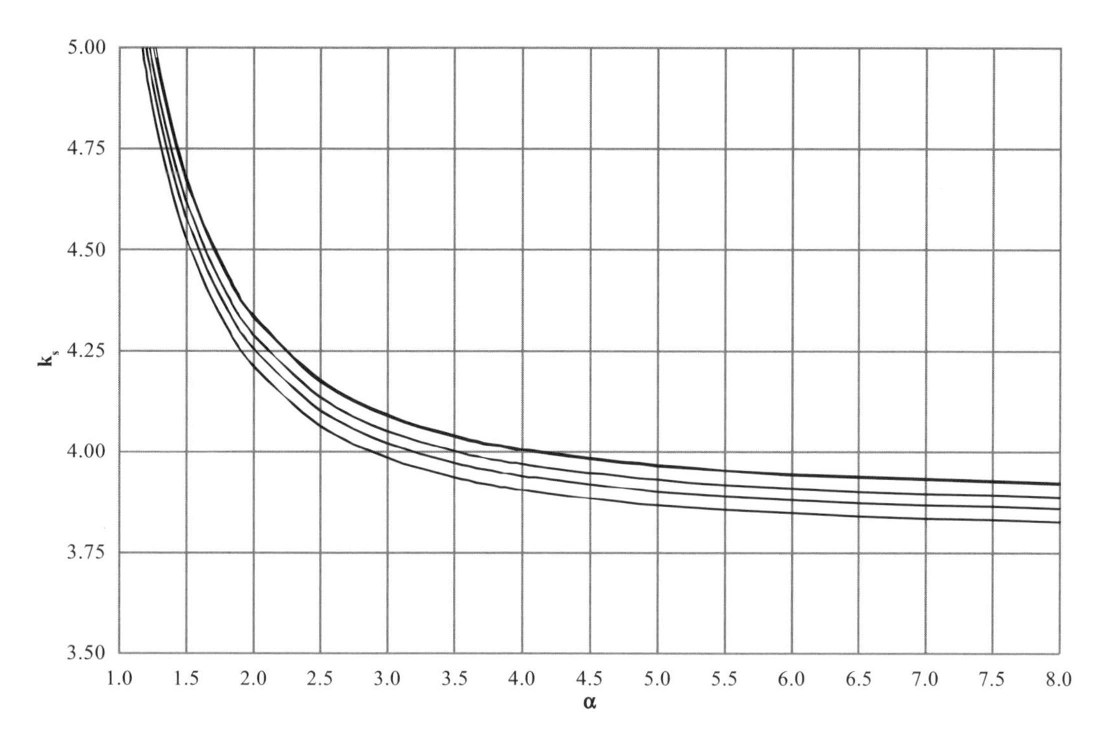

Figure 3. Shear buckling coefficient distribution for different values of t/b and $\gamma = -0.50$.

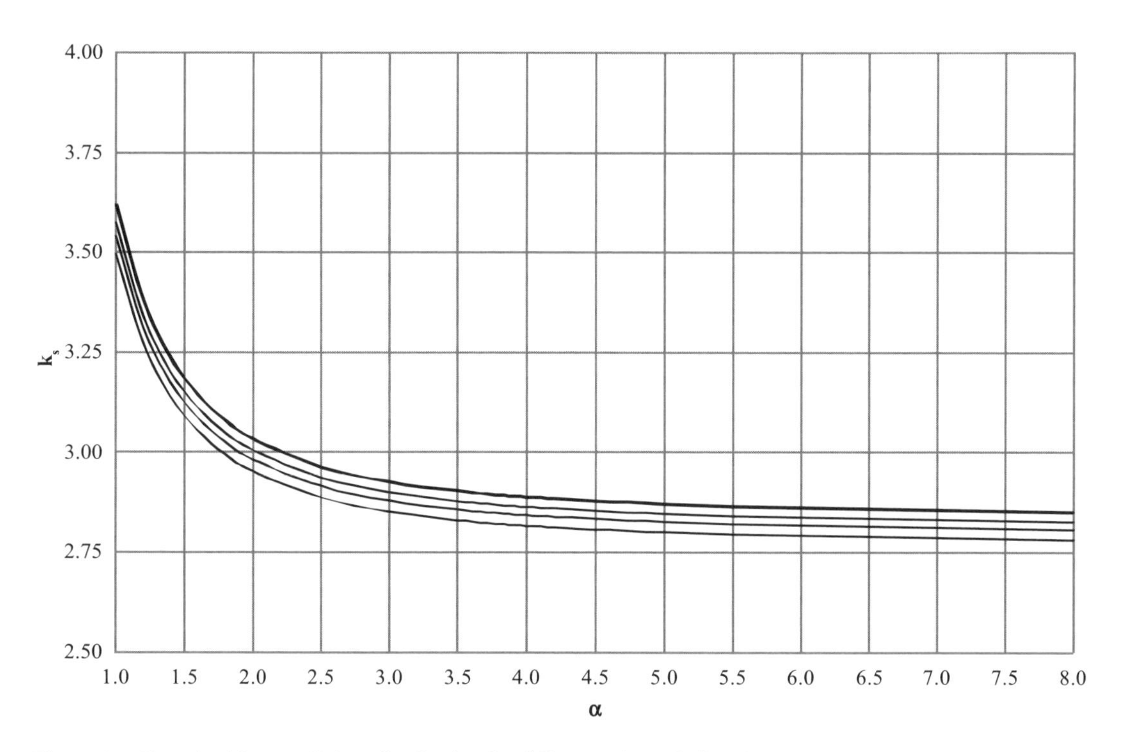

Figure 4. Shear buckling coefficient distribution for different values of t/b and $\gamma = -1.00$.

the other ones to the new obtained formulas. It is noticed that increasing the ratio t/b the shear buckling coefficients always decrease.

The convergence of solution, in terms of shear buckling coefficients, has also been investigated varying the number of harmonics in both x and y directions. It has been found that $M = N = 35$ harmonics are generally sufficient to obtain sufficiently accurate results. Tables 1 and 2 show the convergence of solution for plates under pure shear ($\gamma = 0$) with $t/b = 0$ and $t/b = 0.05$, respectively.

Table 1. Convergence of solution for plates under pure shear with $t/b = 0$.

α	M = N = 5	M = N = 10	M = N = 20	M = N = 30	M = N = 35
0.1	764.500	540.019	538.830	538.788	538.783
0.2	142.797	138.339	138.260	138.254	138.254
0.4	37.935	37.717	37.706	37.706	37.706
0.6	18.985	18.952	18.949	18.949	18.949
0.8	12.159	12.137	12.135	12.135	12.135
1.0	9.343	9.326	9.325	9.325	9.325
1.2	7.999	7.985	7.984	7.984	7.984
1.4	7.301	7.289	7.288	7.287	7.287
1.6	6.921	6.909	6.907	6.907	6.907
1.8	6.701	6.689	6.688	6.688	6.688
2.0	6.561	6.547	6.546	6.546	6.546
2.5	6.070	6.035	6.033	6.033	6.033
3.0	5.863	5.842	5.840	5.840	5.840
3.5	5.765	5.738	5.734	5.734	5.734
4.0	5.665	5.628	5.625	5.625	5.625
5.0	5.712	5.534	5.530	5.530	5.530
6.0	5.954	5.485	5.479	5.479	5.479
7.0	6.306	5.443	5.440	5.440	5.440
8.0	6.703	5.421	5.415	5.415	5.415
∞	13.364	6.759	5.351	5.350	5.350

Table 2. Convergence of solution for plates under pure shear with $t/b = 0.05$.

α	M = N = 5	M = N = 10	M = N = 20	M = N = 30	M = N = 35
0.1	265.539	157.643	138.411	138.252	138.250
0.2	90.197	83.090	82.979	82.967	82.965
0.4	32.092	31.887	31.867	31.864	31.864
0.6	17.437	17.391	17.385	17.384	17.384
0.8	11.472	11.444	11.440	11.440	11.440
1.0	8.926	8.905	8.902	8.902	8.902
1.2	7.694	7.677	7.675	7.675	7.675
1.4	7.052	7.037	7.035	7.035	7.035
1.6	6.702	6.688	6.686	6.686	6.686
1.8	6.501	6.487	6.485	6.485	6.485
2.0	6.371	6.355	6.354	6.354	6.354
2.5	5.898	5.863	5.861	5.861	5.861
3.0	5.707	5.686	5.685	5.685	5.685
3.5	5.612	5.583	5.580	5.579	5.579
4.0	5.521	5.482	5.479	5.479	5.479
5.0	5.579	5.392	5.389	5.389	5.389
6.0	5.823	5.346	5.341	5.340	5.340
7.0	6.171	5.308	5.306	5.306	5.306
8.0	6.564	5.287	5.281	5.281	5.281
∞	13.111	6.621	5.221	5.219	5.219

4 NUMERICAL APPLICATIONS

In the following several buckling analyses of steel platings, under pure shear and shear with uniaxial compression, are presented, in order to make a comparison with the relevant results obtained by some FE eigenvalue buckling analyses, carried out by ANSYS.

In the FE analysis a mesh with a mean element length of 0.01 m has been adopted, choosing the 4-node finite strain SHELL181, suitable for analyzing thin to moderately thick structures and well-suited for linear, large rotation, and/or large strain nonlinear applications.

The analyzed panels, in high strength steel with $E = 2.06E11$ Pa and $\nu = 0.3$, are:

1. Case 1: a = 1.00 m; b = 1.00 m; t = 10-20-30 mm;
2. Case 2: a = 3.00 m; b = 1.00 m; t = 10-20-30 mm;
3. Case 3: a = 8.00 m; b = 1.00 m; t = 10-20-30 mm.

In Tables 3, 4 and 5 the Euler forces per unit of length N_{xy} are shown for different values of the ratio t/b for plates under pure shear. Tables 6, 7 and 8 show the same results for plates under shear plus uniaxial compression with $\gamma = -0.50$, while in Tables 9, 10 and 11 the same results are presented assuming $\gamma = -1.0$.

The above presented numerical applications show that in almost all cases the classical thin plate theory slightly overestimates the critical shear forces per unit of length, as regards the relevant

Table 3. Case 1—$\alpha = 1$—$\gamma = 0$.

t/b	ANSYS (A)	Thin plate (B)	Thick plate (C)	$\frac{B-A}{A}$	$\frac{C-A}{A}$
–	kN/m	kN/m	kN/m	%	%
0.01	1727	1739	1736	0.69	0.52
0.02	13639	13912	13817	2.00	1.31
0.03	45291	46952	46232	3.67	2.08

Table 4. Case 2—$\alpha = 3$—$\gamma = 0$.

t/b	ANSYS (A)	Thin plate (B)	Thick plate (C)	$\frac{B-A}{A}$	$\frac{C-A}{A}$
–	kN/m	kN/m	kN/m	%	%
0.01	1083	1077	1076	–0.55	–0.65
0.02	8609	8616	8578	0.08	–0.36
0.03	28850	29078	28791	0.79	–0.20

Table 5. Case 3—$\alpha = 8$—$\gamma = 0$.

t/b	ANSYS (A)	Thin plate (B)	Thick plate (C)	$\frac{B-A}{A}$	$\frac{C-A}{A}$
–	kN/m	kN/m	kN/m	%	%
0.01	1005	1006	1005	0.10	0.00
0.02	7960	8047	8015	1.09	0.69
0.03	26630	27158	26917	1.98	1.08

Table 6. Case 1—$\alpha = 1$—$\gamma = -0.50$.

t/b	ANSYS (A)	Thin plate (B)	Thick plate (C)	$\frac{B-A}{A}$	$\frac{C-A}{A}$
–	kN/m	kN/m	kN/m	%	%
0.01	1027	1053	1052	2.53	2.43
0.02	8217	8426	8376	2.54	1.94
0.03	27732	28439	28054	2.55	1.16

Table 7. Case 2—$\alpha = 3$—$\gamma = -0.50$.

t/b	ANSYS (A)	Thin plate (B)	Thick plate (C)	$\frac{B-A}{A}$	$\frac{C-A}{A}$
–	kN/m	kN/m	kN/m	%	%
0.01	765	761	761	–0.52	–0.52
0.02	6060	6092	6066	0.53	0.10
0.03	20242	20559	20365	1.57	0.61

Table 8. Case 3—$\alpha = 8$—$\gamma = -0.50$.

t/b	ANSYS (A)	Thin plate (B)	Thick plate (C)	$\frac{B-A}{A}$	$\frac{C-A}{A}$
–	kN/m	kN/m	kN/m	%	%
0.01	720	730	729	1.39	1.25
0.02	5706	5841	5818	2.37	1.96
0.03	19063	19712	19539	3.40	2.50

Table 9. Case 1—$\alpha = 1$—$\gamma = -1.00$.

t/b	ANSYS (A)	Thin plate (B)	Thick plate (C)	$\frac{B-A}{A}$	$\frac{C-A}{A}$
–	kN/m	kN/m	kN/m	%	%
0.01	642	673	672	4.83	4.67
0.02	5075	5387	5358	6.15	5.58
0.03	16953	18181	17962	7.24	5.95

Table 10. Case 2—$\alpha = 3$—$\gamma = -1.00$

t/b	ANSYS (A)	Thin plate (B)	Thick plate (C)	$\frac{B-A}{A}$	$\frac{C-A}{A}$
–	kN/m	kN/m	kN/m	%	%
0.01	539	545	544	1.11	0.93
0.02	4132	4358	4340	5.47	5.03
0.03	14639	14707	14572	0.46	–0.46

Table 11. Case 3—$\alpha = 80$—$\gamma = -1.00$.

t/b	ANSYS (A)	Thin plate (B)	Thick plate (C)	$\frac{B-A}{A}$	$\frac{C-A}{A}$
–	kN/m	kN/m	kN/m	%	%
0.01	533	531	530	–0.38	–0.56
0.02	4233	4247	4230	0.33	–0.07
0.03	14141	14334	14208	1.36	0.47

results obtained by the FE analyses carried out by ANSYS. Obviously, the percentage errors, obtained applying the classical plate theory, increase with the ratio *t*/*b*.

The new formulas, that take into account the shear effects, permit to obtain more accurate results, with percentage errors that are about one half of the relevant ones obtained applying the thin plate theory. Obviously, in this case, despite of the results obtained by the classical theory, the percentage errors are almost constant when the adimensional ratio *t*/*b* increases. Furthermore, the proposed formulas permit to obtain critical values of the shear forces that are always lower than those ones obtained applying the classical thin plate theory and so may be utilized, on the safety side, with good confidence.

5 CONCLUSIONS

In the paper the refined plate theory proposed by Shimpi (2006) has been applied to the buckling analysis of simply supported plates under the combined action of shear and shear with uniaxial stresses. New formulas, that take into account the shear effects by means of two corrective functions, have been derived. Some curves, showing the decrease of the buckling coefficients, as function of the plate slenderness α and the adimensional ratio *t*/*b*, are also presented for plates under shear and shear with uniaxial stresses.

Some numerical applications have been carried out for steel platings, in order to compare the theoretical buckling forces per unit of length, obtained applying the classical thin plate theory and the refined one, with the relevant results obtained some eigenvalue buckling analyses carried out by ANSYS. It has been found that the refined theory always furnishes, respect to the classical one, results closer to the values obtained by the FE analysis.

Some new project formulas that introduce in the classical ones two terms, as function of the adimensional ratio *t*/*b* are also presented for plates under pure shear and shear with uniaxial compression.

REFERENCES

E. Reissner, 1945. The effect of transverse shear deformation on the bending of elastic plates, *Journal of Applied Mechanics Vol. 12* (Transactions ASME 67), pp. 69–77.

H. Tai, S. Kim & J. Lee, 2007. Buckling analysis of plates using two variable refined plate theory, *Proceedings of Pacific Structural Steel Conference 2007, Steel Structures in Natural Hazards*, Wairakei, New Zeeland, 13–16 March, 2007.

J.N. Reddy, 1984. A refined non linear theory of plates with transverse shear deformation, *International Journal of Solid and Structures Vol. 20*, pp. 881–896.

M. Levinson, 1980. An accurate simple theory of the statics and dynamics of elastic plates, *Mechanics Research Communications Vol. 7*, pp. 343–350.

M. Stein & J. Neff, 1947. *NACA Tech. Note.*

O.F. Hughes, 1988. *Ship structural design: a Rationally-Based Computer Aided Optimization Approach*, SNAME Edition.

R.P. Shimpi, H.G. Patel, 2006. A two variable refined plate theory for orthotropic plate analysis, *International Journal of Solid and Structures (43)*, pp. 6783–6799.

R.R. Mindlin, 1951. Influence of rotary inertia and shear on flexural motions of isotropic, elastic plates, *Journal of Applied Mechanics Vol. 18* (Transactions ASME 73), pp. 31–38.

S. Bergman & H. Reissner, 1932. *Z. Flugtech. Motorluftsch., Vol. 23*, p. 6.

S.P. Timoshenko & J.N. Goodier, 1951. *Theory of Elasticity*, Mc-Graw-Hill International Book Company.

S.P. Timoshenko & J.M. Gere, 1963. *Theory of elastic stability*, Mc-Graw-Hill International Book Company, 17th edition.

S.P. Timoshenko & S. Woinowsky-Krieger, 1959. *Theory of Plates and Shells*, Mc-Graw-Hill International Book Company.

Sustainable Maritime Transportation and Exploitation of Sea Resources – Rizzuto & Guedes Soares (eds)
© 2012 Taylor & Francis Group, London, ISBN 978-0-415-62081-9

A decomposition algorithm for large scale surrogate models

Serdar A. Koroglu & Ahmet Ergin
Faculty of Naval Architecture and Ocean Engineering, Istanbul Technical University, Maslak, Istanbul, Turkey

ABSTRACT: In engineering designs, it is necessary to deal with large numbers of parameters and constrains. However, finite computer resources do not allow all the parameters to be taken into consideration. In recent years, surrogate models were employed to increase the computational efficiency. This approach not only increases the computational speed, but it also provides the means for global optimization methods. Unfortunately, the surrogate models have practical limitations. It is very difficult, if not-impossible, to construct a surrogate model with more than a very limited number of parameters. Since the construction of surrogate models requires increasing number of samples with increasing number of parameters, this makes the application of the method infeasible. To overcome these limitations, a decomposition approach is proposed for structural designs having a simple geometrical pattern with repetition. The application of the algorithm to a stiffened panel structure is presented.

1 INTRODUCTION

Current design practice utilizes finite element method with an optimization algorithm to reach a cost effective solution. Computational power requirement of FEM in optimization revealed a need for alternative models such as meta-models (or surrogate models) performing fast and accurate enough solutions within a limited design space. These models provide the means for optimization routines to search using less number of computations in compared with the conventional methods.

Unfortunately, the surrogate models can only be employed with a limited number of design parameters so they are not scalable. Furthermore, due to this limitation, the parametric model cannot be reused for an alternative design. That means wasting a large amount of resources that is not tolerable for everyone.

The proposed framework targets these issues and presents a solution which is fast in global search analyses (like genetic algorithms), scalable, reusable and flexible enough to search much larger design spaces than ever. Since the safety of the model is ensured in the creation phase of the model, it is even safer for inexperienced users.

The application of the algorithm to a stiffened panel structure is presented in a later section. It is based on a test framework that is focused only on the decomposition concept. The results obtained are encouraging to continue with the works suggested in the final discussion section.

2 CURRENT DESIGN PRACTICE

A design in shipbuilding or aerospace industry starts with a specification list including dimensions, carrying capacity, speed, etc. Major design decisions are made at this initial stage. However, the decisions are generally based on a selection process from old but well established designs. When the ship structural engineers are informed about the particulars of an upcoming ship project, they immediately draw a rough picture of the structural design in their mind. This is understandable because of it is too risky being so much creative for million dollar projects. The risk is mainly related to the lack of having an adequate parametric representation and tools to effectively analyze and optimize these parameters. By no means, the structural optimization is impossible. This can be achieved with a limited scope that is determined by the available computational resources. Venkataraman and Haftka (2004) discuss these complexity related limitations on different aspects of design work.

Recently, researchers have focused on the methods adopting alternative approximate methods, instead of finite element models, for increasing the efficiency of the analyses. These approximate methods actually do not replace FEM, instead they generate accurate enough and faster models, based on data generated by FEM simulations. These models can be used globally as well as locally depending on the intention of the designer. When the model is the only reference employed for an optimization routine, it needs to be a global model. Instead, if the model is used to increase the convergence speed of the search routine, it needs to be a local model. More about the global and local models can be found in Barthelemy and Haftka (1993).

The surrogate models have been adopted by various researchers. For instance, Papadrakakis et al. (1998) used the artificial neural networks with evolution strategies for the optimization of a connecting rod and frame structure. Gholizadeh et al. (2008), on the other hand, employed surrogate models to simulate the frequency constraints of a structural optimization problem. A detailed discussion of surrogate models can be found in Simpson et al. (2001).

The faster models have also their own limitations. Although they converge faster in comparison with finite element models, the maximum number of parameters that can be employed, is still the main obstacle for the scalability of the method. Besides the generated model strictly fits the specified problem at hand. Therefore, the model cannot be reused in another simulation.

3 THE FRAMEWORK

By adopting a smart building block concept, the proposed framework provides a scalable and reusable model. As shown in Figure 1, the smart block is a parametric stiffened panel structure. The smart block model is flexible enough to represent a broad spectrum of geometry, material properties and loading types. These blocks are then combined together to form a complex and large scale structure which suits the designers need. However, it should be noted that, as in Patnaik et al. (2000), it is not only a sub-structuring approach simplifying a complex large scale structure, but it also allows the usage of surrogate models for higher dimensional problems.

The algorithm developed searches for an equilibrium condition of the structure adopted under a specific load combination. After the equilibrium condition obtained, the information of interest such as stresses, displacements, etc., can be extracted.

The correct displacement configuration of the building blocks is sought by searching through the possible positions of interface nodes'. The search is finished when the interface forces are in balance. A sample structure defined in terms of the building blocks can be seen in Figure 2.

3.1 *Surrogate-free testing method*

The surrogate model should be created before any building block concept is implemented. However, if the main aim is just to test the concept, it is possible to replace the surrogate model with a finite element model. Usage of FEM in place of a surrogate model, provides all the functionality and the results obtained set the upper bound for performance, since no approximate model can outperform a finite element model.

3.2 *Equilibrium search algorithm*

The algorithm developed uses a simple fact that each sub-structure applies forces and moments to their neighboring sub-structures, and these forces and moments are equal in magnitude, but opposite in direction. It should be noted that this condition can only be realized when the structural system is in equilibrium. The sum of all the forces and moments acting at the interface nodes should be zero. Therefore, the problem is solved by using the following equation.

$$f(x_1, x_2, ..., x_n) = 0 \quad (1)$$

where x_1, x_2 etc. represents the displacement values at the interface nodes, and f is a function representing the total force or moment at a

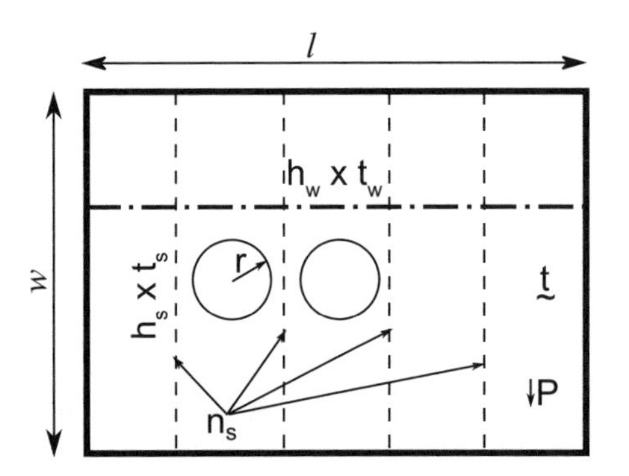

Figure 1. An example parametric building block.

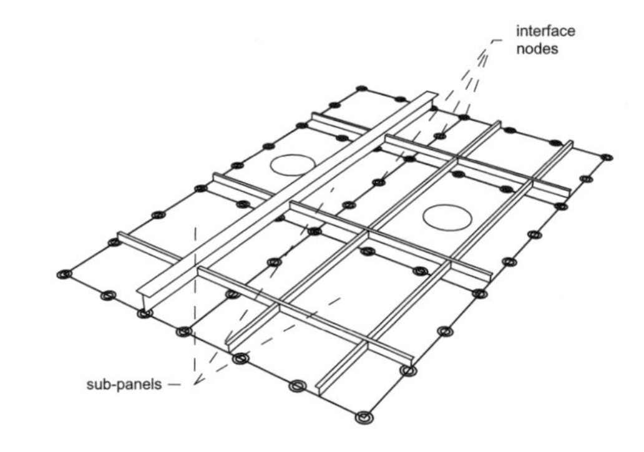

Figure 2. An example structure composed of building blocks.

particular interface node. For practical reasons, the problem is organized such that the number of functions adopted in the solution is minimized.

It is clear that the solution of the problem becomes more difficult with increasing number of interface nodes. In order to overcome this difficulty, the proposed algorithm balances the forces/moments independently for each interface node.

The proposed algorithm is detailed in the following steps:

1. The panel is analyzed to obtain the actual displacements. These calculations are used to check the estimates.
2. In order to find a proper initial set of values, the sub-panels are analyzed separately, and they are assumed free along the interface edges. The average value at each interface node is used as the initial BC for the following analyzes.
3. The panels are analyzed separately again. The boundary values at the interface nodes are now substituted from the last calculations step, instead of free BC.
4. After step 3, the reaction forces/moments are obtained for each panel. These reaction forces/moments guide the search for determining BC for the next calculation step. A crude updating algorithm is applied for two iterations, and it is presented Algorithm 1.

Algorithm 1 Pseudo Code of the First Updating Scheme

```
for all interface node do
  for all displacement BC do
    Take 2 reaction force and the common (inter-
    face edge) displacement BC
    if they are both negative or positive then
      Shift the displacement BC towards negative/
      positive direction, shifting magnitude
      is taken equal to a multiple of the magni-
      tude of the previous BC
    else if they have opposite signs then
      Shift displacement BC towards the BC
      which has larger magnitude, shifting mag-
      nitude is now taken as only a fraction of the
      previous BC
    end if
  end for
end for
```

The algorithm does not take into account the oscillations of the results separately and collectively, and it does not use an updated magnitude information from the previous searches.

5. After the first 5 or 6 iterations, Algorithm 2 is applied addressing the issues listed above.

Algorithm 2 Pseudo Code of the Second Updating Scheme

```
for all interface node do
  for all displacement BC do
    Take 2 reaction force and the common dis-
    placement BC of both current and previous it-
    eration
    Make a linear interpolation using current and
    previous data with common BC is taken as in-
    dependent variable, and sum of reaction forces
    as dependent variable.
    Determine the next common BC when de-
    pendent variable equals to zero (which occurs
    when two reaction forces have opposite sign
    and equal magnitude)
  end for
end for
```

The iteration process is stopped after a predetermined iteration number reached (generally not more than 100 iterations). By using the Algorithm 2, the solution approaches a local minimum and no further improvement of the results can be obtained. Hence, a random disturbance is applied periodically to enable the algorithm to escape from the local minima. The disturbance can be applied as follows:

- Every n iterations (usually 20), the BC is disturbed with uniform random number proportional to its magnitude (usually 20%).
- Every k, l, m iterations, the BC is disturbed with a gaussian random variable having mean = 0 and standard deviation proportional to its magnitude (usually k = 25, l = 10, m = 3; stdev is for k = 190% of BC, l = 40% of BC, m = 1% of BC) and every j iterations the disturbance is reduced (for example by 50%).

4 TEST PROBLEM

The algorithm proposed was tested for a simple stiffened panel structure, and the geometric properties and loading conditions are given in Figure 3 and Table 1. For the structure adopted for the calculations, one interface with 9 nodes was considered. Each interface nodal point has 6 degrees of freedom. Therefore, the problem is to minimize a black box function in order to find the values of 64 variables (degrees of freedom).

As a testbed, finite element sofware package ANSYS, Inc. (2009) was utilized.

The calculations are presented in Figures 4, 5, 6 and 7, 8, 9 respectively for the displacements and interface forces/moments. The calculations are presented as a function of number of iteration together with the actual value in the figures. The figures present the results for some sample

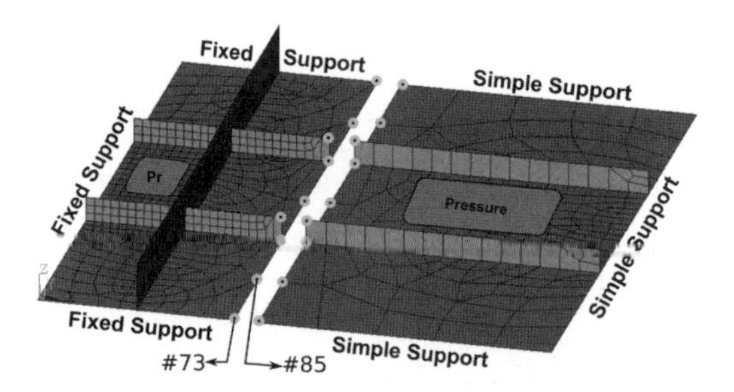

Figure 3. Test problem configuration.

Table 1. Panel properties and loading condition.

Property	Value
Overall panel width	5 m
Overall panel length	6 m
Shorter stiffener height	0.5 m
2 longer stiffener height	0.25 m
Thickness (for both panel and stiffeners)	10 mm
Uniform pressure	5 kN/m^2

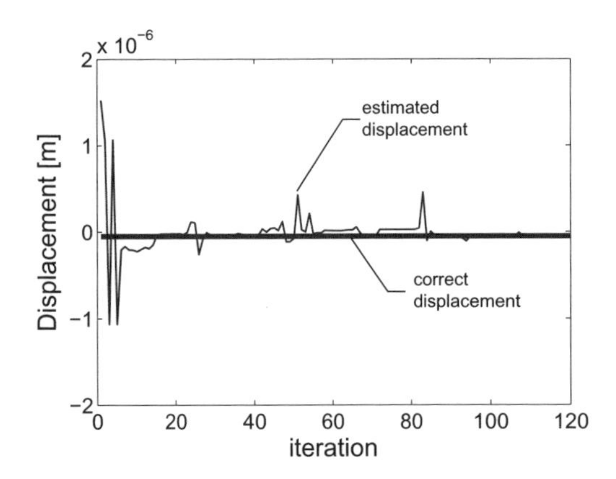

Figure 4. Displacement along y dir. of node 73.

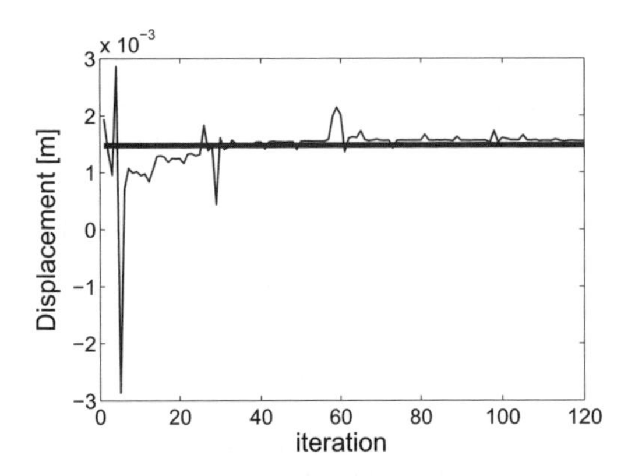

Figure 5. Displacement along z dir. of node 85.

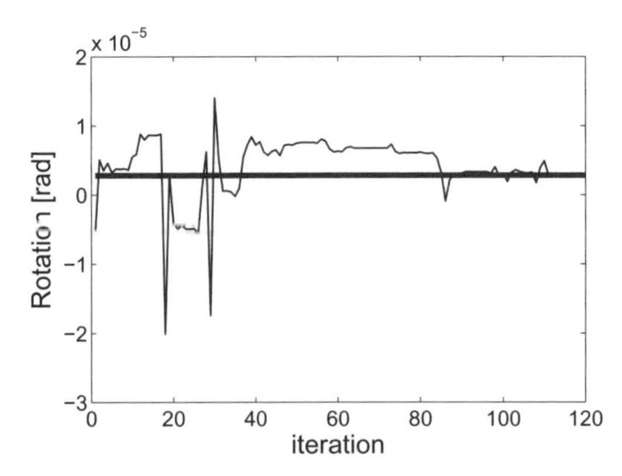

Figure 6. Rotation around z axis of node 85.

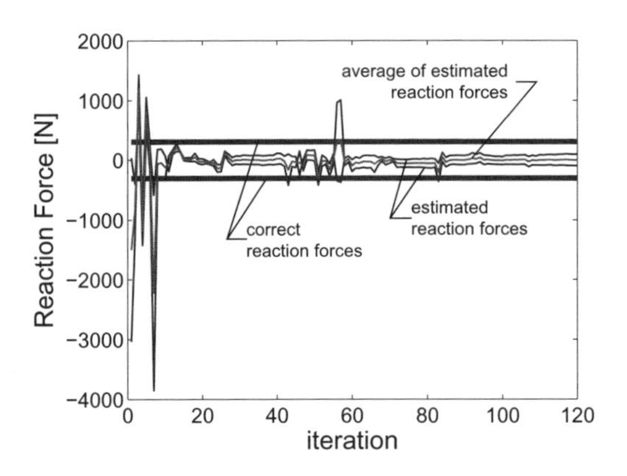

Figure 7. Reaction forces along y dir. of node 73.

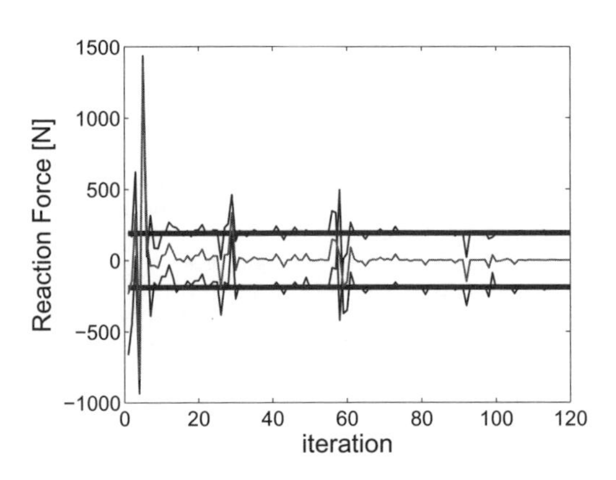

Figure 8. Reaction forces along z dir. of node 85.

interface nodes and for chosen associated degrees of freedoms. In Figures 7–9, the interface nodal reaction forces/moments are presented for two neighboring interface nodes.

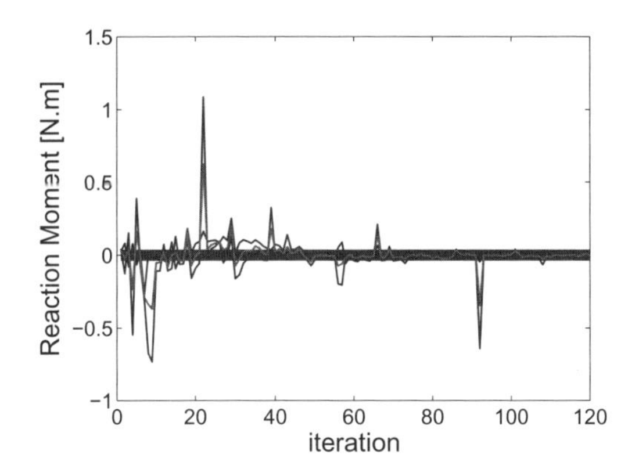

Figure 9. Reaction moments around z axis of node 85.

5 CONCLUSIONS

The search shows a promising trend around the actual values. Unfortunately, some unwanted surges were observed. However, when the stress levels are compared, it is observed that the errors range between 2% and 15%. These differences are expected since no convergence criterion was used and no attempt was made to use a selection procedure. Therefore, more accurate estimates are possible by implementing those criterion and procedures.

6 DISCUSSIONS

Although some useful information can be obtained by using the proposed algorithm, the convergence is not maintained for all the degrees of freedom. In order to improve the convergence, a feedback mechanism should be implemented by taking into consideration the conditions of neighboring nodes.

However, without a fully developed surrogate model, it is not possible to apply the algorithm to a realistic and complex engineering problem. As a key to success, the representation of building blocks deserves much more attention than any other issues such as surrogate modeling, sampling techniques, handling the samples, etc.

REFERENCES

ANSYS, Inc. (2009). *Structural analysis guide: Release 12.0.1.*

Barthelemy, J.F.M. and Haftka, R.T. (1993). Recent advances in approximation concepts for optimum structural design. *Optimization of Large Structural Systems, Vols 1 and 2 231*, 235–256.

Gholizadeh, S., Salajegheh, E. and Torkzadeh, P. (2008). Structural optimization with frequency constraints by genetic algorithm using wavelet radial basis function neural network. *Journal of Sound and Vibration 312*(1–2), 316–331.

Papadrakakis, M., Lagaros, N.D. and Tsom-panakis, Y. (1998). Structural optimization using evolution strategies and neural networks. *Computer Methods in Applied Mechanics and Engineering 156*(1–4), 309–333.

Patnaik, S.N., Coroneos, R.M. and Hopkins, D.A. (2000). Substructuring for structural optimization in a parallel processing environment. *Computer-Aided Civil and Infrastructure Engineering 15*(3), 209–226.

Simpson, T.W., Peplinski, J.D. Koch, P.N. and Allen, J.K. (2001). Metamodels for computer-based engineering design: survey and recommendations. *Engineering with Computers 17*(2), 129–150.

Venkataraman, S. and Haftka, R.T. (2004). Structural optimization complexity: what has moore's law done for us? *Structural and Multidisciplinary Optimization 28*(6), 375–387.

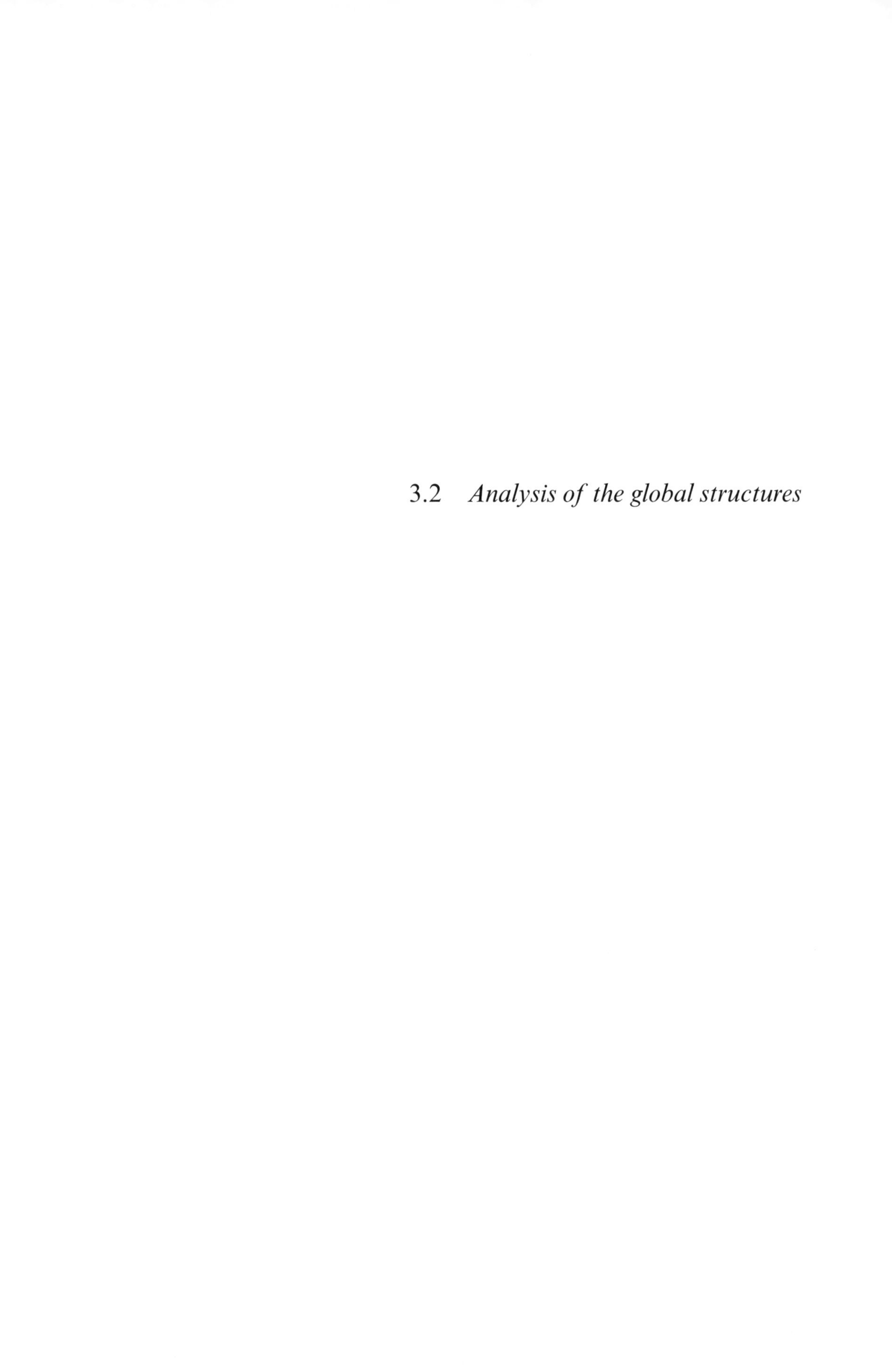

3.2 *Analysis of the global structures*

Sustainable Maritime Transportation and Exploitation of Sea Resources – Rizzuto & Guedes Soares (eds)
 ISBN 978-0-415-62081-9

Box girder strength under pure bending: Comparison of experimental and numerical results

Plamen I. Nikolov
Trouble Shooting Team Ltd, Varna, Bulgaria

ABSTRACT: The progressive collapse behavior of box girder models are studied by the Finite Element Method (FEM) using component and global approaches. By the global FEM method four-point bending is simulated. The nonlinear module of the commercial FEM system COSMOS/M is used. Initial deflections and welding residual stresses are included. Reasonable welding residual stresses are obtained by initial thermal loading and unloading. It is shown that the residual deflections resulting from the thermal stage do not affect the collapse behavior, hence the assumed initial deflections in the buckling modes are not changed. Unloading and reloading in the pre-collapse range are carried out to simulate the welding stress relief in experiments. The ultimate strength is not affected, suggesting that residual stresses are not relieved. Toward an improved efficiency of the global FEM method half-breadth models are validated.

1 INTRODUCTION

The ability to predict accurately the ultimate strength of a ship's hull girder when subjected to longitudinal bending is an important aspect of ship structural design. In this connection extensive research works have been performed on buckling/plastic collapse strength of structural elements and their assemblages, theoretically and experimentally.

As result of demand for an efficient method, Caldwell (1965) suggested the 'component approach' to evaluate the ultimate strength. It was followed by Smith (1977) who was the first to use average stress-average strain relationships of individual elements for progressive collapse analysis of a hull girder cross section. Since then different methods for derivation of the stress-strain relationships have been developed (Yao & Nikolov 1992, Gordo & Guedes Soares 1996).

In parallel, experimental results on progressive collapse behaviour of structural members and systems have been also obtained for validation of calculation methods and understanding the collapse behavior of new systems. Box type specimens have been made and four-point bend tests have been carried out to meet the needs of reliably validation of the simplified numerical methods for progressive collapse behavior of a hull girder cross section (Nishihara 1983, Dow 1991, Gordo & Guedes Soares 2004).

The FEM derivation of stress-strain curves of ordinary stiffener elements composing a real hull girder cross section is becoming affordable and reveals possibilities for improvement of the 'component' approach by enabling complex deflection shapes to be considered (Nikolov 2009). Stress-strain curves of larger structures allow complex interaction between the stiffeners to be accounted for.

Recently, the progress in computer performance is encouraging complex nonlinear FEM simulations of the collapse behavior of larger structures (Luís et al. 2007, Nikolov 2008, Paik et al. 2010). Very often results by FEM analyses are used for validation of the assumptions in the simplified methods. When comparing FEM results with experimental ones any judgment on the validity of either method is difficult. The reasons are mostly related to uncertainties in the specimens imperfections (initial deflections and welding residual stresses) as well as to uncertainties in the initial adjustments of experimental setup and measuring devices.

In the previous study (Nikolov, 2009) different simplified methods and experimental results on the collapse behavior box girders subjected to four-point bending and panels under compression were compared and the validity of the simplified methods was discussed, as well as some disagreements in the experimental results.

The present study is an extension by including the recently reported experimental results on four-point bending of a box type specimen by Saad-Eldeen et al. (2010) denoted as IST. The nonlinear module of the FEM system COSMOS/M is used. The specimens and experimental setup are modeled in full extent. The initial deflections and the welding residual stresses are included. Reasonable residual stresses are obtained by applying thermal loading and unloading before the bending load. When bending load is applied, unloading and reloading in the pre-collapse region is applied to simulate the experimentally intended relief of the welding residual stresses. The validity of assumptions in the numerical methods and in the experiments are discussed. Experimental results obtained by Nishihara (1983) on specimens denoted as MST-3 and MST-4 are also included.

2 TEST SETUP AND SPECIMENS DATA

The subject tests setup and specimens are illustrated in Figure 1. All the collapse tests were carried out on steel girder models under four-point bending. MST-4 and MST-3 are one-span specimens whereas IST is three-span specimen. All the longitudinals stiffening the specimens are flat bars.

The particulars and scantlings of the box girder models are summarized in Table 1. MST-4 and MST-3 differ each other only in the thickness used.

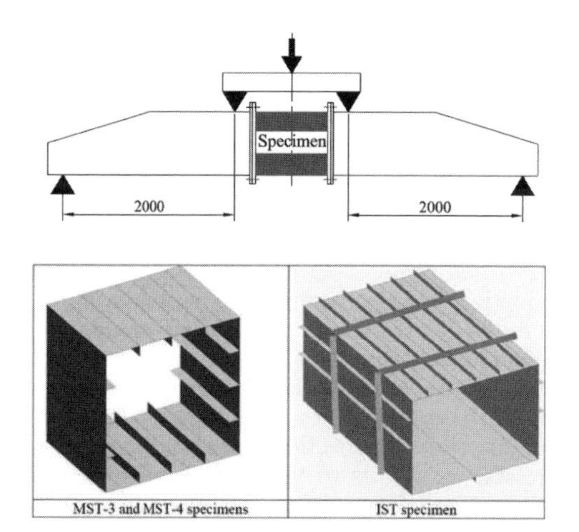

Figure 1. Test set-up and box girder specimens.

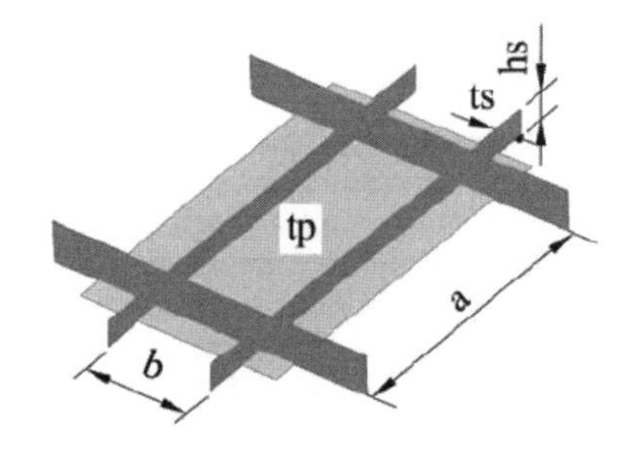

Figure 2. Stiffened panel nomenclature.

Table 1. Specimens data.

		MST-4	MST-3	IST
a	mm	540	540	400
b	mm	180	180	150
t_P	mm	4.35	3.05	4.09
t_S	mm	4.35	3.05	4.35
h_S	Mm	50	50	25
σ_Y	N/mm²	264	287	292
E	GPa	208	207	206
ν	–	0.281	0.277	0.277
β	–	1.47	2.20	1.38
EI	MNm²	258	180	173.1

The IST specimen has been subjected to initial corrosion explaining the thickness variations over the specimen (reduced from the original thickness of 4.5 mm). The thickness of the bottom plating is 3.75 mm. The thicknesses of the PS and SB side plating are also different but are represented by the average one of 3.9 mm.

3 ASSUMED IMPERFECTIONS

Residual deflections and stresses are always present in the welded structures such as the specimens, and they play important role in the structural behavior. However, no measurements seem to be performed on the initial imperfections in the subject specimens.

IST specimen has been subjected to initial unloading and reloading in the pre-collapse region intending to relief the welding residual stress. The validity of this is checked by FEM simulation in Section 6.6. As shown there, the ultimate strength is not affected, suggesting that the welding residual stresses are not relieved and need to be accounted for in the numerical simulations.

The initial deflections and welding residual stresses are explicitly defined in the FEM methods. The complex initial deflections in a real structure are usually idealized in the numerical methods as composed by two components: the plate buckling mode component and the stiffener beam-column buckling mode component. It is known that the panel buckling mode deflection component plays the most important role in the plating collapse behavior. This deflection component is expressed by Equation 1 and is illustrated in Figure 3.

The assumed stiffener initial deflections in beam-column mode is expressed by Equation 2 and illustrated in Figure 4. The double-span model is applicable to continuous stiffened structures since it enables the interaction between the

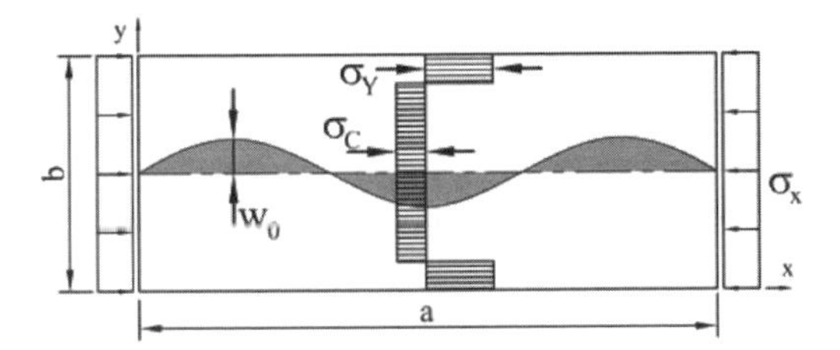

Figure 3. Idealization of the plate imperfections.

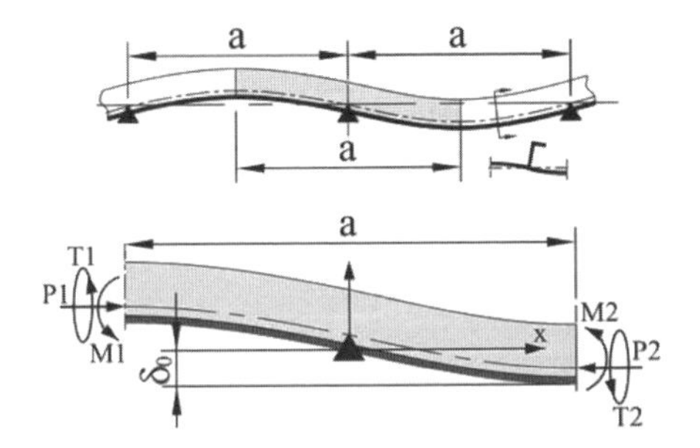

Figure 4. Stiffener double-span model.

spans to be taken into account. The interaction is complex due to the fact that the attached plating in the two spans is in different sides of stiffener bending. Plate-induced failure rises in the span where the panel is in the compression side of bending. Stiffener-induced failure rises in the other span.

$$w_p = w_0 \sin\frac{m\pi x}{a} \sin\frac{\pi y}{b} \tag{1}$$

$$w_s = \delta_0 \sin\frac{\pi x}{a} \tag{2}$$

The welding residual stresses are also idealized when accounted for in the numerical methods. Mostly, the representation by tensile and compressive blocks in equilibrium is used, as illustrated in Figure 3. Analogously, the residual stresses in the transverse direction are represented.

The assumed magnitudes of the imperfections in the present study are summarized below.

$\sigma_c = 0.2\sigma_Y$
$w_0 = 0.025t$
$\delta_0 = 0.0025a$

Alternative calculations without residual stresses are performed attempting to provide upper level of the predictions.

The material is assumed elastic-perfectly plastic. Further details on the modeling assumptions are given in Sections 4 & 5.

4 COMPONENT APPROACH

The progressive collapse behaviour of a longitudinally stiffened girder is simulated by considering the collapse behaviour of the individual elements (components) composing the cross-section. The subdivision of the cross-sections of MST-3 and MST-4 specimens into components is illustrated in Figure 5a, where the components are shrank to be identified. The only interaction between the components follows from the condition that the plane cross-section remains plane and bending take place with respect to the instantaneous neutral axis as shown in Figure 5b.

The nonlinear behaviour of each component is idealized by average stress-average strain relationship which takes in to account the considered collapse modes by the method used. For a given curvature, the axial deformation of each component is calculated and through the average stress-average strain relationship, the stress in that component is evaluated. The moment-curvature relationship is found by incremental procedure.

For the elements which may undergo buckling (stiffeners with attached plating or plate panels between girders) the stress-strain curves are derived considering possible failure modes: plate failure between stiffeners, beam column buckling, torsional buckling and web local buckling. In the previous comparative study (Nikolov, 2009) three different methods to derive the stress-strain relationships of components were used: formula based, analytical and nonlinear FEM.

In the present study the nonlinear module of commercial FEM system COSMOS/M is used to derive the stress-strain curves of the stiffener elements. The model of longitudinally stiffened plating is shown in Figure 6. It comprises two-span, two-bay model. The continuity of the structure is represented by boundary conditions of symmetry

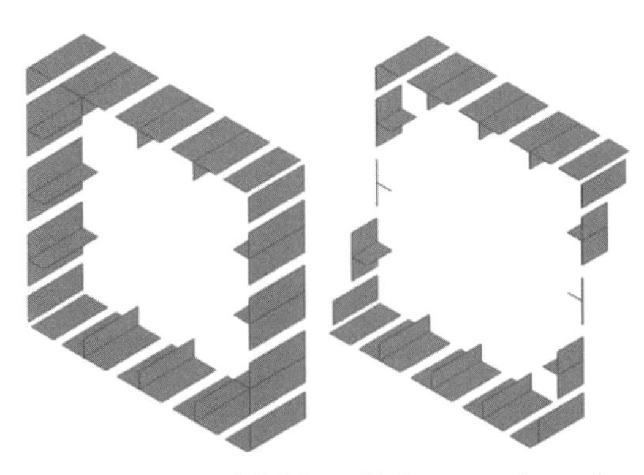

(a) Cross-section subdivision (b) Component interaction

Figure 5. Assumptions in the component approach.

at model sides. The unloaded edges are coupled and free to move.

No elements are provided for the transverse but the deflections in its web plane are restricted.

The assumed initial deflections are also illustrated in that figure as composed by the plate buckling mode component and the stiffener beam-column buckling mode component (magnified 100 and 10 times, respectively, for a better illustration).

The material is assumed to be elastic-perfectly plastic. The compressive loading is applied by forced displacement.

The assumed welding residual stresses are approximately obtained by applying thermal loading and unloading before the compressive load. The thermal load is applied at the nodes connecting the stiffener and the transverse with the plating. With temperature amplitude of T = 2000° and coefficient of thermal expansion α_T = 12e-6, the specified residual stresses are well represented, as illustrated in Figure 7. Tension blocks are observed in the adjacent to the connection line elements of panel and stiffener web. The obtained longitudinal and transverse tension stresses are about 80% and 75% of the yield strength, respectively. In the rest of elements the stresses are compressive, almost uniform with average values in the longitudinal (transverse) panel directions of about 25% (6%) of the tensile stresses, or 20% (5%) of the yield stress. Due to significant plasticity nonlinearity the thermal load stage requires proper temperature increment for a successful solution.

Stress and deflection plots at collapse loads are provided in Figures 8–9 for MST-3 and IST deck stiffeners, respectively. Because of the twice lower stiffener height, the IST collapse mode is a typical beam-column buckling mode. The MST-3 stiffener collapses in torsional buckling mode (MST-4 is similar to MST-3). These collapse modes are also confirmed by the global FEM analyses in Section 6.3.

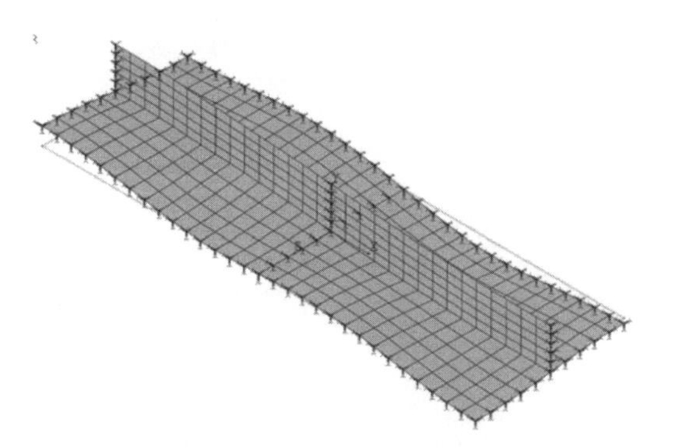

Figure 6. Double-span, double-bay model of stiffener component for FEM calculations.

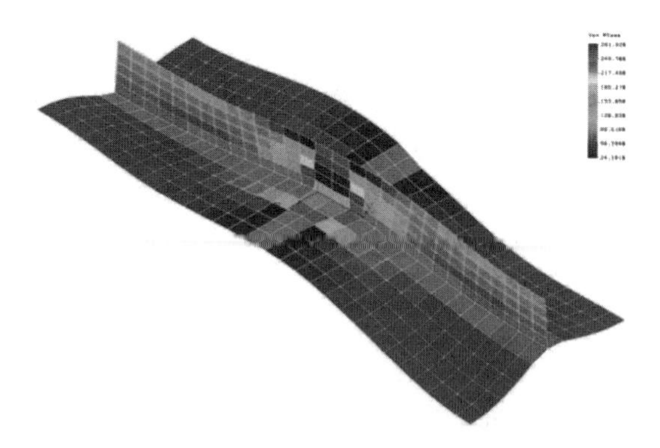

Figure 7. Residual stress and deflections after the thermal stage (MST-3).

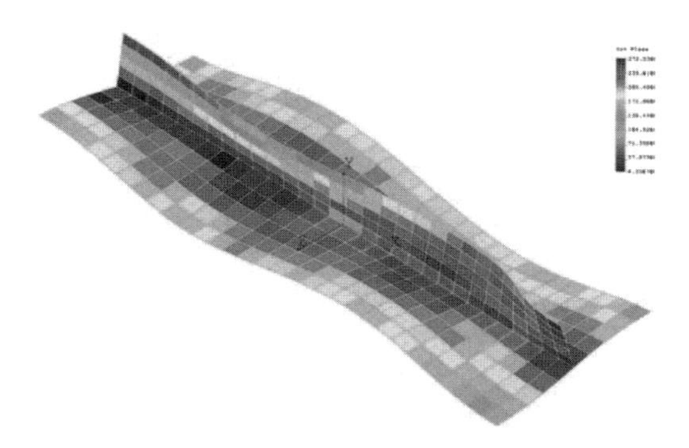

Figure 8. Stress and deflections at collapse load (MST-3).

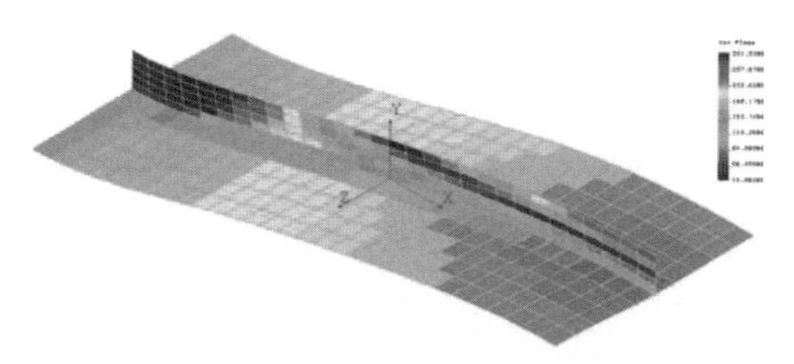

Figure 9. Stress and deflections at collapse load (IST).

The stress distributions at collapse show the relatively low stresses in the heat-treated locations with initially tensile stresses. These portions remain effective in carrying the post-collapse compression.

Relatively uniform are the stresses in the span where the plating is in the tension side of stiffener bending.

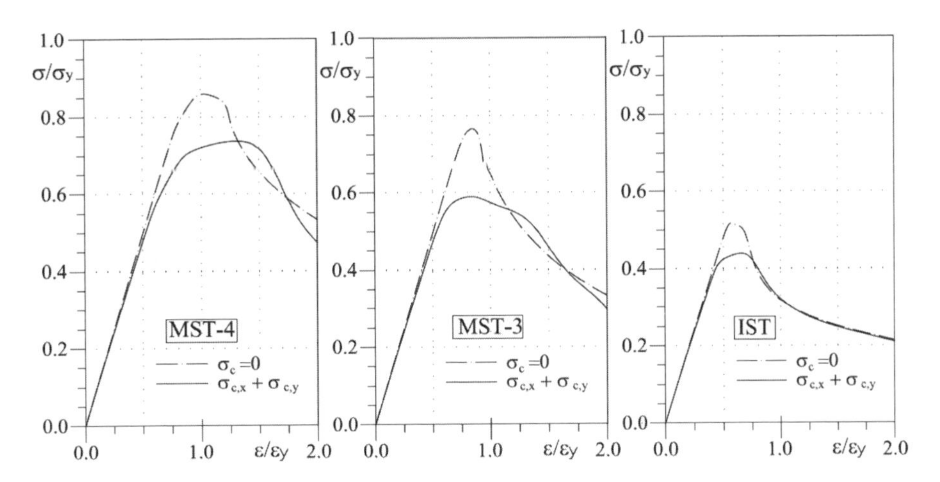

Figure 10. Average stress-average strain relationships of deck stiffener of specimens.

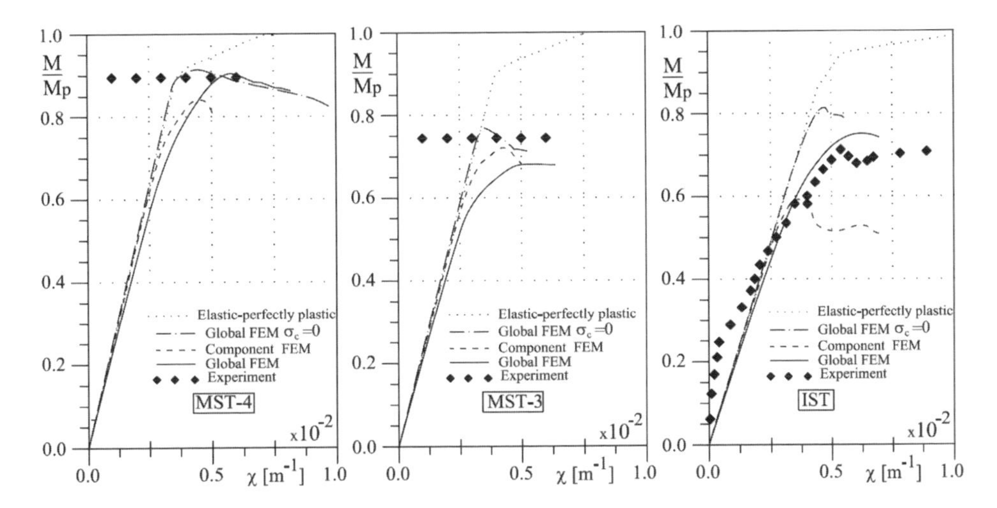

Figure 11. Moment-curvature relationships of box type cross-sections.

5 GLOBAL FEM ANALYSIS (COSMOS/M)

The developed model of the IST specimen including the experimental setup is shown in Figure 12. Because of the symmetry with respect to the mid-girder section, half of the specimen with the adjacent loading arm are represented. Boundary conditions of symmetry are accordingly imposed. Applicability of the conditions of symmetry with respect to the longitudinal plane is validated in Section 6.4.

The mesh of the deck stiffeners and plating is same as that used for the isolated stiffener analyses (see Fig. 6). The size of elements over specimen model is kept almost same. The stiff arms, designated to generate pure bending moment in the specimen, are represented by coarse mesh model.

The model support and loading follow the experimental setup for four-point bending according to Figure 1. The loading is applied by forced displacement.

Model details are provided by the zoomed view in Figure 13. The initial deflections are imposed by modification of the model geometry. The deflections could be visible in that figure if magnified properly (as it was done for the stiffener in Figure 6). The stiffeners deflection magnitude (δ_0) is same for all of them.

The welding residual stresses are approximately obtained by applying thermal loading and unloading before the bending stage. Thermal load is applied at the nodes in connection lines of stiffeners and transverses to plating, as well as to hard corners. The specified residual stresses in Section 3

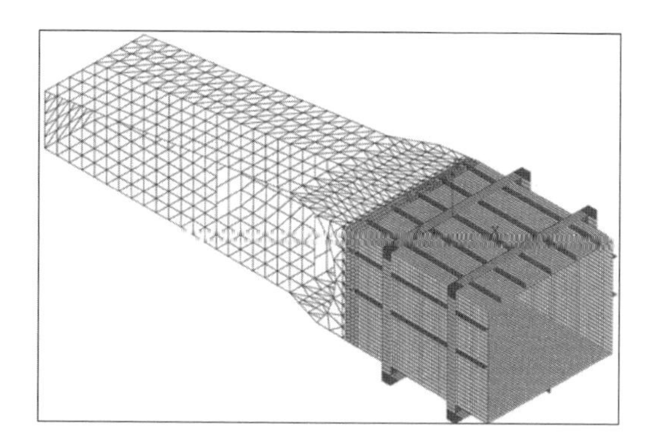

Figure 12. FEM model of IST setup and specimen.

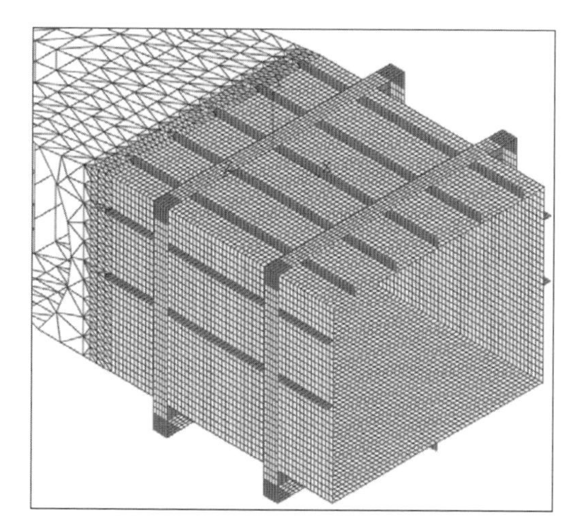

Figure 13. Model details of IST specimen.

are obtained with heating up to temperature of T = 2000° and then cooling back. The coefficient of thermal expansion is set as $\alpha_T = 12e\text{-}6$. Due to significant plasticity nonlinearity proper temperature increment is required for a successful solution. The residual stresses after the thermal stage are illustrated in Figure 14. Using the average (element) Von Mises stresses and shades of gray colors, the block distribution is easily observed. In the tension block the stresses are about 80% of yield stress. In the compression block the average stresses are 25% of the tensile ones. The thermal stage seems reasonable and efficient to be further optimized for application of the residual stresses in the design procedures.

Analogous figures are provided for MST-3 model. Overall view and model details are given in Figures 15 & 16, respectively. The residual stresses after the thermal loading and unloading are shown in Figure 17. These figures are also representative for the MST-4 model.

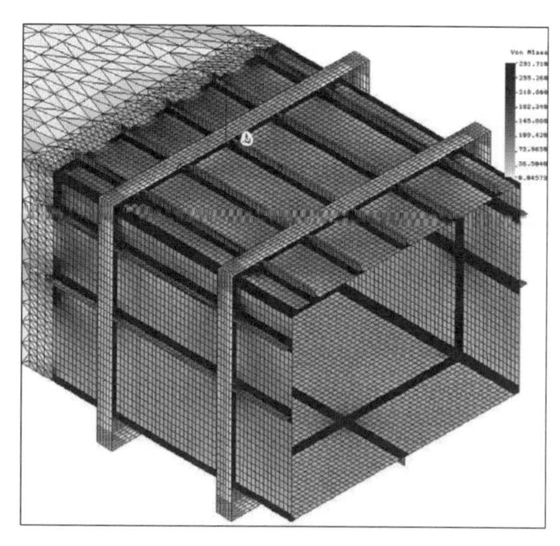

Figure 14. Initial stresses in IST model after thermal loads along longitudinals and transverses.

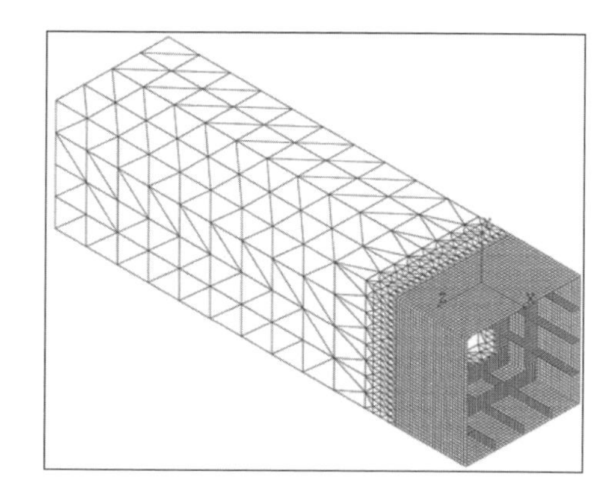

Figure 15. FEM model of MST-3 setup and specimen.

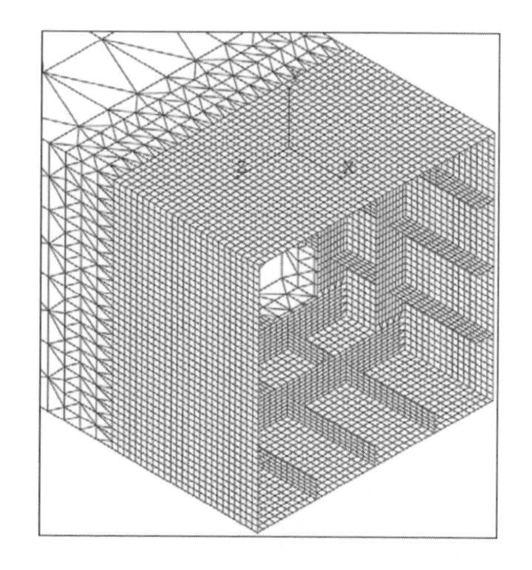

Figure 16. Model details of MST-3 specimen.

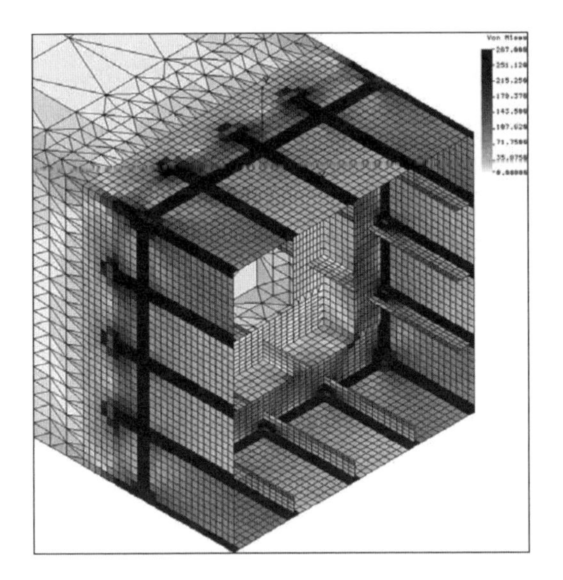

Figure 17. Initial stresses in MST-3 model after thermal loads along longitudinals and transverses.

6 RESULTS AND DISCUSSION

6.1 *Stress-strain relationships of stiffener elements*

The stress-strain relationships for the deck stiffener elements of the 3 specimens are plotted in Figure 10. The stresses and strains are nondimensionalized by the yield stresses and strains at yielding, respectively. The results obtained with considering residual stresses are shown by continuous line whereas the results with neglected welding residual stresses are shown by dashed-doted line. It is well known that the welding residual stresses may reduce significantly the strength and the rigidity. In average, the reduction in strength due to the assumed welding residual stress is about 15% with a maximum reduction of 23% for MST-3 stiffener. In the post-collapse range the influence is not significant, especially for the IST stiffener, where the curves are almost same.

6.2 *Moment-curvature relationships*

Because of nonlinear effects, the curvature is variable along a specimen, although the constant bending moment. It is increasing within failed spans and may be decreasing in the adjacent span after unloading and restoration of the initial deflection shape. In the present study the average curvature over the specimens lengths are used. Calculated and measured moment-curvature relationships of the box girders are compared in Figure 11. The ultimate bending moments are summarized in Table 2.

The experimental results are indicated by symbols. For MST-3 and MST-4 the ultimate bending moments are only shown since the curvatures have not been measured. In general, the measured ultimate bending moments are within the margin of numerical results. Better compliance is observed for MST-4.

The curves traced by dots represent the result obtained by component approach when all elements are assumed to follow the material behavior (hard elements). In this case the influences of the imperfections and buckling effects are excluded and the ultimate strength is equal to the plastic bending moment Mp of the respective cross-section. The bending moments are nondimensionalized by the plastic bending moments, Mp.

The curves obtained by the global FEM analyses with and without residual stresses are plotted by continuous and dashed-doted lines, respectively. The maximum difference between both results is observed at the curvatures where the ultimate strength without residual stresses is attained. After that both curves tend to each other. This effect is very prominent for MST-4 where both curves in the post-collapse range almost coincide.

The results by the component approach considering the welding residual stresses are plotted by dashed lines. In case of MST-4 and IST this method is resulting to the lowest ultimate strength. In case of MST-3 the strength by the global FEM analysis is the lowest one. More analyses are required to clarify the effect of interaction between the components.

The slope of the moment-curvature curve with elastic-perfectly plastic elements (hard elements) prior to first yielding exactly represent the elastic flexural rigidity, EI, of each cross-section (see Table 1). The actual rigidities are usually smaller than the theoretical ones due to presence of initial imperfections and/or elastic buckling. With the assumed imperfections the elastic rigidities are reduced by 8–10%. On contrary, the experimental results for IST, although the presence of imperfections, show initial rigidity exceeding the theoretical one.

Table 2. Summary results on ultimate bending moments, kNm.

	MST-4	MST-3	IST
Mp	1036	791	800
EXPERIMENT	927	589	569
Component FEM	871	570	473
Global FEM	937	539	600
Global FEM ($\sigma_C = 0$)	947	610	651

6.3 *Details on the progressive collapse behaviour*

The stress distribution at collapse load of IST model is illustrated by the Von Mises stresses in Figure 18. The yielding strength is attained at the bottom in tension and at the deck in compression. The stress distribution is more uniform at the bottom, where the initially compressive residual stresses have approached the tensile residual stresses at heat-treated locations. In the deck, subjected to compressive load, the initially tensile stresses at heat-treated locations have been turned into compressive ones but remaining far below the yield stress. Lower are also the stresses in the middle span of the deck grillage where the plating is in the tension side of stiffener bending.

The deflections at collapse load of IST is presented in Figure 19. Because of the small stiffener height, the stiffener height to thickness ratio is far below the Rules required limit, excluding the tripping from the possible collapse modes, but the stiffeners flexural rigidity is relatively low resulting to interframe flexural (beam column) failure mode. This mode is well illustrated in that figure. Permanent deformations in beam-column mode were also observed after the collapse test of IST specimen.

Analogous illustrations for the MST-3 model are provided in Figures 20&21. The side stresses in Figure 20 show the neutral axis shifted downward from its initial position at the mid depth. Tripping is the dominant buckling/collapse mode of MST-3 (and MST-4) as illustrated by the deflections at collapse in Figure 21.

6.4 *Effect of model extent*

In linear problems the symmetry of the structure and the loading (linear symmetry) are utilized to reduce the model extent and calculation time. In nonlinear problems, symmetry of the initial imperfections and collapse modes (nonlinear symmetry) have to be also observed. For models with stiffeners

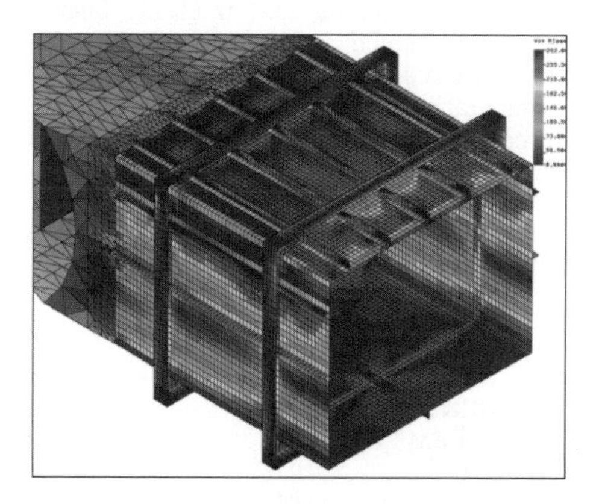

Figure 18. Von Mises stresses at collapse load, IST model.

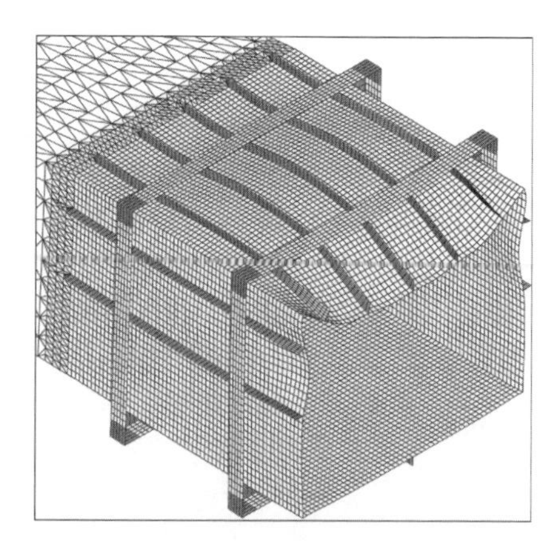

Figure 19. Deflection mode at collapse load of IST model.

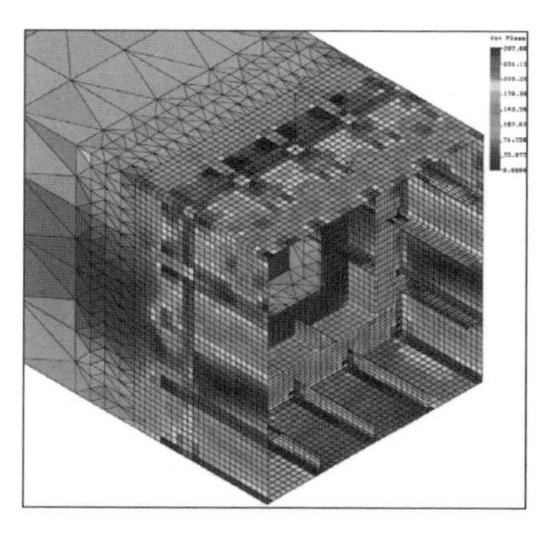

Figure 20. Von Mises stresses at collapse load, MST-3 model.

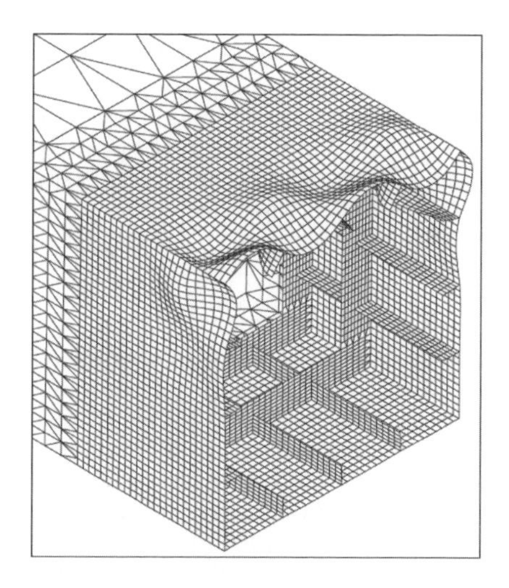

Figure 21. Deflection mode at collapse of MST-3 and MST-4.

in the plane of symmetry, as the studied ones, the symmetry conditions are not valid more because of not allowing alternating plating deflections and stiffener tripping.

To clarify the effect of the model extent on the progressive collapse behaviour, alternative calculations are carried out with half-breadth models and boundary conditions of symmetry in the longitudinal plane. The elements in the longitudinal plane of symmetry are taken with their half thicknesses.

The half-breadth model and the deflections at collapse for MST-3 and MST-4 are shown in Figure 22. Because of the conditions of symmetry the tripping of center stiffener as well as buckling of the attached plating (which are observed for the full-breadth model in Fig. 21) are not allowed. This leads to strengthening of the center stiffener and overestimation of the strength in some extent.

The strengthening effect due to the assumed conditions of symmetry in the center plane on the ultimate strength is 2.6% for MST-3 and 0.7% for MST-4. For IST, of which collapse mode is column buckling, the strengthening effect is negligible.

It is noted that with the increase of number of stiffeners, the effect of a single stiffener on the overall girder behavior would be reduced so as the error imposed by such model simplification.

6.5 *Effect of initial stress components*

As are result of the equilibrium in the thermal stage, together with the welding residual stresses, residual deflections are also obtained. The residual deflections after the thermal stage in IST model are shown in Figure 23. Different are the deflections when the thermal load is applied on both stiffeners and transverses or only to stiffeners. Nevertheless, the girder collapse behaviour in both cases is almost the same as it follows by comparing the respective moment-curvature curves. Although more research is needed, it may be concluded that the initial deflections in buckling modes, which play the main role in the collapse behavior, are not changed by the thermal load.

The half-breadth model is used for these calculations.

It is noted that the residual stresses in the transverse direction may have more significant effect when transverse or lateral loads are included.

6.6 *Behavior after unloading and reloading*

During the experiment the IST specimen has been subjected to initial loading and unloading in two cycles up to 36–38% of the collapse load intending to relief the welding residual stress. To clarify the effect of this, FEM simulation is carried out on the IST half-breadth model. Welding residual stresses along the stiffeners and transverses are included.

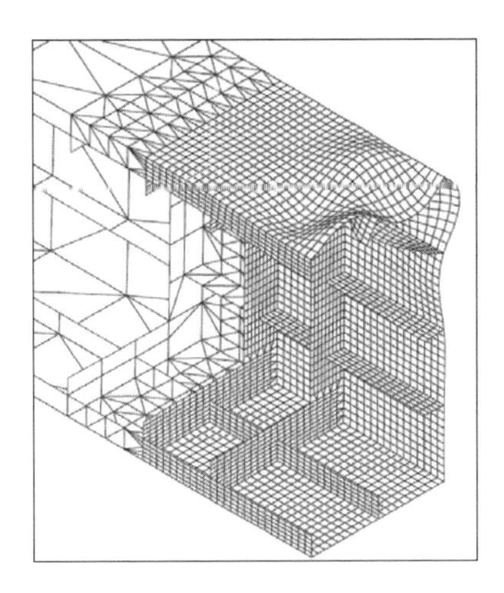

Figure 22. Deflection mode at collapse of MST-3 and MST-4 half-breadth models.

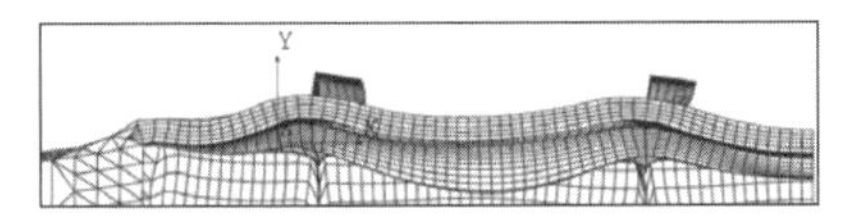

(a) thermal loads applied along the longitudinals and transverses ($\sigma c,x + \sigma c,y$)

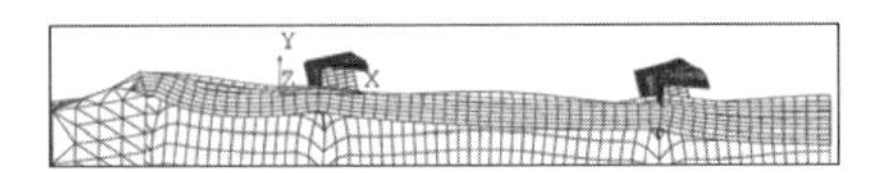

(b) thermal loads along longitudinals only ($\sigma c,x$)

Figure 23. Residual deflections in the deck structure of IST model after the thermal stage.

The moment-curvature curve is shown in Figure 24. Almost linear part on the moment—curvature curve is observed up to 60% of the collapse load. It is well known that in the elastic range the behavior is completely reversible—the unloading path is completely the same with the loading path. To have a noticeable effect, the starting point of unloading is taken as 94% of the collapse load. This point is in the range of significant plasticity nonlinearity as may be observed from the slopes of the moment-curvature curve. The unloading follows different path with residual deflections at the zero load. When the bending load is applied again, the behavior is reversible up to the starting point of unloading, and after that it follows the same behavior under increasing curvature as that without unloading. In other words, the ultimate strength is not affected by reversing load in the pre-collapse range. The residual stresses, affecting the ultimate strength, are also not relieved.

Such behavior is also observed in the experimental moment-deflection curve but at a point of loading in the linear range. One of

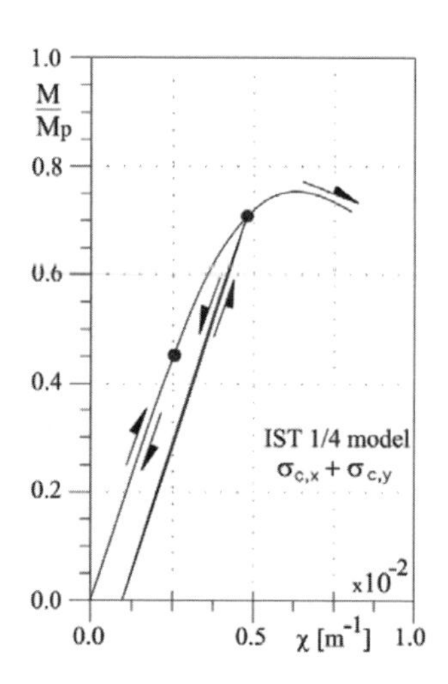

Figure 24. Moment-curvature relationship of IST model with unloading and reloading.

the reasons may be initial adjustment of the test setup.

7 CONCLUSIONS

The progressive collapse behavior of three box girder models were studied by simplified method and global FEM approach. The simplified method is based on the component approach with stress-strain curves of stiffener components derived by FEM. In the global FEM approach four-point bending was simulated. The nonlinear module of the commercial FEM system COSMOS/M was used in the study. Initial deflections and welding residual stresses are included. Reasonable welding residual stresses were obtained by initial thermal loading and unloading.

Uncertainties related with the residual deflections after the thermal stage and the residual stresses after unloading and reloading in the pre-collapse range were clarified. It was shown that the residual deflections resulting from thermal stage do not affect the collapse behavior, hence do not change the assumed initial deflections in buckling mode. Unloading and reloading in the pre-collapse range were carried out to simulate the welding stress relief in one of tests. The ultimate strength was not affected, suggesting that residual stresses were not relieved. In this connection, the welding residual stresses in the specimens have to be specified and taken into account in the calculations.

The efficiency of the global method was improved by considering half-breadth models with conditions of symmetry in the center plane, although the asymmetry of the deflection collapse mode. The half-breadth models were validated and the obtained error from the approximation was 2.6%. Attributed to a single element, the error will reduce with the increase of number of elements involved.

ACKNOWLEDGEMENT

This work is a contribution to the activities of the MARSTRUCT VIRTUAL INSTITUTE, (www.marstruct-vi.com) in particular its Technical Subcommittees 2.3 on Ultimate Strength.

REFERENCES

Caldwell, J.B. 1965. Ultimate Longitudinal Strength, Trans. RINA, Vol. 107, pp. 411–430.

Dow, R.S. 1991. Testing and Analysis of a 1/3-scale Welded Steel Frigate Model, Proceedings of International Conference on Advances in Marine Structures, Vol. 2, Scotland, pp. 749–773.

Gordo, J.M., Guedes Soares, C. & Faulkner, D. 1996. Approximate Assessment of the Ultimate Longitudinal Strength of the Hull Girder, *J. Ship Research*, Vol. 40, No. 1, pp. 60–69.

Gordo, J.M. & Guedes Soares, C. 2004. Experimental Evaluation of the Ultimate Bending Moment of a Box Girder, Marine Systems & Ocean Technology, 1:1, 33–46.

Luís, R.M., Guedes Soares, C. & Nikolov, P.I. 2007, Collapse strength of longitudinal plate assemblies with dimple imperfections. *Advancements in Marine Structures,* P. Das & C. Guedes Soares eds., Taylor & Francis Group, London, UK, pp. 207–215.

Nikolov, P.I. 2008. Collapse strength of damaged plating, *Proc. Int. Offshore & Polar Engineering Conference*, Estoril, Portugal.

Nikolov, P.I. 2009. Box girder strength under pure bending: comparison of experimental and numerical results. Proc. 13th Congress of International Maritime Association of the Mediterranean, Istanbul, Turkey, Vol. I, pp 79–86.

Nishihara, S. 1983. Analysis of Ultimate Strength of Stiffened Rectangular Plate (4 th Report)—On the Ultimate Bending Moment of Ship Hull Girder-, *J. Soc. Naval Arch. of Japan*, vol. 154, pp. 367–375.

Paik, J.K., Kim, D.K. & Kim, M.S. 2010. Effect of Lateral Pressure on the Progressive Hull Collapse Behaviour of a Suezmax-class Double-hull Oil Tanker Subject to Vertical Bending Moments, Proc. of 11th International Symposium on Practical Design of Ships and Other Floating Structures, Rio de Janeiro, pp. 1154–1165.

Saad-Eldeen, S., Garbatov, Y. & Guedes, C. Soares 2010. Experimental Assessment of the Ultimate Strength of a Box Girder Subjected to Four-Point Bending Moment, Proc. of 11th International Symposium on Practical Design of Ships and Other Floating Structures, Rio de Janeiro, pp. 1134–1143.

Smith, C.S. 1977. Influence of Local Compressive Failure on Ultimate Longitudinal Strength of a Ship's Hull, *Proc. PRADS*, A-10, Tokyo, Japan, pp. 72–79.

Yao, T. & Nikolov, P.I. 1992. Progressive Collapse Analysis of a Ship's Hull under Longitudinal Bending (2nd Report), *J. Soc. Naval Arch. of Japan*, vol. 172.

Sustainable Maritime Transportation and Exploitation of Sea Resources – Rizzuto & Guedes Soares (eds)
 ISBN 978-0-415-62081-9

A new approximate method to evaluate the ultimate strength of ship hull girder

G.T. Tayyar & E. Bayraktarkatal
ITU Naval Architecture and Ocean Engineering Faculty, Istanbul, Turkey

ABSTRACT: The most typical consequence of hull girder collapse is the breaking of the hull into two parts as a result of extreme vertical bending moment exceeding the ultimate hull girder strength. Therefore, the ultimate strength of ship hull girder should be known even at the earliest stages of ship design process. The purpose of this study is to develop a fast and reliable calculation model in order to determine the ultimate strength of ship hull girders. The paper describes a new iterative numerical approach for strength of stiffened plates, where the cross-sections remain plane assumption is imposed and the deflection curve is determined by using the curvature values directly against solving by differential equations. The deflection curve is taken as the assembly of chains of circular arcs and performed by Smith's method. The hull girder ultimate strength of a VLCC is analyzed using the proposed simplified method based on iterative approach.

1 INTRODUCTION

The ultimate strength of complicated structures such as ships should be calculated by taking into account material and geometric nonlinearities as well as initial deflections. However, such nonlinear analysis is still a tedious task due to the required large amount of computational time. Although there are several available methods for computing the ultimate strength of structural components or entire structural systems, there is still a demand for methods that can simplify the analysis and easier to operate (Yao et al., 2006).

A new method based on the load shortening relationship is presented here, which can be applied to stiffened plates of the hull girder when ultimate longitudinal strength analysis is performed by the Smith's method. In the Smith's method, a cross section is divided into small elements composed of stiffener(s) and attached plating. A progressive collapse analysis is performed assuming that a plane cross section remains plane and each element behaves according to its load shortening relationship. It is reported that the shape of load-shortening relationship largely affects the ultimate hull girder strength when the Smith's method is applied (Yao 2003); hence, it is important to determine load shortening relationship of element.

Ultimate strength of plates and stiffened plates is the most fundamental source of strength for marine structures, and a great deal of progress has been achieved in the past. To evaluate the ultimate strength of structural members and systems, it is necessary to perform structural analysis considering the influences of both buckling and yielding, which is called the elasto-plastic large deflection analysis. When compressive stresses are developed in the structural members such as columns, plates, stiffened plates etc- under external loads, buckling takes place when the compressive stresses reach a certain critical value. In general, lateral deflection rapidly grows after buckling, which reduces axial/in-plane rigidity of the buckled structural members. Due to lateral deflection, bending stresses are developed in addition to the axial/in-plane stresses (Yao et al., 2006).

The reduction of the load-carrying capacity beyond the ultimate strength of a stiffened palate is generally very rapid because of the progress of the overall buckling deformation, which is normally accompanied by the localization of plastic deformation either in the plate or in the stiffener. It is thus quite important to assess the progressive collapse behavior of structural systems, such as hull girder collapse, that not only the ultimate strength but also the post-ultimate strength behavior is investigated for stiffened panels (Paik et al., 2009).

The accuracy of the ultimate strength computations of the ship cross section is closely related with the strength analysis of the stiffened plates, which in turn depends on the accuracy of the deflection curve. Therefore the proposed method is focused on the mathematical modeling of the deflection curve.

The proposed method is shows how to determine the deflection curve with the curvature values due to moment values. By the help of deflection curve geometry load-shortening diagrams can be easily formed and performed in the Smith's Method to determine ultimate strength.

2 METHOD

In accordance with the second order theory, the axial/in-plane stresses rely on accurate modeling of the deflection curve. Besides using curvature value in differential equations of deflection curve, curvature value also represents the deflection geometrically. In this study, the radius of curvature is directly used in the deflection geometry and the true deflection curve is obtained numerically in an iterative way. The main motivation of this approach is reducing the computational and modeling times, and obtaining more precise solutions. There was no paper available on the determining the deflection curves with the use of curvature values geometrically and are taken as the assembly of chains of circular arcs within the publications recorded by "ISI-Web of Science".

The deflection curve of the structure can be easily modeled and determined by using the curvature values, even if the material or geometrical nonlinearities take effect. Furthermore, the ultimate strength is obtained with a single numerical procedure, regardless of the structure being in elastic or plastic region.

2.1 *Deflection curve*

When the strength of the structure is considered as the equilibrium of internal and external forces, the major input are moment and curvature. The relation between moment and curvature was first settled by James Bernoulli, and the differential equations defining the curvature were obtained by calculus of variations (Heyman 1998).

Considering the equilibrium of a stiffened plate cross section, the internal forces that are distributed over the cross section are statically equivalent to the external bending moment. The distribution of these internal forces over the cross section is obtained by considering the deformation of the stiffened plate (Timoshenko 1948). Deformation of the stiffened plate is examined by using the well-known assumption that the plane sections remain plane and normal to the axis.

The strains of the longitudinal fibers are proportional to the distance z from the neutral surface and inversely proportional to the radius of curvature as shown in Figure 1. The stress in any fiber is proportional to its distance from the neutral axis. The position of the neutral axis and the radius of curvature can be determined from the condition that the forces distributed over any cross section

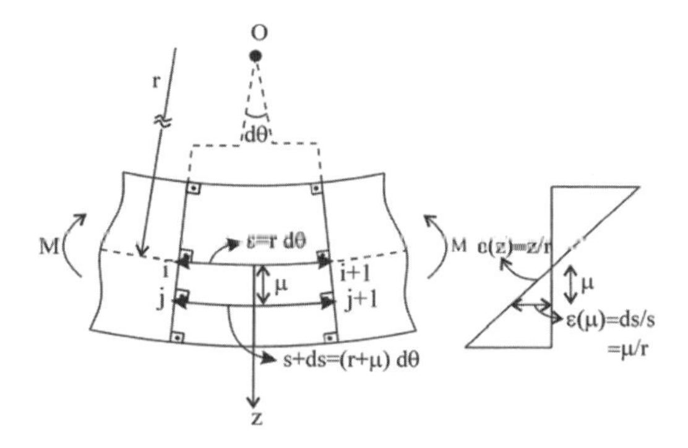

Figure 1. Strains of longitudinal fibers and curvature.

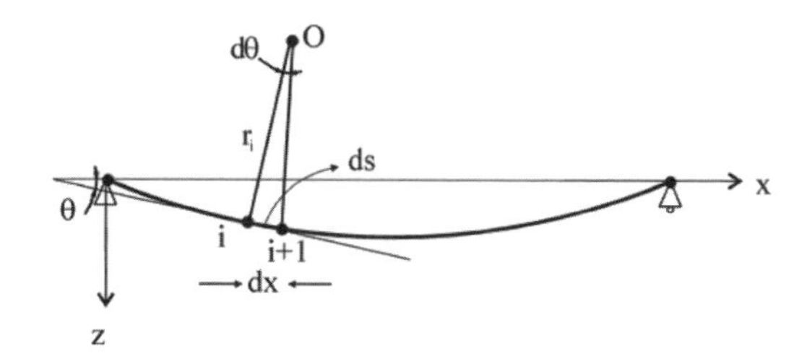

Figure 2. Curvature and slope angle.

gives rise to a resisting couple which balances the external couple M.

Due to the fact that all such forces distributed over the cross section represent a system equivalent to a couple, the resultant of these forces equal zero. Briefly, the relation between the curvature and moments due to external forces can be obtained. The effect of shearing force on the curvature is usually small and neglected (Timoshenko 1948).

To derive an expression for the relation between the curvature and the shape of the curve, two adjacent points i and $i + 1$, ds apart on the deflection curve is considered (Fig. 2). The angle that the tangents at i makes with the x axis is denoted by θ, that the angle between the normal to the curve at i and $i + 1$ is $d\theta$. The intersection point O of these normal gives the center of curvature and defines r, the radius of the curvature (Timoshenko 1948).

The curvature is defined as the rate of change of slope across the curvature axis. By taking the segment sufficiently small, it is assumed that the change of curvature across the segment is negligible. In this case, keeping the slope constant across the segment, a curve with a constant radius will be formed. Curvature of various points on a deflection curve can be obtained by according moment values. In this case, deflection curve will be composed of arcs.

The curve is represented by a sequence of circular arcs, within a user-specified tolerance.

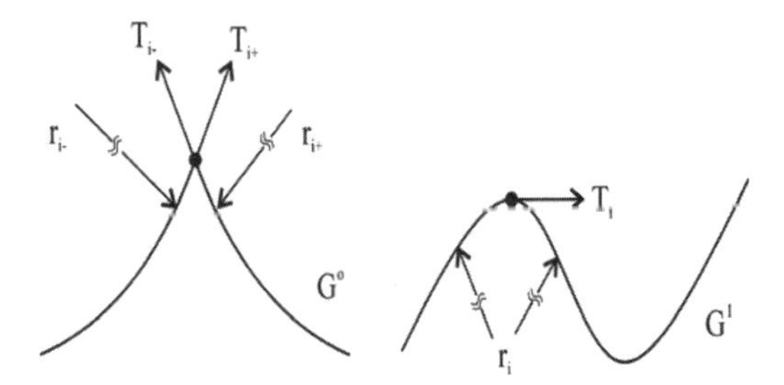

Figure 3. Geometric continuities; zero and first.

By the proposed method, the relation between circular arcs is established.

2.2 *Continuity conditions*

The smoothness of the curve is defined by the geometric continuity G^0, where the curve is continuous, by G^1, where curve and unit tangent vector are continuous, and by G^2, where curve, unit tangent vector and the curvature are continuous (Fig. 3). The loss of continuity of the second and higher derivative is not as important as one might suppose (Bolton 1975).

2.3 *Curve modeling by using curvature*

Radii of arcs are not sufficient for deflection calculations, thus center of the arcs are unknown. In order to appoint the position of center points of arcs, are required to join with G^1 continuity, that have a common unit tangent vector (Barsky & DeRose 1989). This common unit tangent vector also indicates a common unit normal vector between connected bi-arcs.

Arc centers are placed on the same line that passes through the common point of consecutive arcs. So, it is possible to locate the center of the next arc. Figure 4 shows deflection curve obtained by this procedure.

The angle between the point i and z axis is the slope angle θ_i. The angle between the following points' unit normal vectors is center angle of arc, $\mathrm{d}\theta_i$. The angle of the next control point θ_{i+1} is obtained as the sum $\theta_i + \mathrm{d}\theta_i$ as shown in Figure 5. As a result, slope is given by

$$\theta_{i+1} = \mathrm{d}\theta_i + \theta_i \tag{1}$$

This procedure can be simplified by choosing the slope of the starting point is equal to zero, indicating that the moment is at its extremum value.

Figure 6 shows an arc located between points *i, i + 1*, where the center point is O_i, radius of curvature is r_i, arc length is d*s*, arc center angle $\mathrm{d}\theta_i$ and chord length is $\mathrm{d}c_i$. The horizontal and vertical distance components between the points *i* and *i + 1* are obtained from Equations 5 and 6 in terms of d*s* and *r* by using the relations given in Equations 2–4.

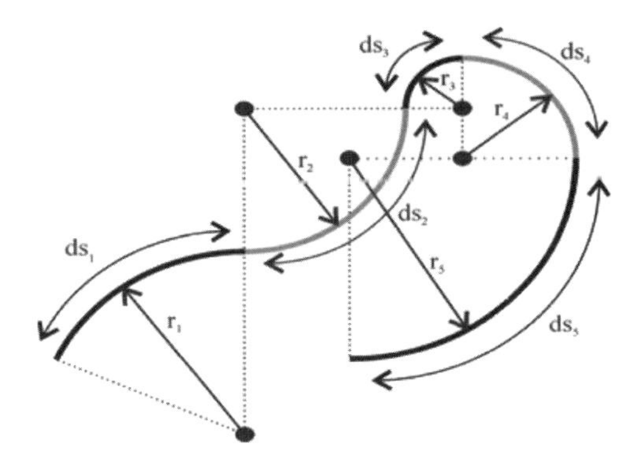

Figure 4. Curve modelling.

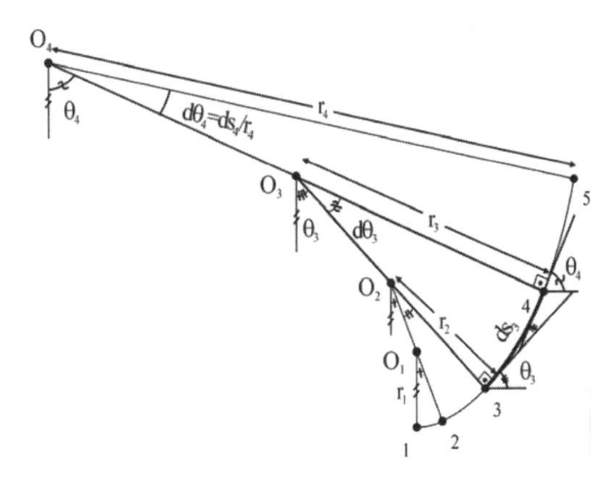

Figure 5. Slope and arc angles.

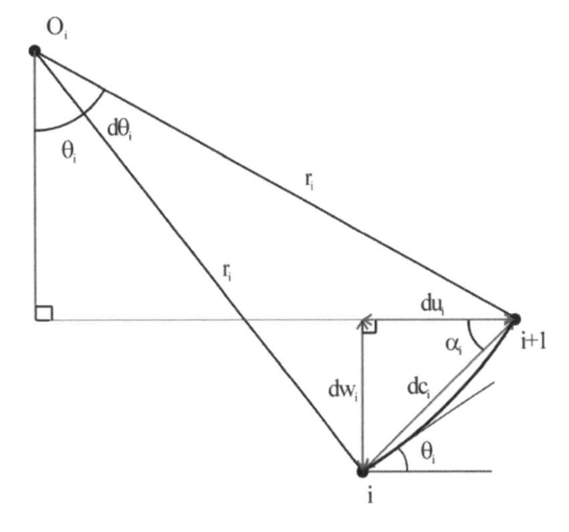

Figure 6. Displacement geometry.

$$\mathrm{d}\theta_i = \mathrm{d}s_i / r_i \tag{2}$$

$$\mathrm{d}c_i = 2\, r_i \sin(\mathrm{d}\theta_i / 2) \tag{3}$$

$$\alpha_i = \theta_i + \frac{\mathrm{d}\theta_i}{2} = \sum_{m=1}^{i-1} \frac{\mathrm{d}s_m}{r_m} + \frac{\mathrm{d}s_i}{2r_i} \tag{4}$$

$$\mathrm{d}w_i = \mathrm{d}c_i \sin(\alpha_i) \tag{5}$$

$$\mathrm{d}u_i = \mathrm{d}w_i / \tan(\alpha_i) \tag{6}$$

The direction of curvature is important in calculating the angle α_i; when it is reversed, θ_i will be decreasing instead of increasing.

It is also possible to use the same approach inversely for computing the initial curvature values. So, there is no need to define the initial deflection with trigonometric series (Tayyar 2011). By adding the initial curvature and the curvature due to external forces the final curvature is obtained, which could be used in equilibrium conditions (Timoshenko 1953).

Although post-collapse strength calculations could be obtained by the proposed method, rigid plastic analyze is preferred in the post-collapse strength calculations for lesser computational times.

A simply supported beam under lateral load is studied and given in appendix in elastic region for a better understanding of the procedure of geometrical curvature curve modelling, with limited segments without considering a convergence.

3 STIFFENED PLATE MEMBERS

When estimating the ultimate strength of a stiffened panel by simplified methods, it is often necessary to accurately evaluate the post-buckling effective width of a local plate panel. The interaction between the plating and stiffener in the buckling behavior must also be carefully considered (Paik et al., 2009). The effective width of the plating between stiffeners is formulated analytically, accounting for applied compressive loads, initial deflections. As the compressive loads increase, the effective width of the buckled plating varies, because it is a function of the applied compressive stresses (Paik et al., 2001). However, most simplified methods assume that the effective width of plating does not depend on the applied compressive loads, and the ultimate effective width of plating typically takes a constant value. If the effective width of plating is treated as a variable, one would find that the equation characterizing the ultimate limit condition for a stiffened panel shows a much higher degree of nonlinearity than it would otherwise (Paik et al., 1999).

The initial shape imperfection and welding residual stresses have a significant influence on the buckling collapse behavior of the plate (Paik et al., 2009). A concept of "effective width" has been used for practical design. In order to calculate effective width, equation 7 is used (von Karman et al., 1932). As can be seen, there is no need for a correction factor for initial deflection, since maximum stress is obtained accurately by the proposed method, unlike the common effective width formulas introduced in literature (Faulkner 1975, Winter 1947). Also, effective width treated as a variable due to the altering maximum in-plane stress of the plate, $\sigma_{\max}$, that can be easily determined within user specified tolerance.

$$\beta = \frac{b}{t}\sqrt{\frac{\sigma_{\max}}{E}} \Rightarrow K = \frac{b_e}{b} = \begin{cases} \dfrac{1.9}{\beta} & \Leftrightarrow \beta > 1 \\ 1 & \Leftrightarrow \beta \leq 1 \end{cases} \tag{7}$$

4 ITERATION PROCEDURE

For each curvature value, the strain of every element is calculated according to the assumption of plane cross-section (Yao 2003), and the stress of the elements is determined from the load-shortening relationship of the elements. The moment of stress on each element is computed about the instantaneous neutral axis. The resultant moment of all elements gives the bending moment of the cross-section for the considered curvature. The location of the instantaneous neutral axis should be determined by iteration or trial and error according to the condition that the sum of the stresses on all elements of the cross-section equals zero (Fig. 7).

The procedure of load-shortening calculations may be summarized as follows.

1. Axial load P is given.
2. Segment number n is given.
3. Effective width is assumed.
4. Determine the elongation of the structure dL.
5. Calculate the segment length (ds = [L-dL]/n).
6. Shortening du is assumed.
7. First point's displacement w_1 is assumed.

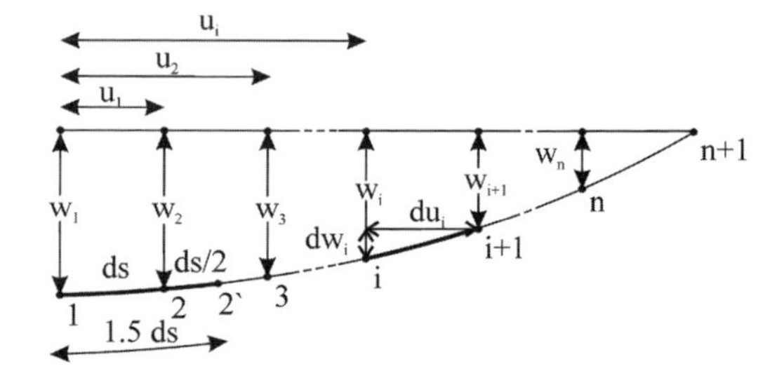

Figure 7. Displacements on the deflection curve.

8. Obtain average moment value of the segment Mi due to reaction forces according: effective width, change in centroid, shortening and axial/in-plane stresses.
9. Obtain Φ_i value from M_i value by using the equilibrium equations.
10. To find the following point location, obtain du_i, dw_i, θ_i, w_{i+1}, u_{i+1}, respectively from equations 2–6 and using ds.
11. To find following segment midpoint location, obtain $du_{i'}$, $dw_{i'}$, $+_{i'}$, $w_{i+1'}$, $u_{i+1'}$, respectively from equations 2–6, and using 1.5 ds.
12. Repeat steps 9–11 till last segment.
13. Go to step 7 and modify until BC's satisfied (For simply supported boundaries $w_{n+1} = 0$). If it is satisfied, proceed to the next step.
14. Determine shortening. If the determined value is close the value in step 6 then move to the next step, otherwise repeat from step 6 by using the determined value.
15. Determine the effective width with equation 8 and repeat the iteration from step 3 until user specified tolerance is obtained.
16. Go to step 2 and increase value of n for convergence.
17. Obtain the load-shortening value du due to P.

5 "ENERGY CONCENTRATION" ANALYSIS

In 1980, The Very Large Crude Carrier *Energy Concentration*, loaded with crude oil, arrived to the port safely, but collapsed during unloading in hogging position. Many studies have been performed concerning this incident lately. Rutherford & Caldwell (1990) indicates that the moment applied at the instance of collapse was 17.94×10^3 MN m. Properties of the *Energy Concentration* can be found in Rutherford & Caldwell (1990). Here, the moment capacity is calculated with the proposed method and comparisons are made with the available results in the open literature.

P-du diagrams of the members are prepared and strength analysis for the whole ship by using progressive collapse analysis. Corrosion effects of the members are also taken into account. It is assumed that welding residual stresses are no longer effective due to the age of the ship (Gordo et al., 1996). Initial deflection value is taken as 5.1. Furthermore, since the draft of the vessel is 12 m, 0.12 N/m pressure is applied on the bottom and side members (Gordo et al., 1996).

The moment capacity developed by the ship section at sagging position is found to be 19.06×10^3 MN m when the initial deflections are ignored. When the initial deflections are considered, the moment capacity decreases to 17.02×10^3 MN m.

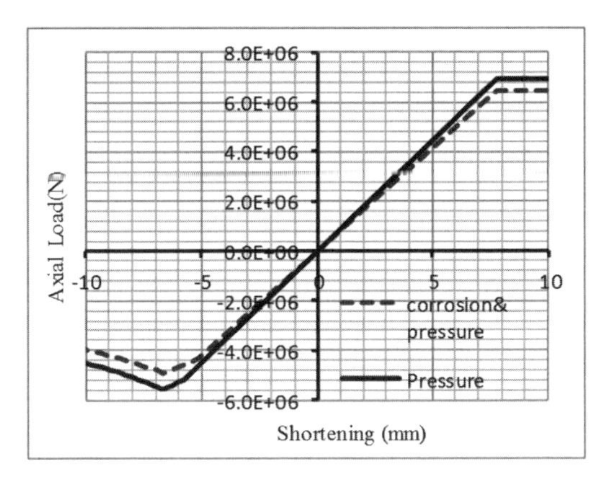

Figure 8. P-du diagram of E102.

Table 1. Results.

	Ultimate strength
Astrup	18.84×10^3 MN m
Chen	20.23×10^3 MN m
Cho	20.09×10^3 MN m
Dow	18.8×10^3 MN m
Masoka	20.01×10^3 MN m
Rigo (1)	18.46×10^3 MN m
Rigo (2)	17.54×10^3 MN m
Yao	19.04×10^3 MN m
ABS	18.22×10^3 MN m
Proposed method	17.02×10^3 MN m

Figure 8 shows P-du diagrams of the member 102, with and without corrosion rate as a result of proposed method (Tayyar 2011).

The predicted moment capacity is compared with other methods in Sun et al. (2005) in Table 1. A reduction of 12% occurred in the ship capacity due to initial deflection by the proposed method. The calculated value is lower than 17.9×10^3 MN m, the moment at the instance of collapse (Tayyar 2011).

6 CONCLUSION

A new solution method is developed for individual ship structure members considering the influences of plate/stiffener interaction, initial deflections and welding residual stress by using curvature geometrically.

With the developed method, progressive collapse analysis is executed by using the solutions of stiffened plate members. The present method is shown to produce reasonably accurate solutions with low modeling and computational times.

This method enables modeling of deflections having complex shapes and their application in strength of stiffened plates. Geometrically using the curvature does not only provide accuracy in deflection calculations but also gives advantages in computation time. Fast and accurate execution of progressive collapse analysis is realized.

ACKNOWLEDGEMENT

The authors express their sincere gratitude to K. Sarioz and B. Ugurlu for their help.

APPENDIX

The equilibrium deflection shape of the beam axis under the lateral load F acting on the middle point of the beam is shown in Figure A1. Because of the symmetric deflection only half of the beam is calculated with using only two segments. Trial deflection at midpoint is z_1 and total shortening of the system is du. Total length of the beam L is 800 mm, area A is 1200 mm^2, inertia moment I is 1440 mm^4, lateral load F is 1000 N and modulus of elasticity E is 200000 N/mm^2.

Trial values: $z_1 = 30$ mm, du = 0 mm, n = 2

Calculations for the first segment:

$$M_{1'} = \frac{F}{2}\left(\frac{L}{2} - \frac{ds}{2}\right) = 150000 \text{ N mm}$$

$$r_1 = \frac{IE}{M_{1'}} = 1920 \text{ mm}$$

$$dw_1 = 2 r_1 \sin\left(\frac{ds}{2r_1}\right) \sin\left(\frac{ds}{2r_1}\right) = 10.41 \text{ mm}$$

$$du_1 = dw_1 / \tan\left(\sum_{m=1}^{i-1} \frac{ds}{r_m} + \frac{ds}{2r_i}\right) = 199.64 \text{ mm}$$

$$z_2 = z_1 - dw_1 = 19.59 \text{ mm}$$

$$x_2 = x_1 + du_1 = 199.64 \text{ mm}$$

$$dw_{1'} = 2 r_1 \sin\left(\frac{1.5ds}{2r_1}\right) sin\left(\frac{1.5ds}{2r_1}\right) = 23.39 \text{ mm}$$

$$du_{1'} = dw_{1'} / \tan\left(\sum_{m=1}^{1-1} \frac{ds}{r_m} + \frac{1.5\,ds}{2r_i}\right) = 298.78 \text{ mm}$$

$$x_{2'} = x_1 + du_{1'} = 298.78 \text{ mm}$$

Now, the average moment for the second segment is 298.78 mm away from the start point. The procedure is continued and Table A.1 describes the first iteration.

As can be seen in Figure A2, the deflection curve of the first iteration does not satisfy the boundary conditions. For the second iteration the assumed deflection for the start point must be revised. After a linear correction, the assumed deflection

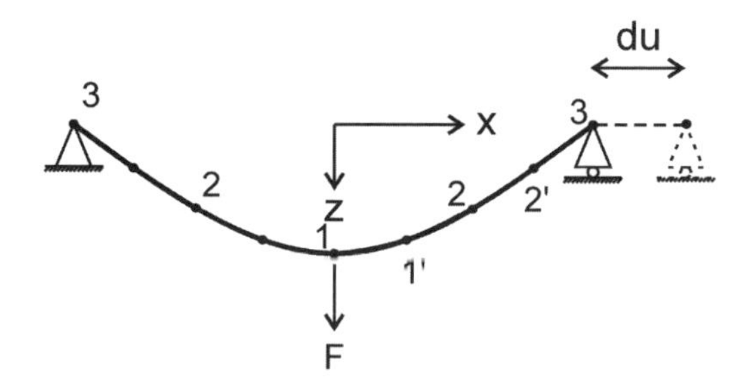

Figure A1. Simply supported beam.

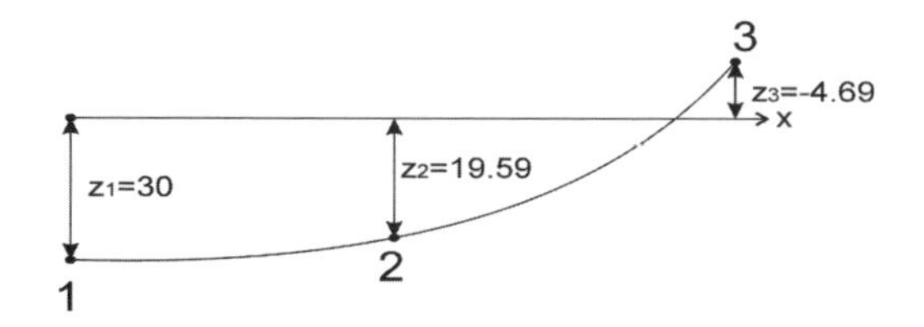

Figure A2. Deflection curve of the first iteration.

Table A.1. First iteration of the calculation.

	M	r_i	dw_i	du_i	x_{i+1}	z_{i+1}
	N mm	mm	mm	mm	mm	mm
1	150000.0	1920	10.41	199.64	199.64	19.59
1'			23.39	298.78	298.78	6.61
2	50609.6	5691	24.29	198.51	398.15	–4.69

Table A.1 Second iteration results.

	M	r_i	dw_i	du_i	x_{i+1}	z_{i+1}
	N mm	mm	mm	mm	mm	mm
1	149075.0	1932	10.34	199.64	199.64	24.35
1'			23.25	298.80	298.80	11.44
2	49677.1	5797	24.09	198.53	398.18	0.25

Table A.2. Third iteration results.

	M	r_i	dw_i	du_i	x_{i+1}	z_{i+1}
	N mm	mm	mm	mm	mm	mm
1	149087.5	1932	10.34	199.64	199.64	24.10
1'			23.25	298.80	298.80	11.69
2	49689.7	5796	24.10	198.53	398.18	0.00

is rearranged to 34.69 mm and shortening value is calculated as 3.7 mm. Until the satisfaction of the boundary conditions, the iteration is repeated as shown in Tables A.2 & A.3.

When the number of segments is increased, the converged maximum deflection becomes 36.71 mm. By neglecting the effect of shortening, the proposed solution gives 37.00 mm, where the analytical solution is 37.01 mm (Tayyar 2011).

REFERENCES

Barsky, B.A. & DeRose, T.D. 1989. Geometric continuity of parametric curves: three equivalent characterizations. *IEEE Computer Graphics & Applications.* 9(6): 60–99.

Bolton, K.M. 1975. Biarc curves. *Computer-Aided Design.* 7(29): 89–92.

Faulkner, D. 1975. A review of effective plating for use in the analysis of stiffened plating in bending and compression. *Journal of Ship Research. 19*(1): 1–17.

Gordo, J.M., Soares, C.G. & Faulkner, D. 1996. Approximate Assessment of the Ultimate Longitudinal Strength of the Hull Girder. *Journal of Ship Research.* 40(1): 60–69.

Heyman, J. 1998. *Structural Analysis A Historical Approach*, Cambridge: Cambridge University Press.

Hu, S.Z. 1993. A Finite Element Assessment of the Buckling Strength Equations of Stiffened Plates, *Ship Structures Symposium*, Virginia.

Hu, Y. Zhang, A. & Sun, J. 2001. Analysis on the ultimate longitudinal strength of a bulk carrier by using a simplified method *Marine Structures* 14:311–330.

Meek, D.S. 2002. Coaxing A Planar Curve To Comply. *Journal Of Computational And Applied Mat*hematics. 140(1): 599–618.

Paik, J., Branner, K., Choo, Y., Czujko, J., Fujikubo, M., Gordo, J., Parmentier, G., Iaccarino, R., O'neil, S., Pasqualino, I., Wang, D., Wang, X. & Zhang, S. 2009. Ultimate Strength, *17th International Ship and Offshore Structures Congress*, Committee III.1, C.D. Jang and S.Y. Hong (Eds), University of Seoul, 1: 375–474.

Paik, J.K., Thayamballi, A.K. & Kim, D.H. 1999. An analytical method for the ultimate compressive strength and effective plating of stiffened panels *Journal of Constructional Steel Research* 49: 43–68.

Paik, J.K., Thamyamballi, A.K., Lee, S.K. & Kang, S.J. 2001. A semi-analytical method for the elastic-plastic large deflection nalysis of welded steel or aluminum plating under combined in-plane and lateral pressure loads. *Thin-Walled Structures* 39: 125–152.

Rutherford, S.E. & Caldwell, J.B. 1990. Ultimate Longitudinal Strength of Ships: A Case Study. *SNAME Transactions.* 98: 441–471.

Sun, H. & Wang, X. 2005. Buckling and Ultimate Strength Assessment of FPSO. *Trans. SNAME*, 113: 634–660.

Tayyar, G.T. 2011 Gemi Kirişinin Nihai Mukavemetinin Tayini (in Turkish) *PhD thesis*, Naval Architecture and Ocean Engineering Faculty, ITU.

Timoshenko, S. 1948. Strength of materials part I elementary theory and problems New York: D. Van Nostrand Company.

Timoshenko, S. 1953. *History of Strength of Materials,* New York: Mc Graw Hill.

von Karman, T., Sechler, E.E. & Donnell, L.H. 1932. Strength of Thin Plates in Compression. *ASME Trans.* 54(5): 53–57.

Winter, G. 1947. Strength of thin steel compression flanges, *ASCE Transactions*. 112: 527–576.

Yao, T., Fujikubo, M. & Khedmati, M. 2000. Progressive Collapse Analysis of a Ships Hull Girder under Longitudinal Bending considering Local Pressure Loads. *Society of Naval Architectures of Japan.*188: 507–515.

Yao, T. 2003. Hull girder strength. *Marine Structures* 16: 1–13.

Yao, T., Brunner, E., Cho, S., Choo, Y., Czujko, J., Estefen, S., Gordo, J., Hess, P., Naar, H., Pu, H., Rigo, P. & Wan, Z. 2006. Ultimate Strength, *16th International Ship and Offshore Structures Congress*, Committee III.1, Frieze, P. & Shenoi, R. (eds), University of Southampton, 1: 359–443.

Sustainable Maritime Transportation and Exploitation of Sea Resources – Rizzuto & Guedes Soares (eds)
 ISBN 978-0-415-62081-9

Torsion dynamic analysis of a ship hull composite model

I. Chirica, E.F. Beznea & D. Boazu
University Dunarea de Jos of Galati, Romania

ABSTRACT: In this paper, the authors are focusing on the torsion dynamic analysis of the ship hull made of composite materials. The numerical analysis (using 2 methods, one of them being a new proposed method) is developed on a typical hull of a container ship. The aim of the work is to analyze the influence of the very large open decks on the dynamic torsion behaviour of the ship hull.

The torsion analysis is performed on a scale model (1:50) of a container ship, made of composite material. The model has the main characteristics: length L = 2.4 m, breadth B = 0.4 m, depth D = 0.2 m.

The first 10 natural frequencies are determined with the proposed numerical method and are compared with the results obtained with FEM package soft COSMOS/M.

1 INTRODUCTION

A growing interest in the foundation of the theory of thin-walled composite beams and of their incorporation in civil and naval constructions, aeronautical, automotive, helicopter and turbo-machinery rotor blades, mechanical, has been developed in the last three decades or so.

In recent years the improved design, fabrication and mechanical performance of low-cost composites has led to increase in the use of composites for large patrol boats, hovercraft, mine hunters and corvettes. Currently, there are all-composite naval ships up to 80–90 m long, and this trend continues. It is predicted that hulls for mid-sized warships, such as frigates that are typically 120–160 m long, may be constructed in composite materials from 2020.

Investigation into the vibrational and the elastic behavior of thin-walled beams with open and closed cross-sections has been carried out since the early works (Vlasov et al., 1961, Timoshenko & Gere 1961).

Monographs written by Kollbrunner & Basler (1969) and Gjelsvik (1981) are useful references for developing new methods of the thin-walled beam theory and its applications. The intensive research works have been made over the years to develop finite element models that can accurately represent the complex flexural-torsional warping deformable response of these structures. Yonghao et al. (1993) present the theory and method of analyzing horizontal-torsion-coupled dynamic behavior of a ship hull with large hatch opening by the transfer matrix method, considering the ship hull as a simplified free-free nonuniform thin-walled beam.

Senjanovic, I. & Fan, Y. (1992) have developed a flexural and torsion theory of thin-walled girders by introducing the concept of effective stiffness and mass parameters as modal quantities. For illustration, analysis of a container ship midship section was performed.

Kim, N.-I. & Kim, M.-Y., (2005) have proposed a general theory for the shear deformable thin-walled beam with non-symmetric open/closed cross-sections. For this purpose, an improved shear deformable beam theory wass developed by introducing Vlasov's assumption and applying Hellinger-Reissner principle, and exact dynamic and static element stiffness matrices were evaluated.

To ensure safe design of a ship's hull, traditionally, the longitudinal strength of the ship hull with length exceeding 60 m must be assessed during the design stage.

The longitudinal failure of ship hulls made of composite materials is usually easier due to the relative low stiffness and relative thin structures. With the trend that the size of ship hull in composite materials is upon large scale, it is becoming necessary to study the longitudinal strength of ship hull in composite materials.

Ship hull structure can be considered as thin-walled structures. Plates and shells have one physical dimension, their thickness, small in comparison with their other two. In thin/thick walled beams all three dimensions are of different order of magnitude. For such structures the wall thickness is small compared with any other characteristic dimension of the cross-section, whereas the linear dimensions of the cross-section are small, compared with the longitudinal dimension.

Ship hulls in composite materials can usually be regarded as assemblies of a series of thin walled stiffened composite panels. Thus, knowing the strength of stiffened composite panels it is possible to estimate the longitudinal strength of ship hulls in composite materials.

Due to their wide applications in civil, aeronautical/aerospace and naval engineering, and due to the increased use in their construction of advanced composite material systems, a comprehensive theory of thin/thick walled beams has to be developed: this is one of the aims of this paper.

The aim of the work is to analyze the influence of the very large open decks on the torsion dynamic behaviour of the ship hull made of composite materials.

2 FINITE MACROELEMENT MODEL OF THIN WALLED BEAM

The methodology proposed to analyze the ship hull dynamic torsion as thin walled beam using macro elements is treated. The outline of the section is considered as polygonal one. The material is orthotropic one. For a straight line portion of cross section outline is corresponding a longitudinal strip plate (Fig. 1). Due to the torsion of the thin walled beam, in the strip plate the stretching-compression, bending and shearing occur. The strip plate is treated as an Euler-Bernoulli plate. The stiffness matrix of the macro-element is obtained by assembling the stiffness matrices of the strips (Chirica et al., 2011).

Two coordinate systems are used:

- global system O_0XYZ having axis O_0X along the torsion centers line of the cross sections of the beam.
- local system attached to each plane k $(F_k^0 x_k y_k z_k)$ having the axis $F_k^0 x_k$ parallel with OX.

The torsion behaviour of the thin-walled beam is depending on the section type. So, the methodology presented in this paper is treating in different way

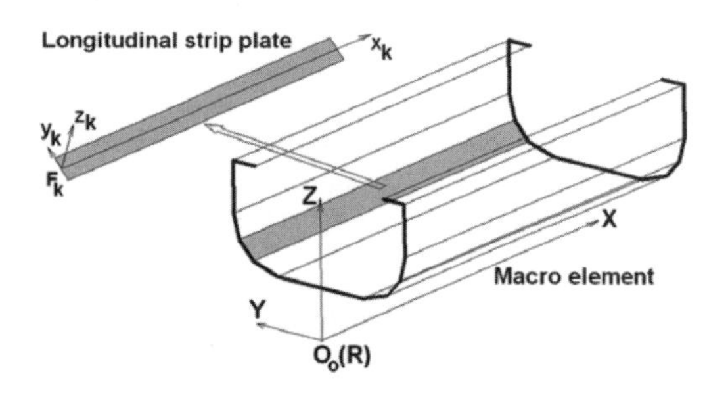

Figure 1. Model of macro-element.

and different hypothesis, depending on the type of cross section: open and closed.

2.1 *Thin-walled theory for open section*

For the cross section type open one, the hypothesis of the Vlasov theory are used:

- The material is linear-elastic, homogeneous, orthotropic generally, having the coordinate system $F_k^o x_k y_k z_k$ as the main orthotropic axis (Fig. 2);
- The shear stresses occurring in the beam cross section are parallel with the median line Γ.
- During the beam deformation the median line Γ does not remain plane. The projection of the median line on the cross section plane remains the same as its initial shape (non-deformed outline hypothesis). For small displacements, the displacement v of the current point F placed on the median line has the equation

$$\mathrm{v}(x,s) = \tilde{r}(s)\varphi(x) \tag{1}$$

- The displacement u along the axis O_oX of the point F is considered as constant on the wall thickness. The displacement u is considered to be in the form

$$u(x,s) = -\omega(s)\varphi'(x) \tag{2}$$

The sectorial coordinate is defined as

$$\omega(s) = \int_{\Gamma} \tilde{r}(s)\,\mathrm{d}s \tag{3}$$

The torsion of the thin-walled beam generates the torsion of the strips and the loading of the strips in their plane.

Using the Equations (1) and (2) for a strip k, it may be written

$$\mathrm{v}_k(x,s) = \tilde{r}_k\,\varphi(x) \tag{4}$$

$$u_k(x,y_k) = -\omega(y_k)\varphi'(x) \tag{5}$$

For the displacement u, due to the tension-compression loading of the strip k (Fig. 3) the approach function is a parabolic one, having the form

$$u_k(\xi) = P_1(\xi)u_i^k + P_2(\xi)u_{ij}^k + P_3(\xi)u_j^k \tag{6}$$

2.2 *Closed section*

For the case of closed section, we assume that u is proportional to the generalized sectorial

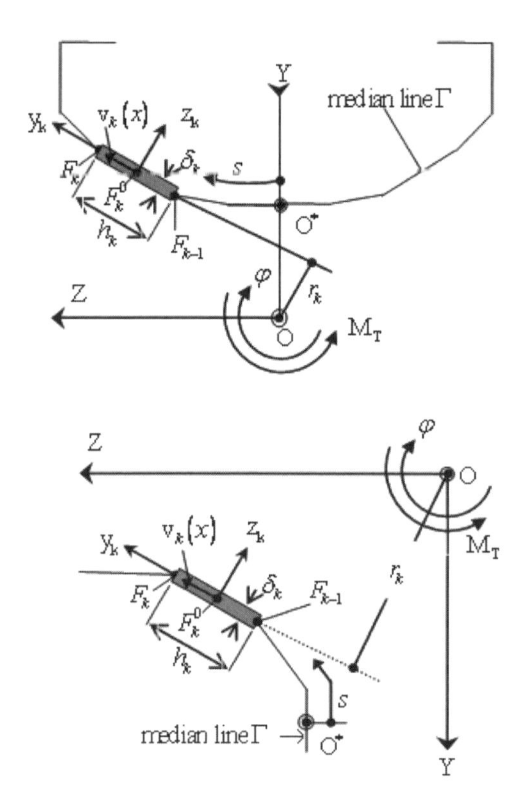

Figure 2. Macro-element with thin-walled beam open section.

co-ordinate $\hat{\omega}$ evaluated to O and O*. Different from classical theory, we assume that u is proportional to the rate of twist

$$u(x,s) = -\hat{\omega}(s)\varphi'(x) \tag{7}$$

The generalized sectorial co-ordinate is defined as

$$\hat{\omega} = \omega - \tilde{\omega} \tag{8}$$

where

$$\omega(s) = \int_0^s r(s)\,\mathrm{d}s,$$

$$\tilde{\omega} = \omega_0 \tilde{s} / \tilde{S}$$

The double of the area surrounded by Γ is

$$\omega_0 = \int_\Gamma r(s)ds$$

$$\tilde{s} = \int_0^s \frac{\mathrm{d}s}{\delta(s)}; \quad \tilde{S} = \int_\Gamma \frac{\mathrm{d}s}{\delta(s)} = \sum_{k=1}^{n} \frac{h_k}{\delta_k}$$

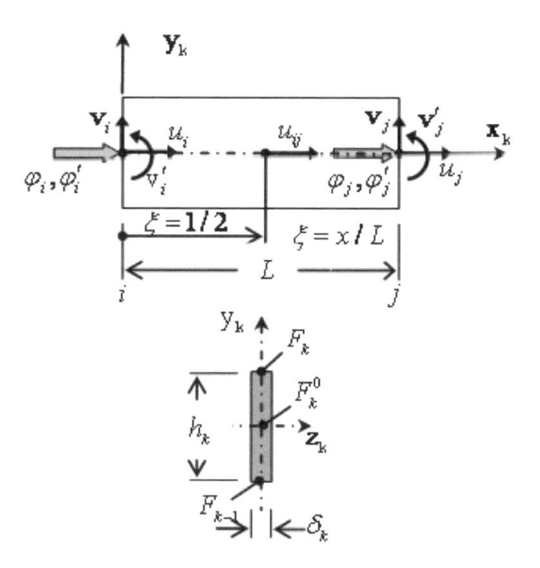

Figure 3. Model of finite strip element.

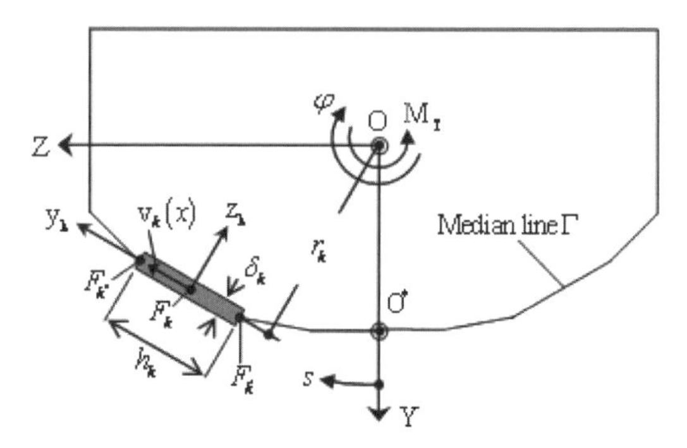

Figure 4. Macro-element with thin-walled beam closed section.

where n is the number of strip-plates.

The torsion loading of the beam generates an in-plane loading of the strip-plate. For each strip-plate, one obtains

$$\mathrm{v}_k(x) = r_k \varphi(x) \tag{9}$$

$$u_k(x, y_k) = -\hat{\omega}(y_k)\varphi'(x) \tag{10}$$

These equations define the displacement field for each stripe-plate. The continuity of the displacement u along the jointing edges between two stripe-plates is embedded in above relation. The linear variation of $\hat{\omega}_k$, the generalized sectorial co-ordinate along the axis y_k (in the reference system $F_k\, x_k\, y_k\, z_k$ associated to stripe-plate k) may be expressed as

$$\hat{\omega} = \hat{\omega}_k + (\hat{\omega}_{k'} - \hat{\omega}_k)\eta \tag{11}$$

where $-1/2 \leq \eta = y_k / h_k < 1/2$. The coordinates $\hat{\omega}_k, \hat{\omega}_{k'}, \hat{\omega}_{k''}$ characterize the points, $F_k, F_{k'}, F_{k''}$. The dependent relation between them is
$\hat{\omega}_k = (\hat{\omega}_{k'} + \hat{\omega}_{k''})/2$

For the longitudinal displacement one obtains

$$u(x, y_k) = -[\, \hat{\omega}_k + (\hat{\omega}_{k''} - \hat{\omega}_{k'})\, \eta \,]\; \varphi'(x) \qquad (12)$$

Using the hypothesis, the strain generated in the stripe-plate k are

$$\gamma_k = \frac{\partial u_k}{\partial y} + \frac{\partial v_k}{\partial x} = \Delta_k \varphi'(x) \qquad (13)$$

$$\varepsilon_k = \frac{\partial u_k}{\partial x} = -[\, \hat{\omega}_k + (\hat{\omega}_{k''} - \hat{\omega}_{k'})\, \eta \,]\; \varphi''(x) \qquad (14)$$

where

$$\Delta_k = \omega_0 / (\tilde{S} \delta_k)$$

Normal stresses σ_k appear in each strip-plate k due to the warping, having the equation

$$\sigma_k(x, y_k) = -E\, \hat{\omega}(y_k)\, \varphi''(x) \qquad (15)$$

In each cross-section, these stresses perform a system of distributed forces in self-equilibrium.

The shear stresses τ_k associates with the deformations γ_k may be determined with the equation

$$\tau_k(x) = G\,\gamma_k = \frac{G\omega_0}{\tilde{S}} \frac{1}{\delta_k} \varphi'(x) \qquad (16)$$

The flow of these stresses, $\tau_k \delta_k$, is constant for each section of thin-walled beam.

The differential equation of the twist angle φ obtained by the Ritz method is

$$E\, I_{\hat{\omega}} \varphi''' - G\, I_T\, \varphi' = -M_T(x) \qquad (17)$$

where

$$I_{\hat{\omega}} = \sum_{k=1}^{n} I_{\hat{\omega}k};\;\; I_{\hat{\omega}k} = h_k \delta_k [\hat{\omega}_k^2 + (\hat{\omega}_{k'} - \hat{\omega}_{k''})^2 / 12]$$

is sectorial moment of inertia,

$$I_T = \omega_0^2 / \tilde{S}$$

is conventional polar moment of inertia.

M_T is the transmitted torque.

The differential equation reveals two components of the transmitted torque:

$M_\gamma = G I_T \varphi'$ – Saint Venant torque

$M_\varepsilon = -E I_{\hat{\omega}} \varphi'''$ – warping torque

The component M_γ of the transmitted torque is the part associated with the strain γ_k and stress τ_k (Saint Venant torsion). The component M_ε is the part of the transmitted torque associate with the shear forces by strip-plates bending generated can be obtained only from equilibrium condition.

For the displacements $\varphi(\xi)$ and $v_k(x)$ polynomial functions (third order) are chosen:

$$\varphi(\xi) = H_1(\xi)\varphi_i + L H_3(\xi)\varphi'_i + H_2(\xi)\varphi_j + L H_4(\xi)\varphi'_j \qquad (18)$$

$$v_k(\xi) = H_1(\xi) v_i^k + L H_3(\xi)\theta_i^k + H_2(\xi) v_j^k + L H_4(\xi)\theta_j^k \qquad (19)$$

For bending and torsion, the well known stiffness and inertia matrices of the beam are used

$$\mathbf{k}_v^k = \frac{EI_k}{L^3}\begin{pmatrix} 12 & 6L & -12 & 6L \\ & 4L^2 & -6L & 2L^2 \\ & symm. & 12 & -6L \\ & & & 4L^2 \end{pmatrix}$$

$$\mathbf{k}_\varphi^k = \frac{GI_{Tk}}{L}\begin{pmatrix} 6/5 & L/10 & -6/5 & L/10 \\ & 2L^2/15 & -L/10 & -L^2/10 \\ & symm. & 6/5 & -L/10 \\ & & & 2L^2/15 \end{pmatrix}$$

$$\mathbf{m}_v^k = \frac{mL}{420}\begin{pmatrix} 156 & 22L & 54 & -13L \\ & 4L^2 & 13L & -3L^2 \\ & symm. & 156 & -22L \\ & & & 4L^2 \end{pmatrix}$$

$$\mathbf{m}_\varphi^k = \frac{\rho I_{Tk} L}{420}\begin{pmatrix} 156 & 22L & 54 & -13L \\ & 4L^2 & 13L & -3L^2 \\ & symm. & 156 & -22L \\ & & & 4L^2 \end{pmatrix}$$

where m is distributed mass per unit length.

Based on the coupling matrices for each strip-beam, the strips' stiffness and inertia matrices for axial and bending loading are assembling to realize the stiffness and inertia matrices of the macroelement. The assembling process was performed by transforming process using the orthogonal process (Ider & Amirouche 1988).

$$\mathbf{K} = \frac{E\, I_\omega}{L^3} \tilde{\mathbf{K}}_{EB} + \frac{E\, I_T}{L} \tilde{\mathbf{K}}_{SV} \qquad (20)$$

where

$$\tilde{\mathbf{K}}_{EB} = \begin{pmatrix} 12 & 6L & -12 & 6L \\ & 4L^2 & -6L & 2L^2 \\ & symm. & 12 & -6L \\ & & & 4L^2 \end{pmatrix}$$

$$\tilde{\mathbf{K}}_{SV} = \begin{pmatrix} 6/5 & L/10 & -6/5 & L/10 \\ & 2L^2/15 & -L/10 & -L^2/30 \\ & symm. & 6/5 & -L/10 \\ & & & 2L^2/15 \end{pmatrix}$$

2.3 *Composite characteristics of the macroelement thin-wall*

In the methodology, the classical thin-walled beam theory for isotropic materials was used. Taking into account the materials characteristics, the orthotropic character of the material was considered.

The equivalent stiffness coefficients for the tension-compression, bending and shearing loading of the strip k are determined.

$$(EA)_k = 2\left(\sum_{i=1}^{ns} E_{i,k} g_{i,k}\right) h_k$$

$$(EI)_k = 2\sum_{i=1}^{ns} E_{i,k} I_{i,k} = \frac{1}{6}\left(\sum_{i=1}^{ns} E_{i,k} g_{i,k}\right) h_k^3$$

$$G_k = \frac{24}{\delta_k^3}\left(\sum_{i=1}^{n_s} G_{i,k} z_{i,k}^2 g_{i,k}\right)$$

$$(GI_T)_k = 8\left(\sum_{i=1}^{ns}\left(G_{i,k} z_{i,k}^2 g_{i,k}\right)\right) h_k$$

$$E_k = \frac{2}{\delta_k}\sum_{i=1}^{ns} E_{i,k} g_{i,k} \tag{21}$$

The Equations (21) are determined according to the Figure 5.

Finally, the results obtained with the proposed methodology for a prismatic hull beam are compared with the ones obtained with analytical solutions.

3 FEM ANALYSIS

A FEM based code was developed. According to the theory presented above the code TORS for static and dynamic numerical analysis may be apply to ship hull behavior loaded especially to torsion. For the present study, a simplified model of a container ship, made of composite materials was analyzed. The mesh is concerning 12 macro-elements (6 macro-elements in the closed parts of the ship model and 6 macro-elements in the open part). Each macro-element, representing a piece of ship model, was modeled with 2D strip elements. The closed type macro-elements concern 16 strip elements. The open type macro-elements were modeled with 12 strip elements (as it is shown in Figure 1). The total number of strip elements (having the same length of 0.2 m) is 228. The outline of the cross section of the model was approached with a polygonal line. The bilge (curved area of the ship model) was meshed with 3 longitudinal strip elements.

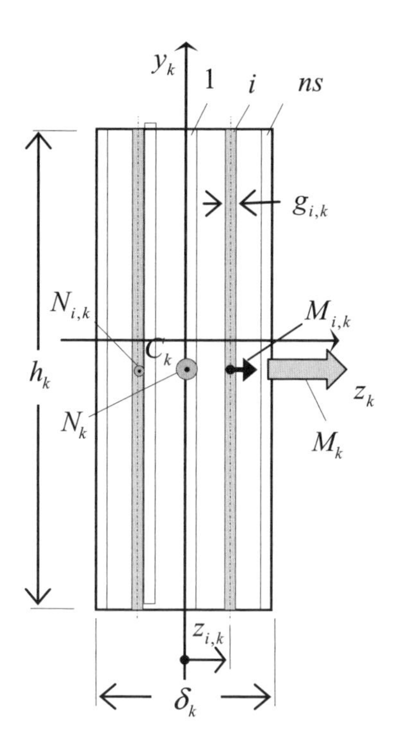

Figure 5. The lay-out of the thin walled.

The results obtained with the code TORS were compared with the ones obtained with COSMOS/M FE soft package. A 3-D model with 4-node SHELL4L composite elements of COSMOS/M was used.

On the beginning, a convergence analysis was performed. Finally, the optimum dimension of the quadrilateral element side (0.02 m) was determined and a number of 18110 elements were used in the mesh model.

The model has the main characteristics: length $L_m = 2.4$ m, breadth $B_m = 0.4$ m, depth $D_m = 0.2$m. The material is E-glass/polyester having the characteristics, determined by experimental tests:

$E_x = 46$ GPa, $E_y = 13$ GPa, $E_z = 13$ GPa,
$G_{xy} = 5$ GPa, $G_{xz} = 5$ GPa, $G_{yz} = 4.6$ GPa,

$\mu_{XY} = 0.3$, $\mu_{YZ} = 0.42$, $\mu_{XZ} = 0.3$
$\rho = 2045$ kg/m^3

The thicknesses' values of the hull shell are: 2 mm for side shell and 3 mm for deck and bulkheads.

The model is a simplified ship hull of a container ship, having the prismatic shell. The model has 3 parts: the ends have closed cross sections, the middle part has open cross section. In the middle part the deck is reduced to two longitudinal strips on sides. To assure the thin walled beam hypothesis (non deformability of the cross section) the model has 13 transversal bulkheads placed at every 0.2 m (Fig. 6).

Due to the fact, the real ship has much stiffened structure in the both end, in the both modeling types (TORS and FEM COSMOS) the model is considered as clamped at the end.

In the Table 1 the natural frequencies obtained so with code TORS and with licensed package soft COSMOS/M are presented. The differences between the results are reasonable.

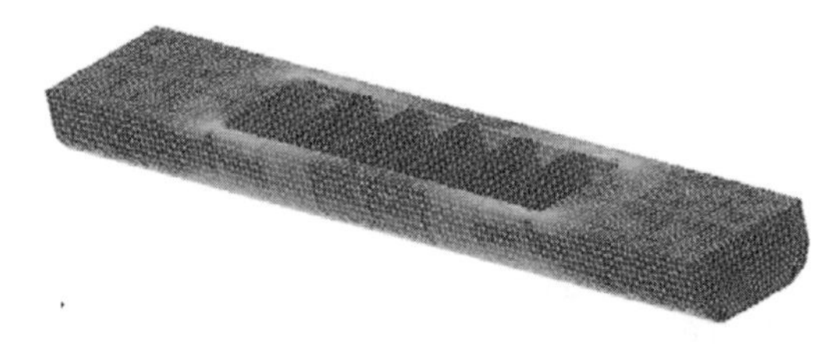

Figure 6. The simplified ship hull model.

Table 1. The first 10 natural frequencies.

No.	TORS	COSMOS/M
	[Hz]	[Hz]
1	309	301
2	329	319
3	353	342
4	368	373
5	416	411
6	429	421
7	451	445
8	479	465
9	519	511
10	550	562

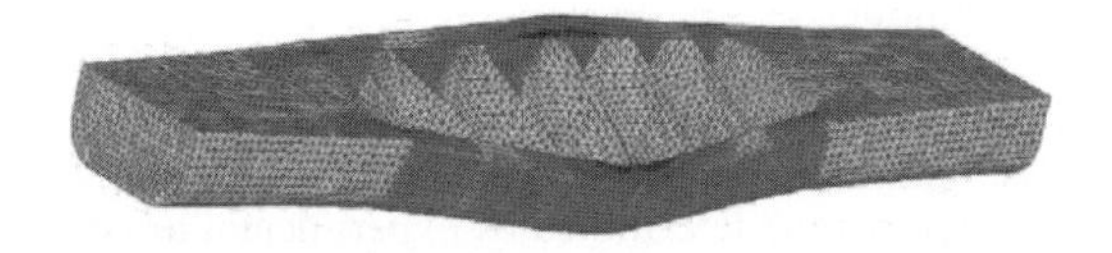

Figure 7. The first torsion modal shape of the ship hull model.

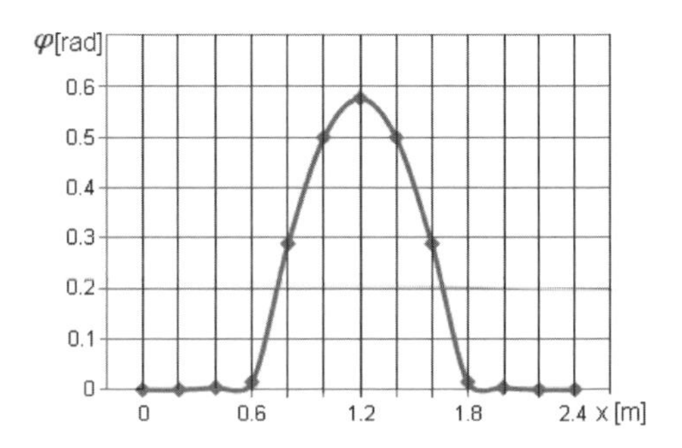

Figure 8. First vibration mode.

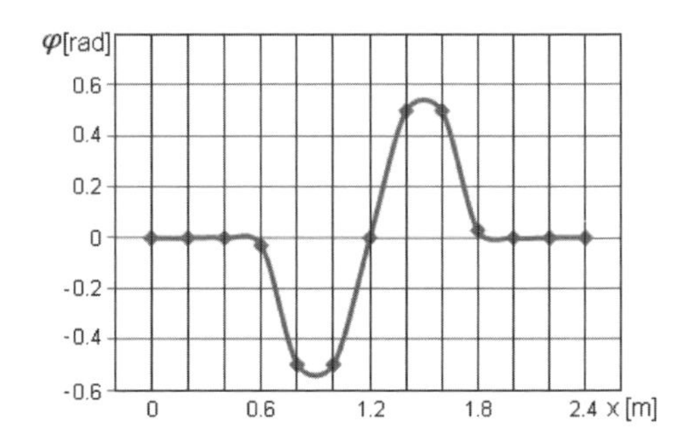

Figure 9 Second vibration mode.

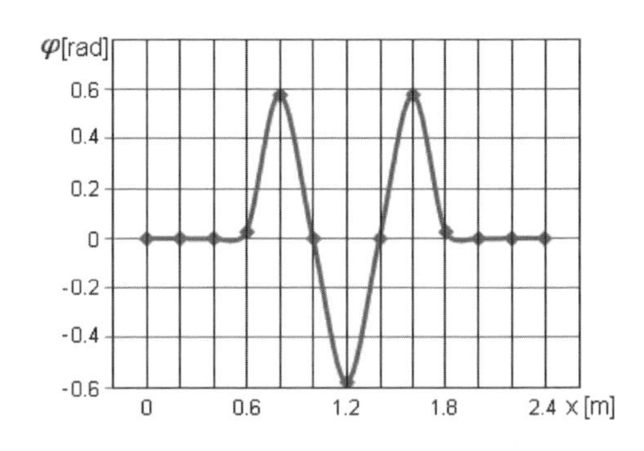

Figure 10. Third vibration mode.

In Figure 7 the first modal shape vibration is shown, where a coupled torsion with lateral bending is observed. The graphical representations of the first 3 vibration modes presented in Figures 8, 9 and 10 give a good information about the differences in torsion stiffness between closed and open section (for the ship hull model the ratio is about 10^4).

4 CONCLUDING REMARKS

The methodology, described in the paper, is concerning torsion static and dynamic response of the ship hull, treated as thin-walled beam. The new methodology is a FEM based macroelement. The results obtained with TORS, a code developed of the methodology based, were compared with those obtained with FEM package soft COSMOS/M.

As it is seen in the graphical representations in Figures 8, 9 and 10 due to very huge torsion stiffness difference between closed and open sections the ship hull ends may be considered as clamped.

The macro-element model is capable of predicting accurate stress state in the static loading, natural frequencies, deflection as well as angle of twist shapes of various configuration including boundary conditions, laminate orientation and type of cross section.

The methodology presented is found to be appropriate and efficient in analyzing flexural-torsion problem of a thin-walled laminated composite beam with a special application in ship design activity.

ACKNOWLEDGEMENT

This work was supported by UEFISCDI, project number PNII—IDEI Code 512/2008 (2009–2011).

REFERENCES

Chirica, I., Musat, S.D., Chirica R. & Beznea, E.F. 2011. Torsional Behaviour of the Ship Hull Composite Model, Computational Materials Science, vol. 50, issue 4: 1381–1386.

Ider, S.K. & Amirouche, F.M.L. 1988. Coordinate Reduction in the Dynamics of Constrained Multibody Systems—A New Approach, *Journal of Applied Mechanics*, December 1988, vol. 55: 899–904.

Kim, N.-I. & Kim, M.-Y. 2005. Exact dynamic/static stiffness matrices of non-symmetric thin walled beams considering coupled shear deformation effects, *Thin-Walled Structures* 43 (2005): 701–734.

Kim, S.B. & Kim, M.Y. 2001. Improved formulation for spatial stability and free vibration of thin-walled tapered beams and space frames, *Eng., Struct.* v. 22:446–458.

Kollbrunner, C.F. & Basler, K. 1969. *Torsion in structures*. Berlin: Springer.

Senjanovic, I. & Fan, Y. 1992. A higher-order theory of thin-walled girders with application to ship structures, *Computers & Structures*, Volume 43, Issue 1: 31–52.

Senjanovic, I., Tomasevic, S. & Grubisic, R. 2007. Coupled Horizontal and Torsional Vibrations of Container Ships, *Brodogradnia*, 58(2007) 4: 365–379.

Timoshenko, S.P. & Gere, J.M. 1961. *Theory of elastic stability*. 2nd ed. New York: McGraw-Hill.

Vlasov, V.Z. 1961. *Thin-Walled Elastic Beams*. 2nd ed. Jerusalem, Israel: Program of Scientific Translations.

Vörös, G.M. 2004. Free Vibration of Thin-Walled Beams, *Periodica Polytechnica Ser. Mech. Eng.*, vol. 48, No. 1: 99–110.

Yonghao, P., Yougang, T. & Xueyong., Y. 1993. The Horizontal-Torsion-Coupled Model Analysis and Dynamic Response Calculation of the Ship Hull with a Large Hatch Openign. *Tianjin: Journal of Tianjin University*: 51–58.

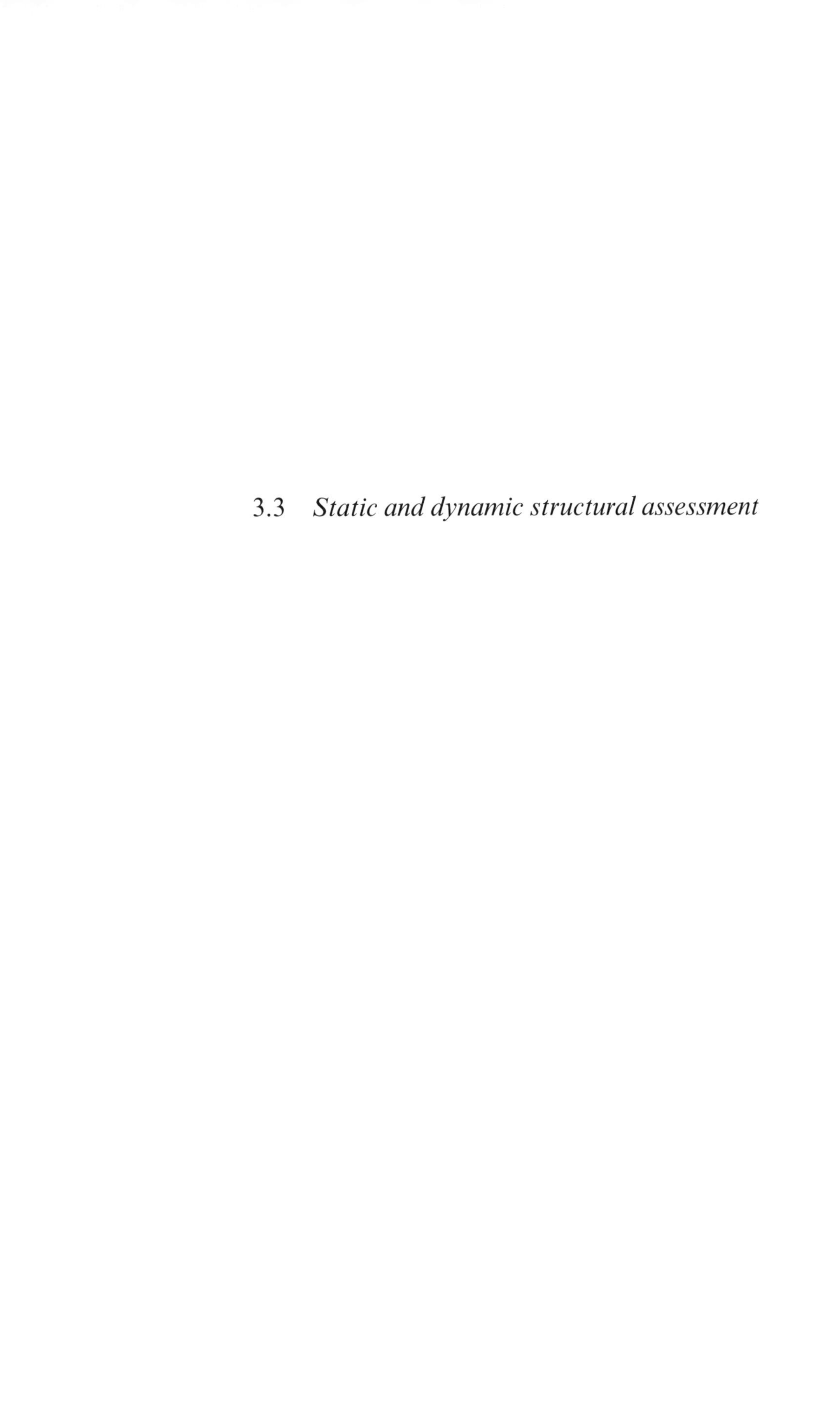

3.3 *Static and dynamic structural assessment*

Sustainable Maritime Transportation and Exploitation of Sea Resources – Rizzuto & Guedes Soares (eds)
© 2012 Taylor & Francis Group, London, ISBN 978-0-415-62081-9

Wave load cases for scantlings of bulk carrier structures

J.A. Jankowski, M. Warmowska & M. Tujakowski
Polish Register of Shipping, Gdańsk, Poland

ABSTRACT: The accuracy of predicting wave-generated stresses in ship structure continues to remain unsatisfactory, affecting safety at sea. Therefore, methods based on simulation of ship motion in waves are being developed, including the simulation of ship structure response to waves, green seas and sloshing. The simulation results are transformed into probability distributions used to determine stresses with assumed probability of their occurrence. The paper presents a method of wave generated stress prediction based on simulation of ship motion, loads and stresses in ship structure in waves.

1 INTRODUCTION

Presently, classification societies require the middle part of a ship structure to be assessed using the Finite Element Method (FEM). This method is well established as it gives satisfactory results of structure response (stress, deflection) to the loads applied. This implies that apposite assessment of ship structure depends on the accuracy of loads applied, mainly the wave loads.

Waving of the sea is a stochastic process and various realizations of the processes (irregular waves) and ship headings to the waves, which can occur with a certain probability in a ship's life, should be taken into account for the assessment of ship structures. This approach determines long-term response of ship structures to waves.

The prediction of wave generated stress level in the structure is of key importance in ship hull safety assessment. The stresses can lead to structure failure in the form of plastic flow, buckling of some structure members or fatigue cracks. The accuracy of wave stresses prediction is still unsatisfactory and therefore continuous development of methods to predict stresses in ship hull structures is necessary.

Severe weather conditions, which randomly occur during a ship's life, have a major impact on sustaining ship structure safety. The "extreme load cases" also depend on ship dimensions, her shape, mass distribution and ship speed. All possible sea states—defined by significant wave height and average zero up-crossing period, as well as all possible ship headings should be analyzed to determine the "extreme load cases". The computations take into account the representation of sea states, corresponding to the scatter diagrams, which determine the probabilities of sea state occurrence (IACS Rec. No. 34).

The classic method for predicting dynamic stresses in hull structures caused by waves is based on a spectral analysis of wave loads. However, this method is based on linear models and the nonlinear effects such as bow flare effect, green water, etc. are disregarded in this method (e.g., Guedes Soares, 2000). Usually the phase shift between, for example, wave generated bending moments, pressures and ship hull accelerations are lost in spectral analysis, which makes it impossible to predict stresses in the structure accurately.

The paper presents a method for predicting stresses in ship structure. According to this method equations of ship motion in irregular waves are solved numerically in time domain taking into account nonlinear effects. Stresses at any point of the hull structure can be computed in time domain using the concept of influence coefficients of wave loads (Jankowski, Bogdaniuk 2007).

However, such an approach is applicable in R&D projects but not in design practice. Therefore, simplifications should be introduced. Studies indicate that long-term response of the structures to sea waving can be approximated by appropriately chosen short-term response (response to one irregular wave). In such approximations the short-term response varies for different types of ship structures.

This paper presents a method of identification of wave load cases—irregular waves, determined by wave parameters and the ship heading, that have a dominant impact on the response of selected bulk carrier structures.

2 PREDICTION OF SHIP STRUCTURE RESPONSE TO WAVES

Seas and oceans are normally divided into distinct areas A_l, $l = 1, 2, \ldots, n$, (Hogben et al., 1986), characterized by the spectral density function. Wave spectrum representing the steady state sea conditions (short-term sea state) depends on the significant wave height H_s and average zero up-crossing period T_o (Ochi, 1998).

The short-term response of the ship to waves (e.g., wave bending moment M_W) is a set of probability distributions (e.g., probability density functions) of the given random variables for one sea state in a given area A_l, for various ship courses etc.

Long-term statistics of the given ship response is the accumulation of response statistics referring to: sea areas A_l, $l = 1, \ldots n$, short-term sea states, ship courses in relation to waves and ship's loading conditions, taking into account the frequencies of their occurrence.

The long-term probability density function $f(y)$ of the ship response y (e.g., wave bending moment M_v as the random variable) can be expressed as:

$$f(y) = \sum_m \sum_l \sum_k \times \left(\int_{-\infty}^{\infty} \int_{-\infty}^{\infty} f_{klm}\left(y \mid (H_s, T_0)\right) g(H_s, T_0) dH_s dT_0 \right) \times p_{kl} p_l p_m \quad (1)$$

where $f(y(H_s, T_0))$ = probability density function of the random variable y in the sea state condition (H_S, T_o) and $g(H_s, T_0)$ is probability density function of of sea state occurrence.

Taking into account the formula determining the conditional distribution, and by approximating (by the relevant sums) the integral occurring in (1), the following formula is obtained:

$$f(y) = \sum_m \sum_l \sum_k \sum_j \sum_i f_{ijklm}\left(y \mid (H_s, T_o)\right) p_{ijl} p_{kl} p_l p_m \quad (2)$$

where f_{ijklm} = the short term probability density function of random variable y; p_m = the probability of the ship's loading condition occurrence (different drafts for different loading conditions); p_l = the probability of ship presence in sea area A_l; p_{kl} = the probability of ship course in relation to waves in sea area A_l (uniform distribution in the interval [0, 2π] is used); p_{ijl} = the probability of the short-term sea state, determined by (H_s, T_o), occurrence in the sea area A_l, $l = 1, \ldots, n$.

The probability distributions of the sea states occurrence are given in the form of a matrix—called scatter diagram, which presents the probabilities p_{ijl} of sea state occurrence in the interval $[H_{si}, H_{si+1}]_l$ x $[T_{oj}, T_{oj+1}]_l$, $i = 1, \ldots, r$, $j = 1, \ldots, s$, $l = 1, \ldots, n$, (Hogben et al., 1986).

The numerical long-term probability density functions (2) of random variable y, representing different ship structure response to waves, is computed, basing on simulations of ship motion and the structure's behavior in irregular waves.

The shipmaster's decisions exercising good seamanship like e.g., speed reduction in high seas and change of course in relation to the waves are projected in the long-term procedure.

3 SIMULATION OF SHIP MOTION IN IRREGULAR WAVES

The simulation of vessel motions in waves is based on numerical solutions of non-linear equations of motion. The non-linear model used is presented in (Jankowski, 2006).

It is assumed that the hydrodynamic forces acting on the ship and defining the equations of its motions can be split into Froude-Krylov forces, diffraction and radiation forces as well as other forces, such as rudder forces and non linear damping.

The Froude-Krylov forces are obtained by integrating the pressure caused by irregular waves, undisturbed by the presence of the ship, over the actual wetted ship surface.

The diffraction forces are determined as a superposition of diffraction forces caused by the harmonic components of the irregular wave. The irregular wave is assumed to be a superposition of harmonic waves. It is assumed that the ship diffracting the waves is in its mean position. This is possible under the assumption that the diffraction phenomenon is described by a linear hydrodynamic problem. Such an approach significantly simplifies calculations because bulky computations can be performed at the beginning of the simulations and the ready solutions can be applied to determine the diffraction forces during the simulation.

The radiation forces are determined by added masses for infinite frequency and by the so-called memory functions given in the form of convolution. The memory functions take into account the disturbance of water, caused by the preceding ship motions, affecting the motion of the ship in the time instant considered (Cumminis, 1962).

The ways of solving 3D hydrodynamic problems and determining forces appearing in the equation of motion are presented in (Jankowski, 2006). The non-linear equations of motion are solved numerically.

4 SIMULATION OF SHIP STRUCTURE RESPONSE TO IRREGULAR WAVES

The stress values in members of ship structure depend on the loads acting on the structure and the geometrical and material features of the structure.

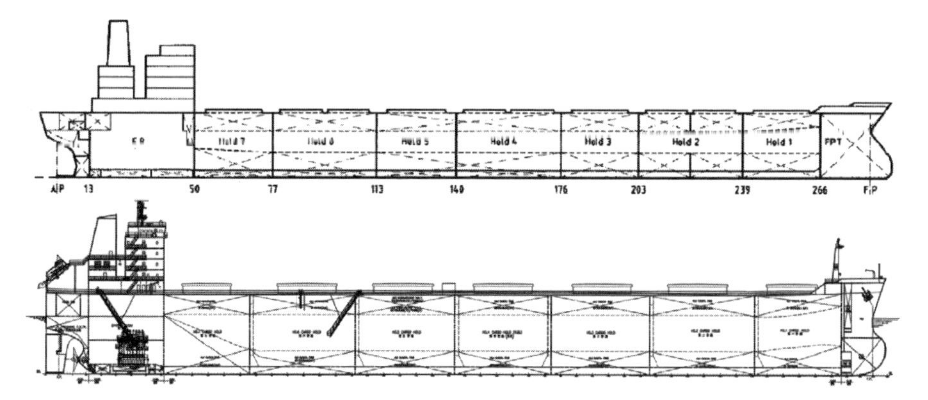

Figure 1. Two panamax bulk carriers of different design covered by the analysis; first built in 1977 and next in 2010.

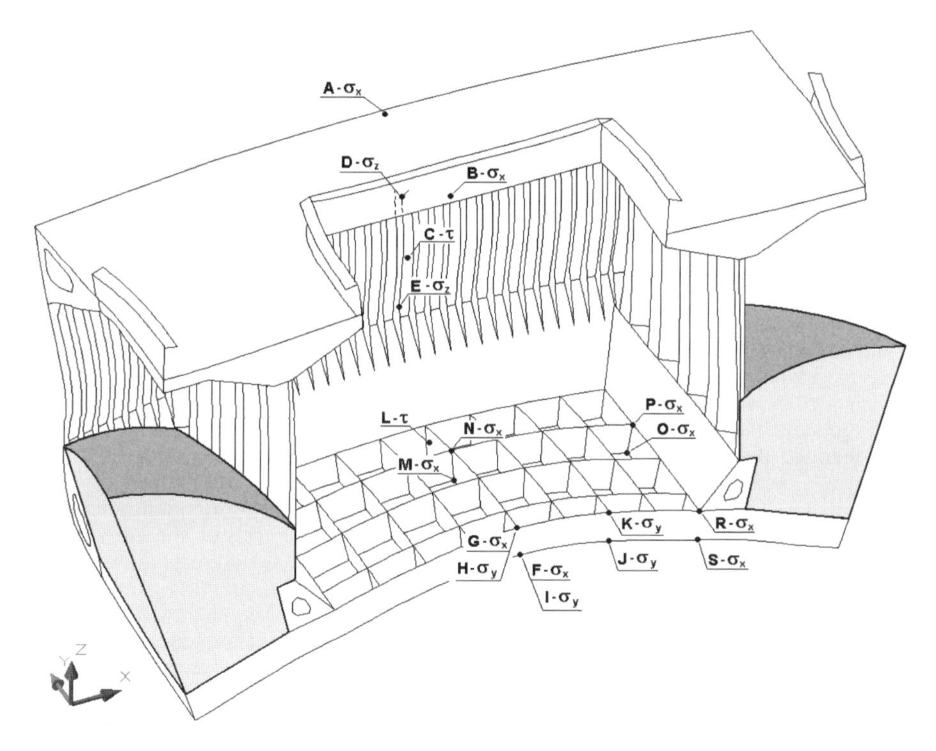

Figure 2. Members of the bulk carrier structure for which stresses were computed.

The structure of two panamax bulk carriers of different designs has been analyzed. The ships are presented in Fig. 1; first a 1977 built ship, and next—a 2010 built ship.

The module of the middle part of each ship hull, embracing three holds, has been modeled to carry out the strength analysis using finite element method (FEM)—Fig. 3. The stresses have been computed for 58 structure members on both ship sides, but only the members for which the computed stresses were significant have been marked in Fig. 2. Stresses have been computed for alternate loading of the ships: the middle hold was empty and the neighboring ones were loaded with heavy cargo (Fig. 2). This cargo rests on inner bottom, bilge tanks and lower stools of the bulkheads.

The impact of the fore and aft parts of the hull, which are not part of the module, is accounted for in the form of the wave vertical and horizontal bending moments, generated at the ends of the cross section $x = x_{aft}$ and $x = x_{fore}$ of the considered hull structure module.

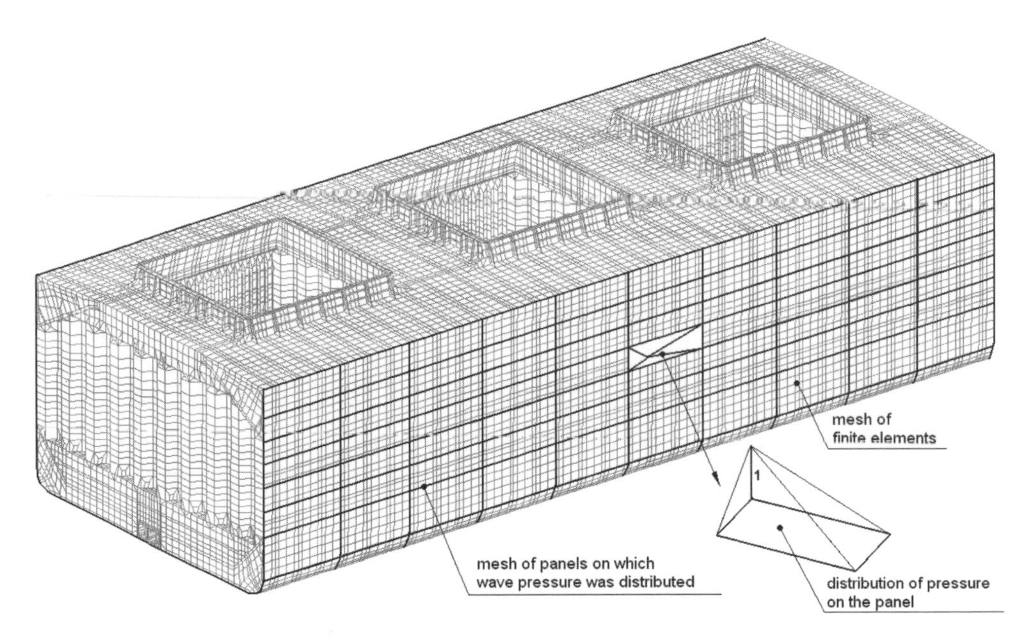

Figure 3. Ship hull module used to compute the influence of coefficients' values, and then to compute the stresses in chosen structure members.

So, the loads of the of the structure of ship module consist of:

- vertical and horizontal bending moments applied at the ends of the ship module;
- pressure generated by water flow on the instantaneous wetted surface of the moving ship (Fig. 4); and
- pressure generated on the inner bottom, bilge tanks and the lower stools of the bulkheads caused by gravitation and inertial forces acting on the cargo.

The simulation (PC computing) of the stresses in the structure members using the hull module (Fig. 3) consisting of approximately 100000 finite elements is practically impossible as it requires:

- fifteen minutes of computations of stresses using FEM for above listed loads in one time step of the simulation;
- the simulations of the stresses in an irregular wave (sea state); it requires computations for thousands of time steps and;
- the long-term computations; it requires the simulations of the stresses for hundreds of sea states.

Therefore, the following method has been adopted to make the simulations of stresses in the structure of two bulk carriers possible:

The stress value at a selected structure member is calculated according to the formula (Jankowski, Bogdaniuk, 2007):

$$\sigma = \sum_i p_{1i} W_{1i} + \sum_k a_{2k} W_{2k} + \sum_l M_l W_{3l} \tag{3}$$

where W_{1i} = the influence coefficients of the wave excited pressure values on wetted surface of the ship; W_{2j} = the influence coefficients of pressure values on inner bottom, bilge tanks and stools, generated a_{2k}; W_{3k} = the influence coefficient values of vertical and horizontal bending moments, generated at the ends of the cross section of the considered hull structure module; p_{1i} = the wave excited pressures on the wetted ship surface (Fig. 4); $a_{2k} = g + a_{vk}$ = gravitational acceleration and acceleration generated by ship motion and acting on the cargo in loaded cargo holds (Fig. 2), transformed in the computations to the pressure $p_{vk} = \rho a_{2k} h$ acting on the inner bottom, stools and bilge tanks, where h is the height (distance) of the cargo surface from the inner bottom, stools and bilge tank plating; M_l = vertical and horizontal bending moments, $l = 1, 2$, generated at the ends of the cross section $x = x_{aft}$ or $x = x_{fore}$ of the considered hull structure module.

The influence coefficients values W_{1i}, W_{2k}, W_{3l} of one structure member for fix i, k, and l are calculated applying FEM to the ship hull module (built of shell, beam and rod finite elements) and applying:

- loads distributed on a smaller number of panels than the number of finite elements (see Fig. 3) to reduce the time of computation.

Figure 4. Wave generated loads on wetted surface of bulk carrier built in 1977; the position of hull module presented in Fig. 3 is marked.

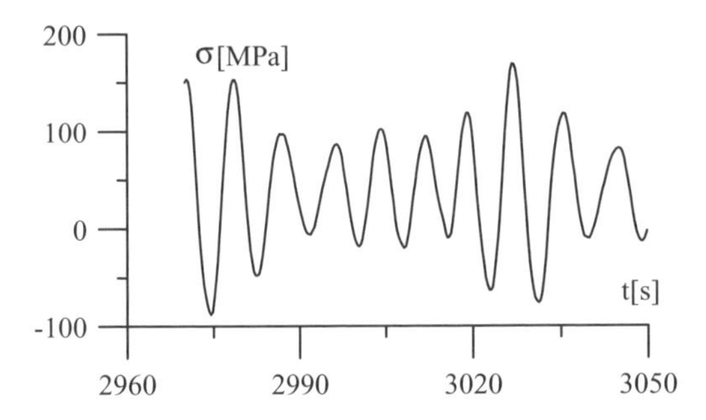

Figure 5. Time history of stresses σ_x in point A shown in Fig. 2, generated by irregular wave determined by: $H_s = 6.5$ m, $T_o = 7.5$ s and $\beta = 120$.

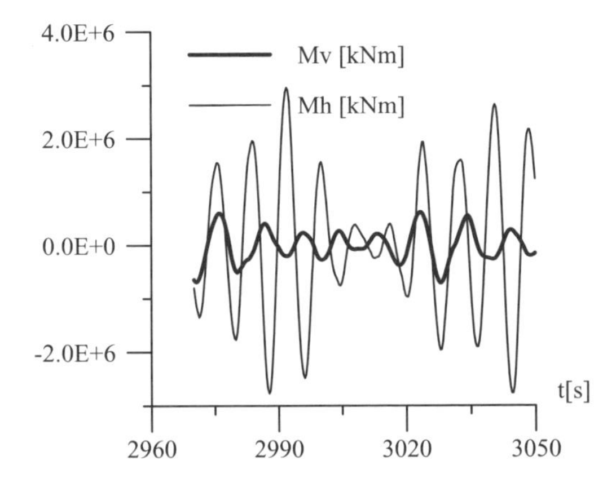

Figure 6. Time history of vertical M_v and horizontal M_h bending moments in the cross section at midship, generated by irregular wave determined by: $H_s = 6.5$ m, $T_o = 7.5$ s and $\beta = 120°$.

- the unit pressure p_{1i}, or p_{2i} at a panel corner and linearly distributed over the panel as shown in Fig. 3, and for unit bending moments M_l $l = 1, 2$ at the ends of the cross section of the hull structure module.

One set of coefficient W_{1i}, W_{2k}, W_{3l} for fix i, k, and l is determined for one panel "corner". Values of these coefficients are computed before performing the simulation of ship dynamics in waves.

The combination of unit pressure distributions for four corners of the panel multiplied by actual pressures results in linear distribution of actual pressure over the panel; further multiplied by influence coefficients values gives stress which is a contribution of such loaded panel to the total stress in the selected structure member. Summing the contributions of stresses, according to formulae (3), from all panels and bending moments gives the stress value in the structure member. This procedure is used in each time step of the simulation.

Computations of the stresses generated by the loaded panel are appropriately organized in the computer program to reduce the time of computations. During the simulation, the actual values of pressures, accelerations and moments: p_{1i}, a_{2k}, M_l, are changing with time. Applying the above presented procedure the stress values in the considered structure member is simulated.

For example, the time history of stresses in the structure member A of the first ship (Fig. 1) are presented in Fig. 5.

An example of the time history of vertical $M_v(x, t)$ and horizontal $M_h(x, t)$ bending moments in the cross section $x = 0$ of the first ship are

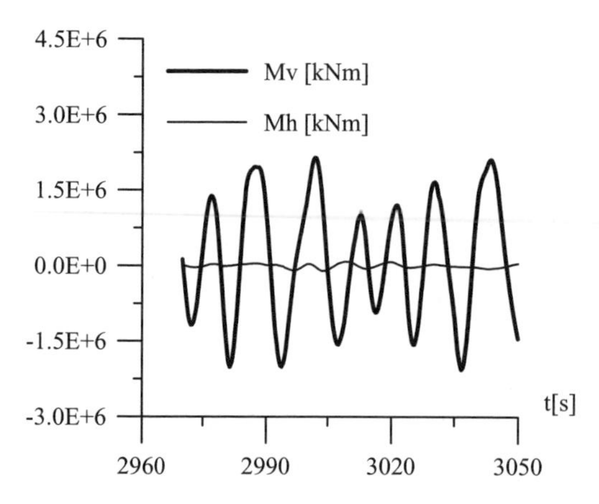

Figure 7. Time history of vertical M_v and horizontal M_h bending moments in the cross section at midship, generated by irregular wave determined by: $H_s = 15.5$ m, $T_o = 11.5$ s and $\beta = 180°$.

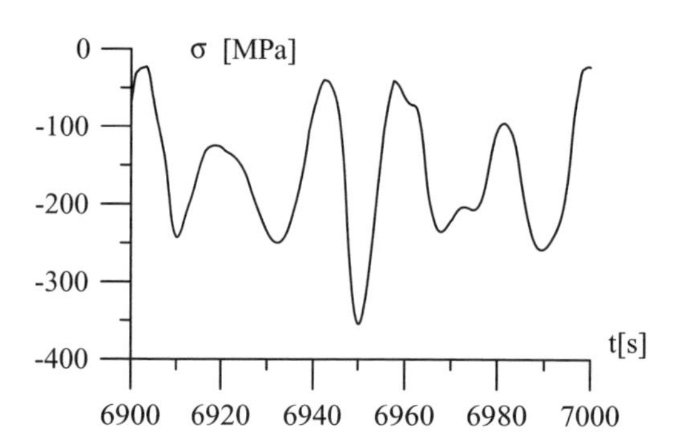

Figure 8. Time history of stresses σ_z in point E shown in Fig. 2, generated by irregular wave determined by: $H_s = 15.5$ m, $T_o = 15.5$ s and $\beta = 180°$.

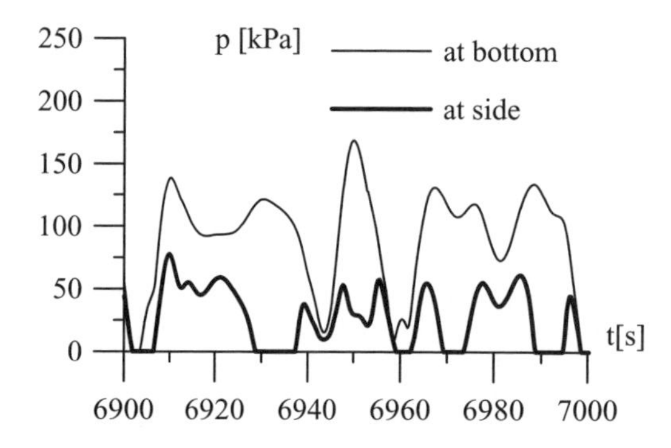

Figure 9. Time history of wave generated pressure p at the bottom and at the side (bold line), generated by irregular wave determined by: $H_s = 15.5$ m, $T_o = 15.5$ s and $\beta = 180°$.

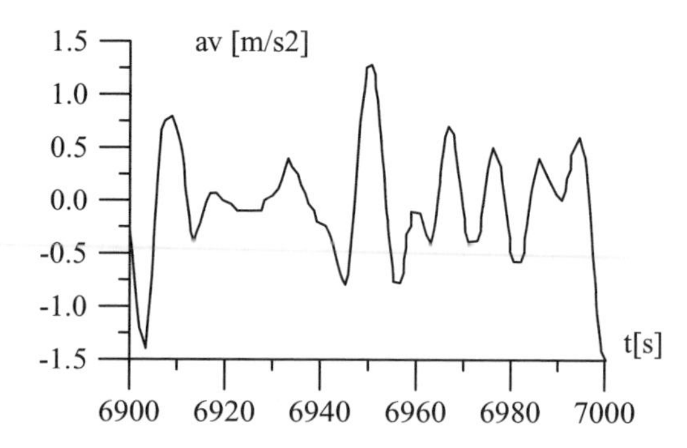

Figure 10. Time history of acceleration a_v in loaded aft hold, generated by irregular wave determined by: $H_s = 15.5$ m, $T_o = 15.5$ s and $\beta = 180°$.

presented in Fig. 6, and Fig. 7. The presented moments are wave generated bending moment.

The next example, the time history of stresses in the structure member E of the first ship (Fig. 1) are presented in Fig. 8.

The time history of wave generated pressure p at the bottom in the ship symmetry plane and at the ship side; and wave vertical acceleration a_v acting on the cargo in the aft hold are shown in Fig. 9 and 10.

The time intervals in the figures, for which the simulations are presented, contain instant t_{max} of maximum/minimum value of σ_{max}. This value is the maximum/minimum value of all sea states (for all simulations).

5 PROBABILITY DISTRIBUTIONS AND THEIR PARAMETERS

The numerical long-term probability density functions of stress σ (as the random variable) in the considered structure member were computed according to formula (2) taking into account the following assumptions:

$p_m = 1$, as only the alternate loading condition of the ship was considered;

$p_l = 1$, as according to the IMO Goal-Based Standards the areas A_l. $l = 1$, should cover only the North Atlantic area;

p_k = uniform probability distribution in the interval [0, 2π], representing the ship course in relation to waves in the sea area A_l was used in the computation of the long-term probability density functions;

p_{ij} = the probability of sea state occurrence in the North Atlantic; IACS Rec. 34 was used in the computations of the long-term probability density functions;

f_{ijk} = the short term probability density function of random variable σ is approximated by a step function obtained by dividing the number of extremes (separately one minimum and one maximum in a stress cycle) belonging to the set interval of stress $\Delta\sigma$ by the total number of extremes occurring in a sea state (represented by one its realizations—an irregular wave) and by the length of the interval $\Delta\sigma$.

$$f_{ij}(\Delta\sigma) = \frac{n}{N|\Delta\sigma|} \tag{4}$$

where n = number of maximums or minimums in the stress interval $\Delta\sigma$; N = total number of maximums or minimums in the time of simulations of one realization of stresses in a chosen structure member in a given sea state; one stress maximum or minimum is per one stress cycle.

The examples of numerical long-term probability density function, computed according to (2), are presented in Fig. 11, Fig. 12 and Fig. 13.

The probability density function of stress σ in the considered structure member enables the computation of its characteristic values, for example, the stress value for the given probability (e.g., 10^{-8}) of exceeding this stress value. The process of computation also enables the selection of:

- the sea state defined by (H_s, T_o) and
- the angle between ship course and wave propagation (Fig. 14);

from all sea states and ship courses accounted for in the simulations that demonstrate the extreme value of stress σ in the considered structure member. They are called the "extreme load cases". All together 640 cases (sea states and ship courses) were used in the computation process .

The results of computations of the characteristic values of long-term probability density function of stress σ in a selected structure members for two bulk carriers of panamax size of different designs

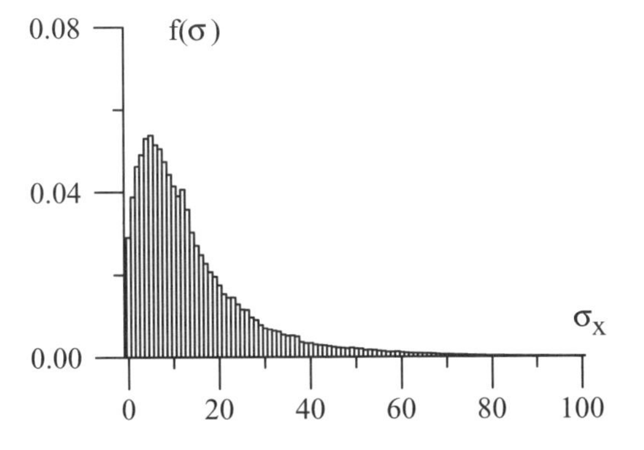

Figure 11. Probability density function of stresses σ_x in structure member A shown in Fig. 2.

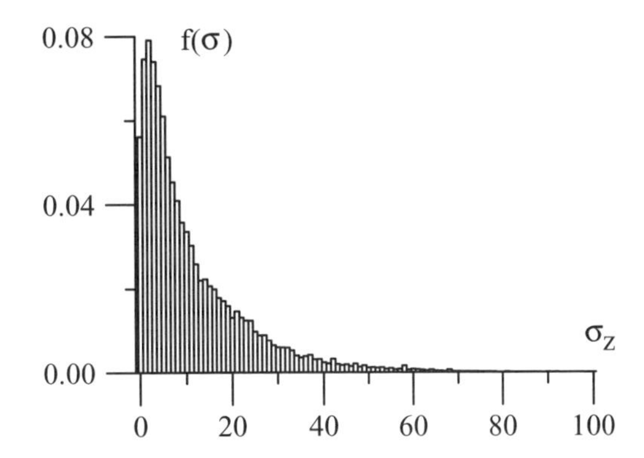

Figure 12. Probability density function of stresses σ_z in structure member E shown in Fig. 2.

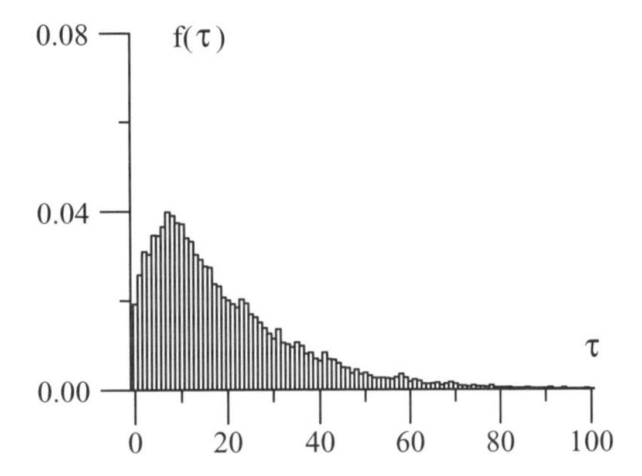

Figure 13. Probability density function of stresses τ in structure member C shown in Fig. 2.

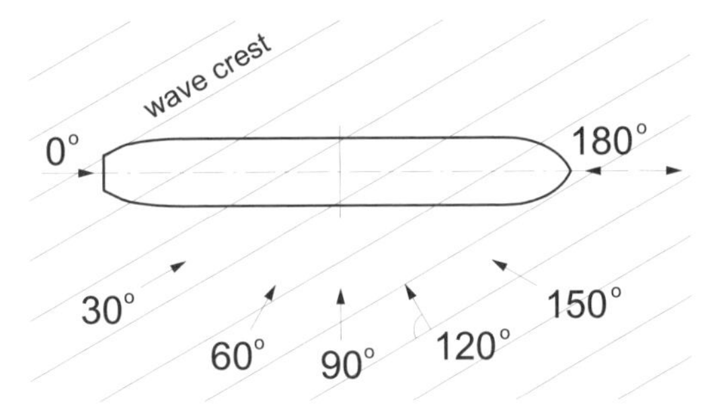

Figure 14. Definition of the angle between the ship course and wave propagation.

are presented in Tables 1 and 2. The "extreme load cases" are also presented in these tables.

The following denotations are used in the tables: σ_{pred} = the stress which can be exceeded with the probability equal to 10^{-8}; σ_{still} = the stress in still water; T_a = the mean period of stress oscillations; β = the angle between the ship course and wave

Table 1. The stresses in selected structure members and load cases for first panamax bulk carrier.

Struct memb.	σ pred [MPa]	σ still [MPa]	T a [s]	β [o]	H s [s]	T o [s]	t max [s]
A	167	41.27	9.74	120	6.5	7.5	3033
B	148	22.73	10.07	180	15.5	11.5	9925
C	−276	−2.37	10.63	180	16.5	12.5	8946
D	−57	−3.3	10.72	180	15.5	11.5	6772
E	−180	−139.68	11.48	180	15.5	15.5	6954
F	−111	−110.56	10.94	180	16.5	12.5	2791
G	95	103.12	10.51	180	16.5	12.5	4773
H	31	50.24	11.84	180	15.5	15.5	2877
I	−68	−73.74	11.91	180	15.5	15.5	9025
J	−72	−62.44	11.91	180	15.5	15.5	9025
K	13	27.4	10.7	180	15.5	15.5	9035
L	−63	−75.09	11.8	180	15.5	15.5	7990
M	−105	−79.33	10.03	180	16.5	12.5	1853
N	119	49.02	9.68	120	6.5	7.5	6804
O	−123	−25.93	9.83	120	6.5	8.5	2217
P	−128	−3.82	9.79	120	6.5	8.5	4260
R	71	5.66	10.47	180	16.5	12.5	1810
S	−83	−25.55	10.65	180	16.5	12.5	2791

Table 2. The stresses in selected structure members and load cases for second panamax bulk carrier.

Struct. memb.	σ pred [MPa]	σ still [MPa]	T a [s]	β [o]	H s [s]	T o [s]	t max [s]
A	233	129.43	9.89	120	6.5	8.5	3332
B	143	110.93	10	180	14.5	9.5	8385
C	193	3.46	10.64	180	14.5	9.5	3265
D	−73	−20.7	10.12	120	6.5	8.5	1073
E	−65	−78.51	10.43	120	6.5	9.5	5991
F	−65	−109.72	11.71	180	15.5	13.5	1477
G	12	28.76	11.15	180	15.5	13.5	2425
H	81	21.64	10.33	180	15.5	11.5	5433
I	−112	−170.06	10.82	180	14.5	9.5	3257
J	−85	−114.87	10.85	180	15.5	11.5	6459
K	−68	−25.2	10.54	180	16.5	12.5	4938
L	41	57.73	11.79	180	15.5	11.5	9062
M	−40	−81.85	11.4	180	15.5	13.5	7635
N	−19	−8.18	11.63	180	14.5	9.5	7346
O	−147	−84.72	10.2	60	6.5	8.5	2536
P	−151	−70.95	10.06	60	6.5	8.5	2536
R	−76	−86.28	10.94	180	16.5	12.5	5265
S	−63	−58.88	10.59	180	15.5	11.5	3353

propagation (Fig. 14); H_s = the significant wave height; T_o = the average zero up-crossing wave period; and t_{max} = the instant of time in which the maximum/minimum stress value $\sigma_{max/min}$ occurs in the sea state defined by (H_s, T_o, β)

The $\sigma_{max/min}$ value is the maximum/minimum value for all studied sea states (all simulations).

6 DISCUSSION OF THE RESULTS AND CONCLUSIONS

The paper presents a method of long-term prediction of stresses in ship structure. This method is based on:

- scatter diagram representing the probability of sea states occurrence in the North Atlantic;
- simulation of ship motion in waves and its structure response (stresses) to the waves;
- assumption that the probability distribution representing the ship course in relation to waves propagation is the uniform probability distribution in the interval [0, 2π];
- assumption that the master of the ship exercises good seamanship.

In the present class rules only the sea area of the North Atlantic—the area featuring the most severe wave conditions, is recommended to use to provide the safety margin.

150 sea states determined by the scatter diagram (IACS Rec. 34) and seven ship courses in relation to waves ($\beta = 0°, 30°, 60°, \ldots, 180°$) are taken into account in the long-term procedures of determining stresses in ship structure (symmetry is taken into account). All together 640 simulations for each of the two panamax type balk carriers (Fig. 1) have been carried out. This number is below 150 multiplied by 7 in the result of the assumed impact of the shipmaster's decision to change ship course in relation to the wave propagation in heavy sea states. For example, it was assumed that in the result of master decision for $\beta = 120°$ the significant wave never exceeds 6.5 m ($H_s <= 6.5$ m) and probability of ship meeting in such waving conditions was modified respectively. Similar modifications were made for other β values, showing that the limiting of H_s depends on β. No limitation on H_s was made for $\beta = 180°$.

However, such an approach cannot be used in design practice. Therefore, a selection of wave load cases, representing the long-term structure response to waves for certain class of ships (e.g., panamax bulk carriers), should be made. The computations presented in Tables 1 and 2 shows that:

- In the first ship (Fig. 1), built in year 1977, the highest stress values occur in the side of the ship: in frame flange (structure member *E*—Fig. 2) $\sigma_z = \sigma_{pred} + \sigma_{still} = -320$ MPa, and in side plate (structure member *C*—Fig. 2) $\tau = -278$ MPa. These values exceed the yield stress equal to 235 MPa and allowable shear stress equal to 135 MPa of the steel used. That ship sank in the year 2000, in the North Atlantic. The sequence of events causing sinking of the ship started from breach of the side.

- In the second ship (Fig. 1), built in year 2010, the highest stress value occurs in the deck plate at side: $\sigma_x = \sigma_{pred} + \sigma_{still} = -362$ MPa. This value exceeds the yield stress equal to 315 MPa of the tensile strength steel applied.

The case of stress in the deck plate at shipside (*A*–Fig. 2) shows that the highest stresses are generated in the sea state ($\beta = 120°$, $H_s = 6.5$ m, $T_o = 8.5$ s). This is the result of contribution of the horizontal bending moment M_h to the stress, much bigger than the contribution of the vertical wave and still water bending moments: $M_v + M_s$. In sea state ($\beta = 180°$, $H_s = 15.5$m, $T_o = 11.5$ s) horizontal bending moment M_h is negligible (see Figs. 6 and 7).

Analyses of the results of the computed strength ranges, presented above, shows that for panamax bulk carriers it is sufficient to take into account only two wave load cases: ($\beta = 120°$, $H_s = 6.5$ m, $T_o = 8.5$ s) and ($\beta = 180°$, $H_s = 15.5$m, $T_o = 11.5$ s).

The sea state ($H_s = 15.5$ m, $T_o = 12$ s) is one of the highest in the scatter diagram applied. If the equivalent regular waves of periods $T_o = 12$ s and $T_o = 8.5$ s are assumed then the length of the wave is equal to $\lambda = 1.56T_o^2/|\cos(\beta)|$, what in the first and the second wave case is approximately equal to the length of panamax bulk carriers (225 m).

However, the equivalent regular wave should only be used, assuming that λ is equal to the length of the ship to determine the parameters of the irregular wave (wave load case). These parameters are used to determine parameters of the irregular waves (T_0, H_s) and next to simulate of ship response to waves (stresses) and to assess the strength of midship module with the use of the Finite Element Method.

The simulation of ship structure response to irregular wave takes into account the shift phase between different loads (pressures, accelerations, bending moments) in a strict manner and do not require the application of combination factors, presented in binding rules of classification societies.

REFERENCES

Cummins, W.E. 1962. Schiffstechnik, Vol. 9, No. 47, *The impulse response function and ship motions*.

Guedes Soares, C. & Toixoira, A.P. 2000. Marine Structures 13, *Structural reliability of two bulk carriers designs*.

Hogben, N., Dacunha, N.M.C. & Oliver, G.F. 1986. BMT, Unwin Brothers Limited, *Global waves statistics*.

IACS Recommendation No. 34., 2000. *Standard wave data*.

Jankowski, J.A. 2006. Technical Report No. 52, Polish Register of Shipping, *Ship facing waves*, (in Polish).

Jankowski, J.A. & Bogdaniuk, M. 2007. Int. Symposium on Maritime Safety, Security and Environmental Protection, *Simulation of wave loads and dynamic stresses in ship hulls in irregular waves*.

Ochi, M.K. 1998. Cambridge Ocean Technology, Series no. 6, Cambridge University Press, *Ocean waves—the stochastic approach*.

Sustainable Maritime Transportation and Exploitation of Sea Resources – Rizzuto & Guedes Soares (eds)

Structural assessment of innovative design of large livestock carrier

J. Andrić, M. Grgić, K. Pirić & V. Žanić
University of Zagreb, Faculty of Mechanical Engineering and Naval Architecture, Zagreb, Croatia

ABSTRACT: The main results of the structural analysis of large livestock carrier have been presented. The most important aspects in the rational structural design of this kind of ships have been underlined. The ship structure is characterized with large openings in the superstructure side shell and with the absence of transverse bulkheads in superstructure part (zone above Deck 6). Ventilations tubes (channels) are integral load carrying part of the structure. Two main structural problems have been evaluated: (1) longitudinal strength, with the appropriate level of superstructure participation in the hull girder bending; (2) transverse/racking strength, where the capability of transverse structure has been evaluated. The strength calculation has been carried out according to RINA Rules using FE coarse mesh approach. Superstructure participation in hull girder longitudinal bending has been evaluated and longitudinal stress distribution over ship height has been analyzed for the prototype and the proposed model. Critical locations (where a fine mesh FE analysis is needed) with the high stress concentrations have been identified. Gain in structural weight has been achieved in the superstructure due to the rational redesign procedure.

1 INTRODUCTION

The extensive superstructure characterizes some ship types: cruise ships, passenger ferries, RoPax ships, car carriers, etc. Examined livestock carrier can be classified as a ship with a strong hull-superstructure interaction. The height of the superstructure is approximately equal to the height of the lower part of the hull and its influence on longitudinal strength of the ship is very important. Thus, it should be taken into consideration during a design and dimensioning of the structural elements.

A term strength deck, which is obvious for ships with a single deck, in this case is obsolete. The design of the structural elements on the assumption that only the lower part of the hull participates in the longitudinal strength is not rational and it could cause an excessive scantlings of structural elements in the lower part of the hull. The assumption that the superstructure is 100% effective and that a classical Bernoulli beam could be applied for estimation of primary stresses also is not valid for this type of the structures. Nonlinear distribution of the primary stresses in the hull is caused by many factors such as reduced shear stiffness of the side shell due to large openings, a shift of the side wall line of the hull or superstructure due to recess, the free end effect of short superstructures, influence of length, breadth and height of the superstructures, etc. For the rational structural design of these kinds of vessels, the correct primary stress distribution amidships along the height of the cross section (lower part of the hull + superstructure) should be determined. Scantlings of structural elements around midship calculated in the early design phase have a huge impact on the further process of the project. The effective superstructure design is very important because of a regulation of weights and vertical position of gravity, due to speed and stability requirements (Andric 2007). Relatively large ship's height (hull + superstructure) could cause a large reduction of the primary stresses in upper lower hull decks if the superstructure only partially participates in longitudinal strength of ship.

Also, this kind of ship is characterized with the absence of transverse bulkheads in the superstructure part, so special care has to be focused also on transverse strength (racking) calculation due to non-symmetrical load cases.

Nowadays, most suitable and accepted method for final checking of structural adequacy of ships with large superstructures is the 3D FEM coarse mesh model of a complete ship.

In the cooperation with the ULJANIK Shipyard, Pula, the University of Zagreb, Faculty of Mechanical Engineering and Naval Architecture has carried out the full ship 3D FEM analysis of the complete hull of Yard number: 486–487 designed in ULJANIK Shipyard (Zanic et al., 2009). Presented research represents continuations of previously presented work, see (Andric et al., 2006), on similar ship already built by ULJANIK (Yard no. 428) and long-term effort in structural design improvement of ULJANIK vessels.

The objective of the analysis was to investigate the structural strength in fulfillment of the requirements for direct calculation of the classification society Registro Italiano Navale (RINA Rules 2008b), including: (1) longitudinal strength analysis, (2) double bottom analysis, (3) transverse strength analysis, (4) detail stress analysis of structure around openings in deck 6, side shell, etc. The main results of those investigations have been summarized through this paper. MAESTRO software (2007) was used for calculation of the structural response of the full ship 3D FEM model.

2 MODEL DESCRIPTION

2.1 *Vessel description*

The vessel is a LIVESTOCK CARRIER 24 000 sq. meters with the following principal dimensions:

Length overall	185.20 m
Length between perpendiculars	169.26 m
Breadth moulded	31.10 m
Depth moulded to DECK no. 6	14.53 m
Depth moulded to DECK no. 10	24.33 m
Draught design moulded	7.70 m
Draught scantling moulded	8.70 m
Deadweight at design draught	10800 t
Deadweight at scantling draught	14700 t
C_B at d = 8.70 m	0.54
Main engine MCR	MAN B&W 7S50 MC-C
Output	116200 kW/127 r.p.m.
Speed trial (90% MCR)	19.8 knots

Ship has 10 decks and a deck house. The pen and access arrangement defines the layout of ordinary and primary pillars in transverse direction. The arrangement is based on *AMSA* (*Australian Maritime Safety Authority*) Rules: the breadth of a single pen is limited to 4.5 m and minimal access breadth between stores for food and water is 0.7 m. Based on these constrains 6 pens (layout: 1 + 2 + 2 + 1) and 3 access are arranged in transverse direction.

The ship is longitudinally framed with the exception of ship sides above Deck 1, some areas within the engine room, fore and aft peak structures. The ship structure is characterized with large openings in superstructure side shell and with the absence of transverse bulkheads in superstructure (zone above Deck 6). Large ventilations tubes (channels) are integral load carrying part of the structure. Two RINA approved material qualities were used:

Steel Grade NSS min. yield stress = 235 N/mm^2
Steel Grade AH36 min. yield stress = 355 N/mm^2

Figure 1 shows an outline of the general arrangement of the vessel.

2.2 *Mathematical model and boundary conditions*

In the view of the non-symmetrical structure and racking loads, the entire full ship FEM model was performed to give deflection, stress and adequacy results and simultaneously provide boundary conditions for further fine mesh analysis of all critical areas. The coarse-mesh, full-asymmetric, global livestock carrier FE model for Yard no. 486 was developed, see Figures 2 and 3. FE modeling details are as follows:

a. Plated areas such as decks, outer shell and bulkheads were represented by the special stiffened shell macroelements. MAESTRO stiffened macroelement uses the NASTRAN type QUAD4 4-node shell elements enhanced with stiffeners in their proper geometrical position regarding axial and bending energy absorption / detailed stress output.
b. Triangular 3 node elements were also applied with appropriate thickness.
c. Smaller primary transverse frames or girders were modeled with bracketed beam macro-elements (where beam theory is applicable).
d. Larger web frames or girders were modeled with elements as ad (a) or (b).

Full ship 3D FEM model had 52 827 nodes and 77 722 macro-elements (stiffened panels and bracketed beams). In general, ventilation ducts (tubes) were modeled with the eight plate segments and are penetrating respective decks with a 'collar' of plate elements. Element thicknesses and properties were implemented from the scantlings shown on the structural plans supplied by the Yard. Net scantling approach was implemented based on complete ship model in accordance with reference (RINA Rules 2008a), (see, Pt. B, Ch. 4, Sec. 2). Manholes (and similar) in the webs of primary

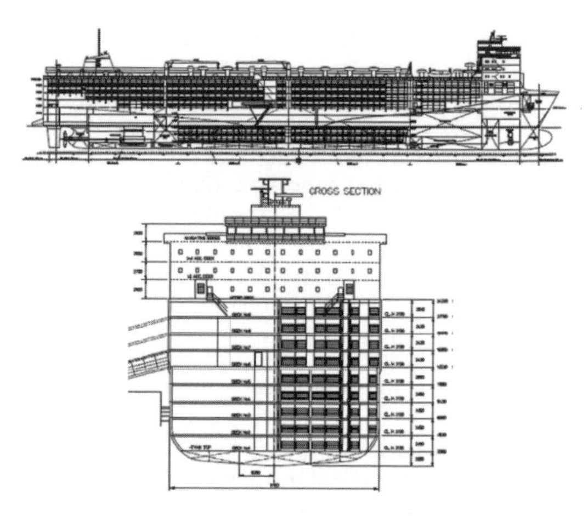

Figure 1. General arrangement of livestock carrier.

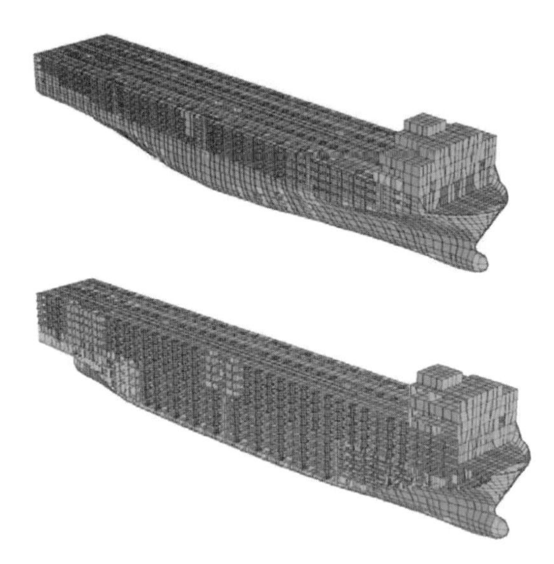

Figure 2. Full ship FEM model of livestock carrier.

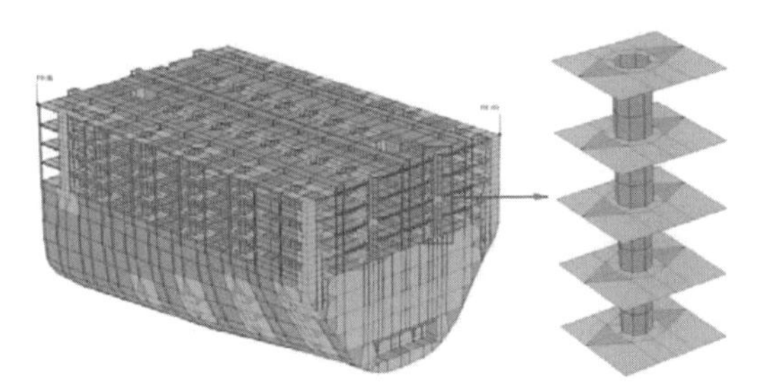

Figure 3. Part of the full ship FEM model—area between Fr. 96 and Fr. 151 and ventilation tubes detail.

supporting members (particularly in double bottom floors) were disregarded (RINA Rules 2008b) and element thickness has been reduced in required proportion.

To prevent rigid body motion of the model and singularity of the free ship stiffness matrix boundary conditions were implemented in accordance with RINA Rules (2008b).

Quasi static balancing of the full ship FEM model was performed using MAESTRO automatic balancing option (changing heave, trim and heel) to achieve permitted reaction force levels on the restrained nodes. Reactions obtained for the balanced ship in all load cases were below RINA requirements.

3 LOADING

One of the characteristics of considered ship are relatively fine hull lines (low block coefficient) and continuous distribution of lightweight which imply that the ship is always in hogging condition at still water i.e. it has excess of buoyancy amidships and excess of weights at the ends. Due to this distribution of static loads, the structure is loaded by very large still water hogging bending moment. A combination of maximum still water bending moment and maximum wave bending moment in hogging condition produces the maximum longitudinal stresses. A combination of minimum still water bending moment in hogging condition and maximum wave bending moment in sagging condition could generate a compression stresses in upper decks. This should be avoided at any cost because the compression stresses could generate panel buckling problems on the upper decks which mostly consist of very thin plates. For this type of ship it is not a case and the structure is always in hogging condition. Nevertheless, the deck structure is designed in such way that stiffened panels withstand compression stress of minimum 30 N/mm^2. The still water shear force distribution is similar to a conventional distribution with the maximum values at positions 0.25 L and 0.75 L from aft perpendicular. The design values of shear force are obtained by summing the maximum value of still water shear force with the maximum value of wave shear force. The shear force could cause large shear stresses on the ship side in the areas where the shear stiffness is decreased. One of the advantages of 3D FEM complete ship models is an avoidance of defining force/displacement boundary conditions at the ends which are necessary for partial ship models. The complete models should be balanced such as the reactions in points where the boundary conditions are placed are negligible. A total bending moment (static and wave) and shear forces are derived from static and dynamic load components distribution. Each loading condition comprises of four sets: (1) light ship weight, (2) deadweight, (3) accelerations and (4) buoyancy loading including dynamic pressures.

Static load of an ideal structure is required to be enlarged and adjusted to the lightweight of a ship according to the trim and stability book (T&S) for considered load case. Based on the model geometry and the scantlings of the elements used, a modeled mass for the hull was generated within the program. Additional masses of machinery and equipment were added to the hull mass by concentrated masses at the appropriate nodes. Those masses include superstructure items (other than directly modeled), wheelhouse, stern ramp, rudder and steering gear, bow and stern thruster, diesel engine and transmissions, boiler, main electric power production, fooders. Cargo loading consists of masses of cattle and food. The distribution of cargo loading was defined by Yard in [6]. Cargo masses were concentrated in mesh nodes of the corresponding decks. The masses of water ballast, fuel oil, fresh water etc. were taken from [6]. These masses were generally placed in appropriate tanks, which were modeled inside the ship structure model. The main engine and massive equipment are defined as concentrated masses on their actual positions. Contents of some smaller tanks were idealized by the concentrated masses. The mass distribution defined above was corrected to achieve the distribution given in Trim and Stability book (2009). Total light ship mass was 11 150 t.

The hull shape of the 3D FEM model slightly differs from the original one, i.e. a difference of a displacement up to 2% are acceptable. A hydrostatic pressure distribution is defined directly by the ship's draught and it is implemented automatically by the software.

Wave induced loads and 3D FEM full ship model implementation

Upon the modeling of wave induced loads it should be mentioned that it is generated with much more uncertainty than the structural model. For a practical implementation of the wave loads on the 3D FEM full ship model there are various methods which use direct calculation of the loads (Payer & Fricke, 1994). Design wave method (RINA Rules, 2008b) uses elements of deterministic approach in defining an equivalent design wave by which the FEM model will be loaded. This method was used in the presented work. The dominated load effect, i.e. the target value by which is loaded the FEM model, is the total vertical bending moment (static + wave) for hogging condition where the wave bending moment is calculated by the Rules.

Equivalent design wave is a regular sinusoidal wave (with Smith's effect included) with the following characteristics: length, height (twice the value of amplitude), phase which defines the position of a crest from the origin. It should be mentioned that due to the high nonlinearity of the hull form (deviation of a side shell from a normal above the waterline) at the ends the real wave bending moment is increased in hogging condition and decreased in sagging condition. Correction factor for hogging condition, regarding linear wave bending moment can be considered for this ship type. The value of vertical wave bending moment calculated according to expression from Requirement S11 is not scaled with calculated correction factor but the total value of the vertical bending moment is reduced according to the RINA requirement (2008b). Acceleration components were also calculated according to Rules (2008a) and implemented in the model as an acceleration vector.

Set of critical load cases is selected aiming to maximize dominant load effects having dominant influence on the strength of some part of the structure, Table 1.

Table 1. Load cases description.

LC	Type of analysis	M_{SW}	M_{wave}	M_{Tot}	M_{FEM}
	MNm		MNm	MNm	MNm
1	Long. Strength	1187	939	1832	1910
2	Long. & Trans. Strength	1187	939	1832	1870
3	Long. Strength	570	−1177	−483	−482
4	Double Bottom	1146	939	1792	1815
5	Double Bottom	1136	939	1782	1760
6	Double Bottom	1136	939	1782	1780
7	Trans. Strength	1136	939	1371	1400
8	Trans. Strength	1136	939	1371	1410

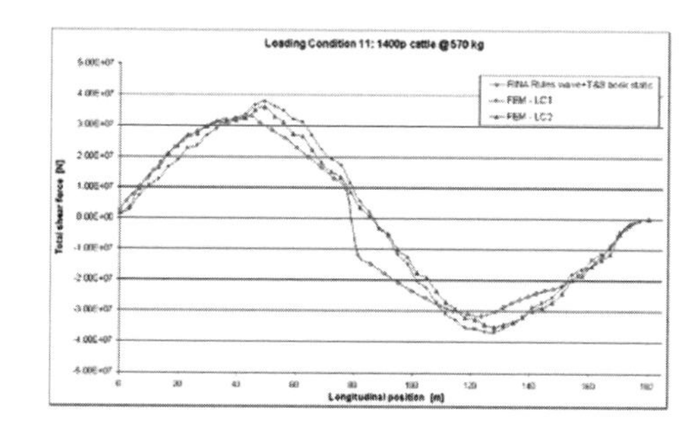

Figure 4. Shear force distributions for LC1 and LC2 and comparison with RINA requirements.

The wave bending moment and the total bending moments were calculated according to RINA Rules (2008a). Implemented distribution of the total vertical bending moment and shear forces shows good agreement with the required/target values, see Figure 4, for LC1 and LC2.

For transverse strength calculation two load cases "b" and "d" were selected. For the transverse strength calculation in the upright condition, the load case LC3 (RINA "b" case) was based on load condition 11 (14000 Cattle). It was chosen according to T&S book due to higher loads on the decks. LC8-IMO full load (Loading Condition 4 according T&S book) was chosen, for the inclined LC1 and LC2 (RINA "d" case), due to high GM that causes the highest transverse acceleration vector.

4 RESULTS

4.1 *Longitudinal strength analysis*

The full ship 3D FE model provided results for the global deformations, the effectiveness of upper decks, the distribution of longitudinal stresses at each level, etc. as well as the boundary conditions for the fine mesh analysis of the critical details. The model used here, for the RINA analysis specified in RINA Rules (2008a), was sufficient for all those purposes. The selected load cases represent the approximation of the extreme condition in accordance with the RINA requests. All conclusions were based on the selected load cases. The behavior of the ship's structure in terms of the global deformation is considered satisfactory from the structural aspect in all loading conditions considered.

Results obtained, confirmed the design assumptions that all continuous superstructure decks participated in the global hull girder bending, see Figures 5 and 6.

The implemented concept (partially effective superstructure) present an opportunity to reduce structural scantlings of superstructure decks (decks above D6) which are affected by the RINA requirements to increase cross section modulus by increased structural scantlings in superstructure decks. Based on RINA request, sensitivity analysis regarding deck plate thicknesses of deck 7 and 8 has been performed. Very low influence was identified. Stress differences are below 10 N/mm^2 due to the increased plate thickness of deck 8, from $t_p = 8$ mm to $t_p = 11$ mm. It has been verified that the level of normal σ_x stresses for proposed lightweight solution satisfied RINA requirements. Maximum normal σ_x stresses were below +160 N/mm^2 (connection of DECK 6 and the outer shell) and −160 N/mm^2 (bottom structure) in region 0.3 L—0.65 L for LC1–3.

Special attention has to be taken to solve the stress concentrations problems that have been identified in several locations:

- Connection of ventilation tubes and pillars with double bottom plating in aft and fore region and connection of ventilation tubes and pillars with deck 6, see Figure 7. Also, rational selection of pillar profiles has to be taken due to bending (and not only axial stress) that occurs in specified position. Cross-like type of profile has been substituted with tube/ rectangular type of profile.
- Side openings in superstructure in fore part of the ship and amidships between deck 6 and 7.
- Large casings openings at deck 6.

4.2 *Double bottom analysis*

The fine mesh model, between Fr. 71 and Fr. 91, was developed and imbedded into global coarse mesh model, see Figure 8.

Two additional RINA load cases "a-crest" were implemented. Specifics of the examined double bottom structure are that pillars and ventilation tubes transfer cargo loading to the double bottom

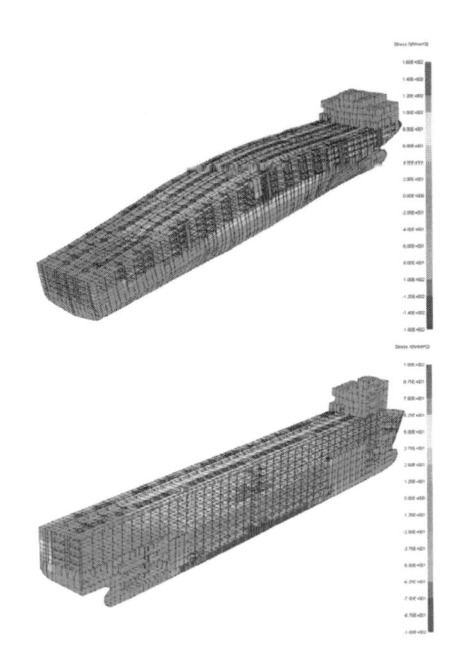

Figure 5. Deformed full and half ship FEM model and normal σ_x stresses for LC2.

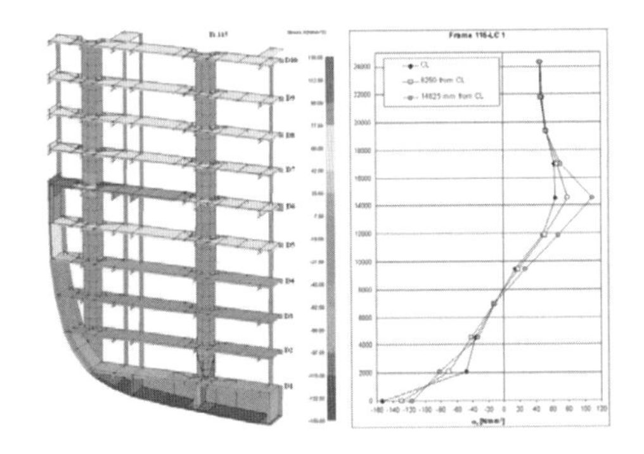

Figure 6. Normal σ_x stresses distribution over ship height at Fr. 115 for LC1.

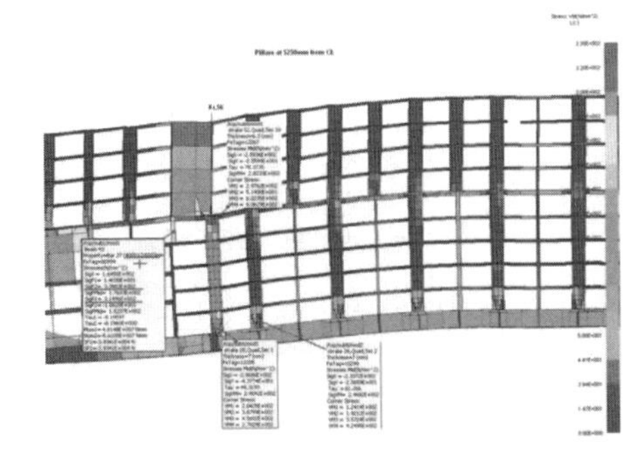

Figure 7. Maximum Von Mises equivalent stresses in ventilation tubes for LC1.

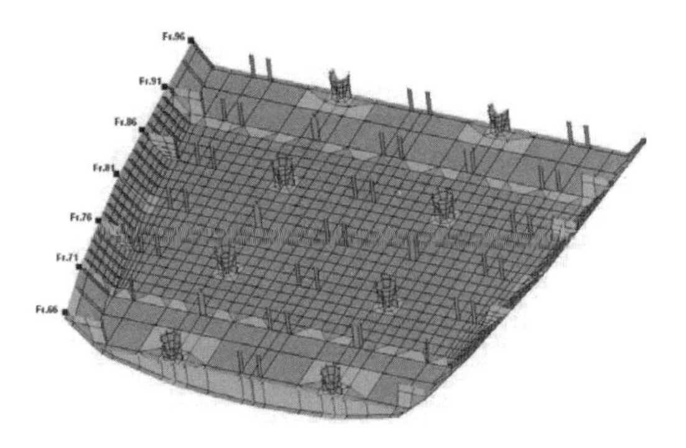

Figure 8. Fine mesh FEM model of double bottom imbedded in global coarse mesh model.

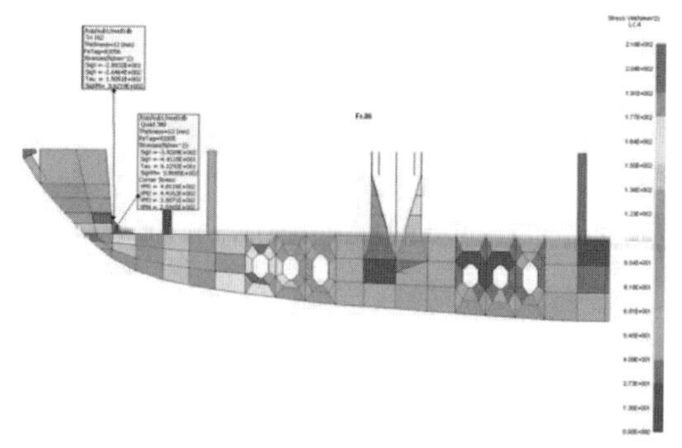

Figure 9. Maximum Von Mises equivalent stresses at Fr. 86 for ballast load case LC4.

and support floors w.r.t to external pressure loading. This fact makes ballast condition critical. Several locations with high stress levels have been identified and evaluated/redesigned:

- Higher stresses were recorded in the bilge region at the intersection of inner bottom plating, floor and partial transverse BHD supporting ventilation pipes (Fr. 76 and Fr. 86, see Figure 9).
- Very high stresses (at the free edge) were recorded in small brackets that connect the floor with the partial TBHD.
- Around manholes (closest to bilge) local stress levels are increased.

The comparable grillage (based on equivalent 3D beam model) calculation of double bottom structure using artificial boundary condition show more conservative results.

4.3 *Transverse strength analysis*

Transverse strength analysis has been performed using presented global coarse mesh model to analyze several critical structural parts.

a. Strength analysis of transverse bulkheads has been performed. The vessel has several waterproof transverse bulkheads from double bottom to D6 (Fr. 51, 101, 151 and 181). Fine mesh model was developed and imbedded into global coarse mesh model around specified transverse bulkheads. Model was balanced to achieve required transverse acceleration vector over the model height. Yield and buckling criteria were evaluated for the inclined "d" type load cases according to RINA Rules (LC7 and LC8).

   Several high stresses areas have been identified on all examined transverse bulkheads mainly in connection with ventilation tubes, partial casing bulkheads and strong web frames above deck 6. Several highly stressed details are given on Figures 10, 11 and 12. Certain changes

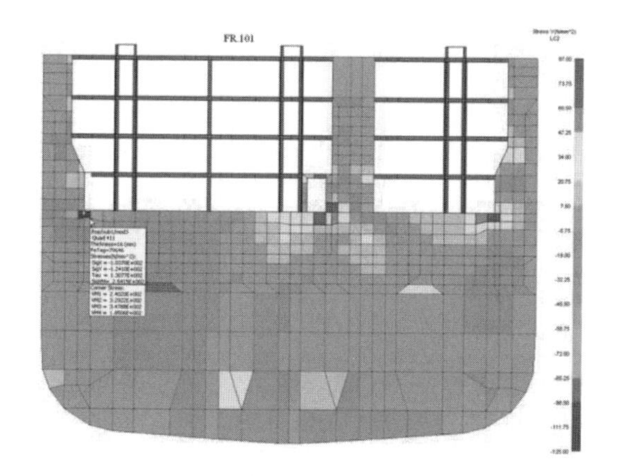

Figure 10. Normal σ_y stresses in transverse bulkhead at Fr. 101 for non-symmetrical load case LC8.

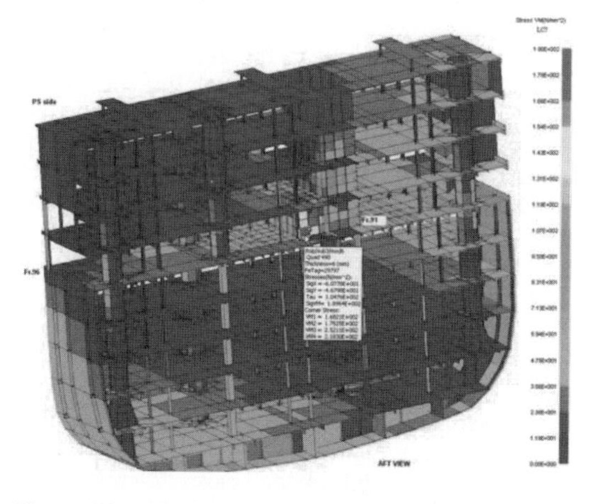

Figure 11. Maximal Von Mises equivalent stresses in partial bulkheads of ventilation trunk for non-symmetrical LC7.

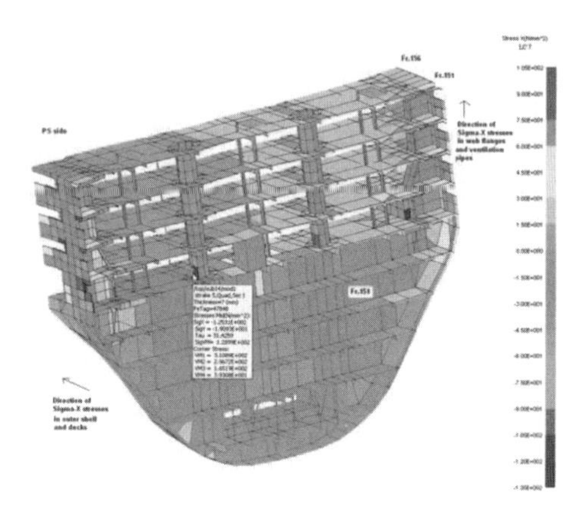

Figure 12. Normal σ_x stresses in transverse bulkhead at Fr. 151 for non-symmetrical load case LC7.

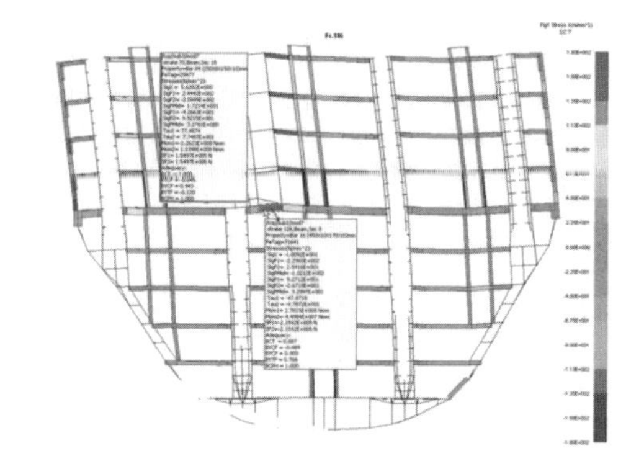

Figure 13. Maximum (axial + bending) stresses in transverse beams flange at Fr. 146 for non-symmetrical LC7.

of plate thickness have been implemented following shipyard suggestions to achieve stress level bellow RINA limits (Pt. B, Ch. 7, Sec. 1–3, 2008a). Detail FEM analysis of critical structural details have been suggested to support local design.

b. Strength of transverse beams and side frames have been evaluated in upright load case (LC2) and inclined load cases (LC7 and LC8), see Table 1. Critical locations have been identified and recommendations to solve identified problems have been suggested. Most critical locations for transverse deck beams are in connection to ventilation trunk bulkheads, ventilation tubes and pillars, see Figure 13.
c. Strength of pillars and ventilation tubes have been evaluated in upright load case (LC2) and inclined load cases (LC7 and LC8), see Table 1. Critical locations regarding achieved stresses have been identified in details. Changes in type of profiles of pillar systems have been suggested. Pillars (with cross-like cross section) due to high participation of bending stresses (in racking and longitudinal strength load cases) should be replaced with rectangular or tube cross section.

4.4 *Fine mesh analysis of critical details*

Several critical structural details have to be evaluated using very fine mesh FE models ($t \times t$). Those details very previously identified in the global-fine mesh model and presented through Ch. 4.1 to 4.3.

- Connection of web frame and transverse bulkhead;
- Connection of ventilation tubes and D6 plating;
- Connection of ventilation trunk and D6 plating;
- Structure around openings at D6 (amidships);
- Side openings above D6-transverse frame design.

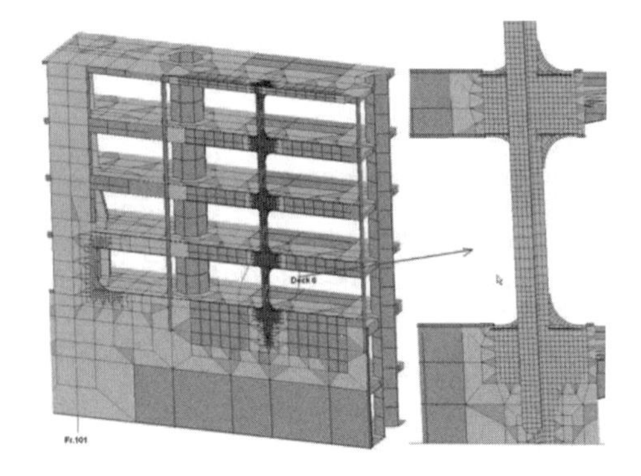

Figure 14. Very fine mesh of superstructure side around Fr. 101 imbedded into global FE model.

Very fine FE mesh (element size $t \times t$) has been developed to support local design of side frames of large side openings above deck 6, see Figure 14. Recommendations for solving those problems have been suggested.

5 CONCLUSIONS

Presented work represents the successful cooperation and joint work of Yard and Faculty design teams as an example of modern procedure in rational structural design. It also represents a progress in developing livestock carrier structural concept as a specific (tailor made) type of vessel offered by ULJANIK shipyard. The conclusions are as follows:

- Only the full ship FEM model is capable to simulate realistic 3D effect of hull/superstructure interaction without the restricting assumptions. Effectiveness of the superstructure in longitudinal strength was analyzed and efficient redesign was developed.

- Sensitivity analysis, based on parametric investigations of different scantlings of the superstructure, has provided the designer with rational arguments regarding benefits/drawbacks of the selected designs.
- Complex transverse strength problems in non-symmetrical load cases, using presented models, can be rapidly solved and provides the head designer with the rational basis for determination of the final design scantlings.
- Pillars have to be rationally designed not only for axial loading but also regarding bending stresses.

Fine mesh and very fine mesh FE analysis was found to be very efficient way for solving the stress concentration problems on previously identified critical locations.

ACKNOWLEDGEMENT

Thanks are due to the long-term support of Croatian Ministry of Science, Education and Sports: projects 120-1201829-1671. Thanks are due to all members of the *OCTOPUS group* (www.fsb.hr/octopus) and to the design team of the *ULJANIK shipyard* (www.uljanik.hr) for fruitful and long term cooperation

REFERENCES

Andric, J. 2007. Decision support methodology for concept design of the ship structures with hull—superstructure interaction, *PhD thesis*, University of Zagreb, Croatia.

Andric, J., Zanic, V. & Grgic, M. 2006. Superstructure Deck Effectiveness in Longitudinal Strength of Livestock Carrier, *XVII. Simpozij Teorija i praksa brodogradnje, Proceedings of SORTA 2006*, pp. 525–540, Opatija, Croatia.

ISSC, 1997. Technical Committee II.1. Quasi-Static Response. Proc. of the 13th International Ship and Offshore Structures Congress Trondheim, Norway.

MAESTRO Version 8.7.6, 2007. Program documentation, DRS Technologies, Stevensville, MD, USA.

Payer, H.G. & Fricke, W. 1994. Rational Dimensioning and Analysis of Complex Ship Structures, *SNAME Transactions*, Vol. 102, pp. 395–417.

RINA rules for classification of ships, newbuildings, 2008a. Part B, Hull and Stability.

RINA RULES, 2008b. Part B, Ch. 7, Appendix 3: Analyses Based On Complete Ship Models, 2008.

Trim and Stability book with Longitudinal Strength Calculation, 2009. Yard no. 486–487, Uljanik, 1 103 102 A.

Zanic, V., Andric, J., Stipcevic, M., Grgic, M. & Piric, K. 2009. Full ship FEM Analysis of Livestock Carrier, *Yard. no. 486–487, Technical Reports ULJ 486- 487/1÷5*, University of Zagreb, Croatia.

Sustainable Maritime Transportation and Exploitation of Sea Resources – Rizzuto & Guedes Soares (eds)
 ISBN 978-0-415-62081-9

A Collision study of a large supply vessel hitting a ship shaped FPSO

Gabriele Notaro, Tom K. Østvold & Eivind Steen
Det Norske Veritas, Norway

Narve Oma & Jon Kippenes
Statoil ASA, Norway

ABSTRACT: Rules and regulations for assessing residual hull strength of ships after accidental collision or grounding events exist but they are all per date linked to voluntary Class Notations outside minimum Class and IACS unified requirements. Within the area of offshore exploration and field operations, focus has been on Accidental Limit State (ALS) scenarios for decades. DNV offshore standards and NORSOK consider collision as a design load case in general which the offshore unit shall resist without jeopardizing the overall safety, w.r.t stability, hull girder strength and pollution. ALS assessments are normally carried out according to simplified plastic collapse models of plates, stiffeners, girders/floors and hull girders etc giving reasonable estimates. However today's commercial non-linear FE programs have reached a high level of robustness and efficiency which makes them very attractive for performing comprehensive assessment of the damages that can be related to the residual hull girder strength and damaged stability. Such a tool is used in the present study simulating a collision between an Offshore Service Vessel (OSV) and a ship shaped FPSO. The focus is on bow impact scenarios estimating when the OSV hits and penetrates into the FPSO side shell. The variation of energy absorption and damage extent has been studied as a function of different physical parameters.

1 INTRODUCTION

A bow impact collision scenario between a double hull ship-shaped FPSO vessel and a large modern OSV has been analysed by using the Non-Linear finite element tool ABAQUS/Explicit. Rules and regulation covering such subject are given by DNV (2008a; 2008b; 2010) and NORSOK (2004; 2007).

The calculations are based on conservative assumptions with an infinitely rigid striking OSV bow hitting the FPSO normal to the side shell at the mid-span between two frames/bulkheads.

The impact elevation relative to the base line depends on the considered loading condition. In this study three different vertical conditions have been analysed, considering the possibility that the relative motion of the vessels in waves can cause different impact scenarios. Two impact speeds have been considered in the assessment and the effect of a 45 degrees impact angle has in addition been analysed for worst damage.

A very refined cargo hold model of the FPSO has been modeled in ABAQUS and the colliding supply vessel is represented as a rigid body. The impact forces and response in the hull girder are evaluated on an initially stress free hull girder, i.e. the loads and deflection induced by the operating condition are not accounted for.

Two different calculation methods have been applied for the investigated cases. In the first approach, the collision has been analyzed considering a case where the striking vessel is forced to move with constant speed during the impact. This case is used to estimate an energy absorption curve and, based on an energy consideration, the final indentation and damage can be estimated for different impact speeds.

The second approach accounts for the natural speed reduction of the OSV during the collision as given by Newton's second law of motion and is used as validation of the forced indentation approach.

2 FINITE ELEMENT MODEL

A detailed model of the mid-ship area of the FPSO has been created in ABAQUS, Figure 1.

The FPSO will be impacted by the OSV, which is considered as a rigid body. This is a conservative assumption since all the impact energy is only absorbed by deformation of the struck vessel.

2.1 *The struck vessel-FPSO*

A detailed model of the struck vessel, including all structural details such as man holes, stiffeners,

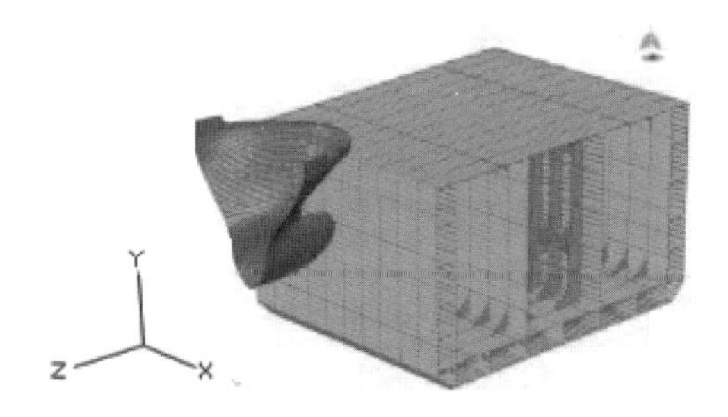

Figure 1. View of the ABAQUS model.

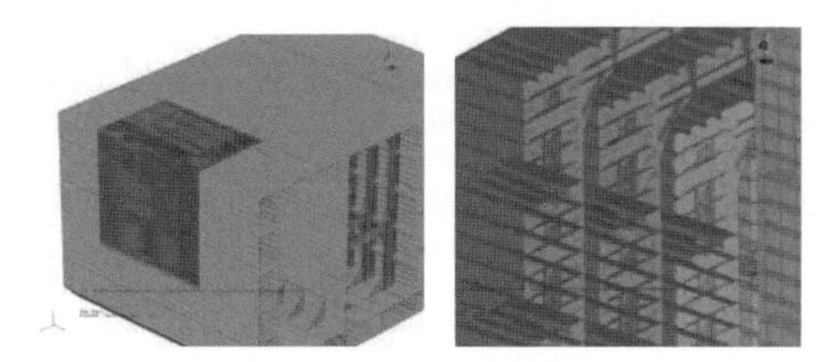

Figure 2. Fine mesh area and model detail.

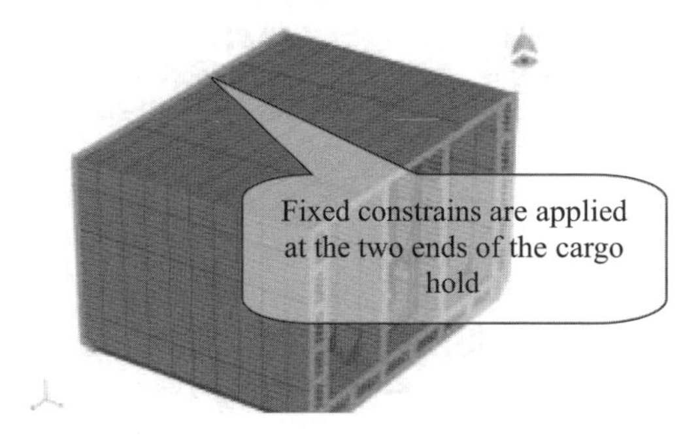

Figure 3. Boundary condition applied to the cargo hold.

flanges and brackets is required to describe the impact resistance, Figure 2. The mesh refinement has to be fine enough to describe correctly the local displacements and capture the folding, buckling and rupture of the impacted members. A typical mesh size in the impact area has to be in the order of 4–5 times the thickness.

The cargo hold has been constrained at the two ends as indicated in Figure 3.

The horizontal global bending effect is not included since it is expected that the collision phenomenon is mainly governed by local plastic behaviour and the overall hull deflection induced by the impact is considered to be of minor importance.

The scantlings are modelled according to the "as built" drawings, i.e. no corrosions margins are deducted. Figure 5 shows a part of the mid-ship section in the impact area. The width of the double side is 3 m.

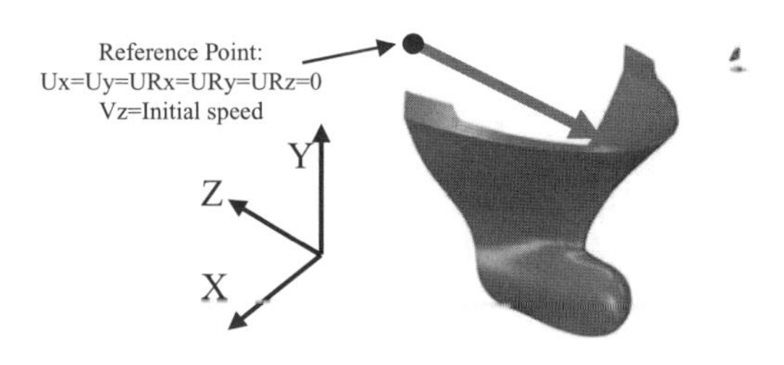

Figure 4. Boundary condition applied to the striking vessel.

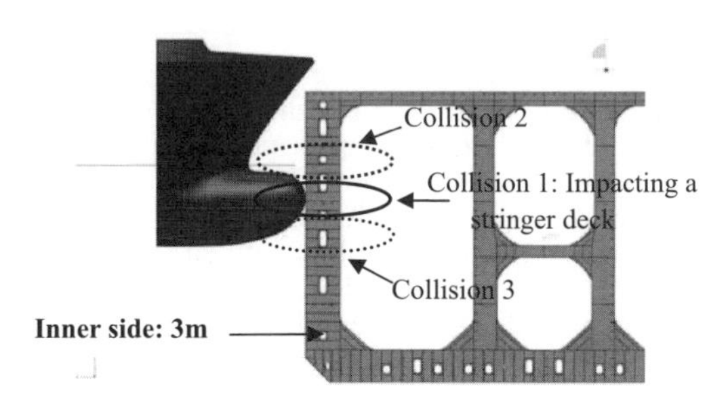

Figure 5. Vertical impact locations.

2.2 *The striking vessel- OSV*

The colliding vessel is represented by a rigid bow. The mass of the vessel and its added mass are explicitly considered only in the cases when the speed reduction during the collision is accounted for. The boundary conditions, including the initial speed, are applied to a reference point that describes the motion of the rigid body. The OSV is constrained to move along a line and the external kinematics, such as vessel rotations, are neglected.

The initial speed and the total mass, including the added mass, will define the initial kinetic energy that has to be dissipated as strain energy during the collision, Equation 1. According to DNV (2008a) the kinetic energy used to define the impact load should not be taken less than 11MJ. The basis for this figure is a supply vessel of 5000 tons moving with a speed of 2 m/s. For bow and stern impacts the added mass is specified in DNV (2008a) and NORSOK (2004) as 10% of the vessel displacement.

$$K_{OSV} = \frac{1}{2}(m_{OSV} + m_{add})v_{OSV}^{2} \quad (1)$$

In the present assessment an OSV with a bulbous bow and a displacement of 8200 tons has been considered. This gives the collision energies summarised in Table 1.

Table 1. Initial kinetic energy-OSV.

Mass [ton]	Added Mass [ton]	Speed [mm/s]	Kin. Energy [MJ]
8 200	820	2000	18
8 200	820	4000	72

3 COLLISION SCENARIOS

The OSV is assumed to hit the FPSO in the middle of two transverse frames. This position is expected to represent the minimum collision resistance of the ship side and will hence ensure that the analyses predict the most severe indentation for the considered initial impact energy.

The vertical location depends on the assumed draft, the loading condition and the relative vertical motions between the two vessels.

Three different vertical impact locations have been considered in this assessment as illustrated in Figure 5:

- Design condition-Full Load, representing the still water condition (*Collision* 1).
- Upper condition-representing a full load condition and relative motions due to waves; the impact location is shifted upwards (*Collision* 2)
- Lower condition—Representing an intermediate loading condition (*Collision* 3).

The effect of the waves on the ship motions is not explicitly accounted for. Instead a likely range of ship impact speeds (ca. 2 to 4 m/s) are considered and the corresponding damages are identified.

90 and 45 impact angles are considered in scenario 2.

4 MATERIAL MODEL AND FRACTURE CRITERIA APPLIED TO THE STRUCK VESSEL

The energy absorption in the FPSO is limited by the local buckling of the structural elements exposed to compressive stresses and fracture for the structural members exposed to tensile stresses. Therefore the FE model and the material model have to be suitable to describe the two behaviours.

When the impacted structural member is exposed to compressive stresses, it has to deform plastically from the onset of yield with subsequent large folding and buckling deformation of the structure. This is achieved by using a very refined mesh in the impact area.

The fracture is assumed to take place when the equivalent tensile plastic strain, Simulia (ABAQUS 6.9) exceeds a critical value (ε_C) defined by the user. The fracture criterion is important to establish. This topic is widely discussed in literature e.g. Hagbart & Amdahl (2008), Ehlers & Varsta (2009) and Tornquist (2003). In the present work the applied material curve has a tri-linear behaviour, and the strain hardening has been taken according to the minimum material requirements specified in DNV (2009).

By using the models for damage and failure simulation for ductile metals included in ABAQUS it is possible to model the material degradation and the progressive reduction in the load carrying capacity. The failure is simulated by linear degradation of the material stiffness over a failure elongation δ (Figure 6), after reaching the necking strain ε_{neck}. Necking is not captured by the mesh refinement used in the collision analysis model and it is therefore necessary to specify the strain at the onset of necking in addition to the elongation δ.

The elongation δ is for practical purposes confined to the area where necking occurs, and is therefore independent of the dimensions of the considered structural element. Hence the fracture criteria will also ensure mesh independent results as long as the mesh is fine enough to capture the stress distribution.

DNV (2009) specifies a minimum failure elongation of 22% over a measurement length of 50 mm. This includes the elongation associated with necking. To determine a failure elongation δ and necking strain consistent with the material specification a tensile test has been simulated using a very fine FE model capable of capturing the necking behaviour of the material. Figure 7 illustrates the considered material curve and correspondent the response with the onset of necking.

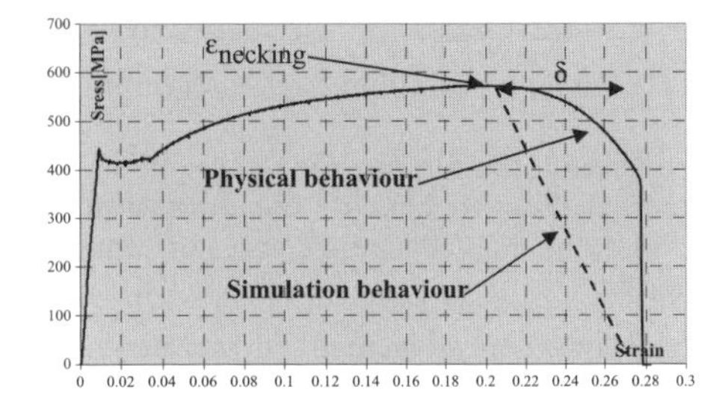

Figure 6. Illustration of material response curve.

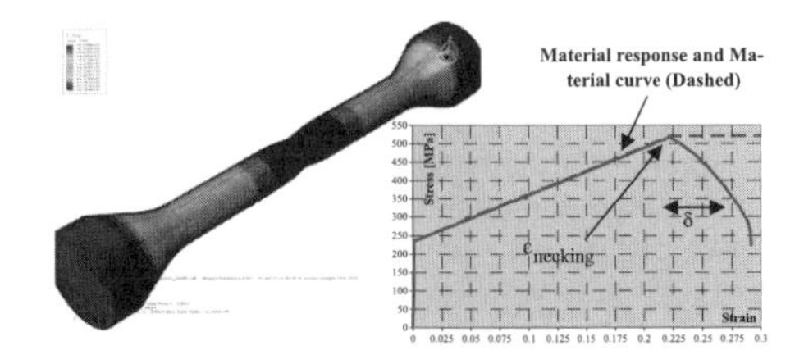

Figure 7. FE tensile test-Material curve and material response.

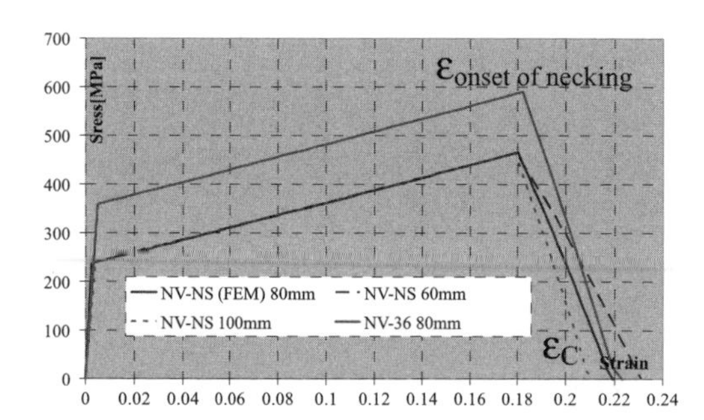

Figure 8. Local stress-strain response relationship.

The determined failure point is defined as the elongation at which the load carrying capacity is reduced to zero (ε_C). This will typically be in the range of 22% while the onset of necking starts at 18% as shown in Figure 8.

The established ABAQUS failure criterion defines the damage initiation when the equivalent plastic strain in an element reaches approximately 18%. After this point the material degradation starts (necking) and the load carrying capacity of the modelled material is linearly reduced until total fracture. When the strain reaches 22%, the element has failed and consequently is deleted from the hull structural model.

Figure 8 shows the expected mesh dependence of the failure criterion on a local level. The load carrying capacity is reduced to zero when the elongation after onset of necking reaches 3.2 mm, which results in a different critical strain for different element sizes.

5 ANALYSIS METHOD AND THEORETICAL CONSIDERATIONS

Two different solution methods are applied in the collision analyses to simulate the collision mechanism and the related energy transfer; these are defined as:

- *The constant speed method*: In this method the striking vessel is forced into the FPSO with a constant speed. A force is applied to the OSV in order to keep the speed constant.
- *The natural speed reduction method*: In this method the striking vessel is moving with an initial speed towards the struck vessel. The speed will decrease during the collision process due to the reaction forces developed in the deforming FPSO structure.

The natural speed variation approach simulates the dynamic response in real time, predicting the time required to stop the impacting OSV from a given initial speed and assessing the corresponding damage level in the FPSO ship side. The constant speed approach is a simplified method introduced for the purpose of possibly justifying using a quasi-static approach for assessing the damage level of the FPSO ship side valid for any initial collision speed. The different approaches are discussed in the subsequent chapters.

5.1 *Energy consideration*

The collision energy dissipated as strain energy may depend on the type of offshore installation as indicated in NORSOK (2004). Within the context of this paper the impact energy is taken as the initial kinetic energy of the OSV.

By running a constant speed case, a force is applied to the OSV and a corresponding external work (W_E) is done on the system. This work is absorbed by the FPSO structure as deformation energy and is used to estimate the impact energy absorption curve shown in Figure 9.

The energy absorption curves can be used to estimate the indentation of the striking vessel into the side of the FPSO. The dashed red line in Figure 9 represents the kinetic energy of the OSV previous to the impact. The OSV will stop when the absorbed energy is equal to the initial kinetic energy, i.e. when the external work (W_E) is equal to the kinetic energy (K_I). This is indicated in Figure 9.

When the speed variation is accounted for, the striking vessel is approaching the FPSO with a prescribed initial speed which, in this specific case, is 2 m/s. During the collision, the retardation of the colliding vessel is proportional to the generated impact forces, according to Newton's second law. In this method, no external work is executed on the system, but the absorbed energy can be monitored by considering the internal energy plus the frictional dissipation. These two contributions are approximately equal to the external work executed in the constant speed method as indicated in Figure 9.

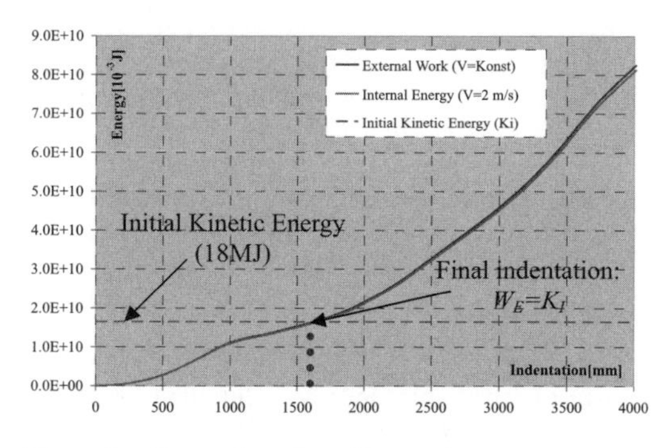

Figure 9. External work and internal energy.

The kinetic energy is mainly absorbed by the deformations of the struck vessel structure. A small fraction of the impact kinetic energy is dissipated by frictional forces during the sliding as illustrated in Figure 10, where the red line represents the frictional dissipation. Figure 10 shows that the kinetic energy reduction is balanced by the increasing internal energy and frictional dissipations. The total energy in the system remains constant.

5.2 *Impact load*

The impact load can be directly monitored by considering the resulting contact force. In the constant speed method, these forces are equal to the applied force required to maintain the constant OSV's speed. The impact force is a direct output from ABAQUS, but the dynamic of the system creates large high frequency vibration in the response, with high and unphysical peaks. A method for removing theses peaks is described in the following.

The work done by the impact forces is defined as follow, Equation 2:

$$W = \int \overline{F} \circ \overline{dU} = \int F_X du_X + \int F_Y du_Y + \int F_Z du_Z \qquad (2)$$

Since $du_X = du_Y = 0$ on the reference point, the only component to execute work into the system is F_Z. It follows that the impact force can be determined by differentiating the work with respect to indentation. An example of this application is indicated in Figure 11, which is obtained from a constant speed case. The impact forces calculated from the external work and from the internal energy are very similar. This finding will be used to monitor the impact forces when the speed variation is accounted for.

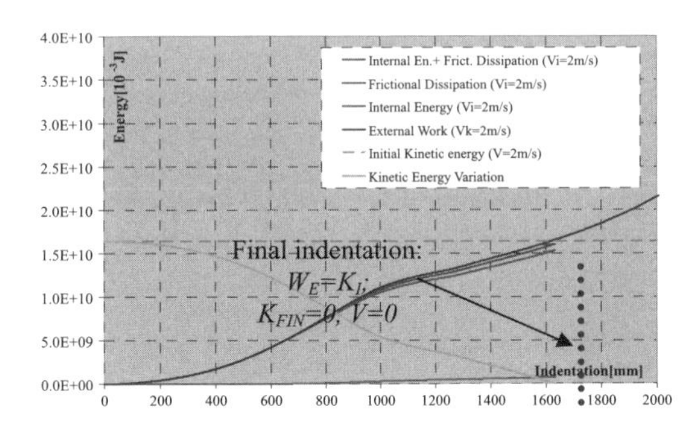

Figure 10. Energy Balance.

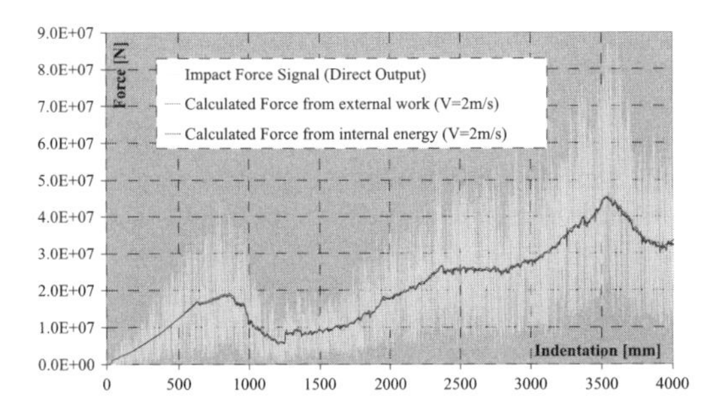

Figure 11. Vibrations in the impact force signal.

6 COLLISION MECHANICS

The impact force evaluated as indicated in 5.2 can be related to the physics of the collision.

The time variation of the impact force in Figure 12 can be associated with plastic deformation and fracture of the ship side and to the progressive damages illustrated in Figure 13 to Figure 16. In this section a constant speed analysis for collision scenario 1 is presented.

The side structure subjected to the impact starts to deform. The impact force increases until local buckling occurs in the stringer deck, Figure 13 and the stiffness of the structure changes (knuckle at 0.3 s in Figure 12). When the strain due to tensile stresses reaches a critical value, the side shell fractures as shown in Figure 14, the stiffness is reduced and the impact force decreases (Knuckle at 0.45 s in Figure 12). The impact force at the side shell failure is 18 MN. After the instant when the main deck is impacted, at 0.6 s indicated in Figure 15-a, the force increases until large plastic deformation

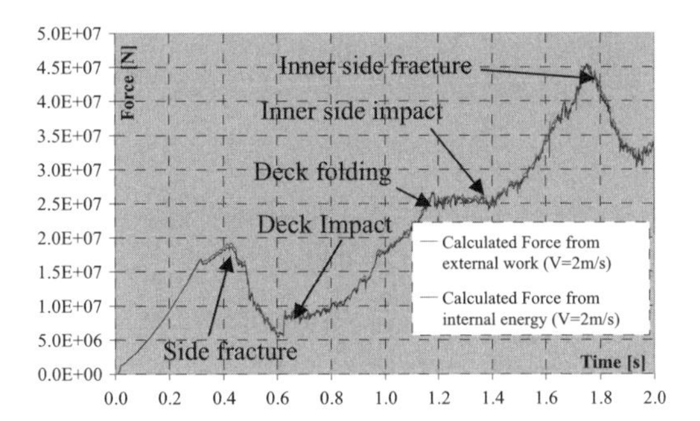

Figure 12. Impact force as function of time (Constant Speed).

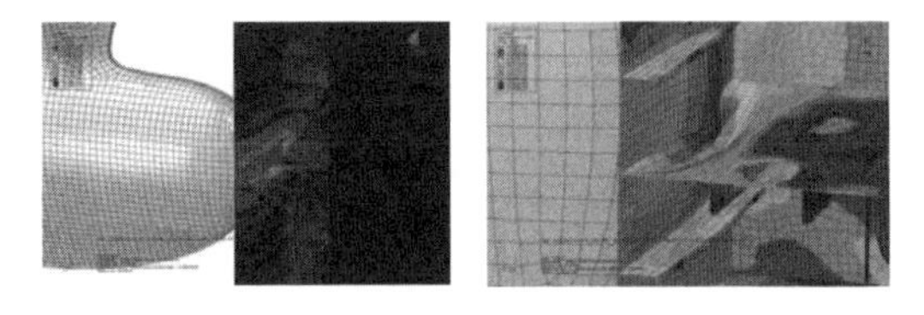

Figure 13. (a,b) Plastic deformation of side structure (0–0.4 s).

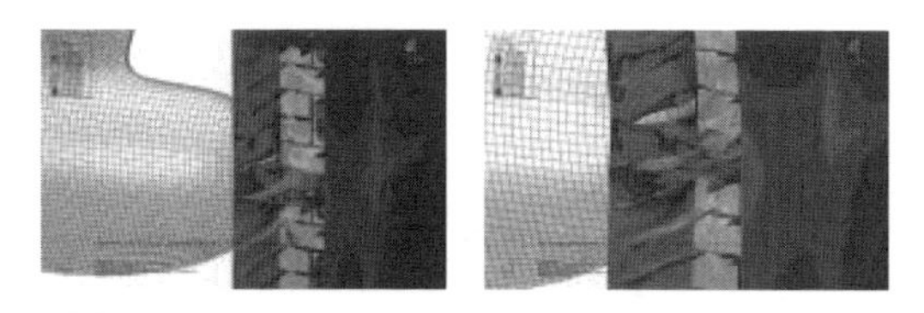

Figure 14. (a, b) Side shell and failure propagation (0.4 s).

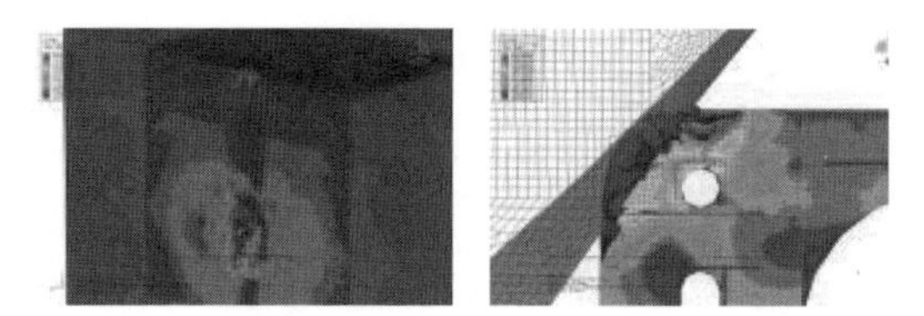

Figure 15. (a, b) Contact between striking bow and deck (0.6 s).

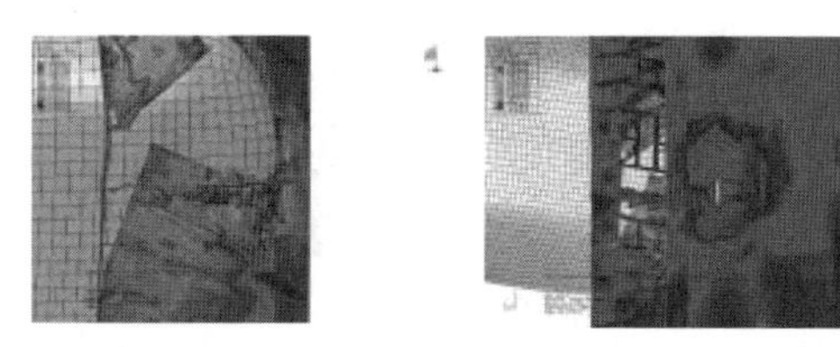

Figure 16. (a, b) Indent and fracture of the inner side.

and folding of the deck structure will limit the energy absorption, Figure 15-b. Once the striking vessel indented the inner side, this will deform and absorb energy until it fractures (Knuckle at 1.75 s in Figure 12). The impact force required to rupture the double side is 45 MN.

7 COMPARISONS BETWEEN THE TWO ANALYSIS METHODS

By introducing the mass of the OSV and removing the constant speed constraint, the speed of the OSV decreases in time due to the collision forces. The relation between the impact force and the speed is given by the second Newton's law, Figure 17.

With an initial speed of 2 m/s (corresponding to an impact energy of 18 MJ) the vessel stops after approximately 1.7 s with a final indentation of 1760 mm. When the initial speed is 4 m/s (approximately 72 MJ), the energy is dissipated after 1.45 s and the final indentation is 3690 mm.

It is found that the impact force and the absorbed energy are uniquely defined by the indentation; i.e. the constant speed approach predicts the same response as the variable speed method. This can be observed by plotting the impact forces evaluated

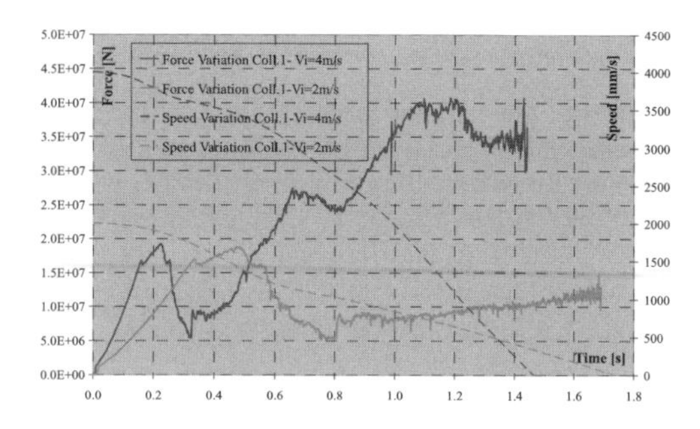

Figure 17. Collision 1, Force and speed variation in time.

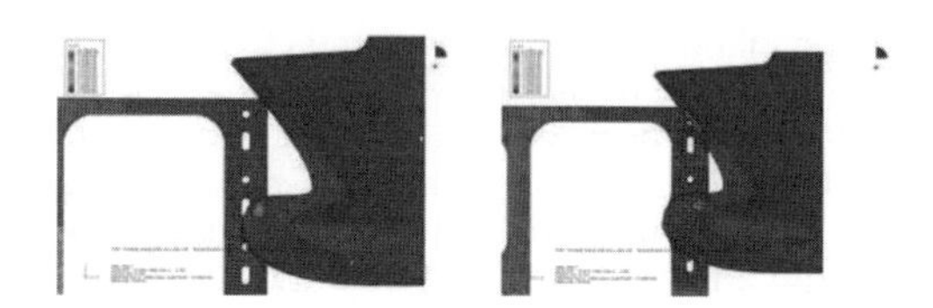

Figure 18. (a, b) Collision 1, final Indentation with 2 and 4 m/s initial speed.

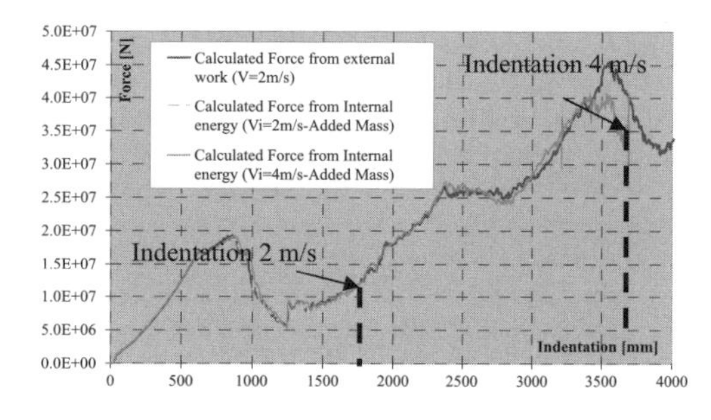

Figure 19. Force versus indentation curve in the different analysis method.

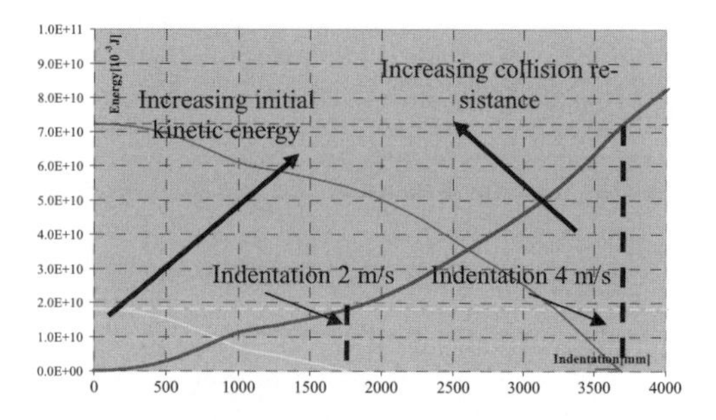

Figure 20. Indentation found from energy curves.

for the different cases against the indentation, Figure 19.

Therefore to assess the impact forces and the final damage for each initial kinetic energy level, it is sufficient to solve the problem using the constant speed approach. The final indentation is thus found from the energy-indentation curves valid for a specific impact scenario and obtained by the quasi-static method, as shown in Figure 20. This means that dynamic effects are found to be negligible for the analysed range of speeds, i.e. from 1 to 4 m/s.

8 RESULTS FOR THE ANALYSED SCENARIOS

The resistance to the impact has been evaluated for different vertical impact locations. Independent of the initial speed, the collision scenario 2 shows the lowest collision resistance.

As can be seen from Figure 21, the initial stiffness is approximately the same for all the assessed cases. The energy absorption and the corresponding collision resistance are higher for collision 3. For Collision 2, after onset of failure of the side shell, there are no structural members absorbing energy until the inner longitudinal bulkhead is reached, Figure 22-a, b). This is reflected into a flatter energy absorption curve and lower impact forces. Hence a deeper indentation is required to stop the striking vessel.

After approximately 3 m, the main deck is also impacted, Figure 22-b. The impact forces starts to increase again and the energy curve gets steeper.

As observed before, the forces versus indentations curves is uniquely defined for a specific

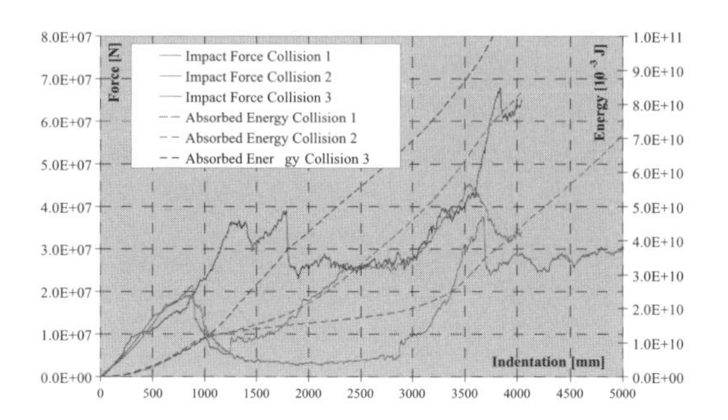

Figure 21. Impact forces and absorbed energy for different impact location (Constant speed analyses).

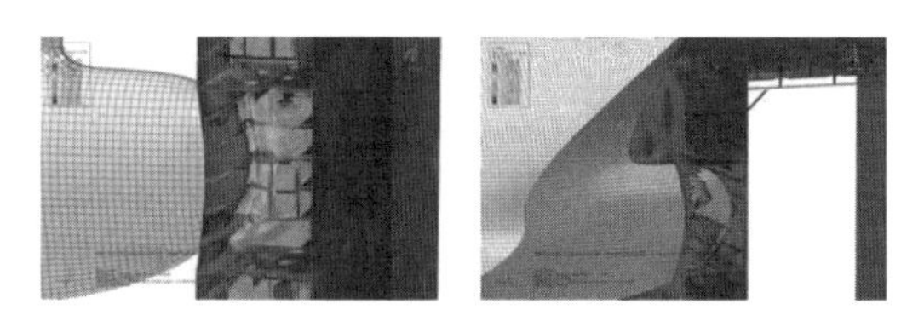

Figure 22. (a, b) Collision 2, Plastic deformation of the side structure and impact at the deck and inner side.

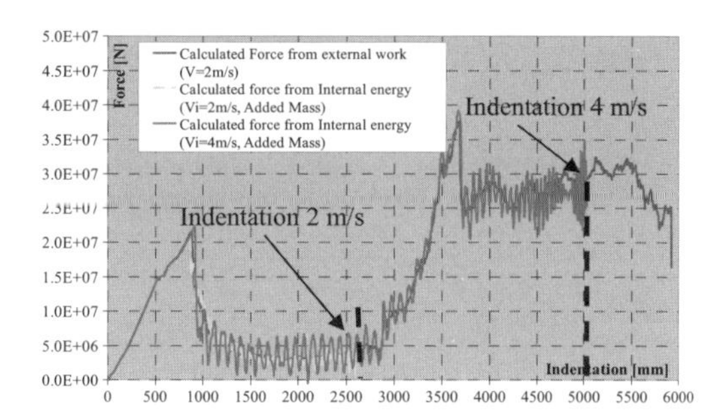

Figure 23. Collision 2, Impact force versus indentation.

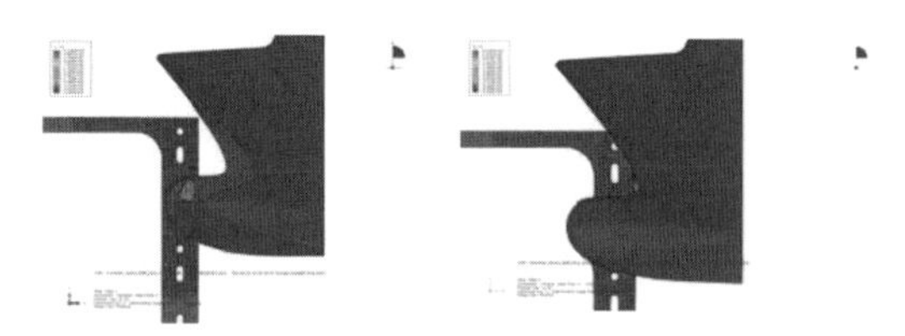

Figure 24. (a, b) Collision 2, Final indentation for 2 and 4 m/s.

Figure 25. (a, b) Collision 3 deformation and fracture.

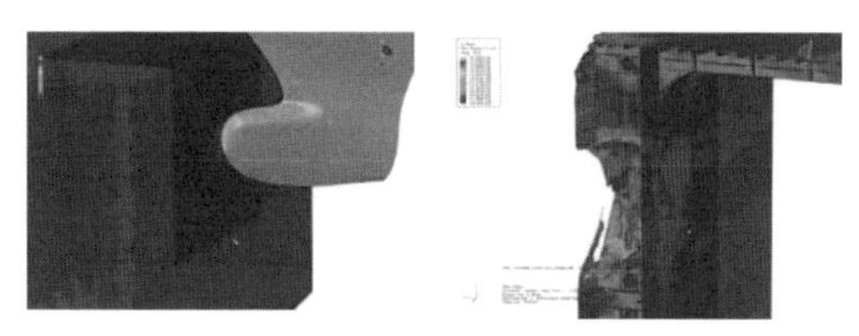

Figure 26. (a, b) Plastic deformation and failure of side staructure.

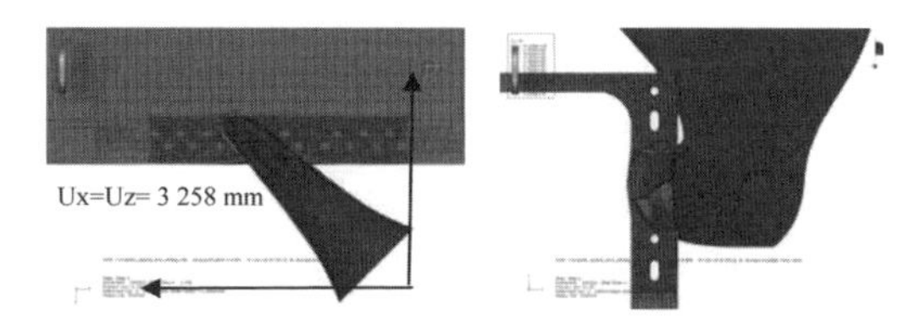

Figure 27. (a, b) 45 degrees Damages with 4 m/s speed.

Table 2. Results summary.

Scenario	Total mass [ton]	Speed [mm/s]	Kinetic impact energy [MJ]	Energy to reach the inner side (3 m indentation) [MJ]	Final indentation [mm]	Critical speed (3 m indentation) [mm/s]	Rupture to inner side
Collision 1	9020	2000	18.0	46.0	1 760	3 192	NO
	9020	4000	72.2	46.0	3 690	3 192	YES
Collision 2	9020	2000	18.0	20.1	2 645	2 110	NO
	9020	4000	72.2	20.1	5 100	2 110	YES
Collision 3	9020	2000	18.0	69.2	1 240	3916	NO
	9020	4000	72.2	69.2	3 190	3916	YES
45 Deg	9020	2000	18.0	57.2	1 340	3 561	NO
	9020	4000	72.2	57.2	3 260	3 561	YES

impact location and is independent of the initial speed or analysis method, as shown in Figure 23.

For collision scenario 2, the final indentation with initial speed of 2 m/s (approximately 18MJ impact energy) is 2.6 m while for 4 m/s (approximately 72 MJ) it is close to 5.1 m.

In collision scenario 3 the impact starts at the sheer strake, which is characterised by higher scantling and high tensile steel. Deformations are induced in a wider area of the ship side as shown in Figure 25. The impact force necessary to determine the rupture is therefore higher.

As shown in Figure 21, the absorbed energy for a given indentation is larger in this case compared to the other cases. The indentation required to fracture at the bulb elevation is 1.25 m. Once the side shell has failed also at the main deck level, the force versus indentation curve follows the trend as recorded for collision scenario 1.

When the 45 degrees impact direction is considered, it is found that the resulting damage extends also in longitudinal direction compared to the 90 degrees condition. The corresponding indentation perpendicular to the ship-shell is therefore lower than in the 90 degrees cases (Figure 26 and Figure 27).

Table 2 resumes the analysed collision cases, and includes the initial impact energies and an assessment of the kinetic energy required to reach the inner side of the vessel. Analogously a "critical speed" can be defined for each scenario as the one required to penetrate 3 m into the ship side.

9 CONCLUSIONS

Today regulatory authorities like DNV (2008a, b–2009) and NORSOK (2004, 2007), give explicit requirements to Accidental Load Scenarios (ALS) in general and for collisions between OSV and floating/fixed offshore installations in particular. To avoid possible penetration of a cargo tank, the side structure of the unit shall be capable of absorbing the energy of a vessel collision with an annual probability of 10^{-4} or at least an impact energy of 11 MJ for bow collision (14 MJ for side collision). This requirement is based on an OSV vessel with 5000 tons displacement and an impact speed of 2 m/s. On the other hand, the average size of today's OSV operating in the North Sea is increasing and a number of accidents characterised by impact energy higher than 11–14 MJ has been observed, PSA Norway. This increases the interest around the collision scenarios characterised by higher impact energies, as addressed in this study.

For ship-shaped FPSO units, DNV (2008b) does not give any explicit strength requirements for collisions with OSV's, i.e. only a damaged stability check is mandatory.

This work is presenting an energy based methodology using advance Non-Linear finite element tools targeted towards assessing likely collision damages as a result of impacts between a ship-shaped FPSO and one of the largest supply vessels used in offshore service today.

The bulbous bow striking OSV has a displacement of approximately 8200 tons. The FPSO vessel has a displacement in the range of 115000–145000 tons depending on the loading conditions.

Focusing on scenario 2, Figure 21 shows that an impact energy of 20MJ (2.1 m/s) is sufficient to reach the inner longitudinal bulkhead.

This scenario is the most severe since the OSV will hit high up in the ship-side and the collision resistance of the deck structure will not be activated before the bulb has penetrated approximately 2.8 m into the double hull. However, the inner longitudinal bulkhead will also deform during the collision contributing to the energy absorption until it fractures. If this additional capacity is accounted for, the speed required to rapture the inner longitudinal bulkhead is 2.9 m/s, corresponding to an impact energy of 38 MJ (Figure 21). Thus, it is shown that there will be some reserve of capacity from the instant when the inner side is starting deforming until it is punctured.

The present study shows that likely collision scenarios for OSV's operating close to a FPSO may lead to penetration and rupture of the side shell. This may be critical for single skin vessels with respect to environmental consequences. However, for the examined double skin FPSO, the likelihood of penetrating and rupture the cargo tanks is considered to be rather small with reference to the collision scenarios defined in existing design standards. An impact speed of around 2.9 m/s is required for an OSV having a displacement of 8200 tons. This corresponds to an initial kinetic energy of 38 MJ.

In this study, a bulbous bow has been used. A different bow shape, as raked bow or X-bow will lead to different deformations and damages. It is noted that the indicated conclusion are referring to a specific FPSO design, characterised by certain scantlings and structural arrangement.

It is thought that the collision damages assessed and methodology reported herein will provide useful background for formulating consistent Rules and Regulations in the future.

The presented study is part of a wider project, which aim's is to assess how the collision damages influence on the residual strength of the hull girder. Stern and side collisions are so far not analysed as they are considered to be less critical from a structural integrity point of view.

The structural integrity of the hull girder will be addressed during future studies and the findings will be compared to previous assessments carried out on the subject, ref. Notaro et al. (2010) and Russo et al. (2009).

ACKNOWLEDGEMENT

The authors would like to thank DNV and Statoil ASA for founding of this project. We would also like to thank our colleagues in DNV for discussions.

REFERENCES

DNV 2008. Offshore Standard, DNV-OS-A101 Safety Principles and Arrangements, October 2008.

DNV 2008b. Offshore Standard, DNV-OS-C102 Structural Design of Offshore Ships, October 2008.

DNV 2009. Offshore Standard, DNV-OS-B101-Metallic Materials, April 2009.

DNV 2010. Recommended Practice, DNV-RP-C204 Design against Accidental Loads, October 2010.

Ehlers, S. & Varsta, P, March, 2009. Strain and Stress Relation for non-linear Finite Element Simulations.

Hagbart, S. & Amdahl, J. 2008. On the resistance to penetration of stiffened plates, Part I-Experiments.

NORSOK STANDARD 2004. October N-004 Rev.2,—Design of steel structures.

NORSOK STANDARD 2007. September, N-003, Edition 2.

Notaro, G, Kippenes, J., Amlashi, H. & Russo, M., Steen, E. PRADS 2010. Conference, Rio de Janeiro 2010-Residual Hull Girder Strength of Ships with Collision and Grounding Damages.

PSA-Norwegian Petroleum Safety Authority, Risk of Collision with Visiting Vessels.

Russo, M., Renoud, F., Steen, E., Kippenes, J. & Oma, N. OMAE 2009. Conference, Honolulu 2009—Residual Strength of a FPSO vessel After Collision Damages.

Simulia, ABAQUS/CAE 6.9-1 User's Manual.

Tornqvist, R. PhD Thesis, June 2003. Design of Crashworthy Ship Structures.

3.4 *Materials*

Sustainable Maritime Transportation and Exploitation of Sea Resources – Rizzuto & Guedes Soares (eds)
 ISBN 978-0-415-62081-9

A statistical study on the material properties of shipbuilding steels

E. VanDerHorn
American Bureau of Shipping (ABS), Technology, Houston, USA

G. Wang
American Bureau of Shipping (ABS), Greater China Division, Shanghai, China

ABSTRACT: The Structural Reliability Approach (SRA) is regarded as a rational method for evaluating the structural safety of ships. One of the key components in a SRA is modeling the probabilistic nature of the material properties. This topic has been extensively studied and the resulting probabilistic models have been established. During the past few decades, there have been advances in steel manufacturing technologies. For reliability purposes, it is important to understand whether the statistical characteristics of steel material properties have changed significantly as a result of these trends.

This paper presents a study on selected mechanical properties of shipbuilding steels from five steel manufacturers in the United States and Asia. Specifically, the statistics of the material's yield stress, ultimate tensile stress, impact strength and percent elongation were studied. The specified minimum values required by current industry practice for each of these material variables provide a set of unified requirements for all steel manufacturers, while the mean value, standard deviation and probability density functions form a basis for comparison between steel manufacturers and relate directly to the overall steel quality. Based on this study, a set of new probabilistic parameters is proposed to represent the steels' mechanical properties. Comparisons are also made between these proposed parameters and established probabilistic models.

1 INTRODUCTION

The use of the Structural Reliability Approach (SRA) in the design of ships and marine structures requires detailed and accurate probabilistic characteristics of material properties. The probabilistic models that are used to represent the random behavior of these material properties must be developed from extensive experimental data. Interest in these SRA methods is not recent, and there has been substantial work done to establish the statistical characteristics of the relevant material properties. A majority of this work, specifically relating to the materials used in the marine industry, was conducted several decades ago. Since then, there have been advancements in steel manufacturing technologies and the development of more advanced quality control methods. Due to these improvements, it is necessary to collect more recent data on shipbuilding material properties and evaluate if these advancements have affected the statistical models for material properties which are currently being applied with the SRA for the design of marine structures.

The focus of this paper is to present a statistical analysis of more recent material properties' data and to evaluate the results with respect to the current statistical models based on previous studies. The most relevant material properties for SRA are the material's yield strength and tensile strength, which are the primary focus of this paper's analysis. The statistical results for the material's percent elongation and Charpy impact energy are also presented.

This study introduces a statistical analysis of shipbuilding steels from five steel manufacturers in the United States and Asia. This data was collected over the past four years, and represents over 140,000 samples. This allows for a comprehensive evaluation of shipbuilding steel material properties as manufactured with the improved production and quality methods, providing a basis for comparison to the established probabilistic models.

2 MATERIAL PROPERTIES

2.1 *Summary of data*

The data in this analysis was collected from five different steel manufacturers between 2004 and 2009, one from the United States and four from Asia. These material tests followed the testing requirements of the American Bureau of Shipping (ABS). See ABS Steel Vessel Rules, Part 2 (2010).

Several material properties were investigated, including yield strength, ultimate tensile strength, percent elongation and impact strength. This analysis considers samples of various shipbuilding steel grades, including mild (A, B, D & E) and high-strength steels (HT 32, HT 36 and HT 40) as defined by International Association of Classification Societies (IACS) requirements. There were also sufficient data samples to separate out high-strength steels which had undergone a thermo-mechanical (TM) rolling process. The data includes the ratio of the mean and rule values, the coefficient of variation (COV), and the number of samples.

2.2 *Summary of previous studies*

Additionally, this paper is interested in the comparison between the statistical models that are traditionally assumed and models developed from this recent data. Literature review indicated that despite extensive literature on the subject of material properties, very little of this literature contains statistical data on shipbuilding steels. This conclusion was also reached by Mansour et al. (1984) and the collection of statistical data presented in his paper is one of the most comprehensive sources with more than 60,000 samples. However, not all of this data is applicable for analysis, since the material type was not reported, which is required to determine the mean to rule value ratio. Other sources of material property statistical information include Caldwell (1972), Galambos & Ravindra (1978) and Guedes Soares (1988). Due to the lack of published statistical data, many of these papers contain overlapping data.

The American Iron and Steel Institute (AISI) performed a study on the variation in steel tensile properties in 1974, but the material type was not reported. Suwan (2003) performed another study on behalf of AISI which included material grades.

Additional data was collected in the early 1990's by the Ship Structures Committee (SSC) in an effort to establish a permanent and easy-to-access repository of material statistics for the marine industry (Kaufman et al., 1991). This is the most recent statistical properties that have been summarized with relation to the marine industry (Hess et al., 2002).

Table 1 is a summary of the yield stress statistics found in the literature.

The ratio of the mean value to the rule value establishes the estimated mean value of the data with respect to the IACS rule-defined nominal value. Reviewing these ratios can then provide a basis for proposing the mean value for a statistical model with respect to the nominal rule value.

Table 1. Statistical information for steel yield stress from literature.

Reference	Steel grade	Mean /Rule value *	COV
Caldwell (1972)	–	–	0.066
AISI (1974)		–	0.066
Galambos & Ravindra (1978)	flange	1.05	0.100
	web	1.10	0.110
Mansour et al. (1984)	–	1.10	0.089
Guedes Soares (1988)	–	–	0.100
	–	–	0.080
Hess et al. (2002)	OS	1.11	0.068
	HT	1.22	0.089
Suwan (2003)	A572	1.16	0.064
	A588	1.15	0.046

*IACS rule value of yield stress:
235 MPa for ordinary steel (OS)
315 MPa for high tensile (HT) 32 steel
355 MPa for high tensile (HT) 36 steel
390 MPa for high tensile (HT) 40 steel.

The published results for other tensile test properties, such as the tensile strength and percent elongation, are limited since they are not used often in SRA calculations. Table 2 shows the statistical data for the tensile stress as given in literature.

There is a significant amount of data available for the Charpy impact energy for steels; however, this data is typically reported at varying temperatures to identify the ductile/brittle transition temperature. There is very little work discussing the variability of Charpy impact energy results at a given temperature. Some discussion was provided by Todinov et al. (2000), but this work is primarily concerned with the statistical behavior of welds. The most comprehensive studies were performed by AISI in 1979 and 1989, which include statistics on the variability of the Charpy impact energy for different metals at three different testing temperatures. Suwan (2003) presented a similar study on the Charpy impact energy in his study performed for AISI. Table 3 shows a summary of the statistical information collected from these studies.

2.3 *Yield strength*

The yield strength is the material strength variable used when determining hull girder strength for SRA. As such, statistics on the variability of the steel's yield stress are the most prominent in the available literature. The data collected for yield stress in earlier studies contain substantial variations in the results. When reviewing the collection of data presented by Mansour et al. (1984), Galambos & Ravindra (1978) suggested that numerical analysis of the yield stress was probably

Table 2. Statistical information for tensile stress from literature.

Reference	Steel grade	Mean /Rule value *	COV
AISI (1974)	–	–	0.035
Mansour et al. (1984)	–	–	0.068
Hess et al. (2002)	OS	1.07	0.064
	HT	1.09	0.048
Suwan (2003)	A572	1.18	0.049
	A588	1.17	0.033

*IACS rule value of tensile stress:
400–520 MPa for ordinary steel (OS)
440–590 MPa for high tensile (HT) 32 steel
490–620 MPa for high tensile (HT) 36 steel
510–650 MPa for high tensile (HT) 40 steel.

Table 3. Statistical information for charpy impact energy from literature.

Reference	Steel grade	Mean /Rule value	COV
AISI (1979)	A572	–	0.354
	A516	–	0.322
	A537	–	0.318
AISI (1989)	A572	–	0.428
	A588	–	0.482
Suwan (2003)	A572	–	0.549
	A588	–	0.457

of limited value due to the different measurement methods. This is due to the lack of uniformity in test method and the sensitivity of the yield strength to the strain rate. This study attempts to minimize this lack of uniformity, as all the steel manufacturers followed ABS class rule requirements for specimen testing. The rule requirements provide a strain rate for testing. This limits the variability between the tests; however, the strain rate for the tests was not reported.

Table 4 provides the statistical information from this study.

The weighted average of the COV's for the data presented is 0.0647. This is a 27% improvement on the weighted average of 0.089 found by Mansour et al. (1984). This improvement is consistent in the data from all the steel makers in this study. It can possibly be attributed to the improved manufacturing and quality control processes that have been developed over the past several decades. It may also be attributed to more uniform testing procedures. For reliability approaches, the COV is in many cases taken as 10%, which can be viewed as conservative when compared to the presented data.

The traditional assumption is that the steel's mean yield strength is 1.1 times the class-defined

Table 4. Statistical information for yield strength for shipbuilding steels (2004–2009).

Reference	Steel grade	Mean /Rule value	COV	No. of tests
Steel Maker I*	A	1.228	0.0465	8365
	AH32	1.204	0.0438	7038
	AH36	1.188	0.0389	3041
	AH40	1.196	0.0434	367
Steel Maker II(*,**)	AH32-TM	1.224	0.0595	325
	DH32-TM	1.225	0.0611	72
	EH32-TM	1.428	0.0447	188
	AH36-TM	1.150	0.0819	165
	DH36-TM	1.199	0.0755	178
	EH36-TM	1.303	0.0598	220
Steel Maker III	AH36	1.170	0.0634	2789
	DH36	1.194	0.0544	587
Steel Maker IV	AH36	1.199	0.0666	185
	DH36	1.167	0.0858	304
	EH36	1.045	0.0566	120
Steel Maker V*	A	1.287	0.0768	41,225
	B	1.293	0.0626	2941
	D	1.336	0.0837	1082
	E	1.346	0.0722	1514
	AH32	1.182	0.0648	36,342
	DH32	1.231	0.0588	2057
	EH32	1.205	0.0580	1626
	AH36	1.154	0.0578	9619
	DH36	1.150	0.0612	4294
	EH36	1.121	0.0687	6038
	AH32-TM	1.342	0.0579	4370
	DH32-TM	1.316	0.0557	420
	EH32-TM	1.322	0.0665	980
	AH36-TM	1.274	0.0547	3127
	DH36-TM	1.240	0.0634	920
	EH36-TM	1.268	0.0657	2264

*TM—Thermo-mechanical rolling
**Data previously reviewed by Guo (2010).

nominal value. As shown in Table 4, the mean value of the steel's yield stress is, on average, about 1.23 times the nominal values. The data suggests that this may vary based on the steel type. Separating the data by steel type, it can be seen that mild steel is on average 1.28 times the rule nominal value, high-strength steel is 1.18 times the rule nominal value and high-strength steels manufactured using thermo-mechanical rolling is 1.30 times the rule nominal value.

Based on the data presented, it is proposed that the mean value be taken as 1.15 times the nominal rule value for mild steels and high-strength steels undergoing a thermo-mechanical rolling process and 1.10 times the nominal value for high-strength steels. The corresponding COV for all steel grades is proposed be as 8%. These assumptions take into account the improvements in the steel properties while still being conservative when compared with the data in Table 4.

Previous work by DNV (1992) notes that a log-normal distribution is appropriate for describing the yield strength and this assumption was supported by the work of Hess et al. (2002). Goodness-of-fit tests were performed on the given data using Easy-Fit, a software for distribution fitting. Based on the Anderson-Darling criteria, it was confirmed that the log-normal distribution is an applicable model for describing the yield strength of steel. It is therefore recommended that the log-normal PDF be used to represent each of the steel grades' yield strength with their respective statistics suggested above. A sample distribution can be seen in Figure 1.

Using this statistical model, the distributions of the data from this study can be compared with the statistical information provided in the literature. Figure 2 shows a sample comparison between three mild steel distributions. Two distributions are presented from the current data (Steel Maker I and Steel Maker V), which can be compared with a distribution derived from the data in the literature (Mansour et al., 1984). The plot of the probability density functions (PDF) supports the observation that there has been a measurable increase in both the quality and yield strength of shipbuilding steels. The most noticeable improvement is the increase in the yield strength seen in both of the modern steels (Steel Maker I and Steel Maker V).

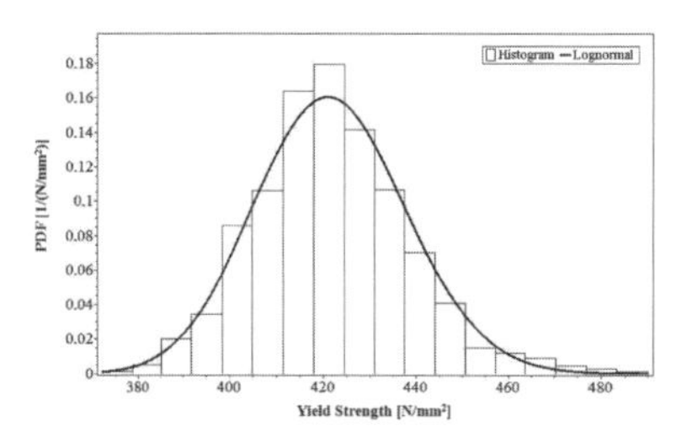

Figure 1. Distribution fitting for yield strength data using EasyFit software.

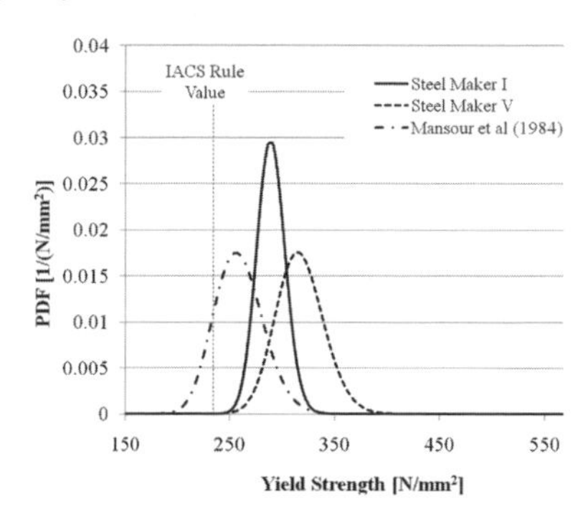

Figure 2. Comparison between mild steel yield strength probability density functions.

2.4 *Tensile strength*

The tensile strength is also a useful property in design, as it provides information on the material's post-yield behavior. The ultimate tensile strength is also much less sensitive to the strain rate than the yield strength, which makes comparisons of various data sources more uniform. There are a very limited numbers of studies which present data on the ultimate tensile strength. Mansour et al. (1984) reviews the results from several tests but with very few samples.

Table 5 provides the statistical information from this study.

Table 5. Statistical information for tensile stress for shipbuilding steels (2004–2009).

Reference	Steel grade	Mean /Rule value	COV	No. of tests
Steel Maker I	A	1.095	0.0204	8365
	AH32	1.134	0.0246	7038
	AH36	1.094	0.0293	3041
	AH40	1.110	0.0300	367
Steel Maker II*	AH32-TM	1.186	0.0272	325
	DH32-TM	1.183	0.0344	72
	EH32-TM	1.218	0.0306	188
	AH36-TM	1.103	0.0407	165
	DH36-TM	1.126	0.0451	178
	EH36-TM	1.133	0.0326	220
Steel Maker III	AH36	1.163	0.0340	2789
	DH36	1.128	0.0526	587
Steel Maker IV	AH36	1.126	0.011	185
	DH36	1.123	0.0355	304
	EH36	1.053	0.0287	120
Steel Maker V*	A	1.129	0.0281	41,225
	B	1.111	0.0340	2941
	D	1.131	0.0247	1082
	E	1.137	0.0216	1514
	AH32	1.202	0.0354	36,342
	DH32	1.200	0.0340	2057
	EH32	1.210	0.0319	1626
	AH36	1.110	0.0266	9619
	DH36	1.110	0.0239	4294
	EH36	1.116	0.0296	6038
	AH32-TM	1.163	0.0361	4370
	DH32-TM	1.168	0.0377	420
	EH32-TM	1.182	0.0455	980
	AH36-TM	1.110	0.0430	3127
	DH36-TM	1.112	0.0469	920
	EH36-TM	1.128	0.0503	2264

*TM—Thermo-mechanical rolling.

As seen in Table 5, the average of the mean values of the steel's tensile strength is 1.15 times the nominal rule value. The weighted average of the COVs for the presented data was 0.031. This is a 54% improvement on the weighted average of 0.068 found by Mansour et al. (1984). Based on the reported data, it is suggested that the steel's tensile strength be taken as 1.05 times the nominal rule value, with a COV of 0.050. This is consistent with the assumptions made by Atua et al. (1996). This assumption can be considered conservative when compared with the data in Table 2.

Goodness-of-fit tests, performed using EasyFit, suggest that the normal distribution is appropriate for describing the tensile strength. This is consistent with the results found by Hess et al. (2002). A sample distribution can be seen in Figure 3.

Using this statistical model, the distributions of the data from this study can be compared with the statistical information provided in the literature. Figure 4 shows a sample comparison between three mild steel distributions. Two distributions are presented from the current data (Steel Maker I and Steel Maker V), which can be compared with a distribution derived from the data in the literature (Hess et al., 2002). The plot of the probability density functions supports the observation that there

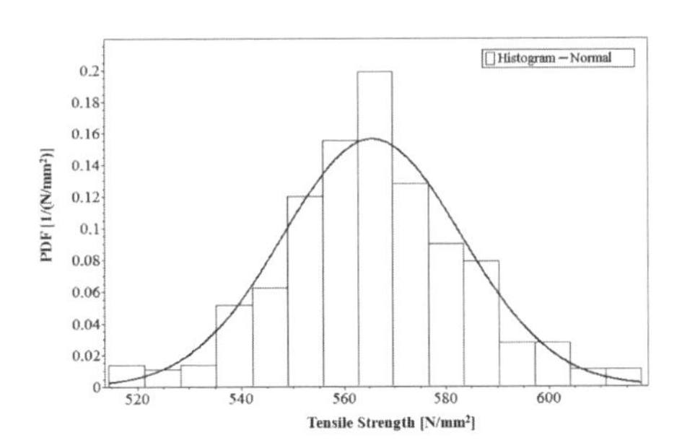

Figure 3. Distribution fitting for tensile strength data using EasyFit software.

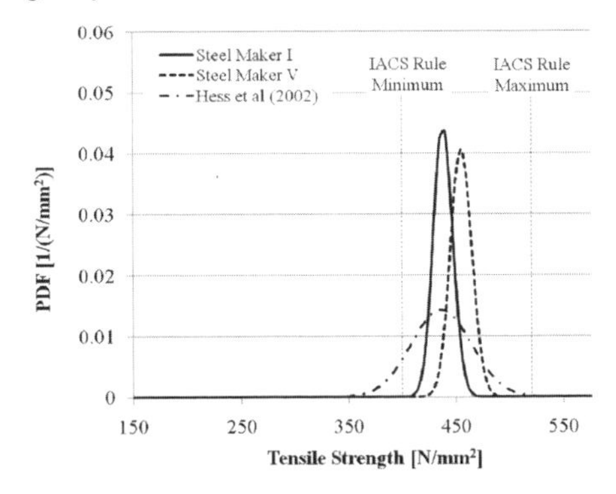

Figure 4. Comparison between mild steel tensile strength probability density functions.

has been a measurable increase in both the quality and tensile strength of shipbuilding steels. The increase in the tensile strength appears to be very slight, but there is a noticeable increase in quality control for both of the modern steels as indicated by the reduce spread in the distributions.

2.5 *Percent elongation*

The percent elongation is less important in overall structural strength calculations, but it is important to ensure that the material's ductility meets the class- defined nominal values. The literature search revealed no collection of statistical data for the percent elongation, so the presented data is not compared to any existing source.

Table 6 provides the statistical information from this study.

Table 6. Statistical information for percent elongation for shipbuilding steels (2004–2009).

Reference	Steel grade	Mean/Rule value	COV	No. of tests
Steel Maker I	A	1.323	0.0694	8365
	AH32	1.186	0.0663	7038
	AH36	1.119	0.0762	3041
	AH40	1.080	0.1606	367
Steel Maker II*	AH32-TM	1.172	0.0958	325
	DH32-TM	1.177	0.1352	72
	EH32-TM	1.138	0.1007	188
	AH36-TM	1.213	0.1143	165
	DH36-TM	1.149	0.1098	178
	EH36-TM	1.200	0.1122	220
Steel Maker III	AH36	1.278	0.0929	2789
	DH36	1.342	0.1058	587
Steel Maker IV	AH36	1.028	0.1021	185
	DH36	1.036	0.0867	304
	EH36	1.337	0.1218	120
Steel Maker V*	A	1.577	0.0835	41,225
	B	1.586	0.0825	2941
	D	1.580	0.0695	1082
	E	1.658	0.1080	1514
	AH32	1.382	0.0705	36,342
	DH32	1.378	0.0669	2057
	EH32	1.413	0.0722	1626
	AH36	1.417	0.0779	9619
	DH36	1.339	0.0894	4294
	EH36	1.429	0.0960	6038
	AH32-TM	1.273	0.1157	4370
	DH32-TM	1.220	0.1216	420
	EH32-TM	1.296	0.1110	980
	AH36-TM	1.209	0.1242	3127
	DH36-TM	1.171	0.1136	920
	EH36-TM	1.196	0.1128	2264

*TM—Thermo-mechanical rolling.

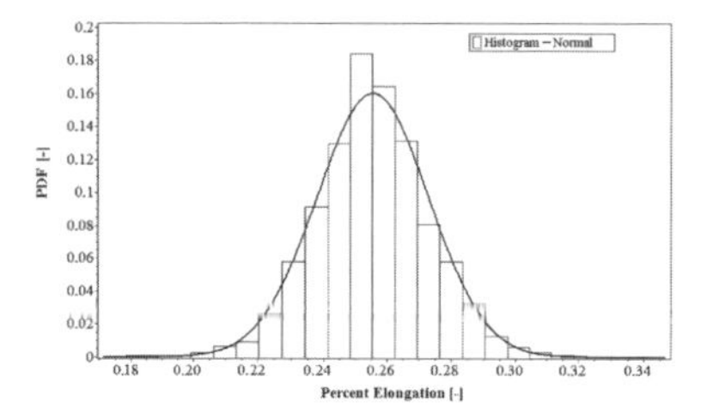

Figure 5. Distribution fitting for percent elongation data using EasyFit software.

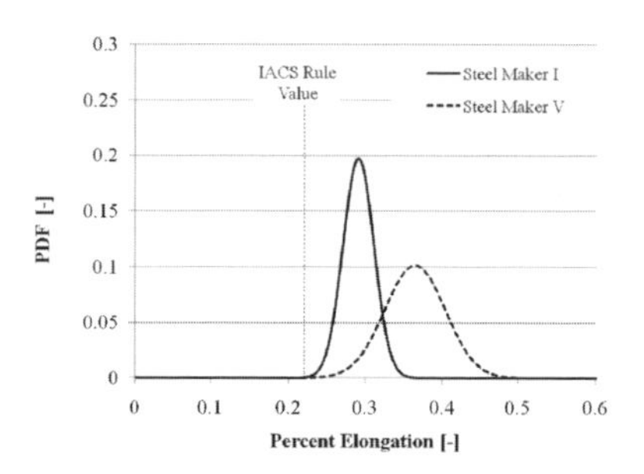

Figure 6. Comparison between mild steel percent elongation probability density functions.

The percent elongation has a weighted COV of 0.082 and the average of the mean values is 1.41 times the nominal rule value. As previously mentioned, there is no existing data to compare these results with. However, compared with the other two material properties presented, the percent elongation seems to display larger variations in measurements. It appears that the manufacturer has an effect on the data and a sensitivity analysis should be performed to determine the significance of this effect.

Goodness-of-fit tests, performed using EasyFit, suggest that the normal distribution is appropriate for describing the percent elongation. A sample distribution can be seen in Figure 5.

Figure 6 shows a sample comparison between two mild steel distributions (Steel Maker I and Steel Maker V). While there was no comparison data provided in the literature, it is observed that similar to the yield strength and tensile strength, the percent elongation probability density functions for modern steels lie visibly above the IACS rule value and have good quality as indicated by the narrow shape of the distributions.

2.6 *Charpy impact energy*

There is a large amount of literature and statistical data for the Charpy impact energy. However, most of the data collected is for the determination of the ductile/brittle transition temperature. There is very little previous work investigating the statistical models for Charpy impact energy tests conducted at a given temperature. This is likely due to the sensitivity of impact energy to both temperature and material thickness. This leads to data which is inherently scattered. Since the data presented in this paper was collected based on class requirements for testing, the testing temperature is set by class requirements, however, this is not possible to verify as the testing temperatures for the data were not reported. There was also substantial variability in the reported thicknesses being tested, even among data collected from the same steel manufacturer. Most of the data was identified by a range of thicknesses, which typically spanned more than one of the class defined thickness categories. For this study, the data was compiled and evaluated compared to the thickest defined category rule nominal value.

Table 7. Statistical information for impact energy for shipbuilding steels (2004–2009).

Reference	Steel grade	Mean/Rule value	COV	No. of tests
Steel Maker I	A	6.561	0.1829	45
	AH32	4.500	0.1444	4927
	AH36	3.660	0.1787	2067
	AH40	5.145	0.1187	348
Steel Maker II*	AH32-TM	5.938	0.1203	325
	DH32-TM	4.249	0.1820	72
	EH32-TM	8.294	0.1438	258
	AH36-TM	4.389	0.1267	165
	DH36-TM	5.784	0.1689	178
	EH36-TM	7.051	0.1049	654
Steel Maker III	AH36	2.752	0.2725	2650
	DH36	3.210	0.4526	565
Steel Maker IV	AH36	2.580	0.3291	403
	DH36	2.250	0.4194	408
	EH36	5.425	0.2209	134
Steel Maker V*	A	4.436	0.5231	10810
	B	3.885	0.3623	1371
	D	5.379	0.2286	1634
	E	5.193	0.2343	3405
	AH32	3.880	0.2229	61319
	DH32	4.161	0.2790	2690
	EH32	3.422	0.3209	2833
	AH36	4.149	0.2279	14520
	DH36	4.069	0.2808	6134
	EH36	3.912	0.3492	10341
	AH32-TM	7.060	0.1317	4355
	DH32-TM	6.704	0.1300	421
	EH32-TM	6.198	0.1854	952
	AH36-TM	6.377	0.1288	3120
	DH36-TM	5.966	0.1495	921
	EH36-TM	5.622	0.1772	2262

*TM—Thermo-mechanical rolling.

The data presented by AISI (1979, 1989) and Suwan (2003) has a weighted average COV of 0.398. The impact energy data in Table 7 has a weighted COV of 0.255. As previously mentioned, the data for Charpy impact energy is inherently scattered, and this can be seen in the large COV values for both the presented data and data found in literature. This scatter was potentially amplified in the presented data by the large thickness ranges (5 mm–80 mm) included in the data. Despite the large thickness ranges represented, there was still a 36% reduction in the COV for the presented data when compared to previously published data.

Class requirements for the impact properties of the materials are based on three thickness ranges. Due to the wide range of thicknesses presented in the data, the mean values were compared with the largest rule nominal value, corresponding with the 70 mm–100 mm thickness range. The average of the mean values was found to be 4.24 times this nominal rule value. Due to the sensitivity of the impact energy results to temperature, thickness and material type, more detailed testing is required to develop a series of statistical models for a selection of a combination of these variables.

3 CONCLUSIONS

The uncertainty in material properties must be accurately characterized to understand their impact on overall structural strength, particularly their statistical models for use in SRA. The statistical models used in SRA must be well defined and supported by adequate testing data. While there is substantial literature discussing material properties, there is a relatively limited literature reporting statistical data. This has led to the use of statistical models based on test data from as far back as 1950. The most recent published test data for shipbuilding steels was performed by the SSC in the early 1990's.

The data presented in this paper indicates that the improvements made in steel manufacturing processes and quality control methods has resulted in adjustments to the existing statistical parameters. All of the material properties showed consistent reductions in the COV values as compared to previously published data. This reduction in the COV values was also consistent among all of the steel manufacturers, indicating that this trend is likely applicable for all steel produced in modern steel mills. Additional data collected from steel makers in other parts of the world would confirm this trend.

Table 8 presents a summary of the proposed probabilistic characteristics for the selected properties.

Table 8. Summary of proposed distribution parameters for steel material properties.

		Statistical information		
Variable	Steel type	Mean	COV	Distribution
Yield Strength (σ_y)*	Mild	1.15 σ_y	0.080	Log-normal
	HS	1.10 σ_y	0.080	
	HS-TM	1.15 σ_y	0.080	
Tensile Strength (σ_u)*	Mild	1.05 σ_u	0.050	Normal**
	HS			
	HS-TM			
Percent Elongation (e)*	Mild	1.30 e	0.150	Normal**
	HS	1.20 e		
	HS-TM	1.10 e		

*σ_y, σ_u, and e represent IACS nominal rule values
**Truncated at zero.

The improvements in steel manufacturing and quality control processes did not alter the underlying types of probability distributions, but did affect the accompanying parameters. It was recommended to increase the assumed mean value for the yield strength of mild steels and high-strength steels which have undergone a thermo-mechanical rolling process. There were also noted decreases in the variability of the data for all the selected material properties as indicated by the smaller COV values.

It is important to note that this data only applies to material for new-build ships. Future work should be focused on collecting data and modifying the statistical models to account for aging ships. While not currently used in SRA, future work should also consider developing probabilistic models for the Charpy impact energy data.

ACKNOWLEDGEMENT

The views expressed herein are the opinions of the author and not necessarily those of ABS. The authors would like to express their appreciation to ABS management, especially R. Basu, for their support of this project. We would also like to thank N. Chen, E. Khoo, M. Lee and L. Ivanov for valuable discussions and suggestions during the development of this paper.

REFERENCES

ABS (2010). Rules for Building and Classing Steel Vessels.

American Iron and Steel Institute (AISI), "The Variation of Product Analysis and Tensile Properties, Carbon steel Plates and Wide Flange Shapes," *Contributions to the Metallurgy of Steel*, September 1974.

American Iron and Steel Institute (AISI), "The Variations of Charpy V-Notch Impact Test Properties in Steel Plates," *Contributions to the Metallurgy of Steel*, January 1979.

American Iron and Steel Institute (AISI), "The Variations of Charpy V-Notch Impact Test Properties in Steel Plates," *Contributions to the Metallurgy of Steel*, July 1989.

Atua, K., Assakkaf, I.A. & Ayyub, B.M. 1996. "Statistical Characteristics of Strength and Load Random Variables of Ship Structures," *Probabilistic Mechanics and Structural Reliability, Proceeding of the Seventh Specialty Conference,* Worcester, Massachusetts, USA.

Caldwell, J.B. (1972). Design-construction. *De Ingenieur*, 49.

"EasyFit v5.5," MathWave Technologies, Available from http://mathwave.com/donwloads.html, 2010.

Galambos, T.V. & Ravindra, T.V. 1978. "Properties of Steel for Use in LRFD," *Journal of Structural Division*, ASCE, Vol. 104, No. ST9.

Guedes Soares, C. (1988). Uncertainty modeling in plate buckling. *Structural Safety*, 5: 17–34.

Guo, J. "Reliability-Based Inspection Planning with Application to Deck Structure thickness Measurement of Corroded Aging Tankers." Thesis. The University of Michigan, 2010. Print.

Hess, P.E., Bruchman, D.D., Assakkaf, I. & Ayyub, B.M. (2002). "Uncertainties in Material Strength. Geometric and Load Variables", *ASNE Naval Engineers Journal*, 114:2, 139–165.

Kaufman, J.G. & Prager, M. 1991. "Marine Structural Steel Toughness Data Bank." *Ship Structural Committee report SSC-352*, Volumes 1–4.

Mansour, A.E., Jan, H.Y., Zigelman, C.I., Chen, Y.N. & Harding, S.J. 1984. "Implementation of Reliability Methods to Marine Structures," *Trans. Society of Naval Architects and Marine Engineers*, Vol. 92, 11–20.

"Structural Reliability Analysis of Marine Structures", *Classification Note No. 30.6*, Det Norkse Veritas 1992.

Suwan, S., Manuel, L. & Frank, K.H. Statistical Analysis of Structural Plate Mechanical Properties. *Rep. no. 03-1*. 2003.

Todinov, M.T., Novovic, M., Bowen, P. & Knott, J.F. Modeling the impact energy in the ductile/brittle transition region of C-Mn multi-run welds, *Materials Science & Engineering A*, A287 (2000) 116–124.

Sustainable Maritime Transportation and Exploitation of Sea Resources – Rizzuto & Guedes Soares (eds)
© 2012 Taylor & Francis Group, London, ISBN 978-0-415-62081-9

Mechanical characteristics of the shipbuilding stainless steel at sub-zero temperatures

J. Kodvanj, K. Žiha, B. Ljubenkov & A. Bakić
University of Zagreb, Faculty of Mechanical Engineering and Naval Architecture, Ivana Lucica Zagreb, Croatia

ABSTRACT: Ships for liquefied natural gas are of increasing interest for extensive research of the classification societies, scientific institutes and shipyards. Since the liquefied gas is transferred in the cargo tanks at –165°C, the mechanical characteristics of the materials for ship building are faced with high-level demands on low temperatures. Therefore this paper brings forward the results of mechanical testing of shipbuilding steel on low temperatures. Mechanical characteristics of material are examined in the Laboratory for experimental mechanics at the Faculty of Mechanical Engineering and Naval Architecture of the University of Zagreb by static tensile test. The specimens are tested in cooled chambers on 4 different temperatures from 22°C to –165°C. Values of the tensile strength, deformation, total uniform elongation, breaking elongation and modulus of elasticity are measured during each test. The conclusion contains remarks about mechanical characteristics of the stainless steel at sub-zero temperatures. The plans for future research include low temperature fatigue and toughness testing.

1 INTRODUCTION

Constant development of the world economy demands huge neccesities for different types of energy. Among them, gases are significant and ecologically very useful item.

The gases could be transported by pipes and in special tanks on ships. Transport by ships started 60 years ago with very strong level of reliability and security because untill today no accident on ships carrying liquified gas has signed. During the transport, temperature of liqiufied gas is –165 °C. Therefore, the ships are under strong demands for mechanical properties of the material, construction, welding procedure and insulation details. Also, ships carrying liquified gas are still interesting for research in scientiffic institutes and clasification societies.

2 CLASSIFICATION SOCIETY DEMANDS FOR SHIPS CARRYING LIQUIFIED GAS

In 1973 IMO (International Maritime Organization) proposed a issue under the title IGC Code 'International Code for Construction and Equipment os Ships carrying Liquified Gases in Bulk' which is accepted and used in the world. The purposes of these codes is to provide an international standard for the safe transport by sea in bulk of liquefied gases and certain other substances, by prescribing the design and construction standards of ships involved in such transport and the equipment they should carry so as to minimize the risk to the ship, its crew and to the environment, having regard to the nature of the products involved.

IGC Code is a part of the SOLAS regulations under the title 'International Convertion for the Safety of Life at Sea'—Chapter VII, part C.

The ships carrying liquified gasesmust fullfil rules and regulations determined with documents:

- International Gas Carrier Code 'IGC Code'—SOLAS—Chapter VII, part C
- Special rules of the Clasification Societies
- Four special demands of the American Coast Guard.

Rules and regulations of the Croatian Register of Shipping are part of the international rules which are concerned on materials and products predicted for assembling and outfitting on ships, off shore objects or other maritime constructions.

Due to cargo type and design temperature, material of tank have to be carefully chosen. It has to be acceptable for a service on sub zero temperatures with acceptable value of the total uniform elongation. The chemical structure of the material has to be approved by classiffication society and the mechanical propreties have to be according allowable values.

Table 1 contains information about steel materials accepted for sub zero temperatures. Fine grained steels with nominal yield strength up to

Table 1. Steel materials for sub-zero temperatures.

Grade	Standard/standard designation	Minimum design temperature [°C]
Fine grained structural steels with nominal yield strength up to 355 N/mm^2	DIN 17102	–45
Nickel steels:	DIN 17280	
0,5% Ni	13 Mn Ni 6 3	–55
1,5% Ni	14 Ni Mn 6	–60
3,5% Ni	10 Ni 14	–90
5% Ni	12 Ni 19	–105
9% Ni	X 8 Ni 9	–165
Austenitic steels	DIN 17440 1.4306 (AISI 304 L) 1.4404 (AISI 316 L) 1.4541 (AISI 321) 1.4550 (AISI 347)	–165

355 N/mm^2 could be used for design tempearture up to –45°C conforming to DIN17102. Nickel steels are acceptable for lowest temperatures conforming to DIN17280 and the austenitic steels are used for design temperature –165°C.

The mechanical properties of materials are tested on low and high temperature according the service conditions. The temperatures changes influence on material structure and the mechanical properties.

Congruatly to service conditions, mechanical properties testing of material is conducting on high and low temperatures. Increase or decrease of temperature influences on a material structure and mechanical properties. If we want to establish influence of temperature change on a mechanical properties of the material, test specimen have to be set in the cold or warm chamber during the tensile test. For material testing at sub zero temperatures, expectations are that values of convencional yield strength and tensile strength increase, value of elongation decreases and value of elastic modulus has no changes.

3 MECHANICAL PROPERTIES TESTING OF THE AUSTENITIC STEEL AT SUB ZERO TEMPERATURES

According to Croatian Register of Shipping demands and data from Table 1, material for tank fabrication of the ships carrying gases is one of the austenitic steels. The steel marked as AISI 316L (X2CrNiMo) or (1.4404) is chosen. The austenitic steels are chrom nickel steels with small precentage of carbon which is in interval from 0.03% to 0.12%. Percentage of chrom in the material is from 12% to 25% and percentage of nickel is from 8% to 25%. This steel has carbon percentage of 0.03% and has totaly austenitic structure on room temperature. The significant characteristics of the steel are good ductility, corrosion resistance and steadiness on high and low temperatures.

The tensile test is carried out on a testing machine at temperatures of –165°C, –80°C, –40°C and at room temperature of 22°C. For each temperature, three tests were done.

Test specimens are prepared according to classification society demands. They could have flat or round shape, but round specimens are easier for testing. Test specimen is illustrated by Figure 1 and characteristic designations are:

- Neck diameter
- Head diameter
- Head length
- Original gauge length
- Parallel test length
- Total specimen length
- Transition radius

The specimen neck diameter is determined with material properties and maximum force of the testing machine. The dimensions of the specimen, used in this test are shown in Table 2.

Tests are carried out on the tensile machine Messphysik Beta 50–5 with maximum force of 50 kN, which is shown by Figure 2. The testing machine is connected to a computer. Before the test starting, testing parameters like type, velocity or neck diameter are defined in the computer. The computer is also used to follow testing and prepare measurement results in form of tables and graphs. In this case, specimens are set in chamber, shown by Figure 3, which is cooled on a proper temperature. The specimen must be two hours on that temperature before the test begins. The liquefied nitrogen is used for cooling and proper temperature is regulated and controlled by special device.

For determining the mechanical properties of materials by the static tensile tests it is necessary to use precise measuring devices such as extensometers to measure the elongation of the specimen.

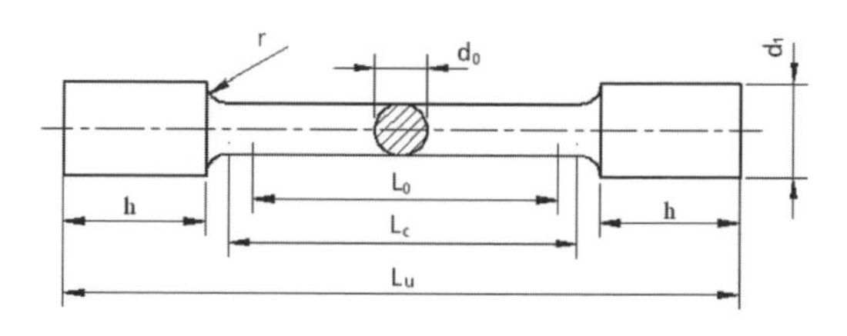

Figure 1. Testing specimen.

Table 2. Dimensions of the testing specimen.

Designation	d_o	d_1	L_o	L_c	L_u	r	h
Dimension mm	8	12	40	48	115	10	30

Figure 2. The testing machine Messphysik Beta 50–5.

Figure 3. The chamber for testing.

In practice, different types of extensometers are used such as strain gauges, clip on extensometers or, more recently, video extensometers and laser speckle extensometers. Here, laser spackle extensometer is used which advantages over other types are as follows:

- Non-contact measurement,
- Large measuring range,
- There is no need for markers on the specimen
- Measurement can be conducted through the glass of the chamber

Laser speckle extensometer consists of two CCD cameras and two laser diode with power of 3mW as shown in Figure 4. During operation the laser extensometer coherent laser beam illuminates the specimen surface, shown by Figure 5, and each camera captures the reflection light.

When coherent laser light reflect from the optically rough surface laser speckle effect appears as a black and white pixels stochastically distributed

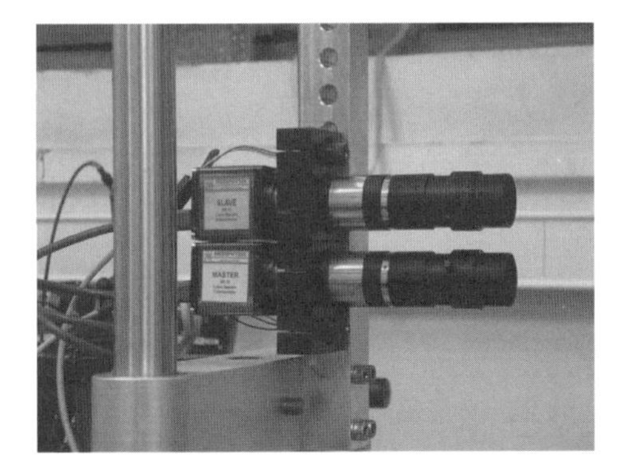

Figure 4. Laser speckle extensometer.

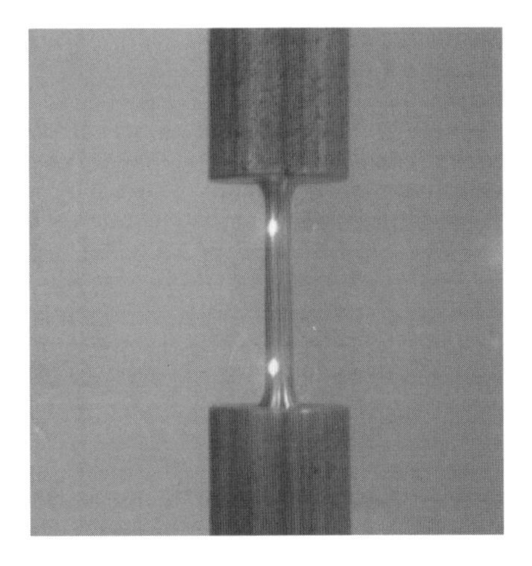

Figure 5. The laser light on the testing specimen surface.

in the picture. The surface of the specimen and therefore the speckle pattern change comparatively slowly during loading. Due to this fact the video processor is able to find an initially stored reference-pattern in consecutive images and measure the distance this pattern has moved in the meantime. Applying this procedure iteratively from image to image in real-time, strain within the distance of the two cameras can be measured.

Static tensile testing of mechanical properties of materials is carried out at temperatures from −165°C, −80°C, −40°C and at room temperature of 22°C. For each temperature, three specimens were tested and the results are shown on the 'stress-strain' diagram and in the tables.

The table contains the serial number, designation, specimen diameter, value of elastic modulus E, the conventional yield strength $R_{p0.2}$, tensile strength R_m, strain ε_m and breaking elongation A. The last row of the table presents mean values of the three test results. The results of material properties testing at the temperature of −80°C are shown in Figure 6 and Table 3.

Test results at the temperature −80°C as well as the results at the other temperatures show good repeatability and low standard deviation.

Comparative stress-strain diagram at low temperatures is shown in Figure 7. Only one representative curve at each temperature is presented in order to better results insight. Numerical values of the results are shown in Table 4.

Test results are within expectations. Lowering the temperature values of yield stress R_e and tensile strength R_m increase while the elongation A and strain ε m decrease. Values of elastic modulus E are approximately equal regardless of the temperature change and it can be seen in Figure 7 that the Hooks line for all curves have the same slope.

Figures 8, 9 and 10 show a particular material property regards to test temperature. A more detailed analysis of measurement results revealed that the tensile strength slightly increases in the range from room temperature down to −40°C after which the increase becomes more pronounced. As opposite to these results, lowering the temperature elongation A slight fall in the range

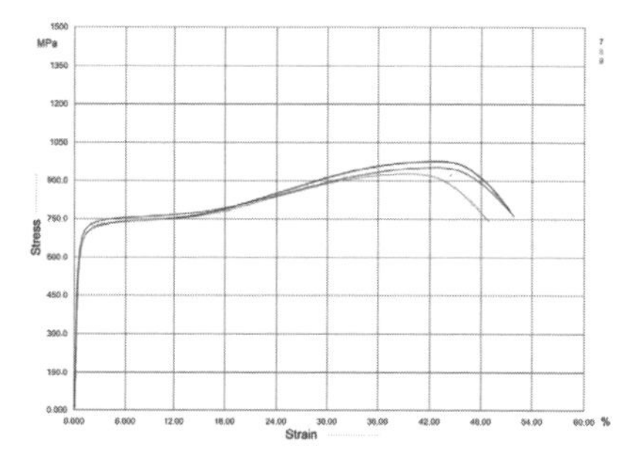

Figure 6. 'Stress-Strain' diagram of tensile test at temperature −80°C.

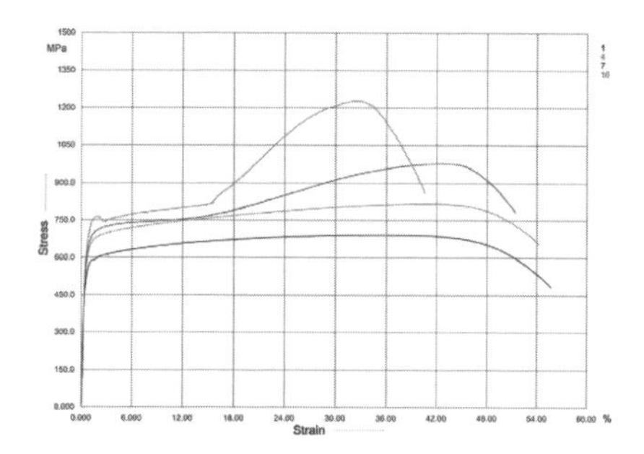

Figure 7. Comparative diagram of tensile tests at sub-zero temperatures.

Table 3. The results of the material mechanical characteristics testing at the temperature −80°C.

Ordinal number	Specimen label	d (mm)	E (MPa)	$R_{p0.2}$ (MPa)
1.	E-7-M80	7,88	195,2	572,6
2.	E-8-M80	7,93	203,6	561,8
3.	E-9-M80	7,95	212,1	580,0
	Mean val.	7,92	203,6	571,5

Ordinal number	Specimen label	Rm (MPa)	ε_m (%)	A (%)
1.	E-7-M80	976,7	42,14	50,94
2.	E-8-M80	927,2	39,14	48,41
3.	E-9-M80	951,3	42,34	51,46
	Mean val.	951,7	41,21	50,27

Table 4. The results of the material characteristics testing at sub-zero temperatures.

Ordinal number	Specimen label	d (mm)	E (MPa)	$R_{p0.2}$ (MPa)
1.	E-1-sobna	7,98	204,4	485,4
2.	E-4-M-40	7,99	198,5	568,5
3.	E-7-M80	7,88	195,2	572,6
4.	E-10-M165	6,85	203,2	603,4

Ordinal number	Specimen label	Rm (MPa)	ε_m (%)	A (%)
1.	E-1-sobna	688,9	35,79	55,44
2.	E-4-M40	815,9	40,30	53,82
3.	E-7-M80	976,7	42,14	50,94
4.	E-10-M165	1227,7	31,68	37,18

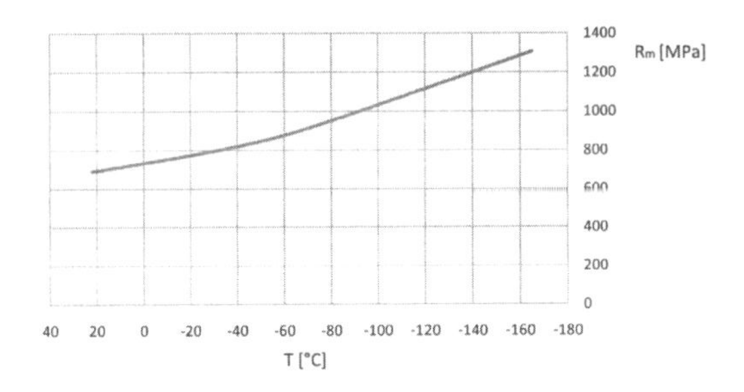

Figure 8. The values of tensile strength regarding temperature.

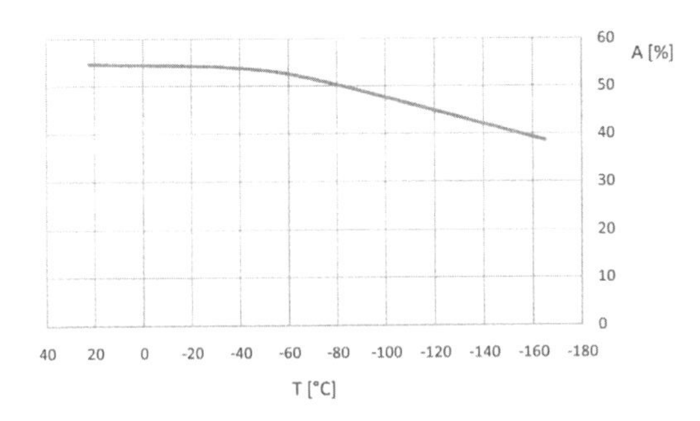

Figure 9. The values of elongation regarding temperature.

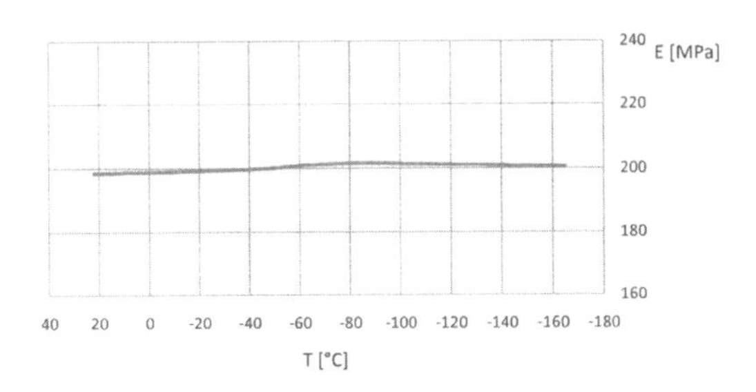

Figure 10. The values of elastic modulus regarding temperature.

from room temperature to –80°C after which the decline becomes more pronounced while the value of elastic modulus E can be concluded that is not dependent on temperature. For further considerations suggest the need for additional cycling test of materials, especially in the temperature range from –80°C and –165°C.

4 CONCLUSION

The paper presented the results of mechanical testing of low temperature effects on shipbuilding steels in use for building LNG carriers. The specimens are tested in cooled chambers on 4 different temperatures from 22°C to –165°C in the Laboratory for experimental mechanics at the Faculty of Mechanical Engineering and Naval Architecture of the University of Zagreb by static tensile test.

The tensile testing on specimens provided by Shipbuilding Industry in Split, Croatia, indicated significant changes of material properties in the range of temperatures from –80°C to –165°C degrees centigrade. By lowering the temperature below –80°C degrees the yield stress R_e and the tensile strength R_m are increasing. In the same time, the fracture elongation A and strains ε_m are decreasing. The modulus of elasticity E is not sensitive to low temperature changes. Change of the mechanical properties of material is determined especially in the zone of large plastic deformation suggesting additional tests.

The experimenting planned in the future will tackle low temperature toughness and low temperature fatigue possibly with constant and variable stress amplitudes.

REFERENCES

Croatian Register of Shipping 2009. Rules for classifications of ships, Part 25—Metallic materials.

Franz, M. 1998. Mechanical properties of materials (in Croatian), Faculty of Mechanical Engineering and Naval Architecture, Zagreb, Croatia.

International Maritime Organization 1986. IGC Code—International for the construction and equipment of ships carrying liquefied gases in bulk.

Ljubičić, P. 2010. Mechanical properties of steel at sub zero temperatures (in Croatian), master thesis, Faculty of Mechanical Engineering and Naval Architecture, Zagreb, Croatia.

Sustainable Maritime Transportation and Exploitation of Sea Resources – Rizzuto & Guedes Soares (eds)
 ISBN 978-0-415-62081-9

Buckling behavior of FRP sandwich panels made by hand layup and vacuum bag infusion procedure

Marco Gaiotti & Cesare M. Rizzo
DINAEL, Faculty of Engineering, University of Genova, Italy

ABSTRACT: This paper deals with strength of composite sandwich panels typically used in pleasure boats and naval industry. The focus is on the influence of fabrication processes onto the mechanical behavior, namely onto buckling strength. Both, experimental and numerical analyses were carried out in the Ship Structures Laboratory of the University of Genova on rectangular specimens, whose dimensions and geometry were conceived to induce buckling collapse, being this one a typical failure mode under compressive loading for thin skins sandwich. Tests on both hand layup fabricated and infusion made panels were carried out measuring compressive load vs. shortening and lateral deflection. Videos of the tests were taken and used to gather some measurements and to assess failure time and location. Experimental results were compared with different numerical models, developed considering large deformations in the PVC core under buckling conditions. Various finite element types and different implementation were considered to model complex interactions between core and skins. The conclusions of this paper aim to provide designers of pleasure and naval crafts with some hints about different strategies to build up cost-effective numerical models for predicting buckling and post-buckling behavior of sandwich hulls, taking advantage of calibrations obtained through experimental testing.

1 INTRODUCTION

Pleasure boat and naval industry have been extending the field of application of composite materials in the past decades to exploit the benefits allowed by novel fabrication technologies with the aim to build highly performing hulls. In particular, the application of sandwich composite laminates allows optimizing the strength to weight ratio, which is a simple but suitable index to evaluate the efficiency of the structural design of a craft.

While mechanical properties of composite materials, including sandwich, can be estimated by analytical formulations like the well known Classical Laminates Theory (CLT), depending on constituent materials properties, content of reinforcing material, construction methods, etc. (see e.g. Greene 1999), the obtained estimates are often rather different from the data achieved through experimental testing.

As a matter of facts, classification societies prefer empirical formulations for their rules rather than the application of the CLT and always require some testing sampled from each hull built aiming at demonstrating that the material characterization is in agreement with properties assumed in the scantling calculations.

Notwithstanding the above, due to the high complexity of the composite materials, which are indeed continuously improved and upgraded, mechanical characterization for some specific loading condition and for particular laminates is still lacking. Even more if the very limited experimental data freely available in open literature are considered.

In the present work, the results of an experimental campaign on sandwich laminates is presented. Tests were carried out aiming at demonstrating whether the infusion fabrication technology applied to sandwich laminates having a PVC core and fiberglass skins offers higher strength than traditional hand lay-up lamination under compressive loading.

Actually, the sandwich laminate was specifically conceived to investigate its buckling behavior and to validate numerical models, whose definition is indeed another and equally important aim of the work.

Kardomateas (2005) suggested a wide range validy elastic solution to the buckling of wide sandwich panels formulated as eigen boundary problem; by the way Hohe et al. (2005) showed the limits of a linear solution due to the transversal compressibility of the core and proposed a geometrically nonlinear FE model.

As reported by Sorensen et al. (2009), buckling is a very critical condition for composite laminates, especially for sandwich laminates. It can generate sudden and unexpected failures due to instability and, in case of particular geometrical features, a very limited post-buckling stiffness is shown.

Gaiotti & Rizzo (2011) recently analyzed the buckling behavior of the bottom structure of a typical single skin composite craft, which may be subject to compressive loading due to global loads.

Gaiotti et al. (2011) compared different finite element (FE) modeling strategies for buckling behavior of single skin composites.

Structural behavior of typical marine composite sandwiches is considered in the present paper taking advantage of experimental data as well as skills acquired in the above mentioned analyses of single skin laminates: two rectangular sandwich panels were built, the first one using the infusion technique and the other one by traditional handy lay-up.

Specimen sets were obtained from both panels and tested in order to assess the different buckling collapse behavior. In the first stage, three long span specimens for each fabrication method were tested to evaluate the post-buckling failure and corresponding finite element models (FEM) were built up and validated using experimental results. Thereafter a second series of shorter span specimens were cut out to study the in-plane compressive failure and modeled applying the FEM as well, as described in the following.

The development of numerical models is a challenging task in these cases because of the coupling between the low stiffness thick core with the rather thin fiberglass skins. Different modeling techniques are presented and compared with the experimental tests in order to analyze different modeling strategies in the light of the experimental results.

2 DESCRIPTION OF SPECIMENS

Two rectangular [1000 × 450 mm] sandwich panels were purposely made for the tests. The selected lamination sequence has been assumed as representative of typical stacking sequences of shipbuilding laminates, even if with a limited number of layers.

On the other hand, one of the aim of the study is the calibration of numerical models for buckling behavior of sandwiches. Therefore, thin skins were deemed suitable for this aim.

Fiberglass layers in an epoxy matrix are coupled to a 1 cm thick PVC core. Table 1 and Figure 1 summarize the stacking sequence and shows the layers fabrics, directions of the reinforcement fibers and relevant specific weight (per unit area or volume):

Table 1. Stacking sequence of sandwich.

Layer #	Material	Weave type	Orientation angle	Weight [g/m^2]	Thick [mm]
1.	fiberglass	Twill	0°/90°	200	0.15
2.		Biax	±45°	600	0.45
	PVC				
3.		–	–	75 g/m^3	10.0
4.	fiberglass	Biax	±45°	600	0.45
5.		Twill	0°/90°	200	0.15

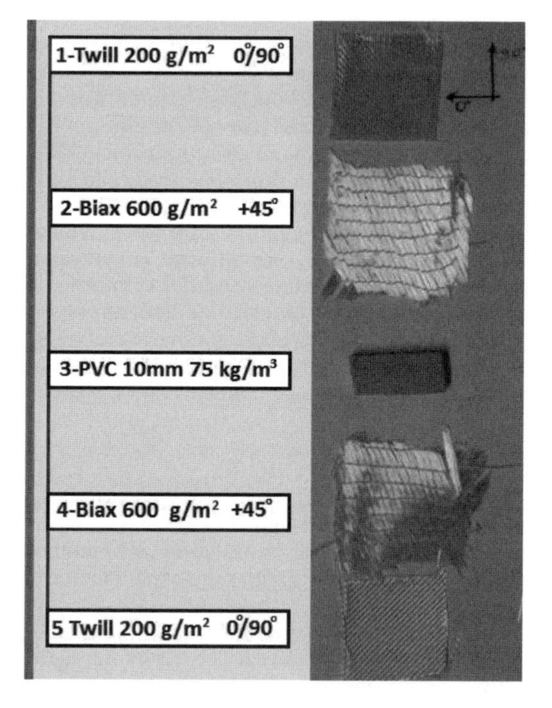

Figure 1. Stacking sequence representation.

The panel built following a traditional hand lay-up method was named "Panel A", while the one infused in a single shot of resin was named "Panel B".

In the first stage of the tests, three [450 × 150 mm] rectangular specimens were cut out from each panel, paying attention to avoid those parts of the panels showing imperfections at visual examination, e.g. visible fiber/matrix decohesions, localized delamination between skins and core, etc. Examples of visible imperfections are shown in Figure 2 while specimens are depicted in Figure 3.

Specimens were named A2, A3, A4 and B2, B3, B4, being obvious the meaning of symbols. The thicknesses of the upper and lower skins, and the global thickness of the sandwich were carefully measured at regular 50 mm steps using a digital caliper on all edges of the six specimens: no significant

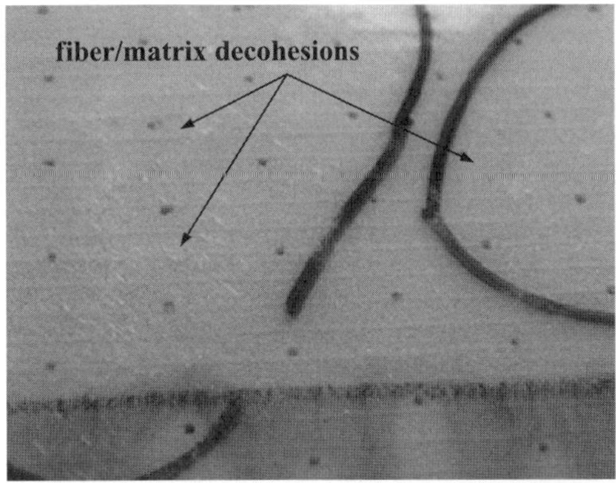

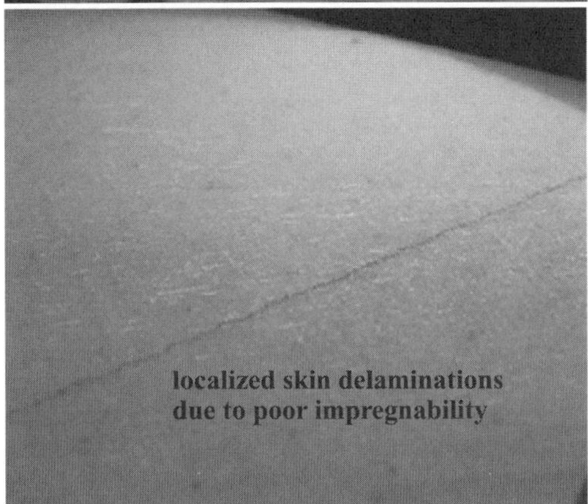

Figure 2. Examples of visible imperfections in the panels.

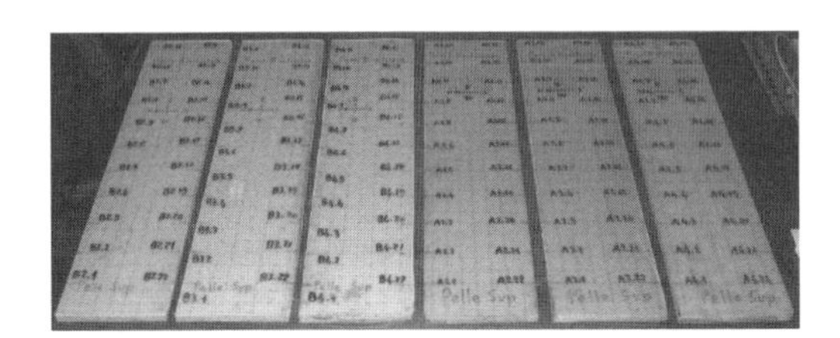

Figure 3. 3 + 3 rectangular specimens cut from Panel A and Panel B, thickness gauging points are shown.

differences were observed between the two fabrication methods (see Table 2), thus indicating a good manufacturing of the hand lay-up lamination.

Thicknesses of skins higher than nominal ones are noted for both fabrication methods. The upper skins are thicker due to gravity effects. In facts, the panels were laminated horizontally on a working plane. The core thickness appears slightly lower than the nominal one.

Thereafter, another set of specimens was obtained for the second phase of experimental

Table 2. Thickness statistics.

		Nominal [mm]	Average [mm]	Standard dev. [mm]
Panel A, hand lay-up	Upper skin	0.60	1.03	0.10
	PVC core	10.0	9.42	0.20
	Lower skin	0.60	0.94	0.12
Panel B infusion	Upper skin	0.60	1.21	0.14
	PVC core	10.0	9.36	0.19
	Lower skin	0.60	0.99	0.12

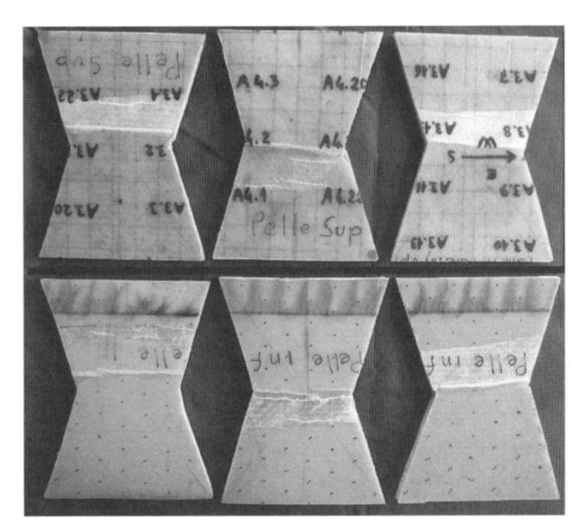

Figure 4. Specimens used for the short beam compressive tests.

testing. Rectangular specimens [200 × 150 mm] were constrained in order to leave a 100 mm span. Such short span was selected in order to produce a non-buckling failure and was appropriately shaped to avoid failure at clamped edges due to local compression of clamping devices and/or due to local bending effects.

The geometry of the samples is reported in Figure 4 and further details are explained in the following Figure 7. It is worth noting that, in order to induce the failure at midspan, a rounded notch was shaped in the specimens. However, this do not impair the scope of the comparative tests since the same geometry was maintained for all specimens.

Namely, three short beam specimens were cut out from each of the original panels. Similarly to the above mentioned ones they were named A2s, A3s, A4s, B2s, B3s, B4s, being obvious the meaning of the symbols.

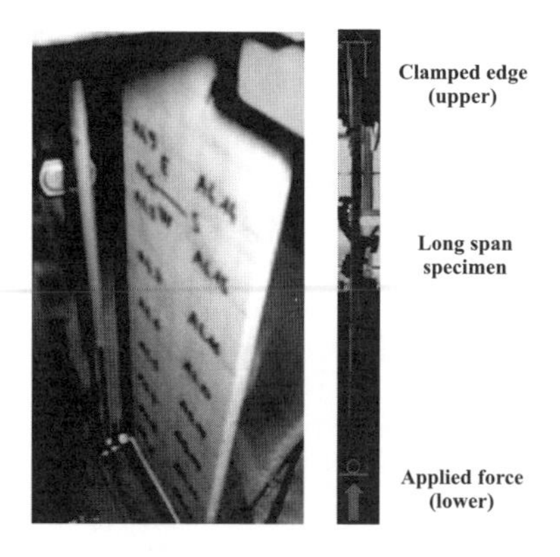

Figure 5. Buckling test equipment (long span tests).

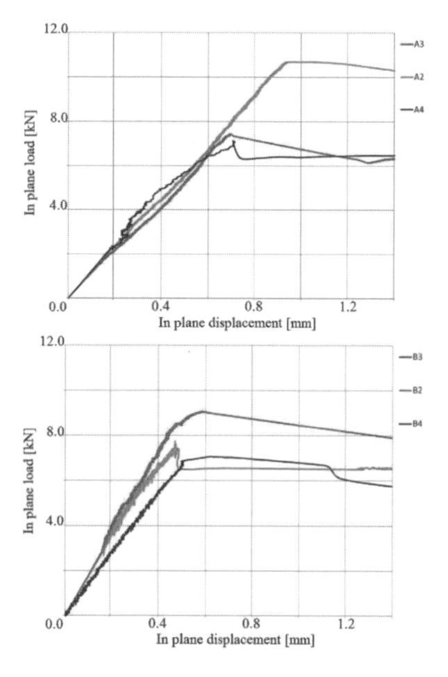

Figure 6. Load vs. displacement of long specimens tests.

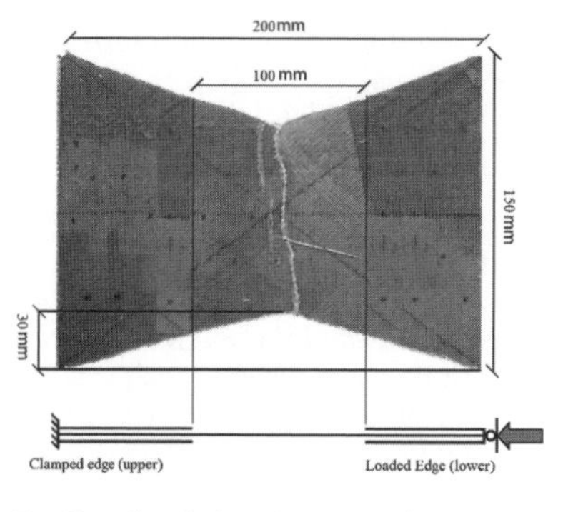

Figure 7. Details of short beam specimens and testing.

3 BUCKLING TESTS OF LONG SPECIMENS

Compressive tests were carried out on long span specimens (A2, A3, A4 and B2, B3, B4) by clamping one short edge of the specimens using a steel profile having a C-shaped cross section, as shown in Figure 5.

The other short edge of specimens was simply supported on the head of the hydraulic actuator applying the force, therefore rotations are free on the lower edge of the specimen. Such boundary condition is due to the articulated actuator head and does not impair the aim of the work. Actually, the paper focuses on the influence of the manufacturing process onto the material strength and in developing suitable numerical modeling of sandwich buckling collapse. Accounting for the above, the imposed boundary condition is not essential.

The hydraulic pressure in the actuator, easily converted into applied load, and the in plane displacements, were measured using a pressure gauge and a pair of displacement sensors, placed along the longer edges of the specimens.

Signals were acquired at sampling frequency $f = 20$ Hz. The rather high acquisition frequency was selected to catch even dynamic and/or impulsive effects taking place during the post buckling phase of the tests, if any.

As expected, after an initial linear load vs. in plane displacement relationship, the panel starts moving out of its plane, introducing a stiffness reduction due to the progressive failure of the sandwich skins in way of the clamped edge. This phenomenon was clearly observed in five out of six tested specimens. Basically, the failure is caused by the large and sudden out of plane displacements taking place in the post buckling equilibrium configuration, leading to high stress concentrations.

However, beside the failures in way of clamped edges appearing during the collapse tests, in all tested specimens cracks first extended into the core starting from the mid span of the skin in compression and eventually leading to global collapse. The tensioned skin was practically undamaged at the end of the experimental tests.

Table 3 and Figure 6 shows the experimental results of the hand lay-up specimens (up) and of the infusion made specimens (down).

From tests results the following considerations can be drawn out:

- A fairly linear relationship is confirmed up to collapse load by all specimens,
- The ultimate load does not seem to be depending on the fabrication method as for all specimens

Table 3. collapse load/displacement of long span specimens.

Specimen	Failure load [kN]	Displace-ment [mm]
A2	10.8	0.92
A3	7.3	0.68
A4	7.1	0.70
B2	7.9	0.47
B3	6.9	0.49
B4	9.1	0.57

the collapse load is about 7 kN, except for specimen A2 and B3 that show slightly higher strength,
- The infusion made specimens are approximately 10% stiffer in terms of elastic compressive modulus.

4 SHORT BEAM COMPRESSIVE TESTS

The previously described short beam specimens were compressive tested similarly to the long span specimens, as shown in Figure 4. Figure 7 shows geometrical details of the specimens and of the clamping devices applying boundary conditions during the tests. While the specimen length was 200 mm, only 100 mm free span was left. Moreover, a V-shaped notch was created as shown in Figure 7, aiming at inducing failure at mid span due to compressive loading without buckling effects. In fact, the scope of the tests was to compare the different failure modes of the specimens obtained by two different fabrication procedures, if any.

The following Table 4 summarizes the results of the short beam tests.

During the tests two different failure modes occurred:

- Early skin de-bonding followed by core failure, observed in half cases and shown in Figure 7(1), at collapse load of about 12 kN,
- Sudden skin fracture leading to global failure, observed in remaining half cases and shown in Figure 7(2), at collapse load of about 14 kN.

At this stage of the research, no dependency from the fabrication method was observed on the failure mode; early failure are instead initiated by local de-bonding on the interface between the fiberglass skins and the PVC core.

If de-bonding of skins occurs, then a typical core local failure is observed followed by the collapse of the sandwich. As far as the adhesion is effective, the specimen sustain the external loads until the collapse under compression of one of the skins.

Table 4. collapse load/displacement of short beam specimens.

Specimen	Failure load [kN] / mode	Displace-ment [mm]
A1s	12.0 / (1)	0.13
A2s	12.3 / (1)	0.14
A3s	13.9 / (2)	0.12
B1s	14.1 / (2)	0.14
B2s	15.2 / (2)	0.14
B3s	12.1 / (1)	0.16

5 NUMERICAL MODELS

5.1 *Long span specimens*

Different modeling strategies were applied to simulate the structural behavior of the sandwich laminate under compressive loading: the most challenging task is the correct simulation of the transversal shrinkage effect (Poisson's effect) of the sandwich core, which is an isotropic foam material.

Testing showed that this effect is negligible in a typical static loading condition but becomes significant in non linear buckling analyses, as the ones presented in this work. Moreover, the bonding between core and skins need to be properly modeled.

A linearized (Eigenvalue) buckling analysis was initially carried out for every test carried out, in order to identify the buckling mode shapes. Shapes of the buckling modes were used to generate initial imperfections in the subsequent non linear collapse buckling simulation. The 1st mode shape was indeed observed during tests. An out of plane displacement of 0.1 mm, corresponding to 1% of the total nominal thickness of the laminate, was assigned to the node having the maximum out of plane component of the Eigen-vector, as described by Sorensen et al. (2009). This is a common assumption for such kind of simulations. A compressive total load corresponding to the critical load obtained by the linearized buckling analysis increased by 50% was applied in 100 identical load steps, aiming to obtain a fair in-plane load vs in-plane displacement relationship.

The following FE models, having nominal dimensions and thicknesses, were run and obtained results were compared to the experimental tests, providing validation of the FE modeling strategies:

A. A traditional 2D layered shell model was initially considered: this model completely ignore the Poisson's effect of the core, modeled as a layer of the element, thus leading to an over-estimated critical load and to a post buckling stiffness much greater than the one of the experimental tests, as shown in Figure 10. In this case as well as in the following ones, layered MITC elements were used as available in the ADINA software. Further details about the element formulation can be found in ADINA, 2008.

B. Owing the limits of the above mentioned simulation, the FGRP skins were later modeled as separated (layered) shell elements and connected each other, node by node, by rigid links on 5 DoF leaving the out of plane translation free and thus allowing the shrinkage of core material). This model correctly estimated the critical load, but in the pre-buckling phase the stiffness was under-estimated because longitudinal stiffness of the core was completely missing.
C. To overcome the above approximations, a combined rigid-links/springs model was developed and studied: five out of six degrees of freedom of each node of the skin shell elements were constrained through rigid links, while a spring element couples the transversal displacements. However, this model does not produce any significant improvements, especially if its complexity is taken into account. The load displacements curve is almost coincident with the one obtained through the previously described model B. Moreover, the longitudinal stiffness of the core material is still missing.
D. Aures further attempt was performed by modeling a grid of shell elements in between the shell elements idealizing the skins, as shown in Figure 9D; the mechanical properties of the shell elements used to built up such grid are set-up to restore the core stiffness of the sandwich also in longitudinal direction. In practice, the sum of the thicknesses of the shell elements representing the core is equal to the width/length of the core of the specimen while the material properties are the ones of the core. This model better agrees with the experimental curves, but the matching cannot be considered satisfactory.

Due to the difficulties introduced by coupling the thick but soft PVC core to the thin but stiff fiberglass skins, none of the simplified 2D shell FE model previously considered seems to be validated by the experimental tests.

A further step is thus needed to improve the numerical analysis, i.e. the natural choice to model the core by using 3D solid elements coupled to 2D shell elements for the skins. Figure 9E shows the FE model while the corresponding results are reported in Figure 10.

However, a problem suddenly arises when attempting to link together different elements because of the different number of degrees of freedom: three translational DOF for the solid elements nodes and six degrees of freedom for the shell nodes. Hence a direct coupling is difficult and it is impossible to correctly transfer the rotational DOF of the skins elements to the core ones.

To overcome this issue, the facing nodes on skins and core, were constrained via rigid links coupling only the three translational DOF. This modeling assumption leaves the shell nodes free to rotate independently from the core, but if the mesh is sufficiently refined, this approximation may be acceptable, being rotation of skins nodes limited.

This latter model matches very well with the experimental results, even in the post-buckling phase, where the specimens clamped edge do not suffer early failure. Of course, modeling is more time consuming and computations requires more computer resources also because of DOF coupling between shell and solid elements.

Figure 10 summarizes the experimental/ numerical comparisons reporting also a curve obtained taking into account the average value of in plane load vs. in plane displacement for the tests carried out with the specimens built by handy lay-up and for tests of infused specimens.

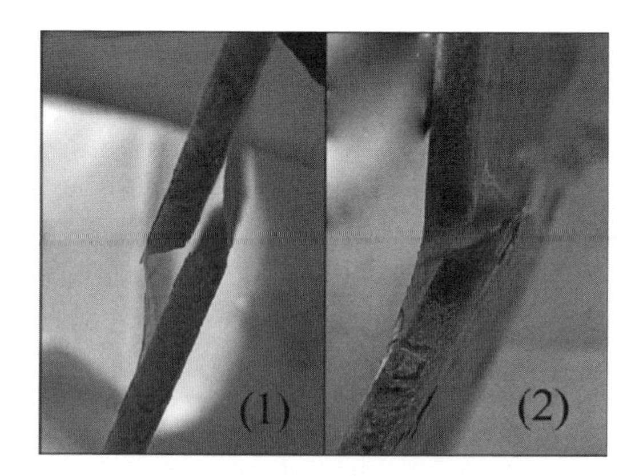

Figure 8. Different failure modes observed during short beam tests.

5.2 *Short beam specimens*

It is rather difficult if not practically impossible to correctly simulate the short beam tests by using simplified FE models matching the needs of the widely applied structural analyses of pleasure crafts.

Actually tests highlighted that the failure mode is basically governed by the efficiency of the adhesion between skins and core.

Therefore, it is required to explicitly model the bonding layer as well as the skins and the core, hence increasing dramatically the complexity of

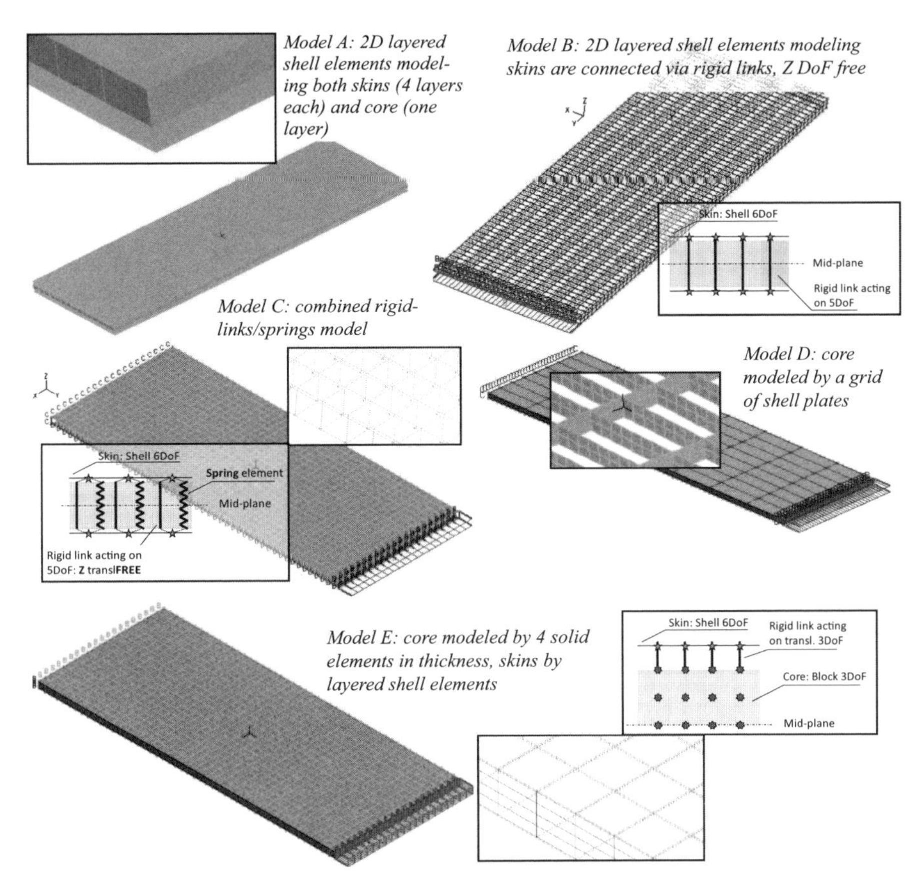

Figure 9. FEM models built and run for the long span specimens buckling analyses.

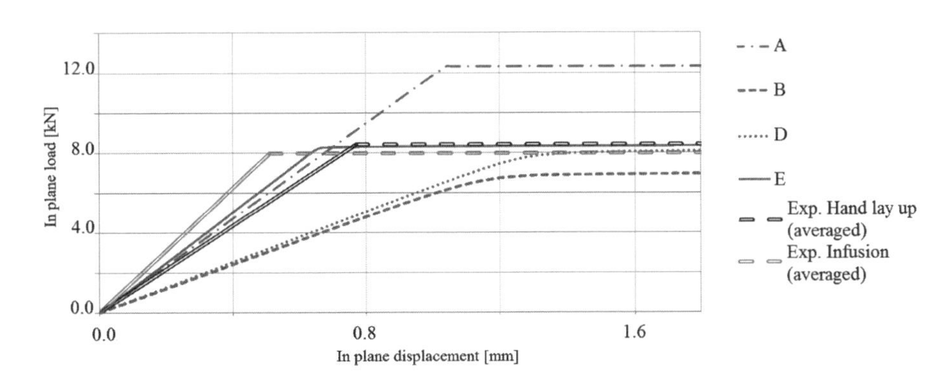

Figure 10. Results of the FEM analyses (a selected experimental results is shown for comparison purposes).

the analyses and being necessary in addition to properly calibrate the necessary input data, e.g. by means of lap shear tests assessing the adhesive layer.

However, a simple FE model was built and run with twofold aim: on one side it was deemed interesting to estimate the stress concentration factor due to the V-notch, on the other side the deviation from experimental testing results of a typical FE model which assumes a perfect bonding between the skins and the core can be estimated.

The FE model is shown in Figure 11, where the longitudinal component of the stress is depicted having applied a compressive load of 15 kN.

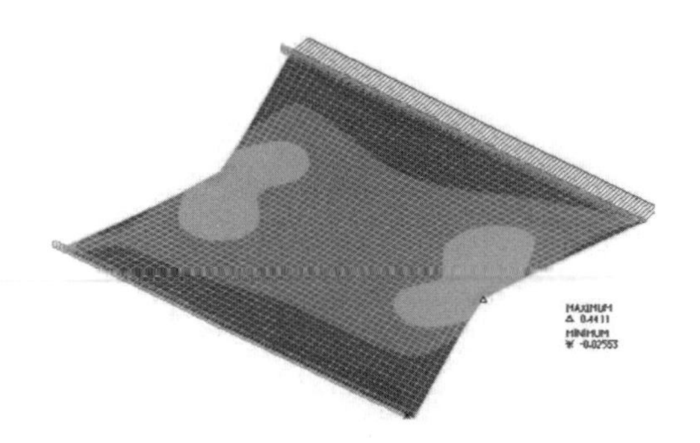

Figure 11. FE model of short beam specimen (load: 15 kN, equivalent stress field according to Tsai-Wu criterion).

Such load is almost the maximum failure load occurred during short beam tests.

The stress concentration factor was found to be about 3.9, assuming an equivalent isotropic material to simulate the laminate. Actually, failures due to boundary effects at edges of laminates are well known. However, it is pointed out that in this case the failure occurred as shown in Figure 8, i.e. involving the whole cross section of the sandwich due to the compressive load.

Despite the compressive load applied in the FE calculation was sufficient to fail the specimens during testing, the most stressed layer of the laminate, i.e. the ±45° layer, is only stressed by an equivalent stress value less than half the one obtained by using the widely applied Tsai-Wu failure criterion (see e.g. Greene, 1999); namely the acting stress is only 44% of the failure criterion stress.

6 CONCLUSIONS

In the present paper two different fabrication technologies for sandwich panels are compared to assess strength of laminates under compressive loads.

The experimental sets of specimens were designed in order to show both buckling and ultimate compressive failure.

No significant difference has been noted depending on the fabrication method from the tests in terms of ultimate strength, but different failure modes were observed depending on localized adhesion strength (see Figure 8).

A parallel work is carried out to obtain a reliable FE model to predict buckling behavior of sandwich panels with low stiffness and soft compressible core: five modeling approaches were tested starting from a basic 2D shell model up to a 3D model with newly proposed shell to solid coupling constraints.

The final result can be considered satisfactory and the experimental validation leads to consider the strategy of modeling the core of the sandwich by using solid elements appropriately coupled with the skins as a quite reliable one and worth of further investigations.

Another important consideration arising from the different models presented in this paper is that the transversal stresses are not negligible in buckling problems where low core stiffness modifies the geometrical shape of the tested panel.

ACKNOWLEDGEMENTS

The authors wish to thank Mr G. Spano who built sandwich panels for the tests and the technicians of the Ship Structures Laboratory of the University of Genova, Mr Stelvio Musicò and Mr Massimo Fiaschi, for their invaluable support during the tests.

REFERENCES

ADINA 2008. Theory and modelling guide v.8.6, ADINA R & D, Inc., Watertown, Massachusset, USA.

Gaiotti, M. & Rizzo, C.M. 2011, An analytical/numerical study on buckling behaviour of typical composite top hat stiffened panels, *Ships and Offshore Structures* (in press).

Gaiotti, M., Rizzo, C.M., Branner, K. & Berring, P. 2011. Finite elements modeling of delaminations in composite laminates, In: Advances in Marine Structures, C. Guedes Soares, W. Fricke Ed.s, *Proc.s MARSTRUCT 2011 Congress*, 28–30 March 2011, ISBN: 9780415677714.

GL 2009. Draft Rules & Guidelines, Part I-3-1 Special crafts, Germanisher Lloyd, Hamburg.

Greene, E. 1999. Marine Composites. Annapolis: Eric Greene Associates Inc.

Hohe, J., Librescu, L. & Yong Oh, S. 2005. Effect of Transverse Core Compressibility on Dynamic Buckling of Sandwich Structures. Sandwich Structures 7: Advancing with Sandwich Structures and Materials, Part 5, 527–536, DOI: 10.1007/1-4020-3848-8_53.

Kardomateas, G. 2005. Global Buckling of Wide Sandwich Panels with Orthotropic Phases: An Elasticity Solution. Sandwich Structures 7: Advancing with Sandwich Structures and Materials 2005, Part 2, 57–66, DOI: 10.1007/1-4020-3848-8_5.

Shenoi, R.A. & Wellicome, J.F. ed.s. 2005. Composite materials in maritime structures. Cambridge: University Press. UK, ISBN-13: 978-0521451543.

Sørensen, B.F., Branner, K., Lund, E., Wedel-Heinen, J. & Garm, J.H. 2009. Improved Design of large wind turbine blade of fibre composites (phase 3). *Summary Report, Risø-R-1699(EN)*, Risø National Laboratory for Sustainable Energy, Denmark.

3.5 *Degradation/Defects in structures*

Sustainable Maritime Transportation and Exploitation of Sea Resources – Rizzuto & Guedes Soares (eds)
© 2012 Taylor & Francis Group, London, ISBN 978-0-415-62081-9

Uncertainty in corrosion wastage prediction of oil tankers

P. Jurišić
Croatian Register of Shipping, Split, Croatia

J. Parunov & K. Žiha
University of Zagreb, Faculty of Mechanical Engineering and Naval Architecture, Zagreb, Croatia

ABSTRACT: The paper describes investigation of the global and local corrosion wastage of three single-hull oil tankers built in eighties. Analysis of data is based on existing thickness measurements of hull elements from Croatian Register of Shipping (CRS) file, gauged on periodic dry-docking and close-up surveys of ships in service after 10, 15 and 20 years. Hull Girder Section Modulus (HGSM) and local corrosion wastage of main deck plates and longitudinals are determined as a function of time taking into account the lifetime of the protective coatings. The available theoretical non-linear corrosion propagation model is fitted to the measurements after 10 and 15 years in order to predict the corrosion loss at 20 years. Uncertainty of the corrosion propagation model is then determined by the statistical analysis of differences between predicted and measured corrosion losses at 20 years.

1 INTRODUCTION

Damages to ships due to corrosion are very likely and the possibility of accident increases with the aging of ships. Statistics reveal that corrosion is the number one cause for marine casualties in old ships. The consequences of corrosion wastage can be very serious in some circumstances. Figure 1 shows the main deck area of 20-year-old tanker where deck plates and longitudinals suffered various degrees of corrosion. In some locations, the web plates of deck longitudinals are totally wasted away. This may cause loss of support of deck plates from longitudinals. Thus, the unsupported span of the deck plate may increase decreasing the plate buckling strength. In heavy seas, elasto-plastic collapse of the main deck structure may occur that eventually could lead to the collapse of the hull girder and the loss of the ship.

Experts in the maritime industry most seriously take into consideration the corrosion wastage as one of the very important degradation factors for ship structural strength. Several theoretical models have been proposed to monitor and predict the wastage condition of a structural components. (Paik et al., 2003, Garbatov et al., 2007, Gua et al., 2008, Melchers 2008).

As a consequence of corrosion wastage of the longitudinally effective plating and longitudinals, Hull Girder Section Modulus (HGSM), as a fundamental measure of the ship longutudinal strength, also deteriorate over the time. There is a clear tendency nowadays to adopt theoretical non-linear models in order to predict the long-term corrosion wastage progression and associated loss of HGSM (Wang et al., 2008).

Such direct approach for corrosion progression could be useful tool for classification societies and ship owners in order to predict long-term behaviour of hull structure and to decide if the renewal of the hull structure is necessary and when would be the optimal time for the repair (Garbatov & Guedes Soares 2010). Furthermore, such direct approach has potential to facilitate application of more accurate computational methods in design and analysis of oil tankers (Parunov et al., 2007, 2008, 2010).

Unfortunately, corrosion progression models that are used today are based on different

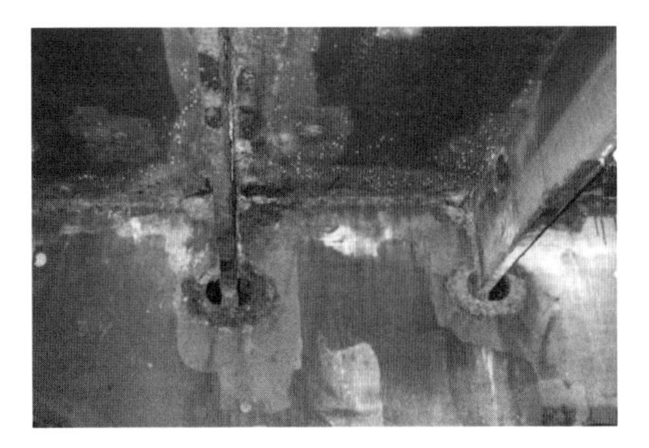

Figure 1. Heavily corroded under-deck of 20 years old oil tanker.

assumptions and methodologies and consequently large uncertainty are associated to their application. Differences arise because of the uncertainties in the thickness measurements and the complexity of the corrosion mechanism (Matsukura et al., 2011).

The aim of the present paper is to investigate practical applicability of the currently used long-term non-linear corrosion wastage predictions. This is done in a way that corrosion losses of three ships after 20 years in service are predicted based on thickness measurements after 10 and 15 years. The actual measurements after 20 years are then compared to the theoretical prediction and statistical measures of uncertainty are determined. Firstly, the assessment of HGSM loss is performed, following by the assessment of local corrosion of main deck plates and longitudinals. For the local corrosion analysis, total number of 6567 measured data are used.

Findings of the study can be used for assessment of the ship hull fitness for service and for planning ship hull inspection as well as for reducing maintenance costs of oil tankers in service.

2 DESCRIPTION OF SHIPS

The paper describes investigation of corrosion wastage of three oil tankers with single-hull structure built in eighties. The whole cargo area is made of mild steel for two sister ships, while one ship has high tensile steel in bottom and deck areas and mild steel in neutral axis area including longitudinal bulkheads.

Central tanks along cargo hold areas are cargo oil tanks, while wing tanks can serve as ballast or cargo oil tanks, see Fig. 2.

3 ASSESSMENT OF HGSM LOSS

The assessment of the HGSM loss is performed based on the following procedure. Firstly, the as-built HGSM is calculated. Then, thickness of structural elements (plates and longitudinals) contributing to the longitudinal strength are modified according to the results of thickness measurements from Croatian Register of Shipping

Table 1. Main characteristic of three single-hull tankers.

		Ship no. 1		Ships no. 2 & 3	
Length	*Lpp*	237	m	205	m
Breadth	*B*	42	m	48	m
Depth	*D*	20.5	m	19	m
Draught	*T*	13	m	13	m
Deadweight	*DWT*	78000	dwt	80000	dwt

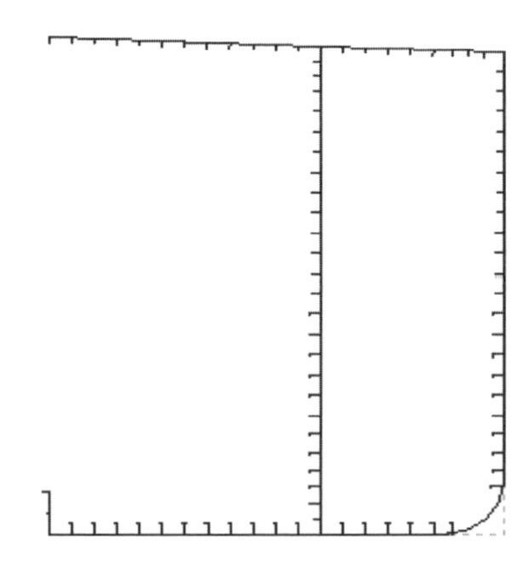

Figure 2. Midship section of single-hull oil tankers.

(CRS) file. Gauging records were performed on periodic dry-dockings and close-up surveys of ships in service after 10, 15 and 20 years. Corrosion model of the ships was performed for the transverse sections with combination of central tanks as cargo oil tanks and wing tanks as ballast tanks.

The aging effect is measured by the HGSM loss, which is the ratio of the as-gauged HGSM over the as built:

$$R(t) = 1\text{-}HGSM(\text{as-gauged at year } t) / HGSM(\text{as-built}) \quad (1)$$

Results for measured R(t) for three ships after 10, 15 and 20 years are presented in the Table 2.

HGSM loss is determined as a function of time taking into account the lifetime of protective coatings. The equation for the HGSM loss after t years of ship service used in the present paper is proposed by Wang et al. (2008):

$$R(t) = C(t-t_0)^I \quad (1)$$

where R(t) is the HGSM loss at age t, while t_0 is the year when HGSM starts to deviate from the as-built condition. C and index I are constants to be determined according to the data set.

In order to investigate usefulness of the prediction model, curves according to the equation 1 are fitted through measurement points of HGSM loss at 10 and 15 years. Since there are three unknown variables in equation (1) (t_0, C and I), while only two available calibration points (at 10 and 15 years), t_0 has to be assumed. Resulting parameters of the equation (1) are presented in the Table 3, while corresponding diagrams are shown in the Figure 3.

It appears from Figure 3 that HGSM loss at 20 years for ship no. 1 is significantly underestimated by the prediction curve. Prediction of HGSM loss in 20 years for ship no. 2 is quite good.

Table 2. Measured R(t) for three ships.

Year	R(t)-Ship 1	R(t)-Ship 2	R(t)-Ship 3
10	0.0151	0.0145	0.0161
15	0.0225	0.0290	0.0445
20	0.0410	0.0470	0.0592

Table 3. Parameters of Equation (1) for different ships based on measurement after 10 and 15 years.

Ship No.	C	t_0, years	I
1	0.60	5	0.58
2	0.44	6	0.86
3	0.92	8	0.81

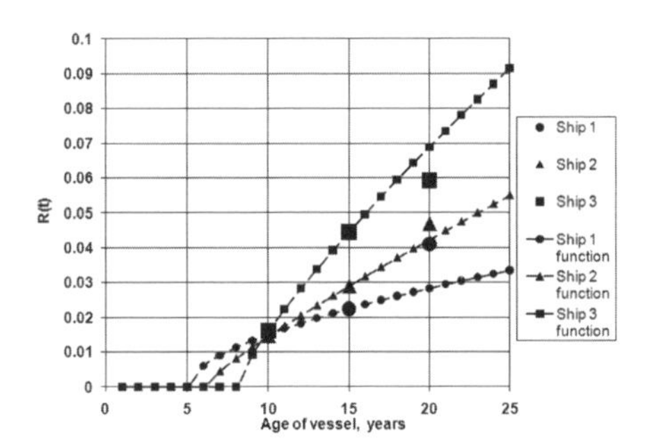

Figure 3. Measured and predicted HGSM losses for different ships.

Curve is almost linear from time of coating breakdown (6 years). Results for ship no. 3 indicate that the coating lifetime for that ship could be the longest (8 years), but after the breakdown of the coating corrosion progression is very fast. Prediction curve for that ship would overestimate actual HGSM loss at 20 years.

It should be mentioned that there is a large uncertainty associated to HGSM loss as the corrosion loss is different for each transverse section of the ship hull. It may be assumed, however, that surveyors of analysed ships measured sections with representative (average) corrosion losses. More research is required to improve reliability of long-term predictions of corrosion losses.

4 ASSESSMENT OF THE LOCAL CORROSION WASTAGE

The progression of the local corrosion wastage is performed separately for plates and longitudinals of the main deck in ballast and cargo tanks. For the local corrosion wastage, equation 1 takes following form (Gua et al., 2008):

$$C(t) = \alpha(t-t_0)^\beta \quad (2)$$

where $C(t)$ is the corrosion wastage at age t; t_0 is the year when thickness of the plates starts to deviate from the as-built condition; α and β are constants that can be determined according to the measurement data.

Total number of measurements of deck plates and longitudinals used in the present study reads 6567. 2135 measurements are used for ship no. 1, 2079 for ship no. 2 while 2353 for ship no. 3. Similar as for HGSM, curves are fitted through measurement points of HGSM loss at 10 and 15 years. Since there are three unknown variables in equation (2) (t_0, α and β), while only two available calibration points (at 10 and 15 years), t_0 has to be assumed.

Fitting of the equation 2 is performed for the mean corrosion diminution thickness and also for the extreme wastage, corresponding to 5% exceeding probability. To estimate extreme corrosion wastage, probability distribution of corrosion wastage is fitted to the each data set measured at year t of the ship life. The Weibull distribution is frequently used for representing the corrosion wastage at certain ship age (Parunov et al., 2007, Gua et al., 2008). Such good fitting of the Weibull distribution is confirmed also in the present study, as may be seen from the Figure 4, representing corrosion wastage of the deck plates in the cargo tanks at year 15 for ship no. 2.

Results of the analysis for the mean and extreme corrosion wastage of the main deck plates in cargo tanks are presented in Figures 5 and 6 respectively.

It appears from Figure 5 that the corrosion model overpredicts mean corrosion wastage for ship no. 2 while underestimate measured corrosion for ships nos. 1 and 3. Coating lifetime for mean corrosion is assumed to be 6.5 years.

One may notice from Figure 6 that the extreme corrosion after 20 years is predicted well for ship no. 3. For ship no. 1 and ship no. 2 corrosion after 20 years is underestimated and overestimated respectively. Prediction curve is almost linear for

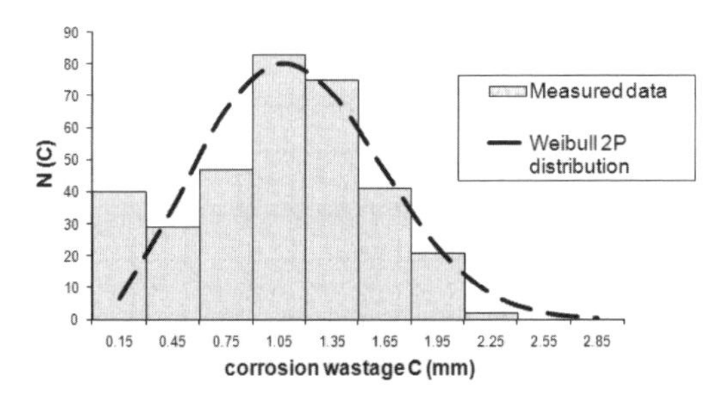

Figure 4. Weibull probability density function of the corrosion wastage of deck plating in cargo tanks.

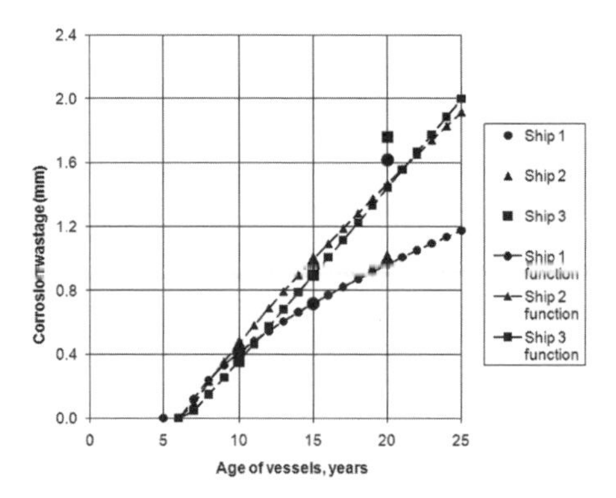

Figure 5. Measured and predicted mean corrosion wastage of the main deck plates in cargo tanks.

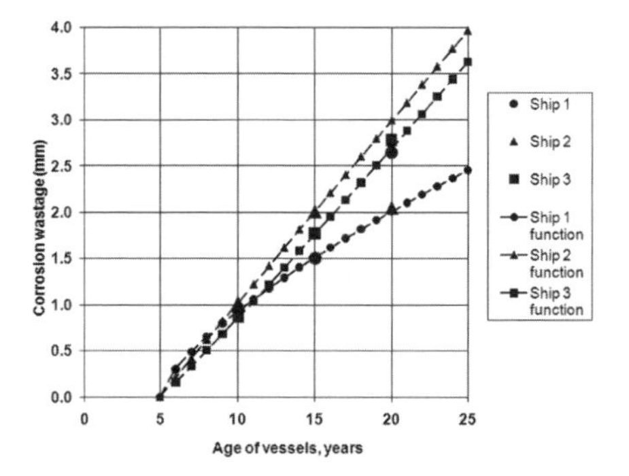

Figure 6. Measured and predicted extreme corrosion wastage of the main deck plates in cargo tanks.

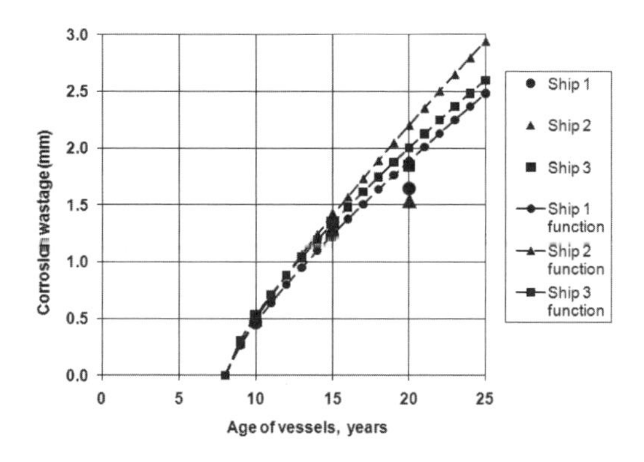

Figure 7. Measured and predicted mean corrosion wastage of the main deck longitudinals in cargo tanks.

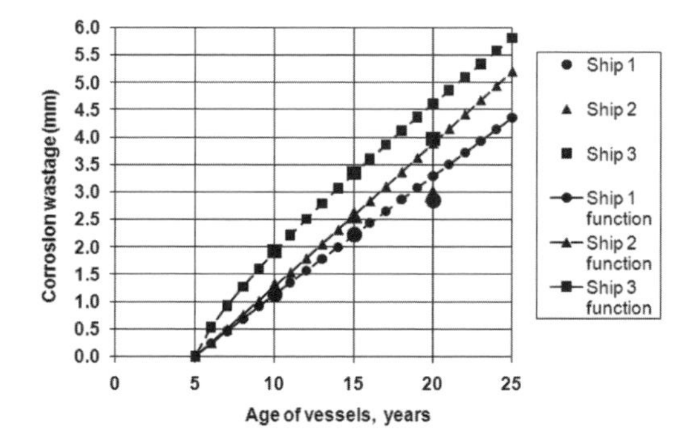

Figure 8. Measured and predicted extreme corrosion wastage of the main deck longitudinals in cargo tanks.

ships no. 2 and 3 while coating breakdown time is estimated to 5 years for all three ships.

It is interesting to notice from both Figures 5 and 6 that for ship no. 2 corrosion deterioration after 20 years is only slightly higher comparing to corrosion wastage after 15 years. For ship no. 1, in contrary, difference between corrosion wastage for 20 and 15 years is unexpectedly large, indicating that in that case concave curve would be more appropriate for corrosion process modelling (Paik et al., 2003).

Results of the analysis for the mean and extreme corrosion wastage of the main deck longitudinals in cargo tanks are presented in Figures 7 and 8 respectively. All three prediction models overestimate corrosion wastage after 20 years. That is valid for both mean and extreme corrosion wastage. The trend noticed for deck plates, that the corrosion rate is reduced between years 15 and 20 for ship no. 2, is evident also for deck longitudinals. By comparing results for deck plates and longitudinals, one may conclude that corrosion wastage of deck longitudinals is larger comparing to deck plates. Thus, for 20 years, extreme corrosion wastage of deck longitudinals is between 3 mm and 4 mm, while for deck plates it is between 2 mm and 3 mm. Coating breakdown time for deck longitudinals is estimated to be 8 and 5 years for mean and extreme corrosion respectively.

Results of the analysis for the mean and extreme corrosion wastage of the main deck plates in ballast tanks are presented in Figure 9 and 10 respectively. For ship no. 1, similar as for cargo tank deck plates, difference between corrosion wastage for 20 and 15 years is quite large, indicating the concave type of the corrosion process. The tendency of the reduced corrosion rate for the period between 15 and 20 years, already noticed for cargo tanks, is evident also for ballast tanks of ship no. 2. Consequently, the corrosion wastage after 20 years is systematically overestimated for that vessel. For ship no. 3, predicted corrosion agrees excellent with measurements for both mean end extreme corrosion wastage.

By comparing Figures 9 and 10 with Figures 5 and 6, it appears that corrosion wastage is higher

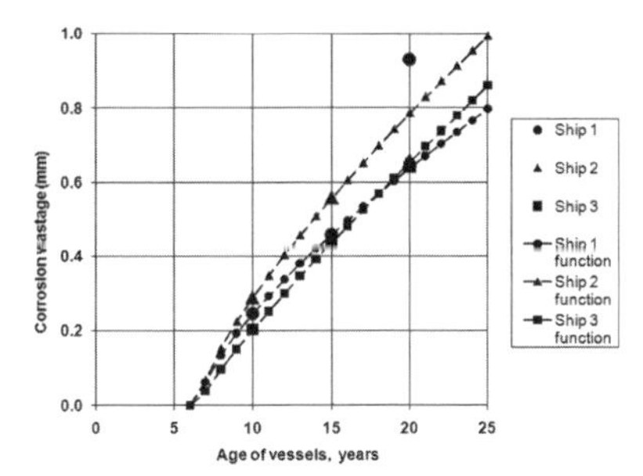

Figure 9. Measured and predicted mean corrosion wastage of the main deck plates in ballast tanks.

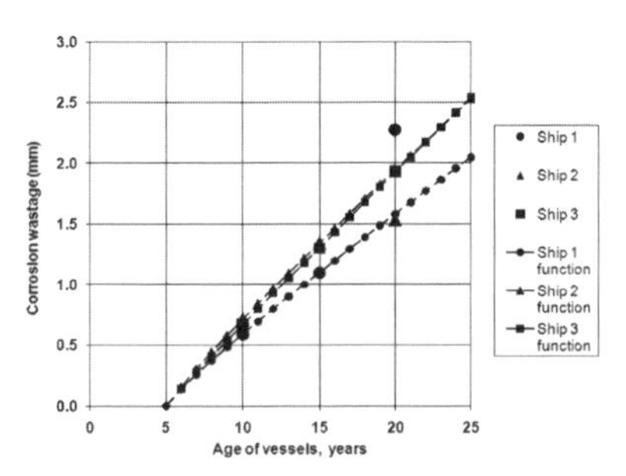

Figure 10. Measured and predicted extreme corrosion wastage of the main deck plates in ballast tanks.

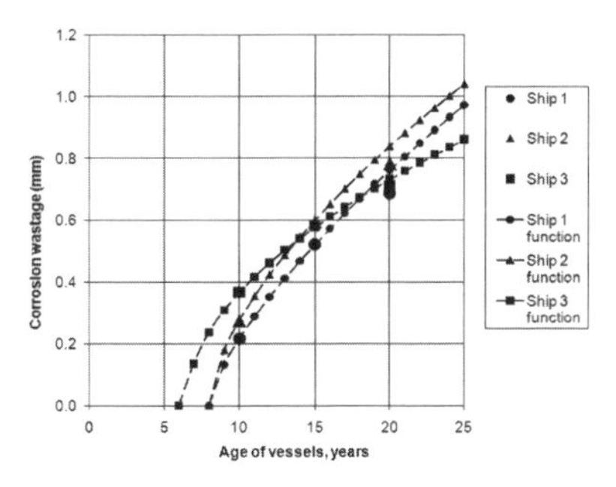

Figure 11. Measured and predicted extreme corrosion wastage of the main deck longitudinals in ballast tanks.

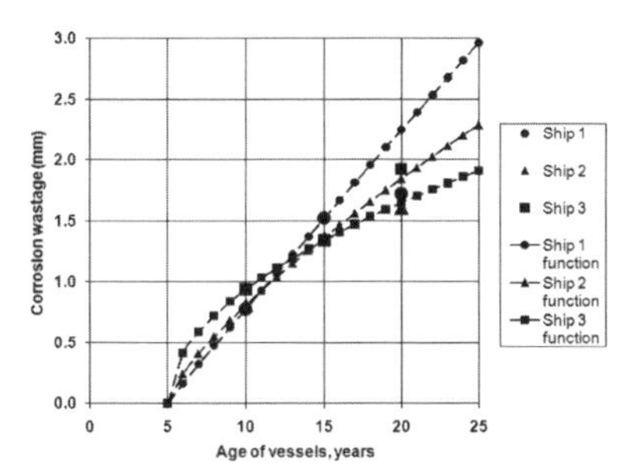

Figure 12. Measured and predicted extreme corrosion wastage of the main deck longitudinals in ballast tanks.

for deck plates in cargo tanks than in ballast tanks. Coating breakdown time is estimated to be the same as for deck plates in cargo tanks, i.e., 6.5 and 5 years for mean and extreme corrosion respectively.

Results of the analysis for the mean and extreme corrosion wastage of the main deck longitudinals in ballast tanks are presented in Figures 11 and 12 respectively. Corrosion prediction model for mean values slightly overestimates measured data after 20 years. For extreme values, prediction model overestimate measured values for ships nos. 1 and 2, while underestimates for ship no. 3. Generally, measured corrosion for deck longitudinals in ballast tanks is approximately the same as for deck plating in ballast tanks. Corrosion breakdown time for deck longitudinals in ballast tanks reads 5 years for extreme corrosion, while 6.5 and 8 years for mean corrosion wastage.

5 UNCERTAINTY OF THE LOCAL CORROSION PROGRESSION MODEL

In order to quantify uncertainty of the corrosion propagation model used in the study, results for the predicted and measured mean and extreme corrosion wastage are presented in Tables 4-6 and subjected to further statistical analysis.

The average ratio of the measured and predicted corrosion wastage read 1.01 and 0.96 for mean end extreme wastage respectively, while the corresponding standard deviations read 0.30 and 0.24. This indicates that the model almost has no bias, but uncertainty of predictions is rather large.

If only cargo tanks are considered, than average ratio measured/predicted corrosion wastage read 1.01 and 0.92 for mean end extreme wastage respectively, while corresponding standard deviation read 0.38 and 0.23.

If only ballast tanks are considered, than average ratio measured/predicted corrosion wastage read 1.01 and 1.00 for mean end extreme wastage respectively, while corresponding standard deviation read 0.22 and 0.26. Therefore, the agreement between predictions and measurements is better for ballast than for cargo tanks.

Table 4. Comparison of predicted and measured corrosion wastage for ship no. 1, mm.

Element	Area	Mean (predicted)	Mean (measured)	Extreme (predicted)	Extreme (measured)
Plates	Cargo	0.96	1.62	2.01	2.65
	Ballast	0.64	0.93	1.58	2.27
Long.	Cargo	1.89	1.65	3.29	2.85
	Ballast	0.76	0.69	2.25	1.72

Table 5. Comparison of predicted and measured corrosion wastage for ship no. 2, mm.

Element	Area	Mean (predicted)	Mean (measured)	Extreme (predicted)	Extreme (measured)
Plates	Cargo	1.47	1.01	2.99	2.05
	Ballast	0.79	0.66	1.95	1.54
Long.	Cargo	2.20	1.54	3.90	2.98
	Ballast	0.84	0.78	1.84	1.61

Table 6. Comparison of predicted and measured corrosion wastage for ship no. 3, mm.

Element	Area	Mean (predicted)	Mean (measured)	Extreme (predicted)	Extreme (measured)
Plates	Cargo	1.45	1.76	2.69	2.78
	Ballast	0.65	0.64	1.93	1.93
Long.	Cargo	2.00	1.85	4.62	3.97
	Ballast	0.73	0.71	1.65	1.92

6 CONCLUSIONS

Analysis of data is based on existing thickness measurements of hull elements of three single-hull oil tankers built in eighties. Corrosion wastage is gauged on periodic dry-docking and close-up surveys of ships in service after 10, 15 and 20 years. The main idea of the paper was to fit available non-linear corrosion prediction models to corrosion measurements after 10 and 15 years and then to compare with measurements after 20 years, aiming to investigate the uncertainty of the long-term corrosion prediction models. Following conclusions may be drawn:

1. estimate of the hull girder section modulus loss after 20 years is satisfactory for 2 out of 3 studied ships. Coating lifetime is between 5 and 8.5 years.
2. corrosion wastage in cargo tanks is larger than in ballast tanks, while deck longitudinals in cargo tanks experienced the largest corrosion wastage of all structural elements analysed
3. non-linear long-term local corrosion progression model calibrated based on measurements after 10 and 15 years in average predicts very well corrosion loss after 20 years. However, uncertainty of such prediction, measured by standard deviation, is rather large.
4. it is also found that prediction model is more accurate and reliable in ballast than in cargo tanks.

Measured extreme corrosion wastage after 20 years of service is well below corrosion addition of 4 mm prescribed in newly developed Common Structural Rules (CSR) for Double-hull Oil Tankers (ABS et al., 2006).

REFERENCES

ABS, DNV, LLOYD'S REGISTER 2006. Common Structural Rules for Double Hull Oil Tankers.

Garbatov, Y., Guedes Soares, C. & Wang, G. 2007. Non-Linear Corrosion Wastage of Deck Plates of Ballast and Cargo Holds of Tankers. *Journal of Offshore Mechanics and Arctic Engineering*. 129(1):48–55.

Garbatov, Y. & Guedes Soares, C. 2010. Risk based maintenance of deteriorated ship structures accounting for historical data. In: *Advanced Ship Design for Pollution Prevention*. ed. Guedes Soares & Parunov. Taylor & Francis Group. London. UK. 131–147.

Guo, et al. 2008. Time-varying ultimate strength of aging tanker deck plate considering corrosion effect. *Marine Structures* 21. 402–419.

Matsukura, T. et al. 2011. A study on long-term prediction of corrosion wastage. In: *Advances in Marine Structures*. Guedes Soares & Fricke (eds). Taylor & Francis Group. London. 699–705.

Melchers, R.E. 2008. Development of new applied models for steel corrosion in marine applications including shipping. *Ships and oddfshore Structure*., 3:2, 135–144.

Paik, J.K., Wang, G., Thayamballi, A.K., Lee, J.M. & Park, Y.I. 2003. Time-dependent risk assessment of aging ships accounting for general/pit corrosion, fatiguecracking and local denting damage. *Transactions SNAME* (2003). Vol. 111. 159–197.

Parunov, J., Žiha, K. & Mage, P. 2008. Corrosion wastage of oil tankers—A case study of an aged ship. In: *Maritime Industry, Ocean Engineering and Coastal Resources*. Guedes Soares & Kolev (ed.). Taylor & Francis Group. London. 271–276.

Parunov, J., Mage, P. & Guedes Soares, C. 2008. Hull girder reliability of an aged oil tanker, *Proceedings of the 27th International Conference on Offshore Mechanics and Artic Engineering*. ASME. Estoril. Portugal, Paper no. 57183.

Parunov, J., Žiha, K., Mage, P. & Jurišić, P. 2010. Hull girder fatigue of corroding oil tanker. In: *Advanced Ship Design for Pollution Prevention*, ed. Guedes Soares & Parunov. Taylor & Francis Group. London. UK. 149–154.

Wang, et al. 2008. A statistical investigation of time-variant hull girder strength of aging ships and coating life. *Marine Structures* 21. 240–256.

Sustainable Maritime Transportation and Exploitation of Sea Resources – Rizzuto & Guedes Soares (eds)
© 2012 Taylor & Francis Group, London, ISBN 978-0-415-62081-9

Influence of weld toe shape and material models on the ultimate strength of a slightly corroded box girder

S. Saad-Eldeen, Y. Garbatov & C. Guedes Soares
Centre for Marine Technology and Engineering (CENTEC), Instituto Superior Técnico, Technical University of Lisbon, Portugal

ABSTRACT: The objective of this paper is to analyze the ultimate strength of a slightly corroded box girder subjected to four-point loading resulting in a constant vertical bending moment along the box girder, using non-linear finite element method. A series of nonlinear collapse analyses have been performed and the ultimate strength has been analyzed using different material models and also different weld toe shapes. Different elasto-plastic material models have been developed accounting for the residual stresses effect and post-buckling behavior. Comparisons between numerical and experimental results have been performed for the slightly corroded box showing a very good agreement.

1 INTRODUCTION

The ability to evaluate the ultimate strength of a ship's hull subjected to longitudinal bending moment is an important aspect of ship structural design.

The linear elastic theory has been employed to predict the longitudinal strength of the ship hull for years. According to this theory, the maximum bending moment that the hull cross-section can withstand is equal to the bending moment corresponding to the first yield, that is, the bending moment when the maximum stress on the hull cross-section reaches the yield stress of the material. In design practice, an allowable stress is used instead of the yield stress, which corresponds to a safety factor against yielding.

However, many studies in the last decades have revealed that to estimate the longitudinal strength of the ship hull several factors have to be accounted for: (1) various possible failure modes including buckling, (2) progressive and interactive behavior of the failure of structural members, (3) redistribution of stresses on the hull cross-section and (4) residual strength of structural members after buckling and even after collapse. By considering these factors, the maximum bending moment that the hull cross-section can withstand is designated by the ultimate longitudinal strength, which represents the maximum load-carrying capacity of the ship hull under longitudinal bending.

The ultimate longitudinal strength assessment is a nonlinear problem in which both the nonlinearity related to material behavior and geometry is involved (Hu, et al., 2001).

Manevich (2007) based on the analysis of plastic buckling problems for bars and plates subjected to compressive load, concluded that in certain slenderness range the hardening does not increase the critical stresses noticeably.

A series of finite element analyses were conducted to simulate the behavior of the tested box girders. Hansen (1996) performed four finite element analyses with different considerations for a cross-section selected from the models tested by Nishihara (1984). The finite element model used in the analyses was built up by 4-noded shell element with 5 layers in the thickness. The model was loaded incrementally with force vector at the end of the load section in order to ensure the overall bending.

Qi, et al. (2005) performed a series of nonlinear finite element analyses for the ultimate strength of tested box girders simulating a large surface ship, frigate and double hull tanker. The analyses demonstrated a good agreement with the test results. Moreover, a comparison has been done to calculate the ultimate hull girder strength for a large double hull tanker using different methods.

Nikolov (2009) performed comparison between the experimental results and different simplified methods of five different box girders used for ultimate strength tests. The stress-strain and moment-curvature relationships were obtained. The comparison showed that there is significant difference between the numerical and in some experimental results.

Saad-Eldeen, et al. (2011a) performed a series of nonlinear finite element analyses for an intact box girder. The most appropriate parameter description for the developed finite element model by varying both the initial imperfection (shape and amplitude) and mesh size that are one of the most important governing parameters affecting the ultimate strength and post-collapse behavior has been defined. The results achieved based on the developed finite element model demonstrated a good agreement with the experimentally defined three-linear trend of the moment-curvature relationship of the intact box girder.

All of the mentioned experimental works were dealing with intact structures. The step forward is the study of ultimate strength of aged box girders performed by Saad-Eldeen, et al. (2010a), where a test of a slightly corroded multispan stiffened box girder representing a midship section of ship hull has been carried out. The box girder was subjected to four-point loading resulting to a uniform bending moment along the specimen. The ultimate bending moment of the tested box girder has been compared with empirical formulae showing a good agreement.

A continuation of the previous study is the one conducted by Saad-Eldeen, et al. (2011b) where a box girder made of mild steel subjected to moderate corrosion was tested and by Saad-Eldeen, et al. (2011c) where a severely corroded box girder was also tested under same conditions. It has been concluded that the load carrying capacity and ultimate bending moment are highly affected by corrosion deterioration of plating and the material properties changes.

The study presented here uses the finite element model defined for the intact box-girder reported by Saad-Eldeen et al. (2011a). The aim of the analysis performed here is to compare the experimental results of the slightly corroded box girder as reported by Saad-Eldeen, et al. (2010a) with the one achieved by finite element method, employing commercial code ANSYS, based on different material and weld toe models.

2 HULL GIRDER STRENGTH ASSESSMENT

The hull girder ultimate strength is defined as the maximum bending capacity of the hull girder beyond which the hull will collapse. Hull girder failure is controlled by the ultimate strength of structural elements, considering buckling and yielding.

The principal parameters governing the buckling strength are the plate and column slenderness defined as:

$$\text{plate slenderness, } \beta = \frac{b}{t}\sqrt{\frac{\sigma_y}{E}} \tag{1}$$

$$\text{column slenderness, } \lambda = \frac{a}{r}\sqrt{\frac{\sigma_y}{E}} \tag{2}$$

where t is the plate thickness, b is the plate width between stiffeners, a is the stiffener span between frames I is the moment of inertia and A is the cross-sectional area of the stiffener including the associated plate, σ_y is the yield stress, E is the Young modulus and r is the radii of gyration of the cross-section of the stiffener with an associated plate:

$$r = \sqrt{\frac{I}{A}} \tag{3}$$

2.1 *Box girder configurations*

The analysed box girder consists of three bays. The deck panel is stiffened with five longitudinal flat bars with a spacing of 150 mm. The side panel is stiffened with two stiffeners on a distance of 300 and 500 mm respectively and the bottom panel was stiffened with one stiffener in the middle, as may be seen Figure 1. The geometry configurations of the analysed box girder are given in Table 1.

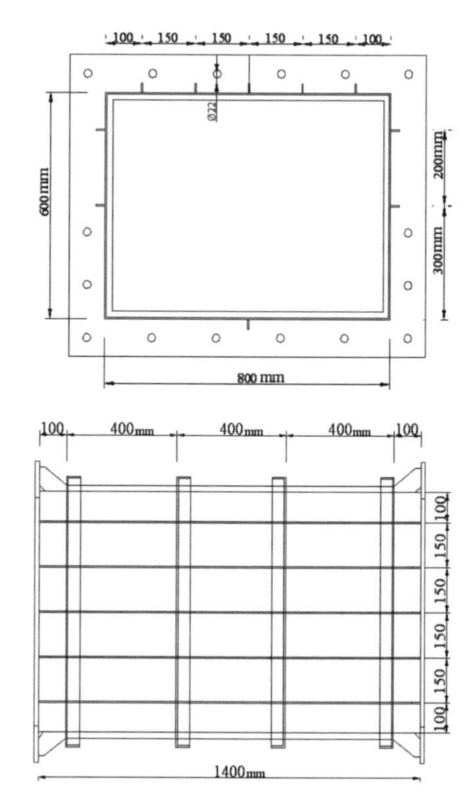

Figure 1. Box girder geometry.

Table 1. Principal characteristics of initial corroded box girder.

Item	Dimensions	units
Deck plating	4.09	mm
Port side plating	3.95	mm
Starboard side plating	3.85	mm
Bottom plating	3.75	mm
Stiffeners	25 × 4.35	mm
Web frames	50 × 50 × 6.14	mm
Brackets	80 × 100 × 3.91	mm

Figure 2. Experimental set up.

The plate and column slenderness are estimated as 1.23 and 0.71 respectively. The ratio of the stiffener height to thickness is 5.73. Saad-Eldeen, et al. (2010b) reported that the slightly corroded box girder matches the 0.2 year service life without accounting for coating life.

2.2 *Experimental scheme*

The box girder has been mounted between two stiff supporting arms, using bolt connections. The box girder is subjected to four-point loading producing a vertical bending moment. The bottom part is subjected to tension and the upper part, deck, is under compression. The bending moment is kept constant along the box girder, between bolt connections. As can be seen in Figure 2, there are four points for transmitting the loading, two are located at the supports of the arms and two are on the frontier between box-girder and supporting arms. To avoid the shear effect, the load is subjected on a certain distance away from the ends of the box girder (the connections between the box girder and the supporting arms). The load is applied by a 700 kN hydraulic jack and it is transmitted to the box girder through a horizontal beam, see Figure 2.

3 FINITE ELEMENT MODEL

Numerical analyses of the ultimate strength for a slightly corroded box girder are performed based on general nonlinear finite element commercial code—ANSYS. The finite element analyses utilize the full Newton–Raphson equilibrium iteration scheme. The large deformation option was activated to solve the geometric and material nonlinearities and to pass through the extreme points. The automatic time stepping features are employed allowing the program to determine appropriate load steps.

The geometry of the box-girder structure is modelled in the way as the real one used during the ultimate strength test without any simplifications (see Figure 7). Shell elements were used to generate the entire FE model. The shell element, SHELL 93 is defined by eight nodes, four nodal thicknesses, with six degrees of freedom at each node: translations in the nodal x, y, and z directions and rotations about the nodal x, y, and z axes. The deformation shapes are quadratic in both in-plane directions. The element has plasticity, stress stiffening, large deflection, and large strain capabilities.

3.1 *Material definition*

The material used in the model was of low carbon steel with the yield stress, σ_y of 235 [MPa], the Young's modulus, E of 206 [GPa] and the Poisson's ratio, υ of 0.3.

Four material models are used in the finite element analyses to invistigate the effect of each model on the ultiamte sterength and the post-collapse behavior of the box girder.

3.1.1 *Elastic-prefectly plastic model*

For the elastic-perfectly plastic (EP) stress-strain model, see Figure 3, at stresses below the yield stress, σ_y, the material behaviour is linear with a tangent modulus $E_{T0} = \sigma_y / \varepsilon_y$. At the level of the yield stress, the material flows plastically without strain hardening. When the material is unloaded by reducing the stress below the yield stress, it behaves elastically in a manner unaffected by the plastic flow.

3.1.2 *Elastic-prefectly plastic with hardening model*

In the case of elastic-perfectly plastic with hardening model (EP+H) (see Figure 4), once the yield stress is reached, the stress continues to increase achieving plastic deformation.

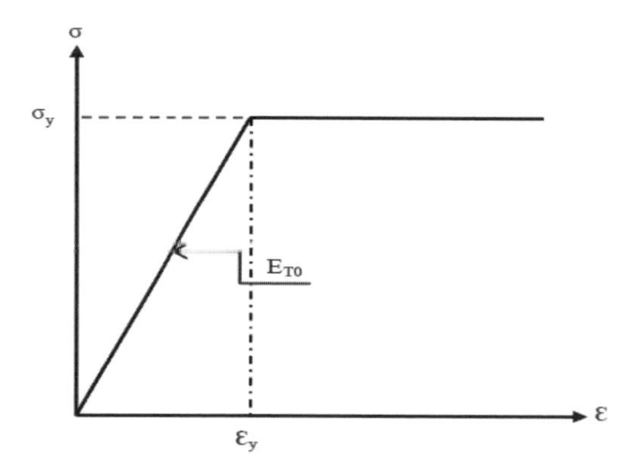

Figure 3. Elasto-perfectly plastic stress–strain model (EP).

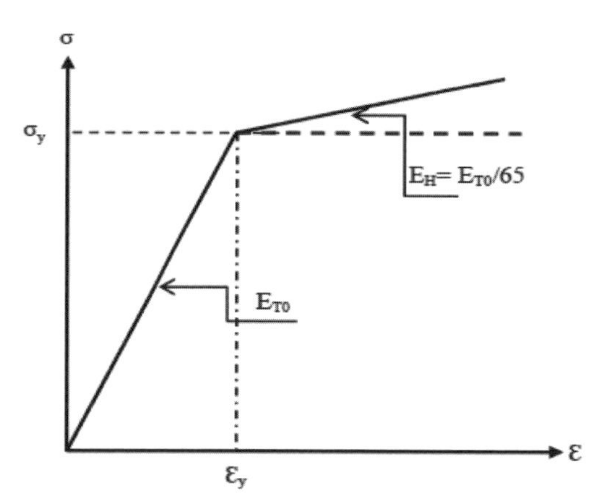

Figure 4. Elasto-perfectly plastic stress–strain with hardening model (EP+H).

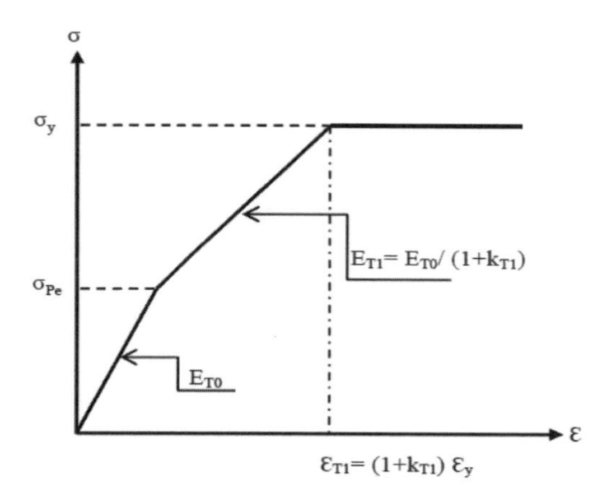

Figure 5. Elasto-plastic modified stress–strain model (EPM).

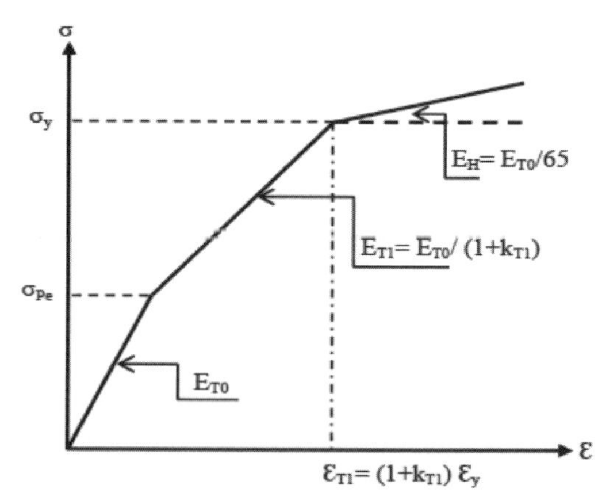

Figure 6. Elasto-plastic modified stress–strain with hardening model (EPM+H).

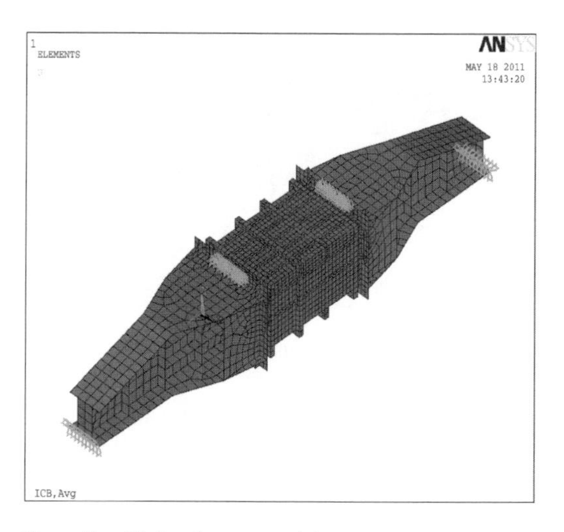

Figure 7. Finite element model.

It is evident that, strain-hardening effect has some influence on the nonlinear behaviour of strucutres. The degree of such an influence is a function of many factors including plate slenderness. The strain-hardening rate of $E_H = E_{T0}/65$ is used. The strain-hardening tangen modulus has been chosen based on a large number of elastic–plastic large deflection analyses reported by Khedmati, et al. (2011).

3.1.3 *Elastico-plastic modified model*

The elasto-plastic modified (EPM) stress strain model, as shown in Figure 5, is a modification of the idealized elastic-plastic model by introducing a second tangent modulus, E_{T1}, starting from the point up to which the stress and strain are linearly, proportional, σ_{Pe}. The second tangent modulus is calcualted as $E_{T1} = E_{T0}/(1 + k_{T1})$.

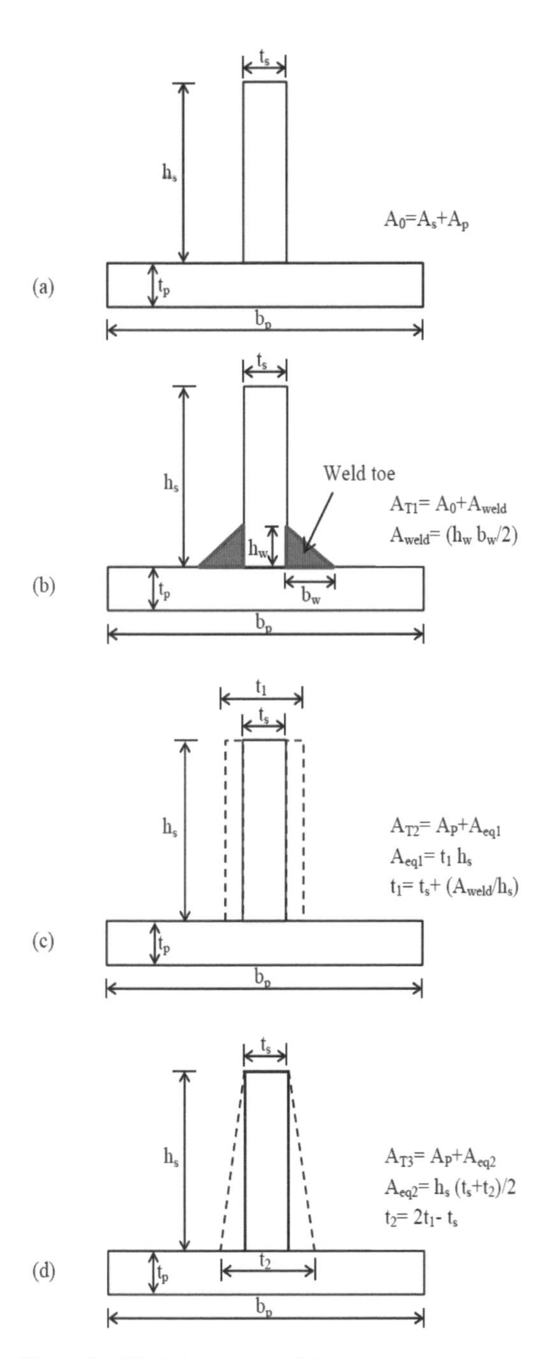

Figure 8. Weld shape toe models.

where E_{T0} is the first tangent modulus, the Young modulus, and k_{T1} is a coefficient adopted to account for the effect of the existing residual stresses in the real structure. The finite element model does not account explicitly for residual stresses.

3.1.4 *Elastico-plastic modified with hardening model*

The elasto-plastic modified stress-strain model (EPM+H), is a combination between the modified elasto-plastic and the strain hardening models, as may be seen in Figure 6.

3.2 *Load application and boundary conditions*

The box girder is subjected to vertical loading producing a constant pure vertical bending moment along the box-girder length. The load is generated by imposed vertical displacement acting on the two heavy plates, which are located outside of analyzed box-girder on the supporting arms connection (see Figure 7). The displacement load is applied by small increments to ensure that the analysis would closely follow the structural load-response curve. The both ends are simply supported, constrained from translations in the vertical and transverse direction. The translation in the longitudinal direction is only constrained at the one of the ends. No rotation is prevented.

3.3 *Weld toe modeling*

Fillet welds are used to build up the box girder welding the stiffeners, web frames etc. to the attached plate, as may be seen in Figure 8 (b).

To account for the presence of the weld toe in the finite element analysis two weld-toe models are used.

The first one models the weld-toe by introducing an extra thickness along the stiffener height, h_s as shown in Figure 8 (c). The added area equals the original weld toe area, A_{weld} Figure 8 (b). The newly defined equivalent thickness of stiffener, t_1 is calculated as $t_1 = t_s\, A_{weld} / h_s$.

The second weld-toe model is defined by increasing the thickness of stiffener at the connecting node of the stiffener and the plate. The newly defined thickness at the connecting node, t_2, can be calculated as given in Figure 8 (d).

4 FINITE ELEMENT MODELLING

The aim of the nonlinear analyses presented in this section is to investigate the effect of different material and weld-toe models on the post-buckling behavior as well as the ultiamte bending moment.

4.1 *Coarse mesh size*

In order to obtain reasonable results, and to find out the appropriate element size to be used for the finite element model a systematic finite element analyses have been carried out by Saad-Eldeen, et al. (2011a) for an intact box girder having the same configurations of the analyzed box girder here. It has been concluded that for this particular box girder ultimate strength analysis, the appropriate finite element size is 5 [cm] as may be seen from Figure 9.

It can be also observed that, there is an inflection point at the level of the element size of 5 [cm], which refers to the change in the gradient of the ultimated bending moment behavior, therefore, for the present FE analysis, the most appropriate element size is chosen as 5 [cm].

4.2 *Initial imperfections*

In the present study, the plate initial imperfection is modeled based on real initial imperfection measurements as reported by Saad-Eldeen, et al. (2011d) and presented in Figure 10 as:

$$w_{i,IC}(z) = w_{io,IC}\left|\sin\left(\frac{\pi z}{a}\right)\right| \tag{4}$$

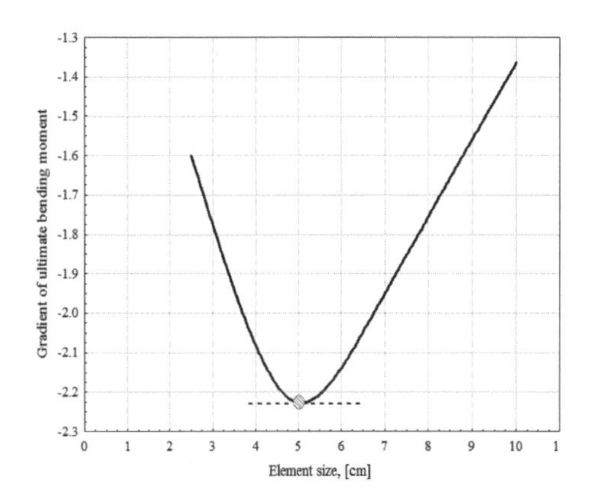

Figure 9. Gradient of ultimate bending moment vs. element size.

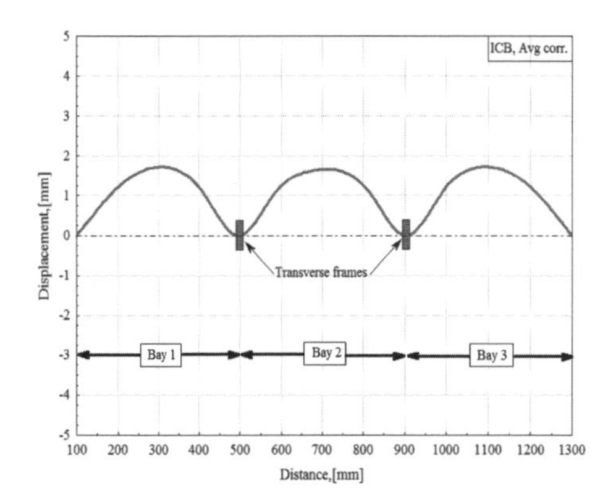

Figure 10. Initial imperfections shape.

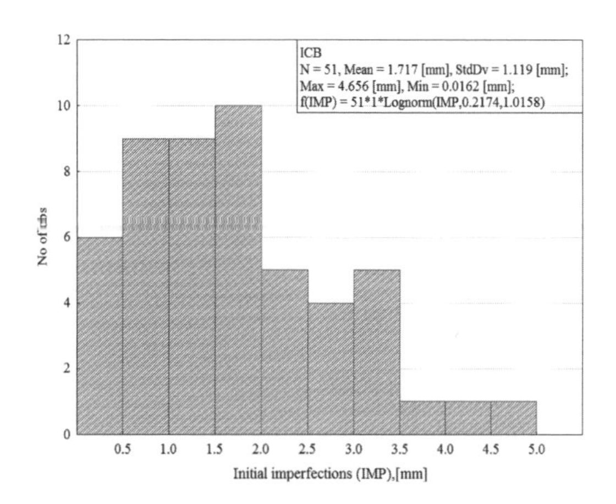

Figure 11. Initial imperfection amplitudes.

where w_0 is the mean value of the amplitude of the imperfections. The mean value and standard deviation for the measured imperfections are 1.717 and 1.119 [mm], respectively, as can be seen in Figure 11. These imperfections have been generated in the finite element model by changing the vertical position of element nodes without inducing any additional stresses.

5 FINITE ELEMENT RESULTS

Two groups of nonlinear FE analyses have been conducted. The first group deals with the effect of different material models on the post-collapse behavior and the ultimate bending moment of the box girder. Four different material models have been used to analyze behavior of the initial corroded box girder under vertical bending moment.

The moment-curvature relationship for each case is plotted in Figure 12. For the two models, without hardening (EP and EPM), it is noticed that with the use of the EP model, the box girder showed a linear relationship until reaching the collapse point, which makes the behavior in the post-buckling far away from the experimental results. The box girder registers a higher ultimate bending moment than the experimentally achieved one as can be seen in Figure 12 and Table 2, with 8.27%.

With the application of the elasto-plastic modified model (EPM), the box girder registered an ultimate bending moment bigger than the experimentally achieved one with 5.94%, which is less than the one obtained by the (EP) model, see Table 2.

In the post-buckling region, the behavior of the box girder using the EPM model, is much more closer to the experimental results, therefore, it is reasonable to have a second tangent modulus with

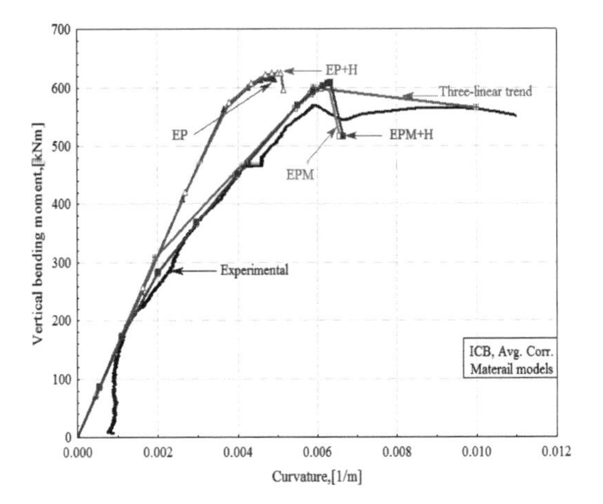

Figure 12. Moment-curvature relationship as a function of different material models.

Table 2. Ultimate bending moment as a function of material models.

	BM [kNm]	Exp [%]	H[%]
Exp	568.94	-	
EP	616.00	8.27%	1.26
EP+H	623.74	9.63%	
EPM	602.75	5.94%	1.00
EPM+H	608.75	7.00%	

an appropriate inclination to cover both effect of the remaining residual stresses and the nonlinear behavior of the box-girder structure.

For the analyzed box girder, the appropriate second tangent modulus is calculated as:

$$E_{T1} = E_{T0} / (1 + k_{T1}) \quad (5)$$

where k_{T1} equals 0.6.

It can be concluded from Table 2 that, as a result of the presence of the second tangent modulus, the ultimate bending moment is reduced by 2.15% with respect to the EP model. Furthermore, the behavior of the box girder in the post-buckling region showed a good agreement with respect to the experimental results as may be seen from Figure 12.

In the other two models (EP+H and EPM+H), which contains hardening, the difference is little, 1.26 and 1.00%, with respect to the models without hardening for EP and EPM, respectively, as given in Table 2 and Figure 12.

Therefore, it can be concluded that for a plate slenderness value of 1.23 and a given hardening rate, the effect of hardening is not essential, which is consistent with the conclusion derived by Manevich (2007) for the plate elements under compression.

The second group of analyses is investigating the effect of using different weld-toe models on the box girder behavior, see Figure 8. Two models have been identified in Section 4.3. The finite element results for each of the model studied are presented in Figure 13 with the use of the EP material model and in Figure 14 for the EPM model.

As given in Figure 13, in case of the box girder without weld toe, the obtained ultimate bending moment is bigger than the experimental results with 8.27% (see Table 3), and the tangent flexural rigidity is equal to the experimental one.

With the use of the weld toe, for the model no. 1, in which the weld toe was uniformly distributed along the height of the stiffener, the ultimate bending moment increased with 5.85% with respect to the model without weld toe.

For the weld toe model no. 2, in which the weld toe was modeled by increasing the thickness at the connecting node of the plate with the stiffener. It clear that, (see Figure 13), there is little difference between the model no. 2 and the first one, 0.41%, see Table 3.

The little difference is as a result of the smaller stiffness that is added to the stiffeners with the use of the weld toe model no. 2, in which the stiffness was not uniformly distributed along the stiffener height. Generally, with the use of the EP material model with or without weld toe modeling, the behavior of the box in the post-buckling region and the ultimate bending moment is far away from the experimental results as can be seen in Figure 13 and Table 3.

With the application of the two weld toe models with the use of the EPM material model, the

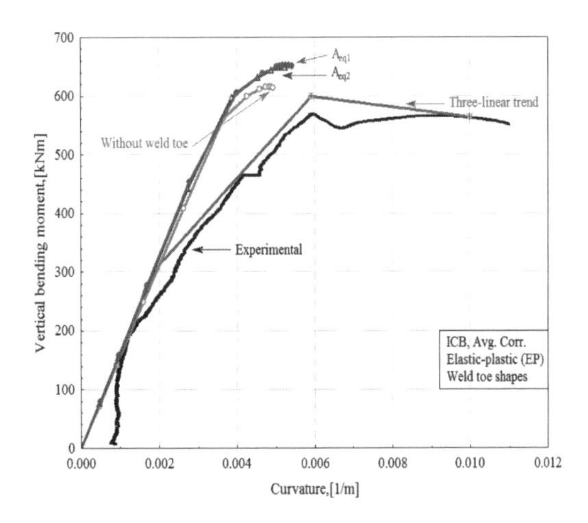

Figure 13. Moment-curvature relationship as a function of different weld toe shape models with EP material.

Table 3. Ultimate bending moment as a function of weld toe shape models and EP material.

	BM [kNm]	Exp [%]	A_{eq} [%]
Exp	568.94	–	
EP	616.003	8.27%	
EP, A_{eq1}	652.057	14.61%	0.41
EP, A_{eq2}	649.415	14.14%	

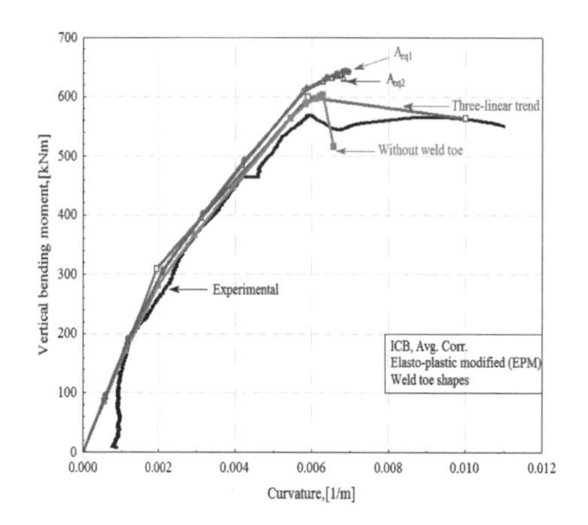

Figure 14. Moment-curvature relationship as a function of different weld toe models with EPM material.

Table 4. Ultimate bending moment as a function of weld toe shape models and EPM material.

	BM [kNm]	Exp [%]	A_{eq}[%]
Exp	568.94	–	
EPM	602.75	5.94%	
EPM,A_{eq1}	643.54	13.11%	0.83
EPM,A_{eq2}	638.20	12.17%	

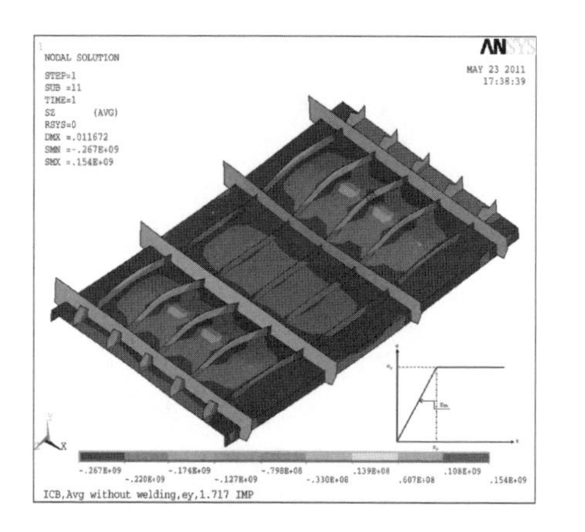

Figure 15. Stress distribution (longitudinal direction), EP material model.

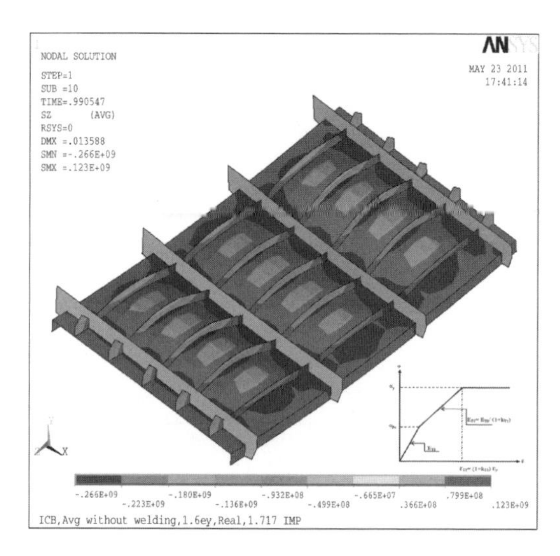

Figure 16. Stress distribution (longitudinal direction), EPM material model.

difference between the weld toe models is the almost twice the difference using EP material from ultimate bending moment point of view, 0.83%, as tabulated in Table 4.

It can be concluded that, the elasto-plastic modified material model (EPM) without hardening and weld toe modeling showed the best agreement with the experimental results of the slightly corroded box girder and for the global behavior of ultimate bending moment.

Figure 15 and Figure 16 show the stress distribution in the longitudinal direction on the deck panel of the slightly corroded box girder, with both EP and EPM material models.

As can be observed from Figure 15 (EP material model) the middle bay (plating and stiffeners) is subjected to higher stresses, which is bigger than the one for the EPM model, see Figure 16. This demonstrates that with the use of the EP material model, the box girder is able to carry more load due to the fact that in the EP model the effect of residual stresses is not accounted for, and on the contrary for the EPM model in which the second tangent modulus is introduced to account for the residual stresses effect.

Moreover, the deck, side connection for the EP model is subjected to higher stresses rather than in EPM due to the same reason, and in both models, the transverse frames are not subjected to higher stresses because they are much stiffer.

6 CONCLUSIONS

A series of nonlinear FE analyses for a slightly corroded box girder have been carried out.

Two groups of analyses have been performed, the first one dealt with different material models and the second one with different weld toe shape models. With the presence of the strain hardening, the difference is small, 1.26 and 1.00%, with respect to the models without hardening for EP and EPM, respectively.

Accounting for the weld toe using the weld shape model no. 1, and weld toe shape model no. 2 the ultimate bending moment increases with 5.9 and 5.4% with respect to the EP model without welding respectively.

With the application of the two weld toe models using the EP and EPM material models, the difference between the weld toe models using the EPM is twice the difference using EP, 0.8%.

With the use of the elastic-perfectly plastic material model with or without hardening and weld toe, the behavior of the box girder is far away from the experimental results.

It has been recognized that the variations in the weld toe shape considered in the two weld toe models is limited on the ultimate strength.

As a result of the introduced second tangent modulus, the ultimate bending moment is reduced by 2.15% with respect to the EP model. An elasto-plastic modified material model EPM has been developed accounting for residual stresses effect and post-buckling behavior demonstrated a very good agreement with the experimental results.

ACKNOWLEDGMENTS

The work reported here is a contribution to the activities of the MARSTRUCT VIRTUAL INSTITUTE, (www.marstruct-vi.com) in particular its Technical Subcommittee 2.3 on Ultimate Strength.

The first author has been funded by the Portuguese Foundation for Science and Technology (Fundação para a Ciência e Tecnologia—FCT) under contract SFRH /BD/ 46790 /2008.

REFERENCES

Hansen, A.M., 1996, Strength of Midship Sections, *Marine Structures*, 9, pp. 471–494.

Hu, Y., Zhanga, A. and Sunb, J., 2001, Analysis on The Ultimate Longitudinal Strength of A Bulk Carrier by Using Simplified Method, *Marine Structures*, 14, pp. 311–330.

Khedmati, M.R., Roshanali, M.M. and Esmaeil, M., 2011, Strength of Steel Plates with Both-sides Randomly Distributed with Corrosion Wastage under Uniaxial compression, *Thin-walled Structures*, 49, pp. 325–342.

Manevich, A.I., 2007, Effect of Strain Hardening on the Buckling of Structural Members and Design Codes Recommendations, *Thin-walled Structures*, 45, pp. 810–815.

Nikolov, P.I., 2009, Box Girder Strength under Pure Bending:Comparison of Experimental and Numerical Results, *Proceedings of the 13th Congress of International Maritime Association of the Mediterranean (IMAM'09)*, İstanbul, Turkey, ITU, Faculty of Naval Architecture and Ocean Engineering, Instanbul, pp. 79–86.

Nishihara, S., 1984, Ultimate Longitudinal Strength of Midship Cross Section, *Naval Architecture & Ocean Engineering*, 22, pp. 200–214.

Qi, E., Cui, W. and Wan, Z., 2005, Comparative Study on the Utimate Hull girder Strength of Large Double Hull Tankres, *Marine Structures*, 18, pp. 227–249.

Saad-Eldeen, S., Garbatov, Y. and Guedes Soares, C., 2010a, Experimental Assessment of the Ultimate Strength of a Box Girder Subjected to four-point Bending Moment, *Proceedings of the 11th International Symposium on Practical Design of Ships and other Floating Structures (PRADS2010)*, Rio de Janeiro, Brasil, pp. 1134:1143.

Saad-Eldeen, S., Garbatov, Y. and Guedes Soares, C., 2010b, Corrosion Dependent Ultimate Strength Assessment of Aged Box Girders Based on Experimental Results *[Submitted for publication]*.

Saad-Eldeen, S., Garbatov, Y. and Guedes Soares, C., 2011a, FE Parameters Estimation and Analysis of Ultimate Strength of Box Girder. In: *Maritime Technology and Engineering [In Press]*. C. Guedes Soares et al., editors. London, UK: Taylor & Francis Group.

Saad-Eldeen, S., Garbatov, Y. and Guedes Soares, C., 2011b, Compressive Strength Assessment of a Moderately Corroded Box Girder, *Marine System and Ocean Technology [In press]*.

Saad-Eldeen, S., Garbatov, Y. and Guedes Soares, C., 2011c, Experimental Assessment of the Ultimate Strength of a Box Girder Subjected to Severe Corrosion, *Marine Structures [In Press]*.

Saad-Eldeen, S., Garbatov, Y. and Guedes Soares, C., 2011d, Analysis of Plate Deflections during Ultimate Strength Experiments of Corroded Box Girders, *[Submitted for publication]*.

Sustainable Maritime Transportation and Exploitation of Sea Resources – Rizzuto & Guedes Soares (eds)

N-SIF based fatigue assessment of hopper knuckle details

Claas Fischer & Wolfgang Fricke
Institut für Konstruktion und Festigkeit von Schiffen, Hamburg University of Technology, Germany

Cesare M. Rizzo
DINAEL, Facoltà di Ingegneria, Università di Genova, Italy

ABSTRACT: A somewhat innovative fatigue assessment approach is applied to a typical shipbuilding structural detail: the hopper knuckle. This detail was benchmarked in the Technical Committee III. 2 report (Fatigue & Fracture) of the last ISSC 2009 Congress and significant scatter among results obtained by various fatigue assessment procedures was found, even when considering specific corrections of the commonly applied approaches for such type of welded joints. This paper aims at providing verification of a Notch Stress Intensity Factor (N-SIF) based approach in a case currently of concern within the shipping community, namely the Strain Energy Density (SED) approach, that is a promising method for fatigue assessment of welded structural components proposed in open literature. Comparisons with experimental and numerical available results are provided with results obtained applying other approaches, highlighting advantages and disadvantages of the SED approach in the captioned case.

1 INTRODUCTION

It is commonly recognized that it is impossible for a physical model to account for all fatigue influencing parameters, thus several approximate models have been conceived for practical fatigue assessments.

Nowadays, fatigue design of ship structures is based on Finite Element (FE) analysis using shell elements, where welds are generally not included in the models. Rather, an estimate of an appropriate stress component at the failure point, which is assumed to govern fatigue, is obtained by conventional extrapolations from stress values in the neighborhood according to the structural stress approach. Hence, it is avoided to deal with stress singularity of notch tip. Solid element models are able to explicitly include the geometry of weld seams and sub-modeling techniques are sometimes applied to evaluate structural stresses assessing the effect of weld geometry onto stress field. In these cases 1-mm fictitious rounding of notches can also be applied, according to the Radaj's worst case suggestions for the notch stress approach (see e.g., Hobbacher 2009).

Nevertheless, robust fatigue assessment methods for ship structures, accounting for the stress field in close vicinity of the notch tip and for the stress singularity at weld notches, are more and more required. In fact, benchmark studies of Technical Committee III. 2 report (Fatigue & Fracture) of the last ISSC 2009 Congress are mostly related to hopper knuckle details in bulk carriers and tankers (see Figure 1), showing that the widely applied 0.5t/1.5t linear extrapolation technique according to the structural (hot-spot) stress approach also suggested by rules of classification societies tends to overestimate the target stress by 38% in average. Different methods were proposed to correct the structural approach for hopper knuckles, e.g., Lotsberg et al. (2008), Osawa et al. (2010).

In this paper, in addition to usual approaches, the fatigue strength of hopper knuckles is assessed according to a Notch Stress Intensity Factor (N-SIF) based approach, that in principle allows accounting for stress field singularity as well as notch opening angle and weld geometry, inclusive of plate thickness (size effect). However, N-SIF based approaches are not yet recognised as a procedure

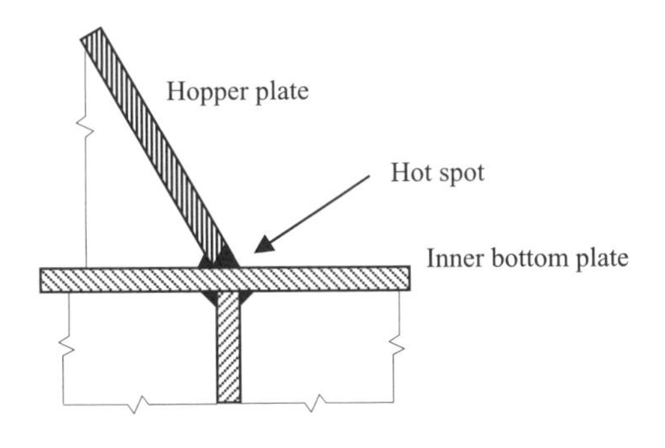

Figure 1. Section of a typical hopper knuckle detail.

for fatigue life assessment in the various industrial fields nor they were agreed to be included in any regulation or standard. At the same time, a variant of the approach, namely the strain energy density (SED) approach, seems relatively straightforward to be implemented in current structural analyses of ship structures, which are more and more based on the finite element method (FEM).

2 N-SIF BASED APPROACHES

While information about N-SIF based fatigue assessment approaches can be found in Rizzo (2010), including a rather comprehensive literature review, a very brief advice is given in the following.

Local approaches for fatigue strength assessment are categorized in the well-known sketch reported by Radaj et al. (2006) on the basis of the parameter believed to govern the fatigue strength behaviour. N-SIF based approaches can be regarded as in between the notch stress or notch strain based approaches and the relatively more complex crack propagation approaches, see Figure 2.

Indeed, by studying the relevant literature, it clearly appears that the theoretical principles of the N-SIF based fatigue approaches originate in the second half of the last century, when analytical formulations for the stress fields in way of crack tips and notch corners were derived dealing with stress singularities. In turn, such works were based on the fracture mechanics theory, analyzing a special case of sharp notch, i.e., the crack. In fact, a crack may be regarded as a notch having null opening angle and null radius at the tip.

Analytical solutions showed that stresses at notch corners are proportional to the term $r^{(\lambda-1)}$ in a polar reference system centered at notch tip, where r is the radial distance from notch tip. It is well known that $(1-\lambda) = 0.5$ in case of cracks but it is lower for open notches, depending on notch angle and loading conditions. In case of 135° notch opening angle, i.e., the usually assumed weld

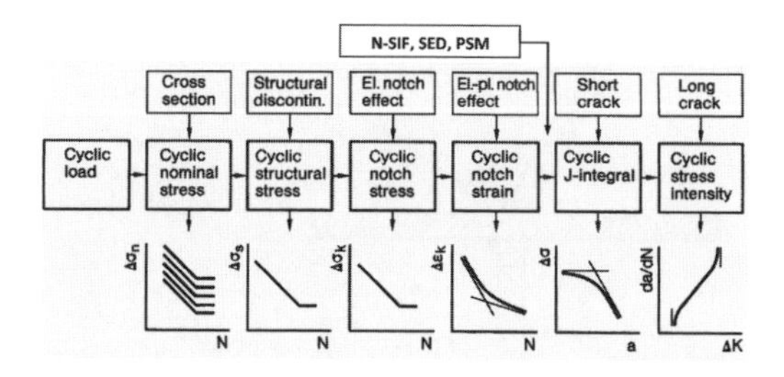

Figure 2. Overview of local fatigue assessment approaches.

toe angle, and loading mode I conditions, $(1-\lambda)$ becomes 0.326.

In short, it was demonstrated that the stress field in way of notches is governed by a parameter analogous to the well-known stress intensity factor (SIF) for cracks, i.e. the N-SIF defined as:

$$K_i^N = \sqrt{2\pi} \cdot \lim_{r \to 0^+} \left[\sigma_i \, r^{(1-\lambda_i)} \right] \tag{1}$$

where $i = 1, 2, 3$ indicates the loading modes I, II or III of cracks and notches (see e.g., Rizzo 2010).

The N-SIF was proven to be a suitable parameter for fatigue assessment of welded joints (Lazzarin & Tovo, 1998). However, the need of micrometric FE mesh size in way of the failure point for its estimate is one of the main disadvantages of this approach.

In more recent years, another approach evolved from the N-SIF one, based on the SED in a control volume and lately, another method based on the N-SIF concept was proposed, allowing relatively straightforward fatigue assessments of welded joints if well calibrated in advance, i.e., the Peak Stress Method (PSM).

While the PSM method takes as fatigue parameter the elastic peak stress evaluated in a point by means of a finite element analysis at the notch corner using mesh size and element type appropriately calibrated with N-SIF results, the SED approach is based on the averaged value of the strain energy computed in a relatively small volume of material.

For high-cycle fatigue of welded joints, assuming an open notch with null radius at the weld toe, Lazzarin and Zambardi (2001) proposed the SED over a small cylindrical volume with a characteristic radius R_0 surrounding the notch tip as the fatigue assessment parameter. In case of two-dimensional (2D) problems the volume degenerates into a sector. The radius over which the SED needs to be computed is thought as a material property and has been derived from Beltrami's failure criterion considering stress intensity factors of cracks under mode I loading condition and high cycle ($N = 5 \times 10^6$) fatigue strength data. $R_0 = 0.28$ mm results in a conservative approach for steel made arc welded joints with failures at both toe and root of the weld seam.

Based on the relationship between the SED in the mentioned control volume and the N-SIFs, a unified S-N curve independent of the welded detail can be drawn. It was shown that fatigue data of failures originating from weld roots and weld toes of a number of welded joints under various loading conditions can be summarized in a convincingly narrow single scatter band (Berto & Lazzarin, 2009). Figure 3 shows the scatter band of fatigue strength data involving fractures initiated from weld toe as well as weld root of the analyzed joints. The slope of the regression is in this case $m = 1.5$, i.e., half of the usual slope of the S-N curves.

Fatigue strength at 2 millions cycles is expressed in terms of SED range as $\Delta W = 0.058$ Nmm/mm^3, at 97.7% survival probability.

It is worth noting that the SED approach allows naturally including effects of mixed mode loading conditions. Moreover, the problem of the dimensional variability of N-SIF at different opening notch angles is overcome as SED dimensions are independent from notch opening angle while N-SIF ones depend by definition on length powered to ($1-\lambda$).

However, mainly 2D problems were considered to define the control volume but the captioned case is a three dimensional one. Here, Lazzarin et al. (2008) suggest a depth of the volume approximately equal to the characteristic radius in order to take account for the secondary stress gradient along the weld seam.

It was demonstrated that the N-SIF K_1^N (loading mode I) is directly related to SED if the contribution relevant to loading mode II can be considered null (e.g., because of symmetry of the stress field) or negligible (large opening angle resulting in a non-singular stress distribution). In such conditions, it is possible to reconvert a posteriori the local strain energy density averaged over a larger control radius $R^* > R_0$ to the correct control volume by the expression:

$$\overline{W}(R_0) = \overline{W}(R^*) \cdot \left(\frac{R_0}{R^*}\right)^{2(\lambda-1)} \tag{2}$$

Hence, the number of degrees of freedom in the computational model can be reduced as the control volume contains less elements.

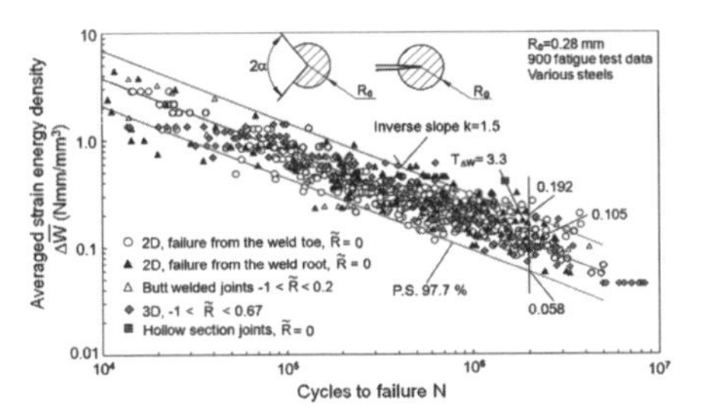

Figure 3. Fatigue strength vs averaged SED for abt. 900 fatigue test data of steel joints (Berto & Lazzarin 2009).

3 TEST CASE AND FE MODELS

The test case presented in the following was earlier analyzed by Osawa et al. (2010) and also included in the ISSC 2009 benchmark study. Experimental values of structural stresses are available along with proposals for the correction of usual procedures to extrapolate the structural stress for fatigue assessment. Figure 4 shows information of the specimen tested in Japan while Figure 5 shows the

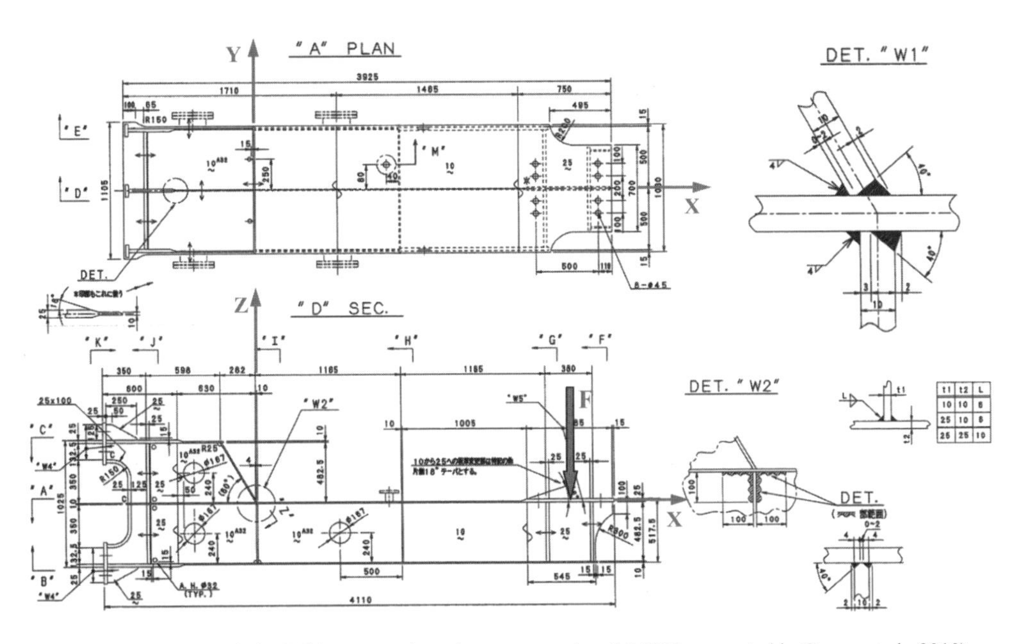

Figure 4. Size and shape of the BC lower stool used as test case (model 6010 presented in Osawa et al., 2010).

FE model built in ANSYS™ environment. The hot spot is the intersection of the inner bottom plate, lower stool plate, floor and center girder of a bulk carrier (BC).

The structure is fixed at the left end (at the stool side) as a cantilever beam and loaded by a vertical force F of 250 kN on the opposite side. The distance to the intersection point of the plates is 2520 mm which results in a nominal stress of 106 MPa at the cross section by considering full section modulus. Additionally, 0.2 MPa internal pressure of tanks (surrounding the hopper knuckle) is also accounted for causing additional local bending.

Shell elements were used to model the structure in a traditional way apart from the refinement zone where various modeling strategies were applied as shown in Figures 5–6.

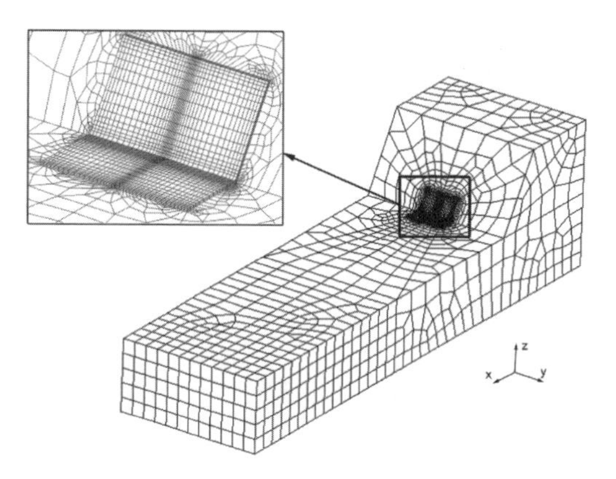

Figure 5. Finite element model built in ANSYS™ environment.

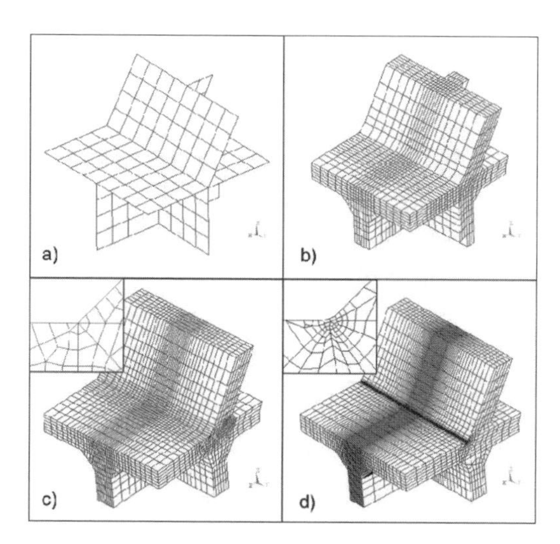

Figure 6. Different modeling strategies for the refinement zone.

A first refinement with shell elements aimed at obtaining stresses at nodes for extrapolations according to the usual structural stress approach. The element size is equal to the plate thickness and no welds were modeled as shown in Figure 6a.

Solid elements were then used for various refinements. The rotational degrees of freedom (DOF) of the shell elements were coupled to the solid elements by means of shell elements with a fictitious thickness placed onto the thickness faces of the solid elements at the boundaries of the refinement zone and perpendicular to the elements representing the plates. The thickness of the fictitious shell elements arranged in the mid-plane of the plates was defined equal to that of actual plates.

All weld seams of the structure in the refinement zone were modeled in the solid element models as shown in Figure 6.

The mesh refinement useful to apply the structural stress approach is shown in Figure 6b. Figure 6c and Figure 6d show the mesh used for the SED approach with $R^* = 1$ mm and the effective notch stress approach with an effective notch root radius of $\rho_{ref} = 1$ mm respectively.

4 STRUCTURAL STRESS APPROACHES

Conventional structural stress approach according to the Recommendations of the International Institute of Welding (IIW, Hobbacher, 2009) was applied, using both 8 nodes shell elements (SHELL281 in ANSYS™) and 20 nodes solid elements (SOLID95) in the refinement zone.

In addition, the correction method proposed by Osawa et al. (2010) was applied to obtain the structural hot-spot stress (HSS) from the shell model, i.e. the extrapolation points are shifted away from the shell-plane intersection line by:

$$\Delta = \frac{t_v}{2}\operatorname{cosec}\Phi - \frac{t_h}{2}\operatorname{cotg}\Phi = 2.89\,\text{mm} \quad (3)$$

being $t_v = t_h = 10$ mm the margin plate and the inner bottom thickness and $\Phi = 60°$ the stool angle. The shell stresses have not been corrected considering the effect of the transversal bending deformation of the main plate caused by pressure loading as mentioned by Osawa et al. (2010) because it was verified that in this case the HSS is practically unchanged.

The obtained results are summarized in Table 1 considering FAT 90 as the fatigue strength for the structural stress approach of load carrying welds while Figure 7 shows the X-stress field at hot spot, X being the direction transverse to the weld.

Calculated stress field of the described FE models are in substantial agreement with the one stated

Table 1. HSS and corresponding fatigue life using FAT 90.

	HSS [MPa]	HSS/FAT	N
Shell, centre line	342.0	3.80	36,448
Shell, 5 mm beneath	320.3	3.56	44,386
Osawa correction	288.5	3.21	60,467
Solid (0.4t/1.0t)	261.6	2.91	81,436

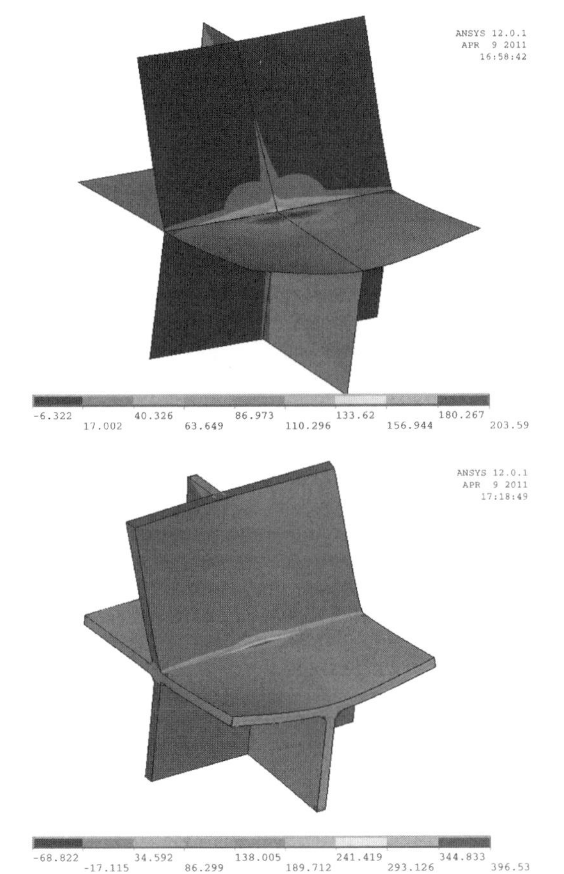

Figure 7. X-stress field at hot spot for shell and solid models (deformation with magnification).

by Osawa et al. (2010). It is worth again noting that weld seams are only modeled in the solid element models as previously described.

According to IIW Recommendations, 0.5 t/1.5 t linear extrapolations for relatively coarse shell meshes show exaggerated structural stresses if the extrapolation path is taken on the inner bottom top surface along the centre floor intersection line. However, the path 5mm beneath the centre line, i.e., taking stresses in the centre of the elements at side of the centre line (as actually proposed in the IIW recommendations), provides lower but still overestimated HSS.

It is noted that also Osawa et al. (2010), used the element center stress to estimate the stresses at extrapolation points but they used 4-nodes elements in lieu of the 8-nodes shell elements applied in the present study, resulting in a HSS of about 300.7 MPa when conventional IIW 0.5t/1.5t linear extrapolation procedure is applied and in a HSS of 257.6 MPa when their proposed correction is considered by shifting the extrapolation points.

As far as extrapolation of structural stresses from results of solid models is concerned, it should be noted that in this case the mesh should be considered relatively fine since six solid elements over the plate thickness were used, having a face size on the plate surface of 0.2 t × 0.2 t = 2 × 2mm. Therefore, according to Hobbacher (2009), the 0.4t/1.0 t linear extrapolation should be followed. Table 1 reports the obtained results applying this method.

HSS extrapolations using solid elements models reported by Osawa et al. (2010) are in this case slightly different, being HSS = 228 MPa. In fact, they used 8 nodes instead of 20 nodes solid elements and a finer mesh, i.e., a minimum element size of t/8 in each direction. Moreover, they followed a 0.5t/1.5t HSS extrapolation rule.

It is noted that IIW recommends a quadratic extrapolation in cases with steep, non-linear stress distributions, which may explain the differences between structural stresses obtained using shell and solid elements.

Fricke and Weißenborn (2004) performed parametric calculations with simplified FE shell models (no scallops) varying hopper angle α, double bottom height h, floor spacing b, inner bottom and hopper plate thickness t, with or without horizontal extension plate in tank (Osawa's model shows an extension plate, see Figure 4). The stress concentration factors K_s are summarized in Figure 8. Considering the geometry of the test case (b/h~1, h/t~52

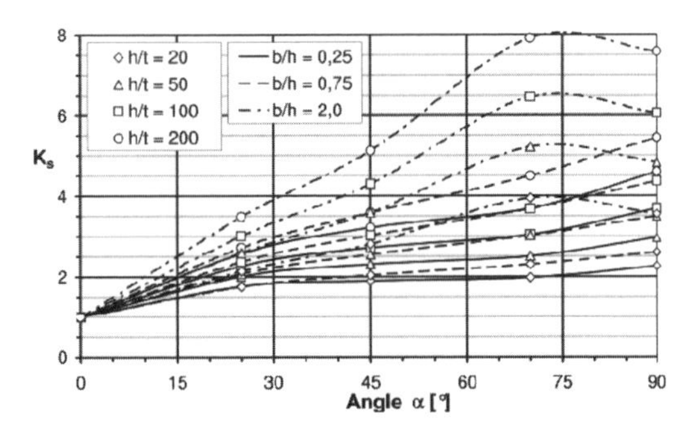

Figure 8. Parametric diagram of stress concentration factors K_S (Fricke & Weißenborn, 2004).

and $\alpha = 60°$), K_s is about 3.25 and therefore the HSS becomes approximately 344.5 MPa.

The DNV Classification Note 30.7 suggests a stress concentration factor $K_s = 7$, even if direct analysis is recommended.

5 NOTCH STRESS APPROACH

For comparison purposes, the notch stress approach according to the 1-mm fictitious notch radius at the weld toe proposed by Radaj was carried out (see e.g., Radaj et al., 2006 and Fricke, 2008).

The solid model was modified as shown in Figure 6d to obtain the required 1 mm fictitious notch rounding at weld toe. Figure 9 shows the principal stress field in the refined zone as the 1st principal stress is recommended to be used in this case.

The calculated notch principal stress in way of the curved surface (hot spot) is 751.4 MPa, leading to a fatigue life of about $N = 5.37 \times 10^4$ cycles if the usual S-N curve (FAT 225) is considered, being the stress concentration factor SCF = 3.34.

6 SED CALCULATIONS

The SED was estimated by modeling different control volumes. Cylindrical volumes having different radii and centered along the weld toe was defined and meshed with 5 elements around the circular sector for solid model shown in Figure 6c.

Eq. 2 was applied to scale the obtained SED values to the radius $R_0 = 0.28$ mm for control volumes having $R^* = 0.62$ mm and $R^* = 1$ mm. Results are in very good agreement, as shown in Figure 10. The plot is symmetric with respect to the center girder of the double bottom as expected and SED is rapidly vanishing towards the sides.

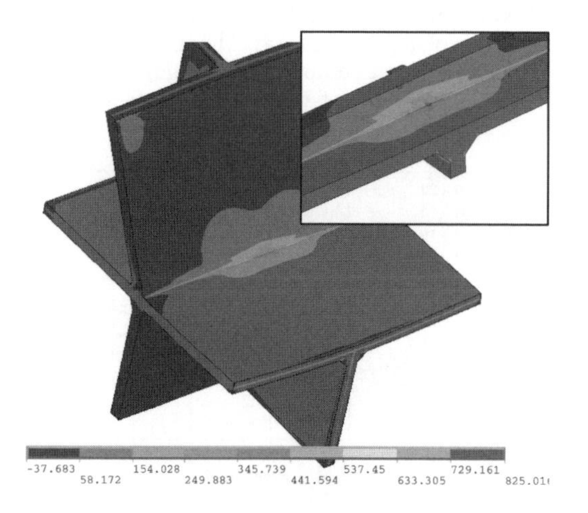

Figure 9. First principal stress field at hot spot of the FE model for the notch stress approach.

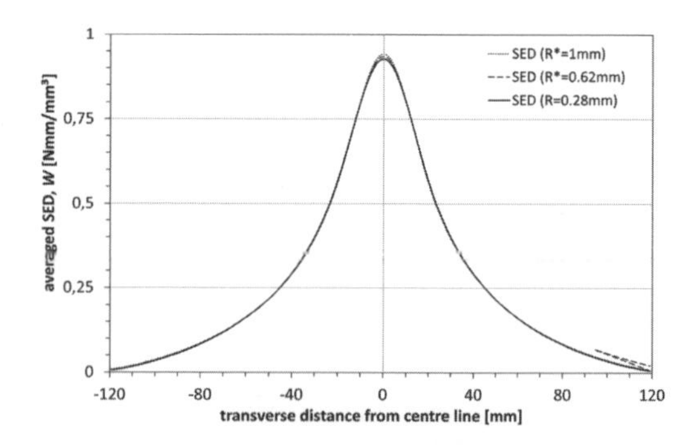

Figure 10. Distribution of SED along weld toe in transversal direction.

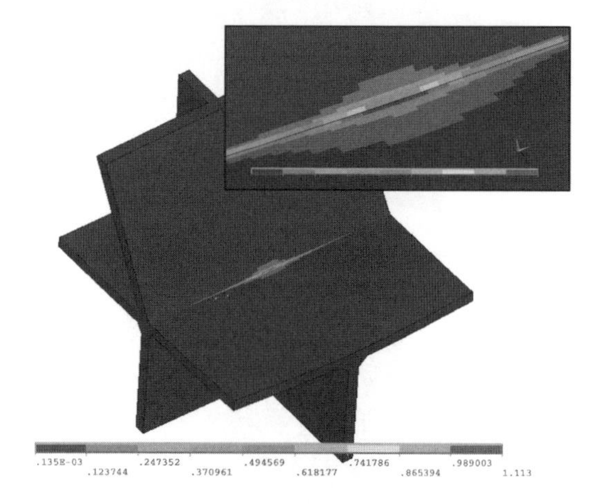

Figure 11. SED field in the refined zone ($R^* = 1$ mm).

Considering data reported in Figure 3, fatigue life of the analyzed detail under constant pulsating load can be estimated as $N \sim 3.14 \times 10^4$ cycles being the computed SED $\Delta W \sim 0.925$ Nmm/mm³.

Assuming linear elastic material behavior, the square root of the ratio between the acting SED and its characteristic value for fatigue assessment can be calculated for comparison purposes as:

$$\frac{S}{S_{char}} = \sqrt{\frac{SED}{SED_{char}}} = \sqrt{\frac{0.925}{0.058}} = 4.00 \tag{4}$$

7 DISCUSSION

Very few independent applications of the N-SIF based approaches for fatigue assessment of welded structural details were carried out up to now (Fischer et al., 2010, 2011). In these papers the comparison was carried out in terms of local stresses.

However, due to the relatively complex 3D geometry and substantial differences among applied methods, in this case the comparison is carried out directly in terms of fatigue lives N and ratio between actual and characteristic fatigue strength values S/S_{char}. It should be kept in mind that fatigue lives are more sensitive to changes than stresses.

Unfortunately, no fatigue tests results are available and only local stress measurements of static tests were reported by Osawa et al. (2010), which are in good agreement with the presented results (see Figure 12, where measured stresses are corrected by the factor 0.92 to match the computed stresses).

The comparison is therefore carried out only among results obtained applying different local approaches for fatigue strength assessment in Table 2.

The comparison among results obtained by the widely applied IIW procedures, including the Osawa's correction proposal, shows some scatter of fatigue lives.

Extrapolations of stresses of solid element models appear to be less conservative while the evaluations obtained from the shell models are in better agreement with the results of the effective notch stress approach.

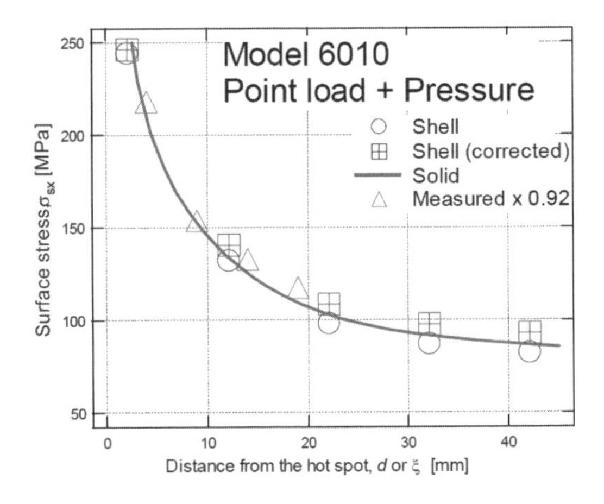

Figure 12. Measured and computed surface stresses approaching the hot spot (Osawa et al., 2010).

Table 2. Comparison of fatigue life and strength results.

	HSS (IIW)		HSS	1-mm	
	Shell	Solid	Osawa	radius	SED
N	44,386	81,436	60.467	53,688	31,379
S/S_{char}	3.56	2.91	3.21	3.34	4.00

Noticeably, the parametric calculations by Fricke and Weißenborn (2004) provide quick and fairly consistent estimates of the structural stress.

It also appears that results from the solid elements models are influenced by mesh size and weld seams modeling, thus leaving some space for interpretation of practical application of the structural stress approach.

The SED approach appears to be the most conservative one. It has been verified that weld seams modeling has a rather large effect on the computation of SED (up to 50% if e.g. only transversal weld seams are modeled).

As a matter of fact, the SED is not a point parameter and therefore it accounts for the stress/strain field of the volume surrounding the hot spot, including the effect of stress/strain components in all directions. Hence, local stiffness of the modeled structure and boundary conditions in way of the hot spot are of paramount importance, especially in geometrically complex three-dimensional cases like the captioned one.

It should be pointed out that, even in the case of a relatively geometrically complex detail, the control volume can be enlarged in order to obtain a conveniently controllable mesh to compute the SED and then the computed value can be easily converted to the volume having $R_0 = 0.28$ mm radius because the contribution of loading mode II to SED is negligible.

On the other hand, appropriate modeling of welded seams is time consuming but necessary and need proper expertise of the FE analyst.

8 CONCLUSIONS

While N-SIF based approaches are going to be recognized as sound engineering technique for fatigue assessment of welded structural components, this paper aimed at providing verification in the case of a typical ship structural detail that is currently of concern within the shipping community.

Certainly, fatigue tests are necessary to finally draw conclusions regarding the effectiveness of the analyzed assessment approaches. However, some useful hints can be summarized as follows:

- The structural stress approach leads to quite scattered results depending on the FE modeling strategy and extrapolation technique;
- The shell element modeling according to IIW Recommendations is quite conservative and correction procedures have been proposed in open literature (Osawa et al., 2010, Lotsberg et al.,

2008) and included in recent classification societies rules (IACS, 2010);

- The solid element modeling, which was assumed as a target for the correction procedures proposed in open literature, is less conservative and relevant results appear in good agreement with measurements of local stresses from static tests;
- Even structural stress evaluated by means of solid element meshes depends on element size and extrapolation technique;
- The results of the effective notch stress approach are in between the ones of the IIW structural stress approach using shell elements models and the relevant proposed correction;
- It appears that appropriate modeling of weld seams is significant for both the notch stress approach and the SED approach;
- The SED approach appears to be the most conservative one; however, it should be recognized that the analyzed detail is a three-dimensional one while 2D details were mainly assessed to obtain the fatigue strength curve of Figure 3 and that local stress approaches assume as the fatigue governing parameter one component of the stress calculated using a fictitious extrapolation while SED naturally accounts for all components as well as the stress field singularity at the notch tip;
- SED calculations using different control volumes are in very good agreement: once the volume is properly defined, SED estimate appears mesh independent, thus avoiding interpretations of the calculation procedure.

Accounting for the above, the SED approach can be considered a very promising fatigue strength assessment method for ship structures, also because it can eventually be automated applying FEM submodeling techniques able to transfer appropriate boundary conditions from large and coarse shell element models of ship structures to smaller and refined solid element models, in a similar way of the test case presented in this paper.

ACKNOWLEDGEMENTS

The grant of the Alexander von Humboldt Stiftung (Germany), funding the Fellowship of Dr. C. Rizzo at the Hamburg University of Technology, is gratefully acknowledged.

REFERENCES

Berto, F. & Lazzarin, P. 2009. A review of the volume-based strain energy density approach applied to V-notches and welded structures, *Theor. Applied Fract. Mech.* 52:183–194.

Fischer, C., Düster, A. & Fricke, W. 2011. Different finite element refinement strategies for the computation of the strain energy density in a welded joint, In *Proc. MARSTRUCT 2011*, Hamburg, Germany.

Fischer, C., Feltz O., Fricke, W. & Lazzarin, P. 2010. Application of the Notch stress intensity and crack propagation approaches to weld toe and root fatigue, *IIW-Doc. XIII-2337-10/XV-1354-10*, International Institute of Welding, Paris.

Fricke, W. 2008. Guideline for the Fatigue Assessment by Notch Stress Analysis for Welded Structures, *IIW-Doc. XIII-2240-08/XV-1289-08*, International Institute of Welding, Paris.

Fricke, W. & Weißenborn, C. 2004. Simplified stress concentration factors for fatigue-critical details in the ship structure (in German). *Schiffbauforschung* 43, pp. 39–50.

Hobbacher, H. 2009. Recommendations for fatigue design of welded joints and components, *IIW doc.1823-07*, Welding Research Council Bulletin 520, New York.

IACS 2010. Common Structural Rules for Bulk Carrier, Consolidated Effective as of 1 July 2010, International Association of Classification Societies, London, www.iacs.org.uk.

ISSC 2009. Report of Committee III.2 Fatigue and Fracture, In *Proc.s 17th Int. Ship and Offshore Structures Congress*, Seoul National University, Korea.

Lazzarin, P. & Tovo, R. 1998. A notch intensity factor approach to the stress analysis of welds. *Fatigue Fract. Eng. Mater. Struct.* 21, 1089–103.

Lazzarin, P. & Zambardi, R. 2001. A finite-volume-energy based approach to predict the static and fatigue behavior of components with sharp V-shaped notches. *Int. J. Fract.* 112, 275–298.

Lazzarin, P., Berto, F., Gomez, F.J. & Zappalorto, M. 2008. Some advantages derived from the use of the strain energy density over a control volume in fatigue strength assessments of welded joints. *Int. J. Fat.*, 30, 1345–1357.

Lotsberg, I., Rundhaug, T.A., Thorkildsen, H., Bøe, Å. & Lindemark, T. 2008. A procedure for fatigue design of web-stiffened cruciform connections, Ships *and Offshore Structures* 3(2):113–126.

Osawa, N., Yamamoto, N., Fukuoka, T., Sawamura, J., Nagai, H. & Maeda, S. 2010. Development of a Structural Hot Spot Stress Estimation Technique Based on Shell FE Analysis for Ship Structural Details under the Real Load, In *Proc.s 11th Int. Symp. on Practical Design of Ships and Other Floating Structures*, Rio de Janeiro, Brasil.

Radaj, D., Sonsino, C.M., Fricke, W. 2006. Fatigue assessment of welded joints by local approaches (Second edition), Woodhead Publishing Ltd, Abington, Cambridge UK, ISBN-13: 978 1 85573 948 2.

Rizzo, C.M. 2010. Application of advanced notch stress approaches to assess fatigue strength of ship structural details: literature review, Report 655, *Schriftenreihe Schiffbau*, Technische Universität Hamburg-Harburg.

Sustainable Maritime Transportation and Exploitation of Sea Resources – Rizzuto & Guedes Soares (eds)

Health monitoring of marine vehicles structures by phase-based FBG detectors

N. Roveri & A. Carcaterra
Department of Mechanics and Aeronautics, University of Rome, 'La Sapienza', Rome, Italy

M. Platini
ACE System Snc., Rome, Italy

ABSTRACT: In the context of a new project named SEALAB, aimed at conceiving an innovative high speed marine vehicle, a new system for monitoring the structural response and potential damages in the ship structures is under development. This system permits the real-time self-monitoring of the structure using a new system of sensor based on optical fibers equipped with FBG (Fiber-Bragg-Gratings) sensors. These fibers are applied using different technologies. For traditional materials, like aluminum made structures, the fibers are simply glued along the system. For composite materials, the fibers are embedded directly into the matrix.

A novel HHT-based method (Hilbert-Huang Transform) for damage detection of panel structures under impulsive or travelling load is proposed. The technique is able to identify the presence and the location of the damage along the panels. The measured data are processed by the HHT technique, and none a priori information is needed about the response of the undamaged structure. Damage location is revealed by direct inspection of the first instantaneous frequency, which presents a sharp crest in correspondence of the damaged section.

1 INTRODUCTION

This paper proposes a new technology for continuous structural monitoring and damage detection in ships structures. Namely the system combines a novel algorithm for damage identification together with an optical sensors based on FBG (Fiber Bragg Grating). The system is particularly appealing when using the optical fibers directly embedded into the composite material.

The development of this new technology is part of a larger project developed at the University "La Sapienza" named SEALAB focused on the concept design and building of an innovative high speed marine vehicle. The new challenging vehicle is a technology platform with many new technologies hosted on board. The vehicle presents innovation in several field, involving the whole architecture of the vehicle, the propulsion, transmission and control systems, the shock attenuation devices and the structural real-time monitoring and damage detection system based on optical technology; a prototype is shown in Fig. 1.

A new architecture of the vehicle has been developed, being it equipped with three types of control surfaces closed in the same loop: submerged foils, aerodynamic surfaces and a special interface system equipped with an electromechanical suspension device that contributes to the vehicle lift, large mitigation of the of water-impact phenomena, and to the trim-keeping at high speed (about 200 km/h).

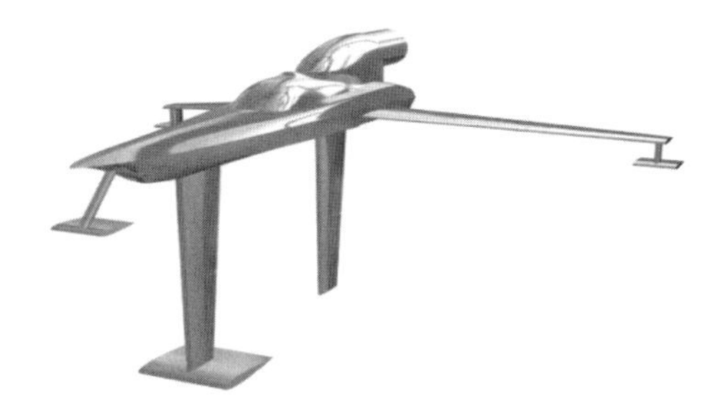

Figure 1. Concept design stage of the high-speed-vehicle SEALAB. Univeristy La Sapienza, Rome, Italy.

The characteristics of navigation of SEALAB implies in general severe structural load conditions, characterized both by large amplitude vibrations, as well as by fatigue phenomena. In fact, violent impacts of the structure on the water surface, frequently identified as slamming, produce dangerous overloads on the ship structures (Carcaterra 2000, 2001, 2004), (Iafrati 2000), as shown in Fig. 2.

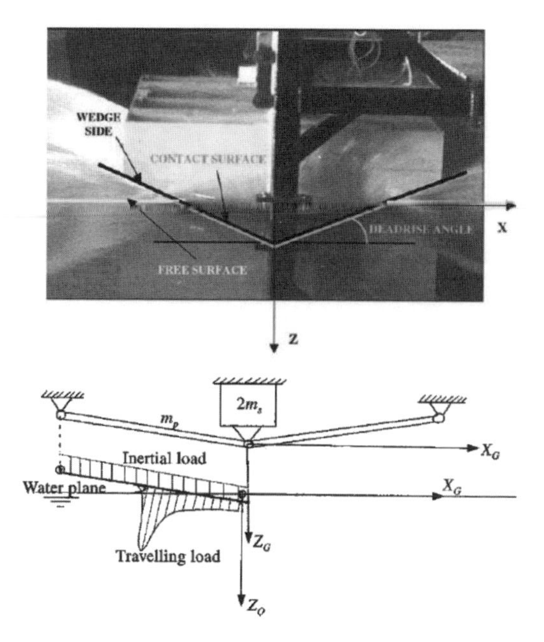

Figure 2. Evidence of the travelling pressure load associated to the wetted part of an impacting hull.

Travelling loads are in this case typically generated by the contact region between water and structure, where the water-front travels along the hull and is characterized by a large peak of pressure close to the boundary of the wetted region. A novel approach to the structural monitoring has been indeed identified as one of the strategy topic in the frame of SEALAB, and the present paper reports on the advances on this subject.

FBG stands for Fiber-Bragg-Grating and is a smart sensor located along an optical fiber. The measurement system scheme is represented in Fig. 3, whose main characteristics can be shortly summarized as follows. A light pulse is generated by an optical source with a suitable spectrum and injected into an optical waveguide. The light beam travels without any significant spectrum modification along the waveguide if it is homogeneous, that is indeed the condition along the optical fiber, up to the section where a FBG sensor is placed. FBG sensor is a short portion of the optical fiber characterized by periodic modification of the refraction index, produced by a set of reflecting parallel plane surfaces normal to the axis of the fiber. In this sense it acts as a resonating structure, characterized by a typical bandwidth. The part of the light beam that belongs to this bandwidth is reflected back, while the transmitted beam is depurated by the reflected part of the spectrum.

When the FBG sensor is integral with the structure it is monitoring, the local strain of the

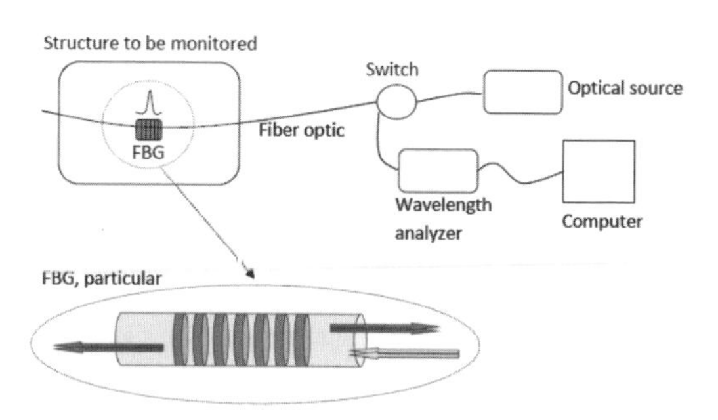

Figure 3. Measuring system of dynamic strain with a fiber optic Bragg grating sensor.

structure induces an equal strain in the fiber producing a modification in the typical distance between the reflecting planes of the FBG sensor. Thus, the characteristic bandwidth of the FBG sensor is modified as a consequence of the applied strain, and the reflected part of the light beam spectrum is also modified. This implies that a suitable analysis of the spectrum of the light reflected by the FBG sensor, allows to identify the characteristic modification of the distance between the reflecting surfaces of the FBG sensor, e.g. its strain can be identified.

In the last decade there has been increasing interest in developing structural health monitoring systems, based on the strain measurement in ship hulls; some applications using FBG sensors have been reported (Johnson 1999), (Gifford 2003), (Li 2006). The novelty of the proposed method relies on the way the acquired signal is analyzed. In fact, in the past years damage identification in engineering structures has been carried out basically through the use of the dynamic response. Most of the identification techniques rely on Fourier spectral analysis and are restricted to linear systems, while engineering structures are often nonlinear and their response is non-stationary. Furthermore modal properties, such as eigenfrequencies, mode shapes, etc., are not very sensitive to incipient structural damage.

An innovative signal processing for structural health monitoring, based on the so-called Empirical Mode Decomposition (EMD), is presented in this paper.

The EMD permits to decompose the acquired signal into a set of basis functions, called Implicit Mode Functions (IMF), which describe the vibratory response of the system. The IMF is a complete, adaptive and nearly orthogonal representation of the analyzed signal, each implicit mode function is almost monocomponent, thus the IMF can determine all the instantaneous frequencies of the non-stationary signal, even of nonlinear structures. The empirically derived basis functions are processed through the Hilbert Transform or AM/FM demodulation, to obtain the analytic signals, namely to obtain amplitude, phase, and instantaneous frequency. A time–frequency spectrum, known as Hilbert–Huang Transformation (HHT), is obtained, which enlighten unique features of the vibratory response of structures.

An improved HHT is here adopted to process time-series data from 1-D structures with and without structural damage. The location and size of damage are identified monitoring the Hilbert—Huang spectrum along the structure. The method shows a potential attractiveness for efficient structural damage diagnostic and health monitoring.

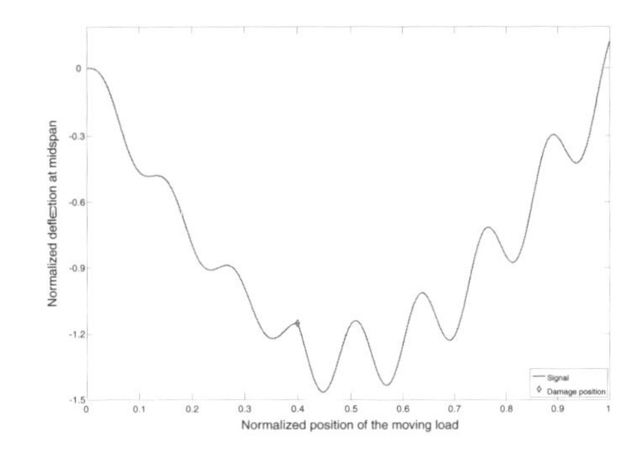

Figure 5. Normalized forced response at midspan of the damaged beam versus Vt/L; ◊ is for the damage position.

2 PRELIMINARY REMARKS

Some preliminary insights into the appearance of damage effects in the beam response provides hints for the identification technique completely described in section 4. A damaged simply supported beam with an open crack is considered, the system is modeled as a two-span beam, each one has a constant density ρ, cross section $A = bh$, where b and h are the cross section width and height, respectively, and uniform bending stiffness EI; each segment obeys to the Euler—Bernoulli beam theory. The crack is located at L_1 and its depth is d, the beam is subjected to a concentrated load P, moving with a constant velocity V, as shown in Fig. 4 .

The analytical response is obtained by the use of the transfer matrix method (Lin 2002) and is plotted in Fig. 5, for $x = 0.5L$, $E = 2.00 \times 10^{11}$ N/m^2, $\rho = 7800$ Kg/m^3, $L = 20$ m, $h = b = 0.2$ m, $L_1 = 0.4L$, $d = 0.5h$, $P = -1000$ N/m and $V = 10$ km/h; the y axis has been normalized by the static deflection of an intact beam,

$$w_{static}(0.5L) = \frac{2PL^3}{\pi^4 EI}.$$

The signal in Fig. 5 can be thought as a sum of two harmonic waves, the amplitude of the higher frequency wave increases as the wave crosses the damaged section. It is possible to prove indeed the explicit solution is the sum of two harmonic waves plus a trend, before the load reaches the crack the harmonic terms can be approximated as follows:

$$w_{H1} \propto \frac{1}{\omega_1^2 - \omega_v^2}\left[sin(\omega_v t) - \frac{\omega_v}{\omega_1} sin(\omega_1 t) \right], \quad t \le \frac{L_1}{V} \tag{1}$$

where $\omega_v = V\sqrt[4]{\rho A/EI}\sqrt{\omega_1}$ is the loading frequency, in the example above $\omega_v = \omega_1/16$. The analysis of Eq. (1) suggests that:

- the *low* frequency wave, or the steady state term, is controlled by the loading frequency ω_v;
- the *high* frequency wave, or the transient term, is controlled by the fundamental frequency ω_1;
- the amplitude ratio between the high and low frequency wave is equal to ω_v/ω_1, it means, when the speed of the moving load is smaller than the critical velocity, the amplitude of the high frequency wave is attenuated.

In the same way, after the load has reached the crack, the harmonic terms can be approximated as follows:

$$w_{H2} \propto \frac{1}{\omega_1^2 - \omega_v^2}\left[sin\left(\omega_v t - \left(\varphi_v + \frac{\pi}{2} \right) \right) + \right.$$
$$\left. -sin\left(\omega_1 t - \left(\varphi_1 + \frac{\pi}{2} \right) \right) \right], \quad \frac{L_1}{V} < t < \frac{L}{V} \tag{2}$$

where $\varphi_v = \omega_v L_1/V$, $\varphi_1 = \omega_1 L_1/V$. The previous equation enlightens:

- there are only two harmonic waves controlled, as for Eq. (1), by the fundamental and the loading frequencies;
- the amplitude ratio between the high and low frequency wave is now unitary;
- a phase shift across the damaged section appears.

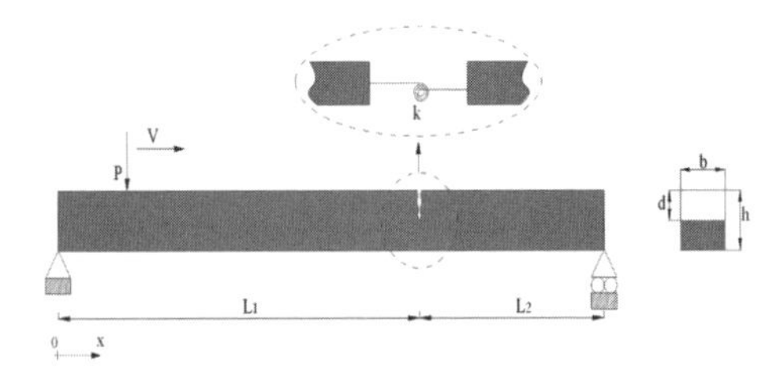

Figure 4. Simply-supported beam with an open crack, subject to a moving load.

The dynamic response can be thought as a sum of a low frequency wave w_v plus a high frequency wave w_1 with piecewise continuous amplitude:

$$w_H(0.5L,t) \propto w_v(t) + \left(\frac{\omega_v}{\omega_1} + c\Pi_T(t-t_d)\right) w_1(t) \quad (3)$$

where $\Pi_T(t-t_d)$ is a shifted rectangular pulse, whose duration and time shift are $2T$ and t_d, with $t_d = (L+L_1)/2V$ and $T = (L-L_1)/2V$. $c = (\omega_1 - \omega_v)/\omega_1$ controls the wave amplitude after the damaged section and tends to zero as $\omega_v \rightarrow \omega_1$. Note the amplitude of the high frequency wave is piecewise continuous. It increases as the moving load has reached the damaged section, as it is shown in Fig. 5. We take advantage of this phenomenon in sections 4 and 5, where the damage position is identified by the knowledge of the dynamic response of the beam, evaluated only at midspan.

3 SHORT OVERVIEW OF THE HILBERT-HUANG TRANSFORM

In 1998, Huang *et al.* (Huang 1998) proposed an innovative technique for analyzing nonlinear and nonstationary signals, called Hilbert-Huang transform (HHT). The method contains two fundamental elements: the empirical mode decomposition (EMD) and the Hilbert spectrum. By the EMD the signal is decomposed into a finite number of intrinsic mode functions (IMFs). Accordingly to Huang, an intrinsic mode is a function which satisfies the two conditions:

1. in the whole data set, the number of extrema and the number of zero crossings must either be equal or differ at most by one;
2. at any point, the mean value of the envelope defined by the local maxima and the envelope defined by local minima is zero.

The name intrinsic mode is referred to the oscillation modes embedded in the vibration data. The properties above make an IMF monocomponent, thus each IMF has a well-behaved Hilbert transform.

3.1 *The empirical mode decomposition method*

Accordingly to the properties 1) and 2) in the last section, to generate the family of IMFs the so-called sifting process is used (Huang 1998):

a. let $s(t)$ the signal to be analyzed; once the extrema are identified, all the local maxima are connected together by a cubic spline function: this is the upper envelope $s_{max}(t)$. This procedure is iterated for all the local minima to define the lower envelope $s_{min}(t)$.
b. The envelopes mean $m_1(t) = (s_{max}(t) - s_{min}(t))/2$ is determined. $m_1(t)$ is subtracted to the original signal to generate the first estimate of the first IMF: $h_1(t) = s(t) - m_1(t)$. This process can be iterated to obtain a better estimate of the first intrinsic mode function. Considering $h_1(t)$as the new input signal $s(t)$ the operations above are repeated until a suitable stopping criterion is satisfied.
c. The original signal remains decomposed into the first IMF c_1 and a residual $r_1 : s(t) = c_1 + r_1$. The residual $r_1 = s(t) - c_1$ has to be considered as the new input signal $s(t)$, repeat the previous steps to obtain the second IMFs and so on. The sifting process stops when the last IMF, called residue, has no more than one extremum: the original signal can be represented as a summation of the IMFs, plus the residue.

It should be noticed, due to the sifting process, the first IMF contains the finest scale, or the shortest period component, while the last IMF contains the largest scale: from the first to the last IMF, the period components are in ascending order.

3.2 *The Hilbert transform and the Hilbert spectrum*

Once the family of IMFs is determined, the Hilbert transform is used to obtain the instantaneous frequencies. Let $x(t) = c_i(t)$ the *i*-th IMF, its Hilbert transform is $y(t) = 1/\pi P \int x(\tau)/t - \tau \, d\tau$, in which P indicates the Cauchy principal value of the singular integral. The analytic signal in polar form is $z(t) = a(t)e^{j\varphi(t)}$, where the instantaneous amplitude and phase are $a(t) = \sqrt{x^2(t) + y^2(t)}$ and $\varphi(t) = atan(y(t)/x(t))$, respectively. The instantaneous frequency is defined as:

$$\omega(t) = \frac{d\varphi(t)}{dt} \quad (4)$$

4 DAMAGE DETECTION USING HILBERT-HUANG TRANSFORM

The presence of a crack in simply supported beam structures excited by a moving load causes a frequency modulation, which can be detected by an innovative method, based on the Hilbert-Huang transform, as explained ahead.

To understand how the proposed technique works in detail and to take advantage of the HHT capabilities, it is crucial to understand (i) how the EMD is used for damage data processing of $w(t)$, and (ii) how the damage presence and location can be extracted from the Hilbert transform of the IMF family. The backbone of the method is articulated as follows:

a. data acquisition: the response $w(t)$ is measured by a single-point sensor placed at midspan;
b. data decomposition: $w(t)$ is decomposed into a family of IMFs through the EMD method, as described in section 3.1;
c. data transformation: the Hilbert transform is applied to each IMF, and the analytic signals are extracted, as in section 3.2;
d. data analysis: the damage position is determined from the analysis of the instantaneous frequency curves, as explained below.

As explained, $w(t)$ can be regarded as the sum of two signals, a low frequency harmonic wave w_v and a high frequency harmonic wave w_1 with an amplitude modulation. The wave form (3) is already a sum of two IMFs, since the harmonic terms verify the properties 1) and 2) enunciated in section 3. Thus, in this simplified analysis, where $w(t)$ is approximated by Eq. (3), there is no need for invoking the EMD method, as it will be done indeed in the simulations of section 5. It is worth to point out the proposed procedure is not limited to linear systems, as in this case.

The Hilbert transform of Eq. (3) produces:

$$H\left[w_H(t)\right] = H\left[w_v(t)\right] + (\omega_v/\omega_1)H\left[w_1(t)\right] + cH\left[\Pi_T(t-t_d)w_1(t)\right] \tag{5}$$

Since the Hilbert transformation of a sinusoidal function only introduces a phase shift of $-\pi/2$, the first and the second term on the r.h.s. do not cause any modulation of the instantaneous frequency. Let focus the attention on the third term. If the Fourier spectra of $\Pi_T(t-t_d)$ and $w_1(t)$ do not overlap, the modulation theorem can be applied (Hahn 1995). Thus: $F[\Pi_T(t-t_d)] = e^{-j\omega t_d} 2T sinc(T\omega/\pi)$, where $sinc(t) = sin(\pi t)/\pi t$ is the normalized *sinc* function, with spectrum roughly within the bandwidth $B_\Pi = [0, f_\Pi]$, where $f_\Pi = \pi/T$. The limit frequency f_Π depends on the damage position and on the speed of the moving load: $f_\Pi = 2V/(L-L_1)$. If $f_1 > f_\Pi$ the modulation theorem holds and $H[\Pi_T(t-t_d)w_1(t)] \approx H[w_1(t)]$. In this case the identification of the damage position from inspection of the IF curve becomes difficult, because of the absence of any significant modulation in it.

On the contrary, if $f_1 < f_\Pi$ the spectra overlap, and it is possible to prove eq. (5) can be approximated by:

$$H\left[w_H(t)\right] \approx c\Omega w_1(t)\left[e^{j\omega_\Omega(t-t_d+T)} sinc\left(2\Omega(t-t_d+T)\right)\right] \tag{6}$$

where: $c = (\omega_1 - \omega_v)/\omega_1$, $\Omega = \pi(f_\Pi - f_1)$, $\omega_\Omega = \pi(f_1 + f_\Pi)$.

Eq. (6) leads to the following conclusions:

- Damage effect: as the moving load crosses the damaged section, a peak occurs in the Hilbert transform (6), and then in the IF curve of the high frequency wave. The induced peak is a crest since $c > 0$. The damage induces two, antithetic, effects: (i) a small global effect, due to the natural frequencies shift, and (ii) a local one, due to an intra-wave modulation which produces a sharp crest in the IF curve;
- Influence of the damage position: since Ω increases as $L_1 \to L$, the closer the damage position to the end of the beam, the sharper and higher the peak in the IF curve.
- Influence of the velocity of the moving load: Ω is proportional to V: the higher V the sharper the peak. On the other hand, the EMD process becomes unable to separate the two frequency components as soon as the forced response shows a single extremum. As a rule, the forced response has at least two extrema when $L/V \geq 3/(4f_1)$ thus, there exists an upper bound for the velocity, $V_{EMD} = 4Lf_1/3$: above that, the frequency components cannot be separated and the damage position cannot be accurately identified.

On the basis of the previous arguments, if (i) $f_1 < f_\Pi$ (peak sharpness) and (ii) $V < V_{EMD}$ (EMD efficiency), then the location of the damage is identified from the position of the highest crest in the IF curve. The condition (i) requires the system response due to the damage has a wideband spectrum, while the condition (ii) is needed for an efficient decomposition of the frequency components. These two conditions can be summarized as:

$$0.5(L-L_1)f_1 < V < 4Lf_1/3 \tag{7}$$

providing the velocity range of applicability of the proposed identification technique.

The previous points suggest, when allowed by service conditions, the moving load should perform two runs on the beam, along opposite directions, to maximize the identification capability for damages located near the endpoints.

5 NUMERICAL VALIDATION AND DISCUSSION

To evaluate the performance of the proposed technique, for the detection of damage in beam structures, some numerical examples are presented.

The aim is to evaluate the sensitivity of the method to:

- the crack depth d;
- the crack location L_1;
- the velocity of the moving load V;
- the ambient noise.

Preliminary, identification is operated in absence of ambient noise. The crack position and depth are chosen as follow: $L_1 = 0.4L$ and $d = 0.4h$, respectively, and the speed of the moving load is $V = 10$ km/h. The fundamental frequency is $f_1 = 1.126$ Hz, being 1.6% smaller than the undamaged case, the moving load frequency is $f_v = 0.069$ Hz while the limit frequency is $f_{\Pi} = 1.454$ Hz, being satisfied the condition $f_1 < f_{\Pi}$ for the damage detection from the IF curve.

Fig. 6 shows the implicit mode functions obtained from the empirical mode decomposition method. As expected, there are only two IMFs. The first is the high frequency wave while the residue is the low frequency wave, discussed in sections 2 and 4. The slope discontinuity induced by the damage is still observable in the IMF curves.

The first subplot in Fig. 7 shows the instantaneous frequency of the first IMF, which makes small

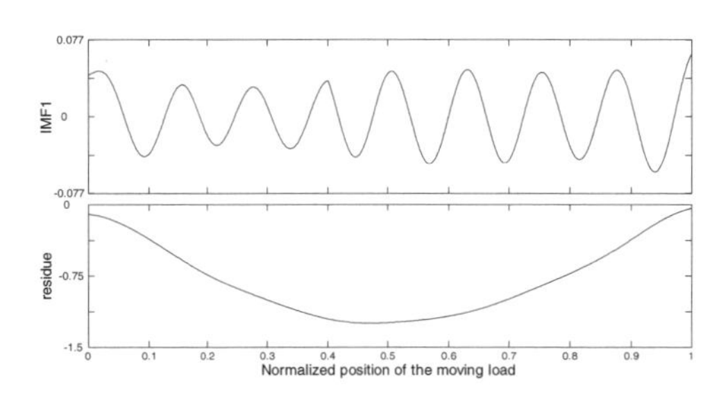

Figure 6. From the subplot on top it is shown, the first IMF, and the residue, versus Vt/L.

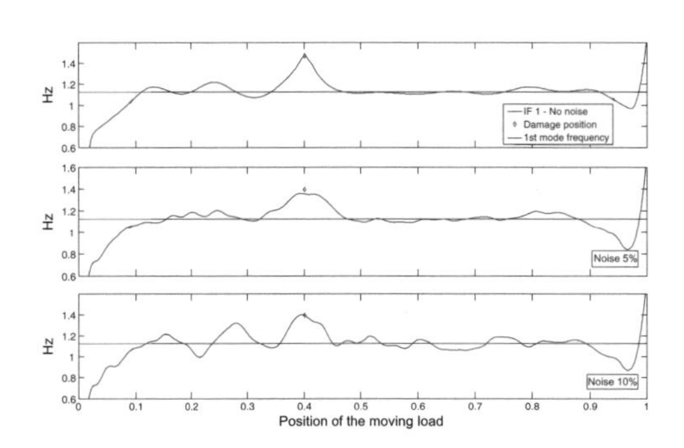

Figure 7. From the subplot on top it is shown, the first IF evaluated in absence of noise, with 5% and 10% of noise level, respectively, versus Vt/L.—is for the fundamental frequency, ◊ is for the damage position.

oscillations around the fundamental frequency. The results are in good agreement with theoretical arguments presented in section 4:

- the figure shows the highest crest, in the first IF curve, is located at $0.4L$, i.e., at the position of the damaged section;
- the presence of a crest instead of a dip, is a first glance surprising and somehow counterintuitive considering the damage decreases the natural frequency of the beam. An explanation for this is found in section 4, this effect happens because of the vibration amplitude increasing across the damaged section;
- end effects, in the form of large swings near the ends, are clearly visible in each IF curve. They are caused by the cubic spline fitting, implemented to generate the envelope curves in the EMD method. For this reason the data near the endpoints of each IF curve, i.e., lower than 0.1 or higher than 0.9, are not included into the identification process.

The influence of ambient noise is considered, adding a white noise to the analytic forced response: $w_{noise} = w + N_{lev} RMS(w)\eta$, where N_{lev} is the noise level, $RMS(w)$ is the root mean square of the analytic forced response and η is the normal distribution vector with zero mean and unitary rms.

Fig. 7 shows, even if the noise introduces a modulation in the instantaneous frequency plots, the damage position can be still correctly identified from the location of the highest crest, which confirms the reliability of the proposed method even with moderately polluted measurements.

The influence of the damage location on producing frequency modulation is considered, Fig. 8 shows the mean absolute percentage error (MAPE) of the estimated damage position L_1: $\mathrm{MAPE} = 100\sum_{i=1}^{10} |L_1 - L_{1i}|/L_1$, where L_{1i} is the i-th sample of the estimated damage position, four

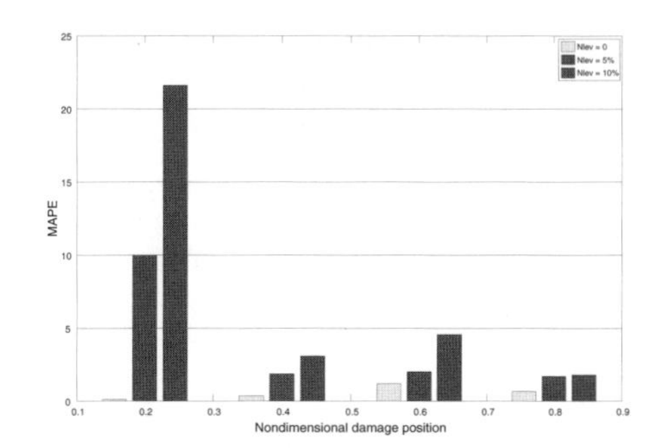

Figure 8. MAPE versus the nondimensional damage position, L1/L, for tree noise levels: 0%, 5%, 10%.

positions of the damage have been considered: $L_1 = 0.2L$, $0.4L$, $0.6L$ and $0.8L$. It appears:

- the error in the case of absence of ambient noise is due to the EMD process;
- when the damage is close to the load entering sections of the beam, i.e. $L_1 = 0.2L$, the frequency ratio f_{Π}/f_1, is close to unity, thus the frequency modulation induced by the sudden change of amplitude is weak, as explained in section 4. This makes the identification difficult in case of noisy ambient;
- when the damage is close to the endpoint of the beam, i.e., $L_1 = 0.8L$, the frequency ratio f_{Π}/f_1, is much higher than unity, the change of amplitude induces a sharp crest in the IF curve, and the damage position can be easily identified, even for moderate noise level;
- the damaged section is accurately identified, since the error remains below 5% for all the damage positions and noise levels, except for $L_1 = 0.2L$.

The sensitivity of the method to the crack depth is now investigated. Fig. 9 shows the MAPE evaluated for four different damage depths, i.e., $d = 0.2h$, $0.3h$, $0.4h$, $0.5h$, when the damaged section is located at $L_I = 0.4L$. The proposed technique fails to identify the damage location for $d = 0.2h$ in the presence of ambient noise, since the error is higher than 50%.

For the other crack depths the location of the damaged section is accurate, errors are indeed below 4%.

Finally, the influence of the velocity of the moving load is studied. The considered velocities are $V = 20$, 40 and 50 km/h; the damaged section is located at $L_I = 0.4L$, its crack depth is $d = 0.4h$, the noise level is 5%. In these conditions the limit velocity is $V_{EMD} \approx 30$ km/h. Fig. 10 shows the smoothed normalized forced responses at midspan and the corresponding first IFs.

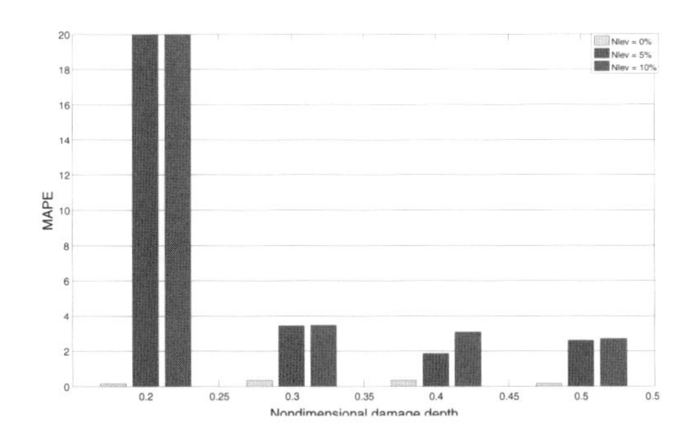

Figure 9. MAPE versus the nondimensional crack depth, d/h, for tree noise levels: 0%, 5%, 10%.

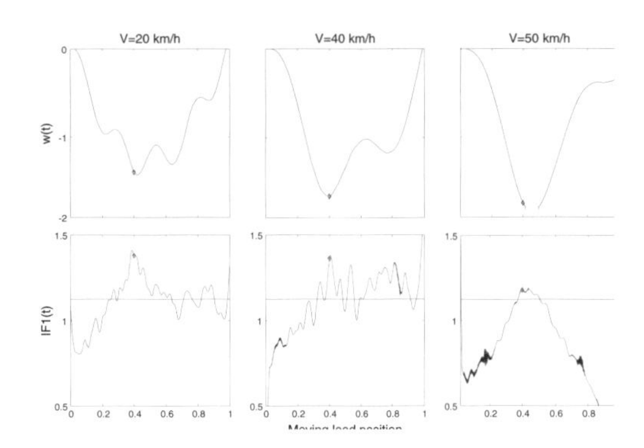

Figure 10. Subplots on top show the smoothed forced responses, subplots on bottom show the corresponding first IF, evaluated for different velocities versus Vt/L;—is for the fundamental frequency, ◊ is for the damage position.

As the velocity increases, the load frequency tends to the fundamental one and the number of extrema of the analyzed signal decreases, making the two wave components difficult to separate. The figure shows the presence of spurious local maxima in the IF curves, for $V = 40$ and 50 km/h. The damaged section is barely detectable from these plots, as anticipated in section 4 since $V > V_{EMD}$.

As a final remark, it is worth to point out the proposed technique fails to identify the damage position when $V > V_{EMD}$ and the forced response is well approximated by a linear model. However, since the amplitude of the forced response increases as the speed of the moving load, the stress-strain relationship may go beyond the linear limit, giving rise to nonlinear deformation and intra-wave frequency modulation, especially when the load crosses the damaged location. In this condition, the inspection of the instantaneous frequency curves should still reveal the position of the damage.

6 CONCLUDING REMARKS

A technique, based on the Hilbert-Huang transform, for monitoring the health safety of beam structures under moving load has been here proposed. It is shown the forced response is the sum of a low and high frequency waves, when the speed of the moving load is smaller than the critical velocity. The presence of a crack produces two antithetic effects: one is global and reduces the natural frequencies of the whole system, one is local, inducing an intra-wave frequency modulation. As the moving load crosses the cracked section, the instantaneous frequency of the high frequency wave has indeed a sharp crest. The identification procedure relies on determining the highest crest in the first instantaneous frequency curve. The performance of the proposed technique are evaluated for different damage locations, crack depths and velocities of the moving load. Results are quite accurate and capable to identify moderately damaged sections, with crack depth about 25% of the section height, they are not very sensitive to ambient noise, the technique shows good performance for velocities far from the critical velocity. Results are influenced by the damage position and, more deeply, by the speed of the moving load: it is found indeed a velocity range of applicability for the proposed method. There exists a lower bound, below that only minor modulations on the instantaneous frequency curves are induced by the damage. Similarly, there exists an upper bound, above that the EMD becomes unable to separate the frequency components of the signal. Outside the range of applicability, results are very sensitive to ambient noise and numerical errors. Further studies are in progress on direct data measurements and nonlinear models of 1-D and 2-D damaged structures, (i) to establish the applicability of the technique for velocities approaching the critical speed, and (ii) to extend the technique to bidimensional structures.

REFERENCES

Carcaterra A. & Ciappi E. (2000a). *Prediction of the Compressible Stage Slamming Force on Rigid and Elastic System Impacting over the Water Surface*. Nonlinear Dynamics, 21(2), 2000, 193–220.

Carcaterra A., Ciappi E., Iafrati A. & Campana E.F. (2000b). *Shock Spectral Analysis of Elastic Systems Impacting on the Water Surface.* Journal of Sound and Vibration, Academic Press, 229(3), 2000, 579–605.

Carcaterra A., Ciappi E. & Mariani R. (2001). *Experimental evidence of critical phenomena in structure-water impact.* In Fluid Structure Interaction, editor K. Chakrabarti, USA, & C.A. Brebbia, UK, 2001.

Carcaterra A., Ciappi E. (2004). *Hydrodynamic shock of elastic structures impacting on the water: theory and experiments.* Journal of Sound and Vibration, 271, 2004, 411–439.

Chance, J. *et al.* (1994). *A simplified approach to the numerical and experimental modeling of the dynamics of a cracked beam.* In: The 12th International Modal Analysis Conference, Honolulu, USA, 1994, pp. 778–785.

Gifford D.K., Childers B.A., Duncan R.G., Jackson A.C., Shaw S. & Schwienberg B. (2003). *Structural integrity monitoring of aircraft panels using a distributed Bragg grating sensing technique.* Smart Sensor Technol Measure Syst, 5050, 2003, 358–66.

Hahn S. (1995). Hilbert transforms in signal processing. Artech House, 1995.

Huang N.E. et al. (1998). *The empirical mode decomposition and the Hilbert spectrum for nonlinear and nonstationary time series analysis.* Proc. R. Soc. London A, 454, 1998, pp. 903–995.

Iafrati A., Carcaterra A., Ciappi E. & Campana E.F. (2000). *Hydroelastic Analysis of a Simple Oscillator Impacting the Free Surface.* Journal of Ship Research, 44(4), 2000, 278–289.

Johnson G.A., Pran K., Wang G., Havsgård G.B. & Vohra S.T. (1999). *Structural monitoring of a composite hull air cushion catamaran with a multi-channel fiber Bragg grating sensor system.* Structural Health Monitoring, 2000, 190–198.

Li H.C.H., Herszberg I., Davis C.E., Mouritz A.P. & Galea, S.C. (2006) *Health monitoring of marine composite structural joints using fibre optic sensors.* Compos Struct, 75(1–4), 2006, 321–327.

Lin H.P., Chang S.C. & Wu J.D. (2002). *Beam vibrations with an arbitrary number of cracks.* Journal of Sound and Vibration, 258(1), 2002, 987–999.

4 *Design & construction*

4.1 *Conceptual design*

Sustainable Maritime Transportation and Exploitation of Sea Resources – Rizzuto & Guedes Soares (eds)
© 2012 Taylor & Francis Group, London, ISBN 978-0-415-62081-9

Multiattribute decision making methodology for handymax tanker design in the concept design phase—A case study

V. Žanić & P. Prebeg
University of Zagreb, Faculty of Mechanical Engineering and Naval Architecture, Zagreb, Croatia

P. Cudina
BRODOSPLIT Shipyard Ltd., Design Office, Split, Croatia

ABSTRACT: An improved methodology of decision making for ship concept design, with the related design procedures and mathematical models, is presented. The mathematical model includes optimization of the main newbuilding characteristics as well as the optimization of its commercial effects. The Multiattribute decision making method is applied and the ship design analysis model is inbuilt into the decision making shell DeMak. Each attribute is defined by the membership grade function based on fuzzy set theory. The consistent relative significance of attributes is obtained as an eigenvector of the subjective decision making matrix. Inadequate information about the configuration of the feasible designs' subspace is overcome by sequential adaptive generation of design points in the feasible region (via e.g. adaptive Monte Carlo method, Particle Swarm Optimization or Genetic Algorithms). The concept of nondominated designs is used for filtering Pareto solutions. In practical applications to the general design of cargo ships, the developed design procedure and the associated mathematical models for the multi-attribute design synthesis have resulted in considerable improvements of the vessel's quality measures. It makes this design approach superior to traditional ship design and ranks it among promising approaches for those ship types in the concept design phase. The application of the method is presented for the concept design of Handymax Product Tanker.

1 INTRODUCTION

The objective of the paper is to present the design methodology capable of supporting the decision making process in the concept design phase, as implemented in the OCTOPUS system, together with the example of its application. Implementation of methodology in form of the interactive design shell is elaborated and tested on general concept design of Handymax Product tanker. Other ship design examples are presented in Zanic et al. (2010) and for structural design in Zanic et al. (2007).

The framework, developed on the basis of the novel methodology, is equipped with a quality output graphics offering the designer interactive possibilities for the design analysis and synthesis. The methodology enables the designer (or a design team) to rely on their own subjective estimation of the importance of particular design attributes, to analyse their interrelations, to make synthesis of the whole design and finally to reveal some profitable behavioural patterns in the design space.

When selecting the final design solution, the designer has at his disposal the sufficient amount of information and possibilities which enable creation of a comprehensive picture of the design: the quality of satisfying the conditions of each design attribute (cost, etc); the relation of design parameters with corresponding attributes in other design solutions and finally the information on what should be considered with special attention in further phases of the design development.

The methodology also enables the selection of a final design in co-operation with the other project stakeholders e.g., ship-owner, since technical and commercial data for all nondominated designs are available.

Therefore the efficient design procedure relies on the concept of nondominance, greatly decreasing the number of design variants to be explored, and methodology of subjective selection of the preferred design among them. Brief description of these two underlying basic concepts is given first (based upon ref. Zanic et al., 2009).

1.1 *Concept of nondominance and Pareto frontier*

The subspace of nondominated or Pareto optimal or efficient designs can be identified when designer's preference structure is applied to the feasible designs (points **y**—see Appendix A1 for definitions) in the attribute space (denoted $\mathbf{Y}^{\geq}$). They are images of the feasible designs (points **x**) in design space $\mathbf{X}^{\geq}$. Only Pareto optimal designs (usually only a small fraction of feasible designs) are of interest to designer since they dominate all other feasible designs.

Preference is a binary relation stating that design $\mathbf{y}^i$ is preferred to design $\mathbf{y}^j$. The "better set" can be defined with respect to given design $\mathbf{y}^0$ if all its elements are preferred to $\mathbf{y}^0$. Conversely, the "worse set" can be formed containing all designs that are worse than $\mathbf{y}^0$ in all attributes i.e. dominated by it.

Finally, the set of non-dominated designs is defined as a set of designs that have no "better set", hence they are not dominated by any design. Alternatively, design is non-dominated if it is better than any other feasible design in at least one objective.

1.2 *Subjectivity in decision making*

Realistic decision *making* must include subjectivity of decision makers (different stake holders will exhibit different preferences), particularly for novel designs. Objectives and measures of the design quality are best represented in the attribute space but the decision making is complicated due to lack of metric. Inclusion of subjectivity is basic to realistic decision-making, opening the possibility of realistic distance norms (metrics). It implies revealing of stakeholders' preferences:

- Subjective comparison of various designs can be performed using fuzzy functions $U_i(y_i)$, e.g., Novak (1998), Grubisic et al., (1997). Membership grade (satisfaction level) value $\mu_i = U_i(y_i)$ has range 0–1. In some problems (e.g., vibration) function $U_i()$ may consist of e.g., series of the inverse bell shaped functions centered at specified values. Concept is widely used in Decision Support problems (DSP) for inter attribute preferences.
- For determination of subjective importance of different attributes via weighting factors w_i, the AHP (Saaty) method, Yu (1985), using the bi-attribute preferences is used as default for intra attribute preferences.
- Combination of subjectivities for each attribute can be achieved e.g., as a $u_i(y_i) = w_i\ U_i(y_i)$.

Subjective metric space can now be formed with the introduction of metric (distance measure) since all attribute values $m_i = u_i(y_i)$ are normalized and scaled to their relative importance.

Value and utility function sets (**v** and **u**), which serve for the final selection of designs, contain values obtained from different attribute functions and include the subjectivity of designer and others involved in decision making. *Distance norms* (L_p metrics), w.r.t. given target ship (e.g., ideal m^*), are commonly used as value functions **v** (see Appendix A1 and the application of L_p for p = 2 in Sec. 3.4).

2 METHODOLOGY FOR THE CONCEPT DESIGN OF HANDYMAX PRODUCT TANKER

In general, the optimization based design process includes: (1) problem identification; (2) formulation of the Decision Support Problem (DSP) methodology and (3) the problem solution (4) sensitivity/robustness assessment, Grubisic et al. (1997). In the sequel, steps 1–3 are elaborated for the general design of Handymax tanker:

2.1 *Ship design problem identification*

Identification of DSP implies (see Appendix 1 for nomenclature and definitions):

- Selection of the ship design variable set **x** and sets of design criteria functions **c** (including constraints **g** and attributes **a**)
- Determination of the ship design objectives values **y** and corresponding measures of robustness.

Mathematical definition of design problem in the OCTOPUS DESIGNER environment implies definition of the ship design parameters, design quality measures ('merits', key performance indicators) and the corresponding structure of sets used for the efficient design description and calculation. In order to define the design task and the targets of the design process, the following design descriptors **d** need to be defined:

Design variables **x** *and/or fixed parameters* $\mathbf{d}^-$,

a. main dimensions (basic design variables):
   - length between perpendiculars L_{pp} (m)
   - scantling draught d_s (m)
   - breadth B (m)
   - block coefficient C_B (–)
b. index of the main engine type I_{ME} (–);
c. design task should be fulfilled within particular parameter limits:
   - deadweight DW (t)
   - volume of cargo space V_{car} (m^3)
   - required trial speed v_{tr} (kn).
d. parameter denoted "specific voluminosity" of the ship $\kappa = V_{car} / (L_{pp}\ B\ D)$ (–)
e. parameters for estimating the influence of the use of high tensile steel on the mass reduction of the steel structure,

f. parameters required for the selection of main engine (max. power of a particular engine MCR_i),
g. parameters required for the cost calculation of materials (costs of feasible main engines C_{MEi}, average unit cost of steel c_{st} and costs of other materials and equipment C_{fix}),
h. parameters required for the calculation of labour costs (productivity of the shipyard P_{cGT}, unit hourly wage V_L and other costs C_{oc}).

Design constraint set $\mathbf{g} = \{g^{minmax}, g^{ratios}\}$:

Design constraints are defined in two ways:

a. min/max values of basic design variables:
   - length: $L_{pp\,min}$, $L_{pp\,max}$
   - scantling draught: $d_{s\,min}$, $d_{s\,max}$
   - breadth: B_{min}, B_{max}
   - block coefficient: $C_{B\,min}$, $C_{B\,max}$
b. limit values of ratios between ship main dimensions:
   - length/breadth: $(L_{pp}/B)_{min}$, $(L_{pp}/B)_{max}$
   - length/scant. draught: $(L_{pp}/d_s)_{min}$, $(L_{pp}/d_s)_{max}$
   - length/depth: $(L_{pp}/D)_{min}$, $(L_{pp}/D)_{max}$
   - breadth/scant. draught: $(B/d_s)_{min}$, $(B/d_s)_{max}$

Dependent design properties–design attributes set **a**

Dependent design properties (attributes) depend upon the values of particular design variables and parameters. They are listed as follows:

a. mass of the steel structure W_{st} (t)
b. cost of materials C_M (USD)
c. cost of labour C_L (USD)
d. cost of ship C_{NB} (USD)
e. obtained deadweight DW (t)
f. obtained volume of cargo space V_{car} (m^3)
g. obtained trial speed v_{tr} (kn)

Design objectives set **o**

Design objectives depend primarily on the type of the ship and its special purposes. Potential design objectives can be the following:

a. minimize the weight of the steel structure
b. minimize the power of the main engine
c. minimize the cost of the material
d. minimize the cost of labour
e. minimize the cost of new building
f. minimize the own mass of the ship
g. maximize the stability
h. maximize the speed (at given CSR)

The design solution quality is estimated by the quality of satisfying particular design objectives or design attributes—by a multi-objective (MODM) or multi-attribute (MADM) decision making (synthesis) of the design. In this paper a multi-attribute synthesis of the design, Yu (1985), is applied in which each design characteristic y_i is enhanced with its own fuzzy function U_i, while the interrelations between importance of the particular design characteristics are given by consistent weights of Saaty's AHP method.

2.2 *Formulation of the Decision Support Problem (DSP) methodology and problem solution*

DSP methodology can be efficiently formulated after the basic characteristics of the design requirements and designer's preferences are revealed. It involves:

- DSP manipulation into equivalent but mathematically more convenient form,
- Selection of solution strategy (e.g., optimization technique) for the manipulated problem,
- Development of the final selection method for the generated design variants,
- Sensitivity/uncertainty analysis (application of function ROB (π-module) in preparation for Handymax vessel).

DSP solution requires practical implementation of selected methodology through two basic calculation (mathematical) models comprising of systems/sets of modules:

- Design analysis model (AM) is used for technical (performance, response, safety) and economical (cost) evaluations. For many engineering problems the mathematical model may be defined via six (generic) meta-systems of which two basic ones (**Φ, ε**) provide physical (ship model) and environmental (ship economy model) definitions of the problem and other four (**ρ, α, π, Ω**). are behavioral systems for modeling of ship performance functions, adequacy w.r.t. given constraints, reliability/robustness to uncertain (stochastic) design parameters and evaluation of quality measures/attribute values. For ship problems the implemented modules were developed by general designers, e.g., Trincas et al. (1994), Grubisic et al. (1997) and Cudina (2008), were used. Similar models were developed by e.g., Belamaric et al. (1999) or Jang et al. (2000).
- Synthesis model (OCTOPUS DESIGNER) includes the interactive decision-making shell with design utilities: design definition modules (**Δ**) and optimization and sensitivity solvers (**Σ**), jointly denoted DeMak, and databases, visualization, filtering and selection modules (**Γ**), jointly denoted DeView. General formulation was presented by Vanderplaats (1999) and numerous examples for structural sub problems were given in Zanic et al. (2000).

DeMak enables selection between MODM and MADM approaches. The latter approach implies:

- *Phase I*: generation, evaluation and filtering of non-dominated ship designs $\mathbf{D}^{k,ND}$ within the Cartesian space **Y**. It is done using solver modules **Σ** of the OCTOPUS DESIGNER.
- *Phase II*: the final selection procedure in the metric space **M** and selection set **L** using the OCTOPUS DESIGNER's interactive modules **Γ**.

In this way, problems of discrete variables and disjoint domain, which prohibit application of many analysis methods (used in MODM), becomes irrelevant. MADM approach is particularly efficient in ship concept design phase and in design of its subsystems in the following preliminary design phase. Implementation of the procedure into OCTOPUS Designer implies integration of user defined design model and user independent parts of the design environment.

The main components of OCTOPUS DESIGNER are DeMakGUI, DeMakMain and DeModel. DeMakGUI and DeMakMain are problem (i.e., analysis model) independent.

DeModel component wraps given User Model component **Ψ** (e.g., Handymax tanker general ship design in this paper). It gives prescribed interface from input modules **Φ, ε** to calculation modules in **Ψ**. This enables communication between User Model and User Model Independent components.

DeMakMain is the main component that encapsulates functionalities necessary for solving DSP Problems. Currently five **Σ** modules (optimization algorithms) are included in OCTOPUS: SLP, ES-AMC, FFE, MOGA and MOPSO (see Appendix A1).

DeMak graphical user interface (GUI) (Fig. 1) enables designer's flexible communication with

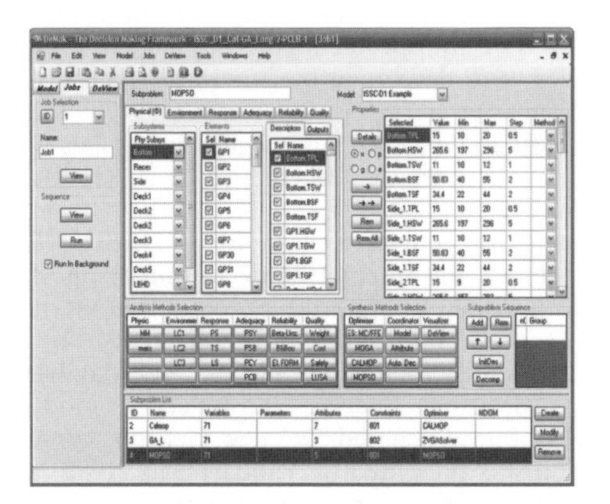

Figure 1. Optimization job navigation panel.

design tools in the form of interactive input definitions (**Δ**) and visualizations.

The upper part of the Figure 1 is used for design problem identification (definition of sets **x, g, a**), while middle part is used for selection of modules (**Φ, ε, ρ, α, π, Ω, Σ** and **Δ**) for the Decision Support Problem (DSP) procedure and the problem solution. Lower part is used for display of selected designs.

The visualization of the output and interactive design selection are transferred to the *DeView* tool (**Γ**).

3 CASE STUDY: CONCEPT DESIGN OF HANDYMAX PRODUCT TANKER

3.1 *Overview of modern Handymax tankers*

For further comparison with the results obtained using proposed design procedure the overview of modern built ships is given in Table 1.

3.2 *Design procedure and design model*

In the sequel, design procedure which enables use of:

- newly developed tools (**Δ, Σ, Γ**) for optimization
- analytical modules **Φ, ε** and $\mathbf{\Psi}^{1-4}$ developed for cargo ship design analysis (see Figure 2), is described. Basic calculation blocks are also briefly described below and in Appendices 2–4. Note that the ships with full forms share a common design procedure (Cudina 2008, Zanic et al., 2009, Belamaric et al., 1999) and the developed design model may be used as template for design of other ship types.

Task 1: Designs in subspace $X^{\geq}$: constraint verification

Main dimensions of a ship L_{pp}, B, d_s, C_B are varied within the design area in given steps: L_{step}, B_{step}, $d_{s\ step}$, C_{Bstep}. Each combination of design variables should be verified with respect to the design constraints set **g**.

Table 1. Modern Handymax tankers.

Shipyard	Mizushima	Onomichi	Shin Kurus.	Brodosplit	Brodotrogir
L_{oa} (m)	182.0	182.5	179.88	183.4	182.5
L_{pp} (m)	174.0	172.0	172.0	175.0	174.8
B (m)	32.20	32.20	32.20	32.0	32.20
D (m)	17.80	19.10	18.70	17.95	17.50
d_s (m)	12.65	12.65	12.0	12.0	12.2
DW (t)	48338	47131	45908	44881	47400
GT	27185	28534	28077	27533	27526
Cargo tank vol. (m^3)	52180	53609	53562	55926	53100
Main engine	6RTA48T	6S50MC	6UEC60LA	6S50MC	6S50MC
SMCR (kW/rev.)	8160/124	8580/127	9267/110	8240/122	8310/123
CSR (kW/rev.)	7460/122.4	7705/123	7877/104	7415/117.8	7480/118.8
$V_{tr,\ ballast}$ (knots)	15.35	16.28	16.82	16.5	16.5
$V_{service}$ (knots)	14.25	15.0	14.6	14.5	14.7

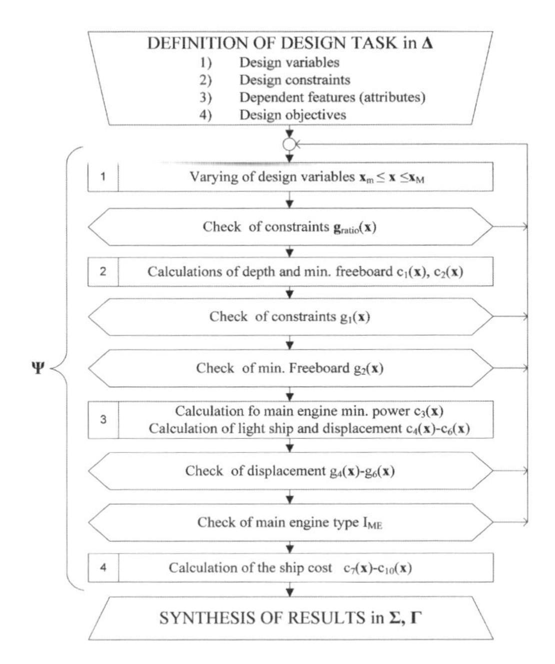

Figure 2. Block diagram of the general design procedure.

Task 2: Ship's depth (c_1) and min freeboard (c_2)
The depth D is obtained by a calculation from input data (required volume of cargo space V_{car} and the specific voluminosity of the ship κ) and the actual combination of design variables L_{pp}, B. Minimum freeboard is calculate by approximate calculation.

Task 3a: Main engine minimum power (c_3)
Approximate calculation of the main engine minimum power is obtained by the method presented in App. A2.

Task 3b: Displacement, light ship, DWT (c_{4-6})
Calculation of the displacement, light ship and deadweight is presented in Appendix A3.

Task 4: Calculation of the ship costs (c_{7-10})
The total cost of a ship C_{NB} comprise the costs of materials C_M, labour C_L and other costs C_{oc}.

Calculation of the cost of the material is carried out by adding the most relevant elements: cost of main engine C_{ME}, cost of steel C_{st} and cost of other material and equipment C_{fix}. The cost of the main engine is given as a fixed value or as a set of fixed values if there is a choice between various main engines. The cost of steel is dependent of the estimated weight of the steel structure W_{st}, ratio between the gross mass of steel and the weight of steel structure W_{gst}/W_{st} and average unit price of steel c_{st}. The cost of other material comprises all other ship's equipment and it can be considered as a fixed value at this design stage.

The cost of labour is presented in Appendix A4. Calculation of "product quantity cGT" is based on the OECD methodology.

Other costs (costs of a classification society, financing, docking, engagement of expert institutions, etc.) can be considered as fixed costs in the initial design stage.

3.3 *Handymax tanker requirements and data sets*

Based on the presented data, the following design requirements are set:

- Deadweight 50,000 t
- Maximum breadth 32.24 m
- Cargo holds capacity 58,000 m^3
- Maximum scantling draught 12.65 m
- Maximum length over all 174.8 m
- Trial speed at the scantling draught 15.0 kn

Design variables and parameters set
Identified design variables and parameters:

- main dimensions L_{pp}, B, d_s and C_B,
- two possible two main engines I_{ME}:

MAN B&W 6S50MC-C7, MCR 9480 kW/127 rpm,
$C_{ME1} = 4.7$ m US $,
MAN B&W 7S50MC-C7, MCR 11060 kW/127 rpm,
$C_{ME2} = 5.2$ m US $,

- main characteristics DW, V_{car} and v_{tr},
- data for the calculation of design attributes:

$\kappa = 0.56$; $f_1 = 3$ (15% of high tensile steel)
$c_{st} = 1000$ US $/t; $W_{gst}/W_{st} = 1.15$
$P_{cGT} = 35$ wh/cGT; $V_L = 30$ US $/wh
$C_{fix} = 16.0$ m US $; $C_{oc} = 4.0$ m US $

Design constraints sets
Constraints $\mathbf{g}^{minmax}$ with stepsize (within the range):

$170 \le L_{pp} \le 174.8$ m	step of 0.2 m
$32.0 \le B \le 32.24$ m	step of 0.04 m
$12.2 \le d_s \le 12.65$ m	step of 0.05 m
$0.82 \le C_B \le 0.845$	step of 0.001

Constraints in $\mathbf{g}^{ratio}$ set for basic ship dimensions:

$5.3 \le L_{pp}/B \le 5.6$; $13.7 \le L_{pp}/d_s \le 14.2$
$2.5 \le B/d_s \le 2.8$; $9.2 \le L_{pp}/D \le 10.0$

Design attributes (direct: y and normalised: m)
Inter and intra attribute preferences ($\mathbf{P}^u$) are listed below via subjective preference matrix with scale 1–9 (Table 2) and Novak's fuzzy functions $U_i(y_i)$ coefficients for the selected design attributes (Table 3).

Table 2. Determined inter-attribute preferences ($P^{uAHP} = [p_i/p_j]$).

	DW	V_{tr}	V_{car}	W_{st}	C_M	C_L	C_{NB}
DW	1	1	1	1/3	1/5	1/5	1/7
V_{tr}	1	1	1	1/3	1/5	1/5	1/7
V_{car}	1	1	1	1/3	1/5	1/5	1/7
W_{st}	3	3	3	1	1/3	1/3	1/5
C_M	5	5	5	3	1	1	1/3
C_L	5	5	5	3	1	1	1/3
C_{NB}	7	7	7	5	3	3	1

9-absolute preference of attribute in row i over one in col. j, 1-equal importance

Table 3. Fuzzy functions coefficients (P^{uNOVAK}).

Attribute	Type of Function	a_1	b_1	c_1	c_2	b_2	a_2
DW·(t)	Ω	49750	49875	50000	50000	50125	50250
v_{tr}·(knots)	Ω	14.7	14.8	15.0	15.0	15.2	15.4
V_{car}·(m^3)	Ω	57800	57900		58000	58100	58200
W_{st}·(t)	Z				8000	8100	8200
C_M·(m·US·$)	Z				22.5	23.0	23.5
C_L·(m·US·$)	Z				22.5	23.0	23.5
C_{NB}·(m·US·$)	Z				48.0	49.0	50.0

a_i, c_i function value at membership grade 0 or 1; b_i is fn.value for m. grade =0.5

In the right and lower part of the Figure 3, numerical data and a graphical representation of fuzzy functions associated to the attributes are shown. Weight of steel structure W_{st}, cost of material C_M, cost of labour C_L and the total cost of newbuilding C_{NB} are Z-shaped and have no lower limits defined since with the presented attributes, the option of "dissatisfaction" with the associated minimal values does not exist. Attributes which are additionally guarantying ship's characteristics (deadweight DW, trial speed v_{tr} and cargo tanks volume V_{car}) have Ω-shaped fuzzy functions with lower limits defined on the level expected to be free of penalties.

In the upper left part of Fig. 3 there is a subjective preference matrix and a graphical representation of attributes subjective weights relevant for decision making process.

Design objectives set o_{1-6}

The following design objectives are identified:

- minimize the weight of steel structure,
- minimize the power of main engine (using catalogue main engines),
- minimize the cost of newbuilding.
- additional objectives used for satisfaction of the constraints within given margins:
  a. Trial speed v_{tr};
  b. Volume of cargo V_{car}
  c. Deadweight DW

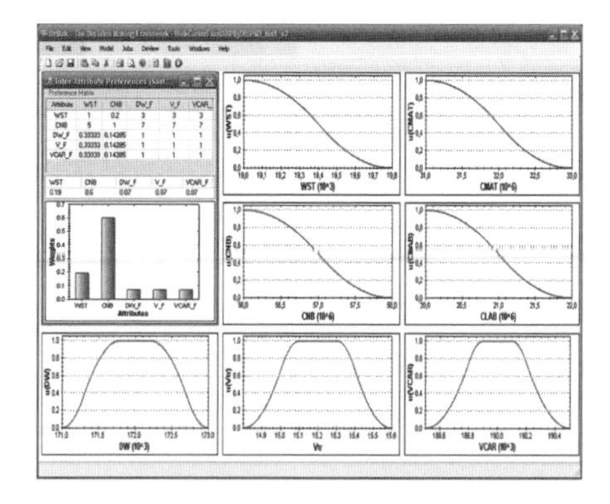

Figure 3. Graphical representation of preferences.

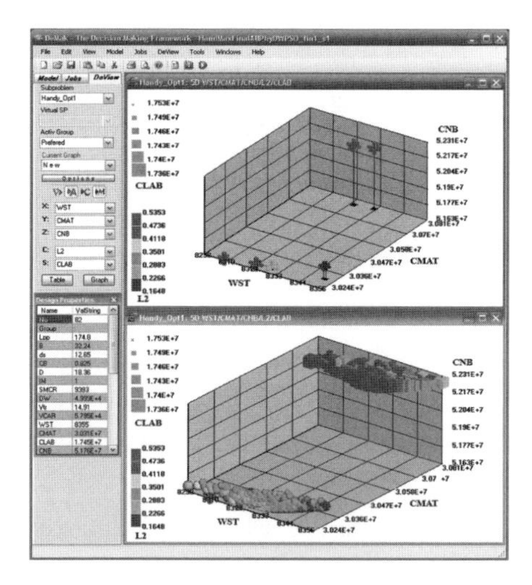

Figure 4. Pareto frontier: Wst-C_{MAT}-C_{NB}-C_{LAB}-L_2.

3.4 *Results and discussion*

The result of the applied optimization process is a hyper surface of nondominated designs (Pareto frontier).

Figure 4 (lower part) shows nondominated designs in a 3D coordinate system: weight of the steel structure Wst, costs of material CM, and costs of newbuilding CNB on the three coordinates. The shape of designation identifies the main engine type (sphere—6S50MC-C, cube—7S50MC-C).

The size of designation represents the cost of labor C,L. The color spectrum represents the remoteness from the target design qualities (e.g. $\mathbf{m}^{ideal}$) using $v = L_2$ metrics (Euclidean distance). Insight in 5D subspaces supports decision making.

The figure illustrates possible combinations of design dimensions for the purpose of analyzing interrelations of particular attributes and their impacts. In this figure, one can also notice two major groups of designs classified according to the selected main engine type (low, left—designs designated by a sphere and characterized by six-cylinder main engines). Designs with smaller main engines are closer to the utopia (colored in azure, blue and green). This means, more or less as in example presented in Cudina et al. (2010), that the cost of newbuilding is dominant and preferred with respect to other design attributes. In this case it leads to a conclusion that the rise in costs of main engine has a greater influence on the total costs of newbuilding than other influential factors. Size of sphere, or cube (cost of labor), is rather uniformly distributed, which leads us to a conclusion that, in this case, the influence of costs of material is stronger than that of costs of labor (process). The difference in the price of main engines at the level of 0.5 m US $ cannot be matched by the difference in the cost of steel and cost of labor. The data referring to any of nondominated designs presented in the figure are in the lower left part. Values which define the presented 5-D space are recapitulated in the upper left part.

Figure 4 (upper part) represents preferred designs in the same coordinate system. Four designs with lowest cost of newbuilding are selected, two designs according to selected metrics which is as close to the utopia (or ideal m*) as possible and two best designs with bigger main engines; all the designs are designated by 3D pluses.

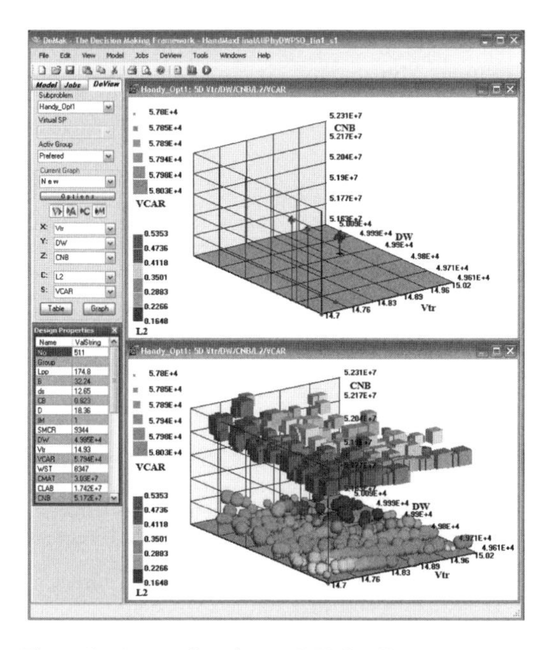

Figure 5. Pareto frontier: v_{tr}-DW-C_{NB}-V_{CAR}-L_2.

In the lower part of Figure 5 the nondominated designs are presented in 3-D space with trial speed v_{tr}, deadweight DW and costs of newbuilding C_{NB} on the three coordinate axes. The shape of designations (cube or sphere) designates the associated main engine (designs designated by a sphere have a 6S50MC-C main engine, and those by a cube a 7S50MC-C). The size of cube, or sphere, represents the volume of cargo tanks V_{CAR} (lower volume—smaller cube or sphere). Thus, a 5-D space of design objectives has been defined. The color spectrum represents the remoteness from the utopia according to L_2 metrics.

The upper part of the figure represents only the preferred designs. These designs are designated by 3-D pluses. The left part of the figure represents the data of a selected design. Nondominated designs presented in this figure are divided into two major groups according to the selected main engine. From this figure, one can conclude that, also in this case, the difference in price of main engines has a major influence on the costs of newbuilding, and that all other elements which are involved in the cost of newbuilding are of minor importance. Since the costs of newbuilding represent the attribute which is preferred with respect to other design attributes, its influence on determining the remoteness from an ideal design („utopia“) is dominant, so that designs with lowest cost are at the same time closest to the ideal, i.e., they are colored in blue in the figure.

Figure 6 gives a representation of subjective selection space **L** with parallel axis presentation. In the lower part there are eight selected preferred designs (vertical axes) identified with design no.

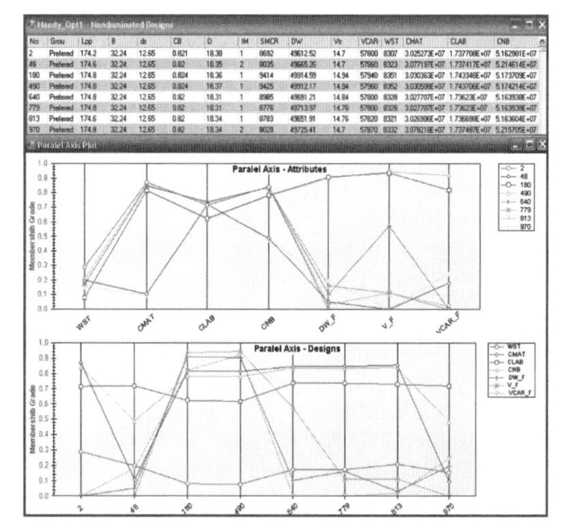

Figure 6. Subjective decision making—parallel axes plot.

Four of eight preferred designs have been selected according to the minimal building cost, two designs have been selected according to the minimal distance from the utopia, and the remaining two are the best among the designs with bigger main engines. Different colours represent subjective satisfaction level m (not attribute value y) for different attributes and all designs. The upper diagram has a reversed situation with the vertical axes representing satisfaction level for particular attributes. Each design is now represented with different colour. A table with alphanumeric data of all preferred designs is given in the upper part of the figure. It can be noted again that the cost of main engine has a major share in the total costs of material, while the differences in the costs of labour are of a lesser degree. The presented diagrams/numbers give the design information required for the final selection of nondominated, subjectively preferred, design. It is possible to select a design with lowest building cost as subjectively preferred design, which is e.g., desired design solution for the Yard.

Design no. 2 was selected as final design. It is presented on the diagram in the upper part of the figure by a red line with little circles and it has the highest value on the ordinate of newbuildings; the diagram in the lower part gives the cost of newbuildings in a light blue line with little triangles with the highest value on the ordinate of design no. 2.

Basic characteristics of the design are as follows:

L_{pp} = 174.2 m v_{tr} = 14.7 kn
B = 32.24 m V_{car} = 57800 m^3
d_s = 12.65 m W_{st} = 8307 t
D = 18.38 m C_M = 30.253 m US $
C_B = 0.821 C_L = 17.377 m US $
DW = 49613 t C_{NB} = 51.630 m US $
Main engine: MAN B&W 6S60MC-C7
$SMCR$ = 8692 kW

As far as the design requirements are concerned, one can notice that the preferred design slightly falls short of the required deadweight, speed and cargo tanks volume. It leads to conclusion that, during the further design development, actions related to cargo holds geometry, all weight groups, hull form and propeller design quality must be performed with special care.

Such quality of the preferred design could have been expected since this solution minimizes the costs of newbuilding and they are given as a dominant attributes in determining preferences among design attributes.

Presented design of Handymax tanker is a preferred design due to its good basic characteristics and especially due to its commercial effects (smaller and cheaper main engine) making it superior to the Table 1 designs.

4 CONCLUSIONS

- The applied optimization method (multi-attribute design synthesis) and the application of the adapted design procedure and associated mathematical models has resulted in an improvement in Handymax product tanker design, which makes this design approach more convenient than the traditional tanker design and ranks it among modern approaches to ship design.
- The adaptation of the mathematical model to particular requirements is very simple; therefore, it is possible to inbuilt the subjective design experience of any shipyard/designer/owner with their particular features, personal experiences and attitudes.
- The methodology also enables the selection of final Handymax design in co-operation with other project stakeholders, e.g., ship-owner, and technical and commercial data for nondominated designs can be available for comparison regarding e.g., life-cycle cost.
- The developed Handymax design module, imbedded in OCTOPUS design environment, can be used as template for the design of a wide range of full form ships in yards and RTD institutions.

ACKNOWLEDGEMENT

Thanks are due to the long-term support of Croatian Ministry of Science, Education and Sport: projects 120-1201829-1671.

REFERENCES

Belamaric, I., Cudina, P. & Ziha, K. (1999): Design Analysis of a New Generation of Suezmax Tankers, J. of Ship Production 15, 53–64.

Cudina, P. (2008): Design Procedure and Mathematical Models in the Concept Design of Tankers and Bulk Carriers, Brodogradnja 59 No. 4, 323–339.

Cudina, P., Zanic, V. & Prebeg P. (2010): Multiattribute Decision Making Methodology in The Concept Design of Tankers and Bulk-Carriers, Proc. of PRADS, Rio de Janeiro.

Grubisic, I., Zanic, V. & Trincas, G. (1997): Sensitivity of Multiattribute Design to Economy Environment: Shortsea Ro-Ro Vessels, Proceedings of VI International Marine Design Conference, Vol. 1, Newcastle, 201–216, Penshaw Press.

Jang, C.D., Yoon, G.J. (2000): Optimum Structural Design of Double Bulk Carriers in Comparison with Conventional Single Hull Types, IMDC 2000, 381–391.

Novak, V. (1989): Fuzzy Sets and their Applications, Adam Hilger, Bristol.
OECD (2007): Organisation for Economic Co-operation and Development (OECD), Directorate for Science, Technology and Industry (STI), Compensated Gross Ton (CGT) System.
OCTOPUS documentation (2010): U. of Zagreb, Faculty of ME and Naval Arch, Zagreb,.
Proceedings of ISSC (2003): Report of Committee IV. 2.
Trincas, G., Zanic, V. & Grubisic, I. (1994): Comprehensive Concept of Fast Ro-Ro Ships by Multiattribute Decision-Making, Proceedings of 5th Int. Marine Design Conference, IMDC'94, Delft.
Vanderplaats, G.N. (1999): Structural Design Optimization Status and Direction, Journal of Aircraft, 36:1, 11–20.
Yu, P.O. (1985): 'Multiple Criteria Decision Making', Plenum Press, New York.
Zanic, V., Jancijev, T. & Andric, J. (2000): Mathematical Models for Analysis and Optimization in Concept and Preliminary Ship Structural Design, IMAM 2000, Naples 3:2, 15–23.
Zanic, V., Prebeg, P. & Kitarovic, S., (2007): "Decision Support Problem Formulation for Structural Concept Design of Ship Structures", MARSTRUCT Conference, Glasgow, pp. 499–512.
Zanic, V., Cudina, P. (2009): Multiattribute Decision Making Methodology in the Concept Design of Tankers and Bulk Carriers, Brodogradnja 60, no. 1, 19–43.

APPENDICES

Appendix A1: Basic design problem notation/definitions:

DESIGN DESCRIPTION

$\mathbf{d}, \mathbf{d}^0, \mathbf{d}^-$ = n-tuple of (all, basic, remaining) design descriptors,
$\mathbf{d} = \{d_i\} = \{\mathbf{x}, \mathbf{d}^-\}$; $d_i = \{d_{i\ \mathrm{MEAN}}$, statistics $_i\}$, $i = 1, \ldots n$
$\mathbf{D}^k$ = design variant $k = \{\mathbf{d}^k, \mathbf{y}^k, \mathbf{m}^k, \mathbf{l}^k\}$
$\mathbf{m}$ = n_y-tuple of normalized attribute values; point in $\mathbf{M}$,
$\mathbf{m}^{ideal} = \{m_i^{\max}\}$; $m_i^{\max}$ is maximal achieved value within feasible designs
$\mathbf{M}$ = metric attribute space spanned by attributes m_i
$\mathbf{l}$ = n_l-tuple of composite attribute values; point $\mathbf{l}^k$ in $\mathbf{L}$
$\mathbf{L}$ = selection space spanned by composite attributes l_i
$\mathbf{x}$ = n_x-tuple of design variables x_i, $i = 1, n_x$; point $\mathbf{x}^k$ in $\mathbf{X}$
$\mathbf{X}, \mathbf{X}^{\geq}, \mathbf{X}^{\mathrm{ND}}$ = design, feasible, nondominated spaces spanned by x_i,
$\mathbf{y}$ = n_y-tuple of design attribute values; point $\mathbf{y}^k$ in $\mathbf{Y}$
$\mathbf{Y}, \mathbf{Y}^{\geq}, \mathbf{Y}^{\mathrm{ND}}$ = attribute spaces spanned by design attributes y_i
$\mathbf{z}$ = set of intermediate results for performance calculations.

MAPPINGS

$\mathbf{c}$ = n_c-tuple of design criteria functions (mappings)
$\mathbf{c} = \{c_i\} = \mathbf{g} \cup \mathbf{a}$
c_i: $(\mathbf{d}, \mathbf{z}) \rightarrow q_i$ (quality measure)
$\mathbf{a}$ = n_a-tuple of design attributes fn's (obtained from $\mathbf{c}$) a_i: $(\mathbf{d}, \mathbf{z}) \rightarrow y_i$
$\mathbf{g}$ = n_g-tuple of design constraint functions (obtained from $\mathbf{c}$)
g_i: $(\mathbf{d}, \mathbf{z}) \rightarrow I_{gi}$ (pass-fail indicator), $\mathbf{X}^{\geq} = \{\mathbf{x} \mid I_{gi}$ = pass, all $i\}$;
$\mathbf{o}$ = n_o-tuple of design objective functions o_i = manipulated a_i, u_i, l_i
$\mathbf{p}$ = n_p-tuple of probabilistically based $\mathbf{c}$-functions e.g.: ROB: $a_i\,(\mathbf{d}, \mathbf{z})$
$\rightarrow$ robustness measure (see grubisic et al. IMDC 1997)
$\mathbf{r}$ = n_r-tuple of design response/performance functions r_i: $\mathbf{d} \rightarrow \mathbf{z}^i$
$\mathbf{u}$ = n_a-tuple of subjectively normalized attribute functions
u_i: $(y_i, \mathbf{P}^u) \rightarrow m_i$; $\mathbf{P}^u$ = designer's inter/intra attribute preference data
$\mathbf{v}$ = n_v-tuple of value fn's v_i: $(\mathbf{m}, \mathbf{P}^v) \rightarrow l_i$ (e.g. $v_i = L_p = \Sigma|\mathbf{m}^*\text{-}\mathbf{m}|^p)^{1/p}$,
$\mathbf{m}^* \subseteq \mathbf{D}^{\mathrm{target}}$); alternatively: $v_i\,(\mathbf{u}\,(\mathrm{ROB}\,(\mathbf{a}\,(\mathbf{d}, \mathbf{z}), \mathbf{P}^u), \mathbf{P}^v) = l_i$.

SETS OF COMPUTATION MODULES

$\boldsymbol{\alpha}$ = adequacy meta-system; subset of modules in the analysis model (AM) containing constraint functions/mappings g_i. Output: I_{gi}
$\boldsymbol{\Gamma}$ = set of synthesis modules in synthesis model (SM) for optimization (using $\mathbf{P}^u$, $\mathbf{P}^v$ data for subjective definition of $\mathbf{u}$ and $\mathbf{v}$), designer interaction with the design process, filtering of designs and visualization of $\mathbf{X}, \mathbf{Y}, \mathbf{M}, \mathbf{L}$ spaces. Output: $\mathbf{m}, \mathbf{l}$.
$\boldsymbol{\Delta}$ = set of modules for the synthesis (optimization) problem definition (selection of variables $\mathbf{x}$ and criteria fn's $\mathbf{c}$).
$\boldsymbol{\varepsilon}$ = environment/economy meta-system (costs, etc.); subset of modules in AM generating descriptors $\mathbf{d}^{\varepsilon} = \mathrm{E}(\mathbf{d}^0) \subseteq \mathbf{d}$
$\boldsymbol{\Phi}$ = ship (physical) meta-system; subset of modules in AM / SM generating descriptors $\mathbf{d}^{\Phi} = \mathrm{F}(\mathbf{d}^0) \subseteq \mathbf{d}$
$\boldsymbol{\Psi}$ = user analysis program block, containing $\{\boldsymbol{\rho}, \boldsymbol{\alpha}, \boldsymbol{\pi}, \boldsymbol{\Omega}\}$ modules dependent on $(\mathbf{d}, \mathbf{z})$. Descriptors $\mathbf{d}$ are generated in the user initialization program block containing $\{\boldsymbol{\Phi}, \boldsymbol{\varepsilon}\}$: Output: $\mathbf{y}, I_{gi}, p_{failure}$, robustness.

π = reliability/robustness meta-system; subset of AM containing: reliability, robustness and risk modules based on **c(d**$_{MEAN}$, **statistics, z)** functions / mappings. Output: $p_{failure}$, robustness measures.
ρ = response/behavior meta-system; subset of AM, containing modules for calculation of functions r_i. Output: **z**.
Σ = set of optimization solvers (Seq. Linear Programming- (SLP), Fractional Factorial Experiments (FFE), Multi Objective Particle Swarm Optimization (MOPSO), Multi Objective Genetic Algorithms(MOGA), Evolution Strategy-Adaptive Monte Carlo (ES-AMC), etc.) generating Pareto frontier $\{\mathbf{x}^k, \mathbf{y}^k\}^{ND}$ by filtering designs in $\mathbf{X}^{\geq}$ U $\mathbf{Y}^{\geq}$ based on objectives **o**.
Ω = design quality meta-system; subset of AM/SM containing functions/mappings a_i. Output: **y**.

Appendix A2: Main engine minimum power

The data base has been created in a way that the SEAKING calculation results of the power delivered to the ship propeller are increased by mechanical losses, and then correlated on the basis of empirical data in using the SEAKING program (developed and published by SSPA) and results obtained at trial sailings. The data base uses the results of approximately 100 ship speed calculations within the following range and expected propeller revolutions of approx. 127 rpm:

$57000 \leq V_D \leq 60000$ m^3 $32.0 \leq B \leq 32.3$ m
$14.8 \leq v_{tr} \leq 15.2$ kn $12.2 \leq d_s \leq 12.65$ m
$170 \leq L_{pp} \leq 175$ m $0.82 \leq C_B \leq 0.845$

The regression analysis results with the approximation function of continuous service rating:

$$CSR = 0.007977\, L_{pp}^{-0.05697}\, B^{1.048}\, d_s^{0.1141}\, C_B^{3.826}\, v_{tr}^{4.251} ((1-0.002913\, L_{pp}/d_s)\ (kW) \quad (A1.1)$$

Calculation of continuous service rating for the purpose of determining the main engine power, if MAN B&W 7S50MC-C is selected, needs to be reduced by 7–8% due to lower efficiency of propulsion system at a higher propeller revolutions.

Selected maximum continuous rating:

$$SMCR = CSR / 0.9\ (kW) \quad (A1.2)$$

Appendix A3: Calculation of the displacement, light ship and deadweight

Displacement is calculated as a product of main dimensions of the ship, i.e., L_{pp}, B, d_s, C_B, and γ_{tot} (sea water density including the influence of ship plating and appendages:

$$\Delta = L_{pp}\, B\, d_s\, C_B\, \gamma_{tot}\ (t) \quad (A2.1)$$

The weight of a ship, according to Cudina (2008), can be divided into three groups: weight of the steel structure W_{st}, weight of machinery equipment W_m and weight of other equipment W_o.

$$LS = W_{st} + W_m + W_o\ (t) \quad (A2.2)$$

$$W_{st} = (1-f_1/100)(0.034\, [L_{pp}\, (B+0.85\, D+0.15\, d_s)]^{1.36} \{ (1 + 0.5\, [(C_B-0.7) + (1-C_B)\, (0.8\, D-d_s)/3\, d_s]\} +350)\ (t) \quad (A2.3)$$

where f_1 (%) is reduction of the total weight of steel structure caused by implementation of high tensile steel.

$$W_m = SMCR\ (860-0.0034\ SMCR)/7350\ (t) \quad (A2.4)$$

$$W_o = (0.29 - L_{pp} / 1620)\, L_{pp}\, B\ (t) \quad (A2.5)$$

Resulted deadweight is calculated as a difference between displacement and light ship:

$$DW = \Delta - LS\ (t) \quad (A2.6)$$

Appendix A4:

Calculation of the cost of labour C_L

$$C_L = cGT\, P_{cGT}\, V_L\ (US\ \$) \quad (A3.1)$$

P_{cGT}—productivity (working hours/cGT)
V_L—unit hourly wage (USD/working hour)
cGT—compensated gross tonnage, according to OECD, defined as:

$$cGT = 48 * GT^{0.57} \quad (A3.2)$$

Gross tonnage:

$$GT = K_1\, V;\ K_1 = 0.2 + 0.02\, \log V \quad (A3.3)$$

Total enclosed ship's volume:

$$V = V_D + V_{cam} + V_{sup} + V_{for}\ (m^3) \quad (A3.4)$$

Ship's volume up to moulded depth D:

$$V_D = L_{pp}\, B\, D\, C_{BD}\ (m^3) \quad (A3.5)$$

Block coefficient is calculated as:

$$C_{BD} = C_B\, [1 + 0.005686\, (D - d_s)]\ (-) \quad (A3.6)$$

Estimated volume of the camber:
$V_{cam} = 1500$ m^3
Estimated volume of the complete superstructure:
$V_{sup} = 5000$ m^3
Estimated volume of the forecastle:
$V_{for} = 500$ m^3

Sustainable Maritime Transportation and Exploitation of Sea Resources – Rizzuto & Guedes Soares (eds)
© 2012 Taylor & Francis Group, London, ISBN 978-0-415-62081-9

Reliability of attribute prediction in small craft concept design

I. Grubisic
University of Zagreb, Zagreb, Croatia

E. Begovic
University Federico II, Naples, Italy

ABSTRACT: A multi-attribute concept design procedure, as developed at the University of Zagreb, is applied to the small craft. Specialized concept design models of several generic types, e.g. mono-hull ferry, catamaran ferry, fishing vessel, fast SAR, patrol and paramilitary craft, were developed and included into a multi-attribute design procedure. In the present paper we address concept design model of the fast mono-hull type service vessels. Calibration of the model is an important task if reliable results are expected. Here we will deal with the reliability of attribute prediction. Practical parametric equations for attribute prediction, based on the previous work, were developed from published empirical methods, from analytical first principle methods, and from the study of selected database vessels. The results of attribute prediction were systematically compared to database values. Best fitting equations are presented in order to make them widely available to small craft designers.

1 INTRODUCTION

1.1 *Background*

A multi-attribute concept design procedure is established as an aid in concept design of generic small craft type defined as: fast mono-hull of round bilge or hard chine form, with transom stern, with steel, aluminum or composite structure.

The procedure operates on the generic concept design model that is structured in advance and calibrated for the ship type in question.

1.2 *Concept design model*

At the level of concept design a parametric model is used. Direct "first principle" approach is not an option since hull form is defined only via dimension ratios and form coefficients, since design is not yet defined in enough detail.

In this paper we will demonstrate the concept design model for fast service vessels, named "SERVES". The types of vessels considered are e.g. patrol, police, SAR, customs, crew transfer, passenger, yacht, etc., consistent with database (Table 1).

1.3 *Multi-attribute design procedure*

For many years design optimization searched for unique optimum based on single criterion (e.g. minimal displacement). Development of operations research techniques leads to the possibility of having many optima. When dealing with multi-dimensional space of design attributes, Pareto approach selects only non-dominated ones by an objective procedure (Grubisic & Begovic, 2003). Selecting preferred design from the set of Pareto optimal designs has to be performed subjectively. This is why design is still dealing with human artistic talents. The idea is not to replace human designer by computer procedure but to use multi-attribute procedure to solve all objectively solvable conflicts and leave the rest (the most difficult ones) to the designer.

Each generic model consists of a number of modules that are responsible for predicting numerical values of design attributes. Comparison of the design attribute values is fundamental for decision making in the design space, through the process of elimination of dominated designs, resulting in multi-dimensional Pareto frontier. Thorough visualization of the design space helps the designer in decision-making focused on Pareto optimal designs only.

Table 1. Types and number of vessel types in database.

Type of vessel	N	Type of vessel	N
CARGO	2	PATROL	70
CREW	5	PAX	12
FIRE	11	PILOT	20
FISH	3	RESEARCH	3
MEDIC	1	SAR	15
MIL	19	WORK	13
MYACHT	22	Total	196

Design model is the core of the multi-attribute design procedure but design decisions are made outside the model. The black box approach would describe the design model as a transformer of input parameters into output attributes. Reliability of attribute prediction by the model is of paramount importance for the applicability of the whole procedure. Therefore, it is in the focus our interest in this paper.

1.4 *Parameters and attributes*

Design parameters define the problem (in this case, a ship) while design attributes describe the capabilities of the ship and serve as a basis in decision-making process.

Design parameters in our model comprise hull dimensions and coefficients, propulsion system, structural material, required capacities, etc.

Design attributes comprise maximal attainable speed, sailing range, sea-keeping attributes, cost from shipyard, etc.

1.5 *Variables and constraints*

Sub set of design parameters is varied while searching for optimal solutions, i.e., variables, e.g., length, beam, draft, depth, etc. Variables are controlled outside the model by the procedure or in some cases manually by the designer.

In the procedure of predicting design attributes the model generates many intermediate attributes that may not be required for selection purposes but that may be subject to restrictions. These attributes are used as constraints to eliminate undesirable designs (e.g., insufficient stability).

2 PARAMETRIC DESIGN MODEL

Parametric model was developed based on the database of 196 vessels (Table 1). The model is composed of parametric relations that predict attribute values based on input parameters or other attributes that were previously defined by the same model. Parametric equations are in greater part result of regression analysis of database vessels. Variability of the relations is described by standard deviation of the sample vessels.

Introducing standard deviation in the estimation process serves two purposes: firstly it demonstrates the variability of the particular parameter; secondly, since parameters are entered into relation that predicts attributes (or constraints) their variability may be estimated too.

Standard deviation is given for most parametric relations. It is common practice to estimate any parameter value to be within one standard deviation each side from the predicted value, e.g.:

$$L_{WL} \cdot 0,01 \cdot (100 - \sigma) < L_{WL} < L_{WL} \cdot 0,01 \cdot (100 + \sigma) \quad (1)$$

Sometimes it may be extended to two or even three standard deviations each way. In the practical implementation of the model in addition to the predicted value the upper and lower suggested values are given as guidance to designer. The procedure is used in two ways;

By imputing parameters of the known vessel it is possible to compare predicted values of attributes and so gain insight into reliability of vessel data.

By modifying suggested values of parameters it is possible to observe consistency and sensitivity of the model to input variation. At the same time a concept design of the vessel may be developed manually (without optimization).

When model is tested to designer's satisfaction it can be transferred into multi-attribute design procedure.

3 "SERVES" DESIGN MODEL

Design model "SERVES" is developed using attribute prediction methods (such as resistance, propeller, weights etc.) and parametric relations developed from database.

Database includes fast vessels having some of the generic characteristics needed for the service i.e. search and rescue, survey, pilot, patrol and other military and paramilitary service vessels, fire fighting, environmental and pollution control, etc. Some fast yachts and sport boats are also included in the database.

Formulations of the parametric relations used in the design model are developed from the available analytic and heuristic estimation methods for powering, sea-keeping, weight, cost estimating, etc. and are calibrated to fit database values.

For the purpose of testing the reliability of the model it was found convenient to program the formulae in spreadsheet. The advantage of spreadsheet approach is simple adjustments and additions to the procedure and quick graphical and tabular output that is easy to control.

Some important parts of the model are presented in sequel.

4 HULL PARAMETERS PREDICTION

Small vessels design starts from a set of requirements. The first one is usually the hull length (ISO 8666). Other parameters are related to length. When these parameters are selected, a number of further parameters are suggested and the design proceeds.

4.1 *Hull principal dimensions*

$$L_{WL} = 0{,}932 \cdot L_H - 0{,}576 \quad \sigma = 4{,}191\% \tag{2}$$

$$B_M = 0{,}932 \cdot L_H^{0{,}567} \quad \sigma = 11{,}216\% \tag{3}$$

Some parameters may be estimated using more than one relation. In this case the mean value is used. Beam at the section of maximal immersed area:

$$B_X = 0{,}900 \cdot B_M - 0{,}053 \quad \sigma = 3{,}967\% \tag{4}$$

$$B_X = 0{,}834 \cdot L_{WL}^{0{,}589} \quad \sigma = 11{,}331\% \tag{5}$$

Draft at the section of maximal immersed area:

$$T_X = 0{,}166 \cdot B_X^{1{,}214} \quad \sigma = 25{,}068\% \tag{6}$$

$$T_X = 0{,}128 \cdot L_{WL}^{0{,}725} \quad \sigma = 24{,}034\% \tag{7}$$

Depth at the section of maximal immersed area:

$$D_X = 0{,}497 \cdot L_{WL}^{0{,}564} \quad \sigma = 16{,}579\% \tag{8}$$

$$D_X = 0{,}548 \cdot B_M^{0{,}940} \quad \sigma = 17{,}342\% \tag{9}$$

$$D_X = 2{,}493 \cdot T_X^{0{,}582} \quad \sigma = 19{,}568\% \tag{10}$$

Transom immersion at rest:

$$T_{T(FNT=5)} = 0{,}7 \cdot \frac{V_{MAX}^2}{962} \quad \sigma = 36\% \tag{11}$$

Full load displacement may be estimated from three relations:

$$W_{FL} = 0{,}058 \cdot L_{WL}^{2{,}256} \quad \sigma = 21{,}9\% \tag{12}$$

$$W_{FL} = 0{,}624 \cdot \left(L_{WL} \cdot B_X \cdot T_X\right)^{0{,}931} \quad \sigma = 21{,}31\% \tag{13}$$

$$W_{FL} = 0{,}459 \cdot \rho_{SW} \cdot L_{WL} \cdot B_X \cdot F_N^{-0{,}44} \quad \sigma = 28{,}33\% \tag{14}$$

Displacement volume is:

$$\nabla = \frac{W_{FL}}{\rho_{SW}} \tag{15}$$

Basic parameters defining hull dimensions are input by the designer within suggested minimal and maximal values.

4.2 *Hull form coefficients*

Hull form parameters developed by Begovic 1998 are used to define input for further hydrodynamic estimates.

$$C_P = 0{,}384 + 0{,}565 \cdot C_B \tag{16}$$

$$C_X = 0{,}29 + 0{,}918 \cdot C_B \tag{17}$$

Water plane related coefficients are found from:

$$C_{WP} = 0{,}47 \cdot C_P + 0{,}467 \tag{18}$$

$$C_{IL} = 1{,}66 \cdot C_{WP} - 0{,}73 \tag{19}$$

$$C_{IT} = 1{,}316 \cdot C_{WP} - 0{,}394 \tag{20}$$

Deadrise angle at the position of longitudinal center of buoyancy is found from relation (21):

$$\beta_X = 103{,}6 - 127 \cdot C_X \tag{21}$$

Centers of buoyancy and flotation are defined as:

$$X_{CB} = 63{,}21 - 28{,}44 \cdot C_P \tag{22}$$

$$X_{CF} = 24{,}19 - 0{,}396 \cdot X_{CB} \tag{23}$$

Height of center of buoyancy above base line is founded on the modification of Papmel's formula:

$$\overline{KB} = 0{,}961 \cdot T_X \cdot \left(1{,}048 - \frac{C_B}{C_B + C_{WP}}\right) \tag{24}$$

Immersed transom area is determined in static condition. A factor of 0,7 is applied to compensate for transom immersion due to dynamic trim:

$$\frac{A_T}{A_X} = 3{,}604 \cdot C_P - 1{,}941 \tag{25}$$

$$\frac{T_T}{T_X} = 0{,}028 + 0{,}986 \cdot \frac{A_T}{A_X} \tag{26}$$

Wetted surface is estimated by Taylor's formula where the coefficient is found by Begovic 1998.

$$C_S = 2{,}61 + \frac{\frac{B_X}{T_X} \cdot \left(\frac{B_X}{T_X} - 0{,}244\right)}{81} \tag{27}$$

$$S_{WS} = C_S \cdot \sqrt{L_{WL} \cdot \nabla} \tag{28}$$

All the relations are programmed in spreadsheet and used as guidance. A Townsin formula is found useful here although it was developed for large ships. The predicted value is compared to the block coefficient defined by volume, length, beam and draft that are input values.

$$C_B = 0,7 + 0,125 \cdot \tan^1\left(\frac{23 - 100 \cdot F_N}{4}\right) \quad (29)$$

Speed related Froude's numbers are found from:

$$F_N = \frac{0,5144 \cdot V_{MAX}}{\sqrt{9,80665 \cdot L_{WL}}} \quad (30)$$

$$F_{N\nabla} = \frac{0,5144 \cdot V_{MAX}}{\sqrt{9,80665 \cdot \sqrt[3]{\nabla}}} \quad (31)$$

5 SEA KEEPING MODEL

Sea keeping is obviously very important for "SERVES" vessels. There are human tolerances that should not be exceeded if effective operation of the vessel is expected. Estimating allowable crew exposure time, i.e. measuring sea-keeping quality of the vessel is solved by estimating vertical accelerations and subsequently estimating allowable crew exposure time according to ISO 2631/3.

5.1 *Acceleration prediction*

Acceleration prediction method, as amended by Lloyd's Register 1996, is used to predict an average of 1/100 of the maximal vertical acceleration (expressed in g) at the centre of gravity of the vessel in head sea:

$$a_{1/100} = 0,0015 \cdot \tau \cdot L_1 \cdot (H_1 + 0,084) \cdot (5 - 0,1 \cdot \beta_X) \cdot \Gamma^2 \quad (34)$$

Recalculated to the RMS value by equation we obtain RMS acceleration prediction as:

$$a_{RMS} = \frac{\tau \cdot L_1 \cdot (H_1 + 0,084) \cdot (5 - 0,1 \cdot \beta_X) \cdot \Gamma^2}{3737} \quad (32)$$

Where:

Where: $\Gamma = V_{MAX}/\sqrt{L_{WL}}$; $L_1 = L_{WL}/B_x \cdot B_X{}^3/\Delta$;

and: $L_{WL}/B_X \geq 3$; $H_1 = H_{1/3}/B_X \geq 0,2$

$\beta_X \leq 30^\circ$ bottom deadrise angle at X_{CG};
$\tau \geq 3^\circ$ trim angle (temporarily set at 3 degrees)

The RMS acceleration is transformed to other statistics by the expression (33):

$$a_{1/N} = a_{RMS} \cdot (1 + \ln N) \quad (33)$$

We can now calculate a_{RMS} acceleration from $a_{1/100}$ by (34):

$$a_{RMS} = \frac{a_{1/100}}{5,605} \quad (34)$$

5.2 *Human tolerance to vertical acceleration*

Criterion of the human tolerance to vertical accelerations ISO 2631/3, is used to estimate maximal allowable time for the crew exposure to such accelerations. The tolerable exposure time in hours is approximated by the following relations for respective encounter frequency ranges:

$$0,1 \leq f_E \leq 0,315 \qquad t = \frac{0,5}{a_{RMS}{}^2} \quad (35)$$

$$0,315 < f_E < 0,63 \qquad t = \frac{14.486 \cdot f_E - 4.0631}{a_{RMS}{}^2} \quad (36)$$

Encounter frequency f_E in H_Z in head sea for the estimated modal period T_0 is defined as.

$$f_E = \frac{1}{T_0} - \frac{v \cdot \cos(\mu \cdot \pi/180)}{T_0{}^2 \cdot \frac{g}{2 \cdot \pi}} \quad (37)$$

The modal period in seconds for the Adriatic Sea is estimated by the following expression:

$$T_0 = 4,45 \cdot H_{1/3}{}^{0,406} \quad (38)$$

6 FREQUENCY OF WAVE HEIGHTS

Probabilities of exceeding significant wave heights in the Adriatic Sea are approximated the by simple Equation (39) and shown in Figure 1:

$$f(H_{1/3}) = \frac{103,13}{1 + \left(\frac{1 + H_{1/3}}{2}\right)^5} \ [\%] \quad (39)$$

7 FREEBOARD

By analyzing freeboards at fore perpendicular from database vessels, the relation of the general exponential type was found applicable with coefficients as given in Table 2:

$$F_{FP} = a \cdot L_{WL}{}^b \quad (40)$$

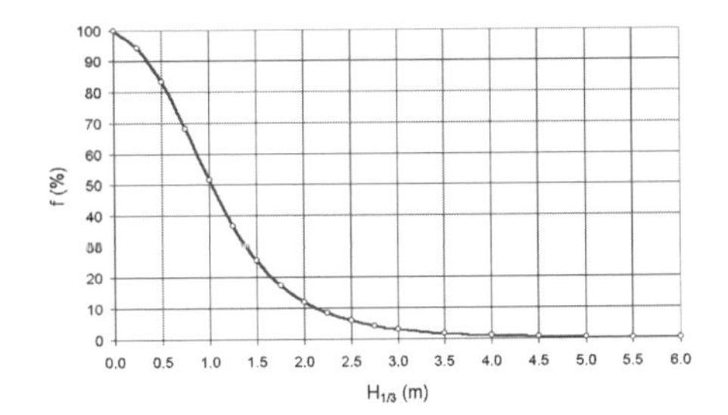

Figure 1. Probabilities of exceeding H1/3 in the Adriatic Sea.

Table 2. Freeboard at fore perpendicular.

Freeboard at F.P. (m)		
Navigating area	*a*	*b*
Protected waters	0,136	0,774
Coastal waters	0,206	0,707
Open ocean	0,329	0,642
Severe sea conditions	0,460	0,616

8 WEIGHT PREDICTION

Weight prediction method was studied in more details in Grubisic & Begovic 2009. Therefore, here only brief presentation is appropriate. Rule length and cubic number, respectively, are set as a basis for weight prediction.

$$L = 0{,}5 \cdot (L_H + L_{WL}) \qquad (41)$$

$$C_N = L \cdot B_M \cdot D_X \qquad (42)$$

Shell surfaces are estimated by:

$$S_1 = 2{,}8 \cdot \sqrt{W_{FL} \cdot L} \qquad (43)$$

$$S_2 = 1{,}09 \cdot (2 \cdot L + B_M) \cdot (D_X - T_X) \qquad (44)$$

$$S_3 = 0{,}823 \cdot L \cdot B_M \qquad (45)$$

$$S_4 = 0{,}6 \cdot N_{WTB} \cdot B_M \cdot D_X \qquad (46)$$

Reduced surface is:

$$S_R = S_1 + 0{,}73 \cdot S_2 + 0{,}69 \cdot S_3 + 0{,}65 \cdot S_4 \qquad (47)$$

Hull structure weight numeral (analogous to Watson 1998) is corrected for displacement and for draft to depth ratio:

$$E_S = 2{,}75 \cdot S_R \cdot \left(0{,}292 + \frac{\nabla}{L_{WL}^2 - 15{,}8}\right) \cdot \left(\frac{T_X}{D_X}\right)^{0{,}244} \qquad (48)$$

Hull material may be aluminum alloy or composite (FRP). Respective hull weight prediction is given by (49) and (50):

$$(W_{100})_{FRP} = (0{,}0135 \cdot G_f \cdot S_f - 0{,}0034) \cdot E_S^{1{,}33} \qquad (49)$$

$$(W_{100})_{Al} = (0{,}002 + 0{,}0064 \cdot G_f \cdot S_f) \cdot E_S^{1{,}33} \qquad (50)$$

Deckhouse weight is estimated by a form of cubic number for average deckhouse dimensions. Specific weight varies very little and may be taken as q = 21 kg/m^3 for aluminum and q = 23 kg/m^3 for FRP deckhouse structure, respectively.

$$W_{150} = q_{DH} \cdot L_{DH} \cdot B_{DH} \cdot H_{DH} \qquad (51)$$

Propulsion machinery weight is divided in two parts. Weight of propulsion motor with appropriate gearbox is known since we start by defining candidate propulsion engine in advance. Therefore, engine weight (wet) including gears is defined by (52),

$$W_{250} = 1{,}07 \cdot N_{PR} \cdot W_{EGB} \qquad (52)$$

The rest of propulsion machinery is related to ship size as given by (53)

$$W_{200-250} = \frac{C_N^{0{,}94}}{46} \qquad (53)$$

Electrical power system weight is:

$$W_{300} = \frac{C_N^{1{,}24}}{592} \qquad (54)$$

Electronic system weight is:

$$W_{400} = \frac{L^{2{,}254}}{1887} \qquad (55)$$

Auxiliary machinery weight is:

$$W_{500} = \frac{(L \cdot B_M)^{1{,}784}}{1295} \qquad (56)$$

Outfit weight is:

$$W_{600} = \frac{L^{2{,}132}}{103} \qquad (57)$$

Special systems weight is:

$$W_{700} = \frac{C_N^{1{,}422}}{3000} \qquad (58)$$

Light ship weight is obtained by summation of individual weights to which a margin is added. Here margin is presented as a single weight but it also can be distributed to individual component weights.

$$\begin{aligned} W_{LS} &= W_{100} + W_{150} + W_{200-250} + W_{250} \\ &+ W_{300} + W_{400} + W_{500} \\ &+ W_{600} + W_{700} + W_{MAR} \end{aligned} \tag{60}$$

Deadweight consists of crew and effects, fresh water, fuel oil and payload, as given by (61), while full load displacement is given by (62).

$$W_{DWT} = W_{PL} + W_{FO} + W_{FW} + W_{CR} \tag{61}$$

$$W_{FL} = W_{LS} + W_{DWT} \tag{62}$$

Since the full load displacement is known it is possible to solve equality constraint (i.e. Archimedes law) by defining residual weight found by subtracting all deadweight items, but one, from full load weight. In this way a fuel oil weight is chosen as residual weight (procedure was first suggested by Brower & Walker 1986).

$$W_{FO} = W_{FL} - \left(W_{LS} + W_{PL} + W_{FO} + W_{CR}\right) \tag{63}$$

From this fuel oil weight as defined by (63), the endurance range will be estimated and later used as attribute in decision-making.

9 RESISTANCE PREDICTION MODEL

Resistance prediction method was developed by Begovic 1998 from the data of 186 models of systematic series of fast craft. Resistance in calm sea condition follows an ITTC-57 approach.

Regression equation was developed after thorough investigation of possible solutions. Residual resistance is estimated by Equation (64) while appropriate coefficients are given in Table 6.

$$\begin{aligned} 10^3 \cdot C_R &= a_1 \cdot (M) + a_2 \cdot (M)^{a_0} + a_3 \cdot \frac{B_X}{T_X} + a_4 \cdot \frac{T_X}{B_X} \\ &+ a_5 \cdot C_P + a_6 \cdot \frac{1}{C_P} + a_7 \cdot C_X + a_8 \cdot \frac{1}{C_X} \\ &+ a_9 \cdot \frac{A_T}{A_X} + a_{10} \cdot C_S + a_{11} \cdot X_{CB} \end{aligned} \tag{64}$$

Friction part is estimated by ITTC-57 formula with correlation addition.

$$C_F = \frac{0,075}{\log\left(R_N - 2\right)^2} \tag{65}$$

$$C_A = 0,0004 \tag{66}$$

$$C_T = C_F + C_R + C_A \tag{67}$$

$$R_T = 0,5 \cdot \rho \cdot v^2 \cdot S_{WS} \cdot C_T \tag{68}$$

Wetted surface is given by (28). The effective power is given by Equation (69).

$$P_E = 0,5 \cdot \rho_{SW} \cdot v^3 \cdot S_{WS} \cdot C_T \tag{69}$$

10 PROPULSION PARAMETRIC MODEL

Different propulsion solutions are possible and choice has to be made in order to produce realistic but not too complicated parametric model.

The non-cavitating propeller on inclined shaft (FPP) propulsion system is selected (Figure 2):

First estimate of propeller diameter may be done by a simple formula:

$$D_P = 12,54 \cdot \frac{\left(\dfrac{P_B}{V_{\text{MAX}}}\right)^{0,25}}{\left(\dfrac{N_{ENG}}{G_R}\right)^{0,50}} \tag{70}$$

Propeller diameter is design parameter and so is the engine power (from the engine catalogue). The following value is calculated in order to eliminate revolutions from the equation:

$$\frac{K_Q}{J^3} = \frac{P_D}{2 \cdot \pi \cdot \rho_{SW} \cdot V_A^{\ 3} \cdot D_P^2} \tag{71}$$

Consequently the propeller is capable of producing thrust of:

$$T_{PR} = 2 \cdot \pi \cdot J_0^2 \cdot \eta_0 \cdot \rho_{SW} \cdot n_0^2 \cdot D_P^4 \cdot \left(\frac{K_Q}{J^3}\right) \tag{72}$$

Together the propellers are capable of overcoming resistance of:

$$R_{PR} = N_{PR} \cdot T_{PR} \cdot (1 - t) \tag{73}$$

Speed equilibrium is found from the intersection point of resistance and thrust force curves respectively. Overcoming the resistance hump is

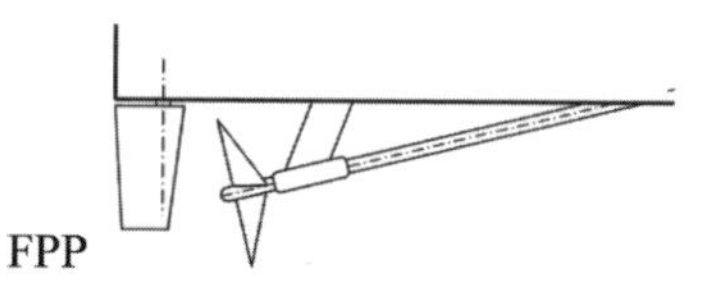

Figure 2. Propulsion configuration in the model.

controlled at the same time. Cavitation is checked by limiting maximal propeller load.

Optimal revolution rate is found from the relation:

$$n_0 = \frac{V_A}{J_0 \cdot D_P} \tag{74}$$

Since the purpose of multi-attribute optimization is to find optimal solutions, it is decided to use only data for optimal propellers. Therefore, parameters of the optimal propellers (i.e. having best efficiency η_0) are defined as dependent of K_Q/J^3. This is the so called "marine engineer's approach".

10.1 *Optimal fixed pitch propeller*

Optimal FPP is selected from the WB 5-75 methodical series of propellers. The equations for the optimal advance coefficient, optimal efficiency and optimal pitch ratio, respectively, are:

$$J_0 = 0{,}1353 \cdot \exp\left(0{,}62949 \cdot \frac{K_Q}{J^3}\right) \cdot \left(\frac{K_Q}{J^3}\right)^{-0{,}55706} \tag{75}$$

$$\eta_0 = 0{,}63 \cdot \exp\left(-2{,}393 \cdot \frac{K_Q}{J^3}\right) \cdot \left(\frac{K_Q}{J^3}\right)^{-0{,}05101} \tag{76}$$

$$\left(\frac{P}{D}\right)_0 = 0{,}1885 \cdot \exp\left(0{,}9183 \cdot \frac{K_Q}{J^3}\right) \cdot \left(\frac{K_Q}{J^3}\right)^{-0{,}5068} \tag{77}$$

Cavitation is checked by an adapted Keller's formula:

$$\frac{A_E}{A_0} = \frac{0{,}2855 \cdot T_{PR}}{D_P{}^2 \cdot (10159 + 1025 \cdot D_P)} \tag{78}$$

This approach to propeller design is greatly simplified. Also, it assumes that all propellers are 5-bladed and optimal. The advantage of this approach is that estimation is much more realistic than approximate regression formula.

11 EQUILIBRIUM SPEED

Equilibrium speed is found by intersecting required propulsive power curve from resistance and curve of propulsive power provided by propulsion system (Figure 3) and Table 6.

If the model is applied to testing known vessel the equilibrium speed when compared to known speed is in most cases within 1 knot either side.

12 EXAMPLE OF THE APPLICATION

Application of the model to design of service vessel is presented in short. When program is started designer enters parameter values as guided by the lower and upper suggested values from parametric relations with one standard deviation. The model calculates attribute values that may be inspected by designer and parameters adjusted in desired direction. Propulsion engine is treated as parameter, since it is selected from a catalogue

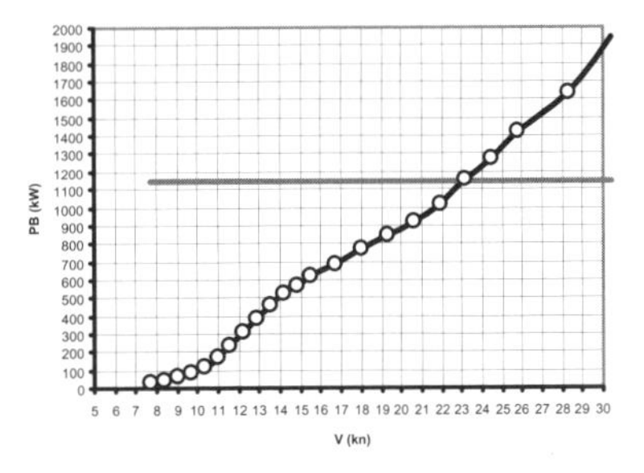

Figure 3. Speed prediction for given power.

Table 3. Design parameters.

Design parameters				Suggested range		
				mean-σ	mean	mean+σ
Hull length, ISO 8666	L_H =	19,80	m	17,03	17,78	18,52
Length on DWL	L_{WL} =	17,92	m	17,03	17,78	18,52
Beam molded	B_M =	5,49	m	4,50	5,07	5,63
Beam at DWL m. sec.	B_X =	4,62	m	4,05	4,73	5,08
Draft at DWL m. sec.	T_X =	1,00	m	0,79	1,05	1,33
Draft at transom	T_T =	0,73	m	0,27	0,42	0,57
Depth amidships	D_X =	2,86	m	2,01	2,58	3,19
Full load displ.	W_{FL} =	45,18	t	28,83	39,10	51,62
Propeller diameter	D_P =	0,86	m	0,76	0,85	0,92
Service speed	V_S =	20,0	kn	–	–	–
Service type(LR SSC)	G_f =	1,20	–			
Service area(LR SSC)	S_f =	0,75	–			
Significant wave hight	H_S =	1,00	m			

Table 4. Propulsion system parameters.

Propulsion system			
Number of propulsors	N_{PR} =	3	
Engine make		DD	
Engine model		8V-92TI	
Rating (A,B,C,D,E)		E	
Brake power (1 engine)	P_{MCR} =	380	kW
Engine revolutions	n_{RPM} =	2300	rpm
Spec. fuel consumption at MCR	SFC =	225	g/kWh
Dry engine weight	W_{ENG} =	1871	kg
Reduction ratio:	R_{GB} =	2,140	
Gearbox weight	W_{GB} =	0	kg
Prop. revolutions	n_{PROP} =	1074	rpm
Weight (eng.+gearbox)	W_{EGB} =	1871	kg

Table 5. Design attributes.

Design attributes			
Max. speed at full load	V_{MAX} =	23,1	kn
Range at V_{MAX} and full load	R_{VMAX} =	198	NM
Free rolling period	T_R =	5,522	s
Average vert. acceleration at CG	a_{vcg} =	0,202	g
Average MSI amidships	MSI =	5,356	%
Cost of new vessel	C_{NEW} =	1,12	10^6EUR
Range at service speed	R_{VS} =	235	NM

Table 6. Coefficients in resistance prediction equation.

FN	a_0	(M)	$10^3(M)^a{}_0$	B_X/T_X	T_X/B_X	C_P	$1/C_P$	C_X	$1/C_X$	A_T/A_X	CS	XCB
0,300	−2,0033	−0,0302	0,1160	−0,2554	−2,2155	9,5150	−4,4203	0,9043	0,6740	0,1026	0,3102	1,2057
0,325	−2,1363	−0,1545	0,1299	−0,2169	−1,7831	13,2405	−2,4283	−2,4336	−0,5496	0,2162	0,2026	1,5535
0,350	−2,2091	−0,2648	0,1435	−0,2218	−2,2652	17,5872	0,6158	−6,1572	−2,2303	0,2420	0,1888	−0,7701
0,375	−2,3607	−0,3829	0,1972	−0,2112	−2,8829	19,3376	2,9282	−8,0751	−3,0708	0,3759	0,0319	−1,5740
0,400	−2,6227	−0,5824	0,3426	−0,1712	−3,1400	18,3604	4,6144	−7,0937	−2,4535	0,7070	−0,5766	−1,7701
0,425	−2,9501	−0,7867	0,7429	−0,2030	−3,6079	14,3357	5,0395	−3,9873	−0,7411	0,9529	−1,0032	−0,5568
0,450	−3,2556	−0,8459	1,7045	−0,2561	−4,2613	11,0344	4,7138	−1,6198	0,7122	0,8623	−1,2437	0,1638
0,475	−3,4680	−0,7567	3,1456	−0,2785	−4,2412	10,0301	4,2966	−2,0030	0,6803	0,6249	−1,3483	2,4205
0,500	−3,6144	−0,6287	4,7066	−0,3232	−4,0263	10,7938	4,4846	−3,7698	−0,2700	0,2803	−1,2244	3,1624
0,525	−3,6866	−0,4828	5,8084	−0,3647	−3,7365	11,7352	4,7535	−5,5669	−1,3682	−0,1246	−1,0204	3,0753
0,550	−3,7140	−0,3604	6,2901	−0,4222	−4,1540	11,8956	4,7434	−6,2087	−1,8846	−0,4675	−0,8470	2,7206
0,575	−3,7109	−0,2990	6,1513	−0,4210	−4,0295	11,2606	4,4412	−5,7545	−1,7466	−0,7423	−0,8773	2,4430
0,600	−3,6819	−0,2832	5,5494	−0,4048	−3,8506	10,2773	4,0439	−4,9607	−1,3513	−0,9304	−0,9174	2,5033
0,650	−3,5869	−0,1892	4,2117	−0,3528	−2,6535	11,4413	4,6059	−7,8489	−3,2987	−1,0453	−0,7941	5,7932
0,700	−3,5050	−0,1564	3,2362	−0,3291	−2,3234	8,2838	3,4592	−6,5322	−2,6076	−0,7594	−0,6514	7,7023
0,750	−3,4223	−0,1131	2,5297	−0,2335	−1,6660	6,2947	3,1352	−5,0405	−1,9353	−0,5998	−0,6994	4,9903
0,800	−3,3539	−0,1102	2,0077	−0,2224	−1,5763	3,8169	2,2106	−2,8089	−1,0797	−0,7109	−0,4811	4,0024
0,850	−3,2810	−0,1053	1,5927	−0,2047	−1,6509	2,4737	1,7671	−2,1359	−0,6994	−0,6380	−0,4611	4,8189
0,900	−3,2587	−0,0447	1,4585	−0,1272	−0,9127	−4,2954	0,1523	−1,8058	−0,8161	0,4907	−0,0122	13,8171
0,950	−3,2224	−0,0579	1,2518	−0,1391	−1,1900	−4,7218	−0,0114	−1,7593	−0,8418	0,4201	0,1479	14,7206
1,000	−3,1866	−0,0466	1,1347	−0,1514	−1,2209	−4,2391	0,3471	−2,7290	−1,3391	0,3523	0,2537	15,2100
1,100	−2,8618	0,0145	0,6633	−0,3661	−2,5015	−0,2994	0,8506	−3,2656	−2,2907	−1,2979	1,5258	5,8639
1,200	−2,7672	0,0456	0,6012	−0,3622	−1,8096	−2,9093	−0,4715	−4,0406	−2,9293	−2,9151	2,4136	12,2215

(Continued)

Table 6. (*Continued*).

Fnv	vms	vkn	Rn	C_F	C_R	C_T	R_T	P_E	QPC	P_D	P_B	dif	cross
0,6757	3,977	7,731	5,99E+07	2,25E–03	5,78E–03	8,03E–03	4941	19,7	0,55	35,7	1140	1104,3	0,00
0,7320	4,308	8,375	6,49E+07	2,22E–03	6,22E–03	8,44E–03	6095	26,3	0,55	47,7	1140	1092,3	0,00
0,7883	4,640	9,019	6,99E+07	2,20E–03	6,75E–03	8,94E–03	7494	34,8	0,55	63,2	1140	1076,8	0,00
0,8446	4,971	9,663	7,49E+07	2,17E–03	7,67E–03	9,84E–03	9463	47,0	0,55	85,5	1140	1054,5	0,00
0,9009	5,303	10,307	7,99E+07	2,15E–03	9,20E–03	1,14E–02	12429	65,9	0,55	119,8	1140	1020,2	0,00
0,9572	5,634	10,952	8,49E+07	2,13E–03	1,13E–02	1,34E–02	16584	93,4	0,55	169,9	1140	970,1	0,00
1,0136	5,965	11,596	8,99E+07	2,12E–03	1,36E–02	1,58E–02	21816	130,1	0,55	236,6	1140	903,4	0,00
1,0699	6,297	12,240	9,49E+07	2,10E–03	1,56E–02	1,77E–02	27260	171,7	0,55	312,1	1140	827,9	0,00
1,1262	6,628	12,884	9,99E+07	2,08E–03	1,67E–02	1,88E–02	32133	213,0	0,55	387,3	1140	752,7	0,00
1,1825	6,960	13,529	1,05E+08	2,07E–03	1,71E–02	1,91E–02	36086	251,1	0,55	456,6	1140	683,4	0,00
1,2388	7,291	14,173	1,10E+08	2,06E–03	1,69E–02	1,89E–02	39138	285,4	0,55	518,8	1140	621,2	0,00
1,2951	7,622	14,817	1,15E+08	2,04E–03	1,62E–02	1,82E–02	41257	314,5	0,55	571,8	1140	568,2	0,00
1,3514	7,954	15,461	1,20E+08	2,03E–03	1,52E–02	1,73E–02	42542	338,4	0,55	615,2	1140	524,8	0,00
1,4640	8,617	16,750	1,30E+08	2,01E–03	1,32E–02	1,52E–02	43995	379,1	0,55	689,3	1140	450,7	0,00
1,5766	9,280	18,038	1,40E+08	1,99E–03	1,16E–02	1,35E–02	45409	421,4	0,55	766,1	1140	373,9	0,00
1,6893	9,942	19,326	1,50E+08	1,97E–03	1,02E–02	1,21E–02	46655	463,9	0,55	843,4	1140	296,6	0,00
1,8019	10,605	20,615	1,60E+08	1,95E–03	9,00E–03	1,09E–02	47920	508,2	0,55	924,0	1140	216,0	0,00
1,9145	11,268	21,903	1,70E+08	1,93E–03	8,08E–03	1,00E–02	49476	557,5	0,55	1013,6	1140	126,4	23,07
2,0271	11,931	23,192	1,80E+08	1,92E–03	7,68E–03	9,59E–03	53147	634,1	0,55	1152,9	1140	–12,9	0,00
2,1397	12,594	24,480	1,90E+08	1,90E–03	7,10E–03	9,00E–03	55559	699,7	0,55	1272,2	1140	–132,2	0,00
2,2523	13,257	25,769	2,00E+08	1,89E–03	6,75E–03	8,64E–03	59071	783,1	0,55	1423,8	1140	–283,8	0,00
2,4776	14,582	28,345	2,20E+08	1,86E–03	5,60E–03	7,46E–03	61775	900,8	0,55	1637,9	1140	–497,9	0,00
2,7028	15,908	30,922	2,40E+08	1,84E–03	5,27E–03	7,12E–03	70082	1114,9	0,55	2027,0	1140	–887,0	0,00
													V_{MAX} =23,07

of candidate engines. Usually designer has quite limited choice of engines and preparing such catalogue is not difficult task. Since results are obtained quickly it is possible to try all candidate engines in short time. A design of fisheries inspection vessel for Adriatic Sea is shown in sequel (Tables 3, 4 and 5):

Design attributes as predicted by the model contain maximal speed that is found in Table 6.

The model quickly predicts attributes for a number of propulsion configurations, different structural materials, etc.

13 CONCLUSIONS

The proposed model of fast service vessels, "SERVES", may be used in several ways:

1. Control of vessel data before inclusion into the database, in order to check for possible errors of data from literature.
2. Concept design of new vessel by manually adjusting design parameters and judging design attributes.
3. Main purpose of the model is inclusion into multi-attribute design procedure. To this end it is possible to test the reliability of the model by comparison to the database vessels.
4. Finally, reliability of all prediction formulas and respective empirical or semi empirical formulas and methods may be tested and quick adaptations made.

The model is programmed in MS EXCEL in order to make it widely available and useful to small design offices where many design calculations are performed in spreadsheet form. The version that is included in the multi-attribute platform is reprogrammed in FORTRAN to make it compatible to the rest of our software.

14 NOMENCLATURE

A_T/A_X	Transom area ratio
a_{vcg}	Average vertical acceleration amidships (g)
B_X	Beam at DWL at section of max. immersed area (m)
B_M	Hull moulded beam
β_X	Dead rise angle amidships (^{0})
C_{NEW}	Acquisition cost of new vessel in 10^6 EUR
∇	Displacement volume (m^3)
D_P	Propeller diameter (m)
D_X	Hull depth (m)
$H_{1/3}$	Significant wave height (m)
MSI	Averaged motion sickness incidence (%)
L_H	Length over deck (ISO 8666) (m)
L_{WL}	Length on DWL at full load (m)
P_B	Total installed power (kW)
R_{VMAX}	Sailing range at maximal speed (NM)
R_{VS}	Sailing range at service speed (NM)
T_R	Free rolling period (s)
V_{MAX}	Maximal speed at full load (kn)
V_S	Service speed at full load (kn)
W_{FL}	Full load displacement (t)

REFERENCES

Begovic, E. 1998. Hydromechanical Module in the Multi-Criteria Design Model of Fast Ships, Master of science thesis, University of Zagreb, Faculty of Mechanical Engineering and Naval Architecture, Zagreb, Croatia (in Croatian).

Brower, K.S. & Walker, W.W. Ship Design Computer Programs -an Interpolative Technique, Naval Engineers Journal, May 1986, pp. 74–87.

CATERPILLAR Application Guidelines for Marine Propulsion Engines, http://www.cat.com/cda/layout?m=37601&x=7.

Grubisic, I. et al. 1999. *Study of the Development of the Maritime Administration Fleet*, Croatian Ministry of Maritime Affairs (in Croatian), Zagreb.

Grubisic, I. & Begovic, E. 2003. Multi-Attribute Concept Design Model of Patrol, Rescue and Antiterrorist Craft, *Proceedings of FAST 2003.*

Grubisic, I. & Begovic, E. 2009. Upgrading weight prediction in small craft concept design, Proceedings of the IMAM-2009, Istanbul.

Lloyd's Register, 1996. Rules and Regulations for the Classification of Special Service Craft, London.

Parsons, M.G. 2003. Parametric Design, in Ship Design and Construction, T. Lamb, Ed., Society of Naval Architects and Marine Engineers, New Jersey, pp. 11–1 to 11–48.

Schneekluth & Bertram, 1998. *Ship Design for Efficiency and Economy*, Butterworth-Hainemann.

Watson, D.G.M. 1998. *Practical Ship Design*, Elsevier Science & Technology Books.

Sustainable Maritime Transportation and Exploitation of Sea Resources – Rizzuto & Guedes Soares (eds)
© 2012 Taylor & Francis Group, London, ISBN 978-0-415-62081-9

A parametric representation of fair hull shapes by means of splines in tension

N. Del Puppo & G. Contento
Department of Mechanical Engineering and Naval Architecture, University of Trieste, Italy

ABSTRACT: The paper presents the development and application of a design tool for the parametric representation of hull shapes by means of splines in tension. The excellent fairness properties of these curves make their use extremely useful, mainly in the frame of an automatic generation of hull shapes for the optimization of given properties, among others the hydrodynamic performances. In the present implementation, these curves are built and handled in a multi-platform application called ParCAD, so that the user can define macro-parameters of the entire hull—lengths, areas, volumes, angles, global or local coefficients, ...—or only some parts of it and control the hull curves shape in a fancy way making use of the ParCAD Graphical User Interface. In this paper a fore-body Ro/pax and an entire sailing yacht hull parametric models are also presented and discussed.

1 INTRODUCTION

In the past decade, the opportunities given by state-of-art parametric CAD, coupled with more and more accurate CFD tools and with the availability of cheaper and increasingly performing computers, have made the virtual hydrodynamic design of ships more attractive. The amount of available literature on this subject is huge, see for instance Tahara et al. (2008), for a thorough discussion. Besides the fluid dynamic design that involves generally fair body shapes (ships, aircrafts, cars, ...), the parametric CAD concept is applied to a variety of engineering applications where the body surface is given by a combination of primitive geometric shapes, i.e., straight lines, planes, circles, cylinders, ... In those cases, it becomes rather easy to handle the primitive shapes controlling directly their main parameters.

This is not the case for the surfaces of most ships and their appendages. Indeed, provided the overall constraints (length, displaced volume, ...) are satisfied, the main parameters governing the hull shape are generally indirect properties of fair curves (area, geometric center, slopes, ...), see Lackenby (1950).

A large amount of papers and practical/research applications of the parametric CAD concept for the automatic design of hull shapes have appeared, among others Harries & Schulze (1997), Harries (1998), Harries & Abt (1998), Abt et al. (2001), Percival et al. (2001), Valorani et al. (2003), Mancuso (2006), Perez-Arribas (2006), Abt & Harries (2007), Perez et al. (2007), Zhang et al. (2008).

Part of the success of these applications was related to the capability of the software to produce hull surfaces that implement the concept of fairness.

The concept of fairness of a curve and the method to "measure" the fairness have been widely treated in the literature, including the attempt to define the local and global fairness with links to the aesthetic aspect too. A comprehensive discussion is given by Roulier & Rando (1994).

The splines in tension were first introduced by Schweikert (1966). These curves, based on a family of hyperbolic functions, behave smoothly through the data points with a minimum number of inflection points or oscillations. Moreover by varying the tension, the interpolated curve can change shape from that of a cubic spline to that of a polyline with still sharp but rounded corners. Cline (1974) has given a slightly different mathematical formulation based on a normalization of the tension. Renka (1993) has proposed a software package based on these curves.

The scope of the present work is to develop an object-oriented parametric modeler for the representation of hull shapes making use of the splines in tension. The main advantages expected from the mathematical model as implemented are given by the fairness and robustness properties of the curves. As far as the implementation of the whole package is concerned, it has been intended to get the maximum flexibility in terms of definition by the user of the control curves of the hull shape and of the parameters, easy maintenance and extensibility of the software core, GUI and batch running modes.

2 MATHEMATICAL MODEL

2.1 *Splines in tension*

Given a set of points (x_i, y_i), $i=1,N$ such that $x_{i+1} > x_i$, a spline in uniform tension is defined as follows:

$$y(x) = f_i(x) \text{ for } \begin{cases} x_i \le x < x_{i+1}, & i=1,N-2 \\ x_i \le x \le x_{i+1}, & i=N-1 \end{cases} \tag{1}$$

The following properties hold for $f_i(x)$ over $x_i \le x \le x_{i+1}$:

- $y_i = f_i(x_i)$
- $f_i \in C^2[x_1, x_N]$
- at most one inflection point is allowed over the interval $x_i \le x \le x_{i+1}$
- the quantity $f_i''(x_i) - \sigma^2 \cdot f_i(x)$ is required to vary linearly over each interval $x_i \le x \le x_{i+1}$, $i=N-1$ where σ is the so called "tension", i.e.

$$f_i''(x) - \sigma^2 \cdot f_i(x) = \left[f_i''(x_i) - \sigma^2 y_i\right]\frac{(x_{i+1}-x)}{h_i} + \left[f_i''(x_{i+1}) - \sigma^2 y_{i+1}\right]\frac{(x-x_i)}{h_i} \tag{2}$$

where $h_i = x_{i+1} - x_i$.

Matching the first and second order derivatives at the inner given points x_i, $i=2,N-1$ and prescribing first order derivatives (slopes) at the extremes x_1, x_N gives a tri-diagonal system of N equations in N unknowns $f_i''(x_i)$.

Alternatively the slopes at the end points can be left free, computing them according to a second order polynomial distribution over the intervals $x_3 - x_1$ and $x_N - x_{N-2}$, then prescribing those values at x_1, x_N and solving the system.

According to the previous definitions and requirements, the spline in tension and its derivatives over the interval $x_i \le x \le x_{i+1}$ become:

$$\begin{aligned} f_i(x) = &\left[\frac{f_i''(x_i)}{\sigma^2}\right] \cdot \frac{sinh(\sigma \cdot (x_{i+1}-x))}{sinh(\sigma \cdot h_i)} \\ &+ \left[y_i - \frac{f_i''(x_i)}{\sigma^2}\right] \cdot \frac{(x_{i+1}-x)}{h_i} \\ &+ \left[\frac{f_i''(x_{i+1})}{\sigma^2}\right] \cdot \frac{sinh(\sigma \cdot (x-x_i))}{sinh(\sigma \cdot h_i)} \\ &+ \left[y_{i+1} - \frac{f_i''(x_{i+1})}{\sigma^2}\right] \cdot \frac{(x-x_i)}{h_i} \end{aligned} \tag{3}$$

$$\begin{aligned} f_i'(x) = &-\left[\frac{f_i''(x_i)}{\sigma}\right] \cdot \frac{cosh(\sigma \cdot (x_{i+1}-x))}{sinh(\sigma \cdot h_i)} \\ &- \left[y_i - \frac{f_i''(x_i)}{\sigma^2}\right] \cdot \frac{1}{h_i} \\ &+ \left[\frac{f_i''(x_{i+1})}{\sigma}\right] \cdot \frac{cosh(\sigma \cdot (x-x_i))}{sinh(\sigma \cdot h_i)} \\ &+ \left[y_{i+1} - \frac{f_i''(x_{i+1})}{\sigma^2}\right] \cdot \frac{1}{h_i} \end{aligned} \tag{4}$$

$$\begin{aligned} f_i''(x) = &f_i''(x_i) \cdot \frac{sinh(\sigma \cdot (x_{i+1}-x))}{sinh(\sigma \cdot h_i)} \\ &+ f_i''(x_{i+1}) \cdot \frac{sinh(\sigma \cdot (x-x_i))}{sinh(\sigma \cdot h_i)} \end{aligned} \tag{5}$$

for $x_i \le x \le x_{i+1}$, $i=N-1$.

The equations above depend nonlinearly on the tension σ so that the result of interpolation may depend on the scale used. To avoid this, Cline (1974) has proposed to normalize the tension by the average horizontal distance between the given points as follows:

$$\sigma^* = \sigma \cdot \frac{(x_N - x_1)}{(N-1)} \tag{6}$$

The obtained function sums up the concept of fairness of a curve, with a minimum number of inflection points or oscillations that can be observed in a cubic spline (Fig. 1).

It must be observed also that:

- for $\sigma \to 0$, the curve converges to a cubic spline
- for $\sigma \to \infty$, the curve converges to a piecewise linear curve (polygonal).

Once the spline in tension through the given points is obtained, the following geometric quantities can be computed:

$$Area = \sum_{i=1}^{N-1} \int_{x_i}^{x_{i+1}} f_i(x)dx \tag{7}$$

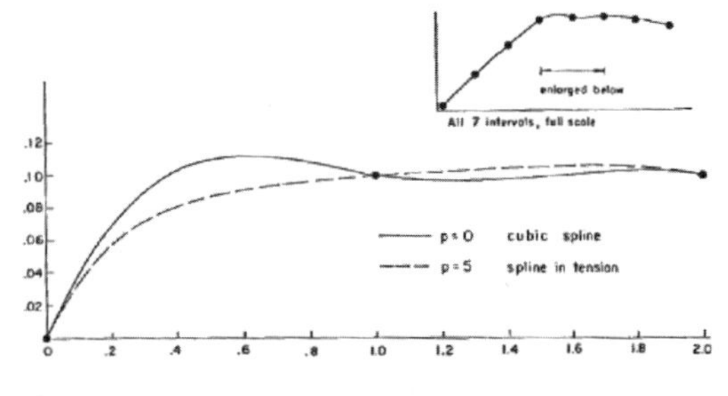

Figure 1. Behaviour of a cubic spline and a spline in tension (after Schweikert 1966).

$$x_C = \frac{\sum_{i=1}^{N-1} \int_{x_i}^{x_{i+1}} f_i(x) \cdot x \ dx}{Area} \quad (8)$$

$$y_C = \frac{\sum_{i=1}^{N-1} \int_{x_i}^{x_{i+1}} f_i(x)^2 dx}{2 \cdot Area} \quad (9)$$

$$arclength = \sum_{i=1}^{N-1} \int_{x_i}^{x_{i+1}} \sqrt{1 + f_i{'}(x)^2} dx \quad (10)$$

$$curv_i(x) = \frac{f_i{''}(x)}{(1 + f_i{'}(x)^2)^{1.5}} \quad (11)$$

$$\overline{\Delta curv} = \frac{\sum_{i=1}^{N-1} \int_{x_i}^{x_{i+1}} abs(curv_i(x+dx) - curv_i(x))dx}{arclength} \quad (12)$$

The parameters in Equations 7–12 represent the area under the curve (*Area*), the coordinates of geometric centre (x_C,y_C), the arclength of the curve (*arclength*) and local curvature (*curv*). $\Delta curv$ represents the curvature overall absolute change along the curve, here used as an index of fairness.

2.2 *Parametric form of the spline in tension*

The form of the spline in tension presented above is not suitable for any arbitrary set of given points, i.e., closed curves or even curves with tangent parallel to the Y axis at some point, extremes included cannot be represented by Equations 3–5.

The problem can be simply overcome by introducing a parametric form of the spline in tension, typically adopting the curvilinear abscissa as independent variable.

Here we define the additional variable s such that:

$$\begin{cases} s_i = 0, \ i = 1 \\ s_i = s_{i-1} + \sqrt{(x_i - x_{i-1})^2 + (y_i - y_{i-1})^2}, \ i = 2, N \end{cases} \quad (13)$$

where s is simply the arc-length of the polygonal passing through the given points, used as a good approximation of the curvilinear abscissa.

We can then use two separate interpolation functions $fx_i(s), fy_i(s)$ with $s \in [s_i, s_{i+1}]$.

In case the slopes α_1, α_N (with respect to the x-axis) are prescribed at one or both end points $(x_1, y_1), (x_N, y_N)$, then we simply have

$$\begin{cases} fx_i{'}(s_i) = \cos(\alpha_i) \\ fy_i{'}(s_i) = \sin(\alpha_i), \qquad i = 1, N \end{cases} \quad (14)$$

According to this parametric form $fx(s), fy(s)$, the parameters defined in Eq. (7–12) can now be computed as:

$$Area = 0.5 \left[\sum_{i=1}^{N-1} \int_{s_i}^{s_{i+1}} \lfloor fy_i(s) \cdot fx_i{'}(s) - fx_i(s) \cdot fy_i{'}(s) \rfloor ds + y_N \cdot x_N - y_1 \cdot x_1 \right] \quad (15)$$

$$x_C = \frac{1}{3 \cdot Area} \left[\sum_{i=1}^{N-1} \int_{s_i}^{s_{i+1}} [fy_i(s) \cdot fx_i{'}(s) - fx_i(s) \cdot fy_i{'}(s)] \cdot fx_i(s) ds + y_N \cdot x_N^2 - y_1 \cdot x_1^2 \right] \quad (16)$$

$$y_C = \frac{1}{3 \cdot Area} \left[\sum_{i=1}^{N-1} \int_{s_i}^{s_{i+1}} [fy_i(s) \cdot fx_i{'}(s) - fx_i(s) \cdot fy_i{'}(s)] \cdot fy_i(s) ds + 0.5 \cdot y_N^2 \cdot x_N - 0.5 \cdot y_1^2 \cdot x_1 \right] \quad (17)$$

$$arclength = \sum_{i=1}^{N-1} \int_{s_i}^{s_{i+1}} ds \quad (18)$$

$$curv_i(s) = \frac{fx_i{'}(s) \cdot fy_i{''}(s) - fy_i{'}(s) \cdot fx_i{''}(s)}{\left(fx_i{'}(s)^2 + fy_i{'}(s)^2 \right)^{1.5}} \quad (19)$$

$$\overline{\Delta curv} = \frac{\sum_{i=1}^{N-1} \int_{s_i}^{s_{i+1}} abs\left(curv_i(s+ds) - curv_i(s) \right) \cdot ds}{arclength} \quad (20)$$

Equations 1 to 20 can be used to obtain an interpolating curve through a set of given points, with or without end slopes.

2.3 *Curves with targets*

The purpose of the work is now to obtain a curve that satisfies a given set of geometric targets (parameters). A sub-set of these targets is trivial (start and end points, start and end slopes) whereas the rest of the targets depend on the shape of the curve.

Specifically the full set of targets adopted in this work is given in Table 1.

The prescribed area can be given with respect to an assigned matching pair given by a coordinate axis and a projection plane. The number of targets can be conveniently reduced, depending on the specific application, where *a*) and *e*) are mandatory, *b*) to *d*) are optional but at least one of them must be prescribed (free target).

When the options *c*) or *d*) are activated, the number of free targets becomes four at most, including *e*). In this case, an iterative process is started such that the minimum of the function

S is searched, where S is defined as the follows (subscript T stands for "*Target*")

$$S = \sum_{i=1}^{N_T} \eta_i^2 \tag{21}$$

N_T is the number of free targets $(0 \le N_T \le 4)$ and η_i are the normalized targets defined as follows:

$$\begin{aligned} \eta_1 &= (Area - Area_T) / Area_T \\ \eta_2 &= (x_C - x_{C_T}) / d \\ \eta_3 &= (y_C - y_{C_T}) / d \\ \eta_4 &= \overline{\Delta curv_T} \end{aligned} \tag{22}$$

If N_T is set to 0, then the curve is simply required to fulfil the constraints *a*), *b*) given in Table 1.

If N_T is either than 0, an iteration process is started in which the spline in tension must fulfil the constraints. The S minimum (see Equation 21) is found letting the shape of the curve being controlled by a number of N_{CP} additional inner control points the spline passes through. These additional points are free to move along the y direction only.

Assuming that the free coordinates of these additional control points are $y_{CP}(i), i = 1, N_{CP}$, then the minimum of S is given by the set of N_{CP} equations:

$$\frac{\partial S}{\partial y_{CP}(i)} = 0, \quad i = 1, N_{CP} \tag{23}$$

Since the system of Equations 23 is nonlinear, we use the Levenberg-Marquardt technique (Levenberg 1944, Marquardt 1963) to obtain the set of N_{CP} coordinates of the additional inner control points.

Table 1. Description of the targets applied to the spline in tension.

Criterion	Description
a	the curve has to start and end on two given points P_B, P_E
b	prescribed at the start point P_B and end point P_E the curve can have one or both slopes α_B, α_E
c	the curve can have a prescribed (target) area ($Area_T$)
d	the curve can have prescribed (target) coordinates of the geometric centre (x_{CT}, y_{CT})
e	the curve must fulfill the fairness criterion of minimum, i.e. $\Delta curv_T = 0$

3 PACKAGE IMPLEMENTATION

The mathematical model here presented was implemented in ParCAD, an object-oriented parametric modeller developed by the authors and written in C++.

ParCAD was intended to satisfy the following requirements:

1. easy maintenance and great extensibility of the software core
2. user-interactive basic geometric entities definition
3. form-parameters, control curves and constraints definition
4. platform independent
5. GUI and batch running modes.

The first requirement was fulfilled developing a generic and object-oriented kernel in C++ in which, besides the spline in tension, some basic geometric entities (i.e., points, lines, circles, interpolation curves and so on) were defined to be used in the parametric model.

The spline in tension with constraints, as described in section 2.3, was implemented in a C++ class called *tspline*.

The *tspline* class comes with a set of public member functions that can be used to define (see Table 2) or get (see Table 3) curve properties.

In this way we ensure the private data (curve points and derivatives) are protected, giving public access only to the curve settings and properties by means of class public member functions.

The full set of points on the curve is computed only once the prescribed geometric constraints are set. This is done by calling the class public member function *compute()*. In this way the user decides when it is time to compute the curve reducing

Table 2. Description of the tspline class member functions used to define the curve parameters.

Criterion	Description
a	define the curve extreme points (P_B, P_E)
b	define the curve slopes (α_B, α_E) at the start (PB) and at the end (P_E) points
c	define the curve target area $Area_T$ related to an assigned matching pair given by a coordinate axis and a projection plane
d	define the curve target geometric centre (x_{CT}, y_{CT}) with regards to an assigned matching pair given by a coordinate axis and a projection plane
e	$\Delta curv_T = 0$ is always fulfilled without explicit input from the user

considerably the computational effort. The member function *compute()* is linked to the Levenberg Marquardt optimization library to obtain the curve that fulfils the constraints given in Table 1.

The remaining requirements of the implementation were satisfied making use of the Qt Nokia libraries. The parametric models are generated interpreting a script file by means of a QScriptEngine object. A set of functions were integrated in ParCAD to allow the script engine instantiate the core geometric entities dynamically. The model script can be defined either using a simple text editor or the built in graphical script editor that comes with a smart code completion which helps the user to define a parametric model. Besides, the script engine is able to interpret the JavaScript language so that it is possible to manage form parameters and geometric constraints using JavaScript loops and conditional constructs.

ParCAD was initially intended to be a fast geometry builder to be used in optimization processes, therefore it is also able to read script files and generate models in a command line modality.

Table 3. Description of the tspline class member functions used to get curve properties.

Request	Description
a	get the curve point coordinates for a given
b	get the curve point coordinates for which the intersection at the given elevation and axis is met
c	get the curvilinear abscissa parameter value for which the intersection at the given elevation and axis is met
d	get the curve slopes for a given curvilinear abscissa value
e	get a complete points set belonging to the curve
f	get and display the curvature over a points set belonging to the curve

The GUI 3D view was developed overloading the Qt QGLWidget class (an OpenGL wrapper widget), so that the external dependences to other libraries were reduced to the minimum necessary, easing code portability.

Finally it has to be noted that the Qt libraries allow the coder to write highly portable GUI; as a matter of fact it was possible to generate working binary packages for most common Operating Systems without weigh down the code with platform dependent directives.

4 TEST CASES

The sample cases proposed are a sailing yacht and a Ro/pax hull.

For the sailing yacht, the parametric design has been applied to the entire hull whereas for the Ro/pax the parametric design has been limited to the forebody only.

4.1 *Sailing yacht*

Figure 2 shows a hull preliminary design sketch.

With reference to Figures 3a, b, the sailing yacht hull shape is basically controlled by 5 longitudinal

Figure 2. Sketch of the hull of the sailing yacht from the preliminary design.

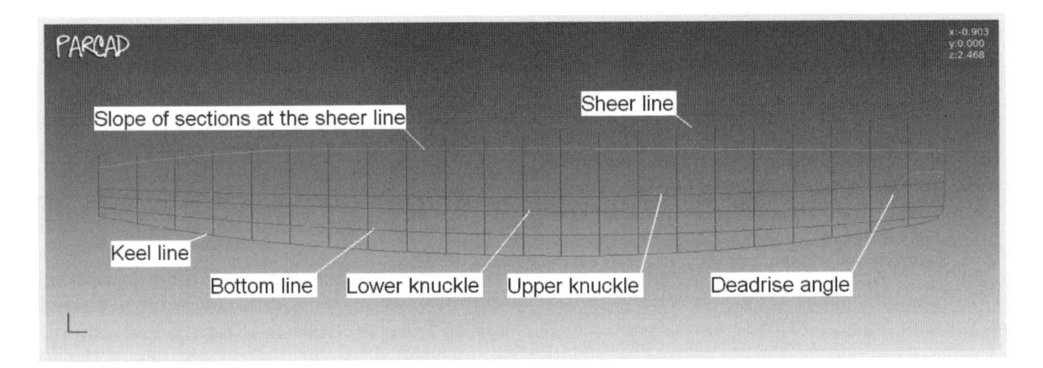

Figure 3a. Longitudinal view and control curves.

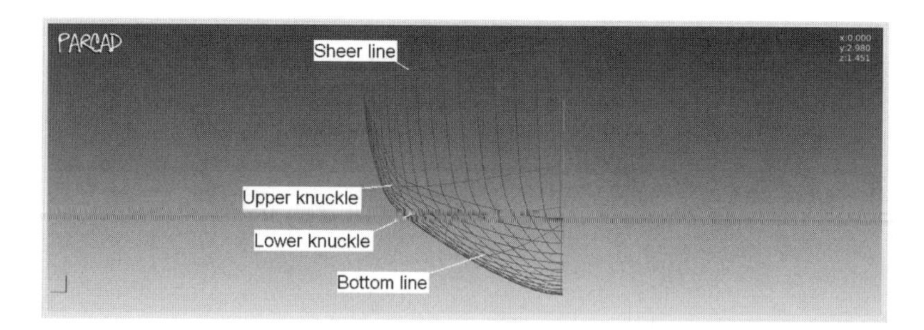

Figure 3b. Body plan and control curves.

Table 4. Sailing yacht design variables.

Item	Reference Axis/Plane	Description
1.	X	Length between perpendiculars
2.	Z	Canoe body draft
3.	X	Max draft longitudinal position
4.	Y	Deck max beam
5.	X	Deck max beam longitudinal position
6.	X	Overhang at stern
7.	X	Overhang at stem
8.	Y	Deck beam at stem
9.	X-Z	Stem slope
10.	X-Z	Keel line slope at AP
11.	Y-Z	Sheer/deck line coordinates at the aft- most station
12.	Y-Z	Bottom line coordinates at the aft-most station
13.	Y-Z	Lower knuckle line coordinates at the aft-most station
14.	Y-Z	Upper knuckle line coordinates at the aft-most station
15.	Z	Bottom line coordinates at stem
16.	Z	Lower knuckle line coordinates at stem
17.	Z	Upper knuckle line coordinates at stem
18.	Y	Bottom line max beam
19.	Y	Lower knuckle line max beam
20.	Y	Upper knuckle line max beam
21.	Z	Bottom line max draft
22.	Z	Lower knuckle line max draft
23.	Z	Upper knuckle line max draft
24.	X	Bottom line max beam long. position
25.	X	Lower knuckle line max beam long position
26.	X	Upper knuckle line max beam long. position
27.	X	Bottom line max draft long. position
28.	X	Lower knuckle line max draft long. position
29.	X	Upper knuckle line max draft long. position

curves (pink lines in Figs. 3a, b), namely the keel line, the sheer line and three intermediate control lines, two of them (upper and lower knuckle) define the beginning and end position of a kind of rounded knuckle.

Moreover, two additional curves (green lines in Figs. 3a, b) control the deadrise angle and the slope of each cross station at the sheer line respectively.

Table 4 describes the design parameters used to generate the entire hull.

For this test case, the shape of the keel line (Fig. 4) is simply controlled by the length between perpendiculars, slope at the aft perpendicular, max draft, longitudinal position of the max draft, "artistic" shape of the stem profile. One could add the subtended area and/or its centre position.

Similarly the remaining longitudinal curves are controlled by few parameters, start and end point coordinates (some are derived from the intersection with the keel line, stem profile and the aft-most station), max depth and max width, longitudinal position of max depth and max width separately. The slopes at one or both end points can be prescribed as additional parameters.

Figure 5 show the effect on the body plan due to some changes mainly on the knuckles and on the bottom line. Specifically, with reference to Table 4, the items that have been changed are #8, 12, 13, 14, 18, 19, 20, 22, 23.

4.2 *Ro/pax ship*

The interest is here focused on the parametric representation of the bulb shape and related changes of the forebody. The hull shape is considered in the final form from stern to the station 18.

The design variables of the bulb investigated in this test are reported in Table 5 and partly shown in Figure 6.

In order to link fairly the aft part of the hull with the bulb, 6 additional control curves are added from the station 18 to the FP station. End points and slopes at the end stations of these control curves are derived automatically from the aft body and the bulb respectively, thus avoiding introducing additional parameters.

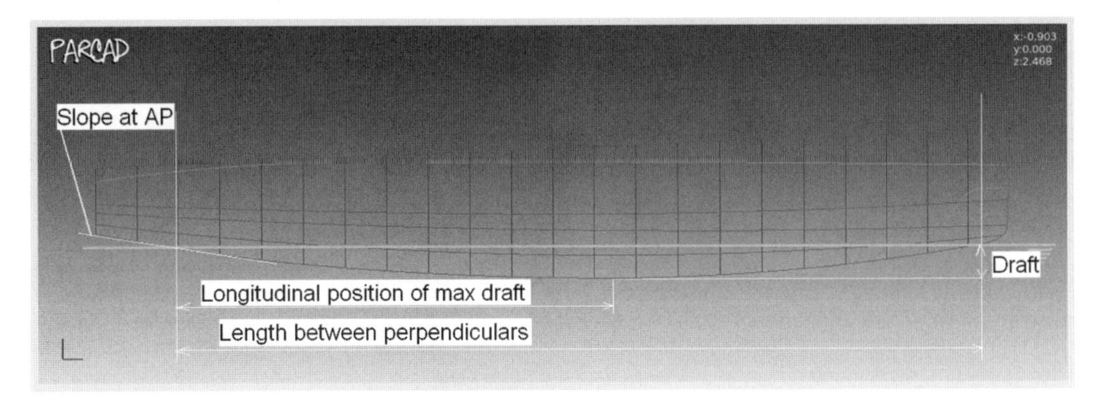

Figure 4. Parameters governing the shape of the keel line.

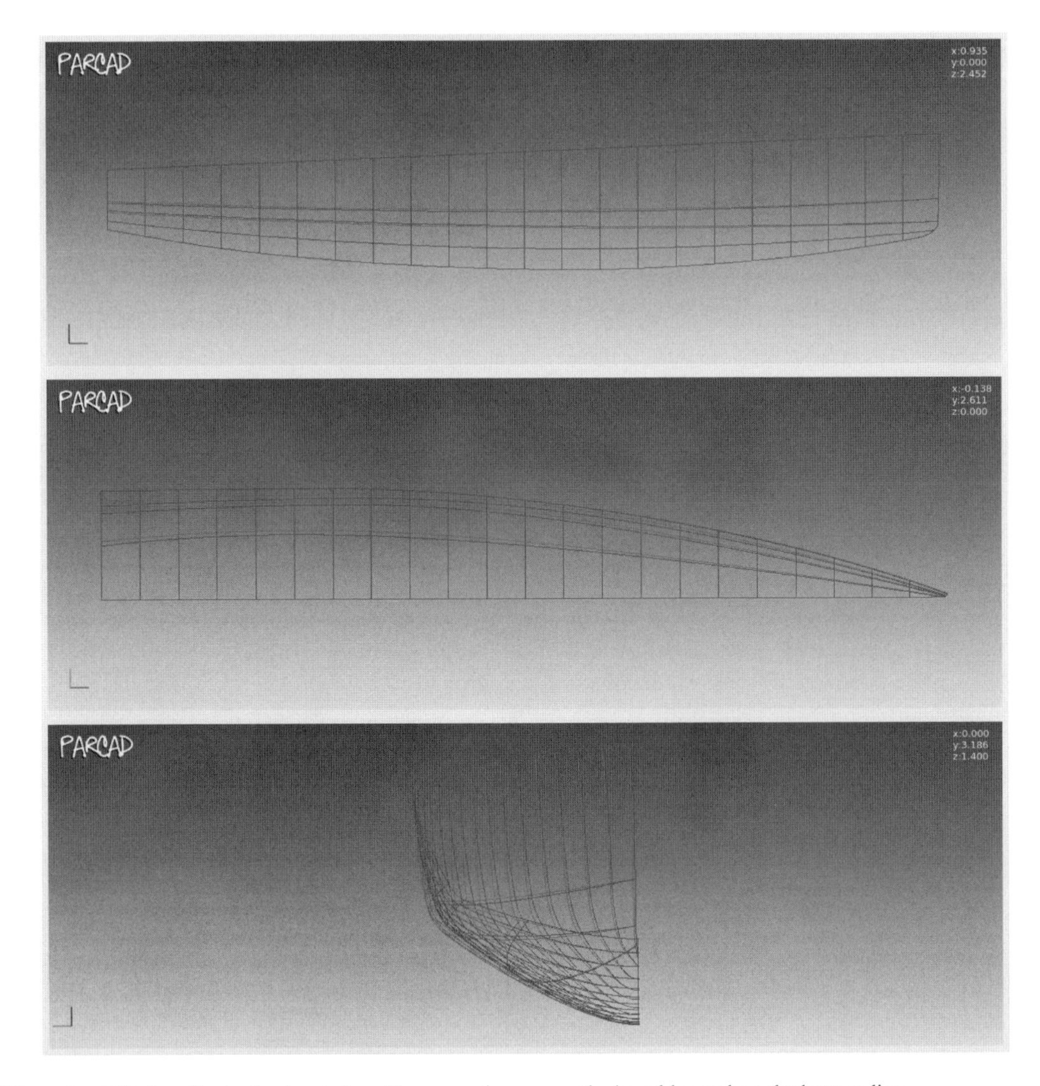

Figure 5. Effect on the body plan induced by some changes on the knuckles and on the bottom line.

Figure 7 and Figure 8 show the results on the fore-body hull shape due to two different sets of parameters (Set A, B). Specifically, with reference to Table 5, the items that have been changed are #1, 2, 5, 6 and #5, 15, 17, for the set A and B respectively.

5 DISCUSSION AND CONCLUSIONS

In this work, a design tool for the parametric representation of hull shapes by means of splines in tension has been developed and presented. The mathematical properties of these curves guarantee fair hull shapes with a minimum number of inflection points even in the presence of a strong local change of the curvature and allow a good flexibility in handling complex geometries. For those cases when the subtended area and/or its centre are prescribed, the curves are defined by means of a non-linear least square process, fulfilling systematically the minimum mean curvature. Slopes at the end points can be added further on as form parameters. The curves can be used also for interpolation, exhibiting the same fairness properties.

Table 5. Bulb design variables.

Item	Reference Axis/Plane	Description
1.	X	Length
2.	Z	Bulb (TIP) elevation at LOA
3.	Z	Upper bound curve elevation at FP
4.	Z	Lower bound curve elevation at FP
5.	Y	Max width at FP
6.	Z	Max width elevation at FP
7.	X-Z	Upper bound curve slope at FP
8.	X-Z	Lower bound curve slope at FP
9.	X-Z	Max width elevation curve slope at FP
10.	X-Z	Max width elevation curve slope at TIP
11.	X-Y	Max width curve slope at FP
12.	X-Y	Max width curve slope at TIP
13.	X-Z	Side area of the bulb upper part
14.	X-Z	Side area of the bulb lower part
15.	X-Y	Bulb waterplane area coefficient
16.	Y-Z	Sections slope at the upper bound curve
17.	Y-Z	Sections slope at the lower bound curve

Above the extremely good behavior of the curves in terms of small curvature variation, a further advantage in using splines in tension regards the control of the shape of the curves that is based on the physical points they pass through whereas B-splines use control points that do not belong to the curve/surface and thus their coordinates might have an ambiguous geometric meaning when used to manipulate the entire curve in a parametric model.

A drawback in our implementation is that the representation of the hull is given by plane control curves (not surfaces) but on the other hand these curves are controlled by higher level curves in the hierarchy, for instance the sectional area curve, the curve of the slope of the cross sections at the center plane, the curves of the max beam and max beam height and so on.

The results obtained in the test cases presented and in other hull forms not shown here, confirm the fairness quality of the curves and the flexibility and robustness of the whole tool to represent a variety of body shapes—including ship appendages—by means of physically based parameters.

The tool is intrinsically suited to be included in a hull form optimization process.

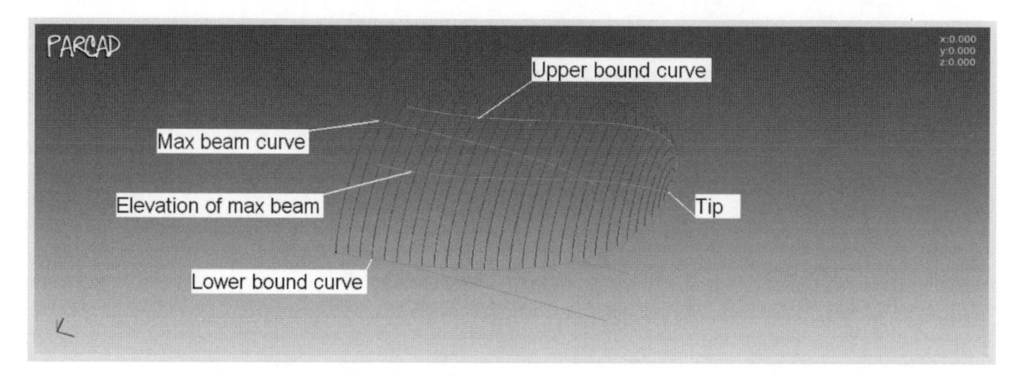

Figure 6. Some curves governing the shape of the bulb.

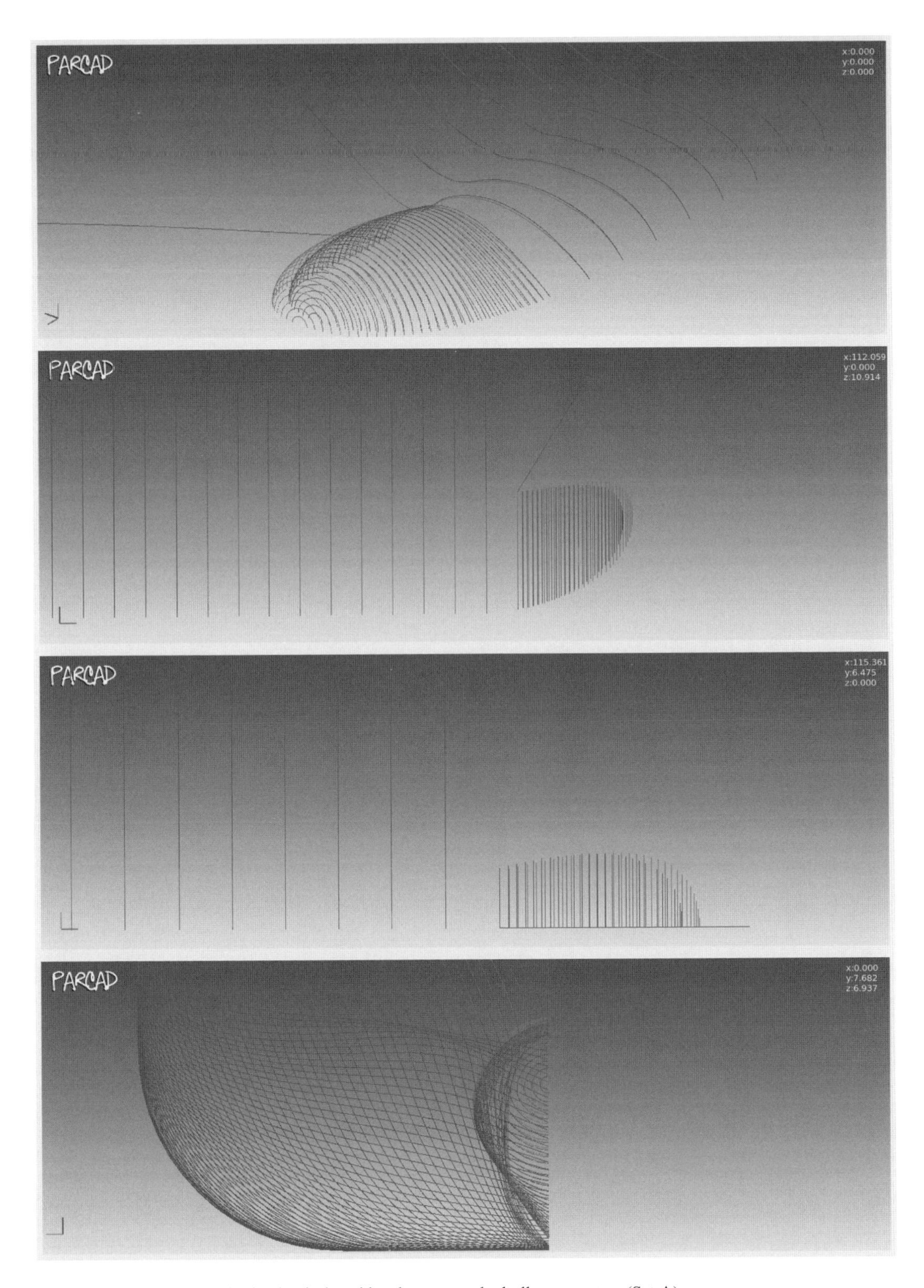

Figure 7. Effects on the body plan induced by changes to the bulb parameters (Set A).

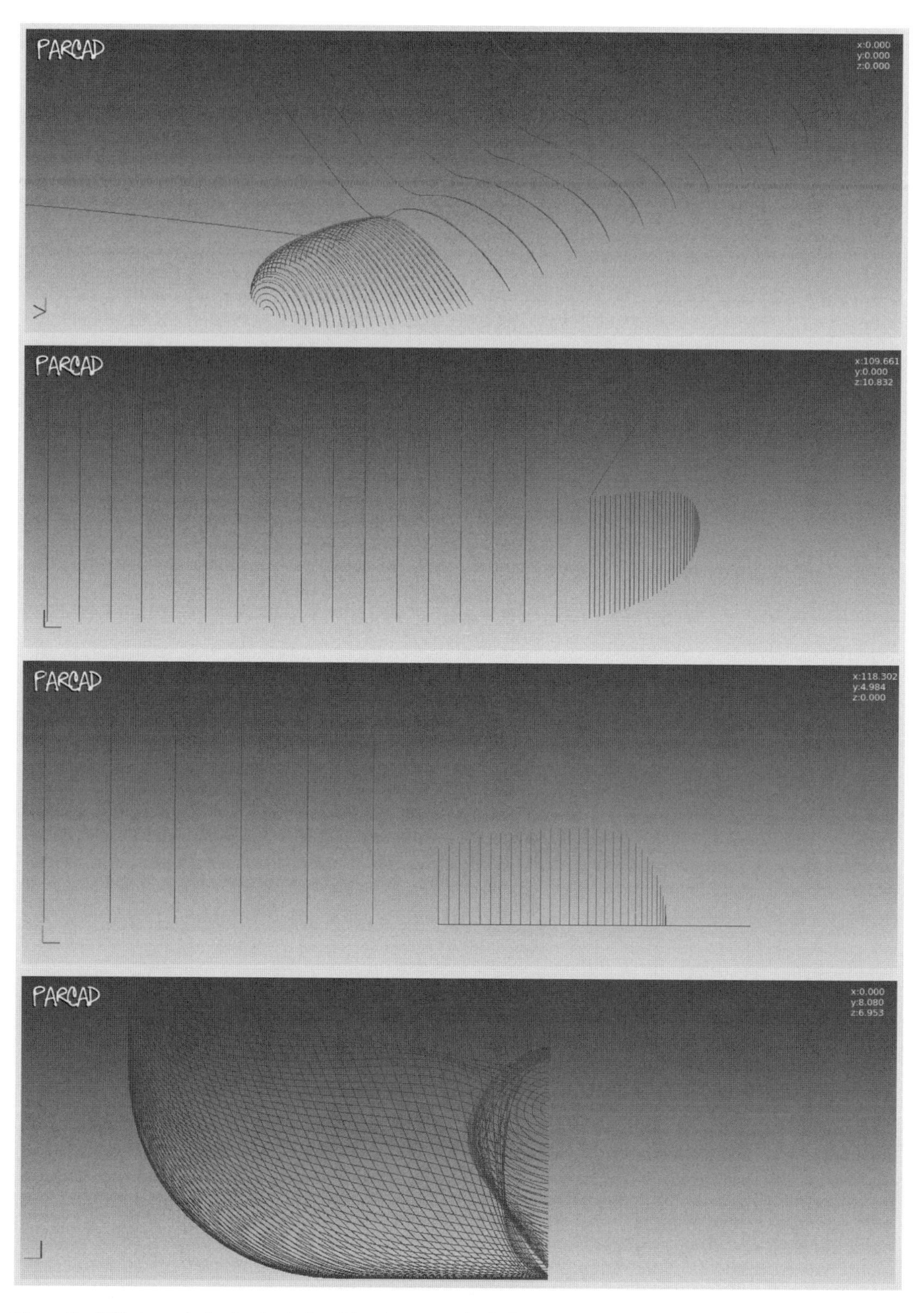

Figure 8. Effects on the body plan induced by changes to the bulb parameters (Set B).

ACKNOWLEDGEMENTS

The Maurizio Cossutti Yacht Design is acknowledged for providing the body plan of the sailing vessel. This work was partly performed in the context of the project OpenSHIP, supported by Regione Friuli Venezia Giulia POR FESR 2007 2013 Obiettivo competitività regionale e occupazione.

REFERENCES

Abt, C., Bade, SD., Birk, L. & Harries, S. (2001). Parametric Hull Form Design—A Step Towards One Week Ship Design. In: Proceedings of 8th International Symposium on Practical Design of Ships and Other Floating Structures (PRADS 2001). Shanghai.

Abt, C. & Harries, S. (2007). FRIENDSHIP Framework-integrating ship-design modeling, simulation and optimization, The Naval Architect, RINA.

Cline, A.K. (1974). Scalar- and Planar- Valued Curve Fitting Using Splines Under Tension. Communications of the Association of Computing Machinery—Numerical Mathematics 17(4):218–228.

Harries, S. (1998). Parametric design and hydrodynamic optimization of ship hull forms. Ph.D. Thesis, Berlin, Germany: Technische Universität.

Harries, S. & Abt, C. (1998). Parametric curve design applying fairness criteria. In: Proceedings of the International Workshop on Creating Fair and Shape-Preserving Curves and Surfaces, Network Fairshape. Berlin/Potsdam, Germany, pp. 67–77.

Harries, S. & Schulze, D. (1997). Numerical Investigation of a Systematic Model Series for the Design of Fast Monohulls, In: Proceedings of the 4th International Conference on Fast Sea Transportation (FAST'97), Sydney, Australia, Vol. 1.

Lackenby, H. (1950). On the systematic geometrical variation of ship forms. Transactions INA 289–316.

Levenberg, K. (1944). A Method for the Solution of Certain Non-Linear Problems in Least Squares, The Quarterly of Applied Mathematics 2:164–168.

Mancuso, A. (2006). Parametric design of sailing hull shapes. Ocean Engineering 33(4):234–246.

Marquardt, D. (1963). An Algorithm for Least-Squares Estimation of Nonlinear Parameters, SIAM Journal on Applied Mathematics 11:431–441.

Percival, S., Hendrix, D. & Noblesse, F. (2001). Hydrodynamic optimization of ship hull forms. Applied Ocean Research 23:337–355.

Perez-Arribas, F., Suarez-Suarez, J.A. & Fernandez-Jambrina, L. (2006). Auto surface modeling of a ship hull. Computer-Aided Design 38:584–594.

Perez, F., Suarez, J.A., Clemente, J.A. & Souto, A. (2007). Geometric modelling of bulbous bows with the use of non-uniform rational B-spline surfaces. Journal of Marine Science and Technology 12:83–94.

Renka, R.J. (1993). TSPACK: Tension Spline Curve-Fitting Package, Association of Computing Machinery. Transactions on Mathematical Software 19(1):81–94.

Roulier, J. & Rando, T. (1994). Measures of Fairness for Curves and Surfaces. In Designing Fair Curves and Surfaces, ed. N.S. Sapidis, Society for Industrial and Applied Mathematics, pp. 75–122.

Schweikert, D.G. (1966). An interpolation curve using a spline in tension. Journal of Mathematics and Physics 45:312–317.

Tahara, Y., Peri, D., Campana, E.F. & Stern, F. (2008). Computational fluid dynamics-based multiobjective optimization of a surface combatant using a global optimization method. Journal of Marine Science and Technology 13:95–116.

Valorani, M., Peri, D. & Campana, E.F. (2003). Sensitivity analysis methods to design optimal ship hulls. Optimization and Engineering, 4:337–364.

Zhang, P., Zhu, D. & Leng, W. (2008). Parametric Approach to Design of Hull Forms, Journal of Hydrodynamics 20(6):804–810.

Sustainable Maritime Transportation and Exploitation of Sea Resources – Rizzuto & Guedes Soares (eds)
 ISBN 978-0-415-62081-9

A performance study of the buoyancy-control type ballast-free ship

M. Arai
Yokohama National University, Yokohama, Japan

S. Suzuki
Modec, Tokyo, Japan

ABSTRACT: This paper describes the concept of the ballast-free ship proposed by Yokohama National University and the results of studies carried out to examine its performance. In the proposed method, seawater outside a ship is taken in from intake vents at the ship bottom, and the seawater flows through the buoyancy control tanks of the ship. After circulating in the buoyancy control tank, the seawater is discharged from the exit vents. To accelerate the water circulation in the buoyancy control tanks, the intake and exit vents are shaped to produce a pressure difference between them during the voyage, i.e., to give high pressure at the intake and low pressure at the exit vents. An effective flow is generated by this pressure difference to ventilate the water inside the buoyancy control tank without any additional devices to accelerate it. The seawater inside the tanks causes the ship to lose its buoyancy and secure sufficient draft to navigate in the lightweight condition. Since the water in the tanks is always displaced by fresh seawater from outside the ship, the water in the tank has the same contents as the local seawater where the ship is navigating, thus preventing the problem of introducing invasive marine species in ballast water into new environments. In this study, we carried out a towing test using a 7-meter-long model ship and confirmed that the increase of the ship's resistance caused by the bottom vents was less than 1 percent of the total ship resistance. Detailed numerical simulations were carried out to study the relationship among the shape of the vents, seawater exchange performance and power increase. Based on the obtained optimized shapes of the vents, we propose a practical open and close mechanism of the bottom vents as well.

1 INTRODUCTION

Ships' ballast water is considered to be one of the main responsible factors causing marine ecological problems worldwide. To cope with this problem, the International Maritime Organization (IMO) adopted the International Convention for the Control and Management of Ship's Ballast Water and Sediments in 2004, and enactment of the convention is expected in the near future (IMO, GloBallast).

After the enactment of this convention, all ocean-going vessels will be required to satisfy the ballast water emission standard. Therefore, development and installation of the ballast water treatment equipment that is in accord with the standard have been hastened. However, a complicated processing system is necessary for the treatment of marine organisms of divergent sizes and shapes and other characteristics. It is also difficult to secure sufficient space for the treatment system in ships, especially in the case of retrofitting. In the worst case, decreased capacity of cargo hold and/or drastic structural modification of the hull may be necessary. In addition, there is a high possibility of causing increased transportation costs due to the increase of electric power and chemicals that are necessary for the ballast water treatment systems.

Approaching this ecological problem from a different viewpoint, Arai et al. (2009a-e) and Kora and Arai (2007) proposed a ballast-free system. With the proposed system, during the lightweight voyage of a ship, seawater circulates in the buoyancy control tanks, and the contents of the seawater in the tanks is the same as the local seawater outside the ship. Thus the migration of marine species from the origin ports to the destination ports is effectively prevented. The proposed system uses a simple mechanism in which circulation of the water in the tank is generated by using the pressure difference between the intake and exit vents caused by the vents' shape. Since installation of the intake and exit vents on the ship bottom is not difficult and the conventional ballast water tanks can be converted to the buoyancy control tanks without serious structural modification, the proposed system is effective not only for use in newly built ships but also in existing ships.

In this paper, we show the results of studies on the relationship between the shape of the intake and exit vents and the water circulation performance, which is one of the key points of the proposed system. We also discuss the increase of ship resistance due to the installation of the vents on the ship bottom. Finally, we propose a few open and close mechanisms of the vents based on the research results.

2 OUTLINE OF THE BUOYANCY-CONTROL TYPE BALLAST-FREE SHIP

Fig. 1 shows the operation of the proposed ballast-free ship. With this method, when the cargoes are discharged at the port, seawater is taken into the buoyancy control tanks using the ship's piping system to secure the necessary draft. During this operation, intake and exit vents on the bottom of the ship are all closed. After the navigation starts, the vents are opened, and circulation of water inside the tanks is generated due to the pressure difference between the intake and exit vents. This pressure difference is caused by the ship's advance speed and the shapes of the intake and exit vents. In some types of ship, particularly in the case when a large volume of water is required to control the buoyancy, it may be advantageous to set the ceiling of the buoyancy control tanks higher than the ship's draft line. In that case we can realize the loading of water in the buoyancy control tanks by closing the air pipe and opening the intake and exit vents after seawater is fully loaded in the tank. Since the seawater inside the tank is suspended from the tank ceiling in this case, care should be taken not to generate too low pressure

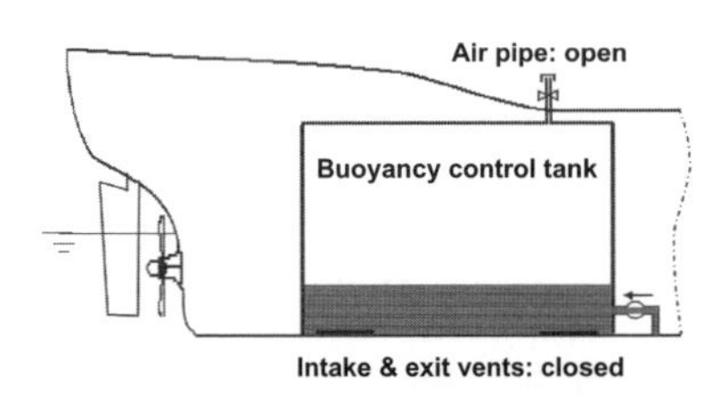

(a) Filling process of the buoyancy control tank

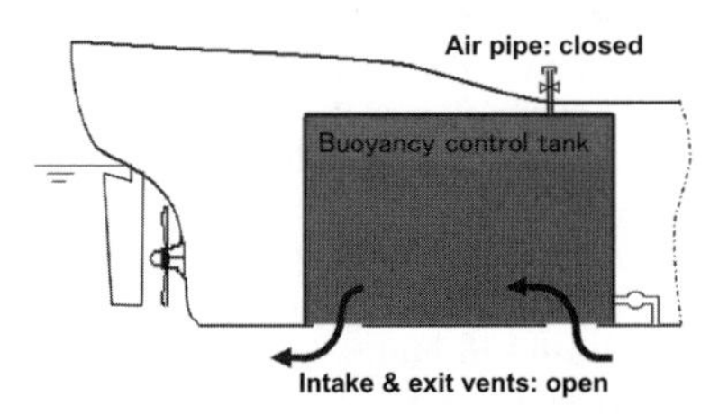

(b) Water circulation in a lightweight condition

Figure 1. Operation of the proposed ballast-free ship.

at the top part of the tank to prevent vaporization of the seawater. The vertical distance between the tank ceiling and the draft line should be less than 10 meters to prevent vaporization.

By the operation described above, the ship can keep sufficient draft for navigation, similar to the conventional ship using seawater in ballast tanks. The intake and exit vents are open during the lightweight voyage, and the water circulation continues. By this mechanism, the water inside the tanks is efficiently replaced by the seawater outside the ship, and the contents of the water in the tanks are kept identical to those of local seawater. At the destination port, the water inside the tank, which has become the local seawater there, is discharged to load the cargo onto the ship.

During the full load voyage, the vents are closed and the buoyancy-control tanks are kept empty, so the ship can secure sufficient buoyancy, the same as conventional ships.

The operations just described make it possible for the ships to prevent the immigration of marine species from the origin ports to the destination ports without the need for serious ship structural modification, loss of cargo capacity, cost increase and so on.

3 SHAPES OF BOTTOM VENTS

In this system, water circulation in the buoyancy control tank is induced by the pressure difference between inside and outside the tank around the opening vents. Therefore, the vent's shape becomes an important design factor. Fig. 2 shows an example of the vents arrangement on the ship's bottom and shapes of intake and exit vents.

In this chapter we will explain the fundamental mechanism of the generation of circulating flow inside the tank. Various shapes of intake vents can be considered. In Fig. 2, a concave-shaped intake vent is shown. In the dented part of the concave, a relatively high pressure inside the vent occurs due to the flow outside the tank. On the other hand, if we install a convex-shaped appendage at the exit vent, a low pressure is generated below the appendage. Thus, the high pressure at the intake pushes the seawater into the buoyancy control tank, and the low pressure at the exit sucks the water out of the tank. These phenomena can be easily explained by Bernoulli's theorem.

One possible side effect is the increase of the ship resistance due to the installation of the appendages on the ship bottom. We therefore performed a towing test of a model ship with bottom appendages to examine the additional resistance, i.e., to determine whether increased engine power was

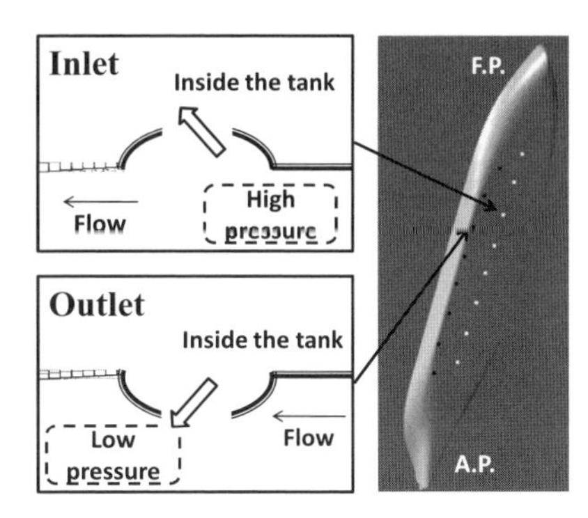

Figure 2. Example of bottom vents on a ship hull.

required to maintain the service speed. In addition, numerical analyses were conducted to evaluate the power increase and water circulation performance of the actual ship as well as to verify the accuracy of the numerical method by the comparison with model test results.

4 MODEL TESTS OF THE EFFECT OF VENTS ON THE RESISTANCE

4.1 *Conditions of model tests*

Fig. 3 illustrates the tanker type model ship we utilized in our tests, with the appendages that simulate the shape of the vents. Twenty appendages, 10 for each side, were installed on the bottom of the model ship near the bilge areas.

Ship resistance was measured by towing the model ship in a towing test basin. Size of the appendage was determined with reference to the results of numerical analysis of water circulation (Arai et al., 2009c, 2009d). Table 1 shows the examined size of the appendages. Shape 1 has the basic size that we expect will be suitable for actual ships. Shape 2 is 1.5 times taller than Shape 1. Shape 3 has double the diameter of Shape 1. Shape 2 and Shape 3 have larger appendages than we would expect to actually implement (i.e., Shape 1), since we wanted to observe clear effect of appendages whose effect had been anticipated to be very small and thus challenging to observe. Those appendages have the shape of paraboloid of revolution and are 1/23.4 scale models of actual ones. The root diameter of Shape 1 was designed to be the size that can be placed between the longitudinal stiffeners of the ship structure. Also, their height was confirmed to be the height at which the effect on the laminar sublayer of the model ship is not serious.

The conditions for the model tests corresponded to the full-load conditions of an actual ship. Inflow and outflow to and from the vents were neglected.

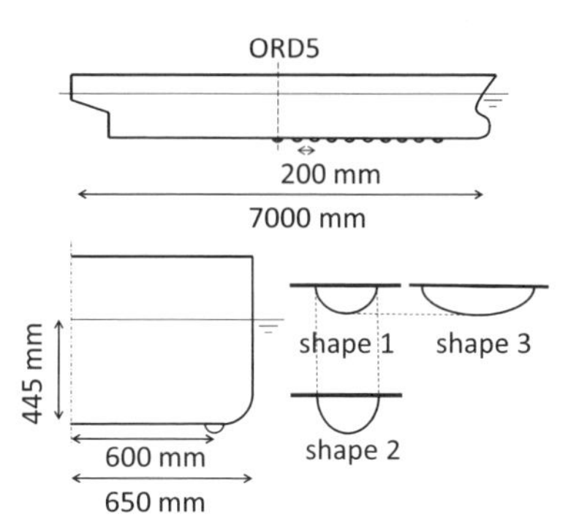

Figure 3. Model ship and appendages simulating the vent system.

Table 1. Size of the bottom vents.

	Model ship		Real ship	
	Diameter[mm]	Appendage height [mm]	Diameter[mm]	Appendage height [mm]
shape1	30.3	7.6	800	200
shape2	30.3	11.4	800	300
shape3	60.5	11.4	1600	300

For the naked hull without appendages and for Shape 2, we also measured the resistance in the lightweight condition for comparison. We tested 5 different speed cases and examined the effect of ship speed.

4.2 *Model test results*

The residuary resistance coefficient (γ_R) for the model with three different appendages and for the naked hull are compared in Fig. 4.
where

$$\gamma_R = R_R / \rho \nabla^{\frac{2}{3}} V^2, \tag{1}$$

R_R: residuary resistance,
ρ: density of water,
∇: displacement of model ship, and
V: model ship speed.

Froude Number Fn = 0.175 corresponds to the design service speed of the ship. The measured residuary resistance coefficient for the full-load condition at around Fn = 0.175 increased in the order of Shape 2, Shape 3 and Shape 1. Compared to the naked hull, the coefficient for Shape 1, which is expected to be used for an actual ship, increased very little in the full-load condition.

Fig. 5 indicates the ratio of measured resistance increase caused by the appendages to the total resistance. The resistance increase of Shape 1 in the full-load condition is 0.5% around the service speed. We did not measure the resistance of

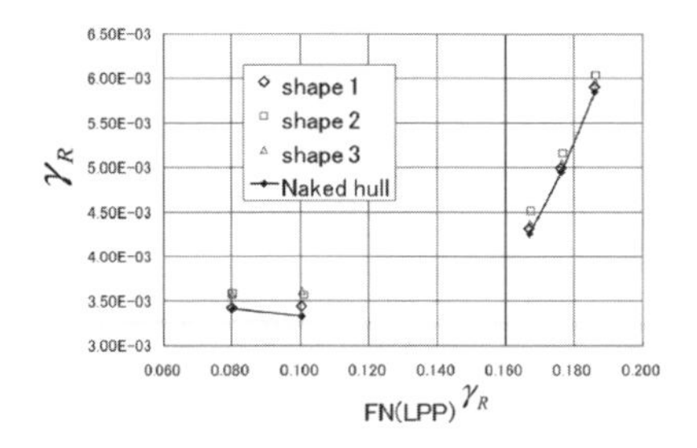

Figure 4. Measured residuary resistance by towing tests.

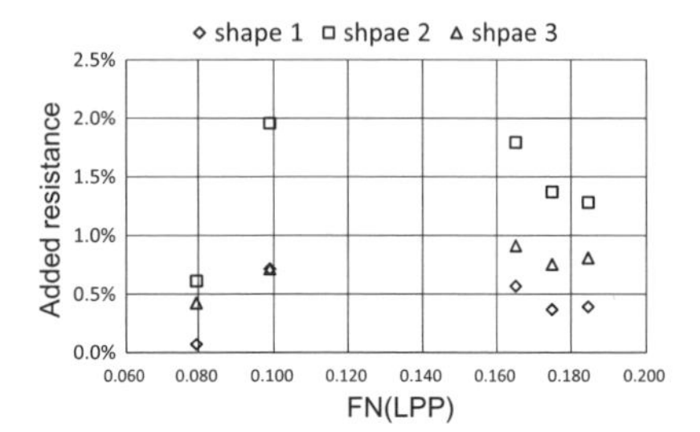

Figure 5. Measured added resistance.

Shape 1 in the lightweight condition; however, if we compare the data of Shape 1 and Shape 2 in the full-load condition, we may estimate the resistance increase of Shape 1 in the lightweight condition to be less than 1%. We note that in the lightweight condition the outflow from the exit vents changes the flow around the appendage and may change the resistance value. This effect will be discussed in Section 6.3.

Based on the test results, we estimate the power increase ratio of the actual ship by the three-dimensional extrapolation method (or Hughes method). The power increase will be compared with numerical results in Chapter 5. In this paper, the power increase ratio will be evaluated by dividing the necessary ship power for the ship with the appendages by the power for the naked hull. The power discussed in this paper is the effective power at the service speed (14.5 knots). We assumed that the appendages are sufficiently small compared to the ship hull such that the wave-making resistance by the appendages can be ignored.

5 NUMERICAL ANALYSIS OF THE FLOW AROUND THE APPENDAGES

5.1 *Numerical method*

To examine the change of the resistance due to the shape variation of the appendages on the ship bottom, we carried out numerical analysis for the appendages that were used in the model tests shown in Chapter 4. In this chapter, we will show the numerical analysis for the appendages without openings. Here, we will compare the results of numerical analysis with measured data shown in Chapter 4 and discuss the accuracy and appropriateness of the numerical method.

For the numerical analysis, we used ANSYS CFX 12.0, which is based on a Finite Volume Method, to solve the flow. To model the turbulent boundary layer, the Shear Stress Transport (SST) Model was applied.

5.2 *Preliminary analysis*

Velocity distribution in the turbulent boundary layer influences the resistance of the appendages. Therefore, we need to reproduce the turbulent boundary layer in the numerical analysis. As a preliminary examination we reproduced the development of the turbulent boundary layer on a plate whose length is equal to the model ship length. The local flow around the appendages was calculated by giving the velocity component, the turbulent kinematic energy and the dissipation rate of the kinematic energy at the boundary of the computational domain. We confirmed by this preliminary analysis that the velocity distribution in the boundary layer satisfied the law of the wall and also that the wall shear stress satisfied Schoenherr's formula.

To examine the effect of the appendages located upstream on the flow around the downstream appendage, we carried out a numerical analysis of the flow around the appendages that were lined up in a series in the direction of the inflow. From the numerical simulation we determined that the resistance of each appendage was almost the same as the result for a single appendage that was set in a flow. Therefore, we concluded that we can practically examine the flow around a series of appendages by analyzing the flow around a single appendage.

Results of a full model and a half model, in which only one side about the center line of the appendage was modeled, were also compared with regard to the vortex generation and resistance value, and we confirmed that the half model provides sufficient results.

5.3 *Condition of the numerical analysis*

Fig. 6 shows the computational domain and boundary conditions for a single appendage. A half of an appendage was modeled. Fig. 6 illustrates the appendage and the ship bottom upside down. The fully developed boundary layer obtained from the preliminary analysis mentioned in Section 5.2 was applied at the upper boundary. The law of the wall was applied on the ship bottom and the

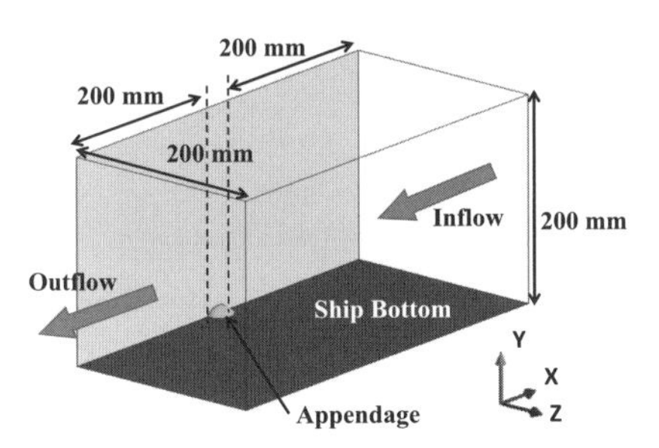

Figure 6. Computational domain and boundary conditions.

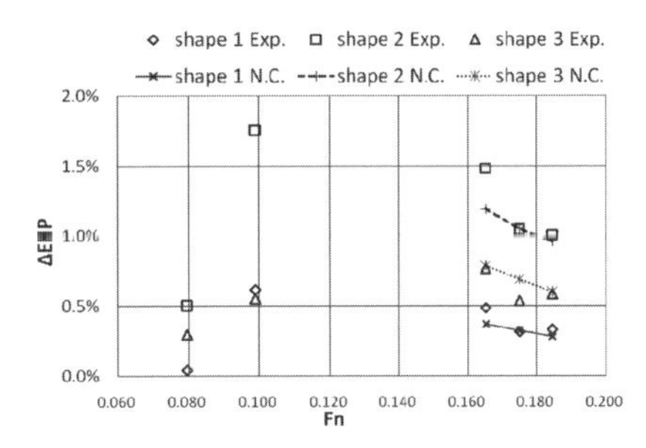

Figure 7. Comparison between measured and calculated results.

appendage surfaces. Symmetric condition was applied at the boundary opposite to the ship bottom and also at the side boundary.

As for the computational cells, hexahedron ones were used near the appendage and tetrahedron ones were used in the far field. The number of nodes on the surface of the half model was 4,000. Viscosity and density of the fluid were set corresponding to the condition of the towing tests.

5.4 *Consistency with the model test*

We will discuss the method we used to confirm the consistency between the model test and the numerical analysis. It is not appropriate to compare the measured ship resistance directly with the result of numerical simulation, since the measured resistance includes the effect of the ship's hull shape. In the analysis of the model test result, we applied a three-dimensional extrapolation method, where components of the resistance are separated and a form factor is used, to estimate the power increase of the actual ship. It is impossible to apply the same procedure to the numerical results to obtain the power increase of the actual ship. Therefore, we estimated the viscous pressure resistance component from the obtained resistance by the numerical analysis and used it to estimate the power increase of the actual ship.

Fig. 7 shows the power increase ratio of the actual ship obtained by the model experiments and by the numerical simulations. We compared the data in the ship's service speed range where the pressure drag component is comparatively large to the total resistance. As shown in Fig. 7, the power increase ratio becomes large in the order of Shape 2, Shape 3 and Shape 1. Moreover, the difference between the measured and calculated results is sufficiently small, less than 0.1% on average, and also the difference becomes smaller according to the increase of Froude number. This tendency is considered to occur because the pressure drag component becomes dominant according to the increase of Froude number. These results indicate that we can discuss the relative effect of the shape change of the appendage by using numerical analysis.

5.5 *Flow around the appendage*

Fig. 8 shows the flow around the bottom appendages obtained by the numerical analysis. The appendages are shown upside down in the figure. Generation of an eddy behind the appendage is clearly observed in the case of Shape 2, in which the power increase is most remarkable both in the model experiment, and the numerical analysis shown in Section 5.4.

In the case of Shape 2, flow separation starts further forward than in the other two cases, and it causes a larger area of dead water and a larger pressure resistance. Although Shape 3 has the largest projection area with respect to the flow, the increase of the power is less than that of Shape 2. This is because the inclination at the root of the appendage connected to the ship bottom is smaller in this case, and therefore the pressure at the stagnation point is smaller and also the separation occurs relatively farther back than in the case of the other two shapes.

5.6 *Effect of the shape of the appendage*

In this section we will summarize the power increase caused by the appendage. From the discussions so far, we know that the power increase around the ship's service speed was less than 1%, except in the case of Shape 2. The power increase of Shape 1 was about 0.5%. Here the power of the ship with appendages was compared with that of the naked hull. Actual ships do have some appendages such as bilge keels below the draft line. For example, if we estimate the viscous resistance of the standard bilge keels by using Shoenherr's formula and convert it to the power increase of the ship, it becomes about 0.7%. That is to say, the power increase caused by the twenty vents of Shape 1 that were set on the bottom of the ship is less than the power increase caused by the bilge keels.

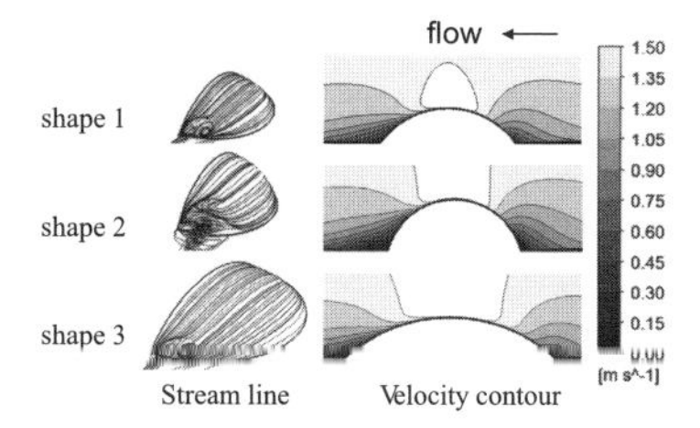

Figure 8. Flow around the bottom appendages.

6 DESIGN OF THE SHAPE OF APPENDAGES

6.1 *Suction performance analysis*

Fig. 9 illustrates the computational model used to study the suction performance around the appendage acting as an exit vent. We assumed the symmetrical relationship and modeled only a half of the fluid domain along the center line of the appendage to the flow. The figure shows a vent and the adjacent fluid domain inside the buoyancy control tank. The figure is illustrated upside down with respect to the actual vent. The domain inside the tank is a half of a hemisphere, and it has a radius equal to about 3 times the diameter of the appendage. An open or continuous boundary condition that does not have any effect on the flow inside the hemisphere domain was set for the surface of the hemisphere. This computational domain shown in Fig. 9 is connected to the whole analysis domain outside the tank shown in Fig. 6. The far side of the open boundary is not analyzed, but the suction effect at the exit vent caused by the outside flow induces the inflow at the open boundary on the surface of the hemisphere.

The shapes of analyzed vents are shown in Fig. 10. The extreme height and width of each appendage are as assumed to be the same as the Shape 1 shown in Fig. 3, except Type B. Type B has the same open area as Type A but has a larger diameter at the foot of the appendage that connects smoothly to the ship's bottom surface. The external shapes of Type E and Type G are the same, but the internal structures are different, as shown in Fig. 10. The opening of Type F is shifted aft compared to that of Type A.

6.2 *Results of numerical analysis*

The figures in the middle column of Fig. 10 show the velocity vector on the centerline plane of the appendages, and the figures in the right column show the stream lines of the seawater that flow out from the vents. Table 2 shows the appendage sizes and openings together with the obtained efficiency index of the appendages. In this study, we define the following index for the evaluation of the

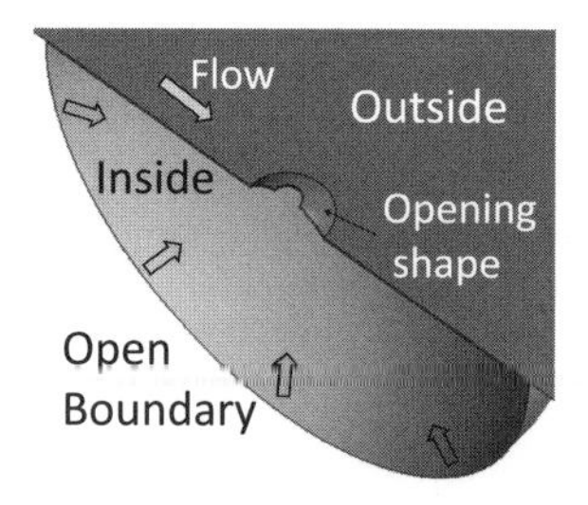

Figure 9. Computational domain for the suction performance analysis.

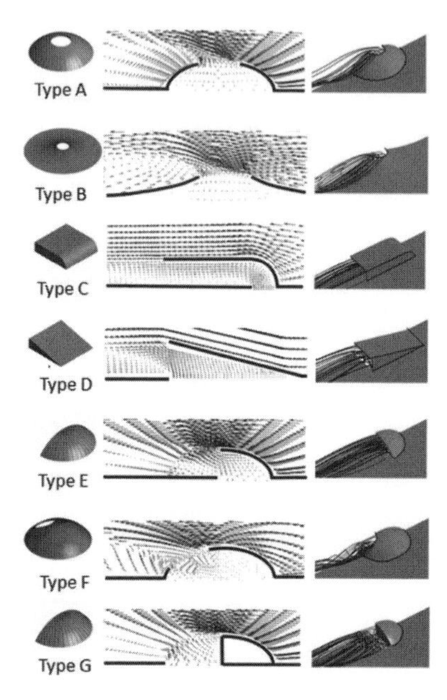

Figure 10. Flow around the various appendages with openings.

appendage, since the goal is to find an appendage with good suction efficiency and low resistance:

$$C = M \cdot V / F \tag{2}$$

where

C: The efficiency index used to evaluate the shape of the appendage,

M [kg/s]: The mass of fluid that flows in at the artificial open boundary defined inside the buoyancy control tank adjacent to the vent (see Fig. 9),

V [m/s]: The external flow speed, which is the same as the ship speed and

F [N]: The resistance (sum of viscous resistance and pressure resistance).

According to the index shown in the right column of Table 2, Type E and Type G seem to be superior. However, this result is obtained from the comparison in model scale. Discussion in the case of the full scale will be given in the next section.

Table 2. Appendage sizes and suction performance.

	Top Height [mm]	Diameter [mm] or Breadth [mm]	M・V/F
Type A	7.6	30.3	0.5
Type B	7.6	*	0.6
Type C	7.6	00.0	1.5
Type D	7.6	30.3	4.7
Type E	7.6	30.3	8.2
Type F	7.6	30.3	1.8
Type G	7.6	30.3	9.6

6.3 *Discussion for the performance of the vents in full scale*

In this section we examine the resistance and seawater suction performance by the numerical analysis. Table 3 shows the conditions of the numerical analysis. Power increase will be estimated by the method based on the non-dimensional pressure drag discussed in Chapter 5. Seawater suction performance will be estimated by the flux of seawater entering the domain adjacent to the exit through the open boundary. A non-dimensional value of the flux divided by the ship speed, water density and open area of the vent is used. The result will be expressed by the ratio of the sucked-out water volume to the tank volume. This ratio indicates the seawater exchange capacity of the vent. The horizontal axis of Fig. 11 shows the power increase of the actual ship, and the vertical axis shows the seawater exchange capacity in 200 nautical miles of the voyage. Desirable performance is in the upper left part of Fig. 11, where there is less power increase and good seawater exchange performance. The variation in the same vent type in the horizontal direction is caused by the difference of ship speed condition (see Table 3). Each of the data shown in Fig. 11 is obtained as the suction performance of a vent itself. In the case of an actual ship, the flow inside the tank is affected by the shape of the intake vent and internal structures as well.

We summarize the findings of the examination as follows:

- The power increase of the Type A vent, which has an opening on the top of the appendage, is about 10 times larger than that of the similar appendage without the opening shown in Chapter 5. In the case of Type A, the outward flow from the opening shifts the location of the separation point more forward than that of the appendage without a top opening.
- As a whole, in the cases of vent types that have a small power increase, the flow from the opening does not prevent the external flow and both flows merge smoothly.
- The resistance increase of types C, D, E and G is small when the seawater is coming out from the vents, but in the case of the full-load condition the openings are closed and a dead water zone will be created behind the appendages. A relatively high increase of resistance can be predicted.
- The acceleration of the outflow from Type E is similar to that from Type G, since the external shapes of the appendages are similar. However, the suction performance of those two types is strikingly different because of the difference in the opening shapes.
- In the cases of Type A and Type B, although the shape of the base part is largely different, the power increase is almost the same. On the other hand, in the case of Type F, whose opening position is shifted backward with respect to the external flow, the power increase becomes about 1/4 of that of Type A. We can say that the position of the opening on the appendage has a large effect on the power increase.
- It is clear from the comparison of the results of Type C and Type D that the blunt forward shape of the appendage causes a greater power increase.

Table 3. Conditions for the numerical analysis.

Ship type	Oil tanker
Ship length	185 m
Total volume of buoyancy control tanks	15,000 m^3
Number of exit vents	20
Operation distance of the system	200 n. mile
Ship speed in Froude number	0.165, 0.175, 0.185

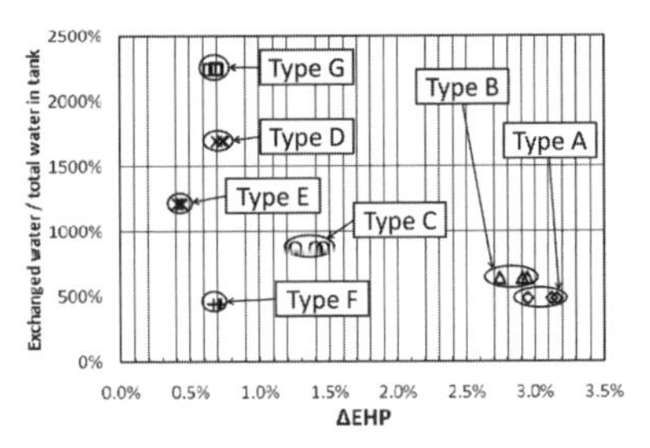

Figure 11. Relation between exchanged water volume and increase of effective horse power (EHP).

7 OPEN AND CLOSE METHOD OF THE VENTS OF THE APPENDAGE

From the examination made in the previous chapter we obtained basic information about the effect of the shape of the appendage and the position of the opening of the vent. In this chapter, we will examine the practical mechanism of the vents that realizes the effective shape in the actual system.

We have discussed the importance of the shape of the exit vents, since the shape of the appendage changes the external flow locally in the lightweight

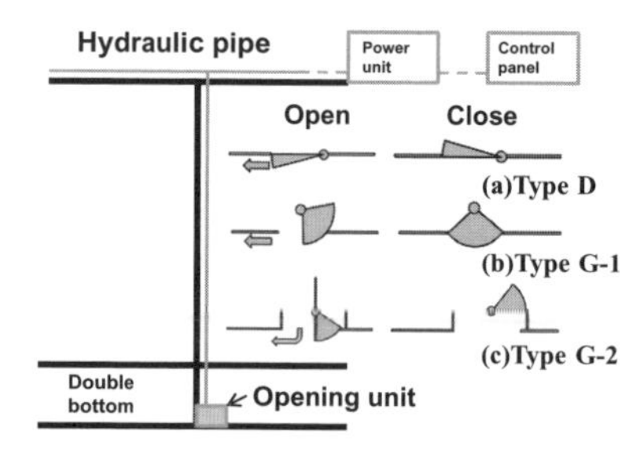

Figure 12. Open and close system on the bottom vents.

condition. In the full-load condition, a smooth ship surface that is similar to the naked hull is preferable. Therefore, it may be useful to discuss the appendage shape and open and close mechanism of the vents. In this chapter we will show the result of examination for Type D and Type G, both of which we think are strong candidates for the actual vents.

Fig. 12 shows a few examples of the mechanisms designed for Type D and Type G. Fig. 12(a) illustrates a mechanism for Type D, in which a structure having a triangular section is joined to the ship hull by a hinge at the apex of the triangle. In this system, the triangular structure can be rotated, and it realizes the Type D in the lightweight condition and the flat surface in the full-load condition. The system shown in Fig. 12(b) can be the Type G in the lightweight condition, and it becomes the appendage without opening (i.e., Shape 1) shown in Chapter 5 in the full-load condition. The system shown in Fig. 12(c) is a variant of that of Fig. 12(b). The mechanism of Fig. 12(c) is similar to that of a butterfly valve.

An ordinary hydraulic system can be used for the open and close mechanism of these vents. Therefore, the installation of the mechanisms shown in this chapter does not require a new type of power unit. It is also possible to apply this ballast-free system to the retrofiting of existing ships by supplying the vent system in a unit that can be installed between longitudinal stiffeners.

8 CONCLUSIONS

We carried out model experiments and numerical analyses on the flow around the vents of a buoyancy-control type ballast-free ship. We made the following conclusions with regard to the power increase and the suction performance of seawater of the several appendages:

1. In the case of installing a sufficient number of the appendages to the ship, the power increase by those appendages is less than 1% in the full-load condition.
2. In this study we compared the compatibility of the model experiments and numerical analyses and showed that we can reasonably evaluate the performance of the various shaped appendages on the ship bottom by the numerical analysis. By using the numerical analysis we can optimize the shape of appendages.
3. We examined the performance of the appendages with openings and found that in the system with superior performance, the water sucked from the buoyancy control tank merges smoothly with the external flow.
4. We proposed a practical mechanism by which the power increase will be reasonably small under both full-load and lightweight conditions.

ACKNOWLEDGEMENT

We would like to thank Prof. K. Suzuki of Yokohama National University and Dr. T. Ohmori of IHI Research Institute for their fruitful discussions and support for the model experiments.

REFERENCES

Arai, M. & Kora, K. 2009a. Ballast-free ships: a new concept for resolving the ballast water management problem. *10th International Marine Design Conference,* Trontheim, Norway: 138–147.

Arai, M., et al. 2009b. A ballast-free ship concept to protect the global marine environment. *GreenTech: Marine Science and Technology for Green Shipping,* Glasgow, UK: 78–87.

Arai, M. & Suzuki, K. 2009c. Experimental and numerical study of the performance of a buoyancy control-type ballast-free ship. *International Symposium on Ship Design and Construction 2009—The Environmentally Friendly Ship,* Tokyo, Japan.

Arai, M., et al. 2009d. Proposal for a ballast-free ship and studies of its performance. *13th Congress of International Marine Association of Mediterranean,* Istanbul, Turkey: 799–805.

Arai, M. & Suzuki, S. 2009e. Study on structural strength of buoyancy control tanks for ballast-free ships. *The 23rd Asian-Pacific Technical Exchange and Advisory Meeting on Marine Structures,* Kaohsiung, Taiwan: 486–492.

GloBallast Partnership. Global Ballast Water Management Programme. *http://globallast.imo.org/*

International Maritime Organization, Ballast Water Management Conventions. *ISBN 92-801-0033-5.*

Kora, K. & Arai, M. 2007. Study of a new non-ballast ship. *21st Asian-Pacific Technical Exchange and Advisory Meeting on Marine Structures,* Yokohama, Japan: 127–133.

Lloyd's Register. 2007. Ballast water treatment technology-Current Status.

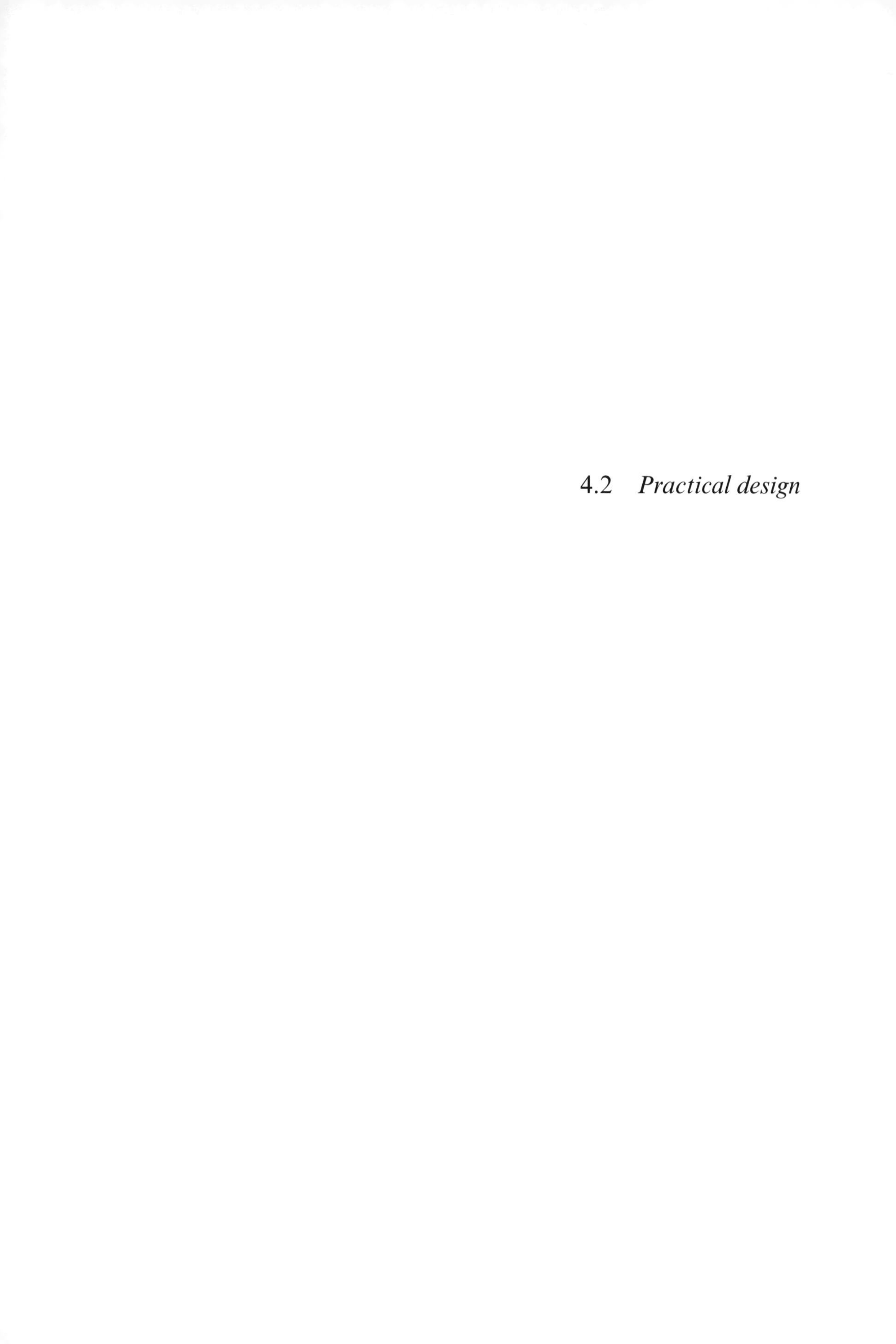

4.2 *Practical design*

Sustainable Maritime Transportation and Exploitation of Sea Resources – Rizzuto & Guedes Soares (eds)

Conditions of Russian water transport and perspectives of river shipbuilding

Gennadiy V. Egorov
Marine Engineering Bureau, Odessa, Ukraine

ABSTRACT: The analysis of a water transport role in economy of Russia, capacities and typical transportations on the mixed navigation vessels is executed, the statistics on composition and conditions of river and mixed river-sea navigation vessels is resulted. Necessity of updating of fleet is proved. The achieved rates of building new vessels are analyzed and future prospects for building river and mixed river-sea navigation vessels are shown. Same time requirements for 2015–2020 periods are of about 350 transport vessels and more than 400 auxiliary vessels for Russian water transport branch.

1 INTRODUCTION

Historically formation of a transport network in Russia developed on the basis of the rivers; at 19th-20th centuries railways were added. From the middle of 20th century river transport became an additional and supplementing mean of cargo transportation. USSR State plan calculations showed that it is enough to have good railway network arranged through whole European part of the country; then river transportations will provide up to 20–25% of additional transportations of mass cargoes and transportation to the areas where railway is absent due to non-profitability.

It is interesting to note, that the best achievement of the USSR river transportation was of 582.3 million tons of cargoes in 1988. The cargo turnover reached 242 billion t·km (in 1985). Average transportation distance of 1 ton of a cargo was of 474 km in 1980, and of 381 km in 1990. For comparison, in 2004 the cargo turnover of Russia river transport was of 87.6 billion t·km with average transportation distance of 640 km (Egorov 2010).

River transport is one of the most safe types of transport. According to the data of the special European Community committee that studies problems of European internal waterways development, it is marked that damages owing to river transport accidents occur 178 times less often than owing to hard vehicles accidents and 13 times less often than owing to railway accidents.

It is a power-saving transport type with a low level of atmosphere and water pollution. For 1 ton of cargo the railway locomotive consumes 8 times bigger energy than river transport vessel; for hard vehicle this factor is 26.

2 THE PURPOSE OF THE PAPER

The analysis of a water transport role in economy of Russia, capacities and typical transportations on the mixed navigation vessels, statistics on composition and conditions of river and mixed river-sea navigation vessels for a substantiation of fleet updating necessity.

3 MAIN MATERIAL STATEMENT

Russian river transport serves 26 autonomous republics, lands, national areas and 42 regions of Russian Federations. Also it provides export/import cargo transportation (mixed river-sea transportation and voyages from mouth river and low depth marine ports of Russia). Extent of internal waterways is of 101.6 thousand km (see Figure 1).

Cargo transportation dynamics by Russian internal water transport is given in the Table 1 (beginning from 2000) (Egorov 2010).

In 2000 for the first time during decade obvious jump in the national economy was reached; some of its reflections can be observed in the parameters of work of internal water transport that was in the lead on growth rates in 2000 and 2001 (113.6% and 106.2% of previous year level accordingly).

For the period 2000–2007 these transportations capacity constantly increased. As a whole growth was of 130%. Thus the greatest achievements are marked on non-metallic building cargoes (160%), grain and other cargoes (140%), timber and metal (120%). Simultaneously small decrease is observed for oil and oil products (–17%), coal (–43%), fertilizers (–14%). Transportations of cement, iron ore and mixed fodders were stabilized at same level.

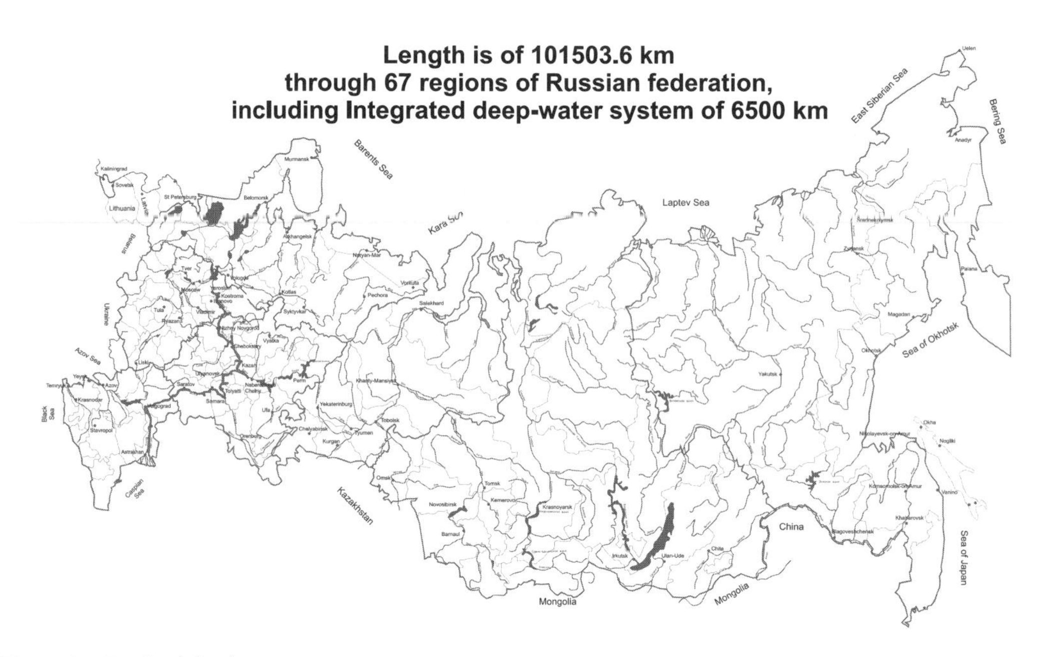

Figure 1. Russian inland water system.

Table 1. Capacity of cargo transportation by Russian internal water transport during 2000–2008 (million tons).

	Year							
Cargo type	2000	2001	2002	2003	2004	2006	2007	2008
Total	116.8	124.0	100.2	100.1	112.9	139.2	152.4	151.0
Including:								
Oil	11.8	13.4	14.5	19.0	18.8	13.9	9.75	9.8
Timber rafting	2.5	2.3	1.9	2.1	2.2	1.7	3.54	2.9
Dry cargoes	102.5	108.4	83.9	78.9	91.9	123.6	139.07	138.3
Including:								
Grain	1.6	2.1	3.9	2.4	2.7	3.3	2.34	0.8
Mixed fodders	0.16	0.3	0.2	0.2	0.3	0.27	0.171	0.04
Coal and coke	3.3	3.3	2.7	3.1	3.1	3.4	2.67	1.9
Timber on vessels	6.0	6.2	6.0	5.3	5.4	6.0	7.3	5.2
Ferrous metals	3.5	2.8	2.5	2.5	3.4	2.8	4.19	2.0
Ores	0.13	0.2	0.35	0.34	0.57	0.81	0.48	0.14
Building	63.5	66.1	56.7	52.6	62.4	93.3	103.5	111.9
Cement	0.1	0.1	0.07	0.15	0.11	0.17	0.42	0.4
Fertilizes	2.9	3.5	3.1	2.9	3.1	2.36	2.46	0.9
Other	21.3	23.6	10.6	9.5	10.9	11.2	15.6	15.0

During 2007 navigation river shipping companies transported 152.4 million tons of cargoes (9.5% more than in 2006). During 2007 In river ports 225.5 million tons of various cargoes were handled (17.6% more than in 2006). These figures include transshipment of export cargoes (17.5 million tons), import cargoes (1.4 million tons) and internal cargoes (205.6 million tons). Thus growth is of 21.7% for export cargoes, 14.3% for import cargoes and 17.3% for internal cargoes.

According to statistics of Russian Ministry of Transport, capacity of transportations by river transport grew till 2009. During 2009 crisis an appreciable decreasing of transportation capacity took place. Mostly it was specified by reduction of mineral-building cargoes transportations.

Growth of capacities of export-import cargoes transportation emphasizes special interest to the mixed river-sea navigation vessels (RSNAV), as to the most effective and profitable component of Russia river transport.

A new stage of development of RSNAV has begun at the end of 80's—the beginning of 90's. As well as earlier, the reasons of new splash in interest to these vessels had purely economical side. On the one hand, active country capitalization has begun, private shipowners have appeared. On the other hand river shipping companies acquired more rights. These river shipping companies did not show themselves actively in the international transportations earlier and possessed a great quantity of "source material" (vessels of inland navigation) which potentially could be conversed in RSNAV. All these changes have occurred on a background of destruction of the centralized system of foreign-economic activity and sharp breaking of cargo parties (down to 1000–5000 t). RSNAV were fitted for work in framework of new economic conditions as well as possible.

At the beginning of 90s RSNAV owned by national shipowners had Significant advantages comparing with sea vessels (including rather smaller average age and residual cost). All these have allowed RSNAV to take strongly place in the transport services market, that earlier was belonged to sea vessels of close carrying capacity.

Growth of amount of RSNAV and restricted sea regions vessels with Russian maritime Register of shipping (RS) classes due to modernization of river vessels was rather rough and reached up to hundred units per year. As a result at the beginning of 2003 the part of RSNAV reached 60% from the number of transport vessels with RS class.

More significant transportations were carried out during the period from 1995 till 2005; their capacities reached about 30 million tons. Since 2006 these transportations were reduced (formally) down to 12 million tons according to change of accounting methodology. This difference was transported by the same RSNAV from sea ports and so was has been related to achievements of sea transport.

Last years crude oil and oil products began to prevail in the structure of export-import transportations; their part grew from 13–18% up to 36–46% of total amount. Dry cargos' part is of 54–64% correspondingly. The basic foreign trade dry cargoes are the grain and regrinding products (18%), timber in vessels (18%), chemical and mineral fertilizers (16%).

Total foreign transportations capacity in prospect till 2025 (Egorov 2010), are given in Table 2 below.

On may see that foreign cargo transportation will increase by 70% till 2020 in comparison with 2008. Thus oil and oil products transportations will increase from 5.7 up to 8.0 million tons (by 40%); dry cargoes transportations will increase from 6.6 up to 10.0 million tons (by 50%).

The basic mass cargoes will stay grain, timber in vessels, ferrous metals, chemical and mineral fertilizers which will reach about 70% in total amount of dry cargoes.

During 2009 the cargo transportations capacity by RSNAV has increased for 26.4% (15.6 comparing with 12.3 million tons); bulk cargoes transportation increased for 49.9% and dry cargoe transportation increased for 6.6%. The dry cargoes transportations capacity has increased due to ferrous metals transportations (+33.7%, up to 1.84 million tons), other dry cargoes (+38.6%, up to 2.42 million tons), chemical and mineral fertilizers (+25.6%, up to 1.1 million tons) and grain (+26.9%, up to 0.93 million tons).

Table 2. Dynamics of cargo import-export transportation through Russia by mixed river-sea navigation vessels, thousand tons.

	Actual data			Forecast		
	2006	2007	2008	2015	2020	2025
Total	20955.2	12047.2	12333.6	18000.0	20000.0	25000.0
Including:						
Oil and oil products	8011.5	4356.3	5672.3	8000.0	8000.0	10000.0
Dry cargoes	12943.7	7690.9	6661.3	10000.0	12000.0	15000.0
Including:						
– grain and regrind products	2467.3	1117.3	732.9	1100.0	2500.0	2800.0
– timber on vessels	2369.5	1408.4	1475.4	2000.0	2000.0	2500.0
– ferrous metals	1853.1	989.9	1378.1	2000.0	2000.0	2200.0
– chemical and mineral Fertilizes	1717.5	1127.5	875.8	1500.0	1700.0	2000.0

Today the earnings of the shipping companies that work on internal Russian waterways are about 870 million of US dollars. The sum of about 360 million of US dollars corresponds to the trans-loading and service companies that work on the rivers.

River transport market is growing; its efficiency is bigger than for car and rail way transportations. OSC "Russian Railways" charge rates growth provides new clients inflow for river transportation. Besides Russian rail way transport transfer capability has known limit, and it becomes reached in some causes. Especially this takes place in summer seasons when additional passenger trains are provided for south directions.

Though even big river shipping companies have faced difficult or will face them in the nearest time. Cargo transportation structure changes, so carriers need for fleet update. Only very big companies are able to effect such changes. But Russian shipbuilding industry has no capacity to fulfill such orders in short time.

More than 80% of vessels which are assumed for building for Russian shipowners also are of RSNAV type, and changing of this parity in nearest years is not expected. (Egorov 2008).

2119 transport vessels of RSNAV type and of restricted marine region navigation (including 1188 ones that effect international voyages) are under supervision of only Russian classification societies (as for January, 2010).

In addition to foreign transportations, SMRSN provide execution of the most important task of north regions supply.

Roads cargo transfer from RSNAV to marine vessels, which is widely applied in the south and the north-west OF Russian Federation is another modern direction of RSNAV usage.

In world marine practice (USA, England, Denmark, Norway, Egypt, etc.) such operations are named as scheme of "ship-to-ship" or STS operations.

The STS scheme of oil reloading from river-sea going vessels to marine vessels with use of storage tankers gives an opportunity to form larger cargo parties for marine tankers with displacement of 100–150 thousand tons. Regulations for road reloading complexes are developed in order to provide safety of STS operations. These Regulation required strict fulfillment of the Russian ecological legislation and usage of worldwide experience that reduces to a minimum risks of environmental contamination.

Sulfur, grain and mineral fertilizers also are trans-loaded through road reloading complexes. The steady export channel of Russian products which is bringing stable income in the budget of Russian Federation (see Table 3).

Table 3. Capacity of cargo transshipment at road trans-loading complexes of marine ports (2007 navigation data).

Port name	Cargo type	Capacity, thousand tons
Caucasus, Kerch	heavy oil	2715
	sulphur	2845
	grain	897
	fertilizes	187
Kronstadt, St. Petersburg	heavy oil	3717
	fertilizes	590
New port	oil	250
Tiksi	oil	50

At present time age structure of inland and mixed river-sea vessels cannot be accepted as satisfactory (by August of 2010 there are 15072 such vessel in Russia).

Age structure of fleet is characterized as follows due to Russian River Registry of shipping (RRR) data (see Table 4):

- mean age of self propelled dry cargo vessels is of 35.5 years;
- mean age of non-self propelled dry cargo vessels is of 30 years;
- mean age of self propelled tankers is of 37 years;
- mean age of non-self propelled tankers is of 29 years;
- mean age of passenger vessels is of 32 years;
- mean age of tug boats that provide service of non-self propelled dry cargo vessels and tankers is of 33.2 years.

Number, average age and general technical condition of mostly widespread projects of river and river-sea cargo vessels are shown in the Table 5.

From all 13476 cargo vessels 1599 ones have technical condition estimation "bad" and 410 ones have estimation "suitable limited".

Average age for prj. 05074 vessels ("Volzhskiy type") is of 21.1 years (6 ones), for prj. 565 vessels ("Volgo-Don" type) is of 35.9 years (41 ones), for prj. 507B vessels ("Volgo-Don" type) is of 40 years (55 ones). From 100 vessels of these projects 6 ones (6%) in a unsatisfactory technical condition.

Hull ageing is the main factor that determine technical condition of "Volgo-Don" vessels. At the majority of "Volgo-Don" vessels the life-time of the main engines is spend.

Average age for 124 tankers of "Volgneft" type is as follows: 42.4 years for prj. 338/550 (16 ones) and 35.3 years for prj. 1557/550A (108 ones). Among them 18 vessels have estimation "bad" (141.5%). The main problem of "Volgoneft" type

Table 4. Transport vessels age (August of 2010) due to RRR data.

Vessels' type	Age groups					Total
	less than 10 years	10–20 years	21–30 years	31–40 years	More than 40 years	
Vessels' distribution due to age groups, (pcs)						
Self propelled dry cargo	17	40	394	296	485	1232
Non-self propelled dry cargo	80	418	2131	1535	717	4881
Self propelled tanker	4	22	120	243	306	695
Non-self propelled tanker	43	131	344	233	145	896
Passenger	199	114	417	295	531	1556
Tugboat	52	349	2193	1781	1437	5812
Total	395	1074	5599	4383	3621	15072
Fleet's age structure, (%)						
Self propelled dry cargo	1.38	3.25	31.98	24.03	39.37	100
Non-self propelled dry cargo	1.64	8.56	43.66	31.45	14.69	100
Self propelled tanker	0.58	3.17	17.27	34.96	44.03	100
Non-self propelled tanker	4.80	14.62	38.39	26.00	16.18	100
Passenger	12.79	7.33	26.80	18.96	34.13	100
Tugboat	0.89	6.00	37.73	30.64	24.72	100
Total	2.62	7.13	37.15	29.08	24.02	100

vessels is double bottom height that doesn't satisfy MARPOL requirements.

Updating ship power stations of tankers is actual. It should include replacement of the main and auxiliary engines, such as electric equipment and fire protection.

For all tankers intensive hull construction corrosion is observed; capacity of repair-renewal works annually grows correspondingly. But these repairs of growing capacity do not cover actual needs and vessels are forwarded into operation with minimal safety factors which does not suffice for a five years' cycle between classifications surveys.

From 4881 dry-cargo barges with average age of 29.9 years 13% have "bad" or "suitable limited" estimations of technical condition. The basic problem for these vessels is maintenance of technical condition by hull repair, including in docks (on slipways). Decision of this problem can extend dry-cargo barges operation for the nearest 10–15 years.

From 896 oil barges with average age of 29.1 year 12% have "bad" or "suitable limited" estimations of technical condition.

Design operation term for vessels is of 25–35 years. Thus in 5–10 years more than 50% of vessel that are under operation at this moment this moment should be discarded. This should lead to a collapse of river transportations capacity.

The basic criteria of the future mass discard of river vessels in the Russian Federation are as follows:

- extreme physical ageing; some river vessels are under operation for 40 years and more (dry-cargo vessels "Shestaya pyatiletka", "Okskiy", "Kaliningrad", "Volgo-Don" (prj. 507B), tankers "Volgoneft" (prj. 558/550), "GES", etc.);
- big capital investments in maintenance of vessel's suitable technical condition for passing classification survey (documents are valid for 5 years under condition of annual confirmation).
- as a result of an expense for repair and confirmation of classification documents aren't compensated during 4–5 years for vessel's directions and cargoes; so repair and class confirmation became economically inexpedient;
- situation when the further operation of a vessel threatens safety of navigation and it is connected to high risks of accident;
- situation when growth of the operational expenses due to vessel's maintenance (e.g., fuel, oil, spare parts, materials, insurance, etc.), makes her further operation unprofitable.

Accordingly to above mentioned one may conclude that forthcoming fleet discarding will put essential and practically irreplaceable damage to

Table 5. Number, average age and general technical condition of mostly widespread projects of "old" river and river-sea cargo vessels (August 2010).

Type, project, Dw	Number of vessels with RRR class	Average age, years	Number of vessels with estimation "bad" and "suitable limited"
Self propelled dry cargo vessels			
Volzhskiy, 05074, 5100 t	6	21.1	–
Volgo-Don, 1565, 5100 t	41	35.9	2
Volgo-Don, 507B, 5210 t	55	40.0	4
Omskiy, 1743, 3070 t	7	25.5	1
Volgo-Balt, 2–95, 3140 t	7	38.2	1
Kaliningrad, 21–88, 21–89, 2200 t	45	45.3	11
Shestaya pyatilrtka, 576, 2050 t	90	49.2	16
STK, 326, 326.1, 1540 t	11	25.6	–
Okskiy, 559, 559B, 559M, 1740 t	38	40.0	9
Okskiy, P97, 1900 t	14	29.6	1
Self propelled tankers			
Volgoneft, 1577/550A, 4875 t	108	35.4	17
Volgoneft, 558/550, 4900 t	16	42.4	1
Volgoflot, 05074T, 5210 t	9	26.3	–
Volgo-Don, 507B, 1565T, 5210 t	4	42.9	–
Nefterudovoz, 1553, 1570, 3345/2855 t	9	36.8	–
GES, 576T, 1820 t	7	49.8	1
Leneneft, 621, 3390 t	9	24.3	1
Leneneft, P77, 2890 t	40	32.3	6
Non-self propelled dry cargo vessels			
3020, 3040, 4640/4000 t	8	7.7	–
Volzhskiy, 05074, 5100 t	12	24.4	2
16800, 2500 t	100	25.1	66
16801, 2600 t	100	23.4	10
P-56, 2800 t	504	30.3	50
P-79, 3750 t	51	27.5	2
P-85, 2500 t	146	27.3	15
1787, 1787U, 3600 t	89	34.1	16
459, 1700 t	38	43.1	11
1653B, 600 t	42	42.4	16
P–89, 1000 t	89	33.5	17
942, 450 t	647	33.3	65
943	178	33.4	28
81210, 250 t	177	18.7	8
183, 200 t	752	29.9	84
Non-self propelled tankers			
16800H, 3000 t	37	19.5	7
05074H, 4800 t	8	26.5	–
P-27, 4600 t	45	35.2	4
P-43, 9200 t	24	32.9	–
1635T*, 2000 t	10	10.1	–
P-93, 400 t	37	31.7	1
P-63, 200 t	99	28.4	16

*Marine Engineering Bureau project.

Table 6. Main projects of mixed river–sea navigation vessels and marine restricted regions vessels.

	Vessels with RRR class			Vessels with RS class			
Type's name, project	M–SP	M–PR	M–PR	R3–RSN	R2–RSN	R2	R1
New vessels							
Nadezhda, 006RSD02*	–	–	–	–	1	–	–
Karelia, 005RSD03*	–	–	–	–	1	11	–
Caspian Express, 003RSD04*, 003RSD04/ALB02, 003RSD04/ALB03	–	–	–	–	–	9	3
Palmali Trader, 006RSD05*	–	–	–	–	–	8	–
Chelsea, 005RSD06*	–	–	–	–	7	–	–
Tanais, 007RSD07*	–	–	–	–	1	–	–
Azov XL, RSD12*	–	–	–	–	–	4	–
Euro cruiser, RSD17*	–	–	–	–	–	–	5
St. Nikolay, RSD20*	–	–	–	–	–	1	–
Sparta, DCV25*	–	–	–	–	–	–	1
Scala, DCV26*	–	–	–	–	–	–	1
Saxona, DCV27*	–	–	–	–	–	–	1
Rusich, 00101	–	–	–	–	–	–	8
Yuzhniy Bug, 17620	–	–	–	–	–	–	9
Existing vessels							
Onego, 10522, 10523, 10535, 199/200	–	–	–	–	–	–	18
Don, 30205	–	–	–	–	–	–	4
Volga, 19610	–	–	–	–	–	3	33
Volga, 19611	–	–	–	–	–	–	5
Volgo-Don, 507B	3	–	45	11	9	–	–
Volgo-Don, 1565, M1565A	12	5	10	17	11	–	–
Volzhskiy, 05074M, 05074A	4	–	–	14	26	3	–
Sibirskiy, 292	9	–	–	6	3	–	–
Sibirskiy, 0225	1	–	–	–	12	–	–
Volgo-Balt, 791	2	–	–	1	–	–	–
Volgo-Balt, 2–95A	7	–	–	54	9	–	–
Amur, 92–040	–	–	–	1	38	–	–
Sormovskiy, 1557	1	–	–	–	54	1	–
Sormovskiy, 614	–	–	–	–	6	–	–
Sormovskiy, 488A	–	–	–	–	5	20	–
Omskiy, 1743, 1743.1, 1743.3, 1743.7	5	–	–	14	85	4	–
Slavutich, Д 080M	–	–	–	8	2	–	–
Baltiyskiy, 613	–	–	–	–	1	10	–
Baltiyskiy, 620	–	–	–	–	–	1	–
Baltiyskiy, 781	1	–	–	–	5	–	–
Morskoy, 1810, 1814	–	–	–	–	4	–	–
Ladoga, П–787, 285, 289	–	–	–	–	3	14	–
STK, 326, 326.1	3	6	–	–	33	–	–
Refrigerator, 037	–	–	–	–	5	–	–
ST, 19620	–	–	–	–	35	–	–
ST, P168M	–	7	–	–	2	–	–
ST, 191	–	–	–	–	11	–	–
Kishinev, 1572	–	–	–	–	–	–	12
Yakutiya, 1576	–	–	–	–	–	–	3
Vasliy Shukshin, 1588, 15881, 15882	–	–	–	–	–	–	12
Nevskiy, P32	–	19	–	1	–	–	–
Kaliningrad, 21–88	4	8	14	4	2	–	–
Kaliningrad, 21–89	–	4	1	–	–	–	–
Shestaya pyatilrtka, 576	–	1	20	–	–	–	–
Fin.1000	4	7	–	1	–	–	–

* Marine Engineering Bureau project.

Table 7. Main projects of mixed river-sea navigation tankers and marine restricted regions tankers.

	Vessels with RRR class			Vessels with RS class			
Type's name, project	M–SP	M–PR	M–SP	M–PR	M–SP	M–PR	M–SP
New vessels							
Armada, 005RST01*	–				–	10	–
New Armada, RST22*	–	–	–	–	–	7	–
Aston Trader, RST09*	–	–	–	–	3	–	–
Roschem, RST14*	–	–	–	–	3	–	–
Eco Mariner, 001RST02*	–	–	–	1	–	–	–
00201L	–	–	–	–	–	–	2
00210, 00215, 00230	–	–	–	–	–	–	7
19612	–	–	–	–	–	–	5
19614	–	–	–	–	14	–	–
19619	–	–	–	–	–	–	9
17103	–	–	–	–	–	2	–
VHX591	–	–	–	–	1	–	–
Existing vessels							
Volgoneft, 558/550	–	3	10	–	–	–	–
Volgoneft, 1577/550A	31	38	14	5	10	–	–
Lenaneft, 621, 621.1	9	–	–	10	10	–	–
Lenaneft, 630, 630.1	–	–	–	–	7	–	–
Lenaneft, P77	7	27	–	4	–	–	–
Nefterudovoz, 1553, 1570	3	–	–	–	31	–	–
Bunkerovschik, 610	–	1	–	–	1	8	9
Oleg Koshevoy, 1677M, 16773, 16776	–	–	–	–	–	–	24
Volgoflot, 05074T	–	–	9	–	–	–	–
Volgo-Don, 1565, 507B, 507A	–	4	–	–	–	–	–
Irkutsk, GES, 576TM after modernization	–	4	–	–	–	–	–

* Marine Engineering Bureau project.

internal river transportation and mixed river-sea transportation.

The basis of vessels of mixed river-sea navigation and marine restricted regions navigation is made from the projects specified in Tables 6 and 7.

RRR and RS class notations of these vessels are specified there also. All new generation vessels (i.e., built in XXI century) have RS of R2-RSN, R2, R1.

Mass series' vessels built in the USSR are placed into RS and RRR classes as a result of re-classification carried out in 90s.

For example, dry-cargo vessels of "Omskiy" type in the majority has been constructed with RRR class notation "O-PR"; nowadays they have RS class notation of R2-RSN (85 ones) or R3-RSN (14 ones); some of them have RRR class notation "M-SP" (5 ones).

Tankers of "Volgoneft" type in the majority has been constructed with RRR class notation "M-PR"; nowadays they have RS class notation of R2-RSN (10 ones) and R3-RSN (5 ones); or RRR class notation "M-SP" (31 ones). 38 ones stayed at RRR class notation "O-PR" and 14 ones have changed to more weak RRR class notation "O-PR" due to bad technical condition.

The tendency to move vessels with RS classes under RRR supervision is observed also. But there are only few cases and basically they are explained by unsatisfactory condition of these vessels. Shipowners return vessels "to the river" in order neither discard vessel nor repair vessel by required RS capacities.

RSNAV transport problems that are already decided or put again are so crucial, that demand providing of reliable transportations and acceptance of measures on reduction of RSNAV operation risk.

The statistics data evidently show that the existing RSNAV fleet was built basically in 70s–80s years of the previous century.

Operation term for RSNAV was set at designing and was usually of 25–35 years under condition of keeping design restrictions due to regions and seasons of navigation (Egorov & Avtutov 2009).

Table 8. Actual updating of Russian marine restricted regions fleet during period 2002—August, 2010.

Vesuresel's type	Number of vessels	Purpose
Tanker (oil product carrier), prj. 005RST01* "Armada" type, Dw 6500/4700 t (see Figure 2)	10	Export-import transportation
Tanker (oil product carrier), prj. RST22* "New Armada" type, Dw 7000/4600 t	7	Export-import transportation
Tanker (oil product carrier), prj. 19612 "SFAT" type, Dw 8000/4420 t	5	Export-import transportation
Tanker (oil product carrier), prj. 19614 "Nizhniy Novgorod" type, Dw 5600/5100 t	17	Export-import transportation
Tanker (oil product carrier), prj. 0201L "Lukoil" type, Dw 6600/3640 t	10	Export-import transportation
Dry cargo vessel, prj. 005RSD03* "Карелия" type, Dw 5500/3340 t	12	Export-import transportation
Dry cargo vessel, prj. RSD17* "Euro Cruiser" type, Dw 6354 t	5	Export-import transportation
Dry cargo vessel, prj. DCV33* Dw 4570 t	6	Export-import transportation
Dry cargo vessel, prj. 003RSD04* with modifications "Caspian Express" type, Dw 3756/2584 t	12	Export-import transportation
Dry cargo vessel, prj. 006RSD05* "Palmali Trader" type, Dw 6970/4580 t	8	Export-import transportation
Dry cargo vessel, prj. RSD19* "Khazar" type, Dw 7004/4596 t	4	Export-import transportation
Dry cargo vessel, prj. 006RSD02* "Nadezhda" type, Dw 7078/4680 t	1	Export-import transportation
Dry cargo vessel, prj. 007RSD07* "Tanais" type, Dw 7215/4778 t	1	Export-import transportation
Dry cargo vessel, prj. RSD20* "St. Nikolay" type, Dw 6862/4280 t	1	Export-import transportation
Dry cargo vessel, prj. 01010 "Valday" type, Dw 5010/3800 t	4	Export-import transportation
Dry cargo vessel, prj. 00101 "Rusich" type, Dw 5190/3855 t	12	Export-import transportation
Dry cargo vessel, prj. RSD12* "Azov XL" type, Dw 8048 t	4	Export-import transportation
Dry cargo vessel, prj. 005RSD06* with modifications "Chelsea" type, Dw 5827/5080 t	8	Export-import transportation
Oil barges, prj. 004ROB05* Dw 4324/3897 t (see Figure 3)	7	Internal transportations within M-PR class
Oil barges, prj. 2731 with modifications Dw 4500/3700 t	9	Internal transportations within M-PR class
Dry cargo barges, prj. 03020, 03040 Dw 5000/4130 t	7	Internal transportations within M-PR class
Dry cargo barges, prj. 82260 Dw 2000 t	18	Internal transportations within M-PR class
Tugboats, prj. 07521 1400 b.h.p.	2	Internal transportations within M-PR class

*Marine Engineering Bureau project.

Figure 2. New Russian river-sea oil tanker.

Figure 3. New Russian river barge.

Operation conditions For RSNAV that work at the European part of the former USSR have changed hardly during last 10–15 years (essential increase of the time of stay in sea conditions at variable loadings). Consequently, from the point of ageing and fatigue hull resources have spend faster than it was supposed at design. As a result there are unprecedented earlier capacities of construction replacement during repairs and actual reduction of time between vessels' dockings.

Clearly, that the cardinal decision of the problem can be provided only by building new RSNAV and river vessels.

There is sub-program "Internal water transport" included in Federal Purpose Plan "Modernization of Russian transport system (2002–2010)" (FPP). In accordance with this sub-program at stage I (2002–2005) it was supposed to build 127 cargo vessels with total carrying capacity of 367 thousand tons and modernize 35 cargo vessels.

Table 9. Forecast of required number of new vessels of river and mixed river-sea vessels and coasters for the nearest 5–10 years.

Vessel's type	Number	Remarks
Multipurpose dry cargo vessel, Dw 7000–8500 tons	7	FPP*
Multipurpose dry cargo vessel, Dw 7000–8500 tons	16	Private companies
Multipurpose dry cargo vessel, Dw 4000–5000 tons	5	FPP*
Multipurpose dry cargo vessel, Dw 4000–5000 tons	12–20	Private companies
River passenger vessels		
Passenger vessel for 212 persons	5	FPP*
Transport river and mixed river-sea vessels		
Mixed river-sea dry cargo vessel Dw about 7000–8500 tons	20–30	Private companies
Mixed river-sea tanker, "Volgo-Don max" type. Dw about 7000 tons	70	Private companies
Dry cargo vessel of "O-PR" class, "Volgomax" type. Dw about 5400 tons	49	FPP*
Tanker of "O-PR" class, "Volgomax" type. Dw about 5400 tons	36	FPP*
Tanker of "M-SP" class, "Lena" type. Dw about 3400 tons	7	FPP*
Dry cargo vessel of "M-SP" class, "Lena" type. Dw about 3400 tons	7	
River and mixed river-sea tugboats and pushers	20–30	
Dry cargo barges, including for pushing trains with marine couplings	80–100	
Oil barges, including for pushing trains with marine couplings	20–30	
Auxiliary boats for internal water ways		
Dredgers and suction-tube dredges	22	FPP*
Way ones of 0.2–1.0 MWt	250	FPP*
Surveying ones of 0.15–0.30 MWt	32	FPP*
Runabout ones of 0.15–0.45 MWt	32	FPP*
Ecological ones of 0.15–0.30 MWt	16	FPP*
Patrol ones of 0.46 MWt	115	FPP*

*Federal purpose plan "Development of transport system of Russian Federation (2010–2015)" (FPP 2008).

At stage II (2006–2010) it was supposed to build 195 cargo vessels with total carrying capacity of 639 thousand tons and modernize 21 cargo vessels.

170 vessels of mixed river-sea sailing type and marine restricted regions type were built de facto from 2002 till August 2010. Mostly they were designated for export/import transportation (see Table 8). From 10 to 25 vessels were launched for operation annually.

It is interesting to note, that 86 vessels from specified in the Table 8, are built due to projects of the Marine Engineering Bureau.

4 CONCLUSION

Design of new vessel projects for transportation of different cargoes in European part of Russian inland waterways and building of such vessels is strategically important problem in view of Russian export/import relations development and strengthening Russian geopolitical position.

Forecast of required number of vessels is given in the Table 9 (Zakharov & Egorov 2009).

1. Most of all (about 150 vessels) there are required dry cargo vessels and tankers which have to change old "Volgo-Don" and "Volgoneft" type vessels on the lines which are oriented for transportation of raw cargoes from Russian river ports to trans-shipping complexes at Gulf of Finland and Kerch Straits. New vessels have to differ in quality from existing ones which ideology was developed in 50-s of last century in such directions as high efficiency, ecology and reliability. This designs have to "solve" problem moments such transport queues under Neva bridges and under Rostov bridge.
2. Transition to qualitatively new level of transportation organization for water transport is needed. Due to experience of highly developed USA river transport this problem can be solved by more wider usage of pushing tug-barge trains (TBT) either of classical river type (domestic couplings of O-200, UDR-100 and other types) or marine types with couplings of Japanese, Finnish or American types. Thus it is necessary to understand clearly, that simple creation of TBT project isn't enough.

   Effective TBT operation demands radical change of transportation organization (i.e., using of so-called "rotator" when on one tug-pusher serves 2–3 barges). Small draught TBT usage is especially interesting to Russian East where appreciable degradation of way conditions is observed last years.
3. Russian shipbuilding yards (Krasnoe Sormovo, Onega Yard, Oka Yard, Neevsk Shipbuilding Yard, Nobel Brothers Yard, Volgograd Yard, Zelenodolsk Yard, Kama Yard) have experience of building such vessels at this century. Their present abilities let to produce annually not bigger than 25 (35 in perspective) of mostly demanded "Volgo-Don Max" type vessels.
4. Same time requirements for 2015–2020 periods are of about 350 transport vessels and more than 400 auxiliary vessels for Russian water transport branch.

The main conclusion: for building of necessary quantity of river and river-sea vessels attraction not only Russian, but also foreign shipyards is required.

REFERENCES

Egorov, G.V. 2008. Substantiation of necessity of mixed river-sea navigation vessels keeping // *Herald of the Odesa National Maritime Univ*. Odesa, ONMU. Vol. 25. 5–26.

Egorov, G.V. 2010. Transportation by national water transport, river fleet conditions and new shipbuilding prospect // *Marine Exchange*. Vol. 4 (34). 20–26.

Egorov, G.V. & Avtutov, N.V. 2009. Prospect of existing mixed river-sea navigation vessels // *Transactions of CSII named by acad. A.N.Krylov*. Vol. 46 (330). 17–28.

FPP. 2008. Federal purpose plan "Development of transport system of Russian Federation (2010–2015)". *The ministry of transport of the Russian Federation*.

Zakharov, I.E. & Egorov, G.V. 2009. Estimation of Russian neediness for new vessels // *Marine Fleet*. Vol. 2. 42–49.

Sustainable Maritime Transportation and Exploitation of Sea Resources – Rizzuto & Guedes Soares (eds)
 ISBN 978-0-415-62081-9

Shallow draught ice-breaking river tug boat design

Gennadiy V. Egorov & Nickolay V. Avtutov
Marine Engineering Bureau, Odessa, Ukraine

ABSTRACT: Existing fleet of Russian river tug boats and pusher tugs continue steaadily grow old. Though necessity of the new tug boats is very high in the river ports due to extremely decrepitude of existing fleet.

In the most of Russian river ports the auxiliary fleet operation is accomplished with a lot of problems; main ones are: long period of winter stay with significant low outdoor air temperature (less than 50ºC); freezing into ice of vessel's hull and screw-rudder units on full-draught depth; every year ice drifting that is cause of significant damages or even loss of vessels or port facilities.

Main principles of river tug boat design for Russia are as follows: operation must be foreseen for low air temperature; construction and equipment of the screw-rudder unit must allow operation of tug boat with low negative outdoor temperatures, full freezing of hull into the ice; tug boat hull must have enforced construction for operation in ice condition; tug boat weight has to be not over than 100 t—due to necessity of vessel' lift on the berth for winter stay; maximal draught should not exceed 1.80 m (due to depth of places of winter stay and operation ways depths).

1 INTRODUCTION

In the sector of river and mixed river-sea transportation shipowners actively invest finances into building of self propelled dry cargo and oil transport vessels as well as non-self propelled barges that provide direct cargo transportation and thus "earn" profit.

Existing tug and tug-pusher river fleet continues becoming out of date steadily. There were 55 vessels with the age up to 10 years, 110 vessels with the age from 10 to 15 years, 972 vessels with the age from 15 to 20 years, 2517 vessels with the age from 20 to 30 years and 3119 vessels with the age more than 30 years with the Russian River Register (RRR) class at the beginning of 2008. Unfortunately for the nearest time there are no plans due to significant investments for buying new representatives of auxiliary fleet.

Though in river ports neediness of new tugs are enough high due to significant dilapidation of existing fleet. One of such ports is river port of Dudinka that is included into Arctic branch of OJS "Mining & Smelting Complex "Norilskiy Nickel" (Egorov 2008).

Design of shallow draught river tugs was described in publications of 70–80-s of last century (Gorbunov *& all* 1990, Grinbaum 1980, Grinbaum *& all* 1980, Lesyukov 1975, Pavlenko *& all* 1985, Zaitsev 1972). Since that time vessels of such type weren't designed.

In the process of new generation of tugs creation the greatest interest is lines optimization on the basis of numerical modeling (Egorov *& all* 2007); usage of present-day technique in the field of weight-dimension characteristics diesel engines that let to increase tug power while keeping tug dimensions; decrease available operational risks by involving rational design of hull constructions (Egorov 2007).

2 AIM OF THE PAPER

Justification of Definition characteristics of river tug that is effective for shallow water and river ice conditions. Prj. TG04 tug designed by Marine Engineering Bureau for the port of Dudinka was taken as an example.

3 MAIN MATERIAL STATEMENT

Port of Dudinka is situated on the right shore of Yenisei River ate distance of 230 km from river's mouth (see Figure 1). This is departmental port that was build in order to provide needs of Norilsk Mining & Smelting Complex.

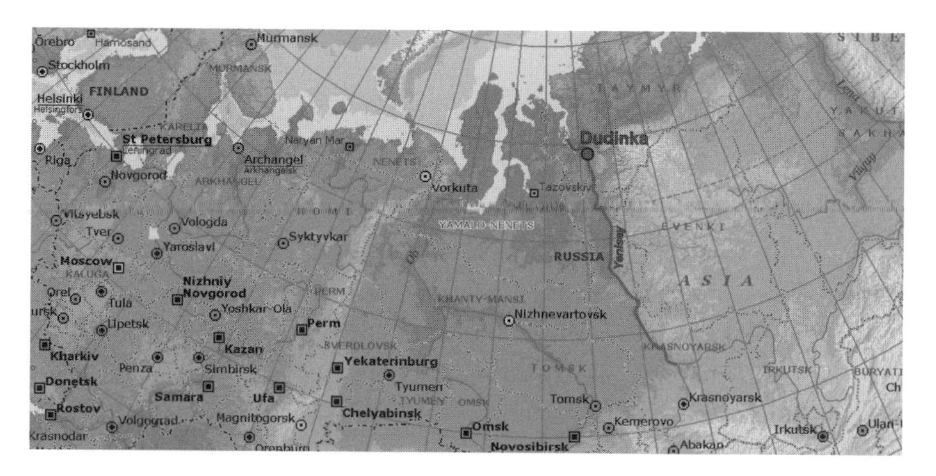

Figure 1. Position of port Dudinka.

Auxiliary fleet operation in the port of Dudinka is connected with a lot of problems; main of them as follows:

- short navigational period (from May till October), so efficient fault-free vessels' work is required;
- long period of the winter stay with extremely low outer air temperature (below –50°C) and freezing into ice for deepness of full draught (including propulsion complex);
- annual ice-drift that may be accompanied with significant damages or even perish of vessels and port facilities (see Figure 2).

Thereafter river tug for severe Arctic conditions has to fit requirements as follows:

- tug has to be design for operation at extremely low outer air temperature;
- equipment and construction of the propulsion complex have to permit to put tug into operation at extremely low temperature from condition of full into-ice freezing;
- tug's hull should be of reinforced type for operation in the conditions of freeze basin;
- tug's weight should not exceed 100 t for providing ability to raise tug onto berth for the winter stay (crane capacity restriction);
- tug's maximal draught should not exceed 1.80 m (due to depth of the boatyard where fleet is placed for the winter stay and due to way conditions of the Dudinka River).

Tug must fulfill works due to providing port fleet safety during ice shifting, so her overall dimensions should me of minimal type. Tug's pollard pull should be not less than 6 t; that provides ability to service all fleet owned by the Arctic branch of Norilsk Complex.

Figure 2. Annual ice-drift.

Table 1. Comparative characteristics of TG04 tug and her prototypes.

Denomination	TG04 prj.	1427 prj.	P14A prj.
Length overall, m	20.45		31.50
Length due waterline L, m	18.50	18.20	30.40
Breadth due waterline on the middle B, m	6.00	4.20	6.60
Breadth overall, m	6.56	4.40	6.80
Depth on the middle H, m	2.40	2.56	1.80
Draught due waterline d, м	1.80	1.43	1.08
Maximal continuous ME power N, kWt	2×221	2×110	2×166
Pull effort, т	6.50		4.15
Gross Tonnage, reg. t.	84.81		171
Crew	6		11
Type and number of propulsion devices	2 FPP	2 FPP	2 FPP

Main tug characteristics and characteristics of some analog vessels (including tug "Tayezhniy" of prj. 1427 that was substituted by TG04 one) are shown in the Table 1.

Beforehand setting of tug's light weight (100 t) and maximal draught (d = 1.80 m) determines main dimension of the vessel (L × B × H = 18,5 6 × 2,40 m).

Tug's architect-constructive type was selected as classic type for port tugs, notably single deck double screw vessel with upper deck fore recess, with wheel house and ER located middle, with ice-breaking stem (see Figure 3).

Sailing regions are basins and mouth reach of the rivers with sea navigation regime of "O" class. The vessel can be operated at sea roughness with 1% probability wave height no more than 2.0 m.

Registry class is determined in accordance with planned sailing region. Class notation is «✠ O 2,0 (ice 30)» of Russian River Registry (RRR). Ice category is assigned higher that recommended

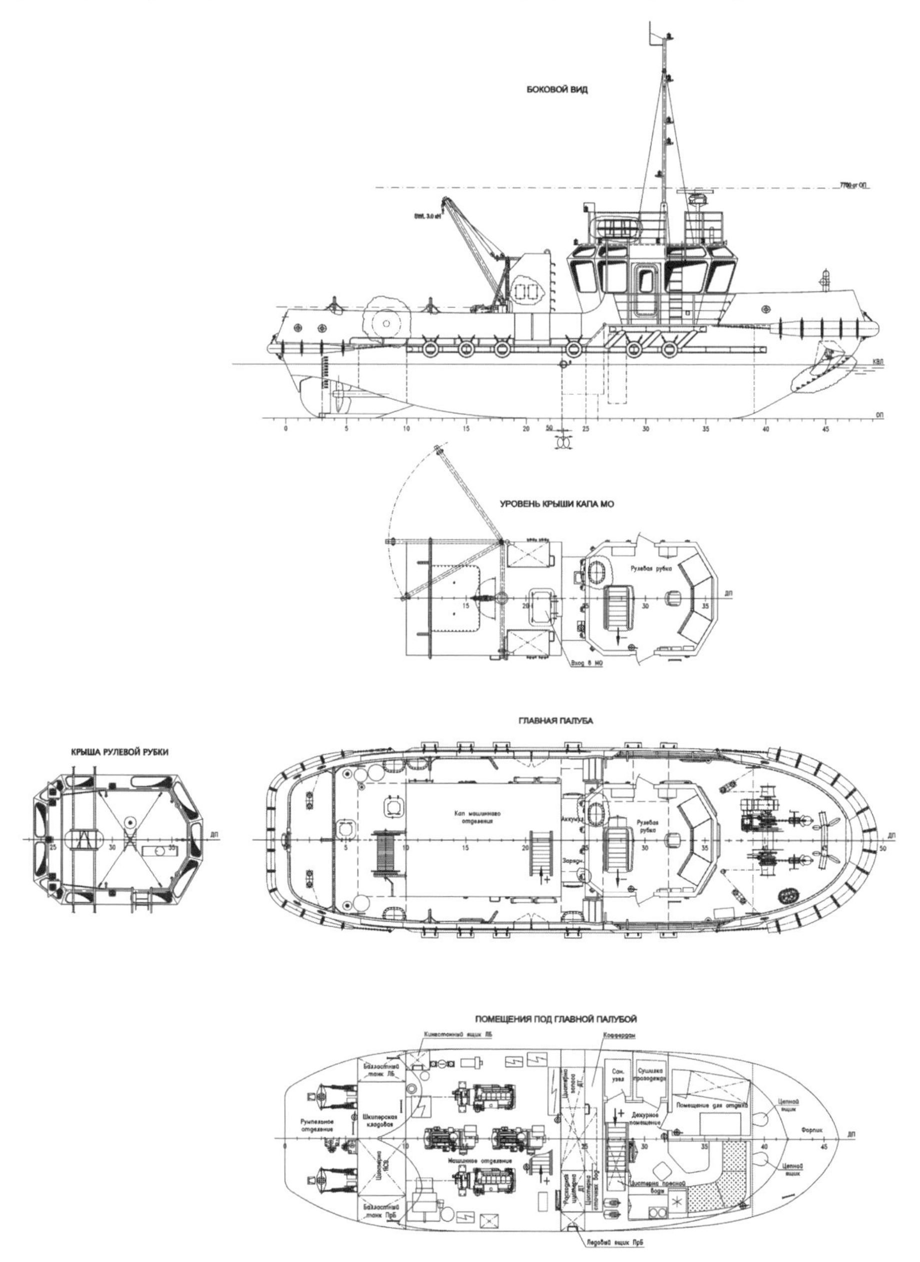

Figure 3. Shallow draught ice-breaking river tug boat general arrangement.

by RRR Rules for vessels of «✠ O 2,0» type due to actual operational conditions at the port of Dudinka.

Tug hull's shapes were determined with help of CFD-modeling (see lines in the Figure 4 and trial in Figure 5).

Same time choice of revolutions and hydrodynamic characteristics of screw propellers (SP) was carried out in order to achieve bollard pull not less than 6.0 t a the project mode.

The project mode means SP work at mooring regime (zero speed) with nominal revolutions and full (100%) loading of main engines (ME).

Initial data for the screw:

- diameter of opened screws $D_p = 1.200$ m;
- number of screws $x = 2$;
- number and power of Main Engines 2×221 kWt.

Screw propellers are places in the light semi-tunnels of the hull in accordance with estimated ice conditions and shallow draught. At the calculated conditions SP are fully immerged, accordingly $D_p/T = 0.667$.

Calculation of optimal revolutions of SP is made in the Table 2.

Revolutions value was varied in the limits 350–420 rpm with the step of 10 rpm during calculation. For each revolutions value pitch ratio was fitted in order to provide 100% load of ME. Correlation between geometric and hydrodynamic characteristics of SP is accepted in accordance with test of 4-blade propellers of "B" series at the Netherlands towing tank.

Blade-area ratio A_E/A_O is accepted of 0.650 on the basis of minimal value $A_E/A_{O\min}$ determination for condition when hard cavitation is absent for each revolutions value. $A_E/A_{O\min}$ value determination is carried out with help of well known Keller formula (1) that sets relations of blade-area with SP pull effort and work conditions.

$$A_E / A_{O\min} = \frac{0.05 + (1.3 + 0.3z) \cdot T_p}{D_p^2 (p_0 + \rho g h - p_v)}, \text{ where:} \quad (1)$$

z—SP blades number,

T_p—SP pulling effort,

$p_0 + \rho gh$—static pressure onto SP axis,

p_v—saturated vapor pressure onto SP axis.

$p_0 - p_v = 99047$ Pa is accepted for calculations, that corresponds water conditions at temperature of 15°C.

ME load due to effective power N_E (Table 2, lines 8 and 9) is calculated accounting efficiency factor of 0.95.

SP bollard pull during mooring regime T_E (Table 2, line 10) is calculated accounting suction; suction coefficient for mooring regime t_{P0} is accepted of 0.100 preliminary.

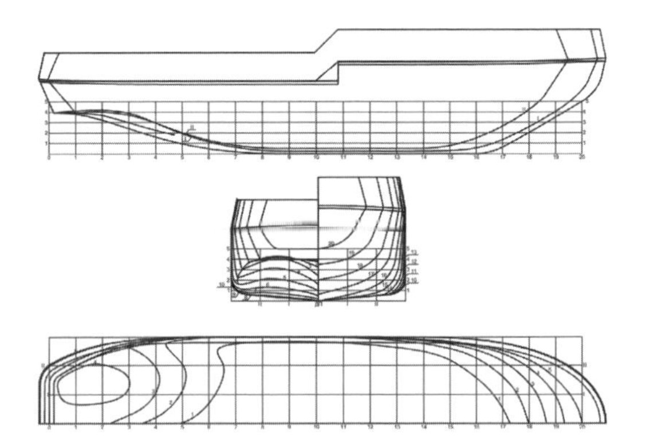

Figure 4. Lines of shallow draught river tug.

Figure 5. Shallow draught ice-breaking river tug boat in trial.

Approximate bollard pull during mooring regime $T_{E\Sigma}$ is given in line 11 of Table 2.

Reduction ratio $i_r = 4{,}409{:}1$ of reduction gears was determined while taking into consideration data from Table 2. This value corresponds to the nominal SP revolutions of 408 rpm.

Table 2. Optimal SP revolutions calculation.

1	n, rpm (preset)	350	360	370	380	390	400	410	420
2	AE/AOmin	0.609	0.615	0.620	0.625	0.630	0.633	0.637	0.640
3	P/DP	1.016	0.973	0.933	0.895	0.860	0.828	0.797	0.768
4	AE/AO	0.650	0.650	0.650	0.650	0.650	0.650	0.650	0.650
5	KT	0.4522	0.4325	0.4137	0.3959	0.3790	0.3629	0.3477	0.3332
6	KQ	0.0677	0.0622	0.0573	0.0529	0.0489	0.0453	0.0421	0.0392
7	TP, kN	0.4522	0.4325	0.4137	0.3959	0.3790	0.3629	0.3477	0.3332
8	NE, kWt	221.0	221.0	221.0	221.0	221.0	221.0	221.0	221.0
9	NE, b.h.p.	300.6	300.6	300.6	300.6	300.6	300.6	300.6	300.6
10	TE, t	2.928	2.962	2.993	3.021	3.046	3.069	3.089	3.106
11	TEΣ, t	5.855	5.924	5.986	6.042	6.092	6.137	6.177	6.213

Bollard pull distribution during mooring regime in accordance with Table 2 data is given in the Figure 2; determined nominal SP revolutions is marked.

One may see in the Figure 6 that pull curve has no maximum; pull increases while RPM increases. Pull curve approaches to some constant value asymptotically.

So when determining optimal RPM one should try to increase it in the limits set by SP cavitation, SP construction and providing of suitable free run abilities of tug. Taking into consideration abovementioned the determination of reduction ratio looks reasonable enough in our case.

SP characteristics of TG04 tug is accepted as follows accounting determined nominal RPM (on the basis of "B" series of SP):

Diameter $D_p = 1.299$ m;
Pitch ratio $P/D_p = 0.803$;
Blade-area ratio $A_E/A_O = 0.650$;
Blade number $z = 4$.

For more detail determination of pull characteristics of the tug it's necessary to take account of interaction between SP and hull.

Wake factor for SP in tunnels for examined vessel's type can be determined by E.E. Pupmel formula:

$$w = 0.11 + \frac{0.16}{x} C_B{}^x \sqrt{\frac{\sqrt[3]{\nabla}}{T}} = 0.247, \text{ where:} \tag{2}$$

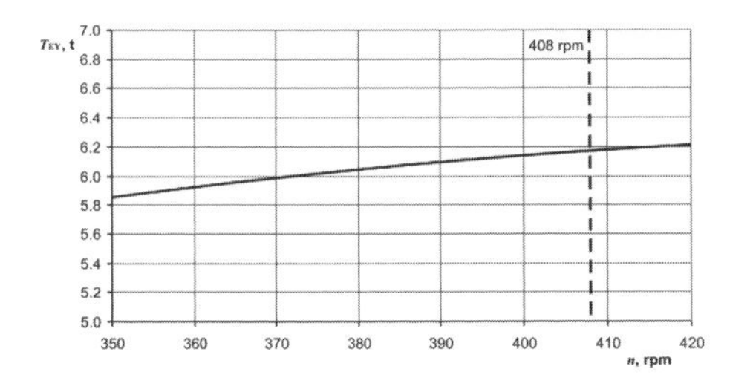

Figure 6. Pull curve.

$x = 2$—SP number;
$C_B = 0.603$—block coefficient;
$\nabla = 120.4$ cub.m—volumetric displacement;
$T = 1.80$ m—draught.

Wake factor is accepted as constant value for various velocities and ME regimes.

Suction coefficient is approximately equal to wake factor for fully immersed SP located in tunnels (in accordance with recommendations (Basin & *all* (1961))). Assume $t_p = w = 0.247$ for tug free run at the speed of 10 kn.

Suction coefficient dependence from ME work regime is set by E.E. Papmel by the next formula:

$$t_p = \frac{t_0}{1 - \dfrac{J}{P_1/D_p}}, \text{ where:} \tag{3}$$

J—SP advanced ratio;

T_1/D_p—pitch ratio for zero thrust; for our case $T_1/D_p = 0.866$.

Using (3) one can defines suction coefficient for mooring regime t_{P0} by inserting known values $t_P = 0.247$, $P_1/D_p = 0.866$ and SP advanced ratio $J = 0.475$ (for speed of 10 kn and nominal RPM). Suction coefficient for mooring regime is of $t_{P0} = 0.110$.

Tug pull capacity data was defined using abovementioned SP characteristics and "hull-SP" interaction coefficients. Corresponding calculations were made for speed range of 0 (mooring regime) to 10 kn and RPM range from 310 to 408. This data is shown in the Table 3 and Figure 7.

For achieving free running speed or speed during towage on the basis of Table 3 or Figure 7 it's necessary to set resistance values for tug and towing object taking into consideration lashing coefficient (depending on the speed value). ME load rises to the maximal capacity (100%) during mooring regime.

Due to calculations bollard pull during mooring regime reaches value of 6.1 t at the nominal ME revolutions.

Table 3. Pulling effort $T_{E\Sigma}$ (t) for different regimes.

n, rpm.	V_S, kn							
	0.00	1.43	2.86	4.29	5.71	7.14	8.57	10.00
408	6.10	5.71	5.28	4.82	4.33	3.80	3.25	2.68
390	5.57	5.20	4.79	4.34	3.87	3.36	2.83	2.27
370	5.02	4.66	4.27	3.84	3.38	2.90	2.39	1.85
350	4.49	4.15	3.78	3.37	2.93	2.47	1.97	1.46
330	3.99	3.67	3.32	2.93	2.51	2.06	1.59	1.10
310	3.52	3.22	2.88	2.51	2.11	1.69	1.24	0.77

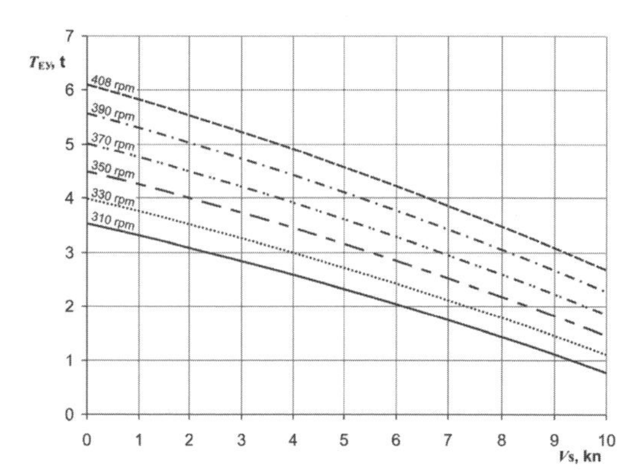

Figure 7. Pulling effort $T_{E\Sigma}$ for different regimes.

Tug's movement is provided by two open cast steel 4-blades fixed pith propellers. Screw propeller's diameter is of 1200 mm, pith ratio is of 0.793, blade-area ratio is of 0.65. Propellers drive from two ME (221 kWt each) is of mechanical type trough reverse gear. Ability of ME work with load of 110% during 1 hour is foreseen.

With the design draught of 1.8 and ME load of 90% tug's speed reaches 10.2 kn (see Figure 8).

Maneuvering characteristics of the tug are provided by two balanced (with double bearings) rudders arranged by sides. Rudder blades are made of streamlined type. Ice "claw" is foreseen for each rudder blade protection against ice operation damages.

For rudder blade turning two independent electro-hydraulic steering engines are installed in the steering gear room. They provide simultaneously (synchronic) and independent rudders putting over to any side. The steering gear ensures putting rudder over from 35° on one side to 35° on the other side at maximum ahead speed within not more than 22 sec.

Tug's docking weight is of 88 t about. Design tug's hull life period of 15 years was accepted in order to achieve such weight.

Vessel's hull is made from D category steel. Hull is set up by transverse framing system; frame spacing is of 600 mm. Floors are located on each frame. Thickness of outer shell, deck and deckhouse bulkheads (from 4 till 9 mm) is accepted due to strength providing and RRR requirements.

Figure 8. Maneuvering in shipyard basin.

Main watertight transverse bulkheads are placed at frs. 6, 10, 23, 29, so divide hull into 5 independent compartments. Bulkheads are of flat type with thickness of 4 or 5 mm. Bulkhead's girders are made from non-symmetrical bulb iron No. 8 and from welded T-profile (4 × 150/8 × 80 mm).

Vessel's stem is of ice-breaking type. It is made from plate with section 30 × 125 mm that is enforced by transverse girders and welded T-profile s6/8 × 80 mm. Side girders is made from non-symmetrical bulb iron No. 8 (frames) and from welded T-profile 5 × 250/8 × 80 mm (web frames and stringers). Floors and bottom stringers are made from welded T-profile with beam web of 5 mm and belt of 8 × 80 mm outside ER and with beam web of 6 mm within ER. Deck girders are made from non-symmetrical bulb iron No. 8 (beams) and from welded T-profile 4 × 200/8 × 80 mm (web beams and carlings). Thickened plates and corresponding enforcements are installed at the places where anchor-mooring and towing equipment is arranged.

Wheel house is arranged on the elevated fore part of upper deck. At this end electrical anchor-mooring-towing winch (pull of 75 kN) and towing bitt that provide ability for different objects canting.

Basic pulling effort of tow winch is of 50 kN for the first layer of drum, maximum pulling effort is of 75 kN for the first layer. Rope capacity of tow winch drum is of 100 m of synthetic rope with diameter of 48 mm.

Vessel's fuel autonomy is of 3.3 days, water autonomy is of 4.8 days. Heavy fuel stores are placed in tanks arranged in deep-tank in area of the ER fore bulkhead. Fuel tanks have no contact with the outside water due to RRR Rules.

Independent fresh water tank is arranged in the middle part of the hull.

Battery box and charging compartment are located in the middle part of the deck.

Engine room with 900 mm height cape, boatswain's locker and steering gear room are located in the aft end of the vessel.

Towing hook (working load about 70 kN) with automatic and remote release is installed onto ER cape for towing operations providing. both synthetic or steel towing rope can be used when working with towing hook.

Boxed band made of 10 mm plate is mounted onto the sides in order to avoid hull damages during towing and canting operations. Additionally cylindrical rubber fender is mounted fore and aft for the same purpose.

Auxiliary power plant consists of two diesel generators of 62 kWt each. Auxiliary power plant provides energy for the vessel and is able to forward energy to the shore or another floating object.

It's foreseen that the vessel can be withdrawn from operation by her freezing in ice for 6 months without ER and everyday quarters warming-up.

Putting into operation after winter staying is carried out with negative outdoor air temperature (down to −20°C) and with positive temperature (provided by electrical air heater) in ER.

Keel of tug-boat "Portoviy 1" (building number is of 701) was laid down on 12.09.08. Tug was launched 27.08.09 and put into operation 17.09.08 (see Figure 9).

Figure 9. Shallow draught ice-breaking river tug boat in Volga river.

Figure 10. Shallow draught ice-breaking river tug boat in Siberia.

Tests carried out on the Volga River in the September, 2009 and first operation 2009–2010 confirmed accepted tug design decisions fully.

4 CONCLUSION

Main characteristics of new river tug boat for the port of Dudinka were reviewed, as follows:

- ability of maneuvering works with river—going vessels and floating cranes;
- perform of auxiliary works such as mooring and in-port operation with barges and vessels;
- vessels' placing in backwater during spring flood;
- ice-breaking works in the port of Dudinka during autumn period;
- placing of non-self propelled fleet for winter stand (placing on a distance from berth in autumn ice with keeping vessel until she freeze in).

The auxiliary fleet of "Mining & Smelting Complex "Norilskiy Nickel" now has the tugboat "Portoviy 1" (TG04 prj.) that is substituted the tug-boat of 1427 prj. (built 1969) it the port of Dudinka (see Figure 10).

It is planned to make decision to build the next vessels of the series TG04 based on results of tug operation in 2010.

REFERENCES

Basin, A.M. & Anfimov, V.N. 1961. Ship's hydrodimamic. L. *River Transport*.

Egorov, G.V. 2007. Design of ships of restricted navigation area based on risk theory. St. Petersburg. *Shipbuiding*.

Egorov, G.V. 2008. Estimation of Russian Federation neediness of new vessels of different types. Perspectives of native shipbuilding // *Shipbuiding and Shiprepair*. Issue 3(29). 42–49.

Egorov, G.V., Stankov, B.N. & Pechenyuk, A.V. 2007. An experience of CFD modeling usage for vessel's propulsion unit design // *Transaction of National Shipbuilding University*. Nikolaev. NSU. Issue 2. 3–11.

Gorbunov, Yu.V., Lyubimov, V.I. & Gamzin, B.P. 1990. Vessels for small rivers. M. *Transport*.

Grinbaum, A.F. 1980. A method to define main characteristics of tug boat // *Shipbuilding*. Issue 12. 5–8.

Grinbaum, A.F., Lobastov, V.P. & Sergeev, I.V. 1987. Tugs, tug pushers and barges for Siberia // *Shipbuilding*. Issue 9. 6–10.

Lesyukov, V.A. 1975. Optimizing of characteristics of inland waterways tugs and tug pushers // *Shipbuiding*. Issue 8. 8–10.

Pavlenko, V.G., Sakhnovskiy, B.M. & Vrublevskaya, V.M. 1985. Cargo transport means for small rivers. L. *Shipbuiding*.

Zaitsev, I.A. 1972. Power stations of tug boats. L. *Shipbuiding*.

Sustainable Maritime Transportation and Exploitation of Sea Resources – Rizzuto & Guedes Soares (eds)
© 2012 Taylor & Francis Group, London, ISBN 978-0-415-62081-9

Feasibility study for adaptation of ITS Etna to MARPOL—Annex I

A. Commander & A. Menna
Surface Ship Design Office, Ship Design and Material Department, Italian Navy General Staff, Rome, Italy

B. Lieutenant & A. Grimaldi
Aux. Systems Office, Ship Design and Material Department, Italian Navy General Staff, Rome, Italy

ABSTRACT: The international agreement MARPOL states, in Annex I (concerning regulations for prevention of pollution by oil), specific requirements for the cargo area of oil tankers. Italian Navy, although MARPOL is not to be applied to war ships, due to its "green policy", has started a lot of studies in order to modify its AOR ships (Auxiliary Oil Replenishers) according to this International threat. This study is about ITS Etna (a single hull AOR, delivered in 1998, with a long-estimated residual operative life) and is divided in two main bodies: in the first one the ITS Etna configuration has been analyzed to verify if the specific requirements stated for oil tankers are satisfied while in the second part a number of solutions have been considered.

1 GENERAL

1.1 *Legal framework*

The international community has been interested in protecting marine and coastal environment since a long time. First international rules, set around early '50. were about different topics: The London Convention (Oilpol 54), for example, concerned the prevention from marine pollution by oil; the Moscow Treaty in 1963 (Partial Test Ban Treaty) stated the prohibition of any nuclear experiment in the marine environment, while the two Conventions of Brussels (Intervention 69 and CLC 1969) were about the intervention in case of oil pollution casualties and about civil responsibility in case of damages due to marine pollution by oil.

The most important convention, regarding marine environment protection, is the International Convention for the Prevention of Pollution from Ships (better known as Marpol), edited in 1973 and modified by the Protocol of 1978 and next amendments. The original Convention, which was signed on 17 February 1973 and entered into force on 2 October 1983, was originally composed of five Annexes concerning the prevention of different forms of marine pollution from ships. In particular:

- Annex I: prevention of pollution by Oil;
- Annex II: prevention of pollution by noxious liquid substances carried in bulk;
- Annex III: prevention of pollution by harmful substances carried in package form;
- Annex IV: prevention of pollution by sewage;
- Annex V: prevention of pollution by garbage.

In 1997 a new Annex, containing regulations for the prevention of air pollution, was added (Annex VI). It entered into force in 2005: since 31 December of the same year, 136 countries, representing about 98% of the world's shipping tonnage, are parties to the Convention. Although the Convention is not to be applied to War Ships (as stated in the article 3) Italian Navy has started, since the late '90, a specific program in order to make its fleet compliant to Marpol rules. In the next paragraphs the requirements stated in the Annex I for Oil tankers will be analyzed.

1.2 *Classification of oil tankers according to Marpol—Annex 1*

According to Regulation 1 of the Annex I of Marpol, modified by later amendments, Oil Tankers are classified depending on:

- Type of cargo;
- Deadweight;
- Date of delivery of the ship.

Referring to the cargo type, a tanker can be classified as an Oil tanker, a Crude Oil tanker, a Product carrier or a Combination carrier as defined by the Regulation 1. Referring to deadweight, defined as "the difference in tones between the displacement of a ship in water of a relative density of 1.025 at the load waterline corresponding to the

Table 1. ITS Etna main features.

Displacement	13,309 t (full load)
Length overall	146.6 m
Beam	21.0 m
Draft	7.25 m
Propulsion system	2 Diesel Engines 12-ZAV-40 S 2 Shafts with variable pitch propellers
Power	17,290 kW
Range (cruise speed)	7600 nm

Table 2. Loading components at full load displacement.

Ballast water	20 t
Fuel oil	760 t
Fresh water	160 t
Kerosene	26 t
Cargo fuel oil	4900 t
Cargo kerosene	1200 t
Lubricating oil	60 t
Sludge	30 t
Crew	40 t
Food	20 t
Ammunitions	98 t
Cargo food	30 t
Spare parts	20 t
Oil in drums	30 t
Liquids in circulation	70 t
Torpedo	125 t
General equipment	15 t
Total (Deadweight)	7604 t

assignment summer freeboard and the lightweight of the ship", the following limits are stated:

- 600 tons and above;
- 5000 tons and above;
- 20,000 tons and above;
- 30,000 tons and above;
- 40,000 tons and above;
- 70,000 tons and above.

Finally, referring to the date of delivery of the ship, Regulation 1 states the following temporal limits:

- Ships delivered on or before 31 December 1979;
- Ships delivered after 31 December 1979;
- Ships delivered on or before 1 June 1982;
- Ships delivered after 1 June 1982;
- Ships delivered before 6 July 1996;
- Ships delivered on or after 6 July 1996;
- Ships delivered on or after 1 February 2002;
- Ships delivered on or after 1 January 2010;
- Ships delivered on or after 1 August 2010.

Figure 1. ITS Etna.

The same Regulation gives more details about each temporal limits, considering the date of building contracts or the beginning of construction works.

1.3 *Classification ITS Etna according to Marpol—Annex 1*

ITS Etna is, according to the military classification, a single hull AOR (Auxiliary Oil Replenisher), built by FINCANTIERI in Riva Trigoso dockyards, and delivered to Italian Navy on 29 July 1998.

The main features of the ship are reported in Table 1.

The full load displacement, at mission start, is 13,309 t while the lightweight is 5705 t. According to the prescriptions reported in the Regulation 1, Deadweight is:

$$DW = 13{,}309 - 5705 = 7604t \tag{1}$$

Loading components, at the full load displacement condition (mission start), are reported in Table 2.

Therefore, on the base of prescriptions of Annex 1, ITS Etna can be classified as an Oil tanker, delivered after 6 July 1996, with a deadweight greater then 5000 t.

2 REQUIREMENTS VERIFICATION

2.1 *Requirements for oil tankers delivered after 6 July 1996 applied to ITS Etna in present configuration*

Marpol states, in Regulation 19 of Annex I, double hull and double bottom requirements for all those oil tankers that, as ITS Etna, have been delivered after 6 July 1996. According to this regulation, for this kind of ships, the entire cargo tank length is to be protected by wing tanks and double bottom unless the criteria reported in paragraph 4 and 5 of the same regulation, are satisfied. Paragraph 4 states that the following condition has to be satisfied for each cargo tank (hydrostatic method):

$$f \cdot h_c \cdot \rho_c \cdot g + p < d_n \cdot \rho_s \cdot g \qquad (2)$$

where f = safety factor = 1.1; h_c = height of cargo in contact with the bottom shell plating in m; ρ_c = maximum cargo density in kg/m^3; d_n = minimum operating draught under any expected loading condition in m; ρ_s = density of sea water in kg/m^3; p = maximum set pressure above atmospheric pressure (gauge pressure) of pressure/vacuum valve provided for the cargo tank in Pa; and g = standard acceleration of gravity = 9.81 m/s^2.

This formula has been checked, for every cargo tank, at all loading conditions of the ship. Results are reported in Table 3.

It's possible to say that, as reported in Table 3, ITS Etna doesn't comply with paragraph 4 of Regulation 19. Paragraph 5 recalls the IMO Resolution MEPC 110 (49) containing guidelines for the approval of alternative methods of design and construction of oil tankers under Regulation 19. This Resolution is based on the philosophy of comparing oil outflow performance in case of collision or stranding of a single hull tanker to that of reference double hull tanker trough the calculation of the so-called pollution-prevention index E given by the formula reported below:

$$E = k_1 \frac{P_0}{P_{0R}} + k_2 \frac{0.01 + O_{MR}}{0.01 + O_M} + k_3 \frac{0.025 + O_{ER}}{0.025 + O_E} \geq 1 \qquad (3)$$

where k_1, k_2 and k_3 are weighting factors; P_O = probability of zero oil outflow for the alternative design; O_M = mean oil outflow parameter for the alternative design; O_E = extreme oil outflow parameter for the alternative design; P_{OR}, O_{MR} and O_{ER} are the corresponding for the reference double hull design of the same cargo oil capacity. Probability of zero outflow parameter (P_0) represents the "probability that no cargo oil will escape in case of

Table 3. Verification of the hydrostatic method at the different loading conditions.

Overload (mission start)	not compliant
Overload (return to port)	not compliant
Full load (mission start)	not compliant
Full load (return to port)	not compliant
Average operative displacement	compliant
On ballast (mission start)	compliant
On ballast (return to port)	compliant
Ready to replenish	not compliant
During replenishment	not compliant
End of replenishment	not compliant

Figure 2. ITS Etna during replenishment operations with French carrier charles de Gaulle.

collision or stranding". Its value may vary from 0 (0%) to 1 (100%). Mean oil outflow is "the sum of all outflow volumes multiplied by their respective probabilities". The corresponding parameter O_M is expressed as "a fraction of the total cargo capacity at 98% of tank filling". At last, Extreme oil outflow is the sum of the outflow volumes characterized by a probability between 0.9 and 1, multiplied by their respective probability. The calculated value is finally multiplied by 10. As previously, the corresponding extreme oil outflow parameter O_E is expressed as "a fraction of the total cargo capacity at 98% of tank filling" The oil outflow parameters, defined as above, have to be calculated, following the probabilistic method, for collision and stranding separately and then combined. Oil outflow parameters for stranding should be calculated referring to 0.0 m tide condition and 2.5 m tide condition and then combined. The combination of the parameters is performed following the instructions contained in the paragraph 5. The damage assumptions for the next analysis of oil outflow, are given by MEPC 110 (49) in terms of density distribution. Values of oil outflow parameters P_{OR}, O_{MR} and O_{ER} are given on deadweight as reported in Table 4.

The probabilistic method above described, has been applied to ITS Etna hypothesizing the lateral and bottom damages displayed in Figure 3 and Figure 4.

Results concerning the bottom damage for 0.0 m tide condition and 2.5 m tide fall condition are reported in Table 5.

The combination of the parameters calculated for side and bottom damage are reported in Table 6.

The pollution-prevention index has been finally calculated, by using (3), referring to the values of P_{OR}, O_{MR} and O_{ER} valid both for ships of 5000 t deadweight and ships of 60,000 t deadweight (as ITS Etna is about 7600 t of deadweight).

As reported in Table 7, in both cases ITS doesn't comply with paragraph 5 of Regulation 19 ($E < 1$).

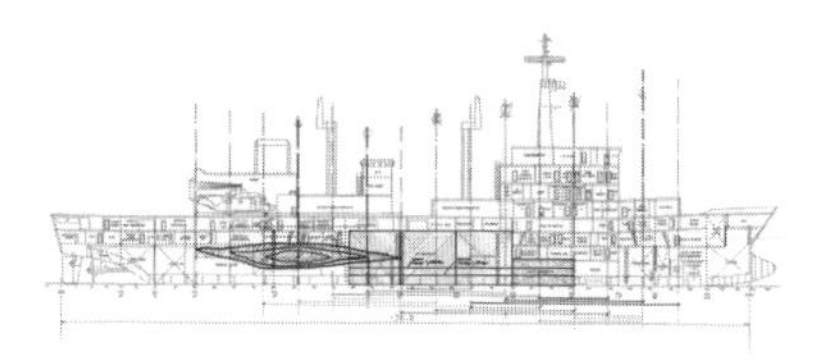

Figure 3. Hypothesis of lateral damage.

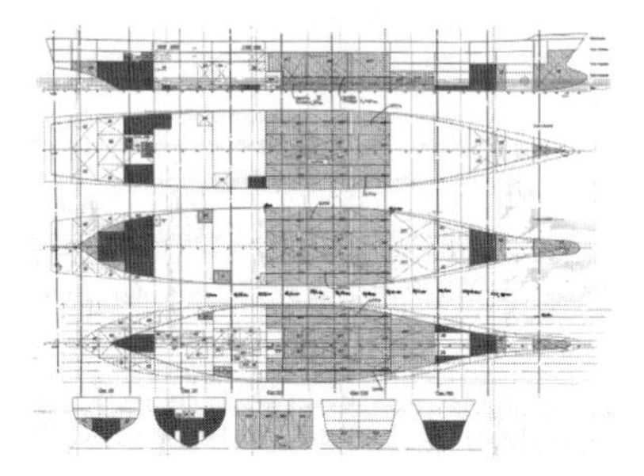

Figure 4. Hypothesis of bottom damage.

Table 4. Reference oil outflow parameters for double hull tanker.

	P_{OR}	O_{MR}	O_{ER}
5000 t	0.81	0.013	0.098
60,000 t	0.81	0.012	0.089
150,000 t	0.79	0.014	0.101
283,000 t	0.77	0.012	0.077

Table 5. Oil outflow parameters for bottom damage.

	0.0 m tide	2.5 m tide
Probability of zero outflow (P_0)	0.519	0.519
Mean outflow (m^3)	84.53	451.31
Extreme outflow (m^3)	327.14	1827.64

Table 6. Oil outflow parameters for combined side and bottom damage.

Probability of zero outflow (P_0)	0.569
Mean outflow (m^3)	282.75
Extreme outflow (m^3)	1098.70
Mean outflow parameter (O_M)	0.039
Extreme outflow parameter (O_E)	0.150

Table 7. Pollution-prevention index E calculated for ships of 5000 t deadweight and 60,000 t deadweight.

E_{5000}	0.614
$E_{60,000}$	0.597

3 PROPOSED SOLUTIONS

3.1 *General*

As previously reported, ITS Etna doesn't comply with prescriptions contained in paragraph 4 and pariagraph 5 of Regulation 19 of Marpol—Annex I. As the verification of the probabilistic method provided by Resolution IMO MEPC 110 (49) depends on cargo tanks volume and position, while the verification of the hydrostatic method (paragraph 4—Regulation 19) depends only on loading conditions, in the next paragraphs the considered design solutions will be described first. After that, the cargo loading instructions for the ship developed to satisfy the hydrostatic method, will be described.

3.2 *Verification of probabilistic method*

In order to satisfy all the requirements contained in MEPC 110 (49) (paragraph 5—Regulation 19—Annex 1) the following design solutions have been analyzed:

- CASE I: insertion of a double bottom, of 3.045 m depth, in all cargo tanks;
- CASE II: insertion of a double bottom, of 3.045 m depth, in all cargo tanks except tanks 1GT and 2 GT (considered out-of-service);
- CASE III: insertion of a double bottom, of 3.045 m, in all cargo tanks and of a longitudinal baffle, at 7.25 m from the center line, in tanks 4GT, 5GT, 6GT, 7GT, 8GT and 9GT;
- CASE IV: insertion of a longitudinal baffle, at 7.25 m from the center line, in tanks 4GT, 5GT, 6GT, 7GT, 8GT and 9GT.

In CASE I oil outflow parameters have been calculated considering a double bottom, of 3.045 m depth, in all cargo tanks. The depth of double bottom has been chosen referring to the maximum vertical damage penetration calculated as indicated in MEPC 110 (49). Results are reported in Table 8.

The Pollution-prevention index has been calculated as in paragraph 2.1.

Although calculated Pollution-prevention Indexes are higher than those calculated for the present configuration of the ship, the requirements stated by MEPC 110 (49) are not satisfied with the designed proposed for CASE I ($E < 1$).

In CASE II oil outflow parameters have been calculated without computing tanks 1GT and 2GT (considered out-of-service). Calculated Pollution-prevention indexes are reported in Table 10.

Even in this case the probabilistic method is not satisfied. CASE III varies from CASE II for the insertion of a longitudinal baffle, at 7.25 m distance from the center line, in tanks 4GT, 5GT,

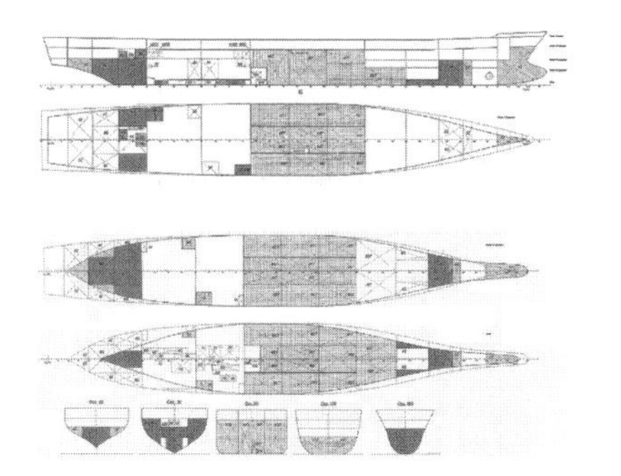

Figure 5. Cargo tanks position.

Table 8. Oil outflow parameters for combined side and bottom damage (CASE I).

	Side damage (40%)
Probability of zero outflow (P_0)	0.649
Mean outflow (m^3)	296.09
Extreme outflow (m^3)	1136.54
	Bottom damage (60%)
Probability of zero outflow (P_0)	1
Mean outflow (m^3)	0
Extreme outflow (m^3)	0
	Combined
Probability of zero outflow (P_0)	0.859
Mean outflow (m^3)	118.43
Extreme outflow (m^3)	454.61
Mean outflow parameter (O_M)	0.023
Extreme outflow parameter (O_E)	0.089

Table 9. Pollution-prevention index E calculated for CASE I.

E_{5000}	0.915
$E_{60,000}$	0.895

Table 10. Pollution-prevention index E calculated for CASE II.

E_{5000}	0.902
$E_{60,000}$	0.883

6GT, 7GT, 8GT and 9GT. The distance of 7.25 m has been chosen referring to the maximum transversal damage penetration calculated as indicated in MEPC 110 (49). The oil outflow parameters have been calculated as in previous cases. Results are reported in Table 11.

Table 11. Oil outflow parameters for combined side and bottom damage (CASE III).

	Side damage (40%)
Probability of zero outflow (P_0)	0.649
Mean outflow (m^3)	163.85
Extreme outflow (m^3)	377.34
	Bottom damage (60%)
Probability of zero outflow (P_0)	1
Mean outflow (m^3)	0
Extreme outflow (m^3)	0
	Combined
Probability of zero outflow (P_0)	0.859
Mean outflow (m^3)	65.54
Extreme outflow (m^3)	150.93
Mean outflow parameter (O_M)	0.013
Extreme outflow parameter (O_E)	0.029

Figure 6. ITS Etna.

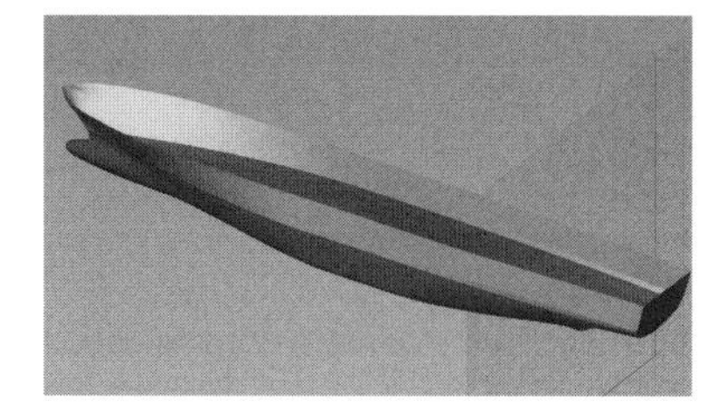

Figure 7. ITS Etna hull.

As made for the other cases, Pollution-prevention indexes have been calculated.

Pollution-prevention indexes calculated for CASE III comply with MEPC 110 (49).

The last case (CASE IV) consists only in the insertion of a longitudinal baffle, in the same tanks of CASE III, without building any double bottom.

Calculated oil outflow parameters for this case are reported in Table 13.

The pollution-prevention indexes calculated for CASE IV are reported in Table 14.

Design outfit proposed in CASE IV complies with prescriptions contained in MEPC 110 (49) and it seems to be more convenient than CASE III design as no double bottom has to be built.

In short, the results gained in this paragraph can be summarized as follows:

- CASE I doesn't comply with MEPC 110 (49);
- CASE II doesn't comply with MEPC 110 (49);
- CASE III complies with MEPC 110 (49);
- CASE IV complies with MEPC 110 (49).

3.3 *Verification of hydrostatic method*

The verification of the hydrostatic method reported in paragraph 4 of Regulation 19, as touched on before, depends basically on cargo fuel level in tanks and not on its weight. According to this and referring to formula (2), it's possible to say that:

- the presence of a double hull doesn't change h_c value for any cargo tank;
- the insertion of baffles inside cargo tanks doesn't influence any factor in formula (2).

Table 12. Pollution-prevention index E calculated for CASE III.

E_{5000}	1.158
$E_{60,000}$	1.124

Table 13. Oil outflow parameters for combined side and bottom damage (CASE IV).

	Side damage (40%)
Probability of zero outflow (P_0)	0.645
Mean outflow (m^3)	226.80
Extreme outflow (m^3)	600.35
	Bottom damage (60%)
Probability of zero outflow (P_0)	0.919
Mean outflow (m^3)	28.55
Extreme outflow (m^3)	140.99
	Combined
Probability of zero outflow (P_0)	0.810
Mean outflow (m^3)	107.85
Extreme outflow (m^3)	324.73
Mean outflow parameter (O_M)	0.014
Extreme outflow parameter (O_E)	0.044

Table 14. Pollution-prevention index E calculated for CASE IV.

E_{5000}	1.048
$E_{60,000}$	1.019

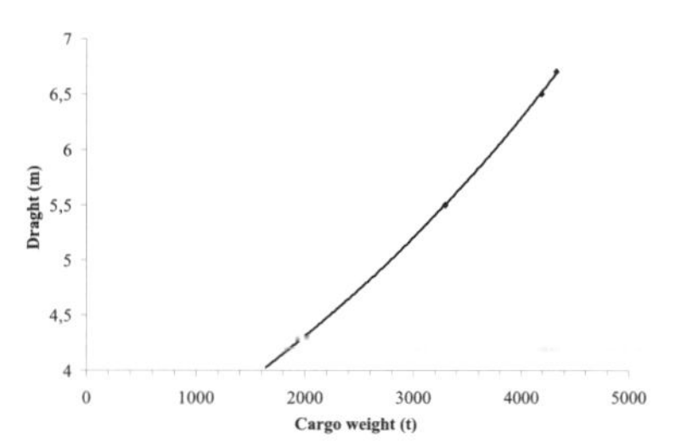

Figure 8. Maximum loading cargo at different draughts.

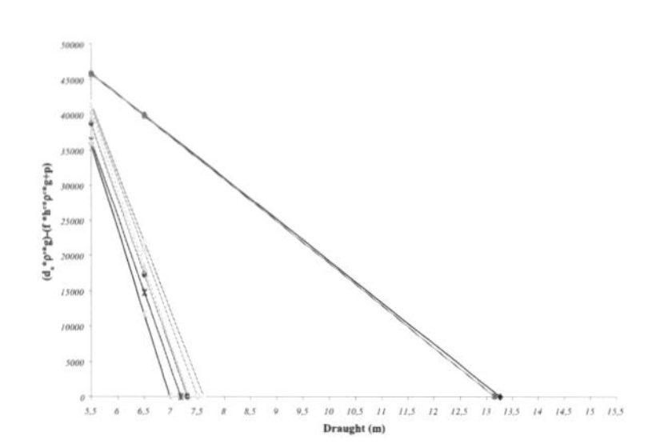

Figure 9. Variation of $(d_n*\rho_s*g)-(f*h_c*\rho_c*g+p)$ for each cargo tank at different draughts.

For these reasons the maximum cargo value has been calculated, for each cargo tank, at different draughts, able to satisfy formula (2) considering fixed all the other loads. Calculations have been reported referring only to CASE IV. This design solution (insertion of a number of longitudinal bulkheads in some tanks) has been considered the most convenient as there is no transported fuel reduction.

The difference between the first and the second part of formula (2) has been calculated, for each cargo tank, at different draughts. Results are displayed in Figure 9, where draught is reported on the x-axis and the correspondent difference $(d_n*\rho_s*g)-(f*h_c*\rho_c*g+p)$ for every cargo tank is reported on y-axis.

4 CONCLUSIONS

ITS Etna, in present configuration, doesn't comply with prescription contained in Annex I concerning Oil Tankers, as it doesn't satisfy the hydrostatic method described in paragraph 4 of Regulation 19 and the probabilistic method described in the IMO Resolution MEPC 110 (49). Four different design layouts have been considered in order to identify which of them complies with the probabilistic method first (as it depends only on cargo tanks disposition and volume) and then with the hydrostatic method (by considering the maximum loading cargo). As demonstrated, only the design layouts proposed in CASE III and CASE IV are able to satisfy Regulation 19 and, at a preliminary analysis, the last one seems to be the most convenient.

Sustainable Maritime Transportation and Exploitation of Sea Resources – Rizzuto & Guedes Soares (eds)
 ISBN 978-0-415-62081-9

Fast all-weather patrol craft for the Mediterranean

Fabio Buzzi & Vincenzo Farinetti
FB Design Srl, Annone Brianza (LC), Italy

ABSTRACT: Twenty-three different Countries lay along the about 2.5 million square kilometres of surface and 46,000 kilometres of coastline of the Mediterranean sea. The Mediterranean has always been a sea of traffic of men and goods. In ancient times all world traffic was here, but even today, every day, almost 1000 ships of 100 tons and up are crossing its waters, and some 300 of them are tankers. Although the Mediterranean were representing a mere 0,8% of the total world water surface, it sees about 20% of the world oil traffic ploughing its waters. Up to recent times several Med areas have seen heavy smuggling of goods, while today law enforcement Agencies have to cope with the clandestine traffic of migrants. Apart from political and security considerations, in these crossings, every year, thousands of migrants lose their lives. This huge traffic, legal and illegal, of dangerous goods, smuggled goods and people requires a gigantic effort in terms of means and resources by the coastal Countries to safeguard security, environment and human lives. Air surveillance and radar coastal surveillance can perfectly cope with the tactical situation, while only direct vessel presence allows interventions for patrol, inspection and major rescue operations. The bulk of this activity should be performed by small to medium craft able to rapidly reach the intervention area, also in adverse weather conditions. This paper deals, inter alia, with the expected characteristics of these types of vessels, in terms of structural strength, payload capacity, speed, sea-keeping behaviour, reliability, availability, maintainability. A new family of all-weather patrol craft has been conceived, built, tested and is currently operating with several Government Agencies within and outside the Mediterranean. These vessels are based on a number of FB Design original patents, relevant to design and manufacturing technology, allowing to reach superior navigation safety, coupled to outstanding performances even in bad weather conditions.

1 INTRODUCTION: GEOGRAPHICAL AND HISTORICAL BACKGROUND

The Mediterranean connects three different Continents and 23 different Countries, not to mention the five Countries surrounding the Black Sea, from those very old like Egypt, Greece or Italy, to the newest like Montenegro or the Palestinian Territories of the Gaza Strip, spanning from the less than 2 square kilometres of Monaco to the almost 2.5 million of Algeria, being this last figure, more or less, equivalent to the total surface of the Mediterranean. The total coastline length of the sea is about 46,000 kilometres, from the 20 kilometres of Bosnia-Herzegovina (even much less in the case of Monaco and Gibraltar), to the 15,000 of Greece.

The Western civilization was born here and the Mediterranean has been a sea of traffic of men and goods since the beginning of known history, when all world traffic was here. But even today about 1000 ships of 100 tons and up are crossing its waters every day, and some 300 of them are tankers: although the Mediterranean were representing less than 1% of the world water surface, about 20% of the world oil traffic is ploughing its waters. Almost 400 million t of oil are transported annually here, with an average of 10 oil spills per year, and it is estimated that some 150,000 t of crude oil are deliberately released into the sea every year. And talking of oil, offshore oil production is expected to increase a lot in the near future.

All this is a terrible threat to the environment and a big challenge for law enforcement and sea protection Agencies.

But there are other issues to be addressed: smuggling of goods, drug trafficking and illegal migrants, looking for better life conditions on the north shores of the sea. More or less all Countries of the Mediterranean are subject to these phenomena, either as origin or as destination. In addition, it must be considered that thousands of these migrants lose their lives at sea every year, so it is not only matter of security, but it is also matter of a huge effort of SAR operations. To cope with all this, all coastal Countries have to make a tremendous effort in terms of means and resources.

2 DIFFERENT ASSETS FOR DIFFERENT TASKS

Different "tools" are best suited to perform all tasks required to ensure safety, law-enforcement and security at sea:

- coastal and airborne radar surveillance can provide information about a wider scenario and directions to other co-operating assets;
- air-planes would allow to rapidly reach the specific spot for visual assessment;
- helicopters can perform search operations in an excellent way and also limited rescue operations;
- large displacement ships are well suited to carry out long range, blue waters, multi-day naval missions, especially in those areas usually called as "outer Mediterranean"; but, for instance, they are not fit to approach small vessels or to pursue and intercept high speed craft;
- fast patrol light craft are the 'work horses' for every day activity made up of rapid intervention, high speed pursuing, vessel interception and boarding, SAR operation, illegal trafficking interdiction, just to mention the main ones. All these mission are usually lasting hours, not days, but they have to be accomplished regardless the weather and sea conditions.

3 OPERATIONAL REQUIREMENTS

The "equation" defining a high speed patrol craft includes a lot of "variables" (the requirements) and finding the right parameters to solve it is not that easy.

The main requirements that such a boat has to meet are:

- speed,
- sea-keeping,
- comfort on board,
- structural strength,
- manoeuvrability,
- sailing safety and course keeping,
- propulsion reliability,
- low acquisition cost,
- low running cost,
- maintainability,
- high availability,
- easy deployment.

3.1 *Speed*

The speed attainable by a craft in water depends upon the resistance offered by the water (and air), the power and efficiency of the propulsion device adopted and the interaction among them. Because of this interaction, it is extremely important to consider the design of the hull and the propulsion device as an integrated system. In addition, when the water surface is rough, it must be considered the increase in resistance and the fact that the propulsion system might be working in less favourable conditions. Therefore to obtain a high speed it is necessary to reduce the resistance and optimize the propulsion system.

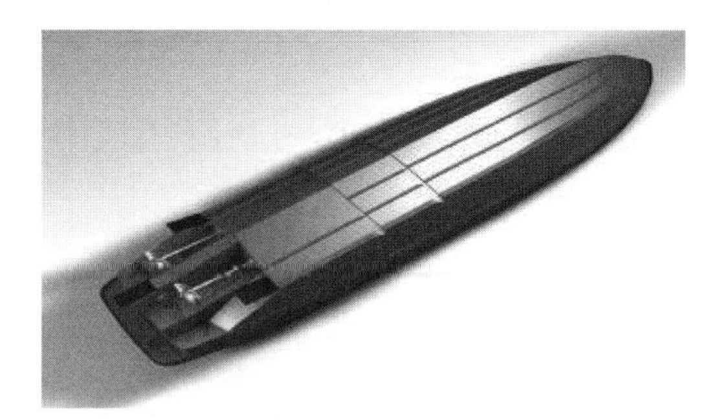

Figure 1. Hard-chine stepped hull with trim tabs and surface propellers.

Hull resistance is made up of frictional and residuary resistance, plus the resistance of appendages. To reduce friction resistance a good portion of the hull should not be in contact with water, which is rather viscous. This goal can be reached through hull ventilation, to replace water with air, and through dynamic lifting forces, i.e. adopting a planing hull, which is reducing also the residuary resistance.

But the planing stage must be reached as soon as possible, not to face a 'hump' in the curve of resistance so high to require a too powerful propulsion plant. Reaching the planing stage can be helped by means of trim tabs.

Appendages can be minimized by adopting a surface propeller drive transmission system. Hence, the optimal solution as regards a high speed monohull should be as indicated here below.

3.2 *Sea-keeping*

Top speed in calm waters is an interesting feature, but what about real operational conditions?

Patrol vessels have to be dependable, and hence they have to be available with almost any weather condition, with reasonable performances.

Some years ago, two similar FB Design boats were competing in the Venice-Montecarlo race. Their underwater bodies were identical, but one was a pure rigid hull, while the second was a RIB. In calm seas the rigid hull was slightly faster, but in waves the RIB had superior performances. Why? Think to conservation of energy. When jumping from wave to wave the rigid hull was losing a good deal of energy at every impact, while the RIB was transforming its kinetic energy into elastic energy by bending its inflated collar that then returned a good part of that energy, like a basket ball bouncing on the floor. Some considerations have to be made regarding the collar based on this concept:

- it has to be positioned at the right height in respect to the water, to be able to perform its proper function;
- it should be of inflatable type, not filled with foam, otherwise it would be just a fender, unable to return any energy;
- the connection of the collar with the hull must be carefully studied to properly cope with the bending cycles to which the collar is subject;
- the shape of the collar in way of the bow has to be suitably streamlined, not to present a sort of "wall" against waves;
- in general the presence of a collar is reducing the beam of the rigid portion of the hull, and hence it is reducing the possible volume for habitability.

Full scale observations have shown that basically only the aft part of the collar is touching the water, being the fore part just a kind of fender. Is it strictly necessary? That's why the STAB® option was introduced: two inflatable sort of sponsons easy to install and remove, just in way of the aft half of the boat, connected to the hull by means of rails. The STAB®s are acting also as roll stabilizers and improve boat lateral stability.

In addition, at every wave encounter, the bow of the boat tends to be submerged. To cope with this the upper works of the boat in way of the bow are suitably shaped to provide dynamic lift and enhanced buoyancy, thus reducing pitching also.

In addition to this, the beam at bow would result wider, allowing easier manoeuvres. Typical resulting boat configuration is shown in Fig. 2.

However, as every-day experience is witnessing, all types of hull are possible, and each of them has a specific field of application. As a kind of general rule we may say that for boats with length in the 20–50 foot range:

- a rigid hull without any inflatable collar is used when calm to moderate sea conditions are expected, high top speed is required and there is no need to perform boarding of other vessels or rescue operations (see Fig. 6—Racing boat)
- a hull with inflatable collar is to be adopted when and where heavy sea conditions are expected, relatively high speed is required to be sustained and boarding and/or rescue operations are required and it is not required to have a large habitability internal volume as needed for long range missions (see Fig. 7—Police boat)
- STAB® fitted boats represent an interesting compromise (see Fig. 2—High-speed patrol boat).

Figure 2. "High speed patrol craft with STAB® option and enhanced lift bow".

3.3 *On-board comfort*

Heavy sea conditions are causing high level of acceleration, up to several g, acting on the boat, and hence on the crew, affecting activity on board, including safety. Crew and transported personnel have to be accommodated in shock absorbing seats, fitted with safety belts. FB Design has devised a full range of shock absorbing seats, with power-adjustable seating support inclination, allowing any position, from fully seated to fully standing.

Figure 3 shows a group of seats designed to withstand impact accelerations up to 12 g, able to allow acceleration reduction up to 80%.

Hot climate or water tightness of cabin of self-righting boats require the adoption of an air conditioning system. Usually an air-conditioning system requires the installation on board of a generating

Figure 3. FB Tecno G12 seat.

Figure 4. FB Condair system.

set or PTOs on reduction gears, but this cannot apply to a boat with out-board propulsion engines. FB Condair is a low-cost, low-weight stand-alone system that could operate on gasoline, or electric power, and hence it is the right solution, even for boats with outboard propulsion engines and no power generating set on board.

3.4 *Structural strength and boat unsinkability*

Usually the structure of patrol boats have to be in accordance with the Rules of a Classification Society defined by the building contract. Very often FB Design boats are complying with RINA Rules for the Italian Coast Guard, but any international Classification Society can be chosen by the Customer.

In general, the strength of FB Design boats is well in excess to Class requirements, thanks to the patented technology adopted to built the boat: the so called Structural Foam®.

Boats built with this technology present a moulded hull and a counter-moulded floor, both laminated in fiberglass, with suitable kevlar reinforcements if required, and divided into six different chambers. These chambers are generated by longitudinal stiffeners previously bonded to the bottom with grp, then glued to the counter-moulded floor with a systems based on aluminum profiles and structural adhesive. After the hull and the counter-moulded interior have been assembled together, a high density polyuretane foam is injected in one shot, through six different points, with resulting filling by capillarity of the six chambers.

In this way the boat has extremely high stiffness and strength and it is unsinkable. Figure 5 here below is showing the main principle of Structural Foam® technology.

FB Design boats are hence extremely robust, as, for instance, demonstrated by "Red FPT" (actually the restored "Cesa 1882", originally built in 1985), winner of the 2010 UIM World Marathon (see Fig. 6), after 25 years of racing activity.

A further interesting feature related to safety is self-righting capability, as shown in Fig. 7 below, during a full-scale self righting test.

While Structural Foam® technology makes this option easy to materialize, other platform issues, like cabin air conditioning mentioned at para 3.3 above, may arise and must be properly addressed.

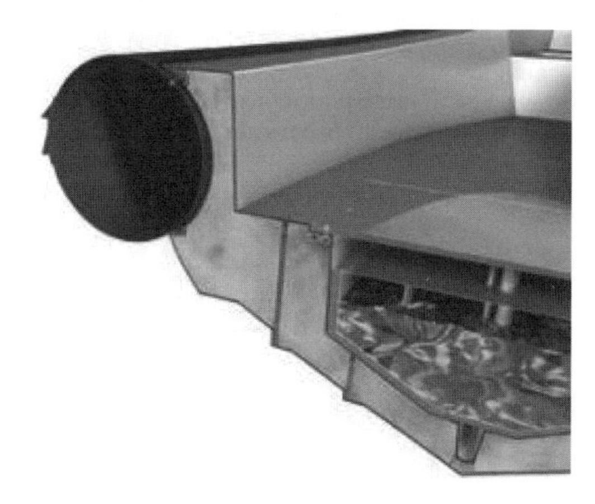

Figure 5. Structural Foam®.

Figure 6. 2010 Cowes-Torquay race.

Figure 7. Full-scale self righting test.

3.5 *Manoeuvrability, course keeping and sailing safety*

Patrol boats might be operating in restricted waters and should be able to easily approach other vessels for inspections or boarding. Therefore manoeuvrability is of paramount importance. FB Design boats are usually adopting Trimax® surface drive systems which present rudders aft of the propellers. In this way the arm of the lifting force of rudders in respect to the center of flotation is maximum, ensuring maximum steering moment.

In addition to this, and as it can be seen in Fig. 6 above, when the boat is impacting on water, the first surface to touch the sea would be the rudder, which is aft of propellers, and hence, in any condition, the resistance of rudders and the thrust of propellers will generate a moment that would keep the course, while the contrary would happen if the first surface (resistance) to hit the water would be fore of propellers.

When sailing in shallow waters, rudders are prone to collide with the sea bed, and this may cause serious damages to the hull. As a standard feature, FB Design boats have safety rudders, fitted with safety pins that, in case of collision, would give way, not causing any damage to the hull. Safety pins can be then replaced in minutes by on-board personnel.

A planing boat, below the hump critical speed, may present a major raise of the bow, causing a dangerous visual reduction. As it was mentioned in para 3.1 above, to reduce the critical speed, a proper trim tab system should be installed. FB GPS 3TAB system features two or three tabs electronically controlled, with angle of attack proportional to the speed of the boat. The system control unit is continuously relating two inputs: the boat speed, through a dedicated GPS antenna, and the actual tab position, electronically monitored. The relationship between speed and tab position is programmed for every type of boat. The central tab (see Fig. 1 above) is aft of propellers, thus ensuring the maximum longitudinal righting moment, to minimize the transition phase.

3.6 *Propulsion reliability*

Obviously, while talking of reliability of a propulsion system, the reliability of the installed engines comes to mind first. But there are many other issues to be properly considered: fuel system, combustion air ducting, shafting, propellers.

The fuel system should be suitably redundant, to avoid, for instance, fuel pollution with water.

Combustion air should be taken as high as possible, as shown in Fig. 8 below, to avoid dangerous water ingestion.

Figure 8. Air intakes above the wheelhouse.

Shafting should be as simple as possible, but allowing a certain trim adjustment.

Propellers, usually of the surface type, should be properly ventilated, to avoid overloading of the engines during the transition phase, when the propellers are completely immersed.

Another issue to be properly considered is safety of propulsion control.

When sailing at high speed in rough sea, with the traditional combined control levers, it may happen to put the gear into reverse, with subsequent disaster. Gears should be operated independently from fuel throttles, through suitably protected switches, as shown in Fig. 9 below.

3.7 *Life-cycle cost considerations*

Lowest life-cycle cost (LCC) is the most straightforward and easy-to-interpret measure of economic evaluation related to any investment. Life-cycle cost analysis (LCCA) is a method for assessing the total cost of ownership. It takes into account all costs of acquiring, owning, and disposing of a system. LCCA is especially useful when project alternatives that fulfill the same performance requirements, but differ with respect to initial costs and operating costs, have to be compared in order to select the one that maximizes net savings.

The different tasks to be assigned to patrol craft may require to have different, specialized boat configurations. This would be in contrast with the basic requirement of low acquisition cost.

To this respect it is advisable to select a yard having a wide range of boats deriving from the same 'family of craft', able to provide economy of scale even in case of small number of different specialized boat configurations, thanks to the modularity of design and components adopted for the actual outfitting of the craft.

This modularity of components would also reduce maintenance cost, thanks to the reduction of spare parts and skilled maintenance personnel, as discussed in paragraph 3.8 here below.

In addition to this, a yard with state-of-the-art technologies for design and construction would

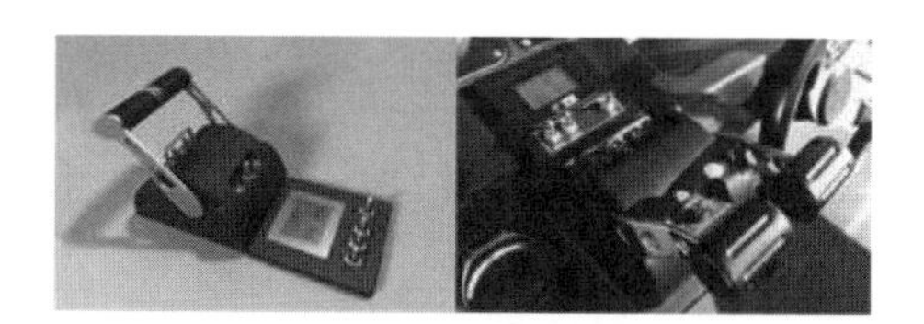

Figure 9. FB Throttle.

allow to quickly and economically define and produce possible modifications to standard designs to perfectly match all Customer requirements.

To minimize running costs it is necessary to rely on:

- a highly efficient hull, allowing to reduce the installed power, and hence fuel consumption, to get the requested performances;
- an optimum seakeeping behavior to reduce stress on hull, systems and crew;
- a rugged structure to avoid hull damages and ensuing expensive repairs and down-times.

3.8 *R.A.M. (Reliability, Availability and Maintainability)*

Should we go by the book, R.A.M. means: set of requirements imposed on a system to ensure that it will be ready for use when required; successfully perform assigned or intended functions, and that can be maintained in its operational state over its specified useful life.

An item is specified, procured, and designed to a functional requirement and it is important that it satisfies this requirement. However it is also desirable that the item would be predictably available and this depends upon its reliability and availability. For items and systems used in critical areas including military equipment, the availability, reliability and maintainability considerations are vital. Reliability can be defined as the ability of an item under given conditions of use, to be retained in, or restored to, a state in which it can perform a required function. The reliability is expressed as a probability (0–1 or 0 to 100%) that the item will not fail in the given time period.

Availability is the ability of an item to be in a state to perform a required function under given conditions at a given instant of time or during a given time interval, assuming that the required external resources are provided. At its simplest level:

$$\text{Availability} = \text{Uptime} / (\text{Downtime} + \text{Uptime})$$

Maintainability represents the ease of restoring the operating condition as soon as possible.

It is important to focus on the concept of 'restoring', not mere 'repairing', meaning by 'restoring' the three phases of: preparation to repair, repair and start-up. For instance, preparation to repair time may be much longer than repair time due to difficulties in reaching the failed component.

Racing experience is extremely important in how to design boats, systems and sub-systems to be easily and rapidly (and hence economically) restored to their intended function.

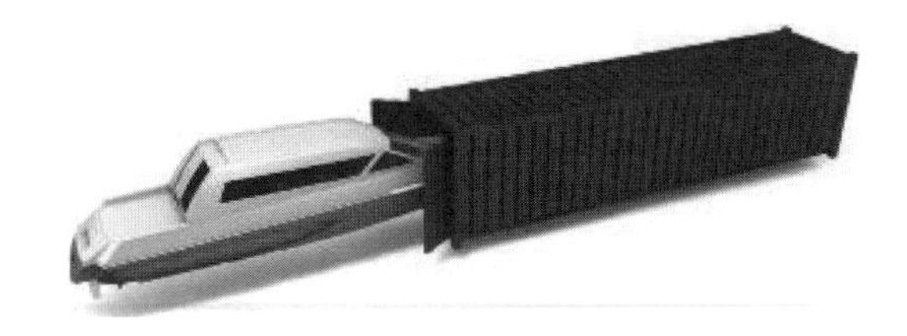

Figure 10. Boat transportation with standard container.

3.9 *Boat deployment*

Countries with consistent coast lines or participating to international task forces may require to safely and quickly deploy their assets.

The safest and most economical way of transportation is by 40' standard container that could be either embarked on board a ship or a truck.

All FB Design RIBs and STABs® up to the 39' size can fit inside a 40' standard container, as shown in Fig. 10 below.

Military air-lift transportation (e.g., with C130 aircraft) might be possible for boats up to 42' of length. To allow easy air transportation special trolleys should be provided.

4 FURTHER CONSIDERATIONS ABOUT BOAT DIMENSIONS AND HULL FORMS

All issues addressed in the above paragraphs are mainly related to high speed planing monohulls, featuring calm sea speed well in excess of 50 knots. Boats with such performances are economically viable if their dimensions are reasonably limited, otherwise the need of installed power would not be cost effective. While a fast patrol craft of 18–20 m and about 25 t of full load displacement could reach 50 kn with a 'manageable' 2 × 1000 HP propulsion package, the same result for a 24 m vessel with some 60 t of full load displacement would require more than a triple power, with corresponding fuel consumption, not to mention dimensions, weight and cost of the propulsion package. From these dimensions and up, i.e., for long-range patrol craft, different performances and different hull configurations should be adopted. Speeds would hardly be beyond 40 kn and hull forms would be displacement hulls, like slender monohulls or catamarans.

Slender monohulls are of simple construction, with their structure withstanding basically longitudinal bending moment and shear only, but they require a powerful propulsion package, usually arranged in a complex 4-shaft configuration.

Catamarans, thanks to their very high double L/B ratio, allow to reach substantial speeds with reduced propulsion power and offer very good transversal stability and a wide platform. In contrast, their construction is more difficult and their structure would be also subject to torsional and transversal bending moments. To successfully cope with such complex loading conditions, catamarans up to some 40 m in length should be built in composite, a technology that allows to generate highly non-isotropic structural details and hence to design and build optimized hulls.

5 SHORT EXAMPLES OF HIGH SPEED CRAFT OPERATING IN MED WATERS

FB 33' RIB—Armed Forces of Malta

Length o.a.: 9.5 m
Propulsion system: 2 × 320 HP inboard
Max speed: 52 kn

FB 38' STAB®—Italian Coast Guard

Length o.a.: 11.4 m
Propulsion system: 2 × 450 HP inboard
Max speed: 50 kn

FB 39' RIB—Albanian police

Length o.a.: 11.98 m
Propulsion system: 2 × 300 HP outboard
Max speed: 50 kn

FB 42' RIB—Greek Navy

Length o.a.: 13.2 m
Propulsion system: 2 × 750 HP inboard
Max speed: over 65 kn

FB 43' SF—Italian Customs Police

Length o.a.: 13.7 m
Propulsion system: 2 × 825 HP inboard
Max speed: over 50 kn

6 CONCLUSIONS

While defense related activity at sea has to be performed by real naval assets and wide strategic scenario picture is offered by air and/or coastal radar surveillance, the bulk of security, environmental protection, SAR and border control missions has to be performed by light high speed craft, having, inter alia, good seakeeping performances.

The most cost effective answer to these requirement is represented by planing monohulls.

Reliability, Availability and Maintainability related issues are of paramount importance to be in position to assure round-o'-clock response, and have to be properly addressed. Endurance ocean races have similar requirements, and hence experience and testing of solutions acquired by the builder in this field is highly valuable.

Planing monohulls have a dimensional limit, in terms of cost effectiveness, beyond which different performances are expected and different hull forms have to be adopted.

Composite catamarans appear to be a good answer to the need of economical blue-waters patrol craft.

Today technology allows to conceive unmanned surface vessels able to perform a wide spectrum of missions in dangerous areas, being these USVs possibly taken to theater, launched, operated and retrieved by a mother ship.

To provide stakeholders with a kind of reference table to select the right vessel for a typical mission, the authors, on the basis of their experience, are proposing the following summary table:

BOAT TYPE VS MAIN MISSION AND DIMENSIONS

MISSION / LENGTH (m)	LOW SPEED PATROL	SAR BOARDING PARTY	HIGH SPEED INTERCEPT	LONG RANGE MISSION	USV
6-10	RHIB	RHIB	—	—	—
10-15	RHIB	RHIB	RIGID HULL STAB ®	—	RIGID HULL STAB ®
15-20	RHIB RIGID HULL	RHIB	RIGID HULL STAB ®	RIGID HULL	—

All experimental results and theoretical data relevant to this paper have not been published yet and presently are only part of internal confidential documents.

Sustainable Maritime Transportation and Exploitation of Sea Resources – Rizzuto & Guedes Soares (eds)
 ISBN 978-0-415-62081-9

IT Navy new multifunctional underwater support vessel programme

Commander M. Cannarozzo & Lieutenant P. Migliore
IT Navy General Staff Submarine Department

ABSTRACT: The designing and developing of a programme with the specific objective to commission a new Multifunctional Underwater Support Vessel comes from the necessity of the IT Navy to replace IT Navy Ship **ANTEO (ARS)** that has reached it's operational life limit. The design criteria are driven by a combination of state-of-the-art commercial Offshore Support Vessel technology, with tight criteria in terms of functionality, ergonomics, safety and living comfort, and with Naval Standards required for naval operations such as helicopter (**Helo**) operations and military communications, all complying with international regulations that will come into force by 2014. The concept is to enable the It Navy to operate with a modern multirole, flexible, deployable unit, in order to support the Navy in it's Underwater Operations mainly for Search and Rescue of Distressed Submarines (Salvage Ship), Supporting the Italian Navy Diving Group (Supply Diving Vessel) and Supporting the Italian Special Forces for training and operational purpose.

1 PROJECT DESCRIPTION

The USSP project born in a strict relation to the necessity of the Italian Navy to provide itself of a modern "**Multifunctional Underwater Support Vessel**" which, in consideration of it's deployment and versatility, allows the Italian Navy to provide a high added value to it's Diving Group. The general objective is to concentrate all subsea support capabilities, including **DISSUB** search and rescue operations, in only one multirole Platform (such as *Rescue and Salvage Ship* **ARS**). To reach this ambition we need to concentrate in this platform a large number of different assets and, more important, integrate everything in a perfect way. At first has been found the type of vessel which best could represent the reference of our project. Has been decided to follow the line of design of the modern Offshore Supply Vessel.

This kind of vessels have the characteristics of modularity, free space on working deck, manoeuvring capacity and stability and fittings which are necessary for the scopes.

Such a vessel class has the particular characteristic of reconfiguring according to the assigned mission; able to operate either as *Supporting Unit* for *Diving Operations* (Supply Diving Vessel—**SDV**).

Now, the Italian Navy can rely for mentioned operations on IT Navy ship "**ANTEO**". It was commissioned in the early 80's with a concept design from the 70's, and although it is provided with high performance equipments.

The useful operational life limit of It Navy "**ANTEO**" (due for 2014 and intended as last reduced efficiency status limit) and the importance for the IT Navy that the program of acquisition of a *Supporting Vessel* for underwater operations (that includes specialized personnel and equipment transportation) will be concluded in a short time, has guided the starting phase to follow next guidelines:

- Strong loading capability and deployment configuration flexibility
- Construction project and line of production already available in the market in order to considerably reduce all related costs.

Making comparison with ITS ANTEO, we can outline the following features:

	New project	ITS ANTEO
Displacement	circa 6000 t	3500 t
V	16 Kn	18 Kn
Autonomy	5000 M	4000 M (14 Kn)
Loa	circa 110 m	99 m
Lbp	circa 103 m	93 m
Bmax	circa 22 m	16 m
H	9 m	5.5 m
Propulsion power	Pod 2 × 3500 Kw	Axial 1 × 4400 Kw
D.P.	Level 2	no
Crew	185–200	145
T.U.P	60 rescuees	no
Crane	100 t a 15 m	25 t

Based on the requirements and capabilities demanded, the general configuration/characteristic of the new **Multifunctional Underwater Support Vessel** represents an "*off shore supply vessel*" already in use in the off- shore under water related jobs industry.

This specific vessel and particularly it's general arrangement are well adapted to perform lifting and laying operations of materials and equipments with the widest range of dimensions and characteristics in all kinds of sea/weather state and conditions.

The new **Multifunctional Underwater Support Vessel** will be able to provide specific dual-use operations as follows:

- Military, maritime and homeland security: Oil pipes and wellheads survey for antiterrorism/sabotage operations, hazardous wreckage search operations, etc.
- Other State/Government/National Organizations Subsea Operations (**Ministero degli Interni**.)

The above mentioned operations are traditionally demanded to The Italian Navy with positive benefits to the whole Nation and can be summarized as follows:

- Search and recovery from the seabed of unestimated archaeological and artistic artefacts.
- Search of hazardous wreckages (environment polluting ships—"poisons ships")
- Support in recovering/removal operations from the seabed.
- Support in cleaning-up operations for sites with a potentially high risk of environmental pollution. (see 2008 Haven tanker accident).

Based on what previously stated, further capabilities are then necessary in order to assure the new USSP with *Underwater Survey* capability and *deep-sea lifting/working* operations capability.

Such capabilities shall be achieved by using a *building block* development concept.

Three different blocks can be considered as follows:

Block 0: Ship in line with proposed **OR**. **ARS/NAI** configuration is intended as a common pool of capabilities necessary for *Submarine Escape and Rescue* (**SMER**) operations and *Diving Operations* and for blocks 1–3 settings availability.

Block 1: Working Class ROV, beyond 3.000 m working depth with possibility to operate for inspections and surveillance activities on hazardous wreckages, recovering and removal operations as well as security inspections on oil conducts, oil pipes and wellheads for antiterrorism/sabotage operations.

Block 2: Watching Class ROV, working depth up to 3.000 m. with possibility to operate along with the capacities for **Block 1,** deployable **AUV** with working depth up to 3.000 m. for researching, localization and identification for sea-depths and underwater structures installed at that depth (Electro ducts, oil ducts, and wellheads)

Figure 1. Preliminary design phase.

Block 3: Reinforcement of diving capability for underwater operations up to 450 m.

All systems and equipments mentioned in Block1 to Block 3 are provided with their respective *Initial Logistic Support* (**SLI**).

2 WHOLE WARSHIP DESIGN LINE AND TASKS

In detail, in order to fulfil *Diving Operations activities* this Unit, through owned or deployable assets/equipments, will be able to accomplish the following task:

- 600 m Submarine Search and Rescue operations with organic SRV (Submarine Rescue Vehicle), Personnel Transfer Under Pressure (TUP) capability in Decompression Chamber up to 60 patients.
- Embarkment and Engagement of other deployable recovery systems in particular US NAVY SRDRS and NSRS.
- EOD/IEDD searching, removal and disposal operations with ship's owned assets.
- Underwater searching and clearance operations for strategic sites such as oil platforms, electro ducts, oil ducts, and underwater conducts.
- General Subsea operations.
- Commanding Unit in National/International scenarios (NATO, UE, Multinational) for subsea operations in relation to SUBMARINE search and rescue activities.
- Supporting Unit for Diving Training Operations.
- Hazardous wreckages, constructions, obstacles search, localization and inspection operations as per **Blocks 1–2.**
- Logistic messing and berthing for crew and, specialized staff on board.

- **Role 1** Medical Treatment facilities (**MTF**) in order to:
  - Fulfil ordinary daily medical checks on Divers
  - First aid and limited time recovery treatment capability for Divers following Subsea operations as specified in specific chapter/sections.
- Due to the reduced number of crew members in relation to the complexity and dimensions of this Unit, all activities related to Ship's conducting operations and technical maintenance will be simplified, therefore all technical equipments must have high flexibility proven reliability and low-cost of employment they must be available in wide and proven option, in line with the most recent personnel and environmental safety and security requirements/regulations.
- All plants and equipments so called essential for Ship's and Personnel's mission and safety must be provided with remote and local diagnosis/monitoring systems.
- All technical plants and equipments and related interface/circuits must be installed in order to guarantee maintenance up to Level 3, all circuits' identification signs must be implemented with bioluminescence materials according to It Navy safety regulations.

It will provide, whether necessary, the following Support capabilities to be embarked as requested:

- State/Government/National Organizations Subsea Operations (Ministero degli Interni).
- Emergency Towing operations.
- Scientific research and experiment operations for underwater environment including oceanographic conditions, magneto-thermic survey.
- Special equipment transportation.

The entire project has been developed according to the principles and rules of international maritime organizations, like:

- NATO Naval safety code
- ANEP 77
- NATO Guidelines for Environmental Factors in NATO Surface Ships (Acoustical, Climatic, Vibration, Colour, Illumination)
- ANEP 25
- ANEP 72 ("design criteria for submarine rescue motherships (moships)").

Obviously ANEP 72, in this case, represents the most important path to follow for developing of project requirements.

3 WHOLE WARSHIP TECHNICAL SPECIFICATIONS

This Unit, in order to fulfil wit the requirements, is designed according to the following tasks:

- Ensure the capability to embark, manoeuvre and operate with Submergence mini sub for rescuing operations up to 600 m depth. The embarked Minisub and all related equipment will be fully deployable;
- Both Minisub and LARS will be compliant with all dissub rescuing and equipment deployment international regulations and specifications and standard transportation requirements;
- The complete Rescue System is designed to be compliant with Voo (Vessel of Opportunity) Ship specifications;
- MULTIFUNCTIONAL UNDERWATER SUPPORT VESSEL Working Deck is designed to integrate any NATO/US Navy Rescue System such as **NSRS** and **SRDRS** system;
- Working Deck Space above 600 sq m. for special equipment loading and lay out. It's designed to fully reflect "Multirole-Ship Concept" avoiding general interference and guaranteeing useful spaces for containerized equipment embarkment.
- Up o 60 rescuees Transfer Under Pressure (**TUP**) capability from Rescue Mnisom to Decompression Chamber on board **USSP**.
- External "moon pool" for **ROV** Launching and Recovering System and several others applications.

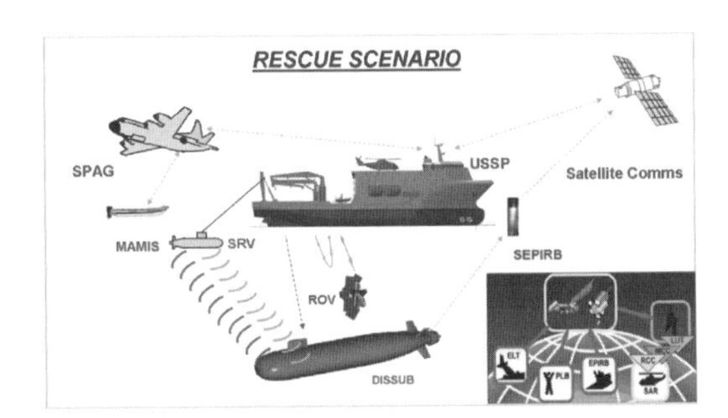

Figure 2. Rescue scenario.

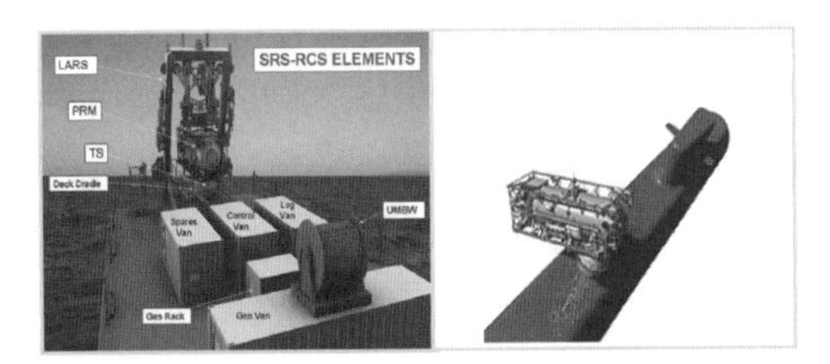

Figure 3. SDRS System.

- Capability for medium/heavy type **AUV** Launching and Recovering System.
- Flying deck with flying operations space engaging medium/heavy type Helicopters (NH 90/EH 101 type);
- Specific areas for Command and Control System for Ship and Underwater activities.
- Specific areas for Command and Control.

Following are listed IT Navy Multifunctional Underwater Support Vessel main characteristics.

a. Propulsion and generation systems

16 knots Continuous maximum speed, 5.000 miles autonomy. It will have efficient manoeuvring capability in confined waters in order to assure a quick rotation from static condition and under movement either with calm and rough sea state and wind speed up to 30 knots. It will be provided with Dynamic Positioning System to guarantee Subsea operations.

b. Integrated diving system

For Subsea Operations this Unit is provided with:

- **600** m. working class **SRV** (Submarine Rescue Vehicle), Aft A-frame Launch and Recovery System. (see attachment D)
- Hyperbaric Treatment Facility (max 7 ATA) for medical treatment (up to 60 patients) separated from the Integrated Hyperbaric Plant.
- **TUP** for **SRV** and similar systems NATO/US Navy up to 5 ATA for transferring of the rescues.
- **C3** System for all Subsea operations with different equipment.
- 300 m. depth internal atmosphere monitoring and ventilation/depressurization system for Distressed Submarine scenario.
- 300 m. Recovery Underwater Bell (SRC- Submarine Rescue Chamber- Mc Cann Bell) and related launch and recovery system.
- Integrated Hyperbaric Facility for diving operations with **SDC** (Submarine Diving Chamber) up to **450 m**. with related Command, Control and Communication System for deep-subsea operations.
- **ADS** (Atmospheric Diving Suit) and related launch and recovery system.
- Arrangement for deployment and Subsea Operations with **NSRS** and **SRDRS** rescue system.

c. Acoustic sensors

Multi beam and Side Scan Sonar capabilities up to 300 m.

- **HIPAP** (High Precision Acoustic Positioning) for Ship's and AUV's tracking and positioning control.
- Underwater Communications Apparatus (Multifrequence Underwater Telephone) compatible with Diving Operations Equipment.
- Acoustic Emission Interceptor and system/equipment for Submarine Emergency Signal localization and other possible emergency acoustic signals.
- 1.000 m. Single beam echo-sounding.
- Shallow-medium water echo-sounding, 150° beam and suitable frequencies for operational deployment up to 1.000 m., **SW/HW** for bathymetric data acquisition, Positioning and Orientation integrated system for Motor Vessel (**POS/MV**), **DGPS** and gyrocompass for bearing/positioning and motion sensors for wave motion compensation.
- Underwater Digital Bathythermograph System
- **MAMIS** System for Internal Submarine Atmosphere Monitoring and Communication.

d. Command & control system

This Unit is arranged with a fully integrated C&C system to perform live management of underwater operation and ship's activities.

The system can be conducted from three different locations:

- Integrated Bridge Console
- Tactical Operation Center
- Underwater Operations Control Room

This system can be integrated into a complex operation tactical web through the Tactical Data Link. The Command Operational Centre is provided with different strategical and tactical plotters, digital tactical displays and communication assets to coordinate external resources.

e. Telecommunications

This Unit is provided with TLC capabilities in order to:

- Guarantee, in National and NATO coordinated operations, communications, data managing and coordination to personnel involved in SaR activities;
- Guarantee coordination of training activities for Divers and SF;

As mentioned above, this system will be fully integrated with C&C System.

f. Integrated defense system

Although it is not a Combatant Ship, this Unit will guarantee a minimum self defense capacity able to contrast conventional and non conventional threat filtered by consecutive lines of defence expressed by the Naval Forces array from any directions.

g. Integrated logistic support

The design approach to the ILS is based on Systems Engineering guidelines. Technical Documents and Logistics Data for the WHOLE WARSHIP are collected in a Common Source

Chart 1. Whole warship characteristics.

Vmax	More than 16 kn
Autonomy	5000 nm 16 kn
Fly deck	NH 90—EH 101
Moon Pool	1 for main deck equipment 1 for SDC
Free main deck space	More than 400 m^2
Rescue Submarine	Minimum 16 pax and deployable LARS
Hospital	Role 1
Crew	Standing 100 pp within 30 DIVERS Accomodations: for 200 pp
Rescue operation max depth	600 m
TUP chambers system	for 60 rescuees
Diving system	450 m—15 divers system

Figure 4. Integrated diving system.

DataBase developed by IT Navy. The aim is to reuse all the project information in a true Life Cycle Management, starting from commissioning trough all the design review phase. Concurrent engineering approach is possible tank to web based applications and the massive use of open document format, based on XML.

More characteristics of the ship are well resumed in the following chart:

4 MAIN PROJECT RELATED ACQUISITIONS

General description
High Power Working Class Underwater Vehicle.

Features:

- Inspection
- Load manipulation and installation
- Heavy weight guide and connection
- Pipe lines connection
- Obstructions cut and disposal

Working ROV

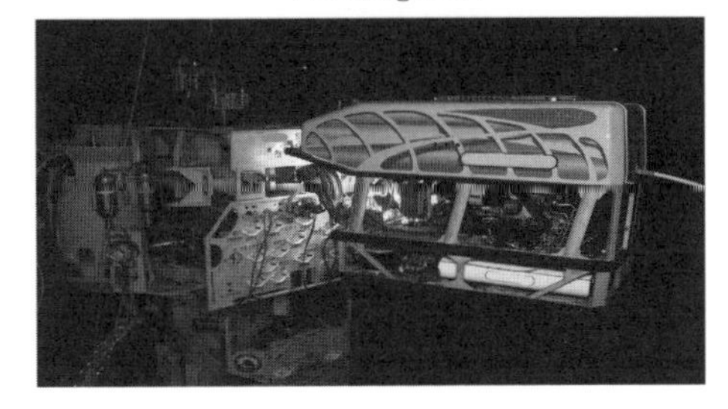

Figure 5. Working ROV example.

Chart 2. Main characteristics.

Power	150–200 HP
Operational depth	3000 m (4000 m option)
Lifting capacity	At least 10 t
Payload capacity	200–300 kg
TMS (Tether Management Sys)	≥ 500 m
Manipulators	Several functions

- Hydraulic power supply for heavy duty tools (saw, drills, torque tool, etc.)
- Other subsea monitoring operations

Standard Equipment:

- nr. 1 20 ft standard container for power control (can be integrated on supporting vessel)
- nr. 1 20 ft standard container for spare parts and maintenance operations
- nr. 1 LARS system
- nr. 1 umbilical control winch
- nr. 1 ROV
- nr. 1 TMS.

Autonomous Underwater Vehicle

General description
Autonomous Underwater Vehicle (**AUV**) is a vehicle without crew (unmanned), autonomous an self propelled for operations in deep water beyond 3.000 m, it can be transported by container, train, high-way, airborne or naval logistic support (**MOSHIP)** main performances are listed below:

- Naval units, airplanes underwater wreckages search, localization and identification
- Archaeological/ historic wreckages underwater search, localization and identification
- Oceanographic, hydro graphic, meteorological data collection—Rapid Environmental Assessment- (**REA**)
- Ground-mine search, localization and classification
- Water chemical analysis

Deployment flexibility

The above listed features are accomplished with a flexible/modular **A**utonomous **U**nderwater **V**ehicle **AUV** able to be reconfigured according to the mission, replacing/integrating different payloads.

All data collected during missions shall bell computer based post processed and analyzed

It shall be able to perform preplanned seabed mapping and exploration missions without requiring specific checks during such operations, random on-going maintenance operations and efficiency checks are minimized, all electronic/ mechanical technical components are commercially available.

The Vehicle (**AUV**) is provided with specific **L**aunch and **R**ecovery **S**ystem **(LARS)** to guarantee lowering and lifting operations up to sea state **4** and **20 ft**. container located **(NOT** permanently secured) in the vicinity of **LARS** for stowage and transportation operations.

Related fittings

AUV is outfitted with related equipment to provide constant tracking (positioning, depth, bearing, speed) and mission diagnosis monitoring (from launch to recovery). It is provided with permanent acoustic tracking system installed on board the Vessel to keep continuous dialogue with the AUV up to 4.000 m depth

Transportation

The vehicle can be airborne along with all it's auxiliary systems.

Project definitions and feasibility studies

In relation to the program have been estimated several Risk Reduction studies. The studies are still under progress and related topics are listed below:

- Naval Architecture
- Propulsion System/ Electric Generation
- Reference Regulations
- Port/Airports/ Transferring Ideal Combinations
- Human Factors
- Logistics
- CBM Maintenance Theoretical Cycle
- CMS
- Security

Specific technical studies

Structural Design
Hoisting system Structural Design
EH101/NH90 Operational Activity Assessment
ROV-UAV Systems

Project definition

Whole Warship
Platform
Combat System
Radiation Hazard
Electromagnetic Interference
Antenna Coverage & Performance

Submarine Rescue Vehicle

A new generation of Submarine Rescue Vehicle (SRV) will be provided to perform primary tasks listed below, it is a 2- crew members vehicle, self propelled and can be airborne, transported by train, high way in support to **MOSHIP** operations.

The main characteristics of the submarine are:

- DISSUB localization up to 600 m by sonar outfitted system
- DISSUB visual inspection
- DISSUB Seat Mating Clearance
- DISSUB Mating Operations
- Emergency Survival Kit Transportation and Delivery
- **15** Rescuees Evacuation from DISSUB an transfer to **MOSHIP**

  evacuation from DISSUB can be performed under several conditions:

  Compartment with atmospheric pressure, high pressure, smoke.
- Decompression medical treatment for rescuees in specific SRV compartment up to **6 ATA**
- 40 inclination
- Underwater Operations with manipulators

General Description

SRV is a **2** compartment vehicle:

- Command compartment, it includes propulsion control system, sonar control, observation bull's eye, navigation control system, communication system, SRV internal atmosphere monitoring system electric equipment and related systems.

Figure 6. Rescue submarine operation.

- Rescue compartment contains rescue equipment/ fittings, atmosphere regeneration system and living space for rescuees.

Radio-electronic equipment

- Navigation control system
- Sonar research control system
- External 360° video control system
- External illumination system
- DISSUB internal atmosphere monitoring system
- SRV internal atmosphere regeneration sys up to 6 ATA

Trasportation

The system can be airborne along with all its necessary auxiliary systems such as deployable A- frame LARS.

5 CONCLUSIONS

The above project is considered the *driver concept* for final *design project* and has been developed by IT NAVY Engineering Department Officers, it represents **It Navy** sensitiveness and attention towards Submarine Department Personneel deployed on board IT Submarines as well as it's commitment to provide and keep a primary position in **SUBSEA** Operations within Mediterranean Sea.

According to a design and construction point of view this project will deeply engage National Naval Shipyard for Military Constructions as well as specialized Industries in design and development of Autonomous Remotely Operated Vehicles for Distress Submarine Rescue Operations and Diving Operations.

With an innovating design and lay out and in full respect of modern Human Factor standards it is outfitted with state-of- the-art platform and diving systems in terms of deployment and technological concept (It is intended for a wide operational use of deployable assets) as follows:

- Mini sub and related LARS
- Electrical propulsion with thrusters for *Station Keeping*
- 100 t crane for lowering/lifting operations
- 450 m Diving operations
- Unmanned Vehicles including WROV for Subsea operations up to 3000 m.
- Complex Command & Control System.

This unit will be certified in compliance with safety and ecological regulations and specifications.

It is oriented according the most recent logistic parameters and it will be supported by a 5 years Temporary Global Support and logistic cycle according to the latest criteria in terms of naval constructions.

Important part of the complex National device for Search and Rescue Operations and **ISMERLO Organization** (*International Submarine Escape and Rescue Liaison Office*) it will accomplish with tasks such as support to US Navy and Multiforce (*NOR-UK-FR*) Rescue Vehicles as well as Vessel of Opportunity (**VOO**) Operations.

Versatile and Multifunction it is due for operational commissioning in 2014 and it will provide dual use activities on behalf of other national and government departments.

Sustainable Maritime Transportation and Exploitation of Sea Resources – Rizzuto & Guedes Soares (eds)
 ISBN 978-0-415-62081-9

A new hydro-oceanographic survey vessel for the Italian Navy

E. Olivo (Cdr)
Italian Navy, La Spezia Naval Shipyard, Italy

ABSTRACT: "Ammiraglio Magnaghi", the oldest and largest survey vessel of Italian Naval Hydrographic Institute, is in service since 1975. Now the Italian Navy needs a replacement vessel with improved research capabilities and seaworthiness to extend the range and capability of research and to better support naval task forces as well as academic institutions. The paper will focus on the process followed through by the Italian Navy during the pre-feasibility study of a new Hydro-Oceanographic Survey Vessel (HOSV) in order to meet all the requirements.

1 INTRODUCTION

ITS Ammiraglio Magnaghi, commissioned in 1975, is near now to the end of its operational life, with levels of efficiency and capabilities no more cost-effective.

At the end of 2009 Navy General Staff wrote a draft version of an operational requirement for a replacement vessel in which was stated the need to provide for the acquisition of a new Hydro-Oceanographic Survey Vessel (HOSV) within 2013, in order to fill the technological gap accumulated in the oceanographic area due to the obsolescence of the ship and its systems.

Key Performance Parameters (KPP) of the new ship will be:

- wide use of merchant-vessel-type solutions, characterized by high reliability and maintainability;
- dynamic positioning capability;
- no environmental impact;
- low acoustic signature;
- good seaworthiness in rough seas.

Considering the above, some critical areas for the definition of the ship requirement were identified:

- regulatory framework;
- hull forms and general arrangements, modeling and simulation;
- propulsion systems;
- noise and vibration;
- manoeuvrability and sea-keeping;
- electromagnetic compatibility of systems and sensors.

At the beginning of 2010 the Navy decided to start the feasibility study of the new ship, launching, at the end of the same year, a series of consultations in the mentioned areas.

The studies, begun in March 2011, are temporally divided over a period of 600 days (two years) and organized into interdependent lots (i.e. the output of a study/lot will be the input of the following one).

In order to supply the necessary start-up of these studies, a in-house pre-feasibility study of the ship has been conducted by the Navy.

2 NEW HOSV GENERAL REQUIREMENTS

2.1 *Mission objectives and main capabilities*

The ship main mission objectives will be to conduct:

- coastal and offshore surveys;
- harbor areas and access routes surveys;
- researches of obstacles, artifacts and relics on the seabed;
- oceanographic and meteorological surveys;
- activities in support of REA (Rapid Environmental Assessment) operations.

As per above, the ship operational capabilities considered fundamental are:

- measuring depths from very shallow waters down to 5000 m (at least);
- the collection of detailed bathymetry of harbors and port access routes;
- exploration and characterization of the seabed down to 5000 m;
- the measurement of oceanographic parameters;
- sampling of sediment;
- measuring meteorological parameters;
- the surveying and geodesy;

2.2 *Regional needs*

The ship will be based in La Spezia (main naval base in northern Italy) and will operate mainly in the the Mediterranean Sea, the Atlantic Ocean (off the African coasts), the Red Sea, the Northern Indian Ocean, the Arabian Sea and the Arabian Gulf (Figure 1).

2.3 *Sea-keeping and station-keeping*

The ship should be able to operate up to sea state (SS) 4 and be able to sail up to SS 6. It will be equipped with a IMO class 2 dynamic positioning (DP) system and should be able to maintain station, on worst heading (crosswind and beam current, acting simultaneously in the same direction), in seas up to SS 4, wind speeds up to 20 knots and current up to 2 knots.

2.4 *Helicopters*

The ship will have a VERTREP area for helicopter operations. The area will be compliant with STANAG 1162 Edition 4 and will allow operations with helos up to EH101.

2.5 *Speed*

The ship should be able to maintain a continuous speed of 15 knots at full load end of life displacement (+4%) in calm sea conditions.

2.6 *Accomodations*

The ship will have accommodations for not less than 110 persons (including 30 scientific personnel). Adopted solutions should be compliant with Navy standards SMM 100 ed. 2003.

There will be a conference room and an exercise room.

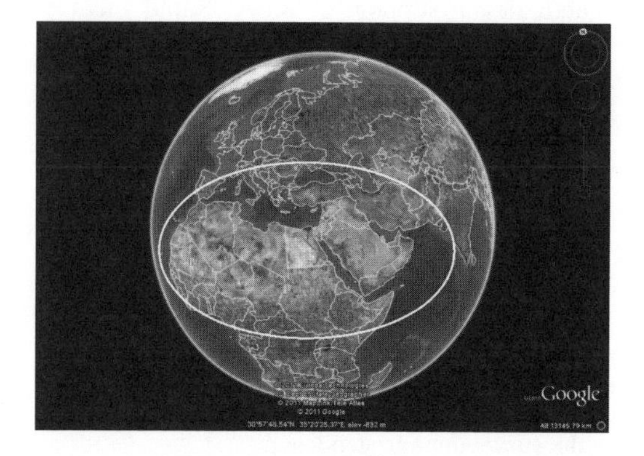

Figure 1. Wide Mediterranean Sea (source: Google Earth).

2.7 *Endurance*

The ship will be capable of remaining at sea for 30 days with the ability to transit for 4 days at cruising speed and to conduct 26 days' station work. Should be able to provide 30 days of hotel service.

2.8 *Range*

At least 7000 miles at 12 knots. The quantity of fuel to be embarked will be assessed according to the need of performing the mission profile.

2.9 *Mission profile*

- 18% in DP operations (~ 1 knots);
- 24% in Towing Ops at low speed (~ 4 knots);
- 43% in Free Run Survey (~ 8 knots);
- 12% at Cruise Speed (~ 12 knots);
- 3% at Max Speed (~ 15 knots).

2.10 *Rate of employment*

The ship will operate 3500 hours per year at sea.

2.11 *Discharges*

The ship will have no environmental impact. This will include the ability, with ship moored or at anchor, to hold any waste on board for 7 days. The ship should be capable to operate wherever and therefore all on-board systems shall comply with IMO MARPOL requirements; standards higher than what expected shall be adopted (i.e.: a double bottom throughout the length of the ship and double-hull for oil and fuel tanks).

2.12 *Safety at sea*

Despite not being armed, the ship will still be a military vessel and thus not subject to any Regulation. However, the ship will be designed and built in accordance to the Code of Safety for Special Purpose Ships, 2008, (SPS Code—Resolution MSC. 266 (84)) except for fire protection and life-saving systems, where Navy standards will be applied.

2.13 *Workboats*

The ship will have two motor boats (Figure 2) for coastal and harbor survey and two RHIBs.

The survey boats will have a length of 9–10 m, width 3 m and maximum draft of 1 m and will be fitted with very shallow water single and multibeam systems.

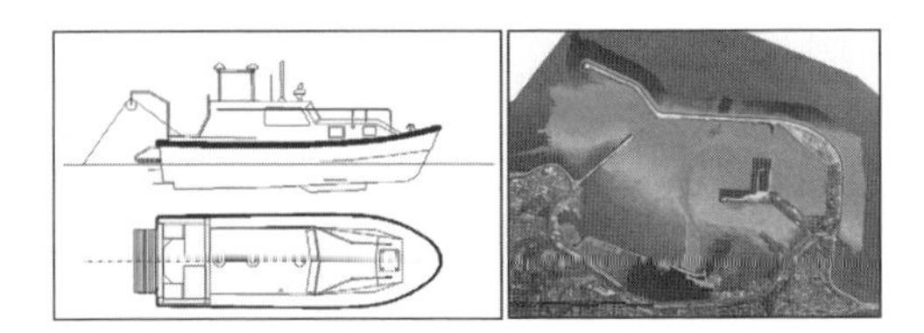

Figure 2. Example of survey motor boats (source: internet).

The RHIBs should be certified as SOLAS Fast Rescue Boats (FRBs).

2.14 *Hydro-oceanographic systems*

The ship will be fitted with a sophisticated suite of acoustic systems, including multi-beam and side scan sonars.

All the eco-sounding systems will be integrated with the other hydro-oceanographic devices.

Most of these sensors will be located in a special appendage called 'gondola' (Figure 3).

The hull-mounted multibeam mapping system (from shallow to deep water [5000 m]), will be used to produce high resolution seafloor maps.

These maps will be used primarily to create a 3D model of the sea bottom, including the identification of the unique seafloor features for further exploration.

The seafloor mapping systems will include also a set of towed side scan sonars for shallow (up to 200 m) and deep water (up to 1500 m) used for search of wrecks and obstacles (optionally with towed magnetometer) and for sea bottom characterization (Figure 4).

The ship will be capable to operate with ROVs and AUVs systems. They will come on board complete of LARS (Launch and recovery system) and will be able to operate up to 3000 m. These latter will be deployable and will be embarked depending on the type of mission that will be given to the ship.

ROVs will be equipped with high-definition cameras and lights, as well as special sensors and manipulators for collecting data and samples.

As might be already noted, particular attention has been given to the concept of mission oriented systems' configuration (MOSC) in order to let the ship to accomplish the mission assigned.

2.15 *Deck working area*

The ship will have a spacious stern working area (540 m^2), wood coated and fitted out with heavy-duty hold-downs (removable twistlock connectors) to provide high flexibility in the accommodation of large and heavy equipment.

Figure 3. Example of gondola (source: internet).

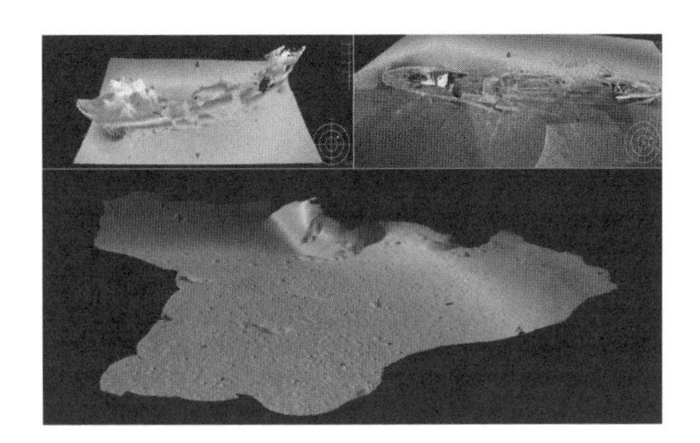

Figure 4. 3D model of wrecks and 3D maps (source: internet).

2.16 *Cranes*

The ship will be fitted with:

- one main crane with a 100 kN lifting capacity, capable of reaching all working deck areas and able to offload vans and heavy equipment. The crane winch shall have an active tension system;
- a second smaller knuckle boom crane, positioned on the opposite side of the main crane, for the launch and recovery of the piston corer system as well as for the transfer of heavy equipment.
- one A frame on the stern, for all the stern-launched instruments and trawls; size and capacity of the frame will be defined with appropriate modeling and simulation studies.
- one CTD frame positioned along side, with a maximum workload of 10 kN.

The fore deck should be equipped at least with one small knuckle crane (10 kN) capable of deploying goods over the side.

2.17 *Winches*

Flexibility in the winch configuration is mandatory as well as the ability to easily remove or install winches on deck and replace winch cable reels.

Winches must be as small as possible but still able to withstand the expected workload.

In general winches shall be multi-purpose and will carry 10,000 m of conducting and towing cables. This may be done with dual drums winches.

Winches supporting stern launched instruments will be assisted by an active heave compensator unit (AHC).

2.18 *Vans*

Up to two 8 ft by 40 ft ISO containers, which may be used for laboratory, storage, or other specialized use, should be hosted onboard. Hookup provision for power, HVAC, fresh and saltwater, compressed air, drains, communications, data and shipboard monitoring systems.

2.19 *Platform stability*

Vessel's stability at zero speed is important. Roll accelerations and velocities can be dangerous for instruments, lifting systems and for the personnel working with them.

Usually this type of vessels have multiple anti-roll tanks (active or passive), each one tuned (i.e. tank dimensions and level of fluid) for different ship loading conditions.

The roll motion of the ship in waves will be defined with appropriate modeling and simulation studies.

In addition to the mentioned systems, it will be assessed the effectiveness of a zero speed stabilizers system.

2.20 *Noise and vibration*

The proper functioning of instruments and non-interference with the environment must be ensured.

Attention should also be paid to noise and vibration levels expected in the cabins and public areas.

The requirement for underwater radiated noise, as well as on board noise and vibration levels, will be defined through a dedicated study.

2.21 *Sick bay*

The ship sick bay will consist of:

- one medical office;
- one medical store;
- one ambulatory equipped for orthopedic surgery;
- one room with four nursed beds and toilet facilities for patients with limited mobility.

The area will be placed on the first deck.

Double doors will facilitate the passage of stretchers.

2.22 *Laboratories*

The ship will be equipped with:

- one oceanography control station of not less than 80 m^2;
- one main laboratory, located on the main deck, below the CTD frame, with direct access outside. The laboratory will be divided into two sections: a 'wet' one in which instruments will be fitted out and samples collected, and a 'dry' one for analysis and data processing.
- one electronic laboratory;
- one REA-METOC room, with direct access to the outside, capable of accommodating up to 8 operators;
- one storage area for hydro-oceanographic equipment and instrumentation, located just below the working area, with direct access through a hatch large enough to allow passage of bulky equipment;
- one room for divers and diving equipment, with direct access to the working area;
- one briefing room capable of hosting press events.

3 COMPARISON WITH OTHER VESSELS

It is useful to compare the characteristics of the planned vessel with others in the worldwide fleet.

These ships are successfully operating for important military hydrographic offices:

HMS ECHO (2002), 90.6 m overall, beam 16.8 m, full load displacement 3500 t, draught at f.l. displ. 5.5 m, diesel-electric propulsion, azimuth thrusters. The ship is operated by the UK Hydrographic Office and is the first of a class of 2 vessels.

USNS PATHFINDER (1994), 100 m overall, beam 17.7 m, full load displacement 4920 t, draught at f.l. displ. 5.8 m, diesel-electric propulsion, azimuth thrusters, 27 scientists. The ship is operated by NAVOCEANO and is the first of a class of seven vessels, the latter of which is currently under construction.

FS POURQUOI PAS? (2005), 107.6 m overall, beam 20 m, full load displacement 6600 t, draught at f.l. displ. 5.8 m, diesel-electric propulsion, 40 scientists. The ship is operated by the *Service Hydrographique et Oceanographique de la Marine (SHOM)*.

SAGAR NIDHI (2003), 103.6 m overall, beam 18 m, full load displacement 4950 t, draught at f.l. displ. 4.3 m, diesel-electric propulsion, azimuth thrusters, 30 scientists. The ship is operated by the Indian National Institute of Ocean Technology (NIOT).

ITS ELETTRA (2002), 93 m overall, beam 15.5 m, full load displacement 2950 t, draught at f.l. displ. 4.9 m, diesel-electric propulsion. The ship is operated by the Italian Navy.

Although the diesel-electric solution appears to be fairly common for this type of ships, the same cannot be said for the propulsors, since all types of solutions (Z-drives, pods, shaft lines) are effectively used.

All ships above, except Sagar Nidhi, do not have a bulbous bow, in order to minimize underwater noise in the fore part of the hull, where echo-sounders are usually set.

ITS Elettra is not a hydrographic vessel, but its hull lines are derived from those of a NATO research vessel and have been studied to minimize noise, as well as maximise sea-keeping qualities.

All ships above, except, again, Sagar Nidhi, are immersed at full load displacement more than 5 m, due to the need of having hull mounted instruments as far as possible from the sea surface in order to minimize noise coming from or reflected by it.

4 NAVY SHIP'S CONCEPT DESIGN

4.1 *Main dimensions*

Length over all	105.6 m
Length between perpendiculars	95 m
Beam	18 m
Full load displacement	4220 t
Draught at f.l. displ.	5.2 m
WL length	97.9 m
WL breadth	18 m
C_B	0.450
C_P	0.515

As most of this type of vessels, the ship will be organized into three main sections:

- a stern working area, where all activities related to launch and recovery of instruments, equipment and research vehicles will be performed and where the containerized modules will be settled;
- a forward area dedicated to VERTERP operations and mooring systems;
- a central part with the superstructure, boarding areas and workboats bays.

Figure 5. Longitudinal view.

According to a good rule of thumb, the gondola with all the hull-mounted echo-sounders is located in the first third of the ship length, as far as possible from the main ship-borne noise sources (engines, propulsors, ...). The bow-thrusters will not interfere with the gondola sensors since they will not operate simultaneously.

4.2 *General description*

As already mentioned, the aft part of the ship is dedicated to the performance of those functions related to the nature of the mission itself: research and survey. In addition to the stern working area, the aft part of the deckhouse houses the laboratories, storage rooms and the divers room. Under the working area, with a hatch for direct access to it, there is a big storage room for heavy and bulky equipment.

The forward part of the ship is where the logistics functions are performed: here you can find the galley, lunch areas, food stores, laundry, gym and wardrooms, In particular, the block associated with the embark and storage of food and the preparation and packaging of meals, develops vertically and is connected via an elevator that reaches the foredeck, next to the VERTREP area. In the foredeck area two small knuckle cranes with a capacity of 10 kN assist the loading and off-loading operations of food provisions from the pier.

Amidships is where the crew work and rest. In the superstructure, the sickbay is located on the main deck, while offices and cabins are on the upper decks. Access through the upper decks is granted by two trunk stairs.

Only exceptions to this topology are the echo-sounders control cabinets room, located at the bottom of the hull, as close as possible to the gondola, and the navigation and oceanography bridges, located on the last upper deck, dominating and controlling the whole ship.

Technical compartments are all placed amidships, below the main deck.

The auxiliary room houses the chillers of the HVAC system, the sewage treatment plant, the desalination system for the production of fresh water and the fuel oil centrifugal separators.

The diesel-generators and the boilers for the production of hot water are located in the engine room.

The after technical compartment is dedicated to the Active Heave Compensation Unit and winches supporting stern launching instruments.

On the corridor deck (2nd deck), over the three compartments just mentioned, there's the

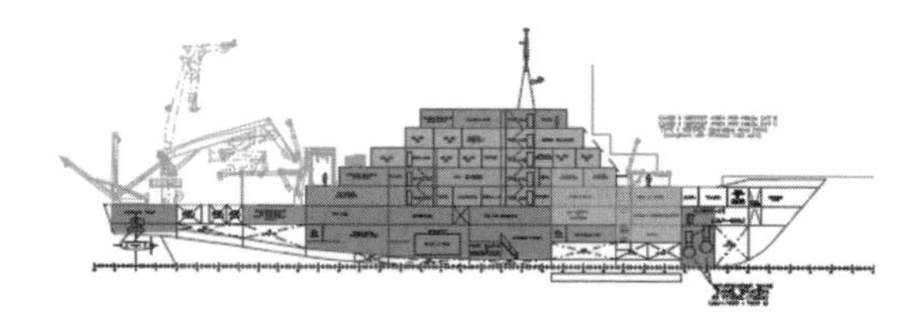

Figure 6. Functional arrangement.

engine control room, as well as workshops, stores and the switchboard rooms. Two watertight sliding doors facilitate the passage through these compartments.

On the main deck, amidships, on both sides, there are the boarding areas with gangways and accommodation ladders, RHIBs and motor boats bays. The mid part of the hull has vertical sides to facilitate workboats launching and recovery operations, as well as ensure the correct positioning of pilot ladders.

4.3 *Hull form*

The comparison with other similar vessels has shown a clear trend towards a propulsion system with azimuth thrusters. Nonetheless the Navy has commissioned a trade-off study between a standard shaft-line based propulsion solution and an azimuthal one.

This has been done since the Navy has already available among its ships a very efficient hull (ITS Elettra), but with a traditional (shaftline based) propulsion.

Therefore, during the pre-feasibility study, Elettra hull lines have been modified in such a manner to host two azimuth thrusters: while the fore body has been kept intact, the aft part of the hull has been completely re-designed with the aim to still have a good balance of volumes and weights, as well as good resistance performances (keeping Elettra as a benchmark).

The result has been quite satisfactory since, according to predictions, the new hull should absorb, at maximum speed, only 7% more power than Elettra and there is enough confidence that this amount of energy can be recovered by the higher propulsive efficiency offered by the azimuth thruster propulsion solutions.

4.4 *Propulsion systems*

As mentioned before, the Navy is looking at a propulsion system based on a diesel-electric solution and the propulsors that will result as the best trade-off among traditional propellers and some typical azimuth thrusters (pods, Z-drives, L-drives, epicycloidal).

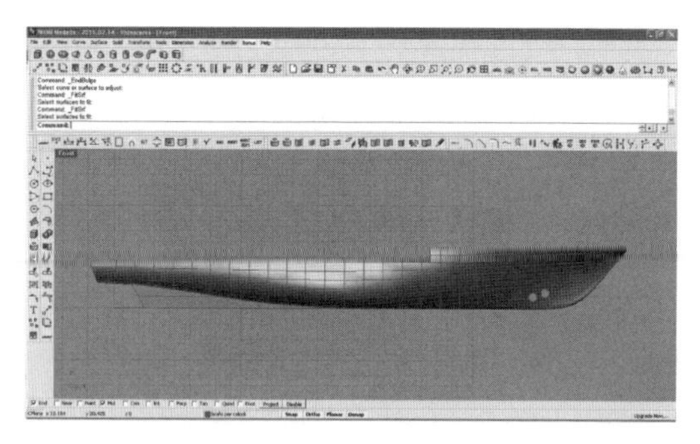

Figure 7. Ship's hull.

A dynamic positioning system compliant with IMO class 2 requirement involves the redundancy of all maneuvering thrusters and control surfaces, if any.

Due to the demanding station keeping requirement, the bow-thrusters are expected to be of not less than 750 kW each.

In station keeping and towing operations the power absorbed by thrusters and equipment (i.e.: AHC unit, winches, scientific instruments, services) is expected to be as much as the power needed for maximum speed.

This means that, according to the mission profile, the power demand will be maximum for almost 45% of the period at sea.

As per above, reliability issues suggest to have a power station composed by four generators at least, one of which always ready in stand-by.

5 THE TRADE-OFF STUDIES

In December 2010 the Navy started a series of consultations to explore more in detail some critical areas of the project.

As already mentioned in the introduction, the areas identified were:

- regulatory framework;
- hull forms and general arrangements, modeling and simulation;
- propulsion systems;
- noise and vibration;
- maneuverability and sea-keeping;
- electromagnetic compatibility of systems and sensors.

In March this year, the first lots of three activities started: regulatory framework definition, hull forms and general arrangements definition and propulsion system definition.

The objective of this very first phase is to define, according to cost-effectiveness parameters, the propulsion system (traditional or azimuthal), the hull

Figure 8. New Italian HOSV.

form and the type of thrusters and to verify the consistency of the solutions with the general arrangements.

The second phase will look more in detail into the solution chosen at the end of the first phase: the electrical system will be designed, hull form will be optimized (both numerically and through model tests), general arrangements will be revised. During this phase some outputs (drawings) will be preliminary approved by a classification society.

In this phase the noise and vibration design will start too (requirements definition, cooperation with general arrangement team regarding noise and vibration management).

A third phase will be dedicated to check and optimize ship maneuverability, resistance (gondola optimization) and sea-keeping qualities (definition of anti-roll system and parametric roll resonance analysis). In this phase, also, design loads of main lifting equipments (i.e.: A frames and cranes) will be valuated through a modeling and simulation analysis.

According to the time schedule approved, the feasibility study should finish by the end of 2012.

6 CONCLUSIONS

The need to replace the main survey vessel of the Naval Hydrographic Institute, ITS 'Ammiraglio Magnaghi', 36 years in service now, with a new modern and affordable ship, has led the Navy to start a thorough study to define operational and technical requirements of the new ship.

Firstly a in-house pre-feasibility study was conducted to identify main items and make ship requirements consistent. Then, a series of consultancies have been activated to support the feasibility study that should finish by the end of 2012.

This will help the Italian Navy to make the right choices and have a new hydro-oceanographic survey vessel able to perform with full satisfaction all missions assigned.

REFERENCES

IMO Resolution MSC.266(84) CODE OF SAFETY FOR SPECIAL PURPOSE SHIPS, 2008, 27th May 2011, <http://www5.imo.org/SharePoint/blastDataHelper.asp/data_id%3D22047/266%2884%29.pdf>

Institut français de recherche pour l'exploitation de la mer (IFREMER), IFREMER Fleet, 27th May 2011: <http://flotte.ifremer.fr/>

Italian Naval Hydrographic Institute, 27th May 2011: <http://www.marina.difesa.it/conosciamoci/comandienti/scientifici/idrografico/Pagine/home.aspx>

Kongsberg Maritime Subsea Division, 27th May 2011, <http://www.km.kongsberg.com/ks/web/nokbg0237.nsf/AllWeb/9DE6D8BDFF41EAA1C125742000444147?OpenDocument>

National Institute of Ocean Technology (NIOT), Vessel Management Cell, Facilities, 27th May 2011, <http://www.niot.res.in/op/vms/vesselmanagement_facilities.php>

Service Hydrographique et Oceanographique de la Marine (SHOM), 27th May 2011: <http://www.shom.fr/>

STANAG 1162 HOS (edition 4)—Vertical replenishment (VERTREP) operating area marking, clearances, and lighting, 27th May 2011, <http://www.tradoc.mil.al/Standartizimi/Downloads/1162Eed04.pdf>

Stato Maggiore Marina, SMM 100 "*Abitabilità* delle Unità Navali della M.M.", edizione Giugno 2003.

Survey Vessels of the World—7th Edition, 2008, Clarkson Research Services Limited, ISBN: 978-1-903352-84-7.

UK Hydrographic Office, Defence Maritime Geospatial Intelligence Centre, 27th May 2011: <http://www.ukho.gov.uk/Defence/Pages/Home.aspx>

UK Royal Navy, Multi-role survey vessels, 27th May 2011: <http://www.royal-navy.mod.uk/operations-and-support/surface-fleet/hydrographic-vessels/multi-role-survey-vessels/index.htm>

US Naval Oceanographic Office (NAVOCEANO), 27th May 2011: <http://www.navmetoccom.navy.mil/>

US Navy's Military Sealift Command (MSC), Oceanographic Survey Ships, 27th May 2011: <http://www.msc.navy.mil/inventory/inventory.asp?var=OceanographicSurveyship>

Sustainable Maritime Transportation and Exploitation of Sea Resources – Rizzuto & Guedes Soares (eds)
© 2012 Taylor & Francis Group, London, ISBN 978-0-415-62081-9

Conversion of conventional barges into floating dry-docks

M.C.T. Reyes, P. Kaleff & Carlos Elias
Universidade Federal do Rio de Janeiro, Brazil

ABSTRACT: The main steps of a float-on or float-off procedure are addressed, focused on barges meant to act as floating dry-docks when fitted with supplemental floatation aids. The operational limitations posed by ballasting, setting the freeboard limits and defining the necessary tank capacities are addressed next and followed by the specific buoyancy and stability issues specific to the variable instantaneous equilibrium states undergone by a barge fitted with floatation aids in the process of laying a supported craft dry or setting it afloat. Since the barge has to be totally sunken prior to hoisting the supported craft or setting it afloat, the design of the supplemental mobile floating aids is the central issue of the proposed procedure. The steps to establish the necessary buoyancy of such floatation aids, their intrinsic stability and their role in the float-on float-off process are treated next followed by considerations regarding implementation as single or double piece elements aiming at the reduction of cross sectional properties. Finally, a case study is presented aimed at illustrating the proposed procedure.

1 INTRODUCTION

Building and maintenance of small craft are generally performed on small launch ways in order to keep costs under control. Since yard operations tend to be less effective on an inclined vessel, a demand for dry dock availability exists but is not implemented due to the costs involved.

In this contribution an alternative to both launch ways and fixed dry-docks is proposed in terms of existing barges adapted to act as float-on and float-off aids in which both the launch operation of a new building as well as the hoisting of craft intended for maintenance may be performed in an expedite and economic manner.

Initially, the main steps of a float-on or float-off procedure are addressed, followed by the operational limitations posed by ballasting, setting the freeboard limits and defining the necessary tank capacities of the existing barge. The buoyancy and stability issues involved are considered next, focused on the variable instantaneous equilibrium states undergone by the barge in the process of laying the supported craft dry or setting it afloat. Since the barge has to be totally sunken prior to hoisting the supported craft or setting it afloat, the design of supplemental mobile floatation aids is the central issue of the proposed procedure. The steps to establish the necessary buoyancy of such floatation aids, their intrinsic stability and their role in the float-on float-off process are treated next followed by considerations regarding implementation as single or double piece elements aiming at the reduction of cross sectional properties. Finally, a case study is presented aimed at illustrating the procedure being proposed.

2 FLOAT-ON AND FLOAT-OFF PROCEDURES

Figure 1 depicts the qualitative steps of a float-on procedure for a barge acting as the supporting structure in association with removable floatation aids. The float-off procedure runs inversely.

Initially, Figure 1a, a barge in parallel floatation and minimum draft is assumed.

Figure 1b, shows the moment, after ballasting, in which the floatation aids are fitted to the barge after the docking blocks have been placed over the deck in order to receive the supported craft. Ideally, the draft should correspond to the minimum allowable barge freeboard.

Subsequently, the barge is ballasted until reaching an immersion sufficient to allow for the supported craft to be towed in and aligned for deposition over the docking blocks. That moment is depicted in Figure 1c.

Once the supported craft touches the upper surface of the docking blocks, the amount of ballast in the barge is reduced until the minimum freeboard for the removal of the floatation aids and further, in case there is still ballast remaining, until the barge's ballast tanks are empty (Figure 1d).

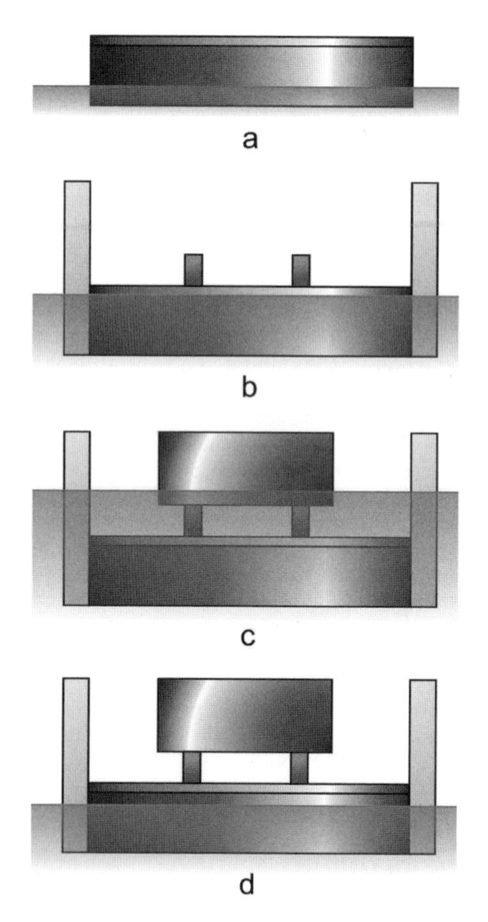

Figure 1. Steps of a float-on procedure for a barge acting as the supporting structure in association with removable floatation aids.

3 HYDROSTATIC CONSIDERATIONS

The practical implementation of procedures like the one mentioned above demands an adequate sizing of the floatation aids and the barge's ballast tanks. The corresponding floatation and hydrostatic stability considerations are presented in what follows.

3.1 *Maximum barge capacity*

The maximum load carrying capacity of the barge and thus the maximal weight of the supported craft, as referenced to Figure 1d, is given by:

$$W_{scm} = \rho \cdot g(V - A_d \cdot fbm) - W_{db} - W_f \quad (1)$$

where V = total volume of the barge; ρ = density of the barge's floatation medium; g = acceleration of gravity; A_d = area of the barge's deck; fbm = barge's free board maximum; W_{db} = weight of the docking bloks; and W_f = weight of the floatation aids.

3.2 *The critical phase of the procedure*

The critical phase of the procedure described above corresponds to the time interval between the instants b and c depicted in Figure 1 or, more precisely, between the instant the supported craft touches down completely on the docking blocks as depicted in Figure 2a and the instant the barge's deck emerges (Figure 2b).

Prior to complete touchdown, the immersed volume of the supported craft contributes both to the elevation of the center of buoyancy and to the increase in waterline area moment of inertia of the ensemble. After touchdown, only the floatation aids provide waterline area and the center of buoyancy of the ensemble moves downwards as the floating aids and docking blocks emerge. On the other hand, once the barge deck emerges, the waterline area assumes its maximum value and stability ceases to be an issue. Since there is no evidence as to which instant of this phase is the most unfavorable, the stability considerations will be developed for a waterline placed at a generic distance x above the barge's deck.

3.3 *Stability requirements*

In the critical phase described above, only the floatation aids and the docking blocks contribute to the restoring moment to inclinations of the ensemble. The effect of the docking blocks will be

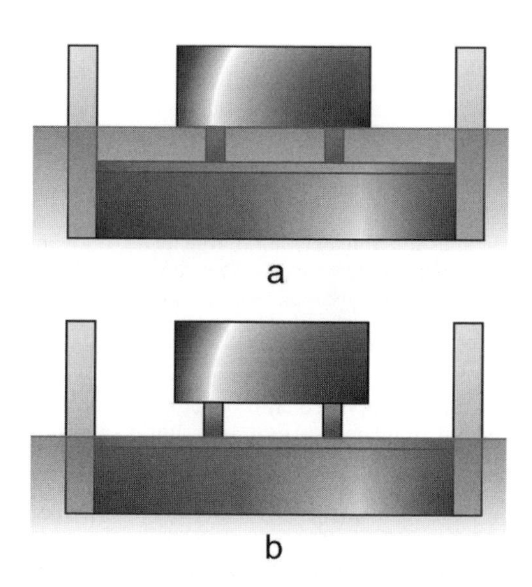

Figure 2. The boundaries of the critical phase of a float-on float-off procedure.

conservatively disregarded since their contribution is less significant than that of the floatation aids due to their proximity to the barge's centerline besides their exact arrangement being operation dependent and thus not precisely describable in general terms.

The moment of inertia of the waterline area of each floatation aid with regard to the central longitudinal axis of the barge, as per Figure 3, may be written as:

$$I = \frac{l_f \cdot b_f^{\,2}}{12} + l_f \cdot b_f \left(\frac{b_f}{2} + \frac{B}{2} \right) \tag{2}$$

where B = beam of the barge; b_f = transverse dimension of each floatation aid; and l_f = longitudinal dimension of each floatation aid.

The corresponding metacentric radius of the ensemble for a waterline placed at a distance x above the barge's deck, as per Figure 4, may be written as (Goldberg, 1988):

$$BM(x) = \frac{4I}{V + x.A_{db} + 4.l_f.b_f\left(D + x\right)} \tag{3}$$

where A_{db} = total horizontal area of the docking blocks; and D = depth of the barge.

The center of buoyancy of the ensemble may be found by considering the individual contributions of the components:

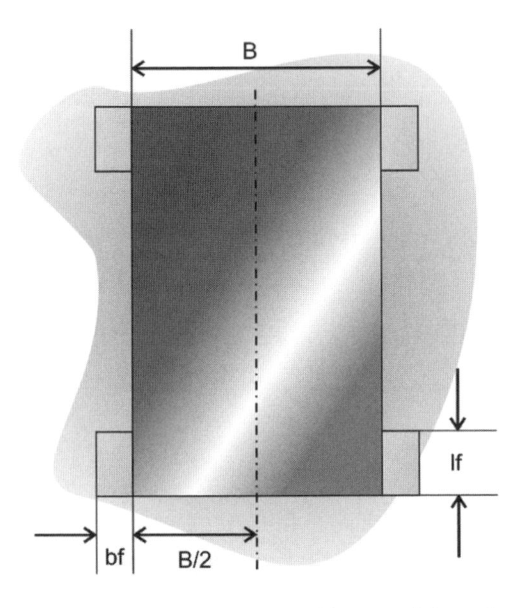

Figure 3. Relevant dimensions of barge and floatation aids for waterline moment of inertia during the critical phase of the float-on float-off procedure.

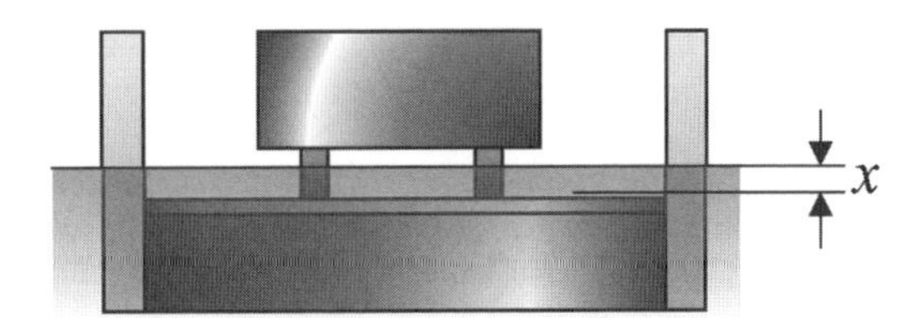

Figure 4. Generic position of an intermediate waterline during the critical phase of the float-on float-off procedure.

$$KB(x) = \frac{VCB.V + (D + x/2)x.A_{db} + l_f.b_f(D + x)(D + x)/2}{V + x.A_{db} + 4.l_f.b_f\left(D + x\right)} \tag{4}$$

where VCB = vertical center of buoyancy of the barge totally submerged.

The height of the ensemble's metacenter may be found by adding Equations (3) and (4). The corresponding value of the metacentric height may be found from that sum by subtracting the height of the center of gravity of the ensemble and taking the surface effect of the barge's ballast tanks into account (Goldberg, 1988):

$$GM(x) = KB(x) + BM(x) - KG(x) - FS(x) \tag{5}$$

with

$$KG(x) = \frac{W \cdot zG + W_{sc}(D + h_{db} + zG_{sc}) + W_f \cdot zG_f}{W + W_{sc} + W_f + W_{db} + W_{bt}} + \frac{W_{db}(D + h_{db}/2) + W_b(x) \cdot zG_b(x)}{W + W_{sc} + W_f + W_{db} + W_b(x)} \tag{6}$$

where W = lightweight of the barge; zG = height of center of gravity of the unloaded barge above barge keel; W_{sc} = weight of the supported craft; h_{db} = height of docking blocks; zG_{sc} = height of center of gravity of the supported craft above barge keel; zG_f = height of center of gravity of the floatation aids above barge keel; W_{db} = weight of the docking blocks; $W_b(x)$ = variable weight of ballast; $zG_b(x)$ = variable height of center of gravity of ballast; and $FS(x)$ = variable free surface effect of the ballast tanks (to be dealt with later).

The cross section of the floatation aids may be determined by trial and error from equations 2 to 6 and a target metacentric height by varying x in adequate steps while simultaneously assessing the content in the ballast tanks as shown in the following.

3.4 *The ballast tanks*

The barge's ballast tanks must be designed to provide the control necessary to an adequate performance in the critical phase of the float-on (float-off) procedure. To that end, it is desirable that one single ballast tank be responsible for the emersion (immersion) control during the critical phase, being thus the only partially filled tank, with the remaining tanks being kept either totally filled or totally empty.

The volume inside the control tank may be written as:

$$Vct = 4 \cdot m \cdot b_f \cdot l_f \cdot h_{db} \tag{7}$$

where m = is a chosen safety margin, larger than unity, inserted to assure that the volume of the control tank alone is sufficient to deal with the emersion (immersion) process during the critical phase.

The total weight of ballast during the critical phase may be written as:

$$W_b(x) = (W_{scm} - W_{sc}) + \rho \cdot g \cdot b_{ct} \cdot l_{ct} \cdot h_{ct}(x) \tag{8}$$

where $h_{ct}(x)$ = variable height of ballast in the control tank; l_{ct} = length of control tank; and b_{ct} = breadth of control tank.

Assuming that the volume corresponding to the safety margin m be equally distributed between the phases prior and following the critical phase:

$$h_{ct}(x) = \frac{(m-1)}{2} D + \frac{(3-m)}{2} \frac{D \,.\, x}{h_{db}} \tag{9}$$

The corresponding height of the center of gravity of the total ballast may be written as:

$$zG_b(x) = \frac{(W_{scm} - W_{sc}) D/2 + \rho \cdot g \cdot b_{ct} \cdot l_{ct} \cdot h_{ct}(x) \cdot h_{ct}(x)/2}{W_b(x)} \tag{10}$$

Finally, the free surface effect, due only to the amount of ballast in the control tank, may be written as (Goldberg, 1988):

$$FS(x) = \frac{\rho \cdot g \cdot l_{ct} \cdot b_{ct}^{3}}{12\left(W + W_{sc} + W_f + W_{db} + W_b(x)\right)} \tag{11}$$

3.5 *Stability of the floatation aids*

Since, in general, the floatation aids will have significant dimensions; it is desirable that their connection to the barge be performed with a minimum of external intervention (like cranes or the like) by floating them into position. To achieve this, the floatation aids must be hydrostatically stable individually and further, be able to receive and expel ballast in order to sink to a level sufficient to allow for an upward connection to the barge. Besides, it is also desirable to keep the vertical interaction forces between the barge and the floatation aids as low as possible, again by adjusting the amount of ballast inside the floatation aids.

Given that their height will in general significantly exceed their cross sectional dimensions, the floatation aids will only remain stable if the height of their center of gravity is sufficiently reduced by adding solid ballast.

In general, the metacentric height of the individual floatation aid may be written as:

$$GM_f = KB_f + BM_f - KG_f - FS_f \tag{12}$$

with

$$KB_f = (D - fbm + h_{fit})/2, \tag{13}$$

$$BM_f = \frac{l_f \cdot b_f^3}{12(D - fbm + h_{fit}) b_f \cdot l_f}, \tag{14}$$

$$KG_f = \frac{W_{lf} \cdot h_f/2 + W_{sb} \cdot zG_{sb} + \rho \cdot g \cdot l_{ft} \cdot b_{ft} \cdot h_{ft} \cdot h_{ft}/2}{W_{lf} + W_{sb} + \rho \cdot g (l_{ft} \cdot b_{ft} \cdot h_{ft})} \tag{15}$$

and

$$FS_f = \frac{\rho \cdot g \cdot l_{ft} \cdot b_{ft}^{3}}{12\left(W_{lf} + W_{sb} + \rho \cdot g (l_{ft} \cdot b_{ft} \cdot h_{ft})\right)} \tag{16}$$

for which h_{fit} = fitting clearance of the device (generally a pair of angle bars) connecting the floatation aids to the barge; W_{lf} = light weight of the floatation aid; W_{sb} = weight of solid ballast in the floatation aid; zG_{sb} = height of center of gravity of solid ballast in the floatation aid; l_{ft} = length of floatation aid ballast tank; b_{ft} = breadth of floatation aid ballast tank; h_{ft} = height of floatation aid ballast tank;

4 SAMPLE APPLICATION

In order to illustrate the proposed procedure, a sample barge and a sample craft to float-on were selected. The main characteristics of the barge, assumed as a perfect parallelepiped, are shown in Table 1 and those of the supported craft in Table 2.

From the sample barge's carrying capacity, a maximum weight of 502 tons for the supported craft was found.

The height of the floatation aids, considering a fitting clearance of 0,3 m, a docking blocks height of 1,0 m and an upper clearance of 0,5 m, was found to be 4,19 m.

With a target metacentric height of 1,0 m (as per ABS 1977), a cross-sectional length (*lf*) of 2,2 m

Table 1. Main characteristics of the supporting barge.

Length	28,80 m
Breadth	11,30 m
Draft (light)	0,60 m
Depth	1,74 m
Light weight	192,89 t
Center of gravity above base line	0,85 m

Table 2. Main characteristics of the supported craft.

Length	24,00 m
Breadth	8,30 m
Draft (docking)	0,80 m
Depth	1,64 m
Light weight	92,29 t
Center of gravity above base line	1,85 m

and a cross-sectional breadth (*bf*) of 1,6 m where found sufficient for each floatation aid.

However, while analyzing the stability of the floatation aids with an individual target metacentric height of 0,15 m, a cross-sectional length of 4,0 m and a cross-sectional breadth of 3,2 m where found necessary.

An additional simulation was performed, to find the maximum height for stability of the individual floatation aids with the minimum cross sectional dimensions necessary to provide stability to the barge ensemble. A value of 2,7 m was found; indicating that two piece floatation aids (with an upper piece of 1,49 m to be lifted into position and attached with external help once the lower piece is floated into position and attached to the barge by purely hydrostatic means) might be a compromise to avoid the excessive cost of a larger cross section.

5 CONCLUSION

It was shown that, with the procedure proposed herein, common barges may be transformed to act as floating dry-docks.

In the practical application presented, the supported craft had a significantly lower weight than the carrying capacity of the barge. This circumstance generated floatation devices with dimensions insufficient to assure their own stability demanding alternative solutions like two-piece devices to keep transverse dimensions to a minimum, but demanding external installation aid or devices with a larger cross section to assure individual stability.

Since the proposed procedure is one of trial and error, it is not possible to assure beforehand whether a given combination of barge and supported craft will result in an optimal solution for the floatation aids. Thus, besides the recommendation to implement the procedure in a worksheet goes the suggestion to run it for a wide range of differing supported craft in order to determine the envelope values of floating aid cross sections ideal for a given barge.

REFERENCES

ABS 1977. Rules for Building and Classing Steel Floating Dry Docks, American Bureau of Shipping.

Goldberg, L.L. 1988. Intact Stability. In Edward V. Lewis (ed), *Principles of Naval Architecture—Second Revision, Vol. I, Ch. 2, 1988*, The Society of Naval Architects and Marine Engineers, New Jersey.

4.3 *Shipyards*

Sustainable Maritime Transportation and Exploitation of Sea Resources – Rizzuto & Guedes Soares (eds)
© 2012 Taylor & Francis Group, London, ISBN 978-0-415-62081-9

Analysis of welded joints as a base for improvement of automated welding

D. Pavletic
Faculty of Engineering, University of Rijeka, Rijeka, Croatia

T. Vidolin
Shipyard MAJ, Rijeka, Croatia

I. Samardzic
Mechanical Engineering Faculty, Josip Juraj Strossmayer University of Osijek, Osijek, Croatia

ABSTRACT: The paper presents results of analysis of welded joint types produced at section assembly of ship hull production. The welded joint analysis, made for a year production at selected shipyard, represents a base for assessment of current level of welding automation achieved in this early phase of ship production. Taking into account main principles of Theory of Constraints and Lean manufacturing, in the paper are examined possible ways of welding automation improvement, based on the results of performed analysis. Authors had analyzed selected shipyard yearly production programme in order to assess automated welding needs and to make suggestion for process improvement. Improvement recommendations take into account modern and proven continuous improvement methodologies.

1 INTRODUCTION

How efficient are installed automated welding lines for a ship hull assemblies welding? Do they have enough capacity? When and how much it will be needed to invest to improve existing capacity?…

To answer some of those questions, and to analyse possible ways of improvements, an analysis of welded joint types, produced at a section assembly of a ship hull production, is carried out. The analysis is based on a yearly production program of selected shipyard. Obtained results give information of how much capacity is needed for current production volume. Based on that information, assessment of suitability of installed automated welding equipment can be made, and possible ways of improvements traced.

2 ANALYSIS OF WELDED JOINT TYPES PRODUCED AT A SECTION ASSEMBLY OF A SHIP HULL PRODUCTION

To assess a demand for the capacity of automated welding lines, the analysis of yearly ships production for selected shipyard is conducted. Based on production of five ship hulls per year, a number of needed subassemblies, panel assemblies and completed panel assemblies are determined (Vidolin 2010).

Two types of ship representatives are selected and that are chemical tanker (in further text denominated as Type A), and oil/chemical tanker (Type B). The main characteristics of selected ships representatives are shown in Table 1.

2.1 *Analysis of panel subassemblies production*

The analysis of panel subassemblies production is conducted for two selected ships types. The analysis has taken into account only simple panel subassemblies, excluding corrugated bulkheads and curved hull subassemblies.

Table 1. A main characteristics of ships' representatives.

	Ship type	
Main characteristics	Chemical tanker (Type A)	Oil/Chemical tanker (Type B)
Length overall, LOA (m)	168,00	182,50
Length between perpendiculars, LPP (m)	160,80	174,80
Breadth, B (m)	26,40	32,20
Depth, moulded to upper deck, H (m)	13,80	17,50
Design draught, D (m)	9,02	12,20
Deadweight, (dwt)	23 400	47 300

As it is shown in Table 2, analysis of panel subassemblies production has resulted with following information:

- Lengths of fillet welds (m),
- Lengths of butt welds (m),
- Number of panel subassemblies prepared for later ship hull construction,
- Number of elements (plates and profiles) in subassemblies,
- Mass of subassemblies

In order to determine the maximum load capacity and the needed capacity of automated welding lines, the analysis of the needed capacity in the most loaded period of the year is conducted. In doing so, two basic scenarios are analyzed, namely:

- Scenario 1—Same sections of both ships are produced at a same time
- Scenario 2—A mid-ship section of one ship and the engine section of the other ship are produced at a same time.

Results of scenarios analysis are shown in Table 3 and Table 4. Comparing these two scenarios it can be concluded that the demand for fillet weld is some 7% higher in Scenario 2. At the same time, the demand for butt welds is 27% higher in Scenario 2, but total demand for butt welds is negligible in comparison to fillet weld demand.

2.2 *Panel assemblies production analysis*

Panel assemblies' production analysis has had an aim to obtain information about:

- Number of panel assemblies produced yearly,
- Masses of panel assemblies,
- Lengths of butt and filet welds.

The analysis of number of panel assemblies yearly production is shown in Table 5, while results of the analysis of weld lengths for two production scenarios are shown on Table 6 and Table 7.

Table 2. Analysis of Type A and Type B ships at a panel subassemblies production.

					Number of elements		Total			
	Group	Length of fillet welds, m	Length of butt welds, m	No. of subass.	Completed panels ass. (no. of plates and profiles)	Other (no. of plates and profiles)	Mass, kg	No. of plates	No. of profiles	No. of elements
Type A	Stern	1114	179	116	18	487	68317	165	340	505
	Engine room	1648	86	251	242	753	123331	416	579	995
	Cargo tanks	27677	1132	2287	2692	7826	1462185	2682	7836	10518
	Bow	1661	33	303	84	675	63612	320	0	759
	Super-structure	3649	276	337	0	1144	123551	368	776	1144
Total for Type A		**35749**	**1706**	**3294**	**3036**	**10885**	**1840996**	**3951**	**9531**	**13921**
Type B	Stern	1070	94	150	66	387	48667	162	291	453
	Engine room	4083	148	666	493	1404	177229	723	1174	1897
	Cargo tanks	15116	257	1772	4672	2446	908539	1817	5301	7118
	Bow	2885	50	417	246	984	117832	439	791	1230
	Super-structure	7697	194	722	0	2397	209442	767	1630	2397
Total for Type B		**30851**	**743**	**3727**	**5477**	**7618**	**1461709**	**3908**	**9187**	**13095**
3 × Type A		107247	5118	9882	9108	32655	5522988	11853	28593	41763
2 × Type B		61702	1486	7454	10954	15236	2923418	7816	18372	26190
Total for five ships		**168949**	**6604**	**17336**	**20062**	**47891**	**8446406**	**19669**	**46965**	**67953**

Table 3. Analysis of weld lengths for scenario 1.

	Length of fillet welds, m	Length of butt welds, m	Total length of welds, m
Total for Type A	4269	82	4351
Total for Type B	8749	261	9010
Total for Scenario 1	**13018**	**343**	**13361**

Table 4. Analysis of weld lengths for scenario 2.

	Length of fillet welds, m	Length of butt welds, m	Total length of welds, m
Total for Type A	13938	437	14375
Total for Type B	–	–	–
Total for Scenario 2	**13938**	**437**	**14375**

Table 5. Analysis of Type A and Type B ships at panel assemblies production lines.

	Group	Number of panels
Type A	Stern	6
	Engine room	34
	Cargo tanks	122
	Bow	10
	Superstructure	0
Total for Type A		**172**
Type B	Stern	2
	Engine room	34
	Cargo tanks	103
	Bow	10
	Superstructure	0
Total for Type B		**149**
3 × Type A		516
2 × Type B		298
Total for five ships		**814**

Table 6. Analysis of weld lengths for scenario 1.

	Length of fillet welds, m	Length of butt welds, m	Total length of welds, m
Total for Type A	6009	829,5	6838,5
Total for Type B	4890	858,5	5748,5
Total for Scenario 1	**10899**	**1688**	**12587**

Table 7. Analysis of weld lengths for scenario 2.

	Length of fillet welds, m	Length of butt welds, m	Total length of welds, m
Total for Type A	5412	1272	6684
Total for Type B	–	–	–
Total for Scenario 2	**5412**	**1272**	**6684**

Table 8. Analysis of Type A and Type B ships at completed panels.

	Group	Number of panels
Type A	Stern	6
	Engine room	29
	Cargo tanks	84
	Bow	10
	Superstructure	0
Total for Type A		**129**
Type B	Stern	2
	Engine room	32
	Cargo tanks	65
	Bow	6
	Superstructure	0
Total for Type B		**105**
3 × Type A		387
2 × Type B		210
Total for five ships		**597**

Table 9. Analysis of weld lengths for scenario 1.

	Length of fillet welds, m	Length of butt welds, m	Total length of welds, m
Total for Type A	2627,6	36,7	2664,3
Total for Type B	5173,3	204,8	5378,1
Total for Scenario 1	**7800,9**	**241,5**	**8042,4**

Table 10. Analysis of weld lengths for scenario 2.

	Length of fillet welds, m	Length of butt welds, m	Total length of welds, m
Total for Type A	8597,9	410	9007,9
Total for Type B	–	–	–
Total for Scenario 2	**8597,9**	**410**	**9007,9**

In the case of panel assemblies, the demand for fillet welds is much higher for Scenario 1, some 50% higher. Also, demand for butt welds is 33% higher for Scenario 1.

2.3 *Analysis of the production of completed panel assemblies*

Analysis of the production of completed panel assemblies had provided information about:

- Number of completed panel assemblies produced yearly,
- Masses of completed panel assemblies,
- Lengths of butt and filet welds.

The analysis of number of completed panel assemblies production per year is shown in Table 8, while results of the analysis of weld lengths for two production scenarios are shown on Table 9 and Table 10.

Length of fillet welds, in the case of completed panel assemblies is 10% greater in Scenario 2. The length of butt welds is again negligible.

3 SUITABILITY OF INSTALLED AUTOMATED WELDING EQUIPMENT

As previously shown, the analysis of ship hull for ships representatives, based on a yearly shipyard production, provides the number of produced panel subassemblies, assemblies and completed assemblies. The data about equipment capacity, in terms of lengths of butt and filler welds, as well as, the number of panel subassemblies, assemblies and completed assemblies, that are welding equipment capable to produce within one month period, are extracted from the technical documentation of the automated welding equipment installed within selected shipyard. Comparing these data with, previously calculated and shown, data for Scenario 1 and Scenario 2, showed that installed automated welding equipment have sufficient capacity and that, at the moment, there are no bottlenecks in the production process. The comparison of the installed capacity and the demand is shown on Table 11.

Although installed welding capacity satisfies current demand, it can be seen that an increase in the production volume will in short time create a lack of capacity, firstly at completed panel assemblies production. Therefore, it is necessary to research a possible ways of capacity increase. Off course, it can be achieved by investing in technology. For example, at subassembly panel production line a high efficiency arc welding process, such is TIME TWIN process, can be installed. By a high efficiency welding process the welding speed will rose from current 40 cm/min, for GMAW process, up to 100 cm/min. Obviously, this will have a strong impact on capacity of welding lines (Milos et al., 2010). An example of welding speeds for 4 mm fillet weld is shown on Table 12.

Further technological improvement can be in a form of additional robot at the robot micro panel line. This improvement can be done by adding an additional robot arm on the existing welding portal or by installing a new portal with one or two robot units. Or, by combination of these two options.

At the completed panel assemblies the capacity can be increased by installing robot welding. This type of welding equipment is specially designed for welding large assemblies/sections.

All above mentioned solutions for improvement demands usually considerable investments, and should be considered for implementation when all other, less resource demanding, solutions are exploited.

Table 11. Analysis of capacity for selected shipyard.

	Installed capacity for selected shipyard		Required capacity for selected production program: Scenario 1		Scenario 2	
	Fillet welds capacity, m	Butt welds capacity, m	Fillet welds requirement, m	Butt welds requirement, m	Fillet welds requirement, m	Butt welds requirement, m
Micro panels (panel subassemblies)	20812	0	13018	343	13938	437
Panel line (panel assemblies)	38016	4752	10899	1688	5412	1275
Completed panel assemblies	8600	0	7800,9	241,5	8597,9	410

Table 12. Welding speeds for different welding processes (fillet weld, a = 4 mm).

Welding process	Welding speed
SMAW	20–25 cm/min
GMAW (Flux cored wire)	25–35 cm/min
SAW (panel line)	120–170 cm/min
TIME TWIN	100–120 cm/min

Bellow will be considered ways of improvement and getting maximum of existing welding equipment.

4 MAIN PRINCIPLES OF THEORY OF CONSTRAINTS AND LEAN MANUFACTURING

To explore possibility for improvement of automated welding in context of selected shipyard and production programme, the two well know improvement methodologies were selected. The one is Theory of Constraints (TOC) and the other is Lean Manufacturing. Main principles of both methodologies are briefly described below.

4.1 *The Theory of Constraints*

The Theory of Constraints assumes that organizations can be measured and controlled by three main measures: throughput, operating expense, and investment. Throughput is value generated through sales. Investment is value the system invests in order to sell its goods. Operating expense is all the values the system spends in order to turn the investment into throughput (Cox et al., 1986).

Theory of Constraints is based on the premise that the rate of goal achievement is limited by at least one constraining process. Only by increasing flow through the constraint can overall throughput be increased (Goldratt 1998).

A constraint is anything that prevents the system from achieving more of its goal. By definition, there are three main types of internal constraints:

- Equipment: The way equipment is currently used limits the ability of the system to produce more salable goods / services.
- People: Lack of skilled people limits the system. Mental models held by people can cause behaviour that becomes a constraint.
- Policy: A written or unwritten policy prevents the system from making more.

Assuming that automated welding lines are constraining process in a ship hull preassemblies and assemblies production, and deciding that people and policy, as possible constraints within process, will not be taken into account at the time, the task is to find ways how to make most of installed welding equipment.

The steps that should be followed in TOC methodology application to selected production process are (Dettmer 1997):

1. Identify the constraint.
   If the improvement of automated welding lines is in focus, based on the TOC principles, the resource that prevents the shipyard to gain more of its goal are automated welding lines/station it selves.
2. Decide how to exploit the constraint.
   In order to get the most capacity out of the constrained processes or equipment, the processes have to be available most of the time. Thus, the process availability has to be assessed and identified tasks that have priority in achieving as higher availability as possible. That certainly involves efficient maintenance. Next thing that has to be put in practice is to process only products that are of acceptable quality. Any product that is of unacceptable quality must be filtered out before they reach the constrained equipment. So, the process quality control has to be stressed out and, if needed, reorganized, embedded in process and made more efficient. The constrained equipment must not wait for other processes. It should be the process/station that is waited for.
3. Subordinate all other processes to above decision.
   To put the constrained equipment to work at its maximum, all other connected processes have to subordinate to needs of constrained equipment. That usually means that reorganization of processes, priorities, buffers and other connected resources must take place.
4. Elevate the constraint (make other major changes needed to break the constraint).
   In case the constrained welding equipment still prevents achieving desirable throughput, some significant technological changes (investment) in process or equipment should be considered. This can, for example, involves outsourcing part of production, or investing into hardware to raise equipment capacity.

4.2 *Lean Manufacturing*

As well as TOC, Lean Manufacturing is also based on several main principles which have to be followed, more or less strictly, to gain benefits from efforts puts into process improvement.

The Lean Manufacturing involves five main steps in implementation. Although these five steps

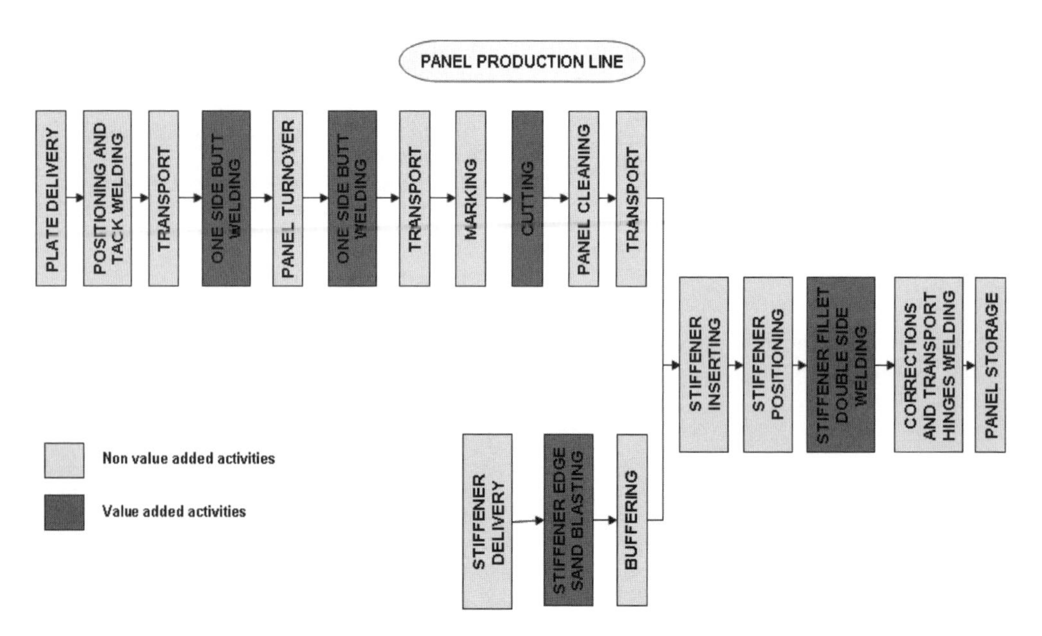

Figure 1. Panel assembly production line value added and non-value added activities.

are quite strait forward, they encompass main Lean principles and it is not always easy to achieve it (Pavletic 2006). At the beginning of most Lean journeys is housekeeping, introduced and accomplished by application of 5S method. By thorough sorting, setting in order and cleaning, which are first three "S" from 5S method, the problems that used to be hidden by sloppy messes are revealed (Rother et al., 1999).

The five-step process:

1. Specify value from the standpoint of the end customer. The customer is willing only to pay for the product he/she order. In contest of construction of ship hull, and simplified, the customer is willing to pay for welded joint of acceptable quality. Of course, only those welded joints that are necessary to keep ship construction elements together. No more and no less.
2. Identify all the steps in the value stream, eliminating whenever possible those steps that do not create value. Value Stream Mapping is a flow of inventory or "the thing" being processed, and flow of the information needed to process it. Each value stream map is primarily focused on the flow of the thing being processed. The welding process is consisting of several activities that are carried out to produce welded assembly. Only few of them are value added activities. All other activities, non value added activities, should be considered as waste, and an effort should be put in to remove or minimize those activities. An example for panel assemblies production line is shown on Figure 1 (Pavletić 2005).
3. Make the value-creating steps occur in tight sequence so the product will flow smoothly toward the customer. At this point also Six Sigma methodology becomes interesting because it is necessary to have high quality to make small batches or, even, single piece flow, and to produce just in time only what the customer demands.
4. As flow is introduced, let customers pull value from the next upstream activity. It should be kept on mind that the goal is not to maximize output but to maximize throughput or goods purchased by a customer.
5. As value is specified, value streams are identified, wasted steps are removed, and flow and pull are introduced, begin the process again and continue it until a state of perfection is reached in which perfect value is created with no waste. By definition, continuous improvement is continuous (Rother 2010). A lean journey always has a next step and never has an end.

5 CONCLUSION

Efficiency of a contemporary commercial ships production greatly depends of efficiency of welding. Although welding as a process is present in almost all phases of ship production, probably the most of it, take its place in welding of ship hull assemblies. To produce welds of acceptable quality, in quantity and time defined by succeeding processes, automation of welding is, probably, today only solutions. Although automated welding systems are technologically sophisticated system, investments in which are usually considerable, they also have its shortcomings, mainly in flexibility. They are designed, selected and installed in process with clearly defined production goals. To change their application area, both in product range or quantity, usually means more investment and technological change on equipment. Before go for that option all other "organizational option" have to be explored. To be able to do that, the needed welding capacity has to be known in all stages of ship hull production. The current, and even future, production range has to be known. The types and quantities of welding joint have to be analysed. Based on such analysis, suitability of installed equipment can be assessed and exploited up to its maximum before an investment in new technology take place. To do so, there are several modern improvement methods, two of them are Theory of Constraints and Lean Manufacturing. By applying main principles of these methods many advantages can be expected, such as decrease of production costs of subassemblies and assemblies, increase in automated welding lines throughput, quality improvement, improvement of process manageability and better working environment.

REFERENCES

Cox, J. & Goldratt, E.M. 1986. *The goal: a process of ongoing improvement,* Croton-on-Hudson, NY, North River Press.

Dettmer, H.W. 1997. *Goldratt's theory of constraints: a systems approach to continuous improvement*, ASQ, Quality Press, Milwoukee, Wisconsin.

Goldratt, E.M. 1998. *Essays on the Theory of Constraints*, Great Barrington, MA, North River Press.

Milos, V., Bilic, D. & Pavletic, D. 2010. *Automated pipes welding*, Engineering Review, Vol. 30–2, Rijeka.

Pavletic, D. 2005. *Lean Welding in Shipbuilding*, 3rd International scientific-professional Conference, 50 years of welding tradition for the future, Slavonski Brod.

Pavletic, D. 2006. *Lean Manufacturing in Shipbuilding*, XVII. SORTA Symposium, Opatija.

Rother, M. 2010. *Toyota KATA*, Mc Graw Hill, New York.

Rother, M. & Shook, J. 1999. *Learning to See*, The Lean Enterprise Institute, Massachusetts.

Vidolin, T. 2010. *Automated welding of ship hull sections*, Graduate thesis, University of Rijeka, Faculty of Engineering, mentor: Assoc. Prof. D. Pavletić, Ph.D.

Sustainable Maritime Transportation and Exploitation of Sea Resources – Rizzuto & Guedes Soares (eds)
© 2012 Taylor & Francis Group, London, ISBN 978-0-415-62081-9

Underwater photogrammetry for 3D modeling of floating objects: The case study of a 19-foot motor boat

F. Menna
3D Optical Metrology Unit, Bruno Kessler Foundation (FBK), Trento, Italy
LTF—DSA, "Parthenope University" Naples, Italy

E. Nocerino, S. Del Pizzo, S. Ackermann & A. Scamardella
LTF—DSA, "Parthenope University" Naples, Italy

ABSTRACT: 3D modeling of floating or semi-submerged objects is a challenging and attractive task for the marine industry especially if the manufacturing of components that have to be replaced or repaired after a damage is necessary or the ship itself has to be converted. Up to now the 3D reverse engineering of ships has required docking operations to carry out a geodetic or photogrammetric survey with high costs for shipowners. In this paper an innovative 3D acquisition method for digital recording of floating objects is presented. The method is based on digital photogrammetry both underwater and terrestrial. Preliminary tests are presented for the case study of a 19-foot motor boat. Two surveys of the boat in floating conditions are carried out and then joined by means of special rigid orientation devices built ad hoc.

1 INTRODUCTION

Nowadays the knowledge and digital representation of boat hull and appendages is a fundamental issue from the naval architect point of view. The 3D digital model is the starting point of the iterative design process (well known as "design spiral") that is followed during the practical design of a new craft. However, retrieving the actual shape of an existing vessel or its part can be also necessary when no product information exists as in the case of digital documentation of maritime heritage (Menna et al., 2011, Wiggenhagen et al., 2004), (Kentley et al., 2007b) or when such information is considered unreliable because of deformation and/or damages occurred over time (Koelman 2010), (Menna et al., 2009), (Menna & Troisi 2007), (Goldan & Kroon 2003). For this purpose, a reverse engineering procedure is needed to reconstruct the shape, dimension, and semantic information of the surveyed object. The measurement technique represents the basis of the reverse engineering workflow and the choice of the survey method depends on several factors (object location, object size, required accuracy, kind of analysis to be performed, etc.) (Remondino & El Hakim 2006) (Remondino et al., 2005), which have to be clearly defined a priori.

For retrieving digital models of marine vehicles or floating objects in general, high precision measurements are usually performed with the object beached or docked. Such operations are highly costly for shipowners as the unit must be set in out of service. Other practical difficulties may arise due to the setting (dry or floating docks) where the survey has to be conducted, often characterized by restricted spaces, disadvantageous environment conditions (e.g., water, wetness, saltiness), etc. In previous works (Menna et al., 2010) (Menna et al., 2009) (Ackermann et al., 2008) (Menna & Troisi 2007), the authors showed that digital photogrammetry is suitable for shipbuilding ashore applications. The photogrammetric technique has been successfully employed for obtaining dense and accurate 3D models of free form surfaces from digital images in accurate, flexible and economical way. Applications of underwater photogrammetry have been also presented; the technique has been proved useful for mapping and retrieving the shape and geometry of objects completely submerged. In particular many published papers are focused on archeological site surveys (Drap et al., 2007), (Canciani et al., 2003) or monitoring of marine fauna populations (Shortis et al., 2007). In the knowledge of the authors up to know there are no scientific publications about the reverse modeling of floating or semi-submerged objects and structures. The measurement of an object below and above the sea level is of great interest for the marine sector especially for manufacturing the replacement elements needed for damage repair or conversion of ships. Usually, the repair of damaged ship hulls starts with a visual inspection for establishing the nature and extent of the damage below and above waterline. This is important for determining the need and the possibility of carrying out repairs in situ, the work method, and the cost of repair for the insurance (Goldan & Kroon 2003). The advantage of recording the shape of a floating or partially submerged structure could be an attractive solution for speeding and reducing the costs of survey and repair operations.

In this contribution, an innovative methodology is presented: a consumer-grade digital camera was employed for retrieving the 3D digital model of a 19-foot pleasure motor boat in floating conditions. By mounting the camera in a low-cost water proof housing, the entire hull (both the parts below and above the sea level) was surveyed. The survey is divided in two photogrammetric surveys: (i) above ("dry") and (ii) below (underwater) the sea level. The two surveys are then joined together by means of ad-hoc orientation devices (which use photogrammetric circular targets). The paper covers the whole workflow of the photogrammetric approach, starting from the camera calibration up to the assessment of the obtainable accuracy and the realization of the digital model suitable for naval architecture purposes. Interesting potentialities of the proposed method are discussed and motivated.

2 DIGITAL PHOTOGRAMMETRY FOR SHIPBUILDING APPLICATIONS

Photogrammetry is a flexible and accurate three-dimensional (3D) measurement technique based on photographs of an object taken from different points of view. Its success became technically and economically evident in the mid 1980s based on analogue large format cameras. Especially for large volume objects with a high number of object points to be measured, close range photogrammetry could exceed the performance of theodolite systems and thus became a standard method for complex 3D measurement tasks (Luhmann 2010).

Once a camera has been calibrated (i.e., geometric parameters of the camera and lens distorsions are known) (Gruen & Beyerv 2001) (Granshaw 1980) the photogrammetric process translates into marking some corresponding object points on the images to determine camera positions and orientations (Exterior Orientation) as well as 3D point coordinates. The recognition of the same object point on two or more images (image correspondences) requires the object surface to have enough texture information (such as natural points and/or edges, etc.). If no features are visible on the images, then artificial targets must be positioned and/or synthetic patterns must be projected or painted on the surface object (Luhmann 2010) (Menna & Troisi 2010) (Menna et al., 2009) (Ackermann et al., 2008). For some applications, i.e., for high automation and accuracy purposes, circular coded targets should be positioned on the object to automatically recognize image correspondences.

Nowadays digital photogrammetry is capable of offering high precision and accuracy levels. The precision of image point measurement can be as high as 1/50 of a pixel, yielding typical measurement precision on the object even better than 1:100000 with respect to the largest object dimension. That corresponds to 0.1 mm for an object of 10 meters (Luhmann 2010).

Over the last few years, the shipbuilding field has seen remarkable experimentation in reverse modeling and re-engineering of ship hulls. In (Koelman 2010) (Goldan & Kroon 2003) CAD applications of photogrammetry for ship repair industry are presented. In (Menna & Troisi, 2010) different low cost techniques, both active and passive, were tested and compared for the 3D reverse engineering of a small screw propeller. In (Menna et al., 2009) close range photogrammetry was used for the digital record of a 24-meter ship hull with 1:40000 relative precision and performing several kind of inspections on its appendages and propellers. In (Ackermann et al., 2008) and (Menna and Troisi 2007) the authors showed the potentialities of image matching algorithms in delivering high density point clouds useful for modeling free form surfaces such as those of towing tank models or sailing boats.

3 DIGITAL PHOTOGRAMMETRY FOR 3D MODELING OF UNDERWATER ENVIRONMENTS

The use of photographs as a mean for visual inspection and analysis of objects located below the sea level has a long history. First underwater photographs taken with primitive camera housings date back to the 1850s, using glass plate as a medium. To see first underwater photogrammetric applications it is necessary to wait until the 1960s when film cameras and television cameras were used in stereo pair configuration (the minimum for 3D measurements) as metrology technique for repairs on marine equipment, archeological site mapping, monitoring of fauna populations and shipwreck surveys (Shortis et al., 2007). The advent of digital era has led to many practical and technical advantages in underwater photogrammetry. Low cost consumer digital cameras mounted in a waterproof housing can deliver very precise measurement whose absolute accuracy is limited only by the turbidity of the water that drastically reduces the overall image contrast, hence the accuracy in image point marking. The accuracy of a photogrammetric system is always related to the calibration of the camera. Utilizing cameras for underwater photogrammetry poses some non trivial modeling problems due to refraction effect and extension of the imaging system into a unit of both camera and protecting housing device.

Two different mathematical models have been proposed for underwater photogrammetry (Telem & Filin 2010) (Shortis et al., 2007):

1. rigorous geometric interpretation of light propagation in multimedia (camera housing-water) also known as *ray tracing* approach (Li et al., 1997);
2. the refractive effect of the different interfaces is absorbed by the camera calibration parameters as if the camera was normally calibrated for terrestrial applications (Harvey & Shortis 1998).

The advantage of the first approach is that the camera can be calibrated out of the water but requires the refractive indices of the air-glass and glass-water interfaces to be assumed or directly measured. Small changes in pressure, temperature and salinity can decrease the accuracy and cannot be eliminated. The second approach has the disadvantage that cameras need to be calibrated underwater. For underwater photogrammetric systems mounted on Remotely Operating Vehicles (ROV) the difficulty becomes obvious especially if surveys have to be carried out at different depths. Shortis et al. (2007) overcame the problem using a special laser array for the calibration of an underwater stereo video system on the field.

In this paper the approach number two for modeling refractive effects of multimedia interfaces has been used.

4 FULL PHOTOGRAMMETRIC SURVEY OF SEMI-SUBMERGED OBJECTS: A CASE STUDY

The case study here presented is part of a wider project called OptiMMA (Optical Metrology for Maritime Applications) involving the Laboratory of Topography and Photogrammetry (LTF) of Parthenope University and the 3DOM optical metrology unit of Bruno Kessler Foundation (FBK). The project is an interdisciplinary work based on optical metrology and 3D reverse engineering techniques in the maritime field for the support of shipbuilding firms, naval architects and designers.

A 19-foot fiberglass boat (Fig. 1) was chosen for testing a new methodology for 3D reverse modeling

Figure 1. The boat used for testing the reverse engineering technique.

Figure 2. Consumer digital camera (left) and waterproof housing (right) used for the photogrammetric survey of the boat.

of floating objects. The study aims to reveal practical and technical issues involved in measuring semi-submerged objects and structures of large size for which dry-docking is usually difficult and expensive. The methodology here presented can be easily extended to larger size boats.

4.1 *Calibration*

The equipment used for the photogrammetric survey consisted in a 7 Mpx CANON A620 (pixel size 2.3 μm) consumer grade digital camera mounted in a dedicated waterproof camera housing (Fig. 2).

A volumetric rigid frame made of aluminum was specifically built for underwater calibrations (Fig. 3—up). It consists of a cross shape with four arms holding four triangular plates. 128 photogrammetric circular coded targets are attached on the frame for high accuracy automatic measurements. The frame measures approximately 530 mm × 530 mm × 200 mm.

Two 1000 mm long aluminum scale bars were also built for scaling photogrammetric underwater measurements. The circular targets of the frame and the scale bars were accurately measured by means of photogrammetric and theodolite measurements in laboratory (Fig. 3—down). The average theoretical precision of target coordinates are $\sigma_X = 0.005$ mm, $\sigma_Y = 0.006$ mm and $\sigma_Z = 0.009$ mm which corresponds to an overall relative precision of 1:100000.

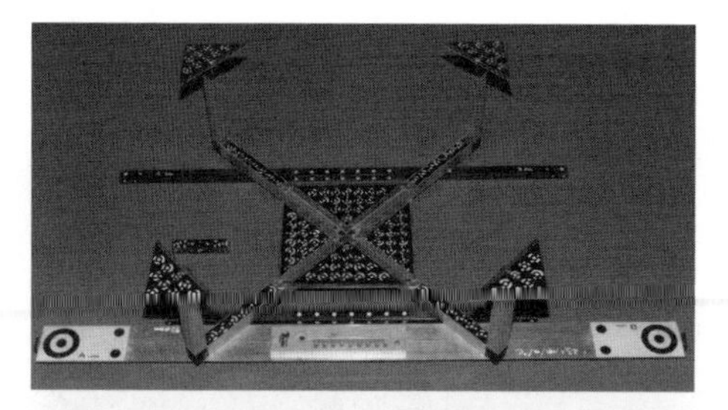

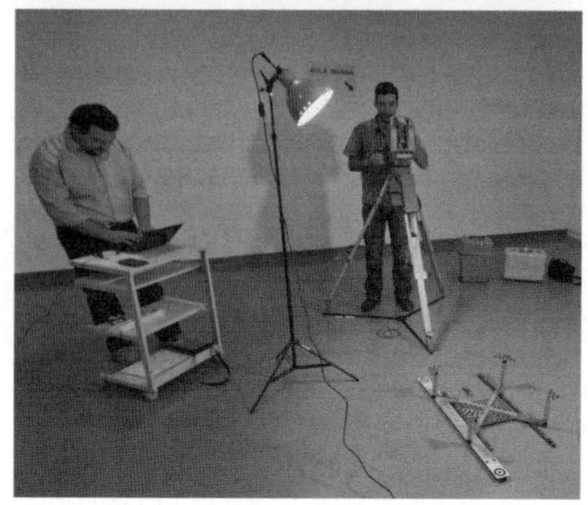

Figure 3. Calibration frame and scale bar (up) arranged for laboratory measurement (down).

The method for 3D modeling of semi-submerged objects herein presented can be summarized as a two-step photogrammetric survey (underwater and above the waterline) followed by a roto-translation of the two separated surveys in a unique reference frame. Four rigid orientation devices (OD) were specifically designed for the photogrammetric survey of floating objects. They consists of an aluminum bar (ca 600 mm long) with two thick Plexiglas plates attached to the extremities (Fig. 4). Each plate has four circular coded targets. The device is intended to be fixed on the surveyed object staying an half (one of the two plates) underwater and the other half above the sea level. The 8 target coordinates for each of the four orientation devices were measured in laboratory with photogrammetric measurements.

Once calibrated, one orientation device allows to join the two surveys executed underwater and above the waterline respectively. In order to guarantee a better solution during the computation of roto-translation parameters a good geometric distribution and redundancy of the orientation devices onto the object is necessary.

The way each OD allows to join the two surveys can be explained as follows: (i) the 4 coded targets on each plate of the OD were previously measured in laboratory (the relative position between the two plates is accurately known); (ii) the coordinates of one plate are measured for example in the underwater survey (Figure 4a), (iii) by means of the measured 4 coded targets, the OD is aligned (roto-translated) in the underwater reference system, then the coordinates of the plate above the sea level are also known with respect to the underwater reference frame; (iv) the coordinates of the emerged plate, known in the underwater system, are used for aligning the survey above the waterline to the underwater survey (Figure 4b,c).

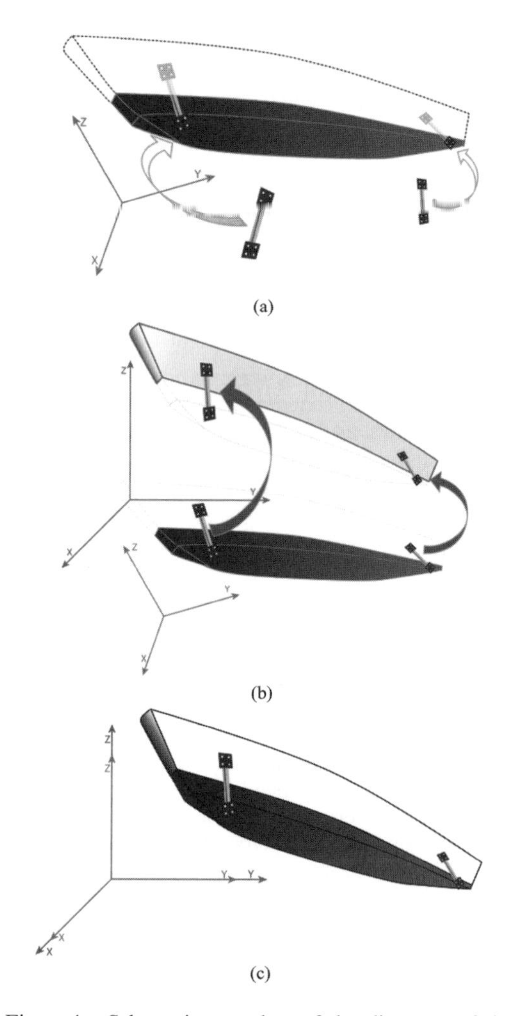

Figure 4. Schematic procedure of the alignment of the two separate surveys above and below the waterline.

In order to investigate the differences in camera calibration parameters between "dry" and underwater calibrations the Canon A620 was mounted in the waterproof camera housing and calibrated in "dry" conditions in laboratory using the calibration frame in Figure 3. Focal length was set to the widest available (having the camera a zoom lens), corresponding to a nominal principal distance of 7.3 mm. The network geometry consisted of 15 convergent poses (average angle between intersecting angles ca 75 degrees) of the calibration frame taken at an average distance of 1 meter. The self calibration procedures in free network solution exposed in (Gruen & Beyerv 2001) (Granshaw 1980) were used. Photo Modeler and Australis software were used to compute the interior and exterior orientation parameters together with the additional lens distortion parameters to model any systematic error. As described in (Wackrow et al., 2007) for the terrestrial case, it is noteworthy that some additional parameters such as decentring distortion parameters (also known as tangential distortions) are not statistically significant for most consumer grade digital cameras. Therefore, ignoring the computation of decentring distortion parameters has no practical effects in terms of reduction of object coordinate precision. On the other hand, in the case of underwater calibrations of video-cameras, (Harvey & Shortis 1998) underline the importance of decentering distortion parameters in absorbing systematic errors whose behaviour cannot be modelled by collinearity equation model. During the experimentation herein proposed, two sets of calibration parameters were computed (with and without decentring distortion parameters). As expressed by (Harvey & Shortis 1998), changes in pressure and temperature and even more the handling of the camera itself produce instability in the camera calibration parameters. For investigating the effect of camera handling during the underwater survey as well as the effects caused by change of pressure and temperature (from the waterline down to 5 meters), two different calibrations were executed at temporal distance of circa 1 hour: (i) shallow water (Fig. 6-left), (ii) at a depth of 4 m (Fig 6-right). The two different depths are the minimum and maximum depths the camera is planned to be used during the underwater survey of the test-boat. For each calibration an average of 16 images were taken with convergent poses (average intersecting angle of ca 85 degrees) at a distance of 1.5 m from the calibration frame.

In Table 1, "dry" and underwater calibration parameters are reported. For each calibration two versions (V1 and V2) are listed: with and without tangential distortion parameters (P1, P2 parameters). For the underwater survey of the boat, calibration parameters without tangential distortions (V2) were used since high statistical correlations (over 97%) were found between principal point position and tangential distortions. Furthermore, some calibration parameters for version V1 such as the principal point position and focal length were not consistent between the two calibrations in shallow water and at 4 m (Table 1). The average ratio between the focal lengths computed in underwater

Figure 5. CAD model of one of the 4 rigid orientation devices.

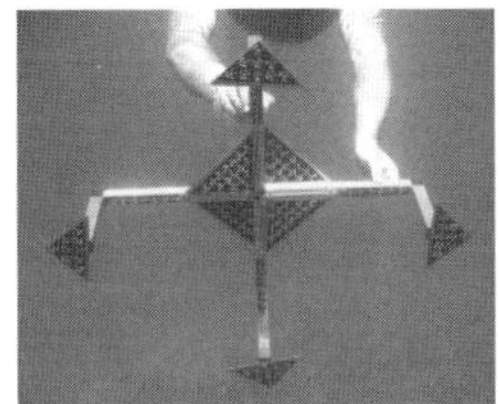

Figure 6. Two images used for the underwater calibrations of the Canon A620 in shallow water (left) and at a depth of 4 m (right) respectively.

Table 1. Calibration parameters of the camera used for the survey of the boat. Both dry and underwater calibrations are reported.

Camera calibration name	DRY_V1	DRY_V2	UW-4m_V1	UW-4m_V2	UW-0.5m_V1	UW-0.5m_V2
Focal length [mm]	7.3175	7.311	9.8299	9.8197	9.8084	9.8113
Principal Point x0 [mm]	−0.0635	−0.066	−0.0344	−0.0606	−0.0807	−0.0632
Principal Point y0 [mm]	−0.051	−0.0648	−0.0173	−0.0478	−0.0544	−0.0609
k1	3.86E-03	3.81E-03	−2.77E-04	−3.12E-04	−3.01E-04	−2.97E-04
k2	−7.47E-05	−7.08E-05	−7.80E-05	−7.45E-05	−7.27E-05	−7.34E-05
k3	0.00E+00	0.00E+00	0.00E+00	0.00E+00	0.00E+00	0.00E+00
P1	−1.25E-05	0.00E+00	−1.25E-04	0.00E+00	8.72E-05	0.00E+00
P2	−6.50E-05	0.00E+00	−1.48E-04	0.00E+00	−3.12E-05	0.00E+00

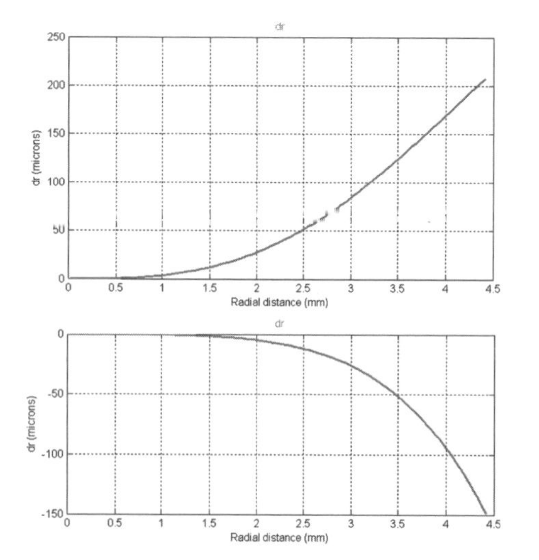

Figure 7. Radial distortion profiles for "dry" calibration (up) and underwater calibrations (down).

and dry calibrations is equal to 1.342 that corresponds to the refractive index of sea water at 26°C and salinity of 38 g/kg. To investigate the accuracy of the digital camera used in these experiments in underwater environment, the 3D coordinates of the calibration frame measured in laboratory were compared to those obtained from the underwater self calibration bundle adjustment. The root mean square error of 3D coordinates measured in underwater calibrations were respectively $\sigma_X = 0.045$ mm, $\sigma_Y = 0.024$ mm, $\sigma_Z = 0.090$ mm.

In Figure 7 radial distortion profiles for versions V2 of Table 1 are plotted for dry (up) and underwater (down) calibrations. For underwater calibrations only one graph is shown since differences between the two calibrations can not be appreciated in the figure.

4.2 *Survey*

The 19-foot boat "Mano 19" was anchored in 6 meters of water along the coast of Procida island in the gulf of Naples. About 50 photogrammetric circular coded targets were stuck both above and below the waterline (Figure 8) and some strips of circular targets were attached along the keel and the stem to determine the boat centreplane. Two aluminum scale bars were attached one below and the other above the waterline.

The four orientation devices were placed aft and fore, symmetrically on the two sides of the boat. Each orientation device has respectively one plate below and above the waterline (Figure 9). Since the boat is made of fiberglass, targets and orientation devices were attached with a special water resistant double-sided tape. The targeting operations required circa 1 hour.

The photogrammetric survey consisted of two set of images of the boat taken underwater and above the waterline trying to keep a good network geometry of camera stations (Fraser 1996). The underwater photogrammetric modelling of boats can be very troublesome since to survey the bottom of the hull, photographs have to be taken pointing the camera up toward the sea surface. In this condition the influence of the dispersion effects of water, the presence of suspension, flare and other optical aberrations reduce sensibly the accuracy of point marking operations on the digital images, hence the precision of 3D point coordinates.

The two sets of images were oriented using Photo Modeler software in two separate projects using coded targets to automate the orientation stage. Table 2 summarizes the main characteristics and results of the two photogrammetric surveys.

To bring the two surveys in a unique reference frame the four orientation devices were used. The procedure consisted in the following steps. First, similarity transformation parameters are computed to bring each of the four OD in the underwater surveys. In this way the 3D coordinates of the 4 points on the plates above the waterline become known in the reference frame of the underwater survey. The operation is repeated for the photogrammetric survey above the waterline hence the 3D coordinates of the 4 points on the plates underwater are known in the reference frame of the "dry" survey. After this operation, the two separate surveys have 32 common points that can be used to compute the similarity transformation parameters to bring the surveys in a unique reference frame. After alignment of the two surveys the standard deviation of

Figure 8. Sticking operations of circular coded targets on the hull of the boat.

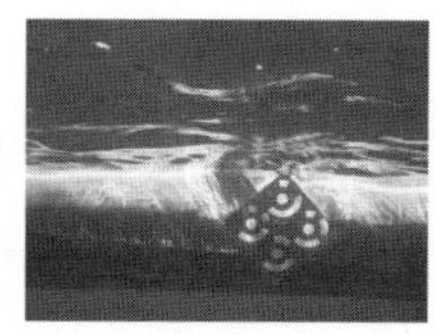

Figure 9. Two photographs of a same orientation device above and below the waterline respectively.

residuals on the plates of the orientation devices were respectively $\sigma_X = 1.1$ mm, $\sigma_Y = 2.1$ mm, $\sigma_Z = 0.9$ mm. From (Figure 9) it is visible that maximum residual exceeds 4 mm (green bars). This behavior is probably due to small movements of the orientation devices caused by movements of the boat. During the survey operation the transit of many local ferries caused waves and frequent roll movements of the boat then, likely, the plates of the orientation devices worked as oars into the water. The roll movement can explain the maximum residuals in y direction (starboard-port axis).

After the joining of the two separate photogrammetric projects, a global bundle adjustment was performed and many manual points, edges and lines were measured on the images to obtain features of interest for the 3D modeling of the boat (Figure 11).

Table 2. Main characteristics and results of the survey.

	Underwater	Above
N° of images	50	40
Average distance [m]	2.5	2.5
Pixel footprint [mm]	0.6	0.8
Average intersecting angle [degrees]	61	67
Theoretical precision σ_X [mm]	0.49	0.41
Theoretical precision σ_Y [mm]	0.57	0.26
Theoretical precision σ_Z [mm]	0.56	0.17
Relative precision	1:6500	1:12000

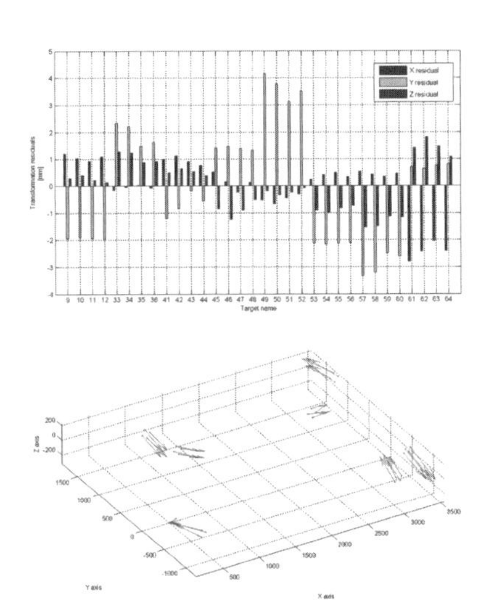

Figure 10. Residuals of the similarity transformation on the targets of the orientation devices.

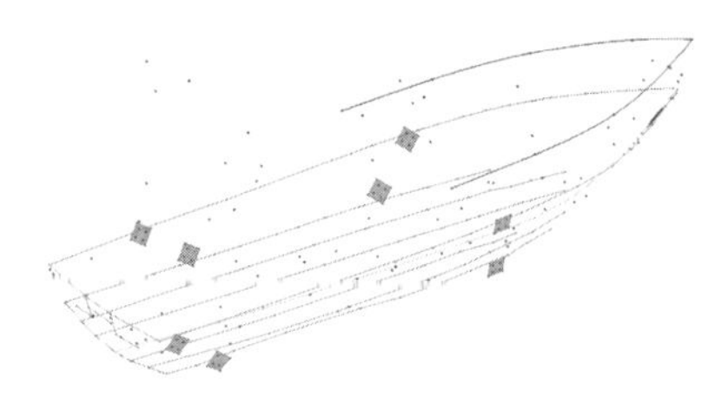

Figure 11. Feature lines and points measured in Photomodeler after the union of the two photogrammetric surveys.

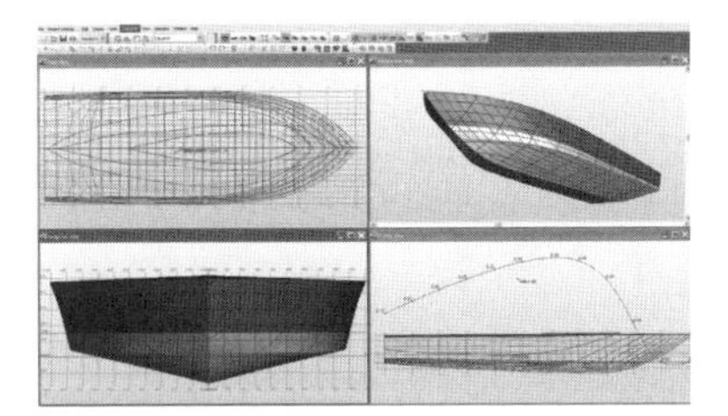

Figure 12. DELFT-ship model of *Manò 19*.

4.3 *3D Modeling*

Feature lines and points were imported in *DELFT-ship* (www.delftship.net) a free hull modeler software that includes hydrostatic calculations.

The 3D photogrammetric data were used as reference for modeling a symmetrical hull composed by subdivision surfaces (Figure 12).

5 CONCLUSIONS

In this contribution a new method for 3D modeling of floating objects has been presented in the case of a 19-foot fiberglass boat. The significance of the method is due to the fact that it does not require the ship to be docked.

Two photogrammetric surveys, underwater and above the sea level respectively were performed with the boat in floating conditions. The two surveys were then joined together by means of ad-hoc orientation devices (calibrated in laboratory), with photogrammetric circular targets. A consumer-grade low cost digital camera mounted in a waterproof housing was employed for retrieving the 3D digital model by means of digital photogrammetry. The whole workflow of dry and underwater camera calibrations, survey operations and 3D modeling of the boat have been analyzed. The results of the experiments are encouraging since sub-millimeter precision was obtained for the two separate photogrammetric surveys and 1-2 millimeter accuracy was obtained from the alignment of the two surveys in a unique reference frame. The larger residuals obtained along the y axis (Figure 10) have shown some instability troubles of the orientation devices probably due to the double-sided tape used to attach them on the hull. It is noteworthy that the characteristics of the boat surveyed in the experiment were really challenging for the tested methodology. In fact, for larger ships made of steel, orientation devices can be attached by means of strong magnets that can assure much more stability. Furthermore, the movements of larger ships are in general smaller than the tested boat.

The case study of the 19-foot boat has been especially interesting for testing the proposed methodology. In fact, dry-docking of vessels of this size is usually easy and inexpensive. In the future, the method will be applied for the reverse engineering of large size vessels.

REFERENCES

Ackermann, S., Menna, F., Scamardella, A. & Troisi, S. 2008. Digital Photogrammetry for High Precision 3D Measurements in Shipbuilding Field. *6th CIRP International Conference on ICME—Intelligent Computation in Manufacturing Engineering*, Naples, Italy, 23–25 July.

Canciani, M., Gambogi, P., Romano, F.G., Cannata, G. & Drap, P. 2003. Low Cost Digital Photogrammetry for Underwater Archaeological Site Survey And Artifact Insertion. The Case Study of the Dolia Wreck In Secche Della Meloria-Livorno-Italia. *International Archives of the Photogrammetry, Remote Sensing and Spatial Information Sciences*, 34 (Part 5/W12), 95–100.

Drap, P., Seinturier, J., Scaradozzi, D., Gambdogi, P., Long L. & Gauch, F. 2007. Photogrammetry for virtual exploration of underwater archeolgical sites. *Proc. XXIth CIPA international Symposium*, Athens, Greece, 1–6 october.

Fraser, C.S. 1996. Network design. *Close-range Photogrammetry and Machine Vision*, Atkinson (Ed.), Whittles Publishing, UK, pp. 256–282.

Goldan, M. & Kroon, R.J.G.A. 2003. As-Built Product Modeling and Reverse Engineering in Shipbuilding Through Combined Digital Photogrammetry and CAD/CAM Technology. *Journal of ship Production*, Vol. 19, No. 2, pp. 98–104.

Granshaw, S.I. 1980. Bundle Adjustment Methods in Engineering Photogrammetry. *Photogrammetric Record*, 10(56), October, pp. 181–207.

Gruen A. & Beyer, H.A. 2001. System calibration through self-calibration. *Calibration and Orientation of Cameras in Computer Vision' Gruen and Huang (Eds.)*, Springer Series in Information Sciences 34, pp. 163–194.

Harvey, E.S. & Shortis, M.R. 1998. Calibration stability of an underwater stereo-video system: Implications for measurement accuracy and precision. *Marine Technology Society Journal*, 32(2): 3–17. http://www.nationalhistoricships.org.uk/

Kentley, E., Stephens, S. & Heighton, M. 2007. Deconstructing Historic Vessels. *Understanding Historic Vessels Volume 2.National Historic Ships.*

Koelman Herbert, J. 2010. Application of a photogrammetry-based system to measure and re-engineer ship hulls and ship parts: An industrial practices-based report. *Computer-Aided Design* 42(8): 731–743.

Lhumann, T. 2010. Close range photogrammetry for industrial applications. *ISPRS Journal of Photogrammetry and Remote Sensing*, Vol. 65, No. 6., pp. 558–569.

Li, R., Li, H., Zou, W., Smith, R.G. & Curran, T.A., 1997. Quantitative photogrammetric analysis of digital underwater video imagery. *IEEE Journal of Oceanic Engineering* 22 (2), 364–375.

Menna, F., Nocerino, E. & Scamardella, A. 2011. Reverse engineering and 3D modelling for digital documentation of maritime heritage. *The International Archives of the Photogrammetry, Remote Sensing and Spatial Information Sciences*, Vol. XXXVIII, Part 5/W16, Trento, Italy, 2–5 March.

Menna, F. & Troisi, S. 2010. Low cost reverse engineering techniques for 3D modelling of propellers. *International Archives of Photogrammetry, Remote Sensing and Spatial Information Sciences*, Vol. XXXVIII, Part 5 Commission V Symposium, Newcastle upon Tyne, UK.

Menna, F., Ackermann, S., Scamardella, A., Troisi, S. & Nocerino, E. 2009. Digital photogrammetry: a useful tool for shipbuilding applications. *13th Congress of Intl. Maritime Assoc. of Mediterranean (IMAM2009)—Towards the Sustainable Marine Technology and Transportation*, Vol. I, Istanbul, Turkey, 12–15 October: 607–614.

Menna, F. & Troisi, S. 2007. Photogrammetric 3D modelling of a boat's hull. *VIII Conference on Optical 3D Measurement Techniques*, Vol. II, pp. 347–354, July 9–12, 2007, Zurich, Switzerland.

Remondino, F. & El-Hakim, S. 2006. Image-based 3D modelling: a review. *The Photogrammetric Record*, Vol. 21(115), September 2006, pp. 269–291.

Remondino, F., Guarnieri, A. & Vettore, A. 2005. 3D modelling of close-range objects: photogrammetry or laser scanning? *Proceedings of SPIE-IS&T Electronic Imaging: Videometrics* VIII, San Jose, California. Vol. 5665, 374 pages: 216–225.

Shortis, M.R., Harvey, E.S. & Seager, J.W. 2007. A review of the status and trends in underwater videometric measurement. Invited paper, *SPIE Conference 6491, Videometrics IX, IS&T/SPIE Electronic Imaging*, San Jose, California, USA, 26 pages.

Telem, G. & Filin, S. 2010. Photogrammetric modeling of underwater environments. *ISPRS Journal 63 of Photogrammetry and Remote Sensing*, Vol. 65, no 5, pp. 433–444.

Wackrow, R., Chandler, J.H. & Bryan, P., 2007: Geometric Consistency and Stability of Consumer-Grade Digital Cameras for Accurate Spatial Measurement. The Photogrammetric Record, vol. 22(118), pp. 121–134. The Remote Sensing and Photogrammetry Society and Blackwell Publishing Ltd.

Wiggenhagen M., Elmhorst A. & Wimann U. 2004. Repeated Object Reconstruction of The Bremen Hanse Cog. Proc of XXth ISPRS Congress, 12–23 July 2004 Istanbul, Turkey.

Sustainable Maritime Transportation and Exploitation of Sea Resources – Rizzuto & Guedes Soares (eds)
© 2012 Taylor & Francis Group, London, ISBN 978-0-415-62081-9

Advanced manufacturing techniques in shipyards

Eda Turan & Ugur Bugra Celebi
Yıldız Technical University, Faculty of Naval Architecture and Maritime,
Naval Architecture and Marine Engineering Department, Yıldız, Istanbul, Turkiye

ABSTRACT: Maritime business has a strategic position in the World and shipbuilding is one of its most significant part. However, the global financial crisis has also affected the shipyards dramatically. Some projects were not managed to be completed while some of them have been cancelled. In spite of this, on the other hand shipyards are trying to make new vessel agreements in order to continue their activities and also go on construction of on-going projects. Due to some projects that have been cancelled/postponed, not only current projects are under risk and pressure of factors such as cost effectiveness, time savings but also new contracts are under risk nowadays. Therefore shipyards should take some precautions to stay competitive in this uncertain situation. One way of delivering the vessels on time and cost effectively is to apply project management system in the shipyards and control its outputs periodically. In addition to this, some shipyards invest on their manufacturing processes in order to increase productivity. In this context, establishing production lines in order to speed up their production capacity, reducing the costs and delivery times by mass production, increasing the profits from each vessel can be listed.

Environmental protection is also a major issue during the shipbuilding period. Usage of robotic welding processes and automated technologies for reducing the wastes and pollutants sourced by the shipyard processes can be beneficial for preserving the environment and improving the production quality.

In this paper, the manufacturing processes of the shipyards are investigated; the benefits of implementing advanced manufacturing techniques in the shipyards considering environmental issues are presented.

Keywords: Shipyards, Manufacturing Techniques, Shipbuilding, Production

1 INTRODUCTION

Shipbuilding is a significant industry in developing countries. Annual vessels' distribution is shown in Fig. 1.

The shipbuilding industry is under the effect of global economical crisis since 2008. Shipyards came across with the huge reduction of new orders as well as cancellation of existing projects. Fig. 2 shows the commercial shipbuilding distribution worldwide.

As a result of the global crisis, some companies stopped their production facilities, while some of them were completely bankrupted. Meanwhile, the companies that have taken preventive precautions regarding these kind of situations still continue their facilities in routine.

The capacity utilization was around 80% in 2008. However this ratio became 50% and also 20% of unemployment was seen in 2009 due to the global crisis in Europe. The export value in Turkey was around 2,646 M$ in 2008, 1,831 M$ in 2009.

One way to cope with the effects of global crisis is adopting effective crisis management systems. In this context, shipyards create crisis management teams and these teams prepare crisis management plans and deal with the events that threaten the company in order to prevent them and get successful outcomes. In some cases, a project management system is applied and the shipyards control

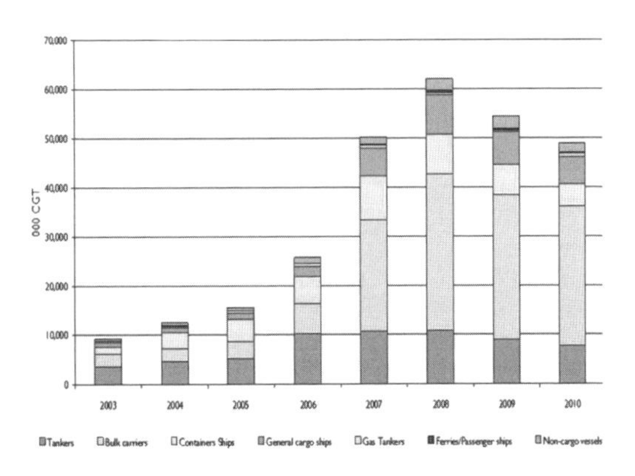

Figure 1. Production index for shipbuilding industry (Ref. 1).

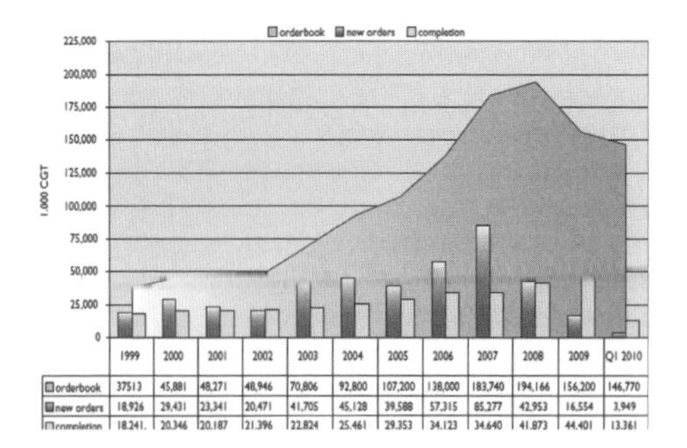

Figure 2. World commercial shipbuilding activity (CESA, 2010).

the outputs periodically. On the other hand, some shipyards that have sufficient capital select to improve their production capabilities through new technologies in order to increase their productivity and stay competitive among their rivals. In this study, conventional manufacturing processes and advanced manufacturing technologies in the shipyards are evaluated and the advantages/disadvantages of adopting new technologies are presented.

There are several production processes which yield hazardous waste and pollutants to the environmental safety and health in a shipyard. Shipyards are often categorized into two basic subdivisions; new shipbuilding division and ship repairing division (Celebi et al., 2010a). The shipbuilding and ship repair industry is the combination of different production processes such as surface preparation, painting and coating, machining and metalworking operations, solvent cleaning and degreasing, welding and cutting operations. The inputs of these processes are various types of products and raw material with an outcome of different forms (solid, liquid, and gaseous) of wastes and pollutants (Akanlar et al., 2009).

The prevention of these waste products at the beginning of the project may be the best solution. Alternatives to reduce the pollution may be to decrease the material inputs, to improve the engineering processes, to reuse the materials, to improve the management practices, to use alternatives to toxic chemicals. Most of the processes such as welding, painting, blasting, fiberglass production that have a direct effect on workers health, i.e., through exposure to VOCs, fumes resulting from burning through base metal and from burning the interior and exterior coatings that are often left in place can cause acute and chronic health problems (Celebi et al., 2010a).

The management of the wastes mentioned above is extremely important for environmental safety and human health. Waste management is the prevention, minimization, reuse, recycling, energy recovery or disposal of waste materials. Source reduction aims to reduce the hazardous materials, pollutants or contaminants released to the environment, prior to recycling, treatment or disposal and minimize the hazards to human health and the environment (Celebi et al., 2010b).

Improvements and innovations outcome from global scientific studies shows their effects on each product. After scientific studies, new production techniques and new products are developed. The push effect of science on the industry is accepted all over the world. The questionnaire of the production processes in the shipbuilding industry by science combines new production techniques, creates more rigid and stable paints, less environmental hazardous blasting material improvements and results with environmentally sensitive ship production (Celebi et al., 2010b).

2 CONVENTIONAL MANUFACTURING TECHNIQUES

The shipbuilding period comprises cutting and marking of steel plates, steel fabrication, assembly of the sections, erection on the slipway, launching of the vessel, sea trials and the delivery of the vessel to the Owner. In general, production type in the shipyards is block fabrication. In this production process, the vessel is composed of several blocks subject to the manufacturing and crane capacities of the shipyard. Each block is assembled in the workshops and after completion of the works and surveys; the blocks are erected on the slipway. The vessel is launched after completion of all blocks' erection on slipway. The outfitting works of the vessel are completed in the sea and the vessels start their sea trials in order to test the efficiency and control whether they meet the requirements of the Owner and comply to the contract items. Consequently, the vessel is delivered to the Owner if the sea trial results can be accepted by the classification societies and the Owners.

Due to the high number of components installed during the production period and equipment testing on the vessel, the shipbuilding industry is more complex compared to other industries. A shipbuilding process is shown in Fig. 3.

According to the data received from a Turkish Shipyard, a chemical carrier which is 8000 DWT has approximately 14 months construction period with average 500 workers per day. 3 months for steel cutting and design approvals; 7 months for completion of the block erections on the slipway and then launching; and 4 months for outfitting works in the sea, sea trials and the delivery procedures that are required in conventional shipbuilding practice.

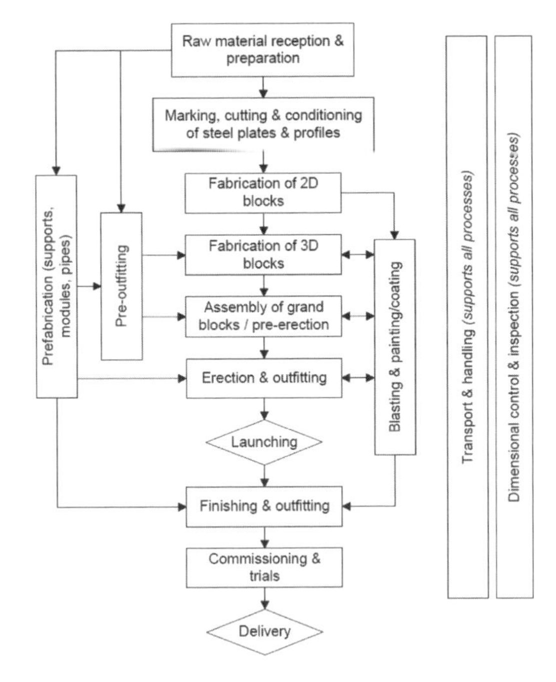

Figure 3. Shipbuilding process (Andritsos, 2000).

In the competitive market situation today, shipyards should seek to decrease the construction periods of the vessels and should stay competitive as per cost, delivery time and quality.

Conventional shipbuilding is a labor-intensive process done manually. Therefore, some defects can be seen in the production period and also delivery times of the vessels can be prolonged. It is possible to reduce the duration of this process, improve the construction quality and decrease the costs with the application of automated technologies and the establishment of panel assembly lines to the shipyards.

3 ADVANCED MANUFACTURING TECHNOLOGIES IN THE SHIPBUILDING INDUSTRY

Maintaining a competitive edge at shipyards requires efficient production. Improvement in cost effectiveness frequently means an increase in automation. Automation in shipbuilding has been applied almost exclusively in steel construction, in particular in cutting and welding.

Maximising the efficiency of shipbuilding essentially means (Andritsos, 2000):

- Assuring the most cost-effective acquisition of the necessary prime materials and components,
- Optimising (in terms of cost and quality) each one of the processes stated above
- Planning the whole sequence of operations in such a way so as to ensure a seamless flow of material and an optimal utilisation of the available resources (labour, equipment, space and time).

The advanced shipbuilding manufacturing automation range is comprised of (Ref. 2):

- Plate and profile prefabrication systems
- Flat-panel production lines
- Double-bottom production lines
- Subassembly production lines (micro, bulkhead)
- Curved section panel lines
- Robotized panel lines or production cells for welding or cutting
- Production logistics and data collection software.

The area required for a panel line installation in the shipyards is dependent on the following items;

- the number of stations which constitute the complete panel line,
- the maximum size of panel likely to be produced,
- turning capability, if using a two-sided welding technique,
- space allowance between panels,
- buffer space for completed panels.

3.1 *Marking and cutting of steel plates*

Steel plates are marked before cutting. The marking methods generally used in the shipyards are powder, inkjet, plasma arc and laser markings.

The location of the steel parts, hull number of the vessel, part number, etc. are also marked on the steel parts in advanced processes.

In general, oxy-fuel, plasma and laser or a combination of them is used for cutting in the shipbuilding industry.

A plate cutting system can be considered complete when it addresses all of the points below (Andritsos and Perez-Prat, 2000):

Figure 4. A production line in a shipyard (Ref. 3).

- Conveyor feed-in and facilities for unloading either plates and small pieces
- Possibility to cut with the bevel for welding
- Integrated marking (identification, part number and block, procedure welding lines, marks for bending.
- Tolerances, change of bevel on contour
- Measurement system /integrated quality control
- Connection to the CAD system
- Multiple torch for simultaneous cutting
- Protection, insonorisation etc.

Cutting of the plates have almost been the first application of automation in the shipbuilding industry. The second application of automation is fabrication of the blocks by automated panel lines.

3.2 *Fabrication of the blocks*

The following types of panel lines are found in the shipyards (Andritsos and Perez-Prat., 2000):

Mini panel line, usable for one plate, working area under the portal(s) about 4×20 m^2

Normal flat panel line, for panels of 10×10 m^2, 15×15 m^2, 20×20 m^2 or larger, depending on the size of ships, the yard is generally building

Double bottom line, usually following the configuration / sequence:

Station 1: Inner floor plates are welded.
Station 2: Floor plates are being put on the panel and welded.
Station 3: Longitudinal floors are put in position and welded.
Station 4: Tank heating pipes and other outfitting elements will be attached.
Station 5: Already bended shell plates will be put on the floor grid and welded.
Station 6: Lifted off the line onto special means of transport for moving to the ring block section building area.

Curved panel line: There are various stations and each station is equipped with a large number of adjustable stanchions. The upper point of all the stanchions forms the shape of the curved panel by means of the software. They usually follow the following configuration:

Station 1: The already shaped plates will be put on them and welded together.
Station 2: Frames will be put on the plate-field and welded by means of robots.
Station 3: Secondary members are being put on the panel and welded as well.
Station 4: As far as practicable outfitting elements will be put on the panel.
Station 5: Lift off the line on special means of transport for moving to the block-building area.

Girder building line, consisting of:

- Crane with beam to put web and girder on the roller—conveyor
- Press roller machine
- Welding robots
- Profile straightening machine
- Discharging conveyor

According to the data received from a Turkish Shipyard which has production line, a 26000 DWT container vessel has approximately 270 working days construction period with an average of 500 workers per day. 55 days for steel cutting, 130 days for pre-manufacturing, 126 days for completion of the block erections on slipway and then launching, 10 days for engine montage, 57 days for outfitting works in the sea, sea trials and the delivery procedures are required in advanced shipbuilding practice.

3.3 *Welding of the blocks*

Welding is one of the major factors in the shipbuilding industry due to the strength of the vessel and also the production quality of the hull construction. Robotic welding is applied in advanced shipbuilding stages to obtain the possible up-time, while in conventional shipbuilding, electrical arc welding is generally done manually.

The arc welding robot is one of the most common functions in the industry today. During this process, electricity jumps from an electrode guided through the seam, to the metal product. This electric arc generates intense heat, enough to melt the metal at the joint. For a robotic arc welding system, a much more controller is also required (Turan et al., 2010).

Robotic welding reduces arc time per weldment by up to 250%. Filler metal usage decreases by up to 60% and it also reduces distortions and improves fit up. Mechanized cutting and welding reduces clean up and improves fit-up. Welding fumes are reduced by up to 60%. Shielding gases usage are reduced by up to 60%. Rework and scrap are reduced considerably. Operator efficiency and productivity also improve. Safety is increased by allowing the operator to stand away from sparks and fumes (Ref. 4)

The stability of the welding process is very sensitive to the main welding parameters such as current, voltage, welding speed, shielding gas and arc length. A small change in the distance between the welding torch and the component being welded or a fault of a welder may produce a considerable variation in the current and in the voltage. However, in robotic welding, the welding defects sourced by the welder can be prevented (Turan et al., 2010).

3.4 *Painting of the blocks*

After completion of the fabrication works, the blocks are transported to the painting workshops.

Painting wastes are known to be the largest category of the hazardous wastes produced in a shipyard. In a typical shipyard painting wastes may account for more than half of the hazardous wastes produced (Celebi et al., 2008).

Painting of the blocks in the closed areas reduces the harmfull effects of shipbuilding practices to the environment. On the other hand, with the automation technologies, the health problems of the workers due to painting can be prevented.

4 ADVANTAGES AND DISADVANTAGES OF ADVANCED MANUFACTURING TECHNOLOGIES

Automation gives advantages such as higher quality, improved capacity and higher productivity in the form of more welded parts per time unit, as well as an improved working environment and work content for the welder.

Automation also improves shipyard's quality and efficiency and enables to lay down more welds in less time.

Other advantages of automation are as follows;

- Reducing the labour-force
- Having free time for the operators
- Inclusion of unmanned working tools to the production period
- Development in the work bench utilization ratio
- Reduction of the need for a high qualified working force in unmanned working conditions
- Reduction of time for work bench preparation
- Inclusion of quick transportation vehicles to the system that enables the engines to maintain their manufacturing processes.
- Development of the operational control
- Reduction of the non-controllable variable number
- Inclusion of the devices to the system that can designate the deviations from the plan and act on these cases.
- Reduction of the need for human communications
- Reduction in the stocks
- Improvement in financial condition
- Having system for Just in Time Production processes
- Reduction in the number of workers.
- Schedule compliance & schedule-driven process improvement
- Faster design-build time
- Improved accuracy control
- Reducing of reworks and skill content.
- More use of automation—welding robots, painting robots, automatic line-heating, automatic welding (Ref. 5).
- Computer-integrated manufacturing—integrated design, planning, procurement etc.
- Operations management—continuous improvement, worker involvement (Ref. 5).
- Other factors: design for production, minimisation of staging, reduction of number of parts (Ref. 5).

In addition to the advantages of advanced manufacturing technologies, there are also some disadvantages. One of the major disadvantage is the unemployment risk due to the workers who are replaced with automatic machines. Another disadvantage of automation is high initial investment costs. On the other hand, the operators of the system should be well-qualified in order to manage appropriately. These disadvantages should also be analysed during decision making.

5 CONCLUSIONS

Conventional shipbuilding techniques and advanced manufacturing technologies in the shipyards are evaluated in this study and the advantages/disadvantages of the advanced manufacturing technologies are put forward.

According to the data received from the shipyards, it is beneficial for the shipyards to apply automation in their production facilities. The authors are aware that it is hard to make investments nowadays due to the global financial crisis. However shipyards can automate some parts of their production instead of making all the system as automated subject to their financial situation and the ratio that they can reserve for new investments.

It is clear from the study that automation improves productivity of the shipyards and prevents reworks. Initial investment cost seems high however in the following years, this investment will yield a benefit and enable the shipyards to build vessels with better quality, less time, less cost and efficiently.

To minimize the hazardous waste materials and to protect the workers health in the long term, traditional production methods should be replaced with alternative new production technologies. These new alternative methods should be defined and shared with top and medium management levels of facilities (Celebi et al., 2010b).

New production technologies to reduce the pollution may be the key to decrease the material inputs, to improve the engineering processes to reuse the materials, to improve the management practices and to use alternative materials to replace toxic chemicals (Celebi et al., 2010b).

In addition, the authors evaluated the manufacturing processes in the shipyards as conventional and advanced systems in this study. Production is compared according to the building times of the vessels. This study can also be evaluated by taking into consideration the invesment costs of the systems and profits from production in the next studies.

ACKNOWLEDGEMENT

The authors wish to thank their colleague Esengül KOCAMAN for her assistance in calculating the container vessel production period and the data for automation technologies in the shipyards.

REFERENCES

Akanlar, F.T., Celebi, U.B. & Vardar, N. 2009. The Importance of Wastewater Treatment in Shipbuilding Industry, *Global Conference On Global Warming-2009 (GCGW-09)* 6–9 July, Istanbul, Turkey.

Andritsos, F. 2000. Integration & robot autonomy: key technologies for competitive shipbuilding, *European Commission, Joint Research Centre Institute for Systems, Informatics & Safety*, TP 210 21020 Ispra (Va), Italy.

Andritsos, F. & Perez-Prat, J. 2000. The Automation and Integration of Production Processes in Shipbuilding, *European Commission Joint Research Centre Institute for systems, Informatics & Safety*, DG Enterprise, unit E.6 Administrative arrangement: 14707-1998-12 A1CA ISP BE.

Celebi, U.B., Akanlar, F.T. & Vardar, N. 2010a. Multimedia Pollutant Sources and Their Effects on the Environment and Waste Management Practice in Turkish Shipyards, *GLOBAL WARMING Green Energy and Technology,* 2010, 579–590, DOI: 10.1007/978-1-4419-1017-2_39.

Celebi, U.B. & Vardar, N. 2008. Investigation of VOC emissions from indoor and outdoor painting processes in shipyards, *Atmospheric Environment* 42 (2008) 5685–5695.

Celebi, U.B., Yildiz, B., Kocal, T. & Turan, E. 2010b. Alternative Automation Technologies To Reduce Waste And Pollutants of Processes In Shipyards, *International Science and Technology Conference*, Proceedings Book, pg. 77. Turkish Republic of Northern Cyprus.

Community of European Shipyards Associations (CESA), 2010. Annual Report 2009–2010.

Turan, E., Kocal, T. & Unlugencoglu, K. 2010. Welding Technologies In Shipbuilding Industry, *International Science and Technology Conference*, Proceedings Book, pg. 875, Turkish Republic of Northern Cyprus.

WEB REFERENCES

http://www.cesa-shipbuilding.org/about_the_industry

Ref. 2: http://www.ship-technology.com/contractors/steel

Ref. 3: http://www.ecvv.com/product/894538.html

Ref. 4: http://www.koike.com/PDF/Portable%20Welding%20and%20Cutting%20Automation.pdf

Ref. 5: http://academic.amc.edu.au/~gthomas/shipproduction/lecture12.pdf

Sustainable Maritime Transportation and Exploitation of Sea Resources – Rizzuto & Guedes Soares (eds)

A supply chain management model for shipyards in Turkey

Ugur Bugra Celebi & Eda Turan
Yıldız Technical University, Faculty of Naval Architecture and Maritime,
Naval Architecture and Marine, Engineering Department, Yıldız, Istanbul, Turkiye

ABSTRACT: Great amount of manpower usage, complex work flows, variety of material usage are the main points to describe shipbuilding industry as heavy industry. Production processes of the shipyards are divided into two main divisions: new building and ship repair industry. Production methods of these two divisions are similar characteristics. New ship construction and ship repairing have many industrial processes in common. Raw material inputs to the shipbuilding and repair industry are primarily steel and other metals, paints and solvents, blasting abrasives, welding machines, welding rods, electrodes, and machine and cutting oils. With over 40 shipyards, Tuzla Bay is one of the areas with the highest density of industrial establishments in Istanbul. Therefore, establishment of supply chain model is extremely important in Turkish Shipyards. Supply chain management in order to meet customer demand, covers the entire chain by suppliers, production facilities, distributors, retailers, and customers of operations planning, implementation and monitoring processes. The purpose of a management process is to maximize the performance and success of the process. Supplier management, technology management and efficient use of resources are very important in shipyard. The usage of new material and new production processes, the improvement on mechanization and automation of processes will reduce the pollutants, decrease the waste outcome and minimize the effects of these hazardous materials on shipyard workers. In this study, an appropriate shipyard supply chain model application is investigated.

Keywords: Supply Chain Management, Shipyards

1 INTRODUCTION

Production processes of shipyards are divided into two main divisions under new shipbuilding and ship repair industry. The shipbuilding and ship repair industry are the combination of different production processes such as surface preparation, painting and coating, machining and metalworking operations, solvent cleaning and degreasing, welding and cutting operations. The inputs of these processes are various types of products and raw material such as primarily steel and other metals, paints and solvents, blasting abrasives, and machine and cutting oils with an outcome of different forms (solid, liquid, and gaseous) of wastes and pollutants as volatile organic compounds (VOCs), particulates (PM), waste solvents, oils and resins, metal bearing sludge and wastewater, waste paint, waste paint chips, and sent abrasives. Especially during welding, blasting and painting processes, Personal Protective Equipment (PPE) usage is required for shipyard workers (Celebi et al., 2009).

In parallel to the global fast improving technologies, nowadays, shipbuilding industry is changing their traditional face with the new technological profile. Shipbuilding and repair industry consist of several branches as shipbuilding for load and passenger transportation, maintenance and repair and ship dismantlement. Shipyards are often categorized into a few basic subdivisions either by type of operations (shipbuilding or ship repairing), by type of ship (commercial or military), and shipbuilding or repairing capacity (first-tier or second-tier). Ships themselves are often classified by their basic dimensions, weight (displacement), load-carrying capacity (deadweight), or their intended service (Akanlar et al., 2009a).

The effect of shipyard on environment and human health can be reduced with technological improvements on shipbuilding industries. The usage of new material and new production processes, the improvement on mechanization and automation of processes will reduce the pollutants, decrease the waste outcome and minimize the effect of these hazardous materials on shipyard workers (Akanlar et al., 2009b).

International shipping can be distinguished from intra-regional and domestic (or coastal) shipping, as well as from freight movements on inland waterways. In terms of mere revenue contribution, the shipping industry can be somewhat simplistically divided between the bulk and liner sectors. The former is broadly characterized by large single shipments of loose cargo in whole ships that are operated on tailor-made voyages and the latter by mixed shipments of containerized cargo in ships that are operated to a regular schedule on pre-defined routes (Cullinane, 2005).

Management's business processes, such as distribution, research and development, operations, or logistics, are heavily impacted by global competition, high-speed information availability, continuously changing business relationships, shorter innovation cycles, and an increasing complexity of products (Larsen et al., 2005).

Supply chain management (SCM) is a new strategic business model aiming to integrate firms' different competencies and critical business processes to increase performance and decrease structural and operational uncertainty at supply chain level. Final aim of SCM is to compete with other supply chains (Unuvara, 2009).

Supply networks where operational control extends well over organizational boundaries have emerged in industries producing relatively complex and customized products with tight profit margins. Products like ships, automobiles and telecommunication systems incorporate complex design and engineering skills that are produced through a tier structured, multi-level supply networks. Efficiency in these networks has stemmed from specialization and cost efficiency in individual value adding operations. Shipbuilding in higher labor cost countries has faced similar transformation toward structured supplier networks, yet in countries with low labor cost integrated shipyards with tens of thousands of workers are common. As labor costs elevate industries are forced to be more efficient and one visible sign of this is the emergence of supplier networks. These networks are comprised of companies that are efficient in creating value in their niche area and their independence of a single value network has diminished and therefore they can better tolerate demand fluctuations from one industry. This is very true with the suppliers that have emerged from the shipbuilding case. This highly cyclical industry having suppliers capable of serving several industrial networks, is less vulnerable to demand fluctuations in one industry. Other competitive industries facing the same battle of cost effectiveness have seen the same evolution taking place in their business environment. Supplier networks with clear tier-structure are less dependent on single value networks, as they are capable to serve other industries and their networks. This was shown to be particularly evident in the shipbuilding supplier networks, where numerous suppliers deliver products and services to a variety of industries other than maritime, including, e.g., civil engineering and construction industries, offshore businesses and machine and equipment industries in general (Hameri, 2005).

Ensuring the coordination while working with several subcontractors is a significant problem of Turkish Shipyards. Worker material flows and procurements are hard to manage. The main purpose of this study is to set a prototype supply chain model for one block of a vessel and the minimization of the problems while working with subcontractors and suppliers. The required raw materials, products, equipments and workers will be present at the construction site on time.

2 SUPPLY CHAIN MANAGEMENT

Supply chain management is to coordinate and manage the material, information and cash flow in the chain of supplier, manufacturer, wholesaler, seller and customer. Supply chain management is also integration of internal and external resources of the company and ensuring to work efficiently. The aim is to increase all values such as production capacity, market sentiment and customer/supplier relationships that constitute all performance of a company. It contributes the customer to get the right product, in right location, time and price and improves the customer satisfaction. The production and marketing activities of the companies will be changed with the application of supply chain management system. In consequence of this, suppliers' integration, just in time production, the stocks approximately zero ratio, the automation in purchasing facilities to be seen in the company and the demands will be met on a regular basis.

Supply chain management consists of managing the material, information and capital flows as seen in Fig. 1. The integration and coordination of these flows internally and externally will also enable the succeed of supply chain management. Today's conditions oblige the companies better control their product prices, hence costs and

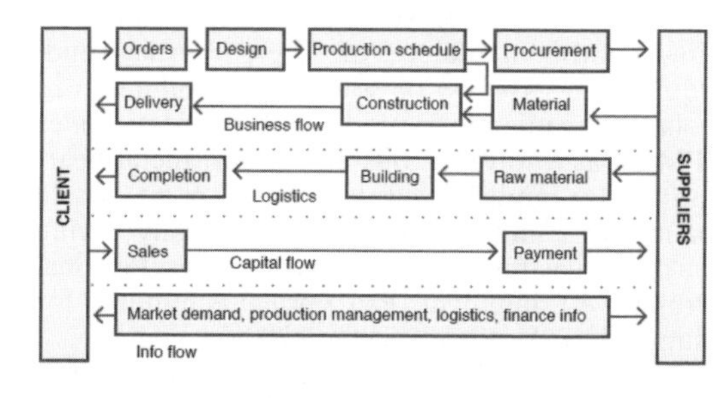

Figure 1. Business flow, material flow, capital flow and information flow (The Link Journal, 2005).

productivities. Performing of this issue occurs not only by improvement of the processes, but also needs the cooperation of seller, customer, distributor and transporter forming the supply chain in mutual trust. Partners' good communications lead to more efficient working conditions. The scope of supply chain management is composed of suppliers, production, storage activities, transportation, distribution of the goods, customer and market conditions. Good flow is carried out in one direction; only from supplier to the customer while, information flow is done in mutually, both from supplier to customer and from customer to supplier.

There are several costs that affect the company in case of any delay or defective manufacturing. These are classified as delivery time costs, storage costs, stocks and inventory costs and quality costs. Long delivery times, low quality control of the manufacturing processes, long storage times, high stock and inventory ratios cause high costs. Low quality control also affects the customer satisfaction. High stock costs result in losses of interest due to the invested capital. Therefore stocks should be defined in accordance with the requirements of production stage.

Supply chain management model of a shipyard improves production processes. Increase in productivity and quality will be obtained with this model. It is also essential for creating a supply chain management model applicable commonly to the shipyards in Tuzla Bay, Turkey. Thus, the investments which were evaluated as it is hard for one shipyard to make with their own facilities in terms of cost and efficiency can be created as whole shipyards in the area to be utilized. The unit prices can also be decreased and the delivery times can be shortened in mass purchases.

2.1 *Selection of the suppliers*

Systematic approach should be applied during selection of the potential supplier companies. Professional competence of the suppliers should be measured both in general aspects and technical criteria. Existing suppliers should also be compared with potential suppliers in order to improve the efficiency. Supply chain circle is shown in Fig. 2.

Companies demonstrating excellence in supply chain management have specific goals with clear objectives against which quantitative metrics can be applied (e.g., "increase the number of inventory turns by 20% this year", or "reduce average production span by 10% this year). Better companies will have specific initiatives involving their suppliers that they can relate directly to such quantitative objectives. Excellence is achieved when companies clearly recognize that specific corporate goals for gains in market share, profitability, speed to market, etc. are not achievable without important contributions by suppliers, are able to think through how their suppliers' performance has to change, and can determine what changes are needed in their supplier management practices to foster that performance (Fleisher et al., 1999).

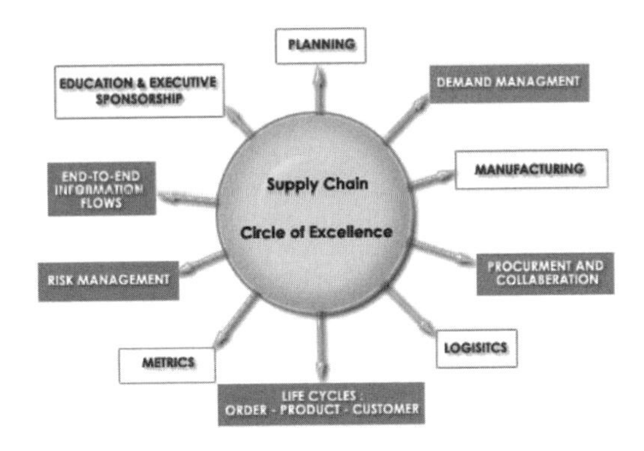

Figure 2. Supply chain circle (Ref. 1).

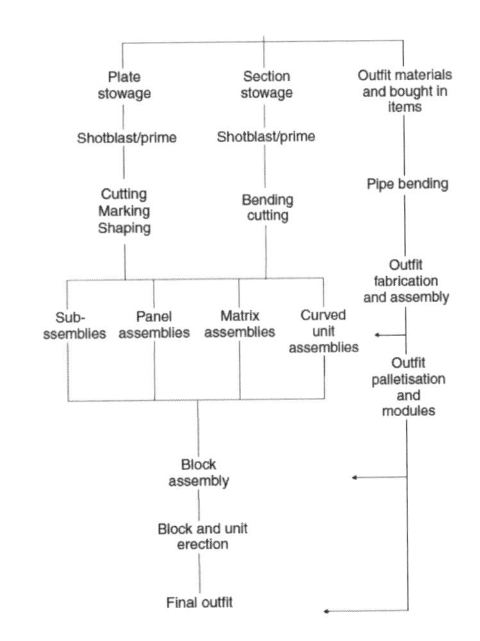

Figure 3. Shipbuilding processes (Eyres, 2007).

2.2 *Workflow and production processes in the shipyards*

Shipbuilding processes are seen in Fig. 3. After procurement of the steel parts, cutting, marking and shaping is done to the shot blasted/primed plates. The parts sent to workshops for assembly.

Steel part are welded on the steel plates in advance. On the other side, outfitting and piping works also start in parallel to the steel construction. Outfitting parts and pipe systems are being mounted on the blocks during block assemblies in order to shorten the shipbuilding processes. After completion of the block assemblies, blocks are erected on slipway. The outfitting works are ongoing on the vessel during slipway stage. The vessel is launched with the completion of the works essential to be finished on slipway. Remaining works continue at the pier of the shipyard until delivery of the vessel to the Owners.

3 SUPPLY CHAIN MANAGEMENT IN THE SHIPYARDS

Supply chain management involves the interaction of different companies, each one performing its value-added activities, aiming to produce a final product. More formally, the supply chain of a company comprises geographically dispersed facilities, where raw materials, intermediate products, or finished products are acquired, transformed, stored, or sold and transportation links that connect facilities along which products flow. The facilities may be operated by the company, or they may be operated by vendors, customers, third-party providers, or other firms. The company's goal is to add value to its products as they pass through its supply chain and to transport them to geographically dispersed markets in the correct quantities, with the correct specifications, at the correct time, and at a competitive cost Maritime and more specifically the ship-repair business, involves the manufacturing activities that are performed by the shipyard and its suppliers and sub-contractors to complete the repair of ships. Each company performs a set of activities that are interrelated and complementary. The execution of these activities involves exchange of information either between the working groups of the shipyard or among the ship-owner with the shipyard and the suppliers and sub-contractors (Xanthakis et al., 2008).

According to Michael Porter's value chain theory, the competitive advantage derives from all the discrete activities of an enterprise, including design, production, marketing and delivery. All these activities contribute to the overall cost of a company. The costs of the shipbuilding value chain are composed of design cost, procurement cost, labor cost and other special expenses. A great part of marine products' costs is determined at the design stage. Design activities are of little expense, but have a great impact on the whole shipbuilding costs. When the design is completed, batch production will follow after a large procurement. The procurement cost is the largest part of the total cost, and with the increase of technical content and value added, the proportion of the procurement cost will rise accordingly. The construction activity lies at the end of the value chain for a shipbuilding enterprise, and its cost involves mainly man-hour expenses and special production expenses. As labor costs rise domestically, the space for decreasing fabrication costs become smaller. The original supply chain is too complex and random, which leads to heavy organization and coordination work in the procurement department of the shipyard and creates additional costs. The change in some ways of the procurement activities will likely increase or decrease the cost of the whole procurement process. For instance, the procurement of good-quality and pretreated steel plates will probably increase the cost of the procurement but simplify the production process, raise the utilization efficiency, reduce the construction cost and optimize the total costs of the value chain. If the procurement department strengthens the management and control of suppliers, it will reduce the cost of internal quality inspection in the shipyard. Shipyards pay more attention to increasing production efficiency and reducing production costs instead of the procurement activities. Shipyard's procurement mainly focuses on the purchase of material and equipments, emphasizes the procurement activity itself, with the price procurement as guideline, and do not attach more importance on the procurement process and lack the philosophy on the supply chain and the modern logistics. The information channel is less advanced without effective method for data processing. The evaluation on suppliers is just like red tape without deep analysis on supplier's costs and capability (The Link Journal, 2005).

Operational sequences and tasks in shipbuilding in the various phases of the shipbuilding supply chain starting from acquisition, project planning/sales, construction/engineering, production, planning/management/processing, materials management, finances, spare part management and after sales/guarantee are in practice integrated to a varying degree and manifold software tools are implemented. The shipbuilding processes are increasingly executed by working groups and in networks. From early design throughout shipbuilding projects simultaneous engineering takes place and involves shipyard service providers, shipyard suppliers and other shipyards. In contrary to the many attempt in the field of process and product modeling, harmonization of processes of the different cooperation partners up to now has not been achieved. (EUROMIND, 2008)

In the shipbuilding supply chain from quotation to delivery the engineering process is central. In the project definition phase the full functionality of the final product is defined, systems are calculated and dimensioned, major components defined and the basic topology configured. In basic design, which follows the project and definition phase, functional and special interfaces are resolved. In this process shipyard suppliers are deeply involved. In detailed design all reference documentation is being prepared in time with the production process. This involves external design offices. The complete design process represents some 10–20% of the total cost of a merchant ship, while 75% of the total costs are already defined in the project definition phase and additional 25% in basic design. These figures show, that the goal must be to display the final product ship and its production processes within short time in sufficient quality so that the later production of the ship can be done according to the calculated conditions (EUROMIND, 2008)

Aiming to improve the efficiency of the procurement, reduce storage and save costs, make analysis from the view of the logistics, attempt to eliminate the unreasonable, the waste and the low efficiency in the production and the procurement activities, making each link reasonable and efficient (The Link Journal, 2005).

3.1 *Benefits of supply chain management*

SCM in shipbuilding lags other industries. The firms exhibiting the most Best Practices still have substantially more progress to make before they will match the leaders in the automotive or aerospace industry sectors. The use of these practices is relatively new in shipbuilding companies, and evidence from other firms suggests that as long as half a decade or more is needed before the Best Practices really take hold. Top management needs to place, or maintain, a priority on supply chain management, and have the vision and stamina to stay the course (Fleischer et al., 1999)

Most SCM approaches can work in shipbuilding. Wide variations in supply chain management philosophies and practices exist in both the foreign and domestic marine industries, as is true of any industry sector. Some companies have instituted, or are instituting, many of the Best Practices, while others appear to have instituted almost none (Fleischer, 1999).

SCM **in shipbuilding is hampered by a lack of consensus on the structure, function and dynamics of the integration of ship production and SCM**. Real progress toward reducing material costs and improving delivery time depends on a deeper understanding of the integration of internal processes and those of suppliers into a "production system." Successful manufacturing firms have learned to design supplier networks that minimize waste and maximize the benefits of supplier knowledge as well as process and material management capabilities. This is only in its infancy in shipbuilding (Fleischer, 1999).

Good practice in SCM leads to business success. Companies in the automotive, aerospace, construction and other industry sectors regarded as leaders in supply chain management are also leaders in important business metrics such as profitability, market share growth, product development time, and production cycle time. These companies attribute a significant part of their business success to their supply chain management philosophies and practices (Fleischer, 1999).

4 AN APPLICATION OF SUPPLY CHAIN MANAGEMENT MODEL IN A SHIPYARD

4.1 *Shipyard/Supplier comparison*

Planning for supply chain management is an important function for any manufacturing company regardless of size or position in the supply chain because every company has customers and suppliers. The complexity of the task, and hence the manner in which it is accomplished, will vary considerably with size and position (Fleischer, 1999).

4.2 *Consolidated purchasing*

Many companies noted for excellence in supply chain management have consolidated the purchasing function. This includes the company's approach to consolidating its supply base in both numbers of suppliers of like or similar items, as well as consolidating multiple purchases of different items from one supplier into something like a single master purchase agreement. It also includes how the purchasing function is consolidated organizationally and the degree of that consolidation (e.g., plant, single geographical area, business unit, multiple geographical areas, and corporation wide) (Fleischer, 1999).

4.3 *Lowest total cost selection*

The lowest total cost idea is that everything involved in supplier selection and management, in receipt and installation of the supplier's product, and support of the end item after delivery that involves the supplier's product represents a cost that should be added to the supplier's product cost to arrive at the total cost associated with that supplier. Competing suppliers are ranked on the basis of total cost, and the one showing the lowest total is selected. Sometimes the term "Best Value" is used for this practice, but many prefer lowest total cost because in has more quantitative connotations. Supplier quality is a key element. Some companies measure quality performance such as numbers of defects by assessing the economic cost to the company, depending not only on the severity of the problem, but also on the when the problem is discovered (Fleischer, 1999).

5 A CASE STUDY: SUPPLY CHAIN MANAGEMENT PROTOTYPE FOR A TYPICAL TURKISH SHIPYARD

After evaluation of the procurement processes of the shipyards, it is seen that especially the shipyards in Tuzla Bay area Istanbul, don't have domestic supplier problems. Istanbul is a metropolitan city and the industry is mainly situated between the cities of Istanbul and Kocaeli. The closeness of the shipyards to these regions is a significant advantage. However, Turkish Shipyards are dependent to abroad for purchasing of main engine, propulsion systems, electronic navigation equipments. Therefore supply and procurement issues should be treated more accurately.

This study makes up a prototype supply chain model that improves the supply chain management of the typical Turkish shipyard. Enhancement of the proposes model and forming a supply chain management model appropriate to the Turkish shipyards is crucial. Testing stage of the prototype is resulted successfully and training is kept on.

5.1 *Existing situation of the shipyard*

Due to there isn't application of supplier management during block constructions in A Shipyard, the shipyard works with high volume of stocks. There are some problems in the procurement and transportation of required materials. In addition, the delays regarding material handlings in the shipyard cause late completion of the blocks. Working with many subcontractors also slows down the processes. Procurement times are being prolonged and complicated due to the subcontractors give their own orders.

5.2 *Supply chain management and process configurations*

The inconveniences in supply chain management in A Shipyard are analyzed. Some meetings were arranged for the observed problems. One improvement is arranging only one supply chain for the subcontractors working block constructions. Thus, material flows were done easier. Working with minimum stock during block construction is a major factor. In order to carry out Just in Time production, the materials and workers of the process should be present when required. This is called just-in time supply chain. The aim is to optimize the supply chain processes with respect to block completion dates. Block completion date is based on for the X block of A Shipyard for this purpose. Calculations were made retrospectively from the block's commencement times and by subtraction of transportation, warehouse periods, procurement and order times were obtained. Block works are prepared in general and detailed in the scope of the study for one block and supply chain model is created for each requirement. Fig. 4 points out construction processes and material requirements of a block. Purchasing processes should be carried out in order to meet these requirements. According to Fig. 4., welded-joints are the most used method in the construction. Therefore, welding machine, electrodes, qualified worker and personal protective equipments should be present for welding of each section. In this purpose, the completion times and other times in intermediate construction stages were calculated for X block in consultation of planning department of the shipyard. Then, the order times of materials such as steel plates, profiles, welding electrodes and paints were calculated in accordance with the calculation of waiting times in warehouse and transportation times subject to latest date. According to subcontractors' man-hours during block construction, specified number of workers has been kept present for the works.

In this study, a significant time-saving was obtained for X block in A Shipyard and block is completed in soonest time of the real completion

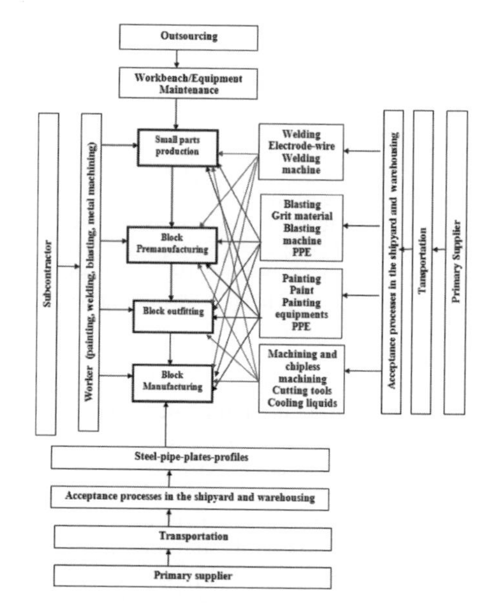

Figure 4. Supply chain management model.

time. In case of investments to the automation technologies in the shipyard, completion of the works will be achieved in shorter periods. So that, the shipyards that effect the environment and human less will also be come out regarding emissions.

The shipbuilding supply chain is a very complex process, especially due to the length of time takes to complete one finished product. It can prove very difficult to successfully manage this type of supply chain without causing a negative impact on the rest of business functions and the overall profit of the company. Building a ship utilizes highly complex processes to design and construct built-to-order products that meet its customer requirements. This process includes the cooperation of all parties, including the customers, the shipbuilder and its suppliers. The process further necessitates a seamless interaction of the suppliers, material management, planning and scheduling, and production. It is this method of these interactions that can either cost the shipbuilder and customer due to rework and rescheduling or allow the shipbuilder to construct the ship at the cost and in the timeframe allotted by the contract and even more desirable, at a profit. Because of the high costs of constructing ships and current practices of shipbuilding, many times the shipbuilder does not make a large profit and sometimes not at all (Ferreira et al., 2010).

6 CONCLUSIONS

Construction processes in the shipyards are very complicated and there are numerous suppliers and subcontractors. Therefore order follow-ups, transportation and storage are difficult factors to manage for all shipyards. In this context, an efficient supply chain management is essential. Thus, they can decrease the costs, delivery times and material handlings/transportations as well as increase the customer satisfaction and their reputability in the sector.

The supply chain management model in Fig. 4 is applied for one block of the shipyard in this study. In case of application of this approach to other blocks (the vessel), the success rate will be much higher. This study attempts the application of the management system to all parts of the vessel.

In order to make a profit, shipyards should create their own supply chain management systems. This is one way of increasing productivity and succeeding on time deliveries. In this paper, shipyards were evaluated based on supply chains. In addition to this approach, it is beneficial for all shipyards to make their organizations in accordance with today's expectations, select their top management from well-qualified managers, give importance to the workers' education and arrange their production facilities in order to stay profitable and competitive in the market.

Shipyards should also give significant importance to the equipments and workbenches' maintenance in addition to process improvements in order to get continuous production.

REFERENCES

Akanlar, F.T., Celebi, U.B. & Vardar, N. 2009a. Alternative Production Processes and New Technologies for Human and Environmentally Responsive Shipyards, *Proceedings of the 13th Congress of Intl. Maritime Assoc. of Mediterranean IMAM 2009, Istanbul, Turkey*, October 12–15, 2009 ISBN Vol. II 978-975-561-357-4: pp. 599–606.

Akanlar, F.T., Celebi, U.B. & Vardar, N. 2009b. New automated technologies in environmentally sensitive shipyards, *Proceedings of the 2nd International CEMEPE & SECOTOX Conference, Mykonos,* June 21–26, 2009 ISBN 978-960-6865-09-1:425–431.

Celebi, U.B., Akanlar, F.T. & Vardar, N. Personal protective equipment to minimize the shipyard production processes health effects on shipyard workers, *Proceedings of the 2nd International CEMEPE & SECOTOX Conference, Mykonos*, June 21–26, 2009 ISBN 978-960-6865-09-1.

Cullinane, K. 2005. Shipping Economics, *Research in Transportation Economics*, Volume 12, 1–17 2005 ISSN: 0739-8859/doi:10.1016/S0739-8859(04)12001-5.

Eyres, D.J. 2007. *Ship Construction, 6th edition*, Butterworth-Heinemann, ISBN: 9-78-0-75-06-8070-7.

Ferreira, S.; Rahman, M.A., Managing Material Flow at the US Shipbuilding Industry.

Fleischer, M., Kohler, R., Lamb, T., Bongiorni, H.B. & Tupper, N. 1999., *Shipbuilding Supply Chain Integration Project* October, 1999 Final Report Mantech Contract F33615-96-C-5511.

Hameri, A. & Paatela, A. 1998. Supply network dynamics as a source of new business, *Int. J. Production Economics* 98 (2005) 41–55.

Link Journal, 2005. Supply Chain Restructuring Based On Integrated Procurement Thelink 2005 Fall 51–61 Ref.1:http://www.intraqq.com/services/mastering_supply_chain.shtml

MD, B. & Sarder Ahad, Ali. 2010. *Proceedings of the 2010 International Conference on Industrial Engineering and Operations Management Dhaka, Bangladesh*, January 9–10, 2010.

Skjøtt-Larsen, T., Kotzab, H. & Grieger, M. Electronic marketplaces and supply chain relationships, *Industrial Marketing Management* 32 (2003) 199–210.

Ünüvara, M. 2009. Resarch On Effect Of Integrated Supply Chain Management Applications To Organizational Structure, *Ege Academic Review* 9 (2) 2009: 559–592.

Xanthakis, M.V., Mourtzis, D. & Chryssolouris, G. 2008. On the information modeling for the electronic operation of supply chains: A maritime case study, S *Robotics and Computer-Integrated Manufacturing* 24 (2008) 140–149. http://www.euromind-project.org/content/view/17/36/ (EUROMIND, 2008).

Sustainable Maritime Transportation and Exploitation of Sea Resources – Rizzuto & Guedes Soares (eds)
 ISBN 978-0-415-62081-9

Can a shipyard work towards lean shipbuilding or agile manufacturing?

D.A. Moura
Federal University of ABC, São Paulo, Brazil

R.C. Botter
University of São Paulo, São Paulo, Brazil

ABSTRACT: Lean production is regarded by many as simply an enhancement of mass production methods, whereas agility implies breaking out of the mass-production mould and producing much more highly customized products—where the customer wants them in any quantity. In a product line context, it amounts to striving for economies of scope, rather than economies of scale ideally serving ever-smaller niche markets, even quantities of one, without the high cost traditionally associated with customization. A lean company may be thought of as a very productive and cost efficient producer of goods or services.

1 INTRODUCTION

1.1 *Toytota Production System (TPS)*

Since the conception of the assembly line and the following development of the Toyota Production System (TPS), efficiency has been a central objective of manufacturing. Lean manufacturing focuses on the systematic elimination of wastes from an organization's operations through a set of synergistic work practices to produce products and services at the rate of demand. Lean manufacturing represents a multifaceted concept that may be grouped together as distinct bundles of organizational practices. A list of bundles of lean practices includes JIT, total quality management, total preventative maintenance, and human resource management, pull, flow, low setup, controlled processes, productive maintenance and involved employees. Lean manufacturing is as a set of practices focused on reduction of wastes and non-value added activities from a firm's manufacturing operations (Yang et al., 2011).

The base of the Toyota Production System (TPS) is to eliminate waste in the system. Therefore work philosophy and a few techniques/tools were inserted in the day-to-day organization to achieve such goal.

The seven types of waste recommended that should be eliminated in TPS are:

- Overproduction
- Transport, which adds no value to the product
- Process, transactions that should not exist
- Waiting time, intermediate stock which generates queue in the process
- Stock, throughout the production process, supply chain and finished products
- Driving, which adds no value to the product;
- Defects, which burden the productive process generating rework, wasted of time, manpower, hours of equipment etc.

1.2 *Agile Manufacturing*

According to Yusuf et al. (1999) agility can be summarized as the use of well-known developed technologies and manufacturing methods. Among them there are Lean Manufacturing, CIM, TQM, MRP II, BPR, Employee Empowerment and OPT. In other words agility is the ability to grow business in competitive markets of continuous and unexpected changes, with rapid response aimed at the consumer/customer valuing the product and service.

- CIM (Computer Integrating Manufacturing);
- TQM (Total Quality Management);
- MRP II (Manufacturing Resources Planning);
- BPR (Business Process Reengineering);
- OPT (Optimized Production Technology).

Agile can be describe as "Ability of an organization to detect changes (which can be opportunities or threats or a combination of both) in its business environment and hence providing focused and rapid responses to its customers and stakeholders by reconfiguring its resources, processes and strategies" (Mathiyakalan, et al., 2005).

An effective integration of response ability and knowledge management in order to rapidly, efficiently and accurately adapt to any unexpected (or unpredictable) change in both proactive and reactive business/customer needs and opportunities without compromising with the cost or the quality of the product/process (Ganguly et al., 2009).

"Ability of a firm to dynamically modify and/or reconfigure individual business processes to accommodate required and potential needs of the firm" (Raschke, David, 2005).

Ability of a firm to redesign their existing processes rapidly and create new processes in a timely fashion in order to be able to take advantage and thrive of the unpredictable and highly dynamic market conditions (Sambamurthy et al., 2003).

"The ability of a firm to excel simultaneously on operations capabilities of quality, delivery, flexibility and cost in a coordinated fashion" (Menor et al., 2001).

The Lean Manufacturing system aims to reduce the lead time for obtaining the components/parts, subsets etc. related to the supply chain, to reduce time of production/processing, to run the process/operation without faults (do it right at the first time) and to eliminate or minimize stocks with high control over the operations, on-time deliveries, increased productivity with efficiency in operations (Shingo).

Research conducted by Iaccoca Institute, Lehigh University, in USA resulted in a report about agility manufacturing. New criteria are (Sharifi, Zhang, 1999):

- Constant changes
- Fast response
- Improved quality
- Social responsibility.

Thus, an agile manufacturing company must have a broad view of new needs in the business environment, skill and ability to deal with turbulence and gain competitive advantage in its businesses.

The four main categories to be an organization in a rapidly changing environment are:

In Fast Response (ability to identify changes and promote rapid responses of reactive and proactive manner):

- Sensitivity to anticipate market changes
- Immediate reaction to changes and insert them into the system
- Absorbing changes.

In Competence (a set of abilities that produces higher productivity, efficiency and effectiveness in operations and processes to the tasks to achieve the goals set by company):

- Have strategic vision
- appropriate technologies or enough technological ability
- Quality of products and services
- Efficiency in costs
- High rate of introduction of new products
- People are trained, certified and involved with the process.
- Efficiency and effectiveness in lean operations
- Internal and external cooperation
- Integration.

In Flexibility (ability to process different products and achieve different goals with the same manufacturing plant):

- Flexibility in the volume of products;
- Flexibility in product models;
- Organizational flexibility;
- Flexible people.

In Quickness (ability to deal with tasks and operations in a shorter time).

- Short time to insert new products in the market
- Fast delivery of products and services
- Fast transaction time

Agile manufacturing encompasses both the concepts of lean and flexible. Also that lean manufacturing is primarily concerned with minimization (if not elimination) of waste through an efficient production process (Ganguly, et al., 2009).

Agile manufacturing means that the production process must be able to respond quickly to changes in information from the market This requires lead time compression in terms of flow of information and material, and the ability, at short notice, to change to a wide variety of products Therefore, the ability to rapidly reconfigure a the production process is essential. In lean manufacturing the ability to change products quickly is also key as any time wasted in changing over to a new product is muda and therefore should be eliminated (Naylor et al., 1999).

To summarize these two characteristics agile manufacturing calls for a high level of rapid reconfiguration and will eliminate as much waste as possible but does not emphasise the elimination of all waste as a prerequisite. Lean manufacturing states that all non-value adding activities, or muda, must be eliminated (Naylor et al., 1999).

Agile manufacturing further requires an all encompassing view, whereas lean production is typically associated only with the factory floor. Agility further embodies such concepts as rapid formation of multi-company alliances or virtual companies to introduce new products to the market. An agile company is primarily characterised as a very fast and efficient learning organisation if it was not first productive and cost efficient (Sharp et al., 1999).

In agile manufacturing, the main features shall be (Yusuf et al., 1999):

- High quality products and highly customized
- Products and services with high added value
- Mobilization of key competences.

- Commitment to social and environmental matters
- Responding to change and uncertainty
- Intern Integration and between companies.

2 THE ENABLERS OF AGILE MANUFACTURING

The enablers of Agile Manufacturing are the strategies, systems, technologies, methodologies and tools that allow the company to become agile. For better understanding, these enablers are classified based on its focus. This classification groups the enablers of Agile Manufacturing, according to the focus on four categories:

- Strategies: Virtual enterprise/virtual manufacturing
  Virtual enterprise is a temporary aggregation of smaller units and its core competencies and associated resources, which gather together to explore business opportunities and act like a single large company. However, as one company is not often able to respond quickly to market needs, the virtual company works for its agility. The subject of virtual enterprises within an agile context is considered vital and indispensable for Agile Manufacturing.
- Integration of supply chain
- Management based on key competences
- Simultaneous Engineering
- Management based on uncertainty and change
- knowledge-based management
- Technologies: Hardware—Tools & Equipment.

To Gunasekaran (1999), Agile Manufacturing requires the rapid shift in product assembly. This is only possible with an adequate structure for the hardware (robots, feeders of flexible parts, module assembly, automated visual inspection, computer-guided vehicles etc.).

- Information Technology: computers and software
  The technology and information systems used in Agile Manufacturing can be divided according to the purposes intended, in:
  Technology and systems dedicated to agile project: CAD, CAM, the computer-aided planning process (CAPP);
  Technologies and systems for the agile production: FMS, CIM.
  Technologies and systems of communication and integration inside and among enterprises MRP, ERP, EDI and electronic commerce.
- CAD (Computer Aided Design);
- CAM (Computer Aided Manufacturing);
- FMS (Flexible Manufacturing System);
- MRP (Material Requirement Planning);
- ERP (Enterprise Resource Planning);
- EDI (Electronic Data Interchange)

Systems: Systems Design

Several techniques and systems are addressed in the literature that support the agile systems design: CAD/CAM, rapid prototyping and QFD are some examples. Regarding the project support systems for Agile Manufacturing, some jobs are worth highlighting:

- QFD (Quality Function Deployment)
- Planning and Control Systems
- Integration of management systems and database.

People

- Continuous improvement
- Commitment of senior management and empowerment
- People multi-qualified, flexible and knowledgeable
- Teamwork and participation
- Training and continuing education.

The main human factors to be considered for an agile manufacturing environment are: continuous improvement, top management commitment and empowerment, use of flexible multi-enabled people, teamwork and participation, training and continuing education.

3 SOME IMPORTANT POINTS TO BECOME LEAN AND/OR AGILE

3.1 *TQM—Total Quality Management*

TQM is something more solid which involves an integrated and shared chain with strategic goals of high performance and quality, aiming at highly competitive markets with sustainable industrial processes and international reference. However, quality program like ISO 9000 does not necessarily guarantee the best quality practices and can not be considered an integrated process throughout the production chain, but it is a first step to check quality.

TQM has the emphasis on continuous improvement of industrial processes, always seeking the feedback system, in order to improve the process and eliminate potential causes of problems. Thus, TQM integrates the suppliers from the development phase of the project, in the quest for continuous improvement with a focus on flawless process, reducing the development time, with operational reliability in the process, and products with no defects according to the specifications of the customer or market, free of processing errors or rework, with a balanced industrial operations, with high productivity and reduced operating costs.

3.2 *Core competencie*

Core competencies are factors that involve collective learning and the way that those values are disseminated in an organization, and how those competences are managed in order to enhance the integration among the agents who seek for competitive advantage of an organization to face competitors.

The core competence of an organization may allow the opening of new markets or be a positive factor to try to keep customers, being an advantage over the competitors when decisions of purchase are made, as well as being an outstanding brand when compared to others. Core competence can make a competitor to have difficulty imitating it (Prahalad, Hamel, 1990).

3.3 *Innovation*

Innovation is a key factor in competitive advantage for an organization. Then, fine tune with the needs of markets is a key factor to promote the competitive edge of companies. Factors such as financial sustainability, ways of relating to their supply chain and customers, reliability and recognized quality of products and service are key points that shall be taken into consideration when making strategic decision for a company to become globally competitive.

Innovation means that industries can gain competitive advantages in their segments. Thus, it is essential that companies make investment as a way to stand out from competitors and gain recognition.

Innovation will require pro-active strategies for anticipating technological and market changes which directly or indirectly affect companies when facing their main competitors. Thus, this process should also be inserted in the supply chain of a client, otherwise it would have difficulties in gaining competitive advantage over the competitor. It is also essential to integrate innovative business strategy of a company and its partners.

3.4 *Advantage in manufacturing*

The competitive advantage in manufacturing shows that the company stands out from its competitors to meet market needs. That means making right is related to the goal of quality performance, making fast relates to Speed, making in time relates to reliability, customization relates to flexibility and making with low cost is related to the objective costs.

The manufacturing strategy, according to, can not be isolated from corporate strategy and should affect and be affected by other areas of business such as Marketing, Finance, Purchasing, Research and Development, Human Resources etc. The authors comment that the manufacturing objectives are expressed in terms of some dimensions of performance used to measure manufacturing strategy, characterized by: cost, quality, flexibility and delivery, as described by Wheelwright (1981).

Technological capability is one of the attributes that can differentiate a company from its competitors. They report that firms that possess technological expertise recognized by the market have an asset difficult to be imitated contributes to the improvement of products, increasing their value and creating a gap in the market among companies that have it and those that still try to achieve. The development of technological capability must be inserted in the strategy defined by the company.

4 SOME EXPERINCES

According to Kim (1997), as South Korea approached the boundaries of technology, activities related to Research and Development (R & D) has become more intense. There was a need for targeted search for relevant information, more interaction between the project team and other departments of the organization like production and marketing, and even with other companies, such as the suppliers, customers, local research institutions, and universities.

One of the policies implemented in Korea was the import of technology and its dissemination to all Korean companies in that segment, aiming to have the largest possible number of Korean companies with knowledge of the new world-leading technologies. Then, Korean companies noticed the need to develop their own technologies, assimilate, adapt and improve the imported technology. For this, there was a need for investment and integration with the areas of research and development (R & D) with the intention of having their own technologies. Therefore, with increasing industrialization, there were government policies focused on increasing research and development (Kim, 1997).

Kim (1997) asserts that the policy aimed at import substitution was critical in creating the demand for foreign technology transfer. The import substitution through protectionism contributed greatly to the transfer of technology from other countries, leveraging various industries and introducing more sophisticated products.

Add to that the export issue, which became the top priority of the Korean government to achieve goals of economic growth. Thus, the government selected strategic industries, both for import substitution and for export promotion.

As a segment changed his condition from not developed to an exporter, the Korean government decreased significantly its protectionism.

The Korean government defined exports target montlhy, and companies were required to achieve that goals being monitored constantly by the Minister of Trade and Industry, directors of the biggest financial institutions, leaders of business associations and representatives of leading exporting companies.

As South Korea was one of the countries that entered the shipbuilding sector much later than its biggest competitors at the time, she had the advantage of the projects best suited their yards, compared to existing in the Asia and Europe. Apart from this, some were designed with huge capacity, exceeding enormously the total capacity of countries considered high-power production for the season. The ability of a single Korean shipyard has already surpassed the total production of a country. In addition to these items, there was the fact that the Korean manpower work more hours per week, compared with European countries, and this has increased the competitiveness of Korean shipbuilding segment of the world.

South Korea has created policies towards the shipbuilding segment that gave sustainability to the sector by promoting the development of technology centers, universities, companies of marine parts, service companies, industrial parks, schools, technical and labor-specialized work, and has focused primarily on the external market. Export was a challenge that has afforded it the policies for the shipbuilding sector and enormous efforts have been made by various actors directly or indirectly related to the country to reach their goals and become globally competitive in that segment.

Both South Korea and Japan have specialized in the production of bulk carriers and tankers focused on mass production, benefiting their production lines because the yards have reduced or eliminated the flexibility offered to the clients, the ship owners, benefiting economy of scale and reducing production costs. Low or no flexibility, high quality, low cost, reduced cycle time for development and production with some innovation/ technology were some of the strategies used by Korean shipyards.

This has seen a huge gain with the learning curve, obtaining a competitive advantage against global competitors. The strategy of South Korea was producing ships different from those produced in Japan, with simpler and cheaper products. Another peculiarity was the planning for the financing focused on exports. There was heavy subsidies in the Korean shipbuilding sector, for insertion of its vessels in various world markets, as well as having strong export policy aimed at solidifying entire structure to make South Korea a country among the most renowned world shipbuilding market.

Japan has established itself in the strategy of cost leadership, according to the model of Porter. With strong participation of several companies related to the sector, with special dedication to factors related to quality control, well-trained manpower able to perform their tasks with the highest quality in the production process, the emphasis for having a classification society qualified and a standardization policy which would help boost the business of shipbuilding.

But soon the focus of Japanese policies shifted to Research and Development, with strong predominance of the critical success factor Innovation according to Bolwijn and Kumpe (1990).

It is critical that a business analyzes the tradeoffs from the manufacturing area, in order that the settings defined in the strategic production can meet the corporate strategies and allow the company to become competitive in highly competitive global markets. Analyzing possible decisions and their alternatives is essential to guide the likely direction to be followed by an organization to promote their competitive advantages in the market.

Japan has guaranteed a minimum production at its shipyards, which contributed to promoting the development of the sector. This program was called Keikaku Zosen. Furthermore, there was a massive investment in automation, to reduce the cost of manpower, and this factor contributed greatly to developing the critical success factor Technology and, thus, Japan is recognized with this competitive advantage ahead the international market of shipbuilding.

Japan has innovated in the production of ships and consequently has increased productivity, but also innovated in the design of vessels. Invested in robotics and in managerial and administrative techniques for controlling the flow of materials and their respective quality.

Another very important factor in the Japanese shipbuilding system was the integration existing in the supply chain among shipyards and their suppliers of ship parts, and there was integration between shipyards and ship owners too, and also between competing shipyards. There was bigger cooperation for product development and technology that would benefit everyone, with government incentives, helping the growth of the local maritime sector. There was the implementation of national policy for promotion of scientific and technological activities involving laboratories, universities, research institutes etc.

Thus, the Japanese were able to get competitive prices globally and even below the market average in the construction of their ships, besides offering special financing conditions for international ship owners to build their ships in shipyards in Japan. For this it was necessary plans, incentive mechanisms and instruments of industrial policy that would involve not only shipbuilding but the chain that was directly or indirectly related to the Japanese shipbuilding industry. For instance: chemical, steel and metallurgic industries, electrical machinery and transport equipment and heavy chemical industry. There was the essential participation of the Ministry of International Trade and Industry to create such industrial policies that ensure sustained growth of the segment.

5 SHIPYARD CAN WORK TOWARDS LEAN SHIPBUILDING OR AGILE MANUFACTURING

In order to work with the production system similar to an automobile assembly plant, a shipyard must acquire most of the parts and components in the form of subsets, available on the market aiming to reducing domestic costs of production.

A key factor in production management is related to the flow of information on the sites, focusing on planning and control of the production process. To make this analogy is relevant to the lean production system with special attention to the Just-In-Time, the resource planning and project management organization.

As the shipbuilding is characterized within the system of production by large projects is essential to focus on managing each activity in order to reduce operating costs, waste and carrying out each task in the correct period without generating stocks.

Integrated information systems are critical to achieving the state-of-the art in various functions of a shipyard. Production features such as cutting boards with numerical control, or the use of automated processes on dedicated production lines, and also functions of planning and control only affect the state of the-art if there are available information systems product, process and resources available and fully integrated.

Concentrating similar production processes identifying families of products that can be manufactured in the same cost centers, using the productive capacity of resources, machinery, equipment, people, in order to generate a continuous flow of operations, without generating intermediate stocks throughout the process production is a prerequisite for entering into the Lean Manufacturing system.

The focus is not to generate batch processing (batch processing), but uniformly according to the needs of each production center, optimizing resources and minimizing or eliminating driving steps, intermediate stock during the production process.

The gain of manufacturing family of products is higher when compared with manufacturing by specialized centers in functions (Liker, Lamb, 2001).

Thus, it is sometimes necessary to duplicate a production center in the layout of a shipyard. It does not mean to double the area that existed initially for this batch operation, but rearrange physically to fill the needs for a continuous production flow. It is often necessary smaller areas and resources with the dismemberment of manufacturing centers that were concentrated.

Eliminating intermediate stocks in the process can provide an enormous gain in physical space for the shipyards. Lean flow allow cost savings in operations and improve efficiency and effectiveness of production, allowing to balance tasks and optimize the use of productive resources.

Reducing or eliminating stock will resulted in the reduction of its costs, involving the supply chain, materials and processes in the physical area, which serve to support the lean production system.

Another relevant factor is the cost of unnecessary drives that are eliminated with the inclusion of a lean production flow.

The problems that arise in the production system will be easier identified and mapped. So, an action plan may be strategically placed to eliminate or minimize them aiming to not to interrupt production.

With the elimination of batch production and the insertion of a lean flow, reducing inventory, an essential factor that will be easily noticed is the quality of manufactured products, as problems related to quality will be easily detected and require quick, efficient and effective solution.

The large batch production does not allow us to understand the problems of quality detected. When they are detected they will have caused more problems along the entire supply chain, manufacturing, increasing costs by increasing waste of resources, time, machine, man-work etc.

The productivity of a company is an important indicator of competitiveness. When production problems are eliminated or reduced to a minimum acceptable, will automatically increase the productivity of the organization by avoiding rework or loss of semi-processed or finished product.

In constructions that operate under a system of large projects with high operational costs, by operations, parts, products, subsets etc. is essential to have quality-assured on the manufacture and also on its supply chain, because production stoppages due to defects can make the final product too much expensive and drive up costs, reducing productivity and competitiveness of a shipyard.

Rework, unnecessary movements, activities that do not add value to the product are factors that minimize the productivity of a company and increase the lead time for implementing the final product, making it uncompetitive compared to its main competitors.

Assured quality of parts, components, assemblies, subassemblies etc. is the backbone of a lean process to eliminate waste and activities that add no value to the final product. Get output with high productivity will require that this concept is widespread in every stage of the production process.

The industrial layout should be efficient and provide operational efficiency by eliminating most unnecessary transport and reducing the operation time in the shipyard.

The implementation of the system 5S's housekeeping is also essential in the whole production system. This type of technic corroborates to increase productivity, to eliminate unnecessary handling or transport, to reduce manufacturing time, to eliminate defects and to improve productivity and strengthen lean production.

Lean production also extends to the supply chain of the shipyards. Receiving materials in time to be processed is important to minimize or eliminate the stocks in the production process.

Receiving the products with assured quality from the supply chain will require that quality control is performed inside the supplier's plant so that the manufacturing system does not stop at the shipyard.

6 CONCLUSIONS

The shipyards must work to minimize or eliminate waste in project and production phases. The integration with the supply chain is essential to develop families of interim products.

The production must fabricated using standard work processes in the same way each time using the same equipment.

To implement agile manufacturing, product design and planning must become very closely integrated with manufacturing, and all bottlenecks in product flow and the flow of engineering information must be minimized. The tight integration between design functions, planning and manufacturing requires precise and sufficiently complete information on all aspects of product, production processes and operations are available. Thus, it is expected that future systems design and planning are closely aligned with the manufacturing technology, and future manufacturing systems will require more complete and more accurate when compared to the information available at this time.

REFERENCES

Bolwijn P.T. & Kumpe, T. 1990. Manufacturing in The 1990s—Productivity, Flexibility and Innovation, Long Range Planning, vol. 23, nº 4, pp. 44–57, UK.

Ganguly, A., Nilchiani, R. & Farr, V.J. 2009. Evaluating agility in corporate enterprises. International Journal of Production Economics, 188, pp. 410–423.

Gunasekaran, A. 1999. Agile manufacturing: a framework for research and development. International Journal of Production Economics, v. 62, n. 1–2, pp. 87–105, May.

Kim, L. 1997. Imitation to innovation: the dynamic of Korea's technological learning, Boston: Harvard Business School Press.

Liker, J. & Lamb, T. 2001. Lean Shipbuilding. The Society of Naval Architects and Marine Engineers. Ship Production Symposium, June 13–15, Ypsilanti, Michigan.

Mathiyakalan, S., Ashrafi, N., Zhang, W., Waage, F., Kuilboer, J.P. & Heimann, D. 2005. Defining business agility: an exploratory study. In: Proceedings of the 16th Information Resources Management Conference, San Diego, CA, May 15–18.

Menor, L.J., Roth, A.V. & Mason, C.H. 2001. Agility in retail banking: a numerical taxonomy of strategic service groups. Manufacturing and Service Operations Management 3 (4), 272–292.

Naylor, J.B., Naim, M.M. & Berry, D. 1999. Leagility: Integrating the lean and agile manufacturing paradigms in the total supply chain. International Journal of Production Economics, 62, pp. 107–118.

Prahalad, C.K. & Hamel, G. 1990. The core competence of the corporation, Harvard Business Review, Vol. 68, No. 3, pp. 79–91.

Raschke, R. & David, J.S. 2005. Business process agility. In: Proceedings of the 11th Americas Conference on Information Systems, Omaha, NE, USA, August, pp. 355–360.

Sambamurthy, V., Bharadwaj, A. & Grover, V. 2003. Shaping agility through digital options: reconceptualizing the role of information technology in contemporary firms. MIS Quarterly 27 (2), 237–263.

Sharifi, H. & Zhang, Z.A. 1999. Methodlogy for achieving agility in manufacturing organizations: an introduction. International Journal of Production Economics, 62, pp. 7–22.

Sharp, J.M., Irani, Z. & Desai, S. 1999. Working towards agile manufacturing in the UK industry. International Journal of Production Economics, 62, pp. 155–169.

Wheelwright, S.C. 1981. Japan; where operations really are strategic. Harvard Business Review. July-August, pp. 67–74.

Yang, M.G.M., Hong, P. & Modi, S.B. 2011. Impact of lean manufacturing and environmental management on business performance: An empirical study of manufacturing firms. International Journal of Production Economics, 129, pp. 251–261.

Yusuf, Y.Y., Sarhadi, M. & Gunasekaran, A. 1999. Agile manufacturing: the drivers, concepts and attributes. International Journal of Production Economics, 62, pp. 33–43.

Sustainable Maritime Transportation and Exploitation of Sea Resources – Rizzuto & Guedes Soares (eds)
© 2012 Taylor & Francis Group, London, ISBN 978-0-415-62081-9

Matching product mix shipyard effectiveness through the design for production concept

D. Kolić, T. Matulja & N. Fafandjel
University of Rijeka, Faculty of Engineering, Rijeka, Croatia

ABSTRACT: The management of most shipyards would prefer to have an order book based on large series production of one type of vessel due to ease of design and minimal product variation. However, it is unrealistic for most shipyards to survive in this way. Therefore a *design for production* methodology for shipyards with product mixes is necessary. In this work, the analysis of design variations and structural configurations of three distinctly different types of ships, a chemical tanker, a car-carrier and a crane barge are presented. The method for determining the technological parameters that raise the production facility compliance level is explained and illustrated. The aim is to adapt all three vessel designs to be virtually equally compliant to the fixed shipyard production facilities as to enable a higher level of shipyard productivity. Finally, *production friendly* design guidelines are developed to be used by ship designers in the future.

1 INTRODUCTION

1.1 *Shipbuilding market and product mixes*

In order for new building shipyards to survive, it is necessary to maintain a full order book. Whereas it is advantageous for shipyards to contract the building a large series of one type of vessel, this is usually not feasible for many shipyards due to the ephemeral market demand. Therefore, in order for the shipyard production facilities to maintain a constant and balanced work load, shipyard management must contract product mixes of vessels. A shipyard with a product mix builds different types of vessels simultaneously. For instance, with the case study presented in this paper, the product mix includes a chemical tanker, a car-carrier and a crane barge. The problem with the product mixes is that often the shipyard facilities are not able to produce the interim products of the different vessel types with equal ease. This results in an unbalanced work load which is not efficient and drives up production costs.

1.2 *Design for production concept*

The *Design For Production* (DFP) concept has been around since the 1970s, first implemented by the consultants A&P Appledore with a Design for Production manual (Lamb, 2003). The National Shipbuilding and Research Program (NSRP) has also taken steps in presenting *Design for Production* research and application in conjunction with industry professionals and consultants such as First Marine International (Dlugokecki et al., 2009). The authors of this paper are introducing the application of the DFP concept in a shipyard through the development of a methodology to be used by future designers along with design guidelines. The interim products chosen for this analysis includes shipbuilding block assembly which lends itself to repetitiveness when the DFP methodology is applied.

2 BACKGROUND

2.1 *Group technology and PWBS*

Traditional shipbuilding design includes creating a design which satisfies the owner and classification society requirements. Likewise, in traditional shipbuilding, the detailed production processes are left to be decided by the various production foremen. It is highly craft based. While DFP also requires the two aforementioned conditions to be met, it goes further by stipulating that the production facility constraints be integrated very early in the design and that production steps, workstations and assembly sequences are decided in a scientific manner (Kolić, 2010). When done properly, this results in a reduction of the number of man-hours and therefore brings down shipyard costs. The application of group technology and a Product oriented Work Breakdown Structure (PWBS) go hand in hand and represent a necessary foundation for any shipyard seriously considering practically applying DFP and improving productivity (Bunch, 2010 & Chirillo, 2010).

2.2 *DFP principles*

The modern shipbuilding process integrates different disciplines while simultaneously complying with owner requests and technical characteristics and quality in the shortest possible time, with minimal costs. In addition, the production process is simplified and standardized as much as possible. The basic principles of DFP include simplicity of construction and choosing building solutions which are compliant to the technological capabilities of the individual shipyard.

Since the purpose of this paper is to analyze interim products from the block assembly process, a DFP compliant design includes the following points (DFP manual, 1999):

- minimal number of elements,
- minimal number of elements which need to be shaped in moulds,
- reduction of the variation of elements,
- reduction of the length of welded connections,
- standardization of elements,
- minimal connecting and adjusting of blocks during assembly,
- elimination of the need for extremely precise joints,
- integration of the hull structure and outfitting,
- elimination of the need for scaffolding,
- securing access for workers to interim products.

The shipyard capabilities adjusted in compliance with interim products implies the following:

- assembly of blocks within the limits of the lifting capacity of the shipyard cranes,
- size of panels, blocks and all interim products to be compliant with the dimensions and capabilities of the workstations,
- capability of using steel plates of maximum size in order to decrease the length of welded joints,
- designing in such as way so that multiple works are undertaken in the workshops and less work is performed on the vessel while on the slipway or along the outfitting pier.

3 SHIPYARD CASE STUDY

3.1 *Shipyard production program mix*

The shipyard production program mix includes the following vessels (3. Maj shipyard archive, 2010):

- 49000/51800 DWT tanker for the transport of oil, oil products and chemicals (Chemical tanker)
- 12300 DWT RO/RO vessel for automobile transport (Car carrier),
- 6300 DWT Deck cargo barge with crane fitted on deck (Crane barge).

The interim product chosen for analysis includes the double bottom block from the parallel mid body. The elements that make up the double bottom block include stiffened panels and built-up panels. A table of design variations is drawn up from the following characteristics (DFP manual, 1999):

- variation of steel plate thickness within a panel,
- orientation of steel plates—direction of straking,
- type of stiffener and its dimensions,
- spacing between stiffeners and number per panel,
- height of transverse web frames,
- spacing and number of webs per panel,
- panel dimensions and mass,
- block mass,
- quality level of steel.

Figures 1 to 3 below show the double bottom block breakdowns for the three vessel types and Tables 1 to 3 show the design variations for these same block types covered in the case study.

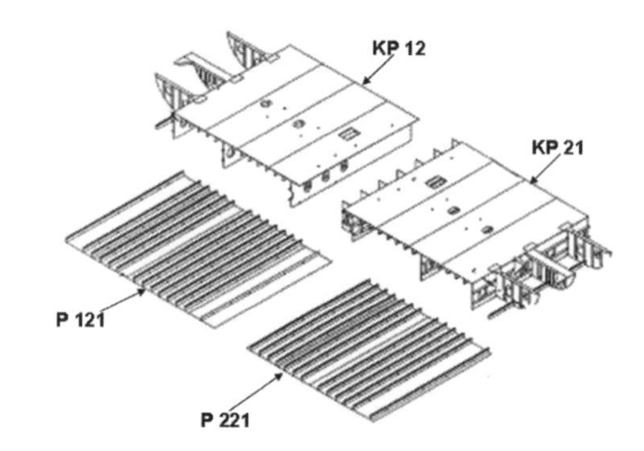

Figure 1. Double bottom block breakdown for chemical tanker.

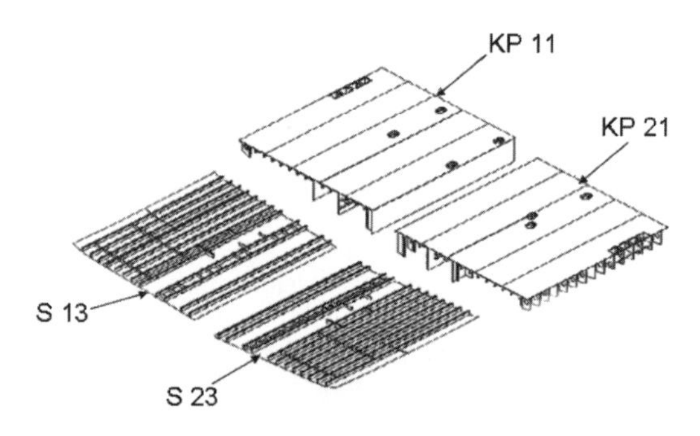

Figure 2. Double bottom block breakdown for the car carrier.

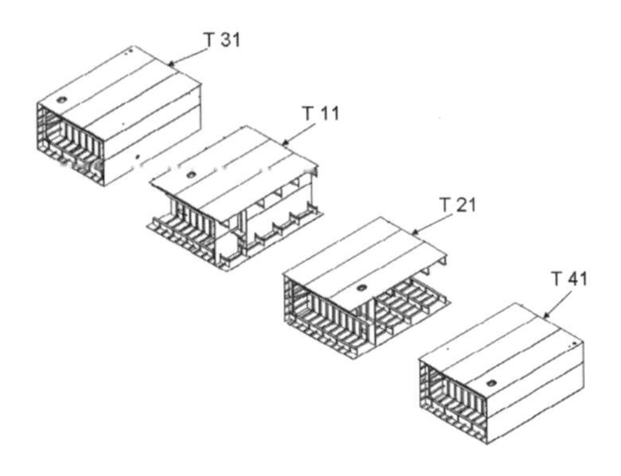

Figure 3. Double bottom block breakdown for the crane barge.

Tables 1 to 3 show the design variations for the main interim products of the double bottom blocks for all the chemical tanker, car carrier and crane barge respectively. Please note the following abbreviations used in the tables.

- P: panel,
- KP: built-up panel
- VT: very large three dimensional block
- HP: Holland profile longitudinal

3.2 *Explanation of tables*

Row 1 shows plate thicknesses which vary between 10 mm and 15 mm. The number of plates per panel

Table 1. Design variations for the chemical tanker double bottom block (3. Maj shipyard archive, 2010, DFP manual, 1999).

		Chemical tanker			
		Double-Bottom			
		Group 3410 - VT02 erection block			
No.	Key areas of variation	KP12 double bottom top	KP22 double bottom top	P121 outer hull bottom	P221 outer hull bottom
1	Plate thickness Number of plates per panel	16 mm 4 plates per panel	16 mm 4 plates per panel	15, 17,5 mm 5 plates per panel	15 mm 4 plates per panel
2	Longitudinal scantlings mm	longitudinals 370 × 13, 2 longitudinal girders 2180 × 16, 2180 × 14,5 tunnel 2180 × 20	longitudinals 370 × 13, bars180 × 13 2 longitudinal girders 2180 × 16, 2180 × 14,5 tunnel 2180 × 20	longitudinals 340 × 14, bar 250 × 16	longitudinals 340 × 14
3	Type of section	HP / plate	HP / bar / plate	HP / bar	HP
4	Longitudinal spacing mm	800	800	800	800
5	No. of longitudinals per panel	12 longitudinals 2 longitudinal girders tunnel	12 longitudinals 1 bar 2 longitudinal girders tunnel	13 longitudinals 1 bar	13 longitudinals
6	Spacing of webs mm	1700/3400	1700/3400	x	x
7	No. of webs per panel	4	4	x	x
8	Depth of webs mm	2180	2180	x	x
9	Panel dimensions mm	11046 × 11998	11046 × 12078	11046 × 14336	11046 × 11876
10	Panel weight t	52 t	55,3 t	27,6 t	27,6 t
11	Block weight t	272 t	272 t	272 t	272 t
12	Steel quality	A	A	A	A
13	Direction of plate straking	longitudinal bow-stern	longitudinal bow-stern	longitudinal bow-stern	longitudinal bow-stern

Table 2. Design variations for the car carrier double bottom block (3. Maj shipyard archive, 2010, DFP manual, 1999).

		Car carrier			
		Double-Bottom block			
		Group 3510 - VT01 Erection block		Group 3511 - VT01 Erection block	
No.	Key areas of variation	KP 11 double bottom top	KP21 double bottom top	KP11 double bottom top	KP21 double bottom top
1	Plate thickness Number of plates per panel	12 mm 5 plates per panel	12 mm 2 plates per panel	12 mm 5 plates per panel	12 mm 5 plates per panel
2	Longitudinal scantlings mm	longitudinals 300 × 11 2 long. girders 2080 × 12, 2080 × 24 bottom centerline girder 2080 × 19	longitudinals 300 × 11 1 longitudinal girder 2080 × 12	longitudinals 300 × 11 2 long. girders 2080 × 12, 2080 × 24 bottom centerline girder 2080 × 19	longitudinals 300 × 11 2 longitudinal girders 2080 × 12, 2080 × 24
3	Type of section	HP / T assembly / plate	HP / T assembly	HP / T assembly / plate	HP / T assembly
4	Longitudinal spacing mm	750	750	750	750
5	No. of longitudinals per panel	11 longitudinals 2 longitudinal girders bottom centerline girder	5 longitudinals 1 longitudinal girder	13 longitudinals 2 longitudinal girders bottom centerline girder	13 longitudinals 2 longitudinal girders
6	Spacing of webs mm	3400	3400	3400	3400
7	No. of webs per panel	4	4	3	3
8	Depth of webs mm	2080	2080	2080	2080
9	Panel dimensions	11746 × 12059	11746 × 4606	12796 × 12059	12796 × 11909
10	Panel weight t	40,8 t	15,1 t	43,9	39,7 t
11	Block weight t	168,5 t	168,5 t	184,7 t	184,7 t
12	Steel quality	A	A	A	A
13	Direction of plate straking	longitudinal bow-stern	longitudinal bow-stern	longitudinal bow-stern	longitudinal bow-stern

varies between 2 and 5. See Tables 1–3. Row 2 shows the longitudinal scantlings or dimensions whereas row 3 describes the type of section such as Holland profile (HP), bar and plate. Row 4 lists the spacing between the longitudinals which is 800 mm for the chemical tanker and 750 mm for the car carrier and crane barge. In row 5, the number of longitudinals varies between 5 and 12 and up to 2 longitudinal girders per panel. For row 6, the spacing of the webs on the built up panels are 1700 or 3400 mm for the chemical tanker and car carrier, while 2000 mm for the crane barge. In row 7, the number of webs per built up panel are 3 or 4 for the chemical tanker and crane barge, while between 1 and 5 for the crane barge. Moving on to row 8, the depth of the webs are 2180 mm for the chemical tanker, 2080 mm for the car carrier, and between 400 and 745 mm for the crane barge. In row 9, the panel dimensions vary between 11000 mm for length and 12059 mm for the beam. In row 10, the panel weight varies between 27,6 tons for panel with a P designation in Table 1 and up to 55,3 tons for built up panels which have a KP designation. The block weight in row 11 is the final block weight prior to

Table 3. Design variations for the crane barge double bottom block (3. Maj shipyard archive, 2010, DFP manual, 1999).

		Crane barge				
		Ring				
		Group 3510—VT01 erection block				
No.	Key areas of variation	KP11 bottom	KP12 deck	KP13 (KP23) longitudinal blkhd	KP14 (KP24) transverse blkhd	KP31 bottom
1	Plate thickness Number of plates per panel	15 mm 4 plates per panel	15 mm 4 plates per panel	12, 15 mm 2 plates per panel	10 mm 2 plates per panel	15 mm 3 plates per panel
2	Longitudinal scantlings	longitudinals 60 × 9	longitudinals 180 × 9	longitudinals 160 × 9 120 × 8	vertical stiffeners	longitudinals 160 × 9
3	Type of section	HP	HP	HP	HP	HP
4	Longitudinal spacing mm	750	750	750	750	750
5	No. of longitudinals per panel	10 longitudinals	10 longitudinals	5 longitudinals	9 vertical. stiffeners	9longitudinals
6	Spacing of webs mm	2000	2000	2000	at half height of the ring 2250 T assembly transverse on stiffening	2000
7	No. of webs per panel	5	5	5	1	5
8	Depth of webs mm	600	600	400	600	735
9	Panel dimensions mm	11000 × 8980	11000 × 8110	11000 × 4500	7485 × 4500	11000 × 7345
10	Panel weight t	16,2 t	15,4 t	6,6 t	3,7 t	14,7 t
11	Block weight t	175,7 t	175,7 t	175,7 t	175,7 t	175,7 t
12	Steel quality	A	A	A	A	A
13	Direction of plate straking	longitudinal bow-stern	longitudinal bow-stern	longitudinal bow-stern	transverse port-starboard	longitudinal bow-stern

erection and includes additional interim products which are not assembled on the panel line or the built up panel line and are therefore not included in this work. The steel quality in row 12 is grade A for all vessel types. Finally, in row 13 the direction of plate straking is longitudinal from bow to stern.

3.3 *Future design guidelines*

The future design guidelines define future characteristics of interim products to be simplified and standardized (Kolić, 2010). The shipyard in this case study has the following constraints of its panel line (3. Maj shipyard archive, 2010 & Kolić, 2010):

- Maximum plate length for the panel line is 14500 mm,
- Maximum panel width is 15200 mm,
- Thickness of plates are between 8–35 mm,
- Maximum mass of panels are 35 tons.

Likewise the built up panel line has the following constraints (3 Maj shipyard archive, 2010):

- Length of panels are between 4000–14500 mm,
- Width of panels are between 4000–14500 mm,
- Girders have a maximum length of 12500 mm, and a maximum height of 3500 mm,
- Maximum mass of 100 tons.

Based upon the given constraints, and tables 1–3, it becomes clear that the performance of the facilities is not used to the fullest extent, since the lengths and widths are lower for most of the panels which is given in row 9 of all the tables. Likewise, the panel weights could also be increased. Additionally, the spacing of the stiffeners is something that could be standardized at 800 mm for all three vessels.

The longitudinals used could also be more standardized which also has a positive effect on the production facilities. A table for bulb profiles could include the following: See Figure 4 and Table 4 below.

Finally the following midship section aids in future designs. See Figure 5.

The midship section above is developed by considering the shipyard facility constraints along with typical design practice. The two main DFP principles of simplification and standardization are the

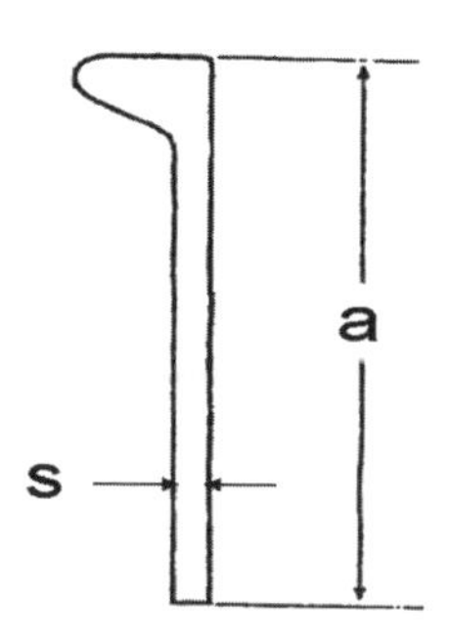

Figure 5. HP longitudinal with height (a) and width (s) defined.

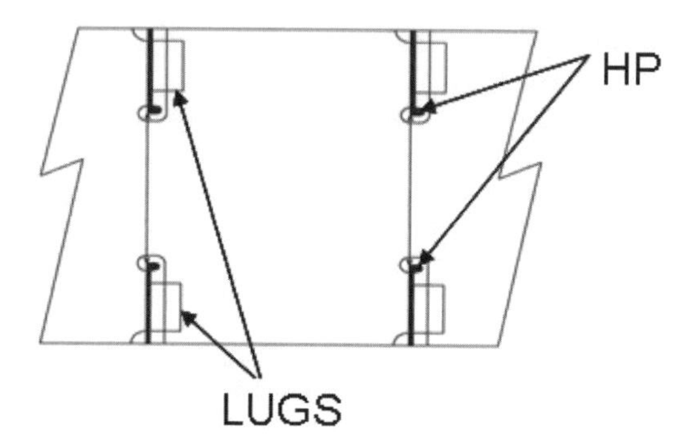

Figure 4. Double bottom with HP longitudinals and lugs defined.

Table 4. Longitudinal dimensions.

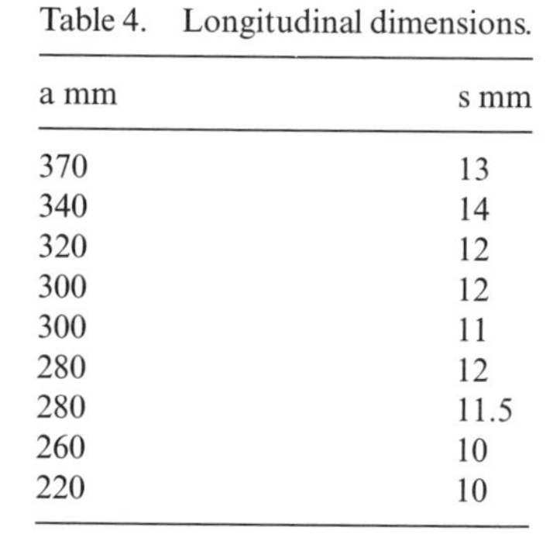

a mm	s mm
370	13
340	14
320	12
300	12
300	11
280	12
280	11.5
260	10
220	10

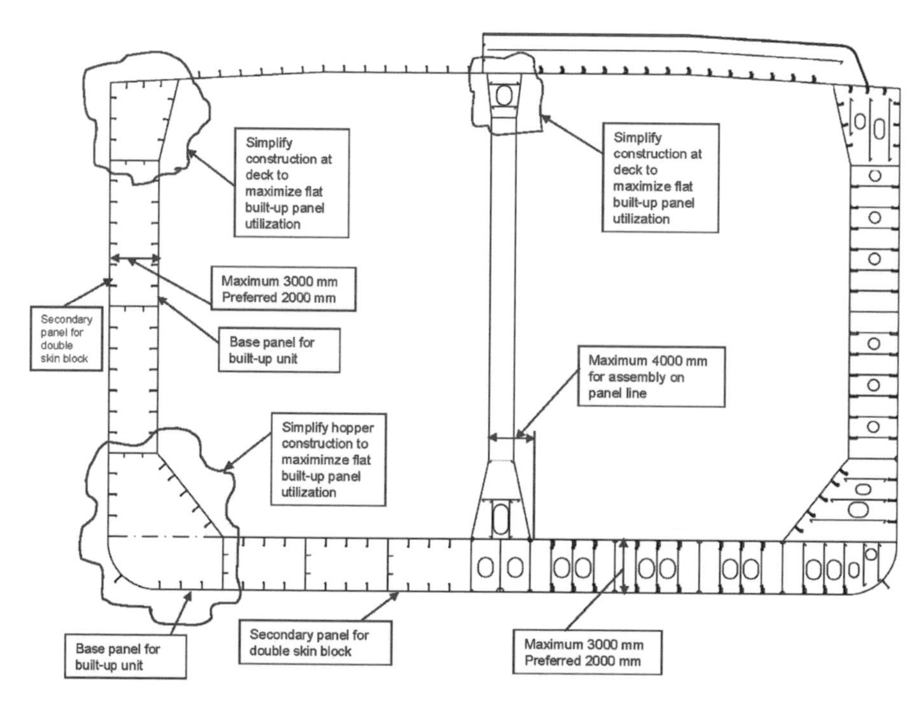

Figure 6. Generic midship section of a chemical tanker with constraints.

underlying points considered in its development. For instance, the preferred double bottom height is 2000 mm, whereas the maximum permissible height is 3000 mm. Likewise recommendations for simplified hopper construction to maximize the use of the flat panel line are illustrated as well. Wherever it is possible to simplify construction and maximize the use of automated production facilities while simultaneously complying with classification society rules and owner requests, this is illustrated in the generic midship section above.

4 CONCLUSIONS

The authors of this paper demonstrated that by breaking down typical interim products of three various types of vessels in tabular form using DFP analysis principles, it is possible to more readily draw up future guidelines which will improve shipyard productivity while maintaining a product mix. The effectiveness of the shipyard facilities is maximized while the unnecessary designs that complicate assembly are minimized and or eliminated. This way costs are reduced and the competitiveness of the shipyard is improved. A shipyard strategy that is geared towards using and further developing guidelines for the design of all interim products, from panels, built up panels, and later into outfitting will further improve shipyard productivity. A strategy of maintaining a product mix along with DFP matching of interim products is an effective path for shipyards to follow. Shipyard management should likewise enforce that interim product matching be performed on all designs including those not done in-house, because only in this way can productivity be maintained and improved.

REFERENCES

Bunch, H.M. 2010. The impact of group technology. *Marine Technology*, 8–9.

Chirillo, L.D. 2010. Development of the PWBS. *Marine Technology,* 72.

Design for Production Manual, 2nd edition, National Shipbuilding Research Program, US Department of the Navy Carderock Division, 1999.

Dlugokecki, V., et al., 2008. Leading the way for mid-tier shipyards to implement design for production methodologies. *Journal of Ship Production* 26(4): 265–272.

Kolić, D., et al., 2010. Proposal for the determination of technological parameters for design rationalization of a shipbuilding production program. *Engineering Review* 30(2): 59–69.

Lamb, T. 2003. *Ship Design and Construction* 1. Jersey City: Society of Naval Architects and Marine Engineers.

3. *Maj shipyard archive*, 2010.

Sustainable Maritime Transportation and Exploitation of Sea Resources – Rizzuto & Guedes Soares (eds)
© 2012 Taylor & Francis Group, London, ISBN 978-0-415-62081-9

Author index